海外优秀数学类教材系列丛书 Pearson

影印版

Elementary Linear Algebra with Applications (Ninth Edition)

初等线性代数及其应用（第9版）

Bernard Kolman
Drexel University

David R. Hill
Temple University

中国教育出版传媒集团
高等教育出版社 · 北京

图字：01-2024-5967号

图书在版编目（CIP）数据

初等线性代数及其应用 ： 第9版 ： 影印版 = Elementary Linear Algebra with Applications (Ninth Edition) ： 英文 / （美）伯纳德·科尔曼，（美）戴维·R. 希尔著. -- 北京 ： 高等教育出版社， 2025. 5. （海外优秀数学类教材系列丛书）. -- ISBN 978-7-04-064530-9

Ⅰ. O151.2

中国国家版本馆 CIP 数据核字第 2025WN8675 号

Chudeng Xianxing Daishu ji Qi Yingyong

策划编辑　刘　荣　　责任编辑　刘　荣　　封面设计　李小璐　　版式设计　徐艳妮

责任校对　吕红颖　　责任印制　耿　轩

出版发行	高等教育出版社	网　　址	http://www.hep.edu.cn
社　　址	北京市西城区德外大街4号		http://www.hep.com.cn
邮政编码	100120	网上订购	http://www.hepmall.com.cn
印　　刷	山东韵杰文化科技有限公司		http://www.hepmall.com
开　　本	889mm×1194mm 1/16		http://www.hepmall.cn
印　　张	40.25		
字　　数	1260千字	版　　次	2025年5月第1版
购书热线	010-58581118	印　　次	2025年5月第1次印刷
咨询电话	400-810-0598	定　　价	102.00元

本书如有缺页、倒页、脱页等质量问题，请到所购图书销售部门联系调换

物 料 号　64530-00

出 版 说 明

本书是“海外优秀数学类教材系列丛书”之一。

该系列教材经国内高校一线教师、专家、学者推荐和评介，遴选 Pearson Education，John Wiley & Sons，McGraw Hill，Cengage Learning 等国外出版公司近几年出版的优秀教材，由高等教育出版社影印出版，原汁原味呈献给读者。

遴选出来的原版教材具有以下特点：

1. 在国际上被广泛使用，在同类教材中具有一定的权威性。
2. 历经多年教学实践检验，取材广泛，内容翔实准确。
3. 教学资源配套完善，方便教师教学、学生学习。
4. 插图精美、丰富，图文并茂，可读性强。
5. 语言简练、流畅，注重可接受性，适合国内学生阅读。

希望该系列教材的出版能为国内教师吸收、借鉴国外优秀教学理念和教学成果，为高等教育提高教学质量、提升人才培养水平做出贡献。

高等教育出版社

2025 年 2 月

谨以此书

献给 Lisa, Stephen 和 Lillie

B. K.

献给 Suzanne

D. R. H.

第 9 版前言

对许多学生而言，线性代数是一门重要的课程。这至少有两个原因。首先，只有少量课程能在数学的其他领域——如多元微积分、微分方程和概率论——以及物理学、生物学、化学、经济学、金融学、心理学、社会学和所有工程领域有如此广泛的应用。其次，这门课程为大二学生提供了学习理解抽象概念的绝佳机会。

本教材介绍大二线性代数的基本思想和计算方法。它包含各种精心挑选的应用，如现实生活中的应用——谷歌（搜索引擎）和全球定位系统（GPS）等。书中还向学生介绍如何理解抽象概念。在讨论线性代数的基本思想时，抽象思想与几何解释和实际计算并重。

本版的新内容

很高兴本教材前八版在过去的 38 年中得到了广泛的接受。在准备本版的过程中，我们仔细考虑了教师和学生的许多建议，以改进书中内容及其呈现方式。令我们特别高兴的是，一些读者在来信中提到，他们原本作为学生使用本教材，现在则是作为教师使用本教材。尽管本次修订改变了不少教材内容，但我们的目标与前八版的保持一致：以学生可以理解的方式呈现线性代数的基本思想。为了实现这一目标，本版包含以下特色：

- 讨论问题已加入章末复习材料，其中的许多问题适合写作项目或小组活动。
- 将原第 2.1 节分成两部分：第 2.1 节“矩阵的阶梯形”和第 2.2 节“线性方程组的求解”，这将为讲述这一重要内容提供教学方法的改进。
- 原第 6 章“行列式”现在变成第 3 章，以便教师更早地讲授这些材料。
- 原第 3.4 节“张成（线性空间）与线性无关”已分成两部分，即第 4.4 节“张成（空间）”和第 4.5 节“线性无关”。由于学生理解这些更抽象的主题时经常有困难，本次修订以较慢的速度呈现这些材料，并提供更多的例子。
- 第 8 章“特征值与特征向量的应用”是本版的修订内容。它由原第 7.3，7.6—7.9 节，原第 7.5 节关于对称图像传输的内容以及原第 8.1，8.2 节组成。
- 将更多的几何材料用于关于对称矩阵对角化和奇异值分解的讨论。
- 第 1.7 节“计算机图形学”已扩充。

- 添加了更多应用，这些应用包括网络和化学平衡方程。
- 扩充了许多论述，以改进教学过程，并增加了更多解释以显示某些材料和结论的重要性。
- 增加了一个简化的讨论，展示线性代数如何应用于全球定位系统（GPS）。
- 增加了更多关于递推关系的材料。
- 介绍了更多不同的向量空间例子。
- 增加了更多讨论线性代数四个基本子空间的材料。
- 添加了更多几何内容。
- 增加了更多图形。
- 增加了更多各层次习题。
- 涉及现实世界数据的习题已更新，以包括最新的数据集。

习题

习题是本教材不可分割的一部分。教材中的许多习题本质上是数值型的，而另一些则是理论型的。本版新增前七章末尾的讨论习题，可用于写作项目或小组活动。许多理论习题和讨论习题，以及一些数值习题，都需要口头讨论解决方案。在这个科技时代，能够仔细而精确地写作尤为重要；解决这类习题有助于提高这项技能。本版包含近 200 个新习题。用特殊符号明确表示前八章需要学生探索的习题，但没有指定任何特定的求解软件。所有奇数号习题的答案都在书末。

呈现方式

我们的经验是，在大二阶段，抽象概念应逐步引入，且应建立在坚实的基础上。因此，我们在研究线性代数的初始阶段，将矩阵视为求解线性方程组时自然出现的一组数字，而求解线性方程组是学生已经熟悉的一个问题。在修订各版时，我们都非常关注改进和完善教材内容的呈现方式。行文时，抽象的概念与对应的几何解释和实际计算并重。附录 C“证明简介”可用于向学生快速介绍数学证明的基础。

覆盖的内容

在使用这本教材时，对于为期一季、每周授课四次的线性代数课程，如果省略选学内容，则可以毫无困难地讲授到特征值和特征向量。改变用于讲授理论材料的时间后，课程的水平和节奏也会相应改变。因此，这本教材可以用于许多不同程度的线性代数课程。

第 1 章讨论矩阵及其性质。该章还简单介绍矩阵变换以及点积在统计学中的应用。**第 2 章**讨论求解线性方程组的方法。**第 3 章**介绍行列式的基本性质及一些应用。**第 4 章**引入一个更抽象的概念——实向量空间，这里会揭示许多自然产生的几何思想中的一部分。例如，我们证明了 n 维实向

量空间与 R^n 同构，后者是所有有序 n 元实数组构成的向量空间，也是所有 $n\times1$ 实矩阵构成的向量空间。由于 R^n 只是 R^2 和 R^3 的简单推广，因此第 4 章开头讨论了 2 维和 3 维空间。这也表明，有限维实向量空间的概念并不像最初引入时看起来那么深奥。**第 5 章**涵盖内积空间，具有很强的几何背景。**第 6 章**讨论矩阵和线性变换，这里考虑维数定理及其在线性方程组求解中的应用。**第 7 章**介绍特征值和特征向量，并彻底解决对称矩阵的对角化问题。**第 8 章**（选学）介绍了特征值和特征向量的一些应用。第 8.3 节"主特征值与主成分分析"强调了线性代数中一些非常有用的结果。授课时可以直接从第 7.2 节转到第 8.4 节"微分方程"，后者展示如何使用线性代数求解微分方程。第 8.5 节"动力系统"给出线性代数在现代应用数学中一个重要领域的应用。该章还讨论实二次型、二次曲线和二次曲面。附录 A 回顾了一些关于集合和函数的非常基本的材料。学生如有需要，可随时查阅该附录。关于复数的附录 B 简单而全面地介绍了复数及其在线性代数中的应用。附录 C 简要介绍了数学证明。

致谢

我们很高兴地感谢前八版的审稿人：已故的 Edward Norman，中佛罗里达大学；Herbert J. Nichol 和已故的 Charles S. Duris，德雷克塞尔大学；Stephen D. Kerr，韦伯州立学院；Norman Lee，鲍尔州立大学；William Briggs，科罗拉多大学；Richard Roth，科罗拉多大学；David Stanford，威廉玛丽学院；David L. Abrahamson，罗德岛学院；Ruth Berger，孟菲斯州立大学；Michael A. Geraghty，艾奥瓦大学；You-Feng Lin，南佛罗里达大学；Lothar Redlin，宾夕法尼亚州立大学阿宾顿分校；Richard Sot，内华达大学雷诺分校；Raymond Southworth，威廉玛丽学院名誉教授；J. Barry Turett，奥克兰大学；Gordon Brown，科罗拉多大学；Matt Insall，密苏里大学罗拉分校；Wolfgang Kappe，纽约州立大学宾厄姆顿分校；Richard P. Kubelka，圣何塞州立大学；James Nation，夏威夷大学；David Peterson，中阿肯色大学；Malcolm J. Sherman，纽约州立大学奥尔巴尼分校；Alex Stanoyevitch，夏威夷大学；Barbara Tabak，布兰迪斯大学；Loring W. Tu，塔夫茨大学；Manfred Kolster，麦克马斯特大学；Daniel Cunningham，布法罗州立学院；Larry K. Chu，迈诺特州立大学；Daniel King，莎拉劳伦斯大学；Kevin Vang，迈诺特州立大学；Avy Soffer，罗格斯大学；David Kaminski，莱斯布里奇大学；Patricia Beaulieu，路易斯安那大学；Will Murray，加利福尼亚州立大学长滩分校，和第 9 版的审稿人：Thalia D. Jeffres，威奇托州立大学；Manuel Lopez，罗切斯特理工学院；Thomas L. Scofield，凯尔文学院；Jim Gehrmann，加利福尼亚州立大学萨克拉门托分校；John M. Erdman，波特兰州立大学；Ada Cheng，凯特琳

大学；Juergen Gerlach，拉德福德大学；Martha Allen，佐治亚学院与州立大学。

这些人的无数建议、评论和批评极大地改进了手稿。

我们感谢 Dennis R. Kletzing，他排版了整个手稿，发现并纠正了手稿中的一些数学错误。很高兴和他一起工作。

我们感谢 Blaise deSesa 仔细检查了所有习题的答案。

我们感谢 William T. Williams 允许我们将他引人注目的画作 Trane 放在本版封面上。

我们也感谢天普大学的 Lilian N. Brady 和 Nina Edelman 批判且仔细地阅读校样；感谢美国和其他国家许多学校的教师和学生与我们分享他们在过去 38 年里使用本教材的经验，并提供有益的建议。

最后，衷心感谢高级执行主编 Scott Disanno、高级编辑 Holly Stark、编辑助理 Jennifer Lonschein 和 Prentice Hall 公司的全体员工在本版的构思、设计、制作及营销阶段所表现出的热情、兴趣与始终如一的合作。和他们一起工作真的很愉快。

B.K.
D.R.H.

致学生

本课程可能与你迄今为止学过的任何其他数学课程不同，至少在两个重要方面如此。首先，这可能是向你介绍抽象数学概念的第一门课程。其次，这是一门很可能对你的职业产生最大影响的数学课程。

与其他数学课程不同的是，本课程不会为你提供解决某些问题的单一计算工具包。相反，我们将展示一种名为线性代数的核心材料，这是通过引入特定的定义并构造用以确定性质和证明定理的流程来实现的。证明定理是一项需要花费时间来掌握的技能，因此我们将非常细心地培养你证明数学结论的能力。我们会逐步向你介绍抽象，并用具体的数值例子和应用充分说明每个抽象概念。虽然你将做很多计算，但是解决大多数问题的目标不仅仅是得到“正确”答案，还要能理解并解释如何得到答案，然后阐释结果。

在日常生活中，线性代数用于解决数学的其他分支、物理学、生物学、化学、工程学、统计学、经济学、金融学、心理学和社会学等领域的问题。使用线性代数的应用包括信息传输、电影和视频特效的开发、声音录制、互联网搜索引擎、全球定位系统（GPS）和经济分析。因此，你将看到线性代数有非常深远的影响。本教材中包含一定数量的应用。如果时间足够，其中的一些可能会在课程中介绍。此外，许多应用可以用于自学。

本教材有四种不同类型的习题。第一类习题是计算型的。这些习题和其中的数字都经过精心挑选，因此几乎所有习题都可以很容易地手算解决。在实际应用中使用线性代数时，你会发现问题的规模要大得多，其中出现的数字并不总是“友好”的。但是这并不是一个问题，因为你几乎肯定会使用功能强大的软件来解决它们。第二类习题是理论型的，其中的一些可能要求你证明一个结论或讨论一个想法。第三类习题提供了使用软件解题的机会。这些习题应该使用计算机和诸如 MATLAB 的软件来解决，其中 MATLAB 是一种强大且基于矩阵的应用程序，并在工业中被广泛使用。第四类习题是讨论型的，可以用作小组项目。在当今世界，仅仅算出答案是不够的；你经常需要准备一份报告，以讨论解决方案、说明其中的步骤并解释结果。这类习题将给你提供这种数学写作的经验。数学也使用文字，而不仅仅使用符号。

如何学好线性代数

- 手拿纸笔，慢慢地读本教材。你可能需要多次阅读某个特定章节。花点时间验证文中标记了“验证”的步骤。
- 及时完成作业。如果只等课堂上教师的解答，你就会错过学习自己解决问题的机会。即使你不能解决一个问题，也要尝试一下，这样当教师在课堂上讲解它时，你会更容易理解解题过程。你可能会发现，与其他学生一起学习课程内容和讨论家庭作业是有益的。
- 确保在你不理解的时候尽快寻求帮助。本课程中的每个抽象概念都基于之前引入的概念，就像先打好地基之后才能盖房子一样。如果对其中任何一个想法感到模糊或者未学习过，你对这门课程的知识掌握程度将不足以让你理解后续内容。
- 利用本教材提供的教学材料。各章每一节末尾都有一个关键术语表，前七章章末都有本章回顾、补充习题、本章测验和讨论习题。奇数号计算题的答案在教材末尾。

我们向你保证，学习线性代数的努力将在其他课程和后续职业生涯中得到充分回报。

我们祝愿你在线性代数学习中取得圆满成功。

Bernard Kolman

David R. Hill

目录

Contents

CHAPTER

1 Linear Equations and Matrices

1.1 Systems of Linear Equations

One of the most frequently recurring practical problems in many fields of study—such as mathematics, physics, biology, chemistry, economics, all phases of engineering, operations research, and the social sciences—is that of solving a system of linear equations. The equation

$$a_1x_1 + a_2x_2 + \cdots + a_nx_n = b, \tag{1}$$

which expresses the real or complex quantity b in terms of the unknowns x_1, x_2, $\ldots$, x_n and the real or complex constants $a_1, a_2, \ldots, a_n$, is called a **linear equation**. In many applications we are given b and must find numbers $x_1, x_2, \ldots, x_n$ satisfying (1).

A **solution** to linear Equation (1) is a sequence of n numbers $s_1, s_2, \ldots, s_n$, which has the property that (1) is satisfied when $x_1 = s_1, x_2 = s_2, \ldots, x_n = s_n$ are substituted in (1). Thus $x_1 = 2$, $x_2 = 3$, and $x_3 = -4$ is a solution to the linear equation

$$6x_1 - 3x_2 + 4x_3 = -13,$$

because

$$6(2) - 3(3) + 4(-4) = -13.$$

More generally, a **system of m linear equations in n unknowns**, $x_1, x_2, \ldots, x_n$, or a **linear system**, is a set of m linear equations each in n unknowns. A linear

Note: Appendix A reviews some very basic material dealing with sets and functions. It can be consulted at any time, as needed.

system can conveniently be written as

$$\begin{aligned} a_{11}x_1 + a_{12}x_2 + \cdots + a_{1n}x_n &= b_1 \\ a_{21}x_1 + a_{22}x_2 + \cdots + a_{2n}x_n &= b_2 \\ &\vdots \\ a_{m1}x_1 + a_{m2}x_2 + \cdots + a_{mn}x_n &= b_m. \end{aligned} \tag{2}$$

Thus the ith equation is

$$a_{i1}x_1 + a_{i2}x_2 + \cdots + a_{in}x_n = b_i.$$

In (2) the a_{ij} are known constants. Given values of $b_1, b_2, \ldots, b_m$, we want to find values of $x_1, x_2, \ldots, x_n$ that will satisfy each equation in (2).

A **solution** to linear system (2) is a sequence of n numbers $s_1, s_2, \ldots, s_n$, which has the property that each equation in (2) is satisfied when $x_1 = s_1, x_2 = s_2, \ldots, x_n = s_n$ are substituted.

If the linear system (2) has no solution, it is said to be **inconsistent**; if it has a solution, it is called **consistent**. If $b_1 = b_2 = \cdots = b_m = 0$, then (2) is called a **homogeneous system**. Note that $x_1 = x_2 = \cdots = x_n = 0$ is always a solution to a homogeneous system; it is called the **trivial solution**. A solution to a homogeneous system in which not all of $x_1, x_2, \ldots, x_n$ are zero is called a **nontrivial solution**.

Consider another system of r linear equations in n unknowns:

$$\begin{aligned} c_{11}x_1 + c_{12}x_2 + \cdots + c_{1n}x_n &= d_1 \\ c_{21}x_1 + c_{22}x_2 + \cdots + c_{2n}x_n &= d_2 \\ &\vdots \\ c_{r1}x_1 + c_{r2}x_2 + \cdots + c_{rn}x_n &= d_r. \end{aligned} \tag{3}$$

We say that (2) and (3) are **equivalent** if they both have exactly the same solutions.

EXAMPLE 1 The linear system

$$\begin{aligned} x_1 - 3x_2 &= -7 \\ 2x_1 + x_2 &= 7 \end{aligned} \tag{4}$$

has only the solution $x_1 = 2$ and $x_2 = 3$. The linear system

$$\begin{aligned} 8x_1 - 3x_2 &= 7 \\ 3x_1 - 2x_2 &= 0 \\ 10x_1 - 2x_2 &= 14 \end{aligned} \tag{5}$$

also has only the solution $x_1 = 2$ and $x_2 = 3$. Thus (4) and (5) are equivalent. ■

To find a solution to a linear system, we shall use a technique called the **method of elimination**; that is, we eliminate some variables by adding a multiple of one equation to another equation. Elimination merely amounts to the development of a new linear system that is equivalent to the original system, but is much simpler to solve. Readers have probably confined their earlier work in this area to linear systems in which $m = n$, that is, linear systems having as many equations as unknowns. In this course we shall broaden our outlook by dealing with systems in which we have $m = n$, $m < n$, and $m > n$. Indeed, there are numerous applications in which $m \neq n$. If we deal with two, three, or four unknowns, we shall often write them as x, y, z, and w. In this section we use the method of elimination as it was studied in high school. In Section 2.2 we shall look at this method in a much more systematic manner.

EXAMPLE 2

The director of a trust fund has \$100,000 to invest. The rules of the trust state that both a certificate of deposit (CD) and a long-term bond must be used. The director's goal is to have the trust yield \$7800 on its investments for the year. The CD chosen returns 5% per annum, and the bond 9%. The director determines the amount x to invest in the CD and the amount y to invest in the bond as follows:

Since the total investment is \$100,000, we must have $x + y = 100{,}000$. Since the desired return is \$7800, we obtain the equation $0.05x + 0.09y = 7800$. Thus, we have the linear system

$$\begin{aligned} x + \quad\; y &= 100{,}000 \\ 0.05x + 0.09y &= \quad 7800. \end{aligned} \tag{6}$$

To eliminate x, we add (-0.05) times the first equation to the second, obtaining

$$0.04y = 2800,$$

an equation having no x term. We have eliminated the unknown x. Then solving for y, we have

$$y = 70{,}000,$$

and substituting into the first equation of (6), we obtain

$$x = 30{,}000.$$

To check that $x = 30{,}000$, $y = 70{,}000$ is a solution to (6), we verify that these values of x and y satisfy *each* of the equations in the given linear system. This solution is the only solution to (6); the system is consistent. The director of the trust should invest \$30,000 in the CD and \$70,000 in the long-term bond. ■

EXAMPLE 3

Consider the linear system

$$\begin{aligned} x - 3y &= -7 \\ 2x - 6y &= \quad 7. \end{aligned} \tag{7}$$

Again, we decide to eliminate x. We add (-2) times the first equation to the second one, obtaining

$$0 = 21,$$

which makes no sense. This means that (7) has no solution; it is inconsistent. We could have come to the same conclusion from observing that in (7) the left side of the second equation is twice the left side of the first equation, but the right side of the second equation is not twice the right side of the first equation. ■

EXAMPLE 4 Consider the linear system

$$\begin{aligned} x + 2y + 3z &= 6 \\ 2x - 3y + 2z &= 14 \\ 3x + y - z &= -2. \end{aligned} \tag{8}$$

To eliminate x, we add (-2) times the first equation to the second one and (-3) times the first equation to the third one, obtaining

$$\begin{aligned} -7y - 4z &= 2 \\ -5y - 10z &= -20. \end{aligned} \tag{9}$$

This is a system of two equations in the unknowns y and z. We multiply the second equation of (9) by $\left(-\frac{1}{5}\right)$, yielding

$$\begin{aligned} -7y - 4z &= 2 \\ y + 2z &= 4, \end{aligned}$$

which we write, by interchanging equations, as

$$\begin{aligned} y + 2z &= 4 \\ -7y - 4z &= 2. \end{aligned} \tag{10}$$

We now eliminate y in (10) by adding 7 times the first equation to the second one, to obtain

$$10z = 30,$$

or

$$z = 3. \tag{11}$$

Substituting this value of z into the first equation of (10), we find that $y = -2$. Then substituting these values of y and z into the first equation of (8), we find that $x = 1$. We observe further that our elimination procedure has actually produced the linear system

$$\begin{aligned} x + 2y + 3z &= 6 \\ y + 2z &= 4 \\ z &= 3, \end{aligned} \tag{12}$$

obtained by using the first equations of (8) and (10) as well as (11). The importance of this procedure is that, although the linear systems (8) and (12) are equivalent, (12) has the advantage that it is easier to solve. ■

EXAMPLE 5 Consider the linear system

$$\begin{aligned} x + 2y - 3z &= -4 \\ 2x + y - 3z &= 4. \end{aligned} \tag{13}$$

Eliminating x, we add (-2) times the first equation to the second equation to get

$$-3y + 3z = 12. \tag{14}$$

We must now solve (14). A solution is

$$y = z - 4,$$

where z can be any real number. Then from the first equation of (13),

$$\begin{aligned} x &= -4 - 2y + 3z \\ &= -4 - 2(z - 4) + 3z \\ &= z + 4. \end{aligned}$$

Thus a solution to the linear system (13) is

$$\begin{aligned} x &= z + 4 \\ y &= z - 4 \\ z &= \text{any real number.} \end{aligned}$$

This means that the linear system (13) has infinitely many solutions. Every time we assign a value to z we obtain another solution to (13). Thus, if $z = 1$, then

$$x = 5, \quad y = -3, \quad \text{and} \quad z = 1$$

is a solution, while if $z = -2$, then

$$x = 2, \quad y = -6, \quad \text{and} \quad z = -2$$

is another solution. ■

These examples suggest that a linear system may have a unique solution, no solution, or infinitely many solutions.

Consider next a linear system of two equations in the unknowns x and y:

$$\begin{aligned} a_1x + a_2y &= c_1 \\ b_1x + b_2y &= c_2. \end{aligned} \tag{15}$$

The graph of each of these equations is a straight line, which we denote by ℓ_1 and ℓ_2, respectively. If $x = s_1$, $y = s_2$ is a solution to the linear system (15), then the point (s_1, s_2) lies on both lines ℓ_1 and ℓ_2. Conversely, if the point (s_1, s_2) lies on both lines ℓ_1 and ℓ_2, then $x = s_1$, $y = s_2$ is a solution to the linear system (15). Thus we are led geometrically to the same three possibilities mentioned previously. See Figure 1.1.

Next, consider a linear system of three equations in the unknowns x, y, and z:

$$\begin{aligned} a_1x + b_1y + c_1z &= d_1 \\ a_2x + b_2y + c_2z &= d_2 \\ a_3x + b_3y + c_3z &= d_3. \end{aligned} \tag{16}$$

The graph of each of these equations is a plane, denoted by P_1, P_2, and P_3, respectively. As in the case of a linear system of two equations in two unknowns,

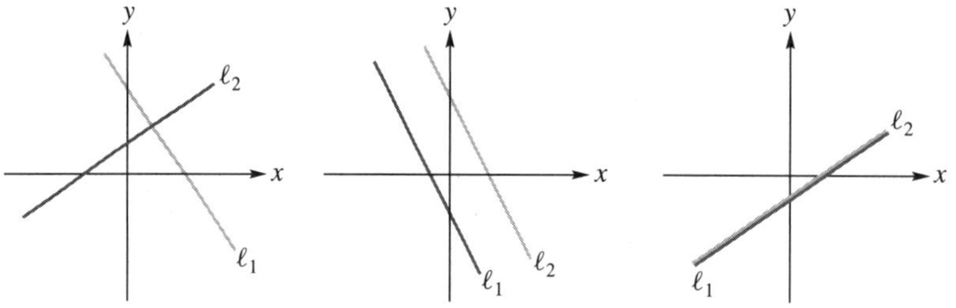

FIGURE 1.1

the linear system in (16) can have infinitely many solutions, a unique solution, or no solution. These situations are illustrated in Figure 1.2. For a more concrete illustration of some of the possible cases, consider that two intersecting walls and the ceiling (planes) of a room intersect in a unique point, a corner of the room, so the linear system has a unique solution. Next, think of the planes as pages of a book. Three pages of a book (held open) intersect in a straight line, the spine. Thus, the linear system has infinitely many solutions. On the other hand, when the book is closed, three pages of a book appear to be parallel and do not intersect, so the linear system has no solution.

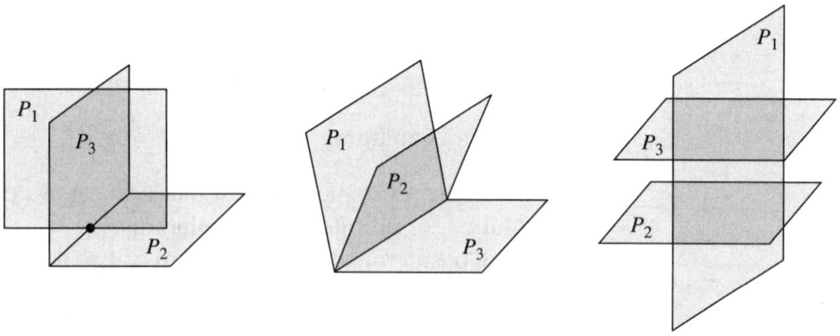

FIGURE 1.2

If we examine the method of elimination more closely, we find that it involves three manipulations that can be performed on a linear system to convert it into an equivalent system. These manipulations are as follows:

1. Interchange the ith and jth equations.
2. Multiply an equation by a nonzero constant.
3. Replace the ith equation by c times the jth equation plus the ith equation, $i \neq j$. That is, replace

$$a_{i1}x_1 + a_{i2}x_2 + \cdots + a_{in}x_n = b_i$$

by

$$(ca_{j1} + a_{i1})x_1 + (ca_{j2} + a_{i2})x_2 + \cdots + (ca_{jn} + a_{in})x_n = cb_j + b_i.$$

It is not difficult to prove that performing these manipulations on a linear system leads to an equivalent system. The next example proves this for the second type of manipulation. Exercises 24 and 25 prove it for the first and third manipulations, respectively.

EXAMPLE 6

Suppose that the ith equation of the linear system (2) is multiplied by the nonzero constant c, producing the linear system

$$\begin{aligned} a_{11}x_1 + a_{12}x_2 + \cdots + a_{1n}x_n &= b_1 \\ a_{21}x_1 + a_{22}x_2 + \cdots + a_{2n}x_n &= b_2 \\ &\vdots \\ ca_{i1}x_1 + ca_{i2}x_2 + \cdots + ca_{in}x_n &= cb_i \\ &\vdots \\ a_{m1}x_1 + a_{m2}x_2 + \cdots + a_{mn}x_n &= b_m. \end{aligned} \tag{17}$$

If $x_1 = s_1, x_2 = s_2, \ldots, x_n = s_n$ is a solution to (2), then it is a solution to all the equations in (17), except possibly to the ith equation. For the ith equation we have

$$c(a_{i1}s_1 + a_{i2}s_2 + \cdots + a_{in}s_n) = cb_i$$

or

$$ca_{i1}s_1 + ca_{i2}s_2 + \cdots + ca_{in}s_n = cb_i.$$

Thus the ith equation of (17) is also satisfied. Hence every solution to (2) is also a solution to (17). Conversely, every solution to (17) also satisfies (2). Hence (2) and (17) are equivalent systems. ■

The following example gives an application leading to a linear system of two equations in three unknowns:

EXAMPLE 7

(Production Planning) A manufacturer makes three different types of chemical products: A, B, and C. Each product must go through two processing machines: X and Y. The products require the following times in machines X and Y:

1. One ton of A requires 2 hours in machine X and 2 hours in machine Y.
2. One ton of B requires 3 hours in machine X and 2 hours in machine Y.
3. One ton of C requires 4 hours in machine X and 3 hours in machine Y.

Machine X is available 80 hours per week, and machine Y is available 60 hours per week. Since management does not want to keep the expensive machines X and Y idle, it would like to know how many tons of each product to make so that the machines are fully utilized. It is assumed that the manufacturer can sell as much of the products as is made.

To solve this problem, we let x_1, x_2, and x_3 denote the number of tons of products A, B, and C, respectively, to be made. The number of hours that machine X will be used is

$$2x_1 + 3x_2 + 4x_3,$$

which must equal 80. Thus we have

$$2x_1 + 3x_2 + 4x_3 = 80.$$

Similarly, the number of hours that machine Y will be used is 60, so we have

$$2x_1 + 2x_2 + 3x_3 = 60.$$

Mathematically, our problem is to find nonnegative values of x_1, x_2, and x_3 so that

$$\begin{aligned} 2x_1 + 3x_2 + 4x_3 &= 80 \\ 2x_1 + 2x_2 + 3x_3 &= 60. \end{aligned}$$

This linear system has infinitely many solutions. Following the method of Example 4, we see that all solutions are given by

$$\begin{aligned} x_1 &= \frac{20 - x_3}{2} \\ x_2 &= 20 - x_3 \\ x_3 &= \text{any real number such that } 0 \le x_3 \le 20, \end{aligned}$$

since we must have $x_1 \ge 0$, $x_2 \ge 0$, and $x_3 \ge 0$. When $x_3 = 10$, we have

$$x_1 = 5, \qquad x_2 = 10, \qquad x_3 = 10$$

while

$$x_1 = \tfrac{13}{2}, \qquad x_2 = 13, \qquad x_3 = 7$$

when $x_3 = 7$. The reader should observe that one solution is just as good as the other. There is no best solution unless additional information or restrictions are given. ■

As you have probably already observed, the method of elimination has been described, so far, in general terms. Thus we have not indicated any rules for selecting the unknowns to be eliminated. Before providing a very systematic description of the method of elimination, we introduce in the next section the notion of a matrix. This will greatly simplify our notational problems and will enable us to develop tools to solve many important applied problems.

Key Terms

Linear equation
Solution of a linear equation
Linear system
Unknowns
Inconsistent system
Consistent system
Homogeneous system
Trivial solution
Nontrivial solution
Equivalent systems
Unique solution
No solution
Infinitely many solutions
Manipulations on linear systems
Method of elimination

1.1 Exercises

In Exercises 1 through 14, solve each given linear system by the method of elimination.

1. $\begin{aligned} x + 2y &= 8 \\ 3x - 4y &= 4 \end{aligned}$

2. $\begin{aligned} 2x - 3y + 4z &= -12 \\ x - 2y + z &= -5 \\ 3x + y + 2z &= 1 \end{aligned}$

3. $\begin{aligned} 3x + 2y + z &= 2 \\ 4x + 2y + 2z &= 8 \\ x - y + z &= 4 \end{aligned}$

4. $\begin{aligned} x + y &= 5 \\ 3x + 3y &= 10 \end{aligned}$

5. $\begin{aligned} 2x + 4y + 6z &= -12 \\ 2x - 3y - 4z &= 15 \\ 3x + 4y + 5z &= -8 \end{aligned}$

6. $\begin{aligned} x + y - 2z &= 5 \\ 2x + 3y + 4z &= 2 \end{aligned}$

7. $\begin{aligned} x + 4y - z &= 12 \\ 3x + 8y - 2z &= 4 \end{aligned}$

8. $\begin{aligned} 3x + 4y - z &= 8 \\ 6x + 8y - 2z &= 3 \end{aligned}$

9. $\begin{aligned} x + y + 3z &= 12 \\ 2x + 2y + 6z &= 6 \end{aligned}$

10. $\begin{aligned} x + y &= 1 \\ 2x - y &= 5 \\ 3x + 4y &= 2 \end{aligned}$

11. $\begin{aligned} 2x + 3y &= 13 \\ x - 2y &= 3 \\ 5x + 2y &= 27 \end{aligned}$

12. $\begin{aligned} x - 5y &= 6 \\ 3x + 2y &= 1 \\ 5x + 2y &= 1 \end{aligned}$

13. $\begin{aligned} x + 3y &= -4 \\ 2x + 5y &= -8 \\ x + 3y &= -5 \end{aligned}$

14. $\begin{aligned} 2x + 3y - z &= 6 \\ 2x - y + 2z &= -8 \\ 3x - y + z &= -7 \end{aligned}$

15. Given the linear system

$$\begin{aligned} 2x - y &= 5 \\ 4x - 2y &= t, \end{aligned}$$

(a) Determine a particular value of t so that the system is consistent.

(b) Determine a particular value of t so that the system is inconsistent.

(c) How many different values of t can be selected in part (b)?

16. Given the linear system

$$\begin{aligned} 3x + 4y &= s \\ 6x + 8y &= t, \end{aligned}$$

(a) Determine particular values for s and t so that the system is consistent.

(b) Determine particular values for s and t so that the system is inconsistent.

(c) What relationship between the values of s and t will guarantee that the system is consistent?

17. Given the linear system

$$\begin{aligned} x + 2y &= 10 \\ 3x + (6 + t)y &= 30, \end{aligned}$$

(a) Determine a particular value of t so that the system has infinitely many solutions.

(b) Determine a particular value of t so that the system has a unique solution.

(c) How many different values of t can be selected in part (b)?

18. Is every homogeneous linear system always consistent? Explain.

19. Given the linear system

$$\begin{aligned} 2x + 3y - z &= 0 \\ x - 4y + 5z &= 0, \end{aligned}$$

(a) Verify that $x_1 = 1$, $y_1 = -1$, $z_1 = -1$ is a solution.

(b) Verify that $x_2 = -2$, $y_2 = 2$, $z_2 = 2$ is a solution.

(c) Is $x = x_1 + x_2 = -1$, $y = y_1 + y_2 = 1$, and $z = z_1 + z_2 = 1$ a solution to the linear system?

(d) Is $3x$, $3y$, $3z$, where x, y, and z are as in part (c), a solution to the linear system?

20. Without using the method of elimination, solve the linear system

$$\begin{aligned} 2x + y - 2z &= -5 \\ 3y + z &= 7 \\ z &= 4. \end{aligned}$$

21. Without using the method of elimination, solve the linear system

$$\begin{aligned} 4x &= 8 \\ -2x + 3y &= -1 \\ 3x + 5y - 2z &= 11. \end{aligned}$$

22. Is there a value of r so that $x = 1$, $y = 2$, $z = r$ is a solution to the following linear system? If there is, find it.

$$\begin{aligned} 2x + 3y - z &= 11 \\ x - y + 2z &= -7 \\ 4x + y - 2z &= 12. \end{aligned}$$

23. Is there a value of r so that $x = r$, $y = 2$, $z = 1$ is a solution to the following linear system? If there is, find it.

$$\begin{aligned} 3x - 2z &= 4 \\ x - 4y + z &= -5 \\ -2x + 3y + 2z &= 9. \end{aligned}$$

24. Show that the linear system obtained by interchanging two equations in (2) is equivalent to (2).

25. Show that the linear system obtained by adding a multiple of an equation in (2) to another equation is equivalent to (2).

26. Describe the number of points that simultaneously lie in each of the three planes shown in each part of Figure 1.2.

27. Describe the number of points that simultaneously lie in each of the three planes shown in each part of Figure 1.3.

28. Let C_1 and C_2 be circles in the plane. Describe the number of possible points of intersection of C_1 and C_2. Illustrate each case with a figure.

29. Let S_1 and S_2 be spheres in space. Describe the number of possible points of intersection of S_1 and S_2. Illustrate each case with a figure.

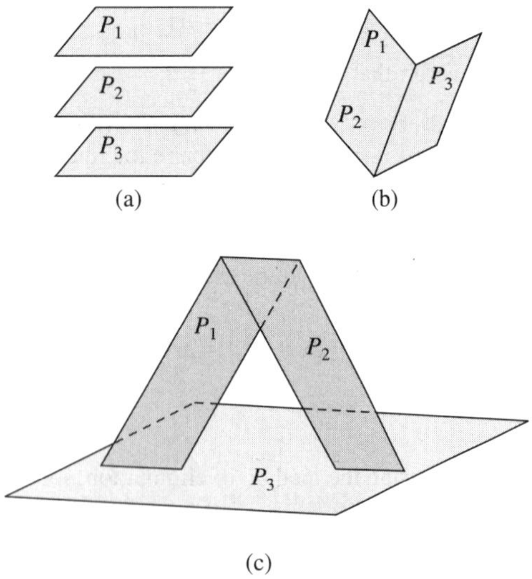

FIGURE 1.3

30. An oil refinery produces low-sulfur and high-sulfur fuel. Each ton of low-sulfur fuel requires 5 minutes in the blending plant and 4 minutes in the refining plant; each ton of high-sulfur fuel requires 4 minutes in the blending plant and 2 minutes in the refining plant. If the blending plant is available for 3 hours and the refining plant is available for 2 hours, how many tons of each type of fuel should be manufactured so that the plants are fully used?

31. A plastics manufacturer makes two types of plastic: regular and special. Each ton of regular plastic requires 2 hours in plant A and 5 hours in plant B; each ton of special plastic requires 2 hours in plant A and 3 hours in plant B. If plant A is available 8 hours per day and plant B is available 15 hours per day, how many tons of each type of plastic can be made daily so that the plants are fully used?

32. A dietician is preparing a meal consisting of foods A, B, and C. Each ounce of food A contains 2 units of protein, 3 units of fat, and 4 units of carbohydrate. Each ounce of food B contains 3 units of protein, 2 units of fat, and 1 unit of carbohydrate. Each ounce of food C contains 3 units of protein, 3 units of fat, and 2 units of carbohydrate. If the meal must provide exactly 25 units of protein, 24 units of fat, and 21 units of carbohydrate, how many ounces of each type of food should be used?

33. A manufacturer makes 2-minute, 6-minute, and 9-minute film developers. Each ton of 2-minute developer requires 6 minutes in plant A and 24 minutes in plant B. Each ton of 6-minute developer requires 12 minutes in plant A and 12 minutes in plant B. Each ton of 9-minute developer requires 12 minutes in plant A and 12 minutes in plant B. If plant A is available 10 hours per day and plant B is available 16 hours per day, how many tons of each type of developer can be produced so that the plants are fully used?

34. Suppose that the three points $(1, -5)$, $(-1, 1)$, and $(2, 7)$ lie on the parabola $p(x) = ax^2 + bx + c$.

(a) Determine a linear system of three equations in three unknowns that must be solved to find a, b, and c.

(b) Solve the linear system obtained in part (a) for a, b, and c.

35. An inheritance of \$24,000 is to be divided among three trusts, with the second trust receiving twice as much as the first trust. The three trusts pay interest annually at the rates of 9%, 10%, and 6%, respectively, and return a total in interest of \$2210 at the end of the first year. How much was invested in each trust?

36. For the software you are using, determine the command that "automatically" solves a linear system of equations.

37. Use the command from Exercise 36 to solve Exercises 3 and 4, and compare the output with the results you obtained by the method of elimination.

38. Solve the linear system

$$\begin{aligned} x + \tfrac{1}{2}y + \tfrac{1}{3}z &= 1 \\ \tfrac{1}{2}x + \tfrac{1}{3}y + \tfrac{1}{4}z &= \tfrac{11}{18} \\ \tfrac{1}{3}x + \tfrac{1}{4}y + \tfrac{1}{5}z &= \tfrac{9}{20} \end{aligned}$$

by using your software. Compare the computed solution with the exact solution $x = \frac{1}{2}$, $y = \frac{1}{3}$, $z = 1$.

39. If your software includes access to a computer algebra system (CAS), use it as follows:

(a) For the linear system in Exercise 38, replace the fraction $\frac{1}{2}$ with its decimal equivalent 0.5. Enter this system into your software and use the appropriate CAS commands to solve the system. Compare the solution with that obtained in Exercise 38.

(b) In some CAS environments you can select the number of digits to be used in the calculations. Perform part (a) with digit choices 2, 4, and 6 to see what influence such selections have on the computed solution.

40. If your software includes access to a CAS and you can select the number of digits used in calculations, do the following: Enter the linear system

$$\begin{aligned} 0.71x + 0.21y &= 0.92 \\ 0.23x + 0.58y &= 0.81 \end{aligned}$$

into the program. Have the software solve the system with digit choices 2, 5, 7, and 12. Briefly discuss any variations in the solutions generated.

1.2 Matrices

If we examine the method of elimination described in Section 1.1, we can make the following observation: Only the numbers in front of the unknowns $x_1, x_2, \ldots, x_n$ and the numbers $b_1, b_2, \ldots, b_m$ on the right side are being changed as we perform the steps in the method of elimination. Thus we might think of looking for a way of writing a linear system without having to carry along the unknowns. Matrices enable us to do this—that is, to write linear systems in a compact form that makes it easier to automate the elimination method by using computer software in order to obtain a fast and efficient procedure for finding solutions. The use of matrices, however, is not merely that of a convenient notation. We now develop operations on matrices and will work with matrices according to the rules they obey; this will enable us to solve systems of linear equations and to handle other computational problems in a fast and efficient manner. Of course, as any good definition should do, the notion of a matrix not only provides a new way of looking at old problems, but also gives rise to a great many new questions, some of which we study in this book.

DEFINITION 1.1

An $m \times n$ **matrix** A is a rectangular array of mn real or complex numbers arranged in m horizontal **rows** and n vertical **columns**:

$$A = \begin{bmatrix} a_{11} & a_{12} & \cdots & a_{1j} & \cdots & a_{1n} \\ a_{21} & a_{22} & \cdots & a_{2j} & \cdots & a_{2n} \\ \vdots & \vdots & & \vdots & & \vdots \\ a_{i1} & a_{i2} & \cdots & a_{ij} & \cdots & a_{in} \\ \vdots & \vdots & & \vdots & & \vdots \\ a_{m1} & a_{m2} & \cdots & a_{mj} & \cdots & a_{mn} \end{bmatrix}. \quad \leftarrow i\text{th row} \qquad (1)$$

$\uparrow$ jth column

The ***i*th row** of A is

$$\begin{bmatrix} a_{i1} & a_{i2} & \cdots & a_{in} \end{bmatrix} \qquad (1 \le i \le m);$$

the ***j*th column** of A is

$$\begin{bmatrix} a_{1j} \\ a_{2j} \\ \vdots \\ a_{mj} \end{bmatrix} \qquad (1 \le j \le n).$$

We shall say that A is ***m* by *n*** (written as $\boldsymbol{m \times n}$). If $m = n$, we say that A is a **square matrix of order *n***, and that the numbers $a_{11}, a_{22}, \ldots, a_{nn}$ form the **main diagonal** of A. We refer to the number a_{ij}, which is in the ith row and jth column of A, as the ***i*, *j*th element** of A, or the **(*i*, *j*) entry** of A, and we often write (1) as

$$A = \begin{bmatrix} a_{ij} \end{bmatrix}.$$

EXAMPLE 1 Let

$$A=\begin{bmatrix}1 & 2 & 3\\ -1 & 0 & 1\end{bmatrix},\quad B=\begin{bmatrix}1+i & 4i\\ 2-3i & -3\end{bmatrix},\quad C=\begin{bmatrix}1\\ -1\\ 2\end{bmatrix},$$

$$D=\begin{bmatrix}1 & 1 & 0\\ 2 & 0 & 1\\ 3 & -1 & 2\end{bmatrix},\quad E=\begin{bmatrix}3\end{bmatrix},\quad F=\begin{bmatrix}-1 & 0 & 2\end{bmatrix}.$$

Then A is a 2×3 matrix with $a_{12}=2$, $a_{13}=3$, $a_{22}=0$, and $a_{23}=1$; B is a 2×2 matrix with $b_{11}=1+i$, $b_{12}=4i$, $b_{21}=2-3i$, and $b_{22}=-3$; C is a 3×1 matrix with $c_{11}=1$, $c_{21}=-1$, and $c_{31}=2$; D is a 3×3 matrix; E is a 1×1 matrix; and F is a 1×3 matrix. In D, the elements $d_{11}=1$, $d_{22}=0$, and $d_{33}=2$ form the main diagonal. ■

For convenience, we focus much of our attention in the illustrative examples and exercises in Chapters 1–6 on matrices and expressions containing only real numbers. Complex numbers make a brief appearance in Chapter 7. An introduction to complex numbers, their properties, and examples and exercises showing how complex numbers are used in linear algebra may be found in Appendix B.

An $n\times 1$ matrix is also called an ***n*-vector** and is denoted by lowercase boldface letters. When n is understood, we refer to n-vectors merely as **vectors**. Vectors are discussed at length in Section 4.1.

EXAMPLE 2

$$\mathbf{u}=\begin{bmatrix}1\\ 2\\ -1\\ 0\end{bmatrix}\text{ is a 4-vector and }\mathbf{v}=\begin{bmatrix}1\\ -1\\ 3\end{bmatrix}\text{ is a 3-vector.}$$ ■

The n-vector all of whose entries are zero is denoted by $\mathbf{0}$.

Observe that if A is an $n\times n$ matrix, then the rows of A are $1\times n$ matrices and the columns of A are $n\times 1$ matrices. The set of all n-vectors with real entries is denoted by R^n. Similarly, the set of all n-vectors with complex entries is denoted by C^n. As we have already pointed out, in the first six chapters of this book we work almost entirely with vectors in R^n.

EXAMPLE 3 **(Tabular Display of Data)** The following matrix gives the airline distances between the indicated cities (in statute miles):

	London	Madrid	New York	Tokyo
London	0	785	3469	5959
Madrid	785	0	3593	6706
New York	3469	3593	0	6757
Tokyo	5959	6706	6757	0

■

EXAMPLE 4 **(Production)** Suppose that a manufacturer has four plants, each of which makes three products. If we let a_{ij} denote the number of units of product i made by plant

j in one week, then the 3×4 matrix

$$\begin{array}{c} \\ \text{Product 1} \\ \text{Product 2} \\ \text{Product 3} \end{array} \begin{array}{c} \begin{array}{cccc} \text{Plant 1} & \text{Plant 2} & \text{Plant 3} & \text{Plant 4} \end{array} \\ \begin{bmatrix} 560 & 360 & 380 & 0 \\ 340 & 450 & 420 & 80 \\ 280 & 270 & 210 & 380 \end{bmatrix} \end{array}$$

gives the manufacturer's production for the week. For example, plant 2 makes 270 units of product 3 in one week. ■

EXAMPLE 5

The windchill table that follows is a matrix.

	°F					
	15	10	5	0	−5	−10
mph						
5	12	7	0	−5	−10	−15
10	−3	−9	−15	−22	−27	−34
15	−11	−18	−25	−31	−38	−45
20	−17	−24	−31	−39	−46	−53

A combination of air temperature and wind speed makes a body feel colder than the actual temperature. For example, when the temperature is 10°F and the wind is 15 miles per hour, this causes a body heat loss equal to that when the temperature is −18°F with no wind. ■

EXAMPLE 6

By a **graph** we mean a set of points called **nodes** or **vertices**, some of which are connected by lines called **edges**. The nodes are usually labeled as $P_1, P_2, \ldots, P_k$, and for now we allow an edge to be traveled in either direction. One mathematical representation of a graph is constructed from a table. For example, the following table represents the graph shown:

	P_1	P_2	P_3	P_4
P_1	0	1	0	0
P_2	1	0	1	1
P_3	0	1	0	1
P_4	0	1	1	0

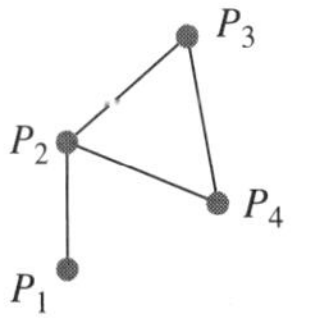

The (i, j) entry $= 1$ if there is an edge connecting vertex P_i to vertex P_j; otherwise, the (i, j) entry $= 0$. The **incidence matrix** $\boldsymbol{A}$ is the $k \times k$ matrix obtained by omitting the row and column labels from the preceding table. The incidence matrix for the corresponding graph is

$$A = \begin{bmatrix} 0 & 1 & 0 & 0 \\ 1 & 0 & 1 & 1 \\ 0 & 1 & 0 & 1 \\ 0 & 1 & 1 & 0 \end{bmatrix}.$$

■

Internet search engines use matrices to keep track of the locations of information, the type of information at a location, keywords that appear in the information, and even the way websites link to one another. A large measure of the effectiveness of the search engine Google© is the manner in which matrices are used to determine which sites are referenced by other sites. That is, instead of directly keeping track of the information content of an actual web page or of an individual search topic, Google's matrix structure focuses on finding web pages that match the search topic, and then presents a list of such pages in the order of their "importance."

Suppose that there are n accessible web pages during a certain month. A simple way to view a matrix that is part of Google's scheme is to imagine an $n \times n$ matrix A, called the "connectivity matrix," that initially contains all zeros. To build the connections, proceed as follows. When you detect that website j links to website i, set entry a_{ij} equal to one. Since n is quite large, in the billions, most entries of the connectivity matrix A are zero. (Such a matrix is called sparse.) If row i of A contains many ones, then there are many sites linking to site i. Sites that are linked to by many other sites are considered more "important" (or to have a higher rank) by the software driving the Google search engine. Such sites would appear near the top of a list returned by a Google search on topics related to the information on site i. Since Google updates its connectivity matrix about every month, n increases over time and new links and sites are adjoined to the connectivity matrix.

In Chapter 8 we elaborate a bit on the fundamental technique used for ranking sites and give several examples related to the matrix concepts involved. Further information can be found in the following sources:

1. Berry, Michael W., and Murray Browne. *Understanding Search Engines—Mathematical Modeling and Text Retrieval*, 2d ed. Philadelphia: Siam, 2005.
2. Moler, Cleve. "The World's Largest Matrix Computation: Google's PageRank Is an Eigenvector of a Matrix of Order 2.7 Billion," MATLAB *News and Notes*, October 2002, pp. 12–13.

Whenever a new object is introduced in mathematics, we must determine when two such objects are equal. For example, in the set of all rational numbers, the numbers $\frac{2}{3}$ and $\frac{4}{6}$ are called equal, although they have different representations. What we have in mind is the definition that a/b equals c/d when $ad = bc$. Accordingly, we now have the following definition:

DEFINITION 1.2 Two $m \times n$ matrices $A = \begin{bmatrix} a_{ij} \end{bmatrix}$ and $B = \begin{bmatrix} b_{ij} \end{bmatrix}$ are **equal** if they agree entry by entry, that is, if $a_{ij} = b_{ij}$ for $i = 1, 2, \ldots, m$ and $j = 1, 2, \ldots, n$.

EXAMPLE 7 The matrices

$$A = \begin{bmatrix} 1 & 2 & -1 \\ 2 & -3 & 4 \\ 0 & -4 & 5 \end{bmatrix} \quad \text{and} \quad B = \begin{bmatrix} 1 & 2 & w \\ 2 & x & 4 \\ y & -4 & z \end{bmatrix}$$

are equal if and only if $w = -1$, $x = -3$, $y = 0$, and $z = 5$. ■

■ Matrix Operations

We next define a number of operations that will produce new matrices out of given matrices. When we are dealing with linear systems, for example, this will enable us to manipulate the matrices that arise and to avoid writing down systems over and over again. These operations and manipulations are also useful in other applications of matrices.

Matrix Addition

DEFINITION 1.3

If $A = \left[a_{ij}\right]$ and $B = \left[b_{ij}\right]$ are both $m \times n$ matrices, then the **sum** $A+B$ is an $m \times n$ matrix $C = \left[c_{ij}\right]$ defined by $c_{ij} = a_{ij} + b_{ij}$, $i = 1, 2, \ldots, m$; $j = 1, 2, \ldots, n$. Thus, to obtain the sum of A and B, we merely add corresponding entries.

EXAMPLE 8

Let

$$A = \begin{bmatrix} 1 & -2 & 3 \\ 2 & -1 & 4 \end{bmatrix} \quad \text{and} \quad B = \begin{bmatrix} 0 & 2 & 1 \\ 1 & 3 & -4 \end{bmatrix}.$$

Then

$$A + B = \begin{bmatrix} 1+0 & -2+2 & 3+1 \\ 2+1 & -1+3 & 4+(-4) \end{bmatrix} = \begin{bmatrix} 1 & 0 & 4 \\ 3 & 2 & 0 \end{bmatrix}.$$

■

EXAMPLE 9

(Production) A manufacturer of a certain product makes three models, A, B, and C. Each model is partially made in factory F_1 in Taiwan and then finished in factory F_2 in the United States. The total cost of each product consists of the manufacturing cost and the shipping cost. Then the costs at each factory (in dollars) can be described by the 3×2 matrices F_1 and F_2:

$$F_1 = \begin{array}{c} \begin{array}{cc} \text{Manufacturing cost} & \text{Shipping cost} \end{array} \\ \left[\begin{array}{cc} 32 & 40 \\ 50 & 80 \\ 70 & 20 \end{array}\right] \begin{array}{l} \text{Model A} \\ \text{Model B} \\ \text{Model C} \end{array} \end{array}$$

$$F_2 = \begin{array}{c} \begin{array}{cc} \text{Manufacturing cost} & \text{Shipping cost} \end{array} \\ \left[\begin{array}{cc} 40 & 60 \\ 50 & 50 \\ 130 & 20 \end{array}\right] \begin{array}{l} \text{Model A} \\ \text{Model B} \\ \text{Model C} \end{array} \end{array}.$$

The matrix $F_1 + F_2$ gives the total manufacturing and shipping costs for each product. Thus the total manufacturing and shipping costs of a model C product are \$200 and \$40, respectively. ■

If $\mathbf{x}$ is an n-vector, then it is easy to show that $\mathbf{x} + \mathbf{0} = \mathbf{x}$, where $\mathbf{0}$ is the n-vector all of whose entries are zero. (See Exercise 16.)

It should be noted that the sum of the matrices A and B is defined only when A and B have the same number of rows and the same number of columns, that is, only when A and B are of the same size.

We now make the convention that when $A + B$ is written, both A and B are of the same size.

The basic properties of matrix addition are considered in the next section and are similar to those satisfied by the real numbers.

Scalar Multiplication

DEFINITION 1.4 If $A = \begin{bmatrix} a_{ij} \end{bmatrix}$ is an $m \times n$ matrix and r is a real number, then the **scalar multiple** of A by r, rA, is the $m \times n$ matrix $C = \begin{bmatrix} c_{ij} \end{bmatrix}$, where $c_{ij} = ra_{ij}$, $i = 1, 2, \ldots, m$ and $j = 1, 2, \ldots, n$; that is, the matrix C is obtained by multiplying each entry of A by r.

EXAMPLE 10 We have

$$-2\begin{bmatrix} 4 & -2 & -3 \\ 7 & -3 & 2 \end{bmatrix} = \begin{bmatrix} (-2)(4) & (-2)(-2) & (-2)(-3) \\ (-2)(7) & (-2)(-3) & (-2)(2) \end{bmatrix}$$

$$= \begin{bmatrix} -8 & 4 & 6 \\ -14 & 6 & -4 \end{bmatrix}. \quad \blacksquare$$

Thus far, addition of matrices has been defined for only two matrices. Our work with matrices will call for adding more than two matrices. Theorem 1.1 in Section 1.4 shows that addition of matrices satisfies the associative property: $A + (B + C) = (A + B) + C$.

If A and B are $m \times n$ matrices, we write $A + (-1)B$ as $A - B$ and call this the **difference between A and B**.

EXAMPLE 11 Let

$$A = \begin{bmatrix} 2 & 3 & -5 \\ 4 & 2 & 1 \end{bmatrix} \quad \text{and} \quad B = \begin{bmatrix} 2 & -1 & 3 \\ 3 & 5 & -2 \end{bmatrix}.$$

Then

$$A - B = \begin{bmatrix} 2-2 & 3+1 & -5-3 \\ 4-3 & 2-5 & 1+2 \end{bmatrix} = \begin{bmatrix} 0 & 4 & -8 \\ 1 & -3 & 3 \end{bmatrix}. \quad \blacksquare$$

Application

Vectors in R^n can be used to handle large amounts of data. Indeed, a number of computer software products, notably, MATLAB®, make extensive use of vectors. The following example illustrates these ideas:

EXAMPLE 12 **(Inventory Control)** Suppose that a store handles 100 different items. The inventory on hand at the beginning of the week can be described by the inventory vector $\mathbf{u}$ in R^{100}. The number of items sold at the end of the week can be described by the 100-vector $\mathbf{v}$, and the vector

$$\mathbf{u} - \mathbf{v}$$

represents the inventory at the end of the week. If the store receives a new shipment of goods, represented by the 100-vector $\mathbf{w}$, then its new inventory would be

$$\mathbf{u} - \mathbf{v} + \mathbf{w}. \quad \blacksquare$$

We shall sometimes use the **summation notation**, and we now review this useful and compact notation.

By $\sum_{i=1}^{n} a_i$ we mean $a_1 + a_2 + \cdots + a_n$. The letter i is called the **index of summation**; it is a dummy variable that can be replaced by another letter. Hence we can write

$$\sum_{i=1}^{n} a_i = \sum_{j=1}^{n} a_j = \sum_{k=1}^{n} a_k.$$

Thus

$$\sum_{i=1}^{4} a_i = a_1 + a_2 + a_3 + a_4.$$

The summation notation satisfies the following properties:

1. $\sum_{i=1}^{n} (r_i + s_i)a_i = \sum_{i=1}^{n} r_i a_i + \sum_{i=1}^{n} s_i a_i$

2. $\sum_{i=1}^{n} c(r_i a_i) = c \sum_{i=1}^{n} r_i a_i$

3. $\sum_{j=1}^{n} \left(\sum_{i=1}^{m} a_{ij} \right) = \sum_{i=1}^{m} \left(\sum_{j=1}^{n} a_{ij} \right)$

Property 3 can be interpreted as follows: The left side is obtained by adding all the entries in each column and then adding all the resulting numbers. The right side is obtained by adding all the entries in each row and then adding all the resulting numbers.

If $A_1, A_2, \ldots, A_k$ are $m \times n$ matrices and $c_1, c_2, \ldots, c_k$ are real numbers, then an expression of the form

$$c_1 A_1 + c_2 A_2 + \cdots + c_k A_k \tag{2}$$

is called a **linear combination** of $A_1, A_2, \ldots, A_k$, and $c_1, c_2, \ldots, c_k$ are called **coefficients**.

The linear combination in Equation (2) can also be expressed in summation notation as

$$\sum_{i=1}^{k} c_i A_i = c_1 A_1 + c_2 A_2 + \cdots + c_k A_k.$$

EXAMPLE 13

The following are linear combinations of matrices:

$$3 \begin{bmatrix} 0 & -3 & 5 \\ 2 & 3 & 4 \\ 1 & -2 & -3 \end{bmatrix} - \frac{1}{2} \begin{bmatrix} 5 & 2 & 3 \\ 6 & 2 & 3 \\ -1 & -2 & 3 \end{bmatrix},$$

$$2 \begin{bmatrix} 3 & -2 \end{bmatrix} - 3 \begin{bmatrix} 5 & 0 \end{bmatrix} + 4 \begin{bmatrix} -2 & 5 \end{bmatrix},$$

$$-0.5 \begin{bmatrix} 1 \\ -4 \\ -6 \end{bmatrix} + 0.4 \begin{bmatrix} 0.1 \\ -4 \\ 0.2 \end{bmatrix}.$$

Using scalar multiplication and matrix addition, we can compute each of these linear combinations. Verify that the results of such computations are, respectively,

$$\begin{bmatrix} -\frac{5}{2} & -10 & \frac{27}{2} \\ 3 & 8 & \frac{21}{2} \\ \frac{7}{2} & -5 & -\frac{21}{2} \end{bmatrix}, \quad \begin{bmatrix} -17 & 16 \end{bmatrix}, \quad \text{and} \quad \begin{bmatrix} -0.46 \\ 0.4 \\ 3.08 \end{bmatrix}.$$

■

EXAMPLE 14

Let

$$\mathbf{p} = \begin{bmatrix} 18.95 \\ 14.75 \\ 8.60 \end{bmatrix}$$

be a 3-vector that represents the current prices of three items at a store. Suppose that the store announces a sale so that the price of each item is reduced by 20%.

(a) Determine a 3-vector that gives the price changes for the three items.
(b) Determine a 3-vector that gives the new prices of the items.

Solution

(a) Since each item is reduced by 20%, the 3-vector

$$-0.20\mathbf{p} = \begin{bmatrix} (-0.20)18.95 \\ (-0.20)14.75 \\ (-0.20)8.60 \end{bmatrix} = \begin{bmatrix} -3.79 \\ -2.95 \\ -1.72 \end{bmatrix} = -\begin{bmatrix} 3.79 \\ 2.95 \\ 1.72 \end{bmatrix}$$

gives the price changes for the three items.
(b) The new prices of the items are given by the expression

$$\mathbf{p} - 0.20\mathbf{p} = \begin{bmatrix} 18.95 \\ 14.75 \\ 8.60 \end{bmatrix} - \begin{bmatrix} 3.79 \\ 2.95 \\ 1.72 \end{bmatrix} = \begin{bmatrix} 15.16 \\ 11.80 \\ 6.88 \end{bmatrix}.$$

Observe that this expression can also be written as

$$\mathbf{p} - 0.20\mathbf{p} = 0.80\mathbf{p}.$$

■

The next operation on matrices is useful in a number of situations.

DEFINITION 1.5

If $A = \begin{bmatrix} a_{ij} \end{bmatrix}$ is an $m \times n$ matrix, then the **transpose** of A, $A^T = \begin{bmatrix} a_{ij}^T \end{bmatrix}$, is the $n \times m$ matrix defined by $a_{ij}^T = a_{ji}$. Thus the transpose of A is obtained from A by interchanging the rows and columns of A.

EXAMPLE 15

Let

$$A = \begin{bmatrix} 4 & -2 & 3 \\ 0 & 5 & -2 \end{bmatrix}, \quad B = \begin{bmatrix} 6 & 2 & -4 \\ 3 & -1 & 2 \\ 0 & 4 & 3 \end{bmatrix}, \quad C = \begin{bmatrix} 5 & 4 \\ -3 & 2 \\ 2 & -3 \end{bmatrix},$$

$$D = \begin{bmatrix} 3 & -5 & 1 \end{bmatrix}, \quad E = \begin{bmatrix} 2 \\ -1 \\ 3 \end{bmatrix}.$$

Then

$$A^T = \begin{bmatrix} 4 & 0 \\ -2 & 5 \\ 3 & -2 \end{bmatrix}, \quad B^T = \begin{bmatrix} 6 & 3 & 0 \\ 2 & -1 & 4 \\ -4 & 2 & 3 \end{bmatrix},$$

$$C^T = \begin{bmatrix} 5 & -3 & 2 \\ 4 & 2 & -3 \end{bmatrix}, \quad D^T = \begin{bmatrix} 3 \\ -5 \\ 1 \end{bmatrix}, \quad \text{and} \quad E^T = \begin{bmatrix} 2 & -1 & 3 \end{bmatrix}.$$ ■

Key Terms

Matrix
Rows
Columns
Size of a matrix
Square matrix
Main diagonal
Element or entry of a matrix
Equal matrices
n-vector (or vector)
R^n, C^n
0, zero vector
Google
Matrix addition
Scalar multiple
Difference of matrices
Summation notation
Index of summation
Linear combination
Coefficients
Transpose

1.2 Exercises

1. Let

$$A = \begin{bmatrix} 2 & -3 & 5 \\ 6 & -5 & 4 \end{bmatrix}, \quad B = \begin{bmatrix} 4 \\ -3 \\ 5 \end{bmatrix},$$

and

$$C = \begin{bmatrix} 7 & 3 & 2 \\ -4 & 3 & 5 \\ 6 & 1 & -1 \end{bmatrix}.$$

(a) What is a_{12}, a_{22}, a_{23}?

(b) What is b_{11}, b_{31}?

(c) What is c_{13}, c_{31}, c_{33}?

2. Determine the incidence matrix associated with each of the following graphs:

(a)

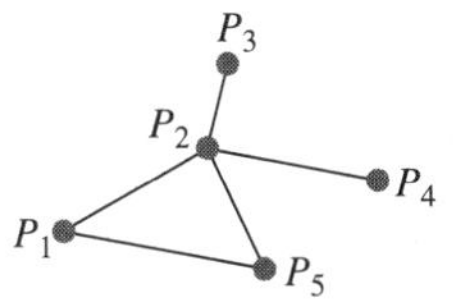

(b)

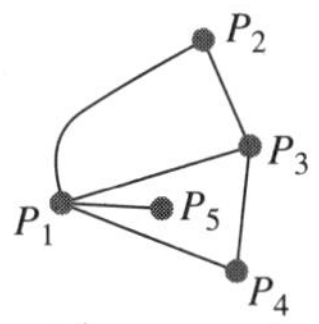

3. For each of the following incidence matrices, construct a graph. Label the vertices $P_1, P_2, \ldots, P_5$.

(a) $A = \begin{bmatrix} 0 & 1 & 0 & 1 & 1 \\ 1 & 0 & 0 & 0 & 1 \\ 0 & 0 & 0 & 1 & 1 \\ 1 & 0 & 1 & 0 & 1 \\ 1 & 1 & 1 & 1 & 0 \end{bmatrix}$

(b) $A = \begin{bmatrix} 0 & 1 & 0 & 0 & 0 \\ 1 & 0 & 1 & 1 & 1 \\ 0 & 1 & 0 & 0 & 0 \\ 0 & 1 & 0 & 0 & 0 \\ 0 & 1 & 0 & 0 & 0 \end{bmatrix}$

4. If

$$\begin{bmatrix} a+b & c+d \\ c-d & a-b \end{bmatrix} = \begin{bmatrix} 4 & 6 \\ 10 & 2 \end{bmatrix},$$

find a, b, c, and d.

5. If

$$\begin{bmatrix} a+2b & 2a-b \\ 2c+d & c-2d \end{bmatrix} = \begin{bmatrix} 4 & -2 \\ 4 & -3 \end{bmatrix},$$

find a, b, c, and d.

In Exercises 6 through 9, let

$$A = \begin{bmatrix} 1 & 2 & 3 \\ 2 & 1 & 4 \end{bmatrix}, \quad B = \begin{bmatrix} 1 & 0 \\ 2 & 1 \\ 3 & 2 \end{bmatrix},$$

$$C = \begin{bmatrix} 3 & -1 & 3 \\ 4 & 1 & 5 \\ 2 & 1 & 3 \end{bmatrix}, \quad D = \begin{bmatrix} 3 & -2 \\ 2 & 4 \end{bmatrix},$$

$$E = \begin{bmatrix} 2 & -4 & 5 \\ 0 & 1 & 4 \\ 3 & 2 & 1 \end{bmatrix}, \quad F = \begin{bmatrix} -4 & 5 \\ 2 & 3 \end{bmatrix},$$

$$\text{and} \quad O = \begin{bmatrix} 0 & 0 & 0 \\ 0 & 0 & 0 \\ 0 & 0 & 0 \end{bmatrix}.$$

6. If possible, compute the indicated linear combination:

(a) $C + E$ and $E + C$ **(b)** $A + B$

(c) $D - F$ **(d)** $-3C + 5O$

(e) $2C - 3E$ **(f)** $2B + F$

7. If possible, compute the indicated linear combination:

(a) $3D + 2F$ **(b)** $3(2A)$ and $6A$

(c) $3A + 2A$ and $5A$

(d) $2(D + F)$ and $2D + 2F$

(e) $(2 + 3)D$ and $2D + 3D$

(f) $3(B + D)$

8. If possible, compute the following:

(a) A^T and $(A^T)^T$

(b) $(C + E)^T$ and $C^T + E^T$

(c) $(2D + 3F)^T$ **(d)** $D - D^T$

(e) $2A^T + B$ **(f)** $(3D - 2F)^T$

9. If possible, compute the following:

(a) $(2A)^T$ **(b)** $(A - B)^T$

(c) $(3B^T - 2A)^T$ **(d)** $(3A^T - 5B^T)^T$

(e) $(-A)^T$ and $-(A^T)$ **(f)** $(C + E + F^T)^T$

10. Is the matrix $\begin{bmatrix} 3 & 0 \\ 0 & 2 \end{bmatrix}$ a linear combination of the matrices $\begin{bmatrix} 1 & 0 \\ 0 & 1 \end{bmatrix}$ and $\begin{bmatrix} 1 & 0 \\ 0 & 0 \end{bmatrix}$? Justify your answer.

11. Is the matrix $\begin{bmatrix} 4 & 1 \\ 0 & -3 \end{bmatrix}$ a linear combination of the matrices $\begin{bmatrix} 1 & 0 \\ 0 & 1 \end{bmatrix}$ and $\begin{bmatrix} 1 & 0 \\ 0 & 0 \end{bmatrix}$? Justify your answer.

12. Let

$$A = \begin{bmatrix} 1 & 2 & 3 \\ 6 & -2 & 3 \\ 5 & 2 & 4 \end{bmatrix} \quad \text{and} \quad I_3 = \begin{bmatrix} 1 & 0 & 0 \\ 0 & 1 & 0 \\ 0 & 0 & 1 \end{bmatrix}.$$

If λ is a real number, compute $\lambda I_3 - A$.

13. If A is an $n \times n$ matrix, what are the entries on the main diagonal of $A - A^T$? Justify your answer.

14. Explain why every incidence matrix A associated with a graph is the same as A^T.

15. Let the $n \times n$ matrix A be equal to A^T. Briefly describe the pattern of the entries in A.

16. If $\mathbf{x}$ is an n-vector, show that $\mathbf{x} + \mathbf{0} = \mathbf{x}$.

17. Show that the summation notation satisfies the following properties:

(a) $\displaystyle\sum_{i=1}^{n} (r_i + s_i)a_i = \sum_{i=1}^{n} r_i a_i + \sum_{i=1}^{n} s_i a_i$

(b) $\displaystyle\sum_{i=1}^{n} c(r_i a_i) = c\left(\sum_{i=1}^{n} r_i a_i\right)$

18. Show that $\displaystyle\sum_{i=1}^{n}\left(\sum_{j=1}^{m} a_{ij}\right) = \sum_{j=1}^{m}\left(\sum_{i=1}^{n} a_{ij}\right)$.

19. Identify the following expressions as true or false. If true, prove the result; if false, give a counterexample.

(a) $\displaystyle\sum_{i=1}^{n} (a_i + 1) = \left(\sum_{i=1}^{n} a_i\right) + n$

(b) $\displaystyle\sum_{i=1}^{n}\left(\sum_{j=1}^{m} 1\right) = mn$

(c) $\displaystyle\sum_{j=1}^{m}\left(\sum_{i=1}^{n} a_i b_j\right) = \left[\sum_{i=1}^{n} a_i\right]\left[\sum_{j=1}^{m} b_j\right]$

20. A large steel manufacturer, who has 2000 employees, lists each employee's salary as a component of a vector $\mathbf{u}$ in R^{2000}. If an 8% across-the-board salary increase has been approved, find an expression involving $\mathbf{u}$ that gives all the new salaries.

21. A brokerage firm records the high and low values of the price of IBM stock each day. The information for a given week is presented in two vectors, $\mathbf{t}$ and $\mathbf{b}$, in R^5, showing the high and low values, respectively. What expression gives the average daily values of the price of IBM stock for the entire 5-day week?

22. For the software you are using, determine the commands to enter a matrix, add matrices, multiply a scalar times a matrix, and obtain the transpose of a matrix for matrices with numerical entries. Practice the commands, using the linear combinations in Example 13.

23. Determine whether the software you are using includes a computer algebra system (CAS), and if it does, do the following:

(a) Find the command for entering a symbolic matrix. (This command may be different than that for entering a numeric matrix.)

(b) Enter several symbolic matrices like

$$A = \begin{bmatrix} r & s & t \\ u & v & w \end{bmatrix} \quad \text{and} \quad B = \begin{bmatrix} a & b & c \\ d & e & f \end{bmatrix}.$$

Compute expressions like $A + B$, $2A$, $3A + B$, $A - 2B$, $A^T + B^T$, etc. (In some systems you must explicitly indicate scalar multiplication with an asterisk.)

24. For the software you are using, determine whether there is a command that will display a graph for an incidence matrix. If there is, display the graphs for the incidence matrices in Exercise 3 and compare them with those that you drew by hand. Explain why the computer-generated graphs need not be identical to those you drew by hand.

1.3 Matrix Multiplication

In this section we introduce the operation of matrix multiplication. Unlike matrix addition, matrix multiplication has some properties that distinguish it from multiplication of real numbers.

DEFINITION 1.6

The **dot product**, or **inner product**, of the n-vectors in R^n

$$\mathbf{a} = \begin{bmatrix} a_1 \\ a_2 \\ \vdots \\ a_n \end{bmatrix} \quad \text{and} \quad \mathbf{b} = \begin{bmatrix} b_1 \\ b_2 \\ \vdots \\ b_n \end{bmatrix}$$

is defined as

$$\mathbf{a} \cdot \mathbf{b} = a_1 b_1 + a_2 b_2 + \cdots + a_n b_n = \sum_{i=1}^{n} a_i b_i .^*$$

The dot product is an important operation that will be used here and in later sections.

EXAMPLE 1

The dot product of

$$\mathbf{u} = \begin{bmatrix} 1 \\ -2 \\ 3 \\ 4 \end{bmatrix} \quad \text{and} \quad \mathbf{v} = \begin{bmatrix} 2 \\ 3 \\ -2 \\ 1 \end{bmatrix}$$

is

$$\mathbf{u} \cdot \mathbf{v} = (1)(2) + (-2)(3) + (3)(-2) + (4)(1) = -6. \quad \blacksquare$$

EXAMPLE 2

Let $\mathbf{a} = \begin{bmatrix} x \\ 2 \\ 3 \end{bmatrix}$ and $\mathbf{b} = \begin{bmatrix} 4 \\ 1 \\ 2 \end{bmatrix}$. If $\mathbf{a} \cdot \mathbf{b} = -4$, find x.

Solution

We have

$$\begin{aligned} \mathbf{a} \cdot \mathbf{b} = 4x + 2 + 6 &= -4 \\ 4x + 8 &= -4 \\ x &= -3. \end{aligned} \quad \blacksquare$$

*The dot product of vectors in C^n is defined in Appendix B.2.

EXAMPLE 3

(Computing a Course Average) Suppose that an instructor uses four grades to determine a student's course average: quizzes, two hourly exams, and a final exam. These are weighted as 10%, 30%, 30%, and 30%, respectively. If a student's scores are 78, 84, 62, and 85, respectively, we can compute the course average by letting

$$\mathbf{w} = \begin{bmatrix} 0.10 \\ 0.30 \\ 0.30 \\ 0.30 \end{bmatrix} \quad \text{and} \quad \mathbf{g} = \begin{bmatrix} 78 \\ 84 \\ 62 \\ 85 \end{bmatrix}$$

and computing

$$\mathbf{w} \cdot \mathbf{g} = (0.10)(78) + (0.30)(84) + (0.30)(62) + (0.30)(85) = 77.1.$$

Thus the student's course average is 77.1. ■

■ Matrix Multiplication

DEFINITION 1.7

If $A = \begin{bmatrix} a_{ij} \end{bmatrix}$ is an $m \times p$ matrix and $B = \begin{bmatrix} b_{ij} \end{bmatrix}$ is a $p \times n$ matrix, then the **product** of A and B, denoted AB, is the $m \times n$ matrix $C = \begin{bmatrix} c_{ij} \end{bmatrix}$, defined by

$$\begin{aligned} c_{ij} &= a_{i1}b_{1j} + a_{i2}b_{2j} + \cdots + a_{ip}b_{pj} \\ &= \sum_{k=1}^{p} a_{ik}b_{kj} \quad (1 \le i \le m,\ 1 \le j \le n). \end{aligned} \tag{1}$$

Equation (1) says that the i, jth element in the product matrix is the dot product of the transpose of the ith row, $\text{row}_i(A)$—that is, $(\text{row}_i(A))^T$—and the jth column, $\text{col}_j(B)$, of B; this is shown in Figure 1.4.

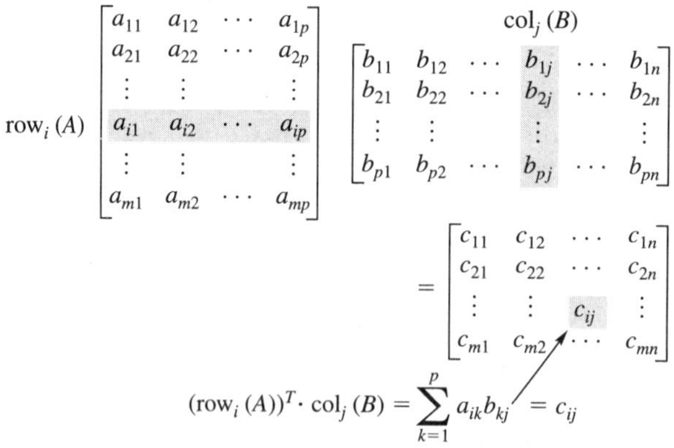

FIGURE 1.4

Observe that the product of A and B is defined only when the number of rows of B is exactly the same as the number of columns of A, as indicated in Figure 1.5.

EXAMPLE 4

Let

$$A = \begin{bmatrix} 1 & 2 & -1 \\ 3 & 1 & 4 \end{bmatrix} \quad \text{and} \quad B = \begin{bmatrix} -2 & 5 \\ 4 & -3 \\ 2 & 1 \end{bmatrix}.$$

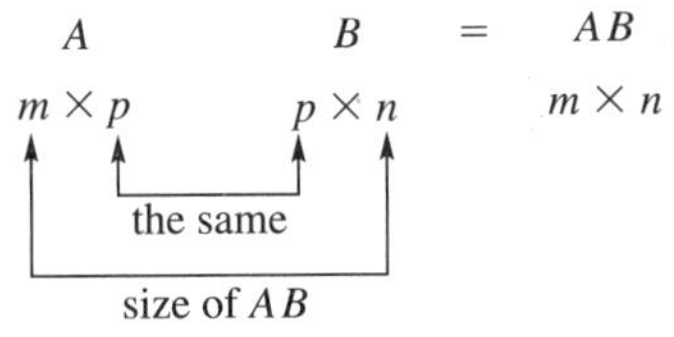

FIGURE 1.5

Then

$$AB = \begin{bmatrix} (1)(-2) + (2)(4) + (-1)(2) & (1)(5) + (2)(-3) + (-1)(1) \\ (3)(-2) + (1)(4) + (4)(2) & (3)(5) + (1)(-3) + (4)(1) \end{bmatrix}$$

$$= \begin{bmatrix} 4 & -2 \\ 6 & 16 \end{bmatrix}.$$

■

EXAMPLE 5

Let

$$A = \begin{bmatrix} 1 & -2 & 3 \\ 4 & 2 & 1 \\ 0 & 1 & -2 \end{bmatrix} \quad \text{and} \quad B = \begin{bmatrix} 1 & 4 \\ 3 & -1 \\ -2 & 2 \end{bmatrix}.$$

Compute the (3, 2) entry of AB.

Solution

If $AB = C$, then the (3, 2) entry of AB is c_{32}, which is $(\text{row}_3(A))^T \cdot \text{col}_2(B)$. We now have

$$(\text{row}_3(A))^T \cdot \text{col}_2(B) = \begin{bmatrix} 0 \\ 1 \\ -2 \end{bmatrix} \cdot \begin{bmatrix} 4 \\ -1 \\ 2 \end{bmatrix} = -5.$$

■

EXAMPLE 6

Let

$$A = \begin{bmatrix} 1 & x & 3 \\ 2 & -1 & 1 \end{bmatrix} \quad \text{and} \quad B = \begin{bmatrix} 2 \\ 4 \\ y \end{bmatrix}.$$

If $AB = \begin{bmatrix} 12 \\ 6 \end{bmatrix}$, find x and y.

Solution

We have

$$AB = \begin{bmatrix} 1 & x & 3 \\ 2 & -1 & 1 \end{bmatrix} \begin{bmatrix} 2 \\ 4 \\ y \end{bmatrix} = \begin{bmatrix} 2 + 4x + 3y \\ 4 - 4 + y \end{bmatrix} = \begin{bmatrix} 12 \\ 6 \end{bmatrix}.$$

Then

$$\begin{aligned} 2 + 4x + 3y &= 12 \\ y &= 6, \end{aligned}$$

so $x = -2$ and $y = 6$.

■

The basic properties of matrix multiplication will be considered in the next section. However, multiplication of matrices requires much more care than their addition, since the algebraic properties of matrix multiplication differ from those satisfied by the real numbers. Part of the problem is due to the fact that AB is defined only when the number of columns of A is the same as the number of rows of B. Thus, if A is an $m \times p$ matrix and B is a $p \times n$ matrix, then AB is an $m \times n$ matrix. What about BA? Four different situations may occur:

1. BA may not be defined; this will take place if $n \neq m$.
2. If BA is defined, which means that $m = n$, then BA is $p \times p$ while AB is $m \times m$; thus, if $m \neq p$, AB and BA are of different sizes.
3. If AB and BA are both of the same size, they may be equal.
4. If AB and BA are both of the same size, they may be unequal.

EXAMPLE 7

If A is a 2×3 matrix and B is a 3×4 matrix, then AB is a 2×4 matrix while BA is undefined. ■

EXAMPLE 8

Let A be 2×3 and let B be 3×2. Then AB is 2×2 while BA is 3×3. ■

EXAMPLE 9

Let

$$A = \begin{bmatrix} 1 & 2 \\ -1 & 3 \end{bmatrix} \quad \text{and} \quad B = \begin{bmatrix} 2 & 1 \\ 0 & 1 \end{bmatrix}.$$

Then

$$AB = \begin{bmatrix} 2 & 3 \\ -2 & 2 \end{bmatrix} \quad \text{while} \quad BA = \begin{bmatrix} 1 & 7 \\ -1 & 3 \end{bmatrix}.$$

Thus $AB \neq BA$. ■

One might ask why matrix equality and matrix addition are defined in such a natural way, while matrix multiplication appears to be much more complicated. Only a thorough understanding of the composition of functions and the relationship that exists between matrices and what are called linear transformations would show that the definition of multiplication given previously is the natural one. These topics are covered later in the book. For now, Example 10 provides a motivation for the definition of matrix multiplication.

EXAMPLE 10

(Ecology) Pesticides are sprayed on plants to eliminate harmful insects. However, some of the pesticide is absorbed by the plant. The pesticides are absorbed by herbivores when they eat the plants that have been sprayed. To determine the amount of pesticide absorbed by a herbivore, we proceed as follows. Suppose that we have three pesticides and four plants. Let a_{ij} denote the amount of pesticide i (in milligrams) that has been absorbed by plant j. This information can be represented by the matrix

$$A = \begin{array}{c} \begin{array}{cccc} \text{Plant 1} & \text{Plant 2} & \text{Plant 3} & \text{Plant 4} \end{array} \\ \left[\begin{array}{cccc} 2 & 3 & 4 & 3 \\ 3 & 2 & 2 & 5 \\ 4 & 1 & 6 & 4 \end{array}\right] \end{array} \begin{array}{l} \text{Pesticide 1} \\ \text{Pesticide 2} \\ \text{Pesticide 3} \end{array}.$$

Now suppose that we have three herbivores, and let b_{ij} denote the number of plants of type i that a herbivore of type j eats per month. This information can be represented by the matrix

$$B = \begin{array}{c} \begin{matrix} \text{Herbivore 1} & \text{Herbivore 2} & \text{Herbivore 3} \end{matrix} \\ \left[\begin{matrix} 20 & 12 & 8 \\ 28 & 15 & 15 \\ 30 & 12 & 10 \\ 40 & 16 & 20 \end{matrix}\right] \begin{matrix} \text{Plant 1} \\ \text{Plant 2} \\ \text{Plant 3} \\ \text{Plant 4} \end{matrix} \end{array}.$$

The (i, j) entry in AB gives the amount of pesticide of type i that animal j has absorbed. Thus, if $i = 2$ and $j = 3$, the (2, 3) entry in AB is

$$\begin{aligned} (\text{row}_2(A))^T \cdot \text{col}_3(B) &= 3(8) + 2(15) + 2(10) + 5(20) \\ &= 174 \text{ mg of pesticide 2 absorbed by herbivore 3.} \end{aligned}$$

If we now have p carnivores (such as a human) who eat the herbivores, we can repeat the analysis to find out how much of each pesticide has been absorbed by each carnivore. ■

It is sometimes useful to be able to find a column in the matrix product AB without having to multiply the two matrices. It is not difficult to show (Exercise 46) that the jth column of the matrix product AB is equal to the matrix product $A\text{col}_j(B)$.

EXAMPLE 11 Let

$$A = \begin{bmatrix} 1 & 2 \\ 3 & 4 \\ -1 & 5 \end{bmatrix} \quad \text{and} \quad B = \begin{bmatrix} -2 & 3 & 4 \\ 3 & 2 & 1 \end{bmatrix}.$$

Then the second column of AB is

$$A\text{col}_2(B) = \begin{bmatrix} 1 & 2 \\ 3 & 4 \\ -1 & 5 \end{bmatrix} \begin{bmatrix} 3 \\ 2 \end{bmatrix} = \begin{bmatrix} 7 \\ 17 \\ 7 \end{bmatrix}.$$

■

Remark If **u** and **v** are n-vectors ($n \times 1$ matrices), then it is easy to show by matrix multiplication (Exercise 41) that

$$\mathbf{u} \cdot \mathbf{v} = \mathbf{u}^T \mathbf{v}.$$

This observation is applied in Chapter 5.

■ The Matrix–Vector Product Written in Terms of Columns

Let

$$A = \begin{bmatrix} a_{11} & a_{12} & \cdots & a_{1n} \\ a_{21} & a_{22} & \cdots & a_{2n} \\ \vdots & \vdots & & \vdots \\ a_{m1} & a_{m2} & \cdots & a_{mn} \end{bmatrix}$$

be an $m \times n$ matrix and let

$$\mathbf{c} = \begin{bmatrix} c_1 \\ c_2 \\ \vdots \\ c_n \end{bmatrix}$$

be an n-vector, that is, an $n \times 1$ matrix. Since A is $m \times n$ and $\mathbf{c}$ is $n \times 1$, the matrix product $A\mathbf{c}$ is the $m \times 1$ matrix

$$A\mathbf{c} = \begin{bmatrix} a_{11} & a_{12} & \cdots & a_{1n} \\ a_{21} & a_{22} & \cdots & a_{2n} \\ \vdots & \vdots & & \vdots \\ a_{m1} & a_{m2} & \cdots & a_{mn} \end{bmatrix} \begin{bmatrix} c_1 \\ c_2 \\ \vdots \\ c_n \end{bmatrix} = \begin{bmatrix} (\text{row}_1(A))^T \cdot \mathbf{c} \\ (\text{row}_2(A))^T \cdot \mathbf{c} \\ \vdots \\ (\text{row}_m(A))^T \cdot \mathbf{c} \end{bmatrix} \tag{2}$$

$$= \begin{bmatrix} a_{11}c_1 + a_{12}c_2 + \cdots + a_{1n}c_n \\ a_{21}c_1 + a_{22}c_2 + \cdots + a_{2n}c_n \\ \vdots \\ a_{m1}c_1 + a_{m2}c_2 + \cdots + a_{mn}c_n \end{bmatrix}.$$

This last expression can be written as

$$c_1 \begin{bmatrix} a_{11} \\ a_{21} \\ \vdots \\ a_{m1} \end{bmatrix} + c_2 \begin{bmatrix} a_{12} \\ a_{22} \\ \vdots \\ a_{m2} \end{bmatrix} + \cdots + c_n \begin{bmatrix} a_{1n} \\ a_{2n} \\ \vdots \\ a_{mn} \end{bmatrix} \tag{3}$$

$$= c_1\text{col}_1(A) + c_2\text{col}_2(A) + \cdots + c_n\text{col}_n(A).$$

Thus the product $A\mathbf{c}$ of an $m \times n$ matrix A and an $n \times 1$ matrix $\mathbf{c}$ can be written as a linear combination of the columns of A, where the coefficients are the entries in the matrix $\mathbf{c}$.

In our study of linear systems of equations we shall see that these systems can be expressed in terms of a matrix–vector product. This point of view provides us with an important way to think about solutions of linear systems.

EXAMPLE 12 Let

$$A = \begin{bmatrix} 2 & -1 & -3 \\ 4 & 2 & -2 \end{bmatrix} \quad \text{and} \quad \mathbf{c} = \begin{bmatrix} 2 \\ -3 \\ 4 \end{bmatrix}.$$

Then the product $A\mathbf{c}$, written as a linear combination of the columns of A, is

$$A\mathbf{c} = \begin{bmatrix} 2 & -1 & -3 \\ 4 & 2 & -2 \end{bmatrix} \begin{bmatrix} 2 \\ -3 \\ 4 \end{bmatrix} = 2\begin{bmatrix} 2 \\ 4 \end{bmatrix} - 3\begin{bmatrix} -1 \\ 2 \end{bmatrix} + 4\begin{bmatrix} -3 \\ -2 \end{bmatrix} = \begin{bmatrix} -5 \\ -6 \end{bmatrix}.$$

■

If A is an $m \times p$ matrix and B is a $p \times n$ matrix, we can then conclude that the jth column of the product AB can be written as a linear combination of the columns of matrix A, where the coefficients are the entries in the jth column of matrix B:

$$\operatorname{col}_j(AB) = A\operatorname{col}_j(B) = b_{1j}\operatorname{col}_1(A) + b_{2j}\operatorname{col}_2(A) + \cdots + b_{pj}\operatorname{col}_p(A).$$

EXAMPLE 13 If A and B are the matrices defined in Example 11, then

$$AB = \begin{bmatrix} 1 & 2 \\ 3 & 4 \\ -1 & 5 \end{bmatrix} \begin{bmatrix} -2 & 3 & 4 \\ 3 & 2 & 1 \end{bmatrix} = \begin{bmatrix} 4 & 7 & 6 \\ 6 & 17 & 16 \\ 17 & 7 & 1 \end{bmatrix}.$$

The columns of AB as linear combinations of the columns of A are given by

$$\operatorname{col}_1(AB) = \begin{bmatrix} 4 \\ 6 \\ 17 \end{bmatrix} = A\operatorname{col}_1(B) = -2\begin{bmatrix} 1 \\ 3 \\ -1 \end{bmatrix} + 3\begin{bmatrix} 2 \\ 4 \\ 5 \end{bmatrix}$$

$$\operatorname{col}_2(AB) = \begin{bmatrix} 7 \\ 17 \\ 7 \end{bmatrix} = A\operatorname{col}_2(B) = 3\begin{bmatrix} 1 \\ 3 \\ -1 \end{bmatrix} + 2\begin{bmatrix} 2 \\ 4 \\ 5 \end{bmatrix}$$

$$\operatorname{col}_3(AB) = \begin{bmatrix} 6 \\ 16 \\ 1 \end{bmatrix} = A\operatorname{col}_3(B) = 4\begin{bmatrix} 1 \\ 3 \\ -1 \end{bmatrix} + 1\begin{bmatrix} 2 \\ 4 \\ 5 \end{bmatrix}.$$

■

■ Linear Systems

Consider the linear system of m equations in n unknowns,

$$\begin{aligned} a_{11}x_1 + a_{12}x_2 + \cdots + a_{1n}x_n &= b_1 \\ a_{21}x_1 + a_{22}x_2 + \cdots + a_{2n}x_n &= b_2 \\ &\vdots \\ a_{m1}x_1 + a_{m2}x_2 + \cdots + a_{mn}x_n &= b_m. \end{aligned} \tag{4}$$

Now define the following matrices:

$$A = \begin{bmatrix} a_{11} & a_{12} & \cdots & a_{1n} \\ a_{21} & a_{22} & \cdots & a_{2n} \\ \vdots & \vdots & & \vdots \\ a_{m1} & a_{m2} & \cdots & a_{mn} \end{bmatrix}, \quad \mathbf{x} = \begin{bmatrix} x_1 \\ x_2 \\ \vdots \\ x_n \end{bmatrix}, \quad \mathbf{b} = \begin{bmatrix} b_1 \\ b_2 \\ \vdots \\ b_m \end{bmatrix}.$$

Then

$$A\mathbf{x} = \begin{bmatrix} a_{11} & a_{12} & \cdots & a_{1n} \\ a_{21} & a_{22} & \cdots & a_{2n} \\ \vdots & \vdots & & \vdots \\ a_{m1} & a_{m2} & \cdots & a_{mn} \end{bmatrix} \begin{bmatrix} x_1 \\ x_2 \\ \vdots \\ x_n \end{bmatrix} = \begin{bmatrix} a_{11}x_1 + a_{12}x_2 + \cdots + a_{1n}x_n \\ a_{21}x_1 + a_{22}x_2 + \cdots + a_{2n}x_n \\ \vdots \\ a_{m1}x_1 + a_{m2}x_2 + \cdots + a_{mn}x_n \end{bmatrix}. \quad (5)$$

The entries in the product $A\mathbf{x}$ at the end of (5) are merely the left sides of the equations in (4). Hence the linear system (4) can be written in matrix form as

$$A\mathbf{x} = \mathbf{b}.$$

The matrix A is called the **coefficient matrix** of the linear system (4), and the matrix

$$\left[\begin{array}{cccc:c} a_{11} & a_{12} & \cdots & a_{1n} & b_1 \\ a_{21} & a_{22} & \cdots & a_{2n} & b_2 \\ \vdots & \vdots & & \vdots & \vdots \\ a_{m1} & a_{m2} & \cdots & a_{mn} & b_m \end{array}\right],$$

obtained by adjoining column $\mathbf{b}$ to A, is called the **augmented matrix** of the linear system (4). The augmented matrix of (4) is written as $\left[A \mid \mathbf{b}\right]$. Conversely, any matrix with more than one column can be thought of as the augmented matrix of a linear system. The coefficient and augmented matrices play key roles in our method for solving linear systems.

Recall from Section 1.1 that if

$$b_1 = b_2 = \cdots = b_m = 0$$

in (4), the linear system is called a **homogeneous system**. A homogeneous system can be written as

$$A\mathbf{x} = \mathbf{0},$$

where A is the coefficient matrix.

EXAMPLE 14 Consider the linear system

$$\begin{aligned} -2x \phantom{{}+3y} + z &= 5 \\ 2x + 3y - 4z &= 7 \\ 3x + 2y + 2z &= 3. \end{aligned}$$

Letting

$$A = \begin{bmatrix} -2 & 0 & 1 \\ 2 & 3 & -4 \\ 3 & 2 & 2 \end{bmatrix}, \quad \mathbf{x} = \begin{bmatrix} x \\ y \\ z \end{bmatrix}, \quad \text{and} \quad \mathbf{b} = \begin{bmatrix} 5 \\ 7 \\ 3 \end{bmatrix},$$

we can write the given linear system in matrix form as

$$A\mathbf{x} = \mathbf{b}.$$

The coefficient matrix is A, and the augmented matrix is

$$\left[\begin{array}{rrr|r} -2 & 0 & 1 & 5 \\ 2 & 3 & -4 & 7 \\ 3 & 2 & 2 & 3 \end{array}\right].$$

■

EXAMPLE 15 The matrix

$$\left[\begin{array}{rrr|r} 2 & -1 & 3 & 4 \\ 3 & 0 & 2 & 5 \end{array}\right]$$

is the augmented matrix of the linear system

$$\begin{aligned} 2x - y + 3z &= 4 \\ 3x \quad\;\; + 2z &= 5. \end{aligned}$$

■

We can express (5) in another form, as follows, using (2) and (3):

$$\begin{aligned} A\mathbf{x} &= \begin{bmatrix} a_{11}x_1 + a_{12}x_2 + \cdots + a_{1n}x_n \\ a_{21}x_1 + a_{22}x_2 + \cdots + a_{2n}x_n \\ \vdots \\ a_{m1}x_1 + a_{m2}x_2 + \cdots + a_{mn}x_n \end{bmatrix} \\ &= \begin{bmatrix} a_{11}x_1 \\ a_{21}x_1 \\ \vdots \\ a_{m1}x_1 \end{bmatrix} + \begin{bmatrix} a_{12}x_2 \\ a_{22}x_2 \\ \vdots \\ a_{m2}x_2 \end{bmatrix} + \cdots + \begin{bmatrix} a_{1n}x_n \\ a_{2n}x_n \\ \vdots \\ a_{mn}x_n \end{bmatrix} \\ &= x_1 \begin{bmatrix} a_{11} \\ a_{21} \\ \vdots \\ a_{m1} \end{bmatrix} + x_2 \begin{bmatrix} a_{12} \\ a_{22} \\ \vdots \\ a_{m2} \end{bmatrix} + \cdots + x_n \begin{bmatrix} a_{1n} \\ a_{2n} \\ \vdots \\ a_{mn} \end{bmatrix} \\ &= x_1 \,\mathbf{col}_1(A) + x_2 \,\mathbf{col}_2(A) + \cdots + x_n \,\mathbf{col}_n(A). \end{aligned}$$

Thus $A\mathbf{x}$ is a linear combination of the columns of A with coefficients that are the entries of $\mathbf{x}$. It follows that the matrix form of a linear system, $A\mathbf{x} = \mathbf{b}$, can be expressed as

$$x_1 \,\mathbf{col}_1(A) + x_2 \,\mathbf{col}_2(A) + \cdots + x_n \,\mathbf{col}_n(A) = \mathbf{b}. \tag{6}$$

Conversely, an equation of the form in (6) always describes a linear system of the form in (4).

EXAMPLE 16 Consider the linear system $A\mathbf{x} = \mathbf{b}$, where the coefficient matrix

$$A = \begin{bmatrix} 3 & 1 & 2 \\ 4 & -5 & 6 \\ 0 & 7 & -3 \\ -1 & 2 & 0 \end{bmatrix}, \quad \mathbf{x} = \begin{bmatrix} x_1 \\ x_2 \\ x_3 \end{bmatrix}, \quad \text{and} \quad \mathbf{b} = \begin{bmatrix} 4 \\ 1 \\ 0 \\ 2 \end{bmatrix}.$$

Writing $A\mathbf{x} = \mathbf{b}$ as a linear combination of the columns of A as in (6), we have

$$x_1 \begin{bmatrix} 3 \\ 4 \\ 0 \\ -1 \end{bmatrix} + x_2 \begin{bmatrix} 1 \\ -5 \\ 7 \\ 2 \end{bmatrix} + x_3 \begin{bmatrix} 2 \\ 6 \\ -3 \\ 0 \end{bmatrix} = \begin{bmatrix} 4 \\ 1 \\ 0 \\ 2 \end{bmatrix}.$$

■

The expression for the linear system $A\mathbf{x} = \mathbf{b}$ as shown in (6), provides an important way to think about solutions of linear systems.

> $A\mathbf{x} = \mathbf{b}$ is consistent if and only if $\mathbf{b}$ can be expressed as a linear combination of the columns of the matrix A.

We encounter this approach in Chapter 2.

Key Terms

Dot product (inner product)
Matrix–vector product
Coefficient matrix
Augmented matrix

1.3 Exercises

In Exercises 1 and 2, compute $\mathbf{a} \cdot \mathbf{b}$.

1. (a) $\mathbf{a} = \begin{bmatrix} 1 \\ 2 \end{bmatrix}$, $\mathbf{b} = \begin{bmatrix} 4 \\ -1 \end{bmatrix}$

(b) $\mathbf{a} = \begin{bmatrix} -3 \\ -2 \end{bmatrix}$, $\mathbf{b} = \begin{bmatrix} 1 \\ -2 \end{bmatrix}$

(c) $\mathbf{a} = \begin{bmatrix} 4 \\ 2 \\ -1 \end{bmatrix}$, $\mathbf{b} = \begin{bmatrix} 1 \\ 3 \\ 6 \end{bmatrix}$

(d) $\mathbf{a} = \begin{bmatrix} 1 \\ 1 \\ 0 \end{bmatrix}$, $\mathbf{b} = \begin{bmatrix} 1 \\ 0 \\ 1 \end{bmatrix}$

2. (a) $\mathbf{a} = \begin{bmatrix} 2 \\ -1 \end{bmatrix}$, $\mathbf{b} = \begin{bmatrix} 3 \\ 2 \end{bmatrix}$

(b) $\mathbf{a} = \begin{bmatrix} 1 \\ -1 \end{bmatrix}$, $\mathbf{b} = \begin{bmatrix} 1 \\ 1 \end{bmatrix}$

(c) $\mathbf{a} = \begin{bmatrix} 1 \\ 2 \\ 3 \end{bmatrix}$, $\mathbf{b} = \begin{bmatrix} -2 \\ 0 \\ 1 \end{bmatrix}$

(d) $\mathbf{a} = \begin{bmatrix} 1 \\ 0 \\ 0 \end{bmatrix}$, $\mathbf{b} = \begin{bmatrix} 1 \\ 0 \\ 0 \end{bmatrix}$

3. Let $\mathbf{a} = \mathbf{b} = \begin{bmatrix} -3 \\ 2 \\ x \end{bmatrix}$. If $\mathbf{a} \cdot \mathbf{b} = 17$, find x.

4. Determine the value of x so that $\mathbf{v} \cdot \mathbf{w} = 0$, where

$$\mathbf{v} = \begin{bmatrix} 1 \\ -3 \\ 4 \\ x \end{bmatrix} \quad \text{and} \quad \mathbf{w} = \begin{bmatrix} x \\ 2 \\ -1 \\ 1 \end{bmatrix}.$$

5. Determine values of x and y so that $\mathbf{v} \cdot \mathbf{w} = 0$ and $\mathbf{v} \cdot \mathbf{u} = 0$, where $\mathbf{v} = \begin{bmatrix} x \\ 1 \\ y \end{bmatrix}$, $\mathbf{w} = \begin{bmatrix} 2 \\ -2 \\ 1 \end{bmatrix}$, and $\mathbf{u} = \begin{bmatrix} 1 \\ 8 \\ 2 \end{bmatrix}$.

6. Determine values of x and y so that $\mathbf{v} \cdot \mathbf{w} = 0$ and $\mathbf{v} \cdot \mathbf{u} = 0$, where $\mathbf{v} = \begin{bmatrix} x \\ 1 \\ y \end{bmatrix}$, $\mathbf{w} = \begin{bmatrix} x \\ -2 \\ 0 \end{bmatrix}$, and $\mathbf{u} = \begin{bmatrix} 0 \\ -9 \\ y \end{bmatrix}$.

7. Let $\mathbf{w} = \begin{bmatrix} \sin\theta \\ \cos\theta \end{bmatrix}$. Compute $\mathbf{w} \cdot \mathbf{w}$.

8. Find all values of x so that $\mathbf{u} \cdot \mathbf{u} = 50$, where $\mathbf{u} = \begin{bmatrix} x \\ 3 \\ 4 \end{bmatrix}$.

9. Find all values of x so that $\mathbf{v} \cdot \mathbf{v} = 1$, where $\mathbf{v} = \begin{bmatrix} \frac{1}{2} \\ -\frac{1}{2} \\ x \end{bmatrix}$.

10. Let $A = \begin{bmatrix} 1 & 2 & x \\ 3 & -1 & 2 \end{bmatrix}$ and $B = \begin{bmatrix} y \\ x \\ 1 \end{bmatrix}$.

If $AB = \begin{bmatrix} 6 \\ 8 \end{bmatrix}$, find x and y.

Consider the following matrices for Exercises 11 through 15:

$$A = \begin{bmatrix} 1 & 2 & 3 \\ 2 & 1 & 4 \end{bmatrix}, \quad B = \begin{bmatrix} 1 & 0 \\ 2 & 1 \\ 3 & 2 \end{bmatrix},$$

$$C = \begin{bmatrix} 3 & -1 & 3 \\ 4 & 1 & 5 \\ 2 & 1 & 3 \end{bmatrix}, \quad D = \begin{bmatrix} 3 & -2 \\ 2 & 5 \end{bmatrix},$$

$$E = \begin{bmatrix} 2 & -4 & 5 \\ 0 & 1 & 4 \\ 3 & 2 & 1 \end{bmatrix}, \quad \textit{and} \quad F = \begin{bmatrix} -1 & 2 \\ 0 & 4 \\ 3 & 5 \end{bmatrix}.$$

11. If possible, compute the following:

(a) AB **(b)** BA **(c)** F^TE

(d) $CB + D$ **(e)** $AB + D^2$, where $D^2 = DD$

12. If possible, compute the following:

(a) $DA + B$ **(b)** EC **(c)** CE

(d) $EB + F$ **(e)** $FC + D$

13. If possible, compute the following:

(a) $FD - 3B$ **(b)** $AB - 2D$

(c) $F^TB + D$ **(d)** $2F - 3(AE)$

(e) $BD + AE$

14. If possible, compute the following:

(a) $A(BD)$ **(b)** $(AB)D$

(c) $A(C + E)$ **(d)** $AC + AE$

(e) $(2AB)^T$ and $2(AB)^T$ **(f)** $A(C - 3E)$

15. If possible, compute the following:

(a) A^T **(b)** $(A^T)^T$

(c) $(AB)^T$ **(d)** B^TA^T

(e) $(C + E)^TB$ and $C^TB + E^TB$

(f) $A(2B)$ and $2(AB)$

16. Let $A = \begin{bmatrix} 1 & 2 & -3 \end{bmatrix}$, $B = \begin{bmatrix} -1 & 4 & 2 \end{bmatrix}$, and $C = \begin{bmatrix} -3 & 0 & 1 \end{bmatrix}$. If possible, compute the following:

(a) AB^T **(b)** CA^T **(c)** $(BA^T)C$

(d) A^TB **(e)** CC^T **(f)** C^TC

(g) B^TCAA^T

17. Let $A = \begin{bmatrix} 2 & 3 \\ -1 & 4 \\ 0 & 3 \end{bmatrix}$ and $B = \begin{bmatrix} 3 & -1 & 3 \\ 1 & 2 & 4 \end{bmatrix}$.

Compute the following entries of AB:

(a) the (1, 2) entry **(b)** the (2, 3) entry

(c) the (3, 1) entry **(d)** the (3, 3) entry

18. If $I_2 = \begin{bmatrix} 1 & 0 \\ 0 & 1 \end{bmatrix}$ and $D = \begin{bmatrix} 2 & 3 \\ -1 & -2 \end{bmatrix}$, compute DI_2 and I_2D.

19. Let

$$A = \begin{bmatrix} 1 & 2 \\ 3 & 2 \end{bmatrix} \quad \text{and} \quad B = \begin{bmatrix} 2 & -1 \\ -3 & 4 \end{bmatrix}.$$

Show that $AB \neq BA$.

20. If A is the matrix in Example 4 and O is the 3×2 matrix every one of whose entries is zero, compute AO.

In Exercises 21 and 22, let

$$A = \begin{bmatrix} 1 & -1 & 2 \\ 3 & 2 & 4 \\ 4 & -2 & 3 \\ 2 & 1 & 5 \end{bmatrix}$$

and

$$B = \begin{bmatrix} 1 & 0 & -1 & 2 \\ 3 & 3 & -3 & 4 \\ 4 & 2 & 5 & 1 \end{bmatrix}.$$

21. Using the method in Example 11, compute the following columns of AB:

(a) the first column **(b)** the third column

22. Using the method in Example 11, compute the following columns of AB:

(a) the second column **(b)** the fourth column

23. Let

$$A = \begin{bmatrix} 2 & -3 & 4 \\ -1 & 2 & 3 \\ 5 & -1 & -2 \end{bmatrix} \quad \text{and} \quad \mathbf{c} = \begin{bmatrix} 2 \\ 1 \\ 4 \end{bmatrix}.$$

Express $A\mathbf{c}$ as a linear combination of the columns of A.

24. Let

$$A = \begin{bmatrix} 1 & -2 & -1 \\ 2 & 4 & 3 \\ 3 & 0 & -2 \end{bmatrix} \quad \text{and} \quad B = \begin{bmatrix} 1 & -1 \\ 3 & 2 \\ 2 & 4 \end{bmatrix}.$$

Express the columns of AB as linear combinations of the columns of A.

25. Let $A = \begin{bmatrix} 2 & -3 & 1 \\ 1 & 2 & 4 \end{bmatrix}$ and $B = \begin{bmatrix} 3 \\ 5 \\ 2 \end{bmatrix}$.

(a) Verify that $AB = 3\mathbf{a}_1 + 5\mathbf{a}_2 + 2\mathbf{a}_3$, where $\mathbf{a}_j$ is the jth column of A for $j = 1, 2, 3$.

(b) Verify that $AB = \begin{bmatrix} (\text{row}_1(A))B \\ (\text{row}_2(A))B \end{bmatrix}$.

26. **(a)** Find a value of r so that $AB^T = 0$, where $A = \begin{bmatrix} r & 1 & -2 \end{bmatrix}$ and $B = \begin{bmatrix} 1 & 3 & -1 \end{bmatrix}$.

(b) Give an alternative way to write this product.

27. Find a value of r and a value of s so that $AB^T = 0$, where $A = \begin{bmatrix} 1 & r & 1 \end{bmatrix}$ and $B = \begin{bmatrix} -2 & 2 & s \end{bmatrix}$.

28. **(a)** Let A be an $m \times n$ matrix with a row consisting entirely of zeros. Show that if B is an $n \times p$ matrix, then AB has a row of zeros.

(b) Let A be an $m \times n$ matrix with a column consisting entirely of zeros and let B be $p \times m$. Show that BA has a column of zeros.

29. Let $A = \begin{bmatrix} -3 & 2 & 1 \\ 4 & 5 & 0 \end{bmatrix}$ with $\mathbf{a}_j$ = the jth column of A, $j = 1, 2, 3$. Verify that

$$A^T A = \begin{bmatrix} \mathbf{a}_1^T\mathbf{a}_1 & \mathbf{a}_1^T\mathbf{a}_2 & \mathbf{a}_1^T\mathbf{a}_3 \\ \mathbf{a}_2^T\mathbf{a}_1 & \mathbf{a}_2^T\mathbf{a}_2 & \mathbf{a}_2^T\mathbf{a}_3 \\ \mathbf{a}_3^T\mathbf{a}_1 & \mathbf{a}_3^T\mathbf{a}_2 & \mathbf{a}_3^T\mathbf{a}_3 \end{bmatrix}.$$

30. Consider the following linear system:

$$\begin{aligned} 2x_1 + 3x_2 - 3x_3 + x_4 + x_5 &= 7 \\ 3x_1 \qquad + 2x_3 \qquad\quad + 3x_5 &= -2 \\ 2x_1 + 3x_2 \qquad - 4x_4 \qquad &= 3 \\ x_3 + x_4 + x_5 &= 5. \end{aligned}$$

(a) Find the coefficient matrix.

(b) Write the linear system in matrix form.

(c) Find the augmented matrix.

31. Write the linear system whose augmented matrix is

$$\left[\begin{array}{cccc|c} -2 & -1 & 0 & 4 & 5 \\ -3 & 2 & 7 & 8 & 3 \\ 1 & 0 & 0 & 2 & 4 \\ 3 & 0 & 1 & 3 & 6 \end{array}\right].$$

32. Write the following linear system in matrix form:

$$\begin{aligned} -2x_1 + 3x_2 &= 5 \\ x_1 - 5x_2 &= 4 \end{aligned}$$

33. Write the following linear system in matrix form:

$$\begin{aligned} 2x_1 + 3x_2 &= 0 \\ 3x_2 + x_3 &= 0 \\ 2x_1 - x_2 &= 0 \end{aligned}$$

34. Write the linear system whose augmented matrix is

(a) $\left[\begin{array}{cccc|c} 2 & 1 & 3 & 4 & 0 \\ 3 & -1 & 2 & 0 & 3 \\ -2 & 1 & -4 & 3 & 2 \end{array}\right]$.

(b) $\left[\begin{array}{cccc|c} 2 & 1 & 3 & 4 & 0 \\ 3 & -1 & 2 & 0 & 3 \\ -2 & 1 & -4 & 3 & 2 \\ 0 & 0 & 0 & 0 & 0 \end{array}\right]$.

35. How are the linear systems obtained in Exercise 34 related?

36. Write each of the following linear systems as a linear combination of the columns of the coefficient matrix:

(a) $\begin{aligned} 3x_1 + 2x_2 + x_3 &= 4 \\ x_1 - x_2 + 4x_3 &= -2 \end{aligned}$

(b) $\begin{aligned} -x_1 + x_2 &= 3 \\ 2x_1 - x_2 &= -2 \\ 3x_1 + x_2 &= 1 \end{aligned}$

37. Write each of the following linear combinations of columns as a linear system of the form in (4):

(a) $x_1 \begin{bmatrix} 2 \\ 0 \end{bmatrix} + x_2 \begin{bmatrix} 1 \\ 3 \end{bmatrix} = \begin{bmatrix} 4 \\ 2 \end{bmatrix}$

(b) $x_1 \begin{bmatrix} 1 \\ 2 \\ -1 \end{bmatrix} + x_2 \begin{bmatrix} 0 \\ 1 \\ 2 \end{bmatrix} + x_3 \begin{bmatrix} 3 \\ 4 \\ 5 \end{bmatrix} + x_4 \begin{bmatrix} 1 \\ 3 \\ 4 \end{bmatrix} = \begin{bmatrix} 2 \\ 5 \\ 8 \end{bmatrix}$

38. Write each of the following as a linear system in matrix form:

(a) $x_1 \begin{bmatrix} 1 \\ 2 \end{bmatrix} + x_2 \begin{bmatrix} 2 \\ 5 \end{bmatrix} + x_3 \begin{bmatrix} 0 \\ 3 \end{bmatrix} = \begin{bmatrix} 1 \\ 1 \end{bmatrix}$

(b) $x_1 \begin{bmatrix} 1 \\ 1 \\ 2 \end{bmatrix} + x_2 \begin{bmatrix} 2 \\ 1 \\ 0 \end{bmatrix} + x_3 \begin{bmatrix} 1 \\ 2 \\ 2 \end{bmatrix} = \begin{bmatrix} 0 \\ 0 \\ 0 \end{bmatrix}$

39. Determine a solution to each of the following linear systems, using the fact that $A\mathbf{x} = \mathbf{b}$ is consistent if and only if $\mathbf{b}$ is a linear combination of the columns of A:

(a) $\begin{bmatrix} 1 & 2 & 1 \\ -3 & 6 & -3 \\ 0 & 1 & -1 \end{bmatrix} \begin{bmatrix} x_1 \\ x_2 \\ x_3 \end{bmatrix} = \begin{bmatrix} 0 \\ 0 \\ 0 \end{bmatrix}$

(b) $\begin{bmatrix} 1 & 2 & 3 & 4 \\ 2 & 3 & 4 & 1 \\ 3 & 4 & 1 & 2 \end{bmatrix} \begin{bmatrix} x_1 \\ x_2 \\ x_3 \\ x_4 \end{bmatrix} = \begin{bmatrix} 20 \\ 20 \\ 20 \end{bmatrix}$

40. Construct a coefficient matrix A so that $\mathbf{x} = \begin{bmatrix} 1 \\ 0 \\ 1 \end{bmatrix}$ is a solution to the system $A\mathbf{x} = \mathbf{b}$, where $\mathbf{b} = \begin{bmatrix} 1 \\ 2 \\ 3 \end{bmatrix}$. Can there be more than one such coefficient matrix? Explain.

41. Show that if $\mathbf{u}$ and $\mathbf{v}$ are n-vectors, then $\mathbf{u} \cdot \mathbf{v} = \mathbf{u}^T \mathbf{v}$.

42. Let A be an $m \times n$ matrix and B an $n \times p$ matrix. What, if anything, can you say about the matrix product AB when

(a) A has a column consisting entirely of zeros?

(b) B has a row consisting entirely of zeros?

43. If $A = \begin{bmatrix} a_{ij} \end{bmatrix}$ is an $n \times n$ matrix, then the **trace** of A, $\text{Tr}(A)$, is defined as the sum of all elements on the main diagonal of A, $\text{Tr}(A) = \sum_{i=1}^{n} a_{ii}$. Show each of the following:

(a) $\text{Tr}(cA) = c\,\text{Tr}(A)$, where c is a real number

(b) $\text{Tr}(A + B) = \text{Tr}(A) + \text{Tr}(B)$

(c) $\text{Tr}(AB) = \text{Tr}(BA)$

(d) $\text{Tr}(A^T) = \text{Tr}(A)$

(e) $\text{Tr}(A^T A) \geq 0$

44. Compute the trace (see Exercise 43) of each of the following matrices:

(a) $\begin{bmatrix} 1 & 0 \\ 2 & 3 \end{bmatrix}$ **(b)** $\begin{bmatrix} 2 & 2 & 3 \\ 2 & 4 & 4 \\ 3 & -2 & -5 \end{bmatrix}$

(c) $\begin{bmatrix} 1 & 0 & 0 \\ 0 & 1 & 0 \\ 0 & 0 & 1 \end{bmatrix}$

45. Show that there are no 2×2 matrices A and B such that $AB - BA = \begin{bmatrix} 1 & 0 \\ 0 & 1 \end{bmatrix}$.

46. **(a)** Show that the jth column of the matrix product AB is equal to the matrix product $A\mathbf{b}_j$, where $\mathbf{b}_j$ is the jth column of B. It follows that the product AB can be written in terms of columns as

$$AB = \begin{bmatrix} A\mathbf{b}_1 & A\mathbf{b}_2 & \cdots & A\mathbf{b}_n \end{bmatrix}.$$

(b) Show that the ith row of the matrix product AB is equal to the matrix product $\mathbf{a}_i B$, where $\mathbf{a}_i$ is the ith row of A. It follows that the product AB can be written in terms of rows as

$$AB = \begin{bmatrix} \mathbf{a}_1 B \\ \mathbf{a}_2 B \\ \vdots \\ \mathbf{a}_m B \end{bmatrix}.$$

47. Show that the jth column of the matrix product AB is a linear combination of the columns of A with coefficients the entries in $\mathbf{b}_j$, the jth column of B.

48. The vector

$$\mathbf{u} = \begin{bmatrix} 20 \\ 30 \\ 80 \\ 10 \end{bmatrix}$$

gives the number of receivers, CD players, speakers, and DVD recorders that are on hand in an audio shop. The vector

$$\mathbf{v} = \begin{bmatrix} 200 \\ 120 \\ 80 \\ 70 \end{bmatrix}$$

gives the price (in dollars) of each receiver, CD player, speaker, and DVD recorder, respectively. What does the dot product $\mathbf{u} \cdot \mathbf{v}$ tell the shop owner?

49. ***(Manufacturing Costs)*** A furniture manufacturer makes chairs and tables, each of which must go through an assembly process and a finishing process. The times required for these processes are given (in hours) by the matrix

$$A = \begin{array}{c} \begin{array}{cc} \text{Assembly process} & \text{Finishing process} \end{array} \\ \left[\begin{array}{cc} 2 & 2 \\ 3 & 4 \end{array}\right] \begin{array}{l} \text{Chair} \\ \text{Table} \end{array} \end{array}.$$

The manufacturer has a plant in Salt Lake City and another in Chicago. The hourly rates for each of the processes are given (in dollars) by the matrix

$$B = \begin{array}{c} \begin{array}{cc} \text{Salt Lake City} & \text{Chicago} \end{array} \\ \left[\begin{array}{cc} 9 & 10 \\ 10 & 12 \end{array}\right] \begin{array}{l} \text{Assembly process} \\ \text{Finishing process} \end{array} \end{array}.$$

What do the entries in the matrix product AB tell the manufacturer?

50. ***(Medicine)*** A diet research project includes adults and children of both sexes. The composition of the participants in the project is given by the matrix

$$A = \begin{array}{c} \begin{array}{cc} \text{Adults} & \text{Children} \end{array} \\ \left[\begin{array}{cc} 80 & 120 \\ 100 & 200 \end{array}\right] \begin{array}{l} \text{Male} \\ \text{Female} \end{array} \end{array}.$$

The number of daily grams of protein, fat, and carbohydrate consumed by each child and adult is given by the matrix

$$B = \begin{array}{c} \begin{array}{ccc} \text{Protein} & \text{Fat} & \text{Carbohydrate} \end{array} \\ \left[\begin{array}{ccc} 20 & 20 & 20 \\ 10 & 20 & 30 \end{array}\right] \begin{array}{l} \text{Adult} \\ \text{Child} \end{array} \end{array}.$$

(a) How many grams of protein are consumed daily by the males in the project?

(b) How many grams of fat are consumed daily by the females in the project?

51. Let $\mathbf{x}$ be an n-vector.

(a) Is it possible for $\mathbf{x} \cdot \mathbf{x}$ to be negative? Explain.

(b) If $\mathbf{x} \cdot \mathbf{x} = 0$, what is $\mathbf{x}$?

52. Let $\mathbf{a}$, $\mathbf{b}$, and $\mathbf{c}$ be n-vectors and let k be a real number.

(a) Show that $\mathbf{a} \cdot \mathbf{b} = \mathbf{b} \cdot \mathbf{a}$.

(b) Show that $(\mathbf{a} + \mathbf{b}) \cdot \mathbf{c} = \mathbf{a} \cdot \mathbf{c} + \mathbf{b} \cdot \mathbf{c}$.

(c) Show that $(k\mathbf{a}) \cdot \mathbf{b} = \mathbf{a} \cdot (k\mathbf{b}) = k(\mathbf{a} \cdot \mathbf{b})$.

53. Let A be an $m \times n$ matrix whose entries are real numbers. Show that if $AA^T = O$ (the $m \times m$ matrix all of whose entries are zero), then $A = O$.

54. Use the matrices A and C in Exercise 11 and the matrix multiplication command in your software to compute AC and CA. Discuss the results.

55. Using your software, compute $B^T B$ and BB^T for

$$B = \begin{bmatrix} 1 & \frac{1}{2} & \frac{1}{3} & \frac{1}{4} & \frac{1}{5} & \frac{1}{6} \end{bmatrix}.$$

Discuss the nature of the results.

1.4 Algebraic Properties of Matrix Operations

In this section we consider the algebraic properties of the matrix operations just defined. Many of these properties are similar to the familiar properties that hold for real numbers. However, there will be striking differences between the set of real numbers and the set of matrices in their algebraic behavior under certain operations—for example, under multiplication (as seen in Section 1.3). The proofs of most of the properties will be left as exercises.

Theorem 1.1 **Properties of Matrix Addition**

Let A, B, and C be $m \times n$ matrices.

(a) $A + B = B + A$.

(b) $A + (B + C) = (A + B) + C$.

(c) There is a unique $m \times n$ matrix O such that

$$A + O = A \tag{1}$$

for any $m \times n$ matrix A. The matrix O is called the $m \times n$ **zero matrix**.

(d) For each $m \times n$ matrix A, there is a unique $m \times n$ matrix D such that

$$A + D = O. \tag{2}$$

We shall write D as $-A$, so (2) can be written as

$$A + (-A) = O.$$

The matrix $-A$ is called the **negative** of A. We also note that $-A$ is $(-1)A$.

Proof

(a) Let

$$A = \begin{bmatrix} a_{ij} \end{bmatrix}, \quad B = \begin{bmatrix} b_{ij} \end{bmatrix},$$
$$A + B = C = \begin{bmatrix} c_{ij} \end{bmatrix}, \quad \text{and} \quad B + A = D = \begin{bmatrix} d_{ij} \end{bmatrix}.$$

We must show that $c_{ij} = d_{ij}$ for all i, j. Now $c_{ij} = a_{ij} + b_{ij}$ and $d_{ij} = b_{ij} + a_{ij}$ for all i, j. Since a_{ij} and b_{ij} are real numbers, we have $a_{ij} + b_{ij} = b_{ij} + a_{ij}$, which implies that $c_{ij} = d_{ij}$ for all i, j.

(c) Let $U = \left[u_{ij} \right]$. Then $A + U = A$ if and only if[†] $a_{ij} + u_{ij} = a_{ij}$, which holds if and only if $u_{ij} = 0$. Thus U is the $m \times n$ matrix all of whose entries are zero: U is denoted by O. ■

The 2×2 zero matrix is

$$O = \begin{bmatrix} 0 & 0 \\ 0 & 0 \end{bmatrix}.$$

EXAMPLE 1

If

$$A = \begin{bmatrix} 4 & -1 \\ 2 & 3 \end{bmatrix},$$

then

$$\begin{bmatrix} 4 & -1 \\ 2 & 3 \end{bmatrix} + \begin{bmatrix} 0 & 0 \\ 0 & 0 \end{bmatrix} = \begin{bmatrix} 4+0 & -1+0 \\ 2+0 & 3+0 \end{bmatrix} = \begin{bmatrix} 4 & -1 \\ 2 & 3 \end{bmatrix}.$$ ■

The 2×3 zero matrix is

$$O = \begin{bmatrix} 0 & 0 & 0 \\ 0 & 0 & 0 \end{bmatrix}.$$

EXAMPLE 2

If $A = \begin{bmatrix} 1 & 3 & -2 \\ -2 & 4 & 3 \end{bmatrix}$, then $-A = \begin{bmatrix} -1 & -3 & 2 \\ 2 & -4 & -3 \end{bmatrix}$. ■

Theorem 1.2 **Properties of Matrix Multiplication**

(a) If A, B, and C are matrices of the appropriate sizes, then

$$A(BC) = (AB)C.$$

(b) If A, B, and C are matrices of the appropriate sizes, then

$$(A + B)C = AC + BC.$$

(c) If A, B, and C are matrices of the appropriate sizes, then

$$C(A + B) = CA + CB. \tag{3}$$

Proof

(a) Suppose that A is $m \times n$, B is $n \times p$, and C is $p \times q$. We shall prove the result for the special case $m = 2$, $n = 3$, $p = 4$, and $q = 3$. The general proof is completely analogous.

Let $A = \left[a_{ij} \right]$, $B = \left[b_{ij} \right]$, $C = \left[c_{ij} \right]$, $AB = D = \left[d_{ij} \right]$, $BC = E = \left[e_{ij} \right]$, $(AB)C = F = \left[f_{ij} \right]$, and $A(BC) = G = \left[g_{ij} \right]$. We must show that $f_{ij} = g_{ij}$ for all i, j. Now

$$f_{ij} = \sum_{k=1}^{4} d_{ik} c_{kj} = \sum_{k=1}^{4} \left(\sum_{r=1}^{3} a_{ir} b_{rk} \right) c_{kj}$$

[†]The connector "if and only if" means that both statements are true or both statements are false. Thus (i) if $A + U = A$, then $a_{ij} + u_{ij} = a_{ij}$; and (ii) if $a_{ij} + u_{ij} = a_{ij}$, then $A + U = A$. See Appendix C, "Introduction to Proofs."

and

$$g_{ij} = \sum_{r=1}^{3} a_{ir} e_{rj} = \sum_{r=1}^{3} a_{ir} \left(\sum_{k=1}^{4} b_{rk} c_{kj} \right).$$

Then, by the properties satisfied by the summation notation,

$$\begin{aligned} f_{ij} &= \sum_{k=1}^{4} (a_{i1} b_{1k} + a_{i2} b_{2k} + a_{i3} b_{3k}) c_{kj} \\ &= a_{i1} \sum_{k=1}^{4} b_{1k} c_{kj} + a_{i2} \sum_{k=1}^{4} b_{2k} c_{kj} + a_{i3} \sum_{k=1}^{4} b_{3k} c_{kj} \\ &= \sum_{r=1}^{3} a_{ir} \left(\sum_{k=1}^{4} b_{rk} c_{kj} \right) = g_{ij}. \end{aligned}$$

The proofs of (b) and (c) are left as Exercise 4. ■

EXAMPLE 3 Let

$$A = \begin{bmatrix} 5 & 2 & 3 \\ 2 & -3 & 4 \end{bmatrix}, \qquad B = \begin{bmatrix} 2 & -1 & 1 & 0 \\ 0 & 2 & 2 & 2 \\ 3 & 0 & -1 & 3 \end{bmatrix},$$

and

$$C = \begin{bmatrix} 1 & 0 & 2 \\ 2 & -3 & 0 \\ 0 & 0 & 3 \\ 2 & 1 & 0 \end{bmatrix}.$$

Then

$$A(BC) = \begin{bmatrix} 5 & 2 & 3 \\ 2 & -3 & 4 \end{bmatrix} \begin{bmatrix} 0 & 3 & 7 \\ 8 & -4 & 6 \\ 9 & 3 & 3 \end{bmatrix} = \begin{bmatrix} 43 & 16 & 56 \\ 12 & 30 & 8 \end{bmatrix}$$

and

$$(AB)C = \begin{bmatrix} 19 & -1 & 6 & 13 \\ 16 & -8 & -8 & 6 \end{bmatrix} \begin{bmatrix} 1 & 0 & 2 \\ 2 & -3 & 0 \\ 0 & 0 & 3 \\ 2 & 1 & 0 \end{bmatrix} = \begin{bmatrix} 43 & 16 & 56 \\ 12 & 30 & 8 \end{bmatrix}.$$ ■

EXAMPLE 4 Let

$$A = \begin{bmatrix} 2 & 2 & 3 \\ 3 & -1 & 2 \end{bmatrix}, \quad B = \begin{bmatrix} 0 & 0 & 1 \\ 2 & 3 & -1 \end{bmatrix}, \quad \text{and} \quad C = \begin{bmatrix} 1 & 0 \\ 2 & 2 \\ 3 & -1 \end{bmatrix}.$$

Then

$$(A + B)C = \begin{bmatrix} 2 & 2 & 4 \\ 5 & 2 & 1 \end{bmatrix} \begin{bmatrix} 1 & 0 \\ 2 & 2 \\ 3 & -1 \end{bmatrix} = \begin{bmatrix} 18 & 0 \\ 12 & 3 \end{bmatrix}$$

and (verify)

$$AC + BC = \begin{bmatrix} 15 & 1 \\ 7 & -4 \end{bmatrix} + \begin{bmatrix} 3 & -1 \\ 5 & 7 \end{bmatrix} = \begin{bmatrix} 18 & 0 \\ 12 & 3 \end{bmatrix}.$$

■

Recall Example 9 in Section 1.3, which shows that AB need not always equal BA. This is the first significant difference between multiplication of matrices and multiplication of real numbers.

Theorem 1.3 **Properties of Scalar Multiplication**

If r and s are real numbers and A and B are matrices of the appropriate sizes, then

(a) $r(sA) = (rs)A$

(b) $(r + s)A = rA + sA$

(c) $r(A + B) = rA + rB$

(d) $A(rB) = r(AB) = (rA)B$

Proof

Exercises 13, 14, 16, and 18. ■

EXAMPLE 5 Let

$$A = \begin{bmatrix} 4 & 2 & 3 \\ 2 & -3 & 4 \end{bmatrix} \quad \text{and} \quad B = \begin{bmatrix} 3 & -2 & 1 \\ 2 & 0 & -1 \\ 0 & 1 & 2 \end{bmatrix}.$$

Then

$$2(3A) = 2\begin{bmatrix} 12 & 6 & 9 \\ 6 & -9 & 12 \end{bmatrix} = \begin{bmatrix} 24 & 12 & 18 \\ 12 & -18 & 24 \end{bmatrix} = 6A.$$

We also have

$$A(2B) = \begin{bmatrix} 4 & 2 & 3 \\ 2 & -3 & 4 \end{bmatrix}\begin{bmatrix} 6 & -4 & 2 \\ 4 & 0 & -2 \\ 0 & 2 & 4 \end{bmatrix} = \begin{bmatrix} 32 & -10 & 16 \\ 0 & 0 & 26 \end{bmatrix} = 2(AB).$$

■

EXAMPLE 6 Scalar multiplication can be used to change the size of entries in a matrix to meet prescribed properties. Let

$$A = \begin{bmatrix} 3 \\ 7 \\ 2 \\ 1 \end{bmatrix}.$$

Then for $k = \frac{1}{7}$, the largest entry of kA is 1. Also if the entries of A represent the volume of products in gallons, for $k = 4$, kA gives the volume in quarts. ■

So far we have seen that multiplication and addition of matrices have much in common with multiplication and addition of real numbers. We now look at some properties of the transpose.

Theorem 1.4 **Properties of Transpose**

If r is a scalar and A and B are matrices of the appropriate sizes, then

(a) $(A^T)^T = A$

(b) $(A+B)^T = A^T + B^T$

(c) $(AB)^T = B^T A^T$

(d) $(rA)^T = rA^T$

Proof

We leave the proofs of (a), (b), and (d) as Exercises 26 and 27.

(c) Let $A = \left[a_{ij}\right]$ and $B = \left[b_{ij}\right]$; let $AB = C = \left[c_{ij}\right]$. We must prove that c^T_{ij} is the (i, j) entry in $B^T A^T$. Now

$$c^T_{ij} = c_{ji} = \sum_{k=1}^{n} a_{jk} b_{ki} = \sum_{k=1}^{n} a^T_{kj} b^T_{ik}$$

$$= \sum_{k=1}^{n} b^T_{ik} a^T_{kj} = \text{the } (i, j) \text{ entry in } B^T A^T. \qquad \blacksquare$$

EXAMPLE 7 Let

$$A = \begin{bmatrix} 1 & 2 & 3 \\ -2 & 0 & 1 \end{bmatrix} \quad \text{and} \quad B = \begin{bmatrix} 3 & -1 & 2 \\ 3 & 2 & -1 \end{bmatrix}.$$

Then

$$A^T = \begin{bmatrix} 1 & -2 \\ 2 & 0 \\ 3 & 1 \end{bmatrix} \quad \text{and} \quad B^T = \begin{bmatrix} 3 & 3 \\ -1 & 2 \\ 2 & -1 \end{bmatrix}.$$

Also,

$$A + B = \begin{bmatrix} 4 & 1 & 5 \\ 1 & 2 & 0 \end{bmatrix} \quad \text{and} \quad (A+B)^T = \begin{bmatrix} 4 & 1 \\ 1 & 2 \\ 5 & 0 \end{bmatrix}.$$

Now

$$A^T + B^T = \begin{bmatrix} 4 & 1 \\ 1 & 2 \\ 5 & 0 \end{bmatrix} = (A+B)^T.$$

$\blacksquare$

EXAMPLE 8 Let

$$A = \begin{bmatrix} 1 & 3 & 2 \\ 2 & -1 & 3 \end{bmatrix} \quad \text{and} \quad B = \begin{bmatrix} 0 & 1 \\ 2 & 2 \\ 3 & -1 \end{bmatrix}.$$

Then

$$AB = \begin{bmatrix} 12 & 5 \\ 7 & -3 \end{bmatrix} \quad \text{and} \quad (AB)^T = \begin{bmatrix} 12 & 7 \\ 5 & -3 \end{bmatrix}.$$

On the other hand,

$$A^T = \begin{bmatrix} 1 & 2 \\ 3 & -1 \\ 2 & 3 \end{bmatrix} \quad \text{and} \quad B^T = \begin{bmatrix} 0 & 2 & 3 \\ 1 & 2 & -1 \end{bmatrix}.$$

Then

$$B^T A^T = \begin{bmatrix} 12 & 7 \\ 5 & -3 \end{bmatrix} = (AB)^T.$$

■

We also note two other peculiarities of matrix multiplication. If a and b are real numbers, then $ab = 0$ can hold only if a or b is zero. However, this is not true for matrices.

EXAMPLE 9 If

$$A = \begin{bmatrix} 1 & 2 \\ 2 & 4 \end{bmatrix} \quad \text{and} \quad B = \begin{bmatrix} 4 & -6 \\ -2 & 3 \end{bmatrix},$$

then neither A nor B is the zero matrix, but $AB = \begin{bmatrix} 0 & 0 \\ 0 & 0 \end{bmatrix}$. ■

If a, b, and c are real numbers for which $ab = ac$ and $a \neq 0$, it follows that $b = c$. That is, we can cancel out the nonzero factor a. However, the cancellation law does not hold for matrices, as the following example shows.

EXAMPLE 10 If

$$A = \begin{bmatrix} 1 & 2 \\ 2 & 4 \end{bmatrix}, \quad B = \begin{bmatrix} 2 & 1 \\ 3 & 2 \end{bmatrix}, \quad \text{and} \quad C = \begin{bmatrix} -2 & 7 \\ 5 & -1 \end{bmatrix},$$

then

$$AB = AC = \begin{bmatrix} 8 & 5 \\ 16 & 10 \end{bmatrix},$$

but $B \neq C$. ■

We summarize some of the differences between matrix multiplication and the multiplication of real numbers as follows: For matrices A, B, and C of the appropriate sizes,

1. AB need not equal BA.
2. AB may be the zero matrix with $A \neq O$ and $B \neq O$.
3. AB may equal AC with $B \neq C$.

In this section we have developed a number of properties about matrices and their transposes. If a future problem involves these concepts, refer to these properties to help solve the problem. These results can be used to develop many more results.

Key Terms

Properties of matrix addition
Zero matrix
Properties of matrix multiplication
Properties of scalar multiplication
Properties of transpose

1.4 Exercises

1. Prove Theorem 1.1(b).

2. Prove Theorem 1.1(d).

3. Verify Theorem 1.2(a) for the following matrices:

$$A = \begin{bmatrix} 1 & 3 \\ 2 & -1 \end{bmatrix}, \quad B = \begin{bmatrix} -1 & 3 & 2 \\ 1 & -3 & 4 \end{bmatrix},$$

$$\text{and} \quad C = \begin{bmatrix} 1 & 0 \\ 3 & -1 \\ 1 & 2 \end{bmatrix}.$$

4. Prove Theorem 1.2(b) and (c).

5. Verify Theorem 1.2(c) for the following matrices:

$$A = \begin{bmatrix} 2 & -3 & 2 \\ 3 & -1 & -2 \end{bmatrix}, \quad B = \begin{bmatrix} 0 & 1 & 2 \\ 1 & 3 & -2 \end{bmatrix},$$

$$\text{and} \quad C = \begin{bmatrix} 1 & -3 \\ -3 & 4 \end{bmatrix}.$$

6. Let $A = \begin{bmatrix} a_{ij} \end{bmatrix}$ be the $n \times n$ matrix defined by $a_{ii} = k$ and $a_{ij} = 0$ if $i \neq j$. Show that if B is any $n \times n$ matrix, then $AB = kB$.

7. Let A be an $m \times n$ matrix and $C = \begin{bmatrix} c_1 & c_2 & \cdots & c_m \end{bmatrix}$ a $1 \times m$ matrix. Prove that

$$CA = \sum_{j=1}^{m} c_j A_j,$$

where A_j is the jth row of A.

8. Let $A = \begin{bmatrix} \cos\theta & \sin\theta \\ -\sin\theta & \cos\theta \end{bmatrix}$.

(a) Determine a simple expression for A^2.

(b) Determine a simple expression for A^3.

(c) Conjecture the form of a simple expression for A^k, k a positive integer.

(d) Prove or disprove your conjecture in part (c).

9. Find a pair of unequal 2×2 matrices A and B, other than those given in Example 9, such that $AB = O$.

10. Find two different 2×2 matrices A such that $A^2 = \begin{bmatrix} 1 & 0 \\ 0 & 1 \end{bmatrix}$.

11. Find two unequal 2×2 matrices A and B such that $AB = \begin{bmatrix} 1 & 0 \\ 0 & 1 \end{bmatrix}$.

12. Find two different 2×2 matrices A such that $A^2 = O$.

13. Prove Theorem 1.3(a).

14. Prove Theorem 1.3(b).

15. Verify Theorem 1.3(b) for $r = 4$, $s = -2$, and $A = \begin{bmatrix} 2 & -3 \\ 4 & 2 \end{bmatrix}$.

16. Prove Theorem 1.3(c).

17. Verify Theorem 1.3(c) for $r = -3$,

$$A = \begin{bmatrix} 4 & 2 \\ 1 & -3 \\ 3 & 2 \end{bmatrix}, \quad \text{and} \quad B = \begin{bmatrix} 0 & 2 \\ 4 & 3 \\ -2 & 1 \end{bmatrix}.$$

18. Prove Theorem 1.3(d).

19. Verify Theorem 1.3(d) for the following matrices:

$$A = \begin{bmatrix} 1 & 3 \\ 2 & -1 \end{bmatrix}, \quad B = \begin{bmatrix} -1 & 3 & 2 \\ 1 & -3 & 4 \end{bmatrix},$$

$$\text{and} \quad r = -3.$$

20. The matrix A contains the weight (in pounds) of objects packed on board a spacecraft on earth. The objects are to be used on the moon where things weigh about $\frac{1}{6}$ as much. Write an expression kA that calculates the weight of the objects on the moon.

21. (a) A is a 360×2 matrix. The first column of A is $\cos 0°, \cos 1°, \ldots, \cos 359°$; and the second column is $\sin 0°, \sin 1°, \ldots, \sin 359°$. The graph of the ordered pairs in A is a circle of radius 1 centered at the origin. Write an expression kA for ordered pairs whose graph is a circle of radius 3 centered at the origin.

(b) Explain how to prove the claims about the circles in part (a).

22. Determine a scalar r such that $A\mathbf{x} = r\mathbf{x}$, where

$$A = \begin{bmatrix} 2 & 1 \\ 1 & 2 \end{bmatrix} \quad \text{and} \quad \mathbf{x} = \begin{bmatrix} 1 \\ 1 \end{bmatrix}.$$

23. Determine a scalar r such that $A\mathbf{x} = r\mathbf{x}$, where

$$A = \begin{bmatrix} 1 & 2 & -1 \\ 1 & 0 & 1 \\ 4 & -4 & 5 \end{bmatrix} \quad \text{and} \quad \mathbf{x} = \begin{bmatrix} -\frac{1}{2} \\ \frac{1}{4} \\ 1 \end{bmatrix}.$$

24. Prove that if $A\mathbf{x} = r\mathbf{x}$ for $n \times n$ matrix A, $n \times 1$ matrix $\mathbf{x}$, and scalar r, then $A\mathbf{y} = r\mathbf{y}$, where $\mathbf{y} = s\mathbf{x}$ for any scalar s.

25. Determine a scalar s such that $A^2\mathbf{x} = s\mathbf{x}$ when $A\mathbf{x} = r\mathbf{x}$.

26. Prove Theorem 1.4(a).

27. Prove Theorem 1.4(b) and (d).

28. Verify Theorem 1.4(a), (b), and (d) for

$$A = \begin{bmatrix} 1 & 3 & 2 \\ 2 & 1 & -3 \end{bmatrix}, \quad B = \begin{bmatrix} 4 & 2 & -1 \\ -2 & 1 & 5 \end{bmatrix},$$
$$\text{and} \quad r = -4.$$

29. Verify Theorem 1.4(c) for

$$A = \begin{bmatrix} 1 & 3 & 2 \\ 2 & 1 & -3 \end{bmatrix} \quad \text{and} \quad B = \begin{bmatrix} 3 & -1 \\ 2 & 4 \\ 1 & 2 \end{bmatrix}.$$

30. Let

$$A = \begin{bmatrix} 2 \\ -1 \\ 3 \end{bmatrix}, \quad B = \begin{bmatrix} 3 \\ -2 \\ -4 \end{bmatrix}, \quad \text{and} \quad C = \begin{bmatrix} -1 \\ 5 \\ 1 \end{bmatrix}.$$

(a) Compute $(AB^T)C$.

(b) Compute B^TC and multiply the result by A on the right. (*Hint*: B^TC is 1×1).

(c) Explain why $(AB^T)C = (B^TC)A$.

31. Determine a constant k such that $(kA)^T(kA) = 1$, where $A = \begin{bmatrix} -2 \\ 1 \\ -1 \end{bmatrix}$. Is there more than one value of k that could be used?

32. Find three 2×2 matrices, A, B, and C such that $AB = AC$ with $B \neq C$ and $A \neq O$.

33. Let A be an $n \times n$ matrix and c a real number. Show that if $cA = O$, then $c = 0$ or $A = O$.

34. Determine all 2×2 matrices A such that $AB = BA$ for any 2×2 matrix B.

35. Show that $(A - B)^T = A^T - B^T$.

36. Let $\mathbf{x}_1$ and $\mathbf{x}_2$ be solutions to the homogeneous linear system $A\mathbf{x} = \mathbf{0}$.

(a) Show that $\mathbf{x}_1 + \mathbf{x}_2$ is a solution.

(b) Show that $\mathbf{x}_1 - \mathbf{x}_2$ is a solution.

(c) For any scalar r, show that $r\mathbf{x}_1$ is a solution.

(d) For any scalars r and s, show that $r\mathbf{x}_1 + s\mathbf{x}_2$ is a solution.

37. Show that if $A\mathbf{x} = \mathbf{b}$ has more than one solution, then it has infinitely many solutions. (*Hint*: If $\mathbf{x}_1$ and $\mathbf{x}_2$ are solutions, consider $\mathbf{x}_3 = r\mathbf{x}_1 + s\mathbf{x}_2$, where $r + s = 1$.)

38. Show that if $\mathbf{x}_1$ and $\mathbf{x}_2$ are solutions to the linear system $A\mathbf{x} = \mathbf{b}$, then $\mathbf{x}_1 - \mathbf{x}_2$ is a solution to the associated homogeneous system $A\mathbf{x} = \mathbf{0}$.

39. Let

$$A = \begin{bmatrix} 6 & -1 & 1 \\ 0 & 13 & -16 \\ 0 & 8 & -11 \end{bmatrix} \quad \text{and} \quad \mathbf{x} = \begin{bmatrix} 10.5 \\ 21.0 \\ 10.5 \end{bmatrix}.$$

(a) Determine a scalar r such that $A\mathbf{x} = r\mathbf{x}$.

(b) Is it true that $A^T\mathbf{x} = r\mathbf{x}$ for the value r determined in part (a)?

40. Repeat Exercise 39 with

$$A = \begin{bmatrix} -3.35 & -3.00 & 3.60 \\ 1.20 & 2.05 & -6.20 \\ -3.60 & -2.40 & 3.85 \end{bmatrix}$$
$$\text{and} \quad \mathbf{x} = \begin{bmatrix} 12.5 \\ -12.5 \\ 6.25 \end{bmatrix}.$$

41. Let $A = \begin{bmatrix} 0.1 & 0.01 \\ 0.001 & 0.0001 \end{bmatrix}$. In your software, set the display format to show as many decimal places as possible, then compute

$$B = 10 * A,$$
$$C = \underbrace{A + A + A + A + A + A + A + A + A + A}_{10 \text{ summands}},$$

and

$$D = B - C.$$

If D is not O, then you have verified that scalar multiplication by a positive integer and successive addition are not the same in your computing environment. (It is not unusual that $D \neq O$, since many computing environments use only a "model" of exact arithmetic, called floating-point arithmetic.)

42. Let $A = \begin{bmatrix} 1 & 1 \\ 1 & 1 \end{bmatrix}$. In your software, set the display to show as many decimal places as possible. Experiment to find a positive integer k such that $A + 10^{-k} * A$ is equal to A. If you find such an integer k, you have verified that there is more than one matrix in your computational environment that plays the role of O.

1.5 Special Types of Matrices and Partitioned Matrices

We have already introduced one special type of matrix O, the matrix all of whose entries are zero. We now consider several other types of matrices whose structures are rather specialized and for which it will be convenient to have special names.

An $n \times n$ matrix $A = \begin{bmatrix} a_{ij} \end{bmatrix}$ is called a **diagonal matrix** if $a_{ij} = 0$ for $i \neq j$. Thus, for a diagonal matrix, the terms *off* the main diagonal are all zero. Note that O is a diagonal matrix. A **scalar matrix** is a diagonal matrix whose diagonal elements are equal. The scalar matrix $I_n = \begin{bmatrix} d_{ij} \end{bmatrix}$, where $d_{ii} = 1$ and $d_{ij} = 0$ for $i \neq j$, is called the $n \times n$ **identity matrix**.

EXAMPLE 1 Let

$$A = \begin{bmatrix} 1 & 0 & 0 \\ 0 & 2 & 0 \\ 0 & 0 & 3 \end{bmatrix}, \quad B = \begin{bmatrix} 2 & 0 & 0 \\ 0 & 2 & 0 \\ 0 & 0 & 2 \end{bmatrix}, \quad \text{and} \quad I_3 = \begin{bmatrix} 1 & 0 & 0 \\ 0 & 1 & 0 \\ 0 & 0 & 1 \end{bmatrix}.$$

Then A, B, and I_3 are diagonal matrices; B and I_3 are scalar matrices; and I_3 is the 3×3 identity matrix. ■

It is easy to show (Exercise 1) that if A is any $m \times n$ matrix, then

$$AI_n = A \quad \text{and} \quad I_m A = A.$$

Also, if A is a scalar matrix, then $A = rI_n$ for some scalar r.

Suppose that A is a square matrix. We now define the powers of a matrix, for p a positive integer, by

$$A^p = \underbrace{A \cdot A \cdot \cdots \cdot A}_{p \text{ factors}}.$$

If A is $n \times n$, we also define

$$A^0 = I_n.$$

For nonnegative integers p and q, the familiar laws of exponents for the real numbers can also be proved for matrix multiplication of a square matrix A (Exercise 8):

$$A^p A^q = A^{p+q} \quad \text{and} \quad (A^p)^q = A^{pq}.$$

It should also be noted that the rule

$$(AB)^p = A^p B^p$$

does not hold for square matrices unless $AB = BA$ (Exercise 9).

An $n \times n$ matrix $A = \begin{bmatrix} a_{ij} \end{bmatrix}$ is called **upper triangular** if $a_{ij} = 0$ for $i > j$. It is called **lower triangular** if $a_{ij} = 0$ for $i < j$. A diagonal matrix is both upper triangular and lower triangular.

EXAMPLE 2 The matrix

$$A = \begin{bmatrix} 1 & 3 & 3 \\ 0 & 3 & 5 \\ 0 & 0 & 2 \end{bmatrix}$$

is upper triangular, and

$$B = \begin{bmatrix} 1 & 0 & 0 \\ 2 & 3 & 0 \\ 3 & 5 & 2 \end{bmatrix}$$

is lower triangular. ■

DEFINITION 1.8 A matrix A with real entries is called **symmetric** if $A^T = A$.

DEFINITION 1.9 A matrix A with real entries is called **skew symmetric** if $A^T = -A$.

EXAMPLE 3 $A = \begin{bmatrix} 1 & 2 & 3 \\ 2 & 4 & 5 \\ 3 & 5 & 6 \end{bmatrix}$ is a symmetric matrix. ■

EXAMPLE 4 $B = \begin{bmatrix} 0 & 2 & 3 \\ -2 & 0 & -4 \\ -3 & 4 & 0 \end{bmatrix}$ is a skew symmetric matrix. ■

We can make a few observations about symmetric and skew symmetric matrices; the proofs of most of these statements will be left as exercises.

It follows from the preceding definitions that if A is symmetric or skew symmetric, then A is a square matrix. If A is a symmetric matrix, then the entries of A are symmetric with respect to the main diagonal of A. Also, A is symmetric if and only if $a_{ij} = a_{ji}$, and A is skew symmetric if and only if $a_{ij} = -a_{ji}$. Moreover, if A is skew symmetric, then the entries on the main diagonal of A are all zero. An important property of symmetric and skew symmetric matrices is the following: If A is an $n \times n$ matrix, then we can show that $A = S + K$, where S is symmetric and K is skew symmetric. Moreover, this decomposition is unique (Exercise 29).

■ Partitioned Matrices

If we start out with an $m \times n$ matrix $A = \begin{bmatrix} a_{ij} \end{bmatrix}$ and then cross out some, but not all, of its rows or columns, we obtain a **submatrix** of A.

EXAMPLE 5 Let

$$A = \begin{bmatrix} 1 & 2 & 3 & 4 \\ -2 & 4 & -3 & 5 \\ 3 & 0 & 5 & -3 \end{bmatrix}.$$

If we cross out the second row and third column, we get the submatrix

$$\begin{bmatrix} 1 & 2 & 4 \\ 3 & 0 & -3 \end{bmatrix}.$$
■

A matrix can be partitioned into submatrices by drawing horizontal lines between rows and vertical lines between columns. Of course, the partitioning can be carried out in many different ways.

EXAMPLE 6

The matrix

$$A = \left[\begin{array}{ccc|cc} a_{11} & a_{12} & a_{13} & a_{14} & a_{15} \\ a_{21} & a_{22} & a_{23} & a_{24} & a_{25} \\ \hline a_{31} & a_{32} & a_{33} & a_{34} & a_{35} \\ a_{41} & a_{42} & a_{43} & a_{44} & a_{45} \end{array}\right] = \begin{bmatrix} A_{11} & A_{12} \\ A_{21} & A_{22} \end{bmatrix}$$

can be partitioned as indicated previously. We could also write

$$A = \left[\begin{array}{cc|cc|c} a_{11} & a_{12} & a_{13} & a_{14} & a_{15} \\ a_{21} & a_{22} & a_{23} & a_{24} & a_{25} \\ \hline a_{31} & a_{32} & a_{33} & a_{34} & a_{35} \\ a_{41} & a_{42} & a_{43} & a_{44} & a_{45} \end{array}\right] = \begin{bmatrix} \widehat{A}_{11} & \widehat{A}_{12} & \widehat{A}_{13} \\ \widehat{A}_{21} & \widehat{A}_{22} & \widehat{A}_{23} \end{bmatrix}, \tag{1}$$

which gives another partitioning of A. We thus speak of **partitioned matrices.** ■

EXAMPLE 7

The augmented matrix (defined in Section 1.3) of a linear system is a partitioned matrix. Thus, if $A\mathbf{x} = \mathbf{b}$, we can write the augmented matrix of this system as $\left[A \mid \mathbf{b}\right]$. ■

If A and B are both $m \times n$ matrices that are partitioned in the same way, then $A + B$ is produced simply by adding the corresponding submatrices of A and B. Similarly, if A is a partitioned matrix, then the scalar multiple cA is obtained by forming the scalar multiple of each submatrix.

If A is partitioned as shown in (1) and

$$B = \left[\begin{array}{cc|cc} b_{11} & b_{12} & b_{13} & b_{14} \\ b_{21} & b_{22} & b_{23} & b_{24} \\ \hline b_{31} & b_{32} & b_{33} & b_{34} \\ b_{41} & b_{42} & b_{43} & b_{44} \\ \hline b_{51} & b_{52} & b_{53} & b_{54} \end{array}\right] = \begin{bmatrix} B_{11} & B_{12} \\ B_{21} & B_{22} \\ B_{31} & B_{32} \end{bmatrix},$$

then by straightforward computations we can show that

$$AB = \left[\begin{array}{c|c} (\widehat{A}_{11}B_{11} + \widehat{A}_{12}B_{21} + \widehat{A}_{13}B_{31}) & (\widehat{A}_{11}B_{12} + \widehat{A}_{12}B_{22} + \widehat{A}_{13}B_{32}) \\ \hline (\widehat{A}_{21}B_{11} + \widehat{A}_{22}B_{21} + \widehat{A}_{23}B_{31}) & (\widehat{A}_{21}B_{12} + \widehat{A}_{22}B_{22} + \widehat{A}_{23}B_{32}) \end{array}\right].$$

EXAMPLE 8 Let

$$A = \left[\begin{array}{cc|cc} 1 & 0 & 1 & 0 \\ 0 & 2 & 3 & -1 \\ \hline 2 & 0 & -4 & 0 \\ 0 & 1 & 0 & 3 \end{array}\right] = \begin{bmatrix} A_{11} & A_{12} \\ A_{21} & A_{22} \end{bmatrix}$$

and let

$$B = \left[\begin{array}{ccc|ccc} 2 & 0 & 0 & 1 & 1 & -1 \\ 0 & 1 & 1 & -1 & 2 & 2 \\ \hline 1 & 3 & 0 & 0 & 1 & 0 \\ -3 & -1 & 2 & 1 & 0 & -1 \end{array}\right] = \begin{bmatrix} B_{11} & B_{12} \\ B_{21} & B_{22} \end{bmatrix}.$$

Then

$$AB = C = \left[\begin{array}{ccc|ccc} 3 & 3 & 0 & 1 & 2 & -1 \\ 6 & 12 & 0 & -3 & 7 & 5 \\ \hline 0 & -12 & 0 & 2 & -2 & -2 \\ -9 & -2 & 7 & 2 & 2 & -1 \end{array}\right] = \begin{bmatrix} C_{11} & C_{12} \\ C_{21} & C_{22} \end{bmatrix},$$

where C_{11} should be $A_{11}B_{11} + A_{12}B_{21}$. We verify that C_{11} is this expression as follows:

$$\begin{aligned} A_{11}B_{11} + A_{12}B_{21} &= \begin{bmatrix} 1 & 0 \\ 0 & 2 \end{bmatrix}\begin{bmatrix} 2 & 0 & 0 \\ 0 & 1 & 1 \end{bmatrix} + \begin{bmatrix} 1 & 0 \\ 3 & -1 \end{bmatrix}\begin{bmatrix} 1 & 3 & 0 \\ -3 & -1 & 2 \end{bmatrix} \\ &= \begin{bmatrix} 2 & 0 & 0 \\ 0 & 2 & 2 \end{bmatrix} + \begin{bmatrix} 1 & 3 & 0 \\ 6 & 10 & -2 \end{bmatrix} \\ &= \begin{bmatrix} 3 & 3 & 0 \\ 6 & 12 & 0 \end{bmatrix} = C_{11}. \end{aligned}$$

■

This method of multiplying partitioned matrices is also known as **block multiplication**. Partitioned matrices can be used to great advantage when matrices exceed the memory capacity of a computer. Thus, in multiplying two partitioned matrices, one can keep the matrices on disk and bring into memory only the submatrices required to form the submatrix products. The products, of course, can be downloaded as they are formed. The partitioning must be done in such a way that the products of corresponding submatrices are defined.

Partitioning of a matrix implies a subdivision of the information into blocks, or units. The reverse process is to consider individual matrices as blocks and adjoin them to form a partitioned matrix. The only requirement is that after the blocks have been joined, all rows have the same number of entries and all columns have the same number of entries.

EXAMPLE 9 Let

$$B = \begin{bmatrix} 2 \\ 3 \end{bmatrix}, \quad C = \begin{bmatrix} 1 & -1 & 0 \end{bmatrix}, \quad \text{and} \quad D = \begin{bmatrix} 9 & 8 & -4 \\ 6 & 7 & 5 \end{bmatrix}.$$

Then we have

$$\begin{bmatrix} B & D \end{bmatrix} = \left[\begin{array}{c|ccc} 2 & 9 & 8 & -4 \\ 3 & 6 & 7 & 5 \end{array}\right], \quad \begin{bmatrix} D \\ C \end{bmatrix} = \left[\begin{array}{ccc} 9 & 8 & -4 \\ 6 & 7 & 5 \\ \hline 1 & -1 & 0 \end{array}\right],$$

and

$$\begin{bmatrix} \begin{bmatrix} D \\ C \end{bmatrix} & C^T \end{bmatrix} = \left[\begin{array}{ccc|c} 9 & 8 & -4 & 1 \\ 6 & 7 & 5 & -1 \\ \hline 1 & -1 & 0 & 0 \end{array}\right].$$

■

Adjoining matrix blocks to expand information structures is done regularly in a variety of applications. It is common for a business to keep monthly sales data for a year in a 1×12 matrix and then adjoin such matrices to build a sales history matrix for a period of years. Similarly, results of new laboratory experiments are adjoined to existing data to update a database in a research facility.

We have already noted in Example 7 that the augmented matrix of the linear system $A\mathbf{x} = \mathbf{b}$ is a partitioned matrix. At times we shall need to solve several linear systems in which the coefficient matrix A is the same, but the right sides of the systems are different, say, **b**, **c**, and **d**. In these cases we shall find it convenient to consider the partitioned matrix $\left[\begin{array}{c|c|c|c} A & \mathbf{b} & \mathbf{c} & \mathbf{d} \end{array}\right]$. (See Section 4.8.)

■ Nonsingular Matrices

We now come to a special type of square matrix and formulate the notion corresponding to the reciprocal of a nonzero real number.

DEFINITION 1.10 An $n \times n$ matrix A is called **nonsingular**, or **invertible**, if there exists an $n \times n$ matrix B such that $AB = BA = I_n$; such a B is called an **inverse** of A. Otherwise, A is called **singular**, or **noninvertible**.

Remark In Theorem 2.11, Section 2.3, we show that if $AB = I_n$, then $BA = I_n$. Thus, to verify that B is an inverse of A, we need verify only that $AB = I_n$.

EXAMPLE 10 Let $A = \begin{bmatrix} 2 & 3 \\ 2 & 2 \end{bmatrix}$ and $B = \begin{bmatrix} -1 & \frac{3}{2} \\ 1 & -1 \end{bmatrix}$. Since $AB = BA = I_2$, we conclude that B is an inverse of A. ■

Theorem 1.5 The inverse of a matrix, if it exists, is unique.

Proof

Let B and C be inverses of A. Then

$$AB = BA = I_n \quad \text{and} \quad AC = CA = I_n.$$

We then have $B = BI_n = B(AC) = (BA)C = I_nC = C$, which proves that the inverse of a matrix, if it exists, is unique. ■

Because of this uniqueness, we write the inverse of a nonsingular matrix A as A^{-1}. Thus

$$AA^{-1} = A^{-1}A = I_n.$$

EXAMPLE 11 Let

$$A = \begin{bmatrix} 1 & 2 \\ 3 & 4 \end{bmatrix}.$$

If A^{-1} exists, let

$$A^{-1} = \begin{bmatrix} a & b \\ c & d \end{bmatrix}.$$

Then we must have

$$AA^{-1} = \begin{bmatrix} 1 & 2 \\ 3 & 4 \end{bmatrix}\begin{bmatrix} a & b \\ c & d \end{bmatrix} = I_2 = \begin{bmatrix} 1 & 0 \\ 0 & 1 \end{bmatrix},$$

so that

$$\begin{bmatrix} a+2c & b+2d \\ 3a+4c & 3b+4d \end{bmatrix} = \begin{bmatrix} 1 & 0 \\ 0 & 1 \end{bmatrix}.$$

Equating corresponding entries of these two matrices, we obtain the linear systems

$$\begin{aligned} a + 2c &= 1 \\ 3a + 4c &= 0 \end{aligned} \quad \text{and} \quad \begin{aligned} b + 2d &= 0 \\ 3b + 4d &= 1. \end{aligned}$$

The solutions are (verify) $a = -2$, $c = \frac{3}{2}$, $b = 1$, and $d = -\frac{1}{2}$. Moreover, since the matrix

$$\begin{bmatrix} a & b \\ c & d \end{bmatrix} = \begin{bmatrix} -2 & 1 \\ \frac{3}{2} & -\frac{1}{2} \end{bmatrix}$$

also satisfies the property that

$$\begin{bmatrix} -2 & 1 \\ \frac{3}{2} & -\frac{1}{2} \end{bmatrix}\begin{bmatrix} 1 & 2 \\ 3 & 4 \end{bmatrix} = \begin{bmatrix} 1 & 0 \\ 0 & 1 \end{bmatrix},$$

we conclude that A is nonsingular and that

$$A^{-1} = \begin{bmatrix} -2 & 1 \\ \frac{3}{2} & -\frac{1}{2} \end{bmatrix}.$$

■

EXAMPLE 12 Let

$$A = \begin{bmatrix} 1 & 2 \\ 2 & 4 \end{bmatrix}.$$

If A^{-1} exists, let

$$A^{-1} = \begin{bmatrix} a & b \\ c & d \end{bmatrix}.$$

Then we must have

$$AA^{-1} = \begin{bmatrix} 1 & 2 \\ 2 & 4 \end{bmatrix}\begin{bmatrix} a & b \\ c & d \end{bmatrix} = I_2 = \begin{bmatrix} 1 & 0 \\ 0 & 1 \end{bmatrix},$$

so that

$$\begin{bmatrix} a+2c & b+2d \\ 2a+4c & 2b+4d \end{bmatrix} = \begin{bmatrix} 1 & 0 \\ 0 & 1 \end{bmatrix}.$$

Equating corresponding entries of these two matrices, we obtain the linear systems

$$\begin{aligned} a + 2c &= 1 \\ 2a + 4c &= 0 \end{aligned} \quad \text{and} \quad \begin{aligned} b + 2d &= 0 \\ 2b + 4d &= 1. \end{aligned}$$

These linear systems have no solutions, so our assumption that A^{-1} exists is incorrect. Thus A is singular. ■

We next establish several properties of inverses of matrices.

Theorem 1.6 If A and B are both nonsingular $n \times n$ matrices, then AB is nonsingular and $(AB)^{-1} = B^{-1}A^{-1}$.

Proof

We have $(AB)(B^{-1}A^{-1}) = A(BB^{-1})A^{-1} = (AI_n)A^{-1} = AA^{-1} = I_n$. Similarly, $(B^{-1}A^{-1})(AB) = I_n$. Therefore AB is nonsingular. Since the inverse of a matrix is unique, we conclude that $(AB)^{-1} = B^{-1}A^{-1}$. ■

Corollary 1.1 If $A_1, A_2, \ldots, A_r$ are $n \times n$ nonsingular matrices, then $A_1A_2\cdots A_r$ is nonsingular and $(A_1A_2\cdots A_r)^{-1} = A_r^{-1}A_{r-1}^{-1}\cdots A_1^{-1}$.

Proof

Exercise 44. ■

Theorem 1.7 If A is a nonsingular matrix, then A^{-1} is nonsingular and $(A^{-1})^{-1} = A$.

Proof

Exercise 45. ■

Theorem 1.8 If A is a nonsingular matrix, then A^T is nonsingular and $(A^{-1})^T = (A^T)^{-1}$.

Proof

We have $AA^{-1} = I_n$. Taking transposes of both sides, we get

$$(A^{-1})^T A^T = I_n^T = I_n.$$

Taking transposes of both sides of the equation $A^{-1}A = I_n$, we find, similarly, that

$$(A^T)(A^{-1})^T = I_n.$$

These equations imply that $(A^{-1})^T = (A^T)^{-1}$. ■

EXAMPLE 13 If

$$A = \begin{bmatrix} 1 & 2 \\ 3 & 4 \end{bmatrix},$$

then from Example 11

$$A^{-1} = \begin{bmatrix} -2 & 1 \\ \frac{3}{2} & -\frac{1}{2} \end{bmatrix} \quad \text{and} \quad (A^{-1})^T = \begin{bmatrix} -2 & \frac{3}{2} \\ 1 & -\frac{1}{2} \end{bmatrix}.$$

Also (verify),

$$A^T = \begin{bmatrix} 1 & 3 \\ 2 & 4 \end{bmatrix} \quad \text{and} \quad (A^T)^{-1} = \begin{bmatrix} -2 & \frac{3}{2} \\ 1 & -\frac{1}{2} \end{bmatrix}.$$

■

Suppose that A is nonsingular. Then $AB = AC$ implies that $B = C$ (Exercise 50), and $AB = O$ implies that $B = O$ (Exercise 51).

It follows from Theorem 1.8 that if A is a symmetric nonsingular matrix, then A^{-1} is symmetric. (See Exercise 54.)

■ Linear Systems and Inverses

If A is an $n \times n$ matrix, then the linear system $A\mathbf{x} = \mathbf{b}$ is a system of n equations in n unknowns. Suppose that A is nonsingular. Then A^{-1} exists, and we can multiply $A\mathbf{x} = \mathbf{b}$ by A^{-1} on the left on both sides, yielding

$$\begin{aligned} A^{-1}(A\mathbf{x}) &= A^{-1}\mathbf{b} \\ (A^{-1}A)\mathbf{x} &= A^{-1}\mathbf{b} \\ I_n\mathbf{x} &= A^{-1}\mathbf{b} \\ \mathbf{x} &= A^{-1}\mathbf{b}. \end{aligned} \tag{2}$$

Moreover, $\mathbf{x} = A^{-1}\mathbf{b}$ is clearly a solution to the given linear system. Thus, if A is nonsingular, we have a unique solution. We restate this result for emphasis:

> If A is an $n \times n$ matrix, then the linear system $A\mathbf{x} = \mathbf{b}$ has the unique solution $\mathbf{x} = A^{-1}\mathbf{b}$. Moreover, if $\mathbf{b} = \mathbf{0}$, then the unique solution to the homogeneous system $A\mathbf{x} = \mathbf{0}$ is $\mathbf{x} = \mathbf{0}$.

If A is a nonsingular $n \times n$ matrix, Equation (2) implies that if the linear system $A\mathbf{x} = \mathbf{b}$ needs to be solved repeatedly for different $\mathbf{b}$'s, we need compute A^{-1} only once; then whenever we change $\mathbf{b}$, we find the corresponding solution $\mathbf{x}$ by forming $A^{-1}\mathbf{b}$. Although this is certainly a valid approach, its value is of a more theoretical rather than practical nature, since a more efficient procedure for solving such problems is presented in Section 2.5.

EXAMPLE 14

Suppose that A is the matrix of Example 11 so that

$$A^{-1} = \begin{bmatrix} -2 & 1 \\ \frac{3}{2} & -\frac{1}{2} \end{bmatrix}.$$

If

$$\mathbf{b} = \begin{bmatrix} 8 \\ 6 \end{bmatrix},$$

then the solution to the linear system $A\mathbf{x} = \mathbf{b}$ is

$$\mathbf{x} = A^{-1}\mathbf{b} = \begin{bmatrix} -2 & 1 \\ \frac{3}{2} & -\frac{1}{2} \end{bmatrix} \begin{bmatrix} 8 \\ 6 \end{bmatrix} = \begin{bmatrix} -10 \\ 9 \end{bmatrix}.$$

On the other hand, if

$$\mathbf{b} = \begin{bmatrix} 10 \\ 20 \end{bmatrix},$$

then

$$\mathbf{x} = A^{-1} \begin{bmatrix} 10 \\ 20 \end{bmatrix} = \begin{bmatrix} 0 \\ 5 \end{bmatrix}.$$

■

■ Application A: Recursion Relation; the Fibonacci Sequence

In 1202, Leonardo of Pisa, also called Fibonacci,* wrote a book on mathematics in which he posed the following problem: A pair of newborn rabbits begins to breed at the age of 1 month, and thereafter produces one pair of offspring per month. Suppose that we start with a pair of newly born rabbits and that none of the rabbits produced from this pair dies. How many pairs of rabbits will there be at the beginning of each month?

At the beginning of month 0, we have the newly born pair of rabbits P_1. At the beginning of month 1 we still have only the original pair of rabbits P_1, which have not yet produced any offspring. At the beginning of month 2 we have the original pair P_1 and its first pair of offspring, P_2. At the beginning of month 3 we have the original pair P_1, its first pair of offspring P_2 born at the beginning of month 2, and its second pair of offspring, P_3. At the beginning of month 4 we have P_1, P_2, and P_3; P_4, the offspring of P_1; and P_5, the offspring of P_2. Let u_n denote the number of pairs of rabbits at the beginning of month n. We see that

$$u_0 = 1, \quad u_1 = 1, \quad u_2 = 2, \quad u_3 = 3, \quad u_4 = 5, \quad u_5 = 8.$$

The sequence expands rapidly, and we get

$$1, 1, 2, 3, 5, 8, 13, 21, 34, 55, 89, 144, \ldots$$

To obtain a formula for u_n, we proceed as follows. The number of pairs of rabbits that are alive at the beginning of month n is u_{n-1}, the number of pairs who were alive the previous month, plus the number of pairs newly born at the beginning of month n. The latter number is u_{n-2}, since a pair of rabbits produces a pair of offspring, starting with its second month of life. Thus

$$u_n = u_{n-1} + u_{n-2}. \tag{3}$$

*Leonardo Fibonacci of Pisa (about 1170–1250) was born and lived most of his life in Pisa, Italy. When he was about 20, his father was appointed director of Pisan commercial interests in northern Africa, now a part of Algeria. Leonardo accompanied his father to Africa and for several years traveled extensively throughout the Mediterranean area on behalf of his father. During these travels he learned the Hindu–Arabic method of numeration and calculation and decided to promote its use in Italy. This was one purpose of his most famous book, *Liber Abaci*, which appeared in 1202 and contained the rabbit problem stated here.

That is, each number is the sum of its two predecessors. The resulting sequence of numbers, called a **Fibonacci sequence**, occurs in a remarkable variety of applications, such as the distribution of leaves on certain trees, the arrangements of seeds on sunflowers, search techniques in numerical analysis, the generation of random numbers in statistics, and others.

To compute u_n by the **recursion relation** (or difference equation) (3), we have to compute $u_0, u_1, \ldots, u_{n-2}, u_{n-1}$. This can be rather tedious for large n. We now develop a formula that will enable us to calculate u_n directly.

In addition to Equation (3), we write

$$u_{n-1} = u_{n-1},$$

so we now have

$$\begin{aligned} u_n &= u_{n-1} + u_{n-2} \\ u_{n-1} &= u_{n-1}, \end{aligned}$$

which can be written in matrix form as

$$\begin{bmatrix} u_n \\ u_{n-1} \end{bmatrix} = \begin{bmatrix} 1 & 1 \\ 1 & 0 \end{bmatrix} \begin{bmatrix} u_{n-1} \\ u_{n-2} \end{bmatrix}. \tag{4}$$

We now define, in general,

$$\mathbf{w}_k = \begin{bmatrix} u_{k+1} \\ u_k \end{bmatrix} \quad \text{and} \quad A = \begin{bmatrix} 1 & 1 \\ 1 & 0 \end{bmatrix} \qquad (0 \le k \le n-1)$$

so that

$$\begin{aligned} \mathbf{w}_0 &= \begin{bmatrix} u_1 \\ u_0 \end{bmatrix} = \begin{bmatrix} 1 \\ 1 \end{bmatrix}, \\ \mathbf{w}_1 &= \begin{bmatrix} u_2 \\ u_1 \end{bmatrix} = \begin{bmatrix} 2 \\ 1 \end{bmatrix}, \ldots, \quad \mathbf{w}_{n-2} = \begin{bmatrix} u_{n-1} \\ u_{n-2} \end{bmatrix}, \quad \text{and} \quad \mathbf{w}_{n-1} = \begin{bmatrix} u_n \\ u_{n-1} \end{bmatrix}. \end{aligned}$$

Then (4) can be written as

$$\mathbf{w}_{n-1} = A\mathbf{w}_{n-2}.$$

Thus

$$\begin{aligned} \mathbf{w}_1 &= A\mathbf{w}_0 \\ \mathbf{w}_2 &= A\mathbf{w}_1 = A(A\mathbf{w}_0) = A^2\mathbf{w}_0 \\ \mathbf{w}_3 &= A\mathbf{w}_2 = A(A^2\mathbf{w}_0) = A^3\mathbf{w}_0 \\ &\vdots \\ \mathbf{w}_{n-1} &= A^{n-1}\mathbf{w}_0. \end{aligned}$$

Hence, to find u_n, we merely have to calculate A^{n-1}, which is still rather tedious if n is large. In Chapter 7 we develop a more efficient way to compute the Fibonacci numbers that involves powers of a *diagonal* matrix. (See the discussion exercises in Chapter 7.)

Key Terms

Diagonal matrix
Identity matrix
Powers of a matrix
Upper triangular matrix
Lower triangular matrix
Symmetric matrix
Skew symmetric matrix
Submatrix
Partitioning
Partitioned matrix
Nonsingular (invertible) matrix
Inverse
Singular (noninvertible) matrix
Properties of nonsingular matrices
Linear system with nonsingular coefficient matrix
Fibonacci sequence

1.5 Exercises

1. (a) Show that if A is any $m \times n$ matrix, then $I_m A = A$ and $AI_n = A$.
   (b) Show that if A is an $n \times n$ scalar matrix, then $A = rI_n$ for some real number r.
2. Prove that the sum, product, and scalar multiple of diagonal, scalar, and upper (lower) triangular matrices is diagonal, scalar, and upper (lower) triangular, respectively.
3. Prove: If A and B are $n \times n$ diagonal matrices, then $AB = BA$.
4. Let
$$A = \begin{bmatrix} 3 & 2 & -1 \\ 0 & -4 & 3 \\ 0 & 0 & 0 \end{bmatrix} \quad \text{and} \quad B = \begin{bmatrix} 6 & -3 & 2 \\ 0 & 2 & 4 \\ 0 & 0 & 3 \end{bmatrix}.$$
   Verify that $A + B$ and AB are upper triangular.
5. Describe all matrices that are both upper and lower triangular.
6. Let $A = \begin{bmatrix} 1 & 2 \\ 3 & -2 \end{bmatrix}$ and $B = \begin{bmatrix} 1 & -1 \\ 2 & 3 \end{bmatrix}$. Compute each of the following:
   (a) A^2 (b) B^3 (c) $(AB)^2$
7. Let $A = \begin{bmatrix} 1 & 0 & -1 \\ 2 & 1 & 1 \\ 3 & 1 & 0 \end{bmatrix}$ and $B = \begin{bmatrix} 0 & 0 & 1 \\ -1 & 1 & 1 \\ 2 & 0 & 1 \end{bmatrix}$.
   Compute each of the following:
   (a) A^3 (b) B^2 (c) $(AB)^3$
8. Let p and q be nonnegative integers and let A be a square matrix. Show that
$$A^p A^q = A^{p+q} \quad \text{and} \quad (A^p)^q = A^{pq}.$$
9. If $AB = BA$ and p is a nonnegative integer, show that $(AB)^p = A^p B^p$.
10. If p is a nonnegative integer and c is a scalar, show that $(cA)^p = c^p A^p$.
11. For a square matrix A and a nonnegative integer p, show that $(A^T)^p = (A^p)^T$.
12. For a nonsingular matrix A and a nonnegative integer p, show that $(A^p)^{-1} = (A^{-1})^p$.
13. For a nonsingular matrix A and nonzero scalar k, show that $(kA)^{-1} = \frac{1}{k}A^{-1}$.
14. (a) Show that every scalar matrix is symmetric.
    (b) Is every scalar matrix nonsingular? Explain.
    (c) Is every diagonal matrix a scalar matrix? Explain.
15. Find a 2×2 matrix $B \neq O$ and $B \neq I_2$ such that $AB = BA$, where $A = \begin{bmatrix} 1 & 2 \\ 2 & 1 \end{bmatrix}$. How many such matrices B are there?
16. Find a 2×2 matrix $B \neq O$ and $B \neq I_2$ such that $AB = BA$, where $A = \begin{bmatrix} 1 & 2 \\ 0 & 1 \end{bmatrix}$. How many such matrices B are there?
17. Prove or disprove: For any $n \times n$ matrix A, $A^T A = AA^T$.
18. (a) Show that A is symmetric if and only if $a_{ij} = a_{ji}$ for all i, j.
    (b) Show that A is skew symmetric if and only if $a_{ij} = -a_{ji}$ for all i, j.
    (c) Show that if A is skew symmetric, then the elements on the main diagonal of A are all zero.
19. Show that if A is a symmetric matrix, then A^T is symmetric.
20. Describe all skew symmetric scalar matrices.
21. Show that if A is any $m \times n$ matrix, then AA^T and $A^T A$ are symmetric.
22. Show that if A is any $n \times n$ matrix, then
    (a) $A + A^T$ is symmetric.
    (b) $A - A^T$ is skew symmetric.
23. Show that if A is a symmetric matrix, then A^k, $k = 2, 3, \ldots$, is symmetric.
24. Let A and B be symmetric matrices.
    (a) Show that $A + B$ is symmetric.
    (b) Show that AB is symmetric if and only if $AB = BA$.

25. **(a)** Show that if A is an upper triangular matrix, then A^T is lower triangular.

(b) Show that if A is a lower triangular matrix, then A^T is upper triangular.

26. If A is a skew symmetric matrix, what type of matrix is A^T? Justify your answer.

27. Show that if A is skew symmetric, then the elements on the main diagonal of A are all zero.

28. Show that if A is skew symmetric, then A^k is skew symmetric for any positive odd integer k.

29. Show that if A is an $n \times n$ matrix, then $A = S + K$, where S is symmetric and K is skew symmetric. Also show that this decomposition is unique. (*Hint*: Use Exercise 22.)

30. Let

$$A = \begin{bmatrix} 1 & 3 & -2 \\ 4 & 6 & 2 \\ 5 & 1 & 3 \end{bmatrix}.$$

Find the matrices S and K described in Exercise 29.

31. Show that the matrix $A = \begin{bmatrix} 2 & 3 \\ 4 & 6 \end{bmatrix}$ is singular.

32. If $D = \begin{bmatrix} 4 & 0 & 0 \\ 0 & -2 & 0 \\ 0 & 0 & 3 \end{bmatrix}$, find D^{-1}.

33. Find the inverse of each of the following matrices:

(a) $A = \begin{bmatrix} 1 & 3 \\ 5 & 2 \end{bmatrix}$ **(b)** $A = \begin{bmatrix} 1 & 2 \\ 2 & 1 \end{bmatrix}$

34. If A is a nonsingular matrix whose inverse is $\begin{bmatrix} 2 & 1 \\ 4 & 1 \end{bmatrix}$, find A.

35. If

$$A^{-1} = \begin{bmatrix} 3 & 2 \\ 1 & 3 \end{bmatrix} \quad \text{and} \quad B^{-1} = \begin{bmatrix} 2 & 5 \\ 3 & -2 \end{bmatrix},$$

find $(AB)^{-1}$.

36. Suppose that

$$A^{-1} = \begin{bmatrix} 1 & 2 \\ 1 & 3 \end{bmatrix}.$$

Solve the linear system $A\mathbf{x} = \mathbf{b}$ for each of the following matrices $\mathbf{b}$:

(a) $\begin{bmatrix} 4 \\ 6 \end{bmatrix}$ **(b)** $\begin{bmatrix} 8 \\ 15 \end{bmatrix}$

37. The linear system $AC\mathbf{x} = \mathbf{b}$ is such that A and C are nonsingular with

$$A^{-1} = \begin{bmatrix} 2 & 1 \\ -1 & 1 \end{bmatrix}, \quad C^{-1} = \begin{bmatrix} 2 & 1 \\ 1 & 2 \end{bmatrix}, \quad \text{and } \mathbf{b} = \begin{bmatrix} 2 \\ 3 \end{bmatrix}.$$

Find the solution $\mathbf{x}$.

38. The linear system $A^2\mathbf{x} = \mathbf{b}$ is such that A is nonsingular with

$$A^{-1} = \begin{bmatrix} 3 & 0 \\ 2 & 1 \end{bmatrix} \quad \text{and} \quad \mathbf{b} = \begin{bmatrix} -1 \\ 2 \end{bmatrix}.$$

Find the solution $\mathbf{x}$.

39. The linear system $A^T\mathbf{x} = \mathbf{b}$ is such that A is nonsingular with

$$A^{-1} = \begin{bmatrix} 4 & 1 \\ 1 & 0 \end{bmatrix} \quad \text{and} \quad \mathbf{b} = \begin{bmatrix} 1 \\ -2 \end{bmatrix}.$$

Find the solution $\mathbf{x}$.

40. The linear system $C^T A\mathbf{x} = \mathbf{b}$ is such that A and C are nonsingular, with

$$A^{-1} = \begin{bmatrix} 2 & 1 \\ 3 & 1 \end{bmatrix}, \quad C^{-1} = \begin{bmatrix} 0 & 2 \\ 1 & 4 \end{bmatrix}, \quad \text{and} \quad \mathbf{b} = \begin{bmatrix} 1 \\ 1 \end{bmatrix}.$$

Find the solution $\mathbf{x}$.

41. Consider the linear system $A\mathbf{x} = \mathbf{b}$, where A is the matrix defined in Exercise 33(a).

(a) Find a solution if $\mathbf{b} = \begin{bmatrix} 3 \\ 4 \end{bmatrix}$.

(b) Find a solution if $\mathbf{b} = \begin{bmatrix} 5 \\ 6 \end{bmatrix}$.

42. Find two 2×2 singular matrices whose sum is nonsingular.

43. Find two 2×2 nonsingular matrices whose sum is singular.

44. Prove Corollary 1.1.

45. Prove Theorem 1.7.

46. Prove that if one row (column) of the $n \times n$ matrix A consists entirely of zeros, then A is singular. (*Hint*: Assume that A is nonsingular; that is, there exists an $n \times n$ matrix B such that $AB = BA = I_n$. Establish a contradiction.)

47. Prove: If A is a diagonal matrix with nonzero diagonal entries $a_{11}, a_{22}, \ldots, a_{nn}$, then A is nonsingular and A^{-1} is a diagonal matrix with diagonal entries $1/a_{11}, 1/a_{22}, \ldots, 1/a_{nn}$.

48. Let $A = \begin{bmatrix} 2 & 0 & 0 \\ 0 & -3 & 0 \\ 0 & 0 & 5 \end{bmatrix}$. Compute A^4.

49. For an $n \times n$ diagonal matrix A whose diagonal entries are $a_{11}, a_{22}, \ldots, a_{nn}$, compute A^p for a nonnegative integer p.

50. Show that if $AB = AC$ and A is nonsingular, then $B = C$.

51. Show that if A is nonsingular and $AB = O$ for an $n \times n$ matrix B, then $B = O$.

52. Let $A = \begin{bmatrix} a & b \\ c & d \end{bmatrix}$. Show that A is nonsingular if and only if $ad - bc \neq 0$.

53. Consider the homogeneous system $A\mathbf{x} = \mathbf{0}$, where A is $n \times n$. If A is nonsingular, show that the only solution is the trivial one, $\mathbf{x} = \mathbf{0}$.

54. Prove that if A is symmetric and nonsingular, then A^{-1} is symmetric.

55. Formulate the method for adding partitioned matrices, and verify your method by partitioning the matrices

$$A = \begin{bmatrix} 1 & 3 & -1 \\ 2 & 1 & 0 \\ 2 & -3 & 1 \end{bmatrix} \quad \text{and} \quad B = \begin{bmatrix} 3 & 2 & 1 \\ -2 & 3 & 1 \\ 4 & 1 & 5 \end{bmatrix}$$

in two different ways and finding their sum.

56. Let A and B be the following matrices:

$$A = \begin{bmatrix} 2 & 1 & 3 & 4 & 2 \\ 1 & 2 & 3 & -1 & 4 \\ 2 & 3 & 2 & 1 & 4 \\ 5 & -1 & 3 & 2 & 6 \\ 3 & 1 & 2 & 4 & 6 \\ 2 & -1 & 3 & 5 & 7 \end{bmatrix}$$

and

$$B = \begin{bmatrix} 1 & 2 & 3 & 4 & 1 \\ 2 & 1 & 3 & 2 & -1 \\ 1 & 5 & 4 & 2 & 3 \\ 2 & 1 & 3 & 5 & 7 \\ 3 & 2 & 4 & 6 & 1 \end{bmatrix}.$$

Find AB by partitioning A and B in two different ways.

57. What type of matrix is a linear combination of symmetric matrices? Justify your answer.

58. What type of matrix is a linear combination of scalar matrices? Justify your answer.

59. The matrix form of the recursion relation

$$u_0 = 0,\ u_1 = 1,\ u_n = 5u_{n-1} - 6u_{n-2}, \quad n \geq 2$$

is written as

$$\mathbf{w}_{n-1} = A\mathbf{w}_{n-2},$$

where

$$\mathbf{w}_{n-2} = \begin{bmatrix} u_{n-1} \\ u_{n-2} \end{bmatrix}, \quad \mathbf{w}_{n-1} = \begin{bmatrix} u_n \\ u_{n-1} \end{bmatrix},$$

$$\text{and} \quad A = \begin{bmatrix} 5 & -6 \\ 1 & 0 \end{bmatrix}.$$

(a) Using

$$\mathbf{w}_0 = \begin{bmatrix} u_1 \\ u_0 \end{bmatrix} = \begin{bmatrix} 1 \\ 0 \end{bmatrix},$$

compute $\mathbf{w}_1$, $\mathbf{w}_2$, and $\mathbf{w}_3$. Then make a list of the terms of the recurrence relation u_2, u_3, u_4.

(b) Express $\mathbf{w}_{n-1}$ as a matrix times $\mathbf{w}_0$.

60. The matrix form of the recursion relation

$$u_0 = 1,\ u_1 = 2,\ u_n = 6u_{n-1} - 8u_{n-2}, \quad n \geq 2$$

is written as

$$\mathbf{w}_{n-1} = A\mathbf{w}_{n-2},$$

where

$$\mathbf{w}_{n-2} = \begin{bmatrix} u_{n-1} \\ u_{n-2} \end{bmatrix}, \quad \mathbf{w}_{n-1} = \begin{bmatrix} u_n \\ u_{n-1} \end{bmatrix},$$

$$\text{and} \quad A = \begin{bmatrix} 6 & -8 \\ 1 & 0 \end{bmatrix}.$$

(a) Using

$$\mathbf{w}_0 = \begin{bmatrix} u_1 \\ u_0 \end{bmatrix} = \begin{bmatrix} 2 \\ 1 \end{bmatrix},$$

compute $\mathbf{w}_1$, $\mathbf{w}_2$, $\mathbf{w}_3$, and $\mathbf{w}_4$. Then make a list of the terms of the recurrence relation u_2, u_3, u_4, u_5.

(b) Express $\mathbf{w}_{n-1}$ as a matrix times $\mathbf{w}_0$.

61. For the software you are using, determine the command(s) or procedures required to do each of the following:

(a) Adjoin a row or column to an existing matrix.

(b) Construct the partitioned matrix

$$\begin{bmatrix} A & O \\ O & B \end{bmatrix}$$

from existing matrices A and B, using appropriate size zero matrices.

(c) Extract a submatrix from an existing matrix.

62. Most software for linear algebra has specific commands for extracting the diagonal, upper triangular part, and lower triangular part of a matrix. Determine the corresponding commands for the software that you are using, and experiment with them.

63. Determine the command for computing the inverse of a matrix in the software you use. Usually, if such a command is applied to a singular matrix, a warning message is displayed. Experiment with your inverse command to determine which of the following matrices are singular:

(a) $\begin{bmatrix} 1 & 2 & 3 \\ 4 & 5 & 6 \\ 7 & 8 & 0 \end{bmatrix}$ **(b)** $\begin{bmatrix} 1 & 2 & 3 \\ 4 & 5 & 6 \\ 7 & 8 & 9 \end{bmatrix}$

(c) $\begin{bmatrix} 1 & 2 & 4 \\ -1 & 1 & -1 \\ 2 & -1 & 3 \end{bmatrix}$

64. If B is the inverse of $n \times n$ matrix A, then Definition 1.10 guarantees that $AB = BA = I_n$. The unstated assumption is that exact arithmetic is used. If computer arithmetic is used to compute AB, then AB need not equal I_n and, in fact, BA need not equal AB. However, both AB and BA should be close to I_n. In your software, use the inverse command (see Exercise 63) and form the products AB and BA for each of the following matrices:

(a) $A = \begin{bmatrix} 1 & \frac{1}{3} \\ 0 & \frac{1}{3} \end{bmatrix}$ **(b)** $A = \begin{bmatrix} \frac{1}{2} & \frac{1}{4} \\ \frac{1}{4} & \frac{1}{2} \end{bmatrix}$

(c) $A = \begin{bmatrix} 1 & \frac{1}{2} & \frac{1}{3} \\ \frac{1}{2} & \frac{1}{3} & \frac{1}{4} \\ \frac{1}{3} & \frac{1}{4} & \frac{1}{5} \end{bmatrix}$

65. In Section 1.1 we studied the method of elimination for solving linear systems $A\mathbf{x} = \mathbf{b}$. In Equation (2) of this section we showed that the solution is given by $\mathbf{x} = A^{-1}\mathbf{b}$, if A is nonsingular. Using your software's command for automatically solving linear systems, and its inverse command, compare these two solution techniques on each of the following linear systems:

(a) $A = \begin{bmatrix} \frac{1}{3} & \frac{2}{3} & \frac{4}{3} \\ 0 & \frac{2}{3} & \frac{4}{3} \\ 0 & 0 & \frac{5}{3} \end{bmatrix}$, $\mathbf{b} = \begin{bmatrix} 2 \\ 2 \\ \frac{10}{3} \end{bmatrix}$

(b) $A = \left[a_{ij} \right]$, where

$$a_{ij} = \frac{1}{i + j - 1},$$

$i, j = 1, 2, \ldots, 10$, and $\mathbf{b}$ = the first column of I_{10}.

66. For the software you are using, determine the command for obtaining the powers $A^2, A^3, \ldots$ of a square matrix A. Then, for

$$A = \begin{bmatrix} 0 & 1 & 0 & 0 & 0 \\ 0 & 0 & 1 & 0 & 0 \\ 0 & 0 & 0 & 1 & 0 \\ 0 & 0 & 0 & 0 & 1 \\ 0 & 0 & 0 & 0 & 0 \end{bmatrix},$$

compute the matrix sequence A^k, $k = 2, 3, 4, 5, 6$. Describe the behavior of A^k as $k \to \infty$.

67. Experiment with your software to determine the behavior of the matrix sequence A^k as $k \to \infty$ for each of the following matrices:

(a) $A = \begin{bmatrix} \frac{1}{2} & \frac{1}{3} \\ \frac{1}{4} & \frac{1}{5} \end{bmatrix}$ **(b)** $A = \begin{bmatrix} 1 & -1 & 0 \\ 0 & 1 & -1 \\ -1 & 0 & 1 \end{bmatrix}$

1.6 Matrix Transformations

In Section 1.2 we introduced the notation R^n for the set of all n-vectors with real entries. Thus R^2 denotes the set of all 2-vectors and R^3 denotes the set of all 3-vectors. It is convenient to represent the elements of R^2 and R^3 geometrically as directed line segments in a rectangular coordinate system.* Our approach in this section is intuitive and will enable us to present some interesting geometric applications in the next section (at this early stage of the course). We return in Section 4.1 to a careful and precise study of 2- and 3-vectors.

The vector

$$\mathbf{x} = \begin{bmatrix} x \\ y \end{bmatrix}$$

in R^2 is represented by the directed line segment shown in Figure 1.6. The vector

$$\mathbf{x} = \begin{bmatrix} x \\ y \\ z \end{bmatrix}$$

*You have undoubtedly seen rectangular coordinate systems in your precalculus or calculus courses.

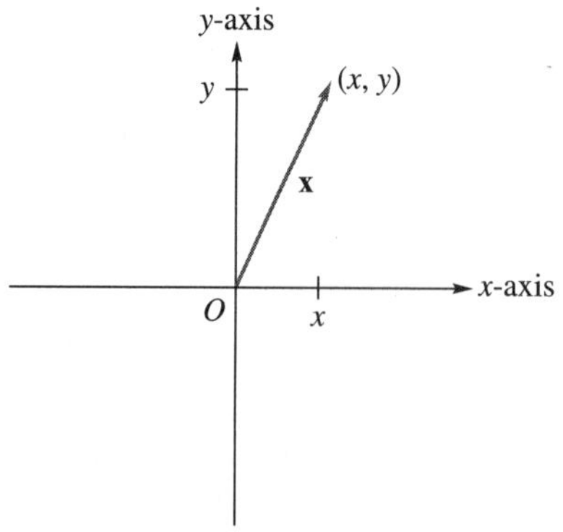

FIGURE 1.6

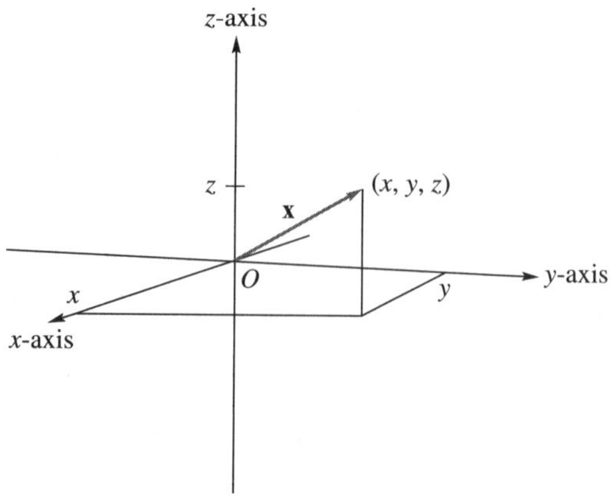

FIGURE 1.7

in R^3 is represented by the directed line segment shown in Figure 1.7.

EXAMPLE 1 Figure 1.8 shows geometric representations of the 2-vectors

$$\mathbf{u}_1 = \begin{bmatrix} 1 \\ 2 \end{bmatrix}, \quad \mathbf{u}_2 = \begin{bmatrix} -2 \\ 1 \end{bmatrix}, \quad \text{and} \quad \mathbf{u}_3 = \begin{bmatrix} 0 \\ 1 \end{bmatrix}$$

in a 2-dimensional rectangular coordinate system. Figure 1.9 shows geometric representations of the 3-vectors

$$\mathbf{v}_1 = \begin{bmatrix} 1 \\ 2 \\ 3 \end{bmatrix}, \quad \mathbf{v}_2 = \begin{bmatrix} -1 \\ 2 \\ -2 \end{bmatrix}, \quad \text{and} \quad \mathbf{v}_3 = \begin{bmatrix} 0 \\ 0 \\ 1 \end{bmatrix}$$

in a 3-dimensional rectangular coordinate system. ■

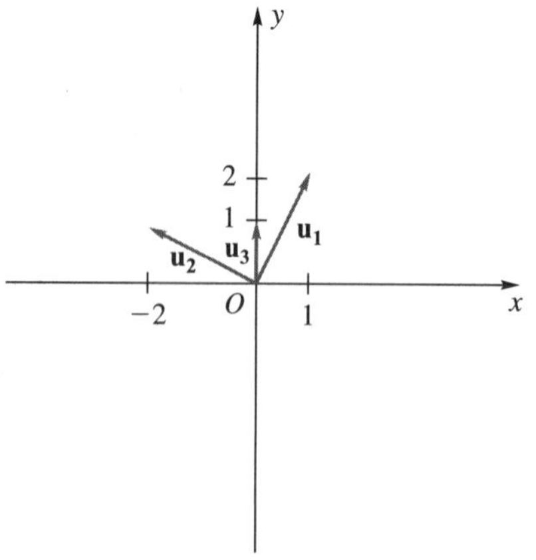

FIGURE 1.8

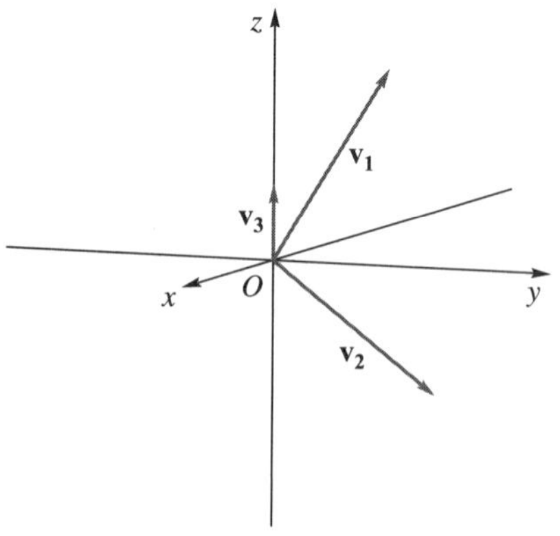

FIGURE 1.9

Functions occur in almost every application of mathematics. In this section we give a brief introduction from a geometric point of view to certain functions mapping R^n into R^m. Since we wish to picture these functions, called matrix transformations, we limit most of our discussion in this section to the situation where m and n have the values 2 or 3. In the next section we give an application of these functions to computer graphics in the plane, that is, for m and n equal to 2. In Chapter 6 we consider in greater detail a more general function, called a linear transformation mapping R^n into R^m. Since every matrix transformation is a linear transformation, we then learn more about the properties of matrix transformations.

Linear transformations play an important role in many areas of mathematics, as well as in numerous applied problems in the physical sciences, the social sciences, and economics.

If A is an $m \times n$ matrix and $\mathbf{u}$ is an n-vector, then the matrix product $A\mathbf{u}$ is an m-vector. A function f mapping R^n into R^m is denoted by $f\colon R^n \to R^m$.[†] A **matrix transformation** is a function $f\colon R^n \to R^m$ defined by $f(\mathbf{u}) = A\mathbf{u}$. The vector $f(\mathbf{u})$ in R^m is called the **image** of $\mathbf{u}$, and the set of all images of the vectors in R^n is called the **range** of f. Although we are limiting ourselves in this section to matrices and vectors with only real entries, an entirely similar discussion can be developed for matrices and vectors with complex entries. (See Appendix B.2.)

EXAMPLE 2

(a) Let f be the matrix transformation defined by

$$f(\mathbf{u}) = \begin{bmatrix} 2 & 4 \\ 3 & 1 \end{bmatrix} \mathbf{u}.$$

The image of $\mathbf{u} = \begin{bmatrix} 2 \\ -1 \end{bmatrix}$ is

$$f(\mathbf{u}) = \begin{bmatrix} 2 & 4 \\ 3 & 1 \end{bmatrix} \begin{bmatrix} 2 \\ -1 \end{bmatrix} = \begin{bmatrix} 0 \\ 5 \end{bmatrix},$$

and the image of $\begin{bmatrix} 1 \\ 2 \end{bmatrix}$ is $\begin{bmatrix} 10 \\ 5 \end{bmatrix}$ (verify).

(b) Let $A = \begin{bmatrix} 1 & 2 & 0 \\ 1 & -1 & 1 \end{bmatrix}$, and consider the matrix transformation defined by

$$f(\mathbf{u}) = A\mathbf{u}.$$

Then the image of $\begin{bmatrix} 1 \\ 0 \\ 1 \end{bmatrix}$ is $\begin{bmatrix} 1 \\ 2 \end{bmatrix}$, the image of $\begin{bmatrix} 0 \\ 1 \\ 3 \end{bmatrix}$ is $\begin{bmatrix} 2 \\ 2 \end{bmatrix}$, and the image of $\begin{bmatrix} -2 \\ 1 \\ 3 \end{bmatrix}$ is $\begin{bmatrix} 0 \\ 0 \end{bmatrix}$ (verify). ■

Observe that if A is an $m \times n$ matrix and $f\colon R^n \to R^m$ is a matrix transformation mapping R^n into R^m that is defined by $f(\mathbf{u}) = A\mathbf{u}$, then a vector $\mathbf{w}$ in R^m is in the range of f only if we can find a vector $\mathbf{v}$ in R^n such that $f(\mathbf{v}) = \mathbf{w}$.

[†]Appendix A, dealing with sets and functions, may be consulted as needed.

EXAMPLE 3

Let $A = \begin{bmatrix} 1 & 2 \\ -2 & 3 \end{bmatrix}$ and consider the matrix transformation defined by $f(\mathbf{u}) = A\mathbf{u}$.

Determine if the vector $\mathbf{w} = \begin{bmatrix} 4 \\ -1 \end{bmatrix}$ is in the range of f.

Solution

The question is equivalent to asking whether there is a vector $\mathbf{v} = \begin{bmatrix} v_1 \\ v_2 \end{bmatrix}$ such that $f(\mathbf{v}) = \mathbf{w}$. We have

$$A\mathbf{v} = \begin{bmatrix} v_1 + 2v_2 \\ -2v_1 + 3v_2 \end{bmatrix} = \mathbf{w} = \begin{bmatrix} 4 \\ -1 \end{bmatrix}$$

or

$$\begin{aligned} v_1 + 2v_2 &= 4 \\ -2v_1 + 3v_2 &= -1. \end{aligned}$$

Solving this linear system of equations by the familiar method of elimination, we get $v_1 = 2$ and $v_2 = 1$ (verify). Thus $\mathbf{w}$ is in the range of f. In particular, if $\mathbf{v} = \begin{bmatrix} 2 \\ 1 \end{bmatrix}$, then $f(\mathbf{v}) = \mathbf{w}$. ■

For matrix transformations where m and n are 2 or 3, we can draw pictures showing the effect of the matrix transformation. This will be illustrated in the examples that follow.

EXAMPLE 4

Let $f\colon R^2 \to R^2$ be the matrix transformation defined by

$$f(\mathbf{u}) = \begin{bmatrix} 1 & 0 \\ 0 & -1 \end{bmatrix}\mathbf{u}.$$

Thus, if $\mathbf{u} = \begin{bmatrix} x \\ y \end{bmatrix}$, then

$$f(\mathbf{u}) = f\left(\begin{bmatrix} x \\ y \end{bmatrix}\right) = \begin{bmatrix} x \\ -y \end{bmatrix}.$$

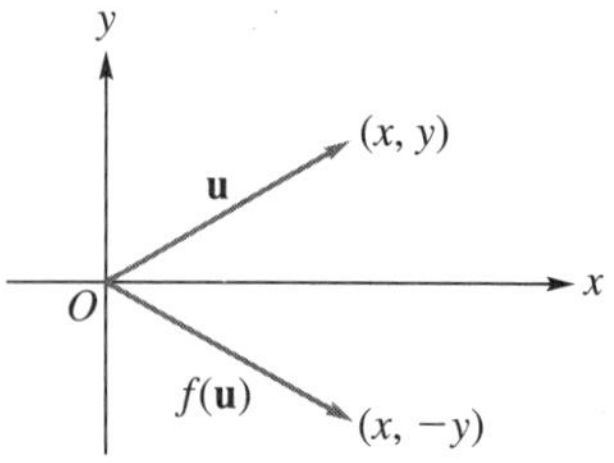

FIGURE 1.10 Reflection with respect to the x-axis.

The effect of the matrix transformation f, called **reflection with respect to the x-axis in R^2**, is shown in Figure 1.10. In Exercise 2 we consider reflection with respect to the y-axis. ■

EXAMPLE 5

Let $f\colon R^3 \to R^2$ be the matrix transformation defined by

$$f(\mathbf{u}) = f\left(\begin{bmatrix} x \\ y \\ z \end{bmatrix}\right) = \begin{bmatrix} 1 & 0 & 0 \\ 0 & 1 & 0 \end{bmatrix}\begin{bmatrix} x \\ y \\ z \end{bmatrix}.$$

Then

$$f(\mathbf{u}) = f\left(\begin{bmatrix} x \\ y \\ z \end{bmatrix}\right) = \begin{bmatrix} x \\ y \end{bmatrix}.$$

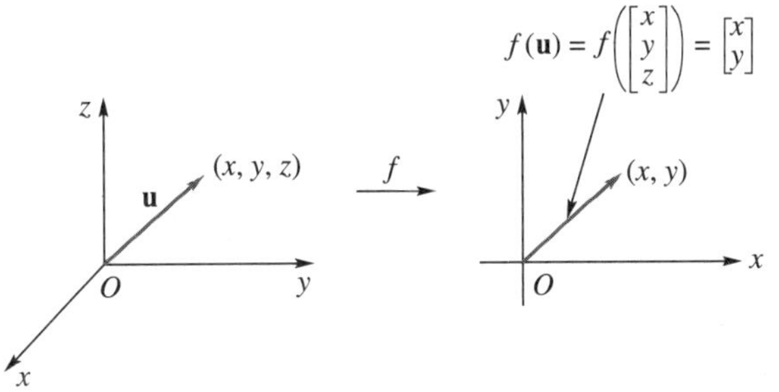

FIGURE 1.11

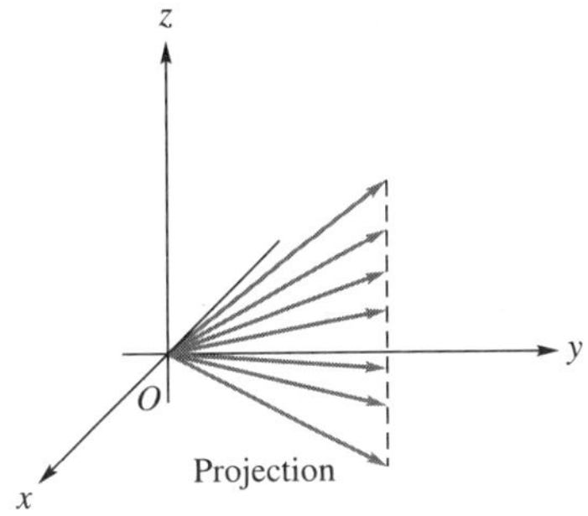

FIGURE 1.12

Figure 1.11 shows the effect of this matrix transformation, which is called **projection into the *xy*-plane**. (*Warning*: Carefully note the axes in Figure 1.11.)

Observe that if

$$\mathbf{v} = \begin{bmatrix} x \\ y \\ s \end{bmatrix},$$

where s is any scalar, then

$$f(\mathbf{v}) = \begin{bmatrix} x \\ y \end{bmatrix} = f(\mathbf{u}).$$

Hence, infinitely many 3-vectors have the same image vector. See Figure 1.12.

Note that the image of the 3-vector $\mathbf{v} = \begin{bmatrix} x \\ y \\ z \end{bmatrix}$ under the matrix transformation $g: R^3 \to R^3$ defined by

$$g(\mathbf{v}) = \begin{bmatrix} 1 & 0 & 0 \\ 0 & 1 & 0 \\ 0 & 0 & 0 \end{bmatrix} \mathbf{v}$$

is $\begin{bmatrix} x \\ y \\ 0 \end{bmatrix}$. The effect of this matrix transformation is shown in Figure 1.13. The picture is almost the same as Figure 1.11. There a 2-vector (the image $f(\mathbf{u})$) lies in the xy-plane, whereas in Figure 1.13 a 3-vector (the image $g(\mathbf{v})$) lies in the xy-plane. Observe that $f(\mathbf{v})$ appears to be the shadow cast by $\mathbf{v}$ onto the xy-plane. ■

EXAMPLE 6

Let $f: R^3 \to R^3$ be the matrix transformation defined by

$$f(\mathbf{u}) = \begin{bmatrix} r & 0 & 0 \\ 0 & r & 0 \\ 0 & 0 & r \end{bmatrix} \mathbf{u},$$

where r is a real number. It is easily seen that $f(\mathbf{u}) = r\mathbf{u}$. If $r > 1$, f is called **dilation**; if $0 < r < 1$, f is called **contraction**. Figure 1.14(a) shows the vector

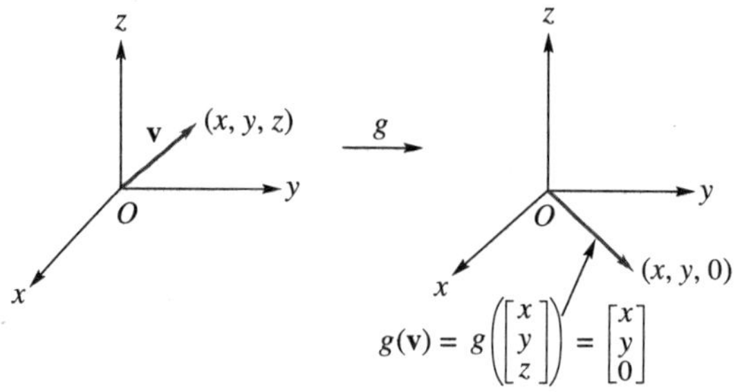

FIGURE 1.13

$f_1(\mathbf{u}) = 2\mathbf{u}$, and Figure 1.14(b) shows the vector $f_2(\mathbf{u}) = \frac{1}{2}\mathbf{u}$. Thus dilation stretches a vector, and contraction shrinks it. Similarly, we can define the matrix transformation $g: R^2 \to R^2$ by

$$g(\mathbf{u}) = \begin{bmatrix} r & 0 \\ 0 & r \end{bmatrix} \mathbf{u}.$$

We also have $g(\mathbf{u}) = r\mathbf{u}$, so again if $r > 1$, g is called **dilation**; if $0 < r < 1$, g is called **contraction**. ■

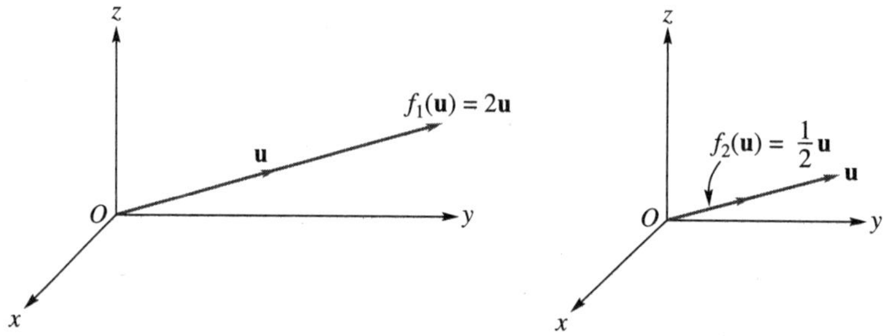

FIGURE 1.14 (a) Dilation: $r > 1$ (b) Contraction: $0 < r < 1$

EXAMPLE 7 A publisher releases a book in three different editions: trade, book club, and deluxe. Each book requires a certain amount of paper and canvas (for the cover). The requirements are given (in grams) by the matrix

$$A = \begin{array}{c} \\ \end{array}\begin{matrix} \text{Trade} & \text{Book club} & \text{Deluxe} \\ \end{matrix}$$

$$A = \begin{bmatrix} 300 & 500 & 800 \\ 40 & 50 & 60 \end{bmatrix} \begin{matrix} \text{Paper} \\ \text{Canvas} \end{matrix}.$$

Let

$$\mathbf{x} = \begin{bmatrix} x_1 \\ x_2 \\ x_3 \end{bmatrix}$$

denote the production vector, where x_1, x_2, and x_3 are the number of trade, book club, and deluxe books, respectively, that are published. The matrix transformation

$f: R^3 \to R^2$ defined by $f(\mathbf{x}) = A\mathbf{x}$ gives the vector

$$\mathbf{y} = \begin{bmatrix} y_1 \\ y_2 \end{bmatrix},$$

where y_1 is the total amount of paper required and y_2 is the total amount of canvas required. ■

EXAMPLE 8

Suppose that we rotate every point in R^2 counterclockwise through an angle ϕ about the origin of a rectangular coordinate system. Thus, if the point P has coordinates (x, y), then after rotating, we get the point P' with coordinates (x', y'). To obtain a relationship between the coordinates of P' and those of P, we let $\mathbf{u}$ be the vector $\begin{bmatrix} x \\ y \end{bmatrix}$, which is represented by the directed line segment from the origin to $P(x, y)$. See Figure 1.15. Also, let θ be the angle made by $\mathbf{u}$ with the positive x-axis.

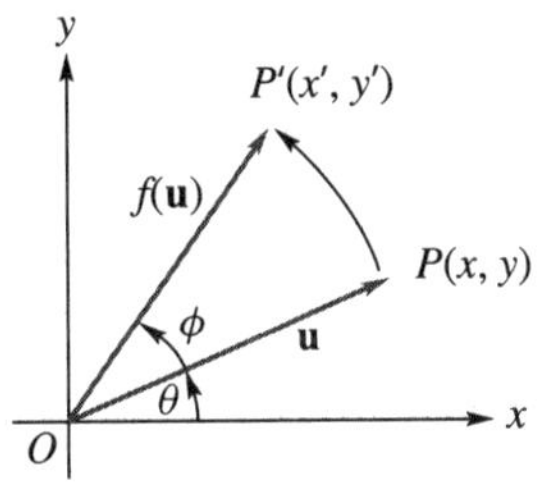

FIGURE 1.15 Rotation

Letting r denote the length of the directed line segment from O to P, we see from Figure 1.15(a) that

$$x = r\cos\theta, \qquad y = r\sin\theta \tag{1}$$

and

$$x' = r\cos(\theta + \phi), \qquad y' = r\sin(\theta + \phi). \tag{2}$$

By the formulas for the sine and cosine of a sum of angles, the equations in (2) become

$$\begin{aligned} x' &= r\cos\theta\cos\phi - r\sin\theta\sin\phi \\ y' &= r\sin\theta\cos\phi + r\cos\theta\sin\phi. \end{aligned}$$

Substituting the expression in (1) into the last pair of equations, we obtain

$$x' = x\cos\phi - y\sin\phi, \qquad y' = x\sin\phi + y\cos\phi. \tag{3}$$

Solving (3) for x and y, we have

$$x = x'\cos\phi + y'\sin\phi \quad \text{and} \quad y = -x'\sin\phi + y'\cos\phi. \tag{4}$$

Equation (3) gives the coordinates of P' in terms of those of P, and (4) expresses the coordinates of P in terms of those of P'. This type of rotation is used to simplify the general equation of second degree:

$$ax^2 + bxy + cy^2 + dx + ey + f = 0.$$

Substituting for x and y in terms of x' and y', we obtain

$$a'x'^2 + b'x'y' + c'y'^2 + d'x' + e'y' + f' = 0.$$

The key point is to choose ϕ so that $b' = 0$. Once this is done (we might now have to perform a translation of coordinates), we identify the general equation of second degree as a circle, ellipse, hyperbola, parabola, or a degenerate form of one of these. This topic is treated from a linear algebra point of view in Section 8.7.

We may also perform this change of coordinates by considering the matrix transformation $f: R^2 \to R^2$ defined by

$$f\left(\begin{bmatrix} x \\ y \end{bmatrix}\right) = \begin{bmatrix} \cos\phi & -\sin\phi \\ \sin\phi & \cos\phi \end{bmatrix} \begin{bmatrix} x \\ y \end{bmatrix}. \tag{5}$$

Then (5) can be written, using (3), as

$$f(\mathbf{u}) = \begin{bmatrix} x\cos\phi - y\sin\phi \\ x\sin\phi + y\cos\phi \end{bmatrix} = \begin{bmatrix} x' \\ y' \end{bmatrix}.$$

It then follows that the vector $f(\mathbf{u})$ is represented by the directed line segment from O to the point P'. Thus, rotation counterclockwise through an angle ϕ is a matrix transformation. ■

Key Terms

Matrix transformation	Image	Dilation
Mapping (function)	Reflection	Contraction
Range	Projection	Rotation

1.6 Exercises

In Exercises 1 through 8, sketch **u** *and its image under each given matrix transformation* f.

1. $f: R^2 \to R^2$ defined by
$$f\left(\begin{bmatrix} x \\ y \end{bmatrix}\right) = \begin{bmatrix} 1 & 0 \\ 0 & -1 \end{bmatrix} \begin{bmatrix} x \\ y \end{bmatrix}; \quad \mathbf{u} = \begin{bmatrix} 2 \\ 3 \end{bmatrix}$$

2. $f: R^2 \to R^2$ **(reflection with respect to the** y**-axis)** defined by
$$f\left(\begin{bmatrix} x \\ y \end{bmatrix}\right) = \begin{bmatrix} -1 & 0 \\ 0 & 1 \end{bmatrix} \begin{bmatrix} x \\ y \end{bmatrix}; \quad \mathbf{u} = \begin{bmatrix} 1 \\ -2 \end{bmatrix}$$

3. $f: R^2 \to R^2$ is a counterclockwise rotation through 30°; $\mathbf{u} = \begin{bmatrix} -1 \\ 3 \end{bmatrix}$

4. $f: R^2 \to R^2$ is a counterclockwise rotation through $\frac{2}{3}\pi$ radians; $\mathbf{u} = \begin{bmatrix} -2 \\ -3 \end{bmatrix}$

5. $f: R^2 \to R^2$ defined by
$$f\left(\begin{bmatrix} x \\ y \end{bmatrix}\right) = \begin{bmatrix} -1 & 0 \\ 0 & -1 \end{bmatrix} \begin{bmatrix} x \\ y \end{bmatrix}; \quad \mathbf{u} = \begin{bmatrix} 3 \\ 2 \end{bmatrix}$$

6. $f: R^2 \to R^2$ defined by
$$f\left(\begin{bmatrix} x \\ y \end{bmatrix}\right) = \begin{bmatrix} 2 & 0 \\ 0 & 2 \end{bmatrix} \begin{bmatrix} x \\ y \end{bmatrix}; \quad \mathbf{u} = \begin{bmatrix} -3 \\ 3 \end{bmatrix}$$

7. $f: R^2 \to R^3$ defined by

$$f\left(\begin{bmatrix} x \\ y \end{bmatrix}\right) = \begin{bmatrix} 1 & 0 \\ 1 & -1 \\ 0 & 0 \end{bmatrix}\begin{bmatrix} x \\ y \end{bmatrix}; \quad \mathbf{u} = \begin{bmatrix} 2 \\ -1 \\ 3 \end{bmatrix}$$

8. $f: R^3 \to R^3$ defined by

$$f\left(\begin{bmatrix} x \\ y \\ z \end{bmatrix}\right) = \begin{bmatrix} 1 & 0 & 1 \\ -1 & 1 & 0 \\ 0 & 0 & 1 \end{bmatrix}\begin{bmatrix} x \\ y \\ z \end{bmatrix}; \quad \mathbf{u} = \begin{bmatrix} 0 \\ -2 \\ 4 \end{bmatrix}$$

In Exercises 9 through 14, let $f: R^2 \to R^3$ be the matrix transformation defined by $f(\mathbf{x}) = A\mathbf{x}$, where

$$A = \begin{bmatrix} 1 & 2 \\ 0 & 1 \\ 1 & 1 \end{bmatrix}.$$

Determine whether each given vector **w** *is in the range of f.*

9. $\mathbf{w} = \begin{bmatrix} 1 \\ -1 \\ 2 \end{bmatrix}$ **10.** $\mathbf{w} = \begin{bmatrix} 1 \\ 1 \\ 1 \end{bmatrix}$ **11.** $\mathbf{w} = \begin{bmatrix} 0 \\ 0 \\ 0 \end{bmatrix}$

12. $\mathbf{w} = \begin{bmatrix} 8 \\ 5 \\ 3 \end{bmatrix}$ **13.** $\mathbf{w} = \begin{bmatrix} 1 \\ 4 \\ 2 \end{bmatrix}$ **14.** $\mathbf{w} = \begin{bmatrix} 1 \\ -1 \\ 1 \end{bmatrix}$

In Exercises 15 through 17, give a geometric description of the matrix transformation $f: R^2 \to R^2$ defined by $f(\mathbf{u}) = A\mathbf{u}$ for each given matrix A.

15. (a) $A = \begin{bmatrix} -1 & 0 \\ 0 & 1 \end{bmatrix}$ **(b)** $A = \begin{bmatrix} 0 & -1 \\ 1 & 0 \end{bmatrix}$

16. (a) $A = \begin{bmatrix} 0 & 1 \\ 1 & 0 \end{bmatrix}$ **(b)** $A = \begin{bmatrix} 0 & -1 \\ -1 & 0 \end{bmatrix}$

17. (a) $A = \begin{bmatrix} 1 & 0 \\ 0 & 0 \end{bmatrix}$ **(b)** $A = \begin{bmatrix} 0 & 0 \\ 0 & 1 \end{bmatrix}$

18. Some matrix transformations f have the property that $f(\mathbf{u}) = f(\mathbf{v})$, when $\mathbf{u} \neq \mathbf{v}$. That is, the images of different vectors can be the same. For each of the following matrix transformations $f: R^2 \to R^2$ defined by $f(\mathbf{u}) = A\mathbf{u}$, find two different vectors **u** and **v** such that $f(\mathbf{u}) = f(\mathbf{v}) = \mathbf{w}$ for the given vector **w**.

(a) $A = \begin{bmatrix} 1 & 2 & 0 \\ 0 & 1 & -1 \end{bmatrix}, \mathbf{w} = \begin{bmatrix} 0 \\ -1 \end{bmatrix}$

(b) $A = \begin{bmatrix} 2 & 1 & 0 \\ 0 & 2 & -1 \end{bmatrix}, \mathbf{w} = \begin{bmatrix} 4 \\ 4 \end{bmatrix}$

19. Let $f: R^2 \to R^2$ be the linear transformation defined by $f(\mathbf{u}) = A\mathbf{u}$, where

$$A = \begin{bmatrix} \cos\phi & -\sin\phi \\ \sin\phi & \cos\phi \end{bmatrix}.$$

For $\phi = 30°$, f defines a counterclockwise rotation by an angle of 30°.

(a) If $T_1(\mathbf{u}) = A^2\mathbf{u}$, describe the action of T_1 on **u**.

(b) If $T_2(\mathbf{u}) = A^{-1}\mathbf{u}$, describe the action of T_2 on **u**.

(c) What is the smallest positive value of k for which $T(\mathbf{u}) = A^k\mathbf{u} = \mathbf{u}$?

20. Let $f: R^n \to R^m$ be a matrix transformation defined by $f(\mathbf{u}) = A\mathbf{u}$, where A is an $m \times n$ matrix.

(a) Show that $f(\mathbf{u} + \mathbf{v}) = f(\mathbf{u}) + f(\mathbf{v})$ for any **u** and **v** in R^n.

(b) Show that $f(c\mathbf{u}) = cf(\mathbf{u})$ for any **u** in R^n and any real number c.

(c) Show that $f(c\mathbf{u} + d\mathbf{v}) = cf(\mathbf{u}) + df(\mathbf{v})$ for any **u** and **v** in R^n and any real numbers c and d.

21. Let $f: R^n \to R^m$ be a matrix transformation defined by $f(\mathbf{u}) = A\mathbf{u}$, where A is an $m \times n$ matrix. Show that if **u** and **v** are vectors in R^n such that $f(\mathbf{u}) = \mathbf{0}$ and $f(\mathbf{v}) = \mathbf{0}$, where

$$\mathbf{0} = \begin{bmatrix} 0 \\ 0 \\ \vdots \\ 0 \\ 0 \end{bmatrix},$$

then $f(c\mathbf{u} + d\mathbf{v}) = \mathbf{0}$ for any real numbers c and d.

22. (a) Let $O: R^n \to R^m$ be the matrix transformation defined by $O(\mathbf{u}) = O\mathbf{u}$, where O is the $m \times n$ zero matrix. Show that $O(\mathbf{u}) = \mathbf{0}$ for all **u** in R^n.

(b) Let $I: R^n \to R^n$ be the matrix transformation defined by $I(\mathbf{u}) = I_n\mathbf{u}$, where I_n is the identity matrix. (See Section 1.5.) Show that $I(\mathbf{u}) = \mathbf{u}$ for all **u** in R^n.

1.7 Computer Graphics (Optional)

We are all familiar with the astounding results being developed with computer graphics in the areas of video games and special effects in the film industry. Computer graphics also play a major role in the manufacturing world. *Computer-aided design* (CAD) is used to create a computer model of a product and then, by subjecting the computer model to a variety of tests (carried out on the computer), changes to the current design can be implemented to obtain an improved design. One of the notable successes of this approach has been in the automobile industry, where the computer model can be viewed from different angles to achieve a most pleasing and popular style and can be tested for strength of components, for roadability, for seating comfort, and for safety in a crash.

In this section we give illustrations of matrix transformations $f: R^2 \to R^2$ that are useful in two-dimensional graphics.

EXAMPLE 1

Let $f: R^2 \to R^2$ be the matrix transformation that performs a reflection with respect to the x-axis. (See Example 4 in Section 1.6.) Then f is defined by $f(\mathbf{v}) = A\mathbf{v}$, where

$$A = \begin{bmatrix} 1 & 0 \\ 0 & -1 \end{bmatrix}.$$

Thus, we have

$$f(\mathbf{v}) = A\mathbf{v} = \begin{bmatrix} 1 & 0 \\ 0 & -1 \end{bmatrix}\begin{bmatrix} x \\ y \end{bmatrix} = \begin{bmatrix} x \\ -y \end{bmatrix}.$$

To illustrate a reflection with respect to the x-axis in computer graphics, let the triangle T in Figure 1.16(a) have vertices

$$(-1, 4), \quad (3, 1), \quad \text{and} \quad (2, 6).$$

To reflect T with respect to the x-axis, we let

$$\mathbf{v}_1 = \begin{bmatrix} -1 \\ 4 \end{bmatrix}, \quad \mathbf{v}_2 = \begin{bmatrix} 3 \\ 1 \end{bmatrix}, \quad \mathbf{v}_3 = \begin{bmatrix} 2 \\ 6 \end{bmatrix}$$

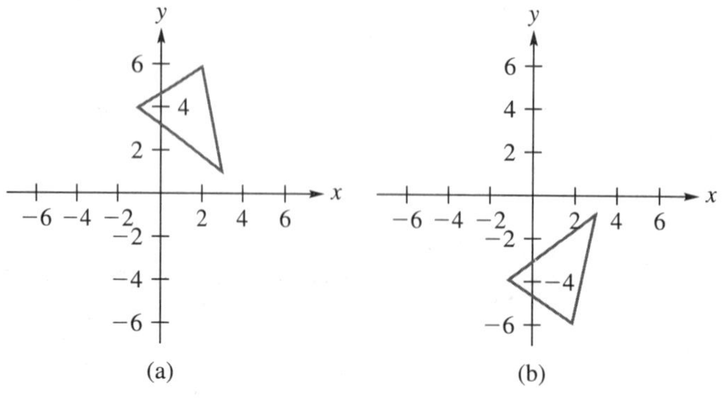

FIGURE 1.16

and compute the images $f(\mathbf{v}_1)$, $f(\mathbf{v}_2)$, and $f(\mathbf{v}_3)$ by forming the products

$$A\mathbf{v}_1 = \begin{bmatrix} 1 & 0 \\ 0 & -1 \end{bmatrix}\begin{bmatrix} -1 \\ 4 \end{bmatrix} = \begin{bmatrix} -1 \\ -4 \end{bmatrix},$$

$$A\mathbf{v}_2 = \begin{bmatrix} 1 & 0 \\ 0 & -1 \end{bmatrix}\begin{bmatrix} 3 \\ 1 \end{bmatrix} = \begin{bmatrix} 3 \\ -1 \end{bmatrix},$$

$$A\mathbf{v}_3 = \begin{bmatrix} 1 & 0 \\ 0 & -1 \end{bmatrix}\begin{bmatrix} 2 \\ 6 \end{bmatrix} = \begin{bmatrix} 2 \\ -6 \end{bmatrix}.$$

These three products can be written in terms of partitioned matrices as

$$A\begin{bmatrix} \mathbf{v}_1 & \mathbf{v}_2 & \mathbf{v}_3 \end{bmatrix} = \begin{bmatrix} -1 & 3 & 2 \\ -4 & -1 & -6 \end{bmatrix}.$$

Thus the image of T has vertices

$$(-1, -4), \quad (3, -1), \quad \text{and} \quad (2, -6)$$

and is displayed in Figure 1.16(b). ■

EXAMPLE 2

The matrix transformation $f\colon R^2 \to R^2$ that performs a reflection with respect to the line $y = -x$ is defined by

$$f(\mathbf{v}) = B\mathbf{v},$$

where

$$B = \begin{bmatrix} 0 & -1 \\ -1 & 0 \end{bmatrix}.$$

To illustrate reflection with respect to the line $y = -x$, we use the triangle T as defined in Example 1 and compute the products

$$B\begin{bmatrix} \mathbf{v}_1 & \mathbf{v}_2 & \mathbf{v}_3 \end{bmatrix} = \begin{bmatrix} 0 & -1 \\ -1 & 0 \end{bmatrix}\begin{bmatrix} -1 & 3 & 2 \\ 4 & 1 & 6 \end{bmatrix} = \begin{bmatrix} -4 & -1 & -6 \\ 1 & -3 & -2 \end{bmatrix}.$$

Thus the image of T has vertices

$$(-4, 1), \quad (-1, -3), \quad \text{and} \quad (-6, -2)$$

and is displayed in Figure 1.17. ■

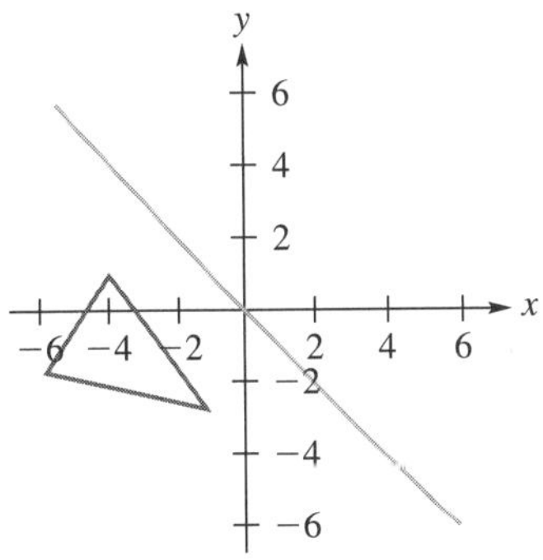

FIGURE 1.17

To perform a reflection with respect to the x-axis on the triangle T of Example 1, followed by a reflection with respect to the line $y = -x$, we compute

$$B(A\mathbf{v}_1), \quad B(A\mathbf{v}_2), \quad \text{and} \quad B(A\mathbf{v}_3).$$

It is not difficult to show that reversing the order of these matrix transformations produces a different image (verify). Thus the order in which graphics transformations are performed is important. This is not surprising, since matrix multiplication, unlike multiplication of real numbers, does not satisfy the commutative property.

EXAMPLE 3

Rotations in a plane have been defined in Example 8 of Section 1.6. A plane figure is rotated counterclockwise through an angle ϕ by the matrix transformation $f: R^2 \to R^2$ defined by $f(\mathbf{v}) = A\mathbf{v}$, where

$$A = \begin{bmatrix} \cos\phi & -\sin\phi \\ \sin\phi & \cos\phi \end{bmatrix}.$$

Now suppose that we wish to rotate the parabola $y = x^2$ counterclockwise through 50°. We start by choosing a sample of points from the parabola, say,

$$(-2, 4), \quad (-1, 1), \quad (0, 0), \quad \left(\tfrac{1}{2}, \tfrac{1}{4}\right), \quad \text{and} \quad (3, 9).$$

[See Figure 1.18(a).] We then compute the images of these points. Thus, letting

$$\mathbf{v}_1 = \begin{bmatrix} -2 \\ 4 \end{bmatrix}, \quad \mathbf{v}_2 = \begin{bmatrix} -1 \\ 1 \end{bmatrix}, \quad \mathbf{v}_3 = \begin{bmatrix} 0 \\ 0 \end{bmatrix}, \quad \mathbf{v}_4 = \begin{bmatrix} \frac{1}{2} \\ \frac{1}{4} \end{bmatrix}, \quad \mathbf{v}_5 = \begin{bmatrix} 3 \\ 9 \end{bmatrix},$$

we compute the products (to four decimal places) (verify)

$$A\begin{bmatrix} \mathbf{v}_1 & \mathbf{v}_2 & \mathbf{v}_3 & \mathbf{v}_4 & \mathbf{v}_5 \end{bmatrix} = \begin{bmatrix} -4.3498 & -1.4088 & 0 & 0.1299 & -4.9660 \\ 1.0391 & -0.1233 & 0 & 0.5437 & 8.0832 \end{bmatrix}.$$

The image points

$$(-4.3498, 1.0391), \quad (-1.4088, -0.1233), \quad (0, 0),$$
$$(0.1299, 0.5437), \quad \text{and} \quad (-4.9660, 8.0832)$$

are plotted, as shown in Figure 1.18(b), and successive points are connected, showing the approximate image of the parabola. ■

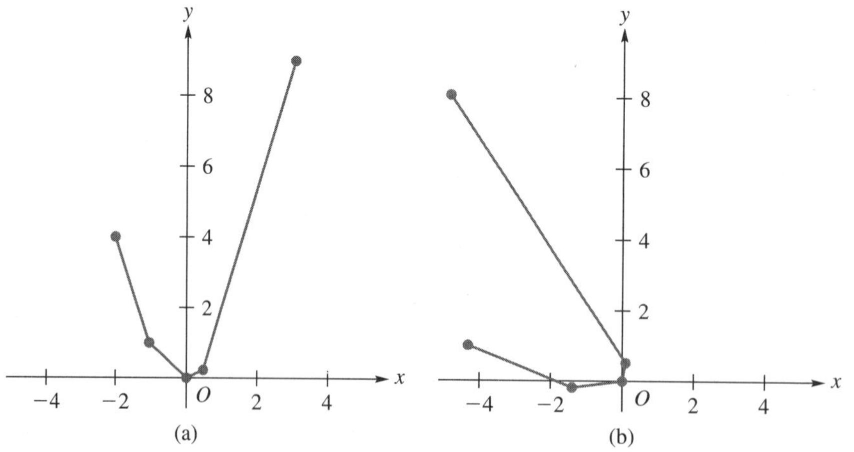

FIGURE 1.18

Rotations are particularly useful in achieving the sophisticated effects seen in arcade games and animated computer demonstrations. For example, to show a wheel spinning, we can rotate the spokes through an angle θ_1, followed by a second rotation through an angle θ_2, and so on. Let the 2-vector $\mathbf{u} = \begin{bmatrix} a_1 \\ a_2 \end{bmatrix}$ represent a spoke of the wheel; let $f: R^2 \to R^2$ be the matrix transformation defined by $f(\mathbf{v}) = A\mathbf{v}$, where

$$A = \begin{bmatrix} \cos\theta_1 & -\sin\theta_1 \\ \sin\theta_1 & \cos\theta_1 \end{bmatrix};$$

and let $g: R^2 \to R^2$ be the matrix transformation defined by $g(\mathbf{v}) = B\mathbf{v}$, where

$$B = \begin{bmatrix} \cos\theta_2 & -\sin\theta_2 \\ \sin\theta_2 & \cos\theta_2 \end{bmatrix}.$$

We represent the succession of rotations of the spoke $\mathbf{u}$ by

$$g(f(\mathbf{u})) = g(A\mathbf{u}) = B(A\mathbf{u}).$$

The product $A\mathbf{u}$ is performed first and generates a rotation of $\mathbf{u}$ through the angle θ_1; then the product $B(A\mathbf{u})$ generates the second rotation. We have

$$B(A\mathbf{u}) = B(a_1\text{col}_1(A) + a_2\text{col}_2(A)) = a_1 B\text{col}_1(A) + a_2 B\text{col}_2(A),$$

and the final expression is a linear combination of column vectors $B\text{col}_1(A)$ and $B\text{col}_2(A)$, which we can write as the product

$$\begin{bmatrix} B\text{col}_1(A) & B\text{col}_2(A) \end{bmatrix} \begin{bmatrix} a_1 \\ a_2 \end{bmatrix}.$$

From the definition of matrix multiplication, $\begin{bmatrix} B\text{col}_1(A) & B\text{col}_2(A) \end{bmatrix} = BA$, so we have

$$B(A\mathbf{u}) = (BA)\mathbf{u}.$$

Thus, instead of applying the transformations in succession, f followed by g, we can achieve the same result by forming the matrix product BA and using it to define a matrix transformation on the spokes of the wheel.

The matrix product BA is given by

$$\begin{aligned} BA &= \begin{bmatrix} \cos\theta_2 & -\sin\theta_2 \\ \sin\theta_2 & \cos\theta_2 \end{bmatrix} \begin{bmatrix} \cos\theta_1 & -\sin\theta_1 \\ \sin\theta_1 & \cos\theta_1 \end{bmatrix} \\ &= \begin{bmatrix} \cos\theta_2\cos\theta_1 - \sin\theta_2\sin\theta_1 & -\cos\theta_2\sin\theta_1 - \sin\theta_2\cos\theta_1 \\ \sin\theta_2\cos\theta_1 + \cos\theta_2\sin\theta_1 & -\sin\theta_2\sin\theta_1 + \cos\theta_2\sin\theta_1 \end{bmatrix}. \end{aligned}$$

Since $g(f(\mathbf{u})) = BA\mathbf{u}$, it follows that this matrix transformation performs a rotation of $\mathbf{u}$ through the angle $\theta_1 + \theta_2$. Thus we have

$$BA = \begin{bmatrix} \cos(\theta_1+\theta_2) & -\sin(\theta_1+\theta_2) \\ \sin(\theta_1+\theta_2) & \cos(\theta_1+\theta_2) \end{bmatrix}.$$

Equating corresponding entries of the two matrix expressions for BA, we have the trigonometric identities for the sine and cosine of the sum of two angles:

$$\begin{aligned} \cos(\theta_1+\theta_2) &= \cos\theta_1\cos\theta_2 - \sin\theta_1\sin\theta_2 \\ \sin(\theta_1+\theta_2) &= \cos\theta_1\sin\theta_2 + \sin\theta_1\cos\theta_2. \end{aligned}$$

See Exercises 16 and 17 for related results.

EXAMPLE 4 A **shear in the x-direction** is the matrix transformation defined by

$$f(\mathbf{v}) = \begin{bmatrix} 1 & k \\ 0 & 1 \end{bmatrix} \mathbf{v},$$

where k is a scalar. A shear in the x-direction takes the point (x, y) to the point $(x + ky, y)$. That is, the point (x, y) is moved parallel to the x-axis by the amount ky.

Consider now the rectangle R, shown in Figure 1.19(a), with vertices

$$(0, 0), \quad (0, 2), \quad (4, 0), \quad \text{and} \quad (4, 2).$$

If we apply the shear in the x-direction with $k = 2$, then the image of R is the parallelogram with vertices

$$(0, 0), \quad (4, 2), \quad (4, 0), \quad \text{and} \quad (8, 2),$$

shown in Figure 1.19(b). If we apply the shear in the x-direction with $k = -3$, then the image of R is the parallelogram with vertices

$$(0, 0), \quad (-6, 2), \quad (4, 0), \quad \text{and} \quad (-2, 2),$$

shown in Figure 1.19(c).

In Exercise 3 we consider shears in the y-direction. ■

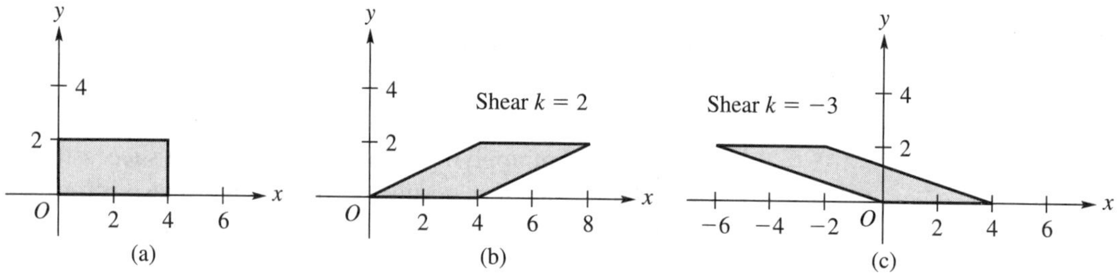

FIGURE 1.19

Other matrix transformations used in two-dimensional computer graphics are considered in the exercises at the end of this section. For a detailed discussion of computer graphics, the reader is referred to the books listed in the Further Readings at the end of this section.

In Examples 1 and 2, we applied a matrix transformation to a triangle, a figure that can be specified by its three vertices. In Example 3, the figure transformed was a parabola, which cannot be specified by a finite number of points. In this case we chose a number of points on the parabola to approximate its shape and computed the images of these approximating points, which, when joined, gave an approximate shape of the parabola.

EXAMPLE 5 Let $f: R^2 \to R^2$ be the matrix transformation called scaling defined by $f(\mathbf{v}) = A\mathbf{v}$, where

$$A = \begin{bmatrix} h & 0 \\ 0 & k \end{bmatrix}$$

with h and k both nonzero. Suppose that we now wish to apply this matrix transformation to a circle of radius 1 that is centered at the origin (the unit circle). Unfortunately, a circle cannot be specified by a finite number of points. However, each point on the unit circle is described by an ordered pair $(\cos\theta, \sin\theta)$, where the angle θ takes on all values from 0 to 2π radians. Thus we now represent an arbitrary point on the unit circle by the vector $\mathbf{u} = \begin{bmatrix} \cos\theta \\ \sin\theta \end{bmatrix}$. Hence the images of the unit circle that are obtained by applying the matrix transformation f are given by

$$f(\mathbf{u}) = A\mathbf{u} = \begin{bmatrix} h & 0 \\ 0 & k \end{bmatrix} \begin{bmatrix} \cos\theta \\ \sin\theta \end{bmatrix} = \begin{bmatrix} h\cos\theta \\ k\sin\theta \end{bmatrix} = \begin{bmatrix} x' \\ y' \end{bmatrix}.$$

We recall that a circle of radius 1 centered at the origin is described by the equation

$$x^2 + y^2 = 1.$$

By Pythagoras's identity, $\sin^2\theta + \cos^2\theta = 1$. Thus, the points $(\cos\theta, \sin\theta)$ lie on the circumference of the unit circle. We now want to write an equation describing the image of the unit circle. We have

$$x' = h\cos\theta \quad \text{and} \quad y' = k\sin\theta,$$

so

$$\frac{x'}{h} = \cos\theta, \quad \frac{y'}{k} = \sin\theta.$$

It then follows that

$$\left(\frac{x'}{h}\right)^2 + \left(\frac{y'}{k}\right)^2 = 1,$$

which is the equation of an ellipse. Thus the image of the unit circle by the matrix transformation f is an ellipse centered at the origin. See Figure 1.20. ■

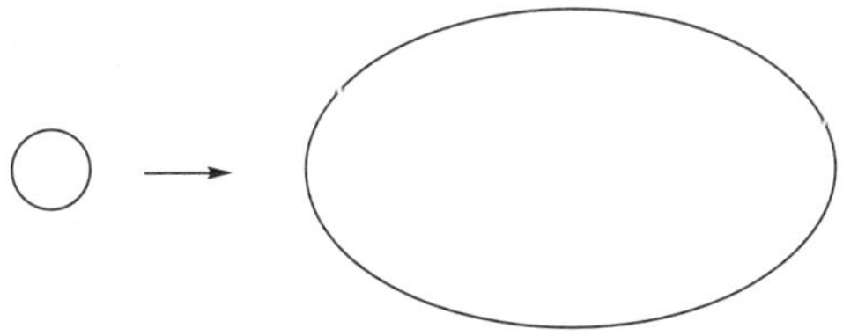

FIGURE 1.20 Unit circle Ellipse

FURTHER READINGS

Cunningham, Steve. *Computer Graphics: Programming, Problem Solving, and Visual Communication*. New Jersey: Prentice Hall, 2007.

Foley, James D., Andries van Dam, Steven K. Feiner, and John F. Hughes. *Computer Graphics: Principles and Practice in C*, 2d ed. Reading, Mass.: Addison Wesley, 1996.

Rogers, D. F., and J. A. Adams. *Mathematical Elements for Computer Graphics*, 2d ed. New York: McGraw-Hill, 1989.

Shirley, Peter, Michael Ashikhmin, Michael Gleicher, Stephen Marschner, Erik Reinhard, Kelvin Sung, William Thompson, and Peter Willemsen. *Fundamentals of Computer Graphics*, 2d ed. Natick, Mass.: A. K. Peters, Ltd., 2005.

Key Terms

Computer graphics
Computer-aided design (CAD)
Shear
Rotation
Contraction
Dilation
Image

1.7 Exercises

1. Let $f: R^2 \to R^2$ be the matrix transformation defined by $f(\mathbf{v}) = A\mathbf{v}$, where

$$A = \begin{bmatrix} -1 & 0 \\ 0 & 1 \end{bmatrix};$$

that is, f is reflection with respect to the y-axis. Find and sketch the image of the rectangle R with vertices $(1, 1)$, $(2, 1)$, $(1, 3)$, and $(2, 3)$.

2. Let R be the rectangle with vertices $(1, 1)$, $(1, 4)$, $(3, 1)$, and $(3, 4)$. Let f be the shear in the x-direction with $k = 3$. Find and sketch the image of R.

3. A **shear in the y-direction** is the matrix transformation $f: R^2 \to R^2$ defined by $f(\mathbf{v}) = A\mathbf{v}$, and

$$A = \begin{bmatrix} 1 & 0 \\ k & 1 \end{bmatrix},$$

where k is a scalar. Let R be the rectangle defined in Exercise 2 and let f be the shear in the y-direction with $k = -2$. Find and sketch the image of R.

4. The matrix transformation $f: R^2 \to R^2$ defined by $f(\mathbf{v}) = A\mathbf{v}$, where

$$A = \begin{bmatrix} k & 0 \\ 0 & k \end{bmatrix},$$

and k is a real number, is called **dilation** if $k > 1$ and **contraction** if $0 < k < 1$. Thus, dilation stretches a vector, whereas contraction shrinks it. If R is the rectangle defined in Exercise 2, find and sketch the image of R for

(a) $k = 4$; **(b)** $k = \frac{1}{4}$.

5. The matrix transformation $f: R^2 \to R^2$ defined by $f(\mathbf{v}) = A\mathbf{v}$, where

$$A = \begin{bmatrix} k & 0 \\ 0 & 1 \end{bmatrix},$$

and k is a real number, is called **dilation in the x-direction** if $k > 1$ and **contraction in the x-direction** if $0 < k < 1$. If R is the unit square and f is dilation in the x-direction with $k = 2$, find and sketch the image of R.

6. The matrix transformation $f: R^2 \to R^2$ defined by $f(\mathbf{v}) = A\mathbf{v}$, where

$$A = \begin{bmatrix} 1 & 0 \\ 0 & k \end{bmatrix}$$

and k is a real number, is called **dilation in the y-direction** if $k > 1$ and **contraction in the y-direction** if $0 < k < 1$. If R is the unit square and f is the contraction in the y-direction with $k = \frac{1}{2}$, find and sketch the image of R.

7. Let T be the triangle with vertices $(5, 0)$, $(0, 3)$, and $(2, -1)$. Find the coordinates of the vertices of the image of T under the matrix transformation f defined by

$$f(\mathbf{v}) = \begin{bmatrix} -2 & 1 \\ 3 & 4 \end{bmatrix} \mathbf{v}.$$

8. Let T be the triangle with vertices $(1, 1)$, $(-3, -3)$, and $(2, -1)$. Find the coordinates of the vertices of the image of T under the matrix transformation defined by

$$f(\mathbf{v}) = \begin{bmatrix} 4 & -3 \\ -4 & 2 \end{bmatrix} \mathbf{v}.$$

9. Let f be the counterclockwise rotation through 60°. If T is the triangle defined in Exercise 8, find and sketch the image of T under f.

10. Let f_1 be reflection with respect to the y-axis and let f_2 be counterclockwise rotation through $\pi/2$ radians. Show that the result of first performing f_2 and then f_1 is not the same as first performing f_1 and then performing f_2.

11. Let A be the singular matrix $\begin{bmatrix} 1 & 2 \\ 2 & 4 \end{bmatrix}$ and let T be the triangle defined in Exercise 8. Describe the image of T under the matrix transformation $f: R^2 \to R^2$ defined by $f(\mathbf{v}) = A\mathbf{v}$. (See also Exercise 21.)

12. Let f be the matrix transformation defined in Example 5. Find and sketch the image of the rectangle with vertices $(0, 0)$, $(1, 0)$, $(1, 1)$, and $(0, 1)$ for $h = 2$ and $k = 3$.

13. Let $f: R^2 \to R^2$ be the matrix transformation defined by $f(\mathbf{v}) = A\mathbf{v}$, where

$$A = \begin{bmatrix} 1 & -1 \\ 2 & 3 \end{bmatrix}.$$

Find and sketch the image of the rectangle defined in Exercise 12.

In Exercises 14 and 15, let f_1, f_2, f_3, and f_4 be the following matrix transformations:

f_1: *counterclockwise rotation through the angle* ϕ

f_2: *reflection with respect to the x-axis*

f_3: *reflection with respect to the y-axis*

f_4: *reflection with respect to the line $y = x$*

14. Let S denote the unit square.

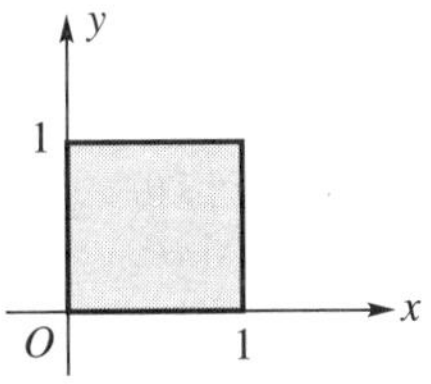

Determine two distinct ways to use the matrix transformations defined on S to obtain the given image. You may apply more than one matrix transformation in succession.

(a)

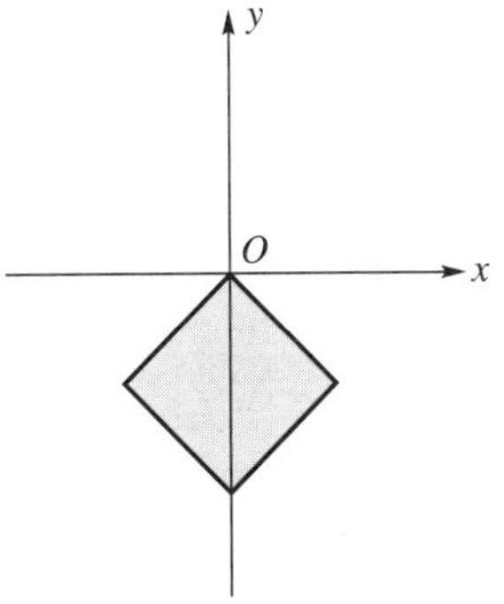

(b)

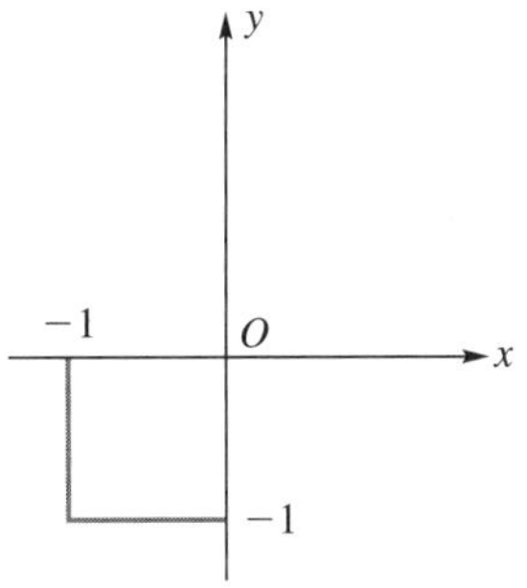

15. Let S denote the triangle shown in the figure.

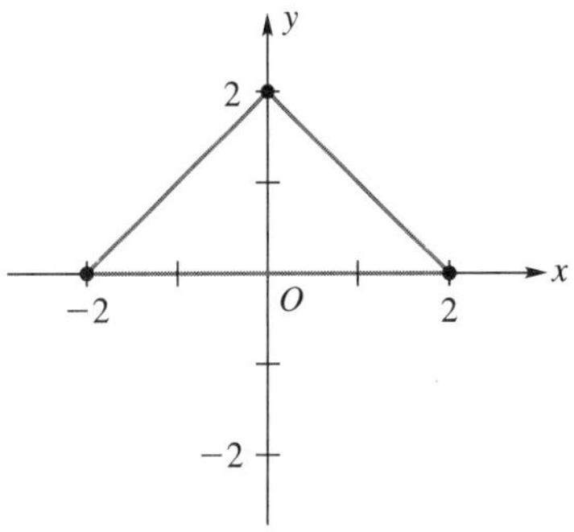

Determine two distinct ways to use the matrix transformations defined on S to obtain the given image. You may apply more than one matrix transformation in succession.

(a)

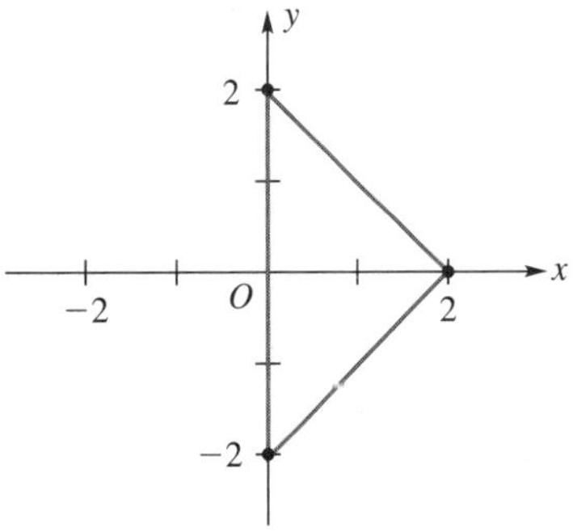

(b)

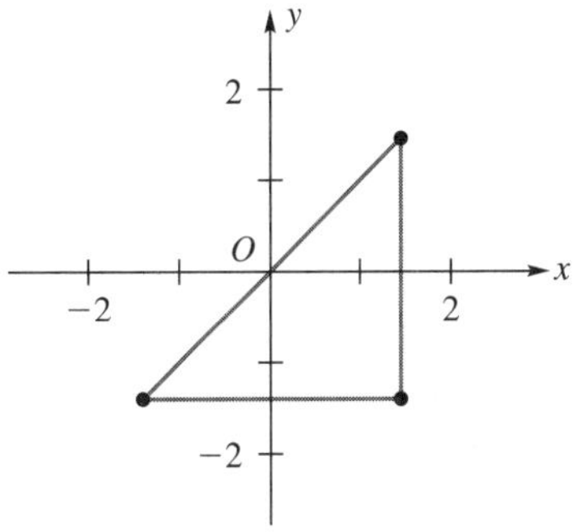

16. Refer to the discussion following Example 3 to develop the double angle identities for sine and cosine by using the matrix transformation $f(f(\mathbf{u})) = A(A\mathbf{u})$, where

$$A = \begin{bmatrix} \cos\theta_1 & -\sin\theta_1 \\ \sin\theta_1 & \cos\theta_1 \end{bmatrix} \quad \text{and} \quad \mathbf{u} = \begin{bmatrix} a_1 \\ a_2 \end{bmatrix}.$$

17. Use a procedure similar to the one discussed after Example 3 to develop sine and cosine expressions for the difference of two angles; $\theta_1 - \theta_2$.

Exercises 18 through 21 require the use of software that supports computer graphics.

18. Define a triangle T by identifying its vertices and sketch it on paper.

(a) Reflect T about the y-axis and record the resulting figure on paper, as Figure 1.

(b) Rotate Figure 1 counterclockwise through 30° and record the resulting figure on paper, as Figure 2.

(c) Reflect T about the line $y = x$, dilate the resulting figure in the x-direction by a factor of 2, and record the new figure on paper, as Figure 3.

(d) Repeat the experiment in part (c), but interchange the order of the matrix transformations. Record the resulting figure on paper, as Figure 4.

(e) Compare Figures 3 and 4.

(f) What does your answer in part (e) imply about the order of the matrix transformations as applied to the triangle?

19. Consider the triangle T defined in Exercise 18. Record T on paper.

(a) Reflect T about the x-axis. Predict the result before execution of the command. Call this matrix transformation L_1.

(b) Reflect the figure obtained in part (a) about the y-axis. Predict the result before execution of the command. Call this matrix transformation L_2.

(c) Record on paper the figure that resulted from parts (a) and (b).

(d) Examine the relationship between the figure obtained in part (b) and T. What single matrix transformation L_3 will accomplish the same result?

(e) Write a formula involving L_1, L_2, and L_3 that expresses the relationship you saw in part (d).

(f) Experiment with the formula in part (e) on several other figures until you can determine whether this formula is correct, in general. Write a brief summary of your experiments, observations, and conclusions.

20. Consider the unit square S and record S on paper.

(a) Reflect S about the x-axis to obtain Figure 1. Now reflect Figure 1 about the y-axis to obtain Figure 2. Finally, reflect Figure 2 about the line $y = -x$ to obtain Figure 3. Record Figure 3 on paper.

(b) Compare S with Figure 3. Denote the reflection about the x-axis as L_1, the reflection about the y-axis as L_2, and the reflection about the line $y = -x$ as L_3. What formula does your comparison suggest when L_1 is followed by L_2, and then by L_3 on S?

(c) If M_i, $i = 1, 2, 3$, denotes the matrix defining L_i, determine the entries of the matrix $M_3M_2M_1$. Does this result agree with your conclusion in part (b)?

(d) Experiment with the successive application of these three matrix transformations on other figures.

21. If your computer graphics software allows you to select any 2×2 matrix to use as a matrix transformation, perform the following experiment: Choose a singular matrix and apply it to a triangle, unit square, rectangle, and pentagon. Write a brief summary of your experiments, observations, and conclusions, indicating the behavior of "singular" matrix transformations.

22. If your software includes access to a computer algebra system (CAS), use it as follows: Let $f(\mathbf{u}) = A\mathbf{u}$ be the matrix transformation defined by

$$A = \begin{bmatrix} \cos\theta_1 & -\sin\theta_1 \\ \sin\theta_1 & \cos\theta_1 \end{bmatrix}$$

and let $g(\mathbf{v}) = B\mathbf{v}$ be the matrix transformation defined by

$$B = \begin{bmatrix} \cos\theta_2 & -\sin\theta_2 \\ \sin\theta_2 & \cos\theta_2 \end{bmatrix}.$$

(a) Find the symbolic matrix BA.

(b) Use CAS commands to simplify BA to obtain the matrix

$$\begin{bmatrix} \cos(\theta_1 + \theta_2) & -\sin(\theta_1 + \theta_2) \\ \sin(\theta_1 + \theta_2) & \cos(\theta_1 + \theta_2) \end{bmatrix}.$$

23. If your software includes access to a computer algebra system (CAS), use it as follows: Let $f(\mathbf{u}) = A\mathbf{u}$ be the matrix transformation defined by

$$A = \begin{bmatrix} \cos\theta & -\sin\theta \\ \sin\theta & \cos\theta \end{bmatrix}.$$

(a) Find the (symbolic) matrix that defines the matrix transformation $f(f(\mathbf{u}))$.

(b) Use CAS commands to simplify the matrix obtained in part (a) so that you obtain the double angle identities for sine and cosine.

24. If your software includes access to a computer algebra system (CAS), use it as follows: Let $f(\mathbf{u}) = A\mathbf{u}$ be the matrix transformation defined by

$$A = \begin{bmatrix} \cos\theta & -\sin\theta \\ \sin\theta & \cos\theta \end{bmatrix}.$$

(a) Find the (symbolic) matrix that defines the matrix transformation $f(f(f(f(\mathbf{u}))))$.

(b) Use CAS commands to simplify the matrix obtained in part (a) so that you obtain the identities for $\sin(4\theta)$ and $\cos(4\theta)$.

1.8 Correlation Coefficient (Optional)

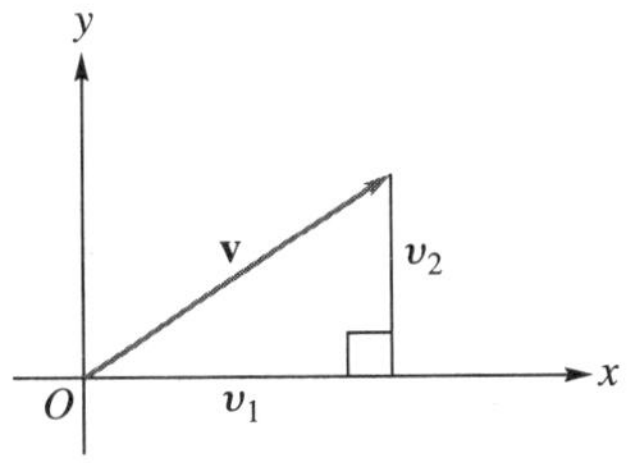

FIGURE 1.21 Length of **v**.

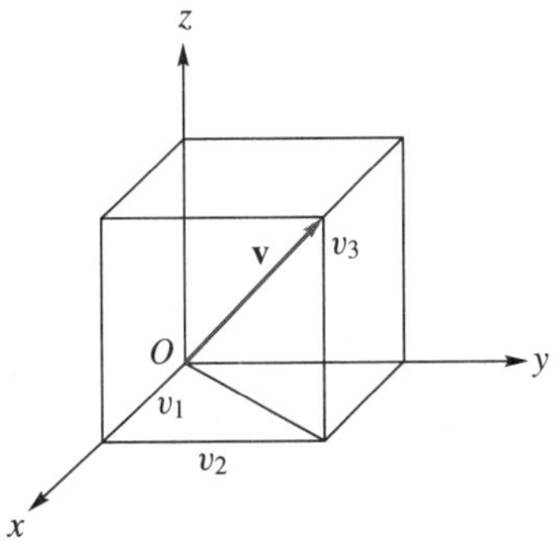

FIGURE 1.22 Length of **v**.

As we noted in Section 1.2, we can use an n-vector to provide a listing of data. In this section we provide a statistical application of the dot product to measure the strength of a linear relationship between two data vectors.

Before presenting this application, we must note two additional properties that vectors possess: length (also known as magnitude) and direction. These notions will be carefully developed in Chapter 4; in this section we merely give the properties without justification.

The length of the n-vector

$$\mathbf{v} = \begin{bmatrix} v_1 \\ v_2 \\ \vdots \\ v_{n-1} \\ v_n \end{bmatrix},$$

denoted as $\|\mathbf{v}\|$, is defined as

$$\|\mathbf{v}\| = \sqrt{v_1^2 + v_2^2 + \cdots + v_{n-1}^2 + v_n^2}. \tag{1}$$

If $n = 2$, the definition given in Equation (1) can be established easily as follows: From Figure 1.21 we see by the Pythagorean theorem that the length of the directed line segment from the origin to the point (v_1, v_2) is $\sqrt{v_1^2 + v_2^2}$. Since this directed line segment represents the vector $\mathbf{v} = \begin{bmatrix} v_1 \\ v_2 \end{bmatrix}$, we agree that $\|\mathbf{v}\|$, the length of the vector $\mathbf{v}$, is the length of the directed line segment. If $n = 3$, a similar proof can be given by applying the Pythagorean theorem twice in Figure 1.22.

It is easiest to determine the direction of an n-vector by defining the angle between two vectors. In Sections 5.1 and 5.4, we define the angle θ between the nonzero vectors $\mathbf{u}$ and $\mathbf{v}$ as the angle determined by the expression

$$\cos\theta = \frac{\mathbf{u}\cdot\mathbf{v}}{\|\mathbf{u}\|\,\|\mathbf{v}\|}.$$

In those sections we show that

$$-1 \le \frac{\mathbf{u}\cdot\mathbf{v}}{\|\mathbf{u}\|\,\|\mathbf{v}\|} \le 1.$$

Hence, this quantity can be viewed as the cosine of an angle $0 \le \theta \le \pi$.

We now turn to our statistical application. We compare two data n-vectors $\mathbf{x}$ and $\mathbf{y}$ by examining the angle θ between the vectors. The closeness of $\cos\theta$ to -1 or 1 measures how near the two vectors are to being parallel, since the angle between parallel vectors is either 0 or π radians. Nearly parallel indicates a strong relationship between the vectors. The smaller $|\cos\theta|$ is, the less likely it is that the vectors are parallel, and hence the weaker the relationship is between the vectors.

Table 1.1 contains data about the ten largest U.S. corporations, ranked by market value for 2004. In addition, we have included the corporate revenue for 2004. All figures are in billions of dollars and have been rounded to the nearest billion.

TABLE 1.1

Corporation	*Market Value (in \$ billions)*	*Revenue (in \$ billions)*
General Electric Corp.	329	152
Microsoft	287	37
Pfizer	285	53
Exxon Mobile	277	271
Citigroup	255	108
Wal-Mart Stores	244	288
Intel	197	34
American International Group	195	99
IBM Corp.	172	96
Johnson & Johnson	161	47

Source: *Time Almanac 2006, Information Please*, Pearson Education, Boston, Mass., 2005.

To display the data in Table 1.1 graphically, we form ordered pairs, (market value, revenue), for each of the corporations and plot this set of ordered pairs. The display in Figure 1.23 is called a **scatter plot**. This display shows that the data are spread out more vertically than horizontally. Hence there is wider variability in the revenue than in the market value.

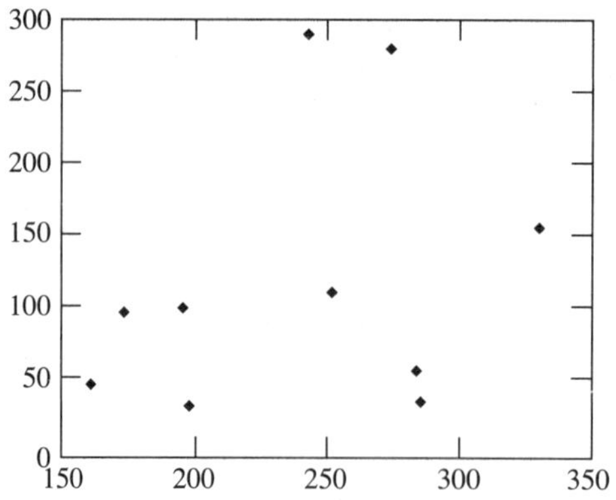

FIGURE 1.23

If we are interested only in how individual values of market value and revenue go together, then we can rigidly translate (shift) the plot so that the pattern of points does not change. One translation that is commonly used is to move the center of the plot to the origin. (If we think of the dots representing the ordered pairs as weights, we see that this amounts to shifting the center of mass to the origin.) To perform this translation, we compute the mean of the market value observations and subtract it from each market value; similarly, we compute the mean of the revenue observations and subtract it from each revenue value. We have (rounded to a whole number)

$$\text{Mean of market values} = 240, \quad \text{mean of revenues} = 119.$$

Subtracting the mean from each observation is called **centering the data**, and the corresponding centered data are displayed in Table 1.2. The corresponding scatter plot of the centered data is shown in Figure 1.24.

TABLE 1.2

Centered Market Value (in $ billions)	*Centered Revenue (in $ billions)*
89	33
47	−82
45	−66
37	152
15	−11
4	169
−43	−85
−45	−20
−68	−23
−79	−72

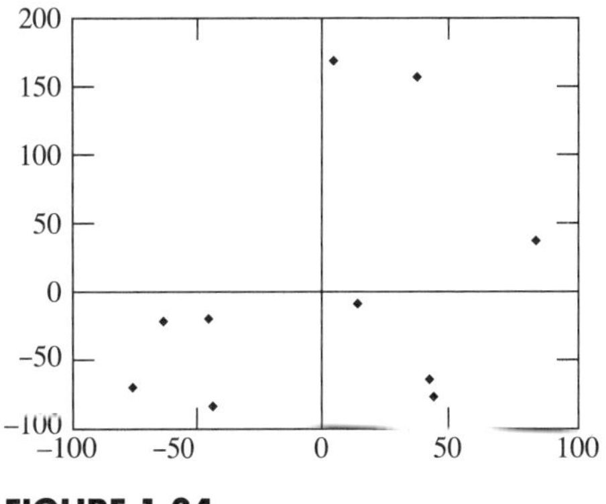

FIGURE 1.24

Note that the arrangement of dots in Figures 1.23 and 1.24 is the same; the scales of the respective axes have changed.

A scatter plot places emphasis on the observed data, not on the variables involved as general entities. What we want is a new way to plot the information that focuses on the variables. Here the variables involved are market value and revenue, so we want one axis for each corporation. This leads to a plot with ten axes, which we are unable to draw on paper. However, we visualize this situation by considering 10-vectors, that is, vectors with ten components, one for each corporation. Thus we define a vector **v** as the vector of centered market values and a vector **w** as the vector of centered revenues:

$$\mathbf{v} = \begin{bmatrix} 89 \\ 47 \\ 45 \\ 37 \\ 15 \\ 4 \\ -43 \\ -45 \\ -68 \\ -79 \end{bmatrix}, \quad \mathbf{w} = \begin{bmatrix} 33 \\ -82 \\ -66 \\ 152 \\ -11 \\ 169 \\ -85 \\ -20 \\ -23 \\ -72 \end{bmatrix}.$$

The best we can do schematically is to imagine **v** and **w** as directed line segments emanating from the origin, which is denoted by **0** (Figure 1.25).

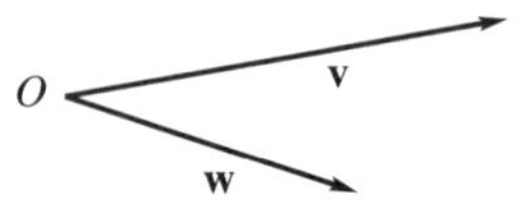

FIGURE 1.25

The representation of the centered information by vectors, as in Figure 1.25, is called a **vector plot**. From statistics, we have the following conventions:

- In a vector plot, the length of a vector indicates the variability of the corresponding variable.
- In a vector plot, the angle between vectors measures how similar the variables are to each other.

The statistical terminology for "how similar the variables are" is "how highly correlated the variables are." Vectors that represent highly correlated variables have either a small angle or an angle close to π radians between them. Vectors that represent uncorrelated variables are nearly perpendicular; that is, the angle between them is near $\pi/2$.

The following chart summarizes the statistical terminology applied to the geometric characteristics of vectors in a vector plot.

Geometric Characteristics	***Statistical Interpretation***
Length of a vector.	Variability of the variable represented.
Angle between a pair of vectors is small.	The variables represented by the vectors are highly positively correlated.
Angle between a pair of vectors is near π.	The variables represented by the vectors are highly negatively correlated.
Angle between a pair of vectors is near $\pi/2$.	The variables represented by the vectors are uncorrelated or unrelated. The variables are said to be perpendicular or orthogonal.

From statistics we have the following measures of a sample of data $\{x_1, x_2, \ldots, x_{n-1}, x_n\}$:

Sample size $= n$, the number of data.

$$\textbf{Sample mean} = \overline{x} = \frac{\sum_{i=1}^{n} x_i}{n}, \text{ the average of the data.}$$

Correlation coefficient: If the n-vectors $\mathbf{x}$ and $\mathbf{y}$ are data vectors where the data have been centered, then the correlation coefficient, denoted by $Cor(\mathbf{x}, \mathbf{y})$, is computed by

$$Cor(\mathbf{x}, \mathbf{y}) = \frac{\mathbf{x} \cdot \mathbf{y}}{\|\mathbf{x}\| \, \|\mathbf{y}\|}.$$

Geometrically, $Cor(\mathbf{x}, \mathbf{y})$ is the cosine of the angle between vectors $\mathbf{x}$ and $\mathbf{y}$.

For the centered data in Table 1.2, the sample size is $n = 10$, the mean of the market value variable is 240, and the mean of the revenue variable is 119. To determine the correlation coefficient for $\mathbf{v}$ and $\mathbf{w}$, we compute

$$Cor(\mathbf{v}, \mathbf{w}) = \cos\theta = \frac{\mathbf{v} \cdot \mathbf{w}}{\|\mathbf{v}\| \, \|\mathbf{w}\|} = 0.2994,$$

and thus

$$\theta = \arccos(0.2994) = 1.2667 \text{ radians} \approx 72.6°.$$

This result indicates that the variables market value and revenue are not highly correlated. This seems to be reasonable, given the physical meaning of the variables from a financial point of view. Including more than the ten top corporations may provide a better measure of the correlation between market value and revenue. Another approach that can be investigated based on the scatter plots is to omit data that seem far from the grouping of the majority of the data. Such data are termed **outliers**, and this approach has validity for certain types of statistical studies.

Figure 1.26 shows scatter plots that geometrically illustrate various cases for the value of the correlation coefficient. This emphasizes that the correlation coefficient is a measure of linear relationship between a pair of data vectors $\mathbf{x}$ and $\mathbf{y}$. The closer all the data points are to the line (in other words, the less scatter), the higher the correlation between the data.

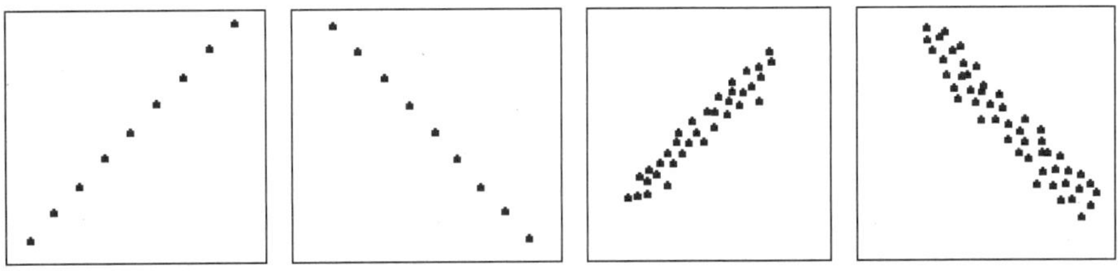

(a) Perfect positive correlation; $Cor(\mathbf{x}, \mathbf{y}) = 1$.
(b) Perfect negative correlation; $Cor(\mathbf{x}, \mathbf{y}) = -1$.
(c) Less than perfect positive correlation; $0 \le Cor(\mathbf{x}, \mathbf{y}) \le 1$.
(d) Less than perfect negative correlation; $-1 \le Cor(\mathbf{x}, \mathbf{y}) \le 0$.

FIGURE 1.26

To compute the correlation coefficient of a set of ordered pairs (x_i, y_i), $i = 1, 2, \ldots, n$, where

$$\mathbf{x} = \begin{bmatrix} x_1 \\ x_2 \\ \vdots \\ x_{n-1} \\ x_n \end{bmatrix} \quad \text{and} \quad \mathbf{y} = \begin{bmatrix} y_1 \\ y_2 \\ \vdots \\ y_{n-1} \\ y_n \end{bmatrix},$$

we use the steps in Table 1.3. The computational procedure in Table 1.3 is called the **Pearson product–moment correlation coefficient** in statistics.

TABLE 1.3

1. Compute the sample means for each data vector:

$$\bar{x} = \frac{\sum_{i=1}^{n} x_i}{n}, \qquad \bar{y} = \frac{\sum_{i=1}^{n} y_i}{n}.$$

2. Determine the centered x-data and the centered y-data as the vectors $\mathbf{x}_c$ and $\mathbf{y}_c$, respectively, where

$$\mathbf{x}_c = \begin{bmatrix} x_1 - \bar{x} & x_2 - \bar{x} & \cdots & x_n - \bar{x} \end{bmatrix}^T$$
$$\mathbf{y}_c = \begin{bmatrix} y_1 - \bar{y} & y_2 - \bar{y} & \cdots & y_n - \bar{y} \end{bmatrix}^T.$$

3. Compute the correlation coefficient as

$$Cor(\mathbf{x}_c, \mathbf{y}_c) = \frac{\mathbf{x}_c \cdot \mathbf{y}_c}{\|\mathbf{x}_c\|\,\|\mathbf{y}_c\|}.$$

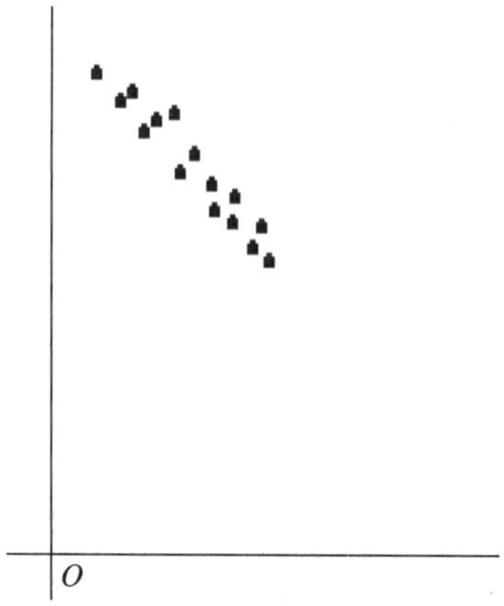

FIGURE 1.27

The correlation coefficient can be an informative statistic in a wide variety of applications. However, care must be exercised in interpreting this numerical estimation of a relationship between data. As with many applied situations, the interrelationships can be much more complicated than they are perceived to be at first glance. Be warned that statistics can be misused by confusing relationship with cause and effect. Here we have provided a look at a computation commonly used in statistical studies that employ dot products and the length of vectors. A much more detailed study is needed to use the correlation coefficient as part of a set of information for hypothesis testing. We emphasize such warnings with the following discussion, adapted from *Misused Statistics*, by A. J. Jaffe and H. F. Spirer (Marcel Dekker, Inc., New York, 1987):

Data involving divorce rate per 1000 of population versus death rate per 1000 of population were collected from cities in a certain region. Figure 1.27 shows a scatter plot of the data. This diagram suggests that divorce rate is highly (negatively) correlated with death rate. Based on this measure of relationship, should we infer that

(i) divorces cause death?
(ii) reducing the divorce rate will reduce the death rate?

Certainly we have not proved any cause-and-effect relationship; hence we must be careful to guard against statements based solely on a numerical measure of relationship.

Key Terms

Dot product
Length of a vector
Direction of a vector
Angle between vectors
Parallel vectors
Perpendicular vectors
Scatter plot
Vector plot
Correlated/uncorrelated variables
Sample size
Sample mean
Correlation coefficient
Outliers

1.8 Exercises

1. The data sets displayed in Figures A, B, C, and D have one of the following correlation coefficients; 0.97, 0.93, 0.88, 0.76. Match the figure with its correlation coefficient.

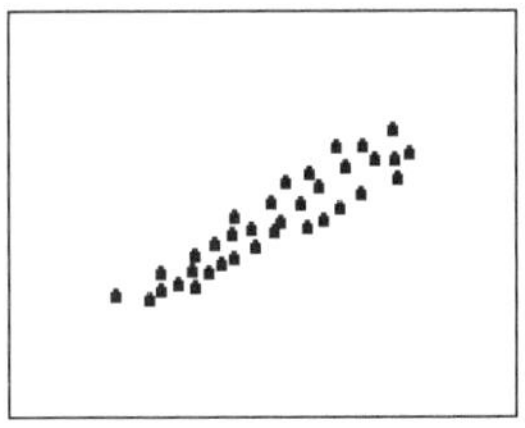

Figure A.

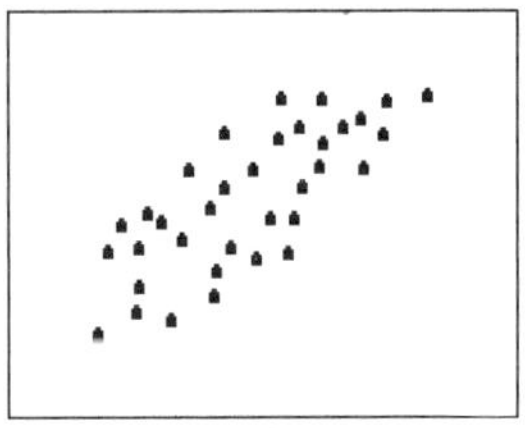

Figure B.

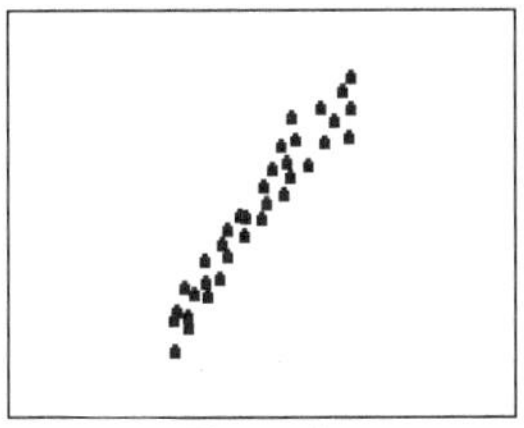

Figure C.

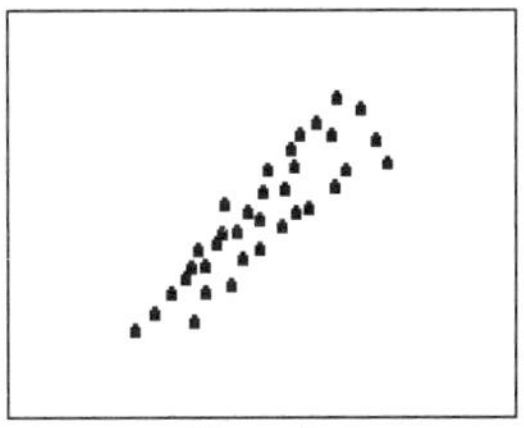

Figure D.

2. A meter that measures flow rates is being calibrated. In this initial test, $n = 8$ flows are sent to the meter and the corresponding meter readings are recorded. Let the set of flows and the corresponding meter readings be given by the 8-vectors $\mathbf{x}$ and $\mathbf{y}$, respectively, where

$$\mathbf{x} = \begin{bmatrix} 1 \\ 2 \\ 3 \\ 4 \\ 5 \\ 6 \\ 7 \\ 8 \end{bmatrix} \quad \text{and} \quad \mathbf{y} = \begin{bmatrix} 1.2 \\ 2.1 \\ 3.3 \\ 3.9 \\ 5.2 \\ 6.1 \\ 6.9 \\ 7.7 \end{bmatrix}.$$

Compute the correlation coefficient between the input flows in $\mathbf{x}$ and the resultant meter readings in $\mathbf{y}$.

3. An experiment to measure the amplitude of a shock wave resulting from the detonation of an explosive charge is conducted by placing recorders at various distances from the charge. (Distances are 100s of feet.) A common arrangement for the recorders is shown in the accompanying figure. The distance of a recorder from the charge and the amplitude of the recorded shock wave are shown in the table. Compute the correlation coefficient between the distance and amplitude data.

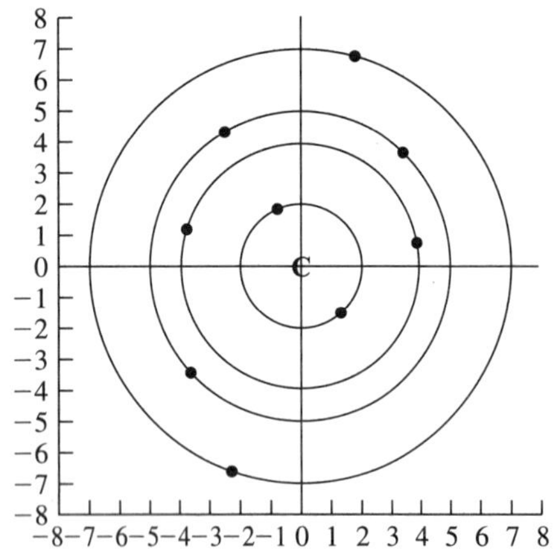

Distance	*Amplitude*
200	12.6
200	19.9
400	9.3
400	9.5
500	7.9
500	7.8
500	8.0
700	6.0
700	6.4

4. An equal number of two-parent families, each with three children younger than ten years old were interviewed in cities of populations ranging from 25,000 to 75,000. Interviewers collected data on (average) yearly living expenses for housing (rental/mortgage payments), food, and clothing. The collected living expense data were rounded to the nearest 100 dollars. Compute the correlation coefficient between the population data and living expense data shown in the following table:

City Population (in 1000s)	*Average Yearly Living Expense (in \$ 100s)*
25	72
30	65
35	78
40	70
50	79
60	85
65	83
75	88

Supplementary Exercises

1. Determine the number of entries on or above the main diagonal of a $k \times k$ matrix when

(a) $k = 2$, **(b)** $k = 3$, **(c)** $k = 4$, **(d)** $k = n$.

2. Let $A = \begin{bmatrix} 0 & 2 \\ 0 & 5 \end{bmatrix}$.

(a) Find a $2 \times k$ matrix $B \neq O$ such that $AB = O$ for $k = 1, 2, 3, 4$.

(b) Are your answers to part (a) unique? Explain.

3. Find all 2×2 matrices with real entries of the form

$$A = \begin{bmatrix} a & b \\ 0 & c \end{bmatrix}$$

such that $A^2 = I_2$.

4. An $n \times n$ matrix A (with real entries) is called a **square root** of the $n \times n$ matrix B (with real entries) if $A^2 = B$.

(a) Find a square root of $B = \begin{bmatrix} 1 & 1 \\ 0 & 1 \end{bmatrix}$.

(b) Find a square root of $B = \begin{bmatrix} 1 & 0 & 0 \\ 0 & 0 & 0 \\ 0 & 0 & 0 \end{bmatrix}$.

(c) Find a square root of $B = I_4$.

(d) Show that there is no square root of

$$B = \begin{bmatrix} 0 & 1 \\ 0 & 0 \end{bmatrix}.$$

5. Let A be an $m \times n$ matrix.

(a) Describe the diagonal entries of $A^T A$ in terms of the columns of A.

(b) Prove that the diagonal entries of $A^T A$ are nonnegative.

(c) When is $A^T A = O$?

6. If A is an $n \times n$ matrix, show that $(A^k)^T = (A^T)^k$ for any positive integer k.

7. Prove that every symmetric upper (or lower) triangular matrix is diagonal.

8. Let A be an $n \times n$ skew symmetric matrix and $\mathbf{x}$ an n-vector. Show that $\mathbf{x}^T A\mathbf{x} = 0$ for all $\mathbf{x}$ in R^n.

9. Let A be an upper triangular matrix. Show that A is nonsingular if and only if all the entries on the main diagonal of A are nonzero.

10. Show that the product of two 2×2 skew symmetric matrices is diagonal. Is this true for $n \times n$ skew symmetric matrices with $n > 2$?

11. Prove that if $\text{Tr}(A^T A) = 0$, then $A = O$.

12. For $n \times n$ matrices A and B, when does $(A+B)(A-B) = A^2 - B^2$?

13. Develop a simple expression for the entries of A^n, where n is a positive integer and

$$A = \begin{bmatrix} 1 & \frac{1}{2} \\ 0 & \frac{1}{2} \end{bmatrix}.$$

14. If $B = PAP^{-1}$, express $B^2, B^3, \ldots, B^k$, where k is a positive integer, in terms of A, P, and P^{-1}.

15. Prove that if A is skew symmetric and nonsingular, then A^{-1} is skew symmetric.

16. Let A be an $n \times n$ matrix. Prove that if $A\mathbf{x} = \mathbf{0}$ for all $n \times 1$ matrices $\mathbf{x}$, then $A = O$.

17. Let A be an $n \times n$ matrix. Prove that if $A\mathbf{x} = \mathbf{x}$ for all $n \times 1$ matrices $\mathbf{x}$, then $A = I_n$.

18. Let A and B be $n \times n$ matrices. Prove that if $A\mathbf{x} = B\mathbf{x}$ for all $n \times 1$ matrices $\mathbf{x}$, then $A = B$.

19. If A is an $n \times n$ matrix, then A is called **idempotent** if $A^2 = A$.

(a) Verify that I_n and O are idempotent.

(b) Find an idempotent matrix that is not I_n or O.

(c) Prove that the only $n \times n$ nonsingular idempotent matrix is I_n.

20. Let A and B be $n \times n$ idempotent matrices. (See Exercise 19.)

(a) Show that AB is idempotent if $AB = BA$.

(b) Show that if A is idempotent, then A^T is idempotent.

(c) Is $A + B$ idempotent? Justify your answer.

(d) Find all values of k for which kA is also idempotent.

21. Let A be an idempotent matrix.

(a) Show that $A^n = A$ for all integers $n \geq 1$.

(b) Show that $I_n - A$ is also idempotent.

22. If A is an $n \times n$ matrix, then A is called **nilpotent** if $A^k = O$ for some positive integer k.

(a) Prove that every nilpotent matrix is singular.

(b) Verify that $A = \begin{bmatrix} 0 & 1 & 1 \\ 0 & 0 & 1 \\ 0 & 0 & 0 \end{bmatrix}$ is nilpotent.

(c) If A is nilpotent, prove that $I_n - A$ is nonsingular. [*Hint*: Find $(I_n - A)^{-1}$ in the cases $A^k = O$, $k = 1, 2, \ldots$, and look for a pattern.]

23. Let $\mathbf{v} = \begin{bmatrix} a \\ b \\ c \\ d \end{bmatrix}$. Determine a vector $\mathbf{w}$ so that $\mathbf{v} \cdot \mathbf{w} = a + b + c + d$. If $\mathbf{v}$ is an n-vector, what is $\mathbf{w}$?

24. Use the result from Exercise 23 to develop a formula for the average of the entries in an n-vector $\mathbf{v} = \begin{bmatrix} v_1 \\ v_2 \\ \vdots \\ v_n \end{bmatrix}$ in terms of a ratio of dot products.

25. For an $n \times n$ matrix A, the main counter diagonal elements are $a_{1n}, a_{2,n-1}, \ldots, a_{n1}$. (Note that a_{ij} is a main counter diagonal element, provided that $i + j = n + 1$.) The sum of the main counter diagonal elements is denoted $\text{Mcd}(A)$, and we have

$$\text{Mcd}(A) = \sum_{i+j=n+1} a_{ij},$$

meaning the sum of all the entries of A whose subscripts add to $n + 1$.

(a) Prove: $\text{Mcd}(cA) = c\,\text{Mcd}(A)$, where c is a real number.

(b) Prove: $\text{Mcd}(A + B) = \text{Mcd}(A) + \text{Mcd}(B)$.

(c) Prove: $\text{Mcd}(A^T) = \text{Mcd}(A)$.

(d) Show by example that $\text{Mcd}(AB)$ need not be equal to $\text{Mcd}(BA)$.

26. An $n \times n$ matrix A is called **block diagonal** if it can be partitioned in such a way that all the nonzero entries are contained in square blocks A_{ii}.

(a) Partition the following matrix into a block diagonal matrix:

$$\begin{bmatrix} 1 & 2 & 0 & 0 \\ 3 & 4 & 0 & 0 \\ 0 & 0 & 1 & 0 \\ 0 & 0 & 2 & 3 \end{bmatrix}.$$

(b) If A is block diagonal, then the linear system $A\mathbf{x} = \mathbf{b}$ is said to be **uncoupled**, because it can be solved by considering the linear systems with coefficient matrices A_{ii} and right sides an appropriate portion of $\mathbf{b}$. Solve $A\mathbf{x} = \mathbf{b}$ by "uncoupling" the linear system when A is the 4×4 matrix of part (a) and

$$\mathbf{b} = \begin{bmatrix} 1 \\ 1 \\ 0 \\ 3 \end{bmatrix}.$$

27. Show that the product of two 2×2 skew symmetric matrices is diagonal. Is this true for $n \times n$ skew symmetric matrices with $n > 2$?

28. Let

$$A = \begin{bmatrix} A_{11} & O \\ O & A_{22} \end{bmatrix}$$

be a block diagonal $n \times n$ matrix. Prove that if A_{11} and A_{22} are nonsingular, then A is nonsingular.

29. Let

$$A = \begin{bmatrix} A_{11} & A_{12} \\ O & A_{22} \end{bmatrix}$$

be a partitioned matrix. If A_{11} and A_{22} are nonsingular, show that A is nonsingular and find an expression for A^{-1}.

*In Exercises 30 through 32, X and Y are $n \times 1$ matrices whose entries are $x_1, x_2, \ldots, x_n$, and $y_1, y_2, \ldots, y_n$, respectively. The **outer product** of X and Y is the matrix product XY^T, which gives the $n \times n$ matrix*

$$\begin{bmatrix} x_1y_1 & x_1y_2 & \cdots & x_1y_n \\ x_2y_1 & x_2y_2 & \cdots & x_2y_n \\ \vdots & \vdots & & \vdots \\ x_ny_1 & x_ny_2 & \cdots & x_ny_n \end{bmatrix}.$$

30. **(a)** Form the outer product of X and Y, where

$$X = \begin{bmatrix} 1 \\ 2 \\ 3 \end{bmatrix} \quad \text{and} \quad Y = \begin{bmatrix} 4 \\ 5 \\ 6 \end{bmatrix}.$$

(b) Form the outer product of X and Y, where

$$X = \begin{bmatrix} 1 \\ 2 \\ 1 \\ 2 \end{bmatrix} \quad \text{and} \quad Y = \begin{bmatrix} -1 \\ 0 \\ 3 \\ 5 \end{bmatrix}.$$

31. Prove or disprove: The outer product of X and Y equals the outer product of Y and X.

32. Prove that $\text{Tr}(XY^T) = X^TY$.

33. Let

$$A = \begin{bmatrix} 1 & 7 \\ 3 & 9 \\ 5 & 11 \end{bmatrix} \quad \text{and} \quad B = \begin{bmatrix} 2 & 4 \\ 6 & 8 \end{bmatrix}.$$

Verify that

$$AB = \sum_{i=1}^{2} \text{outer product of } \text{col}_i(A) \text{ with } \text{row}_i(B).$$

34. Let W be an $n \times 1$ matrix such that $W^TW = 1$. The $n \times n$ matrix

$$H = I_n - 2WW^T$$

is called a **Householder*** **matrix**. (Note that a Householder matrix is the identity matrix plus a scalar multiple of an outer product.)

(a) Show that H is symmetric.

(b) Show that $H^{-1} = H^T$.

ALSTON S. HOUSEHOLDER

*Alston S. Householder (1904–1993) was born in Rockford, Illinois, and died in Malibu, California. He received his undergraduate degree from Northwestern University and his Master of Arts from Cornell University, both in philosophy. He received his Ph.D. in mathematics in 1937 from the University of Chicago. His early work in mathematics dealt with the applications of mathematics to biology. In 1944, he began to work on problems dealing with World War II. In 1946, he became a member of the Mathematics Division of Oak Ridge National Laboratory and became its director in 1948. At Oak Ridge, his interests shifted from mathematical biology to numerical analysis. He is best known for his many important contributions to the field of numerical linear algebra. In addition to his research, Householder occupied a number of posts in professional organizations, served on a variety of editorial boards, and organized the important Gatlinburg conferences (now known as the Householder Symposia), which continue to this day.

35. A **circulant** of order n is the $n \times n$ matrix defined by

$$C = \operatorname{circ}(c_1, c_2, \ldots, c_n) = \begin{bmatrix} c_1 & c_2 & c_3 & \cdots & c_n \\ c_n & c_1 & c_2 & \cdots & c_{n-1} \\ c_{n-1} & c_n & c_1 & \cdots & c_{n-2} \\ \vdots & \vdots & \vdots & & \vdots \\ c_2 & c_3 & c_4 & \cdots & c_1 \end{bmatrix}.$$

The elements of each row of C are the same as those in the previous rows, but shifted one position to the right and wrapped around.

(a) Form the circulant $C = \operatorname{circ}(1, 2, 3)$.

(b) Form the circulant $C = \operatorname{circ}(1, 2, 5, -1)$.

(c) Form the circulant $C = \operatorname{circ}(1, 0, 0, 0, 0)$.

(d) Form the circulant $C = \operatorname{circ}(1, 2, 1, 0, 0)$.

36. Let $C = \operatorname{circ}(c_1, c_2, c_3)$. Under what conditions is C symmetric?

37. Let $C = \operatorname{circ}(c_1, c_2, \ldots, c_n)$ and let $\mathbf{x}$ be the $n \times 1$ matrix of all ones. Determine a simple expression for $C\mathbf{x}$.

38. Verify that for $C = \operatorname{circ}(c_1, c_2, c_3)$, $C^TC = CC^T$.

Chapter Review

True or False

1. A linear system of three equations can have exactly three different solutions.
2. If A and B are $n \times n$ matrices with no zero entries, then $AB \neq O$.
3. If A is an $n \times n$ matrix, then $A + A^T$ is symmetric.
4. If A is an $n \times n$ matrix and $\mathbf{x}$ is $n \times 1$, then $A\mathbf{x}$ is a linear combination of the columns of A.
5. Homogeneous linear systems are always consistent.
6. The sum of two $n \times n$ upper triangular matrices is upper triangular.
7. The sum of two $n \times n$ symmetric matrices is symmetric.
8. If a linear system has a nonsingular coefficient matrix, then the system has a unique solution.
9. The product of two $n \times n$ nonsingular matrices is nonsingular.
10. $A = \begin{bmatrix} 0 & 0 \\ 0 & 1 \end{bmatrix}$ defines a matrix transformation that projects the vector $\begin{bmatrix} x \\ y \end{bmatrix}$ onto the y-axis.

Quiz

1. Find all solutions to the following linear system by the method of elimination:

$$\begin{aligned} 4x + 3y &= -4 \\ 2x - y &= 8 \end{aligned}$$

2. Determine all values of r so that $x = 1$, $y = -1$, $z = r$ is a solution to the following linear system:

$$\begin{aligned} x - 2y + 3z &= 3 \\ 4x + 5y - z &= -1 \\ 6x + y + 5z &= 5 \end{aligned}$$

3. Determine all values of a and b so that

$$\begin{bmatrix} 1 & 2 \\ a & 0 \end{bmatrix} \begin{bmatrix} 3 & b \\ -4 & 1 \end{bmatrix} = \begin{bmatrix} -5 & 6 \\ 12 & 16 \end{bmatrix}.$$

4. Let

$$L = \begin{bmatrix} 2 & 0 & 0 \\ 1 & -2 & 0 \\ a & 1 & 3 \end{bmatrix} \quad \text{and} \quad U = \begin{bmatrix} 1 & 4 & b \\ 0 & -1 & 5 \\ 0 & 0 & c \end{bmatrix}.$$

(a) Determine all values of a so that the (3, 2) entry of LU is 7.

(b) Determine all values of b and c so that the (2, 3) entry of LU is 0.

5. Let $\mathbf{u}$ be a vector in R^2 whose projection onto the x-axis is shown in the figure. Determine the entries of the vector $\mathbf{u}$.

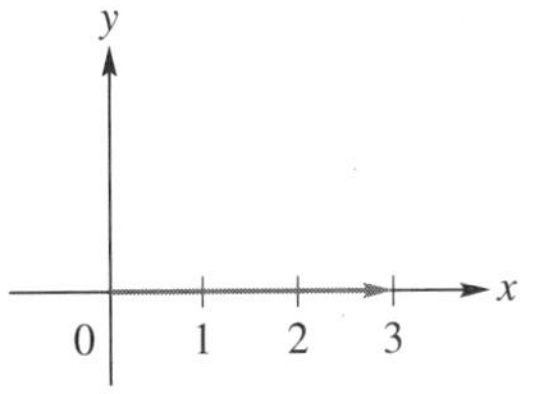

Discussion Exercises

1. In Section 1.5 we briefly introduced recursion relations. We showed how the members of the Fibonacci sequence could be computed by using a matrix form. By successive matrix multiplication we were able to produce 2×1 matrices from which the members of the sequence could be extracted. An alternative approach is to derive a formula that generates any member of the sequence without computing the preceding members. For two-term recursion relations of the form $u_n = au_{n-1} + bu_{n-2}$ we can derive such a formula by solving a 2×2 system of equations as follows: Associated with this recursion relation is a quadratic polynomial of the form $r^2 = ar + b$, called the characteristic polynomial of the recursion relation. Label the roots of this polynomial as r_1 and r_2, with $r_1 \neq r_2$. Then it can be shown that the general term of the recursion relation u_n can be expressed in the form

$$u_n = C_1(r_1)^n + C_2(r_2)^n, \qquad (*)$$

where the constants C_1 and C_2 are determined by solving the system of equations generated by setting $n = 0$ and then $n = 1$ in (*) and using the initial values of u_0 and u_1 given to start the recursion relation.

   (a) Use this approach to find a formula for the nth member of the Fibonacci sequence.

   (b) Since the members of the Fibonacci sequence are integers, explain why the formula from part (a) is rather amazing.

   (c) Given the coefficients a and b of a two-term recursion relation of the form $u_n = au_{n-1} + bu_{n-2}$, construct a general system of equations that can be used to determine the values of the constants C_1 and C_2 in (*).

2. We can use a 3″ by 5″ index card to represent (a portion of) a plane in space. Take three such index cards and cut a slit in each one about halfway through the card along the 5-inch edge. The slit will let you model intersecting a pair of planes by passing the slit in one card through the slit in another card. Each card can be represented by an equation of the form $a_k x + b_k y + c_k z = d_k$ so that for $k = 1, 2, 3$ we have a system of three equations in three unknowns. Using the index cards, configure them so that they represent four different consistent linear systems and four different inconsistent systems. Sketch each configuration and provide a brief description of your diagram. (*Hint*: Two index cards placed exactly one on top of the other represent the same plane.)

3. In Section 1.3 we defined matrix multiplication; if the matrix A is $m \times p$ and the matrix B is $p \times n$, then the product AB is an $m \times n$ matrix. Here we investigate the special case where A is $m \times 1$ and B is $1 \times n$.

   (a) For the case $m = 3$ and $n = 4$, let $A = \begin{bmatrix} 2 \\ 0 \\ 5 \end{bmatrix}$ and $B = \begin{bmatrix} 3 & 4 & 6 & -1 \end{bmatrix}$. Compute AB and carefully describe a pattern of the resulting entries.

   (b) Explain why there will be a row of zeros in AB if the $m \times 1$ matrix A has a zero entry and matrix B is $1 \times n$.

   (c) Suppose that the second entry of the $m \times 1$ matrix A is zero. Describe a resulting pattern of zeros that appears in AA^T.

   (d) Let C be a 5×5 matrix with identical columns. Explain how to generate C by using a product of a 5×1 matrix A and a 1×5 matrix B. (Explicitly indicate the entries of A and B.)

4. In Sections 1.6 and 1.7 we introduced matrix transformations for the manipulation of objects in the plane—for example, dilation, contraction, rotation, reflection, and shear. Another manipulation is called translation, which is like sliding an object over a certain distance. To translate the vector $\mathbf{u} = \begin{bmatrix} x \\ y \end{bmatrix}$ in the plane, we add a fixed vector $\mathbf{a} = \begin{bmatrix} a_1 \\ a_2 \end{bmatrix}$ to it. This action defines a function g mapping R^2 to R^2 given by

$$g(\mathbf{u}) = \mathbf{u} + \mathbf{a}.$$

The function g is called a translation by the vector $\mathbf{a}$. Explain why translations are not matrix transformations for $\mathbf{a} \neq \begin{bmatrix} 0 \\ 0 \end{bmatrix}$. (*Hint*: Use Exercise 20 in Section 1.6.)

5. In Chapter 1 we discussed systems of linear equations in the plane and in space to help connect geometry to the algebraic expressions for the systems. We observed that a pair of lines in the plane could have zero, one, or infinitely many intersection points; similarly, for three planes in space. For nonlinear systems of equations there can be many other sets of intersection points.

   (a) The set of functions $y = f(x) = x^{2n}$, $n = 1, 2, \ldots$, is the family of even power functions. Describe the set of intersection points for any pair of functions in this family.

   (b) For the function $y = f(x) = x^2$, where $x \geq 0$, determine another function $y = g(x)$ so that f and g intersect in exactly four points.

6. People have long been intrigued by magic squares. In the past they were often associated with the supernatural and hence considered to have magical properties. Today they are studied to illustrate mathematical properties and also as puzzles and games. We define a magic square as follows: A magic square is an $n \times n$ matrix of positive integers such that the sum of the entries of each row, each column, and each main diagonal (see the diagram) is equal to the same (magic) constant K.

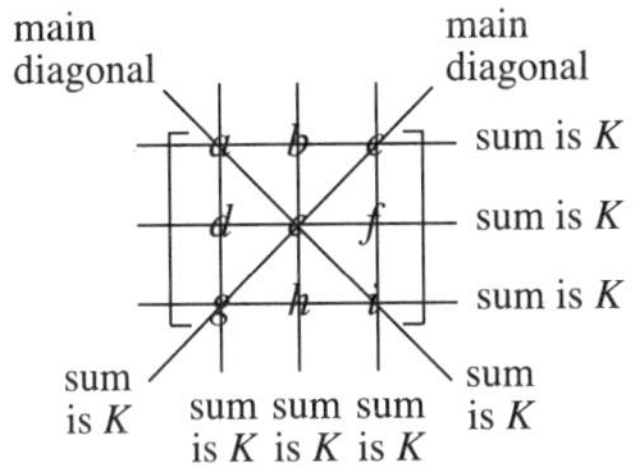

A 3×3 magic square.

The matrix

$$A = \begin{bmatrix} 8 & 1 & 6 \\ 3 & 5 & 7 \\ 4 & 9 & 2 \end{bmatrix}$$

is a magic square with (magic) constant $K = 15$. (Verify.) For 3×3 and other small square matrices A of positive integers, it is rather easy to check whether A is a magic square by observation.

(a) For a 3×3 matrix A, construct a method to check whether A is a magic square by multiplying A by particular matrices and by using an operation defined on matrices. Specifically indicate your strategy and state how each of the portions of the preceding definition are checked.

(b) Briefly discuss how to generalize the technique you devise in part (a) to $n \times n$ matrices.

CHAPTER

2 Solving Linear Systems

2.1 Echelon Form of a Matrix

In Section 1.1 we discussed the method of elimination for solving linear systems, which you studied in high school, and in Section 1.3 we introduced the coefficient matrix and augmented matrix associated with a linear system. In this section we discuss operations on a matrix, which when applied to an augmented matrix can greatly simplify the steps needed to determine the solution of the associated linear system. The operations discussed in this section apply to any matrix, whether or not it is an augmented matrix. In Section 2.2 we apply the constructions developed in this section to the solution of linear systems.

DEFINITION 2.1

An $m \times n$ matrix A is said to be in **reduced row echelon form** if it satisfies the following properties:

(a) All zero rows, if there are any, appear at the bottom of the matrix.

(b) The first nonzero entry from the left of a nonzero row is a 1. This entry is called a **leading one** of its row.

(c) For each nonzero row, the leading one appears to the right and below any leading ones in preceding rows.

(d) If a column contains a leading one, then all other entries in that column are zero.

A matrix in reduced row echelon form appears as a staircase ("echelon") pattern of leading ones descending from the upper left corner of the matrix.

An $m \times n$ matrix satisfying properties (a), (b), and (c) is said to be in **row echelon form**. In Definition 2.1, there may be no zero rows.

A similar definition can be formulated in the obvious manner for **reduced column echelon form** and **column echelon form**.

EXAMPLE 1

The following are matrices in reduced row echelon form, since they satisfy properties (a), (b), (c), and (d):

$$A = \begin{bmatrix} 1 & 0 & 0 & 0 \\ 0 & 1 & 0 & 0 \\ 0 & 0 & 1 & 0 \\ 0 & 0 & 0 & 1 \end{bmatrix}, \quad B = \begin{bmatrix} 1 & 0 & 0 & 0 & -2 & 4 \\ 0 & 1 & 0 & 0 & 4 & 8 \\ 0 & 0 & 0 & 1 & 7 & -2 \\ 0 & 0 & 0 & 0 & 0 & 0 \\ 0 & 0 & 0 & 0 & 0 & 0 \end{bmatrix},$$

and

$$C = \begin{bmatrix} 1 & 2 & 0 & 0 & 1 \\ 0 & 0 & 1 & 2 & 3 \\ 0 & 0 & 0 & 0 & 0 \end{bmatrix}.$$

The matrices that follow are not in reduced row echelon form. (Why not?)

$$D = \begin{bmatrix} 1 & 2 & 0 & 4 \\ 0 & 0 & 0 & 0 \\ 0 & 0 & 1 & -3 \end{bmatrix}, \quad E = \begin{bmatrix} 1 & 0 & 3 & 4 \\ 0 & 2 & -2 & 5 \\ 0 & 0 & 1 & 2 \end{bmatrix},$$

$$F = \begin{bmatrix} 1 & 0 & 3 & 4 \\ 0 & 1 & -2 & 5 \\ 0 & 1 & 2 & 2 \\ 0 & 0 & 0 & 0 \end{bmatrix}, \quad G = \begin{bmatrix} 1 & 2 & 3 & 4 \\ 0 & 1 & -2 & 5 \\ 0 & 0 & 1 & 2 \\ 0 & 0 & 0 & 0 \end{bmatrix}.$$

■

EXAMPLE 2

The following are matrices in row echelon form:

$$H = \begin{bmatrix} 1 & 5 & 0 & 2 & -2 & 4 \\ 0 & 1 & 0 & 3 & 4 & 8 \\ 0 & 0 & 0 & 1 & 7 & -2 \\ 0 & 0 & 0 & 0 & 0 & 0 \\ 0 & 0 & 0 & 0 & 0 & 0 \end{bmatrix}, \quad I = \begin{bmatrix} 1 & 0 & 0 & 0 \\ 0 & 1 & 0 & 0 \\ 0 & 0 & 1 & 0 \\ 0 & 0 & 0 & 1 \end{bmatrix},$$

$$J = \begin{bmatrix} 0 & 0 & 1 & 3 & 5 & 7 & 9 \\ 0 & 0 & 0 & 0 & 1 & -2 & 3 \\ 0 & 0 & 0 & 0 & 0 & 1 & 2 \\ 0 & 0 & 0 & 0 & 0 & 0 & 1 \\ 0 & 0 & 0 & 0 & 0 & 0 & 0 \end{bmatrix}.$$

■

A useful property of matrices in reduced row echelon form (see Exercise 9) is that if A is an $n \times n$ matrix in reduced row echelon form $\neq I_n$, then A has a row consisting entirely of zeros.

We shall now show that every matrix can be put into row (column) echelon form, or into reduced row (column) echelon form, by means of certain row (column) operations.

DEFINITION 2.2

An **elementary row (column) operation** on a matrix A is any one of the following operations:

(a) **Type I:** Interchange any two rows (columns).
(b) **Type II:** Multiply a row (column) by a nonzero number.
(c) **Type III:** Add a multiple of one row (column) to another.

We now introduce the following notation for elementary row and elementary column operations on matrices:

- Interchange rows (columns) i and j, Type I:
$$\mathbf{r}_i \leftrightarrow \mathbf{r}_j \quad (\mathbf{c}_i \leftrightarrow \mathbf{c}_j).$$
- Replace row (column) i by k times row (column) i, Type II:
$$k\mathbf{r}_i \rightarrow \mathbf{r}_i \quad (k\mathbf{c}_i \rightarrow \mathbf{c}_i).$$
- Replace row (column) j by k times row (column) i + row (column) j, Type III:
$$k\mathbf{r}_i + \mathbf{r}_j \rightarrow \mathbf{r}_j \quad (k\mathbf{c}_i + \mathbf{c}_j \rightarrow \mathbf{c}_j).$$

Using this notation, it is easy to keep track of the elementary row and column operations performed on a matrix. For example, we indicate that we have interchanged the ith and jth rows of A as $A_{\mathbf{r}_i \leftrightarrow \mathbf{r}_j}$. We proceed similarly for column operations.

Observe that when a matrix is viewed as the augmented matrix of a linear system, the elementary row operations are equivalent, respectively, to interchanging two equations, multiplying an equation by a nonzero constant, and adding a multiple of one equation to another equation.

EXAMPLE 3 Let

$$A = \begin{bmatrix} 0 & 0 & 1 & 2 \\ 2 & 3 & 0 & -2 \\ 3 & 3 & 6 & -9 \end{bmatrix}.$$

Interchanging rows 1 and 3 of A, we obtain

$$B = A_{\mathbf{r}_1 \leftrightarrow \mathbf{r}_3} = \begin{bmatrix} 3 & 3 & 6 & -9 \\ 2 & 3 & 0 & -2 \\ 0 & 0 & 1 & 2 \end{bmatrix}.$$

Multiplying the third row of A by $\frac{1}{3}$, we obtain

$$C = A_{\frac{1}{3}\mathbf{r}_3 \rightarrow \mathbf{r}_3} = \begin{bmatrix} 0 & 0 & 1 & 2 \\ 2 & 3 & 0 & -2 \\ 1 & 1 & 2 & -3 \end{bmatrix}.$$

Adding (-2) times row 2 of A to row 3 of A, we obtain

$$D = A_{-2\mathbf{r}_2 + \mathbf{r}_3 \rightarrow \mathbf{r}_3} = \begin{bmatrix} 0 & 0 & 1 & 2 \\ 2 & 3 & 0 & -2 \\ -1 & -3 & 6 & -5 \end{bmatrix}.$$

Observe that in obtaining D from A, row 2 of A *did not change.* ■

DEFINITION 2.3

An $m \times n$ matrix B is said to be **row (column) equivalent** to an $m \times n$ matrix A if B can be produced by applying a finite sequence of elementary row (column) operations to A.

EXAMPLE 4

Let

$$A = \begin{bmatrix} 1 & 2 & 4 & 3 \\ 2 & 1 & 3 & 2 \\ 1 & -2 & 2 & 3 \end{bmatrix}.$$

If we add 2 times row 3 of A to its second row, we obtain

$$B = A_{2\mathbf{r}_3+\mathbf{r}_2\to\mathbf{r}_2} = \begin{bmatrix} 1 & 2 & 4 & 3 \\ 4 & -3 & 7 & 8 \\ 1 & -2 & 2 & 3 \end{bmatrix},$$

so B is row equivalent to A.

Interchanging rows 2 and 3 of B, we obtain

$$C = B_{\mathbf{r}_2\leftrightarrow\mathbf{r}_3} = \begin{bmatrix} 1 & 2 & 4 & 3 \\ 1 & -2 & 2 & 3 \\ 4 & -3 & 7 & 8 \end{bmatrix},$$

so C is row equivalent to B.

Multiplying row 1 of C by 2, we obtain

$$D = C_{2\mathbf{r}_1\to\mathbf{r}_1} = \begin{bmatrix} 2 & 4 & 8 & 6 \\ 1 & -1 & 2 & 3 \\ 4 & -3 & 7 & 8 \end{bmatrix},$$

so D is row equivalent to C. It then follows that D is row equivalent to A, since we obtained D by applying three successive elementary row operations to A. Using the notation for elementary row operations, we have

$$D = A_{\substack{2\mathbf{r}_3+\mathbf{r}_2\to\mathbf{r}_2 \\ \mathbf{r}_2\leftrightarrow\mathbf{r}_3 \\ 2\mathbf{r}_1\to\mathbf{r}_1}}.$$

We adopt the convention that the row operations are applied in the order listed. ■

We can readily show (see Exercise 10) that (a) every matrix is row equivalent to itself; (b) if B is row equivalent to A, then A is row equivalent to B; and (c) if C is row equivalent to B and B is row equivalent to A, then C is row equivalent to A. In view of (b), both statements "B is row equivalent to A" and "A is row equivalent to B" can be replaced by "A and B are row equivalent." A similar statement holds for column equivalence.

Theorem 2.1 Every nonzero $m \times n$ matrix $A = \left[a_{ij}\right]$ is row (column) equivalent to a matrix in row (column) echelon form.

Proof

We shall prove that A is row equivalent to a matrix in row echelon form. That is, by using only elementary row operations, we can transform A into a matrix in row echelon form. A completely analogous proof by elementary column operations establishes the result for column equivalence.

We start by looking in matrix A for the first column with a nonzero entry. This column is called the **pivot column**; the first nonzero entry in the pivot column is called the **pivot**. Suppose the pivot column is column j and the pivot occurs in row i. Now interchange, if necessary, rows 1 and i, getting matrix $B = \left[b_{ij}\right]$. Thus the pivot b_{1j} is $\neq 0$. Multiply the first row of B by the reciprocal of the pivot, that is, by $1/b_{1j}$, obtaining matrix $C = \left[c_{ij}\right]$. Note that $c_{1j} = 1$. Now if c_{hj}, $2 \leq h \leq m$, is not zero, then to row h of C we add $-c_{hj}$ times row 1; we do this for each value of h. It follows that the elements in column j, in rows $2, 3, \ldots, m$ of C, are zero. Denote the resulting matrix by D.

Next, consider the $(m-1) \times n$ submatrix A_1 of D obtained by deleting the first row of D. We now repeat this procedure with matrix A_1 instead of matrix A. Continuing this way, we obtain a matrix in row echelon form that is row equivalent to A. ■

EXAMPLE 5

Let

$$A = \begin{bmatrix} 0 & 2 & 3 & -4 & 1 \\ 0 & 0 & 2 & 3 & 4 \\ \textcircled{2} & 2 & -5 & 2 & 4 \\ 2 & 0 & -6 & 9 & 7 \end{bmatrix}.$$

Pivot column (column 1) **Pivot** (circled entry)

Column 1 is the first (counting from left to right) column in A with a nonzero entry, so column 1 is the pivot column of A. The first (counting from top to bottom) nonzero entry in the pivot column occurs in the third row, so the pivot is $a_{31} = 2$. We interchange the first and third rows of A, obtaining

$$B = A_{\mathbf{r}_1 \leftrightarrow \mathbf{r}_3} = \begin{bmatrix} \textcircled{2} & 2 & -5 & 2 & 4 \\ 0 & 0 & 2 & 3 & 4 \\ 0 & 2 & 3 & -4 & 1 \\ 2 & 0 & -6 & 9 & 7 \end{bmatrix}.$$

Multiply the first row of B by the reciprocal of the pivot, that is, by $\dfrac{1}{b_{11}} = \dfrac{1}{2}$, to obtain

$$C = B_{\frac{1}{2}\mathbf{r}_1 \to \mathbf{r}_1} = \begin{bmatrix} 1 & 1 & -\frac{5}{2} & 1 & 2 \\ 0 & 0 & 2 & 3 & 4 \\ 0 & 2 & 3 & -4 & 1 \\ 2 & 0 & -6 & 9 & 7 \end{bmatrix}.$$

Add (-2) times the first row of C to the fourth row of C to produce a matrix D in which the only nonzero entry in the pivot column is $d_{11} = 1$:

$$D = C_{-2\mathbf{r}_1+\mathbf{r}_4\to\mathbf{r}_4} = \begin{bmatrix} 1 & 1 & -\frac{5}{2} & 1 & 2 \\ 0 & 0 & 2 & 3 & 4 \\ 0 & 2 & 3 & -4 & 1 \\ 0 & -2 & -1 & 7 & 3 \end{bmatrix}.$$

Identify A_1 as the submatrix of D obtained by deleting the first row of D: Do not erase the first row of D. Repeat the preceding steps with A_1 instead of A.

$$\begin{matrix} 1 & 1 & -\frac{5}{2} & 1 & 2 \end{matrix}$$

$$A_1 = \begin{bmatrix} 0 & 0 & 2 & 3 & 4 \\ 0 & \textcircled{2} & 3 & -4 & 1 \\ 0 & -2 & -1 & 7 & 3 \end{bmatrix}.$$

Pivot (the circled 2) — **Pivot column** (the second column)

Do $(A_1)_{\mathbf{r}_1\leftrightarrow\mathbf{r}_1}$ to obtain B_1.

$$\begin{matrix} 1 & 1 & -\frac{5}{2} & 1 & 2 \end{matrix}$$

$$B_1 = \begin{bmatrix} 0 & 2 & 3 & -4 & 1 \\ 0 & 0 & 2 & 3 & 4 \\ 0 & -2 & -1 & 7 & 3 \end{bmatrix}.$$

Do $(B_1)_{\frac{1}{2}\mathbf{r}_1\to\mathbf{r}_1}$ to obtain C_1.

$$\begin{matrix} 1 & 1 & -\frac{5}{2} & 1 & 2 \end{matrix}$$

$$C_1 = \begin{bmatrix} 0 & 1 & \frac{3}{2} & -2 & \frac{1}{2} \\ 0 & 0 & 2 & 3 & 4 \\ 0 & -2 & -1 & 7 & 3 \end{bmatrix}.$$

Do $(C_1)_{2\mathbf{r}_1+\mathbf{r}_3\to\mathbf{r}_3}$ to obtain D_1

$$\begin{matrix} 1 & 1 & -\frac{5}{2} & 1 & 2 \end{matrix}$$

$$D_1 = \begin{bmatrix} 0 & 1 & \frac{3}{2} & -2 & \frac{1}{2} \\ 0 & 0 & 2 & 3 & 4 \\ 0 & 0 & 2 & 3 & 4 \end{bmatrix}.$$

Deleting the first row of D_1 yields the matrix A_2. We repeat the procedure with A_2 instead of A. No rows of A_2 have to be interchanged.

$$\begin{array}{ccccc} 1 & 1 & -\frac{5}{2} & 1 & 2 \\ 0 & 1 & \frac{3}{2} & -2 & \frac{1}{2} \end{array}$$

$$A_2 = \begin{bmatrix} 0 & 0 & \textcircled{2} & 3 & 4 \\ 0 & 0 & 2 & 3 & 4 \end{bmatrix} = B_2. \quad \text{Do } (B_2)_{\frac{1}{2}\mathbf{r}_1 \to \mathbf{r}_1} \text{ to obtain } C_2.$$

Pivot — **Pivot column of A_2**

$$\begin{array}{ccccc} 1 & 1 & -\frac{5}{2} & 1 & 2 \\ 0 & 1 & \frac{3}{2} & -2 & \frac{1}{2} \end{array}$$

$$C_2 = \begin{bmatrix} 0 & 0 & 1 & \frac{3}{2} & 2 \\ 0 & 0 & 2 & 3 & 4 \end{bmatrix}.$$

Finally, do $(C_2)_{-2\mathbf{r}_1+\mathbf{r}_2\to\mathbf{r}_2}$ to obtain D_2.

$$\begin{array}{ccccc} 1 & 1 & -\frac{5}{2} & 1 & 2 \\ 0 & 1 & \frac{3}{2} & -2 & \frac{1}{2} \end{array}$$

$$D_2 = \begin{bmatrix} 0 & 0 & 1 & \frac{3}{2} & 2 \\ 0 & 0 & 0 & 0 & 0 \end{bmatrix}.$$

The matrix

$$H = \begin{bmatrix} 1 & 1 & -\frac{5}{2} & 1 & 2 \\ 0 & 1 & \frac{3}{2} & -2 & \frac{1}{2} \\ 0 & 0 & 1 & \frac{3}{2} & 2 \\ 0 & 0 & 0 & 0 & 0 \end{bmatrix}$$

is in row echelon form and is row equivalent to A. ■

When doing hand computations, it is sometimes possible to avoid fractions by suitably modifying the steps in the procedure.

Theorem 2.2 Every nonzero $m \times n$ matrix $A = \left[a_{ij}\right]$ is row (column) equivalent to a unique matrix in reduced row (column) echelon form.

Proof

We proceed as in Theorem 2.1, obtaining matrix H in row echelon form that is row equivalent to A. Suppose that rows $1, 2, \ldots, r$ of H are nonzero and that the leading ones in these rows occur in columns $c_1, c_2, \ldots, c_r$. Then $c_1 < c_2 < \cdots < c_r$. Starting with the last nonzero row of H, we add suitable multiples of this row to all rows above it to make all entries in column c_r above the leading one in row r equal to zero. We repeat this process with rows $r-1, r-2, \ldots,$ and 2, making all entries above a leading one equal to zero. The result is a matrix K in reduced row echelon form that has been derived from H by elementary row operations and is thus row equivalent to H. Since A is row equivalent to H, and H is row equivalent to K, then A is row equivalent to K. An analogous proof can be given to show that A is column equivalent to a matrix in reduced column echelon form. It can be shown, with some difficulty, that there is only one matrix in reduced row echelon form that is row equivalent to a given matrix. For a proof, see K. Hoffman and R. Kunze, *Linear Algebra*, 2d ed. (Englewood Cliffs, N.J.: Prentice-Hall, 1971). ■

Remark It should be noted that a row echelon form of a matrix is not unique.

EXAMPLE 6 Find a matrix in reduced row echelon form that is row equivalent to the matrix A of Example 5.

Solution

We start with the matrix H obtained in Example 5 in row echelon form that is row equivalent to A. We add suitable multiples of each nonzero row of H to zero out all entries above a leading 1. Thus, we start by adding $\left(-\frac{3}{2}\right)$ times the third row of H to its second row:

$$J_1 = H_{-\frac{3}{2}\mathbf{r}_3+\mathbf{r}_2\to\mathbf{r}_2} = \begin{bmatrix} 1 & 1 & -\frac{5}{2} & 1 & 2 \\ 0 & 1 & 0 & -\frac{17}{4} & -\frac{5}{2} \\ 0 & 0 & 1 & \frac{3}{2} & 2 \\ 0 & 0 & 0 & 0 & 0 \end{bmatrix}.$$

Next, we add $\frac{5}{2}$ times the third row of J_1 to its first row:

$$J_2 = (J_1)_{\frac{5}{2}\mathbf{r}_3+\mathbf{r}_1\to\mathbf{r}_1} = \begin{bmatrix} 1 & 1 & 0 & \frac{19}{4} & 7 \\ 0 & 1 & 0 & -\frac{17}{4} & -\frac{5}{2} \\ 0 & 0 & 1 & \frac{3}{2} & 2 \\ 0 & 0 & 0 & 0 & 0 \end{bmatrix}.$$

Finally, we add (-1) times the second row of J_2 to its first row:

$$K = (J_2)_{-1\mathbf{r}_2+\mathbf{r}_1\to\mathbf{r}_1} = \begin{bmatrix} 1 & 0 & 0 & 9 & \frac{19}{2} \\ 0 & 1 & 0 & -\frac{17}{4} & -\frac{5}{2} \\ 0 & 0 & 1 & \frac{3}{2} & 2 \\ 0 & 0 & 0 & 0 & 0 \end{bmatrix}.$$

This is in reduced row echelon form and is row equivalent to A. Alternatively, we can express the reduced row echelon form of A as

$$H_{\substack{-\frac{3}{2}\mathbf{r}_3+\mathbf{r}_2\to\mathbf{r}_2 \\ \frac{5}{2}\mathbf{r}_3+\mathbf{r}_1\to\mathbf{r}_1 \\ -1\mathbf{r}_2+\mathbf{r}_1\to\mathbf{r}_1}}.$$

■

Remark The procedure given here for finding a matrix K in reduced row echelon form that is row equivalent to a given matrix A is not the only one possible. For example, instead of first obtaining a matrix H in row echelon form that is row equivalent to A and then transforming H to reduced row echelon form, we could proceed as follows. First, zero out the entries below a leading 1 and then immediately zero out the entries above the leading 1. This procedure is not as efficient as the procedure given in Example 6.

Key Terms

Elimination method
Reduced row echelon form
Leading one
Row echelon form
Elementary row (column) operation
Row (column) equivalent
Pivot column
Pivot

2.1 Exercises

1. Find a row echelon form of each of the given matrices. Record the row operations you perform, using the notation for elementary row operations.

(a) $A = \begin{bmatrix} -1 & 2 & -5 \\ 2 & -1 & 6 \\ 2 & -2 & 7 \end{bmatrix}$

(b) $A = \begin{bmatrix} 1 & 1 & -1 \\ 3 & 4 & -1 \\ 5 & 6 & -3 \\ -2 & -2 & 2 \end{bmatrix}$

2. Find a row echelon form of each of the given matrices. Record the row operations you perform, using the notation for elementary row operations.

(a) $A = \begin{bmatrix} -1 & 1 & -1 & 0 & 3 \\ -3 & 4 & 1 & 1 & 10 \\ 4 & -6 & -4 & -2 & -14 \end{bmatrix}$

(b) $A = \begin{bmatrix} 1 & 1 & -4 \\ -2 & -1 & 10 \\ 4 & 3 & -12 \end{bmatrix}$

3. Each of the given matrices is in row echelon form. Determine its reduced row echelon form. Record the row operations you perform, using the notation for elementary row operations.

(a) $A = \begin{bmatrix} 1 & 2 & 4 \\ 0 & 1 & -2 \\ 0 & 0 & 1 \end{bmatrix}$

(b) $A = \begin{bmatrix} 1 & 4 & 3 & 5 \\ 0 & 0 & 1 & -4 \\ 0 & 0 & 0 & 1 \\ 0 & 0 & 0 & 0 \end{bmatrix}$

4. Each of the given matrices is in row echelon form. Determine its reduced row echelon form. Record the row operations you perform, using the notation for elementary row operations.

(a) $A = \begin{bmatrix} 1 & 0 & -3 & 2 \\ 0 & 1 & 1 & 1 \\ 0 & 0 & 1 & 2 \\ 0 & 0 & 0 & 0 \end{bmatrix}$

(b) $A = \begin{bmatrix} 1 & 3 & 0 & 2 & 4 \\ 0 & 1 & 0 & 1 & 0 \\ 0 & 0 & 1 & -1 & 0 \end{bmatrix}$

5. Find the reduced row echelon form of each of the given matrices. Record the row operations you perform, using the notation for elementary row operations.

(a) $A = \begin{bmatrix} 1 & 0 & -2 \\ -2 & 1 & 9 \\ 3 & 2 & 4 \end{bmatrix}$

(b) $A = \begin{bmatrix} 1 & 0 & 1 \\ -1 & 2 & -2 \\ 0 & 1 & 0 \\ -2 & 7 & -5 \end{bmatrix}$

6. Find the reduced row echelon form of each of the given matrices. Record the row operations you perform, using the notation for elementary row operations.

(a) $A = \begin{bmatrix} -1 & 2 & -5 \\ 2 & -1 & 6 \\ 2 & -2 & 7 \end{bmatrix}$

(b) $A = \begin{bmatrix} 1 & 1 & -1 \\ 3 & 4 & -1 \\ 5 & 6 & -3 \\ -2 & -2 & 2 \end{bmatrix}$

7. Let x, y, z, and w be nonzero real numbers. Label each of

the following matrices REF if it is in row echelon form, RREF if it is in reduced row echelon form, or N if it is not REF and not RREF:

(a) $\begin{bmatrix} 1 & x & y & 0 \\ 0 & 1 & 0 & z \\ 0 & 0 & w & 1 \end{bmatrix}$ **(b)** $\begin{bmatrix} 1 & x & y & z \\ 0 & 1 & 0 & 0 \\ 0 & 0 & 1 & 0 \\ 0 & 0 & 0 & 1 \end{bmatrix}$

(c) $\begin{bmatrix} 1 & 0 & 0 & x & 0 & 0 \\ 0 & 1 & w & y & 0 & 0 \\ 0 & 0 & 0 & 0 & 1 & 0 \\ 0 & 0 & 0 & 0 & 0 & 1 \end{bmatrix}$

8. Let x, y, z, and w be nonzero real numbers. Label each of the following matrices REF if it is in row echelon form, RREF if it is in reduced row echelon form, or N if it is not REF and not RREF:

(a) $\begin{bmatrix} 1 & 0 & x \\ 0 & 1 & y \\ 0 & 0 & 1 \end{bmatrix}$ **(b)** $\begin{bmatrix} 1 & x & 0 & 0 \\ 0 & 0 & 1 & 0 \\ 0 & 0 & 0 & 1 \\ 0 & 0 & 0 & 0 \end{bmatrix}$

(c) $\begin{bmatrix} 0 & y & 0 \\ 1 & 1 & 0 \\ 0 & 0 & 1 \end{bmatrix}$

9. Let A be an $n \times n$ matrix in reduced row echelon form. Prove that if $A \neq I_n$, then A has a row consisting entirely of zeros.

10. Prove:

(a) Every matrix is row equivalent to itself.

(b) If B is row equivalent to A, then A is row equivalent to B.

(c) If C is row equivalent to B and B is row equivalent to A, then C is row equivalent to A.

11. Let

$$A = \begin{bmatrix} 1 & 2 & -3 & 1 \\ -1 & 0 & 3 & 4 \\ 0 & 1 & 2 & -1 \\ 2 & 3 & 0 & -3 \end{bmatrix}.$$

(a) Find a matrix in column echelon form that is column equivalent to A.

(b) Find a matrix in reduced column echelon form that is column equivalent to A.

12. Repeat Exercise 11 for the matrix

$$\begin{bmatrix} 1 & 2 & 3 & 4 & 5 \\ 2 & 1 & 3 & -1 & 2 \\ 3 & 1 & 2 & 4 & 1 \end{bmatrix}.$$

13. Determine the reduced row echelon form of

$$A = \begin{bmatrix} \cos\theta & \sin\theta \\ -\sin\theta & \cos\theta \end{bmatrix}.$$

2.2 Solving Linear Systems

In this section we use the echelon forms developed in Section 2.1 to more efficiently determine the solution of a linear system compared with the elimination method of Section 1.1. Using the augmented matrix of a linear system together with an echelon form, we develop two methods for solving a system of m linear equations in n unknowns. These methods take the augmented matrix of the linear system, perform elementary row operations on it, and obtain a new matrix that represents an equivalent linear system (i.e., a system that has the same solutions as the original linear system). The important point is that the latter linear system can be solved more easily.

To see how a linear system whose augmented matrix has a particular form can be readily solved, suppose that

$$\left[\begin{array}{ccc|c} 1 & 2 & 0 & 3 \\ 0 & 1 & 1 & 2 \\ 0 & 0 & 1 & -1 \end{array}\right]$$

represents the augmented matrix of a linear system. Then the solution is quickly found from the corresponding equations

$$\begin{aligned} x_1 + 2x_2 \phantom{{}+x_3} &= 3 \\ x_2 + x_3 &= 2 \\ x_3 &= -1 \end{aligned}$$

as

$$\begin{aligned} x_3 &= -1 \\ x_2 &= 2 - x_3 = 2 + 1 = 3 \\ x_1 &= 3 - 2x_2 = 3 - 6 = -3. \end{aligned}$$

The task of this section is to manipulate the augmented matrix representing a given linear system into a form from which the solution can be found more easily.

We now apply row operations to the solution of linear systems.

Theorem 2.3 Let $A\mathbf{x} = \mathbf{b}$ and $C\mathbf{x} = \mathbf{d}$ be two linear systems, each of m equations in n unknowns. If the augmented matrices $\left[A \;\vdots\; \mathbf{b}\right]$ and $\left[C \;\vdots\; \mathbf{d}\right]$ are row equivalent, then the linear systems are equivalent; that is, they have exactly the same solutions.

Proof

This follows from the definition of row equivalence and from the fact that the three elementary row operations on the augmented matrix are the three manipulations on linear systems, discussed in Section 1.1, which yield equivalent linear systems. We also note that if one system has no solution, then the other system has no solution. ■

Recall from Section 1.1 that the linear system of the form

$$\begin{aligned} a_{11}x_1 + a_{12}x_2 + \cdots + a_{1n}x_n &= 0 \\ a_{21}x_1 + a_{22}x_2 + \cdots + a_{2n}x_n &= 0 \\ &\vdots \\ a_{m1}x_1 + a_{m2}x_2 + \cdots + a_{mn}x_n &= 0 \end{aligned} \tag{1}$$

is called a homogeneous system. We can also write (1) in matrix form as

$$A\mathbf{x} = \mathbf{0}. \tag{2}$$

Corollary 2.1 If A and C are row equivalent $m \times n$ matrices, then the homogeneous systems $A\mathbf{x} = \mathbf{0}$ and $C\mathbf{x} = \mathbf{0}$ are equivalent.

Proof

Exercise. ■

We observe that we have developed the essential features of two very straightforward methods for solving linear systems. The idea consists of starting with the linear system $A\mathbf{x} = \mathbf{b}$, then obtaining a partitioned matrix $\left[C \;\vdots\; \mathbf{d}\right]$ in either row echelon form or reduced row echelon form that is row equivalent to the augmented matrix $\left[A \;\vdots\; \mathbf{b}\right]$. Now $\left[C \;\vdots\; \mathbf{d}\right]$ represents the linear system $C\mathbf{x} = \mathbf{d}$, which is quite simple to solve because of the structure of $\left[C \mid \mathbf{d}\right]$, and the set of solutions to this system gives precisely the set of solutions to $A\mathbf{x} = \mathbf{b}$; that is, the linear systems $A\mathbf{x} = \mathbf{b}$ and $C\mathbf{x} = \mathbf{d}$ are equivalent. (See Section 1.1.) The method where $\left[C \mid \mathbf{d}\right]$ is in row echelon form is called **Gaussian elimination**; the method where $\left[C \mid \mathbf{d}\right]$ is in reduced row echelon form is called **Gauss*–Jordan**† **reduction**. Strictly speaking, the original Gauss–Jordan reduction was more along the lines described in the preceding Remark. The version presented in this book is more efficient. In actual practice, neither Gaussian elimination nor Gauss–Jordan reduction is used as much as the method involving the LU-factorization of A that is discussed in Section 2.5. However, Gaussian elimination and Gauss–Jordan reduction are fine for small problems, and we use the latter heavily in this book.

Gaussian elimination consists of two steps:

Step 1. The transformation of the augmented matrix $\left[A \mid \mathbf{b}\right]$ to the matrix $\left[C \mid \mathbf{d}\right]$ in row echelon form using elementary row operations

Step 2. Solution of the linear system corresponding to the augmented matrix $\left[C \mid \mathbf{d}\right]$ using **back substitution**

For the case in which A is $n \times n$, and the linear system $A\mathbf{x} = \mathbf{b}$ has a unique solution, the matrix $\left[C \mid \mathbf{d}\right]$ has the following form:

CARL FRIEDRICH GAUSS

*Carl Friedrich Gauss (1777–1855) was born into a poor working-class family in Brunswick, Germany, and died in Göttingen, Germany, the most famous mathematician in the world. He was a child prodigy with a genius that did not impress his father, who called him a "star-gazer." However, his teachers were impressed enough to arrange for the Duke of Brunswick to provide a scholarship for Gauss at the local secondary school. As a teenager there, he made original discoveries in number theory and began to speculate about non-Euclidean geometry. His scientific publications include important contributions in number theory, mathematical astronomy, mathematical geography, statistics, differential geometry, and magnetism. His diaries and private notes contain many other discoveries that he never published.

An austere, conservative man who had few friends and whose private life was generally unhappy, he was very concerned that proper credit be given for scientific discoveries. When he relied on the results of others, he was careful to acknowledge them; and when others independently discovered results in his private notes, he was quick to claim priority.

In his research Gauss used a method of calculation that later generations generalized to row reduction of matrices and named in his honor, although the method was used in China almost 2000 years earlier.

WILHELM JORDAN

†Wilhelm Jordan (1842–1899) was born in southern Germany. He attended college in Stuttgart and in 1868 became full professor of geodesy at the technical college in Karlsruhe, Germany. He participated in surveying several regions of Germany. Jordan was a prolific writer whose major work, *Handbuch der Vermessungskunde* (*Handbook of Geodesy*), was translated into French, Italian, and Russian. He was considered a superb writer and an excellent teacher. Unfortunately, the Gauss–Jordan reduction method has been widely attributed to Camille Jordan (1838–1922), a well-known French mathematician. Moreover, it seems that the method was also discovered independently at the same time by B. I. Clasen, a priest who lived in Luxembourg. This biographical sketch is based on an excellent article: S. C. Althoen and R. McLaughlin, "Gauss–Jordan reduction: A brief history," *MAA Monthly*, 94 (1987), 130–142.

$$\left[\begin{array}{cccccc|c} 1 & c_{12} & c_{13} & \cdots & c_{1,n-1} & c_{1n} & d_1 \\ 0 & 1 & c_{23} & \cdots & c_{2,n-1} & c_{2n} & d_2 \\ \vdots & \vdots & \vdots & & \vdots & \vdots & \vdots \\ 0 & 0 & 0 & \cdots & 1 & c_{n-1,n} & d_{n-1} \\ 0 & 0 & 0 & \cdots & 0 & 1 & d_n \end{array}\right].$$

(The remaining cases are treated after Example 1.) This augmented matrix represents the linear system

$$\begin{aligned} x_1 + c_{12}x_2 + c_{13}x_3 + \ \dots \ + \quad c_{1n}x_n &= d_1 \\ x_2 + c_{23}x_3 + \ \dots \ + \quad c_{2n}x_n &= d_2 \\ &\vdots \\ x_{n-1} + c_{n-1\,n}x_n &= d_{n-1} \\ x_n &= d_n. \end{aligned}$$

Back substitution proceeds from the nth equation upward, solving for one variable from each equation:

$$\begin{aligned} x_n &= d_n \\ x_{n-1} &= d_{n-1} - c_{n-1\,n}x_n \\ &\vdots \\ x_2 &= d_2 - c_{23}x_3 - c_{24}x_4 - \cdots - c_{2n}x_n \\ x_1 &= d_1 - c_{12}x_2 - c_{13}x_3 - \cdots - c_{1n}x_n. \end{aligned}$$

EXAMPLE 1 The linear system

$$\begin{aligned} x + 2y + 3z &= 9 \\ 2x - \ y + \ z &= 8 \\ 3x \qquad - \ z &= 3 \end{aligned}$$

has the augmented matrix

$$\left[A \mid \mathbf{b}\right] = \left[\begin{array}{ccc|c} 1 & 2 & 3 & 9 \\ 2 & -1 & 1 & 8 \\ 3 & 0 & -1 & 3 \end{array}\right].$$

Transforming this matrix to row echelon form, we obtain (verify)

$$\left[C \mid \mathbf{d}\right] = \left[\begin{array}{ccc|c} 1 & 2 & 3 & 9 \\ 0 & 1 & 1 & 2 \\ 0 & 0 & 1 & 3 \end{array}\right].$$

Using back substitution, we now have

$$\begin{aligned} z &= 3 \\ y &= 2 - z = 2 - 3 = -1 \\ x &= 9 - 2y - 3z = 9 + 2 - 9 = 2; \end{aligned}$$

thus the solution is $x = 2$, $y = -1$, $z = 3$, which is unique. ■

The general case in which A is $m \times n$ is handled in a similar fashion, but we need to elaborate upon several situations that can occur. We thus consider $C\mathbf{x} = \mathbf{d}$, where C is $m \times n$, and $\left[C \mid \mathbf{d}\right]$ is in row echelon form. Then, for example, $\left[C \mid \mathbf{d}\right]$ might be of the following form:

$$\left[\begin{array}{cccccccc|c} 1 & c_{12} & c_{13} & \cdots & \cdots & c_{1,n-2} & c_{1,n-1} & c_{1n} & d_1 \\ 0 & 0 & 1 & c_{24} & \cdots & c_{2,n-2} & c_{2,n-1} & c_{2n} & d_2 \\ \vdots & \vdots & \vdots & \vdots & & \vdots & \vdots & \vdots & \vdots \\ 0 & 0 & 0 & 0 & \cdots & 0 & 1 & c_{k-1,n} & d_{k-1} \\ 0 & 0 & 0 & 0 & \cdots & 0 & 0 & 1 & d_k \\ 0 & 0 & 0 & 0 & \cdots & 0 & 0 & 0 & d_{k+1} \\ \vdots & \vdots & \vdots & \vdots & & \vdots & \vdots & \vdots & \vdots \\ 0 & 0 & 0 & 0 & \cdots & 0 & 0 & 0 & d_m \end{array}\right].$$

This augmented matrix represents the linear system

$$\begin{aligned} x_1 + c_{12}x_2 + c_{13}x_3 + \cdots \qquad\qquad + \quad c_{1n}x_n &= d_1 \\ x_3 + c_{24}x_4 + \cdots + \quad c_{2n}x_n &= d_2 \\ &\vdots \\ x_{n-1} + c_{k-1,n}x_n &= d_{k-1} \\ x_n &= d_k \\ 0x_1 + \quad 0x_2 + \qquad \cdots \qquad + \quad 0x_n &= d_{k+1} \\ &\vdots \\ 0x_1 + \quad 0x_2 + \qquad \cdots \qquad + \quad 0x_n &= d_m. \end{aligned}$$

First, if $d_{k+1} = 1$, then $C\mathbf{x} = \mathbf{d}$ has no solution, since at least one equation is not satisfied. If $d_{k+1} = 0$, which implies that $d_{k+2} = \cdots = d_m = 0$ (since $\left[C \mid \mathbf{d}\right]$ was assumed to be in row echelon form), we then obtain $x_n = d_k$, $x_{n-1} = d_{k-1} - c_{k-1,n}x_n = d_{k-1} - c_{k-1,n}d_k$ and continue using back substitution to find the remaining unknowns corresponding to the leading entry in each row. Of course, in the solution some of the unknowns may be expressed in terms of others that can take on any values whatsoever. This merely indicates that $C\mathbf{x} = \mathbf{d}$ has infinitely many solutions. On the other hand, every unknown may have a determined value, indicating that the solution is unique.

EXAMPLE 2

Let

$$\left[C \mid \mathbf{d}\right] = \left[\begin{array}{ccccc|c} 1 & 2 & 3 & 4 & 5 & 6 \\ 0 & 1 & 2 & 3 & -1 & 7 \\ 0 & 0 & 1 & 2 & 3 & 7 \\ 0 & 0 & 0 & 1 & 2 & 9 \end{array}\right].$$

Then

$$\begin{aligned}
x_4 &= 9 - 2x_5 \\
x_3 &= 7 - 2x_4 - 3x_5 = 7 - 2(9 - 2x_5) - 3x_5 = -11 + x_5 \\
x_2 &= 7 - 2x_3 - 3x_4 + x_5 = 2 + 5x_5 \\
x_1 &= 6 - 2x_2 - 3x_3 - 4x_4 - 5x_5 = -1 - 10x_5 \\
x_5 &= \text{any real number.}
\end{aligned}$$

The system is consistent, and all solutions are of the form

$$\begin{aligned}
x_1 &= -1 - 10r \\
x_2 &= 2 + 5r \\
x_3 &= -11 + r \\
x_4 &= 9 - 2r \\
x_5 &= r, \text{ any real number.}
\end{aligned}$$

Since r can be assigned any real number, the given linear system has infinitely many solutions. ■

EXAMPLE 3 If

$$\left[C \mid \mathbf{d}\right] = \left[\begin{array}{cccc|c} 1 & 2 & 3 & 4 & 5 \\ 0 & 1 & 2 & 3 & 6 \\ 0 & 0 & 0 & 0 & 1 \end{array}\right],$$

then $C\mathbf{x} = \mathbf{d}$ has no solution, since the last equation is

$$0x_1 + 0x_2 + 0x_3 + 0x_4 = 1,$$

which can never be satisfied. ■

When using the Gauss–Jordan reduction procedure, we transform the augmented matrix $\left[A \mid \mathbf{b}\right]$ to $\left[C \mid \mathbf{d}\right]$, which is in reduced row echelon form. This means that we can solve the linear system $C\mathbf{x} = \mathbf{d}$ without back substitution, as the examples that follow show; but of course, it takes more effort to put a matrix in reduced row echelon form than to put it in row echelon form. It turns out that the techniques of Gaussian elimination and Gauss–Jordan reduction, as described in this book, require the same number of operations.

EXAMPLE 4 If

$$\left[C \mid \mathbf{d}\right] = \left[\begin{array}{cccc|c} 1 & 0 & 0 & 0 & 5 \\ 0 & 1 & 0 & 0 & 6 \\ 0 & 0 & 1 & 0 & 7 \\ 0 & 0 & 0 & 1 & 8 \end{array}\right],$$

then the unique solution is

$$\begin{aligned}
x_1 &= 5 \\
x_2 &= 6 \\
x_3 &= 7 \\
x_4 &= 8.
\end{aligned}$$

■

EXAMPLE 5

If

$$\left[C \mid \mathbf{d}\right] = \left[\begin{array}{ccccc|c} 1 & 1 & 2 & 0 & -\frac{5}{2} & \frac{2}{3} \\ 0 & 0 & 0 & 1 & \frac{1}{2} & \frac{1}{2} \\ 0 & 0 & 0 & 0 & 0 & 0 \end{array}\right],$$

then

$$\begin{aligned} x_4 &= \tfrac{1}{2} - \tfrac{1}{2}x_5 \\ x_1 &= \tfrac{2}{3} - x_2 - 2x_3 + \tfrac{5}{2}x_5, \end{aligned}$$

where x_2, x_3, and x_5 can take on any real numbers, so the system has infinitely many solutions. Thus a solution is of the form

$$\begin{aligned} x_1 &= \tfrac{2}{3} - r - 2s + \tfrac{5}{2}t \\ x_2 &= r \\ x_3 &= s \\ x_4 &= \tfrac{1}{2} - \tfrac{1}{2}t \\ x_5 &= t, \end{aligned}$$

where r, s, and t are any real numbers. ■

We now solve a linear system both by Gaussian elimination and by Gauss–Jordan reduction.

EXAMPLE 6

Consider the linear system

$$\begin{aligned} x + 2y + 3z &= 6 \\ 2x - 3y + 2z &= 14 \\ 3x + y - z &= -2. \end{aligned}$$

We form the augmented matrix

$$\left[A \mid \mathbf{b}\right] = \left[\begin{array}{ccc|c} 1 & 2 & 3 & 6 \\ 2 & -3 & 2 & 14 \\ 3 & 1 & -1 & -2 \end{array}\right].$$

Add (-2) times the first row to the second row:

$$\left[A \mid \mathbf{b}\right]_{-2\mathbf{r}_1+\mathbf{r}_2\to\mathbf{r}_2} = \left[\begin{array}{ccc|c} 1 & 2 & 3 & 6 \\ 0 & -7 & -4 & 2 \\ 3 & 1 & -1 & -2 \end{array}\right].$$

Add (-3) times the first row to the third row:

$$\left[A \mid \mathbf{b}\right]_{\substack{-2\mathbf{r}_1+\mathbf{r}_2\to\mathbf{r}_2 \\ -3\mathbf{r}_1+\mathbf{r}_3\to\mathbf{r}_3}} = \left[\begin{array}{ccc|c} 1 & 2 & 3 & 6 \\ 0 & -7 & -4 & 2 \\ 0 & -5 & -10 & -20 \end{array}\right].$$

Multiply the third row by $\left(-\frac{1}{5}\right)$ and interchange the second and third rows:

$$\left[A \mid \mathbf{b}\right]_{\substack{-2\mathbf{r}_1+\mathbf{r}_2\to\mathbf{r}_2\\-3\mathbf{r}_1+\mathbf{r}_3\to\mathbf{r}_3\\-\frac{1}{5}\mathbf{r}_3\to\mathbf{r}_3\\\mathbf{r}_2\leftrightarrow\mathbf{r}_3}} = \left[\begin{array}{ccc|c} 1 & 2 & 3 & 6 \\ 0 & 1 & 2 & 4 \\ 0 & -7 & -4 & 2 \end{array}\right].$$

Add 7 times the second row to the third row:

$$\left[A \mid \mathbf{b}\right]_{\substack{-2\mathbf{r}_1+\mathbf{r}_2\to\mathbf{r}_2\\-3\mathbf{r}_1+\mathbf{r}_3\to\mathbf{r}_3\\-\frac{1}{5}\mathbf{r}_3\to\mathbf{r}_3\\\mathbf{r}_2\leftrightarrow\mathbf{r}_3\\7\mathbf{r}_2+\mathbf{r}_3\to\mathbf{r}_3}} = \left[\begin{array}{ccc|c} 1 & 2 & 3 & 6 \\ 0 & 1 & 2 & 4 \\ 0 & 0 & 10 & 30 \end{array}\right].$$

Multiply the third row by $\frac{1}{10}$:

$$\left[A \mid \mathbf{b}\right]_{\substack{-2\mathbf{r}_1+\mathbf{r}_2\to\mathbf{r}_2\\-3\mathbf{r}_1+\mathbf{r}_3\to\mathbf{r}_3\\-\frac{1}{5}\mathbf{r}_3\to\mathbf{r}_3\\\mathbf{r}_2\leftrightarrow\mathbf{r}_3\\7\mathbf{r}_2+\mathbf{r}_3\to\mathbf{r}_3\\\frac{1}{10}\mathbf{r}_3\to\mathbf{r}_3}} = \left[\begin{array}{ccc|c} 1 & 2 & 3 & 6 \\ 0 & 1 & 2 & 4 \\ 0 & 0 & 1 & 3 \end{array}\right].$$

This matrix is in row echelon form. This means that $z = 3$, and from the second row,

$$y + 2z = 4$$

so that

$$y = 4 - 2(3) = -2.$$

From the first row,

$$x + 2y + 3z = 6,$$

which implies that

$$x = 6 - 2y - 3z = 6 - 2(-2) - 3(3) = 1.$$

Thus $x = 1$, $y = -2$, and $z = 3$ is the solution. This gives the solution by Gaussian elimination.

To solve the given linear system by Gauss–Jordan reduction, we transform the last matrix to $\left[C \mid \mathbf{d}\right]$, which is in reduced row echelon form, by the following steps:

Add (-2) times the third row to the second row:

$$\left[C \mid \mathbf{d}\right]_{-2\mathbf{r}_3+\mathbf{r}_2\to\mathbf{r}_2} = \left[\begin{array}{ccc|c} 1 & 2 & 3 & 6 \\ 0 & 1 & 0 & -2 \\ 0 & 0 & 1 & 3 \end{array}\right].$$

Now add (-3) times the third row to the first row:

$$\left[C \mid \mathbf{d}\right]_{\substack{-2\mathbf{r}_3+\mathbf{r}_2\to\mathbf{r}_2\\-3\mathbf{r}_3+\mathbf{r}_1\to\mathbf{r}_1}} = \left[\begin{array}{ccc|c} 1 & 2 & 0 & -3 \\ 0 & 1 & 0 & -2 \\ 0 & 0 & 1 & 3 \end{array}\right].$$

Finally, add (-2) times the second row to the first row:

$$\left[C \mid \mathbf{d}\right]_{\substack{-2\mathbf{r}_3+\mathbf{r}_2\to\mathbf{r}_2\\ -3\mathbf{r}_3+\mathbf{r}_1\to\mathbf{r}_1\\ -2\mathbf{r}_2+\mathbf{r}_1\to\mathbf{r}_1}} = \left[\begin{array}{ccc|c} 1 & 0 & 0 & 1 \\ 0 & 1 & 0 & -2 \\ 0 & 0 & 1 & 3 \end{array}\right].$$

The solution is $x = 1$, $y = -2$, and $z = 3$, as before. ■

Remarks

1. As we perform elementary row operations, we may encounter a row of the augmented matrix being transformed to reduced row echelon form whose first n entries are zero and whose $n + 1$ entry is not zero. In this case, we can stop our computations and conclude that the given linear system is inconsistent.
2. In both Gaussian elimination and Gauss–Jordan reduction, we can use only row operations. Do not try to use any column operations.

■ Applications

Linear systems arise in a great many applications. In this section we look at several of these.

Quadratic Interpolation

Various approximation techniques in science and engineering use a parabola that passes through three given data points $\{(x_1, y_1), (x_2, y_2), (x_3, y_3)\}$, where $x_i \neq x_j$ for $i \neq j$. We call these **distinct points**, since the x-coordinates are all different. The graph of a quadratic polynomial $p(x) = ax^2+bx+c$ is a parabola, and we use the given data points to determine the coefficients a, b, and c as follows. Requiring that $p(x_i) = y_i$, $i = 1, 2, 3$, gives us three linear equations with unknowns a, b, and c:

$$\begin{aligned} p(x_1) &= y_1 \quad \text{or} \quad ax_1^2 + bx_1 + c = y_1 \\ p(x_2) &= y_2 \quad \text{or} \quad ax_2^2 + bx_2 + c = y_2 \\ p(x_3) &= y_3 \quad \text{or} \quad ax_3^2 + bx_3 + c = y_3. \end{aligned} \tag{3}$$

Let

$$A = \begin{bmatrix} x_1^2 & x_1 & 1 \\ x_2^2 & x_2 & 1 \\ x_3^2 & x_3 & 1 \end{bmatrix}$$

be the coefficient matrix, $\mathbf{v} = \begin{bmatrix} a \\ b \\ c \end{bmatrix}$ and $\mathbf{y} = \begin{bmatrix} y_1 \\ y_2 \\ y_3 \end{bmatrix}$. Then (3) can be written in matrix equation form as $A\mathbf{v} = \mathbf{y}$ whose augmented matrix

$$\left[A \mid \mathbf{y}\right] = \left[\begin{array}{ccc|c} x_1^2 & x_1 & 1 & y_1 \\ x_2^2 & x_2 & 1 & y_2 \\ x_3^2 & x_3 & 1 & y_3 \end{array}\right].$$

We solve this linear system by Gaussian elimination or Gauss–Jordan reduction, obtaining values for a, b, and c. It can be shown that there is a unique solution to this linear system if and only if the points are distinct. The construction of the parabola that matches the points of the given data set is called **quadratic interpolation**, and the parabola is called the **quadratic interpolant**. This process can be generalized to distinct data sets of $n + 1$ points and polynomials of degree n. We illustrate the construction of the quadratic in the following example:

EXAMPLE 7 Find the quadratic interpolant for the three distinct points $\{(1, -5), (-1, 1), (2, 7)\}$.

Solution

Setting up linear system (3), we find that its augmented matrix is (verify)

$$\left[A \mid \mathbf{y}\right] = \left[\begin{array}{ccc|c} 1 & 1 & 1 & -5 \\ 1 & -1 & 1 & 1 \\ 4 & 2 & 1 & 7 \end{array}\right].$$

Solving this linear system, we obtain (verify)

$$a = 5, \quad b = -3, \quad c = -7.$$

Thus the quadratic interpolant is $p(x) = 5x^2 - 3x - 7$, and its graph is given in Figure 2.1. The asterisks represent the three data points. ■

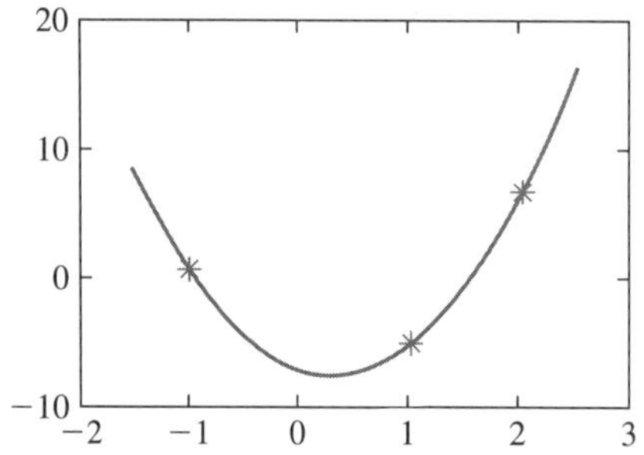

FIGURE 2.1

Temperature Distribution

A simple model for estimating the temperature distribution on a square plate gives rise to a linear system of equations. To construct the appropriate linear system, we use the following information: The square plate is perfectly insulated on its top and bottom so that the only heat flow is through the plate itself. The four edges are held at various temperatures. To estimate the temperature at an interior point on the plate, we use the rule that it is the average of the temperatures at its four compass-point neighbors, to the west, north, east, and south.

EXAMPLE 8 Estimate the temperatures T_i, $i = 1, 2, 3, 4$, at the four equispaced interior points on the plate shown in Figure 2.2.

Solution

We now construct the linear system to estimate the temperatures. The points at which we need the temperatures of the plate for this model are indicated in Figure 2.2 by dots. Using our averaging rule, we obtain the equations

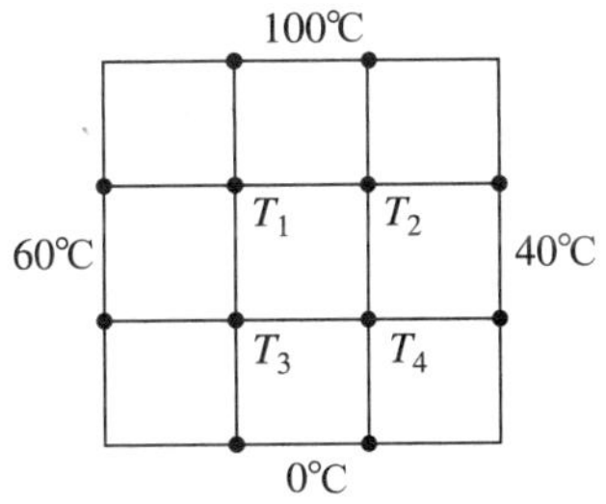

FIGURE 2.2

$$
\begin{aligned}
T_1 &= \frac{60+100+T_2+T_3}{4} &\quad\text{or}\quad & 4T_1 - T_2 - T_3 = 160\\
T_2 &= \frac{T_1+100+40+T_4}{4} &\quad\text{or}\quad & -T_1 + 4T_2 - T_4 = 140\\
T_3 &= \frac{60+T_1+T_4+0}{4} &\quad\text{or}\quad & -T_1 + 4T_3 - T_4 = 60\\
T_4 &= \frac{T_3+T_2+40+0}{4} &\quad\text{or}\quad & - T_2 - T_3 + 4T_4 = 40.
\end{aligned}
$$

The augmented matrix for this linear system is (verify)

$$
\left[A \;\vdots\; \mathbf{b}\right] = \left[\begin{array}{rrrr|r} 4 & -1 & -1 & 0 & 160\\ -1 & 4 & 0 & -1 & 140\\ -1 & 0 & 4 & -1 & 60\\ 0 & -1 & -1 & 4 & 40 \end{array}\right].
$$

Using Gaussian elimination or Gauss–Jordan reduction, we obtain the unique solution (verify)

$$
T_1 = 65^\circ\text{C},\quad T_2 = 60^\circ\text{C},\quad T_3 = 40^\circ\text{C},\quad \text{and}\quad T_4 = 35^\circ\text{C}. \qquad\blacksquare
$$

Global Positioning System

A Global Positioning System (GPS) is a satellite-based global navigation system enabling the user to determine his or her position in 3-dimensional coordinates without the need for further knowledge of navigational calculations. It was developed by the military as a locating utility, and the GPS system operated by the U.S. Department of Defense became operational in 1995. GPS technology has proven to be a useful tool for a wide variety of civilian applications as well and is now available in low-cost units. These units have been incorporated into boats, automobiles, airplanes, and handheld units available for general use such as hiking.

GPS is based on satellite ranging, that is, calculating the distances between a receiver and the position of three or more satellites (four or more if elevation is desired) and then applying some mathematics. Assuming that the positions of the satellites are known, the location of the receiver can be calculated by determining the distance from each of the satellites to the receiver. GPS takes these three or more known references and measured distances and "trilaterates" the position of the receiver. Trilateration is a method of determining the relative position of an object, in this case, orbiting satellites. For GPS calculations, there are three position variables, x, y, and z, together with a fourth variable, t, time. Time must be considered, since the GPS receiver processes signals from the satellites to determine the distances involved. Even though the signals move at the speed of light, there are small time delays for transmission, together with other factors like atmospheric conditions, that must be accounted for to ensure that accurate data are gathered. In this brief discussion of GPS we will use a simplified model to show how linear systems of equations enter into the mathematics that is part of the clever model involved in GPS.

For "real" GPS, we need to think in terms of three dimensions. In this context each satellite is represented by a sphere, and we need four spheres so that the location of the receiver can be determined by computing an intercept of the spheres; that is, a single point of intersection of the spheres. For our discussion, we will think in two dimensions and consider three satellites, each represented by a circle for which the coordinates of the center are known. We will assume that our GPS receiver can determine the distance between its location and the position of the satellite; thus, the radius of each circle is also known. Then, algebraically, we have the equations of three circles as shown in (4), where circle j has center (a_j, b_j) and radius r_j for $j = 1, 2, 3$.

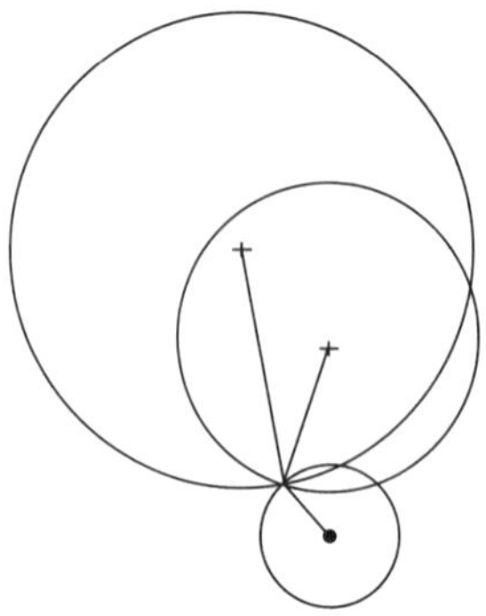

FIGURE 2.3

$$\begin{aligned}(x-a_1)^2+(y-b_1)^2&=r_1^2\\(x-a_2)^2+(y-b_2)^2&=r_2^2\\(x-a_3)^2+(y-b_3)^2&=r_3^2\end{aligned}\tag{4}$$

By construction, the location of the GPS receiver is on the circumference of each circle, so we are guaranteed that there is point (x, y) that satisfies each equation in (4). It is this point that will provide the coordinates of the GPS receiver. In Figure 2.3 we illustrate the system of the (nonlinear) equations in (4). [Why is (4) not a system of linear equations?]

A question that arises is, How do we solve the system of equations in (4), since they are not linear equations? The answer is, We first expand each equation and then eliminate the terms that contain x^2 and y^2 by using algebra. In (5) we show the expansion of each of the equations in (4); note that x^2 and y^2 appear in each equation.

$$\begin{aligned}x^2-2a_1x+a_1^2+y^2-2b_1y+b_1^2&=r_1^2\\x^2-2a_2x+a_2^2+y^2-2b_2y+b_2^2&=r_2^2\\x^2-2a_3x+a_3^2+y^2-2b_3y+b_3^2&=r_3^2\end{aligned}\tag{5}$$

Now we rearrange each equation in (5) to obtain the expressions shown in (6).

$$\begin{aligned}x^2-2a_1x+a_1^2+y^2-2b_1y+b_1^2-r_1^2&=0\\x^2-2a_2x+a_2^2+y^2-2b_2y+b_2^2-r_2^2&=0\\x^2-2a_3x+a_3^2+y^2-2b_3y+b_3^2-r_3^2&=0\end{aligned}\tag{6}$$

Next, set the left side of the first equation in (6) equal to the left side of the second equation in (6) and simplify. Do likewise for the second and third equations in (6). This gives the linear system in x and y in (7).

$$\begin{aligned}-2a_1x+a_1^2-2b_1y+b_1^2-r_1^2&=-2a_2x+a_2^2-2b_2y+b_2^2-r_2^2\\-2a_3x+a_3^2-2b_3y+b_3^2-r_3^2&=-2a_2x+a_2^2-2b_2y+b_2^2-r_2^2\end{aligned}\tag{7}$$

Finally, collect like terms in x and y to get the equations in (8).

$$\begin{aligned}-2(a_1-a_2)x-2(b_1-b_2)y&=(r_1^2-r_2^2)+(a_2^2-a_1^2)+(b_2^2-b_1^2)\\-2(a_3-a_2)x-2(b_3-b_2)y&=(r_3^2-r_2^2)+(a_2^2-a_3^2)+(b_2^2-b_3^2)\end{aligned}\tag{8}$$

To simplify a bit further, we multiply each equation in (8) by -1 and show the matrix formulation for the resulting 2×2 system in (9).

$$\begin{bmatrix} 2(a_1 - a_2) & 2(b_1 - b_2) \\ 2(a_3 - a_2) & 2(b_3 - b_2) \end{bmatrix} \begin{bmatrix} x \\ y \end{bmatrix} = \begin{bmatrix} (r_2^2 - r_1^2) + (a_1^2 - a_2^2) + (b_1^2 - b_2^2) \\ (r_2^2 - r_3^2) + (a_3^2 - a_2^2) + (b_3^2 - b_2^2) \end{bmatrix} \tag{9}$$

So, given the coordinates of the centers and the radii of the three circles, we can determine the location of the GPS receiver in the two-dimensional model.

EXAMPLE 9 The coordinates of the centers and the radii of three circles are shown in Table 2.1. The corresponding system of equations is given in (10) (verify), and its solution is $x = 6$ and $y = 10$ (verify). Thus, the coordinates of the GPS receiver in the two-dimensional system for these three circles is (6, 10).

TABLE 2.1

Circle	*Center*	*Radius*
1	$(-3, 50)$	41
2	$(11, -2)$	13
3	$(13, 34)$	25

$$\begin{bmatrix} -28 & 104 \\ 4 & 72 \end{bmatrix} \begin{bmatrix} x \\ y \end{bmatrix} = \begin{bmatrix} 872 \\ 744 \end{bmatrix}. \tag{10}$$

■

Next we present an approach for GPS in three dimensions. In this case each of the equations in the system that is analogous to those in (4) has the form

$$(x - a_j)^2 + (y - b_j)^2 + (z - c_j)^2 = (\text{distance from receiver to satellite } j)^2 \tag{11}$$

for $j = 1, 2, 3, 4$, where (a_j, b_j, c_j) is the position of the satellite j. The distance from the receiver to satellite j is computed by measuring the time it takes the signal from satellite j to reach the receiver. The satellite contains a clock mechanism that sends the time the signal was sent to the receiver, and we let t be the time the signal was received. Since the signal travels at the speed of light, we can get a good approximation to the distance by using the basic formula distance=speed× elapsed time. Thus there are now four unknowns: x, y, z, and t. We proceed algebraically as we did to get expressions that are analogous to those in (7) and (8). This will yield a system of three equations in four unknowns analogous to the system in (9). We solve this system for x, y, and z in terms of time t. To determine the unknown t, we substitute these expressions for x, y, and z into any of the equations in (11) and then solve the resulting quadratic polynomial for t. Finally, we use the resulting value of t in the expressions for x, y, and z to determine the location of the receiver in three dimensions. This approach uses a system that has infinitely many solutions and then cleverly uses the underlying physical situation to determine the "free" variable t.

The real-life situation is even more complicated than our approach outlined for the three-dimensional case. The satellites are continually moving, so their locations vary with time, inherent errors in time calculations creep in, and a number of other factors introduce more inaccuracies. Highly accurate estimation of the receiver's position is beyond the scope of this course, but there are many books and discussions on the Internet that provide more detailed information. We have presented a basic component of GPS calculations, namely, that linear systems of equations are involved.

Homogeneous Systems

Now we study a homogeneous system $A\mathbf{x} = \mathbf{0}$ of m linear equations in n unknowns.

EXAMPLE 10 Consider the homogeneous system whose augmented matrix is

$$\left[\begin{array}{ccccc|c} 1 & 0 & 0 & 0 & 2 & 0 \\ 0 & 0 & 1 & 0 & 3 & 0 \\ 0 & 0 & 0 & 1 & 4 & 0 \\ 0 & 0 & 0 & 0 & 0 & 0 \end{array}\right].$$

Since the augmented matrix is in reduced row echelon form, the solution is seen to be

$$\begin{aligned} x_1 &= -2r \\ x_2 &= s \\ x_3 &= -3r \\ x_4 &= -4r \\ x_5 &= r, \end{aligned}$$

where r and s are any real numbers. ■

In Example 10 we solved a homogeneous system of m $(= 4)$ linear equations in n $(= 5)$ unknowns, where $m < n$ and the augmented matrix A was in reduced row echelon form. We can ignore any row of the augmented matrix that consists entirely of zeros. Thus let rows $1, 2, \ldots, r$ of A be the nonzero rows, and let the 1 in row i occur in column c_i. We are then solving a homogeneous system of r equations in n unknowns, $r < n$, and in this special case (A is in reduced row echelon form) we can solve for $x_{c_1}, x_{c_2}, \ldots, x_{c_r}$ in terms of the remaining $n - r$ unknowns. Since the latter can take on any real values, there are infinitely many solutions to the system $A\mathbf{x} = \mathbf{0}$; in particular, there is a nontrivial solution. We now show that this situation holds whenever we have $m < n$; A does not have to be in reduced row echelon form.

Theorem 2.4 A homogeneous system of m linear equations in n unknowns always has a nontrivial solution if $m < n$, that is, if the number of unknowns exceeds the number of equations.

Proof

Let B be a matrix in reduced row echelon form that is row equivalent to A. Then the homogeneous systems $A\mathbf{x} = \mathbf{0}$ and $B\mathbf{x} = \mathbf{0}$ are equivalent. As we have just shown, the system $B\mathbf{x} = \mathbf{0}$ has a nontrivial solution, and therefore the same is true for the system $A\mathbf{x} = \mathbf{0}$. ■

We shall use this result in the following equivalent form: If A is $m \times n$ and $A\mathbf{x} = \mathbf{0}$ has only the trivial solution, then $m \geq n$.

EXAMPLE 11

Consider the homogeneous system

$$\begin{aligned} x + \ \ y + z + w &= 0 \\ x \qquad\qquad\ + w &= 0 \\ x + 2y + z \qquad\ &= 0. \end{aligned}$$

The augmented matrix

$$A = \left[\begin{array}{cccc|c} 1 & 1 & 1 & 1 & 0 \\ 1 & 0 & 0 & 1 & 0 \\ 1 & 2 & 1 & 0 & 0 \end{array}\right]$$

is row equivalent to (verify)

$$\left[\begin{array}{cccc|c} 1 & 0 & 0 & 1 & 0 \\ 0 & 1 & 0 & -1 & 0 \\ 0 & 0 & 1 & 1 & 0 \end{array}\right].$$

Hence the solution is

$$\begin{aligned} x &= -r \\ y &= \ \ \ r \\ z &= -r \\ w &= \ \ \ r, \text{ any real number.} \end{aligned}$$

■

■ Application: Chemical Balance Equations

Chemical reactions can be described by equations. The expressions on the left side are called the reactants, and those on the right side are the products, which are produced from the reaction of chemicals on the left. Unlike mathematical equations, the two sides are separated by an arrow, either $\rightarrow$, which indicates that the reactants form the products, or $\leftrightarrow$, which indicates a reversible equation; that is, once the products are formed, they begin to form reactants. A chemical equation is balanced, provided that the number of atoms of each type on the left is the same as the number of atoms of the corresponding type on the right. In Example 12 we illustrate how to construct a homogeneous system of equations whose solution provides appropriate values to balance the atoms in the reactants with those in the products.

EXAMPLE 12

Sodium hydroxide ($NaOH$) reacts with sulfuric acid (H_2SO_4) to form sodium sulfate (Na_2SO_4) and water (H_2O). The chemical equation is

$$NaOH + H_2SO_4 \rightarrow Na_2SO_4 + H_2O.$$

To balance this equation, we insert unknowns, multiplying the chemicals on the left and right to get an equation of the form

$$x\mathrm{NaOH} + y\mathrm{H_2SO_4} \rightarrow z\mathrm{Na_2SO_4} + w\mathrm{H_2O}.$$

Next, we compare the number of sodium (Na), oxygen (O), hydrogen (H), and sulfur (S) atoms on the left side with the numbers on the right. We obtain four linear equations:

$$\begin{aligned} \mathrm{Na}\colon x &= 2z \\ \mathrm{O}\colon x + 4y &= 4z + w \\ \mathrm{H}\colon x + 2y &= 2w \\ \mathrm{S}\colon y &= z \end{aligned}$$

Observe that we made use of the subscripts because they count the number of atoms of a particular element. Rewriting these equations in standard form, we see that we have a homogeneous linear system in four unknowns:

$$\begin{aligned} x \qquad - 2z \qquad &= 0 \\ x + 4y - 4z - \ w &= 0 \\ x + 2y \qquad - 2w &= 0 \\ y - \ z \qquad &= 0. \end{aligned}$$

Writing this system in matrix form, we have the augmented matrix

$$\left[\begin{array}{cccc|c} 1 & 0 & -2 & 0 & 0 \\ 1 & 4 & -4 & -1 & 0 \\ 1 & 2 & 0 & -2 & 0 \\ 0 & 1 & -1 & 0 & 0 \end{array}\right].$$

The reduced row echelon form is

$$\left[\begin{array}{cccc|c} 1 & 0 & 0 & -1 & 0 \\ 0 & 1 & 0 & -\frac{1}{2} & 0 \\ 0 & 0 & 1 & -\frac{1}{2} & 0 \\ 0 & 0 & 0 & 0 & 0 \end{array}\right],$$

and the solution is $x = w$, $y = \frac{1}{2}w$, and $z = \frac{1}{2}w$. Since w can be chosen arbitrarily and we are dealing with atoms, it is convenient to choose values so that all the unknowns are positive integers. One such choice is $w = 2$, which gives $x = 2$, $y = 1$, and $z = 1$. In this case our balanced equation is

$$2\mathrm{NaOH} + \mathrm{H_2SO_4} \rightarrow \mathrm{Na_2SO_4} + 2\mathrm{H_2O}.$$

■

Relationship between Nonhomogeneous Linear Systems and Homogeneous Systems

Let $A\mathbf{x} = \mathbf{b}$, $\mathbf{b} \neq \mathbf{0}$, be a consistent linear system. If $\mathbf{x}_p$ is a particular solution to the given nonhomogeneous system and $\mathbf{x}_h$ is a solution to the associated homogeneous system $A\mathbf{x} = \mathbf{0}$, then $\mathbf{x}_p + \mathbf{x}_h$ is a solution to the given system $A\mathbf{x} = \mathbf{b}$. Moreover, every solution $\mathbf{x}$ to the nonhomogeneous linear system $A\mathbf{x} = \mathbf{b}$ can be written as $\mathbf{x}_p + \mathbf{x}_h$, where $\mathbf{x}_p$ is a particular solution to the given nonhomogeneous system and $\mathbf{x}_h$ is a solution to the associated homogeneous system $A\mathbf{x} = \mathbf{0}$. For a proof, see Exercise 29.

Solving Linear Systems with Complex Entries

Gaussian elimination and Gauss–Jordan reduction can both be used to solve linear systems that have complex entries. The examples that follow show how to solve a linear system with complex entries by using these solution techniques. (For simplicity, we do not show the notation for row operations.) Further illustrations and exercises are given in Appendix B.2.

EXAMPLE 13 Solve the linear system

$$\begin{aligned} (1-i)x + (2+i)y &= 2+2i \\ 2x + (1-2i)y &= 1+3i \end{aligned}$$

by Gaussian elimination.

Solution

The augmented matrix of the given linear system is

$$\left[\begin{array}{cc|c} 1-i & 2+i & 2+2i \\ 2 & 1-2i & 1+3i \end{array}\right].$$

To transform this matrix to row echelon form, we first interchange the two rows (to avoid complicated fractions), obtaining

$$\left[\begin{array}{cc|c} 2 & 1-2i & 1+3i \\ 1-i & 2+i & 2+2i \end{array}\right].$$

Multiply the first row by $\frac{1}{2}$:

$$\left[\begin{array}{cc|c} 1 & \dfrac{1-2i}{2} & \dfrac{1+3i}{2} \\ 1-i & 2+i & 2+2i \end{array}\right].$$

Add $-(1-i)$ times the first row to the second row:

$$\left[\begin{array}{cc|c} 1 & \dfrac{1-2i}{2} & \dfrac{1+3i}{2} \\ 0 & \dfrac{5+5i}{2} & i \end{array}\right].$$

Multiply the second row by $\dfrac{2}{5+5i}$:

$$\left[\begin{array}{cc|c} 1 & \dfrac{1-2i}{2} & \dfrac{1+3i}{2} \\ 0 & 1 & \dfrac{2i}{5+5i} \end{array}\right].$$

Then

$$y = \frac{2i}{5+5i} = \frac{1}{5} + \frac{1}{5}i.$$

Using back substitution, we have

$$x = \frac{1+3i}{2} - \frac{1-2i}{2}y = \frac{1}{5} + \frac{8}{5}i \quad \text{(verify)}.$$ ■

EXAMPLE 14 Solve the linear system whose augmented matrix is

$$\left[\begin{array}{ccc|c} i & 2 & 1-i & 1-2i \\ 0 & 2i & 2+i & -2+i \\ 0 & -i & 1 & -1-i \end{array}\right]$$

by Gauss–Jordan reduction.

Solution

Multiply the first row by $\dfrac{1}{i}$:

$$\left[\begin{array}{ccc|c} 1 & \dfrac{2}{i} & \dfrac{1-i}{i} & \dfrac{1-2i}{i} \\ 0 & 2i & 2+i & -2+i \\ 0 & -i & 1 & -1-i \end{array}\right].$$

Multiply the second row by $\dfrac{1}{2i}$:

$$\left[\begin{array}{ccc|c} 1 & \dfrac{2}{i} & \dfrac{1-i}{i} & \dfrac{1-2i}{i} \\ 0 & 1 & \dfrac{2+i}{2i} & \dfrac{-2+i}{2i} \\ 0 & -i & 1 & -1-i \end{array}\right].$$

Add i times the second row to the third row:

$$\left[\begin{array}{ccc|c} 1 & \dfrac{2}{i} & \dfrac{1-i}{i} & \dfrac{1-2i}{i} \\ 0 & 1 & \dfrac{2+i}{2i} & \dfrac{-2+i}{2i} \\ 0 & 0 & \dfrac{4+i}{2} & \dfrac{-4-i}{2} \end{array}\right].$$

Multiply the third row by $\left(\frac{2}{4+i}\right)$:

$$\left[\begin{array}{ccc|c} 1 & \frac{2}{i} & \frac{1-i}{i} & \frac{1-2i}{i} \\ 0 & 1 & \frac{2+i}{2i} & \frac{-2+i}{2i} \\ 0 & 0 & 1 & -1 \end{array}\right].$$

Add $\left(-\frac{2+i}{2i}\right)$ times the third row to the second row and $\left(-\frac{1-i}{i}\right)$ times the third row to the first row:

$$\left[\begin{array}{ccc|c} 1 & \frac{2}{i} & 0 & \frac{2-3i}{i} \\ 0 & 1 & 0 & 1 \\ 0 & 0 & 1 & -1 \end{array}\right].$$

Add $\left(-\frac{2}{i}\right)$ times the second row to the first row:

$$\left[\begin{array}{ccc|c} 1 & 0 & 0 & -3 \\ 0 & 1 & 0 & 1 \\ 0 & 0 & 1 & -1 \end{array}\right].$$

Hence the solution is $x = -3$, $y = 1$, $z = -1$. ■

Key Terms

Gaussian elimination
Gauss–Jordan reduction
Homogeneous system
Back substitution
Quadratic interpolation
Quadratic interpolant
Global positioning system
Chemical balance equations

2.2 Exercises

1. Each of the given linear systems is in row echelon form. Solve the system.

(a) $\begin{aligned} x + 2y - z &= 6 \\ y + z &= 5 \\ z &= 4 \end{aligned}$ **(b)** $\begin{aligned} x - 3y + 4z + w &= 0 \\ z - w &= 4 \\ w &= 1 \end{aligned}$

2. Each of the given linear systems is in row echelon form. Solve the system.

(a) $\begin{aligned} x + y - z + 2w &= 4 \\ w &= 5 \end{aligned}$ **(b)** $\begin{aligned} x - y + z &= 0 \\ y + 2z &= 0 \\ z &= 1 \end{aligned}$

3. Each of the given linear systems is in reduced row echelon form. Solve the system.

(a) $\begin{aligned} x + y &= 2 \\ z + w &= -3 \end{aligned}$ **(b)** $\begin{aligned} x &= 3 \\ y &= 0 \\ z &= 1 \end{aligned}$

4. Each of the given linear systems is in reduced row echelon form. Solve the system.

(a) $\begin{aligned} x \quad - 2z &= 5 \\ y + z &= 2 \end{aligned}$ **(b)** $\begin{aligned} x &= 1 \\ y &= 2 \\ z - w &= 4 \end{aligned}$

5. Consider the linear system

$$\begin{aligned} x + y + 2z &= -1 \\ x - 2y + z &= -5 \\ 3x + y + z &= 3. \end{aligned}$$

(a) Find all solutions, if any exist, by using the Gaussian elimination method.

(b) Find all solutions, if any exist, by using the Gauss–Jordan reduction method.

6. Repeat Exercise 5 for each of the following linear systems:

(a)
$$\begin{aligned} x + y + 2z + 3w &= 13 \\ x - 2y + z + w &= 8 \\ 3x + y + z - w &= 1 \end{aligned}$$

(b)
$$\begin{aligned} x + y + z &= 1 \\ x + y - 2z &= 3 \\ 2x + y + z &= 2 \end{aligned}$$

(c)
$$\begin{aligned} 2x + y + z - 2w &= 1 \\ 3x - 2y + z - 6w &= -2 \\ x + y - z - w &= -1 \\ 6x + z - 9w &= -2 \\ 5x - y + 2z - 8w &= 3 \end{aligned}$$

In Exercises 7 through 9, solve the linear system, with the given augmented matrix, if it is consistent.

7. (a) $\left[\begin{array}{ccc|c} 1 & 1 & 1 & 0 \\ 1 & 1 & 0 & 3 \\ 0 & 1 & 1 & 1 \end{array}\right]$ **(b)** $\left[\begin{array}{ccc|c} 1 & 2 & 3 & 0 \\ 1 & 1 & 1 & 0 \\ 1 & 1 & 2 & 0 \end{array}\right]$

(c) $\left[\begin{array}{ccc|c} 1 & 2 & 3 & 0 \\ 1 & 1 & 1 & 0 \\ 5 & 7 & 9 & 0 \end{array}\right]$ **(d)** $\left[\begin{array}{ccc|c} 1 & 2 & 3 & 0 \\ 1 & 2 & 1 & 0 \end{array}\right]$

8. (a) $\left[\begin{array}{cccc|c} 1 & 2 & 3 & 1 & 8 \\ 1 & 3 & 0 & 1 & 7 \\ 1 & 0 & 2 & 1 & 3 \end{array}\right]$

(b) $\left[\begin{array}{cccc|c} 1 & 1 & 3 & -3 & 0 \\ 0 & 2 & 1 & -3 & 3 \\ 1 & 0 & 2 & -1 & -1 \end{array}\right]$

9. (a) $\left[\begin{array}{ccc|c} 1 & 2 & 1 & 7 \\ 2 & 0 & 1 & 4 \\ 1 & 0 & 2 & 5 \\ 1 & 2 & 3 & 11 \\ 2 & 1 & 4 & 12 \end{array}\right]$ **(b)** $\left[\begin{array}{ccc|c} 1 & 2 & 1 & 0 \\ 2 & 3 & 0 & 0 \\ 0 & 1 & 2 & 0 \\ 2 & 1 & 4 & 0 \end{array}\right]$

10. Find a 2×1 matrix $\mathbf{x}$ with entries not all zero such that

$$A\mathbf{x} = 4\mathbf{x}, \quad \text{where } A = \begin{bmatrix} 4 & 1 \\ 0 & 2 \end{bmatrix}.$$

[*Hint*: Rewrite the matrix equation $A\mathbf{x} = 4\mathbf{x}$ as $4\mathbf{x} - A\mathbf{x} = (4I_2 - A)\mathbf{x} = \mathbf{0}$, and solve the homogeneous linear system.]

11. Find a 2×1 matrix $\mathbf{x}$ with entries not all zero such that

$$A\mathbf{x} = 3\mathbf{x}, \quad \text{where } A = \begin{bmatrix} 2 & 1 \\ 1 & 2 \end{bmatrix}.$$

12. Find a 3×1 matrix $\mathbf{x}$ with entries not all zero such that

$$A\mathbf{x} = 3\mathbf{x}, \quad \text{where } A = \begin{bmatrix} 1 & 2 & -1 \\ 1 & 0 & 1 \\ 4 & -4 & 5 \end{bmatrix}.$$

13. Find a 3×1 matrix $\mathbf{x}$ with entries not all zero such that

$$A\mathbf{x} = 1\mathbf{x}, \quad \text{where } A = \begin{bmatrix} 1 & 2 & -1 \\ 1 & 0 & 1 \\ 4 & -4 & 5 \end{bmatrix}.$$

14. In the following linear system, determine all values of a for which the resulting linear system has

(a) no solution;

(b) a unique solution;

(c) infinitely many solutions:

$$\begin{aligned} x + y - z &= 2 \\ x + 2y + z &= 3 \\ x + y + (a^2 - 5)z &= a \end{aligned}$$

15. Repeat Exercise 14 for the linear system

$$\begin{aligned} x + y + z &= 2 \\ 2x + 3y + 2z &= 5 \\ 2x + 3y + (a^2 - 1)z &= a + 1. \end{aligned}$$

16. Repeat Exercise 14 for the linear system

$$\begin{aligned} x + y + z &= 2 \\ x + 2y + z &= 3 \\ x + y + (a^2 - 5)z &= a. \end{aligned}$$

17. Repeat Exercise 14 for the linear system

$$\begin{aligned} x + y &= 3 \\ x + (a^2 - 8)y &= a. \end{aligned}$$

18. Let

$$A = \begin{bmatrix} a & b \\ c & d \end{bmatrix} \quad \text{and} \quad \mathbf{x} = \begin{bmatrix} x_1 \\ x_2 \end{bmatrix}.$$

Show that the linear system $A\mathbf{x} = \mathbf{0}$ has only the trivial solution if and only if $ad - bc \neq 0$.

19. Show that $A = \begin{bmatrix} a & b \\ c & d \end{bmatrix}$ is row equivalent to I_2 if and only if $ad - bc \neq 0$.

20. Let $f: R^3 \to R^3$ be the matrix transformation defined by

$$f\left(\begin{bmatrix} x \\ y \\ z \end{bmatrix}\right) = \begin{bmatrix} 4 & 1 & 3 \\ 2 & -1 & 3 \\ 2 & 2 & 0 \end{bmatrix}\begin{bmatrix} x \\ y \\ z \end{bmatrix}.$$

Find x, y, z so that $f\left(\begin{bmatrix} x \\ y \\ z \end{bmatrix}\right) = \begin{bmatrix} 4 \\ 5 \\ -1 \end{bmatrix}$.

21. Let $f: R^3 \to R^3$ be the matrix transformation defined by

$$f\left(\begin{bmatrix} x \\ y \\ z \end{bmatrix}\right) = \begin{bmatrix} 1 & 2 & 3 \\ -3 & -2 & -1 \\ -2 & 0 & 2 \end{bmatrix}\begin{bmatrix} x \\ y \\ z \end{bmatrix}.$$

Find x, y, z so that $f\left(\begin{bmatrix} x \\ y \\ z \end{bmatrix}\right) = \begin{bmatrix} 2 \\ 2 \\ 4 \end{bmatrix}$.

22. Let $f: R^3 \to R^3$ be the matrix transformation defined by

$$f\left(\begin{bmatrix} x \\ y \\ z \end{bmatrix}\right) = \begin{bmatrix} 4 & 1 & 3 \\ 2 & -1 & 3 \\ 2 & 2 & 0 \end{bmatrix}\begin{bmatrix} x \\ y \\ z \end{bmatrix}.$$

Find an equation relating a, b, and c so that we can always compute values of x, y, and z for which

$$f\left(\begin{bmatrix} x \\ y \\ z \end{bmatrix}\right) = \begin{bmatrix} a \\ b \\ c \end{bmatrix}.$$

23. Let $f: R^3 \to R^3$ be the matrix transformation defined by

$$f\left(\begin{bmatrix} x \\ y \\ z \end{bmatrix}\right) = \begin{bmatrix} 1 & 2 & 3 \\ -3 & -2 & -1 \\ -2 & 0 & 2 \end{bmatrix}\begin{bmatrix} x \\ y \\ z \end{bmatrix}.$$

Find an equation relating a, b, and c so that we can always compute values of x, y, and z for which

$$f\left(\begin{bmatrix} x \\ y \\ z \end{bmatrix}\right) = \begin{bmatrix} a \\ b \\ c \end{bmatrix}.$$

Exercises 24 and 25 are optional.

24. **(a)** Formulate the definitions of column echelon form and reduced column echelon form of a matrix.

(b) Prove that every $m \times n$ matrix is column equivalent to a matrix in column echelon form.

25. Prove that every $m \times n$ matrix is column equivalent to a unique matrix in reduced column echelon form.

26. Find an equation relating a, b, and c so that the linear system

$$\begin{aligned} x + 2y - 3z &= a \\ 2x + 3y + 3z &= b \\ 5x + 9y - 6z &= c \end{aligned}$$

is consistent for any values of a, b, and c that satisfy that equation.

27. Find an equation relating a, b, and c so that the linear system

$$\begin{aligned} 2x + 2y + 3z &= a \\ 3x - y + 5z &= b \\ x - 3y + 2z &= c \end{aligned}$$

is consistent for any values of a, b, and c that satisfy that equation.

28. Show that the homogeneous system

$$\begin{aligned} (a - r)x + \qquad dy &= 0 \\ cx + (b - r)y &= 0 \end{aligned}$$

has a nontrivial solution if and only if r satisfies the equation $(a - r)(b - r) - cd = 0$.

29. Let $A\mathbf{x} = \mathbf{b}$, $\mathbf{b} \neq \mathbf{0}$, be a consistent linear system.

(a) Show that if $\mathbf{x}_p$ is a particular solution to the given nonhomogeneous system and $\mathbf{x}_h$ is a solution to the associated homogeneous system $A\mathbf{x} = \mathbf{0}$, then $\mathbf{x}_p + \mathbf{x}_h$ is a solution to the given system $A\mathbf{x} = \mathbf{b}$.

(b) Show that every solution $\mathbf{x}$ to the nonhomogeneous linear system $A\mathbf{x} = \mathbf{b}$ can be written as $\mathbf{x}_p + \mathbf{x}_h$, where $\mathbf{x}_p$ is a particular solution to the given nonhomogeneous system and $\mathbf{x}_h$ is a solution to the associated homogeneous system $A\mathbf{x} = \mathbf{0}$.

[*Hint*: Let $\mathbf{x} = \mathbf{x}_p + (\mathbf{x} - \mathbf{x}_p)$.]

30. Determine the quadratic interpolant to each of the given data sets. Follow the procedure in Example 7.

(a) $\{(0, 2), (1, 5), (2, 14)\}$

(b) $\{(-1, 2), (3, 14), (0, -1)\}$

31. ***(Calculus Required)*** Construct a linear system of equations to determine a quadratic polynomial $p(x) = ax^2 + bx + c$ that satisfies the conditions $p(0) = f(0)$, $p'(0) = f'(0)$, and $p''(0) = f''(0)$, where $f(x) = e^{2x}$.

32. ***(Calculus Required)*** Construct a linear system of equations to determine a quadratic polynomial $p(x) = ax^2 + bx + c$ that satisfies the conditions $p(1) = f(1)$, $p'(1) = f'(1)$, and $p''(1) = f''(1)$, where $f(x) = xe^{x-1}$.

33. Determine the temperatures at the interior points T_i, $i = 1, 2, 3, 4$ for the plate shown in the figure. (See Example 8.)

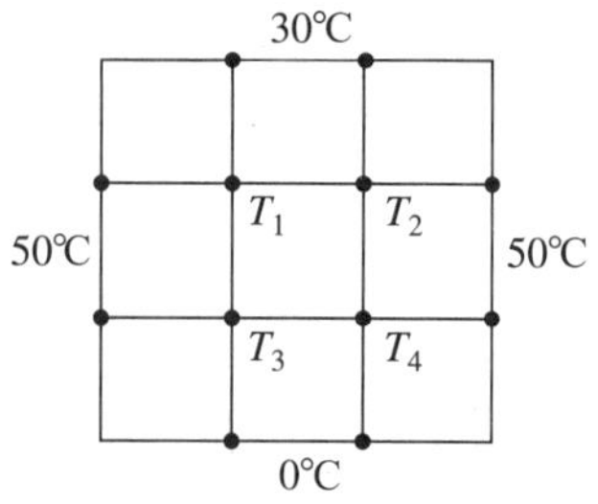

34. Determine the planar location (x, y) of a GPS receiver, using coordinates of the centers and radii for the three circles given in the following tables:

(a)

Circle	*Center*	*Radius*
1	$(-15, 20)$	25
2	$(5, -12)$	13
3	$(9, 40)$	41

(b)

Circle	*Center*	*Radius*
1	$(-10, 13)$	25
2	$(10, -19)$	13
3	$(14, 33)$	41

35. The location of a GPS receiver in a two-dimensional system is $(-4, 3)$. The data used in the calculation are given in the table, except that the radius of the first circle is missing. Determine the value of the missing piece of data.

Circle	*Center*	*Radius*
1	$(-16, 38)$	?
2	$(7, -57)$	61
3	$(32, 80)$	85

36. The location of a GPS receiver in a two-dimensional system is $(6, 8)$. The data used in the calculation are given in the table, except that the radii of circles 1 and 2 are missing. Determine the values of missing pieces of data.

Circle	*Center*	*Radius*
1	$(3, 4)$	?
2	$(10, 5)$	?
3	$(18, 3)$	13

37. Suppose you have a "special edition" GPS receiver for two-dimensional systems that contains three special buttons, labeled C1, C2, and C3. Each button when depressed draws a circle that corresponds to data received from one of three closest satellites. You depress button C1 and then C2. The image on your handheld unit shows a pair of circles that are tangent to each other. What is the location of the GPS receiver? Explain.

38. Rust is formed when there is a chemical reaction between iron and oxygen. The compound that is formed is the reddish brown scales that cover the iron object. Rust is iron oxide whose chemical formula is Fe_2O_3. So a chemical equation for rust is

$$Fe + O_2 \rightarrow Fe_2O_3.$$

Balance this equation.

39. Ethane is a gas similar to methane that burns in oxygen to give carbon dioxide gas and steam. The steam condenses to form water droplets. The chemical equation for this reaction is

$$C_2H_6 + O_2 \rightarrow CO_2 + H_2O.$$

Balance this equation.

In Exercises 40 and 41, solve each given linear system.

40.
$$\begin{aligned} (1-i)x + \quad (2+2i)y &= 1 \\ (1+2i)x + (-2+2i)y &= i \end{aligned}$$

41.
$$\begin{aligned} x + y \qquad &= 3 - i \\ ix + y + \ z &= 3 \\ y + iz &= 3 \end{aligned}$$

In Exercises 42 and 43, solve each linear system whose augmented matrix is given.

42. $$\left[\begin{array}{cc|c} 1-i & 2+2i & i \\ 1+i & -2+2i & -2 \end{array}\right]$$

43. $$\left[\begin{array}{ccc|c} 1 & i & -i & -2+2i \\ 2i & -i & 2 & -2 \\ 1 & 2 & 3i & 2i \end{array}\right]$$

44. Determine whether the software you are using has a command for computing the reduced row echelon form of a matrix. If it does, experiment with that command on some of the previous exercises.

45. Determine whether the software you are using has a command for computing interpolation polynomials, given a set of ordered pairs. If it does, use the command to determine the quadratic interpolant for the data sets in Exercise 30.

46. Determine whether the software you are using has a graphing option as part of a command for computing interpolation polynomials, or if there is an easy way available to graph the interpolant. If it does, use it as follows:

(a) Generate the graphs for the quadratic interpolants for the data sets in Exercise 30. Print out the graphs, and then mark the data points on the graph.

(b) For the data set $\{(0,0),(1,1),(4,2)\}$, generate the quadratic interpolant and graph it over the interval $[0,4]$. Print out the graph and then mark the data points on the graph. This data set is a sample of the function $y = f(x) = \sqrt{x}$. Carefully sketch the graph of f on the printout, making sure it goes through the data points. Briefly discuss the error that would be incurred if you were to evaluate the interpolant at $x = 2$ and $x = 3$ to estimate $\sqrt{2}$ and $\sqrt{3}$, respectively.

2.3 Elementary Matrices; Finding A^{-1}

In this section we develop a method for finding the inverse of a matrix if it exists. To use this method we do not have to find out first whether A^{-1} exists. We start to find A^{-1}; if in the course of the computation we hit a certain situation, then we know that A^{-1} does not exist. Otherwise, we proceed to the end and obtain A^{-1}. This method requires that elementary row operations of types I, II, and III (see Section 2.1) be performed on A. We clarify these notions by starting with the following definition:

DEFINITION 2.4 An $n \times n$ **elementary matrix of type I, type II, or type III** is a matrix obtained from the identity matrix I_n by performing a single elementary row or elementary column operation of type I, type II, or type III, respectively.

EXAMPLE 1 The following are elementary matrices:

$$E_1 = \begin{bmatrix} 0 & 0 & 1 \\ 0 & 1 & 0 \\ 1 & 0 & 0 \end{bmatrix}, \quad E_2 = \begin{bmatrix} 1 & 0 & 0 \\ 0 & -2 & 0 \\ 0 & 0 & 1 \end{bmatrix},$$

$$E_3 = \begin{bmatrix} 1 & 2 & 0 \\ 0 & 1 & 0 \\ 0 & 0 & 1 \end{bmatrix}, \quad \text{and} \quad E_4 = \begin{bmatrix} 1 & 0 & 3 \\ 0 & 1 & 0 \\ 0 & 0 & 1 \end{bmatrix}.$$

Matrix E_1 is of type I—we interchanged the first and third rows of I_3; E_2 is of type II—we multiplied the second row of I_3 by (-2); E_3 is of type III—we added twice the second row of I_3 to the first row of I_3; and E_4 is of type III—we added three times the first column of I_3 to the third column of I_3. ■

Theorem 2.5 Let A be an $m \times n$ matrix, and let an elementary row (column) operation of type I, type II, or type III be performed on A to yield matrix B. Let E be the elementary matrix obtained from I_m (I_n) by performing the same elementary row (column) operation as was performed on A. Then $B = EA$ ($B = AE$).

Proof

Exercise 1. ■

Theorem 2.5 says that an elementary row operation on A can be achieved by premultiplying A (multiplying A on the left) by the corresponding elementary matrix E; an elementary column operation on A can be obtained by postmultiplying A (multiplying A on the right) by the corresponding elementary matrix.

EXAMPLE 2 Let

$$A = \begin{bmatrix} 1 & 3 & 2 & 1 \\ -1 & 2 & 3 & 4 \\ 3 & 0 & 1 & 2 \end{bmatrix}$$

and let $B = A_{-2\mathbf{r}_3+\mathbf{r}_1\to\mathbf{r}_1}$; then

$$B = \begin{bmatrix} -5 & 3 & 0 & -3 \\ -1 & 2 & 3 & 4 \\ 3 & 0 & 1 & 2 \end{bmatrix}.$$

Now let $E = (I_3)_{-2\mathbf{r}_3+\mathbf{r}_1\to\mathbf{r}_1}$; then

$$E = \begin{bmatrix} 1 & 0 & -2 \\ 0 & 1 & 0 \\ 0 & 0 & 1 \end{bmatrix}.$$

We can readily verify that $B = EA$. ■

Theorem 2.6 If A and B are $m \times n$ matrices, then A is row (column) equivalent to B if and only if there exist elementary matrices $E_1, E_2, \ldots, E_k$ such that $B = E_k E_{k-1} \cdots E_2 E_1 A$ $(B = A E_1 E_2 \cdots E_{k-1} E_k)$.

Proof

We prove only the theorem for row equivalence. If A is row equivalent to B, then B results from A by a sequence of elementary row operations. This implies that there exist elementary matrices $E_1, E_2, \ldots, E_k$ such that $B = E_k E_{k-1} \cdots E_2 E_1 A$.

Conversely, if $B = E_k E_{k-1} \cdots E_2 E_1 A$, where the E_i are elementary matrices, then B results from A by a sequence of elementary row operations, which implies that A is row equivalent to B. ■

Theorem 2.7 An elementary matrix E is nonsingular, and its inverse is an elementary matrix of the same type.

Proof

Exercise 6. ■

Thus an elementary row operation can be "undone" by another elementary row operation of the same type.

We now obtain an algorithm for finding A^{-1} if it exists; first, we prove the following lemma:

Lemma 2.1* Let A be an $n \times n$ matrix and let the homogeneous system $A\mathbf{x} = \mathbf{0}$ have only the trivial solution $\mathbf{x} = \mathbf{0}$. Then A is row equivalent to I_n. (That is, the reduced row echelon form of A is I_n.)

Proof

Let B be a matrix in reduced row echelon form that is row equivalent to A. Then the homogeneous systems $A\mathbf{x} = \mathbf{0}$ and $B\mathbf{x} = \mathbf{0}$ are equivalent, and thus $B\mathbf{x} = \mathbf{0}$ also has only the trivial solution. It is clear that if r is the number of nonzero rows of B, then the homogeneous system $B\mathbf{x} = \mathbf{0}$ is equivalent to the homogeneous system whose coefficient matrix consists of the nonzero rows of B and is therefore $r \times n$. Since this last homogeneous system has only the trivial solution, we conclude from Theorem 2.4 that $r \geq n$. Since B is $n \times n$, $r \leq n$. Hence $r = n$, which means that B has no zero rows. Thus $B = I_n$. ■

Theorem 2.8 A is nonsingular if and only if A is a product of elementary matrices.

Proof

If A is a product of elementary matrices $E_1, E_2, \ldots, E_k$, then $A = E_1 E_2 \cdots E_k$. Now each elementary matrix is nonsingular, and by Theorem 1.6, the product of nonsingular matrices is nonsingular; therefore, A is nonsingular.

Conversely, if A is nonsingular, then $A\mathbf{x} = \mathbf{0}$ implies that $A^{-1}(A\mathbf{x}) = A^{-1}\mathbf{0} = \mathbf{0}$, so $I_n\mathbf{x} = \mathbf{0}$ or $\mathbf{x} = \mathbf{0}$. Thus $A\mathbf{x} = \mathbf{0}$ has only the trivial solution. Lemma 2.1 then implies that A is row equivalent to I_n. This means that there exist elementary matrices $E_1, E_2, \ldots, E_k$ such that

$$I_n = E_k E_{k-1} \cdots E_2 E_1 A.$$

It then follows that $A = (E_k E_{k-1} \cdots E_2 E_1)^{-1} = E_1^{-1} E_2^{-1} \cdots E_{k-1}^{-1} E_k^{-1}$. Since the inverse of an elementary matrix is an elementary matrix, we have established the result. ■

Corollary 2.2 A is nonsingular if and only if A is row equivalent to I_n. (That is, the reduced row echelon form of A is I_n.)

Proof

If A is row equivalent to I_n, then $I_n = E_k E_{k-1} \cdots E_2 E_1 A$, where $E_1, E_2, \ldots, E_k$ are elementary matrices. Therefore, it follows that $A = E_1^{-1} E_2^{-1} \cdots E_k^{-1}$. Now the inverse of an elementary matrix is an elementary matrix, and so by Theorem 2.8, A is nonsingular.

Conversely, if A is nonsingular, then A is a product of elementary matrices, $A = E_k E_{k-1} \cdots E_2 E_1$. Now $A = A I_n = E_k E_{k-1} \cdots E_2 E_1 I_n$, which implies that A is row equivalent to I_n. ■

We can see that Lemma 2.1 and Corollary 2.2 imply that if the homogeneous system $A\mathbf{x} = \mathbf{0}$, where A is $n \times n$, has only the trivial solution $\mathbf{x} = \mathbf{0}$, then A is nonsingular. Conversely, consider $A\mathbf{x} = \mathbf{0}$, where A is $n \times n$, and let A be nonsingular. Then A^{-1} exists, and we have $A^{-1}(A\mathbf{x}) = A^{-1}\mathbf{0} = \mathbf{0}$. We also have $A^{-1}(A\mathbf{x}) = (A^{-1}A)\mathbf{x} = I_n\mathbf{x} = \mathbf{x}$, so $\mathbf{x} = \mathbf{0}$, which means that the homogeneous

*A lemma is a theorem that is established for the purpose of proving another theorem.

system has only the trivial solution. We have thus proved the following important theorem:

Theorem 2.9 The homogeneous system of n linear equations in n unknowns $A\mathbf{x} = \mathbf{0}$ has a nontrivial solution if and only if A is singular. (That is, the reduced row echelon form of $A \neq I_n$.) ■

EXAMPLE 3 Let

$$A = \begin{bmatrix} 1 & 2 \\ 2 & 4 \end{bmatrix}.$$

Consider the homogeneous system $A\mathbf{x} = \mathbf{0}$; that is,

$$\begin{bmatrix} 1 & 2 \\ 2 & 4 \end{bmatrix} \begin{bmatrix} x \\ y \end{bmatrix} = \begin{bmatrix} 0 \\ 0 \end{bmatrix}.$$

The reduced row echelon form of the augmented matrix is

$$\left[\begin{array}{cc|c} 1 & 2 & 0 \\ 0 & 0 & 0 \end{array}\right]$$

(verify), so a solution is

$$x = -2r$$
$$y = r,$$

where r is any real number. Thus the homogeneous system has a nontrivial solution, and A is singular. ■

In Section 1.5 we have shown that if the $n \times n$ matrix A is nonsingular, then the system $A\mathbf{x} = \mathbf{b}$ has a unique solution for every $n \times 1$ matrix $\mathbf{b}$. The converse of this statement is also true. (See Exercise 30.)

Note that at this point we have shown that the following statements are equivalent for an $n \times n$ matrix A:

1. A is nonsingular.
2. $A\mathbf{x} = \mathbf{0}$ has only the trivial solution.
3. A is row (column) equivalent to I_n. (The reduced row echelon form of A is I_n.)
4. The linear system $A\mathbf{x} = \mathbf{b}$ has a unique solution for every $n \times 1$ matrix $\mathbf{b}$.
5. A is a product of elementary matrices.

That is, any two of these five statements are pairwise equivalent. For example, statements 1 and 2 are equivalent by Theorem 2.9, while statements 1 and 3 are equivalent by Corollary 2.2. The importance of these five statements being equivalent is that we can always replace any one statement by any other one on the list. As you will see throughout this book, a given problem can often be solved in several alternative ways, and sometimes one procedure is easier to apply than another.

Finding A^{-1}

At the end of the proof of Theorem 2.8, A was nonsingular and

$$A = E_1^{-1}E_2^{-1}\cdots E_{k-1}^{-1}E_k^{-1},$$

from which it follows that

$$A^{-1} = (E_1^{-1}E_2^{-1}\cdots E_{k-1}^{-1}E_k^{-1})^{-1} = E_kE_{k-1}\cdots E_2E_1.$$

This now provides an algorithm for finding A^{-1}. Thus we perform elementary row operations on A until we get I_n; the product of the elementary matrices $E_kE_{k-1}\cdots E_2E_1$ then gives A^{-1}. A convenient way of organizing the computing process is to write down the partitioned matrix $\left[A \mid I_n\right]$. Then

$$\begin{aligned}(E_kE_{k-1}\cdots E_2E_1)\left[A \mid I_n\right] &= \left[E_kE_{k-1}\cdots E_2E_1A \mid E_kE_{k-1}\cdots E_2E_1\right] \\ &= \left[I_n \mid A^{-1}\right].\end{aligned}$$

That is, for A nonsingular, we transform the partitioned matrix $\left[A \mid I_n\right]$ to reduced row echelon form, obtaining $\left[I_n \mid A^{-1}\right]$.

EXAMPLE 4 Let

$$A = \begin{bmatrix} 1 & 1 & 1 \\ 0 & 2 & 3 \\ 5 & 5 & 1 \end{bmatrix}.$$

Assuming that A is nonsingular, we form

$$\left[A \mid I_3\right] = \left[\begin{array}{ccc|ccc} 1 & 1 & 1 & 1 & 0 & 0 \\ 0 & 2 & 3 & 0 & 1 & 0 \\ 5 & 5 & 1 & 0 & 0 & 1 \end{array}\right].$$

We now perform elementary row operations that transform $\left[A \mid I_3\right]$ to $\left[I_3 \mid A^{-1}\right]$; we consider $\left[A \mid I_3\right]$ as a 3×6 matrix, and whatever we do to a row of A we also do to the corresponding row of I_3. In place of using elementary matrices directly, we arrange our computations, using elementary row operations as follows:

$$
\begin{array}{rrr|rrrl}
\multicolumn{3}{c}{A} & \multicolumn{3}{c}{I_3} & \\
1 & 1 & 1 & 1 & 0 & 0 & \text{Apply } -5\mathbf{r}_1 + \mathbf{r}_3 \to \mathbf{r}_3. \\
0 & 2 & 3 & 0 & 1 & 0 & \\
5 & 5 & 1 & 0 & 0 & 1 & \\
\hline
1 & 1 & 1 & 1 & 0 & 0 & \text{Apply } \tfrac{1}{2}\mathbf{r}_2 \to \mathbf{r}_2. \\
0 & 2 & 3 & 0 & 1 & 0 & \\
0 & 0 & -4 & -5 & 0 & 1 & \\
\hline
1 & 1 & 1 & 1 & 0 & 0 & \text{Apply } -\tfrac{1}{4}\mathbf{r}_3 \to \mathbf{r}_3. \\
0 & 1 & \frac{3}{2} & 0 & \frac{1}{2} & 0 & \\
0 & 0 & -4 & -5 & 0 & 1 & \\
\hline
1 & 1 & 1 & 1 & 0 & 0 & \text{Apply } -\tfrac{3}{2}\mathbf{r}_3 + \mathbf{r}_2 \to \mathbf{r}_2 \text{ and} \\
0 & 1 & \frac{3}{2} & 0 & \frac{1}{2} & 0 & -1\mathbf{r}_3 + \mathbf{r}_1 \to \mathbf{r}_1. \\
0 & 0 & 1 & \frac{5}{4} & 0 & -\frac{1}{4} & \\
\hline
1 & 1 & 0 & -\frac{1}{4} & 0 & \frac{1}{4} & \text{Apply } -1\mathbf{r}_2 + \mathbf{r}_1 \to \mathbf{r}_1. \\
0 & 1 & 0 & -\frac{15}{8} & \frac{1}{2} & \frac{3}{8} & \\
0 & 0 & 1 & \frac{5}{4} & 0 & -\frac{1}{4} & \\
\hline
1 & 0 & 0 & \frac{13}{8} & -\frac{1}{2} & -\frac{1}{8} & \\
0 & 1 & 0 & -\frac{15}{8} & \frac{1}{2} & \frac{3}{8} & \\
0 & 0 & 1 & \frac{5}{4} & 0 & -\frac{1}{4} &
\end{array}
$$

Hence

$$
A^{-1} = \begin{bmatrix} \frac{13}{8} & -\frac{1}{2} & -\frac{1}{8} \\ -\frac{15}{8} & \frac{1}{2} & \frac{3}{8} \\ \frac{5}{4} & 0 & -\frac{1}{4} \end{bmatrix}.
$$

We can readily verify that $AA^{-1} = A^{-1}A = I_3$. ■

The question that arises at this point is how to tell when A is singular. The answer is that A is singular if and only if A is row equivalent to matrix B, having at least one row that consists entirely of zeros. We now prove this result.

Theorem 2.10 An $n \times n$ matrix A is singular if and only if A is row equivalent to a matrix B that has a row of zeros. (That is, the reduced row echelon form of A has a row of zeros.)

Proof

First, let A be row equivalent to a matrix B that has a row consisting entirely of zeros. From Exercise 46 of Section 1.5, it follows that B is singular. Now

$B = E_k E_{k-1} \cdots E_1 A$, where $E_1, E_2, \ldots, E_k$ are elementary matrices. If A is nonsingular, then B is nonsingular, a contradiction. Thus A is singular.

Conversely, if A is singular, then A is not row equivalent to I_n, by Corollary 2.2. Thus A is row equivalent to a matrix $B \neq I_n$, which is in reduced row echelon form. From Exercise 9 of Section 2.1, it follows that B must have a row of zeros. ■

This means that in order to find A^{-1}, we do not have to determine, in advance, whether it exists. We merely start to calculate A^{-1}; if at any point in the computation we find a matrix B that is row equivalent to A and has a row of zeros, then A^{-1} does not exist. That is, we transform the partitioned matrix $\left[A \mid I_n\right]$ to reduced row echelon form, obtaining $\left[C \mid D\right]$. If $C = I_n$, then $D = A^{-1}$. If $C \neq I_n$, then C has a row of zeros and we conclude that A is singular.

EXAMPLE 5 Let

$$A = \begin{bmatrix} 1 & 2 & -3 \\ 1 & -2 & 1 \\ 5 & -2 & -3 \end{bmatrix}.$$

To find A^{-1}, we proceed as follows:

	A			I_3		
1	2	−3	1	0	0	Apply $-1\mathbf{r}_1 + \mathbf{r}_2 \to \mathbf{r}_2$.
1	−2	1	0	1	0	
5	−2	−3	0	0	1	
1	2	−3	1	0	0	Apply $-5\mathbf{r}_1 + \mathbf{r}_3 \to \mathbf{r}_3$.
0	−4	4	−1	1	0	
5	−2	−3	0	0	1	
1	2	−3	1	0	0	Apply $-3\mathbf{r}_2 + \mathbf{r}_3 \to \mathbf{r}_3$.
0	−4	4	−1	1	0	
0	−12	12	−5	0	1	
1	2	−3	1	0	0	
0	−4	4	−1	1	0	
0	0	0	−2	−3	1	

At this point A is row equivalent to

$$B = \begin{bmatrix} 1 & 2 & -3 \\ 0 & -4 & 4 \\ 0 & 0 & 0 \end{bmatrix},$$

the last matrix under A. Since B has a row of zeros, we stop and conclude that A is a singular matrix. ■

In Section 1.5 we defined an $n \times n$ matrix B to be the inverse of the $n \times n$ matrix A if $AB = I_n$ and $BA = I_n$. We now show that one of these equations follows from the other.

Theorem 2.11 If A and B are $n \times n$ matrices such that $AB = I_n$, then $BA = I_n$. Thus $B = A^{-1}$.

Proof

We first show that if $AB = I_n$, then A is nonsingular. Suppose that A is singular. Then A is row equivalent to a matrix C with a row of zeros. Now $C = E_k E_{k-1} \cdots E_1 A$, where $E_1, E_2, \ldots, E_k$ are elementary matrices. Then $CB = E_k E_{k-1} \cdots E_1 AB$, so AB is row equivalent to CB. Since CB has a row of zeros, we conclude from Theorem 2.10 that AB is singular. Then $AB = I_n$ is impossible, because I_n is nonsingular. This contradiction shows that A is nonsingular, and so A^{-1} exists. Multiplying both sides of the equation $AB = I_n$ by A^{-1} on the left, we then obtain (verify) $B = A^{-1}$. ■

Remark Theorem 2.11 implies that if we want to check whether a given matrix B is A^{-1}, we need merely check whether $AB = I_n$ *or* $BA = I_n$. That is, we do not have to check both equalities.

Key Terms

Elementary matrix
Inverse matrix
Nonsingular matrix
Singular matrix

2.3 Exercises

1. Prove Theorem 2.5.

2. Let A be a 4×3 matrix. Find the elementary matrix E that, as a premultiplier of A—that is, as EA—performs the following elementary row operations on A:

(a) Multiplies the second row of A by (-2).

(b) Adds 3 times the third row of A to the fourth row of A.

(c) Interchanges the first and third rows of A.

3. Let A be a 3×4 matrix. Find the elementary matrix F that, as a postmultiplier of A—that is, as AF—performs the following elementary column operations on A:

(a) Adds (-4) times the first column of A to the second column of A.

(b) Interchanges the second and third columns of A.

(c) Multiplies the third column of A by 4.

4. Let

$$A = \begin{bmatrix} 1 & 0 & 0 \\ 0 & 1 & 0 \\ -2 & 0 & 1 \end{bmatrix}.$$

(a) Find a matrix C in reduced row echelon form that is row equivalent to A. Record the row operations used.

(b) Apply the same operations to I_3 that were used to obtain C. Denote the resulting matrix by B.

(c) How are A and B related? (*Hint*: Compute AB and BA.)

5. Let

$$A = \begin{bmatrix} 1 & 0 & 1 \\ 2 & 1 & 0 \\ 0 & -1 & 1 \end{bmatrix}.$$

(a) Find a matrix C in reduced row echelon form that is row equivalent to A. Record the row operations used.

(b) Apply the same operations to I_3 that were used to obtain C. Denote the resulting matrix by B.

(c) How are A and B related? (*Hint*: Compute AB and BA.)

6. Prove Theorem 2.7. (*Hint*: Find the inverse of the elementary matrix of type I, type II, and type III.)

7. Find the inverse of $A = \begin{bmatrix} 1 & 3 \\ 2 & 4 \end{bmatrix}$.

8. Find the inverse of $A = \begin{bmatrix} 1 & 2 & 3 \\ 0 & 2 & 3 \\ 1 & 2 & 4 \end{bmatrix}$.

9. Which of the given matrices are singular? For the nonsingular ones, find the inverse.

(a) $\begin{bmatrix} 1 & 3 \\ 2 & 6 \end{bmatrix}$ **(b)** $\begin{bmatrix} 1 & 3 \\ -2 & 6 \end{bmatrix}$

(c) $\begin{bmatrix} 1 & 2 & 3 \\ 1 & 1 & 2 \\ 0 & 1 & 2 \end{bmatrix}$ **(d)** $\begin{bmatrix} 1 & 2 & 3 \\ 1 & 1 & 2 \\ 0 & 1 & 1 \end{bmatrix}$

10. Invert each of the following matrices, if possible:

(a) $\begin{bmatrix} 1 & 2 & -3 & 1 \\ -1 & 3 & -3 & -2 \\ 2 & 0 & 1 & 5 \\ 3 & 1 & -2 & 5 \end{bmatrix}$ **(b)** $\begin{bmatrix} 3 & 1 & 2 \\ 2 & 1 & 2 \\ 1 & 2 & 2 \end{bmatrix}$

(c) $\begin{bmatrix} 1 & 2 & 3 \\ 1 & 1 & 2 \\ 1 & 1 & 0 \end{bmatrix}$ **(d)** $\begin{bmatrix} 2 & 1 & 3 \\ 0 & 1 & 2 \\ 1 & 0 & 3 \end{bmatrix}$

11. Find the inverse, if it exists, of each of the following:

(a) $\begin{bmatrix} 1 & 1 & 1 \\ 1 & 2 & 3 \\ 0 & 1 & 1 \end{bmatrix}$ **(b)** $\begin{bmatrix} 1 & 1 & 1 & 1 \\ 1 & 2 & -1 & 2 \\ 1 & -1 & 2 & 1 \\ 1 & 3 & 3 & 2 \end{bmatrix}$

(c) $\begin{bmatrix} 1 & 1 & 1 & 1 \\ 1 & 3 & 1 & 2 \\ 1 & 2 & -1 & 1 \\ 5 & 9 & 1 & 6 \end{bmatrix}$ **(d)** $\begin{bmatrix} 1 & 2 & 1 \\ 1 & 3 & 2 \\ 1 & 0 & 1 \end{bmatrix}$

(e) $\begin{bmatrix} 1 & 2 & 2 \\ 1 & 3 & 1 \\ 1 & 1 & 3 \end{bmatrix}$

12. Find the inverse, if it exists, of each of the following:

(a) $A = \begin{bmatrix} 1 & 1 & 2 & 1 \\ 0 & -2 & 0 & 0 \\ 1 & 2 & 1 & -2 \\ 0 & 3 & 2 & 1 \end{bmatrix}$

(b) $A = \begin{bmatrix} 1 & 1 & 1 & 1 \\ 1 & 3 & 1 & 2 \\ 1 & 2 & -1 & 1 \\ 5 & 9 & 1 & 6 \end{bmatrix}$

In Exercises 13 and 14, prove that each given matrix A is nonsingular and write it as a product of elementary matrices. (Hint: First, write the inverse as a product of elementary matrices; then use Theorem 2.7.)

13. $A = \begin{bmatrix} 1 & 2 \\ 3 & 4 \end{bmatrix}$ **14.** $A = \begin{bmatrix} 1 & 2 & 3 \\ 0 & 1 & 2 \\ 1 & 0 & 3 \end{bmatrix}$

15. If A is a nonsingular matrix whose inverse is $\begin{bmatrix} 4 & 2 \\ 1 & 1 \end{bmatrix}$, find A.

16. If $A^{-1} = \begin{bmatrix} 1 & 1 & 1 \\ 1 & 1 & 2 \\ 1 & -1 & 1 \end{bmatrix}$, find A.

17. Which of the following homogeneous systems have a nontrivial solution?

(a) $\begin{aligned} x + 2y + 3z &= 0 \\ 2y + 2z &= 0 \\ x + 2y + 3z &= 0 \end{aligned}$

(b) $\begin{aligned} 2x + y - z &= 0 \\ x - 2y - 3z &= 0 \\ -3x - y + 2z &= 0 \end{aligned}$

(c) $\begin{aligned} 3x + y + 3z &= 0 \\ -2x + 2y - 4z &= 0 \\ 2x - 3y + 5z &= 0 \end{aligned}$

18. Which of the following homogeneous systems have a nontrivial solution?

(a) $\begin{aligned} x + y + 2z &= 0 \\ 2x + y + z &= 0 \\ 3x - y + z &= 0 \end{aligned}$

(b) $\begin{aligned} x - y + z &= 0 \\ 2x + y &= 0 \\ 2x - 2y + 2z &= 0 \end{aligned}$

(c) $\begin{aligned} 2x - y + 5z &= 0 \\ 3x + 2y - 3z &= 0 \\ x - y + 4z &= 0 \end{aligned}$

19. Find all value(s) of a for which the inverse of

$$A = \begin{bmatrix} 1 & 1 & 0 \\ 1 & 0 & 0 \\ 1 & 2 & a \end{bmatrix}$$

exists. What is A^{-1}?

20. For what values of a does the homogeneous system

$$\begin{aligned} (a-1)x + 2y &= 0 \\ 2x + (a-1)y &= 0 \end{aligned}$$

have a nontrivial solution?

21. Prove that

$$A = \begin{bmatrix} a & b \\ c & d \end{bmatrix}$$

is nonsingular if and only if $ad - bc \neq 0$. If this condition holds, show that

$$A^{-1} = \begin{bmatrix} \dfrac{d}{ad-bc} & \dfrac{-b}{ad-bc} \\ \dfrac{-c}{ad-bc} & \dfrac{a}{ad-bc} \end{bmatrix}.$$

22. Let

$$A = \begin{bmatrix} 2 & 3 & -1 \\ 1 & 0 & 3 \\ 0 & 2 & -3 \\ -2 & 1 & 3 \end{bmatrix}.$$

Find the elementary matrix that as a postmultiplier of A performs the following elementary column operations on A:

(a) Multiplies the third column of A by (-3).

(b) Interchanges the second and third columns of A.

(c) Adds (-5) times the first column of A to the third column of A.

23. Prove that two $m \times n$ matrices A and B are row equivalent if and only if there exists a nonsingular matrix P such that $B = PA$. (*Hint*: Use Theorems 2.6 and 2.8.)

24. Let A and B be row equivalent $n \times n$ matrices. Prove that A is nonsingular if and only if B is nonsingular.

25. Let A and B be $n \times n$ matrices. Show that if AB is nonsingular, then A and B must be nonsingular. (*Hint*: Use Theorem 2.9.)

26. Let A be an $m \times n$ matrix. Show that A is row equivalent to O if and only if $A = O$.

27. Let A and B be $m \times n$ matrices. Show that A is row equivalent to B if and only if A^T is column equivalent to B^T.

28. Show that a square matrix which has a row or a column consisting entirely of zeros must be singular.

29. **(a)** Is $(A+B)^{-1} = A^{-1} + B^{-1}$?

(b) Is $(cA)^{-1} = \dfrac{1}{c}A^{-1}$?

30. If A is an $n \times n$ matrix, prove that A is nonsingular if and only if the linear system $A\mathbf{x} = \mathbf{b}$ has a unique solution for every $n \times 1$ matrix $\mathbf{b}$.

31. Prove that the inverse of a nonsingular upper (lower) triangular matrix is upper (lower) triangular.

32. If the software you use has a command for computing reduced row echelon form, use it to determine whether the matrices A in Exercises 9, 10, and 11 have an inverse by operating on the matrix $\left[\, A \;\vdots\; I_n \,\right]$. (See Example 4.)

33. Repeat Exercise 32 on the matrices given in Exercise 63 of Section 1.5.

2.4 Equivalent Matrices

We have thus far considered A to be row (column) equivalent to B if B results from A by a finite sequence of elementary row (column) operations. A natural extension of this idea is that of considering B to arise from A by a finite sequence of elementary row *or* elementary column operations. This leads to the notion of equivalence of matrices. The material discussed in this section is used in Section 4.9.

DEFINITION 2.5 If A and B are two $m \times n$ matrices, then A is **equivalent** to B if we obtain B from A by a finite sequence of elementary row or elementary column operations.

As we have seen in the case of row equivalence, we can show (see Exercise 1) that (a) every matrix is equivalent to itself; (b) if B is equivalent to A, then A is equivalent to B; (c) if C is equivalent to B, and B is equivalent to A, then C is equivalent to A. In view of (b), both statements "A is equivalent to B" and "B is equivalent to A" can be replaced by "A and B are equivalent." We can also show that if two matrices are row equivalent, then they are equivalent. (See Exercise 4.)

Theorem 2.12 If A is any nonzero $m \times n$ matrix, then A is equivalent to a partitioned matrix of the form

$$\begin{bmatrix} I_r & O_{r,n-r} \\ O_{m-r,r} & O_{m-r,n-r} \end{bmatrix}^{\dagger}.$$

†Here, $O_{r,n-r}$ is the $r \times (n-r)$ zero matrix; similarly, $O_{m-r,r}$ is the $(m\ -)\, r \times r$ zero matrix, etc.

Proof

By Theorem 2.2, A is row equivalent to a matrix B that is in reduced row echelon form. Using elementary column operations of type I, we get B to be equivalent to a matrix C of the form

$$\begin{bmatrix} I_r & U_{r,n-r} \\ O_{m-r,r} & O_{m-r,n-r} \end{bmatrix},$$

where r is the number of nonzero rows in B. By elementary column operations of type III, C is equivalent to a matrix D of the form

$$\begin{bmatrix} I_r & O_{r,n-r} \\ O_{m-r,r} & O_{m-r,n-r} \end{bmatrix}.$$

From Exercise 1, it then follows that A is equivalent to D. ■

Of course, in Theorem 2.12, r may equal m, in which case there will not be any zero rows at the bottom of the matrix. (What happens if $r = n$? If $r = m = n$?)

Recall from Section 2.1 that we introduced the following notation for elementary column operations:

- Interchange columns i and j: $\mathbf{c}_i \leftrightarrow \mathbf{c}_j$
- Replace column i by k times column i: $k\mathbf{c}_i \rightarrow \mathbf{c}_i$
- Replace column j by k times column i + column j: $k\mathbf{c}_i + \mathbf{c}_j \rightarrow \mathbf{c}_j$

EXAMPLE 1

Let

$$A = \begin{bmatrix} 1 & 1 & 2 & -1 \\ 1 & 2 & 1 & 0 \\ -1 & -4 & 1 & -2 \\ 1 & -2 & 5 & -4 \end{bmatrix}.$$

To find a matrix of the form described in Theorem 2.12, which is equivalent to A, we proceed as follows. Apply row operation $-1\mathbf{r}_1 + \mathbf{r}_2 \rightarrow \mathbf{r}_2$ to obtain

$$\begin{bmatrix} 1 & 1 & 2 & -1 \\ 0 & 1 & -1 & 1 \\ -1 & -4 & 1 & -2 \\ 1 & -2 & 5 & -4 \end{bmatrix} \qquad \text{Apply } 1\mathbf{r}_1 + \mathbf{r}_3 \rightarrow \mathbf{r}_3.$$

$$\begin{bmatrix} 1 & 1 & 2 & -1 \\ 0 & 1 & -1 & 1 \\ 0 & -3 & 3 & -3 \\ 1 & -2 & 5 & -4 \end{bmatrix} \qquad \text{Apply } -1\mathbf{r}_1 + \mathbf{r}_4 \rightarrow \mathbf{r}_4.$$

$$\begin{bmatrix} 1 & 1 & 2 & -1 \\ 0 & 1 & -1 & 1 \\ 0 & -3 & 3 & -3 \\ 0 & -3 & 3 & -3 \end{bmatrix} \qquad \text{Apply } -1\mathbf{r}_3 + \mathbf{r}_4 \rightarrow \mathbf{r}_4.$$

$$\begin{bmatrix} 1 & 1 & 2 & -1 \\ 0 & 1 & -1 & 1 \\ 0 & -3 & 3 & -3 \\ 0 & 0 & 0 & 0 \end{bmatrix} \quad \text{Apply } -\tfrac{1}{3}\mathbf{r}_3 \rightarrow \mathbf{r}_3.$$

$$\begin{bmatrix} 1 & 1 & 2 & -1 \\ 0 & 1 & -1 & 1 \\ 0 & 1 & -1 & 1 \\ 0 & 0 & 0 & 0 \end{bmatrix} \quad \text{Apply } -1\mathbf{r}_2 + \mathbf{r}_3 \rightarrow \mathbf{r}_3.$$

$$\begin{bmatrix} 1 & 1 & 2 & -1 \\ 0 & 1 & -1 & 1 \\ 0 & 0 & 0 & 0 \\ 0 & 0 & 0 & 0 \end{bmatrix} \quad \text{Apply } -1\mathbf{r}_2 + \mathbf{r}_1 \rightarrow \mathbf{r}_1.$$

$$\begin{bmatrix} 1 & 0 & 3 & -2 \\ 0 & 1 & -1 & 1 \\ 0 & 0 & 0 & 0 \\ 0 & 0 & 0 & 0 \end{bmatrix} \quad \text{Apply } -3\mathbf{c}_1 + \mathbf{c}_3 \rightarrow \mathbf{c}_3.$$

$$\begin{bmatrix} 1 & 0 & 0 & -2 \\ 0 & 1 & -1 & 1 \\ 0 & 0 & 0 & 0 \\ 0 & 0 & 0 & 0 \end{bmatrix} \quad \text{Apply } 2\mathbf{c}_1 + \mathbf{c}_4 \rightarrow \mathbf{c}_4.$$

$$\begin{bmatrix} 1 & 0 & 0 & 0 \\ 0 & 1 & -1 & 1 \\ 0 & 0 & 0 & 0 \\ 0 & 0 & 0 & 0 \end{bmatrix} \quad \text{Apply } 1\mathbf{c}_2 + \mathbf{c}_3 \rightarrow \mathbf{c}_3.$$

$$\begin{bmatrix} 1 & 0 & 0 & 0 \\ 0 & 1 & 0 & 1 \\ 0 & 0 & 0 & 0 \\ 0 & 0 & 0 & 0 \end{bmatrix} \quad \text{Apply } -1\mathbf{c}_2 + \mathbf{c}_4 \rightarrow \mathbf{c}_4.$$

$$\begin{bmatrix} 1 & 0 & 0 & 0 \\ 0 & 1 & 0 & 0 \\ 0 & 0 & 0 & 0 \\ 0 & 0 & 0 & 0 \end{bmatrix}.$$

This is the matrix desired. ■

The following theorem gives another useful way to look at the equivalence of matrices:

Theorem 2.13 Two $m \times n$ matrices A and B are equivalent if and only if $B = PAQ$ for some nonsingular matrices P and Q.

Proof

Exercise 5. ■

We next prove a theorem that is analogous to Corollary 2.2.

Theorem 2.14 An $n \times n$ matrix A is nonsingular if and only if A is equivalent to I_n.

Proof

If A is equivalent to I_n, then I_n arises from A by a sequence of elementary row or elementary column operations. Thus there exist elementary matrices $E_1, E_2, \ldots, E_r, F_1, F_2, \ldots, F_s$ such that

$$I_n = E_r E_{r-1} \cdots E_2 E_1 A F_1 F_2 \cdots F_s.$$

Let $E_r E_{r-1} \cdots E_2 E_1 = P$ and $F_1 F_2 \cdots F_s = Q$. Then $I_n = PAQ$, where P and Q are nonsingular. It then follows that $A = P^{-1}Q^{-1}$, and since P^{-1} and Q^{-1} are nonsingular, A is nonsingular.

Conversely, if A is nonsingular, then from Corollary 2.2 it follows that A is row equivalent to I_n. Hence A is equivalent to I_n. ■

Key Terms

Equivalent matrix
Partitioned matrix

2.4 Exercises

1. (a) Prove that every matrix A is equivalent to itself.

   (b) Prove that if B is equivalent to A, then A is equivalent to B.

   (c) Prove that if C is equivalent to B and B is equivalent to A, then C is equivalent to A.

2. For each of the following matrices, find a matrix of the form described in Theorem 2.12 that is equivalent to the given matrix:

   (a) $\begin{bmatrix} 1 & 2 & -1 & 4 \\ 5 & 1 & 2 & -3 \\ 2 & 1 & 4 & 3 \\ 2 & 0 & 1 & 2 \\ 5 & 1 & 2 & 3 \end{bmatrix}$ (b) $\begin{bmatrix} 1 & 2 & 1 \\ 2 & 3 & 1 \\ 2 & 1 & 3 \end{bmatrix}$

   (c) $\begin{bmatrix} 1 & -2 & 1 \\ 2 & 3 & 2 \\ 3 & 1 & 3 \end{bmatrix}$ (d) $\begin{bmatrix} 1 & 3 & -1 & 2 \\ 2 & -4 & -2 & 1 \\ 3 & 1 & 2 & -3 \\ 7 & 3 & -2 & 5 \end{bmatrix}$

3. Repeat Exercise 2 for the following matrices:

   (a) $\begin{bmatrix} 1 & 2 & 3 & -1 \\ 1 & 0 & 2 & 3 \\ 3 & 4 & 8 & 1 \end{bmatrix}$ (b) $\begin{bmatrix} 3 & 4 & 1 \\ 1 & 2 & -2 \\ 5 & 6 & 4 \\ 5 & 8 & -1 \end{bmatrix}$

   (c) $\begin{bmatrix} 2 & 3 & 4 & -1 \\ 1 & 2 & 1 & -1 \\ 2 & -1 & 1 & 1 \\ 4 & 2 & 5 & 0 \\ 4 & 3 & 3 & -1 \end{bmatrix}$ (d) $\begin{bmatrix} 1 & 2 & 3 \\ 1 & -1 & 0 \\ 0 & 1 & 2 \end{bmatrix}$

4. Show that if A and B are row equivalent, then they are equivalent.

5. Prove Theorem 2.13.

6. Let

$$A = \begin{bmatrix} 1 & 2 & 3 \\ 1 & 1 & 2 \\ 0 & 1 & 1 \end{bmatrix}.$$

Find a matrix B of the form described in Theorem 2.12 that is equivalent to A. Also, find nonsingular matrices P and Q such that $B = PAQ$.

7. Repeat Exercise 6 for

$$A = \begin{bmatrix} 1 & -1 & 2 & 3 \\ 2 & -1 & 3 & 1 \\ 4 & -3 & 7 & 7 \\ 0 & -1 & 1 & 5 \end{bmatrix}.$$

8. Let A be an $m \times n$ matrix. Show that A is equivalent to O if and only if $A = O$.

9. Let A and B be $m \times n$ matrices. Show that A is equivalent to B if and only if A^T is equivalent to B^T.

10. For each of the following matrices A, find a matrix $B \neq A$ that is equivalent to A:

(a) $A = \begin{bmatrix} 1 & -2 & 3 & 1 \\ 0 & -1 & 4 & 3 \\ 1 & 0 & -2 & -1 \end{bmatrix}$ **(b)** $A = \begin{bmatrix} 1 & 3 \\ 2 & 6 \end{bmatrix}$

(c) $A = \begin{bmatrix} 1 & 2 & 3 & 4 & 3 \\ 0 & 1 & -2 & 0 & 2 \\ -1 & 3 & 2 & 0 & 1 \end{bmatrix}$

11. Let A and B be equivalent square matrices. Prove that A is nonsingular if and only if B is nonsingular.

2.5 *LU*-Factorization (Optional)

In this section we discuss a variant of Gaussian elimination (presented in Section 2.2) that decomposes a matrix as a product of a lower triangular matrix and an upper triangular matrix. This decomposition leads to an algorithm for solving a linear system $A\mathbf{x} = \mathbf{b}$ that is the most widely used method on computers for solving a linear system. A main reason for the popularity of this method is that it provides the cheapest way of solving a linear system for which we repeatedly have to change the right side. This type of situation occurs often in applied problems. For example, an electric utility company must determine the inputs (the unknowns) needed to produce some required outputs (the right sides). The inputs and outputs might be related by a linear system, whose coefficient matrix is fixed, while the right side changes from day to day, or even hour to hour. The decomposition discussed in this section is also useful in solving other problems in linear algebra.

When U is an upper triangular matrix all of whose diagonal entries are different from zero, then the linear system $U\mathbf{x} = \mathbf{b}$ can be solved without transforming the augmented matrix $\left[\, U \mid \mathbf{b} \,\right]$ to reduced row echelon form or to row echelon form. The augmented matrix of such a system is given by

$$\left[\begin{array}{ccccc|c} u_{11} & u_{12} & u_{13} & \cdots & u_{1n} & b_1 \\ 0 & u_{22} & u_{23} & \cdots & u_{2n} & b_2 \\ 0 & 0 & u_{33} & \cdots & u_{3n} & b_3 \\ \vdots & \vdots & \vdots & & \vdots & \vdots \\ 0 & 0 & 0 & \cdots & u_{nn} & b_n \end{array}\right].$$

The solution is obtained by the following algorithm:

$$\begin{aligned} x_n &= \frac{b_n}{u_{nn}} \\ x_{n-1} &= \frac{b_{n-1} - u_{n-1,n}x_n}{u_{n-1,n-1}} \\ &\;\;\vdots \\ x_j &= \frac{b_j - \displaystyle\sum_{k=n}^{j-1} u_{jk}x_k}{u_{jj}}, \qquad j = n, n-1, \ldots, 2, 1. \end{aligned}$$

This procedure is merely **back substitution**, which we used in conjunction with Gaussian elimination in Section 2.2, where it was additionally required that the diagonal entries be 1.

In a similar manner, if L is a lower triangular matrix all of whose diagonal entries are different from zero, then the linear system $L\mathbf{x} = \mathbf{b}$ can be solved by **forward substitution**, which consists of the following procedure: The augmented matrix has the form

$$\left[\begin{array}{ccccc|c} \ell_{11} & 0 & 0 & \cdots & 0 & b_1 \\ \ell_{21} & \ell_{22} & 0 & \cdots & 0 & b_2 \\ \ell_{31} & \ell_{32} & \ell_{33} & \cdots & 0 & b_3 \\ \vdots & \vdots & \vdots & & \vdots & \vdots \\ \ell_{n1} & \ell_{n2} & \ell_{n3} & \cdots & \ell_{nn} & b_n \end{array}\right],$$

and the solution is given by

$$\begin{aligned} x_1 &= \frac{b_1}{\ell_{11}} \\ x_2 &= \frac{b_2 - \ell_{21}x_1}{\ell_{22}} \\ &\vdots \\ x_j &= \frac{b_j - \sum_{k=1}^{j-1} \ell_{jk}x_k}{\ell_{jj}}, \qquad j = 2, \ldots, n. \end{aligned}$$

That is, we proceed from the first equation downward, solving for one unknown from each equation.

We illustrate forward substitution in the following example:

EXAMPLE 1 To solve the linear system

$$\begin{aligned} 5x_1 \qquad\qquad\qquad &= 10 \\ 4x_1 - 2x_2 \qquad\quad &= 28 \\ 2x_1 + 3x_2 + 4x_3 &= 26 \end{aligned}$$

we use forward substitution. Hence we obtain from the previous algorithm

$$\begin{aligned} x_1 &= \frac{10}{5} = 2 \\ x_2 &= \frac{28 - 4x_1}{-2} = -10 \\ x_3 &= \frac{26 - 2x_1 - 3x_2}{4} = 13, \end{aligned}$$

which implies that the solution to the given lower triangular system of equations is

$$\mathbf{x} = \begin{bmatrix} 2 \\ -10 \\ 13 \end{bmatrix}.$$

■

As illustrated in the discussion at the beginning of Section 2.2 and Example 1 of this section, the ease with which systems of equations with upper or lower triangular coefficient matrices can be solved is quite attractive. The forward substitution and back substitution algorithms are fast and simple to use. These are used in another important numerical procedure for solving linear systems of equations, which we develop next.

Suppose that an $n \times n$ matrix A can be written as a product of a matrix L in lower triangular form and a matrix U in upper triangular form; that is,

$$A = LU.$$

In this case we say that A has an ***LU*-factorization** or an ***LU*-decomposition**. The LU-factorization of a matrix A can be used to efficiently solve a linear system $A\mathbf{x} = \mathbf{b}$. Substituting LU for A, we have

$$(LU)\mathbf{x} = \mathbf{b},$$

or by (a) of Theorem 1.2 in Section 1.4,

$$L(U\mathbf{x}) = \mathbf{b}.$$

Letting $U\mathbf{x} = \mathbf{z}$, this matrix equation becomes

$$L\mathbf{z} = \mathbf{b}.$$

Since L is in lower triangular form, we solve directly for $\mathbf{z}$ by forward substitution. Once we determine $\mathbf{z}$, since U is in upper triangular form, we solve $U\mathbf{x} = \mathbf{z}$ by back substitution. In summary, if an $n \times n$ matrix A has an LU-factorization, then the solution of $A\mathbf{x} = \mathbf{b}$ can be determined by a forward substitution followed by a back substitution. We illustrate this procedure in the next example.

EXAMPLE 2 Consider the linear system

$$\begin{aligned} 6x_1 - 2x_2 - 4x_3 + 4x_4 &= 2 \\ 3x_1 - 3x_2 - 6x_3 + x_4 &= -4 \\ -12x_1 + 8x_2 + 21x_3 - 8x_4 &= 8 \\ -6x_1 \qquad - 10x_3 + 7x_4 &= -43 \end{aligned}$$

whose coefficient matrix

$$A = \begin{bmatrix} 6 & -2 & -4 & 4 \\ 3 & -3 & -6 & 1 \\ -12 & 8 & 21 & -8 \\ -6 & 0 & -10 & 7 \end{bmatrix}$$

has an LU-factorization, where

$$L = \begin{bmatrix} 1 & 0 & 0 & 0 \\ \frac{1}{2} & 1 & 0 & 0 \\ -2 & -2 & 1 & 0 \\ -1 & 1 & -2 & 1 \end{bmatrix} \quad \text{and} \quad U = \begin{bmatrix} 6 & -2 & -4 & 4 \\ 0 & -2 & -4 & -1 \\ 0 & 0 & 5 & -2 \\ 0 & 0 & 0 & 8 \end{bmatrix}$$

(verify). To solve the given system using this LU-factorization, we proceed as follows. Let

$$\mathbf{b} = \begin{bmatrix} 2 \\ -4 \\ 8 \\ -43 \end{bmatrix}.$$

Then we solve $A\mathbf{x} = \mathbf{b}$ by writing it as $LU\mathbf{x} = \mathbf{b}$. First, let $U\mathbf{x} = \mathbf{z}$ and use forward substitution to solve $L\mathbf{z} = \mathbf{b}$:

$$\begin{bmatrix} 1 & 0 & 0 & 0 \\ \frac{1}{2} & 1 & 0 & 0 \\ -2 & -2 & 1 & 0 \\ -1 & 1 & -2 & 1 \end{bmatrix} \begin{bmatrix} z_1 \\ z_2 \\ z_3 \\ z_4 \end{bmatrix} = \begin{bmatrix} 2 \\ -4 \\ 8 \\ -43 \end{bmatrix}.$$

We obtain

$$\begin{aligned} z_1 &= 2 \\ z_2 &= -4 - \tfrac{1}{2}z_1 = -5 \\ z_3 &= 8 + 2z_1 + 2z_2 = 2 \\ z_4 &= -43 + z_1 - z_2 + 2z_3 = -32. \end{aligned}$$

Next we solve $U\mathbf{x} = \mathbf{z}$,

$$\begin{bmatrix} 6 & -2 & -4 & 4 \\ 0 & -2 & -4 & -1 \\ 0 & 0 & 5 & -2 \\ 0 & 0 & 0 & 8 \end{bmatrix} \begin{bmatrix} x_1 \\ x_2 \\ x_3 \\ x_4 \end{bmatrix} = \begin{bmatrix} 2 \\ -5 \\ 2 \\ -32 \end{bmatrix},$$

by back substitution. We obtain

$$\begin{aligned} x_4 &= \frac{-32}{8} = -4 \\ x_3 &= \frac{2 + 2x_4}{5} = -1.2 \\ x_2 &= \frac{-5 + 4x_3 + x_4}{-2} = 6.9 \\ x_1 &= \frac{2 + 2x_2 + 4x_3 - 4x_4}{6} = 4.5. \end{aligned}$$

Thus the solution to the given linear system is

$$\mathbf{x} = \begin{bmatrix} 4.5 \\ 6.9 \\ -1.2 \\ -4 \end{bmatrix}.$$

■

Next, we show how to obtain an LU-factorization of a matrix by modifying the Gaussian elimination procedure from Section 2.2. No row interchanges will be permitted, and we do not require that the diagonal entries have value 1. At the end of this section we provide a reference that indicates how to enhance the LU-factorization scheme presented to deal with matrices where row interchanges are necessary. We observe that the only elementary row operation permitted is the one that adds a multiple of one row to a different row.

To describe the LU-factorization, we present a step-by-step procedure in the next example.

EXAMPLE 3

Let A be the coefficient matrix of the linear system of Example 2.

$$A = \begin{bmatrix} 6 & -2 & -4 & 4 \\ 3 & -3 & -6 & 1 \\ -12 & 8 & 21 & -8 \\ -6 & 0 & -10 & 7 \end{bmatrix}$$

We proceed to "zero out" entries below the diagonal entries, using only the row operation that adds a multiple of one row to a different row.

Procedure	Matrices Used
Step 1. "Zero out" below the first diagonal entry of A. Add $\left(-\frac{1}{2}\right)$ times the first row of A to the second row of A. Add 2 times the first row of A to the third row of A. Add 1 times the first row of A to the fourth row of A. Call the new resulting matrix U_1.	$U_1 = \begin{bmatrix} 6 & -2 & -4 & 4 \\ 0 & -2 & -4 & -1 \\ 0 & 4 & 13 & 0 \\ 0 & -2 & -14 & 11 \end{bmatrix}$
We begin building a lower triangular matrix L_1 with 1's on the main diagonal, to record the row operations. Enter the *negatives of the multipliers* used in the row operations in the first column of L_1, below the first diagonal entry of L_1.	$L_1 = \begin{bmatrix} 1 & 0 & 0 & 0 \\ \frac{1}{2} & 1 & 0 & 0 \\ -2 & * & 1 & 0 \\ -1 & * & * & 1 \end{bmatrix}$
Step 2. "Zero out" below the second diagonal entry of U_1. Add 2 times the second row of U_1 to the third row of U_1. Add (-1) times the second row of U_1 to the fourth row of U_1. Call the new resulting matrix U_2.	$U_2 = \begin{bmatrix} 6 & -2 & -4 & 4 \\ 0 & -2 & -4 & -1 \\ 0 & 0 & 5 & -2 \\ 0 & 0 & -10 & 12 \end{bmatrix}$

Enter the negatives of the multipliers from the row operations below the second diagonal entry of L_1. Call the new matrix L_2.

$$L_2 = \begin{bmatrix} 1 & 0 & 0 & 0 \\ \frac{1}{2} & 1 & 0 & 0 \\ -2 & -2 & 1 & 0 \\ -1 & 1 & * & 1 \end{bmatrix}$$

Step 3. "Zero out" below the third diagonal entry of U_2. Add 2 times the third row of U_2 to the fourth row of U_2. Call the new resulting matrix U_3.

$$U_3 = \begin{bmatrix} 6 & -2 & -4 & 4 \\ 0 & -2 & -4 & -1 \\ 0 & 0 & 5 & -2 \\ 0 & 0 & 0 & 8 \end{bmatrix}$$

Enter the negative of the multiplier below the third diagonal entry of L_2. Call the new matrix L_3.

$$L_3 = \begin{bmatrix} 1 & 0 & 0 & 0 \\ \frac{1}{2} & 1 & 0 & 0 \\ -2 & -2 & 1 & 0 \\ -1 & 1 & -2 & 1 \end{bmatrix}$$

Let $L = L_3$ and $U = U_3$. Then the product LU gives the original matrix A (verify). This linear system of equations was solved in Example 2 by using the LU-factorization just obtained. ■

Remark In general, a given matrix may have more than one LU-factorization. For example, if A is the coefficient matrix considered in Example 2, then another LU-factorization is LU, where

$$L = \begin{bmatrix} 2 & 0 & 0 & 0 \\ 1 & -1 & 0 & 0 \\ -4 & 2 & 1 & 0 \\ -2 & -1 & -2 & 2 \end{bmatrix} \quad \text{and} \quad U = \begin{bmatrix} 3 & -1 & -2 & 2 \\ 0 & 2 & 4 & 1 \\ 0 & 0 & 5 & -2 \\ 0 & 0 & 0 & 4 \end{bmatrix}.$$

There are many methods for obtaining an LU-factorization of a matrix besides the scheme for **storage of multipliers** described in Example 3. It is important to note that if $a_{11} = 0$, then the procedure used in Example 3 fails. Moreover, if the second diagonal entry of U_1 is zero or if the third diagonal entry of U_2 is zero, then the procedure also fails. In such cases we can try rearranging the equations of the system and beginning again or using one of the other methods for LU-factorization. Most computer programs for LU-factorization incorporate row interchanges into the storage of multipliers scheme and use additional strategies to help control roundoff error. If row interchanges are required, then the product of L and U is not necessarily A—it is a matrix that is a permutation of the rows of A. For example, if row interchanges occur when using the **lu** command in MATLAB in the form **[L,U] = lu(A)**, then MATLAB responds as follows: The matrix that it yields as L is not lower triangular, U is upper triangular, and LU is A.

Key Terms

Lower triangular matrix
Upper triangular matrix
Decomposition
Back and forward substitution
LU-factorization (-decomposition)
Storage of multipliers

2.5 Exercises

In Exercises 1 through 4, solve the linear system $A\mathbf{x} = \mathbf{b}$ *with the given LU-factorization of the coefficient matrix A. Solve the linear system by using a forward substitution followed by a back substitution.*

1. $A = \begin{bmatrix} 2 & 8 & 0 \\ 2 & 2 & -3 \\ 1 & 2 & 7 \end{bmatrix}, \quad \mathbf{b} = \begin{bmatrix} 18 \\ 3 \\ 12 \end{bmatrix},$

$L = \begin{bmatrix} 2 & 0 & 0 \\ 2 & -3 & 0 \\ 1 & -1 & 4 \end{bmatrix}, \quad U = \begin{bmatrix} 1 & 4 & 0 \\ 0 & 2 & 1 \\ 0 & 0 & 2 \end{bmatrix}$

2. $A = \begin{bmatrix} 8 & 12 & -4 \\ 6 & 5 & 7 \\ 2 & 1 & 6 \end{bmatrix}, \quad \mathbf{b} = \begin{bmatrix} -36 \\ 11 \\ 16 \end{bmatrix},$

$L = \begin{bmatrix} 4 & 0 & 0 \\ 3 & 2 & 0 \\ 1 & 1 & 1 \end{bmatrix}, \quad U = \begin{bmatrix} 2 & 3 & -1 \\ 0 & -2 & 5 \\ 0 & 0 & 2 \end{bmatrix}$

3. $A = \begin{bmatrix} 2 & 3 & 0 & 1 \\ 4 & 5 & 3 & 3 \\ -2 & -6 & 7 & 7 \\ 8 & 9 & 5 & 21 \end{bmatrix}, \quad \mathbf{b} = \begin{bmatrix} -2 \\ -2 \\ -16 \\ -66 \end{bmatrix},$

$L = \begin{bmatrix} 1 & 0 & 0 & 0 \\ 2 & 1 & 0 & 0 \\ -1 & 3 & 1 & 0 \\ 4 & 3 & 2 & 1 \end{bmatrix},$

$U = \begin{bmatrix} 2 & 3 & 0 & 1 \\ 0 & -1 & 3 & 1 \\ 0 & 0 & -2 & 5 \\ 0 & 0 & 0 & 4 \end{bmatrix}$

4. $A = \begin{bmatrix} 4 & 2 & 1 & 0 \\ -4 & -6 & 1 & 3 \\ 8 & 16 & -3 & -4 \\ 20 & 10 & 4 & -3 \end{bmatrix}, \quad \mathbf{b} = \begin{bmatrix} 6 \\ 13 \\ -20 \\ 15 \end{bmatrix},$

$L = \begin{bmatrix} 1 & 0 & 0 & 0 \\ -1 & 1 & 0 & 0 \\ 2 & -3 & 1 & 0 \\ 5 & 0 & -1 & 1 \end{bmatrix},$

$U = \begin{bmatrix} 4 & 2 & 1 & 0 \\ 0 & -4 & 2 & 3 \\ 0 & 0 & 1 & 5 \\ 0 & 0 & 0 & 2 \end{bmatrix}$

In Exercises 5 through 10, find an LU-factorization of the coefficient matrix of the given linear system $A\mathbf{x} = \mathbf{b}$*. Solve the linear system by using a forward substitution followed by a back substitution.*

5. $A = \begin{bmatrix} 2 & 3 & 4 \\ 4 & 5 & 10 \\ 4 & 8 & 2 \end{bmatrix}, \quad \mathbf{b} = \begin{bmatrix} 6 \\ 16 \\ 2 \end{bmatrix}$

6. $A = \begin{bmatrix} -3 & 1 & -2 \\ -12 & 10 & -6 \\ 15 & 13 & 12 \end{bmatrix}, \quad \mathbf{b} = \begin{bmatrix} 15 \\ 82 \\ -5 \end{bmatrix}$

7. $A = \begin{bmatrix} 4 & 2 & 3 \\ 2 & 0 & 5 \\ 1 & 2 & 1 \end{bmatrix}, \quad \mathbf{b} = \begin{bmatrix} 1 \\ -1 \\ -3 \end{bmatrix}$

8. $A = \begin{bmatrix} -5 & 4 & 0 & 1 \\ -30 & 27 & 2 & 7 \\ 5 & 2 & 0 & 2 \\ 10 & 1 & -2 & 1 \end{bmatrix}, \quad \mathbf{b} = \begin{bmatrix} -17 \\ -102 \\ -7 \\ -6 \end{bmatrix}$

9. $A = \begin{bmatrix} 2 & 1 & 0 & -4 \\ 1 & 0 & 0.25 & -1 \\ -2 & -1.1 & 0.25 & 6.2 \\ 4 & 2.2 & 0.3 & -2.4 \end{bmatrix}, \quad \mathbf{b} = \begin{bmatrix} -3 \\ -1.5 \\ 5.6 \\ 2.2 \end{bmatrix}$

10. $A = \begin{bmatrix} 4 & 1 & 0.25 & -0.5 \\ 0.8 & 0.6 & 1.25 & -2.6 \\ -1.6 & -0.08 & 0.01 & 0.2 \\ 8 & 1.52 & -0.6 & -1.3 \end{bmatrix},$

$\mathbf{b} = \begin{bmatrix} -0.15 \\ 9.77 \\ 1.69 \\ -4.576 \end{bmatrix}$

11. In the software you are using, investigate to see whether there is a command for obtaining an LU-factorization of a matrix. If there is, use it to find the LU-factorization of matrix A in Example 2. The result obtained in your software need not be that given in Example 2 or 3, because there are many ways to compute such a factorization. Also, some software does not explicitly display L and U, but gives a matrix from which L and U can be "decoded." See the documentation on your software for more details.

12. In the software you are using, investigate to see whether there are commands for doing forward substitution or back substitution. Experiment with the use of such commands on the linear systems, using L and U from Examples 2 and 3.

Supplementary Exercises

1. Let

$$A = \begin{bmatrix} 2 & -4 & 0 \\ 0 & 1 & 3 \\ 0 & 0 & 6 \end{bmatrix}.$$

Find a matrix B in reduced row echelon form that is row equivalent to A, using elementary matrices.

2. Find all values of a for which the following linear systems have solutions:

(a) $\begin{aligned} x + 2y + z &= a^2 \\ x + y + 3z &= a \\ 3x + 4y + 7z &= 8 \end{aligned}$ **(b)** $\begin{aligned} x + 2y + z &= a^2 \\ x + y + 3z &= a \\ 3x + 4y + 8z &= 8 \end{aligned}$

3. Find all values of a for which the following homogeneous system has nontrivial solutions:

$$\begin{aligned} (1-a)x \qquad\quad + z &= 0 \\ - ay + z &= 0 \\ y + z &= 0 \end{aligned}$$

4. Find all values of a, b, and c so that the linear system

$$A\mathbf{x} = \begin{bmatrix} a \\ b \\ c \end{bmatrix}$$

is consistent for

$$A = \begin{bmatrix} 1 & 0 & 2 \\ 1 & 1 & -2 \\ 1 & 3 & -10 \end{bmatrix}.$$

5. Let A be an $n \times n$ matrix.

(a) Suppose that the matrix B is obtained from A by multiplying the jth row of A by $k \neq 0$. Find an elementary row operation that, when applied to B, gives A.

(b) Suppose that the matrix C is obtained from A by interchanging the ith and jth rows of A. Find an elementary row operation that, when applied to C, gives A.

(c) Suppose that the matrix D is obtained from A by adding k times the jth row of A to its ith row. Find an elementary row operation that, when applied to D, gives A.

6. Exercise 5 implies that the effect of any elementary row operation can be reversed by another (suitable) elementary row operation.

(a) Suppose that the matrix E_1 is obtained from I_n by multiplying the jth row of I_n by $k \neq 0$. Explain why E_1 is nonsingular.

(b) Suppose that the matrix E_2 is obtained from I_n by interchanging the ith and jth rows of I_n. Explain why E_2 is nonsingular.

(c) Suppose that the matrix E_3 is obtained from I_n by adding k times the jth row of I_n to its ith row. Explain why E_3 is nonsingular.

7. Find the inverse of

$$\begin{bmatrix} 1 & 0 & 1 \\ 1 & 1 & 0 \\ 0 & 1 & 1 \end{bmatrix}.$$

8. Find the inverse of

$$\begin{bmatrix} 1 & a & 0 & 0 \\ 0 & 1 & a & 0 \\ 0 & 0 & 1 & a \\ 0 & 0 & 0 & 1 \end{bmatrix}.$$

9. As part of a project, two students must determine the inverse of a given 10×10 matrix A. Each performs the required calculation, and they return their results A_1 and A_2, respectively, to the instructor.

(a) What must be true about the two results? Why?

(b) How does the instructor check their work without repeating the calculations?

10. Compute the vector $\mathbf{w}$ for each of the following expressions without computing the inverse of any matrix, given that

$$A = \begin{bmatrix} 1 & 0 & -2 \\ 1 & 1 & 0 \\ 0 & 1 & 1 \end{bmatrix}, \quad C = \begin{bmatrix} 1 & 1 & 1 \\ 2 & 3 & 1 \\ 1 & 2 & 1 \end{bmatrix},$$

$$F = \begin{bmatrix} 2 & 1 & 0 \\ -3 & 0 & 2 \\ -1 & 1 & 2 \end{bmatrix}, \quad \mathbf{v} = \begin{bmatrix} 6 \\ 7 \\ -3 \end{bmatrix}:$$

(a) $\mathbf{w} = A^{-1}(C + F)\mathbf{v}$ **(b)** $\mathbf{w} = (F + 2A)C^{-1}\mathbf{v}$

11. Determine all values of s so that

$$A = \begin{bmatrix} 0 & 1 & 2 \\ 2 & 1 & 1 \\ 1 & s & 2 \end{bmatrix}$$

is nonsingular.

12. Determine all values of s so that

$$A = \begin{bmatrix} s & 1 & 0 \\ 1 & s & 1 \\ 0 & 1 & s \end{bmatrix}$$

is nonsingular.

13. Show that the matrix

$$\begin{bmatrix} \cos\theta & \sin\theta \\ -\sin\theta & \cos\theta \end{bmatrix}$$

is nonsingular, and compute its inverse.

14. Let $\mathbf{u}$ and $\mathbf{v}$ be solutions to the homogeneous linear system $A\mathbf{x} = \mathbf{0}$.

(a) Show that $\mathbf{u} + \mathbf{v}$ is a solution.

(b) Show that $\mathbf{u} - \mathbf{v}$ is a solution.

(c) For any scalar r, show that $r\mathbf{u}$ is a solution.

(d) For any scalars r and s, show that $r\mathbf{u} + s\mathbf{v}$ is a solution.

15. Show that if $\mathbf{u}$ and $\mathbf{v}$ are solutions to the linear system $A\mathbf{x} = \mathbf{b}$, then $\mathbf{u} - \mathbf{v}$ is a solution to the associated homogeneous system $A\mathbf{x} = \mathbf{0}$.

16. Justify Remark 1 following Example 6 in Section 2.2.

17. Show that if A is singular and $A\mathbf{x} = \mathbf{b}$, $\mathbf{b} \neq \mathbf{0}$, has one solution, then it has infinitely many. (*Hint*: Use Exercise 29 in Section 2.2.)

Exercises 18 through 20 use material from Section 2.5.

18. Let $A = \begin{bmatrix} 2 & 6 \\ 1 & 1 \end{bmatrix}$, $L = \begin{bmatrix} 2 & 0 \\ t & -2 \end{bmatrix}$, and $U = \begin{bmatrix} 1 & 3 \\ 0 & s \end{bmatrix}$. Find scalars s and t so that $LU = A$.

19. Let

$$A = \begin{bmatrix} 6 & 2 & 8 \\ 9 & 5 & 11 \\ 3 & 1 & 6 \end{bmatrix}, \quad L = \begin{bmatrix} 2 & 0 & 0 \\ t & s & 0 \\ 1 & 0 & -1 \end{bmatrix},$$

and

$$U = \begin{bmatrix} r & 1 & 4 \\ 0 & 2 & -1 \\ 0 & 0 & p \end{bmatrix}.$$

Find scalars r, s, t, and p so that $LU = A$.

20. Let A have an LU-factorization, $A = LU$. By inspecting the lower triangular matrix L and the upper triangular matrix U, explain how to claim that the linear system $A\mathbf{x} = LU\mathbf{x} = \mathbf{b}$ does not have a unique solution.

21. Show that the outer product of X and Y is row equivalent either to O or to a matrix with $n - 1$ rows of zeros. (See Supplementary Exercises 30 through 32 in Chapter 1.)

Chapter Review

True or False

1. Every matrix in row echelon form is also in reduced row echelon form.
2. If the augmented matrices of two linear systems are row equivalent, then the systems have exactly the same solutions.
3. If a homogeneous linear system has more equations than unknowns, then it has a nontrivial solution.
4. The elementary matrix

$$\begin{bmatrix} 0 & 0 & 1 \\ 0 & 1 & 0 \\ 1 & 0 & 0 \end{bmatrix}$$

times the augmented matrix $\left[A \;\vdots\; \mathbf{b} \right]$ of a $3 \times n$ linear system will interchange the first and third equations.

5. The reduced row echelon form of a nonsingular matrix is an identity matrix.
6. If A is $n \times n$, then $A\mathbf{x} = \mathbf{0}$ has a nontrivial solution if and only if A is singular.
7. If an $n \times n$ matrix A can be expressed as a product of elementary matrices, then A is nonsingular.
8. The reduced row echelon form of a singular matrix has a row of zeros.
9. Any matrix equivalent to an identity matrix is nonsingular.
10. If A is $n \times n$ and the reduced row echelon form of $\left[A \;\vdots\; I_n \right]$ is $\left[C \;\vdots\; D \right]$, then $C = I_n$ and $D = A^{-1}$.

Quiz

1. Determine the reduced row echelon form of

$$A = \begin{bmatrix} 1 & 1 & 5 \\ 2 & -2 & -2 \\ -3 & 1 & -3 \end{bmatrix}.$$

2. After some row operations, the augmented matrix of the linear system $A\mathbf{x} = \mathbf{b}$ is

$$\left[C \;\vdots\; \mathbf{d} \right] = \left[\begin{array}{rrrr|r} 1 & -2 & 4 & 5 & -6 \\ 0 & 0 & 1 & 3 & 0 \\ 0 & 0 & 0 & 0 & 0 \\ 0 & 0 & 0 & 0 & 0 \end{array}\right].$$

(a) Is C in reduced row echelon form? Explain.

(b) How many solutions are there for $A\mathbf{x} = \mathbf{b}$?

(c) Is A nonsingular? Explain.

(d) Determine all possible solutions to $A\mathbf{x} = \mathbf{b}$.

3. Determine k so that $A = \begin{bmatrix} 2 & 1 & 4 \\ 1 & -2 & 1 \\ 2 & 6 & k \end{bmatrix}$ is singular.

4. Find all solutions to the homogeneous linear system with coefficient matrix

$$A = \begin{bmatrix} 2 & 0 & 1 \\ 1 & 2 & 1 \\ 1 & 2 & 3 \\ 1 & 0 & 2 \\ 2 & 1 & 4 \end{bmatrix}.$$

5. If $A = \begin{bmatrix} 1 & 2 & 1 \\ 1 & 1 & 1 \\ 2 & 1 & 0 \end{bmatrix}$, then find (if possible) A^{-1}.

6. Let A and B be $n \times n$ nonsingular matrices. Find nonsingular matrices P and Q so that $PAQ = B$.

7. Fill in the blank in the following statement:
If A is ________, then A and A^T are row equivalent.

Discussion Exercises

1. The reduced row echelon form of the matrix A is I_3. Describe all possible matrices A.

2. The reduced row echelon form of the matrix A is

$$\begin{bmatrix} 1 & 2 & 0 \\ 0 & 0 & 1 \\ 0 & 0 & 0 \end{bmatrix}.$$

Find three different such matrices A. Explain how you determined your matrices.

3. Let A be a 2×2 real matrix. Determine conditions on the entries of A so that $A^2 = I_2$.

4. An agent is on a mission, but is not sure of her location. The agent's handheld GPS unit is malfunctioning, but her radio unit is working. The radio unit's battery is so low that she can use it only very briefly. Turning it on, she is able to contact three radio beacons, which give approximate mileage from her position to each beacon. She quickly records the following information: 700 miles from Athens, 1300 miles from Rome, and 900 miles from Sophia. Determine the agent's approximate location. Explain your procedure.

5. The exercises dealing with GPS in Section 2.2 were constructed so that the answers were whole numbers or very close to whole numbers. The construction procedure worked in reverse. Namely, we chose the coordinates (x, y) where we wanted the three circles to intersect and then set out to find the centers and radii of three circles that would intersect at that point. We wanted the centers to be ordered pairs of integers and the radii to have positive integer lengths. Discuss how to use Pythagorean triples of natural numbers to complete such a construction.

6. After Example 9 in Section 2.2, we briefly outlined an approach for GPS in three dimensions that used a set of four equations of the form

$$(x - a_j)^2 + (y - b_j)^2 + (z - c_j)^2 = \text{(distance from the receiver to satellite } j)^2,$$

where the distance on the right side came from the expression "distance = speed × elapsed time." The speed in this case is related to the speed of light. A very nice example of a situation such as this appears in the work of Dan Kalman ("An Underdetermined Linear System for GPS," *The College Mathematics Journal*, vol. 33, no. 5, Nov. 2002, pp. 384–390). In this paper the distance from the satellite to the receiver is expressed in terms of the time t as $0.047(t - \text{satellite to receiver time})$, where 0.047 is the speed of light scaled to earth radius units. Thus the four equations have the form

$$(x - a_j)^2 + (y - b_j)^2 + (z - c_j)^2 =$$
$$0.047^2(t - \text{satellite } j \text{ to receiver time})^2,$$

where (a_j, b_j, c_j) is the location of satellite j, for $j = 1, 2, 3, 4$. For the data in the next table, determine the location (x, y, z) of the GPS receiver on the sphere we call earth. Carefully discuss your steps.

Satellite	*Position* (a_j, b_j, c_j)	*Time it took the signal to go from the satellite to the GPS receiver*
1	(1, 2, 0)	19.9
2	(2, 0, 2)	2.4
3	(1, 1, 1)	32.6
4	(2, 1, 0)	19.9

7. In Gaussian elimination for a square linear system with a nonsingular coefficient matrix, we use row operations to obtain a row equivalent linear system that is upper triangular and then use back substitution to obtain the solution. A crude measure of work involved counts the number of multiplies and divides required to get the upper triangular form. Let us assume that we do not need to interchange rows to have nonzero pivots and that we will not require the diagonal entries of the upper triangular coefficient matrix to be 1's. Hence, we will proceed by using only the row operation $k\mathbf{r}_i + \mathbf{r}_j \to \mathbf{r}_j$ for an appropriate choice of multiplier k. In this situation we can give an expression for the multiplier k that works for each row operation; we have

$$k = \frac{\text{entry to be eliminated}}{\text{pivot}}.$$

Since we are not making the pivots 1, we must count a division each time we use a row operation.

(a) Assume that the coefficient matrix is 5×5. Determine the number of multiplies and divides required to obtain a row equivalent linear system whose coefficient matrix is upper triangular. Do not forget to apply the row operations to the augmented column.

(b) Generalize the result from part (a) to $n \times n$ linear systems. Provide a compact formula for the total number of multiplies and divides required.

8. ***(Network Analysis)*** The central business area of many large cities is a network of one-way streets. Any repairs to these thoroughfares, closing for emergencies and accidents, or civic functions disrupts the normal flow of traffic. For one-way street networks there is a simple rule: Vehicles entering an intersection from a street must also exit the intersection by another street. (We will assume that parking lots and garages are located outside the network.) Thus for each intersection we have an equilibrium equation or, put simply, an input-equals-output equation. After some data collection involving entry and exit volumes at intersections, a city traffic commission can construct network models for traffic flow patterns involving linear systems. The figure shows a street network where the direction of traffic flow is indicated by arrows and the average number of vehicles per hour that enter or exit on a street appears near the street.

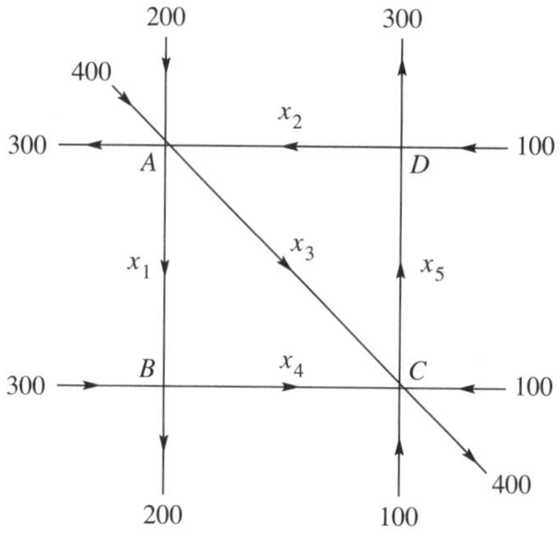

(a) For each intersection A through D, construct an input-equals-output equation. Then rearrange the equations so you can write the system in matrix form, using the coefficients of x_1 in column 1, those for x_2 in column 2, and so on. Determine the reduced row echelon form for the augmented matrix and solve for the unknowns corresponding to leading 1's.

(b) Since each $x_i \geq 0$, determine any restrictions on the unknowns.

(c) Explain what happens in this model if the street from intersection B to C is closed.

9. Solve each of the following matrix equations:

(a) $A\mathbf{x} = \mathbf{x} + \mathbf{b}$,

$$\text{where } A = \begin{bmatrix} 4 & 1 & 0 & 0 \\ 1 & 4 & 1 & 0 \\ 0 & 1 & 4 & 1 \\ 0 & 0 & 1 & 4 \end{bmatrix}, \quad \mathbf{b} = \begin{bmatrix} 2 \\ 1 \\ 0 \\ 2 \end{bmatrix}.$$

(b) $A\mathbf{x} = A^2\mathbf{x} + \mathbf{b}$,

$$\text{where } A = \begin{bmatrix} 3 & 1 & 0 \\ 1 & 3 & 1 \\ 0 & 1 & 3 \end{bmatrix}, \quad \mathbf{b} = \begin{bmatrix} 1 \\ -1 \\ 1 \end{bmatrix}.$$

CHAPTER

3 Determinants

3.1 Definition

In Exercise 43 of Section 1.3, we defined the trace of a square ($n \times n$) matrix $A = \left[a_{ij}\right]$ by $\mathrm{Tr}(A) = \sum_{i=1}^{n} a_{ii}$. Another very important number associated with a square matrix A is the determinant of A, which we now define. Determinants first arose in the solution of linear systems. Although the methods given in Chapter 2 for solving such systems are more efficient than those involving determinants, determinants will be useful for our further study of a linear transformation $L: V \to V$ in Chapter 6. First, we deal briefly with permutations, which are used in our definition of determinant. Throughout this chapter, when we use the term *matrix*, we mean *square matrix*.

DEFINITION 3.1

Let $S = \{1, 2, \ldots, n\}$ be the set of integers from 1 to n, arranged in ascending order. A rearrangement $j_1 j_2 \cdots j_n$ of the elements of S is called a **permutation** of S. We can consider a permutation of S to be a one-to-one mapping of S onto itself.

To illustrate the preceding definition, let $S = \{1, 2, 3, 4\}$. Then 4231 is a permutation of S. It corresponds to the function $f: S \to S$ defined by

$$\begin{aligned} f(1) &= 4 \\ f(2) &= 2 \\ f(3) &= 3 \\ f(4) &= 1. \end{aligned}$$

We can put any one of the n elements of S in first position, any one of the remaining $n - 1$ elements in second position, any one of the remaining $n - 2$ elements in third position, and so on until the nth position can be filled only by the last remaining element. Thus there are $n(n-1)(n-2)\cdots 2\cdot 1 = n!$ (n factorial) permutations of S; we denote the set of all permutations of S by S_n.

EXAMPLE 1 Let $S = \{1, 2, 3\}$. The set S_3 of all permutations of S consists of the $3! = 6$ permutations 123, 132, 213, 231, 312, and 321. The diagram in Figure 3.1(a) can be used to enumerate all the permutations of S. Thus, in Figure 3.1(a) we start out from the node labeled 1 and proceed along one of two branches, one leading to node 2 and the other leading to node 3. Once we arrive at node 2 from node 1, we can go only to node 3. Similarly, once we arrive at node 3 from node 1, we can go only to node 2. Thus we have enumerated the permutations 123, 132. The diagram in Figure 3.1(b) yields the permutations 213 and 231, and the diagram in Figure 3.1(c) yields the permutations 312 and 321.

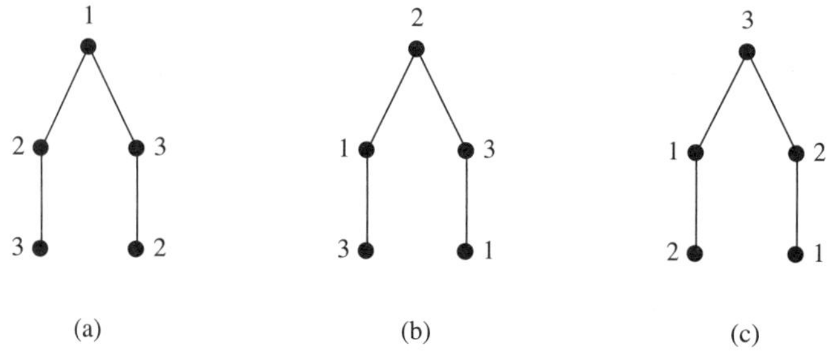

FIGURE 3.1

The graphical method illustrated in Figure 3.1 can be generalized to enumerate all the permutations of the set $\{1, 2, \ldots, n\}$. ■

A permutation $j_1 j_2 \ldots j_n$ is said to have an **inversion** if a larger integer, j_r, precedes a smaller one, j_s. A permutation is called **even** if the total number of inversions in it is even, or **odd** if the total number of inversions in it is odd. If $n \geq 2$, there are $n!/2$ even and $n!/2$ odd permutations in S_n.

EXAMPLE 2 S_1 has only $1! = 1$ permutation: 1, which is even because there are no inversions. ■

EXAMPLE 3 S_2 has $2! = 2$ permutations: 12, which is even (no inversions), and 21, which is odd (one inversion). ■

EXAMPLE 4 In the permutation 4312 in S_4, 4 precedes 3, 4 precedes 1, 4 precedes 2, 3 precedes 1, and 3 precedes 2. Thus the total number of inversions in this permutation is 5, and 4312 is odd. ■

EXAMPLE 5 S_3 has $3! = 3\cdot 2\cdot 1 = 6$ permutations: 123, 231, and 312, which are even, and 132, 213, and 321, which are odd. ■

DEFINITION 3.2 Let $A = \left[a_{ij}\right]$ be an $n \times n$ matrix. The **determinant** function, denoted by **det**, is defined by

$$\det(A) = \sum(\pm)a_{1j_1}a_{2j_2}\cdots a_{nj_n},$$

where the summation is over all permutations $j_1 j_2 \cdots j_n$ of the set $S = \{1, 2, \ldots, n\}$. The sign is taken as $+$ or $-$ according to whether the permutation $j_1 j_2 \cdots j_n$ is even or odd.

In each term $(\pm)a_{1j_1}a_{2j_2}\cdots a_{nj_n}$ of $\det(A)$, the row subscripts are in natural order and the column subscripts are in the order $j_1 j_2 \cdots j_n$. Thus each term in $\det(A)$, with its appropriate sign, is a product of n entries of A, with exactly one entry from each row and exactly one entry from each column. Since we sum over all permutations of S, $\det(A)$ has $n!$ terms in the sum.

Another notation for $\det(A)$ is $|A|$. We shall use both $\det(A)$ and $|A|$.

EXAMPLE 6

If $A = \begin{bmatrix} a_{11} \end{bmatrix}$ is a 1×1 matrix, then $\det(A) = a_{11}$. ■

EXAMPLE 7

If

$$A = \begin{bmatrix} a_{11} & a_{12} \\ a_{21} & a_{22} \end{bmatrix},$$

then to obtain $\det(A)$, we write down the terms $a_{1-}a_{2-}$ and replace the dashes with all possible elements of S_2: The subscripts become 12 and 21. Now 12 is an even permutation and 21 is an odd permutation. Thus

$$\det(A) = a_{11}a_{22} - a_{12}a_{21}.$$

Hence we see that $\det(A)$ can be obtained by forming the product of the entries on the line from left to right and subtracting from this number the product of the entries on the line from right to left.

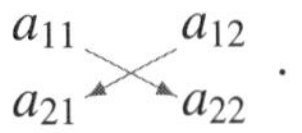

Thus, if $A = \begin{bmatrix} 2 & -3 \\ 4 & 5 \end{bmatrix}$, then $|A| = (2)(5) - (-3)(4) = 22$. ■

EXAMPLE 8

If

$$A = \begin{bmatrix} a_{11} & a_{12} & a_{13} \\ a_{21} & a_{22} & a_{23} \\ a_{31} & a_{32} & a_{33} \end{bmatrix},$$

then to compute $\det(A)$, we write down the six terms $a_{1-}a_{2-}a_{3-}$, $a_{1-}a_{2-}a_{3-}$, $a_{1-}a_{2-}a_{3-}$, $a_{1-}a_{2-}a_{3-}$, $a_{1-}a_{2-}a_{3-}$, $a_{1-}a_{2-}a_{3-}$. All the elements of S_3 are used to replace the dashes, and if we prefix each term by + or − according to whether the permutation is even or odd, we find that (verify)

$$\begin{aligned} \det(A) = a_{11}a_{22}a_{33} + a_{12}a_{23}a_{31} + a_{13}a_{21}a_{32} - a_{11}a_{23}a_{32} \\ - a_{12}a_{21}a_{33} - a_{13}a_{22}a_{31}. \end{aligned} \tag{1}$$

We can also obtain $|A|$ as follows. Repeat the first and second columns of A, as shown next. Form the sum of the products of the entries on the lines from left to right, and subtract from this number the products of the entries on the lines from right to left (verify):

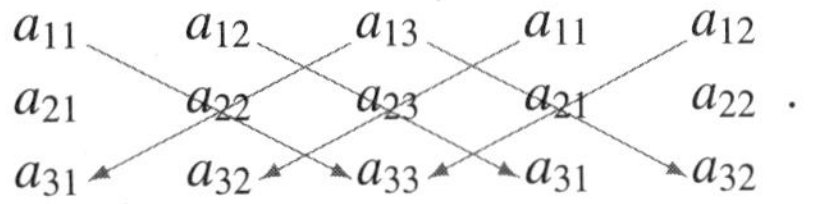

■

EXAMPLE 9 Let

$$A = \begin{bmatrix} 1 & 2 & 3 \\ 2 & 1 & 3 \\ 3 & 1 & 2 \end{bmatrix}.$$

Evaluate $|A|$.

Solution

Substituting in (1), we find that

$$\begin{vmatrix} 1 & 2 & 3 \\ 2 & 1 & 3 \\ 3 & 1 & 2 \end{vmatrix} = (1)(1)(2) + (2)(3)(3) + (3)(2)(1) - (1)(3)(1)$$
$$- (2)(2)(2) - (3)(1)(3) = 6.$$

We could obtain the same result by using the easy method illustrated previously, as follows:

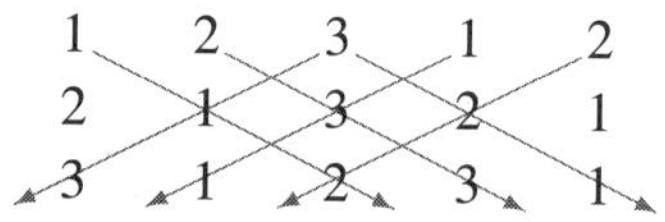

$$|A| = (1)(1)(2) + (2)(3)(3) + (3)(2)(1) - (3)(1)(3) - (1)(3)(1)$$
$$- (2)(2)(2) = 6.$$ ■

Warning The methods used for computing $\det(A)$ in Examples 7–9 do not apply for $n \geq 4$.

It may already have struck the reader that Definition 3.2 is an extremely tedious way of computing determinants for a sizable value of n. In fact, $10! = 3.6288 \times 10^6$ and $20! = 2.4329 \times 10^{18}$, each an enormous number. In Section 3.2 we develop properties of determinants that will greatly reduce the computational effort.

Permutations are studied at some depth in abstract algebra courses and in courses dealing with group theory. As we just noted, we shall develop methods for evaluating determinants other than those involving permutations. However, we do require the following important property of permutations: If we interchange two numbers in the permutation $j_1 j_2 \cdots j_n$, then the number of inversions is either increased or decreased by an odd number.

A proof of this fact can be given by first noting that if two adjacent numbers in the permutation $j_1 j_2 \cdots j_n$ are interchanged, then the number of inversions is either increased or decreased by 1. Thus consider the permutations $j_1 j_2 \cdots j_e j_f \cdots j_n$ and $j_1 j_2 \cdots j_f j_e \cdots j_n$. If $j_e j_f$ is an inversion, then $j_f j_e$ is not an inversion, and the second permutation has one fewer inversion than the first one; if $j_e j_f$ is not an inversion, then $j_f j_e$ is, and so the second permutation has one more inversion than the first. Now an interchange of any two numbers in a permutation $j_1 j_2 \cdots j_n$ can always be achieved by an odd number of successive interchanges of adjacent numbers. Thus, if we wish to interchange j_c and j_k $(c < k)$ and there are s numbers between j_c and j_k, we move j_c to the right by interchanging adjacent numbers, until j_c follows j_k. This requires $s+1$ steps. Next, we move j_k to the left by interchanging adjacent numbers until it is where j_c was. This requires s steps. Thus the total number of adjacent interchanges required is $(s+1)+s=2s+1$, which is always odd. Since each adjacent interchange changes the number of inversions by 1 or -1, and since the sum of an odd number of numbers each of which is 1 or -1, is always odd, we conclude that the number of inversions is changed by an odd number. Thus, the number of inversions in 54132 is 8, and the number of inversions in 52134 (obtained by interchanging 2 and 4) is 5.

Key Terms

Trace
Determinant (det)
Permutation (even/odd)
Inversion

3.1 Exercises

1. Find the number of inversions in each of the following permutations of $S=\{1,2,3,4,5\}$:
(a) 52134 **(b)** 45213 **(c)** 42135

2. Find the number of inversions in each of the following permutations of $S=\{1,2,3,4,5\}$:
(a) 13542 **(b)** 35241 **(c)** 12345

3. Determine whether each of the following permutations of $S=\{1,2,3,4\}$ is even or odd:
(a) 4213 **(b)** 1243 **(c)** 1234

4. Determine whether each of the following permutations of $S=\{1,2,3,4\}$ is even or odd:
(a) 3214 **(b)** 1423 **(c)** 2143

5. Determine the sign associated with each of the following permutations of the column indices of a 5×5 matrix:
(a) 25431 **(b)** 31245 **(c)** 21345

6. Determine the sign associated with each of the following permutations of the column indices of a 5×5 matrix:
(a) 52341 **(b)** 34125 **(c)** 14523

7. (a) Find the number of inversions in the permutation 436215.

(b) Verify that the number of inversions in the permutation 416235, obtained from that in part (a) by interchanging two numbers, differs from the answer in part (a) by an odd number.

8. Evaluate:
(a) $\begin{vmatrix} 2 & -1 \\ 3 & 2 \end{vmatrix}$ **(b)** $\begin{vmatrix} 2 & 1 \\ 4 & 3 \end{vmatrix}$

9. Evaluate:
(a) $\begin{vmatrix} 1 & 2 \\ 2 & 4 \end{vmatrix}$ **(b)** $\begin{vmatrix} 3 & 1 \\ -3 & -1 \end{vmatrix}$

10. Let $A=\left[a_{ij}\right]$ be a 4×4 matrix. Develop the general expression for $\det(A)$.

11. Evaluate:
(a) $\det\left(\begin{bmatrix} 2 & 1 & 3 \\ 3 & 2 & 1 \\ 0 & 1 & 2 \end{bmatrix}\right)$ **(b)** $\begin{vmatrix} 2 & 1 & 3 \\ -3 & 2 & 1 \\ -1 & 3 & 4 \end{vmatrix}$

(c) $\det\left(\begin{bmatrix} 0 & 0 & 0 & 3 \\ 0 & 0 & 4 & 0 \\ 0 & 2 & 0 & 0 \\ 6 & 0 & 0 & 0 \end{bmatrix}\right)$

12. Evaluate:
(a) $\det\left(\begin{bmatrix} 2 & 0 & 0 \\ 0 & -3 & 0 \\ 0 & 0 & 4 \end{bmatrix}\right)$

(b) $\det\left(\begin{bmatrix} 2 & 4 & 5 \\ 0 & -6 & 2 \\ 0 & 0 & 3 \end{bmatrix}\right)$

(c) $\begin{vmatrix} 0 & 0 & 2 & 0 \\ 0 & 3 & 0 & 0 \\ 6 & 0 & 0 & 0 \\ 0 & 0 & 0 & 5 \end{vmatrix}$

13. Evaluate:
(a) $\det\left(\begin{bmatrix} t-1 & 2 \\ 3 & t-2 \end{bmatrix}\right)$

(b) $\begin{vmatrix} t-1 & -1 & -2 \\ 0 & t & 2 \\ 0 & 0 & t-3 \end{vmatrix}$

14. Evaluate:

(a) $\begin{vmatrix} t & 4 \\ 5 & t-8 \end{vmatrix}$

(b) $\det\left(\begin{bmatrix} t-1 & 0 & 1 \\ -2 & t & -1 \\ 0 & 0 & t+1 \end{bmatrix}\right)$

15. For each of the matrices in Exercise 13, find values of t for which the determinant is 0.

16. For each of the matrices in Exercise 14, find values of t for which the determinant is 0.

3.2 Properties of Determinants

In this section we examine properties of determinants that simplify their computation.

Theorem 3.1 If A is a matrix, then $\det(A) = \det(A^T)$.

Proof

Let $A = [a_{ij}]$ and $A^T = [b_{ij}]$, where $b_{ij} = a_{ji}$. We have

$$\det(A^T) = \sum (\pm) b_{1j_1} b_{2j_2} \cdots b_{nj_n} = \sum (\pm) a_{j_1 1} a_{j_2 2} \cdots a_{j_n n}.$$

We can then write $b_{1j_1} b_{2j_2} \cdots b_{nj_n} = a_{j_1 1} a_{j_2 2} \cdots a_{j_n n} = a_{1k_1} a_{2k_2} \cdots a_{nk_n}$, which is a term of $\det(A)$. Thus the terms in $\det(A^T)$ and $\det(A)$ are identical. We must now check that the signs of corresponding terms are also identical. It can be shown, by the properties of permutations discussed in an abstract algebra course,* that the number of inversions in the permutation $k_1 k_2 \ldots k_n$, which determines the sign associated with the term $a_{1k_1} a_{2k_2} \cdots a_{nk_n}$, is the same as the number of inversions in the permutation $j_1 j_2 \ldots j_n$, which determines the sign associated with the term $b_{1j_1} b_{2j_2} \cdots b_{nj_n}$. As an example,

$$b_{13} b_{24} b_{35} b_{41} b_{52} = a_{31} a_{42} a_{53} a_{14} a_{25} = a_{14} a_{25} a_{31} a_{42} a_{53};$$

the number of inversions in the permutation 45123 is 6, and the number of inversions in the permutation 34512 is also 6. Since the signs of corresponding terms are identical, we conclude that $\det(A^T) = \det(A)$. ■

EXAMPLE 1 Let A be the matrix in Example 9 of Section 3.1. Then

$$A^T = \begin{bmatrix} 1 & 2 & 3 \\ 2 & 1 & 1 \\ 3 & 3 & 2 \end{bmatrix}.$$

Substituting in (1) of Section 3.1 (or using the method of lines given in Example 8 of Section 3.1), we find that

$$\begin{aligned} |A^T| &= (1)(1)(2) + (2)(1)(3) + (3)(2)(3) \\ &\quad - (1)(1)(3) - (2)(2)(2) - (3)(1)(3) = 6 = |A|. \end{aligned}$$ ■

*See J. Fraleigh, *A First Course in Abstract Algebra*, 7th ed., Reading, Mass.: Addison-Wesley Publishing Company, Inc., 2003; and J. Gallian, *Contemporary Abstract Algebra*, 5th ed., Lexington, Mass., D. C. Heath and Company, 2002.

Theorem 3.1 will enable us to replace "row" by "column" in many of the additional properties of determinants; we see how to do this in the following theorem:

Theorem 3.2 If matrix B results from matrix A by interchanging two different rows (columns) of A, then $\det(B) = -\det(A)$.

Proof

Suppose that B arises from A by interchanging rows r and s of A, say, $r < s$. Then we have $b_{rj} = a_{sj}$, $b_{sj} = a_{rj}$, and $b_{ij} = a_{ij}$ for $i \neq r, i \neq s$. Now

$$\begin{aligned}\det(B) &= \sum(\pm)b_{1j_1}b_{2j_2}\cdots b_{rj_r}\cdots b_{sj_s}\cdots b_{nj_n}\\ &= \sum(\pm)a_{1j_1}a_{2j_2}\cdots a_{sj_r}\cdots a_{rj_s}\cdots a_{nj_n}\\ &= \sum(\pm)a_{1j_1}a_{2j_2}\cdots a_{rj_s}\cdots a_{sj_r}\cdots a_{nj_n}.\end{aligned}$$

The permutation $j_1j_2\ldots j_s\ldots j_r\ldots j_n$ results from the permutation $j_1j_2\ldots j_r\ldots j_s\ldots j_n$ by an interchange of two numbers, and the number of inversions in the former differs by an odd number from the number of inversions in the latter. This means that the sign of each term in $\det(B)$ is the negative of the sign of the corresponding term in $\det(A)$. Hence $\det(B) = -\det(A)$.

Now let B arise from A by interchanging two columns of A. Then B^T arises from A^T by interchanging two rows of A^T. So $\det(B^T) = -\det(A^T)$, but $\det(B^T) = \det(B)$ and $\det(A^T) = \det(A)$. Hence $\det(B) = -\det(A)$. ■

In the results to follow, proofs will be given only for the rows of A; the proofs for the corresponding column cases proceed as at the end of the proof of Theorem 3.2.

EXAMPLE 2 We have $|A| = \begin{vmatrix} 2 & -1 \\ 3 & 2 \end{vmatrix} = -\begin{vmatrix} 3 & 2 \\ 2 & -1 \end{vmatrix} = \begin{vmatrix} 2 & 3 \\ -1 & 2 \end{vmatrix} = |A^T| = 7.$ ■

Theorem 3.3 If two rows (columns) of A are equal, then $\det(A) = 0$.

Proof

Suppose that rows r and s of A are equal. Interchange rows r and s of A to obtain a matrix B. Then $\det(B) = -\det(A)$. On the other hand, $B = A$, so $\det(B) = \det(A)$. Thus $\det(A) = -\det(A)$, and so $\det(A) = 0$. ■

EXAMPLE 3 We have $\begin{vmatrix} 1 & 2 & 3 \\ -1 & 0 & 7 \\ 1 & 2 & 3 \end{vmatrix} = 0$. (Verify by the use of Definition 3.2.) ■

Theorem 3.4 If a row (column) of A consists entirely of zeros, then $\det(A) = 0$.

Proof

Let the ith row of A consist entirely of zeros. Since each term in Definition 3.2 for the determinant of A contains a factor from the ith row, each term in $\det(A)$ is zero. Hence $\det(A) = 0$. ■

EXAMPLE 4

We have $\begin{vmatrix} 1 & 2 & 3 \\ 4 & 5 & 6 \\ 0 & 0 & 0 \end{vmatrix} = 0$. (Verify by the use of Definition 3.2.) ■

Theorem 3.5 If B is obtained from A by multiplying a row (column) of A by a real number k, then $\det(B) = k\det(A)$.

Proof

Suppose that the rth row of $A = [a_{ij}]$ is multiplied by k to obtain $B = [b_{ij}]$. Then $b_{ij} = a_{ij}$ if $i \neq r$ and $b_{rj} = ka_{rj}$. Using Definition 3.2, we obtain $\det(B)$ as

$$\begin{aligned}\det(B) &= \sum(\pm)b_{1j_1}b_{2j_2}\cdots b_{rj_r}\cdots b_{nj_n}\\ &= \sum(\pm)a_{1j_1}a_{2j_2}\cdots(ka_{rj_r})\cdots a_{nj_n}\\ &= k\left(\sum(\pm)a_{1j_1}a_{2j_2}\cdots a_{rj_r}\cdots a_{nj_n}\right) = k\det(A).\end{aligned}$$ ■

EXAMPLE 5

We have $\begin{vmatrix} 2 & 6 \\ 1 & 12 \end{vmatrix} = 2\begin{vmatrix} 1 & 3 \\ 1 & 12 \end{vmatrix} = (2)(3)\begin{vmatrix} 1 & 1 \\ 1 & 4 \end{vmatrix} = 6(4-1) = 18.$ ■

We can use Theorem 3.5 to simplify the computation of $\det(A)$ by factoring out common factors from rows and columns of A.

EXAMPLE 6

We have

$$\begin{vmatrix} 1 & 2 & 3 \\ 1 & 5 & 3 \\ 2 & 8 & 6 \end{vmatrix} = 2\begin{vmatrix} 1 & 2 & 3 \\ 1 & 5 & 3 \\ 1 & 4 & 3 \end{vmatrix} = (2)(3)\begin{vmatrix} 1 & 2 & 1 \\ 1 & 5 & 1 \\ 1 & 4 & 1 \end{vmatrix} = (2)(3)(0) = 0.$$

Here, we first factored out 2 from the third row and 3 from the third column, and then used Theorem 3.3, since the first and third columns are equal. ■

Theorem 3.6 If $B = [b_{ij}]$ is obtained from $A = [a_{ij}]$ by adding to each element of the rth row (column) of A, k times the corresponding element of the sth row (column), $r \neq s$, of A, then $\det(B) = \det(A)$.

Proof

We prove the theorem for rows. We have $b_{ij} = a_{ij}$ for $i \neq r$, and $b_{rj} = a_{rj} + ka_{sj}$, $r \neq s$, say, $r < s$. Then

$$\begin{aligned}\det(B) &= \sum(\pm)b_{1j_1}b_{2j_2}\cdots b_{rj_r}\cdots b_{nj_n}\\ &= \sum(\pm)a_{1j_1}a_{2j_2}\cdots(a_{rj_r} + ka_{sj_r})\cdots a_{sj_s}\cdots a_{nj_n}\\ &= \sum(\pm)a_{1j_1}a_{2j_2}\cdots a_{rj_r}\cdots a_{sj_s}\cdots a_{nj_n}\\ &\quad + \sum(\pm)a_{1j_1}a_{2j_2}\cdots(ka_{sj_r})\cdots a_{sj_s}\cdots a_{nj_n}.\end{aligned}$$

Now the first term in this last expression is $\det(A)$, while the second is

$$k\left[\sum(\pm)a_{1j_1}a_{2j_2}\cdots a_{sj_r}\cdots a_{sj_s}\cdots a_{nj_n}\right].$$

Note that

$$\sum(\pm)a_{1j_1}a_{2j_2}\cdots a_{sj_r}\cdots a_{sj_s}\cdots a_{nj_n}$$

$$= \begin{vmatrix} a_{11} & a_{12} & \cdots & a_{1n} \\ a_{21} & a_{22} & \cdots & a_{2n} \\ \vdots & \vdots & & \vdots \\ a_{s1} & a_{s2} & \cdots & a_{sn} \\ \vdots & \vdots & & \vdots \\ a_{s1} & a_{s2} & \cdots & a_{sn} \\ \vdots & \vdots & & \vdots \\ a_{n1} & a_{n2} & \cdots & a_{nn} \end{vmatrix} \begin{matrix} \\ \\ \\ \leftarrow r\text{th row} \\ \\ \leftarrow s\text{th row} \\ \\ \\ \end{matrix}$$

$$= 0,$$

because this matrix has two equal rows. Hence $\det(B) = \det(A) + 0 = \det(A)$. ■

EXAMPLE 7

We have

$$\begin{vmatrix} 1 & 2 & 3 \\ 2 & -1 & 3 \\ 1 & 0 & 1 \end{vmatrix} = \begin{vmatrix} 5 & 0 & 9 \\ 2 & -1 & 3 \\ 1 & 0 & 1 \end{vmatrix},$$

obtained by adding twice the second row to the first row. By applying the definition of determinant to the first and second determinant, both are seen to have the value 4. ■

Theorem 3.7 If a matrix $A = \left[a_{ij}\right]$ is upper (lower) triangular, then $\det(A) = a_{11}a_{22}\cdots a_{nn}$; that is, the determinant of a triangular matrix is the product of the elements on the main diagonal.

Proof

Let $A = \left[a_{ij}\right]$ be upper triangular (that is, $a_{ij} = 0$ for $i > j$). Then a term $a_{1j_1}a_{2j_2}\cdots a_{nj_n}$ in the expression for $\det(A)$ can be nonzero only for $1 \le j_1$, $2 \le j_2, \ldots, n \le j_n$. Now $j_1 j_2 \ldots j_n$ must be a permutation, or rearrangement, of $\{1, 2, \ldots, n\}$. Hence we must have $j_1 = 1$, $j_2 = 2$, $\ldots$, $j_n = n$. Thus the only term of $\det(A)$ that can be nonzero is the product of the elements on the main diagonal of A. Hence $\det(A) = a_{11}a_{22}\cdots a_{nn}$.

We leave the proof of the lower triangular case to the reader. ■

Recall that in Section 2.1 we introduced compact notation for elementary row and elementary column operations on matrices. In this chapter, we use the same notation for rows and columns of a determinant:

- Interchange rows (columns) i and j:

$$\mathbf{r}_i \leftrightarrow \mathbf{r}_j \quad (\mathbf{c}_i \leftrightarrow \mathbf{c}_j).$$

- Replace row (column) i by k times row (column) i:

$$k\mathbf{r}_i \to \mathbf{r}_i \quad (k\mathbf{c}_i \to \mathbf{c}_i).$$

- Replace row (column) j by k times row (column) i + row (column) j:

$$k\mathbf{r}_i + \mathbf{r}_j \to \mathbf{r}_j \quad (k\mathbf{c}_i + \mathbf{c}_j \to \mathbf{c}_j).$$

Using this notation, it is easy to keep track of the elementary row and column operations performed on a matrix. For example, we indicate that we have interchanged the ith and jth rows of A as $A_{\mathbf{r}_i \leftrightarrow \mathbf{r}_j}$. We proceed similarly for column operations.

We can now interpret Theorems 3.2, 3.5, and 3.6 in terms of this notation as follows:

$$\begin{aligned} \det(A_{\mathbf{r}_i \leftrightarrow \mathbf{r}_j}) &= -\det(A), \quad i \neq j \\ \det(A_{k\mathbf{r}_i \to \mathbf{r}_i}) &= k\det(A) \\ \det(A_{k\mathbf{r}_i + \mathbf{r}_j \to \mathbf{r}_j}) &= \det(A), \quad i \neq j. \end{aligned}$$

It is convenient to rewrite these properties in terms of $\det(A)$:

$$\begin{aligned} \det(A) &= -\det(A_{\mathbf{r}_i \leftrightarrow \mathbf{r}_j}), \quad i \neq j \\ \det(A) &= \frac{1}{k}\det(A_{k\mathbf{r}_i \to \mathbf{r}_i}), \quad k \neq 0 \\ \det(A) &= \det(A_{k\mathbf{r}_i + \mathbf{r}_j \to \mathbf{r}_j}), \quad i \neq j. \end{aligned}$$

Theorems 3.2, 3.5, and 3.6 are useful in evaluating determinants. What we do is transform A by means of our elementary row or column operations to a triangular matrix. Of course, we must keep track of how the determinant of the resulting matrices changes as we perform the elementary row or column operations.

EXAMPLE 8

Let $A = \begin{bmatrix} 4 & 3 & 2 \\ 3 & -2 & 5 \\ 2 & 4 & 6 \end{bmatrix}$. Compute $\det(A)$.

Solution

We have

$$\begin{aligned} \det(A) &= 2\det(A_{\frac{1}{2}\mathbf{r}_3 \to \mathbf{r}_3}) && \text{Multiply row 3 by } \tfrac{1}{2}. \\ &= 2\det\left(\begin{bmatrix} 4 & 3 & 2 \\ 3 & -2 & 5 \\ 1 & 2 & 3 \end{bmatrix}\right) \\ &= (-1)2\det\left(\begin{bmatrix} 4 & 3 & 2 \\ 3 & -2 & 5 \\ 1 & 2 & 3 \end{bmatrix}_{\mathbf{r}_1 \leftrightarrow \mathbf{r}_3}\right) && \text{Interchange rows 1 and 3.} \end{aligned}$$

$$= -2\det\left(\begin{bmatrix} 1 & 2 & 3 \\ 3 & -2 & 5 \\ 4 & 3 & 2 \end{bmatrix}\right)$$

$$= -2\det\left(\begin{bmatrix} 1 & 2 & 3 \\ 3 & -2 & 5 \\ 4 & 3 & 2 \end{bmatrix}_{\substack{-3\mathbf{r}_1+\mathbf{r}_2\to\mathbf{r}_2 \\ -4\mathbf{r}_1+\mathbf{r}_3\to\mathbf{r}_3}}\right)$$ Zero out below the (1, 1) entry.

$$= -2\det\left(\begin{bmatrix} 1 & 2 & 3 \\ 0 & -8 & -4 \\ 0 & -5 & -10 \end{bmatrix}\right)$$

$$= -2\det\left(\begin{bmatrix} 1 & 2 & 3 \\ 0 & -8 & -4 \\ 0 & -5 & -10 \end{bmatrix}_{-\frac{5}{8}\mathbf{r}_2+\mathbf{r}_3\to\mathbf{r}_3}\right)$$ Zero out below the (2, 2) entry.

$$= -2\det\left(\begin{bmatrix} 1 & 2 & 3 \\ 0 & -8 & -4 \\ 0 & 0 & -\frac{30}{4} \end{bmatrix}\right).$$

Next we compute the determinant of the upper triangular matrix.

$$\det(A) = -2(1)(-8)\left(-\frac{30}{4}\right) = -120 \qquad \text{by Theorem 3.7.}$$

The operations chosen are not the most efficient, but we do avoid fractions during the first few steps. ■

Remark The method used to compute a determinant in Example 8 will be referred to as computation **via reduction to triangular form**.

We can now compute the determinant of the identity matrix I_n: $\det(I_n) = 1$. We can also compute the determinants of the elementary matrices discussed in Section 2.3, as follows.

Let E_1 be an elementary matrix of type I; that is, E_1 is obtained from I_n by interchanging, say, the ith and jth rows of I_n. By Theorem 3.2 we have that $\det(E_1) = -\det(I_n) = -1$. Now let E_2 be an elementary matrix of type II; that is, E_2 is obtained from I_n by multiplying, say, the ith row of I_n by $k \neq 0$. By Theorem 3.5 we have that $\det(E_2) = k\det(I_n) = k$. Finally, let E_3 be an elementary matrix of type III; that is, E_3 is obtained from I_n by adding k times the sth row of I_n to the rth row of I_n ($r \neq s$). By Theorem 3.6 we have that $\det(E_3) = \det(I_n) = 1$. Thus the determinant of an elementary matrix is never zero.

Next, we prove that the determinant of a product of two matrices is the product of their determinants and that A is nonsingular if and only if $\det(A) \neq 0$.

Lemma 3.1 If E is an elementary matrix, then $\det(EA) = \det(E)\det(A)$, and $\det(AE) = \det(A)\det(E)$.

Proof

If E is an elementary matrix of type I, then EA is obtained from A by interchanging two rows of A, so $\det(EA) = -\det(A)$. Also $\det(E) = -1$. Thus $\det(EA) = \det(E)\det(A)$.

If E is an elementary matrix of type II, then EA is obtained from A by multiplying a given row of A by $k \neq 0$. Then $\det(EA) = k\det(A)$ and $\det(E) = k$, so $\det(EA) = \det(E)\det(A)$.

Finally, if E is an elementary matrix of type III, then EA is obtained from A by adding a multiple of a row of A to a different row of A. Then $\det(EA) = \det(A)$ and $\det(E) = 1$, so $\det(EA) = \det(E)\det(A)$.

Thus, in all cases, $\det(EA) = \det(E)\det(A)$. By a similar proof, we can show that $\det(AE) = \det(A)\det(E)$. ■

It also follows from Lemma 3.1 that if $B = E_r E_{r-1} \cdots E_2 E_1 A$, then

$$\begin{aligned}\det(B) &= \det(E_r(E_{r-1} \cdots E_2 E_1 A)) \\ &= \det(E_r)\det(E_{r-1}E_{r-2} \cdots E_2 E_1 A) \\ &\ \ \vdots \\ &= \det(E_r)\det(E_{r-1}) \cdots \det(E_2)\det(E_1)\det(A).\end{aligned}$$

Theorem 3.8 If A is an $n \times n$ matrix, then A is nonsingular if and only if $\det(A) \neq 0$.

Proof

If A is nonsingular, then A is a product of elementary matrices (Theorem 2.8). Thus let $A = E_1 E_2 \cdots E_k$. Then

$$\det(A) = \det(E_1 E_2 \cdots E_k) = \det(E_1)\det(E_2) \cdots \det(E_k) \neq 0.$$

If A is singular, then A is row equivalent to a matrix B that has a row of zeros (Theorem 2.10). Then $A = E_1 E_2 \cdots E_r B$, where $E_1, E_2, \ldots, E_r$ are elementary matrices. It then follows by the observation following Lemma 3.1 that

$$\det(A) = \det(E_1 E_2 \cdots E_r B) = \det(E_1)\det(E_2) \cdots \det(E_r)\det(B) = 0,$$

since $\det(B) = 0$. ■

Corollary 3.1 If A is an $n \times n$ matrix, then $A\mathbf{x} = \mathbf{0}$ has a nontrivial solution if and only if $\det(A) = 0$.

Proof

If $\det(A) \neq 0$, then, by Theorem 3.8, A is nonsingular, and thus $A\mathbf{x} = \mathbf{0}$ has only the trivial solution (Theorem 2.9 in Section 2.3 or by the boxed remark preceding Example 14 in Section 1.5).

Conversely, if $\det(A) = 0$, then A is singular (Theorem 3.8). Suppose that A is row equivalent to a matrix B in reduced row echelon form. By Corollary 2.2 in Section 2.3, $B \neq I_n$, and by Exercise 9 in Section 2.1, B has a row of zeros.

The system $B\mathbf{x} = \mathbf{0}$ has the same solutions as the system $A\mathbf{x} = \mathbf{0}$. Let C_1 be the matrix obtained by deleting the zero rows of B. Then the system $B\mathbf{x} = \mathbf{0}$ has the same solutions as the system $C_1\mathbf{x} = \mathbf{0}$. Since the latter is a homogeneous system of at most $n - 1$ equations in n unknowns, it has a nontrivial solution (Theorem 2.4). Hence the given system $A\mathbf{x} = \mathbf{0}$ has a nontrivial solution. ■

EXAMPLE 9 Let A be a 4×4 matrix with $\det(A) = -2$.

(a) Describe the set of all solutions to the homogeneous system $A\mathbf{x} = \mathbf{0}$.

(b) If A is transformed to reduced row echelon form B, what is B?

(c) Give an expression for a solution to the linear system $A\mathbf{x} = \mathbf{b}$, where

$$\mathbf{b} = \begin{bmatrix} 1 \\ 2 \\ 3 \\ 4 \end{bmatrix}.$$

(d) Can the linear system $A\mathbf{x} = \mathbf{b}$ have more than one solution? Explain.

(e) Does A^{-1} exist?

Solution

(a) Since $\det(A) \neq 0$, by Corollary 3.1, the homogeneous system has only the trivial solution.

(b) Since $\det(A) \neq 0$, by Corollary 2.2 in Section 2.3, A is a nonsingular matrix, and by Theorem 2.2, $B = I_n$.

(c) A solution to the given system is given by $\mathbf{x} = A^{-1}\mathbf{b}$.

(d) No. The solution given in part (c) is the only one.

(e) Yes. ■

Theorem 3.9 If A and B are $n \times n$ matrices, then $\det(AB) = \det(A)\det(B)$.

Proof

If A is nonsingular, then A is row equivalent to I_n. Thus $A = E_k E_{k-1} \cdots E_2 E_1 I_n = E_k E_{k-1} \cdots E_2 E_1$, where $E_1, E_2, \ldots, E_k$ are elementary matrices. Then

$$\det(A) = \det(E_k E_{k-1} \cdots E_2 E_1) = \det(E_k)\det(E_{k-1}) \cdots \det(E_2)\det(E_1).$$

Now

$$\begin{aligned}\det(AB) &= \det(E_k E_{k-1} \cdots E_2 E_1 B) \\ &= \det(E_k)\det(E_{k-1}) \cdots \det(E_2)\det(E_1)\det(B) \\ &= \det(A)\det(B).\end{aligned}$$

If A is singular, then $\det(A) = 0$ by Theorem 3.8. Moreover, if A is singular, then A is row equivalent to a matrix C that has a row consisting entirely of zeros (Theorem 2.10). Thus $C = E_k E_{k-1} \cdots E_2 E_1 A$, so

$$CB = E_k E_{k-1} \cdots E_2 E_1 AB.$$

This means that AB is row equivalent to CB, and since CB has a row consisting entirely of zeros, it follows that AB is singular. Hence $\det(AB) = 0$, and in this case we also have $\det(AB) = \det(A)\det(B)$. ■

EXAMPLE 10 Let

$$A = \begin{bmatrix} 1 & 2 \\ 3 & 4 \end{bmatrix} \quad \text{and} \quad B = \begin{bmatrix} 2 & -1 \\ 1 & 2 \end{bmatrix}.$$

Then

$$|A| = -2 \quad \text{and} \quad |B| = 5.$$

On the other hand, $AB = \begin{bmatrix} 4 & 3 \\ 10 & 5 \end{bmatrix}$, and $|AB| = -10 = |A||B|$. ■

Corollary 3.2 If A is nonsingular, then $\det(A^{-1}) = \dfrac{1}{\det(A)}$.

Proof

Exercise 18. ■

Corollary 3.3 If A and B are similar matrices, then $\det(A) = \det(B)$.

Proof

Exercise 33. ■

The determinant of a sum of two $n \times n$ matrices A and B is, in general, not the sum of the determinants of A and B. The best result we can give along these lines is that if A, B, and C are $n \times n$ matrices all of whose entries are equal except for the kth row (column), and the kth row (column) of C is the sum of the kth rows (columns) of A and B, then $\det(C) = \det(A) + \det(B)$. We shall not prove this result, but will consider an example.

EXAMPLE 11 Let

$$A = \begin{bmatrix} 2 & 2 & 3 \\ 0 & 3 & 4 \\ 0 & 2 & 4 \end{bmatrix}, \quad B = \begin{bmatrix} 2 & 2 & 3 \\ 0 & 3 & 4 \\ 1 & -2 & -4 \end{bmatrix},$$

and

$$C = \begin{bmatrix} 2 & 2 & 3 \\ 0 & 3 & 4 \\ 1 & 0 & 0 \end{bmatrix}.$$

Then $|A| = 8$, $|B| = -9$, and $|C| = -1$, so $|C| = |A| + |B|$. ■

Key Terms

Properties of the determinant
Elementary matrix
Reduction to triangular form

3.2 Exercises

1. Compute the following determinants via reduction to triangular form or by citing a particular theorem or corollary:

(a) $\begin{vmatrix} 3 & 0 \\ 2 & 1 \end{vmatrix}$

(b) $\begin{vmatrix} 2 & 1 \\ 4 & 3 \end{vmatrix}$

(c) $\begin{vmatrix} 4 & 0 & 0 \\ 0 & 2 & 0 \\ 0 & 0 & 3 \end{vmatrix}$

(d) $\begin{vmatrix} 4 & 1 & 3 \\ 2 & 3 & 0 \\ 1 & 3 & 2 \end{vmatrix}$

(e) $\begin{vmatrix} 4 & 2 & 2 & 0 \\ 2 & 0 & 0 & 0 \\ 3 & 0 & 0 & 1 \\ 0 & 0 & 1 & 0 \end{vmatrix}$ (f) $\begin{vmatrix} 4 & 2 & 3 & -4 \\ 3 & -2 & 1 & 5 \\ -2 & 0 & 1 & -3 \\ 8 & -2 & 6 & 4 \end{vmatrix}$

2. Compute the following determinants via reduction to triangular form or by citing a particular theorem or corollary:

(a) $\begin{vmatrix} 2 & -2 \\ 3 & -1 \end{vmatrix}$ (b) $\begin{vmatrix} 4 & 2 & 0 \\ 0 & -2 & 5 \\ 0 & 0 & 3 \end{vmatrix}$

(c) $\begin{vmatrix} 3 & 4 & 2 \\ 2 & 5 & 0 \\ 3 & 0 & 0 \end{vmatrix}$ (d) $\begin{vmatrix} 4 & -3 & 5 \\ 5 & 2 & 0 \\ 2 & 0 & 4 \end{vmatrix}$

(e) $\begin{vmatrix} 4 & 0 & 0 & 0 \\ -1 & 2 & 0 & 0 \\ 1 & 2 & -3 & 0 \\ 1 & 5 & 3 & 5 \end{vmatrix}$

(f) $\begin{vmatrix} 2 & 0 & 1 & 4 \\ 3 & 2 & -4 & -2 \\ 2 & 3 & -1 & 0 \\ 11 & 8 & -4 & 6 \end{vmatrix}$

3. If $\begin{vmatrix} a_1 & a_2 & a_3 \\ b_1 & b_2 & b_3 \\ c_1 & c_2 & c_3 \end{vmatrix} = 3$, find

$$\begin{vmatrix} a_1+2b_1-3c_1 & a_2+2b_2-3c_2 & a_3+2b_3-3c_3 \\ b_1 & b_2 & b_3 \\ c_1 & c_2 & c_3 \end{vmatrix}.$$

4. If $\begin{vmatrix} a_1 & a_2 & a_3 \\ b_1 & b_2 & b_3 \\ c_1 & c_2 & c_3 \end{vmatrix} = -2$, find

$$\begin{vmatrix} a_1-\frac{1}{2}a_3 & a_2 & a_3 \\ b_1-\frac{1}{2}b_3 & b_2 & b_3 \\ c_1-\frac{1}{2}c_3 & c_2 & c_3 \end{vmatrix}.$$

5. If $\begin{vmatrix} a_1 & a_2 & a_3 \\ b_1 & b_2 & b_3 \\ c_1 & c_2 & c_3 \end{vmatrix} = 4$, find

$$\begin{vmatrix} a_1 & a_2 & 4a_3-2a_2 \\ b_1 & b_2 & 4b_3-2b_2 \\ \frac{1}{2}c_1 & \frac{1}{2}c_2 & 2c_3-c_2 \end{vmatrix}.$$

6. Verify that $\det(AB) = \det(A)\det(B)$ for the following:

(a) $A = \begin{bmatrix} 1 & -2 & 3 \\ -2 & 3 & 1 \\ 0 & 1 & 0 \end{bmatrix}$, $B = \begin{bmatrix} 1 & 0 & 2 \\ 3 & -2 & 5 \\ 2 & 1 & 3 \end{bmatrix}$

(b) $A = \begin{bmatrix} 2 & 3 & 6 \\ 0 & 3 & 2 \\ 0 & 0 & -4 \end{bmatrix}$, $B = \begin{bmatrix} 3 & 0 & 0 \\ 4 & 5 & 0 \\ 2 & 1 & -2 \end{bmatrix}$

7. Evaluate:

(a) $\begin{vmatrix} -4 & 2 & 0 & 0 \\ 2 & 3 & 1 & 0 \\ 3 & 1 & 0 & 2 \\ 1 & 3 & 0 & 3 \end{vmatrix}$

(b) $\begin{vmatrix} 2 & 0 & 0 & 0 \\ -5 & 3 & 0 & 0 \\ 3 & 2 & 4 & 0 \\ 4 & 2 & 1 & -5 \end{vmatrix}$

(c) $\begin{vmatrix} t-1 & -1 & -2 \\ 0 & t-2 & 2 \\ 0 & 0 & t-3 \end{vmatrix}$

(d) $\begin{vmatrix} t+1 & 4 \\ 2 & t-3 \end{vmatrix}$

8. Is $\det(AB) = \det(BA)$? Justify your answer.

9. If $\det(AB) = 0$, is $\det(A) = 0$ or $\det(B) = 0$? Give reasons for your answer.

10. Show that if k is a scalar and A is $n \times n$, then $\det(kA) = k^n \det(A)$.

11. Show that if A is $n \times n$ with n odd and skew symmetric, then $\det(A) = 0$.

12. Show that if A is a matrix such that in each row and in each column one and only one element is not equal to 0, then $\det(A) \neq 0$.

13. Show that $\det(AB^{-1}) = \dfrac{\det(A)}{\det(B)}$.

14. Show that if $AB = I_n$, then $\det(A) \neq 0$ and $\det(B) \neq 0$.

15. **(a)** Show that if $A = A^{-1}$, then $\det(A) = \pm 1$.

(b) If $A^T = A^{-1}$, what is $\det(A)$?

16. Show that if A and B are square matrices, then

$$\det\left(\begin{bmatrix} A & O \\ O & B \end{bmatrix}\right) = (\det A)(\det B).$$

17. If A is a nonsingular matrix such that $A^2 = A$, what is $\det(A)$?

18. Prove Corollary 3.2.

19. Show that if A, B, and C are square matrices, then

$$\det\left(\begin{bmatrix} A & O \\ C & B \end{bmatrix}\right) = (\det A)(\det B).$$

20. Show that if A and B are both $n \times n$, then

(a) $\det(A^T B^T) = \det(A)\det(B^T)$;

(b) $\det(A^T B^T) = \det(A^T)\det(B)$.

21. Verify the result in Exercise 16 for $A = \begin{bmatrix} 1 & 2 \\ 3 & 4 \end{bmatrix}$ and $B = \begin{bmatrix} 2 & 1 \\ -3 & 2 \end{bmatrix}$.

22. Use the properties of Section 3.2 to prove that

$$\begin{vmatrix} 1 & a & a^2 \\ 1 & b & b^2 \\ 1 & c & c^2 \end{vmatrix} = (b-a)(c-a)(c-b).$$

(*Hint*: Use factorization.) This determinant is called a **Vandermonde* determinant**.

23. If $\det(A) = 2$, find $\det(A^5)$.

24. Use Theorem 3.8 to determine which of the following matrices are nonsingular:

(a) $\begin{bmatrix} 1 & 2 & 3 \\ 0 & 1 & 2 \\ 2 & -3 & 1 \end{bmatrix}$ **(b)** $\begin{bmatrix} 1 & 2 \\ 3 & 4 \end{bmatrix}$

25. Use Theorem 3.8 to determine which of the following matrices are nonsingular:

(a) $\begin{bmatrix} 1 & 3 & 2 \\ 2 & 1 & 4 \\ 1 & -7 & 2 \end{bmatrix}$

(b) $\begin{bmatrix} 1 & 2 & 0 & 5 \\ 3 & 4 & 1 & 7 \\ -2 & 5 & 2 & 0 \\ 0 & 1 & 2 & -7 \end{bmatrix}$

26. Use Theorem 3.8 to determine all values of t so that the following matrices are nonsingular:

(a) $\begin{bmatrix} t & 1 & 2 \\ 3 & 4 & 5 \\ 6 & 7 & 8 \end{bmatrix}$ **(b)** $\begin{bmatrix} t & 1 & 2 \\ 0 & 1 & 1 \\ 1 & 0 & t \end{bmatrix}$

(c) $\begin{bmatrix} t & 0 & 0 & 1 \\ 0 & t & 0 & 0 \\ 0 & 0 & t & 0 \\ 1 & 0 & 0 & t \end{bmatrix}$

27. Use Corollary 3.1 to find out whether the following homogeneous system has a nontrivial solution (do *not* solve):

$$\begin{aligned} x_1 - 2x_2 + x_3 &= 0 \\ 2x_1 + 3x_2 + x_3 &= 0 \\ 3x_1 + x_2 + 2x_3 &= 0 \end{aligned}$$

28. Repeat Exercise 27 for the following homogeneous system:

$$\begin{bmatrix} 1 & 2 & 0 & 1 \\ 0 & 1 & 2 & 3 \\ 0 & 0 & 1 & 2 \\ 0 & 1 & 2 & -1 \end{bmatrix} \begin{bmatrix} x_1 \\ x_2 \\ x_3 \\ x_4 \end{bmatrix} = \begin{bmatrix} 0 \\ 0 \\ 0 \\ 0 \end{bmatrix}.$$

29. Let $A = \begin{bmatrix} a_{ij} \end{bmatrix}$ be an upper triangular matrix. Prove that A is nonsingular if and only if $a_{ii} \neq 0$ for $i = 1, 2, \ldots, n$.

30. Let A be a 3×3 matrix with $\det(A) = 3$.

(a) What is the reduced row echelon form to which A is row equivalent?

(b) How many solutions does the homogeneous system $A\mathbf{x} = \mathbf{0}$ have?

31. Let A be a 4×4 matrix with $\det(A) = 0$.

(a) Describe the reduced row echelon form matrix to which A is row equivalent.

(b) How many solutions does the homogeneous system $A\mathbf{x} = \mathbf{0}$ have?

32. Let $A^2 = A$. Prove that either A is singular or $\det(A) = 1$.

33. Prove Corollary 3.3.

34. Let $AB = AC$. Prove that if $\det(A) \neq 0$, then $B = C$.

35. Determine whether the software you are using has a command for computing the determinant of a matrix. If it does, verify the computations in Examples 8, 10, and 11. Experiment further by finding the determinant of the matrices in Exercises 1 and 2.

36. Assuming that your software has a command to compute the determinant of a matrix, read the accompanying software documentation to determine the method used. Is the description closest to that in Section 3.1, Example 8 in Section 3.2, or the material in Section 2.5?

---

*Alexandre-Théophile Vandermonde (1735–1796) was born and died in Paris. His father was a physician who encouraged his son to pursue a career in music. Vandermonde followed his father's advice and did not get interested in mathematics until he was 35 years old. His entire mathematical output consisted of four papers. He also published papers on chemistry and on the manufacture of steel. Although Vandermonde is best known for his determinant, it does not appear in any of his four papers. It is believed that someone mistakenly attributed this determinant to him. However, in his fourth mathematical paper, Vandermonde made significant contributions to the theory of determinants. He was a staunch republican who fully backed the French Revolution.

37. **Warning:** Theorem 3.8 assumes that all calculations for $\det(A)$ are done by exact arithmetic. As noted previously, this is usually not the case in software. Hence, computationally, the determinant may not be a valid test for nonsingularity. Perform the following experiment: Let $A = \begin{bmatrix} 1 & 2 & 3 \\ 4 & 5 & 6 \\ 7 & 8 & 9 \end{bmatrix}$. Show that $\det(A)$ is 0, either by hand or by using your software. Next, show by hand computation that $\det(B) = -3\epsilon$, where $B = \begin{bmatrix} 1 & 2 & 3 \\ 4 & 5 & 6 \\ 7 & 8 & 9+\epsilon \end{bmatrix}$. Hence, theoretically, for any $\epsilon \neq 0$, matrix B is nonsingular. Let your software compute $\det(B)$ for $\epsilon = \pm 10^{-k}$, $k = 5, 6, \ldots, 20$. Do the computational results match the theoretical result? If not, formulate a conjecture to explain why not.

3.3 Cofactor Expansion

Thus far we have evaluated determinants by using Definition 3.2 and the properties established in Section 3.2. We now develop a method for evaluating the determinant of an $n \times n$ matrix that reduces the problem to the evaluation of determinants of matrices of order $n-1$. We can then repeat the process for these $(n-1) \times (n-1)$ matrices until we get to 2×2 matrices.

DEFINITION 3.3

Let $A = \left[a_{ij}\right]$ be an $n \times n$ matrix. Let M_{ij} be the $(n-1) \times (n-1)$ submatrix of A obtained by deleting the ith row and jth column of A. The determinant $\det(M_{ij})$ is called the **minor** of a_{ij}.

DEFINITION 3.4

Let $A = \left[a_{ij}\right]$ be an $n \times n$ matrix. The **cofactor** A_{ij} of a_{ij} is defined as $A_{ij} = (-1)^{i+j}\det(M_{ij})$.

EXAMPLE 1

Let

$$A = \begin{bmatrix} 3 & -1 & 2 \\ 4 & 5 & 6 \\ 7 & 1 & 2 \end{bmatrix}.$$

Then

$$\det(M_{12}) = \begin{vmatrix} 4 & 6 \\ 7 & 2 \end{vmatrix} = 8 - 42 = -34, \quad \det(M_{23}) = \begin{vmatrix} 3 & -1 \\ 7 & 1 \end{vmatrix} = 3 + 7 = 10,$$

and

$$\det(M_{31}) = \begin{vmatrix} -1 & 2 \\ 5 & 6 \end{vmatrix} = -6 - 10 = -16.$$

Also,

$$A_{12} = (-1)^{1+2}\det(M_{12}) = (-1)(-34) = 34,$$
$$A_{23} = (-1)^{2+3}\det(M_{23}) = (-1)(10) = -10,$$

and

$$A_{31} = (-1)^{3+1}\det(M_{31}) = (1)(-16) = -16. \quad \blacksquare$$

If we think of the sign $(-1)^{i+j}$ as being located in position (i, j) of an $n \times n$ matrix, then the signs form a checkerboard pattern that has a $+$ in the $(1, 1)$ position. The patterns for $n = 3$ and $n = 4$ are as follows:

$$\begin{bmatrix} + & - & + \\ - & + & - \\ + & - & + \end{bmatrix} \qquad \begin{bmatrix} + & - & + & - \\ - & + & - & + \\ + & - & + & - \\ - & + & - & + \end{bmatrix}$$

$$n = 3 \qquad\qquad n = 4$$

Theorem 3.10 Let $A = \left[a_{ij}\right]$ be an $n \times n$ matrix. Then

$$\det(A) = a_{i1}A_{i1} + a_{i2}A_{i2} + \cdots + a_{in}A_{in}$$
[expansion of $\det(A)$ along the ith row]

and

$$\det(A) = a_{1j}A_{1j} + a_{2j}A_{2j} + \cdots + a_{nj}A_{nj}$$
[expansion of $\det(A)$ along the jth column].

Proof

The first formula follows from the second by Theorem 3.1, that is, from the fact that $\det(A^T) = \det(A)$. We omit the general proof and consider the 3×3 matrix $A = \left[a_{ij}\right]$. From (1) in Section 3.1,

$$\begin{aligned} \det(A) &= a_{11}a_{22}a_{33} + a_{12}a_{23}a_{31} + a_{13}a_{21}a_{32} \\ &\quad - a_{11}a_{23}a_{32} - a_{12}a_{21}a_{33} - a_{13}a_{22}a_{31}. \end{aligned} \tag{1}$$

We can write this expression as

$$\begin{aligned} \det(A) &= a_{11}(a_{22}a_{33} - a_{23}a_{32}) + a_{12}(a_{23}a_{31} - a_{21}a_{33}) \\ &\quad + a_{13}(a_{21}a_{32} - a_{22}a_{31}). \end{aligned}$$

Now,

$$A_{11} = (-1)^{1+1}\begin{vmatrix} a_{22} & a_{23} \\ a_{32} & a_{33} \end{vmatrix} = (a_{22}a_{33} - a_{23}a_{32}),$$

$$A_{12} = (-1)^{1+2}\begin{vmatrix} a_{21} & a_{23} \\ a_{31} & a_{33} \end{vmatrix} = (a_{23}a_{31} - a_{21}a_{33}),$$

$$A_{13} = (-1)^{1+3}\begin{vmatrix} a_{21} & a_{22} \\ a_{31} & a_{32} \end{vmatrix} = (a_{21}a_{32} - a_{22}a_{31}).$$

Hence

$$\det(A) = a_{11}A_{11} + a_{12}A_{12} + a_{13}A_{13},$$

which is the expansion of $\det(A)$ along the first row.

If we now write (1) as

$$\det(A) = a_{13}(a_{21}a_{32} - a_{22}a_{31}) + a_{23}(a_{12}a_{31} - a_{11}a_{32}) + a_{33}(a_{11}a_{22} - a_{12}a_{21}),$$

we can verify that

$$\det(A) = a_{13}A_{13} + a_{23}A_{23} + a_{33}A_{33},$$

which is the expansion of $\det(A)$ along the third column. ■

EXAMPLE 2

To evaluate the determinant

$$\begin{vmatrix} 1 & 2 & -3 & 4 \\ -4 & 2 & 1 & 3 \\ 3 & 0 & 0 & -3 \\ 2 & 0 & -2 & 3 \end{vmatrix},$$

it is best to expand along either the second column or the third row because they each have two zeros. Obviously, the optimal course of action is to expand along the row or column that has the largest number of zeros, because in that case the cofactors A_{ij} of those a_{ij} which are zero do not have to be evaluated, since $a_{ij}A_{ij} = (0)(A_{ij}) = 0$. Thus, expanding along the third row, we have

$$\begin{vmatrix} 1 & 2 & -3 & 4 \\ -4 & 2 & 1 & 3 \\ 3 & 0 & 0 & -3 \\ 2 & 0 & -2 & 3 \end{vmatrix}$$

$$= (-1)^{3+1}(3)\begin{vmatrix} 2 & -3 & 4 \\ 2 & 1 & 3 \\ 0 & -2 & 3 \end{vmatrix} + (-1)^{3+2}(0)\begin{vmatrix} 1 & -3 & 4 \\ -4 & 1 & 3 \\ 2 & -2 & 3 \end{vmatrix}$$

$$+ (-1)^{3+3}(0)\begin{vmatrix} 1 & 2 & 4 \\ -4 & 2 & 3 \\ 2 & 0 & 3 \end{vmatrix} + (-1)^{3+4}(-3)\begin{vmatrix} 1 & 2 & -3 \\ -4 & 2 & 1 \\ 2 & 0 & -2 \end{vmatrix}$$

$$= (+1)(3)(20) + 0 + 0 + (-1)(-3)(-4) = 48.$$ ■

We can use the properties of Section 3.2 to introduce many zeros in a given row or column and then expand along that row or column. Consider the following example:

EXAMPLE 3 We have

$$\begin{vmatrix} 1 & 2 & -3 & 4 \\ -4 & 2 & 1 & 3 \\ 1 & 0 & 0 & -3 \\ 2 & 0 & -2 & 3 \end{vmatrix}_{\mathbf{c}_4+3\mathbf{c}_1\to\mathbf{c}_4} = \begin{vmatrix} 1 & 2 & -3 & 7 \\ -4 & 2 & 1 & -9 \\ 1 & 0 & 0 & 0 \\ 2 & 0 & -2 & 9 \end{vmatrix}$$

$$= (-1)^{3+1}(1)\begin{vmatrix} 2 & -3 & 7 \\ 2 & 1 & -9 \\ 0 & -2 & 9 \end{vmatrix}_{\mathbf{r}_1-\mathbf{r}_2\to\mathbf{r}_1}$$

$$= (-1)^{4}(1)\begin{vmatrix} 0 & -4 & 16 \\ 2 & 1 & -9 \\ 0 & -2 & 9 \end{vmatrix}$$

$$= (-1)^4(1)(8) = 8.$$ ■

■ Application to Computing Areas

Consider the triangle with vertices (x_1, y_1), (x_2, y_2), and (x_3, y_3), which we show in Figure 3.2.

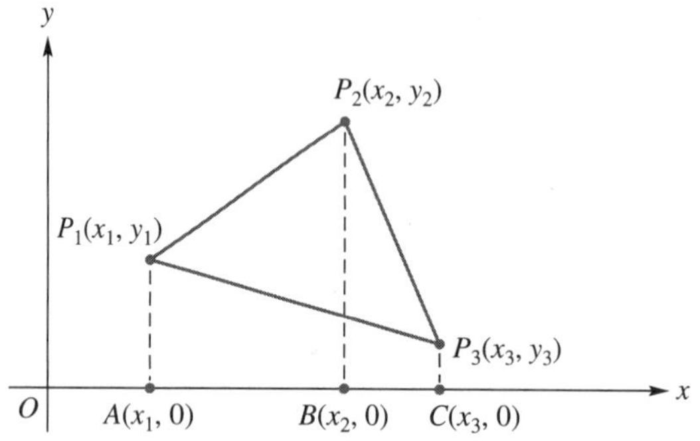

FIGURE 3.2

We may compute the area of this triangle as

$$\text{area of trapezoid } AP_1P_2B + \text{area of trapezoid } BP_2P_3C \\ - \text{area of trapezoid } AP_1P_3C.$$

Now recall that the area of a trapezoid is $\frac{1}{2}$ the distance between the parallel sides of the trapezoid times the sum of the lengths of the parallel sides. Thus

area of triangle $P_1P_2P_3$

$$= \frac{1}{2}(x_2 - x_1)(y_1 + y_2) + \frac{1}{2}(x_3 - x_2)(y_2 + y_3) - \frac{1}{2}(x_3 - x_1)(y_1 + y_3)$$

$$= \frac{1}{2}x_2y_1 - \frac{1}{2}x_1y_2 + \frac{1}{2}x_3y_2 - \frac{1}{2}x_2y_3 - \frac{1}{2}x_3y_1 + \frac{1}{2}x_1y_3$$

$$= -\frac{1}{2}[(x_2y_3 - x_3y_2) - (x_1y_3 - x_3y_1) + (x_1y_2 - x_2y_1)]. \tag{2}$$

The expression in brackets in (2) is the cofactor expansion about the third column of the matrix

$$\begin{bmatrix} x_1 & y_1 & 1 \\ x_2 & y_2 & 1 \\ x_3 & y_3 & 1 \end{bmatrix}.$$

This determinant may be positive or negative, depending upon the location of the points and how they are labeled. Thus, for a triangle with vertices (x_1, y_1), (x_2, y_2), and (x_3, y_3), we have

$$\text{area of triangle} = \frac{1}{2}\left|\det\left(\begin{bmatrix} x_1 & y_1 & 1 \\ x_2 & y_2 & 1 \\ x_3 & y_3 & 1 \end{bmatrix}\right)\right|. \tag{3}$$

(The area is $\frac{1}{2}$ the absolute value of the determinant.)

EXAMPLE 4

Compute the area of the triangle T, shown in Figure 3.3, with vertices $(-1, 4)$, $(3, 1)$, and $(2, 6)$.

Solution

By Equation (3), the area of T is

$$\frac{1}{2}\left|\det\left(\begin{bmatrix} -1 & 4 & 1 \\ 3 & 1 & 1 \\ 2 & 6 & 1 \end{bmatrix}\right)\right| = \frac{1}{2}|17| = 8.5.$$

■

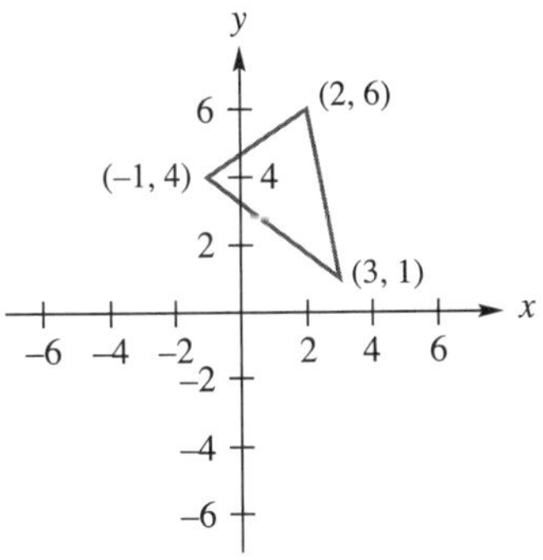

FIGURE 3.3

Suppose we now have the parallelogram shown in Figure 3.4. Since a diagonal divides the parallelogram into two equal triangles, it follows from Equation (3) that

$$\text{area of parallelogram} = \left|\det\left(\begin{bmatrix} x_1 & y_1 & 1 \\ x_2 & y_2 & 1 \\ x_3 & y_3 & 1 \end{bmatrix}\right)\right|. \tag{4}$$

The determinant in (4) can also be evaluated as follows:

$$\det\left(\begin{bmatrix} x_1 & y_1 & 1 \\ x_2 & y_2 & 1 \\ x_3 & y_3 & 1 \end{bmatrix}\right) = \det\left(\begin{bmatrix} x_1 & y_1 & 1 \\ x_2 - x_1 & y_2 - y_1 & 0 \\ x_3 - x_1 & y_3 - y_1 & 0 \end{bmatrix}\right) \qquad \begin{matrix} \mathbf{r}_1 - \mathbf{r}_2 \to \mathbf{r}_2 \\ \mathbf{r}_1 - \mathbf{r}_3 \to \mathbf{r}_3 \end{matrix}$$

$$= \det\left(\begin{bmatrix} x_2 - x_1 & y_2 - y_1 \\ x_3 - x_1 & y_3 - y_1 \end{bmatrix}\right) \qquad \text{Expansion by cofactors about the third column}$$

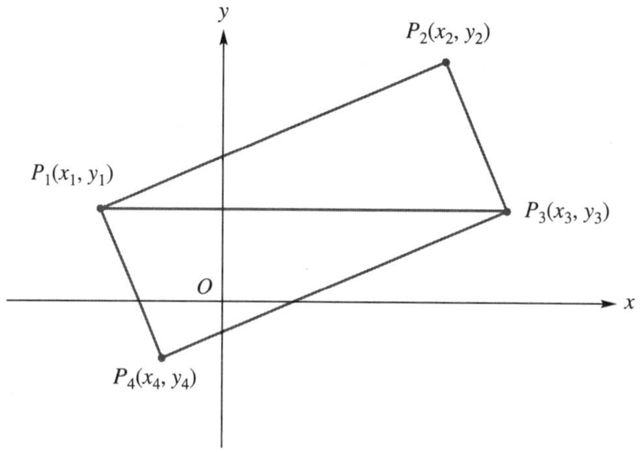

FIGURE 3.4

Hence the absolute value of the determinant of a 2×2 matrix represents the area of a parallelogram.

An arbitrary 2×2 matrix defines a matrix transformation $f\colon R^2 \to R^2$ by

$$f(\mathbf{v}) = A\mathbf{v}$$

for a vector $\mathbf{v}$ in R^2. How does the area of the image of a closed figure, such as a triangle T that is obtained by applying f to T, compare with the area of T? Thus, suppose that T is the triangle with vertices (x_1, y_1), (x_2, y_2), and (x_3, y_3), and let

$$A = \begin{bmatrix} a & b \\ c & d \end{bmatrix}.$$

In Exercise 19, we ask you to first compute the coordinates of the image $f(T)$ and then show that

$$\text{area of } f(T) = |\det(A)| \cdot \text{area of } T.$$

EXAMPLE 5

Compute the area of the parallelogram P with vertices $(-1, 4)$, $(3, 1)$, $(2, 6)$, and $(6, 3)$, shown in Figure 3.5.

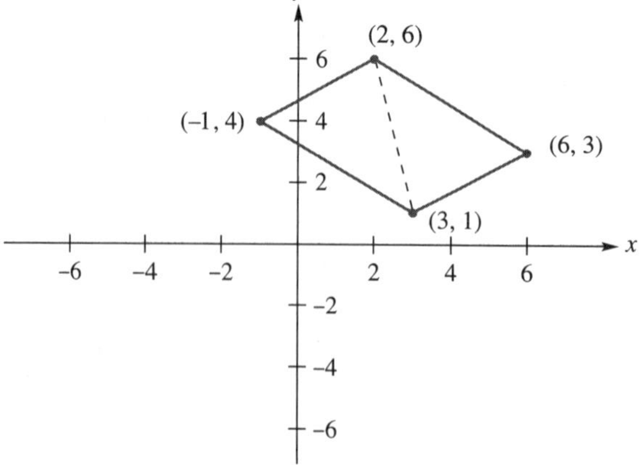

FIGURE 3.5

Solution

The dashed diagonal in Figure 3.5 divides the parallelogram into two equal triangles, whose area has been computed in Example 4. Hence, the area of the parallelogram is 17. ■

EXAMPLE 6 Consider the triangle T defined in Example 4 and let

$$A = \begin{bmatrix} 6 & -3 \\ -1 & 1 \end{bmatrix}.$$

The image of T, using the matrix transformation defined by the matrix A, is the triangle with vertices $(-18, 5)$, $(15, -2)$, and $(-6, 4)$ (verify). See Figure 3.6. The area of this triangle is, by Equation (3),

$$\frac{1}{2}\left|\det\left(\begin{bmatrix} -18 & 5 & 1 \\ 15 & -2 & 1 \\ -6 & 4 & 1 \end{bmatrix}\right)\right| = \frac{1}{2}|51| = 25.5.$$

Since $\det(A) = 3$, we see that $3 \times$ area of triangle T = area of the image. ■

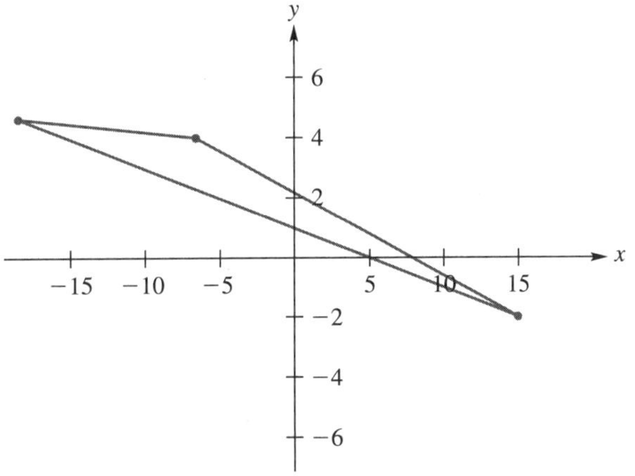

FIGURE 3.6

Remark The result discussed in Example 6 is true in general; that is, if S is a closed figure in 2-space (3-space), A is a matrix of appropriate dimension, and f is the matrix transformation defined by A, then the area of the image $= |\det(A)| \cdot$ area of S (volume of the image $= |\det(A)| \cdot$ volume of S).

Key Terms

Minor	Expansion along a row or column
Cofactor	Area of a triangle

3.3 Exercises

1. Let $A = \begin{bmatrix} 1 & 0 & -2 \\ 3 & 1 & 4 \\ 5 & 2 & -3 \end{bmatrix}$. Find the following minors:

(a) $\det(M_{13})$ **(b)** $\det(M_{22})$

(c) $\det(M_{31})$ **(d)** $\det(M_{32})$

2. Let $A = \begin{bmatrix} 2 & -1 & 0 & 3 \\ 1 & 2 & -2 & 4 \\ -1 & 1 & -3 & -2 \\ 0 & 2 & -1 & 5 \end{bmatrix}$. Find the following minors:

(a) $\det(M_{12})$ **(b)** $\det(M_{23})$

(c) $\det(M_{33})$ **(d)** $\det(M_{41})$

3. Let $A = \begin{bmatrix} -1 & 2 & 3 \\ -2 & 5 & 4 \\ 0 & 1 & -3 \end{bmatrix}$. Find the following cofactors:

(a) A_{13} **(b)** A_{21} **(c)** A_{32} **(d)** A_{33}

4. Let $A = \begin{bmatrix} 1 & 0 & 3 & 0 \\ 2 & 1 & -4 & -1 \\ 3 & 2 & 4 & 0 \\ 0 & 3 & -1 & 0 \end{bmatrix}$. Find the following cofactors:

(a) A_{12} **(b)** A_{23} **(c)** A_{33} **(d)** A_{41}

5. Use Theorem 3.10 to evaluate the determinants in Exercise 1(a), (d), and (e) of Section 3.2.

6. Use Theorem 3.10 to evaluate the determinants in Exercise 1(b), (c), and (f) of Section 3.2.

7. Use Theorem 3.10 to evaluate the determinants in Exercise 2(a), (c), and (f) of Section 3.2.

8. Use Theorem 3.10 to evaluate the determinants in Exercise 2(b), (d), and (e) of Section 3.2.

9. Show by a column (row) expansion that if $A = \begin{bmatrix} a_{ij} \end{bmatrix}$ is upper (lower) triangular, then $\det(A) = a_{11}a_{22}\cdots a_{nn}$.

10. If $A = \begin{bmatrix} a_{ij} \end{bmatrix}$ is a 3×3 matrix, develop the general expression for $\det(A)$ by expanding

(a) along the second column;

(b) along the third row.

Compare these answers with those obtained for Example 8 in Section 3.1.

11. Find all values of t for which

(a) $\begin{vmatrix} t-2 & 2 \\ 3 & t-3 \end{vmatrix} = 0;$

(b) $\begin{vmatrix} t-1 & -4 \\ 0 & t-4 \end{vmatrix} = 0.$

12. Find all values of t for which

$$\begin{vmatrix} t-1 & 0 & 1 \\ -2 & t+2 & -1 \\ 0 & 0 & t+1 \end{vmatrix} = 0.$$

13. Let A be an $n \times n$ matrix.

(a) Show that $f(t) = \det(tI_n - A)$ is a polynomial in t of degree n.

(b) What is the coefficient of t^n in $f(t)$?

(c) What is the constant term in $f(t)$?

14. Verify your answers to Exercise 13 with the following matrices:

(a) $\begin{bmatrix} 1 & 2 \\ 3 & 4 \end{bmatrix}$ **(b)** $\begin{bmatrix} 1 & 3 & 2 \\ 2 & -1 & 3 \\ 3 & 0 & 1 \end{bmatrix}$

(c) $\begin{bmatrix} 1 & 1 \\ 1 & 1 \end{bmatrix}$

15. Let T be the triangle with vertices $(3, 3)$, $(-1, -1)$, $(4, 1)$.

(a) Find the area of the triangle T.

(b) Find the coordinates of the vertices of the image of T under the matrix transformation with matrix representation

$$A = \begin{bmatrix} 4 & -3 \\ -4 & 2 \end{bmatrix}.$$

(c) Find the area of the triangle whose vertices are obtained in part (b).

16. Find the area of the parallelogram with vertices $(2, 3)$, $(5, 3)$, $(4, 5)$, $(7, 5)$.

17. Let Q be the quadrilateral with vertices $(-2, 3)$, $(1, 4)$, $(3, 0)$, and $(-1, -3)$. Find the area of Q.

18. Prove that a rotation leaves the area of a triangle unchanged.

19. Let T be the triangle with vertices (x_1, y_1), (x_2, y_2), and (x_3, y_3), and let

$$A = \begin{bmatrix} a & b \\ c & d \end{bmatrix}.$$

Let f be the matrix transformation defined by $f(\mathbf{v}) = A\mathbf{v}$ for a vector $\mathbf{v}$ in R^2. First, compute the vertices of $f(T)$ and the image of T under f, and then show that the area of $f(T)$ is $|\det(A)| \cdot$ area of T.

3.4 Inverse of a Matrix

We saw in Section 3.3 that Theorem 3.10 provides formulas for expanding $\det(A)$ along either a row or a column of A. Thus $\det(A) = a_{i1}A_{i1}+a_{i2}A_{i2}+\cdots+a_{in}A_{in}$ is the expansion of $\det(A)$ along the ith row. It is interesting to ask what $a_{i1}A_{k1} + a_{i2}A_{k2} + \cdots + a_{in}A_{kn}$ is for $i \neq k$, because as soon as we answer this question, we obtain another method for finding the inverse of a nonsingular matrix.

Theorem 3.11 If $A = \left[a_{ij}\right]$ is an $n \times n$ matrix, then

$$\begin{aligned} a_{i1}A_{k1} + a_{i2}A_{k2} + \cdots + a_{in}A_{kn} &= 0 \quad \text{for } i \neq k; \\ a_{1j}A_{1k} + a_{2j}A_{2k} + \cdots + a_{nj}A_{nk} &= 0 \quad \text{for } j \neq k. \end{aligned}$$

Proof

We prove only the first formula. The second follows from the first one by Theorem 3.1.

Consider the matrix B obtained from A by replacing the kth row of A by its ith row. Thus B is a matrix having two identical rows—the ith and kth—so $\det(B) = 0$. Now expand $\det(B)$ along the kth row. The elements of the kth row of B are $a_{i1}, a_{i2}, \ldots, a_{in}$. The cofactors of the kth row are $A_{k1}, A_{k2}, \ldots, A_{kn}$. Thus

$$0 = \det(B) = a_{i1}A_{k1} + a_{i2}A_{k2} + \cdots + a_{in}A_{kn},$$

which is what we wanted to show. ■

This theorem says that if we sum the products of the elements of any row (column) times the corresponding cofactors of any other row (column), then we obtain zero.

EXAMPLE 1 Let $A = \begin{bmatrix} 1 & 2 & 3 \\ -2 & 3 & 1 \\ 4 & 5 & -2 \end{bmatrix}$. Then

$$A_{21} = (-1)^{2+1}\begin{vmatrix} 2 & 3 \\ 5 & -2 \end{vmatrix} = 19,$$

$$A_{22} = (-1)^{2+2}\begin{vmatrix} 1 & 3 \\ 4 & -2 \end{vmatrix} = -14, \quad \text{and} \quad A_{23} = (-1)^{2+3}\begin{vmatrix} 1 & 2 \\ 4 & 5 \end{vmatrix} = 3.$$

Now

$$a_{31}A_{21} + a_{32}A_{22} + a_{33}A_{23} = (4)(19) + (5)(-14) + (-2)(3) = 0,$$

and

$$a_{11}A_{21} + a_{12}A_{22} + a_{13}A_{23} = (1)(19) + (2)(-14) + (3)(3) = 0.$$ ■

We may summarize our expansion results by writing

$$\begin{aligned} a_{i1}A_{k1} + a_{i2}A_{k2} + \cdots + a_{in}A_{kn} &= \det(A) \quad \text{if } i = k \\ &= 0 \quad \text{if } i \neq k \end{aligned}$$

and

$$a_{1j}A_{1k} + a_{2j}A_{2k} + \cdots + a_{nj}A_{nk} = \det(A) \quad \text{if } j = k$$
$$= 0 \quad \text{if } j \neq k.$$

DEFINITION 3.5 Let $A = \left[a_{ij}\right]$ be an $n \times n$ matrix. The $n \times n$ matrix adj A, called the **adjoint** of A, is the matrix whose (i, j)th entry is the cofactor A_{ji} of a_{ji}. Thus

$$\text{adj } A = \begin{bmatrix} A_{11} & A_{21} & \cdots & A_{n1} \\ A_{12} & A_{22} & \cdots & A_{n2} \\ \vdots & \vdots & & \vdots \\ A_{1n} & A_{2n} & \cdots & A_{nn} \end{bmatrix}.$$

Remark It should be noted that the term *adjoint* has other meanings in linear algebra in addition to its use in the preceding definition.

EXAMPLE 2 Let $A = \begin{bmatrix} 3 & -2 & 1 \\ 5 & 6 & 2 \\ 1 & 0 & -3 \end{bmatrix}$. Compute adj A.

Solution

We first compute the cofactors of A. We have

$$A_{11} = (-1)^{1+1} \begin{vmatrix} 6 & 2 \\ 0 & -3 \end{vmatrix} = -18,$$

$$A_{12} = (-1)^{1+2} \begin{vmatrix} 5 & 2 \\ 1 & -3 \end{vmatrix} = 17, \quad A_{13} = (-1)^{1+3} \begin{vmatrix} 5 & 6 \\ 1 & 0 \end{vmatrix} = -6,$$

$$A_{21} = (-1)^{2+1} \begin{vmatrix} -2 & 1 \\ 0 & -3 \end{vmatrix} = -6,$$

$$A_{22} = (-1)^{2+2} \begin{vmatrix} 3 & 1 \\ 1 & -3 \end{vmatrix} = -10, \quad A_{23} = (-1)^{2+3} \begin{vmatrix} 3 & -2 \\ 1 & 0 \end{vmatrix} = -2,$$

$$A_{31} = (-1)^{3+1} \begin{vmatrix} -2 & 1 \\ 6 & 2 \end{vmatrix} = -10,$$

$$A_{32} = (-1)^{3+2} \begin{vmatrix} 3 & 1 \\ 5 & 2 \end{vmatrix} = -1, \quad A_{33} = (-1)^{3+3} \begin{vmatrix} 3 & -2 \\ 5 & 6 \end{vmatrix} = 28.$$

Then

$$\text{adj } A = \begin{bmatrix} -18 & -6 & -10 \\ 17 & -10 & -1 \\ -6 & -2 & 28 \end{bmatrix}.$$

■

Theorem 3.12 If $A = \left[a_{ij}\right]$ is an $n \times n$ matrix, then $A(\text{adj } A) = (\text{adj } A)A = \det(A)I_n$.

Proof

We have

$$A(\text{adj } A) = \begin{bmatrix} a_{11} & a_{12} & \cdots & a_{1n} \\ a_{21} & a_{22} & \cdots & a_{2n} \\ \vdots & \vdots & & \vdots \\ a_{i1} & a_{i2} & \cdots & a_{in} \\ \vdots & \vdots & & \vdots \\ a_{n1} & a_{n2} & \cdots & a_{nn} \end{bmatrix} \begin{bmatrix} A_{11} & A_{21} & \cdots & A_{j1} & \cdots & A_{n1} \\ A_{12} & A_{22} & \cdots & A_{j2} & \cdots & A_{n2} \\ \vdots & \vdots & & \vdots & & \vdots \\ A_{1n} & A_{2n} & \cdots & A_{jn} & \cdots & A_{nn} \end{bmatrix}.$$

The (i, j)th element in the product matrix $A(\text{adj } A)$ is, by Theorem 3.10,

$$\begin{aligned} a_{i1}A_{j1} + a_{i2}A_{j2} + \cdots + a_{in}A_{jn} &= \det(A) \quad \text{if } i = j \\ &= 0 \qquad\quad\ \text{if } i \neq j. \end{aligned}$$

This means that

$$A(\text{adj } A) = \begin{bmatrix} \det(A) & 0 & \cdots & 0 \\ 0 & \det(A) & \cdots & 0 \\ \vdots & \vdots & & \vdots \\ 0 & 0 & \cdots & \det(A) \end{bmatrix} = \det(A)I_n.$$

The (i, j)th element in the product matrix $(\text{adj } A)A$ is, by Theorem 3.10,

$$A_{1i}a_{1j} + A_{2i}a_{2j} + \cdots + A_{ni}a_{nj} = \begin{cases} \det(A) & \text{if } i = j \\ 0 & \text{if } i \neq j. \end{cases}$$

Thus $(\text{adj } A)A = \det(A)I_n$. ■

EXAMPLE 3 Consider the matrix of Example 2. Then

$$\begin{bmatrix} 3 & -2 & 1 \\ 5 & 6 & 2 \\ 1 & 0 & -3 \end{bmatrix} \begin{bmatrix} -18 & -6 & -10 \\ 17 & 10 & -1 \\ -6 & -2 & 28 \end{bmatrix} = \begin{bmatrix} -94 & 0 & 0 \\ 0 & -94 & 0 \\ 0 & 0 & -94 \end{bmatrix}$$

$$= -94 \begin{bmatrix} 1 & 0 & 0 \\ 0 & 1 & 0 \\ 0 & 0 & 1 \end{bmatrix}$$

and

$$\begin{bmatrix} -18 & -6 & -10 \\ 17 & -10 & -1 \\ -6 & -2 & 28 \end{bmatrix} \begin{bmatrix} 3 & -2 & 1 \\ 5 & 6 & 2 \\ 1 & 0 & -3 \end{bmatrix} = -94 \begin{bmatrix} 1 & 0 & 0 \\ 0 & 1 & 0 \\ 0 & 0 & 1 \end{bmatrix}.$$ ■

We now have a new method for finding the inverse of a nonsingular matrix, and we state this result as the following corollary:

Corollary 3.4 If A is an $n \times n$ matrix and $\det(A) \neq 0$, then

$$A^{-1} = \frac{1}{\det(A)}(\text{adj } A) = \begin{bmatrix} \frac{A_{11}}{\det(A)} & \frac{A_{21}}{\det(A)} & \cdots & \frac{A_{n1}}{\det(A)} \\ \frac{A_{12}}{\det(A)} & \frac{A_{22}}{\det(A)} & \cdots & \frac{A_{n2}}{\det(A)} \\ \vdots & \vdots & & \vdots \\ \frac{A_{1n}}{\det(A)} & \frac{A_{2n}}{\det(A)} & \cdots & \frac{A_{nn}}{\det(A)} \end{bmatrix}.$$

Proof

By Theorem 3.12, $A(\text{adj } A) = \det(A)I_n$, so if $\det(A) \neq 0$, then

$$A\left(\frac{1}{\det(A)}(\text{adj } A)\right) = \frac{1}{\det(A)}(A(\text{adj } A)) = \frac{1}{\det(A)}(\det(A)I_n) = I_n.$$

Hence

$$A^{-1} = \frac{1}{\det(A)}(\text{adj } A).$$

■

EXAMPLE 4 Again consider the matrix of Example 2. Then $\det(A) = -94$, and

$$A^{-1} = \frac{1}{\det(A)}(\text{adj } A) = \begin{bmatrix} \frac{18}{94} & \frac{6}{94} & \frac{10}{94} \\ -\frac{17}{94} & \frac{10}{94} & \frac{1}{94} \\ \frac{6}{94} & \frac{2}{94} & -\frac{28}{94} \end{bmatrix}.$$

■

We might note that the method of inverting a nonsingular matrix given in Corollary 3.4 is much less efficient than the method given in Chapter 2. In fact, the computation of A^{-1}, using determinants, as given in Corollary 3.4, becomes too expensive for $n > 4$. We discuss these matters in Section 3.6, where we deal with determinants from a computational point of view. However, Corollary 3.4 is still a useful result on other grounds.

Key Terms

Inverse
Adjoint

3.4 Exercises

1. Verify Theorem 3.11 for the matrix

$$A = \begin{bmatrix} -2 & 3 & 0 \\ 4 & 1 & -3 \\ 2 & 0 & 1 \end{bmatrix}$$

by computing $a_{11}A_{12} + a_{21}A_{22} + a_{31}A_{32}$.

2. Let $A = \begin{bmatrix} 2 & 1 & 3 \\ -1 & 2 & 0 \\ 3 & -2 & 1 \end{bmatrix}$.

   (a) Find adj A.

   (b) Compute $\det(A)$.

   (c) Verify Theorem 3.12; that is, show that

$$A(\text{adj } A) = (\text{adj } A)A = \det(A)I_3.$$

3. Let $A = \begin{bmatrix} 6 & 2 & 8 \\ -3 & 4 & 1 \\ 4 & -4 & 5 \end{bmatrix}$. Follow the directions of Exercise 2.

4. Find the inverse of the matrix in Exercise 2 by the method given in Corollary 3.4.

5. Repeat Exercise 11 of Section 2.3 by the method given in Corollary 3.4. Compare your results with those obtained earlier.

6. Prove that if A is a symmetric matrix, then adj A is symmetric.

7. Use the method given in Corollary 3.4 to find the inverse, if it exists, of

   (a) $\begin{bmatrix} 0 & 2 & 1 & 3 \\ 2 & -1 & 3 & 4 \\ -2 & 1 & 5 & 2 \\ 0 & 1 & 0 & 2 \end{bmatrix}$,

   (b) $\begin{bmatrix} 4 & 2 & 2 \\ 0 & 1 & 2 \\ 1 & 0 & 3 \end{bmatrix}$, **(c)** $\begin{bmatrix} 3 & 2 \\ -3 & 4 \end{bmatrix}$.

8. Prove that if A is a nonsingular upper triangular matrix, then A^{-1} is upper triangular.

9. Use the method given in Corollary 3.4 to find the inverse of

$$A = \begin{bmatrix} a & b \\ c & d \end{bmatrix} \quad \text{if } ad - bc \neq 0.$$

10. Use the method given in Corollary 3.4 to find the inverse of

$$A = \begin{bmatrix} 1 & a & a^2 \\ 1 & b & b^2 \\ 1 & c & c^2 \end{bmatrix}.$$

[*Hint*: See Exercise 22 in Section 3.2, where $\det(A)$ is computed.]

11. Use the method given in Corollary 3.4 to find the inverse of

$$A = \begin{bmatrix} 4 & 0 & 0 \\ 0 & -3 & 0 \\ 0 & 0 & 2 \end{bmatrix}.$$

12. Use the method given in Corollary 3.4 to find the inverse of

$$A = \begin{bmatrix} 4 & 1 & 2 \\ 0 & -3 & 3 \\ 0 & 0 & 2 \end{bmatrix}.$$

13. Prove that if A is singular, then adj A is singular. [*Hint*: First show that if A is singular, then $A(\text{adj } A) = O$.]

14. Prove that if A is an $n \times n$ matrix, then $\det(\text{adj } A) = [\det(A)]^{n-1}$.

15. Assuming that your software has a command for computing the inverse of a matrix (see Exercise 63 in Section 1.5), read the accompanying software documentation to determine the method used. Is the description closer to that in Section 2.3 or Corollary 3.4? See also the comments in Section 3.6.

3.5 Other Applications of Determinants

We can use the results developed in Theorem 3.12 to obtain another method for solving a linear system of n equations in n unknowns. This method is known as **Cramer's rule**.

Theorem 3.13 **Cramer's* Rule**

Let

$$\begin{aligned} a_{11}x_1 + a_{12}x_2 + \cdots + a_{1n}x_n &= b_1 \\ a_{21}x_1 + a_{22}x_2 + \cdots + a_{2n}x_n &= b_2 \\ &\vdots \\ a_{n1}x_1 + a_{n2}x_2 + \cdots + a_{nn}x_n &= b_n \end{aligned}$$

be a linear system of n equations in n unknowns, and let $A = \left[a_{ij}\right]$ be the coefficient matrix so that we can write the given system as $A\mathbf{x} = \mathbf{b}$, where

$$\mathbf{b} = \begin{bmatrix} b_1 \\ b_2 \\ \vdots \\ b_n \end{bmatrix}.$$

If $\det(A) \neq 0$, then the system has the unique solution

$$x_1 = \frac{\det(A_1)}{\det(A)}, \quad x_2 = \frac{\det(A_2)}{\det(A)}, \quad \ldots, \quad x_n = \frac{\det(A_n)}{\det(A)},$$

where A_i is the matrix obtained from A by replacing the ith column of A by $\mathbf{b}$.

Proof

If $\det(A) \neq 0$, then, by Theorem 3.8, A is nonsingular. Hence

$$\mathbf{x} = \begin{bmatrix} x_1 \\ x_2 \\ \vdots \\ x_n \end{bmatrix} = A^{-1}\mathbf{b} = \begin{bmatrix} \frac{A_{11}}{\det(A)} & \frac{A_{21}}{\det(A)} & \cdots & \frac{A_{n1}}{\det(A)} \\ \frac{A_{12}}{\det(A)} & \frac{A_{22}}{\det(A)} & \cdots & \frac{A_{n2}}{\det(A)} \\ \vdots & \vdots & & \vdots \\ \frac{A_{1i}}{\det(A)} & \frac{A_{2i}}{\det(A)} & \cdots & \frac{A_{ni}}{\det(A)} \\ \vdots & \vdots & & \vdots \\ \frac{A_{1n}}{\det(A)} & \frac{A_{2n}}{\det(A)} & \cdots & \frac{A_{nn}}{\det(A)} \end{bmatrix} \begin{bmatrix} b_1 \\ b_2 \\ \vdots \\ b_n \end{bmatrix}.$$

This means that

$$x_i = \frac{A_{1i}}{\det(A)}b_1 + \frac{A_{2i}}{\det(A)}b_2 + \cdots + \frac{A_{ni}}{\det(A)}b_n \qquad \text{for } i = 1, 2, \ldots, n.$$

GABRIEL CRAMER

*Gabriel Cramer (1704–1752) was born in Geneva, Switzerland, and lived there all his life. Remaining single, he traveled extensively, taught at the Académie de Calvin, and participated actively in civic affairs.

The rule for solving systems of linear equations appeared in an appendix to his 1750 book, *Introduction à l'analyse des lignes courbes algébriques*. It was known previously by other mathematicians, but was not widely known or clearly explained until its appearance in Cramer's influential work.

Now let

$$A_i = \begin{bmatrix} a_{11} & a_{12} & \cdots & a_{1,i-1} & b_1 & a_{1,i+1} & \cdots & a_{1n} \\ a_{21} & a_{22} & \cdots & a_{2,i-1} & b_2 & a_{2,i+1} & \cdots & a_{2n} \\ \vdots & \vdots & & \vdots & \vdots & \vdots & & \vdots \\ a_{n1} & a_{n2} & \cdots & a_{n,i-1} & b_n & a_{n,i+1} & \cdots & a_{nn} \end{bmatrix}.$$

If we evaluate $\det(A_i)$ by expanding along the cofactors of the ith column, we find that

$$\det(A_i) = A_{1i}b_1 + A_{2i}b_2 + \cdots + A_{ni}b_n.$$

Hence

$$x_i = \frac{\det(A_i)}{\det(A)} \qquad \text{for } i = 1, 2, \ldots, n. \tag{1}$$

■

In the expression for x_i given in Equation (1), the determinant, $\det(A_i)$, of A_i can be calculated by any method desired. It was only in the *derivation* of the expression for x_i that we had to evaluate $\det(A_i)$ by expanding along the ith column.

EXAMPLE 1

Consider the following linear system:

$$\begin{aligned} -2x_1 + 3x_2 - x_3 &= 1 \\ x_1 + 2x_2 - x_3 &= 4 \\ -2x_1 - x_2 + x_3 &= -3. \end{aligned}$$

We have $|A| = \begin{vmatrix} -2 & 3 & -1 \\ 1 & 2 & -1 \\ -2 & -1 & 1 \end{vmatrix} = -2$. Then

$$x_1 = \frac{\begin{vmatrix} 1 & 3 & -1 \\ 4 & 2 & -1 \\ -3 & -1 & 1 \end{vmatrix}}{|A|} = \frac{-4}{-2} = 2,$$

$$x_2 = \frac{\begin{vmatrix} -2 & 1 & -1 \\ 1 & 4 & -1 \\ -2 & -3 & 1 \end{vmatrix}}{|A|} = \frac{-6}{-2} = 3,$$

and

$$x_3 = \frac{\begin{vmatrix} -2 & 3 & 1 \\ 1 & 2 & 4 \\ -2 & -1 & -3 \end{vmatrix}}{|A|} = \frac{-8}{-2} = 4.$$

■

We note that Cramer's rule is applicable only to the case in which we have n equations in n unknowns and the coefficient matrix A is nonsingular. If we have to solve a linear system of n equations in n unknowns whose coefficient matrix is singular, then we must use the Gaussian elimination or Gauss–Jordan reduction methods as discussed in Section 2.2. Cramer's rule becomes computationally inefficient for $n \geq 4$, so it is better, in general, to use the Gaussian elimination or Gauss–Jordan reduction methods.

Note that at this point we have shown that the following statements are equivalent for an $n \times n$ matrix A:

1. A is nonsingular.
2. $A\mathbf{x} = \mathbf{0}$ has only the trivial solution.
3. A is row (column) equivalent to I_n.
4. The linear system $A\mathbf{x} = \mathbf{b}$ has a unique solution for every $n \times 1$ matrix $\mathbf{b}$.
5. A is a product of elementary matrices.
6. $\det(A) \neq 0$.

Key Terms

Cramer's rule

3.5 Exercises

1. If possible, solve the following linear systems by Cramer's rule:

$$\begin{aligned} 2x_1 + 4x_2 + 6x_3 &= 2 \\ x_1 \qquad\quad + 2x_3 &= 0 \\ 2x_1 + 3x_2 - x_3 &= -5. \end{aligned}$$

2. Repeat Exercise 1 for the linear system

$$\begin{bmatrix} 1 & 1 & 1 & -2 \\ 0 & 2 & 1 & 3 \\ 2 & 1 & -1 & 2 \\ 1 & -1 & 0 & 1 \end{bmatrix} \begin{bmatrix} x_1 \\ x_2 \\ x_3 \\ x_4 \end{bmatrix} = \begin{bmatrix} -4 \\ 4 \\ 5 \\ 4 \end{bmatrix}.$$

3. Solve the following linear system for x_3, by Cramer's rule:

$$\begin{aligned} 2x_1 + x_2 + x_3 &= 6 \\ 3x_1 + 2x_2 - 2x_3 &= -2 \\ x_1 + x_2 + 2x_3 &= -4. \end{aligned}$$

4. Repeat Exercise 5 of Section 2.2; use Cramer's rule.

5. Repeat Exercise 1 for the following linear system:

$$\begin{aligned} 2x_1 - x_2 + 3x_3 &= 0 \\ x_1 + 2x_2 - 3x_3 &= 0 \\ 4x_1 + 2x_2 + x_3 &= 0. \end{aligned}$$

6. Repeat Exercise 6(b) of Section 2.2; use Cramer's rule.

7. Repeat Exercise 1 for the following linear systems:

$$\begin{aligned} 2x_1 + 3x_2 + 7x_3 &= 0 \\ -2x_1 \qquad\quad - 4x_3 &= 0 \\ x_1 + 2x_2 + 4x_3 &= 0. \end{aligned}$$

3.6 Determinants from a Computational Point of View

In Chapter 2 we discussed three methods for solving a linear system: Gaussian elimination, Gauss–Jordan reduction, and LU-factorization. In this chapter, we have presented one more way: Cramer's rule. We also have two methods for inverting a nonsingular matrix: the method developed in Section 2.3, which uses elementary matrices; and the method involving determinants, presented in Section 3.4. In this section we discuss criteria to be considered when selecting one or another of these methods.

In general, if we are seeking numerical answers, then any method involving determinants can be used for $n \leq 4$. Gaussian elimination, Gauss–Jordan reduction, and LU-factorization all require approximately $n^3/3$ operations to solve the linear system $A\mathbf{x} = \mathbf{b}$, where A is an $n \times n$ matrix. We now compare these methods with Cramer's rule, when A is 25×25, which in the world of real applications is a small problem. (In some applications A can be as large as $100{,}000 \times 100{,}000$.)

If we find $\mathbf{x}$ by Cramer's rule, then we must first obtain $\det(A)$. Suppose that we compute $\det(A)$ by cofactor expansion, say,

$$\det(A) = a_{11}A_{11} + a_{21}A_{21} + \cdots + a_{n1}A_{n1},$$

where we have expanded along the first column of A. If each cofactor A_{ij} is available, we need 25 multiplications to compute $\det(A)$. Now each cofactor A_{ij} is the determinant of a 24×24 matrix, and it can be expanded along a particular row or column, requiring 24 multiplications. Thus the computation of $\det(A)$ requires more than $25 \times 24 \times \cdots \times 2 \times 1 = 25!$ (approximately 1.55×10^{25}) multiplications. Even if we were to use a supercomputer capable of performing *ten trillion* (1×10^{12}) multiplications *per second* (3.15×10^{19} *per year*), it would take *49,000 years* to evaluate $\det(A)$. However, Gaussian elimination takes approximately $25^3/3$ multiplications, and we obtain the solution in less than *one second*. Of course, $\det(A)$ can be computed in a much more efficient way, by using elementary row operations to reduce A to triangular form and then using Theorem 3.7. (See Example 8 in Section 3.2.) When implemented this way, Cramer's rule will require approximately n^4 multiplications for an $n \times n$ matrix, compared with $n^3/3$ multiplications for Gaussian elimination. The most widely used method in practice is LU-factorization because it is cheapest, especially when we need to solve many linear systems with different right sides.

The importance of determinants obviously does not lie in their computational use; determinants enable us to express the inverse of a matrix and the solutions to a system of n linear equations in n unknowns by means of *expressions* or *formulas*. The other methods mentioned previously for solving a linear system, and the method for finding A^{-1} by using elementary matrices, have the property that we cannot write a *formula* for the answer; we must proceed algorithmically to obtain the answer. Sometimes we do not need a numerical answer, but merely an expression for the answer, because we may wish to further manipulate the answer—for example, integrate it.

Supplementary Exercises

1. Compute $|A|$ for each of the following:

(a) $A = \begin{bmatrix} 2 & 3 & 4 \\ 1 & 2 & 4 \\ 4 & 3 & 1 \end{bmatrix}$

(b) $A = \begin{bmatrix} 2 & 1 & 0 \\ 1 & 2 & 1 \\ 0 & 1 & 2 \end{bmatrix}$

(c) $A = \begin{bmatrix} 2 & 1 & -1 & 2 \\ 2 & -3 & -1 & 4 \\ 1 & 3 & 2 & -3 \\ 1 & -2 & -1 & 1 \end{bmatrix}$

(d) $A = \begin{bmatrix} 2 & 1 & 0 & 0 \\ 1 & 2 & 1 & 0 \\ 0 & 1 & 2 & 1 \\ 0 & 0 & 1 & 2 \end{bmatrix}$

2. Find all values of t for which $\det(tI_3 - A) = 0$ for each of the following:

(a) $A = \begin{bmatrix} 1 & 4 & 0 \\ 0 & 4 & 0 \\ 0 & 0 & 1 \end{bmatrix}$

(b) $A = \begin{bmatrix} 2 & -2 & 0 \\ -3 & 1 & 0 \\ 0 & 0 & 3 \end{bmatrix}$

(c) $A = \begin{bmatrix} 0 & 1 & 0 \\ 0 & 0 & 1 \\ 6 & -11 & 6 \end{bmatrix}$

(d) $A = \begin{bmatrix} 0 & 1 & 0 \\ 0 & 0 & 1 \\ 3 & 1 & -3 \end{bmatrix}$

3. Show that if $A^n = O$ for some positive integer n (i.e., if A is a nilpotent matrix), then $\det(A) = 0$.

4. Using only elementary row or elementary column operations and Theorems 3.2, 3.5, and 3.6 (do not expand the determinants), verify the following:

(a) $\begin{vmatrix} a-b & 1 & a \\ b-c & 1 & b \\ c-a & 1 & c \end{vmatrix} = \begin{vmatrix} a & 1 & b \\ b & 1 & c \\ c & 1 & a \end{vmatrix}$

(b) $\begin{vmatrix} 1 & a & bc \\ 1 & b & ca \\ 1 & c & ab \end{vmatrix} = \begin{vmatrix} 1 & a & a^2 \\ 1 & b & b^2 \\ 1 & c & c^2 \end{vmatrix}$

5. Show that if A is an $n \times n$ matrix, then $\det(AA^T) \geq 0$.

6. Prove or disprove that the determinant function is a linear transformation of M_{nn} into R^1.

7. Show that if A is a nonsingular matrix, then adj A is nonsingular and

$$(\operatorname{adj} A)^{-1} = \frac{1}{\det(A)} A = \operatorname{adj}(A^{-1}).$$

8. Prove that if two rows (columns) of the $n \times n$ matrix A are proportional, then $\det(A) = 0$.

9. Let Q be the $n \times n$ real matrix in which each entry is 1. Show that $\det(Q - nI_n) = 0$.

10. Let A be an $n \times n$ matrix with integer entries. Prove that A is nonsingular and A^{-1} has integer entries if and only if $\det(A) = \pm 1$.

11. Let A be an $n \times n$ matrix with integer entries and $\det(A) = \pm 1$. Show that if $\mathbf{b}$ has all integer entries, then every solution to $A\mathbf{x} = \mathbf{b}$ consists of integers.

Chapter Review

True or False

1. $\det(A + B) = \det(A) + \det(B)$
2. $\det(A^{-1}B) = \dfrac{\det(B)}{\det(A)}$
3. If $\det(A) = 0$, then A has at least two equal rows.
4. If A has a column of all zeros, then $\det(A) = 0$.
5. A is singular if and only if $\det(A) = 0$.
6. If B is the reduced row echelon form of A, then $\det(B) = \det(A)$.
7. The determinant of an elementary matrix is always 1.
8. If A is nonsingular, then $A^{-1} = \dfrac{1}{\det(A)} \operatorname{adj}(A)$.
9. If T is a matrix transformation from $R^2 \to R^2$ defined by $A = \begin{bmatrix} 5 & 2 \\ 2 & 2 \end{bmatrix}$, then the area of the image of a closed plane figure S under T is six times the area of S.
10. If all the diagonal elements of an $n \times n$ matrix A are zero, then $\det(A) = 0$.
11. $\det(AB^TA^{-1}) = \det B$.
12. $\dfrac{1}{c}(\det cA) = \det(A)$.

Quiz

1. Let A, B, and C be 2×2 matrices with $\det(A) = 3$, $\det(B) = -2$, and $\det(C) = 4$. Compute $\det(6A^T BC^{-1})$.
2. Prove or disprove: For a 3×3 matrix A, if B is the matrix obtained by adding 5 to each entry of A, then $\det(B) = 5 + \det(A)$.
3. Let B be the matrix obtained from A after the row operations $2\mathbf{r}_3 \to \mathbf{r}_3$, $\mathbf{r}_1 \leftrightarrow \mathbf{r}_2$, $4\mathbf{r}_1 + \mathbf{r}_3 \to \mathbf{r}_3$, and $-2\mathbf{r}_1 + \mathbf{r}_4 \to \mathbf{r}_4$ have been performed. If $\det(B) = 2$, find $\det(A)$.
4. Compute the determinant of

$$A = \begin{bmatrix} 1 & 2 & 1 & 0 \\ 0 & 1 & 2 & 3 \\ -2 & 0 & -1 & -1 \\ 3 & 0 & 0 & -1 \end{bmatrix}$$

by using row operations to obtain upper triangular form.

5. Let A be a lower triangular matrix. Prove that A is singular if and only if some diagonal entry of A is zero.
6. Evaluate $\det(A)$ by using expansion along a row or column, given

$$A = \begin{bmatrix} 0 & 1 & 2 & 0 \\ 1 & 3 & 4 & 5 \\ -2 & 0 & -1 & 1 \\ 1 & -1 & 0 & 0 \end{bmatrix}.$$

7. Use the adjoint to compute A^{-1} for

$$A = \begin{bmatrix} 2 & 3 & 1 \\ 0 & 1 & 1 \\ 2 & -1 & -2 \end{bmatrix}.$$

8. Solve the linear system $A\mathbf{x} = \mathbf{b}$ by using Cramer's rule, given

$$A = \begin{bmatrix} 3 & 4 & 2 \\ 1 & 2 & 2 \\ 3 & 0 & 1 \end{bmatrix} \quad \text{and} \quad \mathbf{b} = \begin{bmatrix} 1 \\ -1 \\ 4 \end{bmatrix}.$$

Discussion Exercises

1. Show that $\det(A)$ is $(a^4 - b^4)/(a - b)$, where

$$A = \begin{bmatrix} a+b & ab & 0 \\ 1 & a+b & ab \\ 0 & 1 & a+b \end{bmatrix}$$

for a and b any real numbers and $a \neq b$, carefully explaining all steps in your proof.

2. Let

$$A = \begin{bmatrix} a+b & ab & 0 & 0 \\ 1 & a+b & ab & 0 \\ 0 & 1 & a+b & ab \\ 0 & 0 & 1 & a+b \end{bmatrix}.$$

   (a) Formulate a conjecture for $\det(A)$. Explain how you arrived at your conjecture.

   (b) Prove your conjecture. (If you have access to software that incorporates a computer algebra system, then use it in your proof.)

3. Let a_j and b_j, $j = 1, 2, 3, 4$, be any real numbers and c_1 and d_1 be any real numbers. Prove that $\det(A) = 0$, where

$$A = \begin{bmatrix} a_1 & a_2 & a_3 & a_4 \\ b_1 & b_2 & b_3 & b_4 \\ c_1 & 0 & 0 & 0 \\ d_1 & 0 & 0 & 0 \end{bmatrix}.$$

Carefully explain all steps in your proof.

4. Determine all values of x so that $\det(A) = 0$, where

$$A = \begin{bmatrix} x & 1 \\ 2 & x \end{bmatrix}.$$

5. Let

$$A = \begin{bmatrix} x & 1 & 2 \\ 1 & x & 1 \\ 1 & 1 & 1 \end{bmatrix}.$$

   (a) Determine $p(x) = \det(A)$.

   (b) Graph $y = p(x)$. If the graph has any x-intercepts, determine the value of $\det(A)$ for those values.

6. Let $P(x_1, y_1)$ and $Q(x_2, y_2)$ be two points in the plane. Prove that the equation of the line through P and Q is given by $\det(A) = 0$, where

$$A = \begin{bmatrix} x & y & 1 \\ x_1 & y_1 & 1 \\ x_2 & y_2 & 1 \end{bmatrix}.$$

7. Let A_n be the $n \times n$ matrix of the form

$$\begin{bmatrix} x & 1 & 0 & 0 & 0 & \cdots & \cdots & 0 \\ 1 & x & 1 & 0 & 0 & \cdots & \cdots & 0 \\ 0 & 1 & x & 1 & 0 & \cdots & \cdots & 0 \\ \vdots & \ddots & \ddots & \ddots & \ddots & \ddots & & \vdots \\ \vdots & & \ddots & \ddots & \ddots & \ddots & \ddots & \vdots \\ \vdots & & & \ddots & \ddots & \ddots & \ddots & 0 \\ 0 & \cdots & \cdots & \cdots & 0 & 1 & x & 1 \\ 0 & \cdots & \cdots & \cdots & \cdots & 0 & 1 & x \end{bmatrix}.$$

Show that for $n \geq 3$,

$$\det(A_n) = x \det(A_{n-1}) - \det(A_{n-2})$$

$\left(\textit{Note}: A_1 = \begin{bmatrix} x \end{bmatrix}, A_2 = \begin{bmatrix} x & 1 \\ 1 & x \end{bmatrix}.\right)$

CHAPTER

4 Real Vector Spaces

In Sections 1.2, 1.3, and 1.6, we have given brief glimpses of 2-vectors and 3-vectors and of some of their properties from an intuitive, somewhat informal point of view. In this chapter we first develop the notion of 2-vectors and 3-vectors along with their properties very carefully and systematically. A good mastery of this basic material will be helpful in understanding n-vectors and in the study of Section 4.2, where a more general notion of vector will be introduced. Moreover, n-vectors and this more general concept of vector will be used in many parts of the book.

4.1 Vectors in the Plane and in 3-Space

In many applications we deal with measurable quantities such as pressure, mass, and speed, which can be completely described by giving their magnitude. They are called **scalars** and will be denoted by lowercase *italic* letters such as c, d, r, s, and t. There are many other measurable quantities, such as velocity, force, and acceleration, which require for their description not only magnitude, but also a sense of direction. These are called **vectors**, and their study comprises this chapter. Vectors will be denoted by lowercase **boldface** letters, such as **u**, **v**, **x**, **y**, and **z**. The reader may already have encountered vectors in elementary physics and in calculus.

■ Vectors in the Plane

We draw a pair of perpendicular lines intersecting at a point O, called the **origin**. One of the lines, the ***x*-axis**, is usually taken in a horizontal position. The other line, the ***y*-axis**, is then taken in a vertical position. The x- and y-axes together are called **coordinate axes** (Figure 4.1), and they form a **rectangular coordinate system**, or a **Cartesian** (after René Descartes*) **coordinate system**. We now choose a point

*René Descartes (1596–1650) was one of the best-known scientists and philosophers of his day; he was considered by some to be the founder of modern philosophy. After completing a university degree in law, he turned to the private study of mathematics, simultaneously pursuing interests in Parisian

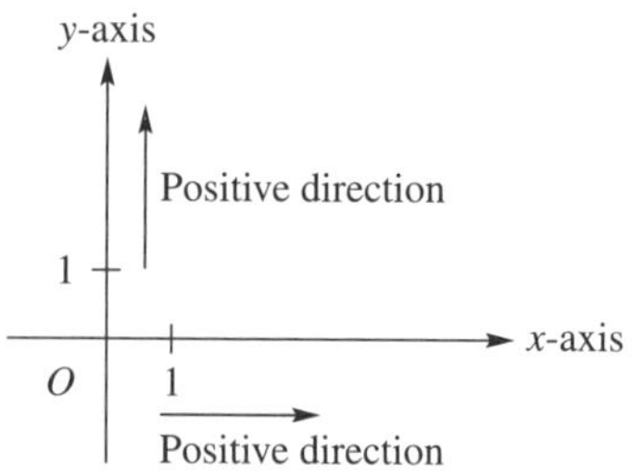

FIGURE 4.1

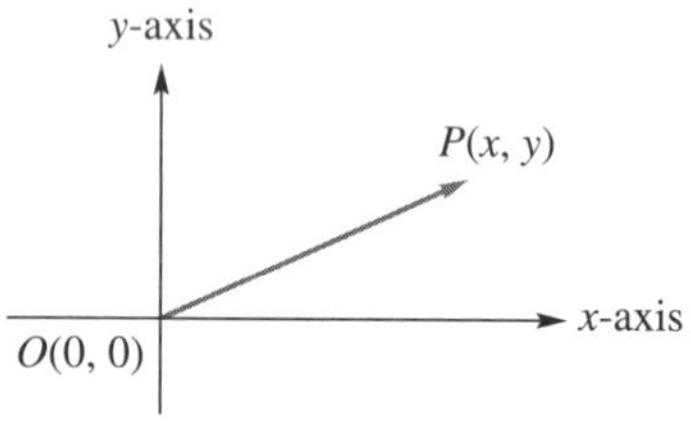

FIGURE 4.2

on the x-axis to the right of O and a point on the y-axis above O to fix the units of length and positive directions on the x- and y-axes. Frequently, but not always, these points are chosen so that they are both equidistant from O—that is, so that the same unit of length is used for both axes.

With each point P in the plane we associate an ordered pair (x, y) of real numbers, its **coordinates**. Conversely, we can associate a point in the plane with each ordered pair of real numbers. Point P with coordinates (x, y) is denoted by $P(x, y)$, or simply by (x, y). The set of all points in the plane is denoted by R^2; it is called **2-space**.

Consider the 2×1 matrix

$$\mathbf{x} = \begin{bmatrix} x \\ y \end{bmatrix},$$

where x and y are real numbers. With $\mathbf{x}$ we associate the directed line segment with the initial point the origin and terminal point $P(x, y)$. The directed line segment from O to P is denoted by $\overrightarrow{OP}$; O is called its **tail** and P its **head**. We distinguish tail and head by placing an arrow at the head (Figure 4.2). A directed line segment has a **direction**, indicated by the arrow at its head. The **magnitude** of a directed line segment is its length. Thus a directed line segment can be used to describe force, velocity, or acceleration. Conversely, with the directed line segment $\overrightarrow{OP}$ with tail $O(0, 0)$ and head $P(x, y)$ we can associate the matrix

$$\begin{bmatrix} x \\ y \end{bmatrix}.$$

DEFINITION 4.1

A **vector in the plane** is a 2×1 matrix

$$\mathbf{x} = \begin{bmatrix} x \\ y \end{bmatrix},$$

where x and y are real numbers, called the **components** (or **entries**) of $\mathbf{x}$. We refer to a vector in the plane merely as a **vector** or as a **2-vector**.

Thus, with every vector, we can associate a directed line segment and, conversely, with every directed line segment we can associate a vector. Frequently, the notions of directed line segment and vector are used interchangeably, and a directed line segment is called a **vector**.

RENÉ DESCARTES

night life and in the military, volunteering for brief periods in the Dutch, Bavarian, and French armies. The most productive period of his life was 1628–1648, when he lived in Holland. In 1649 he accepted an invitation from Queen Christina of Sweden to be her private tutor and to establish an Academy of Sciences there. Unfortunately, he did not carry out this project, since he died of pneumonia in 1650.

In 1619 Descartes had a dream in which he realized that the method of mathematics is the best way for obtaining truth. However, his only mathematical publication was *La Géométrie*, which appeared as an appendix to his major philosophical work, *Discours de la méthode pour bien conduire sa raison, et chercher la vérité dans les sciences* (*Discourse on the Method of Reasoning Well and Seeking Truth in the Sciences*). In *La Géométrie* he proposes the radical idea of doing geometry algebraically. To express a curve algebraically, one chooses any convenient line of reference and, on the line, a point of reference. If y represents the distance from any point of the curve to the reference line and x represents the distance along the line to the reference point, there is an equation relating x and y that represents the curve. The systematic use of "Cartesian" coordinates described in this section was introduced later in the seventeenth century by mathematicians carrying on Descartes's work.

Since a vector is a matrix, the vectors

$$\mathbf{u} = \begin{bmatrix} x_1 \\ y_1 \end{bmatrix} \quad \text{and} \quad \mathbf{v} = \begin{bmatrix} x_2 \\ y_2 \end{bmatrix}$$

are said to be **equal** if $x_1 = x_2$ and $y_1 = y_2$. That is, two vectors are equal if their respective components are equal.

EXAMPLE 1

The vectors

$$\begin{bmatrix} a+b \\ 2 \end{bmatrix} \quad \text{and} \quad \begin{bmatrix} 3 \\ a-b \end{bmatrix}$$

are equal if

$$\begin{aligned} a + b &= 3 \\ a - b &= 2, \end{aligned}$$

which means (verify) that $a = \frac{5}{2}$ and $b = \frac{1}{2}$. ■

Frequently, in physical applications it is necessary to deal with a directed line segment $\overrightarrow{PQ}$ from the point $P(x, y)$ (not the origin) to the point $Q(x', y')$, as shown in Figure 4.3(a). Such a directed line segment will also be called a **vector in the plane**, or simply a **vector** with **tail** $P(x, y)$ and **head** $Q(x', y')$. The **components** of such a vector are $x' - x$ and $y' - y$. Thus the vector $\overrightarrow{PQ}$ in Figure 4.3(a) can also be represented by the vector

$$\begin{bmatrix} x' - x \\ y' - y \end{bmatrix}$$

with tail O and head $P''(x' - x, y' - y)$. Two such vectors in the plane will be called **equal** if their respective components are equal. Consider the vectors $\overrightarrow{P_1Q_1}$, $\overrightarrow{P_2Q_2}$, and $\overrightarrow{P_3Q_3}$ joining the points $P_1(3, 2)$ and $Q_1(5, 5)$, $P_2(0, 0)$ and $Q_2(2, 3)$, $P_3(-3, 1)$ and $Q_3(-1, 4)$, respectively, as shown in Figure 4.3(b). Since they all have the same components, they are equal.

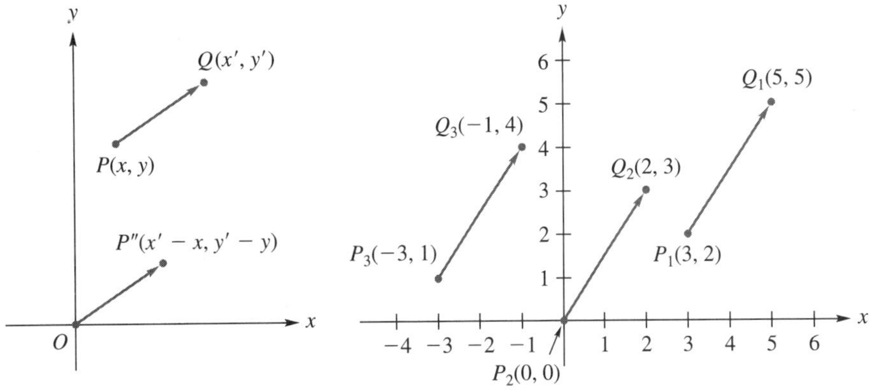

(a) Different directed line segments representing the same vector.

(b) Vectors in the plane.

FIGURE 4.3

Moreover, the head $Q_4(x_4', y_4')$ of the vector

$$\overrightarrow{P_4Q_4} = \begin{bmatrix} 2 \\ 3 \end{bmatrix} = \overrightarrow{P_2Q_2}$$

with tail $P_4(-5, 2)$ can be determined as follows. We must have $x_4' - (-5) = 2$ and $y_4' - 2 = 3$, so that $x_4' = 2 - 5 = -3$ and $y_4' = 3 + 2 = 5$. Similarly, the tail $P_5(x_5, y_5)$ of the vector

$$\overrightarrow{P_5Q_5} = \begin{bmatrix} 2 \\ 3 \end{bmatrix}$$

with head $Q_5(8, 6)$ is determined as follows. We must have $8 - x_5 = 2$ and $6 - y_5 = 3$, so that $x_5 = 8 - 2 = 6$ and $y_5 = 6 - 3 = 3$.

With each vector

$$\mathbf{x} = \begin{bmatrix} x \\ y \end{bmatrix}$$

we can also associate the unique point $P(x, y)$; conversely, with each point $P(x, y)$ we associate the unique vector

$$\begin{bmatrix} x \\ y \end{bmatrix}.$$

Hence we also write the vector $\mathbf{x}$ as (x, y). Of course, this association is carried out by means of the directed line segment $\overrightarrow{OP}$, where O is the origin and P is the point with coordinates (x, y) (Figure 4.2).

Thus the plane may be viewed both as the set of all points or as the set of all vectors. For this reason, and depending upon the context, we sometimes take R^2 as the set of all ordered pairs (x, y) and sometimes as the set of all 2×1 matrices

$$\begin{bmatrix} x \\ y \end{bmatrix}.$$

DEFINITION 4.2

Let

$$\mathbf{u} = \begin{bmatrix} u_1 \\ u_2 \end{bmatrix} \quad \text{and} \quad \mathbf{v} = \begin{bmatrix} v_1 \\ v_2 \end{bmatrix}$$

be two vectors in the plane. The **sum** of the vectors $\mathbf{u}$ and $\mathbf{v}$ is the vector

$$\mathbf{u} + \mathbf{v} = \begin{bmatrix} u_1 + v_1 \\ u_2 + v_2 \end{bmatrix}.$$

Remark Observe that vector addition is a special case of matrix addition.

EXAMPLE 2

Let $\mathbf{u} = \begin{bmatrix} 2 \\ 3 \end{bmatrix}$ and $\mathbf{v} = \begin{bmatrix} 3 \\ -4 \end{bmatrix}$. Then

$$\mathbf{u} + \mathbf{v} = \begin{bmatrix} 2 + 3 \\ 3 + (-4) \end{bmatrix} = \begin{bmatrix} 5 \\ -1 \end{bmatrix}.$$

See Figure 4.4. ■

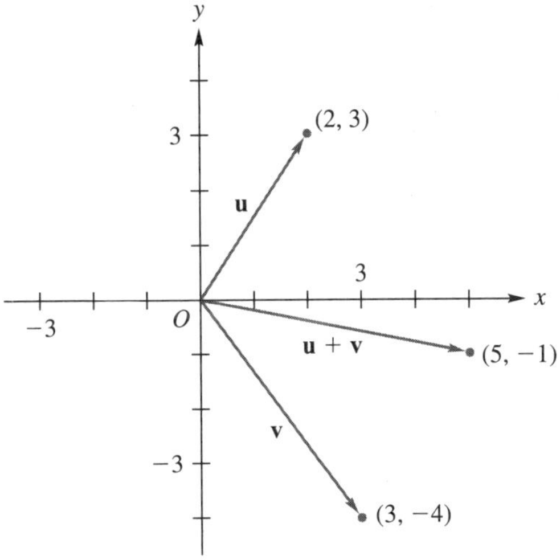

FIGURE 4.4

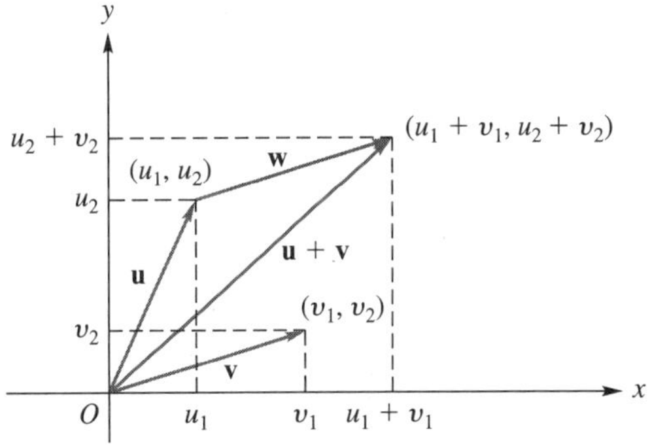

FIGURE 4.5 Vector addition.

We can interpret vector addition geometrically, as follows. In Figure 4.5 the directed line segment **w** is parallel to **v**, it has the same length as **v**, and its tail is the head (u_1, u_2) of **u**, so its head is $(u_1 + v_1, u_2 + v_2)$. Thus the vector with tail O and head $(u_1 + v_1, u_2 + v_2)$ is $\mathbf{u} + \mathbf{v}$. We can also describe $\mathbf{u} + \mathbf{v}$ as the diagonal of the parallelogram defined by **u** and **v**, as shown in Figure 4.6.

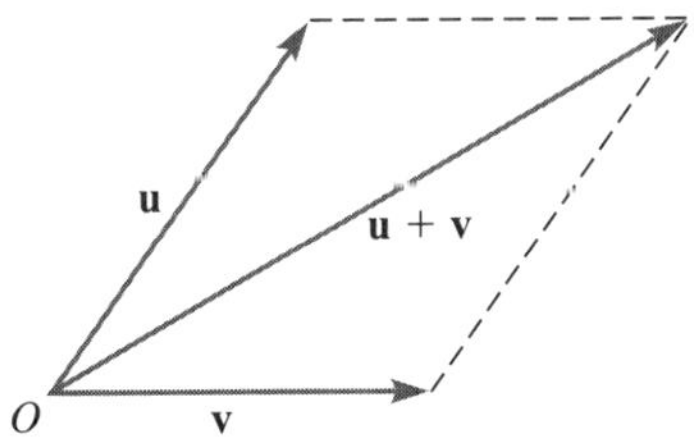

FIGURE 4.6 Vector addition.

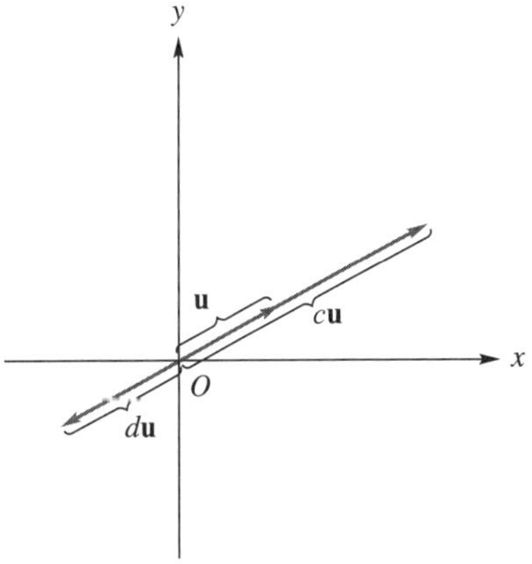

FIGURE 4.7 Scalar multiplication.

DEFINITION 4.3

If $\mathbf{u} = \begin{bmatrix} u_1 \\ u_2 \end{bmatrix}$ is a vector and c is a scalar (a real number), then the **scalar multiple** $c\mathbf{u}$ of **u** by c is the vector $\begin{bmatrix} cu_1 \\ cu_2 \end{bmatrix}$. Thus the scalar multiple $c\mathbf{u}$ is obtained by multiplying each component of **u** by c. If $c > 0$, then $c\mathbf{u}$ is in the same direction as **u**, whereas if $d < 0$, then $d\mathbf{u}$ is in the opposite direction (Figure 4.7).

EXAMPLE 3 If $c = 2$, $d = -3$, and $\mathbf{u} = \begin{bmatrix} 2 \\ -3 \end{bmatrix}$, then

$$c\mathbf{u} = 2\begin{bmatrix} 2 \\ -3 \end{bmatrix} = \begin{bmatrix} 2(2) \\ 2(-3) \end{bmatrix} = \begin{bmatrix} 4 \\ -6 \end{bmatrix}$$

and

$$d\mathbf{u} = -3\begin{bmatrix} 2 \\ -3 \end{bmatrix} = \begin{bmatrix} (-3)(2) \\ (-3)(-3) \end{bmatrix} = \begin{bmatrix} -6 \\ 9 \end{bmatrix},$$

which are shown in Figure 4.8. ■

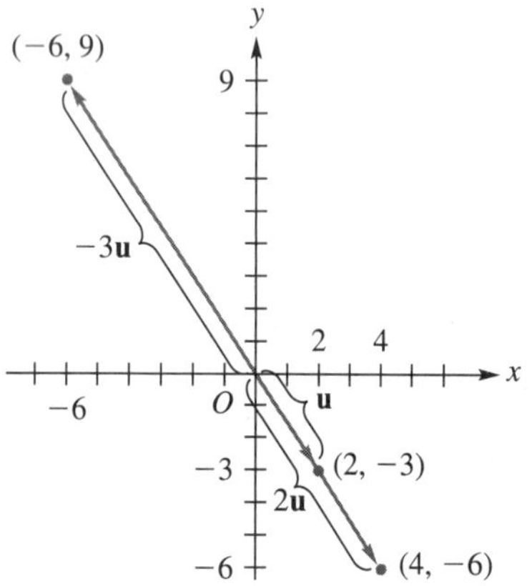

FIGURE 4.8

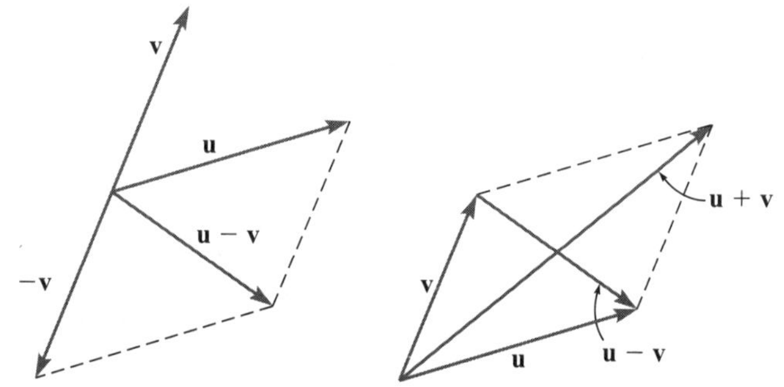

(a) Difference between vectors. (b) Vector sum and vector difference.

FIGURE 4.9

The vector

$$\begin{bmatrix} 0 \\ 0 \end{bmatrix}$$

is called the **zero vector** and is denoted by $\mathbf{0}$. If $\mathbf{u}$ is any vector, it follows that (Exercise 21)

$$\mathbf{u} + \mathbf{0} = \mathbf{u}.$$

We can also show (Exercise 22) that

$$\mathbf{u} + (-1)\mathbf{u} = \mathbf{0},$$

and we write $(-1)\mathbf{u}$ as $-\mathbf{u}$ and call it the **negative** of $\mathbf{u}$. Moreover, we write $\mathbf{u} + (-1)\mathbf{v}$ as $\mathbf{u} - \mathbf{v}$ and call it the **difference between u and v**. It is shown in Figure 4.9(a). Observe that while vector addition gives one diagonal of a parallelogram, vector subtraction gives the other diagonal [see Figure 4.9(b)].

■ Vectors in Space

The foregoing discussion of vectors in the plane can be generalized to vectors in space, as follows. We first fix a **coordinate system** by choosing a point, called the **origin**, and three lines, called the **coordinate axes**, each passing through the origin so that each line is perpendicular to the other two. These lines are individually called the x-, y-, and z-axes. On each of these axes we choose a point fixing the units of length and positive directions on the coordinate axes. Frequently, but not always, the same unit of length is used for all the coordinate axes. In Figure 4.10 we show two of the many possible coordinate systems.

The coordinate system shown in Figure 4.10(a) is called a **right-handed coordinate system**; the one in Figure 4.10(b) is called **left-handed**. A right-handed system is characterized by the following property: If we curl the fingers of the right hand in the direction of a 90° rotation from the positive x-axis to the positive y-axis, then the thumb will point in the direction of the positive z-axis. (See Figure 4.11.)

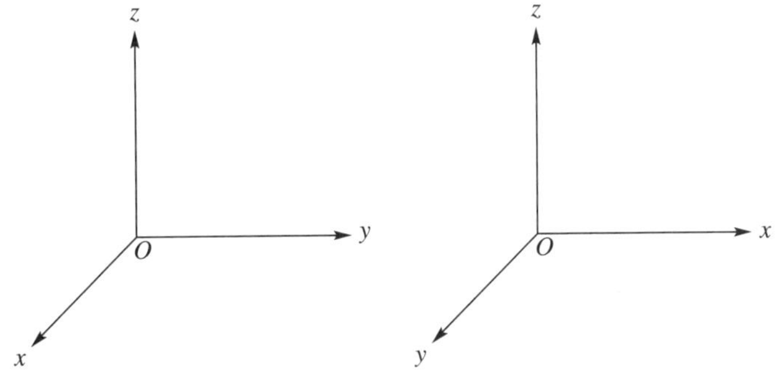

(a) Right-handed coordinate system. (b) Left-handed coordinate system.

FIGURE 4.10

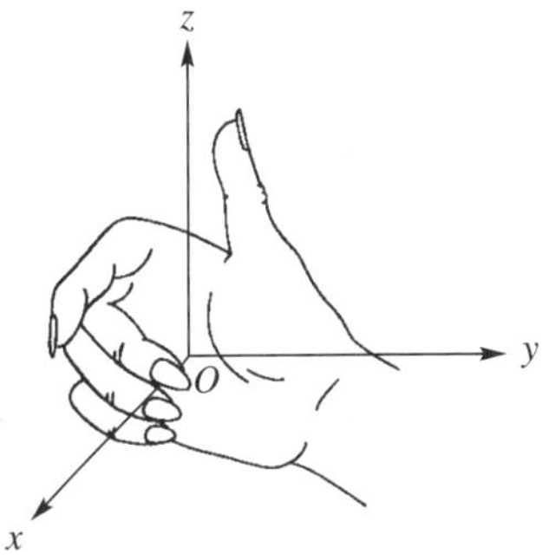

FIGURE 4.11

If we rotate the x-axis counterclockwise toward the y-axis, then a right-hand screw will move in the positive z-direction (see Figure 4.11).

With each point P in space we associate an ordered triple (x, y, z) of real numbers, its coordinates. Conversely, we can associate a point in space with each ordered triple of real numbers. The point P with coordinates x, y, and z is denoted by $P(x, y, z)$, or simply by (x, y, z). The set of all points in space is called **3-space** and is denoted by R^3.

A **vector in space**, or **3-vector**, or simply a **vector**, is a 3×1 matrix

$$\mathbf{x} = \begin{bmatrix} x \\ y \\ z \end{bmatrix},$$

where x, y, and z are real numbers, called the **components** of vector $\mathbf{x}$. Two vectors in space are said to be **equal** if their respective components are equal.

As in the plane, with the vector

$$\mathbf{x} = \begin{bmatrix} x \\ y \\ z \end{bmatrix}$$

we associate the directed line segment $\overrightarrow{OP}$, whose tail is $O(0, 0, 0)$ and whose head is $P(x, y, z)$; conversely, with each directed line segment we associate the vector **x** [see Figure 4.12(a)]. Thus we can also write the vector **x** as (x, y, z). Again, as in the plane, in physical applications we often deal with a directed line segment $\overrightarrow{PQ}$, from the point $P(x, y, z)$ (not the origin) to the point $Q(x', y', z')$, as shown in Figure 4.12(b). Such a directed line segment will also be called a **vector in R^3**, or simply a **vector** with tail $P(x, y, z)$ and head $Q(x', y', z')$. The components of such a vector are $x' - x$, $y' - y$, and $z' - z$. Two such vectors in R^3 will be called **equal** if their respective components are equal. Thus the vector $\overrightarrow{PQ}$ in Figure 4.12(b) can also be represented by the vector $\begin{bmatrix} x' - x \\ y' - y \\ z' - z \end{bmatrix}$ with tail O and head $P''(x' - x, y' - y, z' - z)$.

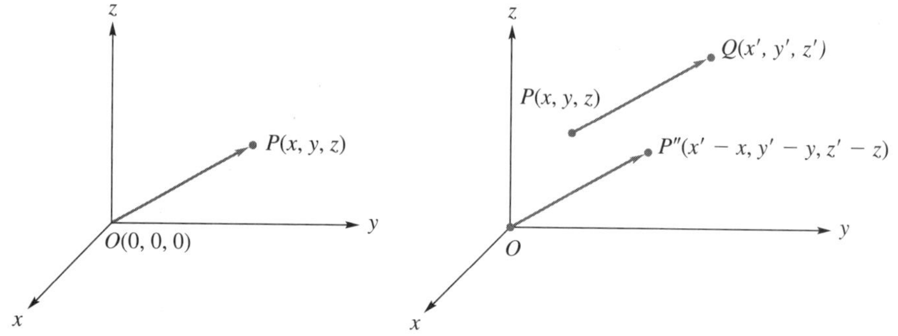

FIGURE 4.12 (a) A vector in R^3. (b) Different directed line segments representing the same vector.

If $\mathbf{u} = \begin{bmatrix} u_1 \\ u_2 \\ u_3 \end{bmatrix}$ and $\mathbf{v} = \begin{bmatrix} v_1 \\ v_2 \\ v_3 \end{bmatrix}$ are vectors in R^3 and c is a scalar, then the **sum** $\mathbf{u} + \mathbf{v}$ and the **scalar multiple** $c\mathbf{u}$ are defined, respectively, as

$$\mathbf{u} + \mathbf{v} = \begin{bmatrix} u_1 + v_1 \\ u_2 + v_2 \\ u_3 + v_3 \end{bmatrix} \quad \text{and} \quad c\mathbf{u} = \begin{bmatrix} cu_1 \\ cu_2 \\ cu_3 \end{bmatrix}.$$

The sum is shown in Figure 4.13, which resembles Figure 4.5, and the scalar multiple is shown in Figure 4.14, which resembles Figure 4.8.

EXAMPLE 4 Let

$$\mathbf{u} = \begin{bmatrix} 2 \\ 3 \\ -1 \end{bmatrix} \quad \text{and} \quad \mathbf{v} = \begin{bmatrix} 3 \\ -4 \\ 2 \end{bmatrix}.$$

Compute: (a) $\mathbf{u} + \mathbf{v}$; (b) $-2\mathbf{u}$; (c) $3\mathbf{u} - 2\mathbf{v}$.

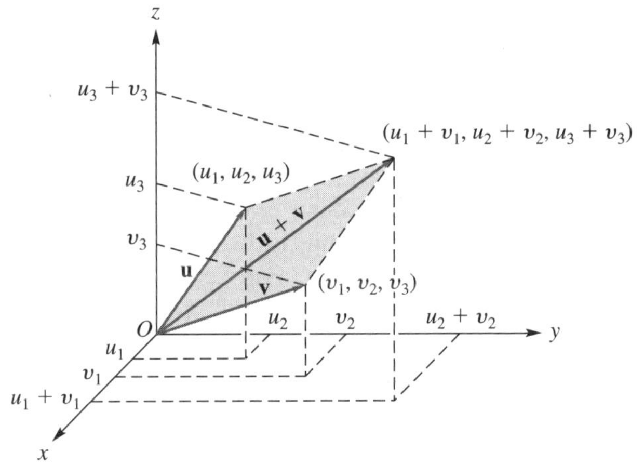

FIGURE 4.13 Vector addition.

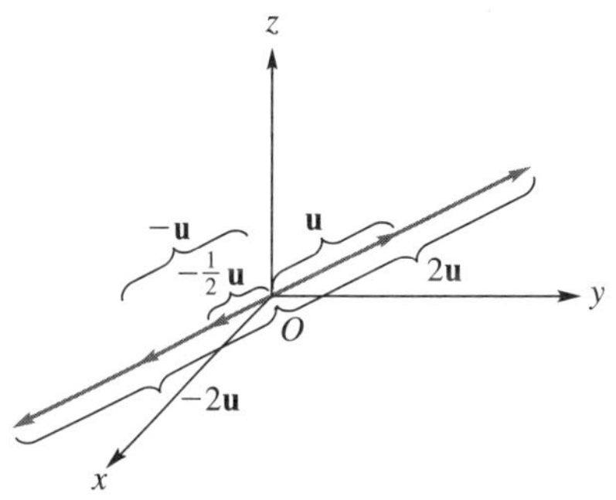

FIGURE 4.14 Scalar multiplication.

Solution

(a) $\mathbf{u} + \mathbf{v} = \begin{bmatrix} 2+3 \\ 3+(-4) \\ -1+2 \end{bmatrix} = \begin{bmatrix} 5 \\ -1 \\ 1 \end{bmatrix}$

(b) $-2\mathbf{u} = \begin{bmatrix} -2(2) \\ -2(3) \\ -2(-1) \end{bmatrix} = \begin{bmatrix} -4 \\ -6 \\ 2 \end{bmatrix}$

(c) $3\mathbf{u} - 2\mathbf{v} = \begin{bmatrix} 3(2) \\ 3(3) \\ 3(-1) \end{bmatrix} - \begin{bmatrix} 2(3) \\ 2(-4) \\ 2(2) \end{bmatrix} = \begin{bmatrix} 0 \\ 17 \\ -7 \end{bmatrix}$ ■

The **zero** vector in R^3 is denoted by $\mathbf{0}$, where

$$\mathbf{0} = \begin{bmatrix} 0 \\ 0 \\ 0 \end{bmatrix}.$$

The vector $\mathbf{0}$ has the property that if $\mathbf{u}$ is any vector in R^3, then

$$\mathbf{u} + \mathbf{0} = \mathbf{u}.$$

The **negative** of the vector $\mathbf{u} = \begin{bmatrix} u_1 \\ u_2 \\ u_3 \end{bmatrix}$ is the vector $-\mathbf{u} = \begin{bmatrix} -u_1 \\ -u_2 \\ -u_3 \end{bmatrix}$, and

$$\mathbf{u} + (-\mathbf{u}) = \mathbf{0}.$$

Observe that we have defined a vector in the plane as an ordered pair of real numbers, or as a 2×1 matrix. Similarly, a vector in space is an ordered triple of real numbers, or a 3×1 matrix. However, in physics we often treat a vector as a directed line segment. Thus we have three very different representations of a vector, and we can then ask why all three are valid. That is, why are we justified in referring to an ordered pair of real numbers, a 2×1 matrix, and a directed line segment by the same name, "vector"?

To answer this question, we first observe that, mathematically speaking, the only thing that concerns us is the behavior of the object we call "vector." It turns out that all three objects behave, from an algebraic point of view, in exactly the same manner. Moreover, many other objects that arise naturally in applied problems behave, algebraically speaking, as do the aforementioned objects. To a mathematician, this is a perfect situation. For we can now abstract those features that all such objects have in common (i.e., those properties that make them all behave alike) and define a new structure. The great advantage of doing this is that we can now talk about properties of all such objects at the same time without having to refer to any one object in particular. This, of course, is much more efficient than studying the properties of each object separately. For example, the theorem presented next summarizes the properties of addition and scalar multiplication for vectors in the plane and in space. Moreover, this theorem will serve as the model for the generalization of the set of all vectors in the plane or in space to a more abstract setting.

Theorem 4.1 If $\mathbf{u}$, $\mathbf{v}$, and $\mathbf{w}$ are vectors in R^2 or R^3, and c and d are real scalars, then the following properties are valid:

(a) $\mathbf{u}+\mathbf{v}=\mathbf{v}+\mathbf{u}$

(b) $\mathbf{u}+(\mathbf{v}+\mathbf{w})=(\mathbf{u}+\mathbf{v})+\mathbf{w}$

(c) $\mathbf{u}+\mathbf{0}=\mathbf{0}+\mathbf{u}=\mathbf{u}$

(d) $\mathbf{u}+(-\mathbf{u})=\mathbf{0}$

(e) $c(\mathbf{u}+\mathbf{v})=c\mathbf{u}+c\mathbf{v}$

(f) $(c+d)\mathbf{u}=c\mathbf{u}+d\mathbf{u}$

(g) $c(d\mathbf{u})=(cd)\mathbf{u}$

(h) $1\mathbf{u}=\mathbf{u}$

Proof

(a) Suppose that $\mathbf{u}$ and $\mathbf{v}$ are vectors in R^2 so that

$$\mathbf{u}=\begin{bmatrix} u_1 \\ u_2 \end{bmatrix} \quad \text{and} \quad \mathbf{v}=\begin{bmatrix} v_1 \\ v_2 \end{bmatrix}.$$

Then

$$\mathbf{u}+\mathbf{v}=\begin{bmatrix} u_1+v_1 \\ u_2+v_2 \end{bmatrix} \quad \text{and} \quad \mathbf{v}+\mathbf{u}=\begin{bmatrix} v_1+u_1 \\ v_2+u_2 \end{bmatrix}.$$

Since the components of $\mathbf{u}$ and $\mathbf{v}$ are real numbers, $u_1+v_1=v_1+u_1$ and $u_2+v_2=v_2+u_2$. Therefore,

$$\mathbf{u}+\mathbf{v}=\mathbf{v}+\mathbf{u}.$$

A similar proof can be given if $\mathbf{u}$ and $\mathbf{v}$ are vectors in R^3.

Property (a) can also be established geometrically, as shown in Figure 4.15. The proofs of the remaining properties will be left as exercises. Remember, they can all be proved by either an algebraic or a geometric approach. ■

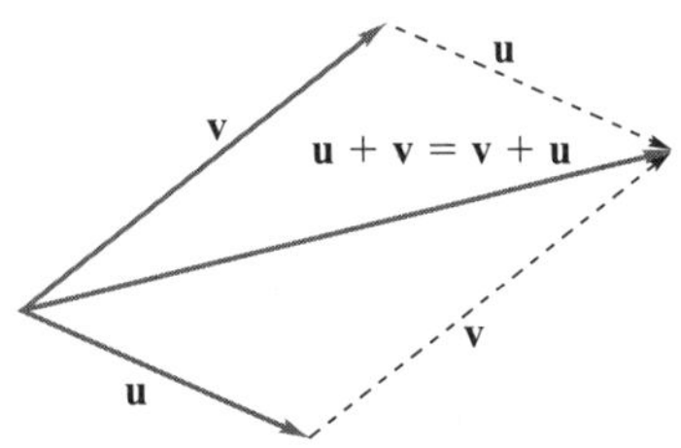

FIGURE 4.15 Vector addition.

Key Terms

Vectors
Rectangular (Cartesian) coordinate system
Coordinate axes
x-axis, y-axis, z-axis
Origin
Coordinates
2-space, R^2
Tail of a vector
Head of a vector
Directed line segment
Magnitude of a vector
Vector in the plane
Components of a vector
Equal vectors
Scalar multiple of a vector
Vector addition
Zero vector
Difference of vectors
Right- (left-) handed coordinate system
3-space, R^3
Vector in space

4.1 Exercises

1. Sketch a directed line segment in R^2, representing each of the following vectors:

(a) $\mathbf{u} = \begin{bmatrix} -2 \\ 3 \end{bmatrix}$ **(b)** $\mathbf{v} = \begin{bmatrix} 3 \\ 4 \end{bmatrix}$

(c) $\mathbf{w} = \begin{bmatrix} -3 \\ -3 \end{bmatrix}$ **(d)** $\mathbf{z} = \begin{bmatrix} 0 \\ -3 \end{bmatrix}$

2. Determine the head of the vector $\begin{bmatrix} -2 \\ 5 \end{bmatrix}$ whose tail is $(-3, 2)$. Make a sketch.

3. Determine the tail of the vector $\begin{bmatrix} 2 \\ 6 \end{bmatrix}$ whose head is $(1, 2)$. Make a sketch.

4. Determine the tail of the vector $\begin{bmatrix} 2 \\ 4 \\ -1 \end{bmatrix}$ whose head is $(3, -2, 2)$.

5. For what values of a and b are the vectors $\begin{bmatrix} a-b \\ 2 \end{bmatrix}$ and $\begin{bmatrix} 4 \\ a+b \end{bmatrix}$ equal?

6. For what values of a, b, and c are the vectors $\begin{bmatrix} 2a-b \\ a-2b \\ 6 \end{bmatrix}$ and $\begin{bmatrix} -2 \\ 2 \\ a+b-2c \end{bmatrix}$ equal?

In Exercises 7 and 8, determine the components of each vector $\overrightarrow{PQ}$.

7. **(a)** $P(1, 2)$, $Q(3, 5)$

(b) $P(-2, 2, 3)$, $Q(-3, 5, 2)$

8. **(a)** $P(-1, 0)$, $Q(-3, -4)$

(b) $P(1, 1, 2)$, $Q(1, -2, -4)$

In Exercises 9 and 10, find a vector whose tail is the origin that represents each vector $\overrightarrow{PQ}$.

9. **(a)** $P(-1, 2)$, $Q(3, 5)$

(b) $P(1, 1, -2)$, $Q(3, 4, 5)$

10. **(a)** $P(2, -3)$, $Q(-2, 4)$

(b) $P(-2, -3, 4)$, $Q(0, 0, 1)$

11. Compute $\mathbf{u} + \mathbf{v}$, $\mathbf{u} - \mathbf{v}$, $2\mathbf{u}$, and $3\mathbf{u} - 2\mathbf{v}$ if

(a) $\mathbf{u} = \begin{bmatrix} 2 \\ 3 \end{bmatrix}$, $\mathbf{v} = \begin{bmatrix} -2 \\ 5 \end{bmatrix}$;

(b) $\mathbf{u} = \begin{bmatrix} 0 \\ 3 \end{bmatrix}$, $\mathbf{v} = \begin{bmatrix} 3 \\ 2 \end{bmatrix}$;

(c) $\mathbf{u} = \begin{bmatrix} 2 \\ 6 \end{bmatrix}$, $\mathbf{v} = \begin{bmatrix} 3 \\ 2 \end{bmatrix}$.

12. Compute $\mathbf{u} + \mathbf{v}$, $2\mathbf{u} - \mathbf{v}$, $3\mathbf{u} - 2\mathbf{v}$, and $\mathbf{0} - 3\mathbf{v}$ if

(a) $\mathbf{u} = \begin{bmatrix} 1 \\ 2 \\ 3 \end{bmatrix}$, $\mathbf{v} = \begin{bmatrix} 2 \\ 0 \\ 1 \end{bmatrix}$;

(b) $\mathbf{u} = \begin{bmatrix} 2 \\ -1 \\ 4 \end{bmatrix}$, $\mathbf{v} = \begin{bmatrix} 1 \\ 2 \\ -3 \end{bmatrix}$;

(c) $\mathbf{u} = \begin{bmatrix} 1 \\ 0 \\ -1 \end{bmatrix}$, $\mathbf{v} = \begin{bmatrix} -1 \\ 1 \\ 4 \end{bmatrix}$.

13. Let

$$\mathbf{u} = \begin{bmatrix} 2 \\ 3 \\ -1 \end{bmatrix}, \quad \mathbf{v} = \begin{bmatrix} -1 \\ 2 \\ 4 \end{bmatrix}, \quad \mathbf{w} = \begin{bmatrix} 0 \\ 1 \\ -1 \end{bmatrix},$$

$c = -2$, and $d = 3$. Compute each of the following:

(a) $\mathbf{u} + \mathbf{v}$

(b) $c\mathbf{u} + d\mathbf{w}$

(c) $\mathbf{u} + \mathbf{v} + \mathbf{w}$

(d) $c\mathbf{u} + d\mathbf{v} + \mathbf{w}$

14. Let

$$\mathbf{x} = \begin{bmatrix} 1 \\ 2 \end{bmatrix}, \quad \mathbf{y} = \begin{bmatrix} -3 \\ 4 \end{bmatrix},$$

$$\mathbf{z} = \begin{bmatrix} r \\ 4 \end{bmatrix}, \quad \text{and} \quad \mathbf{u} = \begin{bmatrix} -2 \\ s \end{bmatrix}.$$

Find r and s so that

(a) $\mathbf{z} = 2\mathbf{x}$, **(b)** $\frac{3}{2}\mathbf{u} = \mathbf{y}$, **(c)** $\mathbf{z} + \mathbf{u} = \mathbf{x}$.

15. Let

$$\mathbf{x} = \begin{bmatrix} 1 \\ -2 \\ 3 \end{bmatrix}, \quad \mathbf{y} = \begin{bmatrix} -3 \\ 1 \\ 3 \end{bmatrix}, \quad \mathbf{z} = \begin{bmatrix} r \\ -1 \\ s \end{bmatrix},$$

$$\text{and} \quad \mathbf{u} = \begin{bmatrix} 3 \\ t \\ 2 \end{bmatrix}.$$

Find r, s, and t so that

(a) $\mathbf{z} = \frac{1}{2}\mathbf{x}$, **(b)** $\mathbf{z} + \mathbf{u} = \mathbf{x}$, **(c)** $\mathbf{z} - \mathbf{x} = \mathbf{y}$.

16. If possible, find scalars c_1 and c_2 so that

$$c_1 \begin{bmatrix} 1 \\ -2 \end{bmatrix} + c_2 \begin{bmatrix} 3 \\ -4 \end{bmatrix} = \begin{bmatrix} -5 \\ 6 \end{bmatrix}.$$

17. If possible, find scalars c_1, c_2, and c_3 so that

$$c_1 \begin{bmatrix} 1 \\ 2 \\ -3 \end{bmatrix} + c_2 \begin{bmatrix} -1 \\ 1 \\ 1 \end{bmatrix} + c_3 \begin{bmatrix} -1 \\ 4 \\ -1 \end{bmatrix} = \begin{bmatrix} 2 \\ -2 \\ 3 \end{bmatrix}.$$

18. If possible, find scalars c_1 and c_2, not both zero, so that

$$c_1 \begin{bmatrix} 1 \\ 2 \end{bmatrix} + c_2 \begin{bmatrix} 3 \\ 4 \end{bmatrix} = \begin{bmatrix} 0 \\ 0 \end{bmatrix}.$$

19. If possible, find scalars c_1, c_2, and c_3, not all zero, so that

$$c_1 \begin{bmatrix} 1 \\ 2 \\ -1 \end{bmatrix} + c_2 \begin{bmatrix} 1 \\ 3 \\ -2 \end{bmatrix} + c_3 \begin{bmatrix} 3 \\ 7 \\ -4 \end{bmatrix} = \begin{bmatrix} 0 \\ 0 \\ 0 \end{bmatrix}.$$

20. Let

$$\mathbf{i} = \begin{bmatrix} 1 \\ 0 \\ 0 \end{bmatrix}, \quad \mathbf{j} = \begin{bmatrix} 0 \\ 1 \\ 0 \end{bmatrix}, \quad \text{and} \quad \mathbf{k} = \begin{bmatrix} 0 \\ 0 \\ 1 \end{bmatrix}.$$

Find scalars c_1, c_2, and c_3 so that any vector $\mathbf{u} = \begin{bmatrix} r \\ s \\ t \end{bmatrix}$ can be written as $\mathbf{u} = c_1\mathbf{i} + c_2\mathbf{j} + c_3\mathbf{k}$.

21. Show that if $\mathbf{u}$ is a vector in R^2 or R^3, then $\mathbf{u} + \mathbf{0} = \mathbf{u}$.

22. Show that if $\mathbf{u}$ is a vector in R^2 or R^3, then

$$\mathbf{u} + (-1)\mathbf{u} = \mathbf{0}.$$

23. Prove part (b) and parts (d) through (h) of Theorem 4.1.

24. Determine whether the software you use supports graphics. If it does, experiment with plotting vectors in R^2. Usually, you must supply coordinates for the head and tail of the vector and then tell the software to connect these points. The points in Exercises 7(a) and 8(a) can be used this way.

25. Assuming that the software you use supports graphics (see Exercise 24), plot the vector

$$\mathbf{v} = \begin{bmatrix} 3 \\ 4 \end{bmatrix}$$

on the same coordinate axes for each of the following:

(a) $\mathbf{v}$ is to have head (1, 1).

(b) $\mathbf{v}$ is to have head (2, 3).

26. Determine whether the software you use supports three-dimensional graphics, that is, plots points in R^3. If it does, experiment with plotting points and connecting them to form vectors in R^3.

4.2 Vector Spaces

A useful procedure in mathematics and other disciplines involves classification schemes. That is, we form classes of objects based on properties they have in common. This allows us to treat all members of the class as a single unit. Thus, instead of dealing with each distinct member of the class, we can develop notions that apply to each and every member of the class based on the properties that they have in common. In many ways this helps us work with more abstract and comprehensive ideas.

Linear algebra has such a classification scheme that has become very important. This is the notion of a **vector space**. A vector space consists of a set of objects and two operations on these objects that satisfy a certain set of rules. If we have a vector space, we will automatically be able to attribute to it certain properties that hold for all vector spaces. Thus, upon meeting some new vector space, we will not have to verify everything from scratch.

The name "vector space" conjures up the image of directed line segments from the plane, or 3-space, as discussed in Section 4.1. This is, of course, where the name of the classification scheme is derived from. We will see that matrices and n-vectors will give us examples of vector spaces, but other collections of objects, like polynomials, functions, and solutions to various types of equations, will also be vector spaces. For particular vector spaces, the members of the set of objects and the operations on these objects can vary, but the rules governing the properties satisfied by the operations involved will always be the same.

DEFINITION 4.4

A **real vector space** is a set V of elements on which we have two operations $\oplus$ and $\odot$ defined with the following properties:

(a) If $\mathbf{u}$ and $\mathbf{v}$ are any elements in V, then $\mathbf{u} \oplus \mathbf{v}$ is in V. (We say that V is **closed** under the operation $\oplus$.)

(1) $\mathbf{u} \oplus \mathbf{v} = \mathbf{v} \oplus \mathbf{u}$ for all $\mathbf{u}$, $\mathbf{v}$ in V.

(2) $\mathbf{u} \oplus (\mathbf{v} \oplus \mathbf{w}) = (\mathbf{u} \oplus \mathbf{v}) \oplus \mathbf{w}$ for all $\mathbf{u}$, $\mathbf{v}$, $\mathbf{w}$ in V.

(3) There exists an element $\mathbf{0}$ in V such that $\mathbf{u} \oplus \mathbf{0} = \mathbf{0} \oplus \mathbf{u} = \mathbf{u}$ for any $\mathbf{u}$ in V.

(4) For each $\mathbf{u}$ in V there exists an element $-\mathbf{u}$ in V such that $\mathbf{u} \oplus -\mathbf{u} = -\mathbf{u} \oplus \mathbf{u} = \mathbf{0}$.

(b) If $\mathbf{u}$ is any element in V and c is any real number, then $c \odot \mathbf{u}$ is in V (i.e., V is closed under the operation $\odot$).

(5) $c \odot (\mathbf{u} \oplus \mathbf{v}) = c \odot \mathbf{u} \oplus c \odot \mathbf{v}$ for any $\mathbf{u}$, $\mathbf{v}$ in V and any real number c.

(6) $(c + d) \odot \mathbf{u} = c \odot \mathbf{u} \oplus d \odot \mathbf{u}$ for any $\mathbf{u}$ in V and any real numbers c and d.

(7) $c \odot (d \odot \mathbf{u}) = (cd) \odot \mathbf{u}$ for any $\mathbf{u}$ in V and any real numbers c and d.

(8) $1 \odot \mathbf{u} = \mathbf{u}$ for any $\mathbf{u}$ in V.

The elements of V are called **vectors**; the elements of the set of real numbers R are called **scalars**. The operation $\oplus$ is called **vector addition**; the operation $\odot$ is called **scalar multiplication**. The vector $\mathbf{0}$ in property (3) is called a **zero vector**. The vector $-\mathbf{u}$ in property (4) is called a **negative of u**. It can be shown (see Exercises 19 and 20) that $\mathbf{0}$ and $-\mathbf{u}$ are unique.

If we allow the scalars to be complex numbers, we obtain a **complex vector space**. More generally, the scalars can be members of a field* F, and we obtain a vector space over F. Such spaces are important in many applications in mathematics and the physical sciences. We provide a brief introduction to complex vector

*A field is an algebraic structure enjoying the arithmetic properties shared by the real, complex, and rational numbers. Fields are studied in detail in an abstract algebra course.

spaces in Appendix B. However, in this book most of our attention will be focused on real vector spaces.

In order to specify a vector space, we must be given a set V and two operations $\oplus$ and $\odot$ satisfying all the properties of the definition. We shall often refer to a real vector space merely as a **vector space**. Thus a "vector" is now an element of a vector space and no longer needs to be interpreted as a directed line segment. In our examples we shall see, however, how this name came about in a natural manner. We now consider some examples of vector spaces, leaving it to the reader to verify that all the properties of Definition 4.4 hold.

EXAMPLE 1 Consider R^n, the set of all $n \times 1$ matrices

$$\begin{bmatrix} a_1 \\ a_2 \\ \vdots \\ a_n \end{bmatrix}$$

with real entries. Let the operation $\oplus$ be matrix addition and let the operation $\odot$ be multiplication of a matrix by a real number (scalar multiplication).

By the use of the properties of matrices established in Section 1.4, it is not difficult to show that R^n is a vector space by verifying that the properties of Definition 4.4 hold. Thus the matrix $\begin{bmatrix} a_1 \\ a_2 \\ \vdots \\ a_n \end{bmatrix}$, as an element of R^n, is now called an n-vector, or merely a *vector*. We have already discussed R^2 and R^3 in Section 4.1. See Figure 4.16 for geometric representations of R^2 and R^3. Although we shall see later that many geometric notions, such as length and the angle between vectors, can be defined in R^n for $n > 3$, we cannot draw pictures in these cases. ■

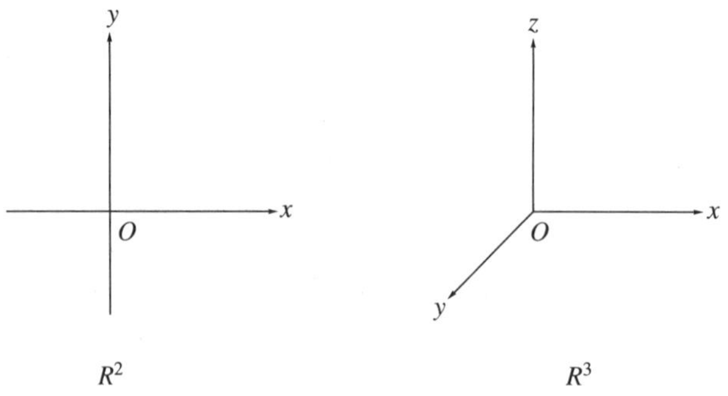

FIGURE 4.16

EXAMPLE 2 The set of all $m \times n$ matrices with matrix addition as $\oplus$ and multiplication of a matrix by a real number as $\odot$ is a vector space (verify). We denote this vector space by M_{mn}. ■

EXAMPLE 3 The set of all real numbers with $\oplus$ as the usual addition of real numbers and $\odot$ as the usual multiplication of real numbers is a vector space (verify). In this case the real numbers play the dual roles of both vectors and scalars. This vector space is essentially the case with $n = 1$ of Example 1. ■

EXAMPLE 4

Let R_n be the set of all $1 \times n$ matrices $\begin{bmatrix} a_1 & a_2 & \cdots & a_n \end{bmatrix}$, where we define $\oplus$ by

$$\begin{bmatrix} a_1 & a_2 & \cdots & a_n \end{bmatrix} \oplus \begin{bmatrix} b_1 & b_2 & \cdots & b_n \end{bmatrix} = \begin{bmatrix} a_1 + b_1 & a_2 + b_2 & \cdots & a_n + b_n \end{bmatrix}$$

and we define $\odot$ by

$$c \odot \begin{bmatrix} a_1 & a_2 & \cdots & a_n \end{bmatrix} = \begin{bmatrix} ca_1 & ca_2 & \cdots & ca_n \end{bmatrix}.$$

Then R_n is a vector space (verify). This is just a special case of Example 2. ■

EXAMPLE 5

Let V be the set of all 2×2 matrices with trace equal to zero; that is,

$$A = \begin{bmatrix} a & b \\ c & d \end{bmatrix} \text{ is in } V \quad \text{provided} \quad \mathrm{Tr}(A) = a + d = 0.$$

(See Section 1.3, Exercise 43 for the definition and properties of the trace of a matrix.) The operation $\oplus$ is standard matrix addition, and the operation $\odot$ is standard scalar multiplication of matrices; then V is a vector space. We verify properties (a), (3), (4), (b), and (7) of Definition 4.4. The remaining properties are left for the student to verify.

Let

$$A = \begin{bmatrix} a & b \\ c & d \end{bmatrix} \quad \text{and} \quad B = \begin{bmatrix} r & s \\ t & p \end{bmatrix}$$

be any elements of V. Then $\mathrm{Tr}(A) = a + d = 0$ and $\mathrm{Tr}(B) = r + p = 0$. For property (a), we have

$$A \oplus B = \begin{bmatrix} a + r & b + s \\ c + t & d + p \end{bmatrix}$$

and

$$\mathrm{Tr}(A \oplus B) = (a + r) + (d + p) = (a + d) + (r + p) = 0 + 0 = 0,$$

so $A \oplus B$ is in V; that is, V is closed under the operation $\oplus$. To verify property (3), observe that the matrix

$$\begin{bmatrix} 0 & 0 \\ 0 & 0 \end{bmatrix}$$

has trace equal to zero, so it is in V. Then it follows from the definition of $\oplus$ that property (3) is valid in V, so $\begin{bmatrix} 0 & 0 \\ 0 & 0 \end{bmatrix}$ is the zero vector, which we denote as $\mathbf{0}$. To verify property (4), let A, as given previously, be an element of V and let

$$C = \begin{bmatrix} -a & -b \\ -c & -d \end{bmatrix}.$$

We first show that C is in V:

$$\mathrm{Tr}(C) = (-a) + (-c) = -(a + c) = 0.$$

Then we have

$$A \oplus C = \begin{bmatrix} a & b \\ c & d \end{bmatrix} + \begin{bmatrix} -a & -b \\ -c & -d \end{bmatrix} = \begin{bmatrix} 0 & 0 \\ 0 & 0 \end{bmatrix} = \mathbf{0},$$

so $C = -A$. For property (b), let k be any real number. We have

$$k \odot A = \begin{bmatrix} ka & kb \\ kc & kd \end{bmatrix}$$

and

$$\text{Tr}(k \odot A) = ka + kd = k(a + d) = 0,$$

so $k \odot A$ is in V; that is, V is closed under the operation $\odot$. For property (7), let k and m be any real numbers. Then

$$k \odot (m \odot A) = k \odot \begin{bmatrix} ma & mb \\ mc & md \end{bmatrix} = \begin{bmatrix} kma & kmb \\ kmc & kmd \end{bmatrix}$$

and

$$(km) \odot A = \begin{bmatrix} kma & kmb \\ kmc & kmd \end{bmatrix}.$$

It follows that $k \odot (m \odot A) = (km) \odot A$. ■

EXAMPLE 6

Another source of examples are sets of polynomials; therefore, we recall some well-known facts about such functions. A **polynomial** (in t) is a function that is expressible as

$$p(t) = a_n t^n + a_{n-1} t^{n-1} + \cdots + a_1 t + a_0,$$

where $a_0, a_1, \ldots, a_n$ are real numbers and n is a nonnegative integer. If $a_n \neq 0$, then $p(t)$ is said to have **degree *n***. Thus the degree of a polynomial is the highest power of a term having a nonzero coefficient; $p(t) = 2t + 1$ has degree 1, and the constant polynomial $p(t) = 3$ has degree 0. The **zero polynomial**, denoted by $\mathbf{0}$, has no degree. We now let P_n be the set of all polynomials of degree $\leq n$ together with the zero polynomial. If $p(t)$ and $q(t)$ are in P_n, we can write

$$p(t) = a_n t^n + a_{n-1} t^{n-1} + \cdots + a_1 t + a_0$$

and

$$q(t) = b_n t^n + b_{n-1} t^{n-1} + \cdots + b_1 t + b_0.$$

We define $p(t) \oplus q(t)$ as

$$p(t) \oplus q(t) = (a_n + b_n)t^n + (a_{n-1} + b_{n-1})t^{n-1} + \cdots + (a_1 + b_1)t + (a_0 + b_0).$$

If c is a scalar, we also define $c \odot p(t)$ as

$$c \odot p(t) = (ca_n)t^n + (ca_{n-1})t^{n-1} + \cdots + (ca_1)t + (ca_0).$$

We now show that P_n is a vector space.

Let $p(t)$ and $q(t)$, as before, be elements of P_n; that is, they are polynomials of degree $\leq n$ or the zero polynomial. Then the previous definitions of the operations $\oplus$ and $\odot$ show that $p(t) \oplus q(t)$ and $c \odot p(t)$, for any scalar c, are polynomials of degree $\leq n$ or the zero polynomial. That is, $p(t) \oplus q(t)$ and $c \odot p(t)$ are in P_n so that (a) and (b) in Definition 4.4 hold. To verify property (1), we observe that

$$q(t) \oplus p(t) = (b_n + a_n)t^n + (b_{n-1} + a_{n-1})t^{n-1} + \cdots + (b_1 + a_1)t + (a_0 + b_0),$$

and since $a_i + b_i = b_i + a_i$ holds for the real numbers, we conclude that $p(t) \oplus q(t) = q(t) \oplus p(t)$. Similarly, we verify property (2). The zero polynomial is the element $\mathbf{0}$ needed in property (3). If $p(t)$ is as given previously, then its negative, $-p(t)$, is

$$-a_n t^n - a_{n-1}t^{n-1} - \cdots - a_1 t - a_0.$$

We shall now verify property (6) and will leave the verification of the remaining properties to the reader. Thus

$$\begin{aligned}
(c+d) \odot p(t) &= (c+d)a_n t^n + (c+d)a_{n-1}t^{n-1} + \cdots + (c+d)a_1 t \\
&\quad + (c+d)a_0 \\
&= ca_n t^n + da_n t^n + ca_{n-1}t^{n-1} + da_{n-1}t^{n-1} + \cdots + ca_1 t \\
&\quad + da_1 t + ca_0 + da_0 \\
&= c(a_n t^n + a_{n-1}t^{n-1} + \cdots + a_1 t + a_0) \\
&\quad + d(a_n t^n + a_{n-1}t^{n-1} + \cdots + a_1 t + a_0) \\
&= c \odot p(t) \oplus d \odot p(t).
\end{aligned}$$

■

Remark We show later that the vector space P_n behaves algebraically in exactly the same manner as R^{n+1}.

For each natural number n, we have just defined the vector space P_n of all polynomials of degree $\leq n$ together with the zero polynomial. We could also consider the space P of *all* polynomials (of any degree), together with the zero polynomial. Here P is the mathematical union of all the vector spaces P_n. Two polynomials $p(t)$ of degree n and $q(t)$ of degree m are added in P in the same way as they would be added in P_r, where r is the maximum of the two numbers m and n. Then P is a vector space (Exercise 6).

As in the case of ordinary real-number arithmetic, in an expression containing both $\odot$ and $\oplus$, the $\odot$ operation is performed first. Moreover, the familiar arithmetic rules, when parentheses are encountered, apply in this case also.

EXAMPLE 7

Let V be the set of all real-valued continuous functions defined on R^1. If f and g are in V, we define $f \oplus g$ by $(f \oplus g)(t) = f(t) + g(t)$. If f is in V and c is a scalar, we define $c \odot f$ by $(c \odot f)(t) = cf(t)$. Then V is a vector space, which is denoted by $C(-\infty, \infty)$. (See Exercise 13.) ■

EXAMPLE 8

Let V be the set of all real multiples of exponential functions of the form e^{kx}, where k is any real number. Define vector addition $\oplus$ as

$$c_1 e^{kx} \oplus c_2 e^{mx} = c_1 c_2 e^{(k+m)x}$$

and scalar multiplication $\odot$ as

$$r \odot c_1 e^{kx} = rc_1 e^{kx}.$$

From the definitions of $\oplus$ and $\odot$ we see that V is closed under both operations. It can be shown that properties (1) and (2) hold. As for property (3), we have $e^{0x} = 1$, so for any vector $c_1 e^{kx}$ in V, we have $c_1 e^{kx} \oplus 1 = 1 \oplus c_1 e^{kx} = c_1 e^{kx}$. Hence the zero vector in this case is the number 1. Next we consider property (4). For any vector $c_1 e^{kx}$ in V, we need to determine another vector $c_2 e^{mx}$ in V so that

$$c_1 e^{kx} \oplus c_2 e^{mx} = c_1 c_2 e^{(k+m)x} = 1.$$

We have $0e^{kx} = 0$ is in V; however, there is no vector in V that we can add to 0 (using the definition of $\oplus$) to get 1; thus V with the given operations $\oplus$ and $\odot$ is not a vector space. ■

EXAMPLE 9

Let V be the set of all real numbers with the operations $\mathbf{u} \oplus \mathbf{v} = \mathbf{u} - \mathbf{v}$ ($\oplus$ is ordinary subtraction) and $c \odot \mathbf{u} = c\mathbf{u}$ ($\odot$ is ordinary multiplication). Is V a vector space? If it is not, which properties in Definition 4.4 fail to hold?

Solution

If $\mathbf{u}$ and $\mathbf{v}$ are in V, and c is a scalar, then $\mathbf{u} \oplus \mathbf{v}$ and $c \odot \mathbf{u}$ are in V, so that (a) and (b) in Definition 4.4 hold. However, property (1) fails to hold, since

$$\mathbf{u} \oplus \mathbf{v} = \mathbf{u} - \mathbf{v} \quad \text{and} \quad \mathbf{v} \oplus \mathbf{u} = \mathbf{v} - \mathbf{u},$$

and these are not the same, in general. (Find $\mathbf{u}$ and $\mathbf{v}$ such that $\mathbf{u} - \mathbf{v} \neq \mathbf{v} - \mathbf{u}$.) Also, we shall let the reader verify that properties (2), (3), and (4) fail to hold. Properties (5), (7), and (8) hold, but property (6) does not hold, because

$$(c + d) \odot \mathbf{u} = (c + d)\mathbf{u} = c\mathbf{u} + d\mathbf{u},$$

whereas

$$c \odot \mathbf{u} \oplus d \odot \mathbf{u} = c\mathbf{u} \oplus d\mathbf{u} = c\mathbf{u} - d\mathbf{u},$$

and these are not equal, in general. Thus V is not a vector space. ■

EXAMPLE 10

Let V be the set of all ordered triples of real numbers (x, y, z) with the operations $(x, y, z) \oplus (x', y', z') = (x', y + y', z + z')$; $c \odot (x, y, z) = (cx, cy, cz)$. We can readily verify that properties (1), (3), (4), and (6) of Definition 4.4 fail to hold. For example, if $\mathbf{u} = (x, y, z)$ and $\mathbf{v} = (x', y', z')$, then

$$\mathbf{u} \oplus \mathbf{v} = (x, y, z) \oplus (x', y', z') = (x', y + y', z + z'),$$

whereas

$$\mathbf{v} \oplus \mathbf{u} = (x', y', z') \oplus (x, y, z) = (x, y' + y, z' + z),$$

so property (1) fails to hold when $x \neq x'$. Also,

$$\begin{aligned}(c + d) \odot \mathbf{u} &= (c + d) \odot (x, y, z) \\ &= ((c + d)x, (c + d)y, (c + d)z) \\ &= (cx + dx, cy + dy, cz + dz),\end{aligned}$$

whereas

$$\begin{aligned} c \odot \mathbf{u} \oplus d \odot \mathbf{u} &= c \odot (x, y, z) \oplus d \odot (x, y, z) \\ &= (cx, cy, cz) \oplus (dx, dy, dz) \\ &= (dx, cy + dy, cz + dz), \end{aligned}$$

so property (6) fails to hold when $cx \neq 0$. Thus V is not a vector space. ■

EXAMPLE 11 Let V be the set of all integers; define $\oplus$ as ordinary addition and $\odot$ as ordinary multiplication. Here V is not a vector, because if $\mathbf{u}$ is any nonzero vector in V and $c = \sqrt{3}$, then $c \odot \mathbf{u}$ is not in V. Thus (b) fails to hold. ■

> To verify that a given set V with two operations $\oplus$ and $\odot$ is a real vector space, we must show that it satisfies all the properties of Definition 4.4. The first thing to check is whether (a) and (b) hold, for if either of these fails, we do not have a vector space. If both (a) and (b) hold, it is recommended that (3), the existence of a zero element, be verified next. Naturally, if (3) fails to hold, we do not have a vector space and do not have to check the remaining properties.

The following theorem presents some useful properties common to all vector spaces:

Theorem 4.2 If V is a vector space, then

(a) $0 \odot \mathbf{u} = \mathbf{0}$ for any vector $\mathbf{u}$ in V.

(b) $c \odot \mathbf{0} = \mathbf{0}$ for any scalar c.

(c) If $c \odot \mathbf{u} = \mathbf{0}$, then either $c = 0$ or $\mathbf{u} = \mathbf{0}$.

(d) $(-1) \odot \mathbf{u} = -\mathbf{u}$ for any vector $\mathbf{u}$ in V.

Proof

(a) We have

$$0 \odot \mathbf{u} = (0 + 0) \odot \mathbf{u} = 0 \odot \mathbf{u} + 0 \odot \mathbf{u} \tag{1}$$

by (6) of Definition 4.4. Adding $-0 \odot \mathbf{u}$ to both sides of Equation (1), we obtain by (2), (3), and (4) of Definition 4.4,

$$0 \odot \mathbf{u} = \mathbf{0}.$$

Parts (b) and (c) are left as Exercise 21.

(d) $(-1) \odot \mathbf{u} \oplus \mathbf{u} = (-1) \odot \mathbf{u} \oplus 1 \odot \mathbf{u} = (-1 + 1) \odot \mathbf{u} = 0 \odot \mathbf{u} = \mathbf{0}$. Since $-\mathbf{u}$ is unique, we conclude that

$$(-1) \odot \mathbf{u} = -\mathbf{u}.$$ ■

In the examples in this section we have introduced notation for a number of sets that will be used in the exercises and examples throughout the rest of this book. The following table summarizes the notation and the descriptions of the set:

R^n	the set of $n \times 1$ matrices
R_n	the set of $1 \times n$ matrices
M_{mn}	the set of $m \times n$ matrices
P_n	the set of all polynomials of degree n or less together with the zero polynomial
P	the set of all polynomials
$C(-\infty, \infty)$	the set of all real-valued continuous functions with domain all real numbers

We also want to emphasize that if V is a vector space, its elements are called vectors regardless of the form of the elements of V.

Key Terms

Real vector space	Closed	Polynomial	M_{mn}
Vectors	Zero vector	Degree of a polynomial	P_n
Scalars	Negative of a vector	Zero polynomial	P
Vector addition	Complex vector space	R^n	$C(-\infty, \infty)$
Scalar multiplication	Field	R_n	

4.2 Exercises

1. Let V be the set of all polynomials of (exactly) degree 2 with the definitions of addition and scalar multiplication as in Example 6.

(a) Show that V is not closed under addition.

(b) Is V closed under scalar multiplication? Explain.

2. Let V be the set of all 2×2 matrices $A = \begin{bmatrix} a & b \\ c & d \end{bmatrix}$ such that the product $abcd = 0$. Let the operation $\oplus$ be standard addition of matrices and the operation $\odot$ be standard scalar multiplication of matrices.

(a) Is V closed under addition?

(b) Is V closed under scalar multiplication?

(c) What is the zero vector in the set V?

(d) Does every matrix A in V have a negative that is in V? Explain.

(e) Is V a vector space? Explain.

3. Let V be the set of all 2×2 matrices $A = \begin{bmatrix} a & b \\ 2b & d \end{bmatrix}$. Let the operation $\oplus$ be standard addition of matrices and the operation $\odot$ be standard scalar multiplication of matrices.

(a) Is V closed under addition?

(b) Is V closed under scalar multiplication?

(c) What is the zero vector in the set V?

(d) Does every matrix A in V have a negative that is in V? Explain.

(e) Is V a vector space? Explain.

4. Let V be the set of all 2×1 matrices $\mathbf{v} = \begin{bmatrix} v_1 \\ v_2 \end{bmatrix}$ with integer entries such that $|v_1 + v_2|$ is even. For example,

$$\begin{bmatrix} 2 \\ 4 \end{bmatrix}, \quad \begin{bmatrix} 0 \\ -8 \end{bmatrix}, \quad \begin{bmatrix} 3 \\ 5 \end{bmatrix}, \quad \text{and} \quad \begin{bmatrix} 0 \\ 0 \end{bmatrix}$$

all belong to V. Let the operation $\oplus$ be standard addition of matrices and the operation $\odot$ be standard scalar multiplication of matrices. Is V a vector space? Explain.

5. Prove in detail that R^n is a vector space.

6. Show that P, the set of all polynomials, is a vector space.

In Exercises 7 through 11, the given set together with the given operations is not a vector space. List the properties of Definition 4.4 that fail to hold.

7. The set of all positive real numbers with the operations of $\oplus$ as ordinary addition and $\odot$ as ordinary multiplication.

8. The set of all ordered pairs of real numbers with the operations

$$(x, y) \oplus (x', y') = (x + x', y + y')$$

and

$$r \odot (x, y) = (x, ry).$$

9. The set of all ordered triples of real numbers with the operations

$$(x, y, z) \oplus (x', y', z') = (x + x', y + y', z + z')$$

and

$$r \odot (x, y, z) = (x, 1, z).$$

10. The set of all 2×1 matrices $\begin{bmatrix} x \\ y \end{bmatrix}$, where $x \leq 0$, with the usual operations in R^2.

11. The set of all ordered pairs of real numbers with the operations $(x, y) \oplus (x', y') = (x + x', y + y')$ and $r \odot (x, y) = (0, 0)$.

12. Let V be the set of all positive real numbers; define $\oplus$ by $\mathbf{u} \oplus \mathbf{v} = \mathbf{uv}$ ($\oplus$ is ordinary multiplication) and define $\odot$ by $c \odot \mathbf{v} = \mathbf{v}^c$. Prove that V is a vector space.

13. Let V be the set of all real-valued continuous functions. If f and g are in V, define $f \oplus g$ by

$$(f \oplus g)(t) = f(t) + g(t).$$

If f is in V, define $c \odot f$ by $(c \odot f)(t) = cf(t)$. Prove that V is a vector space. (This is the vector space defined in Example 7.)

14. Let V be the set consisting of a single element $\mathbf{0}$. Let $\mathbf{0} \oplus \mathbf{0} = \mathbf{0}$ and $c \odot \mathbf{0} = \mathbf{0}$. Prove that V is a vector space.

15. (***Calculus Required***) Consider the differential equation $y'' - y' + 2y = 0$. A solution is a real-valued function f satisfying the equation. Let V be the set of all solutions to the given differential equation; define $\oplus$ and $\odot$ as in Exercise 13. Prove that V is a vector space. (See also Section 8.5.)

16. Let V be the set of all positive real numbers; define $\oplus$ by $\mathbf{u} \oplus \mathbf{v} = \mathbf{uv} - 1$ and $\odot$ by $c \odot \mathbf{v} = \mathbf{v}$. Is V a vector space?

17. Let V be the set of all real numbers; define $\oplus$ by $\mathbf{u} \oplus \mathbf{v} = \mathbf{uv}$ and $\odot$ by $c \odot \mathbf{u} = c + \mathbf{u}$. Is V a vector space?

18. Let V be the set of all real numbers; define $\oplus$ by $\mathbf{u} \oplus \mathbf{v} = 2\mathbf{u} - \mathbf{v}$ and $\odot$ by $c \odot \mathbf{u} = c\mathbf{u}$. Is V a vector space?

19. Prove that a vector space has only one zero vector.

20. Prove that a vector $\mathbf{u}$ in a vector space has only one negative, $-\mathbf{u}$.

21. Prove parts (b) and (c) of Theorem 4.2.

22. Prove that the set V of all real-valued functions is a vector space under the operations defined in Exercise 13.

23. Prove that $-(-\mathbf{v}) = \mathbf{v}$.

24. Prove that if $\mathbf{u} \oplus \mathbf{v} = \mathbf{u} \oplus \mathbf{w}$, then $\mathbf{v} = \mathbf{w}$.

25. Prove that if $\mathbf{u} \neq \mathbf{0}$ and $a \odot \mathbf{u} = b \odot \mathbf{u}$, then $a = b$.

26. Example 6 discusses the vector space P_n of polynomials of degree n or less. Operations on polynomials can be performed in linear algebra software by associating a row matrix of size $n+1$ with polynomial $p(t)$ of P_n. The row matrix consists of the coefficients of $p(t)$, using the association

$$p(t) = a_n t^n + a_{n-1} t^{n-1} + \cdots + a_1 t + a_0$$
$$\rightarrow \begin{bmatrix} a_n & a_{n-1} & \cdots & a_1 & a_0 \end{bmatrix}.$$

If any term of $p(t)$ is explicitly missing, a zero is used for its coefficient. Then the addition of polynomials corresponds to matrix addition, and multiplication of a polynomial by a scalar corresponds to scalar multiplication of matrices. With your software, perform each given operation on polynomials, using the matrix association as just described. Let $n = 3$ and

$$p(t) = 2t^3 + 5t^2 + t - 2, \quad q(t) = t^3 + 3t + 5.$$

(a) $p(t) + q(t)$ **(b)** $5p(t)$ **(c)** $3p(t) - 4q(t)$

4.3 Subspaces

In this section we begin to analyze the structure of a vector space. First, it is convenient to have a name for a subset of a given vector space that is itself a vector space with respect to the same operations as those in V. Thus we have a definition.

DEFINITION 4.5

Let V be a vector space and W a nonempty subset of V. If W is a vector space with respect to the operations in V, then W is called a **subspace** of V.

It follows from Definition 4.5 that to verify that a subset W of a vector space V is a subspace, one must check that (a), (b), and (1) through (8) of Definition 4.4 hold. However, the next theorem says that it is enough to merely check that (a) and (b) hold to verify that a subset W of a vector space V is a subspace. Property (a) is called the **closure** property for $\oplus$, and property (b) is called the **closure** property for $\odot$.

Theorem 4.3 Let V be a vector space with operations $\oplus$ and $\odot$ and let W be a nonempty subset of V. Then W is a subspace of V if and only if the following conditions hold:

(a) If $\mathbf{u}$ and $\mathbf{v}$ are any vectors in W, then $\mathbf{u} \oplus \mathbf{v}$ is in W.

(b) If c is any real number and $\mathbf{u}$ is any vector in W, then $c \odot \mathbf{u}$ is in W.

Proof

If W is a subspace of V, then it is a vector space and (a) and (b) of Definition 4.4 hold; these are precisely (a) and (b) of the theorem.

Conversely, suppose that (a) and (b) hold. We wish to show that W is a subspace of V. First, from (b) we have that $(-1) \odot \mathbf{u}$ is in W for any $\mathbf{u}$ in W. From (a) we have that $\mathbf{u} \oplus (-1) \odot \mathbf{u}$ is in W. But $\mathbf{u} \oplus (-1) \odot \mathbf{u} = \mathbf{0}$, so $\mathbf{0}$ is in W. Then $\mathbf{u} \oplus \mathbf{0} = \mathbf{u}$ for any $\mathbf{u}$ in W. Finally, properties (1), (2), (5), (6), (7), and (8) hold in W because they hold in V. Hence W is a subspace of V. ■

Examples of subspaces of a given vector space occur frequently. We investigate several of these. More examples will be found in the exercises.

EXAMPLE 1

Every vector space has at least two subspaces, itself and the subspace $\{\mathbf{0}\}$ consisting only of the zero vector. (Recall that $\mathbf{0} \oplus \mathbf{0} = \mathbf{0}$ and $c \odot \mathbf{0} = \mathbf{0}$ in any vector space.) Thus $\{\mathbf{0}\}$ is closed for both operations and hence is a subspace of V. The subspace $\{\mathbf{0}\}$ is called the **zero subspace** of V. ■

EXAMPLE 2

Let P_2 be the set consisting of all polynomials of degree ≤ 2 and the zero polynomial; P_2 is a subset of P, the vector space of all polynomials. To verify that P_2 is a *subspace* of P, show it is closed for $\oplus$ and $\odot$. In general, the set P_n consisting of all polynomials of degree $\leq n$ and the zero polynomial is a subspace of P. Also, P_n is a subspace of P_{n+1}. ■

EXAMPLE 3

Let V be the set of all polynomials of degree exactly $= 2$; V is a *subset* of P, the vector space of all polynomials; but V is not a *subspace* of P, because the sum of the polynomials $2t^2 + 3t + 1$ and $-2t^2 + t + 2$ is not in V, since it is a polynomial of degree 1. (See also Exercise 1 in Section 4.2.) ■

EXAMPLE 4

Which of the following subsets of R^2 with the usual operations of vector addition and scalar multiplication are subspaces?

(a) W_1 is the set of all vectors of the form $\begin{bmatrix} x \\ y \end{bmatrix}$, where $x \geq 0$.

(b) W_2 is the set of all vectors of the form $\begin{bmatrix} x \\ y \end{bmatrix}$, where $x \geq 0$, $y \geq 0$.

(c) W_3 is the set of all vectors of the form $\begin{bmatrix} x \\ y \end{bmatrix}$, where $x = 0$.

Solution

(a) W_1 is the right half of the xy-plane (see Figure 4.17). It is not a subspace of

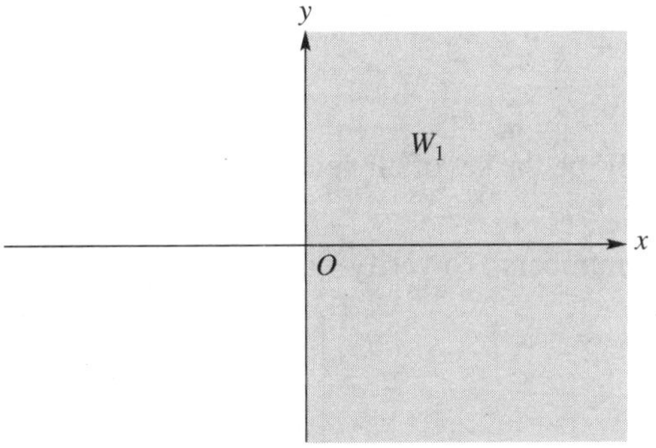

FIGURE 4.17

R^2, because if we take the vector $\begin{bmatrix} 2 \\ 3 \end{bmatrix}$ in W_1, then the scalar multiple

$$-3 \odot \begin{bmatrix} 2 \\ 3 \end{bmatrix} = \begin{bmatrix} -6 \\ -9 \end{bmatrix}$$

is not in W_1, so property (b) in Theorem 4.3 does not hold.

(b) W_2 is the first quadrant of the xy-plane. (See Figure 4.18.) The same vector and scalar multiple as in part (a) shows that W_2 is not a subspace.

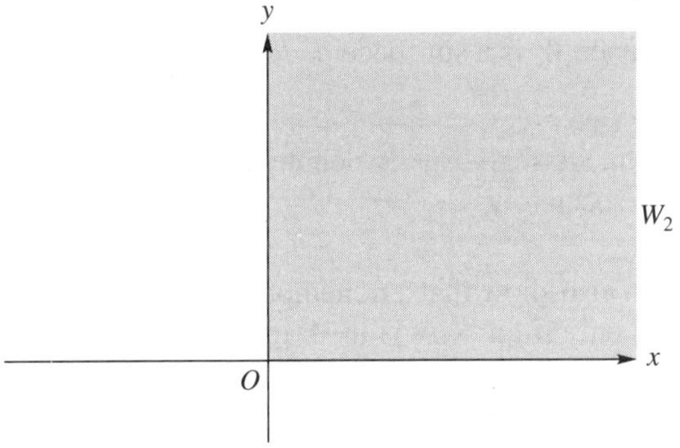

FIGURE 4.18

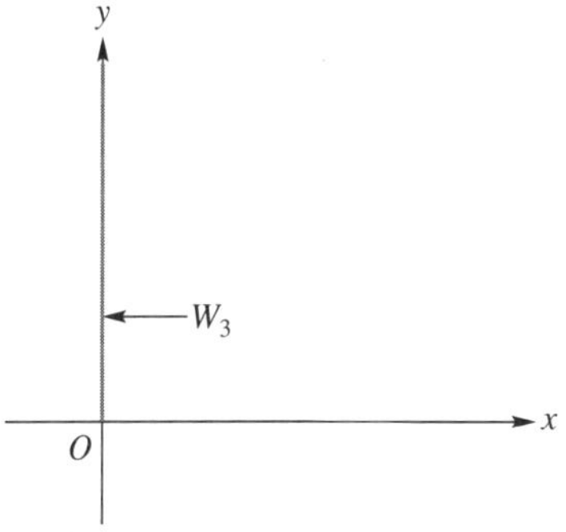

FIGURE 4.19

(c) W_3 is the y-axis in the xy-plane (see Figure 4.19). To see whether W_3 is a subspace, let

$$\mathbf{u} = \begin{bmatrix} 0 \\ b_1 \end{bmatrix} \quad \text{and} \quad \mathbf{v} = \begin{bmatrix} 0 \\ b_2 \end{bmatrix}$$

be vectors in W_3. Then

$$\mathbf{u} \oplus \mathbf{v} = \begin{bmatrix} 0 \\ b_1 \end{bmatrix} + \begin{bmatrix} 0 \\ b_2 \end{bmatrix} = \begin{bmatrix} 0 \\ b_1 + b_2 \end{bmatrix},$$

which is in W_3, so property (a) in Theorem 4.3 holds. Moreover, if c is a scalar, then

$$c \odot \mathbf{u} = c \odot \begin{bmatrix} 0 \\ b_1 \end{bmatrix} = \begin{bmatrix} 0 \\ cb_1 \end{bmatrix},$$

which is in W_3 so property (b) in Theorem 4.3 holds. Hence W_3 is a subspace of R^2. ■

EXAMPLE 5 Let W be the set of all vectors in R^3 of the form $\begin{bmatrix} a \\ b \\ a+b \end{bmatrix}$, where a and b are any real numbers. To verify Theorem 4.3(a) and (b), we let

$$\mathbf{u} = \begin{bmatrix} a_1 \\ b_1 \\ a_1 + b_1 \end{bmatrix} \quad \text{and} \quad \mathbf{v} = \begin{bmatrix} a_2 \\ b_2 \\ a_2 + b_2 \end{bmatrix}$$

be two vectors in W. Then

$$\mathbf{u} \oplus \mathbf{v} = \begin{bmatrix} a_1 + a_2 \\ b_1 + b_2 \\ (a_1 + b_1) + (a_2 + b_2) \end{bmatrix} = \begin{bmatrix} a_1 + a_2 \\ b_1 + b_2 \\ (a_1 + a_2) + (b_1 + b_2) \end{bmatrix}$$

is in W, for W consists of all those vectors whose third entry is the sum of the first two entries. Similarly,

$$c \odot \begin{bmatrix} a_1 \\ b_1 \\ a_1 + b_1 \end{bmatrix} = \begin{bmatrix} ca_1 \\ cb_1 \\ c(a_1 + b_1) \end{bmatrix} = \begin{bmatrix} ca_1 \\ cb_1 \\ ca_1 + cb_1 \end{bmatrix}$$

is in W. Hence W is a subspace of R^3. ■

Henceforth, we shall usually denote $\mathbf{u} \oplus \mathbf{v}$ and $c \odot \mathbf{u}$ in a vector space V as $\mathbf{u} + \mathbf{v}$ and $c\mathbf{u}$, respectively.

We can also show that a nonempty subset W of a vector space V is a subspace of V if and only if $c\mathbf{u} + d\mathbf{v}$ is in W for any vectors $\mathbf{u}$ and $\mathbf{v}$ in W and any scalars c and d.

EXAMPLE 6 A simple way of constructing subspaces in a vector space is as follows. Let $\mathbf{v}_1$ and $\mathbf{v}_2$ be fixed vectors in a vector space V and let W be the set of all vectors in V of the form

$$a_1\mathbf{v}_1 + a_2\mathbf{v}_2,$$

where a_1 and a_2 are any real numbers. To show that W is a subspace of V, we verify properties (a) and (b) of Theorem 4.3. Thus let

$$\mathbf{w}_1 = a_1\mathbf{v}_1 + a_2\mathbf{v}_2 \quad \text{and} \quad \mathbf{w}_2 = b_1\mathbf{v}_1 + b_2\mathbf{v}_2$$

be vectors in W. Then

$$\mathbf{w}_1 + \mathbf{w}_2 = (a_1\mathbf{v}_1 + a_2\mathbf{v}_2) + (b_1\mathbf{v}_1 + b_2\mathbf{v}_2) = (a_1 + b_1)\mathbf{v}_1 + (a_2 + b_2)\mathbf{v}_2,$$

which is in W. Also, if c is a scalar, then

$$c\mathbf{w}_1 = c(a_1\mathbf{v}_1 + a_2\mathbf{v}_2) = (ca_1)\mathbf{v}_1 + (ca_2)\mathbf{v}_2$$

is in W. Hence W is a subspace of V. ■

The construction carried out in Example 6 for two vectors can be performed for more than two vectors. We now consider the following definition:

DEFINITION 4.6

Let $\mathbf{v}_1, \mathbf{v}_2, \ldots, \mathbf{v}_k$ be vectors in a vector space V. A vector $\mathbf{v}$ in V is called a **linear combination** of $\mathbf{v}_1, \mathbf{v}_2, \ldots, \mathbf{v}_k$ if

$$\mathbf{v} = a_1\mathbf{v}_1 + a_2\mathbf{v}_2 + \cdots + a_k\mathbf{v}_k = \sum_{j=1}^{k} a_j\mathbf{v}_j$$

for some real numbers $a_1, a_2, \ldots, a_k$.

Remark Summation notation was introduced in Section 1.2 for linear combinations of matrices, and properties of summation appeared in the Exercises for Section 1.2.

Remark Definition 4.6 was stated for a finite set of vectors, but it also applies to an infinite set S of vectors in a vector space using corresponding notation for infinite sums.

EXAMPLE 7

In Example 5 we showed that W, the set of all vectors in R^3 of the form $\begin{bmatrix} a \\ b \\ a+b \end{bmatrix}$, where a and b are any real numbers, is a subspace of R^3. Let

$$\mathbf{v}_1 = \begin{bmatrix} 1 \\ 0 \\ 1 \end{bmatrix} \quad \text{and} \quad \mathbf{v}_2 = \begin{bmatrix} 0 \\ 1 \\ 1 \end{bmatrix}.$$

Then every vector in W is a linear combination of $\mathbf{v}_1$ and $\mathbf{v}_2$, since

$$a\mathbf{v}_1 + b\mathbf{v}_2 = \begin{bmatrix} a \\ b \\ a+b \end{bmatrix}.$$

■

EXAMPLE 8

In Example 2, P_2 was the vector space of all polynomials of degree 2 or less and the zero polynomial. Every vector in P_2 has the form $at^2 + bt + c$, so each vector in P_2 is a linear combination of t^2, t, and 1. ■

In Figure 4.20 we show the vector $\mathbf{v}$ as a linear combination of the vectors $\mathbf{v}_1$ and $\mathbf{v}_2$.

EXAMPLE 9

In R^3 let

$$\mathbf{v}_1 = \begin{bmatrix} 1 \\ 2 \\ 1 \end{bmatrix}, \quad \mathbf{v}_2 = \begin{bmatrix} 1 \\ 0 \\ 2 \end{bmatrix}, \quad \text{and} \quad \mathbf{v}_3 = \begin{bmatrix} 1 \\ 1 \\ 0 \end{bmatrix}.$$

The vector

$$\mathbf{v} = \begin{bmatrix} 2 \\ 1 \\ 5 \end{bmatrix}$$

is a linear combination of $\mathbf{v}_1$, $\mathbf{v}_2$, and $\mathbf{v}_3$ if we can find real numbers a_1, a_2, and a_3 so that

$$a_1\mathbf{v}_1 + a_2\mathbf{v}_2 + a_3\mathbf{v}_3 = \mathbf{v}.$$

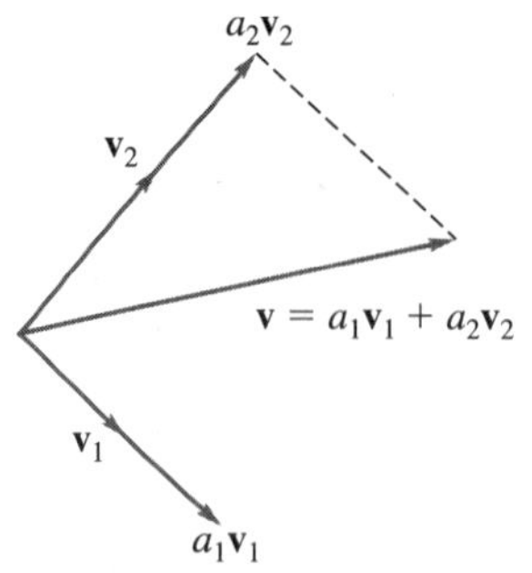

FIGURE 4.20 Linear combination of two vectors.

Substituting for $\mathbf{v}$, $\mathbf{v}_1$, $\mathbf{v}_2$, and $\mathbf{v}_3$, we have

$$a_1 \begin{bmatrix} 1 \\ 2 \\ 1 \end{bmatrix} + a_2 \begin{bmatrix} 1 \\ 0 \\ 2 \end{bmatrix} + a_3 \begin{bmatrix} 1 \\ 1 \\ 0 \end{bmatrix} = \begin{bmatrix} 2 \\ 1 \\ 5 \end{bmatrix}.$$

Equating corresponding entries leads to the linear system (verify)

$$\begin{aligned} a_1 + a_2 + a_3 &= 2 \\ 2a_1 + a_3 &= 1 \\ a_1 + 2a_2 &= 5. \end{aligned}$$

Solving this linear system by the methods of Chapter 2 gives (verify) $a_1 = 1$, $a_2 = 2$, and $a_3 = -1$, which means that $\mathbf{v}$ is a linear combination of $\mathbf{v}_1$, $\mathbf{v}_2$, and $\mathbf{v}_3$. Thus

$$\mathbf{v} = \mathbf{v}_1 + 2\mathbf{v}_2 - \mathbf{v}_3.$$

Figure 4.21 shows $\mathbf{v}$ as a linear combination of $\mathbf{v}_1$, $\mathbf{v}_2$, and $\mathbf{v}_3$. ■

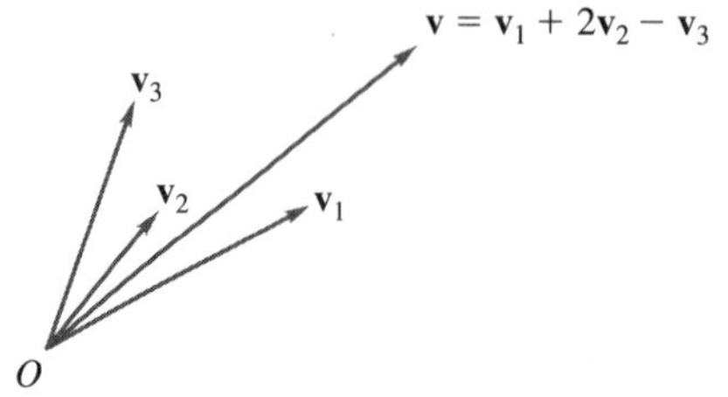

FIGURE 4.21

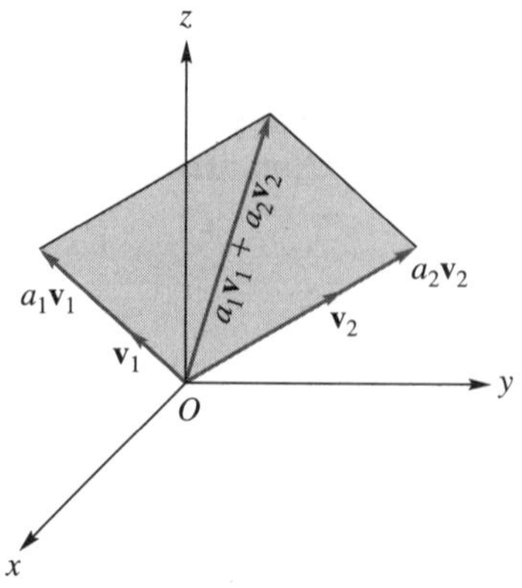

FIGURE 4.22

In Figure 4.22 we represent a portion of the set of all linear combinations of the noncollinear vectors $\mathbf{v}_1$ and $\mathbf{v}_2$ in R^2. The entire set of all linear combinations of the vectors $\mathbf{v}_1$ and $\mathbf{v}_2$ is a plane that passes through the origin and contains the vectors $\mathbf{v}_1$ and $\mathbf{v}_2$.

EXAMPLE 10 In Section 2.2 we observed that if A is an $m \times n$ matrix, then the homogeneous system of m equations in n unknowns with coefficient matrix A can be written as

$$A\mathbf{x} = \mathbf{0},$$

where $\mathbf{x}$ is a vector in R^n and $\mathbf{0}$ is the zero vector. Thus the set W of all solutions is a subset of R^n. We now show that W is a subspace of R^n (called the **solution space** of the homogeneous system, or the **null space** of the matrix A) by verifying (a) and (b) of Theorem 4.3. Let $\mathbf{x}$ and $\mathbf{y}$ be solutions. Then

$$A\mathbf{x} = \mathbf{0} \quad \text{and} \quad A\mathbf{y} = \mathbf{0}.$$

Now

$$A(\mathbf{x} + \mathbf{y}) = A\mathbf{x} + A\mathbf{y} = \mathbf{0} + \mathbf{0} = \mathbf{0},$$

so $\mathbf{x} + \mathbf{y}$ is a solution. Also, if c is a scalar, then

$$A(c\mathbf{x}) = c(A\mathbf{x}) = c\mathbf{0} = \mathbf{0},$$

so $c\mathbf{x}$ is a solution. Thus W is closed under addition and scalar multiplication of vectors and is therefore a subspace of R^n. ■

It should be noted that the set of all solutions to the linear system $A\mathbf{x} = \mathbf{b}$, $\mathbf{b} \neq \mathbf{0}$, is not a subspace of R^n (see Exercise 23).

We leave it as an exercise to show that the subspaces of R^1 are $\{\mathbf{0}\}$ and R^1 itself (see Exercise 28). As for R^2, its subspaces are $\{\mathbf{0}\}$, R^2, and any set consisting of all scalar multiples of a nonzero vector (Exercise 29), that is, any line passing through the origin. Exercise 43 in Section 4.6 asks you to show that all the subspaces of R^3 are $\{\mathbf{0}\}$, R^3 itself, and any line or plane passing through the origin.

■ Lines in R^3

As you will recall, a line in the xy-plane, R^2, is often described by the equation $y = mx + b$, where m is the slope of the line and b is the y-intercept [i.e., the line intersects the y-axis at the point $P_0(0, b)$]. We may describe a line in R^2 in terms of vectors by specifying its direction and a point on the line. Thus, let $\mathbf{v}$ be the vector giving the direction of the line, and let $\mathbf{u}_0 = \begin{bmatrix} 0 \\ b \end{bmatrix}$ be the position vector of the point $P_0(0, b)$ at which the line intersects the y-axis. Then the line through P_0 and parallel to $\mathbf{v}$ consists of the points $P(x, y)$ whose position vector $\mathbf{u} = \begin{bmatrix} x \\ y \end{bmatrix}$ satisfies (see Figure 4.23)

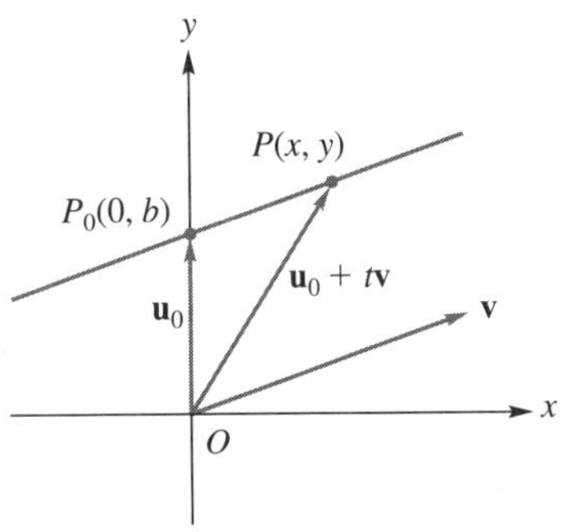

FIGURE 4.23

$$\mathbf{u} = \mathbf{u}_0 + t\mathbf{v}, \qquad -\infty < t < +\infty.$$

We now turn to lines in R^3. In R^3 we determine a line by specifying its direction and one of its points. Let

$$\mathbf{v} = \begin{bmatrix} a \\ b \\ c \end{bmatrix}$$

be a nonzero vector in R^3. Then the line ℓ_0 through the origin and parallel to $\mathbf{v}$ consists of all the points $P(x, y, z)$ whose position vector $\mathbf{u} = \begin{bmatrix} x \\ y \\ z \end{bmatrix}$ is of the form $\mathbf{u} = t\mathbf{v}$, $-\infty < t < \infty$ [see Figure 4.24(a)]. It is easy to verify that the line ℓ_0 is a subspace of R^3. Now let $P_0(x_0, y_0, z_0)$ be a point in R^3, and let $\mathbf{u}_0 = \begin{bmatrix} x_0 \\ y_0 \\ z_0 \end{bmatrix}$ be the position vector of P_0. Then the line ℓ through P_0 and parallel to $\mathbf{v}$ consists of the points $P(x, y, z)$ whose position vector, $\mathbf{u} = \begin{bmatrix} x \\ y \\ z \end{bmatrix}$ satisfies [see Figure 4.24(b)]

$$\mathbf{u} = \mathbf{u}_0 + t\mathbf{v}, \qquad -\infty < t < \infty. \tag{1}$$

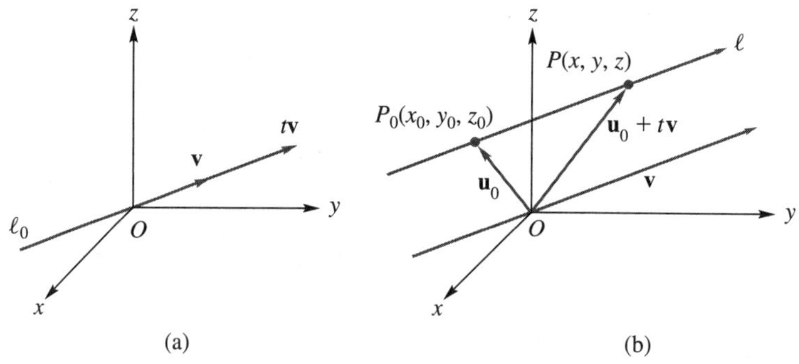

FIGURE 4.24 Line in R^3.

Equation (1) is called a **parametric equation** of ℓ, since it contains the parameter t, which can be assigned any real number. Equation (1) can also be written in terms of the components as

$$\begin{aligned} x &= x_0 + ta \\ y &= y_0 + tb \qquad -\infty < t < \infty, \\ z &= z_0 + tc, \end{aligned}$$

which are called **parametric equations** of ℓ.

EXAMPLE 11 Parametric equations of the line passing through the point $P_0(-3, 2, 1)$ and parallel to the vector $\mathbf{v} = \begin{bmatrix} 2 \\ -3 \\ 4 \end{bmatrix}$ are

$$\begin{aligned} x &= -3 + 2t \\ y &= 2 - 3t \qquad -\infty < t < \infty. \\ z &= 1 + 4t \end{aligned}$$

■

EXAMPLE 12 Find parametric equations of the line ℓ through the points $P_0(2, 3, -4)$ and $P_1(3, -2, 5)$.

Solution

The desired line is parallel to the vector $\mathbf{v} = \overrightarrow{P_0P_1}$. Now

$$\mathbf{v} = \begin{bmatrix} 3 - 2 \\ -2 - 3 \\ 5 - (-4) \end{bmatrix} = \begin{bmatrix} 1 \\ -5 \\ 9 \end{bmatrix}.$$

Since P_0 is on the line, we can write the following parametric equations of ℓ:

$$\begin{aligned} x &= 2 + t \\ y &= 3 - 5t \qquad -\infty < t < \infty. \\ z &= -4 + 9t \end{aligned}$$

Of course, we could have used the point P_1 instead of P_0. In fact, we could use any point on the line in a parametric equation for ℓ. Thus a line can be represented in infinitely many ways in parametric form. ■

Key Terms

Subspace
Zero subspace
Subset
Closure property
Linear combination
Null (solution) space
Parametric equation

4.3 Exercises

1. The set W consisting of all the points in R^2 of the form (x, x) is a straight line. Is W a subspace of R^2? Explain.

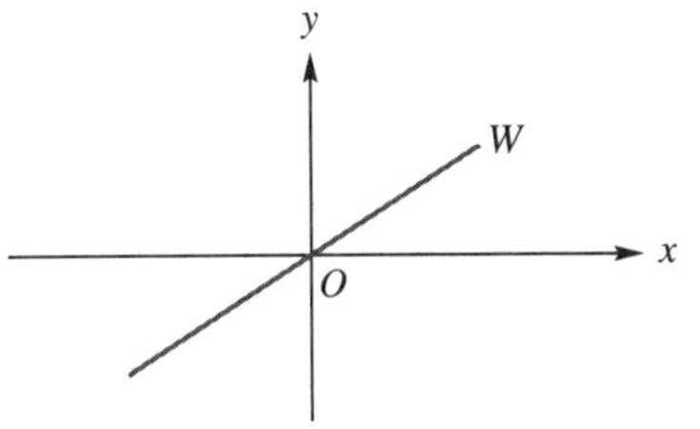

2. Let W be the set of all points in R^3 that lie in the xy-plane. Is W a subspace of R^3? Explain.

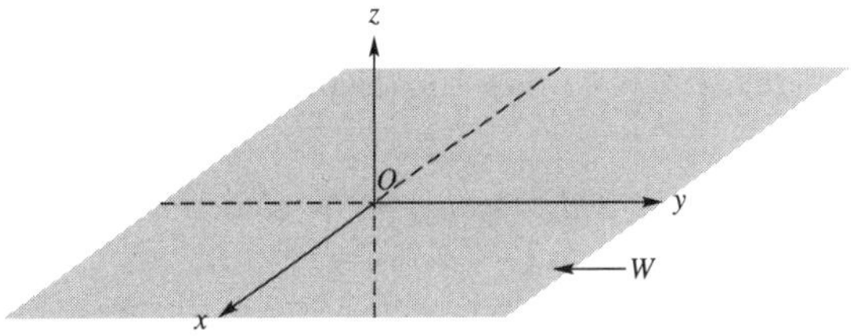

3. Consider the circle in the xy-plane centered at the origin whose equation is $x^2 + y^2 = 1$. Let W be the set of all vectors whose tail is at the origin and whose head is a point inside or on the circle. Is W a subspace of R^2? Explain.

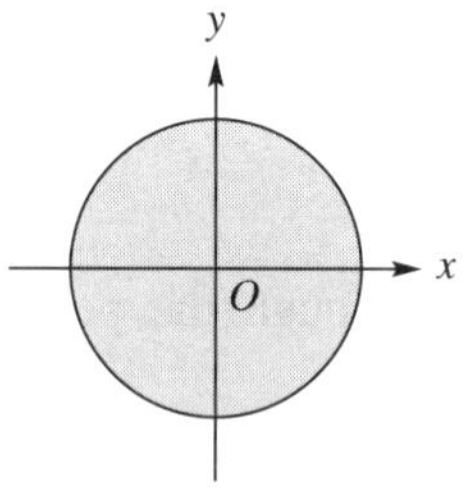

4. Consider the unit square shown in the accompanying figure. Let W be the set of all vectors of the form $\begin{bmatrix} x \\ y \end{bmatrix}$, where $0 \le x \le 1$, $0 \le y \le 1$. That is, W is the set of all vectors whose tail is at the origin and whose head is a point inside or on the square. Is W a subspace of R^2? Explain.

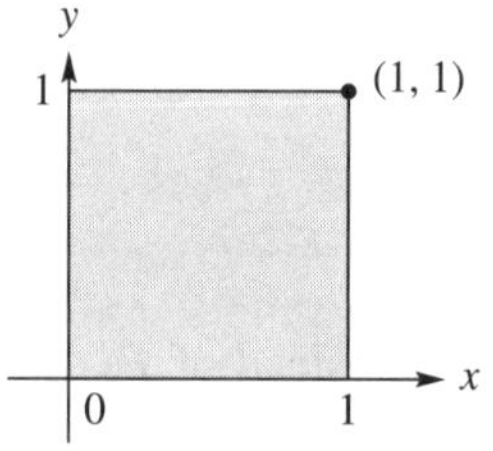

In Exercises 5 and 6, which of the given subsets of R^3 are subspaces?

5. The set of all vectors of the form

(a) $\begin{bmatrix} a \\ b \\ 1 \end{bmatrix}$ **(b)** $\begin{bmatrix} a \\ b \\ a+2b \end{bmatrix}$ **(c)** $\begin{bmatrix} a \\ 0 \\ 0 \end{bmatrix}$

(d) $\begin{bmatrix} a \\ b \\ c \end{bmatrix}$, where $a + 2b - c = 0$

6. The set of all vectors of the form

(a) $\begin{bmatrix} a \\ b \\ 0 \end{bmatrix}$ **(b)** $\begin{bmatrix} a \\ b \\ c \end{bmatrix}$, where $a > 0$

(c) $\begin{bmatrix} a \\ a \\ c \end{bmatrix}$ **(d)** $\begin{bmatrix} a \\ b \\ c \end{bmatrix}$, where $2a - b + c = 1$

In Exercises 7 and 8, which of the given subsets of R_4 are subspaces?

7. **(a)** $\begin{bmatrix} a & b & c & d \end{bmatrix}$, where $a - b = 2$

(b) $\begin{bmatrix} a & b & c & d \end{bmatrix}$, where $c = a + 2b$ and $d = a - 3b$

(c) $\begin{bmatrix} a & b & c & d \end{bmatrix}$, where $a = 0$ and $b = -d$

8. **(a)** $\begin{bmatrix} a & b & c & d \end{bmatrix}$, where $a = b = 0$

(b) $\begin{bmatrix} a & b & c & d \end{bmatrix}$, where $a = 1$, $b = 0$, and $a + d = 1$

(c) $\begin{bmatrix} a & b & c & d \end{bmatrix}$, where $a > 0$ and $b < 0$

In Exercises 9 and 10, which of the given subsets of the vector space, M_{23}, of all 2×3 matrices are subspaces?

9. The set of all matrices of the form

(a) $\begin{bmatrix} a & b & c \\ d & 0 & 0 \end{bmatrix}$, where $b = a + c$

(b) $\begin{bmatrix} a & b & c \\ d & 0 & 0 \end{bmatrix}$, where $c > 0$

(c) $\begin{bmatrix} a & b & c \\ d & e & f \end{bmatrix}$, where $a = -2c$ and $f = 2e + d$

10. **(a)** $\begin{bmatrix} a & b & c \\ d & e & f \end{bmatrix}$, where $a = 2c + 1$

(b) $\begin{bmatrix} 0 & 1 & a \\ b & c & 0 \end{bmatrix}$

(c) $\begin{bmatrix} a & b & c \\ d & e & f \end{bmatrix}$, where $a + c = 0$ and $b + d + f = 0$

11. Let W be the set of all 3×3 matrices whose trace is zero. Show that S is a subspace of M_{33}.

12. Let W be the set of all 3×3 matrices of the form

$$\begin{bmatrix} a & 0 & b \\ 0 & c & 0 \\ d & 0 & e \end{bmatrix}.$$

Show that W is a subspace of M_{33}.

13. Let W be the set of all 2×2 matrices

$$A = \begin{bmatrix} a & b \\ c & d \end{bmatrix}$$

such that $a + b + c + d = 0$. Is W a subspace of M_{22}? Explain.

14. Let W be the set of all 2×2 matrices

$$A = \begin{bmatrix} a & b \\ c & d \end{bmatrix}$$

such that $A\mathbf{z} = \mathbf{0}$, where $\mathbf{z} = \begin{bmatrix} 1 \\ 1 \end{bmatrix}$. Is W a subspace of M_{22}? Explain.

In Exercises 15 and 16, which of the given subsets of the vector space P_2 are subspaces?

15. The set of all polynomials of the form

(a) $a_2t^2 + a_1t + a_0$, where $a_0 = 0$

(b) $a_2t^2 + a_1t + a_0$, where $a_0 = 2$

(c) $a_2t^2 + a_1t + a_0$, where $a_2 + a_1 = a_0$

16. **(a)** $a_2t^2 + a_1t + a_0$, where $a_1 = 0$ and $a_0 = 0$

(b) $a_2t^2 + a_1t + a_0$, where $a_1 = 2a_0$

(c) $a_2t^2 + a_1t + a_0$, where $a_2 + a_1 + a_0 = 2$

17. Which of the following subsets of the vector space M_{nn} are subspaces?

(a) The set of all $n \times n$ symmetric matrices

(b) The set of all $n \times n$ diagonal matrices

(c) The set of all $n \times n$ nonsingular matrices

18. Which of the following subsets of the vector space M_{nn} are subspaces?

(a) The set of all $n \times n$ singular matrices

(b) The set of all $n \times n$ upper triangular matrices

(c) The set of all $n \times n$ skew symmetric matrices

19. (***Calculus Required***) Which of the following subsets are subspaces of the vector space $C(-\infty, \infty)$ defined in Example 7 of Section 4.2?

(a) All nonnegative functions

(b) All constant functions

(c) All functions f such that $f(0) = 0$

(d) All functions f such that $f(0) = 5$

(e) All differentiable functions

20. (***Calculus Required***) Which of the following subsets are subspaces of the vector space $C(-\infty, \infty)$ defined in Example 7 of Section 4.2?

(a) All integrable functions

(b) All bounded functions

(c) All functions that are integrable on $[a, b]$

(d) All functions that are bounded on $[a, b]$

21. Show that P is a subspace of the vector space $C(-\infty, \infty)$ defined in Example 7 of Section 4.2.

22. Prove that P_2 is a subspace of P_3.

23. Show that the set of all solutions to the linear system $A\mathbf{x} = \mathbf{b}$, $\mathbf{b} \neq \mathbf{0}$, is not a subspace of R^n.

24. If A is a nonsingular matrix, what is the null space of A?

25. Show that every vector in R^3 of the form

$$\begin{bmatrix} t \\ -t \\ t \end{bmatrix},$$

for t any real number, is in the null space of the matrix

$$A = \begin{bmatrix} 1 & 2 & 1 \\ -1 & 0 & 1 \\ 2 & 6 & 4 \end{bmatrix}.$$

26. Let $\mathbf{x}_0$ be a fixed vector in a vector space V. Show that the set W consisting of all scalar multiples $c\mathbf{x}_0$ of $\mathbf{x}_0$ is a subspace of V.

27. Let A be an $m \times n$ matrix. Is the set W of all vectors $\mathbf{x}$ in R^n such that $A\mathbf{x} \neq \mathbf{0}$ a subspace of R^n? Justify your answer.

28. Show that the only subspaces of R^1 are $\{\mathbf{0}\}$ and R^1 itself.

29. Show that the only subspaces of R^2 are $\{\mathbf{0}\}$, R^2, and any set consisting of all scalar multiples of a nonzero vector.

30. Determine which of the following subsets of R^2 are subspaces:

(a) V_1

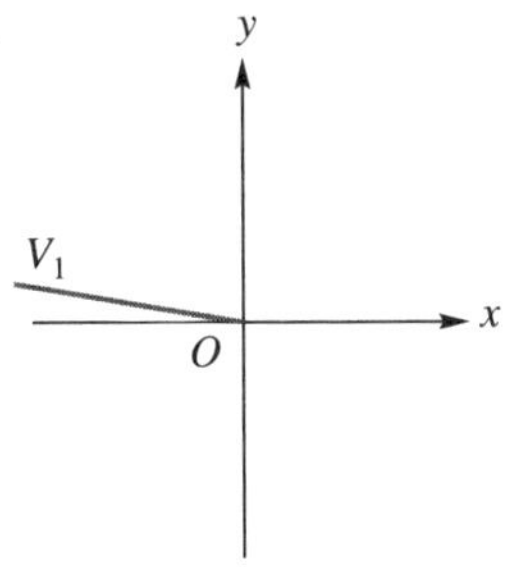

(b) V_2

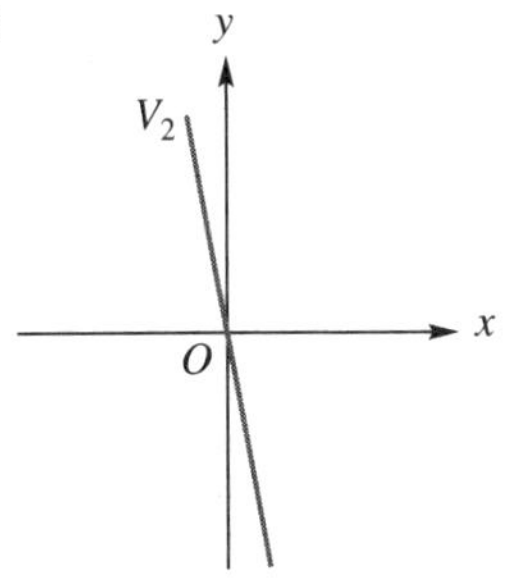

(c) V_3

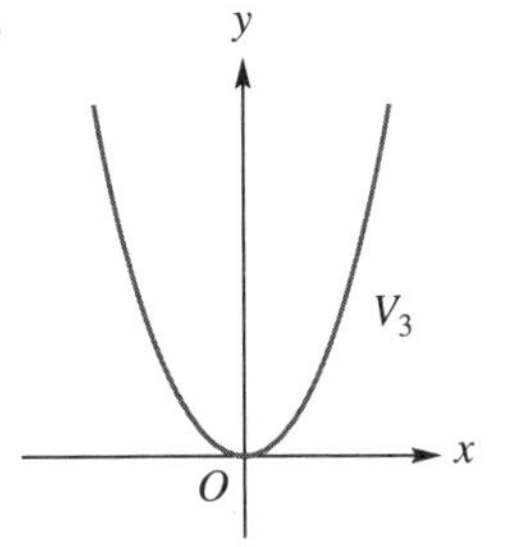

(d) V_4

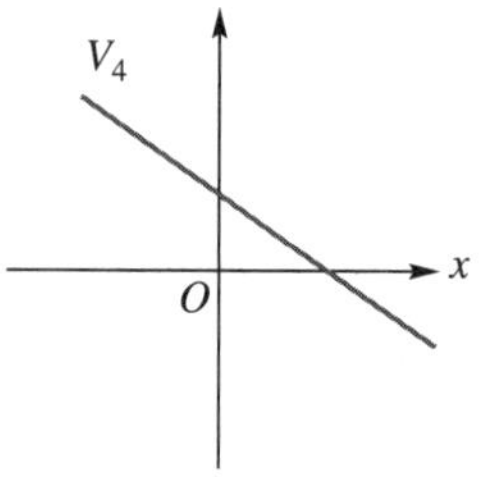

31. The set W of all 2×3 matrices of the form

$$\begin{bmatrix} a & b & c \\ a & 0 & 0 \end{bmatrix},$$

where $c = a + b$, is a subspace of M_{23}. Show that every vector in W is a linear combination of

$$\mathbf{w}_1 = \begin{bmatrix} 1 & 0 & 1 \\ 1 & 0 & 0 \end{bmatrix} \quad \text{and} \quad \mathbf{w}_2 = \begin{bmatrix} 0 & 1 & 1 \\ 0 & 0 & 0 \end{bmatrix}.$$

32. The set W of all 2×2 symmetric matrices is a subspace of M_{22}. Find three 2×2 matrices $\mathbf{v}_1$, $\mathbf{v}_2$, and $\mathbf{v}_3$ so that every vector in W can be expressed as a linear combination of $\mathbf{v}_1$, $\mathbf{v}_2$, and $\mathbf{v}_3$.

33. Which of the following vectors in R^3 are linear combinations of

$$\mathbf{v}_1 = \begin{bmatrix} 4 \\ 2 \\ -3 \end{bmatrix}, \quad \mathbf{v}_2 = \begin{bmatrix} 2 \\ 1 \\ -2 \end{bmatrix}, \quad \text{and} \quad \mathbf{v}_3 = \begin{bmatrix} -2 \\ -1 \\ 0 \end{bmatrix}?$$

(a) $\begin{bmatrix} 1 \\ 1 \\ 1 \end{bmatrix}$ **(b)** $\begin{bmatrix} 4 \\ 2 \\ -6 \end{bmatrix}$ **(c)** $\begin{bmatrix} -2 \\ -1 \\ 1 \end{bmatrix}$ **(d)** $\begin{bmatrix} -1 \\ 2 \\ 3 \end{bmatrix}$

34. Which of the following vectors in R_4 are linear combinations of

$$\mathbf{v}_1 = \begin{bmatrix} 1 & 2 & 1 & 0 \end{bmatrix}, \quad \mathbf{v}_2 = \begin{bmatrix} 4 & 1 & -2 & 3 \end{bmatrix},$$
$$\mathbf{v}_3 = \begin{bmatrix} 1 & 2 & 6 & -5 \end{bmatrix}, \quad \mathbf{v}_4 = \begin{bmatrix} -2 & 3 & -1 & 2 \end{bmatrix}?$$

(a) $\begin{bmatrix} 3 & 6 & 3 & 0 \end{bmatrix}$ **(b)** $\begin{bmatrix} 1 & 0 & 0 & 0 \end{bmatrix}$

(c) $\begin{bmatrix} 3 & 6 & -2 & 5 \end{bmatrix}$ **(d)** $\begin{bmatrix} 0 & 0 & 0 & 1 \end{bmatrix}$

35. **(a)** Show that a line ℓ_0 through the origin of R^3 is a subspace of R^3.

(b) Show that a line ℓ in R^3 not passing through the origin is not a subspace of R^3.

36. State which of the following points are on the line

$$\begin{aligned} x &= 3 + 2t \\ y &= -2 + 3t \qquad -\infty < t < \infty \\ z &= 4 + 3t \end{aligned}$$

(a) $(1, 1, 1)$ **(b)** $(1, -1, 0)$

(c) $(1, 0, -2)$ **(d)** $\left(4, -\frac{1}{2}, \frac{5}{2}\right)$

37. State which of the following points are on the line

$$\begin{aligned} x &= 4 - 2t \\ y &= -3 + 2t \qquad -\infty < t < \infty \\ z &= 4 - 5t \end{aligned}$$

(a) $(0, 1, -6)$ **(b)** $(1, 2, 3)$

(c) $(4, -3, 4)$ **(d)** $(0, 1, -1)$

38. Find parametric equations of the line through $P_0(x_0, y_0, z_0)$ parallel to $\mathbf{v}$.

(a) $P_0(3, 4, -2)$, $\mathbf{v} = \begin{bmatrix} 4 \\ -5 \\ 2 \end{bmatrix}$

(b) $P_0(3, 2, 4)$, $\mathbf{v} = \begin{bmatrix} -2 \\ 5 \\ 1 \end{bmatrix}$

39. Find parametric equations of the line through the given points.

(a) $(2, -3, 1)$, $(4, 2, 5)$

(b) $(-3, -2, -2)$, $(5, 5, 4)$

40. Numerical experiments in software *cannot* be used to verify that a set V with two operations $\oplus$ and $\odot$ is a vector space or a subspace. Such a verification must be done "abstractly" to take into account all possibilities for elements of V. However, numerical experiments can yield counterexamples which show that V is not a vector space or not a subspace. Use your software to verify that each of the following is *not* a subspace of M_{22}, with the usual operations of addition of matrices and scalar multiplication:

(a) The set of symmetric matrices with the $(1, 1)$ entry equal to 3

(b) The set of matrices whose first column is $\begin{bmatrix} 0 & 1 \end{bmatrix}^T$

(c) The set of matrices $\begin{bmatrix} a & b \\ c & d \end{bmatrix}$ such that $ad - bc \neq 0$

41. A linear combination of vectors $\mathbf{v}_1, \mathbf{v}_2, \ldots, \mathbf{v}_k$ in R^n with coefficients $a_1, \ldots, a_k$ is given algebraically, as in Definition 4.6, by

$$\mathbf{v} = a_1\mathbf{v}_1 + a_2\mathbf{v}_2 + \cdots + a_k\mathbf{v}_k.$$

In software we can compute such a linear combination of

columns by a matrix multiplication $\mathbf{v} = A\mathbf{c}$, where

$$A = \begin{bmatrix} \mathbf{v}_1 & \mathbf{v}_2 & \cdots & \mathbf{v}_k \end{bmatrix} \quad \text{and} \quad \mathbf{c} = \begin{bmatrix} a_1 \\ a_2 \\ \vdots \\ a_k \end{bmatrix}.$$

That is, matrix A has $\text{col}_j(A) = \mathbf{v}_j$ for $j = 1, 2, \ldots, k$. Experiment with your software with such linear combinations.

(a) Using $\mathbf{v}_1$, $\mathbf{v}_2$, $\mathbf{v}_3$ from Example 9, compute

$$5\mathbf{v}_1 - 2\mathbf{v}_2 + 4\mathbf{v}_3.$$

(b) Using $\mathbf{v}_j$, where $\mathbf{v}_1 = 2t^2 + t + 2$, $\mathbf{v}_2 = t^2 - 2t$, $\mathbf{v}_3 = 5t^2 - 5t + 2$, and $\mathbf{v}_4 = -t^2 - 3t - 2$, compute

$$3\mathbf{v}_1 - \mathbf{v}_2 + 4\mathbf{v}_3 + 2\mathbf{v}_4.$$

(See also Exercise 46 in Section 1.3.)

42. In Exercise 41, suppose that the vectors were in R_n. Devise a procedure that uses matrix multiplication for forming linear combinations of vectors in R_n.

4.4 Span

Thus far we have defined a mathematical structure called a real vector space and noted some of its properties. In Example 6 in Section 4.3 we showed that the set of all linear combinations of two fixed vectors $\mathbf{v}_1$ and $\mathbf{v}_2$ from a vector space V gave us a subspace W of V. We further observed that the only real vector space having a finite number of vectors in it is the vector space whose only vector is $\mathbf{0}$. For if $\mathbf{v} \neq \mathbf{0}$ is a vector in a vector space V, then by Exercise 25 in Section 4.2, $c \odot \mathbf{v} \neq c' \odot \mathbf{v}$, where c and c' are distinct real numbers, so V has infinitely many vectors in it. Also, from Example 1 in Section 4.3 we see that every vector space V has the zero subspace $\{\mathbf{0}\}$, which has only finitely many vectors in it.

From the preceding discussion we have that, except for the vector space $\{\mathbf{0}\}$, a vector space V will have infinitely many vectors. However, in this section and several that follow we show that many real vector spaces V studied here contain a finite subset of vectors that can be used to completely describe every vector in V. It should be noted that, in general, there is more than one subset that can be used to describe every vector in V. We now turn to a formulation of these ideas. Remember that we will denote vector addition $\mathbf{u} \oplus \mathbf{v}$ and scalar multiplication $c \odot \mathbf{u}$ in a vector space V as $\mathbf{u} + \mathbf{v}$ and $c\mathbf{u}$, respectively.

Linear combinations play an important role in describing vector spaces. In Example 6 in Section 4.3 we observed that the set of all possible linear combinations of a pair of vectors in a vector space V gives us a subspace. We have the following definition to help with such constructions:

DEFINITION 4.7 If $S = \{\mathbf{v}_1, \mathbf{v}_2, \ldots, \mathbf{v}_k\}$ is a set of vectors in a vector space V, then the set of all vectors in V that are linear combinations of the vectors in S is denoted by

$$\text{span } S \quad \text{or} \quad \text{span } \{\mathbf{v}_1, \mathbf{v}_2, \ldots, \mathbf{v}_k\}.$$

Remark Definition 4.7 is stated for a finite set of vectors, but it also applies to an infinite set S of vectors in a vector space.

EXAMPLE 1 Consider the set S of 2×3 matrices given by

$$S = \left\{ \begin{bmatrix} 1 & 0 & 0 \\ 0 & 0 & 0 \end{bmatrix}, \begin{bmatrix} 0 & 1 & 0 \\ 0 & 0 & 0 \end{bmatrix}, \begin{bmatrix} 0 & 0 & 0 \\ 0 & 1 & 0 \end{bmatrix}, \begin{bmatrix} 0 & 0 & 0 \\ 0 & 0 & 1 \end{bmatrix} \right\}.$$

Then span S is the set in M_{23} consisting of all vectors of the form

$$a\begin{bmatrix}1&0&0\\0&0&0\end{bmatrix}+b\begin{bmatrix}0&1&0\\0&0&0\end{bmatrix}+c\begin{bmatrix}0&0&0\\0&1&0\end{bmatrix}+d\begin{bmatrix}0&0&0\\0&0&1\end{bmatrix}$$
$$=\begin{bmatrix}a&b&0\\0&c&d\end{bmatrix},\text{ where } a, b, c, \text{ and } d \text{ are real numbers.}$$

That is, span S is the subset of M_{23} consisting of all matrices of the form

$$\begin{bmatrix}a&b&0\\0&c&d\end{bmatrix},$$

where a, b, c, and d are real numbers. ■

EXAMPLE 2

(a) Let $S = \{t^2, t, 1\}$ be a subset of P_2. We have span $S = P_2$. (See Example 8 in Section 4.3.)

(b) Let

$$S=\left\{\begin{bmatrix}2\\0\\0\end{bmatrix},\begin{bmatrix}0\\-1\\0\end{bmatrix},\begin{bmatrix}0\\0\\0\end{bmatrix}\right\}$$

be a subset of R^3. Span S is the set of all vectors in R^3 of the form $\begin{bmatrix}a\\b\\0\end{bmatrix}$, where a and b are any real numbers. (Verify.)

(c) In Figure 4.22 in Section 4.3 span$\{\mathbf{v}_1, \mathbf{v}_2\}$ is the plane that passes through the origin and contains $\mathbf{v}_1$ and $\mathbf{v}_2$. ■

The following theorem is a generalization of Example 6 in Section 4.3:

Theorem 4.4 Let $S = \{\mathbf{v}_1, \mathbf{v}_2, \ldots, \mathbf{v}_k\}$ be a set of vectors in a vector space V. Then span S is a subspace of V.

Proof

Let

$$\mathbf{u}=\sum_{j=1}^{k} a_j\mathbf{v}_j \quad\text{and}\quad \mathbf{w}=\sum_{j=1}^{k} b_j\mathbf{v}_j$$

for some real numbers $a_1, a_2, \ldots, a_k$ and $b_1, b_2, \ldots, b_k$. We have

$$\mathbf{u}+\mathbf{w}=\sum_{j=1}^{k} a_j\mathbf{v}_j+\sum_{j=1}^{k} b_j\mathbf{v}_j=\sum_{j=1}^{k}(a_j+b_j)\mathbf{v}_j,$$

using Exercise 17 (a) and (b) in Section 1.2. Moreover, for any real number c,

$$c\mathbf{u}=c\left(\sum_{j=1}^{k} a_j\mathbf{v}_j\right)=\sum_{j=1}^{k}(ca_j)\mathbf{v}_j.$$

Since the sum $\mathbf{u}+\mathbf{w}$ and the scalar multiple $c\mathbf{u}$ are linear combinations of the vectors in S, then span S is a subspace of V. ■

EXAMPLE 3 Let $S = \{t^2, t\}$ be a subset of the vector space P_2. Then span S is the subspace of all polynomials of the form $at^2 + bt$, where a and b are any real numbers. ■

EXAMPLE 4 Let

$$S = \left\{ \begin{bmatrix} 1 & 0 \\ 0 & 0 \end{bmatrix}, \begin{bmatrix} 0 & 0 \\ 0 & 1 \end{bmatrix} \right\}$$

be a subset of the vector space M_{22}. Then span S is the subspace of all 2×2 diagonal matrices. ■

In order to completely describe a vector space V, we use the concept of span as stated in the following definition:

DEFINITION 4.8 Let S be a set of vectors in a vector space V. If every vector in V is a linear combination of the vectors in S, then the set S is said to **span** V, or V is spanned by the set S; that is, span $S = V$.

Remark If span $S = V$, S is called a **spanning set** of V. A vector space can have many spanning sets. In our examples we used sets S containing a finite number of vectors, but some vector spaces need spanning sets with infinitely many vectors. See Example 5.

EXAMPLE 5 Let P be the vector space of all polynomials. Let $S = \{1, t, t^2, \dots\}$; that is, the set of all (nonnegative integer) powers of t. Then span $S = P$. Every spanning set for P will have infinitely many vectors. ■

Remark The majority of vector spaces and subspaces in the examples and exercises in this book will have spanning sets with finitely many vectors.

Another type of question that we encounter is; For a given subset S of a vector space V, is the vector $\mathbf{w}$ of V in span S? We will show in the following examples that for sets S with a finite number of vectors we can answer such questions by solving a linear system of equations:

EXAMPLE 6 In R^3, let

$$\mathbf{v}_1 = \begin{bmatrix} 2 \\ 1 \\ 1 \end{bmatrix} \quad \text{and} \quad \mathbf{v}_2 = \begin{bmatrix} 1 \\ -1 \\ 3 \end{bmatrix}.$$

Determine whether the vector

$$\mathbf{v} = \begin{bmatrix} 1 \\ 5 \\ -7 \end{bmatrix}$$

belongs to span$\{\mathbf{v}_1, \mathbf{v}_2\}$.

Solution

If we can find scalars a_1 and a_2 such that

$$a_1\mathbf{v}_1 + a_2\mathbf{v}_2 = \mathbf{v},$$

then $\mathbf{v}$ belongs to span $\{\mathbf{v}_1, \mathbf{v}_2\}$. Substituting for $\mathbf{v}_1$, $\mathbf{v}_2$, and $\mathbf{v}$, we have

$$a_1 \begin{bmatrix} 2 \\ 1 \\ 1 \end{bmatrix} + a_2 \begin{bmatrix} 1 \\ -1 \\ 3 \end{bmatrix} = \begin{bmatrix} 1 \\ 5 \\ -7 \end{bmatrix}.$$

This expression corresponds to the linear system whose augmented matrix is (verify)

$$\left[\begin{array}{cc|c} 2 & 1 & 1 \\ 1 & -1 & 5 \\ 1 & 3 & -7 \end{array}\right].$$

The reduced row echelon form of this system is (verify)

$$\left[\begin{array}{cc|c} 1 & 0 & 2 \\ 0 & 1 & -3 \\ 0 & 0 & 0 \end{array}\right],$$

which indicates that the linear system is consistent, $a_1 = 2$, and $a_2 = 3$. Hence $\mathbf{v}$ belongs to span $\{\mathbf{v}_1, \mathbf{v}_2\}$. ■

EXAMPLE 7

In P_2, let

$$\mathbf{v}_1 = 2t^2 + t + 2, \quad \mathbf{v}_2 = t^2 - 2t, \quad \mathbf{v}_3 = 5t^2 - 5t + 2, \quad \mathbf{v}_4 = -t^2 - 3t - 2.$$

Determine whether the vector

$$\mathbf{v} = t^2 + t + 2$$

belongs to span $\{\mathbf{v}_1, \mathbf{v}_2, \mathbf{v}_3, \mathbf{v}_4\}$.

Solution

If we can find scalars a_1, a_2, a_3, and a_4 so that

$$a_1\mathbf{v}_1 + a_2\mathbf{v}_2 + a_3\mathbf{v}_3 + a_4\mathbf{v}_4 = \mathbf{v},$$

then $\mathbf{v}$ belongs to span $\{\mathbf{v}_1, \mathbf{v}_2, \mathbf{v}_3, \mathbf{v}_4\}$. Substituting for $\mathbf{v}_1$, $\mathbf{v}_2$, $\mathbf{v}_3$, and $\mathbf{v}_4$, we have

$$\begin{aligned} a_1(2t^2 + t + 2) + a_2(t^2 - 2t) + a_3(5t^2 - 5t + 2) + a_4(-t^2 - 3t - 2) \\ = t^2 + t + 2, \end{aligned}$$

or

$$\begin{aligned} (2a_1 + a_2 + 5a_3 - a_4)t^2 + (a_1 - 2a_2 - 5a_3 - 3a_4)t + (2a_1 + 2a_3 - 2a_4) \\ = t^2 + t + 2. \end{aligned}$$

Now two polynomials agree for all values of t only if the coefficients of respective powers of t agree. Thus we get the linear system

$$\begin{aligned} 2a_1 + a_2 + 5a_3 - a_4 &= 1 \\ a_1 - 2a_2 - 5a_3 - 3a_4 &= 1 \\ 2a_1 \qquad + 2a_3 - 2a_4 &= 2. \end{aligned}$$

To determine whether this system of linear equations is consistent, we form the augmented matrix and transform it to reduced row echelon form, obtaining (verify)

$$\left[\begin{array}{cccc|c} 1 & 0 & 1 & -1 & 0 \\ 0 & 1 & 3 & 1 & 0 \\ 0 & 0 & 0 & 0 & 1 \end{array}\right],$$

which indicates that the system is inconsistent; that is, it has no solution. Hence $\mathbf{v}$ does not belong to span $\{\mathbf{v}_1, \mathbf{v}_2, \mathbf{v}_3, \mathbf{v}_4\}$. ■

Remark In general, to determine whether a specific vector $\mathbf{v}$ belongs to span S, we investigate the consistency of an appropriate linear system.

EXAMPLE 8

Let V be the vector space R^3. Let

$$\mathbf{v}_1 = \begin{bmatrix} 1 \\ 2 \\ 1 \end{bmatrix}, \quad \mathbf{v}_2 = \begin{bmatrix} 1 \\ 0 \\ 2 \end{bmatrix}, \quad \text{and} \quad \mathbf{v}_3 = \begin{bmatrix} 1 \\ 1 \\ 0 \end{bmatrix}.$$

To find out whether $\mathbf{v}_1$, $\mathbf{v}_2$, $\mathbf{v}_3$ span V, we pick any vector $\mathbf{v} = \begin{bmatrix} a \\ b \\ c \end{bmatrix}$ in V (a, b, and c are arbitrary real numbers) and determine whether there are constants a_1, a_2, and a_3 such that

$$a_1\mathbf{v}_1 + a_2\mathbf{v}_2 + a_3\mathbf{v}_3 = \mathbf{v}.$$

This leads to the linear system (verify)

$$\begin{aligned} a_1 + a_2 + a_3 &= a \\ 2a_1 + a_3 &= b \\ a_1 + 2a_2 &= c. \end{aligned}$$

A solution is (verify)

$$a_1 = \frac{-2a + 2b + c}{3}, \quad a_2 = \frac{a - b + c}{3}, \quad a_3 = \frac{4a - b - 2c}{3}.$$

Thus $\mathbf{v}_1$, $\mathbf{v}_2$, $\mathbf{v}_3$ span V. This is equivalent to saying that

$$\text{span}\,\{\mathbf{v}_1, \mathbf{v}_2, \mathbf{v}_3\} = R^3.$$ ■

EXAMPLE 9

Let V be the vector space P_2. Let $\mathbf{v}_1 = t^2 + 2t + 1$ and $\mathbf{v}_2 = t^2 + 2$. Does $\{\mathbf{v}_1, \mathbf{v}_2\}$ span V?

Solution

Let $\mathbf{v} = at^2 + bt + c$ be any vector in V, where a, b, and c are any real numbers. We must find out whether there are constants a_1 and a_2 such that

$$a_1\mathbf{v}_1 + a_2\mathbf{v}_2 = \mathbf{v},$$

or

$$a_1(t^2 + 2t + 1) + a_2(t^2 + 2) = at^2 + bt + c.$$

Thus

$$(a_1 + a_2)t^2 + (2a_1)t + (a_1 + 2a_2) = at^2 + bt + c.$$

Equating the coefficients of respective powers of t, we get the linear system

$$\begin{aligned} a_1 + a_2 &= a \\ 2a_1 &= b \\ a_1 + 2a_2 &= c. \end{aligned}$$

Transforming the augmented matrix of this linear system, we obtain (verify)

$$\left[\begin{array}{cc|c} 1 & 0 & 2a - c \\ 0 & 1 & c - a \\ 0 & 0 & b - 4a + 2c \end{array}\right].$$

If $b - 4a + 2c \neq 0$, then the system is inconsistent and there is no solution. Hence $\{\mathbf{v}_1, \mathbf{v}_2\}$ does not span V. ■

EXAMPLE 10 Consider the homogeneous linear system $A\mathbf{x} = \mathbf{0}$, where

$$A = \begin{bmatrix} 1 & 1 & 0 & 2 \\ -2 & -2 & 1 & -5 \\ 1 & 1 & -1 & 3 \\ 4 & 4 & -1 & 9 \end{bmatrix}.$$

From Example 10 in Section 4.3, the set of all solutions to $A\mathbf{x} = \mathbf{0}$ forms a subspace of R^4. To determine a spanning set for the solution space of this homogeneous system, we find that the reduced row echelon form of the augmented matrix is (verify)

$$\left[\begin{array}{cccc|c} 1 & 1 & 0 & 2 & 0 \\ 0 & 0 & 1 & -1 & 0 \\ 0 & 0 & 0 & 0 & 0 \\ 0 & 0 & 0 & 0 & 0 \end{array}\right].$$

The general solution is then given by

$$x_1 = -r - 2s, \quad x_2 = r, \quad x_3 = s, \quad x_4 = s,$$

where r and s are any real numbers. In matrix form we have that any member of the solution space is given by

$$\mathbf{x} = r \begin{bmatrix} -1 \\ 1 \\ 0 \\ 0 \end{bmatrix} + s \begin{bmatrix} -2 \\ 0 \\ 1 \\ 1 \end{bmatrix}.$$

Hence $\begin{bmatrix} -1 \\ 1 \\ 0 \\ 0 \end{bmatrix}$ and $\begin{bmatrix} -2 \\ 0 \\ 1 \\ 1 \end{bmatrix}$ span the solution space. ■

Key Terms

Span of a set
Set S spans vector space V
V is spanned by the set S
Consistent system
Inconsistent system
Homogeneous system

4.4 Exercises

1. For each of the following vector spaces, give two different spanning sets:
   (a) R^3 (b) M_{22} (c) P_2
2. In each part, explain why the set S is not a spanning set for the vector space V.
   (a) $S = \{t^3, t^2, t\}$, $V = P_3$
   (b) $S = \left\{\begin{bmatrix}1\\0\end{bmatrix}, \begin{bmatrix}0\\0\end{bmatrix}\right\}$, $V = R^2$
   (c) $S = \left\{\begin{bmatrix}1 & 0\\0 & 1\end{bmatrix}, \begin{bmatrix}0 & 1\\1 & 0\end{bmatrix}\right\}$, $V = M_{22}$
3. In each part, determine whether the given vector $p(t)$ in P_2 belongs to span $\{p_1(t), p_2(t), p_3(t)\}$, where
$$p_1(t) = t^2 + 2t + 1, \quad p_2(t) = t^2 + 3, \quad \text{and} \quad p_3(t) = t - 1.$$
   (a) $p(t) = t^2 + t + 2$
   (b) $p(t) = 2t^2 + 2t + 3$
   (c) $p(t) = -t^2 + t - 4$
   (d) $p(t) = -2t^2 + 3t + 1$
4. In each part, determine whether the given vector A in M_{22} belongs to span $\{A_1, A_2, A_3\}$, where
$$A_1 = \begin{bmatrix}1 & -1\\0 & 3\end{bmatrix}, \quad A_2 = \begin{bmatrix}1 & 1\\0 & 2\end{bmatrix}, \quad \text{and} \quad A_3 = \begin{bmatrix}2 & 2\\-1 & 1\end{bmatrix}.$$
   (a) $A = \begin{bmatrix}5 & 1\\-1 & 9\end{bmatrix}$ (b) $A = \begin{bmatrix}-3 & -1\\3 & 2\end{bmatrix}$
   (c) $A = \begin{bmatrix}3 & -2\\3 & 2\end{bmatrix}$ (d) $A = \begin{bmatrix}1 & 0\\2 & 1\end{bmatrix}$
5. Which of the following vectors span R_2?
   (a) $\begin{bmatrix}1 & 2\end{bmatrix}, \begin{bmatrix}-1 & 1\end{bmatrix}$
   (b) $\begin{bmatrix}0 & 0\end{bmatrix}, \begin{bmatrix}1 & 1\end{bmatrix}, \begin{bmatrix}-2 & -2\end{bmatrix}$
   (c) $\begin{bmatrix}1 & 3\end{bmatrix}, \begin{bmatrix}2 & -3\end{bmatrix}, \begin{bmatrix}0 & 2\end{bmatrix}$
   (d) $\begin{bmatrix}2 & 4\end{bmatrix}, \begin{bmatrix}-1 & 2\end{bmatrix}$
6. Which of the following sets of vectors span R^4?
   (a) $\left\{\begin{bmatrix}1\\-1\\2\\0\end{bmatrix}, \begin{bmatrix}0\\1\\1\\1\end{bmatrix}\right\}$
   (b) $\left\{\begin{bmatrix}3\\2\\1\\0\end{bmatrix}, \begin{bmatrix}1\\2\\-1\\0\end{bmatrix}, \begin{bmatrix}0\\0\\0\\1\end{bmatrix}\right\}$
   (c) $\left\{\begin{bmatrix}3\\2\\-1\\2\end{bmatrix}, \begin{bmatrix}4\\0\\0\\2\end{bmatrix}, \begin{bmatrix}3\\2\\-1\\2\end{bmatrix}, \begin{bmatrix}5\\6\\-3\\2\end{bmatrix}, \begin{bmatrix}0\\4\\-2\\-1\end{bmatrix}\right\}$
   (d) $\left\{\begin{bmatrix}1\\1\\0\\0\end{bmatrix}, \begin{bmatrix}1\\2\\-1\\1\end{bmatrix}, \begin{bmatrix}0\\0\\1\\-1\end{bmatrix}, \begin{bmatrix}2\\1\\2\\-1\end{bmatrix}\right\}$
7. Which of the following sets of vectors span R_4?
   (a) $\begin{bmatrix}1 & 0 & 0 & 1\end{bmatrix}, \begin{bmatrix}0 & 1 & 0 & 0\end{bmatrix}, \begin{bmatrix}1 & 1 & 1 & 1\end{bmatrix}, \begin{bmatrix}1 & 1 & 1 & 0\end{bmatrix}$
   (b) $\begin{bmatrix}1 & 2 & 1 & 0\end{bmatrix}, \begin{bmatrix}1 & 1 & -1 & 0\end{bmatrix}, \begin{bmatrix}0 & 0 & 0 & 1\end{bmatrix}$
   (c) $\begin{bmatrix}6 & 4 & -2 & 4\end{bmatrix}, \begin{bmatrix}2 & 0 & 0 & 1\end{bmatrix}, \begin{bmatrix}3 & 2 & -1 & 2\end{bmatrix}, \begin{bmatrix}5 & 6 & -3 & 2\end{bmatrix}, \begin{bmatrix}0 & 4 & -2 & -1\end{bmatrix}$
   (d) $\begin{bmatrix}1 & 1 & 0 & 0\end{bmatrix}, \begin{bmatrix}1 & 2 & -1 & 1\end{bmatrix}, \begin{bmatrix}0 & 0 & 1 & 1\end{bmatrix}, \begin{bmatrix}2 & 1 & 2 & 1\end{bmatrix}$
8. Which of the following sets of polynomials span P_2?
   (a) $\{t^2 + 1, t^2 + t, t + 1\}$
   (b) $\{t^2 + 1, t - 1, t^2 + t\}$
   (c) $\{t^2 + 2, 2t^2 - t + 1, t + 2, t^2 + t + 4\}$
   (d) $\{t^2 + 2t - 1, t^2 - 1\}$
9. Do the polynomials $t^3 + 2t + 1$, $t^2 - t + 2$, $t^3 + 2$, $-t^3 + t^2 - 5t + 2$ span P_3?

10. Does the set

$$S = \left\{ \begin{bmatrix} 1 & 1 \\ 0 & 0 \end{bmatrix}, \begin{bmatrix} 0 & 0 \\ 1 & 1 \end{bmatrix}, \begin{bmatrix} 1 & 0 \\ 0 & 1 \end{bmatrix}, \begin{bmatrix} 0 & 1 \\ 1 & 1 \end{bmatrix} \right\}$$

span M_{22}?

11. Find a set of vectors spanning the solution space of $A\mathbf{x} = \mathbf{0}$, where

$$A = \begin{bmatrix} 1 & 0 & 1 & 0 \\ 1 & 2 & 3 & 1 \\ 2 & 1 & 3 & 1 \\ 1 & 1 & 2 & 1 \end{bmatrix}.$$

12. Find a set of vectors spanning the null space of

$$A = \begin{bmatrix} 1 & 1 & 2 & -1 \\ 2 & 3 & 6 & -2 \\ -2 & 1 & 2 & 2 \\ 0 & -2 & -4 & 0 \end{bmatrix}.$$

13. The set W of all 2×2 matrices A with trace equal to zero is a subspace of M_{22}. Let

$$S = \left\{ \begin{bmatrix} 0 & 1 \\ 0 & 0 \end{bmatrix}, \begin{bmatrix} 0 & 0 \\ 1 & 0 \end{bmatrix}, \begin{bmatrix} 1 & 0 \\ 0 & -1 \end{bmatrix} \right\}.$$

Show that span $S = W$.

14. The set W of all 3×3 matrices A with trace equal to zero is a subspace of M_{33}. (See Exercise 11 in Section 4.3.) Determine a subset S of W so that span $S = W$.

15. The set W of all 3×3 matrices of the form

$$\begin{bmatrix} a & 0 & b \\ 0 & c & 0 \\ d & 0 & e \end{bmatrix}$$

is a subspace of M_{33}. (See Exercise 12 in Section 4.3.) Determine a subset S of W so that span $S = W$.

16. Let T be the set of all matrices of the form $AB - BA$, where A and B are $n \times n$ matrices. Show that span T is not M_{nn}. (*Hint*: Use properties of trace.)

17. Determine whether your software has a command for finding the null space (see Example 10 in Section 4.3) of a matrix A. If it does, use it on the matrix A in Example 10 and compare the command's output with the results in Example 10. To experiment further, use Exercises 11 and 12.

4.5 Linear Independence

In Section 4.4 we developed the notion of the span of a set of vectors together with spanning sets of a vector space or subspace. Spanning sets S provide vectors so that any vector in the space can be expressed as a linear combination of the members of S. We remarked that a vector space can have many different spanning sets and that spanning sets for the same space need not have the same number of vectors. We illustrate this in Example 1.

EXAMPLE 1 In Example 5 of Section 4.3 we showed that the set W of all vectors of the form

$$\begin{bmatrix} a \\ b \\ a+b \end{bmatrix},$$

where a and b are any real numbers, is a subspace of R^3. Each of the following sets is a spanning set for W (verify):

$$S_1 = \left\{ \begin{bmatrix} 1 \\ 0 \\ 1 \end{bmatrix}, \begin{bmatrix} 0 \\ 1 \\ 1 \end{bmatrix}, \begin{bmatrix} 3 \\ 2 \\ 5 \end{bmatrix} \right\}$$

$$S_2 = \left\{ \begin{bmatrix} 1 \\ 0 \\ 1 \end{bmatrix}, \begin{bmatrix} 0 \\ 1 \\ 1 \end{bmatrix}, \begin{bmatrix} 0 \\ 0 \\ 0 \end{bmatrix}, \begin{bmatrix} 2 \\ 0 \\ 2 \end{bmatrix} \right\} \qquad S_3 = \left\{ \begin{bmatrix} 1 \\ 0 \\ 1 \end{bmatrix}, \begin{bmatrix} 0 \\ 1 \\ 1 \end{bmatrix} \right\}$$

■

We observe that the set S_3 is a more "efficient" spanning set, since each vector of W is a linear combination of two vectors, compared with three vectors when using S_1 and four vectors when using S_2. If we can determine a spanning set for a vector space V that is minimal, in the sense that it contains the fewest number of vectors, then we have an efficient way to describe every vector in V.

In Example 1, since the vectors in S_3 span W and S_3 is a subset of S_1 and S_2, it follows that the vector

$$\begin{bmatrix} 3 \\ 2 \\ 5 \end{bmatrix}$$

in S_1 must be a linear combination of the vectors in S_3, and similarly, both vectors

$$\begin{bmatrix} 0 \\ 0 \\ 0 \end{bmatrix} \quad \text{and} \quad \begin{bmatrix} 2 \\ 0 \\ 2 \end{bmatrix}$$

in S_2 must be linear combinations of the vectors in S_3. Observe that

$$3\begin{bmatrix} 1 \\ 0 \\ 1 \end{bmatrix} + 2\begin{bmatrix} 0 \\ 1 \\ 1 \end{bmatrix} = \begin{bmatrix} 3 \\ 2 \\ 5 \end{bmatrix},$$

$$0\begin{bmatrix} 1 \\ 0 \\ 1 \end{bmatrix} + 0\begin{bmatrix} 0 \\ 1 \\ 1 \end{bmatrix} = \begin{bmatrix} 0 \\ 0 \\ 0 \end{bmatrix},$$

$$2\begin{bmatrix} 1 \\ 0 \\ 1 \end{bmatrix} + 0\begin{bmatrix} 0 \\ 1 \\ 1 \end{bmatrix} = \begin{bmatrix} 2 \\ 0 \\ 2 \end{bmatrix}.$$

In addition, for set S_1 we observe that

$$3\begin{bmatrix} 1 \\ 0 \\ 1 \end{bmatrix} + 2\begin{bmatrix} 0 \\ 1 \\ 1 \end{bmatrix} - 1\begin{bmatrix} 3 \\ 2 \\ 5 \end{bmatrix} = \begin{bmatrix} 0 \\ 0 \\ 0 \end{bmatrix},$$

and for set S_2 we observe that

$$0\begin{bmatrix} 1 \\ 0 \\ 1 \end{bmatrix} + 0\begin{bmatrix} 0 \\ 1 \\ 1 \end{bmatrix} - 1\begin{bmatrix} 0 \\ 0 \\ 0 \end{bmatrix} = \begin{bmatrix} 0 \\ 0 \\ 0 \end{bmatrix},$$

$$2\begin{bmatrix} 1 \\ 0 \\ 1 \end{bmatrix} + 0\begin{bmatrix} 0 \\ 1 \\ 1 \end{bmatrix} - 1\begin{bmatrix} 2 \\ 0 \\ 2 \end{bmatrix} = \begin{bmatrix} 0 \\ 0 \\ 0 \end{bmatrix}.$$

It follows that if span $S = V$, and there is a linear combination of the vectors in S with coefficients not all zero that gives the zero vector, then some subset of S is also a spanning set for V.

Remark The preceding discussion motivates the next definition. In the preceding discussion, which is based on Example 1, we used the observation that S_3 was a subset of S_1 and of S_2. However, that observation is a special case which need not apply when comparing two spanning sets for a vector space.

DEFINITION 4.9

The vectors $\mathbf{v}_1, \mathbf{v}_2, \ldots, \mathbf{v}_k$ in a vector space V are said to be **linearly dependent** if there exist constants $a_1, a_2, \ldots, a_k$, not all zero, such that

$$\sum_{j=1}^{k} a_j \mathbf{v}_j = a_1\mathbf{v}_1 + a_2\mathbf{v}_2 + \cdots + a_k\mathbf{v}_k = \mathbf{0}. \tag{1}$$

Otherwise, $\mathbf{v}_1, \mathbf{v}_2, \ldots, \mathbf{v}_k$ are called **linearly independent**. That is, $\mathbf{v}_1, \mathbf{v}_2, \ldots, \mathbf{v}_k$ are linearly independent if, whenever $a_1\mathbf{v}_1 + a_2\mathbf{v}_2 + \cdots + a_k\mathbf{v}_k = \mathbf{0}$,

$$a_1 = a_2 = \cdots = a_k = 0.$$

If $S = \{\mathbf{v}_1, \mathbf{v}_2, \ldots, \mathbf{v}_k\}$, then we also say that the set S is **linearly dependent** or **linearly independent** if the vectors have the corresponding property.

It should be emphasized that for any vectors $\mathbf{v}_1, \mathbf{v}_2, \ldots, \mathbf{v}_k$, Equation (1) always holds if we choose all the scalars $a_1, a_2, \ldots, a_k$ equal to zero. The important point in this definition is whether it is possible to satisfy (1) with at least one of the scalars different from zero.

Remark Definition 4.9 is stated for a finite set of vectors, but it also applies to an infinite set S of a vector space, using corresponding notation for infinite sums.

Remark We connect Definition 4.9 to "efficient" spanning sets in Section 4.6.

To determine whether a set of vectors is linearly independent or linearly dependent, we use Equation (1). Regardless of the form of the vectors, Equation (1) yields a homogeneous linear system of equations. It is always consistent, since $a_1 = a_2 = \cdots = a_k = 0$ is a solution. However, the main idea from Definition 4.9 is whether there is a nontrivial solution.

EXAMPLE 2

Determine whether the vectors

$$\mathbf{v}_1 = \begin{bmatrix} 3 \\ 2 \\ 1 \end{bmatrix}, \quad \mathbf{v}_2 = \begin{bmatrix} 1 \\ 2 \\ 0 \end{bmatrix}, \quad \mathbf{v}_3 = \begin{bmatrix} -1 \\ 2 \\ -1 \end{bmatrix}$$

are linearly independent.

Solution

Forming Equation (1),

$$a_1 \begin{bmatrix} 3 \\ 2 \\ 1 \end{bmatrix} + a_2 \begin{bmatrix} 1 \\ 2 \\ 0 \end{bmatrix} + a_3 \begin{bmatrix} -1 \\ 2 \\ -1 \end{bmatrix} = \begin{bmatrix} 0 \\ 0 \\ 0 \end{bmatrix},$$

we obtain the homogeneous system (verify)

$$\begin{aligned} 3a_1 + a_2 - a_3 &= 0 \\ 2a_1 + 2a_2 + 2a_3 &= 0 \\ a_1 - a_3 &= 0. \end{aligned}$$

The corresponding augmented matrix is

$$\left[\begin{array}{rrr|r} 3 & 1 & -1 & 0 \\ 2 & 2 & 2 & 0 \\ 1 & 0 & -1 & 0 \end{array}\right],$$

whose reduced row echelon form is (verify)

$$\left[\begin{array}{rrr|r} 1 & 0 & -1 & 0 \\ 0 & 1 & 2 & 0 \\ 0 & 0 & 0 & 0 \end{array}\right].$$

Thus there is a nontrivial solution

$$\begin{bmatrix} k \\ -2k \\ k \end{bmatrix}, \quad k \neq 0 \text{ (verify)},$$

so the vectors are linearly dependent. ■

EXAMPLE 3 Are the vectors $\mathbf{v}_1 = \begin{bmatrix} 1 & 0 & 1 & 2 \end{bmatrix}$, $\mathbf{v}_2 = \begin{bmatrix} 0 & 1 & 1 & 2 \end{bmatrix}$, and $\mathbf{v}_3 = \begin{bmatrix} 1 & 1 & 1 & 3 \end{bmatrix}$ in R_4 linearly dependent or linearly independent?

Solution

We form Equation (1),

$$a_1\mathbf{v}_1 + a_2\mathbf{v}_2 + a_3\mathbf{v}_3 = \mathbf{0},$$

and solve for a_1, a_2, and a_3. The resulting homogeneous system is (verify)

$$\begin{aligned} a_1 + a_3 &= 0 \\ a_2 + a_3 &= 0 \\ a_1 + a_2 + a_3 &= 0 \\ 2a_1 + 2a_2 + 3a_3 &= 0. \end{aligned}$$

The corresponding augmented matrix is (verify)

$$\left[\begin{array}{rrr|r} 1 & 0 & 1 & 0 \\ 0 & 1 & 1 & 0 \\ 1 & 1 & 1 & 0 \\ 2 & 2 & 3 & 0 \end{array}\right],$$

and its reduced row echelon form is (verify)

$$\left[\begin{array}{rrr|r} 1 & 0 & 0 & 0 \\ 0 & 1 & 0 & 0 \\ 0 & 0 & 1 & 0 \\ 0 & 0 & 0 & 0 \end{array}\right].$$

Thus the only solution is the trivial solution $a_1 = a_2 = a_3 = 0$, so the vectors are linearly independent. ■

EXAMPLE 4 Are the vectors

$$\mathbf{v}_1 = \begin{bmatrix} 2 & 1 \\ 0 & 1 \end{bmatrix}, \quad \mathbf{v}_2 = \begin{bmatrix} 1 & 2 \\ 1 & 0 \end{bmatrix}, \quad \mathbf{v}_3 = \begin{bmatrix} 0 & -3 \\ -2 & 1 \end{bmatrix}$$

in M_{22} linearly independent?

Solution

We form Equation (1),

$$a_1 \begin{bmatrix} 2 & 1 \\ 0 & 1 \end{bmatrix} + a_2 \begin{bmatrix} 1 & 2 \\ 1 & 0 \end{bmatrix} + a_3 \begin{bmatrix} 0 & -3 \\ -2 & 1 \end{bmatrix} = \begin{bmatrix} 0 & 0 \\ 0 & 0 \end{bmatrix},$$

and solve for a_1, a_2, and a_3. Performing the scalar multiplications and adding the resulting matrices gives

$$\begin{bmatrix} 2a_1 + a_2 & a_1 + 2a_2 - 3a_3 \\ a_2 - 2a_3 & a_1 + a_3 \end{bmatrix} = \begin{bmatrix} 0 & 0 \\ 0 & 0 \end{bmatrix}.$$

Using the definition for equal matrices, we have the linear system

$$\begin{aligned} 2a_1 + a_2 \phantom{{}- 3a_3} &= 0 \\ a_1 + 2a_2 - 3a_3 &= 0 \\ a_2 - 2a_3 &= 0 \\ a_1 \phantom{{}+ 2a_2} + a_3 &= 0. \end{aligned}$$

The corresponding augmented matrix is

$$\left[\begin{array}{ccc|c} 2 & 1 & 0 & 0 \\ 1 & 2 & -3 & 0 \\ 0 & 1 & -2 & 0 \\ 1 & 0 & 1 & 0 \end{array}\right],$$

and its reduced row echelon form is (verify)

$$\left[\begin{array}{ccc|c} 1 & 0 & 1 & 0 \\ 0 & 1 & -2 & 0 \\ 0 & 0 & 0 & 0 \\ 0 & 0 & 0 & 0 \end{array}\right].$$

Thus there is a nontrivial solution

$$\begin{bmatrix} -k \\ 2k \\ k \end{bmatrix}, \quad k \neq 0 \text{ (verify)},$$

so the vectors are linearly dependent. ■

EXAMPLE 5

To find out whether the vectors $\mathbf{v}_1 = \begin{bmatrix} 1 & 0 & 0 \end{bmatrix}$, $\mathbf{v}_2 = \begin{bmatrix} 0 & 1 & 0 \end{bmatrix}$, and $\mathbf{v}_3 = \begin{bmatrix} 0 & 0 & 1 \end{bmatrix}$ in R_3 are linearly dependent or linearly independent, we form Equation (1),

$$a_1\mathbf{v}_1 + a_2\mathbf{v}_2 + a_3\mathbf{v}_3 = \mathbf{0},$$

and solve for a_1, a_2, and a_3. Since $a_1 = a_2 = a_3 = 0$ (verify), we conclude that the given vectors are linearly independent. ■

EXAMPLE 6

Are the vectors $\mathbf{v}_1 = t^2 + t + 2$, $\mathbf{v}_2 = 2t^2 + t$, and $\mathbf{v}_3 = 3t^2 + 2t + 2$ in P_2 linearly dependent or linearly independent?

Solution

Forming Equation (1), we have (verify)

$$\begin{aligned} a_1 + 2a_2 + 3a_3 &= 0 \\ a_1 + a_2 + 2a_3 &= 0 \\ 2a_1 + 2a_3 &= 0, \end{aligned}$$

which has infinitely many solutions (verify). A particular solution is $a_1 = 1$, $a_2 = 1$, $a_3 = -1$, so

$$\mathbf{v}_1 + \mathbf{v}_2 - \mathbf{v}_3 = \mathbf{0}.$$

Hence the given vectors are linearly dependent. ■

EXAMPLE 7

Consider the vectors

$$\mathbf{v}_1 = \begin{bmatrix} 1 \\ 2 \\ -1 \end{bmatrix}, \quad \mathbf{v}_2 = \begin{bmatrix} 1 \\ -2 \\ 1 \end{bmatrix}, \quad \mathbf{v}_3 = \begin{bmatrix} -3 \\ 2 \\ -1 \end{bmatrix}, \quad \text{and} \quad \mathbf{v}_4 = \begin{bmatrix} 2 \\ 0 \\ 0 \end{bmatrix}$$

in R^3. Is $S = \{\mathbf{v}_1, \mathbf{v}_2, \mathbf{v}_3, \mathbf{v}_4\}$ linearly dependent or linearly independent?

Solution

Setting up Equation (1), we are led to the homogeneous system

$$\begin{aligned} a_1 + a_2 - 3a_3 + 2a_4 &= 0 \\ 2a_1 - 2a_2 + 2a_3 &= 0 \\ -a_1 + a_2 - a_3 &= 0, \end{aligned}$$

of three equations in four unknowns. By Theorem 2.4, we are assured of the existence of a nontrivial solution. Hence S is linearly dependent. In fact, two of the infinitely many solutions are

$$\begin{aligned} &a_1 = 1, \quad a_2 = 2, \quad a_3 = 1, \quad a_4 = 0; \\ &a_1 = 1, \quad a_2 = 1, \quad a_3 = 0, \quad a_4 = -1. \end{aligned}$$ ■

EXAMPLE 8

Determine whether the vectors

$$\begin{bmatrix} -1 \\ 1 \\ 0 \\ 0 \end{bmatrix} \quad \text{and} \quad \begin{bmatrix} -2 \\ 0 \\ 1 \\ 1 \end{bmatrix}$$

found in Example 10 in Section 4.4 as spanning the solution space of $A\mathbf{x} = \mathbf{0}$ are linearly dependent or linearly independent.

Solution

Forming Equation (1),

$$a_1 \begin{bmatrix} -1 \\ 1 \\ 0 \\ 0 \end{bmatrix} + a_2 \begin{bmatrix} -2 \\ 0 \\ 1 \\ 1 \end{bmatrix} = \begin{bmatrix} 0 \\ 0 \\ 0 \\ 0 \end{bmatrix},$$

we obtain the homogeneous system

$$\begin{aligned} -a_1 - 2a_2 &= 0 \\ a_1 + 0a_2 &= 0 \\ 0a_1 + a_2 &= 0 \\ 0a_1 + a_2 &= 0, \end{aligned}$$

whose only solution is $a_1 = a_2 = 0$ (verify). Hence the given vectors are linearly independent. ■

We can use determinants to determine whether a set of n vectors in R^n or R_n is linearly independent.

Theorem 4.5 Let $S = \{\mathbf{v}_1, \mathbf{v}_2, \ldots, \mathbf{v}_n\}$ be a set of n vectors in R^n (R_n). Let A be the matrix whose columns (rows) are the elements of S. Then S is linearly independent if and only if $\det(A) \neq 0$.

Proof

We shall prove the result for columns only; the proof for rows is analogous.

Suppose that S is linearly independent. Then it follows that the reduced row echelon form of A is I_n. Thus, A is row equivalent to I_n, and hence $\det(A) \neq 0$. Conversely, if $\det(A) \neq 0$, then A is row equivalent to I_n. Now assume that the rows of A are linearly dependent. Then it follows that the reduced row echelon form of A has a zero row, which contradicts the earlier conclusion that A is row equivalent to I_n. Hence, the rows of A are linearly independent. ■

EXAMPLE 9 Is $S = \{[1 \;\; 2 \;\; 3], [0 \;\; 1 \;\; 2], [3 \;\; 0 \;\; -1]\}$ a linearly independent set of vectors in R^3?

Solution

We form the matrix A whose rows are the vectors in S:

$$A = \begin{bmatrix} 1 & 2 & 3 \\ 0 & 1 & 2 \\ 3 & 0 & -1 \end{bmatrix}.$$

Since $\det(A) = 2$ (verify), we conclude that S is linearly independent. ■

Theorem 4.6 Let S_1 and S_2 be finite subsets of a vector space and let S_1 be a subset of S_2. Then the following statements are true:

(a) If S_1 is linearly dependent, so is S_2.

(b) If S_2 is linearly independent, so is S_1.

Proof

Let

$$S_1 = \{\mathbf{v}_1, \mathbf{v}_2, \ldots, \mathbf{v}_k\} \quad \text{and} \quad S_2 = \{\mathbf{v}_1, \mathbf{v}_2, \ldots, \mathbf{v}_k, \mathbf{v}_{k+1}, \ldots, \mathbf{v}_m\}.$$

We first prove (a). Since S_1 is linearly dependent, there exist $a_1, a_2, \ldots, a_k$, not all zero, such that

$$a_1\mathbf{v}_1 + a_2\mathbf{v}_2 + \cdots + a_k\mathbf{v}_k = \mathbf{0}.$$

Then

$$a_1\mathbf{v}_1 + a_2\mathbf{v}_2 + \cdots + a_k\mathbf{v}_k + 0\mathbf{v}_{k+1} + \cdots + 0\mathbf{v}_m = \mathbf{0}. \tag{2}$$

Since not all the coefficients in (2) are zero, we conclude that S_2 is linearly dependent.

Statement (b) is the contrapositive of statement (a), so it is logically equivalent to statement (a). If we insist on proving it, we may proceed as follows. Let S_2 be linearly independent. If S_1 is assumed as linearly dependent, then S_2 is linearly dependent, by (a), a contradiction. Hence, S_1 must be linearly independent. ■

At this point, we have established the following results:

- The set $S = \{\mathbf{0}\}$ consisting only of $\mathbf{0}$ is linearly dependent, since, for example, $5\mathbf{0} = \mathbf{0}$, and $5 \neq 0$.
- From this it follows that if S is any set of vectors that contains $\mathbf{0}$, then S must be linearly dependent.
- A set of vectors consisting of a single nonzero vector is linearly independent (verify).
- If $\mathbf{v}_1, \mathbf{v}_2, \ldots, \mathbf{v}_k$ are vectors in a vector space V and any two of them are equal, then $\mathbf{v}_1, \mathbf{v}_2, \ldots, \mathbf{v}_k$ are linearly dependent (verify).

We consider next the meaning of linear independence in R^2 and R^3. Suppose that $\mathbf{v}_1$ and $\mathbf{v}_2$ are linearly dependent in R^2. Then there exist scalars a_1 and a_2, not both zero, such that

$$a_1\mathbf{v}_1 + a_2\mathbf{v}_2 = \mathbf{0}.$$

If $a_1 \neq 0$, then $\mathbf{v}_1 = \left(-\dfrac{a_2}{a_1}\right)\mathbf{v}_2$. If $a_2 \neq 0$, then $\mathbf{v}_2 = \left(-\dfrac{a_1}{a_2}\right)\mathbf{v}_1$. Thus one of the vectors is a multiple of the other. Conversely, suppose that $\mathbf{v}_1 = a\mathbf{v}_2$. Then

$$1\mathbf{v}_1 - a\mathbf{v}_2 = \mathbf{0},$$

and since the coefficients of $\mathbf{v}_1$ and $\mathbf{v}_2$ are not both zero, it follows that $\mathbf{v}_1$ and $\mathbf{v}_2$ are linearly dependent. Thus $\mathbf{v}_1$ and $\mathbf{v}_2$ are linearly dependent in R^2 if and only if one of the vectors is a multiple of the other [Figure 4.25(a)]. Hence two vectors in R^2 are linearly dependent if and only if they both lie on the same line passing through the origin [Figure 4.25(a)].

FIGURE 4.25 (a) Linearly dependent vectors in R^2. (b) Linearly independent vectors in R^2.

Suppose now that $\mathbf{v}_1$, $\mathbf{v}_2$, and $\mathbf{v}_3$ are linearly dependent in R^3. Then we can write

$$a_1\mathbf{v}_1 + a_2\mathbf{v}_2 + a_3\mathbf{v}_3 = \mathbf{0},$$

where a_1, a_2, and a_3 are not all zero, say, $a_2 \neq 0$. Then

$$\mathbf{v}_2 = \left(-\frac{a_1}{a_2}\right)\mathbf{v}_1 - \left(\frac{a_3}{a_2}\right)\mathbf{v}_3,$$

which means that $\mathbf{v}_2$ is in the subspace W spanned by $\mathbf{v}_1$ and $\mathbf{v}_3$.

Now W is either a plane through the origin (when $\mathbf{v}_1$ and $\mathbf{v}_3$ are linearly independent) or a line through the origin (when $\mathbf{v}_1$ and $\mathbf{v}_3$ are linearly dependent), or $W = \{\mathbf{0}\}$. Since a line through the origin always lies in a plane through the origin, we conclude that $\mathbf{v}_1$, $\mathbf{v}_2$, and $\mathbf{v}_3$ all lie in the same plane through the origin. Conversely, suppose that $\mathbf{v}_1$, $\mathbf{v}_2$, and $\mathbf{v}_3$ all lie in the same plane through the origin. Then either all three vectors are the zero vector, all three vectors lie on the same line through the origin, or all three vectors lie in a plane through the origin spanned by two vectors, say, $\mathbf{v}_1$ and $\mathbf{v}_3$. Thus, in all these cases, $\mathbf{v}_2$ is a linear combination of $\mathbf{v}_1$ and $\mathbf{v}_3$:

$$\mathbf{v}_2 = c_1\mathbf{v}_1 + c_3\mathbf{v}_3.$$

Then

$$c_1\mathbf{v}_1 - 1\mathbf{v}_2 + c_3\mathbf{v}_3 = \mathbf{0},$$

which means that $\mathbf{v}_1$, $\mathbf{v}_2$, and $\mathbf{v}_3$ are linearly dependent. Hence three vectors in R^3 are linearly dependent if and only if they all lie in the same plane passing through the origin [Figure 4.26(a)].

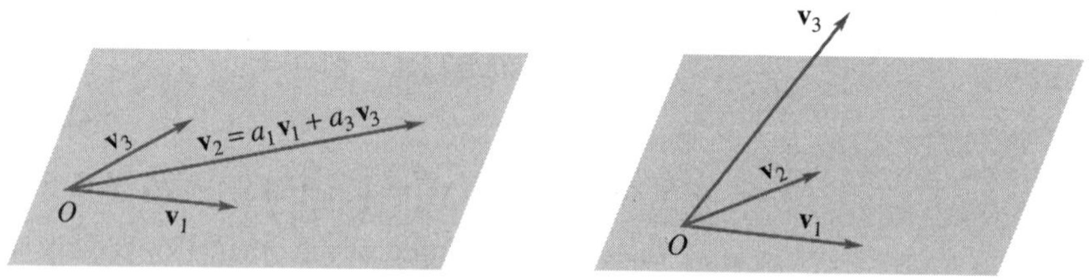

FIGURE 4.26 (a) Linearly dependent vectors in R^3. (b) Linearly independent vectors in R^3.

More generally, let $\mathbf{u}$ and $\mathbf{v}$ be nonzero vectors in a vector space V. We can show (Exercise 18) that $\mathbf{u}$ and $\mathbf{v}$ are linearly dependent if and only if there is a scalar k such that $\mathbf{v} = k\mathbf{u}$. Equivalently, $\mathbf{u}$ and $\mathbf{v}$ are linearly independent if and only if neither vector is a multiple of the other.

Theorem 4.7 The nonzero vectors $\mathbf{v}_1, \mathbf{v}_2, \ldots, \mathbf{v}_n$ in a vector space V are linearly dependent if and only if one of the vectors $\mathbf{v}_j$ $(j \geq 2)$ is a linear combination of the preceding vectors $\mathbf{v}_1, \mathbf{v}_2, \ldots, \mathbf{v}_{j-1}$.

Proof

If $\mathbf{v}_j$ is a linear combination of $\mathbf{v}_1, \mathbf{v}_2, \ldots, \mathbf{v}_{j-1}$, that is,

$$\mathbf{v}_j = a_1\mathbf{v}_1 + a_2\mathbf{v}_2 + \cdots + a_{j-1}\mathbf{v}_{j-1},$$

then

$$a_1\mathbf{v}_1 + a_2\mathbf{v}_2 + \cdots + a_{j-1}\mathbf{v}_{j-1} + (-1)\mathbf{v}_j + 0\mathbf{v}_{j+1} + \cdots + 0\mathbf{v}_n = \mathbf{0}.$$

Since at least one coefficient, -1, is nonzero, we conclude that $\mathbf{v}_1, \mathbf{v}_2, \ldots, \mathbf{v}_n$ are linearly dependent.

Conversely, suppose that $\mathbf{v}_1, \mathbf{v}_2, \ldots, \mathbf{v}_n$ are linearly dependent. Then there exist scalars, $a_1, a_2, \ldots, a_n$, not all zero, such that

$$a_1\mathbf{v}_1 + a_2\mathbf{v}_2 + \cdots + a_n\mathbf{v}_n = \mathbf{0}.$$

Now let j be the largest subscript for which $a_j \neq 0$. If $j \geq 2$, then

$$\mathbf{v}_j = -\left(\frac{a_1}{a_j}\right)\mathbf{v}_1 - \left(\frac{a_2}{a_j}\right)\mathbf{v}_2 - \cdots - \left(\frac{a_{j-1}}{a_j}\right)\mathbf{v}_{j-1}.$$

If $j = 1$, then $a_1\mathbf{v}_1 = \mathbf{0}$, which implies that $\mathbf{v}_1 = \mathbf{0}$, a contradiction of the hypothesis that none of the vectors is the zero vector. Thus one of the vectors $\mathbf{v}_j$ is a linear combination of the preceding vectors $\mathbf{v}_1, \mathbf{v}_2, \ldots, \mathbf{v}_{j-1}$. ■

EXAMPLE 10 Let $V = R_3$ and also $\mathbf{v}_1 = \begin{bmatrix} 1 & 2 & -1 \end{bmatrix}$, $\mathbf{v}_2 = \begin{bmatrix} 1 & -2 & 1 \end{bmatrix}$, $\mathbf{v}_3 = \begin{bmatrix} -3 & 2 & -1 \end{bmatrix}$, and $\mathbf{v}_4 = \begin{bmatrix} 2 & 0 & 0 \end{bmatrix}$. We find (verify) that

$$\mathbf{v}_1 + \mathbf{v}_2 + 0\mathbf{v}_3 - \mathbf{v}_4 = \mathbf{0},$$

so $\mathbf{v}_1$, $\mathbf{v}_2$, $\mathbf{v}_3$, and $\mathbf{v}_4$ are linearly dependent. We then have

$$\mathbf{v}_4 = \mathbf{v}_1 + \mathbf{v}_2 + 0\mathbf{v}_3.$$ ■

Remarks

1. We observe that Theorem 4.7 does not say that *every* vector $\mathbf{v}$ is a linear combination of the preceding vectors. Thus, in Example 10, we also have $\mathbf{v}_1 + 2\mathbf{v}_2 + \mathbf{v}_3 + 0\mathbf{v}_4 = \mathbf{0}$. We cannot solve, in this equation, for $\mathbf{v}_4$ as a linear combination of $\mathbf{v}_1$, $\mathbf{v}_2$, and $\mathbf{v}_3$, since its coefficient is zero.
2. We can also prove that if $S = \{\mathbf{v}_1, \mathbf{v}_2, \ldots, \mathbf{v}_k\}$ is a set of vectors in a vector space V, then S is linearly dependent if and only if one of the vectors in S is a linear combination of all the other vectors in S (see Exercise 19). For instance, in Example 10,
$$\mathbf{v}_1 = -\mathbf{v}_2 - 0\mathbf{v}_3 + \mathbf{v}_4; \quad \mathbf{v}_2 = -\tfrac{1}{2}\mathbf{v}_1 - \tfrac{1}{2}\mathbf{v}_3 - 0\mathbf{v}_4.$$
3. Observe that if $\mathbf{v}_1, \mathbf{v}_2, \ldots, \mathbf{v}_k$ are linearly independent vectors in a vector space, then they must be distinct and nonzero.

The result shown next is used in Section 4.6, as well as in several other places. Suppose that $S = \{\mathbf{v}_1, \mathbf{v}_2, \dots, \mathbf{v}_n\}$ spans a vector space V, and $\mathbf{v}_j$ is a linear combination of the preceding vectors in S. Then the set

$$S_1 = \{\mathbf{v}_1, \mathbf{v}_2, \dots, \mathbf{v}_{j-1}, \mathbf{v}_{j+1}, \dots, \mathbf{v}_n\},$$

consisting of S with $\mathbf{v}_j$ deleted, also spans V. To show this result, observe that if $\mathbf{v}$ is any vector in V, then, since S spans V, we can find scalars $a_1, a_2, \dots, a_n$ such that

$$\mathbf{v} = a_1\mathbf{v}_1 + a_2\mathbf{v}_2 + \cdots + a_{j-1}\mathbf{v}_{j-1} + a_j\mathbf{v}_j + a_{j+1}\mathbf{v}_{j+1} + \cdots + a_n\mathbf{v}_n.$$

Now if

$$\mathbf{v}_j = b_1\mathbf{v}_1 + b_2\mathbf{v}_2 + \cdots + b_{j-1}\mathbf{v}_{j-1},$$

then

$$\begin{aligned}\mathbf{v} &= a_1\mathbf{v}_1 + a_2\mathbf{v}_2 + \cdots + a_{j-1}\mathbf{v}_{j-1} + a_j(b_1\mathbf{v}_1 + b_2\mathbf{v}_2 + \cdots + b_{j-1}\mathbf{v}_{j-1}) \\ &\quad + a_{j+1}\mathbf{v}_{j+1} + \cdots + a_n\mathbf{v}_n \\ &= c_1\mathbf{v}_1 + c_2\mathbf{v}_2 + \cdots + c_{j-1}\mathbf{v}_{j-1} + c_{j+1}\mathbf{v}_{j+1} + \cdots + c_n\mathbf{v}_n,\end{aligned}$$

which means that span $S_1 = V$.

EXAMPLE 11 Consider the set of vectors $S = \{\mathbf{v}_1, \mathbf{v}_2, \mathbf{v}_3, \mathbf{v}_4\}$ in R^4, where

$$\mathbf{v}_1 = \begin{bmatrix}1\\1\\0\\0\end{bmatrix}, \quad \mathbf{v}_2 = \begin{bmatrix}1\\0\\1\\0\end{bmatrix}, \quad \mathbf{v}_3 = \begin{bmatrix}0\\1\\1\\0\end{bmatrix}, \quad \text{and} \quad \mathbf{v}_4 = \begin{bmatrix}2\\1\\1\\0\end{bmatrix},$$

and let $W =$ span S. Since

$$\mathbf{v}_4 = \mathbf{v}_1 + \mathbf{v}_2,$$

we conclude that $W =$ span S_1, where $S_1 = \{\mathbf{v}_1, \mathbf{v}_2, \mathbf{v}_3\}$. ■

Key Terms

Consistent system
Inconsistent system
Homogeneous system
Linearly independent set
Linearly dependent set

4.5 Exercises

1. Show that

$$S = \left\{ \begin{bmatrix}2\\1\\3\end{bmatrix}, \begin{bmatrix}3\\-1\\2\end{bmatrix}, \begin{bmatrix}10\\0\\10\end{bmatrix} \right\}$$

is a linearly dependent set in R^3.

2. Show that

$$S = \left\{ \begin{bmatrix}1\\2\\1\end{bmatrix}, \begin{bmatrix}0\\1\\1\end{bmatrix}, \begin{bmatrix}2\\0\\1\end{bmatrix} \right\}$$

is a linearly independent set in R^3.

3. Determine whether

$$S = \left\{ \begin{bmatrix} 1 \\ 2 \\ 1 \\ -1 \end{bmatrix}, \begin{bmatrix} 4 \\ 3 \\ 1 \\ 0 \end{bmatrix}, \begin{bmatrix} 2 \\ 0 \\ 1 \\ 3 \end{bmatrix} \right\}$$

is a linearly independent set in R^4.

4. Determine whether

$$S = \{ \begin{bmatrix} 3 & 1 & 2 \end{bmatrix}, \begin{bmatrix} 3 & 8 & -5 \end{bmatrix}, \begin{bmatrix} -3 & 6 & -9 \end{bmatrix} \}$$

is a linearly independent set in R_3.

In Exercises 5 through 8, each given augmented matrix is derived from Equation (1).

5. $\left[\begin{array}{cccc|c} 2 & 1 & 3 & 2 & 0 \\ -1 & 0 & 0 & 1 & 0 \\ 1 & -1 & 2 & 1 & 0 \\ 5 & 1 & 8 & 5 & 0 \end{array}\right]$

Is the set S linearly independent?

6. $\left[\begin{array}{cccc|c} 1 & 0 & 2 & 0 & 0 \\ 0 & 1 & -1 & 0 & 0 \\ 0 & 0 & 0 & 1 & 0 \\ 0 & 0 & 0 & 0 & 0 \end{array}\right]$

Is the set S linearly independent?

7. $\left[\begin{array}{ccc|c} 1 & 2 & 1 & 0 \\ 0 & -1 & 0 & 0 \\ 0 & 0 & 2 & 0 \end{array}\right]$ Is the set S linearly independent?

8. $\left[\begin{array}{c|c} A & O \end{array}\right]$, where A is 5×5 and nonsingular. Is the set S linearly independent?

9. Let $\mathbf{x}_1 = \begin{bmatrix} 2 \\ -1 \\ 1 \end{bmatrix}$, $\mathbf{x}_2 = \begin{bmatrix} 4 \\ -7 \\ -1 \end{bmatrix}$, $\mathbf{x}_3 = \begin{bmatrix} 1 \\ 2 \\ 2 \end{bmatrix}$ belong to the solution space of $A\mathbf{x} = \mathbf{0}$. Is $\{\mathbf{x}_1, \mathbf{x}_2, \mathbf{x}_3\}$ linearly independent?

10. Let $\mathbf{x}_1 = \begin{bmatrix} 1 \\ 2 \\ 0 \\ 1 \end{bmatrix}$, $\mathbf{x}_2 = \begin{bmatrix} 1 \\ 0 \\ -1 \\ 1 \end{bmatrix}$, $\mathbf{x}_3 = \begin{bmatrix} 1 \\ 6 \\ 2 \\ 0 \end{bmatrix}$ belong to the null space of A. Is $\{\mathbf{x}_1, \mathbf{x}_2, \mathbf{x}_3\}$ linearly independent?

11. Which of the given vectors in R_3 are linearly dependent? For those which are, express one vector as a linear combination of the rest.

(a) $\begin{bmatrix} 1 & 1 & 0 \end{bmatrix}, \begin{bmatrix} 0 & 2 & 3 \end{bmatrix}, \begin{bmatrix} 1 & 2 & 3 \end{bmatrix}, \begin{bmatrix} 3 & 6 & 6 \end{bmatrix}$

(b) $\begin{bmatrix} 1 & 1 & 0 \end{bmatrix}, \begin{bmatrix} 3 & 4 & 2 \end{bmatrix}$

(c) $\begin{bmatrix} 1 & 1 & 0 \end{bmatrix}, \begin{bmatrix} 0 & 2 & 3 \end{bmatrix}, \begin{bmatrix} 1 & 2 & 3 \end{bmatrix}, \begin{bmatrix} 0 & 0 & 0 \end{bmatrix}$

12. Consider the vector space M_{22}. Follow the directions of Exercise 11.

(a) $\begin{bmatrix} 1 & 1 \\ 2 & 1 \end{bmatrix}, \begin{bmatrix} 1 & 0 \\ 0 & 2 \end{bmatrix}, \begin{bmatrix} 0 & 3 \\ 2 & 1 \end{bmatrix}, \begin{bmatrix} 4 & 6 \\ 8 & 6 \end{bmatrix}$

(b) $\begin{bmatrix} 1 & 1 \\ 1 & 1 \end{bmatrix}, \begin{bmatrix} 1 & 0 \\ 0 & 2 \end{bmatrix}, \begin{bmatrix} 0 & 1 \\ 0 & 2 \end{bmatrix}$

(c) $\begin{bmatrix} 1 & 1 \\ 1 & 1 \end{bmatrix}, \begin{bmatrix} 2 & 3 \\ 1 & 2 \end{bmatrix}, \begin{bmatrix} 3 & 1 \\ 2 & 1 \end{bmatrix}, \begin{bmatrix} 2 & 2 \\ 1 & 1 \end{bmatrix}$

13. Consider the vector space P_2. Follow the directions of Exercise 11.

(a) $t^2 + 1, t - 2, t + 3$ **(b)** $2t^2 + t, t^2 + 3, t$

(c) $2t^2 + t + 1, 3t^2 + t - 5, t + 13$

14. Let V be the vector space of all real-valued continuous functions. Follow the directions of Exercise 11.

(a) $\cos t, \sin t, e^t$ **(b)** $t, e^t, \sin t$

(c) t^2, t, e^t **(d)** $\cos^2 t, \sin^2 t, \cos 2t$

15. Consider the vector space R^3. Follow the directions of Exercise 11.

(a) $\begin{bmatrix} 1 \\ 0 \\ 0 \end{bmatrix}, \begin{bmatrix} 0 \\ 1 \\ 1 \end{bmatrix}, \begin{bmatrix} 1 \\ 2 \\ -1 \end{bmatrix}$

(b) $\begin{bmatrix} 1 \\ 1 \\ -1 \end{bmatrix}, \begin{bmatrix} 0 \\ 1 \\ 1 \end{bmatrix}, \begin{bmatrix} 1 \\ 1 \\ 1 \end{bmatrix}, \begin{bmatrix} 1 \\ 2 \\ -2 \end{bmatrix}$

(c) $\begin{bmatrix} 1 \\ 0 \\ 0 \end{bmatrix}, \begin{bmatrix} 2 \\ 1 \\ 1 \end{bmatrix}, \begin{bmatrix} -1 \\ 2 \\ 1 \end{bmatrix}$

16. For what values of c are the vectors $\begin{bmatrix} -1 & 0 & -1 \end{bmatrix}$, $\begin{bmatrix} 2 & 1 & 2 \end{bmatrix}$, and $\begin{bmatrix} 1 & 1 & c \end{bmatrix}$ in R_3 linearly dependent?

17. For what values of c are the vectors $t + 3$ and $2t + c^2 + 2$ in P_1 linearly independent?

18. Let $\mathbf{u}$ and $\mathbf{v}$ be nonzero vectors in a vector space V. Show that $\mathbf{u}$ and $\mathbf{v}$ are linearly dependent if and only if there is a scalar k such that $\mathbf{v} = k\mathbf{u}$. Equivalently, $\mathbf{u}$ and $\mathbf{v}$ are linearly independent if and only if neither vector is a multiple of the other.

19. Let $S = \{\mathbf{v}_1, \mathbf{v}_2, \ldots, \mathbf{v}_k\}$ be a set of vectors in a vector space V. Prove that S is linearly dependent if and only if one of the vectors in S is a linear combination of all the other vectors in S.

20. Suppose that $S = \{\mathbf{v}_1, \mathbf{v}_2, \mathbf{v}_3\}$ is a linearly independent set of vectors in a vector space V. Prove that $T = \{\mathbf{w}_1, \mathbf{w}_2, \mathbf{w}_3\}$ is also linearly independent, where $\mathbf{w}_1 = \mathbf{v}_1 + \mathbf{v}_2 + \mathbf{v}_3$, $\mathbf{w}_2 = \mathbf{v}_2 + \mathbf{v}_3$, and $\mathbf{w}_3 = \mathbf{v}_3$.

21. Suppose that $S = \{\mathbf{v}_1, \mathbf{v}_2, \mathbf{v}_3\}$ is a linearly independent set of vectors in a vector space V. Is $T = \{\mathbf{w}_1, \mathbf{w}_2, \mathbf{w}_3\}$, where $\mathbf{w}_1 = \mathbf{v}_1+\mathbf{v}_2$, $\mathbf{w}_2 = \mathbf{v}_1+\mathbf{v}_3$, $\mathbf{w}_3 = \mathbf{v}_2+\mathbf{v}_3$, linearly dependent or linearly independent? Justify your answer.

22. Suppose that $S = \{\mathbf{v}_1, \mathbf{v}_2, \mathbf{v}_3\}$ is a linearly dependent set of vectors in a vector space V. Is $T = \{\mathbf{w}_1, \mathbf{w}_2, \mathbf{w}_3\}$, where $\mathbf{w}_1 = \mathbf{v}_1$, $\mathbf{w}_2 = \mathbf{v}_1 + \mathbf{v}_3$, $\mathbf{w}_3 = \mathbf{v}_1 + \mathbf{v}_2 + \mathbf{v}_3$, linearly dependent or linearly independent? Justify your answer.

23. Show that if $\{\mathbf{v}_1, \mathbf{v}_2\}$ is linearly independent and $\mathbf{v}_3$ does not belong to span $\{\mathbf{v}_1, \mathbf{v}_2\}$, then $\{\mathbf{v}_1, \mathbf{v}_2, \mathbf{v}_3\}$ is linearly independent.

24. Suppose that $\{\mathbf{v}_1, \mathbf{v}_2, \ldots, \mathbf{v}_n\}$ is a linearly independent set of vectors in R^n. Show that if A is an $n \times n$ nonsingular matrix, then $\{A\mathbf{v}_1, A\mathbf{v}_2, \ldots, A\mathbf{v}_n\}$ is linearly independent.

25. Let A be an $m \times n$ matrix in reduced row echelon form. Prove that the nonzero rows of A, viewed as vectors in R_n, form a linearly independent set of vectors.

26. Let $S = \{\mathbf{u}_1, \mathbf{u}_2, \ldots, \mathbf{u}_k\}$ be a set of vectors in a vector space and let $T = \{\mathbf{v}_1, \mathbf{v}_2, \ldots, \mathbf{v}_m\}$, where each $\mathbf{v}_i$, $i = 1, 2, \ldots, m$, is a linear combination of the vectors in S. Prove that

$$\mathbf{w} = b_1\mathbf{v}_1 + b_2\mathbf{v}_2 + \cdots + b_m\mathbf{v}_m$$

is a linear combination of the vectors in S.

27. Let S_1 and S_2 be finite subsets of a vector space and let S_1 be a subset of S_2. If S_2 is linearly dependent, why or why not is S_1 linearly dependent? Give an example.

28. Let S_1 and S_2 be finite subsets of a vector space and let S_1 be a subset of S_2. If S_1 is linearly independent, why or why not is S_2 linearly independent? Give an example.

29. Let A be an $m \times n$ matrix. Associate with A the vector $\mathbf{w}$ in R^{mn} obtained by "stringing out" the columns of A. For example, with

$$A = \begin{bmatrix} 1 & 4 \\ 2 & 5 \\ 3 & 6 \end{bmatrix}$$

we associate the vector

$$\mathbf{w} = \begin{bmatrix} 1 \\ 2 \\ 3 \\ 4 \\ 5 \\ 6 \end{bmatrix}.$$

Determine whether your software has such a command. If it does, use it with the vectors in Example 4, together with your software's reduced row echelon form command, to show that the vectors are linearly dependent. Apply this technique to solve Exercise 12.

30. As noted in the Remark after Example 7 in Section 4.4, to determine whether a specific vector $\mathbf{v}$ belongs to span S, we investigate the consistency of an appropriate nonhomogeneous linear system $A\mathbf{x} = \mathbf{b}$. In addition, to determine whether a set of vectors is linearly independent, we investigate the null space of an appropriate homogeneous system $A\mathbf{x} = \mathbf{0}$. These investigations can be performed computationally, using a command for reduced row echelon form, if available. We summarize the use of a reduced row echelon form command in these cases, as follows: Let RREF(C) represent the reduced row echelon form of matrix C.

(i) $\mathbf{v}$ belongs to span S, provided that RREF$\left(\left[A \mid \mathbf{b}\right]\right)$ contains no row of the form $\left[0 \;\cdots\; 0 \mid *\right]$, where $*$ represents a nonzero number.

(ii) The set of vectors is linearly independent if RREF$\left(\left[A \mid \mathbf{0}\right]\right)$ contains only rows from an identity matrix and possibly rows of all zeros.

Experiment in your software with this approach, using the data given in Example 8 in Section 4.4 and Examples 3, 5, 6, and 7.

31. (**Warning:** The strategy given in Exercise 30 assumes the computations are performed by using exact arithmetic. Most software uses a model of exact arithmetic called floating point arithmetic; hence the use of reduced row echelon form may yield incorrect results in these cases. Computationally, the "line between" linear independence and linear dependence may be blurred.) Experiment in your software with the use of reduced row echelon form for the vectors in R^2 given here. Are they linearly independent or linearly dependent? Compare the theoretical answer with the computational answer from your software.

(a) $\begin{bmatrix} 1 \\ 0 \end{bmatrix}, \begin{bmatrix} 1 \\ 1 \times 10^{-5} \end{bmatrix}$ **(b)** $\begin{bmatrix} 1 \\ 0 \end{bmatrix}, \begin{bmatrix} 1 \\ 1 \times 10^{-10} \end{bmatrix}$

(c) $\begin{bmatrix} 1 \\ 0 \end{bmatrix}, \begin{bmatrix} 1 \\ 1 \times 10^{-16} \end{bmatrix}$

4.6 Basis and Dimension

In this section we continue our study of the structure of a vector space V by determining a set of vectors in V that completely describes V. Here we bring together the topics of span from Section 4.4 and linear independence from Section 4.5. In the case of vector spaces that can be completely described by a finite set of vectors, we prove further properties that reveal more details about the structure of such vector spaces.

■ Basis

DEFINITION 4.10 The vectors $\mathbf{v}_1, \mathbf{v}_2, \ldots, \mathbf{v}_k$ in a vector space V are said to form a **basis** for V if (a) $\mathbf{v}_1, \mathbf{v}_2, \ldots, \mathbf{v}_k$ span V and (b) $\mathbf{v}_1, \mathbf{v}_2, \ldots, \mathbf{v}_k$ are linearly independent.

Remark If $\mathbf{v}_1, \mathbf{v}_2, \ldots, \mathbf{v}_k$ form a basis for a vector space V, then they must be distinct and nonzero.

Remark We state Definition 4.10 for a finite set of vectors, but it also applies to an infinite set of vectors in a vector space.

EXAMPLE 1 Let $V = R^3$. The vectors $\begin{bmatrix} 1 \\ 0 \\ 0 \end{bmatrix}, \begin{bmatrix} 0 \\ 1 \\ 0 \end{bmatrix}, \begin{bmatrix} 0 \\ 0 \\ 1 \end{bmatrix}$ form a basis for R^3, called the **natural basis** or **standard basis**, for R^3. We can readily see how to generalize this to obtain the natural basis for R^n. Similarly, $\begin{bmatrix} 1 & 0 & 0 \end{bmatrix}, \begin{bmatrix} 0 & 1 & 0 \end{bmatrix}, \begin{bmatrix} 0 & 0 & 1 \end{bmatrix}$ is the natural basis for R_3. ■

The natural basis for R^n is denoted by $\{\mathbf{e}_1, \mathbf{e}_2, \ldots, \mathbf{e}_n\}$, where

$$\mathbf{e}_i = \begin{bmatrix} 0 \\ \vdots \\ 0 \\ 1 \\ 0 \\ \vdots \\ 0 \end{bmatrix} \leftarrow i\text{th row;}$$

that is, $\mathbf{e}_i$ is an $n \times 1$ matrix with a 1 in the ith row and zeros elsewhere.

The natural basis for R^3 is also often denoted by

$$\mathbf{i} = \begin{bmatrix} 1 \\ 0 \\ 0 \end{bmatrix}, \quad \mathbf{j} = \begin{bmatrix} 0 \\ 1 \\ 0 \end{bmatrix}, \quad \text{and} \quad \mathbf{k} = \begin{bmatrix} 0 \\ 0 \\ 1 \end{bmatrix}.$$

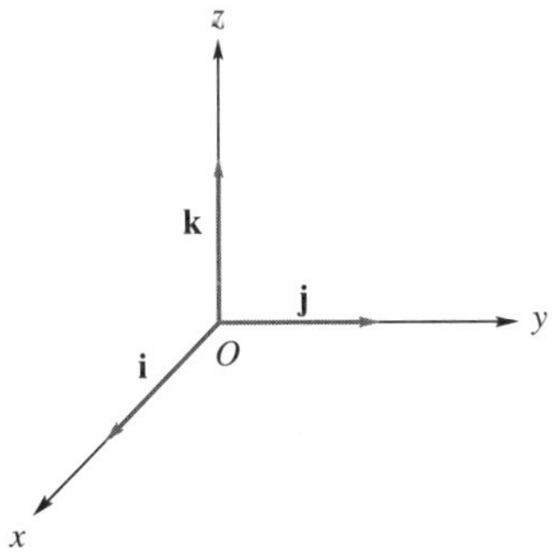

FIGURE 4.27 Natural basis for R^3.

These vectors are shown in Figure 4.27. Thus any vector $\mathbf{v} = \begin{bmatrix} a_1 \\ a_2 \\ a_3 \end{bmatrix}$ in R^3 can be written as

$$\mathbf{v} = a_1\mathbf{i} + a_2\mathbf{j} + a_3\mathbf{k}.$$

EXAMPLE 2 Show that the set $S = \{t^2 + 1, t - 1, 2t + 2\}$ is a basis for the vector space P_2.

Solution

To do this, we must show that S spans V and is linearly independent. To show that it spans V, we take any vector in V, that is, a polynomial $at^2 + bt + c$, and find constants a_1, a_2, and a_3 such that

$$\begin{aligned} at^2 + bt + c &= a_1(t^2 + 1) + a_2(t - 1) + a_3(2t + 2) \\ &= a_1t^2 + (a_2 + 2a_3)t + (a_1 - a_2 + 2a_3). \end{aligned}$$

Since two polynomials agree for all values of t only if the coefficients of respective powers of t agree, we get the linear system

$$\begin{aligned} a_1 \qquad\qquad &= a \\ a_2 + 2a_3 &= b \\ a_1 - a_2 + 2a_3 &= c. \end{aligned}$$

Solving, we have

$$a_1 = a, \quad a_2 = \frac{a + b - c}{2}, \quad a_3 = \frac{c + b - a}{4}.$$

Hence S spans V.

To illustrate this result, suppose that we are given the vector $2t^2 + 6t + 13$. Here, $a = 2$, $b = 6$, and $c = 13$. Substituting in the foregoing expressions for a, b, and c , we find that

$$a_1 = 2, \quad a_2 = -\tfrac{5}{2}, \quad a_3 = \tfrac{17}{4}.$$

Hence

$$2t^2 + 6t + 13 = 2(t^2 + 1) - \tfrac{5}{2}(t - 1) + \tfrac{17}{4}(2t + 2).$$

To show that S is linearly independent, we form

$$a_1(t^2 + 1) + a_2(t - 1) + a_3(2t + 2) = 0.$$

Then

$$a_1t^2 + (a_2 + 2a_3)t + (a_1 - a_2 + 2a_3) = 0.$$

Again, this can hold for all values of t only if

$$\begin{aligned} a_1 \qquad\qquad &= 0 \\ a_2 + 2a_3 &= 0 \\ a_1 - a_2 + 2a_3 &= 0. \end{aligned}$$

The only solution to this homogeneous system is $a_1 = a_2 = a_3 = 0$, which implies that S is linearly independent. Thus S is a basis for P_2. ■

The set of vectors $\{t^n, t^{n-1}, \ldots, t, 1\}$ forms a basis for the vector space P_n called the **natural**, or **standard basis**, for P_n.

EXAMPLE 3

Show that the set $S = \{\mathbf{v}_1, \mathbf{v}_2, \mathbf{v}_3, \mathbf{v}_4\}$, where

$$\mathbf{v}_1 = \begin{bmatrix} 1 & 0 & 1 & 0 \end{bmatrix}, \quad \mathbf{v}_2 = \begin{bmatrix} 0 & 1 & -1 & 2 \end{bmatrix},$$
$$\mathbf{v}_3 = \begin{bmatrix} 0 & 2 & 2 & 1 \end{bmatrix}, \quad \text{and} \quad \mathbf{v}_4 = \begin{bmatrix} 1 & 0 & 0 & 1 \end{bmatrix},$$

is a basis for R_4.

Solution

To show that S is linearly independent, we form the equation

$$a_1\mathbf{v}_1 + a_2\mathbf{v}_2 + a_3\mathbf{v}_3 + a_4\mathbf{v}_4 = \mathbf{0}$$

and solve for a_1, a_2, a_3, and a_4. Substituting for $\mathbf{v}_1$, $\mathbf{v}_2$, $\mathbf{v}_3$, and $\mathbf{v}_4$, we obtain the linear system

$$\begin{aligned} a_1 \qquad\qquad\qquad + a_4 &= 0 \\ a_2 + 2a_3 \qquad &= 0 \\ a_1 - \quad a_2 + 2a_3 \qquad &= 0 \\ 2a_2 + \quad a_3 + a_4 &= 0, \end{aligned}$$

which has as its only solution $a_1 = a_2 = a_3 = a_4 = 0$ (verify), showing that S is linearly independent. Observe that the columns of the coefficient matrix of the preceding linear system are $\mathbf{v}_1^T$, $\mathbf{v}_2^T$, $\mathbf{v}_3^T$, and $\mathbf{v}_4^T$.

To show that S spans R_4, we let $\mathbf{v} = \begin{bmatrix} a & b & c & d \end{bmatrix}$ be any vector in R_4. We now seek constants a_1, a_2, a_3, and a_4 such that

$$a_1\mathbf{v}_1 + a_2\mathbf{v}_2 + a_3\mathbf{v}_3 + a_4\mathbf{v}_4 = \mathbf{v}.$$

Substituting for $\mathbf{v}_1$, $\mathbf{v}_2$, $\mathbf{v}_3$, $\mathbf{v}_4$, and $\mathbf{v}$, we find a solution for a_1, a_2, a_3, and a_4 (verify) to the resulting linear system. Hence S spans R_4 and is a basis for R_4. ■

EXAMPLE 4

The set W of all 2×2 matrices with trace equal to zero is a subspace of M_{22}. Show that the set $S = \{\mathbf{v}_1, \mathbf{v}_2, \mathbf{v}_3\}$, where

$$\mathbf{v}_1 = \begin{bmatrix} 0 & 1 \\ 0 & 0 \end{bmatrix}, \quad \mathbf{v}_2 = \begin{bmatrix} 0 & 0 \\ 1 & 0 \end{bmatrix}, \quad \text{and} \quad \mathbf{v}_3 = \begin{bmatrix} 1 & 0 \\ 0 & -1 \end{bmatrix}$$

is a basis for W.

Solution

To do this, we must show that span $S = W$ and S is linearly independent. To show that span $S = W$, we take any vector $\mathbf{v}$ in W, that is,

$$\mathbf{v} = \begin{bmatrix} a & b \\ c & -a \end{bmatrix},$$

and find constants a_1, a_2, and a_3 such that

$$a_1\mathbf{v}_1 + a_2\mathbf{v}_2 + a_3\mathbf{v}_3 = \mathbf{v}.$$

Substituting for $\mathbf{v}_1$, $\mathbf{v}_2$, $\mathbf{v}_3$, and $\mathbf{v}$ and performing the matrix operations on the left, we obtain (verify)

$$\begin{bmatrix} a_3 & a_1 \\ a_2 & -a_3 \end{bmatrix} = \begin{bmatrix} a & b \\ c & -a \end{bmatrix}.$$

Equating corresponding entries gives $a_1 = b$, $a_2 = c$, and $a_3 = a$, so S spans W. If we replace the vector $\mathbf{v}$ by the zero matrix, it follows in a similar fashion that $a_1 = a_2 = a_3 = 0$, so S is a linearly independent set. Hence S is a basis for W. ■

EXAMPLE 5

Find a basis for the subspace V of P_2, consisting of all vectors of the form $at^2 + bt + c$, where $c = a - b$.

Solution

Every vector in V is of the form

$$at^2 + bt + a - b,$$

which can be written as

$$a(t^2 + 1) + b(t - 1),$$

so the vectors $t^2 + 1$ and $t - 1$ span V. Moreover, these vectors are linearly independent because neither one is a multiple of the other. This conclusion could also be reached (with more work) by writing the equation

$$a_1(t^2 + 1) + a_2(t - 1) = 0,$$

or

$$a_1t^2 + a_2t + (a_1 - a_2) = 0.$$

Since this equation is to hold for all values of t, we must have $a_1 = 0$ and $a_2 = 0$. ■

A vector space V is called **finite-dimensional** if there is a finite subset of V that is a basis for V. If there is no such finite subset of V, then V is called **infinite-dimensional**.

We now establish some results about finite-dimensional vector spaces that will tell about the number of vectors in a basis, compare two different bases, and give properties of bases. First, we observe that if $\{\mathbf{v}_1, \mathbf{v}_2, \dots, \mathbf{v}_k\}$ is a basis for a vector space V, then $\{c\mathbf{v}_1, \mathbf{v}_2, \dots, \mathbf{v}_k\}$ is also a basis when $c \neq 0$ (Exercise 35). Thus a basis for a nonzero vector space is never unique.

Theorem 4.8 If $S = \{\mathbf{v}_1, \mathbf{v}_2, \dots, \mathbf{v}_n\}$ is a basis for a vector space V, then every vector in V can be written in one and only one way as a linear combination of the vectors in S.

Proof

First, every vector $\mathbf{v}$ in V can be written as a linear combination of the vectors in S because S spans V. Now let

$$\mathbf{v} = a_1\mathbf{v}_1 + a_2\mathbf{v}_2 + \cdots + a_n\mathbf{v}_n$$

and

$$\mathbf{v} = b_1\mathbf{v}_1 + b_2\mathbf{v}_2 + \cdots + b_n\mathbf{v}_n.$$

We must show that $a_i = b_i$ for $i = 1, 2, \dots, n$. We have

$$\mathbf{0} = \mathbf{v} - \mathbf{v} = (a_1 - b_1)\mathbf{v}_1 + (a_2 - b_2)\mathbf{v}_2 + \cdots + (a_n - b_n)\mathbf{v}_n.$$

Since S is linearly independent, we conclude that

$$a_i - b_i = 0 \qquad \text{for } i = 1, 2, \ldots, n. \qquad \blacksquare$$

We can also prove (Exercise 44) that if $S = \{\mathbf{v}_1, \mathbf{v}_2, \ldots, \mathbf{v}_n\}$ is a set of nonzero vectors in a vector space V such that every vector in V can be written in one and only one way as a linear combination of the vectors in S, then S is a basis for V.

Even though a nonzero vector space contains an infinite number of elements, a vector space with a finite basis is, in a sense, completely described by a finite number of vectors, namely, by those vectors in the basis.

Theorem 4.9 Let $S = \{\mathbf{v}_1, \mathbf{v}_2, \ldots, \mathbf{v}_n\}$ be a set of nonzero vectors in a vector space V and let $W = \text{span } S$. Then some subset of S is a basis for W.

Proof

Case I If S is linearly independent, then since S already spans W, we conclude that S is a basis for W.

Case II If S is linearly dependent, then

$$a_1\mathbf{v}_1 + a_2\mathbf{v}_2 + \cdots + a_n\mathbf{v}_n = \mathbf{0}, \tag{1}$$

where $a_1, a_2, \ldots, a_n$ are not all zero. Thus some $\mathbf{v}_j$ is a linear combination of the preceding vectors in S (Theorem 4.7). We now delete $\mathbf{v}_j$ from S, getting a subset S_1 of S. Then, by the observation made at the end of Section 4.5, we conclude that $S_1 = \{\mathbf{v}_1, \mathbf{v}_2, \ldots, \mathbf{v}_{j-1}\mathbf{v}_{j+1}, \ldots, \mathbf{v}_n\}$ also spans W.

If S_1 is linearly independent, then S_1 is a basis. If S_1 is linearly dependent, delete a vector of S_1 that is a linear combination of the preceding vectors of S_1 and get a new set S_2 which spans W. Continuing, since S is a finite set, we will eventually find a subset T of S that is linearly independent and spans W. The set T is a basis for W.

Alternative Constructive Proof when V Is R^m or R_m, $n \geq m$. (By the results seen in Section 4.7, this proof is also applicable when V is P_m or M_{pq}, where $n \geq pq$.) We take the vectors in S as $m \times 1$ matrices and form Equation (1). This equation leads to a homogeneous system in the n unknowns $a_1, a_2, \ldots, a_n$; the columns of its $m \times n$ coefficient matrix A are $\mathbf{v}_1, \mathbf{v}_2, \ldots, \mathbf{v}_n$. We now transform A to a matrix B in reduced row echelon form, having r nonzero rows, $1 \leq r \leq m$. Without loss of generality, we may assume that the r leading 1's in the r nonzero rows of B occur in the first r columns. Thus we have

$$B = \begin{bmatrix} 1 & 0 & 0 & \cdots & 0 & b_{1,r+1} & \cdots & b_{1n} \\ 0 & 1 & 0 & \cdots & 0 & b_{2,r+1} & \cdots & b_{2n} \\ 0 & 0 & 1 & \cdots & 0 & b_{3,r+1} & \cdots & b_{3n} \\ \vdots & \vdots & \vdots & & \vdots & \vdots & & \vdots \\ 0 & 0 & 0 & \cdots & 1 & b_{r,r+1} & \cdots & b_{rn} \\ 0 & 0 & 0 & \cdots & 0 & 0 & \cdots & 0 \\ \vdots & \vdots & \vdots & & \vdots & \vdots & & \vdots \\ 0 & 0 & 0 & \cdots & 0 & 0 & \cdots & 0 \end{bmatrix}.$$

Solving for the unknowns corresponding to the leading 1's, we see that $a_1, a_2, \ldots, a_r$ can be solved for in terms of the other unknowns $a_{r+1}, a_{r+2}, \ldots, a_n$.

Thus,

$$\begin{aligned} a_1 &= -b_{1,r+1}a_{r+1} - b_{1,r+2}a_{r+2} - \cdots - b_{1n}a_n \\ a_2 &= -b_{2,r+1}a_{r+1} - b_{2,r+2}a_{r+2} - \cdots - b_{2n}a_n \\ &\;\vdots \\ a_r &= -b_{r,r+1}a_{r+1} - b_{r,r+2}a_{r+2} - \cdots - b_{rn}a_n, \end{aligned} \tag{2}$$

where $a_{r+1}, a_{r+2}, \ldots, a_n$ can be assigned arbitrary real values. Letting

$$a_{r+1} = 1, \quad a_{r+2} = 0, \quad \ldots, \quad a_n = 0$$

in Equation (2) and using these values in Equation (1), we have

$$-b_{1,r+1}\mathbf{v}_1 - b_{2,r+1}\mathbf{v}_2 - \cdots - b_{r,r+1}\mathbf{v}_r + \mathbf{v}_{r+1} = \mathbf{0},$$

which implies that $\mathbf{v}_{r+1}$ is a linear combination of $\mathbf{v}_1, \mathbf{v}_2, \ldots, \mathbf{v}_r$. By the remark made at the end of Section 4.5, the set of vectors obtained from S by deleting $\mathbf{v}_{r+1}$ spans W. Similarly, letting $a_{r+1} = 0, a_{r+2} = 1, a_{r+3} = 0, \ldots, a_n = 0$, we find that $\mathbf{v}_{r+2}$ is a linear combination of $\mathbf{v}_1, \mathbf{v}_2, \ldots, \mathbf{v}_r$ and the set of vectors obtained from S by deleting $\mathbf{v}_{r+1}$ and $\mathbf{v}_{r+2}$ spans W. Continuing in this manner, $\mathbf{v}_{r+3}, \mathbf{v}_{r+4}, \ldots, \mathbf{v}_n$ are linear combinations of $\mathbf{v}_1, \mathbf{v}_2, \ldots, \mathbf{v}_r$, so it follows that $\{\mathbf{v}_1, \mathbf{v}_2, \ldots, \mathbf{v}_r\}$ spans W.

We next show that $\{\mathbf{v}_1, \mathbf{v}_2, \ldots, \mathbf{v}_r\}$ is linearly independent. Consider the matrix B_D that we get by deleting from B all columns not containing a leading 1. In this case, B_D consists of the first r columns of B. Thus

$$B_D = \begin{bmatrix} 1 & 0 & 0 & \cdots & 0 \\ 0 & 1 & 0 & \cdots & 0 \\ 0 & 0 & 1 & \cdots & 0 \\ \vdots & \vdots & \vdots & & \vdots \\ 0 & 0 & 0 & \cdots & 1 \\ 0 & 0 & 0 & \cdots & 0 \\ \vdots & \vdots & \vdots & & \vdots \\ 0 & 0 & 0 & \cdots & 0 \end{bmatrix}.$$

Let A_D be the matrix obtained from A by deleting the columns corresponding to the columns that were deleted in B to obtain B_D. In this case, the columns of A_D are $\mathbf{v}_1, \mathbf{v}_2, \ldots, \mathbf{v}_r$, the first r columns of A. Since A and B are row equivalent, so are A_D and B_D. Then the homogeneous systems

$$A_D\mathbf{x} = \mathbf{0} \quad \text{and} \quad B_D\mathbf{x} = \mathbf{0}$$

are equivalent. Recall now that the homogeneous system $B_D\mathbf{x} = \mathbf{0}$ can be written equivalently as

$$x_1\mathbf{y}_1 + x_2\mathbf{y}_2 + \cdots + x_r\mathbf{y}_r = \mathbf{0}, \tag{3}$$

where $\mathbf{x} = \begin{bmatrix} x_1 \\ x_2 \\ \vdots \\ x_r \end{bmatrix}$ and $\mathbf{y}_1, \mathbf{y}_2, \ldots, \mathbf{y}_r$ are the columns of B_D. Since the columns of B_D form a linearly independent set of vectors in R^m, Equation (3) has only the trivial solution. Hence $A_D\mathbf{x} = \mathbf{0}$ also has only the trivial solution. Thus the columns of A_D are linearly independent. That is, $\{\mathbf{v}_1, \mathbf{v}_2, \ldots, \mathbf{v}_r\}$ is linearly independent. ■

The first proof of Theorem 4.9 leads to a simple procedure for finding a subset T of a set S so that T is a basis for span S. Let $S = \{\mathbf{v}_1, \mathbf{v}_2, \ldots, \mathbf{v}_n\}$ be a set of nonzero vectors in a vector space V. The procedure for finding a subset T of S that is a basis for $W = \text{span } S$ is as follows:

Step 1. Form Equation (1),

$$a_1\mathbf{v}_1 + a_2\mathbf{v}_2 + \cdots + a_n\mathbf{v}_n = \mathbf{0},$$

which we solve for $a_1, a_2, \ldots, a_n$. If these are all zero, then S is linearly independent and is then a basis for W.

Step 2. If $a_1, a_2, \ldots, a_n$ are not all zero, then S is linearly dependent, so one of the vectors in S—say, $\mathbf{v}_j$—is a linear combination of the preceding vectors in S. Delete $\mathbf{v}_j$ from S, getting the subset S_1, which also spans W.

Step 3. Repeat Step 1, using S_1 instead of S. By repeatedly deleting vectors of S, we derive a subset T of S that spans W and is linearly independent. Thus T is a basis for W.

This procedure can be rather tedious, since *every time* we delete a vector from S, we must solve a linear system. In Section 4.9 we present a much more efficient procedure for finding a basis for $W = \text{span } S$, but the basis is *not* guaranteed to be a subset of S. In many cases this is not a cause for concern, since one basis for $W = \text{span } S$ is as good as any other basis. However, there are cases when the vectors in S have some special properties and we want the basis for $W = \text{span } S$ to have the same properties, so we want the basis to be a subset of S. If $V = R^m$ or R_m, the alternative proof of Theorem 4.9 yields a very efficient procedure (see Example 6) for finding a basis for $W = \text{span } S$ consisting of vectors from S.

Let $V = R^m$ or R_m and let $S = \{\mathbf{v}_1, \mathbf{v}_2, \ldots, \mathbf{v}_n\}$ be a set of nonzero vectors in V. The procedure for finding a subset T of S that is a basis for $W = \text{span } S$ is as follows.

Step 1. Form Equation (1),

$$a_1\mathbf{v}_1 + a_2\mathbf{v}_2 + \cdots + a_n\mathbf{v}_n = \mathbf{0}.$$

Step 2. Construct the augmented matrix associated with the homogeneous system of Equation (1), and transform it to reduced row echelon form.

Step 3. The vectors corresponding to the columns containing the leading 1's form a basis T for $W = \text{span } S$.

Recall that in the alternative proof of the theorem we assumed without loss of generality that the r leading 1's in the r nonzero rows of B occur in the first r columns. Thus, if $S = \{\mathbf{v}_1, \mathbf{v}_2, \ldots, \mathbf{v}_6\}$ and the leading 1's occur in columns 1, 3, and 4, then $\{\mathbf{v}_1, \mathbf{v}_3, \mathbf{v}_4\}$ is a basis for span S.

Remark In Step 2 of the foregoing procedure, it is sufficient to transform the augmented matrix to row echelon form.

EXAMPLE 6

Let $V = R_3$ and $S = \{\mathbf{v}_1, \mathbf{v}_2, \mathbf{v}_3, \mathbf{v}_4, \mathbf{v}_5\}$, where $\mathbf{v}_1 = \begin{bmatrix} 1 & 0 & 1 \end{bmatrix}$, $\mathbf{v}_2 = \begin{bmatrix} 0 & 1 & 1 \end{bmatrix}$, $\mathbf{v}_3 = \begin{bmatrix} 1 & 1 & 2 \end{bmatrix}$, $\mathbf{v}_4 = \begin{bmatrix} 1 & 2 & 1 \end{bmatrix}$, and $\mathbf{v}_5 = \begin{bmatrix} -1 & 1 & -2 \end{bmatrix}$. We find that S spans R_3 (verify), and we now wish to find a subset of S that is a basis for R_3. Using the procedure just developed, we proceed as follows:

Step 1. $a_1\begin{bmatrix} 1 & 0 & 1 \end{bmatrix} + a_2\begin{bmatrix} 0 & 1 & 1 \end{bmatrix} + a_3\begin{bmatrix} 1 & 1 & 2 \end{bmatrix} + a_4\begin{bmatrix} 1 & 2 & 1 \end{bmatrix} +$
$a_5\begin{bmatrix} -1 & 1 & -2 \end{bmatrix} = \begin{bmatrix} 0 & 0 & 0 \end{bmatrix}$.

Step 2. Equating corresponding components, we obtain the homogeneous system

$$\begin{aligned} a_1 \qquad\;\; + a_3 + \;\; a_4 - \;\; a_5 &= 0 \\ a_2 + \;\; a_3 + 2a_4 + \;\; a_5 &= 0 \\ a_1 + a_2 + 2a_3 + \;\; a_4 - 2a_5 &= 0. \end{aligned}$$

The reduced row echelon form of the associated augmented matrix is (verify)

$$\left[\begin{array}{ccccc|c} 1 & 0 & 1 & 0 & -2 & 0 \\ 0 & 1 & 1 & 0 & -1 & 0 \\ 0 & 0 & 0 & 1 & 1 & 0 \end{array}\right].$$

Step 3. The leading 1's appear in columns 1, 2, and 4, so $\{\mathbf{v}_1, \mathbf{v}_2, \mathbf{v}_4\}$ is a basis for R_3. ■

Remark In the alternative proof of Theorem 4.9, the order of the vectors in the original spanning set S determines which basis for V is obtained. If, for example, we consider Example 6, where $S = \{\mathbf{w}_1, \mathbf{w}_2, \mathbf{w}_3, \mathbf{w}_4, \mathbf{w}_5\}$ with $\mathbf{w}_1 = \mathbf{v}_5$, $\mathbf{w}_2 = \mathbf{v}_4$, $\mathbf{w}_3 = \mathbf{v}_3$, $\mathbf{w}_4 = \mathbf{v}_2$, and $\mathbf{w}_5 = \mathbf{v}_1$, then the reduced row echelon form of the augmented matrix is (verify)

$$\left[\begin{array}{ccccc|c} 1 & 0 & 0 & 1 & -1 & 0 \\ 0 & 1 & 0 & -1 & 1 & 0 \\ 0 & 0 & 1 & 2 & -1 & 0 \end{array}\right].$$

It then follows that $\{\mathbf{w}_1, \mathbf{w}_2, \mathbf{w}_3\} = \{\mathbf{v}_5, \mathbf{v}_4, \mathbf{v}_3\}$ is a basis for R_3.

We are now about to establish a major result (Corollary 4.1, which follows from Theorem 4.10) of this section, which will tell us about the number of vectors in two different bases.

Theorem 4.10 If $S = \{\mathbf{v}_1, \mathbf{v}_2, \ldots, \mathbf{v}_n\}$ is a basis for a vector space V and $T = \{\mathbf{w}_1, \mathbf{w}_2, \ldots, \mathbf{w}_r\}$ is a linearly independent set of vectors in V, then $r \leq n$.

Proof

Let $T_1 = \{\mathbf{w}_1, \mathbf{v}_1, \ldots, \mathbf{v}_n\}$. Since S spans V, so does T_1. Since $\mathbf{w}_1$ is a linear combination of the vectors in S, we find that T_1 is linearly dependent. Then, by Theorem 4.7, some $\mathbf{v}_j$ is a linear combination of the preceding vectors in T_1. Delete that particular vector $\mathbf{v}_j$.

Let $S_1 = \{\mathbf{w}_1, \mathbf{v}_1, \ldots, \mathbf{v}_{j-1}, \mathbf{v}_{j+1}, \ldots, \mathbf{v}_n\}$. Note that S_1 spans V. Next, let $T_2 = \{\mathbf{w}_2, \mathbf{w}_1, \mathbf{v}_1, \ldots, \mathbf{v}_{j-1}, \mathbf{v}_{j+1}, \ldots, \mathbf{v}_n\}$. Then T_2 is linearly dependent and some vector in T_2 is a linear combination of the preceding vectors in T_2. Since T is linearly independent, this vector cannot be $\mathbf{w}_1$, so it is $\mathbf{v}_i$, $i \neq j$. Repeat this process over and over. Each time there is a new $\mathbf{w}$ vector available from the set T, it is possible to discard one of the $\mathbf{v}$ vectors from the set S. Thus the number r of $\mathbf{w}$ vectors must be no greater than the number n of $\mathbf{v}$ vectors. That is, $r \leq n$. ■

Corollary 4.1 If $S = \{\mathbf{v}_1, \mathbf{v}_2, \ldots, \mathbf{v}_n\}$ and $T = \{\mathbf{w}_1, \mathbf{w}_2, \ldots, \mathbf{w}_n\}$ are bases for a vector space V, then $n = m$.

Proof

Since S is a basis and T is linearly independent, Theorem 4.10 implies that $m \leq n$. Similarly, we obtain $n \leq m$ because T is a basis and S is linearly independent. Hence $n = m$. ■

A vector space or subspace can have many different bases. For example, the natural basis B_1 for R^2 is

$$B_1 = \left\{ \begin{bmatrix} 1 \\ 0 \end{bmatrix}, \begin{bmatrix} 0 \\ 1 \end{bmatrix} \right\},$$

but the set

$$B_2 = \left\{ \begin{bmatrix} 1 \\ 1 \end{bmatrix}, \begin{bmatrix} 1 \\ 2 \end{bmatrix} \right\}$$

is also a basis for R^2. (Verify.) From Theorem 4.8 we have that every vector in R^2 can be written in one and only one way as a linear combination of the vectors in B_1 and in one and only one way as a linear combination of the vectors in B_2. For any vector $\mathbf{v}$ in R^2, where

$$\mathbf{v} = \begin{bmatrix} a \\ b \end{bmatrix},$$

we have

$$\mathbf{v} = a \begin{bmatrix} 1 \\ 0 \end{bmatrix} + b \begin{bmatrix} 0 \\ 1 \end{bmatrix}.$$

However, to express $\mathbf{v}$ as a linear combination of the vectors in B_2, we must find scalars c_1 and c_2 so that

$$c_1 \begin{bmatrix} 1 \\ 1 \end{bmatrix} + c_2 \begin{bmatrix} 1 \\ 2 \end{bmatrix} = \begin{bmatrix} a \\ b \end{bmatrix}.$$

Solving c_1 and c_2 requires the solution of the linear system of equations whose augmented matrix is

$$\left[\begin{array}{cc|c} 1 & 1 & a \\ 1 & 2 & b \end{array}\right] \quad \text{(verify)}.$$

The solution to this linear system is $c_1 = 2a - b$ and $c_2 = b - a$. (Verify.) Thus

$$\mathbf{v} = (2a - b) \begin{bmatrix} 1 \\ 1 \end{bmatrix} + (b - a) \begin{bmatrix} 1 \\ 2 \end{bmatrix}.$$

In this case we observe that the natural basis is more convenient for representing all the vectors in R^2, since the solution of a linear system is not required to determine the corresponding coefficients. In some applications in Chapter 8, a basis other than the natural basis is more convenient. So the choice of basis for a vector space can be important in representing the vectors in that space. We study the representation of a vector in terms of different bases in more detail in Section 4.8, where we discuss the coordinates of a vector with respect to an ordered basis.

■ Dimension

Although a vector space may have many bases, we have just shown that, for a particular vector space V, all bases have the same number of vectors. We can then make the following definition:

DEFINITION 4.11 The **dimension** of a nonzero vector space V is the number of vectors in a basis for V. We often write **dim** V for the dimension of V. We also define the dimension of the trivial vector space $\{\mathbf{0}\}$ to be zero.

EXAMPLE 7 The set $S = \{t^2, t, 1\}$ is a basis for P_2, so $\dim P_2 = 3$. ■

EXAMPLE 8 Let V be the subspace of R_3 spanned by $S = \{\mathbf{v}_1, \mathbf{v}_2, \mathbf{v}_3\}$, where $\mathbf{v}_1 = \begin{bmatrix} 0 & 1 & 1 \end{bmatrix}$, $\mathbf{v}_2 = \begin{bmatrix} 1 & 0 & 1 \end{bmatrix}$, and $\mathbf{v}_3 = \begin{bmatrix} 1 & 1 & 2 \end{bmatrix}$. Thus every vector in V is of the form

$$a_1\mathbf{v}_1 + a_2\mathbf{v}_2 + a_3\mathbf{v}_3,$$

where a_1, a_2, and a_3 are arbitrary real numbers. We find that S is linearly dependent, and $\mathbf{v}_3 = \mathbf{v}_1 + \mathbf{v}_2$ (verify). Thus $S_1 = \{\mathbf{v}_1, \mathbf{v}_2\}$ also spans V. Since S_1 is linearly independent (verify), we conclude that it is a basis for V. Hence $\dim V = 2$. ■

DEFINITION 4.12 Let S be a set of vectors in a vector space V. A subset T of S is called a **maximal independent subset** of S if T is a linearly independent set of vectors that is not properly contained in any other linearly independent subset of S.

EXAMPLE 9 Let V be R^3 and consider the set $S = \{\mathbf{v}_1, \mathbf{v}_2, \mathbf{v}_3, \mathbf{v}_4\}$, where

$$\mathbf{v}_1 = \begin{bmatrix} 1 \\ 0 \\ 0 \end{bmatrix}, \quad \mathbf{v}_2 = \begin{bmatrix} 0 \\ 1 \\ 0 \end{bmatrix}, \quad \mathbf{v}_3 = \begin{bmatrix} 0 \\ 0 \\ 1 \end{bmatrix}, \quad \text{and} \quad \mathbf{v}_4 = \begin{bmatrix} 1 \\ 1 \\ 1 \end{bmatrix}.$$

Maximal independent subsets of S are

$$\{\mathbf{v}_1, \mathbf{v}_2, \mathbf{v}_3\}, \quad \{\mathbf{v}_1, \mathbf{v}_2, \mathbf{v}_4\}, \quad \{\mathbf{v}_1, \mathbf{v}_3, \mathbf{v}_4\}, \quad \text{and} \quad \{\mathbf{v}_2, \mathbf{v}_3, \mathbf{v}_4\}.$$ ■

Corollary 4.2 If the vector space V has dimension n, then a maximal independent subset of vectors in V contains n vectors.

Proof

Let $S = \{\mathbf{v}_1, \mathbf{v}_2, \ldots, \mathbf{v}_k\}$ be a maximal independent subset of V. If span $S \neq V$, then there exists a vector $\mathbf{v}$ in V that cannot be written as a linear combination of $\mathbf{v}_1, \mathbf{v}_2, \ldots, \mathbf{v}_k$. It follows by Theorem 4.7 that $\{\mathbf{v}_1, \mathbf{v}_2, \ldots, \mathbf{v}_k, \mathbf{v}\}$ is a linearly independent set of vectors. However, this contradicts the assumption that S is a maximal independent subset of V. Hence span $S = V$, which implies that set S is a basis for V and $k = n$ by Corollary 4.1. ■

Corollary 4.3 If a vector space V has dimension n, then a minimal spanning set* for V contains n vectors.

Proof

Exercise 38. ■

Although Corollaries 4.2 and 4.3 are, theoretically, of considerable importance, they can be computationally awkward.

From the preceding results, we can make the following observations: If V has dimension n, then any set of $n + 1$ vectors in V is necessarily linearly dependent; also, any set of $n - 1$ vectors in V cannot span V. More generally, we can establish the following results:

Corollary 4.4 If vector space V has dimension n, then any subset of $m > n$ vectors must be linearly dependent.

Proof

Exercise 39. ■

Corollary 4.5 If vector space V has dimension n, then any subset of $m < n$ vectors cannot span V.

Proof

Exercise 40. ■

In Section 4.5, we have already observed that the set $\{\mathbf{0}\}$ is linearly dependent. This is why in Definition 4.11 we defined the dimension of the trivial vector space $\{\mathbf{0}\}$ to be zero.

Thus R^3 has dimension 3, R_2 has dimension 2, and R^n and R_n both have dimension n. Similarly, P_3 has dimension 4 because $\{t^3, t^2, t, 1\}$ is a basis for P_3. In general, P_n has dimension $n + 1$. Most vector spaces considered henceforth in this book are finite-dimensional. Although infinite-dimensional vector spaces are very important in mathematics and physics, their study lies beyond the scope of this book. The vector space P of all polynomials is an infinite-dimensional vector space (Exercise 36).

Section 4.3 included an exercise (Exercise 29) to show that the subspaces of R^2 are $\{\mathbf{0}\}$, R^2 itself, and any line passing through the origin. We can now establish this result by using the material developed in this section. First, we have $\{\mathbf{0}\}$ and

*If S is a set of vectors spanning a vector space V, then S is called a **minimal spanning set** for V if S does not properly contain any other set spanning V.

R^2, the trivial subspaces of dimensions 0 and 2, respectively. The subspace V of R^2 spanned by a vector $\mathbf{v} \neq \mathbf{0}$ is a one-dimensional subspace of R^2; V is a line through the origin. Thus the subspaces of R^2 are $\{\mathbf{0}\}$, R^2, and all the lines through the origin. In a similar way, Exercise 43 at the end of this section asks you to show that the subspaces of R^3 are $\{\mathbf{0}\}$, R^3 itself, and all lines and planes passing through the origin. We now prove a theorem that we shall have occasion to use several times in constructing a basis containing a given set of linearly independent vectors.

Theorem 4.11 If S is a linearly independent set of vectors in a finite-dimensional vector space V, then there is a basis T for V that contains S.

Proof

Let $S = \{\mathbf{v}_1, \mathbf{v}_2, \dots, \mathbf{v}_m\}$ be a linearly independent set of vectors in the n-dimensional vector space V, where $m < n$. Now let $\{\mathbf{w}_1, \mathbf{w}_2, \dots, \mathbf{w}_n\}$ be a basis for V and let $S_1 = \{\mathbf{v}_1, \mathbf{v}_2, \dots, \mathbf{v}_m, \mathbf{w}_1, \mathbf{w}_2, \dots, \mathbf{w}_n\}$. Since S_1 spans V, by Theorem 4.9 it contains a basis T for V. Recall that T is obtained by deleting from S_1 every vector that is a linear combination of the preceding vectors. Since S is linearly independent, none of the $\mathbf{v}_i$ can be linear combinations of other $\mathbf{v}_j$ and thus are not deleted. Hence T will contain S. ■

EXAMPLE 10 Suppose that we wish to find a basis for R_4 that contains the vectors

$$\mathbf{v}_1 = \begin{bmatrix} 1 & 0 & 1 & 0 \end{bmatrix} \quad \text{and} \quad \mathbf{v}_2 = \begin{bmatrix} -1 & 1 & -1 & 0 \end{bmatrix}.$$

We use Theorem 4.11 as follows. First, let $\{\mathbf{e}_1', \mathbf{e}_2', \mathbf{e}_3', \mathbf{e}_4'\}$ be the natural basis for R_4, where

$$\mathbf{e}_1' = \begin{bmatrix} 1 & 0 & 0 & 0 \end{bmatrix}, \quad \mathbf{e}_2' = \begin{bmatrix} 0 & 1 & 0 & 0 \end{bmatrix}, \quad \mathbf{e}_3' = \begin{bmatrix} 0 & 0 & 1 & 0 \end{bmatrix},$$

and

$$\mathbf{e}_4' = \begin{bmatrix} 0 & 0 & 0 & 1 \end{bmatrix}.$$

Form the set $S = \{\mathbf{v}_1, \mathbf{v}_2, \mathbf{e}_1', \mathbf{e}_2', \mathbf{e}_3', \mathbf{e}_4'\}$. Since $\{\mathbf{e}_1', \mathbf{e}_2', \mathbf{e}_3', \mathbf{e}_4'\}$ spans R_4, so does S. We now use the alternative proof of Theorem 4.9 to find a subset of S that is a basis for R_4. Thus we form Equation (1),

$$a_1\mathbf{v}_1 + a_2\mathbf{v}_2 + a_3\mathbf{e}_1' + a_4\mathbf{e}_2' + a_5\mathbf{e}_3' + a_6\mathbf{e}_4' = \begin{bmatrix} 0 & 0 & 0 & 0 \end{bmatrix},$$

which leads to the homogeneous system

$$\begin{aligned} a_1 - a_2 + a_3 \qquad\qquad\qquad &= 0 \\ -a_2 \qquad + a_4 \qquad\qquad &= 0 \\ a_1 - a_2 \qquad\qquad + a_5 \qquad &= 0 \\ a_6 &= 0. \end{aligned}$$

Transforming the augmented matrix to reduced row echelon form, we get (verify)

$$\left[\begin{array}{cccccc|c} 1 & 0 & 0 & 1 & 1 & 0 & 0 \\ 0 & 1 & 0 & 1 & 0 & 0 & 0 \\ 0 & 0 & 1 & 0 & -1 & 0 & 0 \\ 0 & 0 & 0 & 0 & 0 & 1 & 0 \end{array}\right].$$

Since the leading 1's appear in columns 1, 2, 3, and 6, we conclude that $\{\mathbf{v}_1, \mathbf{v}_2, \mathbf{e}'_1, \mathbf{e}'_4\}$ is a basis for R_4 containing $\mathbf{v}_1$ and $\mathbf{v}_2$. ■

It can be shown (Exercise 41) that if W is a subspace of a finite-dimensional vector space V, then W is finite-dimensional and $\dim W \leq \dim V$.

As defined earlier, a given set S of vectors in a vector space V is a basis for V if it spans V and is linearly independent. However, if we are given the *additional* information that the dimension of V is n, we need verify only one of the two conditions. This is the content of the following theorem:

Theorem 4.12 Let V be an n-dimensional vector space.

(a) If $S = \{\mathbf{v}_1, \mathbf{v}_2, \ldots, \mathbf{v}_n\}$ is a linearly independent set of vectors in V, then S is a basis for V.

(b) If $S = \{\mathbf{v}_1, \mathbf{v}_2, \ldots, \mathbf{v}_n\}$ spans V, then S is a basis for V.

Proof

Exercise 45. ■

As a particular application of Theorem 4.12, we have the following: To determine whether a subset S of R^n (R_n) is a basis for R^n (R_n), first count the number of elements in S. If S has n elements, we can use either part (a) or part (b) of Theorem 4.12 to determine whether S is or is not a basis. If S does not have n elements, it is not a basis for R^n (R_n). (Why?) The same line of reasoning applies to any vector space or subspace whose *dimension is known*.

EXAMPLE 11 In Example 6, since $\dim R_3 = 3$ and the set S contains five vectors, we conclude by Theorem 4.12 that S is not a basis for R_3. In Example 3, since $\dim R_4 = 4$ and the set S contains four vectors, it is possible for S to be a basis for R_4. If S is linearly independent *or* spans R_4, it is a basis; otherwise, it is not a basis. Thus we need check only one of the conditions in Theorem 4.12, not both. ■

We now recall that if a set S of n vectors in R^n (R_n) is linearly independent, then S spans R^n (R_n), and conversely, if S spans R^n (R_n), then S is linearly independent. Thus the condition in Theorem 4.5 in Section 4.5 [that $\det(A) \neq 0$] is also necessary and sufficient for S to span R^n (R_n).

Theorem 4.13 Let S be a finite subset of the vector space V that spans V. A maximal independent subset T of S is a basis for V.

Proof

Exercise 46. ■

Key Terms

Basis of a vector space
Natural (standard) basis
Finite-dimensional vector space
Infinite-dimensional vector space
Dimension of a subspace
Maximal independent set
Minimal spanning set

4.6 Exercises

1. Which of the following sets of vectors are bases for R^2?

   (a) $\left\{\begin{bmatrix}1\\3\end{bmatrix}, \begin{bmatrix}1\\-1\end{bmatrix}\right\}$ (b) $\left\{\begin{bmatrix}0\\0\end{bmatrix}, \begin{bmatrix}1\\2\end{bmatrix}, \begin{bmatrix}2\\4\end{bmatrix}\right\}$

   (c) $\left\{\begin{bmatrix}1\\2\end{bmatrix}, \begin{bmatrix}2\\-3\end{bmatrix}, \begin{bmatrix}3\\2\end{bmatrix}\right\}$ (d) $\left\{\begin{bmatrix}1\\3\end{bmatrix}, \begin{bmatrix}-2\\6\end{bmatrix}\right\}$

2. Which of the following sets of vectors are bases for R^3?

   (a) $\left\{\begin{bmatrix}1\\2\\0\end{bmatrix}, \begin{bmatrix}0\\1\\-1\end{bmatrix}\right\}$

   (b) $\left\{\begin{bmatrix}1\\1\\-1\end{bmatrix}, \begin{bmatrix}2\\3\\4\end{bmatrix}, \begin{bmatrix}4\\1\\-1\end{bmatrix}, \begin{bmatrix}0\\1\\-1\end{bmatrix}\right\}$

   (c) $\left\{\begin{bmatrix}3\\2\\2\end{bmatrix}, \begin{bmatrix}-1\\2\\1\end{bmatrix}, \begin{bmatrix}0\\1\\0\end{bmatrix}\right\}$

   (d) $\left\{\begin{bmatrix}1\\0\\0\end{bmatrix}, \begin{bmatrix}0\\2\\-1\end{bmatrix}, \begin{bmatrix}3\\4\\1\end{bmatrix}, \begin{bmatrix}0\\1\\0\end{bmatrix}\right\}$

3. Which of the following sets of vectors are bases for R_4?

   (a) $\{\begin{bmatrix}1 & 0 & 0 & 1\end{bmatrix}, \begin{bmatrix}0 & 1 & 0 & 0\end{bmatrix}, \begin{bmatrix}1 & 1 & 1 & 1\end{bmatrix}, \begin{bmatrix}0 & 1 & 1 & 1\end{bmatrix}\}$

   (b) $\{\begin{bmatrix}1 & -1 & 0 & 2\end{bmatrix}, \begin{bmatrix}3 & -1 & 2 & 1\end{bmatrix}, \begin{bmatrix}1 & 0 & 0 & 1\end{bmatrix}\}$

   (c) $\{\begin{bmatrix}-2 & 4 & 6 & 4\end{bmatrix}, \begin{bmatrix}0 & 1 & 2 & 0\end{bmatrix}, \begin{bmatrix}-1 & 2 & 3 & 2\end{bmatrix}, \begin{bmatrix}-3 & 2 & 5 & 6\end{bmatrix}, \begin{bmatrix}-2 & -1 & 0 & 4\end{bmatrix}\}$

   (d) $\{\begin{bmatrix}0 & 0 & 1 & 1\end{bmatrix}, \begin{bmatrix}-1 & 1 & 1 & 2\end{bmatrix}, \begin{bmatrix}1 & 1 & 0 & 0\end{bmatrix}, \begin{bmatrix}2 & 1 & 2 & 1\end{bmatrix}\}$

4. Which of the following sets of vectors are bases for P_2?

   (a) $\{-t^2 + t + 2, 2t^2 + 2t + 3, 4t^2 - 1\}$

   (b) $\{t^2 + 2t - 1, 2t^2 + 3t - 2\}$

   (c) $\{t^2 + 1, 3t^2 + 2t + 1, 6t^2 + 6t + 3\}$

   (d) $\{3t^2 + 2t + 1, t^2 + t + 1, t^2 + 1\}$

5. Which of the following sets of vectors are bases for P_3?

   (a) $\{t^3 + 2t^2 + 3t, 2t^3 + 1, 6t^3 + 8t^2 + 6t + 4, t^3 + 2t^2 + t + 1\}$

   (b) $\{t^3 + t^2 + 1, t^3 - 1, t^3 + t^2 + t\}$

   (c) $\{t^3 + t^2 + t + 1, t^3 + 2t^2 + t + 3, 2t^3 + t^2 + 3t + 2, t^3 + t^2 + 2t + 2\}$

   (d) $\{t^3 - t, t^3 + t^2 + 1, t - 1\}$

6. Show that the set of matrices

$$\left\{\begin{bmatrix}1 & 1\\0 & 0\end{bmatrix}, \begin{bmatrix}0 & 0\\1 & 1\end{bmatrix}, \begin{bmatrix}1 & 0\\0 & 1\end{bmatrix}, \begin{bmatrix}0 & 1\\1 & 1\end{bmatrix}\right\}$$

   forms a basis for the vector space M_{22}.

In Exercises 7 and 8, determine which of the given subsets forms a basis for R^3. Express the vector

$$\begin{bmatrix}2\\1\\3\end{bmatrix}$$

as a linear combination of the vectors in each subset that is a basis.

7. (a) $\left\{\begin{bmatrix}1\\1\\1\end{bmatrix}, \begin{bmatrix}1\\2\\3\end{bmatrix}, \begin{bmatrix}0\\1\\0\end{bmatrix}\right\}$

   (b) $\left\{\begin{bmatrix}1\\2\\3\end{bmatrix}, \begin{bmatrix}2\\1\\3\end{bmatrix}, \begin{bmatrix}0\\0\\0\end{bmatrix}\right\}$

8. (a) $\left\{\begin{bmatrix}2\\1\\3\end{bmatrix}, \begin{bmatrix}1\\2\\1\end{bmatrix}, \begin{bmatrix}1\\1\\4\end{bmatrix}, \begin{bmatrix}1\\5\\1\end{bmatrix}\right\}$

   (b) $\left\{\begin{bmatrix}1\\1\\2\end{bmatrix}, \begin{bmatrix}2\\2\\0\end{bmatrix}, \begin{bmatrix}3\\4\\-1\end{bmatrix}\right\}$

In Exercises 9 and 10, determine which of the given subsets form a basis for P_2. Express $5t^2 - 3t + 8$ as a linear combination of the vectors in each subset that is a basis.

9. (a) $\{t^2 + t, t - 1, t + 1\}$ (b) $\{t^2 + 1, t - 1\}$

10. (a) $\{t^2 + t, t^2, t^2 + 1\}$ (b) $\{t^2 + 1, t^2 - t + 1\}$

11. Find a basis for the subspace W of R^3 spanned by

$$\left\{\begin{bmatrix}1\\2\\2\end{bmatrix}, \begin{bmatrix}3\\2\\1\end{bmatrix}, \begin{bmatrix}11\\10\\7\end{bmatrix}, \begin{bmatrix}7\\6\\4\end{bmatrix}\right\}.$$

   What is the dimension of W?

12. Find a basis for the subspace W of R_4 spanned by the set of vectors

$$\{\begin{bmatrix}1 & 1 & 0 & -1\end{bmatrix}, \begin{bmatrix}0 & 1 & 2 & 1\end{bmatrix}, \begin{bmatrix}1 & 0 & 1 & -1\end{bmatrix}, \begin{bmatrix}1 & 1 & -6 & -3\end{bmatrix}, \begin{bmatrix}-1 & -5 & 1 & 0\end{bmatrix}\}.$$

What is dim W?

13. Let W be the subspace of P_3 spanned by

$$\{t^3 + t^2 - 2t + 1, t^2 + 1, t^3 - 2t, 2t^3 + 3t^2 - 4t + 3\}.$$

Find a basis for W. What is the dimension of W?

14. Let

$$S = \left\{\begin{bmatrix}1 & 0\\0 & 1\end{bmatrix}, \begin{bmatrix}0 & 1\\1 & 0\end{bmatrix}, \begin{bmatrix}1 & 1\\1 & 1\end{bmatrix}, \begin{bmatrix}-1 & 1\\1 & -1\end{bmatrix}\right\}.$$

Find a basis for the subspace $W = \text{span } S$ of M_{22}.

15. Find all values of a for which

$$\{\begin{bmatrix}a^2 & 0 & 1\end{bmatrix}, \begin{bmatrix}0 & a & 2\end{bmatrix}, \begin{bmatrix}1 & 0 & 1\end{bmatrix}\}$$

is a basis for R_3.

16. Find a basis for the subspace W of M_{33} consisting of all symmetric matrices.

17. Find a basis for the subspace of M_{33} consisting of all diagonal matrices.

18. Let W be the subspace of the space of all continuous real-valued functions spanned by $\{\cos^2 t, \sin^2 t, \cos 2t\}$. Find a basis for W. What is the dimension of W?

In Exercises 19 and 20, find a basis for the given subspaces of R^3 and R^4.

19. **(a)** All vectors of the form $\begin{bmatrix}a\\b\\c\end{bmatrix}$, where $b = a + c$

(b) All vectors of the form $\begin{bmatrix}a\\b\\c\end{bmatrix}$, where $b = a$

(c) All vectors of the form $\begin{bmatrix}a\\b\\c\end{bmatrix}$, where $2a + b - c = 0$

20. **(a)** All vectors of the form $\begin{bmatrix}a\\b\\c\end{bmatrix}$, where $a = 0$

(b) All vectors of the form $\begin{bmatrix}a+c\\a-b\\b+c\\-a+b\end{bmatrix}$

(c) All vectors of the form $\begin{bmatrix}a\\b\\c\end{bmatrix}$, where $a - b + 5c = 0$

21. Find a basis for the subspace of P_2 consisting of all vectors of the form $at^2 + bt + c$, where $c = 2a - 3b$.

22. Find a basis for the subspace of P_3 consisting of all vectors of the form $at^3 + bt^2 + ct + d$, where $c = a - 2d$ and $b = 5a + 3d$.

In Exercises 23 and 24, find the dimensions of the given subspaces of R_4.

23. **(a)** All vectors of the form $\begin{bmatrix}a & b & c & d\end{bmatrix}$, where $d = a + b$

(b) All vectors of the form $\begin{bmatrix}a & b & c & d\end{bmatrix}$, where $c = a - b$ and $d = a + b$

24. **(a)** All vectors of the form $\begin{bmatrix}a & b & c & d\end{bmatrix}$, where $a = b$

(b) All vectors of the form

$$\begin{bmatrix}a+c & a-b & b+c & -a+b\end{bmatrix}$$

25. Find the dimensions of the subspaces of R^2 spanned by the vectors in Exercise 1.

26. Find the dimensions of the subspaces of R^3 spanned by the vectors in Exercise 2.

27. Find the dimensions of the subspaces of R_4 spanned by the vectors in Exercise 3.

28. Find a basis for R^3 that includes

(a) the vector $\begin{bmatrix}1\\0\\2\end{bmatrix}$;

(b) the vectors $\begin{bmatrix}1\\0\\2\end{bmatrix}$ and $\begin{bmatrix}0\\1\\3\end{bmatrix}$.

29. Find a basis for P_3 that includes the vectors $t^3 + t$ and $t^2 - t$.

30. Find a basis for M_{23}. What is the dimension of M_{23}? Generalize to M_{mn}.

31. Find the dimension of the subspace of P_2 consisting of all vectors of the form $at^2 + bt + c$, where $c = b - 2a$.

32. Find the dimension of the subspace of P_3 consisting of all vectors of the form $at^3 + bt^2 + ct + d$, where $b = 3a - 5d$ and $c = d + 4a$.

33. Give an example of a two-dimensional subspace of R^4.

34. Give an example of a two-dimensional subspace of P_3.

35. Prove that if $\{\mathbf{v}_1, \mathbf{v}_2, \ldots, \mathbf{v}_k\}$ is a basis for a vector space V, then $\{c\mathbf{v}_1, \mathbf{v}_2, \ldots, \mathbf{v}_k\}$, for $c \neq 0$, is also a basis for V.

36. Prove that the vector space P of all polynomials is not finite-dimensional. [*Hint*: Suppose that $\{p_1(t), p_2(t), \ldots, p_k(t)\}$ is a finite basis for P. Let $d_j = \text{degree } p_j(t)$. Establish a contradiction.]

37. Let V be an n-dimensional vector space. Show that any $n + 1$ vectors in V form a linearly dependent set.

38. Prove Corollary 4.3.

39. Prove Corollary 4.4.

40. Prove Corollary 4.5.

41. Show that if W is a subspace of a finite-dimensional vector space V, then W is finite-dimensional and $\dim W \leq \dim V$.

42. Show that if W is a subspace of a finite-dimensional vector space V and $\dim W = \dim V$, then $W = V$.

43. Prove that the subspaces of R^3 are $\{\mathbf{0}\}$, R^3 itself, and any line or plane passing through the origin.

44. Let $S = \{\mathbf{v}_1, \mathbf{v}_2, \ldots, \mathbf{v}_n\}$ be a set of nonzero vectors in a vector space V such that every vector in V can be written in one and only one way as a linear combination of the vectors in S. Prove that S is a basis for V.

45. Prove Theorem 4.12.

46. Prove Theorem 4.13.

47. Suppose that $\{\mathbf{v}_1, \mathbf{v}_2, \ldots, \mathbf{v}_n\}$ is a basis for R^n. Show that if A is an $n \times n$ nonsingular matrix, then

$$\{A\mathbf{v}_1, A\mathbf{v}_2, \ldots, A\mathbf{v}_n\}$$

is also a basis for R^n. (*Hint*: See Exercise 24 in Section 4.5.)

48. Suppose that $\{\mathbf{v}_1, \mathbf{v}_2, \ldots, \mathbf{v}_n\}$ is a linearly independent set of vectors in R^n and let A be a singular matrix. Prove or disprove that $\{A\mathbf{v}_1, A\mathbf{v}_2, \ldots, A\mathbf{v}_n\}$ is linearly independent.

49. Find a basis for the subspace W of all 3×3 matrices with trace equal to zero. What is $\dim W$?

4.7 Homogeneous Systems

In Example 12 in Section 2.2 we have seen how the solution of chemical balance equations requires the solution of a homogeneous linear system of equations. Indeed, homogeneous systems play a central role in linear algebra. This will be seen in Chapter 7, where the foundations of the subject are all integrated to solve one of the major problems occurring in a wide variety of applications. In this section we deal with several problems involving homogeneous systems that will arise in Chapter 7. Here we are able to focus our attention on these problems without being distracted by the additional material in Chapter 7.

Consider the homogeneous system

$$A\mathbf{x} = \mathbf{0},$$

where A is an $m \times n$ matrix. As we have already observed in Example 10 of Section 4.3, the set of all solutions to this homogeneous system is a subspace of R^n. An extremely important problem, which will occur repeatedly in Chapter 7, is that of finding a basis for this solution space. To find such a basis, we use the method of Gauss–Jordan reduction presented in Section 2.2. Thus we transform the augmented matrix $\left[A \;\vdots\; \mathbf{0}\right]$ of the system to a matrix $\left[B \;\vdots\; \mathbf{0}\right]$ in reduced row echelon form, where B has r nonzero rows, $1 \leq r \leq m$. Without loss of generality, we may assume that the leading 1's in the r nonzero rows occur in the first r

columns. If $r = n$, then

$$[B \mid \mathbf{0}] = \left[\begin{array}{ccccc:c} 1 & 0 & \cdots & 0 & 0 & 0 \\ 0 & 1 & \cdots & 0 & 0 & 0 \\ \vdots & \vdots & & \vdots & \vdots & \vdots \\ 0 & 0 & \cdots & 0 & 1 & 0 \\ 0 & 0 & \cdots & 0 & 0 & 0 \\ \vdots & \vdots & & \vdots & \vdots & \vdots \\ 0 & 0 & \cdots & 0 & 0 & 0 \end{array}\right] \quad \begin{array}{l} \left.\right\} r = n \\ \left.\right\} m \end{array}$$

(the first n columns are marked n; the first $r = n$ rows are marked $r = n$; all rows are marked m)

and the only solution to $A\mathbf{x} = \mathbf{0}$ is the trivial one. The solution space has no basis, and its dimension is zero.

If $r < n$, then

$$[B \mid \mathbf{0}] = \left[\begin{array}{cccccccc:c} 1 & 0 & 0 & \cdots & 0 & b_{1,r+1} & \cdots & b_{1n} & 0 \\ 0 & 1 & 0 & \cdots & 0 & b_{2,r+1} & \cdots & b_{2n} & 0 \\ 0 & 0 & 1 & \cdots & 0 & b_{3,r+1} & \cdots & b_{3n} & \vdots \\ \vdots & \vdots & \vdots & & \vdots & \vdots & & \vdots & \\ 0 & 0 & 0 & \cdots & 1 & b_{r,r+1} & \cdots & b_{rn} & 0 \\ 0 & 0 & 0 & \cdots & 0 & 0 & \cdots & 0 & 0 \\ \vdots & \vdots & \vdots & & \vdots & \vdots & & \vdots & \vdots \\ 0 & 0 & 0 & \cdots & 0 & 0 & \cdots & 0 & 0 \end{array}\right] \quad \begin{array}{l} \left.\right\} r \\ \left.\right\} m. \end{array}$$

(the first n columns are marked n; the first r rows are marked r; all rows are marked m)

Solving for the unknowns corresponding to the leadings 1's, we have

$$\begin{aligned} x_1 &= -b_{1,r+1}x_{r+1} - b_{1,r+2}x_{r+2} - \cdots - b_{1n}x_n \\ x_2 &= -b_{2,r+1}x_{r+1} - b_{2,r+2}x_{r+2} - \cdots - b_{2n}x_n \\ &\vdots \\ x_r &= -b_{r,r+1}x_{r+1} - b_{r,r+2}x_{r+2} - \cdots - b_{rn}x_n, \end{aligned}$$

where $x_{r+1}, x_{r+2}, \ldots, x_n$ can be assigned arbitrary real values s_j, $j = 1, 2, \ldots, p$, and $p = n - r$. Thus

$$\mathbf{x} = \begin{bmatrix} x_1 \\ x_2 \\ \vdots \\ x_r \\ x_{r+1} \\ x_{r+2} \\ \vdots \\ x_n \end{bmatrix} = \begin{bmatrix} -b_{1,r+1}s_1 - b_{1,r+2}s_2 - \cdots - b_{1n}s_p \\ -b_{2,r+1}s_1 - b_{2,r+2}s_2 - \cdots - b_{2n}s_p \\ \vdots \\ -b_{r,r+1}s_1 - b_{r,r+2}s_2 - \cdots - b_{rn}s_p \\ s_1 \\ s_2 \\ \vdots \\ s_p \end{bmatrix}$$

$$= s_1 \begin{bmatrix} -b_{1,r+1} \\ -b_{2,r+1} \\ \vdots \\ -b_{r,r+1} \\ 1 \\ 0 \\ 0 \\ \vdots \\ 0 \\ 0 \end{bmatrix} + s_2 \begin{bmatrix} -b_{1,r+2} \\ -b_{2,r+2} \\ \vdots \\ -b_{r,r+2} \\ 0 \\ 1 \\ 0 \\ \vdots \\ 0 \\ 0 \end{bmatrix} + \cdots + s_p \begin{bmatrix} -b_{1n} \\ -b_{2n} \\ \vdots \\ -b_{rn} \\ 0 \\ 0 \\ 0 \\ \vdots \\ 0 \\ 1 \end{bmatrix}.$$

Since $s_1, s_2, \ldots, s_p$ can be assigned arbitrary real values, we make the following choices for these values:

$$\begin{array}{lllll} s_1 = 1, & s_2 = 0, & \ldots, & & s_p = 0, \\ s_1 = 0, & s_2 = 1, & \ldots, & & s_p = 0, \\ & & \vdots & & \\ s_1 = 0, & s_2 = 0, & \ldots, & s_{p-1} = 0, & s_p = 1. \end{array}$$

These yield the solutions

$$\mathbf{x}_1 = \begin{bmatrix} -b_{1,r+1} \\ -b_{2,r+1} \\ \vdots \\ -b_{r,r+1} \\ 1 \\ 0 \\ 0 \\ \vdots \\ 0 \\ 0 \end{bmatrix}, \quad \mathbf{x}_2 = \begin{bmatrix} -b_{1,r+2} \\ -b_{2,r+2} \\ \vdots \\ -b_{r,r+2} \\ 0 \\ 1 \\ 0 \\ \vdots \\ 0 \\ 0 \end{bmatrix}, \ldots, \quad \mathbf{x}_p = \begin{bmatrix} -b_{1n} \\ -b_{2n} \\ \vdots \\ -b_{rn} \\ 0 \\ 0 \\ 0 \\ \vdots \\ 0 \\ 1 \end{bmatrix}.$$

Since

$$\mathbf{x} = s_1\mathbf{x}_1 + s_2\mathbf{x}_2 + \cdots + s_p\mathbf{x}_p,$$

we see that $\{\mathbf{x}_1, \mathbf{x}_2, \dots, \mathbf{x}_p\}$ spans the solution space of $A\mathbf{x} = \mathbf{0}$. Moreover, if we form the equation

$$a_1\mathbf{x}_1 + a_2\mathbf{x}_2 + \cdots + a_p\mathbf{x}_p = \mathbf{0},$$

its coefficient matrix is the matrix whose columns are $\mathbf{x}_1, \mathbf{x}_2, \dots, \mathbf{x}_p$. If we look at rows $r+1, r+2, \dots, n$ of this matrix, we readily see that

$$a_1 = a_2 = \cdots = a_p = 0.$$

Hence $\{\mathbf{x}_1, \mathbf{x}_2, \dots, \mathbf{x}_p\}$ is linearly independent and forms a basis for the solution space of $A\mathbf{x} = \mathbf{0}$ (the null space of A).

The procedure for finding a basis for the solution space of a homogeneous system $A\mathbf{x} = \mathbf{0}$, or the null space of A, where A is $m \times n$, is as follows:

Step 1. Solve the given homogeneous system by Gauss–Jordan reduction. If the solution contains no arbitrary constants, then the solution space is $\{\mathbf{0}\}$, which has no basis; the dimension of the solution space is zero.

Step 2. If the solution $\mathbf{x}$ contains arbitrary constants, write $\mathbf{x}$ as a linear combination of vectors $\mathbf{x}_1, \mathbf{x}_2, \dots, \mathbf{x}_p$ with $s_1, s_2, \dots, s_p$ as coefficients:

$$\mathbf{x} = s_1\mathbf{x}_1 + s_2\mathbf{x}_2 + \cdots + s_p\mathbf{x}_p.$$

Step 3. The set of vectors $\{\mathbf{x}_1, \mathbf{x}_2, \dots, \mathbf{x}_p\}$ is a basis for the solution space of $A\mathbf{x} = \mathbf{0}$; the dimension of the solution space is p.

Remark In Step 1, suppose that the matrix in reduced row echelon form to which $\left[A \;\vdots\; \mathbf{0}\right]$ has been transformed has r nonzero rows (also, r leading 1's). Then $p = n - r$. That is, the dimension of the solution space is $n - r$. Moreover, a solution $\mathbf{x}$ to $A\mathbf{x} = \mathbf{0}$ has $n - r$ arbitrary constants.

If A is an $m \times n$ matrix, we refer to the dimension of the null space of A as the **nullity** of A, denoted by nullity A.

EXAMPLE 1

Find a basis for and the dimension of the solution space W of the homogeneous system

$$\begin{bmatrix} 1 & 1 & 4 & 1 & 2 \\ 0 & 1 & 2 & 1 & 1 \\ 0 & 0 & 0 & 1 & 2 \\ 1 & -1 & 0 & 0 & 2 \\ 2 & 1 & 6 & 0 & 1 \end{bmatrix} \begin{bmatrix} x_1 \\ x_2 \\ x_3 \\ x_4 \\ x_5 \end{bmatrix} = \begin{bmatrix} 0 \\ 0 \\ 0 \\ 0 \\ 0 \end{bmatrix}.$$

Solution

Step 1. To solve the given system by the Gauss–Jordan reduction method, we transform the augmented matrix to reduced row echelon form, obtaining (verify)

$$\left[\begin{array}{ccccc|c} 1 & 0 & 2 & 0 & 1 & 0 \\ 0 & 1 & 2 & 0 & -1 & 0 \\ 0 & 0 & 0 & 1 & 2 & 0 \\ 0 & 0 & 0 & 0 & 0 & 0 \\ 0 & 0 & 0 & 0 & 0 & 0 \end{array}\right].$$

Every solution is of the form (verify)

$$\mathbf{x} = \begin{bmatrix} -2s - t \\ -2s + t \\ s \\ -2t \\ t \end{bmatrix}, \tag{1}$$

where s and t are any real numbers.

Step 2. Every vector in W is a solution and is therefore of the form given by Equation (1). We can thus write every vector in W as

$$\mathbf{x} = s \begin{bmatrix} -2 \\ -2 \\ 1 \\ 0 \\ 0 \end{bmatrix} + t \begin{bmatrix} -1 \\ 1 \\ 0 \\ -2 \\ 1 \end{bmatrix}. \tag{2}$$

Since s and t can take on any values, we first let $s = 1$, $t = 0$, and then let $s = 0$, $t = 1$, in Equation (2), obtaining as solutions

$$\mathbf{x}_1 = \begin{bmatrix} -2 \\ -2 \\ 1 \\ 0 \\ 0 \end{bmatrix} \quad \text{and} \quad \mathbf{x}_2 = \begin{bmatrix} -1 \\ 1 \\ 0 \\ -2 \\ 1 \end{bmatrix}.$$

Step 3. The set $\{\mathbf{x}_1, \mathbf{x}_2\}$ is a basis for W. Moreover, $\dim W = 2$. ■

The following example illustrates a type of problem that we will be solving often in Chapter 7:

EXAMPLE 2

Find a basis for the solution space of the homogeneous system $(\lambda I_3 - A)\mathbf{x} = \mathbf{0}$ for $\lambda = -2$ and

$$A = \begin{bmatrix} -3 & 0 & -1 \\ 2 & 1 & 0 \\ 0 & 0 & -2 \end{bmatrix}.$$

Solution

We form $-2I_3 - A$:

$$-2\begin{bmatrix} 1 & 0 & 0 \\ 0 & 1 & 0 \\ 0 & 0 & 1 \end{bmatrix} - \begin{bmatrix} -3 & 0 & -1 \\ 2 & 1 & 0 \\ 0 & 0 & -2 \end{bmatrix} = \begin{bmatrix} 1 & 0 & 1 \\ -2 & -3 & 0 \\ 0 & 0 & 0 \end{bmatrix}.$$

This last matrix is the coefficient matrix of the homogeneous system, so we transform the augmented matrix

$$\left[\begin{array}{ccc|c} 1 & 0 & 1 & 0 \\ -2 & -3 & 0 & 0 \\ 0 & 0 & 0 & 0 \end{array}\right]$$

to reduced row echelon form, obtaining (verify)

$$\left[\begin{array}{ccc|c} 1 & 0 & 1 & 0 \\ 0 & 1 & -\frac{2}{3} & 0 \\ 0 & 0 & 0 & 0 \end{array}\right].$$

Every solution is then of the form (verify)

$$\mathbf{x} = \begin{bmatrix} -s \\ \frac{2}{3}s \\ s \end{bmatrix},$$

where s is any real number. Then every vector in the solution can be written as

$$\mathbf{x} = s\begin{bmatrix} -1 \\ \frac{2}{3} \\ 1 \end{bmatrix},$$

so $\left\{\begin{bmatrix} -1 \\ \frac{2}{3} \\ 1 \end{bmatrix}\right\}$ is a basis for the solution space. ■

Another important problem that we have to solve often in Chapter 7 is illustrated in the following example:

EXAMPLE 3

Find all real numbers λ such that the homogeneous system $(\lambda I_2 - A)\mathbf{x} = \mathbf{0}$ has a nontrivial solution for

$$A = \begin{bmatrix} 1 & 5 \\ 3 & -1 \end{bmatrix}.$$

Solution

We form $\lambda I_2 - A$:

$$\lambda\begin{bmatrix} 1 & 0 \\ 0 & 1 \end{bmatrix} - \begin{bmatrix} 1 & 5 \\ 3 & -1 \end{bmatrix} = \begin{bmatrix} \lambda - 1 & -5 \\ -3 & \lambda + 1 \end{bmatrix}.$$

The homogeneous system $(\lambda I_2 - A)\mathbf{x} = \mathbf{0}$ is then

$$\begin{bmatrix} \lambda - 1 & -5 \\ -3 & \lambda + 1 \end{bmatrix}\begin{bmatrix} x_1 \\ x_2 \end{bmatrix} = \begin{bmatrix} 0 \\ 0 \end{bmatrix}.$$

It follows from Corollary 3.1 in Section 3.2 that this homogeneous system has a nontrivial solution if and only if

$$\det\left(\begin{bmatrix} \lambda - 1 & -5 \\ -3 & \lambda + 1 \end{bmatrix}\right) = 0,$$

that is, if and only if

$$\begin{aligned}(\lambda - 1)(\lambda + 1) - 15 &= 0\\ \lambda^2 - 16 &= 0\\ \lambda = 4 \quad \text{or} \quad \lambda &= -4.\end{aligned}$$

Thus, when $\lambda = 4$ or -4, the homogeneous system $(\lambda I_2 - A)\mathbf{x} = \mathbf{0}$ for the given matrix A has a nontrivial solution. ■

■ Relationship between Nonhomogeneous Linear Systems and Homogeneous Systems

We have already noted in Section 4.3 that if A is $m \times n$, then the set of all solutions to the linear system $A\mathbf{x} = \mathbf{b}$, $\mathbf{b} \neq \mathbf{0}$, is not a subspace of R^n. The following example illustrates a geometric relationship between the set of all solutions to the nonhomogeneous system $A\mathbf{x} = \mathbf{b}$, $\mathbf{b} \neq \mathbf{0}$, and the associated homogeneous system $A\mathbf{x} = \mathbf{0}$.

EXAMPLE 4 Consider the linear system

$$\begin{bmatrix} 1 & 2 & -3 \\ 2 & 4 & -6 \\ 3 & 6 & -9 \end{bmatrix} \begin{bmatrix} x_1 \\ x_2 \\ x_3 \end{bmatrix} = \begin{bmatrix} 2 \\ 4 \\ 6 \end{bmatrix}.$$

The set of all solutions to this linear system consists of all vectors of the form

$$\mathbf{x} = \begin{bmatrix} 2 - 2r + 3s \\ r \\ s \end{bmatrix}$$

(verify), which can be written as

$$\mathbf{x} = \begin{bmatrix} 2 \\ 0 \\ 0 \end{bmatrix} + r \begin{bmatrix} -2 \\ 1 \\ 0 \end{bmatrix} + s \begin{bmatrix} 3 \\ 0 \\ 1 \end{bmatrix}.$$

The set of all solutions to the associated homogeneous system is the two-dimensional subspace of R^3 consisting of all vectors of the form

$$\mathbf{x} = r \begin{bmatrix} -2 \\ 1 \\ 0 \end{bmatrix} + s \begin{bmatrix} 3 \\ 0 \\ 1 \end{bmatrix}.$$

This subspace is a plane Π_1 passing through the origin; the set of all solutions to the given nonhomogeneous system is a plane Π_2 that does not pass through the origin and is obtained by shifting Π_1 parallel to itself. This situation is illustrated in Figure 4.28. ■

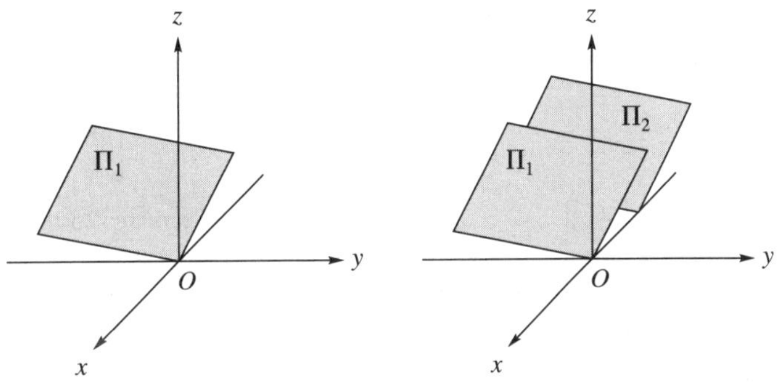

FIGURE 4.28 Π_1 is the solution space to $A\mathbf{x} = \mathbf{0}$. Π_2 is the set of all solutions to $A\mathbf{x} = \mathbf{b}$.

The following result, which is important in the study of differential equations, was presented in Section 2.2 and its proof left to Exercise 29(b) of that section.

If $\mathbf{x}_p$ is a particular solution to the nonhomogeneous system $A\mathbf{x} = \mathbf{b}$, $\mathbf{b} \neq \mathbf{0}$, and $\mathbf{x}_h$ is a solution to the associated homogeneous system $A\mathbf{x} = \mathbf{0}$, then $\mathbf{x}_p + \mathbf{x}_h$ is a solution to the given system $A\mathbf{x} = \mathbf{b}$. Moreover, every solution $\mathbf{x}$ to the nonhomogeneous linear system $A\mathbf{x} = \mathbf{b}$ can be written as $\mathbf{x}_p + \mathbf{x}_h$, where $\mathbf{x}_p$ is a particular solution to the given nonhomogeneous system and $\mathbf{x}_h$ is a solution to the associated homogeneous system $A\mathbf{x} = \mathbf{0}$. Thus, in Example 4,

$$\mathbf{x}_p = \begin{bmatrix} 2 \\ 0 \\ 0 \end{bmatrix} \quad \text{and} \quad \mathbf{x}_h = r\begin{bmatrix} -2 \\ 1 \\ 0 \end{bmatrix} + s\begin{bmatrix} 3 \\ 0 \\ 1 \end{bmatrix},$$

where r and s are any real numbers.

Key Terms

Homogeneous system
Solution (null) space
Dimension
Nullity
Arbitrary constants

4.7 Exercises

1. Let

$$A = \begin{bmatrix} 2 & -1 & -2 \\ -4 & 2 & 4 \\ -8 & 4 & 8 \end{bmatrix}.$$

(a) Find the set of all solutions to $A\mathbf{x} = \mathbf{0}$.

(b) Express each solution as a linear combination of two vectors in R^3.

(c) Sketch these vectors in a three-dimensional coordinate system to show that the solution space is a plane through the origin.

2. Let

$$A = \begin{bmatrix} 1 & 1 & -2 \\ -2 & -2 & 4 \\ -1 & -1 & 2 \end{bmatrix}.$$

(a) Find the set of all solutions to $A\mathbf{x} = \mathbf{0}$.

(b) Express each solution as a linear combination of two vectors in R^3.

(c) Sketch these vectors in a three-dimensional coordinate system to show that the solution space is a plane through the origin.

In Exercises 3 through 10, find a basis for and the dimension of the solution space of the given homogeneous system.

3. $\begin{aligned} x_1 + x_2 + x_3 + x_4 &= 0 \\ 2x_1 + x_2 - x_3 + x_4 &= 0 \end{aligned}$

4. $\begin{bmatrix} 1 & -1 & 1 & -2 & 1 \\ 3 & -3 & 2 & 0 & 2 \end{bmatrix} \begin{bmatrix} x_1 \\ x_2 \\ x_3 \\ x_4 \\ x_5 \end{bmatrix} = \begin{bmatrix} 0 \\ 0 \end{bmatrix}$

5. $\begin{aligned} x_1 + 2x_2 - x_3 + 3x_4 &= 0 \\ 2x_1 + 2x_2 - x_3 + 2x_4 &= 0 \\ x_1 \qquad + 3x_3 + 3x_4 &= 0 \end{aligned}$

6. $\begin{aligned} x_1 - x_2 + 2x_3 + 3x_4 + 4x_5 &= 0 \\ -x_1 + 2x_2 + 3x_3 + 4x_4 + 5x_5 &= 0 \\ x_1 - x_2 + 3x_3 + 5x_4 + 6x_5 &= 0 \\ 3x_1 - 4x_2 + x_3 + 2x_4 + 3x_5 &= 0 \end{aligned}$

7. $\begin{bmatrix} 1 & 2 & 1 & 2 & 1 \\ 1 & 2 & 2 & 1 & 2 \\ 2 & 4 & 3 & 3 & 3 \\ 0 & 0 & 1 & -1 & -1 \end{bmatrix} \begin{bmatrix} x_1 \\ x_2 \\ x_3 \\ x_4 \\ x_5 \end{bmatrix} = \begin{bmatrix} 0 \\ 0 \\ 0 \\ 0 \end{bmatrix}$

8. $\begin{bmatrix} 1 & 0 & 2 \\ 2 & 1 & 3 \\ 3 & 1 & 2 \end{bmatrix} \begin{bmatrix} x_1 \\ x_2 \\ x_3 \end{bmatrix} = \begin{bmatrix} 0 \\ 0 \\ 0 \end{bmatrix}$

9. $\begin{bmatrix} 1 & 2 & 2 & -1 & 1 \\ 0 & 2 & 2 & -2 & -1 \\ 2 & 6 & 2 & -4 & 1 \\ 1 & 4 & 0 & -3 & 0 \end{bmatrix} \begin{bmatrix} x_1 \\ x_2 \\ x_3 \\ x_4 \\ x_5 \end{bmatrix} = \begin{bmatrix} 0 \\ 0 \\ 0 \\ 0 \end{bmatrix}$

10. $\begin{bmatrix} 1 & 2 & -3 & -2 & 1 & 3 \\ 1 & 2 & -4 & 3 & 3 & 4 \\ -2 & -4 & 6 & 4 & -3 & 2 \\ 0 & 0 & -1 & 5 & 1 & 9 \\ 1 & 2 & -3 & -2 & 0 & 7 \end{bmatrix} \begin{bmatrix} x_1 \\ x_2 \\ x_3 \\ x_4 \\ x_5 \\ x_6 \end{bmatrix} = \begin{bmatrix} 0 \\ 0 \\ 0 \\ 0 \\ 0 \end{bmatrix}$

In Exercises 11 and 12, find a basis for the null space of each given matrix A.

11. $A = \begin{bmatrix} 1 & 2 & 3 & -1 \\ 2 & 3 & 2 & 0 \\ 3 & 4 & 1 & 1 \\ 1 & 1 & -1 & 1 \end{bmatrix}$

12. $A = \begin{bmatrix} 1 & -1 & 2 & 1 & 0 \\ 2 & 0 & 1 & -1 & 3 \\ 5 & -1 & 3 & 0 & 3 \\ 4 & -2 & 5 & 1 & 3 \\ 1 & 3 & -4 & -5 & 6 \end{bmatrix}$

In Exercises 13 through 16, find a basis for the solution space of the homogeneous system $(\lambda I_n - A)\mathbf{x} = \mathbf{0}$ for the given scalar λ and given matrix A.

13. $\lambda = 1,\ A = \begin{bmatrix} 3 & 2 \\ 1 & 2 \end{bmatrix}$

14. $\lambda = -3,\ A = \begin{bmatrix} -4 & -3 \\ 2 & 3 \end{bmatrix}$

15. $\lambda = 1,\ A = \begin{bmatrix} 0 & 0 & 1 \\ 1 & 0 & -3 \\ 0 & 1 & 3 \end{bmatrix}$

16. $\lambda = 3,\ A = \begin{bmatrix} 1 & 1 & -2 \\ -1 & 2 & 1 \\ 0 & 1 & -1 \end{bmatrix}$

In Exercises 17 through 20, find all real numbers λ such that the homogeneous system $(\lambda I_n - A)\mathbf{x} = \mathbf{0}$ has a nontrivial solution.

17. $A = \begin{bmatrix} 2 & 3 \\ 2 & -3 \end{bmatrix}$

18. $A = \begin{bmatrix} 3 & 0 \\ 2 & -2 \end{bmatrix}$

19. $A = \begin{bmatrix} 0 & 0 & 0 \\ 0 & 1 & -1 \\ 1 & 0 & 0 \end{bmatrix}$

20. $A = \begin{bmatrix} -2 & 0 & 0 \\ 0 & -2 & -3 \\ 0 & 4 & 5 \end{bmatrix}$

In Exercises 21 and 22, determine the solution to the linear system $A\mathbf{x} = \mathbf{b}$ and write it in the form $\mathbf{x} = \mathbf{x}_p + \mathbf{x}_h$.

21. $A = \begin{bmatrix} 1 & 2 \\ 2 & 4 \end{bmatrix},\ \mathbf{b} = \begin{bmatrix} -2 \\ -4 \end{bmatrix}$

22. $A = \begin{bmatrix} 1 & -1 & 2 \\ 1 & 1 & 2 \\ 1 & -3 & 2 \end{bmatrix},\ \mathbf{b} = \begin{bmatrix} 1 \\ -1 \\ 3 \end{bmatrix}$

23. Let $S = \{\mathbf{x}_1, \mathbf{x}_2, \ldots, \mathbf{x}_k\}$ be a set of solutions to a homogeneous system $A\mathbf{x} = \mathbf{0}$. Show that every vector in span S is a solution to $A\mathbf{x} = \mathbf{0}$.

24. Show that if the $n \times n$ coefficient matrix A of the homogeneous system $A\mathbf{x} = \mathbf{0}$ has a row or column of zeros, then $A\mathbf{x} = \mathbf{0}$ has a nontrivial solution.

25. **(a)** Show that the zero matrix is the only 3×3 matrix whose null space has dimension 3.

(b) Let A be a nonzero 3×3 matrix and suppose that $A\mathbf{x} = \mathbf{0}$ has a nontrivial solution. Show that the dimension of the null space of A is either 1 or 2.

26. Matrices A and B are $m \times n$, and their reduced row echelon forms are the same. What is the relationship between the null space of A and the null space of B?

4.8 Coordinates and Isomorphisms

Coordinates

If V is an n-dimensional vector space, we know that V has a basis S with n vectors in it; thus far we have not paid much attention to the order of the vectors in S. However, in the discussion of this section we speak of an **ordered basis** $S = \{\mathbf{v}_1, \mathbf{v}_2, \ldots, \mathbf{v}_n\}$ for V; thus $S_1 = \{\mathbf{v}_2, \mathbf{v}_1, \ldots, \mathbf{v}_n\}$ is a different ordered basis for V.

If $S = \{\mathbf{v}_1, \mathbf{v}_2, \ldots, \mathbf{v}_n\}$ is an ordered basis for the n-dimensional vector space V, then by Theorem 4.8 every vector $\mathbf{v}$ in V can be uniquely expressed in the form

$$\mathbf{v} = a_1\mathbf{v}_1 + a_2\mathbf{v}_2 + \cdots + a_n\mathbf{v}_n,$$

where $a_1, a_2, \ldots, a_n$ are real numbers. We shall refer to

$$\left[\mathbf{v}\right]_S = \begin{bmatrix} a_1 \\ a_2 \\ \vdots \\ a_n \end{bmatrix}$$

as the **coordinate vector of v with respect to the ordered basis *S***. The entries of $\left[\mathbf{v}\right]_S$ are called the **coordinates of v with respect to *S***.

EXAMPLE 1 Consider the vector space P_1 and let $S = \{t, 1\}$ be an ordered basis for P_1. If $\mathbf{v} = p(t) = 5t - 2$, then $\left[\mathbf{v}\right]_S = \begin{bmatrix} 5 \\ -2 \end{bmatrix}$ is the coordinate vector of $\mathbf{v}$ with respect to the ordered basis S. On the other hand, if $T = \{t+1, t-1\}$ is the ordered basis, we have $5t - 2 = \frac{3}{2}(t + 1) + \frac{7}{2}(t - 1)$, which implies that

$$\left[\mathbf{v}\right]_T = \begin{bmatrix} \frac{3}{2} \\ \frac{7}{2} \end{bmatrix}. \qquad \blacksquare$$

Notice that the coordinate vector $\left[\mathbf{v}\right]_S$ depends upon the order in which the vectors in S are listed; a change in the order of this listing may change the coordinates of $\mathbf{v}$ with respect to S.

EXAMPLE 2 Consider the vector space R^3 and let $S = \{\mathbf{v}_1, \mathbf{v}_2, \mathbf{v}_3\}$ be an ordered basis for R^3, where

$$\mathbf{v}_1 = \begin{bmatrix} 1 \\ 1 \\ 0 \end{bmatrix}, \quad \mathbf{v}_2 = \begin{bmatrix} 2 \\ 0 \\ 1 \end{bmatrix}, \quad \text{and} \quad \mathbf{v}_3 = \begin{bmatrix} 0 \\ 1 \\ 2 \end{bmatrix}.$$

If

$$\mathbf{v} = \begin{bmatrix} 1 \\ 1 \\ -5 \end{bmatrix},$$

compute $\left[\mathbf{v}\right]_S$.

Solution

To find $\left[\mathbf{v}\right]_S$, we need to find the constants a_1, a_2, and a_3 such that

$$a_1\mathbf{v}_1 + a_2\mathbf{v}_2 + a_3\mathbf{v}_3 = \mathbf{v},$$

which leads to the linear system whose augmented matrix is (verify)

$$\left[\begin{array}{ccc|c} 1 & 2 & 0 & 1 \\ 1 & 0 & 1 & 1 \\ 0 & 1 & 2 & -5 \end{array}\right], \tag{1}$$

or equivalently,

$$\left[\begin{array}{ccc|c} \mathbf{v}_1 & \mathbf{v}_2 & \mathbf{v}_3 & \mathbf{v} \end{array}\right].$$

Transforming the matrix in (1) to reduced row echelon form, we obtain the solution (verify)

$$a_1 = 3, \quad a_2 = -1, \quad a_3 = -2,$$

so

$$\left[\mathbf{v}\right]_S = \begin{bmatrix} 3 \\ -1 \\ -2 \end{bmatrix}.$$

■

In Example 5 of Section 1.7 we showed that the matrix transformation $f: R^2 \to R^2$ defined by $f(\mathbf{v}) = A\mathbf{v}$, where

$$A = \begin{bmatrix} h & 0 \\ 0 & k \end{bmatrix},$$

with h and k nonzero, maps the unit circle to an ellipse centered at the origin. (See Figure 1.20 in Section 1.7.) Using techniques from Chapter 8, we can show that for any 2×2 matrix A with real entries, the matrix transformation $f: R^2 \to R^2$ defined by $f(\mathbf{v}) = A\mathbf{v}$ maps the unit circle into an ellipse centered at the origin. With a general 2×2 matrix the ellipse may be rotated so that its major and minor axes are not parallel to the coordinate axes. (See Figure 4.29.)

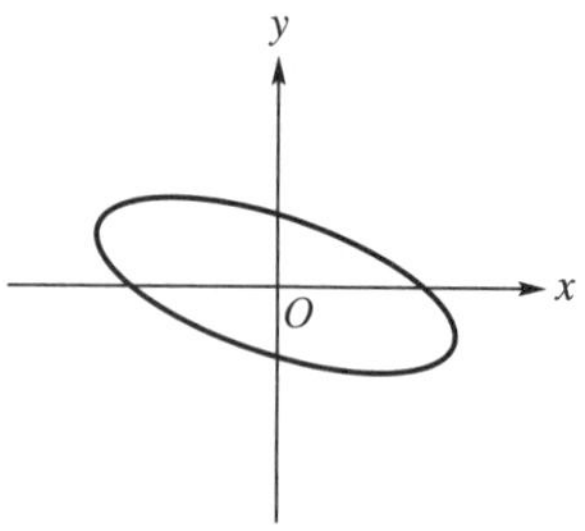

FIGURE 4.29

Any point on the unit circle has coordinates $x = \cos(\theta)$, $y = \sin(\theta)$, and the vector $\mathbf{v}$ from the origin to (x, y) is a linear combination of the natural basis for R^2; that is,

$$\mathbf{v} = \cos(\theta)\begin{bmatrix} 1 \\ 0 \end{bmatrix} + \sin(\theta)\begin{bmatrix} 0 \\ 1 \end{bmatrix}.$$

The coordinates of the image of the point (x, y) by the matrix transformation f are computed from

$$f(\mathbf{v}) = A\mathbf{v} = A\begin{bmatrix} \cos(\theta) \\ \sin(\theta) \end{bmatrix}.$$

If we take n successive images of the unit circle, using the matrix transformation f, then the coordinates of the point (x, y) are computed from the matrix product

$$A^n \begin{bmatrix} \cos(\theta) \\ \sin(\theta) \end{bmatrix}$$

and can be quite complicated expressions. However, a change of basis that uses particular properties of the matrix A can greatly simplify the computation of such coordinates. We illustrate this in Example 3. In Chapter 7 we explain how to determine such bases.

EXAMPLE 3 Let the matrix transformation $f: R^2 \to R^2$ be defined by $f(\mathbf{v}) = A\mathbf{v}$, where

$$A = \begin{bmatrix} 0.97 & 0.12 \\ 0.03 & 0.88 \end{bmatrix}.$$

The image of the unit circle by this matrix transformation is shown in Figure 4.30.

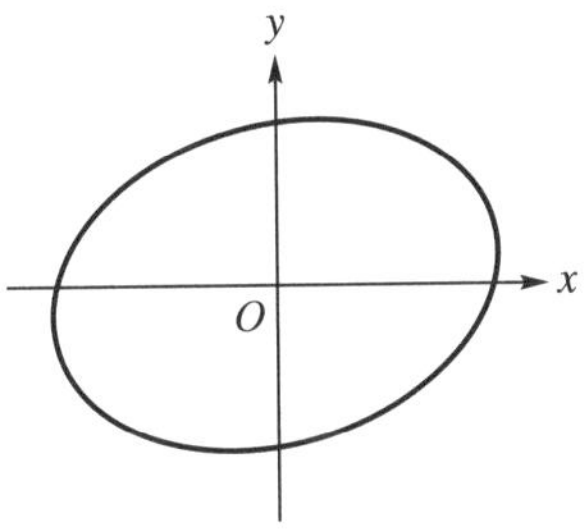

FIGURE 4.30

Let

$$\mathbf{v}_1 = \begin{bmatrix} 4 \\ 1 \end{bmatrix} \quad \text{and} \quad \mathbf{v}_2 = \begin{bmatrix} 1 \\ -1 \end{bmatrix}.$$

Observe that

$$A\mathbf{v}_1 = \mathbf{v}_1 \quad \text{and} \quad A\mathbf{v}_2 = 0.85\mathbf{v}_2;$$

furthermore, that $\{\mathbf{v}_1, \mathbf{v}_2\}$ is a linearly independent set (verify) and hence a basis for R^2. (Explain.) It follows that a vector $\mathbf{v}$ from the origin to a point $(x, y) = (\cos(\theta), \sin(\theta))$ on the unit circle is a linear combination of $\mathbf{v}_1$ and $\mathbf{v}_2$, and we have (verify)

$$\mathbf{v} = \begin{bmatrix} \cos(\theta) \\ \sin(\theta) \end{bmatrix} = \frac{\cos(\theta) + \sin(\theta)}{5}\mathbf{v}_1 + \frac{\cos(\theta) - 4\sin(\theta)}{5}\mathbf{v}_2.$$

Then $f(\mathbf{v})$ is given by

$$\begin{aligned} A\mathbf{v} &= \frac{\cos(\theta) + \sin(\theta)}{5}A\mathbf{v}_1 + \frac{\cos(\theta) - 4\sin(\theta)}{5}A\mathbf{v}_2 \\ &= \frac{\cos(\theta) + \sin(\theta)}{5}\mathbf{v}_1 + \frac{\cos(\theta) - 4\sin(\theta)}{5}0.85\mathbf{v}_2, \end{aligned}$$

so the coordinates of $f(\mathbf{v})$ with respect to the $\{\mathbf{v}_1, \mathbf{v}_2\}$ basis are

$$\left(\frac{\cos(\theta) + \sin(\theta)}{5}, \frac{\cos(\theta) - 4\sin(\theta)}{5}0.85\right).$$

The coordinates of the image (x, y) on the unit circle with respect to the $\{\mathbf{v}_1, \mathbf{v}_2\}$ basis after n applications of the matrix transformation f are given by

$$\left(\frac{\cos(\theta) + \sin(\theta)}{5}, \frac{\cos(\theta) - 4\sin(\theta)}{5}(0.85)^n\right),$$

which is quite easy to compute, since we need not perform successive matrix products. ■

Let $S = \{\mathbf{v}_1, \mathbf{v}_2, \ldots, \mathbf{v}_n\}$ be an ordered basis for a vector space V and let $\mathbf{v}$ and $\mathbf{w}$ be vectors in V such that

$$\left[\mathbf{v}\right]_S = \left[\mathbf{w}\right]_S.$$

We shall now show that $\mathbf{v} = \mathbf{w}$. Let

$$\left[\mathbf{v}\right]_S = \begin{bmatrix} a_1 \\ a_2 \\ \vdots \\ a_n \end{bmatrix} \quad \text{and} \quad \left[\mathbf{w}\right]_S = \begin{bmatrix} b_1 \\ b_2 \\ \vdots \\ b_n \end{bmatrix}.$$

Since $\left[\mathbf{v}\right]_S = \left[\mathbf{w}\right]_S$, we have

$$a_i = b_i, \qquad i = 1, 2, \ldots, n.$$

Then

$$\mathbf{v} = a_1\mathbf{v}_1 + a_2\mathbf{v}_2 + \cdots + a_n\mathbf{v}_n$$

and

$$\begin{aligned} \mathbf{w} &= b_1\mathbf{v}_1 + b_2\mathbf{v}_2 + \cdots + b_n\mathbf{v}_n \\ &= a_1\mathbf{v}_1 + a_2\mathbf{v}_2 + \cdots + a_n\mathbf{v}_n \\ &= \mathbf{v}. \end{aligned}$$

Of course, if $\mathbf{v}$ and $\mathbf{w}$ are vectors in the vector space V and $\mathbf{v} = \mathbf{w}$, then $\left[\mathbf{v}\right]_S = \left[\mathbf{w}\right]_S$, where S is an ordered basis for V.

Thus we have that there is a one-to-one correspondence between a vector $\mathbf{v}$ of a vector space V and the coordinate vector $\left[\mathbf{v}\right]_S$ of $\mathbf{v}$ with respect to a fixed ordered basis S for V.

The choice of an ordered basis and the consequent assignment of a coordinate vector for every $\mathbf{v}$ in V enables us to "picture" the vector space. We illustrate this notion by using Example 1. Choose a fixed point O in the plane R^2, and draw any two arrows $\mathbf{w}_1$ and $\mathbf{w}_2$ from O that depict the basis vectors t and 1 in the ordered basis $S = \{t, 1\}$ for P_1 (see Figure 4.31). The directions of $\mathbf{w}_1$ and $\mathbf{w}_2$ determine two lines, which we call the $\boldsymbol{x_1}$**- and** $\boldsymbol{x_2}$**-axes**, respectively. The positive direction on the x_1-axis is in the direction of $\mathbf{w}_1$; the negative direction on the x_1-axis is

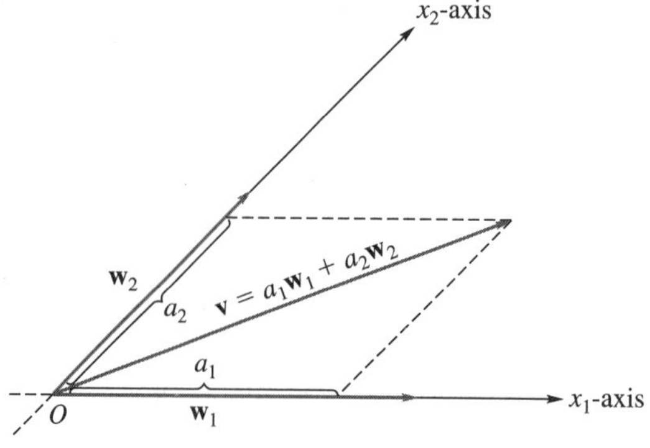

FIGURE 4.31

along $-\mathbf{w}_1$. Similarly, the positive direction on the x_2-axis is in the direction of $\mathbf{w}_2$; the negative direction on the x_2-axis is along $-\mathbf{w}_2$. The lengths of $\mathbf{w}_1$ and $\mathbf{w}_2$ determine the scales on the x_1- and x_2-axes, respectively. If $\mathbf{v}$ is a vector in P_1, we can write $\mathbf{v}$, uniquely, as $\mathbf{v} = a_1\mathbf{w}_1 + a_2\mathbf{w}_2$. We now mark off a segment of length $|a_1|$ on the x_1-axis (in the positive direction if a_1 is positive and in the negative direction if a_1 is negative) and draw a line through the endpoint of this segment parallel to $\mathbf{w}_2$. Similarly, mark off a segment of length $|a_2|$ on the x_2-axis (in the positive direction if a_2 is positive and in the negative direction if a_2 is negative) and draw a line through the endpoint of this segment parallel to $\mathbf{w}_1$. We draw a directed line segment from O to the point of intersection of these two lines. This directed line segment represents $\mathbf{v}$.

Isomorphisms

If $\mathbf{v}$ and $\mathbf{w}$ are vectors in an n-dimensional vector space V with an ordered basis $S = \{\mathbf{v}_1, \mathbf{v}_2, \ldots, \mathbf{v}_n\}$, then we can write $\mathbf{v}$ and $\mathbf{w}$, uniquely, as

$$\mathbf{v} = a_1\mathbf{v}_1 + a_2\mathbf{v}_2 + \cdots + a_n\mathbf{v}_n, \quad \mathbf{w} = b_1\mathbf{v}_1 + b_2\mathbf{v}_2 + \cdots + b_n\mathbf{v}_n.$$

Thus with $\mathbf{v}$ and $\mathbf{w}$ we associate $\left[\mathbf{v}\right]_S$ and $\left[\mathbf{w}\right]_S$, respectively, elements in R^n:

$$\begin{aligned} \mathbf{v} &\to \left[\mathbf{v}\right]_S \\ \mathbf{w} &\to \left[\mathbf{w}\right]_S. \end{aligned}$$

The sum $\mathbf{v} + \mathbf{w} = (a_1 + b_1)\mathbf{v}_1 + (a_2 + b_2)\mathbf{v}_2 + \cdots + (a_n + b_n)\mathbf{v}_n$, which means that with $\mathbf{v} + \mathbf{w}$ we associate the vector

$$\left[\mathbf{v} + \mathbf{w}\right]_S = \begin{bmatrix} a_1 + b_1 \\ a_2 + b_2 \\ \vdots \\ a_n + b_n \end{bmatrix} = \left[\mathbf{v}\right]_S + \left[\mathbf{w}\right]_S.$$

Therefore,

$$\mathbf{v} + \mathbf{w} \to \left[\mathbf{v} + \mathbf{w}\right]_S = \left[\mathbf{v}\right]_S + \left[\mathbf{w}\right]_S.$$

That is, when we add $\mathbf{v}$ and $\mathbf{w}$ in V, we add their associated coordinate vectors $\left[\mathbf{v}\right]_S$ and $\left[\mathbf{w}\right]_S$ to obtain the coordinate vector $\left[\mathbf{v} + \mathbf{w}\right]_S$ in R^n associated with $\mathbf{v} + \mathbf{w}$.

Similarly, if c is a real number, then

$$c\mathbf{v} = (ca_1)\mathbf{v}_1 + (ca_2)\mathbf{v}_2 + \cdots + (ca_n)\mathbf{v}_n,$$

which implies that

$$\left[c\mathbf{v}\right]_S = \begin{bmatrix} ca_1 \\ ca_2 \\ \vdots \\ ca_n \end{bmatrix} = c\left[\mathbf{v}\right]_S.$$

Therefore,

$$c\mathbf{v} \to \left[c\mathbf{v}\right]_S = c\left[\mathbf{v}\right]_S,$$

and thus when $\mathbf{v}$ is multiplied by a scalar c, we multiply $\left[\mathbf{v}\right]_S$ by c to obtain the coordinate vector in R^n associated with $c\mathbf{v}$.

This discussion suggests that, from an algebraic point of view, V and R^n behave rather similarly. We now clarify this notion.

Let L be a function mapping a vector space V into a vector space W. Recall that L is **one-to-one** if $L(\mathbf{v}_1) = L(\mathbf{v}_2)$, for $\mathbf{v}_1, \mathbf{v}_2$ in V, implies that $\mathbf{v}_1 = \mathbf{v}_2$. Also, L is **onto** if for each $\mathbf{w}$ in W there is at least one $\mathbf{v}$ in V for which $L(\mathbf{v}) = \mathbf{w}$.* Thus the mapping $L\colon R^3 \to R^2$ defined by

$$L\left(\begin{bmatrix} a_1 \\ a_2 \\ a_3 \end{bmatrix}\right) = \begin{bmatrix} a_1 + a_2 \\ a_1 \end{bmatrix}$$

is onto. To see this, suppose that $\mathbf{w} = \begin{bmatrix} b_1 \\ b_2 \end{bmatrix}$; we seek $\mathbf{v} = \begin{bmatrix} a_1 \\ a_2 \\ a_3 \end{bmatrix}$ such that

$$L(\mathbf{v}) = \begin{bmatrix} a_1 + a_2 \\ a_1 \end{bmatrix} = \mathbf{w} = \begin{bmatrix} b_1 \\ b_2 \end{bmatrix}.$$

Thus we obtain the solution: $a_1 = b_2$, $a_2 = b_1 - b_2$, and a_3 is arbitrary. However, L is not one-to-one, for if $\mathbf{v}_1 = \begin{bmatrix} 1 \\ 2 \\ 3 \end{bmatrix}$ and $\mathbf{v}_2 = \begin{bmatrix} 1 \\ 2 \\ 4 \end{bmatrix}$, then

$$L(\mathbf{v}_1) = L(\mathbf{v}_2) = \begin{bmatrix} 3 \\ 1 \end{bmatrix}, \quad \text{but} \quad \mathbf{v}_1 \neq \mathbf{v}_2.$$

DEFINITION 4.13

Let V be a real vector space with operations $\oplus$ and $\odot$, and let W be a real vector space with operations $\boxplus$ and $\boxdot$. A one-to-one function L mapping V onto W is called an **isomorphism** (from the Greek *isos*, meaning "the same," and *morphos*, meaning "structure") of V onto W if

(a) $L(\mathbf{v} \oplus \mathbf{w}) = L(\mathbf{v}) \boxplus L(\mathbf{w})$ for $\mathbf{v}$, $\mathbf{w}$ in V;

(b) $L(c \odot \mathbf{v}) = c \boxdot L(\mathbf{v})$ for $\mathbf{v}$ in V, c a real number.

In this case we say that **V is isomorphic to W.**

It also follows from Definition 4.13 that if L is an isomorphism of V onto W, then

$$L(a_1 \odot \mathbf{v}_1 \oplus a_2 \odot \mathbf{v}_2 \oplus \cdots \oplus a_k \odot \mathbf{v}_k) = a_1 \boxdot L(\mathbf{v}_1) \boxplus a_2 \boxdot L(\mathbf{v}_2) \boxplus \cdots \boxplus a_k \boxdot L(\mathbf{v}_k),$$

where $\mathbf{v}_1, \mathbf{v}_2, \ldots, \mathbf{v}_k$ are vectors in V and $a_1, a_2, \ldots, a_k$ are scalars [see Exercise 27(c)].

Remark A function L mapping a vector space V into a vector space W satisfying properties (a) and (b) of Definition 4.13 is called a linear transformation. These functions will be studied in depth in Chapter 6. Thus an isomorphism of a vector

*See Appendix A for further discussion of one-to-one and onto functions.

space V onto a vector space W is a linear transformation that is one-to-one and onto.

As a result of Theorem 4.15, to follow, we can replace the expressions "V is isomorphic to W" and "W is isomorphic to V" by "V and W are isomorphic."

Isomorphic vector spaces differ only in the nature of their elements; their algebraic properties are identical. That is, if the vector spaces V and W are isomorphic, under the isomorphism L, then for each $\mathbf{v}$ in V there is a unique $\mathbf{w}$ in W so that $L(\mathbf{v}) = \mathbf{w}$ and, conversely, for each $\mathbf{w}$ in W there is a unique $\mathbf{v}$ in V so that $\mathbf{w} = L(\mathbf{v})$. If we now replace each element of V by its image under L and replace the operations $\oplus$ and $\odot$ by $\boxplus$ and $\boxdot$, respectively, we get precisely W. The most important example of isomorphic vector spaces is given in the following theorem:

Theorem 4.14 If V is an n-dimensional real vector space, then V is isomorphic to R^n.

Proof

Let $S = \{\mathbf{v}_1, \mathbf{v}_2, \ldots, \mathbf{v}_n\}$ be an ordered basis for V, and let $L: V \to R^n$ be defined by

$$L(\mathbf{v}) = \left[\mathbf{v}\right]_S = \begin{bmatrix} a_1 \\ a_2 \\ \vdots \\ a_n \end{bmatrix},$$

where $\mathbf{v} = a_1\mathbf{v}_1 + a_2\mathbf{v}_2 + \cdots + a_n\mathbf{v}_n$.

We show that L is an isomorphism. First, L is one-to-one. Let

$$\left[\mathbf{v}\right]_S = \begin{bmatrix} a_1 \\ a_2 \\ \vdots \\ a_n \end{bmatrix} \quad \text{and} \quad \left[\mathbf{w}\right]_S = \begin{bmatrix} b_1 \\ b_2 \\ \vdots \\ b_n \end{bmatrix}$$

and suppose that $L(\mathbf{v}) = L(\mathbf{w})$. Then $\left[\mathbf{v}\right]_S = \left[\mathbf{w}\right]_S$, and from our earlier remarks it follows that $\mathbf{v} = \mathbf{w}$.

Next, L is onto, for if $\mathbf{w} = \begin{bmatrix} b_1 \\ b_2 \\ \vdots \\ b_n \end{bmatrix}$ is a given vector in R^n and we let

$$\mathbf{v} = b_1\mathbf{v}_1 + b_2\mathbf{v}_2 + \cdots + b_n\mathbf{v}_n,$$

then $L(\mathbf{v}) = \mathbf{w}$.

Finally, L satisfies Definition 4.13(a) and (b). Let $\mathbf{v}$ and $\mathbf{w}$ be vectors in V such that $\left[\mathbf{v}\right]_S = \begin{bmatrix} a_1 \\ a_2 \\ \vdots \\ a_n \end{bmatrix}$ and $\left[\mathbf{w}\right]_S = \begin{bmatrix} b_1 \\ b_2 \\ \vdots \\ b_n \end{bmatrix}$. Then

$$L(\mathbf{v} + \mathbf{w}) = \left[\mathbf{v} + \mathbf{w}\right]_S = \left[\mathbf{v}\right]_S + \left[\mathbf{w}\right]_S = L(\mathbf{v}) + L(\mathbf{w})$$

and

$$L(c\mathbf{v}) = \left[c\mathbf{v}\right]_S = c\left[\mathbf{v}\right]_S = cL(\mathbf{v}),$$

as we saw before. Hence V and R^n are isomorphic. ■

Another example of isomorphism is given by the vector spaces discussed in the review section at the beginning of this chapter: R^2, the vector space of directed line segments emanating from a point in the plane and the vector space of all ordered pairs of real numbers. There is a corresponding isomorphism for R^3.

Some important properties of isomorphisms are given in Theorem 4.15.

Theorem 4.15 **(a)** Every vector space V is isomorphic to itself.

(b) If V is isomorphic to W, then W is isomorphic to V.

(c) If U is isomorphic to V and V is isomorphic to W, then U is isomorphic to W.

Proof

Exercise 28. [Parts (a) and (c) are not difficult to show; (b) is slightly harder and will essentially be proved in Theorem 6.7.] ■

The following theorem shows that all vector spaces of the same dimension are, algebraically speaking, alike, and conversely, that isomorphic vector spaces have the same dimensions:

Theorem 4.16 Two finite-dimensional vector spaces are isomorphic if and only if their dimensions are equal.

Proof

Let V and W be n-dimensional vector spaces. Then V and R^n are isomorphic and W and R^n are isomorphic. From Theorem 4.15 it follows that V and W are isomorphic.

Conversely, let V and W be isomorphic finite-dimensional vector spaces; let $L\colon V \to W$ be an isomorphism. Assume that $\dim V = n$, and let $S = \{\mathbf{v}_1, \mathbf{v}_2, \ldots, \mathbf{v}_n\}$ be a basis for V.

We now prove that the set $T = \{L(\mathbf{v}_1), L(\mathbf{v}_2), \ldots, L(\mathbf{v}_n)\}$ is a basis for W. First, T spans W. If $\mathbf{w}$ is any vector in W, then $\mathbf{w} = L(\mathbf{v})$ for some $\mathbf{v}$ in V. Since S is a basis for V, $\mathbf{v} = a_1\mathbf{v}_1 + a_2\mathbf{v}_2 + \cdots + a_n\mathbf{v}_n$, where the a_i are uniquely determined real numbers, so

$$\begin{aligned} L(\mathbf{v}) &= L(a_1\mathbf{v}_1 + a_2\mathbf{v}_2 + \cdots + a_n\mathbf{v}_n) \\ &= L(a_1\mathbf{v}_1) + L(a_2\mathbf{v}_2) + \cdots + L(a_n\mathbf{v}_n) \\ &= a_1L(\mathbf{v}_1) + a_2L(\mathbf{v}_2) + \cdots + a_nL(\mathbf{v}_n). \end{aligned}$$

Thus T spans W.

Now suppose that

$$a_1L(\mathbf{v}_1) + a_2L(\mathbf{v}_2) + \cdots + a_nL(\mathbf{v}_n) = \mathbf{0}_W.$$

Then $L(a_1\mathbf{v}_1 + a_2\mathbf{v}_2 + \cdots + a_n\mathbf{v}_n) = \mathbf{0}_W$. From Exercise 29(a), $L(\mathbf{0}_V) = \mathbf{0}_W$. Since L is one-to-one, we get $a_1\mathbf{v}_1 + a_2\mathbf{v}_2 + \cdots + a_n\mathbf{v}_n = \mathbf{0}_V$. Since S is linearly independent, we conclude that $a_1 = a_2 = \cdots = a_n = 0$, which means that T is linearly independent. Hence T is a basis for W, and $\dim W = n$. ■

As a consequence of Theorem 4.16, the spaces R^n and R^m are isomorphic if and only if $n = m$. (See Exercise 30.) Moreover, the vector spaces P_n and R^{n+1} are isomorphic. (See Exercise 32.)

We can now establish the converse of Theorem 4.14, as follows:

Corollary 4.6 If V is a finite-dimensional vector space that is isomorphic to R^n, then $\dim V = n$.

Proof

This result follows from Theorem 4.16. ■

If $L\colon V \to W$ is an isomorphism, then since L is a one-to-one onto mapping, it has an inverse L^{-1}. (This will be shown in Theorem 6.7.) It can be shown that $L^{-1}\colon W \to V$ is also an isomorphism. (This will also be essentially shown in Theorem 6.7.) Moreover, if $S = \{\mathbf{v}_1, \mathbf{v}_2, \ldots, \mathbf{v}_n\}$ is a basis for V, then $T = L(S) = \{L(\mathbf{v}_1), L(\mathbf{v}_2), \ldots, L(\mathbf{v}_n)\}$ is a basis for W, as we have seen in the proof of Theorem 4.16.

As an example of isomorphism, we note that the vector spaces P_3, R_4, and R^4 are all isomorphic, since each has dimension four.

We have shown in this section that the idea of a finite-dimensional vector space, which at first seemed fairly abstract, is not so mysterious. In fact, such a vector space does not differ much from R^n in its algebraic behavior.

■ Transition Matrices

We now look at the relationship between two coordinate vectors for the same vector $\mathbf{v}$ with respect to different bases. Thus, let $S = \{\mathbf{v}_1, \mathbf{v}_2, \ldots, \mathbf{v}_n\}$ and $T = \{\mathbf{w}_1, \mathbf{w}_2, \ldots, \mathbf{w}_n\}$ be two ordered bases for the n-dimensional vector space V. If $\mathbf{v}$ is any vector in V, then

$$\mathbf{v} = c_1\mathbf{w}_1 + c_2\mathbf{w}_2 + \cdots + c_n\mathbf{w}_n \quad \text{and} \quad \left[\mathbf{v}\right]_T = \begin{bmatrix} c_1 \\ c_2 \\ \vdots \\ c_n \end{bmatrix}.$$

Then

$$\begin{aligned} \left[\mathbf{v}\right]_S &= \left[c_1\mathbf{w}_1 + c_2\mathbf{w}_2 + \cdots + c_n\mathbf{w}_n\right]_S \\ &= \left[c_1\mathbf{w}_1\right]_S + \left[c_2\mathbf{w}_2\right]_S + \cdots + \left[c_n\mathbf{w}_n\right]_S \\ &= c_1\left[\mathbf{w}_1\right]_S + c_2\left[\mathbf{w}_2\right]_S + \cdots + c_n\left[\mathbf{w}_n\right]_S. \end{aligned} \tag{2}$$

Let the coordinate vector of $\mathbf{w}_j$ with respect to S be denoted by

$$\left[\mathbf{w}_j\right]_S = \begin{bmatrix} a_{1j} \\ a_{2j} \\ \vdots \\ a_{nj} \end{bmatrix}.$$

The $n \times n$ matrix whose jth column is $\left[\mathbf{w}_j\right]_S$ is called the **transition matrix from the *T*-basis to the *S*-basis** and is denoted by $P_{S\leftarrow T}$. Then Equation (2) can be written in matrix form as

$$[\mathbf{v}]_S = P_{S\leftarrow T}[\mathbf{v}]_T. \tag{3}$$

Thus, to find the transition matrix $P_{S\leftarrow T}$ from the T-basis to the S-basis, we first compute the coordinate vector of each member of the T-basis with respect to the S-basis. Forming a matrix with these vectors as columns arranged in their natural order, we obtain the transition matrix. Equation (3) says that the coordinate vector of $\mathbf{v}$ with respect to the basis S is the transition matrix $P_{S\leftarrow T}$ times the coordinate vector of $\mathbf{v}$ with respect to the basis T. Figure 4.32 illustrates Equation (3).

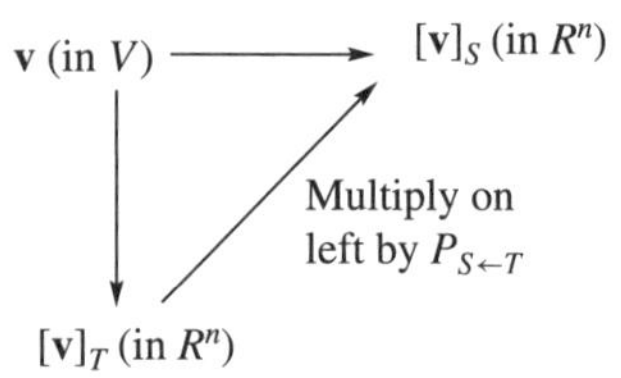

FIGURE 4.32

EXAMPLE 4

Let V be R^3 and let $S = \{\mathbf{v}_1, \mathbf{v}_2, \mathbf{v}_3\}$ and $T = \{\mathbf{w}_1, \mathbf{w}_2, \mathbf{w}_3\}$ be ordered bases for R^3, where

$$\mathbf{v}_1 = \begin{bmatrix} 2 \\ 0 \\ 1 \end{bmatrix}, \quad \mathbf{v}_2 = \begin{bmatrix} 1 \\ 2 \\ 0 \end{bmatrix}, \quad \mathbf{v}_3 = \begin{bmatrix} 1 \\ 1 \\ 1 \end{bmatrix}$$

and

$$\mathbf{w}_1 = \begin{bmatrix} 6 \\ 3 \\ 3 \end{bmatrix}, \quad \mathbf{w}_2 = \begin{bmatrix} 4 \\ -1 \\ 3 \end{bmatrix}, \quad \mathbf{w}_3 = \begin{bmatrix} 5 \\ 5 \\ 2 \end{bmatrix}.$$

(a) Compute the transition matrix $P_{S\leftarrow T}$ from the T-basis to the S-basis.

(b) Verify Equation (3) for $\mathbf{v} = \begin{bmatrix} 4 \\ -9 \\ 5 \end{bmatrix}$.

Solution

(a) To compute $P_{S\leftarrow T}$, we need to find a_1, a_2, a_3 such that

$$a_1\mathbf{v}_1 + a_2\mathbf{v}_2 + a_3\mathbf{v}_3 = \mathbf{w}_1,$$

which leads to a linear system of three equations in three unknowns, whose augmented matrix is

$$\left[\begin{array}{ccc|c} \mathbf{v}_1 & \mathbf{v}_2 & \mathbf{v}_3 & \mathbf{w}_1 \end{array}\right].$$

That is, the augmented matrix is

$$\left[\begin{array}{ccc|c} 2 & 1 & 1 & 6 \\ 0 & 2 & 1 & 3 \\ 1 & 0 & 1 & 3 \end{array}\right].$$

Similarly, we need to find b_1, b_2, b_3 and c_1, c_2, c_3 such that

$$b_1\mathbf{v}_1 + b_2\mathbf{v}_2 + b_3\mathbf{v}_3 = \mathbf{w}_2$$
$$c_1\mathbf{v}_1 + c_2\mathbf{v}_2 + c_3\mathbf{v}_3 = \mathbf{w}_3.$$

These vector equations lead to two linear systems, each of three equations in three unknowns, whose augmented matrices are

$$\left[\begin{array}{ccc|c} \mathbf{v}_1 & \mathbf{v}_2 & \mathbf{v}_3 & \mathbf{w}_2 \end{array}\right] \quad \text{and} \quad \left[\begin{array}{ccc|c} \mathbf{v}_1 & \mathbf{v}_2 & \mathbf{v}_3 & \mathbf{w}_3 \end{array}\right],$$

or specifically,

$$\left[\begin{array}{ccc|c} 2 & 1 & 1 & 4 \\ 0 & 2 & 1 & -1 \\ 1 & 0 & 1 & 3 \end{array}\right] \quad \text{and} \quad \left[\begin{array}{ccc|c} 2 & 1 & 1 & 5 \\ 0 & 2 & 1 & 5 \\ 1 & 0 & 1 & 2 \end{array}\right].$$

Since the coefficient matrix of all three linear systems is $\begin{bmatrix} \mathbf{v}_1 & \mathbf{v}_2 & \mathbf{v}_3 \end{bmatrix}$, we can transform the three augmented matrices to reduced row echelon form simultaneously by transforming the partitioned matrix

$$\left[\begin{array}{ccc|c|c|c} \mathbf{v}_1 & \mathbf{v}_2 & \mathbf{v}_3 & \mathbf{w}_1 & \mathbf{w}_2 & \mathbf{w}_3 \end{array}\right]$$

to reduced row echelon form. Thus, we transform

$$\left[\begin{array}{ccc|c|c|c} 2 & 1 & 1 & 6 & 4 & 5 \\ 0 & 2 & 1 & 3 & -1 & 5 \\ 1 & 0 & 1 & 3 & 3 & 2 \end{array}\right]$$

to reduced row echelon form, obtaining (verify)

$$\left[\begin{array}{ccc|c|c|c} 1 & 0 & 0 & 2 & 2 & 1 \\ 0 & 1 & 0 & 1 & -1 & 2 \\ 0 & 0 & 1 & 1 & 1 & 1 \end{array}\right],$$

which implies that the transition matrix from the T-basis to the S-basis is

$$P_{S\leftarrow T} = \begin{bmatrix} 2 & 2 & 1 \\ 1 & -1 & 2 \\ 1 & 1 & 1 \end{bmatrix}.$$

(b) If $\mathbf{v} = \begin{bmatrix} 4 \\ -9 \\ 5 \end{bmatrix}$, then expressing $\mathbf{v}$ in terms of the T-basis, we have (verify)

$$\mathbf{v} = \begin{bmatrix} 4 \\ -9 \\ 5 \end{bmatrix} = 1\begin{bmatrix} 6 \\ 3 \\ 3 \end{bmatrix} + 2\begin{bmatrix} 4 \\ -1 \\ 3 \end{bmatrix} - 2\begin{bmatrix} 5 \\ 5 \\ 2 \end{bmatrix}.$$

So $\begin{bmatrix} \mathbf{v} \end{bmatrix}_T = \begin{bmatrix} 1 \\ 2 \\ -2 \end{bmatrix}$. Then

$$\begin{bmatrix} \mathbf{v} \end{bmatrix}_S = P_{S\leftarrow T}\begin{bmatrix} \mathbf{v} \end{bmatrix}_T = \begin{bmatrix} 2 & 2 & 1 \\ 1 & -1 & 2 \\ 1 & 1 & 1 \end{bmatrix}\begin{bmatrix} 1 \\ 2 \\ -2 \end{bmatrix} = \begin{bmatrix} 4 \\ -5 \\ 1 \end{bmatrix}.$$

If we compute $\begin{bmatrix} \mathbf{v} \end{bmatrix}_S$ directly, we find that

$$\mathbf{v} = \begin{bmatrix} 4 \\ -9 \\ 5 \end{bmatrix} = 4\begin{bmatrix} 2 \\ 0 \\ 1 \end{bmatrix} - 5\begin{bmatrix} 1 \\ 2 \\ 0 \end{bmatrix} + 1\begin{bmatrix} 1 \\ 1 \\ 1 \end{bmatrix}, \quad \text{so } \begin{bmatrix} \mathbf{v} \end{bmatrix}_S = \begin{bmatrix} 4 \\ -5 \\ 1 \end{bmatrix}.$$

Hence,

$$\left[\mathbf{v}\right]_S = P_{S\leftarrow T}\left[\mathbf{v}\right]_T. \quad \blacksquare$$

We next want to show that the transition matrix $P_{S\leftarrow T}$ from the T-basis to the S-basis is nonsingular. Suppose that $P_{S\leftarrow T}\left[\mathbf{v}\right]_T = \mathbf{0}_{R^n}$ for some $\mathbf{v}$ in V. From Equation (3) we have

$$P_{S\leftarrow T}\left[\mathbf{v}\right]_T = \left[\mathbf{v}\right]_S = \mathbf{0}_{R^n}.$$

If $\mathbf{v} = b_1\mathbf{v}_1 + b_2\mathbf{v}_2 + \cdots + b_n\mathbf{v}_n$, then

$$\begin{bmatrix} b_1 \\ b_2 \\ \vdots \\ b_n \end{bmatrix} = \left[\mathbf{v}\right]_S = \mathbf{0}_{R^n} = \begin{bmatrix} 0 \\ 0 \\ \vdots \\ 0 \end{bmatrix},$$

so

$$\mathbf{v} = 0\mathbf{v}_1 + 0\mathbf{v}_2 + \cdots + 0\mathbf{v}_n = \mathbf{0}_V.$$

Hence, $\left[\mathbf{v}\right]_T = \mathbf{0}_{R^n}$. Thus the homogeneous system $P_{S\leftarrow T}\mathbf{x} = \mathbf{0}$ has only the trivial solution; it then follows from Theorem 2.9 that $P_{S\leftarrow T}$ is nonsingular. Of course, we then also have

$$\left[\mathbf{v}\right]_T = P_{S\leftarrow T}^{-1}\left[\mathbf{v}\right]_S.$$

That is, $P_{S\leftarrow T}^{-1}$ is then the transition matrix from the S-basis to the T-basis; the jth column of $P_{S\leftarrow T}^{-1}$ is $\left[\mathbf{v}_j\right]_T$.

Remark In Exercises 39 through 41 we ask you to show that if S and T are ordered bases for the vector space R^n, then

$$P_{S\leftarrow T} = M_S^{-1}M_T,$$

where M_S is the $n \times n$ matrix whose jth column is $\mathbf{v}_j$ and M_T is the $n \times n$ matrix whose jth column is $\mathbf{w}_j$. This formula implies that $P_{S\leftarrow T}$ is nonsingular, and it is helpful in solving some of the exercises in this section.

EXAMPLE 5

Let S and T be the ordered bases for R^3 defined in Example 4. Compute the transition matrix $Q_{T\leftarrow S}$ from the S-basis to the T-basis directly and show that $Q_{T\leftarrow S} = P_{S\leftarrow T}^{-1}$.

Solution

$Q_{T\leftarrow S}$ is the matrix whose columns are the solution vectors to the linear systems obtained from the vector equations

$$\begin{aligned} a_1\mathbf{w}_1 + a_2\mathbf{w}_2 + a_3\mathbf{w}_3 &= \mathbf{v}_1 \\ b_1\mathbf{w}_1 + b_2\mathbf{w}_2 + b_3\mathbf{w}_3 &= \mathbf{v}_2 \\ c_1\mathbf{w}_1 + c_2\mathbf{w}_2 + c_3\mathbf{w}_3 &= \mathbf{v}_3. \end{aligned}$$

As in Example 4, we can solve these linear systems simultaneously by transforming the partitioned matrix

$$\left[\mathbf{w}_1 \quad \mathbf{w}_2 \quad \mathbf{w}_3 \;\vdots\; \mathbf{v}_1 \;\vdots\; \mathbf{v}_2 \;\vdots\; \mathbf{v}_3\right]$$

to reduced row echelon form. That is, we transform

$$\left[\begin{array}{ccc|c|c|c} 6 & 4 & 5 & 2 & 1 & 1 \\ 3 & -1 & 5 & 0 & 2 & 1 \\ 3 & 3 & 2 & 1 & 0 & 1 \end{array}\right]$$

to reduced row echelon form, obtaining (verify)

$$\left[\begin{array}{ccc|c|c|c} 1 & 0 & 0 & \frac{3}{2} & \frac{1}{2} & -\frac{5}{2} \\ 0 & 1 & 0 & -\frac{1}{2} & -\frac{1}{2} & \frac{3}{2} \\ 0 & 0 & 1 & -1 & 0 & 2 \end{array}\right],$$

so

$$Q_{T \leftarrow S} = \begin{bmatrix} \frac{3}{2} & \frac{1}{2} & -\frac{5}{2} \\ -\frac{1}{2} & -\frac{1}{2} & \frac{3}{2} \\ -1 & 0 & 2 \end{bmatrix}.$$

Multiplying $Q_{T \leftarrow S}$ by $P_{S \leftarrow T}$, we find (verify) that $Q_{T \leftarrow S} P_{S \leftarrow T} = I_3$, so we conclude that $Q_{T \leftarrow S} = P^{-1}_{S \leftarrow T}$. ■

EXAMPLE 6

Let V be P_1, and let $S = \{\mathbf{v}_1, \mathbf{v}_2\}$ and $T = \{\mathbf{w}_1, \mathbf{w}_2\}$ be ordered bases for P_1, where

$$\mathbf{v}_1 = t, \quad \mathbf{v}_2 = t - 3, \quad \mathbf{w}_1 = t - 1, \quad \mathbf{w}_2 = t + 1.$$

(a) Compute the transition matrix $P_{S \leftarrow T}$ from the T-basis to the S-basis.

(b) Verify Equation (3) for $\mathbf{v} = 5t + 1$.

(c) Compute the transition matrix $Q_{T \leftarrow S}$ from the S-basis to the T-basis and show that $Q_{T \leftarrow S} = P^{-1}_{S \leftarrow T}$.

Solution

(a) To compute $P_{S \leftarrow T}$, we need to solve the vector equations

$$\begin{aligned} a_1\mathbf{v}_1 + a_2\mathbf{v}_2 &= \mathbf{w}_1 \\ b_1\mathbf{v}_1 + b_2\mathbf{v}_2 &= \mathbf{w}_2 \end{aligned}$$

simultaneously by transforming the resulting partitioned matrix (verify)

$$\left[\begin{array}{cc|c|c} 1 & 1 & 1 & 1 \\ 0 & -3 & -1 & 1 \end{array}\right]$$

to reduced row echelon form. The result is (verify)

$$\left[\begin{array}{cc|c|c} 1 & 0 & \frac{2}{3} & \frac{4}{3} \\ 0 & 1 & \frac{1}{3} & -\frac{1}{3} \end{array}\right],$$

so

$$P_{S \leftarrow T} = \begin{bmatrix} \frac{2}{3} & \frac{4}{3} \\ \frac{1}{3} & -\frac{1}{3} \end{bmatrix}.$$

(b) If $\mathbf{v} = 5t + 1$, then expressing $\mathbf{v}$ in terms of the T-basis, we have (verify)

$$\mathbf{v} = 5t + 1 = 2(t - 1) + 3(t + 1),$$

so $\left[\mathbf{v}\right]_T = \begin{bmatrix} 2 \\ 3 \end{bmatrix}$. Then

$$\left[\mathbf{v}\right]_S = P_{S \leftarrow T} \left[\mathbf{v}\right]_T = \begin{bmatrix} \frac{2}{3} & \frac{4}{3} \\ \frac{1}{3} & -\frac{1}{3} \end{bmatrix} \begin{bmatrix} 2 \\ 3 \end{bmatrix} = \begin{bmatrix} \frac{16}{3} \\ -\frac{1}{3} \end{bmatrix}.$$

Computing $\left[\mathbf{v}\right]_S$ directly, we find that

$$\mathbf{v} = 5t + 1 = \tfrac{16}{3}t - \tfrac{1}{3}(t - 3), \quad \text{so } \left[\mathbf{v}\right]_S = \begin{bmatrix} \frac{16}{3} \\ -\frac{1}{3} \end{bmatrix}.$$

Hence

$$\left[\mathbf{v}\right]_S = P_{S \leftarrow T} \left[\mathbf{v}\right]_T .$$

(c) The transition matrix $Q_{T \leftarrow S}$ from the S-basis to the T-basis is derived (verify) by transforming the partitioned matrix

$$\left[\begin{array}{rr|r|r} 1 & 1 & 1 & 1 \\ -1 & 1 & 0 & -3 \end{array}\right]$$

to reduced row echelon form, yielding (verify)

$$\left[\begin{array}{rr|r|r} 1 & 0 & \frac{1}{2} & 2 \\ 0 & 1 & \frac{1}{2} & -1 \end{array}\right].$$

Hence,

$$Q_{T \leftarrow S} = \begin{bmatrix} \frac{1}{2} & 2 \\ \frac{1}{2} & -1 \end{bmatrix}.$$

Multiplying $Q_{T \leftarrow S}$ by $P_{S \leftarrow T}$, we find that $Q_{T \leftarrow S} P_{S \leftarrow T} = I_2$ (verify), so we conclude that $Q_{T \leftarrow S} = P_{S \leftarrow T}^{-1}$. ■

Key Terms

Ordered basis
Coordinates of a vector relative to an ordered basis
Coordinates
One-to-one function
Onto function
Isomorphism
Isomorphic vector spaces
Transition matrix

4.8 Exercises

In Exercises 1 through 6, compute the coordinate vector of $\mathbf{v}$ *with respect to each given ordered basis S for V.*

1. V is R^2, $S = \left\{\begin{bmatrix}1\\0\end{bmatrix}, \begin{bmatrix}0\\1\end{bmatrix}\right\}$, $\mathbf{v} = \begin{bmatrix}3\\-2\end{bmatrix}$.

2. V is R_3, $S = \{\begin{bmatrix}1 & -1 & 0\end{bmatrix}, \begin{bmatrix}0 & 1 & 0\end{bmatrix}, \begin{bmatrix}1 & 0 & 2\end{bmatrix}\}$, $\mathbf{v} = \begin{bmatrix}2 & -1 & -2\end{bmatrix}$.

3. V is P_1, $S = \{t+1, t-2\}$, $\mathbf{v} = t+4$.

4. V is P_2, $S = \{t^2 - t + 1, t + 1, t^2 + 1\}$, $\mathbf{v} = 4t^2 - 2t + 3$.

5. V is M_{22}, $S = \left\{\begin{bmatrix}1 & 0\\0 & 0\end{bmatrix}, \begin{bmatrix}0 & 0\\1 & 0\end{bmatrix}, \begin{bmatrix}0 & 1\\0 & 0\end{bmatrix}, \begin{bmatrix}0 & 0\\0 & 1\end{bmatrix}\right\}$,
$\mathbf{v} = \begin{bmatrix}1 & 0\\-1 & 2\end{bmatrix}$.

6. V is M_{22},
$$S = \left\{\begin{bmatrix}1 & -1\\0 & 0\end{bmatrix}, \begin{bmatrix}0 & 1\\1 & 0\end{bmatrix}, \begin{bmatrix}1 & 0\\0 & -1\end{bmatrix}, \begin{bmatrix}1 & 0\\-1 & 0\end{bmatrix}\right\},$$
$$\mathbf{v} = \begin{bmatrix}1 & 3\\-2 & 2\end{bmatrix}.$$

In Exercises 7 through 12, compute the vector $\mathbf{v}$ *if the coordinate vector* $[\mathbf{v}]_S$ *is given with respect to each ordered basis S for V.*

7. V is R^2, $S = \left\{\begin{bmatrix}2\\1\end{bmatrix}, \begin{bmatrix}-1\\1\end{bmatrix}\right\}$, $[\mathbf{v}]_S = \begin{bmatrix}1\\2\end{bmatrix}$.

8. V is R_3, $S = \{\begin{bmatrix}0 & 1 & -1\end{bmatrix}, \begin{bmatrix}1 & 0 & 0\end{bmatrix}, \begin{bmatrix}1 & 1 & 1\end{bmatrix}\}$,
$$[\mathbf{v}]_S = \begin{bmatrix}-1\\1\\2\end{bmatrix}.$$

9. V is P_1, $S = \{t, 2t - 1\}$, $[\mathbf{v}]_S = \begin{bmatrix}-2\\3\end{bmatrix}$.

10. V is P_2, $S = \{t^2 + 1, t + 1, t^2 + t\}$, $[\mathbf{v}]_S = \begin{bmatrix}3\\-1\\-2\end{bmatrix}$.

11. V is M_{22},
$$S = \left\{\begin{bmatrix}-1 & 0\\1 & 0\end{bmatrix}, \begin{bmatrix}2 & 2\\0 & 1\end{bmatrix}, \begin{bmatrix}1 & 2\\-1 & 3\end{bmatrix}, \begin{bmatrix}0 & 0\\2 & 3\end{bmatrix}\right\},$$
$$[\mathbf{v}]_S = \begin{bmatrix}2\\1\\-1\\3\end{bmatrix}.$$

12. V is M_{22},
$$S = \left\{\begin{bmatrix}1 & -2\\0 & 0\end{bmatrix}, \begin{bmatrix}-1 & 3\\0 & 1\end{bmatrix}, \begin{bmatrix}1 & 0\\0 & 0\end{bmatrix}, \begin{bmatrix}0 & -1\\1 & 0\end{bmatrix}\right\},$$
$$[\mathbf{v}]_S = \begin{bmatrix}0\\1\\0\\2\end{bmatrix}.$$

In Example 3 we showed that an appropriate choice of basis could greatly simplify the computation of the values of a sequence of the form $A\mathbf{v}$, $A^2\mathbf{v}$, $A^3\mathbf{v}$, ... *Exercises 13 and 14 require an approach similar to that in Example 3.*

13. Let
$$S = \{\mathbf{v}_1, \mathbf{v}_2\} = \left\{\begin{bmatrix}1\\-1\end{bmatrix}, \begin{bmatrix}1\\-2\end{bmatrix}\right\},$$
$$A = \begin{bmatrix}-0.85 & -0.55\\1.10 & 0.80\end{bmatrix}, \quad \text{and} \quad \mathbf{v} = \begin{bmatrix}2\\6\end{bmatrix}.$$

(a) Show that S is a basis for R^2.

(b) Find $[\mathbf{v}]_S$.

(c) Determine a scalar λ_1 such that $A\mathbf{v}_1 = \lambda_1\mathbf{v}_1$.

(d) Determine a scalar λ_2 such that $A\mathbf{v}_2 = \lambda_2\mathbf{v}_2$.

(e) Use the basis S and the results of parts (b) through (d) to determine an expression for $A^n\mathbf{v}$ that is a linear combination of $\mathbf{v}_1$ and $\mathbf{v}_2$.

(f) As n increases, describe the limiting behavior of the sequence $A\mathbf{v}$, $A^2\mathbf{v}$, $A^3\mathbf{v}$, ..., $A^n\mathbf{v}$, ...

14. Let
$$S = \{\mathbf{v}_1, \mathbf{v}_2\} = \left\{\begin{bmatrix}1\\-1\end{bmatrix}, \begin{bmatrix}-2\\3\end{bmatrix}\right\},$$
$$A = \begin{bmatrix}-1 & -2\\3 & 4\end{bmatrix}, \quad \text{and} \quad \mathbf{v} = \begin{bmatrix}4\\3\end{bmatrix}.$$

Follow the directions for (a) through (f) in Exercise 13.

15. Let $S = \left\{\begin{bmatrix}1\\2\end{bmatrix}, \begin{bmatrix}0\\1\end{bmatrix}\right\}$ and $T = \left\{\begin{bmatrix}1\\1\end{bmatrix}, \begin{bmatrix}2\\3\end{bmatrix}\right\}$ be ordered bases for R^2. Let $\mathbf{v} = \begin{bmatrix}1\\5\end{bmatrix}$ and $\mathbf{w} = \begin{bmatrix}5\\4\end{bmatrix}$.

(a) Find the coordinate vectors of $\mathbf{v}$ and $\mathbf{w}$ with respect to the basis T.

(b) What is the transition matrix $P_{S\leftarrow T}$ from the T- to the S-basis?

(c) Find the coordinate vectors of $\mathbf{v}$ and $\mathbf{w}$ with respect to S, using $P_{S\leftarrow T}$.

(d) Find the coordinate vectors of $\mathbf{v}$ and $\mathbf{w}$ with respect to S directly.

(e) Find the transition matrix $Q_{T \leftarrow S}$ from the S- to the T-basis.

(f) Find the coordinate vectors of $\mathbf{v}$ and $\mathbf{w}$ with respect to T, using $Q_{T \leftarrow S}$. Compare the answers with those of (a).

16. Let

$$S = \left\{ \begin{bmatrix} 1 \\ 0 \\ 1 \end{bmatrix}, \begin{bmatrix} -1 \\ 0 \\ 0 \end{bmatrix}, \begin{bmatrix} 0 \\ 1 \\ 2 \end{bmatrix} \right\}$$

and

$$T = \left\{ \begin{bmatrix} -1 \\ 1 \\ 0 \end{bmatrix}, \begin{bmatrix} 1 \\ 2 \\ -1 \end{bmatrix}, \begin{bmatrix} 0 \\ 1 \\ 0 \end{bmatrix} \right\}$$

be ordered bases for R^3. Let

$$\mathbf{v} = \begin{bmatrix} 1 \\ 3 \\ 8 \end{bmatrix} \quad \text{and} \quad \mathbf{w} = \begin{bmatrix} -1 \\ 8 \\ -2 \end{bmatrix}.$$

Follow the directions for (a) through (f) in Exercise 15.

17. Let $S = \{t^2 + 1, t - 2, t + 3\}$ and $T = \{2t^2 + t, t^2 + 3, t\}$ be ordered bases for P_2. Let $\mathbf{v} = 8t^2 - 4t + 6$ and $\mathbf{w} = 7t^2 - t + 9$. Follow the directions for (a) through (f) in Exercise 15.

18. Let

$$S = \{[1 \quad 1 \quad 1], [1 \quad 2 \quad 3], [1 \quad 0 \quad 1]\}$$

and

$$T = \{[0 \quad 1 \quad 1], [1 \quad 0 \quad 0], [1 \quad 0 \quad 1]\}$$

be ordered bases for R_3. Let

$$\mathbf{v} = [-1 \quad 4 \quad 5] \quad \text{and} \quad \mathbf{w} = [2 \quad 0 \quad -6].$$

Follow the directions for (a) through (f) in Exercise 15.

19. Let

$$S = \left\{ \begin{bmatrix} 1 & 0 \\ 0 & 0 \end{bmatrix}, \begin{bmatrix} 0 & 1 \\ 1 & 0 \end{bmatrix}, \begin{bmatrix} 0 & 2 \\ 0 & 1 \end{bmatrix}, \begin{bmatrix} 0 & 0 \\ 1 & 1 \end{bmatrix} \right\}$$

and

$$T = \left\{ \begin{bmatrix} 1 & 1 \\ 0 & 0 \end{bmatrix}, \begin{bmatrix} 0 & 0 \\ 1 & 0 \end{bmatrix}, \begin{bmatrix} 0 & 0 \\ 0 & 1 \end{bmatrix}, \begin{bmatrix} 1 & 0 \\ 0 & 0 \end{bmatrix} \right\}$$

be ordered bases for M_{22}. Let

$$\mathbf{v} = \begin{bmatrix} 1 & 1 \\ 1 & 1 \end{bmatrix} \quad \text{and} \quad \mathbf{w} = \begin{bmatrix} 1 & 2 \\ -2 & 1 \end{bmatrix}.$$

Follow the directions for (a) through (f) in Exercise 15.

20. Let

$$S = \{[1 \quad -1], [2 \quad 1]\}$$

and

$$T = \{[3 \quad 0], [4 \quad -1]\}$$

be ordered bases for R_2. If $\mathbf{v}$ is in R_2 and $[\mathbf{v}]_T = \begin{bmatrix} 1 \\ 2 \end{bmatrix}$, determine $[\mathbf{v}]_S$.

21. Let $S = \{t + 1, t - 2\}$ and $T = \{t - 5, t - 2\}$ be ordered bases for P_1. If $\mathbf{v}$ is in P_1 and $[\mathbf{v}]_T = \begin{bmatrix} -1 \\ 3 \end{bmatrix}$, determine $[\mathbf{v}]_S$.

22. Let

$$S = \{[-1 \quad 2 \quad 1], [0 \quad 1 \quad 1], [-2 \quad 2 \quad 1]\}$$

and

$$T = \{[-1 \quad 1 \quad 0], [0 \quad 1 \quad 0], [0 \quad 1 \quad 1]\}$$

be ordered bases for R_3. If $\mathbf{v}$ is in R_3 and

$$[\mathbf{v}]_S = \begin{bmatrix} 2 \\ 0 \\ 1 \end{bmatrix},$$

determine $[\mathbf{v}]_T$.

23. If the vector $\mathbf{v}$ in P_2 has the coordinate vector $\begin{bmatrix} 1 \\ 2 \\ 3 \end{bmatrix}$ with respect to the ordered basis $T = \{t^2, t - 1, 1\}$, what is $[\mathbf{v}]_S$ if $S = \{t^2 + t + 1, t + 1, 1\}$?

24. Let $S = \{\mathbf{v}_1, \mathbf{v}_2, \mathbf{v}_3\}$ and $T = \{\mathbf{w}_1, \mathbf{w}_2, \mathbf{w}_3\}$ be ordered bases for R^3, where

$$\mathbf{v}_1 = \begin{bmatrix} 1 \\ 0 \\ 1 \end{bmatrix}, \quad \mathbf{v}_2 = \begin{bmatrix} 1 \\ 1 \\ 0 \end{bmatrix}, \quad \mathbf{v}_3 = \begin{bmatrix} 0 \\ 0 \\ 1 \end{bmatrix}.$$

Suppose that the transition matrix from T to S is

$$P_{S \leftarrow T} = \begin{bmatrix} 1 & 1 & 2 \\ 2 & 1 & 1 \\ -1 & -1 & 1 \end{bmatrix}.$$

Determine T.

25. Let $S = \{\mathbf{v}_1, \mathbf{v}_2\}$ and $T = \{\mathbf{w}_1, \mathbf{w}_2\}$ be ordered bases for P_1, where

$$\mathbf{w}_1 = t, \quad \mathbf{w}_2 = t - 1.$$

If the transition matrix from S to T is $\begin{bmatrix} 2 & 3 \\ -1 & 2 \end{bmatrix}$, determine S.

26. Let $S = \{\mathbf{v}_1, \mathbf{v}_2\}$ and $T = \{\mathbf{w}_1, \mathbf{w}_2\}$ be ordered bases for R^2, where

$$\mathbf{v}_1 = \begin{bmatrix} 1 \\ 2 \end{bmatrix}, \quad \mathbf{v}_2 = \begin{bmatrix} 0 \\ 1 \end{bmatrix}.$$

If the transition matrix from S to T is $\begin{bmatrix} 2 & 1 \\ 1 & 1 \end{bmatrix}$, determine T.

27. Let $S = \{\mathbf{v}_1, \mathbf{v}_2\}$ and $T = \{\mathbf{w}_1, \mathbf{w}_2\}$ be ordered bases for P_1, where

$$\mathbf{w}_1 = t - 1, \quad \mathbf{w}_2 = t + 1.$$

If the transition matrix from T to S is $\begin{bmatrix} 1 & 2 \\ 2 & 3 \end{bmatrix}$, determine S.

28. Prove parts (a) and (c) of Theorem 4.15.

29. Let $L: V \to W$ be an isomorphism of vector space V onto vector space W.

(a) Prove that $L(\mathbf{0}_V) = \mathbf{0}_W$.

(b) Show that $L(\mathbf{v} - \mathbf{w}) = L(\mathbf{v}) - L(\mathbf{w})$.

(c) Show that

$$L(a_1\mathbf{v}_1 + a_2\mathbf{v}_2 + \cdots + a_k\mathbf{v}_k) = a_1L(\mathbf{v}_1) + a_2L(\mathbf{v}_2) + \cdots + a_kL(\mathbf{v}_k).$$

30. Prove that R^n and R^m are isomorphic if and only if $n = m$.

31. Find an isomorphism $L: R_n \to R^n$.

32. Find an isomorphism $L: P_2 \to R^3$. More generally, show that P_n and R^{n+1} are isomorphic.

33. **(a)** Show that M_{22} is isomorphic to R^4.

(b) What is $\dim M_{22}$?

34. Let V be the subspace of the vector space of all real-valued continuous functions that has basis $S = \{e^t, e^{-t}\}$. Show that V and R^2 are isomorphic.

35. Let V be the subspace of the vector space of all real-valued functions that is *spanned* by the set

$$S = \{\cos^2 t, \sin^2 t, \cos 2t\}.$$

Show that V and R_2 are isomorphic.

36. Let V and W be isomorphic vector spaces. Prove that if V_1 is a subspace of V, then V_1 is isomorphic to a subspace W_1 of W.

37. Let $S = \{\mathbf{v}_1, \mathbf{v}_2, \ldots, \mathbf{v}_n\}$ be an ordered basis for the n-dimensional vector space V, and let $\mathbf{v}$ and $\mathbf{w}$ be two vectors in V. Show that $\mathbf{v} = \mathbf{w}$ if and only if $\left[\mathbf{v}\right]_S = \left[\mathbf{w}\right]_S$.

38. Show that if S is an ordered basis for an n-dimensional vector space V, $\mathbf{v}$ and $\mathbf{w}$ are vectors in V, and c is a scalar, then

$$\left[\mathbf{v} + \mathbf{w}\right]_S = \left[\mathbf{v}\right]_S + \left[\mathbf{w}\right]_S$$

and

$$\left[c\mathbf{v}\right]_S = c\left[\mathbf{v}\right]_S.$$

In Exercises 39 through 41, let $S = \{\mathbf{v}_1, \mathbf{v}_2, \ldots, \mathbf{v}_n\}$ and $T = \{\mathbf{w}_1, \mathbf{w}_2, \ldots, \mathbf{w}_n\}$ be ordered bases for the vector space R^n.

39. Let M_S be the $n \times n$ matrix whose jth column is $\mathbf{v}_j$ and let M_T be the $n \times n$ matrix whose jth column is $\mathbf{w}_j$. Prove that M_S and M_T are nonsingular. (*Hint*: Consider the homogeneous systems $M_S\mathbf{x} = \mathbf{0}$ and $M_T\mathbf{x} = \mathbf{0}$.)

40. If $\mathbf{v}$ is a vector in V, show that

$$\mathbf{v} = M_S\left[\mathbf{v}\right]_S \quad \text{and} \quad \mathbf{v} = M_T\left[\mathbf{v}\right]_T.$$

41. **(a)** Use Equation (3) and Exercises 39 and 40 to show that

$$P_{S\leftarrow T} = M_S^{-1}M_T.$$

(b) Show that $P_{S\leftarrow T}$ is nonsingular.

(c) Verify the result in part (a) of Example 4.

42. Let S be an ordered basis for n-dimensional vector space V. Show that if $\{\mathbf{w}_1, \mathbf{w}_2, \ldots, \mathbf{w}_k\}$ is a linearly independent set of vectors in V, then

$$\left\{\left[\mathbf{w}_1\right]_S, \left[\mathbf{w}_2\right]_S, \ldots, \left[\mathbf{w}_k\right]_S\right\}$$

is a linearly independent set of vectors in R^n.

43. Let $S = \{\mathbf{v}_1, \mathbf{v}_2, \ldots, \mathbf{v}_n\}$ be an ordered basis for an n-dimensional vector space V. Show that

$$\left\{\left[\mathbf{v}_1\right]_S, \left[\mathbf{v}_2\right]_S, \ldots, \left[\mathbf{v}_n\right]_S\right\}$$

is an ordered basis for R^n.

4.9 Rank of a Matrix

In this section we obtain another effective method for finding a basis for a vector space V spanned by a given set of vectors $S = \{\mathbf{v}_1, \mathbf{v}_2, \ldots, \mathbf{v}_k\}$. In Section 4.6 we developed a technique for choosing a basis for V that is a subset of S (Theorem 4.9). The method to be developed in this section produces a basis for V that is not guaranteed to be a subset of S. We shall also attach a unique number to a matrix A that we later show gives us information about the dimension of the solution space of a homogeneous system with coefficient matrix A.

DEFINITION 4.14 Let

$$A = \begin{bmatrix} a_{11} & a_{12} & \cdots & a_{1n} \\ a_{21} & a_{22} & \cdots & a_{2n} \\ \vdots & \vdots & & \vdots \\ a_{m1} & a_{m2} & \cdots & a_{mn} \end{bmatrix}$$

be an $m \times n$ matrix. The rows of A, considered as vectors in R_n, span a subspace of R_n called the **row space** of A. Similarly, the columns of A, considered as vectors in R^m, span a subspace of R^m called the **column space** of A.

Theorem 4.17 If A and B are two $m \times n$ row (column) equivalent matrices, then the row (column) spaces of A and B are equal.

Proof

If A and B are row equivalent, then the rows of B are derived from the rows of A by a finite number of the three elementary row operations. Thus each row of B is a linear combination of the rows of A. Hence the row space of B is contained in the row space of A. If we apply the inverse elementary row operations to B, we get A, so the row space of A is contained in the row space of B. Hence the row spaces of A and B are identical. The proof for the column spaces is similar. ■

We can use this theorem to find a basis for a subspace spanned by a given set of vectors. We illustrate this method with the following example:

EXAMPLE 1 Find a basis for the subspace V of R_5 that is spanned by $S = \{\mathbf{v}_1, \mathbf{v}_2, \mathbf{v}_3, \mathbf{v}_4\}$, where

$$\mathbf{v}_1 = \begin{bmatrix} 1 & -2 & 0 & 3 & -4 \end{bmatrix}, \quad \mathbf{v}_2 = \begin{bmatrix} 3 & 2 & 8 & 1 & 4 \end{bmatrix},$$
$$\mathbf{v}_3 = \begin{bmatrix} 2 & 3 & 7 & 2 & 3 \end{bmatrix}, \quad \text{and} \quad \mathbf{v}_4 = \begin{bmatrix} -1 & 2 & 0 & 4 & -3 \end{bmatrix}.$$

Solution

Note that V is the row space of the matrix A whose rows are the given vectors.

$$A = \begin{bmatrix} 1 & -2 & 0 & 3 & -4 \\ 3 & 2 & 8 & 1 & 4 \\ 2 & 3 & 7 & 2 & 3 \\ -1 & 2 & 0 & 4 & -3 \end{bmatrix}.$$

Using elementary row operations, we find that A is row equivalent to the matrix (verify)

$$B = \begin{bmatrix} 1 & 0 & 2 & 0 & 1 \\ 0 & 1 & 1 & 0 & 1 \\ 0 & 0 & 0 & 1 & -1 \\ 0 & 0 & 0 & 0 & 0 \end{bmatrix},$$

which is in reduced row echelon form. The row spaces of A and B are identical, and a basis for the row space of B consists of

$$\mathbf{w}_1 = \begin{bmatrix} 1 & 0 & 2 & 0 & 1 \end{bmatrix}, \quad \mathbf{w}_2 = \begin{bmatrix} 0 & 1 & 1 & 0 & 1 \end{bmatrix}, \quad \text{and} \quad \mathbf{w}_3 = \begin{bmatrix} 0 & 0 & 0 & 1 & -1 \end{bmatrix}.$$

(See Exercise 25 in Section 4.5.) Hence, $\{\mathbf{w}_1, \mathbf{w}_2, \mathbf{w}_3\}$ is also a basis for V. ■

It is not necessary to find a matrix B in reduced row echelon form that is row equivalent to A. All that is required is that we have a matrix B which is row equivalent to A and such that we can easily obtain a basis for the row space of B. Often, we do not have to reduce A all the way to reduced row echelon form to get such a matrix B. We can show that if A is row equivalent to a matrix B which is in row echelon form, then the nonzero rows of B form a basis for the row space of A.

Of course, the basis produced by the procedure used in Example 1 may not be a subset of the given spanning set. The method used in Example 6 of Section 4.6 always gives a basis that is a subset of the spanning set. However, the basis for a subspace V of R^n that is obtained by the procedure used in Example 1 is analogous in its simplicity to the natural basis for R^n. Thus if

$$\mathbf{v} = \begin{bmatrix} a_1 \\ a_2 \\ \vdots \\ a_n \end{bmatrix}$$

is a vector in V and $\{\mathbf{v}_1, \mathbf{v}_2, \ldots, \mathbf{v}_k\}$ is a basis for V obtained by the method of Example 1 where the leading 1's occur in columns $j_1, j_2, \ldots, j_k$, then it can be shown (Exercise 42) that

$$\mathbf{v} = a_{j_1}\mathbf{v}_1 + a_{j_2}\mathbf{v}_2 + \cdots + a_{j_k}\mathbf{v}_k.$$

EXAMPLE 2

Let V be the subspace of Example 1. Given that the vector

$$\mathbf{v} = \begin{bmatrix} 5 & 4 & 14 & 6 & 3 \end{bmatrix}$$

is in V, write $\mathbf{v}$ as a linear combination of the basis determined in Example 1.

Solution

We have $j_1 = 1$, $j_2 = 2$, and $j_3 = 4$, so $\mathbf{v} = 5\mathbf{w}_1 + 4\mathbf{w}_2 + 6\mathbf{w}_3$. ■

Remark The following example illustrates how to use the procedure given in Example 1 to find a basis for a subspace of a vector space that is not R^n or R_n:

EXAMPLE 3 Let V be the subspace of P_4 spanned by $S = \{\mathbf{v}_1, \mathbf{v}_2, \mathbf{v}_3, \mathbf{v}_4\}$, where $\mathbf{v}_1 = t^4 + t^2 + 2t + 1$, $\mathbf{v}_2 = t^4 + t^2 + 2t + 2$, $\mathbf{v}_3 = 2t^4 + t^3 + t + 2$, and $\mathbf{v}_4 = t^4 + t^3 - t^2 - t$. Find a basis for V.

Solution

Since P_4 is isomorphic to R_5 under the isomorphism L defined by

$$L(at^4 + bt^3 + ct^2 + dt + e) = \begin{bmatrix} a & b & c & d & e \end{bmatrix},$$

then $L(V)$ is isomorphic to a subspace W of R_5. (See Exercise 36.) The subspace W is spanned by $\{L(\mathbf{v}_1), L(\mathbf{v}_2), L(\mathbf{v}_3), L(\mathbf{v}_4)\}$, as we have seen in the proof of Theorem 4.16. We now find a basis for W by proceeding as in Example 1. Thus W is the row space of the matrix

$$A = \begin{bmatrix} 1 & 0 & 1 & 2 & 1 \\ 1 & 0 & 1 & 2 & 2 \\ 2 & 1 & 0 & 1 & 2 \\ 1 & 1 & -1 & -1 & 0 \end{bmatrix},$$

and A is row equivalent to (verify)

$$B = \begin{bmatrix} 1 & 0 & 1 & 2 & 0 \\ 0 & 1 & -2 & -3 & 0 \\ 0 & 0 & 0 & 0 & 1 \\ 0 & 0 & 0 & 0 & 0 \end{bmatrix}.$$

A basis for W is therefore $T = \{\mathbf{w}_1, \mathbf{w}_2, \mathbf{w}_3\}$, where $\mathbf{w}_1 = \begin{bmatrix} 1 & 0 & 1 & 2 & 0 \end{bmatrix}$, $\mathbf{w}_2 = \begin{bmatrix} 0 & 1 & -2 & -3 & 0 \end{bmatrix}$, and $\mathbf{w}_3 = \begin{bmatrix} 0 & 0 & 0 & 0 & 1 \end{bmatrix}$. A basis for V is then

$$\left\{L^{-1}(\mathbf{w}_1), L^{-1}(\mathbf{w}_2), L^{-1}(\mathbf{w}_3)\right\} = \left\{t^4 + t^2 + 2t,\ t^3 - 2t^2 - 3t,\ 1\right\}. \quad \blacksquare$$

DEFINITION 4.15 The dimension of the row (column) space of A is called the **row (column) rank** of A.

If A and B are row equivalent, then row rank A = row rank B; and if A and B are column equivalent, then column rank A = column rank B. Therefore, if we start out with an $m \times n$ matrix A and find a matrix B in reduced row echelon form that is row equivalent to A, then A and B have equal row ranks. But the row rank of B is clearly the number of nonzero rows. Thus we have a good method for finding the row rank of a given matrix A.

EXAMPLE 4 Find a basis for the row space of the matrix A defined in the solution of Example 1 that contains only row vectors from A. Also, compute the row rank of A.

Solution

Using the procedure in the alternative proof of Theorem 4.9, we form the equation

$$a_1\begin{bmatrix}1 & -2 & 0 & 3 & -4\end{bmatrix}+a_2\begin{bmatrix}3 & 2 & 8 & 1 & 4\end{bmatrix}+a_3\begin{bmatrix}2 & 3 & 7 & 2 & 3\end{bmatrix}$$
$$+a_4\begin{bmatrix}-1 & 2 & 0 & 4 & -3\end{bmatrix}=\begin{bmatrix}0 & 0 & 0 & 0 & 0\end{bmatrix}$$

$$\left[\begin{array}{rrrr|r}1 & 3 & 2 & -1 & 0\\ -2 & 2 & 3 & 2 & 0\\ 0 & 8 & 7 & 0 & 0\\ 3 & 1 & 2 & 4 & 0\\ -4 & 4 & 3 & -3 & 0\end{array}\right]=\left[A^T \;\middle|\; \mathbf{0}\right]; \tag{1}$$

that is, the coefficient matrix is A^T. Transforming the augmented matrix $\left[A^T \;\middle|\; \mathbf{0}\right]$ in (1) to reduced row echelon form, we obtain (verify)

$$\left[\begin{array}{cccc|c}1 & 0 & \frac{11}{24} & 0 & 0\\ 0 & 1 & 0 & -\frac{49}{24} & 0\\ 0 & 0 & 1 & \frac{7}{3} & 0\\ 0 & 0 & 0 & 0 & 0\\ 0 & 0 & 0 & 0 & 0\end{array}\right]. \tag{2}$$

Since the leading 1's in (2) occur in columns 1, 2, and 3, we conclude that the first three rows of A form a basis for the row space of A. That is,

$$\left\{\begin{bmatrix}1 & -2 & 0 & 3 & -4\end{bmatrix},\begin{bmatrix}3 & 2 & 8 & 1 & 4\end{bmatrix},\begin{bmatrix}2 & 3 & 7 & 2 & 3\end{bmatrix}\right\}$$

is a basis for the row space of A. The row rank of A is 3. ■

EXAMPLE 5 Find a basis for the column space of the matrix A defined in the solution of Example 1, and compute the column rank of A.

Solution 1

Writing the columns of A as row vectors, we obtain the matrix A^T, which when transformed to reduced row echelon form is (as we saw in Example 4)

$$\begin{bmatrix}1 & 0 & 0 & \frac{11}{24}\\ 0 & 1 & 0 & -\frac{49}{24}\\ 0 & 0 & 1 & \frac{7}{3}\\ 0 & 0 & 0 & 0\\ 0 & 0 & 0 & 0\end{bmatrix}.$$

Thus, the vectors $\begin{bmatrix}1 & 0 & 0 & \frac{11}{24}\end{bmatrix}$, $\begin{bmatrix}0 & 1 & 0 & -\frac{49}{24}\end{bmatrix}$, and $\begin{bmatrix}0 & 0 & 1 & \frac{7}{3}\end{bmatrix}$ form a basis for the row space of A^T. Hence the vectors

$$\begin{bmatrix} 1 \\ 0 \\ 0 \\ \frac{11}{24} \end{bmatrix}, \quad \begin{bmatrix} 0 \\ 1 \\ 0 \\ -\frac{49}{24} \end{bmatrix}, \quad \text{and} \quad \begin{bmatrix} 0 \\ 0 \\ 1 \\ \frac{7}{3} \end{bmatrix}$$

form a basis for the column space of A, and we conclude that the column rank of A is 3.

Solution 2

If we want to find a basis for the column space of A that contains only the column vectors from A, we follow the procedure developed in the proof of Theorem 4.9, forming the equation

$$a_1 \begin{bmatrix} 1 \\ 3 \\ 2 \\ -1 \end{bmatrix} + a_2 \begin{bmatrix} -2 \\ 2 \\ 3 \\ 2 \end{bmatrix} + a_3 \begin{bmatrix} 0 \\ 8 \\ 7 \\ 0 \end{bmatrix} + a_4 \begin{bmatrix} 3 \\ 1 \\ 2 \\ 4 \end{bmatrix} + a_5 \begin{bmatrix} -4 \\ 4 \\ 3 \\ -3 \end{bmatrix} = \begin{bmatrix} 0 \\ 0 \\ 0 \\ 0 \end{bmatrix}$$

whose augmented matrix is $\left[A \mid \mathbf{0}\right]$. Transforming this matrix to reduced row echelon form, we obtain (as in Example 1)

$$\left[\begin{array}{ccccc|c} 1 & 0 & 2 & 0 & 1 & 0 \\ 0 & 1 & 1 & 0 & 1 & 0 \\ 0 & 0 & 0 & 1 & -1 & 0 \\ 0 & 0 & 0 & 0 & 0 & 0 \end{array}\right].$$

Since the leading 1's occur in columns 1, 2, and 4, we conclude that the first, second, and fourth columns of A form a basis for the column space of A. That is,

$$\left\{ \begin{bmatrix} 1 \\ 3 \\ 2 \\ -1 \end{bmatrix}, \begin{bmatrix} -2 \\ 2 \\ 3 \\ 2 \end{bmatrix}, \begin{bmatrix} 3 \\ 1 \\ 2 \\ 4 \end{bmatrix} \right\}$$

is a basis for the column space of A. The column rank of A is 3. ■

We may also conclude that if A is an $m \times n$ matrix and P is a nonsingular $m \times m$ matrix, then row rank (PA) = row rank A, for A and PA are row equivalent (Exercise 23 in Section 2.3). Similarly, if Q is a nonsingular $n \times n$ matrix, then column rank (AQ) = column rank A. Moreover, since dimension $R_n = n$, we see that row rank $A \leq n$. Also, since the row space of A is spanned by m vectors, row rank $A \leq m$. Thus row rank $A \leq$ minimum $\{m, n\}$.

In Examples 4 and 5 we observe that the row and column ranks of A are equal. This is always true and is a very important result in linear algebra. We now turn to the proof of this theorem.

Theorem 4.18 The row rank and column rank of the $m \times n$ matrix $A = \begin{bmatrix} a_{ij} \end{bmatrix}$ are equal.

Proof

Let $\mathbf{v}_1, \mathbf{v}_2, \ldots, \mathbf{v}_m$ be the row vectors of A, where

$$\mathbf{v}_i = \begin{bmatrix} a_{i1} & a_{i2} & \cdots & a_{in} \end{bmatrix}, \qquad i = 1, 2, \ldots, m.$$

Let row rank $A = r$ and let the set of vectors $\{\mathbf{w}_1, \mathbf{w}_2, \ldots, \mathbf{w}_r\}$ form a basis for the row space of A, where $\mathbf{w}_i = \begin{bmatrix} b_{i1} & b_{i2} & \cdots & b_{in} \end{bmatrix}$ for $i = 1, 2, \ldots, r$. Now each of the row vectors is a linear combination of $\mathbf{w}_1, \mathbf{w}_2, \ldots, \mathbf{w}_r$:

$$\begin{aligned} \mathbf{v}_1 &= c_{11}\mathbf{w}_1 + c_{12}\mathbf{w}_2 + \cdots + c_{1r}\mathbf{w}_r \\ \mathbf{v}_2 &= c_{21}\mathbf{w}_1 + c_{22}\mathbf{w}_2 + \cdots + c_{2r}\mathbf{w}_r \\ &\vdots \\ \mathbf{v}_m &= c_{m1}\mathbf{w}_1 + c_{m2}\mathbf{w}_2 + \cdots + c_{mr}\mathbf{w}_r, \end{aligned}$$

where the c_{ij} are uniquely determined real numbers. Recalling that two matrices are equal if and only if the corresponding entries are equal, we equate the entries of these vector equations to get

$$\begin{aligned} a_{1j} &= c_{11}b_{1j} + c_{12}b_{2j} + \cdots + c_{1r}b_{rj} \\ a_{2j} &= c_{21}b_{1j} + c_{22}b_{2j} + \cdots + c_{2r}b_{rj} \\ &\vdots \\ a_{mj} &= c_{m1}b_{1j} + c_{m2}b_{2j} + \cdots + c_{mr}b_{rj} \end{aligned}$$

or

$$\begin{bmatrix} a_{1j} \\ a_{2j} \\ \vdots \\ a_{mj} \end{bmatrix} = b_{1j}\begin{bmatrix} c_{11} \\ c_{21} \\ \vdots \\ c_{m1} \end{bmatrix} + b_{2j}\begin{bmatrix} c_{12} \\ c_{22} \\ \vdots \\ c_{m2} \end{bmatrix} + \cdots + b_{rj}\begin{bmatrix} c_{1r} \\ c_{2r} \\ \vdots \\ c_{mr} \end{bmatrix}$$

for $j = 1, 2, \ldots, n$.

Since every column of A is a linear combination of r vectors, the dimension of the column space of A is at most r, or column rank $A \leq r =$ row rank A. Similarly, we get row rank $A \leq$ column rank A. Hence, the row and column ranks of A are equal.

Alternative Proof: Let $\mathbf{x}_1, \mathbf{x}_2, \ldots, \mathbf{x}_n$ denote the columns of A. To determine the dimension of the column space of A, we use the procedure in the alternative proof of Theorem 4.9. Thus we consider the equation

$$c_1\mathbf{x}_1 + c_2\mathbf{x}_2 + \cdots + c_n\mathbf{x}_n = \mathbf{0}.$$

We now transform the augmented matrix, $\left[A \mid \mathbf{0} \right]$, of this homogeneous system to reduced row echelon form. The vectors corresponding to the columns containing the leading 1's form a basis for the column space of A. Thus the column rank of A is the number of leading 1's. But this number is also the number of nonzero rows in the reduced row echelon form matrix that is row equivalent to A, so it is the row rank of A. Thus row rank $A =$ column rank A. ■

Since the row and column ranks of a matrix are equal, we now merely refer to the **rank** of a matrix. Note that rank $I_n = n$. Theorem 2.13 states that A is equivalent to B if and only if there exist nonsingular matrices P and Q such that $B = PAQ$. If A is equivalent to B, then rank A = rank B, for rank B = rank(PAQ) = rank(PA) = rank A.

We also recall from Section 2.4 that if A is an $m \times n$ matrix, then A is equivalent to a matrix $C = \begin{bmatrix} I_r & O \\ O & O \end{bmatrix}$. Now rank A = rank $C = r$. We use these facts to establish the result that if A and B are $m \times n$ matrices of equal rank, then A and B are equivalent. Thus let rank $A = r =$ rank B. Then there exist nonsingular matrices P_1, Q_1, P_2, and Q_2 such that

$$P_1 A Q_1 = \begin{bmatrix} I_r & O \\ O & O \end{bmatrix} = P_2 B Q_2.$$

Then $P_2^{-1} P_1 A Q_1 Q_2^{-1} = B$. Letting $P = P_2^{-1} P_1$ and $Q = Q_1 Q_2^{-1}$, we find that P and Q are nonsingular and $B = PAQ$. Hence A and B are equivalent.

If A is an $m \times n$ matrix, we have defined (see Section 4.7) the nullity of A as the dimension of the null space of A, that is, the dimension of the solution space of $A\mathbf{x} = \mathbf{0}$. If A is transformed to a matrix B in reduced row echelon form, having r nonzero rows, then we know that the dimension of the solution space of $A\mathbf{x} = \mathbf{0}$ is $n - r$. Since r is also the rank of A, we have obtained a fundamental relationship between the rank and nullity of A, which we state in the following theorem:

Theorem 4.19 If A is an $m \times n$ matrix, then rank A + nullity $A = n$. ■

EXAMPLE 6 Let

$$A = \begin{bmatrix} 1 & 1 & 4 & 1 & 2 \\ 0 & 1 & 2 & 1 & 1 \\ 0 & 0 & 0 & 1 & 2 \\ 1 & -1 & 0 & 0 & 2 \\ 2 & 1 & 6 & 0 & 1 \end{bmatrix},$$

as it was defined in Example 1 of Section 4.7. When A is transformed to reduced row echelon form, we get

$$\begin{bmatrix} 1 & 0 & 2 & 0 & 1 \\ 0 & 1 & 2 & 0 & -1 \\ 0 & 0 & 0 & 1 & 2 \\ 0 & 0 & 0 & 0 & 0 \\ 0 & 0 & 0 & 0 & 0 \end{bmatrix}.$$

Then rank $A = 3$ and nullity $A = 2$. This agrees with the result obtained in solving Example 1 of Section 4.7, where we found that the dimension of the solution space of $A\mathbf{x} = \mathbf{0}$ is 2. Thus, Theorem 4.19 has been verified. ■

The following example illustrates geometrically some of the ideas discussed previously:

EXAMPLE 7

Let

$$A = \begin{bmatrix} 3 & -1 & 2 \\ 2 & 1 & 3 \\ 7 & 1 & 8 \end{bmatrix}.$$

Transforming A to reduced row echelon form, we find that

$$\begin{bmatrix} 1 & 0 & 1 \\ 0 & 1 & 1 \\ 0 & 0 & 0 \end{bmatrix} \quad \text{(verify)},$$

so we conclude the following:

- Rank $A = 2$.
- Dimension of row space of $A = 2$, so the row space of A is a two-dimensional subspace of R_3, that is, a plane passing through the origin.

From the reduced row echelon form matrix that A has been transformed to, we see that every solution to the homogeneous system $A\mathbf{x} = \mathbf{0}$ is of the form

$$\mathbf{x} = \begin{bmatrix} -r \\ -r \\ r \end{bmatrix},$$

where r is an arbitrary constant (verify), so the solution space of this homogeneous system, or the null space of A, is a line passing through the origin. Moreover, the dimension of the null space of A, or the nullity of A, is 1. Thus Theorem 4.19 has been verified.

Of course, we already know that the dimension of the column space of A is also 2. We could produce this result by finding a basis consisting of two vectors for the column space of A. Thus, the column space of A is also a two-dimensional subspace of R^3—that is, a plane passing through the origin. These results are illustrated in Figure 4.33. ■

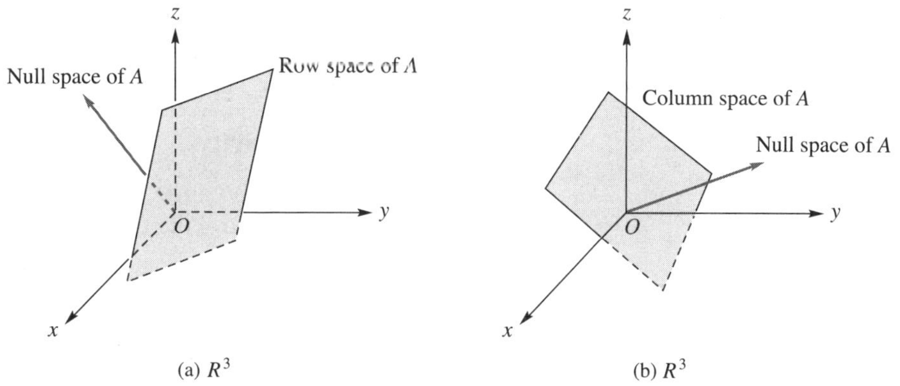

FIGURE 4.33

In Chapter 5 we investigate the relationship between the subspaces associated with a matrix. However, using Example 7, we can make the following observations, which provide a way to visualize the subspaces associated with a matrix:

- Since $A\mathbf{x} = \mathbf{0}$, we have

$$\mathbf{row}_j(A)\mathbf{x} = 0, \quad j = 1, 2, 3,$$

or equivalently,

$$(\mathbf{row}_j(A))^T \cdot \mathbf{x} = 0, \quad j = 1, 2, 3.$$

- If $\mathbf{v}$ is any vector in the row space of A, then $\mathbf{v}$ is a linear combination of the rows of A. Thus we can find scalars c_1, c_2, and c_3 so that

$$\mathbf{v} = c_1\,\mathbf{row}_1(A) + c_2\,\mathbf{row}_2(A) + c_3\,\mathbf{row}_3(A).$$

It follows that

$$\mathbf{v}^T \cdot \mathbf{x} = 0 \quad \text{(verify)}.$$

In Chapter 5 we show that a geometric interpretation of this relationship between the row space of A and the null space of A is expressed by saying these two subspaces are perpendicular (orthogonal). Thus, the plane representing the row space of A and the arrow representing the null space of A intersect in a right angle in Figure 4.33(a).

- The reduced row echelon form of A^T is

$$\begin{bmatrix} 1 & 0 & 1 \\ 0 & 1 & 2 \\ 0 & 0 & 0 \end{bmatrix} \quad \text{(verify)},$$

so the row space of A^T is also a plane in R^3 passing through the origin. (It follows that the row space of A^T is the column space of A.) The null space of A^T consists of all vectors $\mathbf{x}$ of the form (verify)

$$\mathbf{x} = \begin{bmatrix} -r \\ -2r \\ r \end{bmatrix},$$

where r is an arbitrary constant. As in the discussion for the matrix A, the null space of A^T is perpendicular to the row space of A^T. See Figure 4.33(b).

- Observe that the row and null space associated with A are different than those associated with A^T.

Rank and Singularity

The rank of a square matrix can be used to determine whether the matrix is singular or nonsingular. We first prove the following theorem:

Theorem 4.20 If A is an $n \times n$ matrix, then rank $A = n$ if and only if A is row equivalent to I_n.

Proof

If rank $A = n$, then A is row equivalent to a matrix B in reduced row echelon form, and rank $B = n$. Since rank $B = n$, we conclude that B has no zero rows, and this implies (by Exercise 9 of Section 2.1) that $B = I_n$. Hence, A is row equivalent to I_n.

Conversely, if A is row equivalent to I_n, then rank A = rank $I_n = n$. ■

Corollary 4.7 A is nonsingular if and only if rank $A = n$.

Proof

This follows from Theorem 4.20 and Corollary 2.2. ■

From a practical point of view, this result is not too useful, since most of the time we want to know not only whether A is nonsingular, but also its inverse. The method developed in Chapter 1 enables us to find A^{-1}, if it exists, and tells us if it does not exist. Thus we do not have to learn first if A^{-1} exists and then go through another procedure to obtain it.

Corollary 4.8 If A is an $n \times n$ matrix, then rank $A = n$ if and only if $\det(A) \neq 0$.

Proof

By Corollary 4.7 we know that A is nonsingular if and only if rank $A = n$; and by Theorem 3.8, A is nonsingular if and only if $\det(A) \neq 0$. ■

This result also is not very useful from a computational point of view, since it is simpler to find rank A directly than by computing $\det(A)$.

Corollary 4.9 The homogeneous system $A\mathbf{x} = \mathbf{0}$, where A is $n \times n$, has a nontrivial solution if and only if rank $A < n$.

Proof

This follows from Corollary 4.7 and from the fact that $A\mathbf{x} = \mathbf{0}$ has a nontrivial solution if and only if A is singular (Theorem 2.9). ■

Corollary 4.10 Let A be an $n \times n$ matrix. The linear system $A\mathbf{x} = \mathbf{b}$ has a unique solution for every $n \times 1$ matrix $\mathbf{b}$ if and only if rank $A = n$.

Proof

Exercise 43. ■

Let $S = \{\mathbf{v}_1, \mathbf{v}_2, \ldots, \mathbf{v}_n\}$ be a set of n vectors in R_n, and let A be the matrix whose jth row is $\mathbf{v}_j$. It can be shown (Exercise 37) that S is linearly independent if and only if rank $A = n$ and if and only if $\det(A) \neq 0$. Similarly, let $S = \{\mathbf{v}_1, \mathbf{v}_2, \ldots, \mathbf{v}_n\}$ be a set of n vectors in R^n, and let A be the matrix whose jth column is $\mathbf{v}_j$. It can then be shown (Exercise 38) that S is linearly independent if and only if rank $A = n$ and if and only if $\det(A) \neq 0$. Moreover, it follows (Exercise 49) that a set $S = \{\mathbf{v}_1, \mathbf{v}_2, \ldots, \mathbf{v}_n\}$ of vectors in R^n (R_n) spans R^n (R_n) if and only if the rank of the matrix A whose jth column (jth row) is $\mathbf{v}_j$ if and only if $\det(A) \neq 0$.

EXAMPLE 8 Is $S = \{\begin{bmatrix}1 & 2 & 3\end{bmatrix}, \begin{bmatrix}0 & 1 & 2\end{bmatrix}, \begin{bmatrix}3 & 0 & -1\end{bmatrix}\}$ a linearly independent set of vectors in R_3?

Solution

We form the matrix A whose rows are the vectors in S:

$$A = \begin{bmatrix} 1 & 2 & 3 \\ 0 & 1 & 2 \\ 3 & 0 & -1 \end{bmatrix}.$$

Since $\det(A) = 3$ (verify), we conclude that S is linearly independent. The result also follows from the observation that $\det(A) = 2$. ■

EXAMPLE 9

To find out if $S = \{t^2 + t, t + 1, t - 1\}$ is a basis for P_2, we note that P_2 is a three-dimensional vector space isomorphic to R^3 under the mapping $L\colon P_2 \to R^3$ defined by $L(at^2 + bt + c) = \begin{bmatrix} a \\ b \\ c \end{bmatrix}$. Therefore S is a basis for P_2 if and only if $T = \{L(t^2+t), L(t+1), L(t-1)\}$ is a basis for R^3. To decide whether this is so, proceed as follows. Let A be the matrix whose columns are $L(t^2 + t)$, $L(t + 1)$, $L(t - 1)$, respectively. Now

$$L(t^2+t) = \begin{bmatrix} 1 \\ 1 \\ 0 \end{bmatrix}, \quad L(t+1) = \begin{bmatrix} 0 \\ 1 \\ 1 \end{bmatrix}, \quad \text{and} \quad L(t-1) = \begin{bmatrix} 0 \\ 1 \\ -1 \end{bmatrix},$$

so

$$A = \begin{bmatrix} 1 & 0 & 0 \\ 1 & 1 & 1 \\ 0 & 1 & -1 \end{bmatrix}.$$

Since rank $A = 3$ (verify) or $\det(A) = -2$ (verify), we conclude that T is linearly independent. Hence S is linearly independent, and since $\dim P_2 = 3$, S is a basis for P_2. ■

■ Applications of Rank to the Linear System $A\mathbf{x} = \mathbf{b}$

In Corollary 4.9 we have seen that the rank of A provides us with information about the existence of a nontrivial solution to the homogeneous system $A\mathbf{x} = \mathbf{0}$. We now obtain some results that use the rank of A to provide information about the solutions to the linear system $A\mathbf{x} = \mathbf{b}$, where $\mathbf{b}$ is an arbitrary $n \times 1$ matrix. When $\mathbf{b} \neq \mathbf{0}$, the linear system is said to be **nonhomogeneous**.

Theorem 4.21 The linear system $A\mathbf{x} = \mathbf{b}$ has a solution if and only if rank A = rank $\left[A \,\vdots\, \mathbf{b}\right]$, that is, if and only if the ranks of the coefficient and augmented matrices are equal.

Proof

First, observe that if $A = \left[a_{ij}\right]$ is $m \times n$, then the given linear system may be written as

$$x_1 \begin{bmatrix} a_{11} \\ a_{21} \\ \vdots \\ a_{m1} \end{bmatrix} + x_2 \begin{bmatrix} a_{12} \\ a_{22} \\ \vdots \\ a_{m2} \end{bmatrix} + \cdots + x_n \begin{bmatrix} a_{1n} \\ a_{2n} \\ \vdots \\ a_{mn} \end{bmatrix} = \begin{bmatrix} b_1 \\ b_2 \\ \vdots \\ b_m \end{bmatrix}. \tag{3}$$

Suppose now that $A\mathbf{x} = \mathbf{b}$ has a solution. Then there exist values of $x_1, x_2, \ldots, x_n$ that satisfy Equation (3). Thus $\mathbf{b}$ is a linear combination of the columns of A and so belongs to the column space of A. Hence rank A = rank $\left[A \mid \mathbf{b}\right]$.

Conversely, suppose that rank A = rank $\left[A \mid \mathbf{b}\right]$. Then $\mathbf{b}$ is in the column space of A, which means that we can find values of $x_1, x_2, \ldots, x_n$ that satisfy Equation (3). Hence $A\mathbf{x} = \mathbf{b}$ has a solution. ■

EXAMPLE 10 Consider the linear system

$$\begin{bmatrix} 2 & 1 & 3 \\ 1 & -2 & 2 \\ 0 & 1 & 3 \end{bmatrix} \begin{bmatrix} x_1 \\ x_2 \\ x_3 \end{bmatrix} = \begin{bmatrix} 1 \\ 2 \\ 3 \end{bmatrix}.$$

Since rank A = rank $\left[A \mid \mathbf{b}\right] = 3$ (verify), the linear system has a solution. ■

EXAMPLE 11 The linear system

$$\begin{bmatrix} 1 & 2 & 3 \\ 1 & -3 & 4 \\ 2 & -1 & 7 \end{bmatrix} \begin{bmatrix} x_1 \\ x_2 \\ x_3 \end{bmatrix} = \begin{bmatrix} 4 \\ 5 \\ 6 \end{bmatrix}$$

has no solution, because rank $A = 2$ and rank $\left[A \mid \mathbf{b}\right] = 3$ (verify). ■

The following statements are equivalent for an $n \times n$ matrix A:

1. A is nonsingular.
2. $A\mathbf{x} = \mathbf{0}$ has only the trivial solution.
3. A is row (column) equivalent to I_n.
4. For every vector $\mathbf{b}$ in R^n, the system $A\mathbf{x} = \mathbf{b}$ has a unique solution.
5. A is a product of elementary matrices.
6. $\det(A) \neq 0$.
7. The rank of A is n.
8. The nullity of A is zero.
9. The rows of A form a linearly independent set of vectors in R_n.
10. The columns of A form a linearly independent set of vectors in R^n.

Key Terms

Row space of a matrix
Column space of a matrix
Row (column) rank
Rank
Singular/nonsingular matrices
Nonhomogeneous linear system
Augmented matrix

4.9 Exercises

1. Find a basis for the subspace V of R^3 spanned by

$$S = \left\{ \begin{bmatrix} 1 \\ 2 \\ 3 \end{bmatrix}, \begin{bmatrix} 2 \\ 1 \\ 4 \end{bmatrix}, \begin{bmatrix} -1 \\ -1 \\ 2 \end{bmatrix}, \begin{bmatrix} 0 \\ 1 \\ 2 \end{bmatrix}, \begin{bmatrix} 1 \\ 1 \\ 1 \end{bmatrix} \right\}$$

and write each of the following vectors in terms of the basis vectors:

(a) $\begin{bmatrix} 3 \\ 4 \\ 12 \end{bmatrix}$ **(b)** $\begin{bmatrix} 3 \\ 2 \\ 2 \end{bmatrix}$ **(c)** $\begin{bmatrix} 1 \\ 2 \\ 6 \end{bmatrix}$

2. Find a basis for the subspace of P_3 spanned by

$$S = \{t^3 + t^2 + 2t + 1, t^3 - 3t + 1, t^2 + t + 2, t + 1, t^3 + 1\}.$$

3. Find a basis for the subspace of M_{22} spanned by

$$S = \left\{ \begin{bmatrix} 1 & 2 \\ 1 & 1 \end{bmatrix}, \begin{bmatrix} 2 & 1 \\ 3 & 1 \end{bmatrix}, \begin{bmatrix} 0 & 2 \\ 1 & 2 \end{bmatrix}, \begin{bmatrix} 3 & 2 \\ 1 & 4 \end{bmatrix}, \begin{bmatrix} 5 & 0 \\ 0 & -1 \end{bmatrix} \right\}.$$

4. Find a basis for the subspace of R_2 spanned by

$$S = \{[1 \quad 2], [2 \quad 3], [3 \quad 1], [-4 \quad 3]\}.$$

In Exercises 5 and 6, find a basis for the row space of A consisting of vectors that (a) are not necessarily row vectors of A; and (b) are row vectors of A.

5. $A = \begin{bmatrix} 1 & 2 & -1 \\ 1 & 9 & -1 \\ -3 & 8 & 3 \\ -2 & 3 & 2 \end{bmatrix}$

6. $A = \begin{bmatrix} 1 & 2 & -1 & 3 \\ 3 & 5 & 2 & 0 \\ 0 & 1 & 2 & 1 \\ -1 & 0 & -2 & 7 \end{bmatrix}$

In Exercises 7 and 8, find a basis for the column space of A consisting of vectors that (a) are not necessarily column vectors of A; and (b) are column vectors of A.

7. $A = \begin{bmatrix} 1 & -2 & 7 & 0 \\ 1 & -1 & 4 & 0 \\ 3 & 2 & -3 & 5 \\ 2 & 1 & -1 & 3 \end{bmatrix}$

8. $A = \begin{bmatrix} -2 & 2 & 3 & 7 & 1 \\ -2 & 2 & 4 & 8 & 0 \\ -3 & 3 & 2 & 8 & 4 \\ 4 & -2 & 1 & -5 & -7 \end{bmatrix}$

In Exercises 9 and 10, find the row and column ranks of the given matrices.

9. **(a)** $\begin{bmatrix} 1 & 2 & 3 & 2 & 1 \\ 3 & 1 & -5 & -2 & 1 \\ 7 & 8 & -1 & 2 & 5 \end{bmatrix}$

(b) $\begin{bmatrix} 1 & 3 & 2 & 0 & 0 & 1 \\ 2 & 1 & -5 & 1 & 2 & 0 \\ 3 & 2 & 5 & 1 & -2 & 1 \\ 5 & 8 & 9 & 1 & -2 & 2 \\ 9 & 9 & 4 & 2 & 0 & 2 \end{bmatrix}$

10. **(a)** $\begin{bmatrix} 1 & 2 & 3 & 2 & 1 \\ 0 & 5 & 4 & 0 & -1 \\ 2 & -1 & 2 & 4 & 3 \end{bmatrix}$

(b) $\begin{bmatrix} 1 & 1 & -1 & 2 & 0 \\ 2 & -4 & 0 & 1 & 1 \\ 5 & -1 & -3 & 7 & 1 \\ 3 & -9 & 1 & 0 & 2 \end{bmatrix}$

11. Let A be an $m \times n$ matrix in row echelon form. Prove that rank $A =$ the number of nonzero rows of A.

12. For each of the following matrices, verify Theorem 4.18 by computing the row and column ranks:

(a) $\begin{bmatrix} 1 & 2 & 3 \\ -1 & 2 & 1 \\ 3 & 1 & 2 \end{bmatrix}$

(b) $\begin{bmatrix} 1 & -2 & -1 \\ 2 & -1 & 3 \\ 7 & -8 & 3 \end{bmatrix}$ **(c)** $\begin{bmatrix} 1 & -2 & -1 \\ 2 & -1 & 3 \\ 7 & -8 & 3 \\ 5 & -7 & 0 \end{bmatrix}$

In Exercises 13 and 14, compute the rank and nullity of each given matrix and verify Theorem 4.19.

13. **(a)** $\begin{bmatrix} 1 & -1 & 2 & 3 \\ 2 & 6 & -8 & 1 \\ 5 & 3 & -2 & 10 \end{bmatrix}$

(b) $\begin{bmatrix} 1 & 2 & 0 & 3 \\ 3 & 2 & -1 & 0 \\ 2 & -1 & 0 & 1 \end{bmatrix}$

14. (a) $\begin{bmatrix} 1 & 3 & -2 & 4 \\ -1 & 4 & -5 & 10 \\ 3 & 2 & 1 & -2 \\ 3 & -5 & 8 & -16 \end{bmatrix}$

(b) $\begin{bmatrix} 1 & 1 & 1 & 1 \\ 2 & -1 & 0 & 0 \\ 0 & 1 & -1 & 2 \\ 1 & 1 & -1 & 2 \end{bmatrix}$

15. Which of the following matrices are equivalent?

$$A = \begin{bmatrix} 1 & 2 & 1 & 3 \\ 2 & 1 & -4 & -5 \\ 7 & 8 & -5 & -1 \\ 10 & 14 & -2 & 3 \end{bmatrix},$$

$$B = \begin{bmatrix} 1 & 2 & 1 & 3 \\ 2 & 1 & -4 & -5 \\ 1 & 1 & 0 & 0 \\ 0 & 0 & 1 & 1 \end{bmatrix},$$

$$C = \begin{bmatrix} 1 & 5 & 1 & 3 \\ 2 & 1 & 2 & 1 \\ -3 & 0 & 1 & 0 \\ 4 & 7 & -4 & 3 \end{bmatrix},$$

$$D = \begin{bmatrix} 1 & 2 & -4 & 3 \\ 4 & 7 & -4 & 1 \\ 7 & 12 & -4 & -1 \\ 2 & 3 & 4 & -5 \end{bmatrix},$$

$$E = \begin{bmatrix} 4 & 3 & -1 & -5 \\ -2 & -6 & -7 & 10 \\ -2 & -3 & -2 & 5 \\ 0 & -6 & -10 & 10 \end{bmatrix}$$

In Exercises 16 and 17, determine which of the given linear systems are consistent by comparing the ranks of the coefficient and augmented matrices.

16. (a) $\begin{bmatrix} 1 & 2 & 5 & -2 \\ 2 & 3 & -2 & 4 \\ 5 & 1 & 0 & 2 \end{bmatrix}\begin{bmatrix} x_1 \\ x_2 \\ x_3 \\ x_4 \end{bmatrix} = \begin{bmatrix} 0 \\ 0 \\ 0 \end{bmatrix}$

(b) $\begin{bmatrix} 1 & 2 & 5 & -2 \\ 2 & 3 & -2 & 4 \\ 5 & 1 & 0 & 2 \end{bmatrix}\begin{bmatrix} x_1 \\ x_2 \\ x_3 \\ x_4 \end{bmatrix} = \begin{bmatrix} -1 \\ -13 \\ 3 \end{bmatrix}$

17. (a) $\begin{bmatrix} 1 & -2 & -3 & 4 \\ 4 & -1 & -5 & 6 \\ 2 & 3 & 1 & -2 \end{bmatrix}\begin{bmatrix} x_1 \\ x_2 \\ x_3 \\ x_4 \end{bmatrix} = \begin{bmatrix} 1 \\ 2 \\ 2 \end{bmatrix}$

(b) $\begin{bmatrix} 1 & 1 & 1 \\ 1 & -1 & 1 \\ 5 & 1 & 5 \end{bmatrix}\begin{bmatrix} x_1 \\ x_2 \\ x_3 \end{bmatrix} = \begin{bmatrix} 6 \\ 2 \\ 5 \end{bmatrix}$

In Exercises 18 and 19, use Corollary 4.7 to find which of the given matrices are nonsingular.

18. (a) $\begin{bmatrix} 1 & 2 & -3 \\ -1 & 2 & 3 \\ 0 & 8 & 0 \end{bmatrix}$

(b) $\begin{bmatrix} 1 & 2 & -3 \\ -1 & 2 & 3 \\ 0 & 1 & 1 \end{bmatrix}$

19. (a) $\begin{bmatrix} 1 & 1 & 2 \\ -1 & 3 & 4 \\ -5 & 7 & 8 \end{bmatrix}$

(b) $\begin{bmatrix} 1 & 1 & 4 & -1 \\ 1 & 2 & 3 & 2 \\ -1 & 3 & 2 & 1 \\ -2 & 6 & 12 & -4 \end{bmatrix}$

In Exercises 20 and 21, use Corollary 4.8 to find out whether rank $A = 3$ *for each given matrix.*

20. (a) $A = \begin{bmatrix} 1 & 2 & 3 \\ 2 & 1 & 0 \\ -3 & 1 & 2 \end{bmatrix}$

(b) $A = \begin{bmatrix} 1 & 3 & -4 \\ -2 & 1 & 2 \\ -9 & 15 & 0 \end{bmatrix}$

21. (a) $A = \begin{bmatrix} 1 & 0 & 1 \\ 1 & 1 & 0 \\ 2 & 1 & 0 \end{bmatrix}$

(b) $A = \begin{bmatrix} 2 & 3 & -1 \\ 1 & 2 & -2 \\ -1 & -3 & 5 \end{bmatrix}$

In Exercises 22 and 23, use Corollary 4.9 to find which of the given homogeneous systems have a nontrivial solution.

22. (a) $\begin{bmatrix} 1 & 2 & 3 \\ 0 & 1 & 0 \\ 1 & 0 & 3 \end{bmatrix}\begin{bmatrix} x_1 \\ x_2 \\ x_3 \end{bmatrix} = \begin{bmatrix} 0 \\ 0 \\ 0 \end{bmatrix}$

(b) $\begin{bmatrix} 1 & 1 & 2 & -1 \\ 1 & 3 & -1 & 2 \\ 1 & 1 & 1 & 3 \\ 1 & 2 & 1 & 1 \end{bmatrix}\begin{bmatrix} x_1 \\ x_2 \\ x_3 \\ x_4 \end{bmatrix} = \begin{bmatrix} 0 \\ 0 \\ 0 \\ 0 \end{bmatrix}$

23. **(a)** $\begin{bmatrix} 1 & 2 & -1 \\ 2 & -1 & 3 \\ 5 & -4 & 3 \end{bmatrix} \begin{bmatrix} x_1 \\ x_2 \\ x_3 \end{bmatrix} = \begin{bmatrix} 0 \\ 0 \\ 0 \end{bmatrix}$

(b) $\begin{bmatrix} 1 & 2 & 3 \\ -1 & 2 & -1 \\ 1 & 6 & 5 \end{bmatrix} \begin{bmatrix} x_1 \\ x_2 \\ x_3 \end{bmatrix} = \begin{bmatrix} 0 \\ 0 \\ 0 \end{bmatrix}$

In Exercises 24 and 25, find rank A by obtaining a matrix of the form $\begin{bmatrix} I_r & O \\ O & O \end{bmatrix}$ *that is equivalent to A.*

24. **(a)** $A = \begin{bmatrix} 1 & 1 & -2 \\ 1 & 2 & 3 \\ 0 & 1 & 3 \end{bmatrix}$

(b) $A = \begin{bmatrix} 1 & 1 & -2 & 0 & 0 \\ 1 & 2 & 3 & 6 & 7 \\ 2 & 1 & 3 & 6 & 5 \end{bmatrix}$

25. **(a)** $A = \begin{bmatrix} 1 & 1 & -2 \\ 1 & 2 & 3 \\ 3 & 4 & -1 \end{bmatrix}$

(b) $A = \begin{bmatrix} 1 & -1 & 2 & 3 \\ 2 & 2 & 0 & 1 \\ 1 & -5 & 6 & 8 \\ 4 & 0 & 4 & 6 \end{bmatrix}$

In Exercises 26 and 27, use Corollary 4.10 to determine whether the linear system $A\mathbf{x} = \mathbf{b}$ *has a unique solution for every* 3×1 *matrix* $\mathbf{b}$.

26. $A = \begin{bmatrix} 1 & 2 & -2 \\ 0 & 8 & -7 \\ 3 & -2 & 1 \end{bmatrix}$

27. $A = \begin{bmatrix} 1 & -1 & 2 \\ 3 & 2 & 3 \\ 1 & -2 & 1 \end{bmatrix}$

In Exercises 28 through 33, solve using the concept of rank.

28. Is

$$S = \left\{ \begin{bmatrix} 2 \\ 2 \\ 3 \end{bmatrix}, \begin{bmatrix} 1 \\ 0 \\ 2 \end{bmatrix}, \begin{bmatrix} 0 \\ 1 \\ 3 \end{bmatrix} \right\}$$

a linearly independent set of vectors in R^3?

29. Is

$$S = \{[4 \;\; 1 \;\; 2], [2 \;\; 5 \;\; -5], [2 \;\; -1 \;\; 3]\}$$

a linearly independent set of vectors in R_3?

30. Does the set

$$S = \left\{ \begin{bmatrix} 3 \\ 1 \\ 0 \end{bmatrix}, \begin{bmatrix} 0 \\ -1 \\ 1 \end{bmatrix}, \begin{bmatrix} 1 \\ 2 \\ -1 \end{bmatrix} \right\}$$

span R^3?

31. Is $S = \{t^3 + t + 1, 2t^2 + 3, t - 1, 2t^3 - 2t^2\}$ a basis for P_3?

32. Is

$$S = \left\{ \begin{bmatrix} 1 & 1 \\ 0 & 1 \end{bmatrix}, \begin{bmatrix} 0 & 2 \\ 1 & 3 \end{bmatrix}, \begin{bmatrix} 0 & 0 \\ -1 & 2 \end{bmatrix}, \begin{bmatrix} 0 & -2 \\ 0 & 3 \end{bmatrix} \right\}$$

a basis for M_{22}?

33. For what values of c is the set $\{t+3, 2t+c^2+2\}$ linearly independent?

34. **(a)** If A is a 3×4 matrix, what is the largest possible value for rank A?

(b) If A is a 4×6 matrix, show that the columns of A are linearly dependent.

(c) If A is a 5×3 matrix, show that the rows of A are linearly dependent.

35. Let A be a 7×3 matrix whose rank is 3.

(a) Are the rows of A linearly dependent or linearly independent? Justify your answer.

(b) Are the columns of A linearly dependent or linearly independent? Justify your answer.

36. Let A be a 3×5 matrix.

(a) Give *all* possible values for the rank of A.

(b) If the rank of A is 3, what is the dimension of its column space?

(c) If the rank of A is 3, what is the dimension of the solution space of the homogeneous system $A\mathbf{x} = \mathbf{0}$?

37. Let $S = \{\mathbf{v}_1, \mathbf{v}_2, \ldots, \mathbf{v}_n\}$ be a set of n vectors in R_n, and let A be the matrix whose jth row is $\mathbf{v}_j$. Show that S is linearly independent if and only if rank $A = n$.

38. Let $S = \{\mathbf{v}_1, \mathbf{v}_2, \ldots, \mathbf{v}_n\}$ be a set of n vectors in R^n, and let A be the matrix whose jth column is $\mathbf{v}_j$. Show that S is linearly independent if and only if rank $A = n$.

39. Let A be an $n \times n$ matrix. Show that the homogeneous system $A\mathbf{x} = \mathbf{0}$ has a nontrivial solution if and only if the columns of A are linearly dependent.

40. Let A be an $n \times n$ matrix. Show that rank $A = n$ if and only if the columns of A are linearly independent.

41. Let A be an $n \times n$ matrix. Prove that the rows of A are linearly independent if and only if the columns of A span R^n.

42. Let $S = \{\mathbf{v}_1, \mathbf{v}_2, \ldots, \mathbf{v}_k\}$ be a basis for a subspace V of R_n that is obtained by the method of Example 1. If

$$\mathbf{v} = \begin{bmatrix} a_1 & a_2 & \cdots & a_n \end{bmatrix}$$

belongs to V and the leading 1's in the reduced row echelon form from the method in Example 1 occur in columns $j_1, j_2, \ldots, j_k$, then show that

$$\mathbf{v} = a_{j_1}\mathbf{v}_1 + a_{j_2}\mathbf{v}_2 + \cdots + a_{j_k}\mathbf{v}_k.$$

43. Prove Corollary 4.10.

44. Let A be an $m \times n$ matrix. Show that the linear system $A\mathbf{x} = \mathbf{b}$ has a solution for every $m \times 1$ matrix $\mathbf{b}$ if and only if rank $A = m$.

45. Let A be an $m \times n$ matrix with $m \neq n$. Show that either the rows or the columns of A are linearly dependent.

46. Suppose that the linear system $A\mathbf{x} = \mathbf{b}$, where A is $m \times n$, is consistent (i.e., has a solution). Prove that the solution is unique if and only if rank $A = n$.

47. What can you say about the dimension of the solution space of a homogeneous system of 8 equations in 10 unknowns?

48. Is it possible that all nontrivial solutions of a homogeneous system of 5 equations in 7 unknowns be multiples of each other? Explain.

49. Show that a set $S = \{\mathbf{v}_1, \mathbf{v}_2, \ldots, \mathbf{v}_n\}$ of vectors in R^n (R_n) spans R^n (R_n) if and only if the rank of the matrix whose jth column (jth row) is $\mathbf{v}_j$ is n.

50. Determine whether your software has a command for computing the rank of a matrix. If it does, experiment with the command on matrices A in Examples 4 and 5 and Exercises 13 and 14.

51. Assuming that exact arithmetic is used, rank A is the number of nonzero rows in the reduced row echelon form of A. Compare the results by using your rank command and the reduced row echelon form approach on the following matrices:

$$A = \begin{bmatrix} 1 & 1 \\ 0 & 1 \times 10^{-j} \end{bmatrix}, \qquad j = 5, 10, 16.$$

(See Exercise 31 in Section 4.5.)

Supplementary Exercises

1. Let $C[a, b]$ denote the set of all real-valued continuous functions defined on $[a, b]$. If f and g are in $C[a, b]$, we define $f \oplus g$ by $(f \oplus g)(t) = f(t) + g(t)$, for t in $[a, b]$. If f is in $C[a, b]$ and c is a scalar, we define $c \odot f$ by $(c \odot f)(t) = cf(t)$, for t in $[a, b]$.
   (a) Show that $C[a, b]$ is a real vector space.
   (b) Let $W(k)$ be the set of all functions in $C[a, b]$ with $f(a) = k$. For what values of k will $W(k)$ be a subspace of $C[a, b]$?
   (c) Let $t_1, t_2, \ldots, t_n$ be a fixed set of points in $[a, b]$. Show that the subset of all functions f in $C[a, b]$ that have roots at $t_1, t_2, \ldots, t_n$, that is, $f(t_i) = 0$ for $i = 1, 2, \ldots, n$, forms a subspace.

2. In R^4, let W be the subset of all vectors

$$\mathbf{v} = \begin{bmatrix} a_1 \\ a_2 \\ a_3 \\ a_4 \end{bmatrix}$$

that satisfy $a_4 - a_3 = a_2 - a_1$.
   (a) Show that W is a subspace of R^4.
   (b) Show that

$$S = \left\{ \begin{bmatrix} 1 \\ 0 \\ 0 \\ -1 \end{bmatrix}, \begin{bmatrix} 0 \\ 1 \\ 0 \\ 1 \end{bmatrix}, \begin{bmatrix} 1 \\ 1 \\ 1 \\ 1 \end{bmatrix}, \begin{bmatrix} 0 \\ 0 \\ 1 \\ 1 \end{bmatrix} \right\}$$

   spans W.
   (c) Find a subset of S that is a basis for W.
   (d) Express $\mathbf{v} = \begin{bmatrix} 0 \\ 4 \\ 2 \\ 6 \end{bmatrix}$ as a linear combination of the basis obtained in part (c).

3. Consider

$$\mathbf{v}_1 = \begin{bmatrix} 1 \\ 1 \\ -2 \\ 1 \end{bmatrix}, \quad \mathbf{v}_2 = \begin{bmatrix} 1 \\ 5 \\ 2 \\ -1 \end{bmatrix}, \quad \text{and} \quad \mathbf{v}_3 = \begin{bmatrix} 3 \\ 0 \\ 2 \\ 1 \end{bmatrix}.$$

Determine whether each vector $\mathbf{v}$ belongs to span $\{\mathbf{v}_1, \mathbf{v}_2, \mathbf{v}_3\}$.

(a) $\mathbf{v} = \begin{bmatrix} 0 \\ 7 \\ 4 \\ 0 \end{bmatrix}$ **(b)** $\mathbf{v} = \begin{bmatrix} 5 \\ 6 \\ 2 \\ 1 \end{bmatrix}$ **(c)** $\mathbf{v} = \begin{bmatrix} 3 \\ -8 \\ -6 \\ 5 \end{bmatrix}$

4. Let A be a fixed $n \times n$ matrix and let the set of all $n \times n$ matrices B such that $AB = BA$ be denoted by $C(A)$. Is $C(A)$ a subspace of M_{nn}?

5. Let W and U be subspaces of vector space V.

(a) Show that $W \cup U$, the set of all vectors $\mathbf{v}$ that are either in W or in U, is not always a subspace of V.

(b) When is $W \cup U$ a subspace of V?

(c) Show that $W \cap U$, the set of all vectors $\mathbf{v}$ that are in both W and U, is a subspace of V.

6. Prove that a subspace W of R^3 coincides with R^3 if and only if it contains the vectors $\begin{bmatrix} 1 \\ 0 \\ 0 \end{bmatrix}$, $\begin{bmatrix} 0 \\ 1 \\ 0 \end{bmatrix}$, and $\begin{bmatrix} 0 \\ 0 \\ 1 \end{bmatrix}$.

7. Let A be a fixed $m \times n$ matrix and define W to be the subset of all $m \times 1$ matrices $\mathbf{b}$ in R^m for which the linear system $A\mathbf{x} = \mathbf{b}$ has a solution.

(a) Is W a subspace of R^m?

(b) What is the relationship between W and the column space of A?

8. Consider vector space R_2.

(a) For what values of m and b will all vectors of the form $\begin{bmatrix} x & mx + b \end{bmatrix}$ be a subspace of R_2?

(b) For what value of r will the set of all vectors of the form $\begin{bmatrix} x & rx^2 \end{bmatrix}$ be a subspace of R_2?

9. Let W be a nonempty subset of a vector space V. Prove that W is a subspace of V if and only if $r\mathbf{u} + s\mathbf{v}$ is in W for any vectors $\mathbf{u}$ and $\mathbf{v}$ in W and any scalars r and s.

10. Let A be an $n \times n$ matrix and λ a scalar. Show that the set W consisting of all vectors $\mathbf{x}$ in R^n such that $A\mathbf{x} = \lambda\mathbf{x}$ is a subspace of R^n.

11. For what values of a is the vector $\begin{bmatrix} a^2 \\ -3a \\ -2 \end{bmatrix}$ in

$$\text{span}\left\{ \begin{bmatrix} 1 \\ 2 \\ 3 \end{bmatrix}, \begin{bmatrix} 0 \\ 1 \\ 1 \end{bmatrix}, \begin{bmatrix} 1 \\ 3 \\ 4 \end{bmatrix} \right\}?$$

12. For what values of a is the vector $\begin{bmatrix} a^2 \\ a \\ 1 \end{bmatrix}$ in

$$\text{span}\left\{ \begin{bmatrix} 1 \\ 2 \\ 3 \end{bmatrix}, \begin{bmatrix} 1 \\ 1 \\ 1 \end{bmatrix}, \begin{bmatrix} 0 \\ 1 \\ 2 \end{bmatrix} \right\}?$$

13. For what values of k will the set S form a basis for R^6?

$$S = \left\{ \begin{bmatrix} 1 \\ 2 \\ -2 \\ 1 \\ 1 \\ 1 \end{bmatrix}, \begin{bmatrix} 0 \\ 2 \\ 0 \\ 0 \\ 0 \\ 0 \end{bmatrix}, \begin{bmatrix} 0 \\ 5 \\ k \\ 1 \\ 0 \\ 0 \end{bmatrix}, \begin{bmatrix} 0 \\ 2 \\ 1 \\ k \\ 0 \\ 0 \end{bmatrix}, \begin{bmatrix} 0 \\ 1 \\ -2 \\ 4 \\ -3 \\ 1 \end{bmatrix}, \begin{bmatrix} 0 \\ 3 \\ 1 \\ 1 \\ 1 \\ 0 \end{bmatrix} \right\}$$

14. Consider the subspace of R^4 given by

$$W = \text{span}\left\{ \begin{bmatrix} 1 \\ 1 \\ 1 \\ 1 \end{bmatrix}, \begin{bmatrix} 3 \\ 2 \\ 1 \\ 1 \end{bmatrix}, \begin{bmatrix} 1 \\ 2 \\ 3 \\ 1 \end{bmatrix}, \begin{bmatrix} 0 \\ 2 \\ 4 \\ 1 \end{bmatrix} \right\}.$$

(a) Determine a subset S of the spanning set that is a basis for W.

(b) Find a basis T for W that is not a subset of the spanning set.

(c) Find the coordinate vector of $\mathbf{v} = \begin{bmatrix} 3 \\ -1 \\ -5 \\ 0 \end{bmatrix}$ with respect to each of the bases from parts (a) and (b).

15. Prove that if $S = \{\mathbf{v}_1, \mathbf{v}_2, \ldots, \mathbf{v}_k\}$ is a basis for a subspace W of vector space V, then there is a basis for V that includes the set S. (*Hint*: Use Theorem 4.11.)

16. Let $V = \text{span}\,\{\mathbf{v}_1, \mathbf{v}_2\}$, where

$$\mathbf{v}_1 = \begin{bmatrix} 1 \\ 0 \\ 2 \end{bmatrix} \quad \text{and} \quad \mathbf{v}_2 = \begin{bmatrix} 1 \\ 1 \\ 1 \end{bmatrix}.$$

Find a basis S for R^3 that includes $\mathbf{v}_1$ and $\mathbf{v}_2$. (*Hint*: Use the technique developed in the Alternative Constructive Proof of Theorem 4.9.)

17. Describe the set of all vectors $\mathbf{b}$ in R^3 for which the linear system $A\mathbf{x} = \mathbf{b}$ is consistent.

(a) $A = \begin{bmatrix} 1 & -2 & 1 & 0 \\ 2 & 1 & 1 & 2 \\ 1 & -7 & 2 & -2 \end{bmatrix}$

(b) $A = \begin{bmatrix} 1 & 2 & 1 \\ 1 & 3 & 1 \\ 2 & 4 & 3 \end{bmatrix}$

18. Find a basis for the solution space of the homogeneous system $(\lambda I_3 - A)\mathbf{x} = \mathbf{0}$ for each given scalar λ and given matrix A.

(a) $\lambda = 1, A = \begin{bmatrix} 0 & 0 & 1 \\ 1 & 0 & -3 \\ 0 & 1 & 3 \end{bmatrix}$

(b) $\lambda = 3, A = \begin{bmatrix} 1 & 1 & -2 \\ -1 & 2 & 1 \\ 0 & 1 & -1 \end{bmatrix}$

19. Show that rank A = rank A^T for any $m \times n$ matrix A.

20. Let A and B be $m \times n$ matrices that are row equivalent.

(a) Prove that rank A = rank B.

(b) Prove that for $\mathbf{x}$ in R^n, $A\mathbf{x} = \mathbf{0}$ if and only if $B\mathbf{x} = \mathbf{0}$.

21. Let A be $m \times n$ and B be $n \times k$.

(a) Prove that $\text{rank}(AB) \leq \min\{\text{rank } A, \text{rank } B\}$.

(b) Find A and B such that $\text{rank}(AB) < \min\{\text{rank } A, \text{rank } B\}$.

(c) If $k = n$ and B is nonsingular, prove that $\text{rank}(AB) = \text{rank } A$.

(d) If $m = n$ and A is nonsingular, prove that $\text{rank}(AB) = \text{rank } B$.

(e) For nonsingular matrices P and Q, what is $\text{rank}(PAQ)$?

22. For an $m \times n$ matrix A, let the set of all vectors $\mathbf{x}$ in R^n such that $A\mathbf{x} = \mathbf{0}$ be denoted by NS(A), which in Example 10 of Section 4.3 has been shown to be a subspace of R^n, called the null space of A.

(a) Prove that rank $A + \dim \text{NS}(A) = n$.

(b) For $m = n$, prove that A is nonsingular if and only if $\dim \text{NS}(A) = 0$.

23. Let A be an $m \times n$ matrix and B a nonsingular $m \times m$ matrix. Prove that $\text{NS}(BA) = \text{NS}(A)$. (See Exercise 22.)

24. Find $\dim \text{NS}(A)$ (see Exercise 22) for each of the following matrices:

(a) $A = \begin{bmatrix} 1 & 2 & 1 \\ 2 & 0 & 3 \\ 0 & 4 & -1 \end{bmatrix}$

(b) $A = \begin{bmatrix} 1 & 2 & 4 & 1 \\ 2 & 1 & 1 & 1 \\ -1 & 4 & 10 & 1 \end{bmatrix}$

25. Any nonsingular 3×3 matrix P represents a transition matrix from some ordered basis $T = \{\mathbf{w}_1, \mathbf{w}_2, \mathbf{w}_3\}$ to some other ordered basis $S = \{\mathbf{v}_1, \mathbf{v}_2, \mathbf{v}_3\}$. Let

$$P = \begin{bmatrix} 1 & 1 & 1 \\ 0 & 2 & 1 \\ 1 & 0 & 3 \end{bmatrix}.$$

(a) If $\mathbf{v}_1 = \begin{bmatrix} 1 \\ 1 \\ 0 \end{bmatrix}$, $\mathbf{v}_2 = \begin{bmatrix} 1 \\ 0 \\ 1 \end{bmatrix}$, and $\mathbf{v}_3 = \begin{bmatrix} 1 \\ 1 \\ 1 \end{bmatrix}$, find T.

(b) If $\mathbf{w}_1 = \begin{bmatrix} 1 \\ 0 \\ 2 \end{bmatrix}$, $\mathbf{w}_2 = \begin{bmatrix} 2 \\ 0 \\ 1 \end{bmatrix}$, and $\mathbf{w}_3 = \begin{bmatrix} 1 \\ 1 \\ 0 \end{bmatrix}$, find S.

26. In Supplementary Exercises 30 through 32 for Chapter 1, we defined the outer product of two $n \times 1$ column matrices X and Y as XY^T. Determine the rank of an outer product.

27. Suppose that A is an $n \times n$ matrix and that there is no nonzero vector $\mathbf{x}$ in R^n such that $A\mathbf{x} = \mathbf{x}$. Show that $A - I_n$ is nonsingular.

28. Let A be an $m \times n$ matrix. Prove that if A^TA is nonsingular, then rank $A = n$.

29. Prove or find a counterexample to disprove each of the following:

(a) $\text{rank}(A + B) \leq \max\{\text{rank } A, \text{rank } B\}$

(b) $\text{rank}(A + B) \geq \min\{\text{rank } A, \text{rank } B\}$

(c) $\text{rank}(A + B) = \text{rank } A + \text{rank } B$

30. Let A be an $n \times n$ matrix and $\{\mathbf{v}_1, \mathbf{v}_2, \ldots, \mathbf{v}_k\}$ a linearly dependent set of vectors in R^n. Are $A\mathbf{v}_1, A\mathbf{v}_2, \ldots, A\mathbf{v}_k$ linearly dependent or linearly independent vectors in R^n? Justify your answer.

31. Let A be an $m \times n$ matrix. Show that the linear system $A\mathbf{x} = \mathbf{b}$ has at most one solution for every $m \times 1$ matrix $\mathbf{b}$ if and only if the associated homogeneous system $A\mathbf{x} = \mathbf{0}$ has only the trivial solution.

32. Let A be an $m \times n$ matrix. Show that the linear system $A\mathbf{x} = \mathbf{b}$ has at most one solution for every $m \times 1$ matrix $\mathbf{b}$ if and only if the columns of A are linearly independent.

33. What can you say about the solutions to the consistent nonhomogeneous linear system $A\mathbf{x} = \mathbf{b}$ if the rank of A is less than the number of unknowns?

34. Let W_1 and W_2 be subspaces of a vector space V. Let $W_1 + W_2$ be the set of all vectors $\mathbf{v}$ in V such that $\mathbf{v} = \mathbf{w}_1 + \mathbf{w}_2$, where $\mathbf{w}_1$ is in W_1 and $\mathbf{w}_2$ is in W_2. Show that $W_1 + W_2$ is a subspace of V.

35. Let W_1 and W_2 be subspaces of a vector space V with $W_1 \cap W_2 = \{\mathbf{0}\}$. Let $W_1 + W_2$ be as defined in Exercise 34. Suppose that $V = W_1 + W_2$. Prove that every vector in V can be uniquely written as $\mathbf{w}_1 + \mathbf{w}_2$, where $\mathbf{w}_1$ is in W_1 and $\mathbf{w}_2$ is in W_2. In this case we write $V = W_1 \oplus W_2$ and say that V is the **direct sum** of the subspaces W_1 and W_2.

36. Let $S = \{\mathbf{v}_1, \mathbf{v}_2, \ldots, \mathbf{v}_k\}$ be a set of vectors in a vector space V, and let W be a subspace of V containing S. Show that W contains span S.

Chapter Review

True or False

1. In the accompanying figure, $\mathbf{w} = \mathbf{u} - \mathbf{v}$.

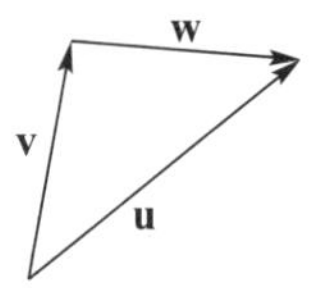

2. If V is a real vector space, then for every vector $\mathbf{u}$ in V, the scalar 0 times $\mathbf{u}$ gives the zero vector in V.
3. Let V be the vector space R^2. Then the set of all vectors with head and tail on the line $y = 2x + 1$ is a subspace of R^2.
4. If W is a subspace of the vector space V, then every linear combination of vectors from V is also in W.
5. The span of a set S is the same as the set of all linear combinations of the vectors in S.
6. Every subspace of R^3 contains infinitely many vectors.
7. If A is an $m \times n$ matrix, then the set of all solutions to $A\mathbf{x} = \mathbf{0}$ is a subspace of R^n.
8. If the set S spans a subspace W of a vector space V, then every vector in W can be written as a linear combination of the vectors from S.
9. Two vectors are linearly dependent, provided that one is a scalar multiple of the other.
10. Any subset of a linearly independent set is linearly dependent.
11. Any set containing the zero vector is linearly independent.
12. A basis for a vector space V is a linearly independent set that spans V.
13. R^n contains infinitely many vectors, so we say it is infinite-dimensional.
14. Two bases for a subspace of R^n must contain the same number of vectors.
15. Every set that spans a subspace contains a basis for the subspace.
16. The dimension of the null space of matrix A is the number of arbitrary constants in the solution to the linear system $A\mathbf{x} = \mathbf{0}$.
17. If a nonhomogeneous 3×3 linear system $A\mathbf{x} = \mathbf{b}$ has a solution $\mathbf{x} = \mathbf{x}_h + \mathbf{x}_p$ where $\mathbf{x}_h$ contains two arbitrary constants, then the set of solutions is a plane that does not go through the origin.
18. If $S = \{\mathbf{v}_1, \mathbf{v}_2, \mathbf{v}_3\}$ is a basis for a vector space V and $\mathbf{u}$ in V has
$$\left[\mathbf{u}\right]_S = \begin{bmatrix} k_1 \\ k_2 \\ k_3 \end{bmatrix},$$
then $\mathbf{u} = k_1\mathbf{v}_1 + k_2\mathbf{v}_2 + k_3\mathbf{v}_3$.
19. rank A is the number of zero rows in the row echelon form of A.
20. If A is 4×4 with rank $A = 4$, then $A\mathbf{x} = \mathbf{b}$ has exactly four solutions.
21. If the $n \times n$ matrix A is singular, then rank $A \leq n - 1$.
22. If A is $n \times n$ with rank $A = n - 1$, then A is singular.

Quiz

1. Let V be the set of all 2×1 real matrices with operations
$$\begin{bmatrix} a \\ b \end{bmatrix} \oplus \begin{bmatrix} c \\ d \end{bmatrix} = \begin{bmatrix} ad \\ bc \end{bmatrix}$$
and
$$k \odot \begin{bmatrix} a \\ b \end{bmatrix} = \begin{bmatrix} ka \\ kb \end{bmatrix}, \quad \text{for } k \text{ real.}$$
Is V a vector space? Explain.
2. Let V be the set of all 2×2 real matrices with operations standard matrix addition and
$$k \odot \begin{bmatrix} a \\ b \end{bmatrix} = \begin{bmatrix} k(a+b) \\ k(a+b) \end{bmatrix}.$$
Is V a vector space? Explain.
3. Is the set of all vectors of the form $\begin{bmatrix} a \\ b \\ -a \end{bmatrix}$, where a and b are any real numbers, a subspace W of R^3? Explain. If it is, find a basis for W.
4. Is the set of all vectors of the form $\begin{bmatrix} a \\ b \\ c \end{bmatrix}$, where a, b, and c are any real numbers with $a + b + c > 0$, a subspace W of R^3? Explain. If it is, find a basis for W.
5. Let W be the set of all vectors $p(t)$ in P_2 such that $p(0) = 0$. Show that W is a subspace of P_2 and find a basis for W.

6. Let

$$S = \left\{ \begin{bmatrix} 2 \\ 1 \\ -1 \end{bmatrix}, \begin{bmatrix} 3 \\ -1 \\ 2 \end{bmatrix}, \begin{bmatrix} -2 \\ 4 \\ -6 \end{bmatrix} \right\}.$$

Determine whether span $S = R^3$.

7. Let

$$S = \left\{ \begin{bmatrix} 1 \\ 2 \\ 0 \end{bmatrix}, \begin{bmatrix} 1 \\ 0 \\ 1 \end{bmatrix}, \begin{bmatrix} 2 \\ 8 \\ -2 \end{bmatrix}, \begin{bmatrix} 3 \\ 0 \\ 0 \end{bmatrix}, \begin{bmatrix} 4 \\ 0 \\ 1 \end{bmatrix} \right\}.$$

Show that span $S = R^3$ and find a basis for R^3 consisting of vectors from S.

8. Find a basis for the null space of

$$A = \begin{bmatrix} 1 & 2 & 1 & 1 \\ 0 & 1 & 0 & 1 \\ 2 & 3 & 2 & 1 \\ 1 & 0 & 1 & -1 \end{bmatrix}.$$

9. Determine a basis for the row space of

$$\begin{bmatrix} 2 & 3 & -2 \\ 1 & -1 & 4 \\ -1 & 2 & -6 \end{bmatrix}.$$

10. If A is 3×3 with rank $A = 2$, show that the dimension of the null space of A is 1.

11. Determine the solution to the linear system $A\mathbf{x} = \mathbf{b}$ and write it in the form $\mathbf{x} = \mathbf{x}_p + \mathbf{x}_h$, for

$$A = \begin{bmatrix} 1 & 2 & -1 & 3 \\ -2 & 2 & -3 & 3 \\ 0 & 6 & -1 & 3 \\ -1 & 4 & 0 & 0 \end{bmatrix} \quad \text{and} \quad \mathbf{b} = \begin{bmatrix} 2 \\ 3 \\ 7 \\ 5 \end{bmatrix}.$$

12. Determine all values of c so that the set $\{t+3, 2t+c^2+2\}$ is linearly independent.

Discussion Exercises

1. Let a_j and b_j, $j = 1, 2, 3, 4, 5$, and c_j, d_j, and e_j, $j = 1, 2$ be any real numbers. Construct the matrix

$$A = \begin{bmatrix} a_1 & a_2 & a_3 & a_4 & a_5 \\ b_1 & b_2 & b_3 & b_4 & b_5 \\ 0 & c_1 & 0 & c_2 & 0 \\ 0 & d_1 & 0 & d_2 & 0 \\ 0 & e_1 & 0 & e_2 & 0 \end{bmatrix}.$$

Discuss how to show that $\det(A) = 0$, without actually computing the determinant.

2. Let k be any real number except 1 and -2. Show that

$$S = \left\{ \begin{bmatrix} 1 \\ 1 \\ k \end{bmatrix}, \begin{bmatrix} k \\ 1 \\ 1 \end{bmatrix}, \begin{bmatrix} 1 \\ k \\ 1 \end{bmatrix} \right\}$$

is a basis for R^3.

3. Let $S = \{\begin{bmatrix} k & 1 & 0 \end{bmatrix}, \begin{bmatrix} 1 & k & 1 \end{bmatrix}, \begin{bmatrix} 0 & 1 & k \end{bmatrix}\}$. Determine all the real values of k so that S is a linearly dependent set.

4. Let W be the set of all $n \times n$ matrices A such that $A^{-1} = A^T$. Prove or disprove that W is a subspace of M_{nn}.

5. An $n \times n$ matrix A is called **involutory** if $A^2 = I_n$. Prove or disprove that the set of all $n \times n$ involutory matrices is a subspace of M_{nn}.

6. In Section 1.5 we discussed the Fibonacci sequence and illustrated its computation by using a recurrence relation that was expressed as a matrix product

$$\mathbf{w}_{n-1} = A^{n-1}\mathbf{w}_0,$$

where

$$\mathbf{w}_0 = \begin{bmatrix} 1 \\ 1 \end{bmatrix} \quad \text{and} \quad A = \begin{bmatrix} 1 & 1 \\ 1 & 0 \end{bmatrix}.$$

Let $S = \{\mathbf{v}_1, \mathbf{v}_2\}$, where

$$\mathbf{v}_1 = \begin{bmatrix} \frac{1+\sqrt{5}}{2} \\ 1 \end{bmatrix} \quad \text{and} \quad \mathbf{v}_2 = \begin{bmatrix} \frac{1-\sqrt{5}}{2} \\ 1 \end{bmatrix}.$$

(a) Show that S is a basis for R^2.

(b) Show that $A\mathbf{v}_1$ is a scalar multiple of $\mathbf{v}_1$, $\lambda_1\mathbf{v}_1$, and determine the scalar λ_1.

(c) Show that $A\mathbf{v}_2$ is a scalar multiple of $\mathbf{v}_2$, $\lambda_2\mathbf{v}_2$, and determine the scalar λ_2.

(d) Find scalars c_1 and c_2 so that $\mathbf{w}_0 = c_1\mathbf{v}_1 + c_2\mathbf{v}_2$.

(e) Form the expression $\mathbf{w}_{n-1} = A^{n-1}\mathbf{w}_0$ by using the result of part (d) and simplify as much as possible.

(f) Use the result from part (e) to obtain an expression for u_n in terms of the powers of λ_1 and λ_2. Explain all the steps in your computation.

CHAPTER

5 Inner Product Spaces

As we noted in Chapter 4, when physicists talk about vectors in R^2 and R^3, they usually refer to objects that have magnitude and direction. However, thus far in our study of vector spaces we have refrained from discussing these notions. In this chapter we deal with magnitude and direction in a vector space.

5.1 Length and Direction in R^2 and R^3

Length

In this section we address the notions of magnitude and direction in R^2 and R^3, and in the next section we generalize these to R^n. We consider R^2 and R^3 with the usual Cartesian coordinate system. The **length**, or **magnitude**, of the vector $\mathbf{v} = \begin{bmatrix} v_1 \\ v_2 \end{bmatrix}$ in R^2, denoted by $\|\mathbf{v}\|$, is by the Pythagorean theorem (see Figure 5.1)

$$\|\mathbf{v}\| = \sqrt{v_1^2 + v_2^2}. \tag{1}$$

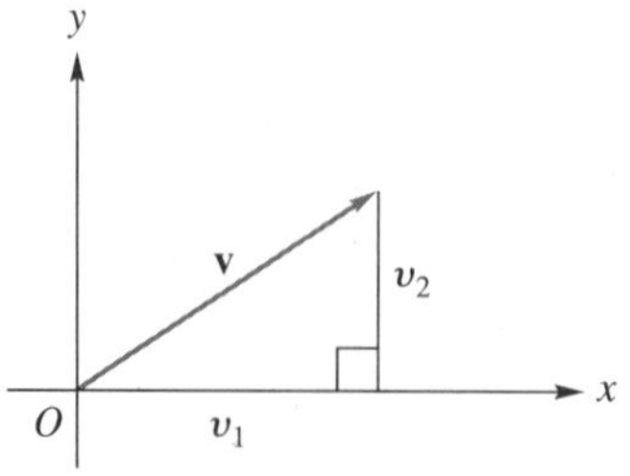

FIGURE 5.1 Length of v.

Note: This chapter may also be covered before Section 7.3, which is where the material it discusses is used.

EXAMPLE 1 If

$$\mathbf{v} = \begin{bmatrix} 2 \\ -5 \end{bmatrix},$$

then, by Equation (1),

$$\|\mathbf{v}\| = \sqrt{(2)^2 + (-5)^2} = \sqrt{4 + 25} = \sqrt{29}. \quad \blacksquare$$

Consider now the points $P_1(u_1, u_2)$ and $P_2(v_1, v_2)$, as shown in Figure 5.2(a). Applying the Pythagorean theorem to triangle P_1RP_2, we find that the distance from P_1 to P_2, the length of the line segment from P_1 to P_2, is given by

$$\sqrt{(v_1 - u_1)^2 + (v_2 - u_2)^2}.$$

If $\mathbf{u} = \begin{bmatrix} u_1 \\ u_2 \end{bmatrix}$ and $\mathbf{v} = \begin{bmatrix} v_1 \\ v_2 \end{bmatrix}$ are vectors in R^2, as shown in Figure 5.2(b), then their heads are at the points $P_1(u_1, u_2)$ and $P_2(v_1, v_2)$, respectively. We then define the distance between the vectors $\mathbf{u}$ and $\mathbf{v}$ as the distance between the points P_1 and P_2. The distance between $\mathbf{u}$ and $\mathbf{v}$ is

$$\sqrt{(v_1 - u_1)^2 + (v_2 - u_2)^2} = \|\mathbf{v} - \mathbf{u}\|, \tag{2}$$

since $\mathbf{v} - \mathbf{u} = \begin{bmatrix} v_1 - u_1 \\ v_2 - u_2 \end{bmatrix}$.

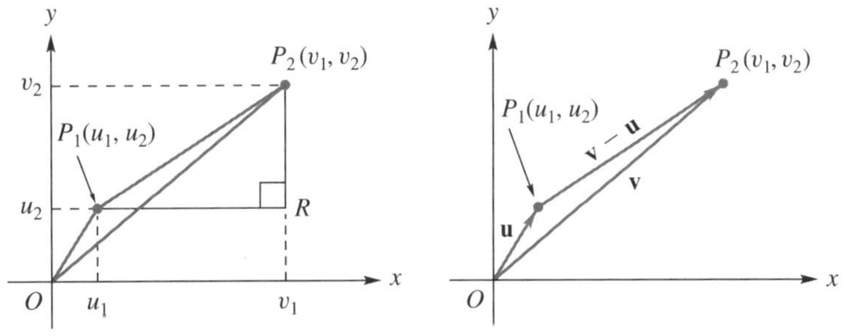

(a) Distance between the points $P_1(u_1, u_2)$ and $P_2(v_1, v_2)$.

(b) Distance between the vectors $\mathbf{u}$ and $\mathbf{v}$.

FIGURE 5.2

EXAMPLE 2 Compute the distance between the vectors

$$\mathbf{u} = \begin{bmatrix} -1 \\ 5 \end{bmatrix} \quad \text{and} \quad \mathbf{v} = \begin{bmatrix} 3 \\ 2 \end{bmatrix}.$$

Solution

By Equation (2), the distance between $\mathbf{u}$ and $\mathbf{v}$ is

$$\|\mathbf{v} - \mathbf{u}\| = \sqrt{(3 + 1)^2 + (2 - 5)^2} = \sqrt{4^2 + (-3)^2} = \sqrt{25} = 5. \quad \blacksquare$$

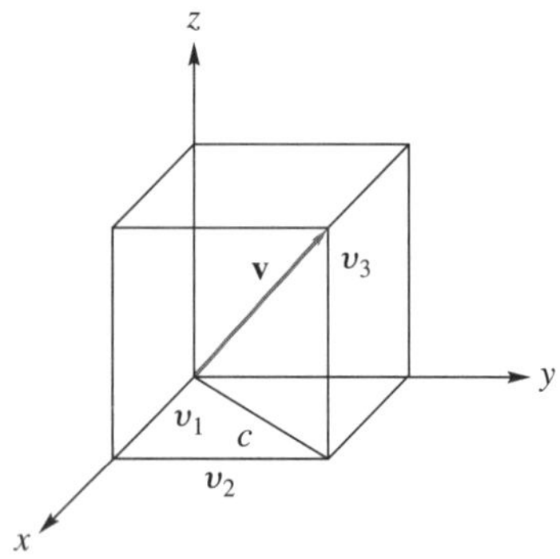

FIGURE 5.3 Length of v.

Now let $\mathbf{v} = \begin{bmatrix} v_1 \\ v_2 \\ v_3 \end{bmatrix}$ be a vector in R^3. Using the Pythagorean theorem twice (Figure 5.3), we obtain the **length** of **v**, also denoted by $\|\mathbf{v}\|$, as

$$\|\mathbf{v}\| = \sqrt{c^2 + v_3^2} = \sqrt{\left(\sqrt{v_1^2 + v_2^2}\right)^2 + v_3^2} = \sqrt{v_1^2 + v_2^2 + v_3^2}. \tag{3}$$

It follows from Equation (3) that the zero vector has length zero. It is easy to show that the zero vector is the only vector whose length is zero.

If $P_1(u_1, u_2, u_3)$ and $P_2(v_1, v_2, v_3)$ are points in R^3, then as in the case for R^2, the distance between P_1 and P_2 is given by

$$\sqrt{(v_1 - u_1)^2 + (v_2 - u_2)^2 + (v_3 - u_3)^2}.$$

Again, as in R^2, if $\mathbf{u} = \begin{bmatrix} u_1 \\ u_2 \\ u_3 \end{bmatrix}$ and $\mathbf{v} = \begin{bmatrix} v_1 \\ v_2 \\ v_3 \end{bmatrix}$ are vectors in R^3, then the distance between **u** and **v** is given by

$$\|\mathbf{v} - \mathbf{u}\| = \sqrt{(v_1 - u_1)^2 + (v_2 - u_2)^2 + (v_3 - u_3)^2}. \tag{4}$$

EXAMPLE 3

Compute the length of the vector

$$\mathbf{v} = \begin{bmatrix} 1 \\ 2 \\ 3 \end{bmatrix}.$$

Solution

By Equation (3), the length of **v** is

$$\|\mathbf{v}\| = \sqrt{1^2 + 2^2 + 3^3} = \sqrt{14}.$$ ■

EXAMPLE 4

Compute the distance between the vectors

$$\mathbf{u} = \begin{bmatrix} 1 \\ 2 \\ 3 \end{bmatrix} \quad \text{and} \quad \mathbf{v} = \begin{bmatrix} -4 \\ 3 \\ 5 \end{bmatrix}.$$

Solution

By Equation (4), the distance between **u** and **v** is

$$\|\mathbf{v} - \mathbf{u}\| = \sqrt{(-4 - 1)^2 + (3 - 2)^2 + (5 - 3)^2} = \sqrt{30}.$$ ■

■ Direction

The direction of a vector in R^2 is given by specifying its angle of inclination, or slope. The direction of a vector **v** in R^3 is specified by giving the cosines of the angles that the vector **v** makes with the positive x-, y-, and z-axes (see Figure 5.4); these are called **direction cosines**.

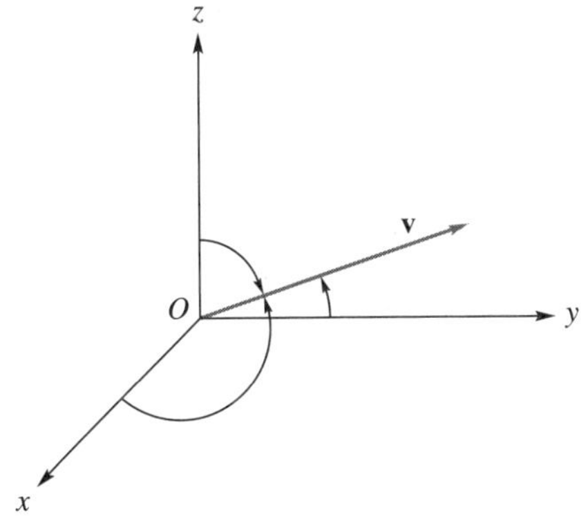

FIGURE 5.4

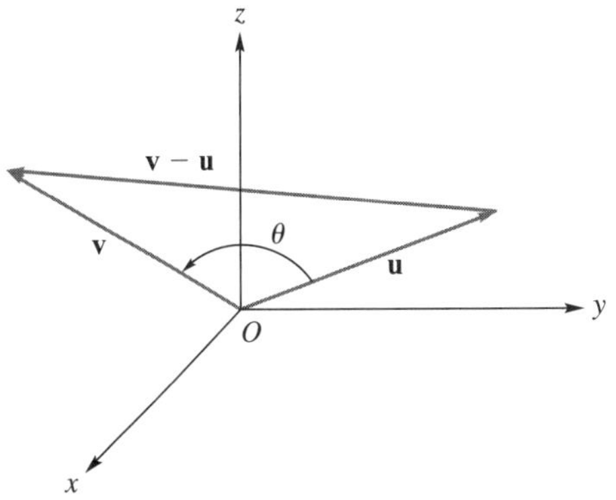

FIGURE 5.5

Instead of dealing with the special problem of finding the cosines of these angles for a vector in R^3, or the angle of inclination for a vector in R^2, we consider the more general problem of determining the **angle** θ, $0 \leq \theta \leq \pi$, between two nonzero vectors in R^2 or R^3. As shown in Figure 5.5, let

$$\mathbf{u} = \begin{bmatrix} u_1 \\ u_2 \\ u_3 \end{bmatrix} \quad \text{and} \quad \mathbf{v} = \begin{bmatrix} v_1 \\ v_2 \\ v_3 \end{bmatrix}$$

be two vectors in R^3. By the law of cosines, we have

$$\|\mathbf{v} - \mathbf{u}\|^2 = \|\mathbf{u}\|^2 + \|\mathbf{v}\|^2 - 2\|\mathbf{u}\|\,\|\mathbf{v}\| \cos\theta.$$

Hence

$$\begin{aligned} \cos\theta &= \frac{\|\mathbf{u}\|^2 + \|\mathbf{v}\|^2 - \|\mathbf{v} - \mathbf{u}\|^2}{2\|\mathbf{u}\|\,\|\mathbf{v}\|} = \frac{(u_1^2 + u_2^2 + u_3^2) + (v_1^2 + v_2^2 + v_3^2)}{2\|\mathbf{u}\|\,\|\mathbf{v}\|} \\ &\quad - \frac{(v_1 - u_1)^2 + (v_2 - u_2)^2 + (v_3 - u_3)^2}{2\|\mathbf{u}\|\,\|\mathbf{v}\|} \\ &= \frac{u_1 v_1 + u_2 v_2 + u_3 v_3}{\|\mathbf{u}\|\,\|\mathbf{v}\|}. \end{aligned}$$

Thus

$$\cos\theta = \frac{u_1 v_1 + u_2 v_2 + u_3 v_3}{\|\mathbf{u}\|\,\|\mathbf{v}\|}, \quad 0 \leq \theta \leq \pi. \tag{5}$$

In a similar way, if $\mathbf{u} = \begin{bmatrix} u_1 \\ u_2 \end{bmatrix}$ and $\mathbf{v} = \begin{bmatrix} v_1 \\ v_2 \end{bmatrix}$ are nonzero vectors in R^2 and θ is the angle between **u** and **v**, then

$$\cos\theta = \frac{u_1 v_1 + u_2 v_2}{\|\mathbf{u}\|\,\|\mathbf{v}\|}, \quad 0 \leq \theta \leq \pi. \tag{6}$$

The zero vector in R^2 or R^3 has no specific direction. The law of cosines expression given previously is true if $\mathbf{v} \neq \mathbf{0}$ and $\mathbf{u} = \mathbf{0}$ for any angle θ. Thus the zero vector can be assigned any direction.

EXAMPLE 5

Let $\mathbf{u} = \begin{bmatrix} 1 \\ 1 \\ 0 \end{bmatrix}$ and $\mathbf{v} = \begin{bmatrix} 0 \\ 1 \\ 1 \end{bmatrix}$. The angle θ between $\mathbf{u}$ and $\mathbf{v}$ is determined by

$$\cos\theta = \frac{(1)(0) + (1)(1) + (0)(1)}{\sqrt{1^2 + 1^2 + 0^2}\,\sqrt{0^2 + 1^2 + 1^2}} = \frac{1}{2}.$$

Since $0 \leq \theta \leq \pi$, $\theta = 60°$. ■

The length of a vector and the cosine of an angle between two nonzero vectors in R^2 or R^3 can be expressed in terms of the dot product, which was defined in Section 1.3. We now recall this definition from a different point of view, to anticipate a generalization of the dot product introduced in the next section.

The **standard inner product**, or **dot product**, on R^2 (R^3) is the function that assigns to each ordered pair of vectors $\mathbf{u} = \begin{bmatrix} u_1 \\ u_2 \end{bmatrix}$, $\mathbf{v} = \begin{bmatrix} v_1 \\ v_2 \end{bmatrix}$ in R^2 ($\mathbf{u} = \begin{bmatrix} u_1 \\ u_2 \\ u_3 \end{bmatrix}$, $\mathbf{v} = \begin{bmatrix} v_1 \\ v_2 \\ v_3 \end{bmatrix}$ in R^3) the number $\mathbf{u} \cdot \mathbf{v}$ defined by

$$u_1v_1 + u_2v_2 \quad \text{in } R^2$$

$$(u_1v_1 + u_2v_2 + u_3v_3 \quad \text{in } R^3).$$

Remark We have already observed in Section 1.3 that if we view the vectors $\mathbf{u}$ and $\mathbf{v}$ in R^2 or R^3 as matrices, then we can write the dot product of $\mathbf{u}$ and $\mathbf{v}$, in terms of matrix multiplication, as $\mathbf{u}^T\mathbf{v}$, where we have ignored the brackets around the 1×1 matrix $\mathbf{u}^T\mathbf{v}$. (See Exercise 41 in Section 1.3.)

If we examine Equations (1) and (3), we see that if $\mathbf{v}$ is a vector in R^2 or R^3, then

$$\|\mathbf{v}\| = \sqrt{\mathbf{v} \cdot \mathbf{v}}. \tag{7}$$

We can also write Equations (5) and (6) for the cosine of the angle θ between two nonzero vectors $\mathbf{u}$ and $\mathbf{v}$ in R^2 and R^3 as

$$\cos\theta = \frac{\mathbf{u} \cdot \mathbf{v}}{\|\mathbf{u}\|\,\|\mathbf{v}\|}, \qquad 0 \leq \theta \leq \pi. \tag{8}$$

It is shown in Section 5.3 that

$$-1 \leq \frac{\mathbf{u} \cdot \mathbf{v}}{\|\mathbf{u}\|\|\mathbf{v}\|} \leq 1.$$

It then follows that two vectors $\mathbf{u}$ and $\mathbf{v}$ in R^2 or R^3 are **orthogonal**, or **perpendicular**, if and only if $\mathbf{u} \cdot \mathbf{v} = 0$.

EXAMPLE 6

The vectors $\mathbf{u} = \begin{bmatrix} 2 \\ -4 \end{bmatrix}$ and $\mathbf{v} = \begin{bmatrix} 4 \\ 2 \end{bmatrix}$ are orthogonal, since

$$\mathbf{u} \cdot \mathbf{v} = (2)(4) + (-4)(2) = 0.$$

See Figure 5.6. ■

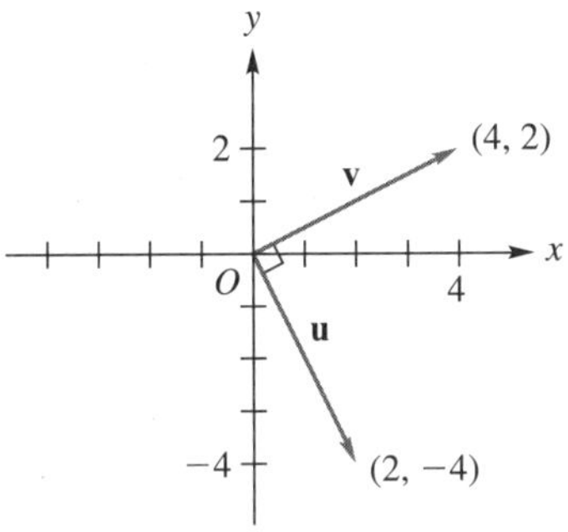

FIGURE 5.6

We note the following properties of the standard inner product on R^2 and R^3 that will motivate our next section.

Theorem 5.1 Let $\mathbf{u}$, $\mathbf{v}$, and $\mathbf{w}$ be vectors in R^2 or R^3, and let c be a scalar. The standard inner product on R^2 and R^3 has the following properties:

(a) $\mathbf{u} \cdot \mathbf{u} \geq 0$; $\mathbf{u} \cdot \mathbf{u} = 0$ if and only if $\mathbf{u} = \mathbf{0}$

(b) $\mathbf{u} \cdot \mathbf{v} = \mathbf{v} \cdot \mathbf{u}$

(c) $(\mathbf{u} + \mathbf{v}) \cdot \mathbf{w} = \mathbf{u} \cdot \mathbf{w} + \mathbf{v} \cdot \mathbf{w}$

(d) $c\mathbf{u} \cdot \mathbf{v} = c(\mathbf{u} \cdot \mathbf{v})$, for any real scalar c

Proof

Exercise 13. ■

■ Unit Vectors

A **unit vector** in R^2 or R^3 is a vector whose length is 1. If $\mathbf{x}$ is any nonzero vector, then the vector

$$\mathbf{u} = \frac{1}{\|\mathbf{x}\|}\mathbf{x}$$

is a unit vector in the direction of $\mathbf{x}$ (Exercise 34).

EXAMPLE 7

Let $\mathbf{x} = \begin{bmatrix} -3 \\ 4 \end{bmatrix}$. Then $\|\mathbf{x}\| = \sqrt{(-3)^2 + 4^2} = 5$. Hence the vector

$$\mathbf{u} = \frac{1}{5}\begin{bmatrix} -3 \\ 4 \end{bmatrix} = \begin{bmatrix} -\frac{3}{5} \\ \frac{4}{5} \end{bmatrix}$$

is a unit vector. Observe that

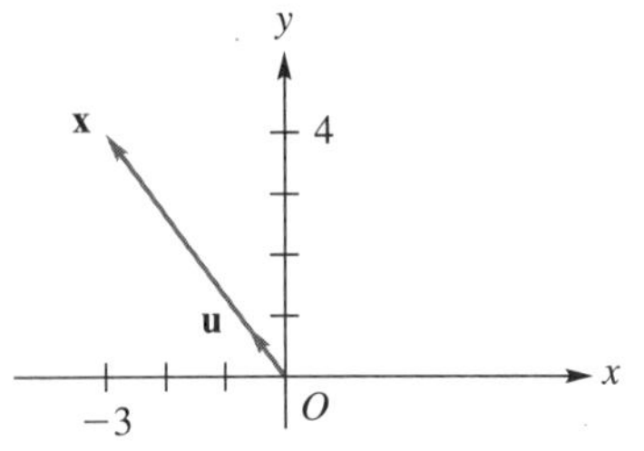

FIGURE 5.7

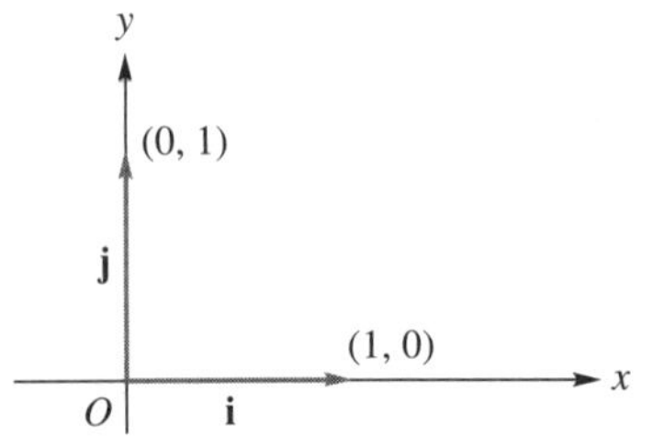

FIGURE 5.8

$$\|\mathbf{u}\| = \sqrt{\left(-\frac{3}{5}\right)^2 + \left(\frac{4}{5}\right)^2} = \sqrt{\frac{9+16}{25}} = 1.$$

Notice also that **u** points in the direction of **x** (Figure 5.7). ■

There are two unit vectors in R^2 that are of special importance. These are $\mathbf{i} = \begin{bmatrix} 1 \\ 0 \end{bmatrix}$ and $\mathbf{j} = \begin{bmatrix} 0 \\ 1 \end{bmatrix}$, the unit vectors along the positive x- and y-axes, respectively, shown in Figure 5.8. Observe that **i** and **j** are orthogonal. Since **i** and **j** form the natural basis for R^2, every vector in R^2 can be written uniquely as a linear combination of the orthogonal vectors **i** and **j**. Thus, if $\mathbf{u} = \begin{bmatrix} u_1 \\ u_2 \end{bmatrix}$ is a vector in R^2, then

$$\mathbf{u} = u_1 \begin{bmatrix} 1 \\ 0 \end{bmatrix} + u_2 \begin{bmatrix} 0 \\ 1 \end{bmatrix} = u_1\mathbf{i} + u_2\mathbf{j}.$$

Similarly, the vectors in the natural basis for R^3,

$$\mathbf{i} = \begin{bmatrix} 1 \\ 0 \\ 0 \end{bmatrix}, \quad \mathbf{j} = \begin{bmatrix} 0 \\ 1 \\ 0 \end{bmatrix}, \quad \text{and} \quad \mathbf{k} = \begin{bmatrix} 0 \\ 0 \\ 1 \end{bmatrix},$$

are unit vectors that are mutually orthogonal. Thus, if $\mathbf{u} = \begin{bmatrix} u_1 \\ u_2 \\ u_3 \end{bmatrix}$ is a vector in R^3, then $\mathbf{u} = u_1\mathbf{i} + u_2\mathbf{j} + u_3\mathbf{k}$.

■ Resultant Force and Velocity

When several forces act on a body, we can find a single force, called the **resultant force**, having an equivalent effect. The resultant force can be determined using vectors. The following example illustrates the method:

EXAMPLE 8

Suppose that a force of 12 pounds is applied to an object along the negative x-axis and a force of 5 pounds is applied to the object along the positive y-axis. Find the magnitude and direction of the resultant force.

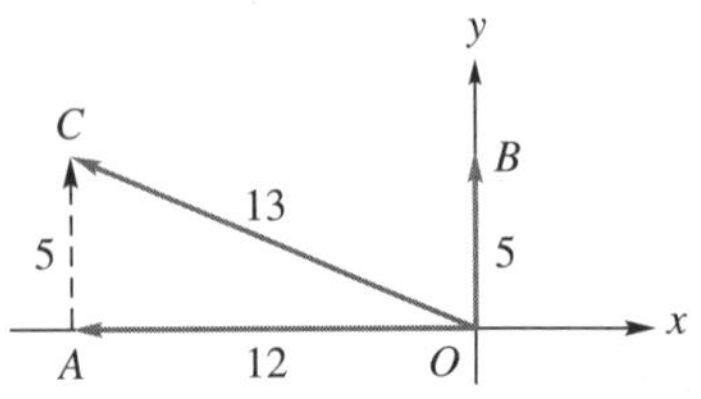

FIGURE 5.9

Solution

In Figure 5.9 we have represented the force along the negative x-axis by the vector $\overrightarrow{OA}$ and the force along the positive y-axis by the vector $\overrightarrow{OB}$. The resultant force is the vector $\overrightarrow{OC} = \overrightarrow{OA} + \overrightarrow{OB}$. Thus the magnitude of the resultant force is 13 pounds, and its direction is as indicated in the figure. ■

Vectors are also used in physics to deal with velocity problems, as the following example illustrates:

EXAMPLE 9

Suppose that a boat is traveling east across a river at a rate of 4 miles per hour while the river's current is flowing south at a rate of 3 miles per hour. Find the resultant velocity of the boat.

Solution

In Figure 5.10 we have represented the velocity of the boat by the vector $\overrightarrow{OA}$ and the velocity of the river's current by the vector $\overrightarrow{OB}$. The resultant velocity is the vector $\overrightarrow{OC} = \overrightarrow{OA} + \overrightarrow{OB}$. Thus the magnitude of the resultant velocity is 5 miles per hour, and its direction is as indicated in the figure. ■

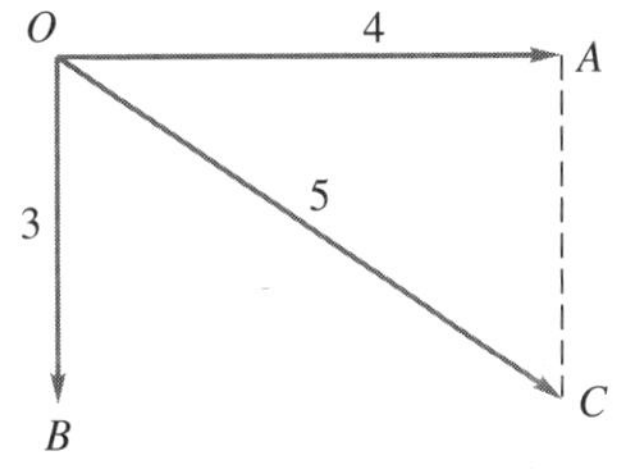

FIGURE 5.10

Key Terms

Length (magnitude) of a vector
Distance between vectors
Direction cosines
Law of cosines
Standard inner product
Dot product
Orthogonal (perpendicular) vectors
Unit vectors

5.1 Exercises

In Exercises 1 and 2, find the length of each vector.

1. **(a)** $\begin{bmatrix} 1 \\ 0 \end{bmatrix}$ **(b)** $\begin{bmatrix} 0 \\ 0 \end{bmatrix}$ **(c)** $\begin{bmatrix} 1 \\ 2 \end{bmatrix}$

2. **(a)** $\begin{bmatrix} 0 \\ -2 \\ 0 \end{bmatrix}$ **(b)** $\begin{bmatrix} -1 \\ -3 \\ -4 \end{bmatrix}$ **(c)** $\begin{bmatrix} 1 \\ -2 \\ 4 \end{bmatrix}$

In Exercises 3 and 4, compute $\|\mathbf{u} - \mathbf{v}\|$.

3. **(a)** $\mathbf{u} = \begin{bmatrix} 1 \\ 0 \end{bmatrix}, \mathbf{v} = \begin{bmatrix} 1 \\ 1 \end{bmatrix}$

(b) $\mathbf{u} = \begin{bmatrix} 0 \\ 0 \end{bmatrix}, \mathbf{v} = \begin{bmatrix} 1 \\ -1 \end{bmatrix}$

4. **(a)** $\mathbf{u} = \begin{bmatrix} 1 \\ 2 \\ 3 \end{bmatrix}, \mathbf{v} = \begin{bmatrix} 4 \\ 5 \\ 6 \end{bmatrix}$

(b) $\mathbf{u} = \begin{bmatrix} -1 \\ -2 \\ -3 \end{bmatrix}, \mathbf{v} = \begin{bmatrix} -4 \\ -5 \\ -6 \end{bmatrix}$

In Exercises 5 and 6, find the distance between **u** *and* **v**.

5. **(a)** $\mathbf{u} = \begin{bmatrix} 1 \\ 2 \end{bmatrix}, \mathbf{v} = \begin{bmatrix} -4 \\ -5 \end{bmatrix}$

(b) $\mathbf{u} = \begin{bmatrix} 1 \\ 2 \end{bmatrix}, \mathbf{v} = \begin{bmatrix} 4 \\ -5 \end{bmatrix}$

6. **(a)** $\mathbf{u} = \begin{bmatrix} -1 \\ -2 \\ -3 \end{bmatrix}, \mathbf{v} = \begin{bmatrix} 4 \\ 5 \\ 6 \end{bmatrix}$

(b) $\mathbf{u} = \begin{bmatrix} 0 \\ 1 \\ -1 \end{bmatrix}, \mathbf{v} = \begin{bmatrix} 1 \\ 2 \\ 0 \end{bmatrix}$

In Exercises 7 and 8, determine all values of c so that each given condition is satisfied.

7. $\|\mathbf{u}\| = 3$ for $\mathbf{u} = \begin{bmatrix} 2 \\ c \\ 0 \end{bmatrix}$

8. $\|\mathbf{u}\| = 1$ for $\mathbf{u} = \begin{bmatrix} \frac{1}{c} \\ \frac{2}{c} \\ -\frac{2}{c} \end{bmatrix}$

9. For each pair of vectors **u** and **v** in Exercise 5, find the cosine of the angle θ between **u** and **v**.

10. For each pair of vectors in Exercise 6, find the cosine of the angle θ between **u** and **v**.

11. For each of the following vectors **v**, find the direction cosines (the cosine of the angles between **v** and the positive x-, y-, and z-axes):

(a) $\mathbf{v} = \begin{bmatrix} 1 \\ 0 \\ 0 \end{bmatrix}$ **(b)** $\mathbf{v} = \begin{bmatrix} 1 \\ 3 \\ 2 \end{bmatrix}$

(c) $\mathbf{u} = \begin{bmatrix} -1 \\ -2 \\ -3 \end{bmatrix}$ (d) $\mathbf{u} = \begin{bmatrix} 4 \\ -3 \\ 2 \end{bmatrix}$

12. Let P and Q be the points in R^3, with respective coordinates $(3, -1, 2)$ and $(4, 2, -3)$. Find the length of the segment PQ.

13. Prove Theorem 5.1.

14. Verify Theorem 5.1 for

$$\mathbf{u} = \begin{bmatrix} 1 \\ 2 \\ 3 \end{bmatrix}, \quad \mathbf{v} = \begin{bmatrix} -2 \\ 4 \\ 3 \end{bmatrix}, \quad \mathbf{w} = \begin{bmatrix} 0 \\ 3 \\ -2 \end{bmatrix},$$

and $c = -3$.

15. Show that in R^2,

(a) $\mathbf{i} \cdot \mathbf{i} = \mathbf{j} \cdot \mathbf{j} = 1$; (b) $\mathbf{i} \cdot \mathbf{j} = 0$.

16. Show that in R^3,

(a) $\mathbf{i} \cdot \mathbf{i} = \mathbf{j} \cdot \mathbf{j} = \mathbf{k} \cdot \mathbf{k} = 1$;

(b) $\mathbf{i} \cdot \mathbf{j} = \mathbf{i} \cdot \mathbf{k} = \mathbf{j} \cdot \mathbf{k} = 0$.

17. Which of the vectors $\mathbf{v}_1 = \begin{bmatrix} 1 \\ 2 \end{bmatrix}$, $\mathbf{v}_2 = \begin{bmatrix} 0 \\ 1 \end{bmatrix}$, $\mathbf{v}_3 = \begin{bmatrix} -2 \\ -4 \end{bmatrix}$, $\mathbf{v}_4 = \begin{bmatrix} -2 \\ 1 \end{bmatrix}$, $\mathbf{v}_5 = \begin{bmatrix} 2 \\ 4 \end{bmatrix}$, and $\mathbf{v}_6 = \begin{bmatrix} -6 \\ 3 \end{bmatrix}$ are

(a) orthogonal? (b) in the same direction?

(c) in opposite directions?

18. Which of the vectors $\mathbf{v}_1 = \begin{bmatrix} 1 \\ -1 \\ -2 \end{bmatrix}$, $\mathbf{v}_2 = \begin{bmatrix} 3 \\ -1 \\ 2 \end{bmatrix}$, $\mathbf{v}_3 = \begin{bmatrix} 2 \\ 4 \\ -1 \end{bmatrix}$, $\mathbf{v}_4 = \begin{bmatrix} \frac{1}{2} \\ 0 \\ \frac{1}{4} \end{bmatrix}$, $\mathbf{v}_5 = \begin{bmatrix} \frac{1}{2} \\ -\frac{1}{2} \\ -1 \end{bmatrix}$, $\mathbf{v}_6 = \begin{bmatrix} -\frac{2}{3} \\ -\frac{4}{3} \\ \frac{1}{3} \end{bmatrix}$ are

(a) orthogonal? (b) in the same direction?

(c) in opposite directions?

19. (***Optional***) Which of the following pairs of lines are perpendicular?

(a) $\begin{aligned} x &= 2 + 2t \\ y &= -3 - 3t \\ z &= 4 + 4t \end{aligned}$ and $\begin{aligned} x &= 2 + t \\ y &= 4 - t \\ z &= 5 - t \end{aligned}$

(b) $\begin{aligned} x &= 3 - t \\ y &= 4 + 4t \\ z &= 2 + 2t \end{aligned}$ and $\begin{aligned} x &= 2t \\ y &= 3 - 2t \\ z &= 4 + 2t \end{aligned}$

20. (***Optional***) Find parametric equations of the line passing through $(3, -1, -3)$ and perpendicular to the line passing through $(3, -2, 4)$ and $(0, 3, 5)$.

21. A ship is being pushed by a tugboat with a force of 300 pounds along the negative y-axis while another tugboat is pushing along the negative x-axis with a force of 400 pounds. Find the magnitude and sketch the direction of the resultant force.

22. Suppose that an airplane is flying with an airspeed of 260 kilometers per hour while a wind is blowing to the west at 100 kilometers per hour. Indicate on a figure the appropriate direction that the plane must follow to fly directly south. What will be the resultant speed?

23. Let points A, B, C, and D in R^3 have respective coordinates $(1, 2, 3)$, $(-2, 3, 5)$, $(0, 3, 6)$, and $(3, 2, 4)$. Prove that $ABCD$ is a parallelogram.

24. Find c so that the vector $\mathbf{v} = \begin{bmatrix} 1 \\ c \end{bmatrix}$ is orthogonal to $\mathbf{w} = \begin{bmatrix} 2 \\ -1 \end{bmatrix}$.

25. Find c so that the vector $\mathbf{v} = \begin{bmatrix} 2 \\ c \\ 3 \end{bmatrix}$ is orthogonal to $\mathbf{w} = \begin{bmatrix} 1 \\ -2 \\ 1 \end{bmatrix}$.

26. If possible, find a, b, and c so that $\mathbf{v} = \begin{bmatrix} a \\ b \\ c \end{bmatrix}$ is orthogonal to both $\mathbf{w} = \begin{bmatrix} 1 \\ 2 \\ 1 \end{bmatrix}$ and $\mathbf{x} = \begin{bmatrix} 1 \\ -1 \\ 1 \end{bmatrix}$.

27. If possible, find a and b so that $\mathbf{v} = \begin{bmatrix} a \\ b \\ 2 \end{bmatrix}$ is orthogonal to both $\mathbf{w} = \begin{bmatrix} 2 \\ 1 \\ 1 \end{bmatrix}$ and $\mathbf{x} = \begin{bmatrix} 1 \\ 0 \\ 1 \end{bmatrix}$.

28. Find c so that the vectors $\begin{bmatrix} c \\ 4 \end{bmatrix}$ and $\begin{bmatrix} 2 \\ 5 \end{bmatrix}$ are parallel.

29. Let θ be the angle between the nonzero vectors $\mathbf{u}$ and $\mathbf{v}$ in R^2 or R^3. Show that if $\mathbf{u}$ and $\mathbf{v}$ are parallel, then $\cos\theta = \pm 1$.

30. Show that the only vector $\mathbf{x}$ in R^2 or R^3 that is orthogonal to every other vector is the zero vector.

31. Prove that if $\mathbf{v}$, $\mathbf{w}$, and $\mathbf{x}$ are in R^2 or R^3 and $\mathbf{v}$ is orthogonal to both $\mathbf{w}$ and $\mathbf{x}$, then $\mathbf{v}$ is orthogonal to every vector in span $\{\mathbf{w}, \mathbf{x}\}$.

32. Let **u** be a fixed vector in R^2 (R^3). Prove that the set V of all vectors **v** in R^2 (R^3) such that **u** and **v** are orthogonal is a subspace of R^2 (R^3).

33. Prove that if c is a scalar and **v** is a vector in R^2 or R^3, then $\|c\mathbf{v}\| = |c|\,\|\mathbf{v}\|$.

34. Show that if **x** is a nonzero vector in R^2 or R^3, then $\mathbf{u} = \frac{1}{\|\mathbf{x}\|}\mathbf{x}$ is a unit vector in the direction of **x**.

35. Let $S = \{\mathbf{v}_1, \mathbf{v}_2, \mathbf{v}_3\}$ be a set of nonzero vectors in R^3 such that any two vectors in S are orthogonal. Prove that S is linearly independent.

36. Prove that for any vectors **u**, **v**, and **w** in R^2 or R^3, we have

$$\mathbf{u}\cdot(\mathbf{v}+\mathbf{w}) = \mathbf{u}\cdot\mathbf{v}+\mathbf{u}\cdot\mathbf{w}.$$

37. Prove that for any vectors **u**, **v**, and **w** in R^2 or R^3 and any scalar c, we have

(a) $(\mathbf{u}+c\mathbf{v})\cdot\mathbf{w} = \mathbf{u}\cdot\mathbf{w}+c(\mathbf{v}\cdot\mathbf{w})$;

(b) $\mathbf{u}\cdot(c\mathbf{v}) = c(\mathbf{u}\cdot\mathbf{v})$;

(c) $(\mathbf{u}+\mathbf{v})\cdot(c\mathbf{w}) = c(\mathbf{u}\cdot\mathbf{w})+c(\mathbf{v}\cdot\mathbf{w})$.

38. Prove that the diagonals of a rectangle are of equal length. [*Hint*: Take the vertices of the rectangle as $(0, 0)$, $(0, b)$, $(a, 0)$, and (a, b).]

39. Prove that the angles at the base of an isosceles triangle are equal.

40. Prove that a parallelogram is a rhombus, a parallelogram with four equal sides, if and only if its diagonals are orthogonal.

41. To compute the dot product of a pair of vectors **u** and **v** in R^2 or R^3, use the matrix product operation in your software as follows: Let U and V be column matrices for vectors **u** and **v**, respectively. Then $\mathbf{u}\cdot\mathbf{v}$ is the product of U^T and V (or V^T and U). Experiment by choosing several pairs of vectors in R^2 and R^3. (Determine whether your software has a particular command for computing a dot product.)

42. Determine whether there is a command in your software to compute the length of a vector. If there is, use it on the vector in Example 3 and then compute the distance between the vectors in Example 4.

43. Assuming that your software has a command to compute the length of a vector (see Exercise 42), determine a unit vector in the direction of **v** for each of the following:

(a) $\mathbf{v} = \begin{bmatrix} 2 \\ 4 \end{bmatrix}$ **(b)** $\mathbf{v} = \begin{bmatrix} 7 \\ 1 \\ 0 \end{bmatrix}$ **(c)** $\mathbf{v} = \begin{bmatrix} 1 \\ 2 \\ -1 \end{bmatrix}$

44. Referring to Exercise 41, how could your software check for orthogonal vectors?

5.2 Cross Product in R^3 (Optional)

In this section we discuss an operation that is meaningful only in R^3. Despite this limitation, it has a number of important applications, some of which we discuss in this section. Suppose that $\mathbf{u} = u_1\mathbf{i}+u_2\mathbf{j}+u_3\mathbf{k}$ and $\mathbf{v} = v_1\mathbf{i}+v_2\mathbf{j}+v_3\mathbf{k}$ and that we want to find a vector $\mathbf{w} = \begin{bmatrix} x \\ y \\ z \end{bmatrix}$ orthogonal (perpendicular) to both **u** and **v**. Thus we want $\mathbf{u}\cdot\mathbf{w} = 0$ and $\mathbf{v}\cdot\mathbf{w} = 0$, which leads to the linear system

$$\begin{aligned} u_1x + u_2y + u_3z &= 0 \\ v_1x + v_2y + v_3z &= 0. \end{aligned} \tag{1}$$

It can be shown that

$$\mathbf{w} = \begin{bmatrix} u_2v_3 - u_3v_2 \\ u_3v_1 - u_1v_3 \\ u_1v_2 - u_2v_1 \end{bmatrix}$$

is a solution to Equation (1) (verify). Of course, we can also write **w** as

$$\mathbf{w} = (u_2v_3 - u_3v_2)\mathbf{i} + (u_3v_1 - u_1v_3)\mathbf{j} + (u_1v_2 - u_2v_1)\mathbf{k}. \tag{2}$$

This vector is called the **cross product** of **u** and **v** and is denoted by $\mathbf{u} \times \mathbf{v}$. Note that the cross product, $\mathbf{u} \times \mathbf{v}$, is a vector, while the dot product, $\mathbf{u}\cdot\mathbf{v}$, is a scalar, or number. Although the cross product is not defined on R^n if $n \neq 3$, it has many applications; we shall use it when we study planes in R^3.

EXAMPLE 1

Let $\mathbf{u} = 2\mathbf{i} + \mathbf{j} + 2\mathbf{k}$ and $\mathbf{v} = 3\mathbf{i} - \mathbf{j} - 3\mathbf{k}$. From Equation (2),

$$\mathbf{u} \times \mathbf{v} = -\mathbf{i} + 12\mathbf{j} - 5\mathbf{k}.$$

■

Let $\mathbf{u}$, $\mathbf{v}$, and $\mathbf{w}$ be vectors in R^3 and c a scalar. The cross product operation satisfies the following properties, whose verification we leave to the reader:

(a) $\mathbf{u} \times \mathbf{v} = -(\mathbf{v} \times \mathbf{u})$

(b) $\mathbf{u} \times (\mathbf{v} + \mathbf{w}) = \mathbf{u} \times \mathbf{v} + \mathbf{u} \times \mathbf{w}$

(c) $(\mathbf{u} + \mathbf{v}) \times \mathbf{w} = \mathbf{u} \times \mathbf{w} + \mathbf{v} \times \mathbf{w}$

(d) $c(\mathbf{u} \times \mathbf{v}) = (c\mathbf{u}) \times \mathbf{v} = \mathbf{u} \times (c\mathbf{v})$

(e) $\mathbf{u} \times \mathbf{u} = \mathbf{0}$

(f) $\mathbf{0} \times \mathbf{u} = \mathbf{u} \times \mathbf{0} = \mathbf{0}$

(g) $\mathbf{u} \times (\mathbf{v} \times \mathbf{w}) = (\mathbf{u} \cdot \mathbf{w})\mathbf{v} - (\mathbf{u} \cdot \mathbf{v})\mathbf{w}$

(h) $(\mathbf{u} \times \mathbf{v}) \times \mathbf{w} = (\mathbf{w} \cdot \mathbf{u})\mathbf{v} - (\mathbf{w} \cdot \mathbf{v})\mathbf{u}$

EXAMPLE 2

It follows from (2) that

$$\mathbf{i} \times \mathbf{i} = \mathbf{j} \times \mathbf{j} = \mathbf{k} \times \mathbf{k} = \mathbf{0},$$
$$\mathbf{i} \times \mathbf{j} = \mathbf{k}, \quad \mathbf{j} \times \mathbf{k} = \mathbf{i}, \quad \mathbf{k} \times \mathbf{i} = \mathbf{j}.$$

Also,

$$\mathbf{j} \times \mathbf{i} = -\mathbf{k}, \quad \mathbf{k} \times \mathbf{j} = -\mathbf{i}, \quad \mathbf{i} \times \mathbf{k} = -\mathbf{j}.$$

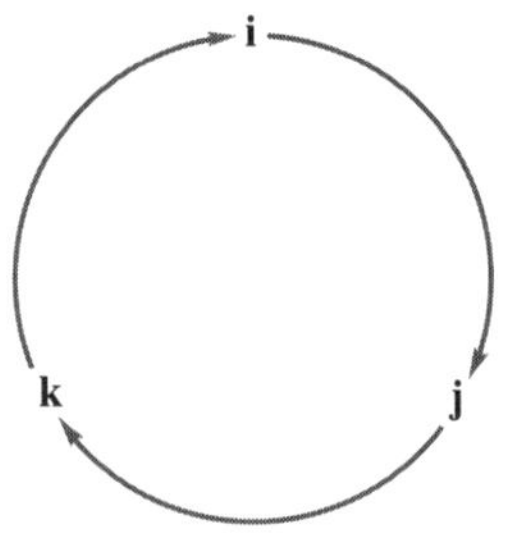

FIGURE 5.11

These rules can be remembered by the method illustrated in Figure 5.11. Moving around the circle in a clockwise direction, we see that the cross product of two vectors taken in the indicated order is the third vector; moving in a counterclockwise direction, we see that the cross product taken in the indicated order is the negative of the third vector. The cross product of a vector with itself is the zero vector. ■

Although many of the familiar properties of the real numbers hold for the cross product, it should be noted that two important properties do not hold. The commutative law does not hold, since $\mathbf{u} \times \mathbf{v} = -(\mathbf{v} \times \mathbf{u})$. Also, the associative law does not hold, since $\mathbf{i} \times (\mathbf{i} \times \mathbf{j}) = \mathbf{i} \times \mathbf{k} = -\mathbf{j}$ while $(\mathbf{i} \times \mathbf{i}) \times \mathbf{j} = \mathbf{0} \times \mathbf{j} = \mathbf{0}$.

We now take a closer look at the geometric properties of the cross product. First, we observe the following additional property of the cross product, whose proof we leave to the reader:

$$(\mathbf{u} \times \mathbf{v}) \cdot \mathbf{w} = \mathbf{u} \cdot (\mathbf{v} \times \mathbf{w}) \qquad \text{(Exercise 7)} \tag{3}$$

EXAMPLE 3

Let $\mathbf{u}$ and $\mathbf{v}$ be as in Example 1, and let $\mathbf{w} = \mathbf{i} + 2\mathbf{j} + 3\mathbf{k}$. Then

$$\begin{aligned} \mathbf{u} \times \mathbf{v} &= -\mathbf{i} + 12\mathbf{j} - 5\mathbf{k} \quad \text{and} \quad (\mathbf{u} \times \mathbf{v}) \cdot \mathbf{w} = 8 \\ \mathbf{v} \times \mathbf{w} &= \ \ 3\mathbf{i} - 12\mathbf{j} + 7\mathbf{k} \quad \text{and} \quad \mathbf{u} \cdot (\mathbf{v} \times \mathbf{w}) = 8, \end{aligned}$$

which illustrates Equation (3). ■

From the construction of $\mathbf{u} \times \mathbf{v}$, it follows that $\mathbf{u} \times \mathbf{v}$ is orthogonal to both $\mathbf{u}$ and $\mathbf{v}$; that is,

$$(\mathbf{u} \times \mathbf{v}) \cdot \mathbf{u} = 0, \tag{4}$$
$$(\mathbf{u} \times \mathbf{v}) \cdot \mathbf{v} = 0. \tag{5}$$

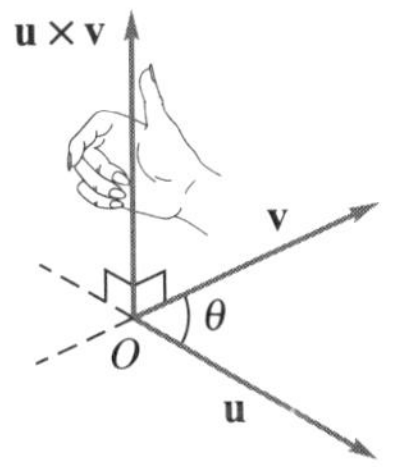

FIGURE 5.12

These equations can also be verified directly by the definitions of $\mathbf{u} \times \mathbf{v}$ and dot product, or by Equation (3) and properties (a) and (e) of the cross product operation. Then $\mathbf{u} \times \mathbf{v}$ is also orthogonal to the plane determined by $\mathbf{u}$ and $\mathbf{v}$. It can be shown that if θ is the angle between $\mathbf{u}$ and $\mathbf{v}$, then the direction of $\mathbf{u} \times \mathbf{v}$ is determined as follows. If we curl the fingers of the right hand in the direction of a rotation through the angle θ from $\mathbf{u}$ to $\mathbf{v}$, then the thumb will point in the direction of $\mathbf{u} \times \mathbf{v}$ (Figure 5.12).

The magnitude of $\mathbf{u} \times \mathbf{v}$ can be determined as follows. From Equation (7) of Section 5.1, it follows that

$$\begin{aligned}
\|\mathbf{u} \times \mathbf{v}\|^2 &= (\mathbf{u} \times \mathbf{v}) \cdot (\mathbf{u} \times \mathbf{v}) \\
&= \mathbf{u} \cdot [\mathbf{v} \times (\mathbf{u} \times \mathbf{v})] && \text{by (3)} \\
&= \mathbf{u} \cdot [(\mathbf{v} \cdot \mathbf{v})\mathbf{u} - (\mathbf{v} \cdot \mathbf{u})\mathbf{v}] && \text{by property (g) for cross product} \\
&= (\mathbf{u} \cdot \mathbf{u})(\mathbf{v} \cdot \mathbf{v}) - (\mathbf{v} \cdot \mathbf{u})(\mathbf{v} \cdot \mathbf{u}) && \text{by (d) and (b) of Theorem 4.1} \\
&= \|\mathbf{u}\|^2 \|\mathbf{v}\|^2 - (\mathbf{u} \cdot \mathbf{v})^2 && \text{by Equation (7) of Section 4.1 and (b) of Theorem 4.1.}
\end{aligned}$$

From Equation (8) of Section 5.1, it follows that

$$\mathbf{u} \cdot \mathbf{v} = \|\mathbf{u}\| \|\mathbf{v}\| \cos\theta,$$

where θ is the angle between $\mathbf{u}$ and $\mathbf{v}$. Hence

$$\begin{aligned}
\|\mathbf{u} \times \mathbf{v}\|^2 &= \|\mathbf{u}\|^2 \|\mathbf{v}\|^2 - \|\mathbf{u}\|^2 \|\mathbf{v}\|^2 \cos^2\theta \\
&= \|\mathbf{u}\|^2 \|\mathbf{v}\|^2 (1 - \cos^2\theta) \\
&= \|\mathbf{u}\|^2 \|\mathbf{v}\|^2 \sin^2\theta.
\end{aligned}$$

Taking square roots, we obtain

$$\|\mathbf{u} \times \mathbf{v}\| = \|\mathbf{u}\| \|\mathbf{v}\| \sin\theta, \qquad 0 \le \theta \le \pi. \tag{6}$$

Note that in (6) we do not have to write $|\sin\theta|$, since $\sin\theta$ is nonnegative for $0 \le \theta \le \pi$. It follows that vectors $\mathbf{u}$ and $\mathbf{v}$ are parallel if and only if $\mathbf{u} \times \mathbf{v} = \mathbf{0}$ (Exercise 9).

We now consider several applications of cross product.

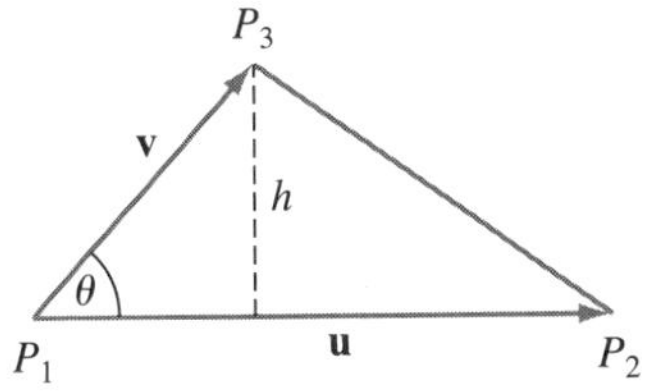

FIGURE 5.13

Area of a Triangle

Consider the triangle with vertices P_1, P_2, and P_3 (Figure 5.13). The area of this triangle is $\frac{1}{2}bh$, where b is the base and h is the height. If we take the segment between P_1 and P_2 to be the base and denote $\overrightarrow{P_1P_2}$ by the vector $\mathbf{u}$, then

$$b = \|\mathbf{u}\|.$$

Letting $\overrightarrow{P_1P_3} = \mathbf{v}$, we find that the height h is given by

$$h = \|\mathbf{v}\| \sin\theta.$$

Hence, by (6), the area A_T of the triangle is

$$A_T = \tfrac{1}{2}\|\mathbf{u}\|\,\|\mathbf{v}\| \sin\theta = \tfrac{1}{2}\|\mathbf{u} \times \mathbf{v}\|.$$

EXAMPLE 4 Find the area of the triangle with vertices $P_1(2, 2, 4)$, $P_2(-1, 0, 5)$, and $P_3(3, 4, 3)$.

Solution

We have

$$\mathbf{u} = \overrightarrow{P_1P_2} = -3\mathbf{i} - 2\mathbf{j} + \mathbf{k}$$
$$\mathbf{v} = \overrightarrow{P_1P_3} = \mathbf{i} + 2\mathbf{j} - \mathbf{k}.$$

Then the area of the triangle A_T is

$$A_T = \tfrac{1}{2}\|(-3\mathbf{i} - 2\mathbf{j} + \mathbf{k}) \times (\mathbf{i} + 2\mathbf{j} - \mathbf{k})\|$$
$$= \tfrac{1}{2}\|-2\mathbf{j} - 4\mathbf{k}\| = \|-\mathbf{j} - 2\mathbf{k}\| = \sqrt{5}.$$ ■

■ Area of a Parallelogram

The area A_P of the parallelogram with adjacent sides $\mathbf{u}$ and $\mathbf{v}$ (Figure 5.14) is $2A_T$, so

$$A_P = \|\mathbf{u} \times \mathbf{v}\|.$$

EXAMPLE 5 If P_1, P_2, and P_3 are as in Example 4, then the area of the parallelogram with adjacent sides $\overrightarrow{P_1P_2}$ and $\overrightarrow{P_1P_3}$ is $2\sqrt{5}$. (Verify.) ■

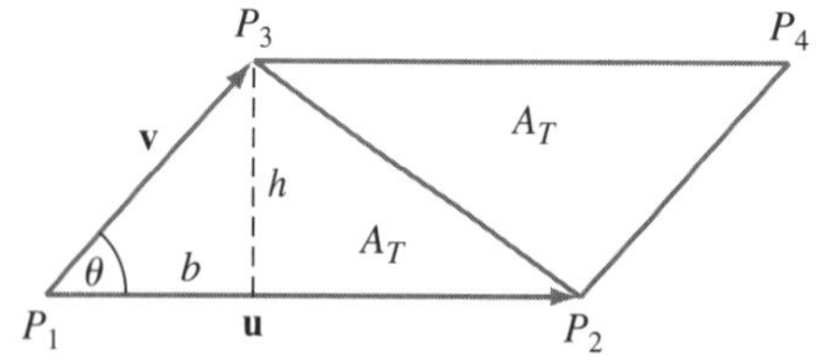

FIGURE 5.14

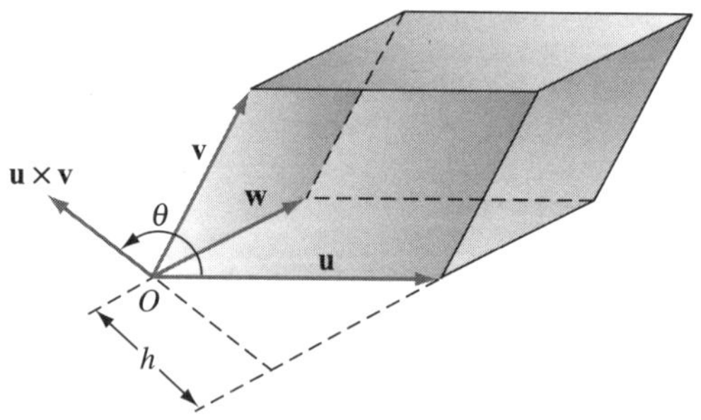

FIGURE 5.15

■ Volume of a Parallelepiped

Consider the parallelepiped with a vertex at the origin and edges $\mathbf{u}$, $\mathbf{v}$, and $\mathbf{w}$ (Figure 5.15). The volume V of the parallelepiped is the product of the area of the face containing $\mathbf{v}$ and $\mathbf{w}$ and the distance h from this face to the face parallel to it. Now

$$h = \|\mathbf{u}\|\,|\cos\theta|,$$

where θ is the angle between $\mathbf{u}$ and $\mathbf{v} \times \mathbf{w}$, and the area of the face determined by $\mathbf{v}$ and $\mathbf{w}$ is $\|\mathbf{v} \times \mathbf{w}\|$. Hence

$$V = \|\mathbf{v} \times \mathbf{w}\| \|\mathbf{u}\| |\cos\theta| = |\mathbf{u} \cdot (\mathbf{v} \times \mathbf{w})|.$$

EXAMPLE 6

Consider the parallelepiped with a vertex at the origin and edges $\mathbf{u} = \mathbf{i} - 2\mathbf{j} + 3\mathbf{k}$, $\mathbf{v} = \mathbf{i} + 3\mathbf{j} + \mathbf{k}$, and $\mathbf{w} = 2\mathbf{i} + \mathbf{j} + 2\mathbf{k}$. Then

$$\mathbf{v} \times \mathbf{w} = 5\mathbf{i} - 5\mathbf{k}.$$

Hence $\mathbf{u} \cdot (\mathbf{v} \times \mathbf{w}) = -10$. Thus the volume V is given by

$$V = |\mathbf{u} \cdot (\mathbf{v} \times \mathbf{w})| = |-10| = 10. \quad ■$$

■ Planes

A plane in R^3 can be determined by specifying a point in the plane and a vector perpendicular to the plane. This vector is called a **normal** to the plane.

To obtain an equation of the plane passing through the point $P_0(x_0, y_0, z_0)$ and having the nonzero vector $\mathbf{v} = a\mathbf{i} + b\mathbf{j} + c\mathbf{k}$ as a normal, we proceed as follows: A point $P(x, y, z)$ lies in the plane if and only if the vector $\overrightarrow{P_0P}$ is perpendicular to $\mathbf{v}$ (Figure 5.16). Thus $P(x, y, z)$ lies in the plane if and only if

$$\mathbf{v} \cdot \overrightarrow{P_0P} = 0. \tag{7}$$

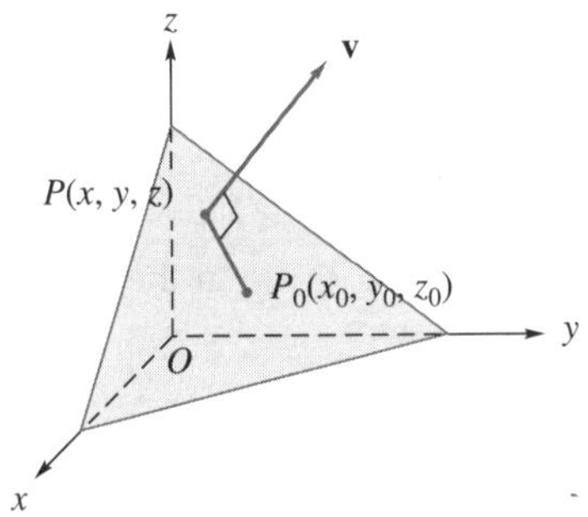

FIGURE 5.16

Since

$$\overrightarrow{P_0P} = (x - x_0)\mathbf{i} + (y - y_0)\mathbf{j} + (z - z_0)\mathbf{k},$$

we can write (7) as

$$a(x - x_0) + b(y - y_0) + c(z - z_0) = 0. \tag{8}$$

EXAMPLE 7

Find an equation of the plane passing through the point $(3, 4, -3)$ and perpendicular to the vector $\mathbf{v} = 5\mathbf{i} - 2\mathbf{j} + 4\mathbf{k}$.

Solution

Substituting in (8), we obtain the equation of the plane as

$$5(x - 3) - 2(y - 4) + 4(z + 3) = 0. \quad ■$$

A plane is also determined by three noncollinear points in it, as we show in the following example:

EXAMPLE 8

Find an equation of the plane passing through the points $P_1(2, -2, 1)$, $P_2(-1, 0, 3)$, and $P_3(5, -3, 4)$.

Solution

The nonparallel vectors $\overrightarrow{P_1P_2} = -3\mathbf{i} + 2\mathbf{j} + 2\mathbf{k}$ and $\overrightarrow{P_1P_3} = 3\mathbf{i} - \mathbf{j} + 3\mathbf{k}$ lie in the plane, since the points P_1, P_2, and P_3 lie in the plane. The vector

$$\mathbf{v} = \overrightarrow{P_1P_2} \times \overrightarrow{P_1P_3} = 8\mathbf{i} + 15\mathbf{j} - 3\mathbf{k}$$

is then perpendicular to both $\overrightarrow{P_1P_2}$ and $\overrightarrow{P_1P_3}$ and is thus a normal to the plane. Using the vector $\mathbf{v}$ and the point $P_1(2, -2, 1)$ in (8), we obtain

$$8(x-2) + 15(y+2) - 3(z-1) = 0 \tag{9}$$

as an equation of the plane. ■

If we multiply out and simplify, (8) can be rewritten as

$$ax + by + cz + d = 0. \tag{10}$$

EXAMPLE 9

Equation (10) of the plane in Example 8 can be rewritten as

$$8x + 15y - 3z + 17 = 0. \tag{11}$$

■

It is not difficult to show (Exercise 24) that the graph of an equation of the form given in (10), where a, b, c, and d are constants (with a, b, and c not all zero), is a plane with normal $\mathbf{v} = a\mathbf{i} + b\mathbf{j} + c\mathbf{k}$; moreover, if $d = 0$, it is a two-dimensional subspace of R^3.

EXAMPLE 10

An alternative solution to Example 8 is as follows. Let the equation of the desired plane be

$$ax + by + cz + d = 0, \tag{12}$$

where a, b, c, and d are to be determined. Since P_1, P_2, and P_3 lie in the plane, their coordinates satisfy (12). Thus we obtain the linear system

$$\begin{aligned} 2a - 2b + c + d &= 0 \\ -a + 3c + d &= 0 \\ 5a - 3b + 4c + d &= 0. \end{aligned}$$

Solving this system, we have (verify)

$$a = \tfrac{8}{17}r, \quad b = \tfrac{15}{17}r, \quad c = -\tfrac{3}{17}r, \quad \text{and} \quad d = r,$$

where r is any real number. Letting $r = 17$, we find that

$$a = 8, \quad b = 15, \quad c = -3, \quad \text{and} \quad d = 17,$$

which yields (11) as in the first solution. ■

EXAMPLE 11

Find parametric equations of the line of intersection of the planes

$$\Pi_1\colon 2x + 3y - 2z + 4 = 0 \quad \text{and} \quad \Pi_2\colon x - y + 2z + 3 = 0.$$

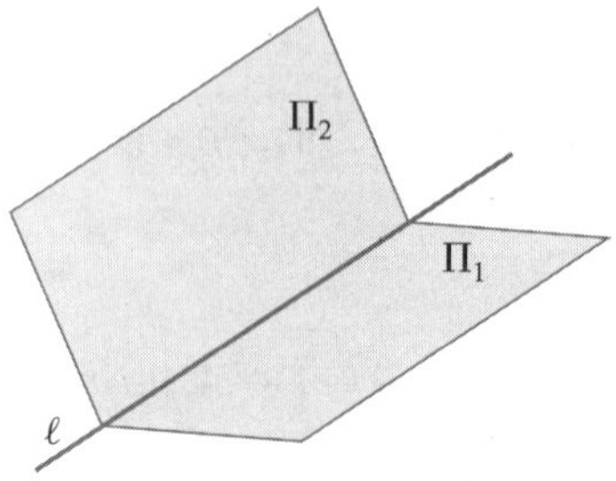

FIGURE 5.17

Solution

Solving the linear system consisting of the equations of Π_1 and Π_2, we get (verify)

$$\begin{aligned} x &= -\tfrac{13}{5} - \tfrac{4}{5}t \\ y &= \tfrac{2}{5} + \tfrac{6}{5}t \qquad -\infty < t < \infty \\ z &= 0 + t \end{aligned}$$

as parametric equations (see Section 4.3) of the line ℓ of intersection of the planes (see Figure 5.17). ■

As we have indicated, the cross product cannot be generalized to R^n. However, we can generalize the notions of length, direction, and standard inner product to R^n in the natural manner, but there are some things to be checked. For example, if we define the cosine of the angle θ between two nonzero vectors $\mathbf{u}$ and $\mathbf{v}$ in R^n as

$$\cos\theta = \frac{\mathbf{u}\cdot\mathbf{v}}{\|\mathbf{u}\|\,\|\mathbf{v}\|},$$

we must check that $-1 \leq \cos\theta \leq 1$; otherwise, it would be misleading to call this fraction $\cos\theta$. (See Section 5.3.) Rather than verify this property for R^n now, we obtain this result in the next section, where we formulate the notion of inner product in any real vector space.

■ Determinants and Cross Product (Optional)

Determinants can be applied to the computation of cross products. Recall that the cross product $\mathbf{u} \times \mathbf{v}$ of the vectors $\mathbf{u} = u_1\mathbf{i} + u_2\mathbf{j} + u_3\mathbf{k}$ and $\mathbf{v} = v_1\mathbf{i} + v_2\mathbf{j} + v_3\mathbf{k}$ in R^3 is

$$\mathbf{u} \times \mathbf{v} = (u_2v_3 - u_3v_2)\mathbf{i} + (u_3v_1 - u_1v_3)\mathbf{j} + (u_1v_2 - u_2v_1)\mathbf{k}.$$

If we formally write the matrix

$$C = \begin{bmatrix} \mathbf{i} & \mathbf{j} & \mathbf{k} \\ u_1 & u_2 & u_3 \\ v_1 & v_2 & v_3 \end{bmatrix},$$

then the determinant of C, evaluated by expanding along the cofactors of the first row, is $\mathbf{u} \times \mathbf{v}$; that is,

$$\mathbf{u} \times \mathbf{v} = \det(C) = \begin{vmatrix} u_2 & u_3 \\ v_2 & v_3 \end{vmatrix}\mathbf{i} - \begin{vmatrix} u_1 & u_3 \\ v_1 & v_3 \end{vmatrix}\mathbf{j} + \begin{vmatrix} u_1 & u_2 \\ v_1 & v_2 \end{vmatrix}\mathbf{k}.$$

Of course, C is not really a matrix, and $\det(C)$ is not really a determinant, but it is convenient to think of the computation in this way.

EXAMPLE 12

If $\mathbf{u} = 2\mathbf{i} + \mathbf{j} + 2\mathbf{k}$ and $\mathbf{v} = 3\mathbf{i} - \mathbf{j} - 3\mathbf{k}$, as in Example 1, then

$$C = \begin{bmatrix} \mathbf{i} & \mathbf{j} & \mathbf{k} \\ 2 & 1 & 2 \\ 3 & -1 & -3 \end{bmatrix},$$

and $\det(C) = \mathbf{u} \times \mathbf{v} = -\mathbf{i} + 12\mathbf{j} - 5\mathbf{k}$, when expanded along its first row. ■

Key Terms

Cross product
Unit vectors, **i**, **j**, **k**
Orthogonal vectors
Parallel vectors
Length of a vector
Area of a triangle
Area of a parallelogram
Volume of a parallelepiped

5.2 Exercises

1. Compute $\mathbf{u} \times \mathbf{v}$.
   (a) $\mathbf{u} = 2\mathbf{i} + 3\mathbf{j} + 4\mathbf{k}$, $\mathbf{v} = -\mathbf{i} + 3\mathbf{j} - \mathbf{k}$
   (b) $\mathbf{u} = \mathbf{i} + \mathbf{k}$, $\mathbf{v} = 2\mathbf{i} + 3\mathbf{j} - \mathbf{k}$
   (c) $\mathbf{u} = \mathbf{i} - \mathbf{j} + 2\mathbf{k}$, $\mathbf{v} = 3\mathbf{i} - 4\mathbf{j} + \mathbf{k}$
   (d) $\mathbf{u} = \begin{bmatrix} 2 \\ -1 \\ 1 \end{bmatrix}$, $\mathbf{v} = -2\mathbf{u}$
2. Compute $\mathbf{u} \times \mathbf{v}$.
   (a) $\mathbf{u} = \mathbf{i} - \mathbf{j} + 2\mathbf{k}$, $\mathbf{v} = 3\mathbf{i} + \mathbf{j} + 2\mathbf{k}$
   (b) $\mathbf{u} = 2\mathbf{i} + \mathbf{j} - 2\mathbf{k}$, $\mathbf{v} = \mathbf{i} + 3\mathbf{k}$
   (c) $\mathbf{u} = 2\mathbf{j} + \mathbf{k}$, $\mathbf{v} = 3\mathbf{u}$
   (d) $\mathbf{u} = \begin{bmatrix} 4 \\ 0 \\ -2 \end{bmatrix}$, $\mathbf{v} = \begin{bmatrix} 0 \\ 2 \\ -1 \end{bmatrix}$
3. Let $\mathbf{u} = \mathbf{i} + 2\mathbf{j} - 3\mathbf{k}$, $\mathbf{v} = 2\mathbf{i} + 3\mathbf{j} + \mathbf{k}$, $\mathbf{w} = 2\mathbf{i} - \mathbf{j} + 2\mathbf{k}$, and $c = -3$. Verify properties (a) through (h) for the cross product operation.
4. Prove properties (a) through (h) for the cross product operation.
5. Let $\mathbf{u} = 2\mathbf{i} - \mathbf{j} + 3\mathbf{k}$, $\mathbf{v} = 3\mathbf{i} + \mathbf{j} - \mathbf{k}$, and $\mathbf{w} = 3\mathbf{i} + \mathbf{j} + 2\mathbf{k}$. Verify Equation (3).
6. Verify that each of the cross products $\mathbf{u} \times \mathbf{v}$ in Exercise 1 is orthogonal to both $\mathbf{u}$ and $\mathbf{v}$.
7. Show that $(\mathbf{u} \times \mathbf{v}) \cdot \mathbf{w} = \mathbf{u} \cdot (\mathbf{v} \times \mathbf{w})$.
8. Verify Equation (6) for the pairs of vectors in Exercise 1.
9. Show that $\mathbf{u}$ and $\mathbf{v}$ are parallel if and only if $\mathbf{u} \times \mathbf{v} = \mathbf{0}$.
10. Show that $\|\mathbf{u} \times \mathbf{v}\|^2 + (\mathbf{u} \cdot \mathbf{v})^2 = \|\mathbf{u}\|^2\|\mathbf{v}\|^2$.
11. Prove the **Jacobi identity**

$$(\mathbf{u} \times \mathbf{v}) \times \mathbf{w} + (\mathbf{v} \times \mathbf{w}) \times \mathbf{u} + (\mathbf{w} \times \mathbf{u}) \times \mathbf{v} = \mathbf{0}.$$

12. Find the area of the triangle with vertices $P_1(1, -2, 3)$, $P_2(-3, 1, 4)$, and $P_3(0, 4, 3)$.
13. Find the area of the triangle with vertices P_1, P_2, and P_3, where $\overrightarrow{P_1P_2} = 2\mathbf{i} + 3\mathbf{j} - \mathbf{k}$ and $\overrightarrow{P_1P_3} = \mathbf{i} + 2\mathbf{j} + 2\mathbf{k}$.
14. Find the area of the parallelogram with adjacent sides $\mathbf{u} = \mathbf{i} + 3\mathbf{j} - 2\mathbf{k}$ and $\mathbf{v} = 3\mathbf{i} - \mathbf{j} - \mathbf{k}$.
15. Find the volume of the parallelepiped with a vertex at the origin and edges $\mathbf{u} = 2\mathbf{i} - \mathbf{j}$, $\mathbf{v} = \mathbf{i} - 2\mathbf{j} - 2\mathbf{k}$, and $\mathbf{w} = 3\mathbf{i} - \mathbf{j} + \mathbf{k}$.
16. Repeat Exercise 15 for $\mathbf{u} = \mathbf{i} - 2\mathbf{j} + 4\mathbf{k}$, $\mathbf{v} = 3\mathbf{i} + 4\mathbf{j} + \mathbf{k}$, and $\mathbf{w} = -\mathbf{i} + \mathbf{j} + \mathbf{k}$.
17. Determine which of the following points are in the plane

$$3(x - 2) + 2(y + 3) - 4(z - 4) = 0:$$

   (a) $(0, -2, 3)$ (b) $(1, -2, 3)$
18. Find an equation of the plane passing through the given point and perpendicular to the given vector.
   (a) $(0, 2, -3)$, $3\mathbf{i} - 2\mathbf{j} + 4\mathbf{k}$
   (b) $(-1, 3, 2)$, $\mathbf{j} - 3\mathbf{k}$
19. Find an equation of the plane passing through the given points.
   (a) $(0, 1, 2)$, $(3, -2, 5)$, $(2, 3, 4)$
   (b) $(2, 3, 4)$, $(1, -2, 3)$, $(-5, -4, 2)$
20. Find parametric equations of the line of intersection of the given planes.
   (a) $2x + 3y - 4z + 5 = 0$ and $-3x + 2y + 5z + 6 = 0$
   (b) $3x - 2y - 5z + 4 = 0$ and $2x + 3y + 4z + 8 = 0$
21. Find an equation of the plane through $(-2, 3, 4)$ and perpendicular to the line through $(4, -2, 5)$ and $(0, 2, 4)$.
22. Find the point of intersection of the line

$$x = 2 - 3t$$
$$y = 4 + 2t \qquad -\infty < t < \infty,$$
$$z = 3 - 5t,$$

and the plane $2x + 3y + 4z + 8 = 0$.
23. Find a line passing through $(-2, 5, -3)$ and perpendicular to the plane $2x - 3y + 4z + 7 = 0$.
24. (a) Show that the graph of an equation of the form given in (10), with a, b, and c not all zero, is a plane with normal $\mathbf{v} = a\mathbf{i} + b\mathbf{j} + c\mathbf{k}$.
   (b) Show that the set of all points on the plane $ax + by + cz = 0$ is a subspace of R^3.
   (c) Find a basis for the subspace given by the plane $2x - 3y + 4z = 0$.
25. Find a basis for the subspace given by the plane $-3x + 2y + 5z = 0$.
26. Let $\mathbf{u} = u_1\mathbf{i} + u_2\mathbf{j} + u_3\mathbf{k}$, $\mathbf{v} = v_1\mathbf{i} + v_2\mathbf{j} + v_3\mathbf{k}$, and $\mathbf{w} = w_1\mathbf{i} + w_2\mathbf{j} + w_3\mathbf{k}$ be vectors in R^3. Show that

$$(\mathbf{u} \times \mathbf{v}) \cdot \mathbf{w} = \begin{vmatrix} u_1 & u_2 & u_3 \\ v_1 & v_2 & v_3 \\ w_1 & w_2 & w_3 \end{vmatrix}.$$

27. Compute each $\mathbf{u} \times \mathbf{v}$ by the method of Example 12.
   (a) $\mathbf{u} = 2\mathbf{i} + 3\mathbf{j} + 4\mathbf{k}$, $\mathbf{v} = -\mathbf{i} + 3\mathbf{j} - \mathbf{k}$

(b) $\mathbf{u} = \mathbf{i} + \mathbf{k}$, $\mathbf{v} = 2\mathbf{i} + 3\mathbf{j} - \mathbf{k}$

(c) $\mathbf{u} = \mathbf{i} - \mathbf{j} + 2\mathbf{k}$, $\mathbf{v} = 3\mathbf{i} - 4\mathbf{j} + \mathbf{k}$

(d) $\mathbf{u} = 2\mathbf{i} + \mathbf{j} - 2\mathbf{k}$, $\mathbf{v} = \mathbf{i} + 3\mathbf{k}$

28. If (x_1, y_1) and (x_2, y_2) are distinct points in the plane, show that

$$\begin{vmatrix} x & y & 1 \\ x_1 & y_1 & 1 \\ x_2 & y_2 & 1 \end{vmatrix} = 0$$

is the equation of the line through (x_1, y_1) and (x_2, y_2). Use this result to develop a test for collinearity of three points.

29. Let $P_i(x_i, y_i, z_i)$, $i = 1, 2, 3$, be three points in 3-space. Show that

$$\begin{vmatrix} x & y & z & 1 \\ x_1 & y_1 & z_1 & 1 \\ x_2 & y_2 & z_2 & 1 \\ x_3 & y_3 & z_3 & 1 \end{vmatrix} = 0$$

is the equation of a plane (see Section 5.2) through points P_i, $i = 1, 2, 3$.

30. Determine whether your software has a command for computing cross products. If it does, check your results in Exercises 1 and 2.

5.3 Inner Product Spaces

In this section we use the properties of the standard inner product or dot product on R^3 listed in Theorem 5.1 as our foundation for generalizing the notion of the inner product to any real vector space. Here, V is an arbitrary vector space, not necessarily finite-dimensional, where the scalars are restricted to real numbers.

DEFINITION 5.1

Let V be a real vector space. An **inner product** on V is a function that assigns to each ordered pair of vectors $\mathbf{u}$, $\mathbf{v}$ in V a real number $(\mathbf{u}, \mathbf{v})$ satisfying the following properties:

(a) $(\mathbf{u}, \mathbf{u}) \geq 0$; $(\mathbf{u}, \mathbf{u}) = 0$ if and only if $\mathbf{u} = \mathbf{0}_V$

(b) $(\mathbf{v}, \mathbf{u}) = (\mathbf{u}, \mathbf{v})$ for any $\mathbf{u}$, $\mathbf{v}$ in V

(c) $(\mathbf{u} + \mathbf{v}, \mathbf{w}) = (\mathbf{u}, \mathbf{w}) + (\mathbf{v}, \mathbf{w})$ for any $\mathbf{u}$, $\mathbf{v}$, $\mathbf{w}$ in V

(d) $(c\mathbf{u}, \mathbf{v}) = c(\mathbf{u}, \mathbf{v})$ for $\mathbf{u}$, $\mathbf{v}$ in V and c a real scalar

From these properties it follows that $(\mathbf{u}, c\mathbf{v}) = c(\mathbf{u}, \mathbf{v})$, because $(\mathbf{u}, c\mathbf{v}) = (c\mathbf{v}, \mathbf{u}) = c(\mathbf{v}, \mathbf{u}) = c(\mathbf{u}, \mathbf{v})$. Also, $(\mathbf{u}, \mathbf{v} + \mathbf{w}) = (\mathbf{u}, \mathbf{v}) + (\mathbf{u}, \mathbf{w})$.

EXAMPLE 1

In Section 1.3 we defined the standard inner product, or dot product, on R^n as the function that assigns to each ordered pair of vectors

$$\mathbf{u} = \begin{bmatrix} u_1 \\ u_2 \\ \vdots \\ u_n \end{bmatrix} \quad \text{and} \quad \mathbf{v} = \begin{bmatrix} v_1 \\ v_2 \\ \vdots \\ v_n \end{bmatrix}$$

in R^n the number, denoted by $(\mathbf{u}, \mathbf{v})$, given by

$$(\mathbf{u}, \mathbf{v}) = u_1v_1 + u_2v_2 + \cdots + u_nv_n.$$

Of course, we must verify that this function satisfies the properties of Definition 5.1. ■

Remarks

1. See Appendix B.2 for the definition of inner product on an arbitrary vector space V where the scalars are restricted to be complex numbers. (Of course, they could be real numbers.)
2. If we view the vectors $\mathbf{u}$ and $\mathbf{v}$ in R^n as $n \times 1$ matrices, then we can write the standard inner product of $\mathbf{u}$ and $\mathbf{v}$ in terms of matrix multiplication as

$$(\mathbf{u}, \mathbf{v}) = \mathbf{u}^T \mathbf{v}, \tag{1}$$

where we have ignored the brackets around the 1×1 matrix $\mathbf{u}^T\mathbf{v}$ (Exercise 39).

EXAMPLE 2

Let V be any finite-dimensional vector space and let $S = \{\mathbf{u}_1, \mathbf{u}_2, \ldots, \mathbf{u}_n\}$ be an ordered basis for V. If

$$\mathbf{v} = a_1\mathbf{u}_1 + a_2\mathbf{u}_2 + \cdots + a_n\mathbf{u}_n$$

and

$$\mathbf{w} = b_1\mathbf{u}_1 + b_2\mathbf{u}_2 + \cdots + b_n\mathbf{u}_n,$$

we define

$$(\mathbf{v}, \mathbf{w}) = \left(\left[\mathbf{v}\right]_S, \left[\mathbf{w}\right]_S\right) = a_1b_1 + a_2b_2 + \cdots + a_nb_n.$$

It is not difficult to verify that this defines an inner product on V (Exercise 4). This definition of $(\mathbf{v}, \mathbf{w})$ as an inner product on V uses the standard inner product on R^n. ■

Example 2 shows that we can define an inner product on any finite-dimensional vector space. Of course, if we change the basis for V in Example 2, we obtain a different inner product.

EXAMPLE 3

Let $\mathbf{u} = \begin{bmatrix} u_1 \\ u_2 \end{bmatrix}$ and $\mathbf{v} = \begin{bmatrix} v_1 \\ v_2 \end{bmatrix}$ be vectors in R^2. We define

$$(\mathbf{u}, \mathbf{v}) = u_1v_1 - u_2v_1 - u_1v_2 + 3u_2v_2.$$

Show that this gives an inner product on R^2.

Solution

We have

$$\begin{aligned}(\mathbf{u}, \mathbf{u}) &= u_1^2 - 2u_1u_2 + 3u_2^2 = u_1^2 - 2u_1u_2 + u_2^2 + 2u_2^2 \\ &= (u_1 - u_2)^2 + 2u_2^2 \geq 0.\end{aligned}$$

Moreover, if $(\mathbf{u}, \mathbf{u}) = 0$, then $u_1 = u_2$ and $u_2 = 0$, so $\mathbf{u} = \mathbf{0}$. Conversely, if $\mathbf{u} = \mathbf{0}$, then $(\mathbf{u}, \mathbf{u}) = 0$. We can also verify (see Exercise 2) the remaining three properties of Definition 5.1. This inner product is, of course, not the standard inner product on R^2. ■

Example 3 shows that on one vector space we may have more than one inner product, since we also have the standard inner product on R^2.

EXAMPLE 4

Let V be the vector space of all continuous real-valued functions on the unit interval $[0, 1]$. For f and g in V, we let $(f, g) = \int_0^1 f(t)g(t)\,dt$. We now verify that this is an inner product on V, that is, that the properties of Definition 5.1 are satisfied.

Using results from calculus, we have for $f \neq 0$, the zero function,

$$(f, f) = \int_0^1 (f(t))^2\,dt \geq 0.$$

Moreover, if $(f, f) = 0$, then $f = 0$. Conversely, if $f = 0$, then $(f, f) = 0$. Also,

$$(f, g) = \int_0^1 f(t)g(t)\,dt = \int_0^1 g(t)f(t)\,dt = (g, f).$$

Next,

$$\begin{aligned}(f+g, h) &= \int_0^1 (f(t)+g(t))h(t)\,dt = \int_0^1 f(t)h(t)\,dt + \int_0^1 g(t)h(t)\,dt \\ &= (f, h) + (g, h).\end{aligned}$$

Finally,

$$(cf, g) = \int_0^1 (cf(t))g(t)\,dt = c\int_0^1 f(t)g(t)\,dt = c(f, g).$$

Thus, for example, if f and g are the functions defined by $f(t) = t + 1$, $g(t) = 2t + 3$, then

$$(f, g) = \int_0^1 (t+1)(2t+3)\,dt = \int_0^1 (2t^2 + 5t + 3)\,dt = \tfrac{37}{6}. \quad ■$$

EXAMPLE 5

Let $V = R_2$; if $\mathbf{u} = \begin{bmatrix} u_1 & u_2 \end{bmatrix}$ and $\mathbf{v} = \begin{bmatrix} v_1 & v_2 \end{bmatrix}$ are vectors in V, we define $(\mathbf{u}, \mathbf{v}) = u_1v_1 - u_2v_1 - u_1v_2 + 5u_2v_2$. The verification that this function is an inner product is entirely analogous to the verification required in Example 3 (Exercise 5). ■

EXAMPLE 6

Let $V = P$; if $p(t)$ and $q(t)$ are polynomials in P, we define

$$(p(t), q(t)) = \int_0^1 p(t)q(t)\,dt.$$

The verification that this function is an inner product is identical to the verification given for Example 4 (Exercise 6). ■

We now show that every inner product on a finite-dimensional vector space V is completely determined, in terms of a given basis, by a certain matrix.

Theorem 5.2 Let $S = \{\mathbf{u}_1, \mathbf{u}_2, \ldots, \mathbf{u}_n\}$ be an ordered basis for a finite-dimensional vector space V, and assume that we are given an inner product on V. Let $c_{ij} = (\mathbf{u}_i, \mathbf{u}_j)$ and $C = \begin{bmatrix} c_{ij} \end{bmatrix}$. Then

(a) C is a symmetric matrix.

(b) C determines $(\mathbf{v}, \mathbf{w})$ for every $\mathbf{v}$ and $\mathbf{w}$ in V.

Proof

(a) Exercise.

(b) If $\mathbf{v}$ and $\mathbf{w}$ are in V, then

$$\mathbf{v} = a_1\mathbf{u}_1 + a_2\mathbf{u}_2 + \cdots + a_n\mathbf{u}_n$$
$$\mathbf{w} = b_1\mathbf{u}_1 + b_2\mathbf{u}_2 + \cdots + b_n\mathbf{u}_n,$$

which implies that

$$\left[\mathbf{v}\right]_S = \begin{bmatrix} a_1 \\ a_2 \\ \vdots \\ a_n \end{bmatrix} \quad \text{and} \quad \left[\mathbf{w}\right]_S = \begin{bmatrix} b_1 \\ b_2 \\ \vdots \\ b_n \end{bmatrix}.$$

The inner product $(\mathbf{v}, \mathbf{w})$ can then be expressed as

$$\begin{aligned}
(\mathbf{v}, \mathbf{w}) &= \left(\sum_{i=1}^{n} a_i\mathbf{u}_i, \mathbf{w}\right) = \sum_{i=1}^{n} a_i(\mathbf{u}_i, \mathbf{w}) \\
&= \sum_{i=1}^{n} a_i\left(\mathbf{u}_i, \sum_{j=1}^{n} b_j\mathbf{u}_j\right) \\
&= \sum_{i=1}^{n} a_i \sum_{j=1}^{n} b_j(\mathbf{u}_i, \mathbf{u}_j) = \sum_{i=1}^{n}\sum_{j=1}^{n} a_i b_j(\mathbf{u}_i, \mathbf{u}_j) \\
&= \sum_{i=1}^{n}\sum_{j=1}^{n} a_i c_{ij} b_j \\
&= \left[\mathbf{v}\right]_S^T C \left[\mathbf{w}\right]_S,
\end{aligned}$$

so

$$(\mathbf{v}, \mathbf{w}) = \left[\mathbf{v}\right]_S^T C \left[\mathbf{w}\right]_S, \tag{2}$$

which means that C determines $(\mathbf{v}, \mathbf{w})$ for every $\mathbf{v}$ and $\mathbf{w}$ in V. ■

Thus the inner product in Equation (2) is the product of three matrices. We next show that the inner product in (2) can also be expressed in terms of a standard inner product on R^n. We first establish the following result for the standard inner product on R^n:

If $A = \left[a_{ij}\right]$ is an $n \times n$ matrix and $\mathbf{x}$ and $\mathbf{y}$ are vectors in R^n, then

$$(A\mathbf{x}, \mathbf{y}) = (\mathbf{x}, A^T\mathbf{y}). \tag{3}$$

Equation (1), together with associativity of matrix multiplication and Theorem 1.4(c), can now be used to prove (3):

$$(A\mathbf{x}, \mathbf{y}) = (A\mathbf{x})^T\mathbf{y} = (\mathbf{x}^T A^T)\mathbf{y} = \mathbf{x}^T(A^T\mathbf{y}) = (\mathbf{x}, A^T\mathbf{y}).$$

Using Equation (1), we can now write (2) as

$$(\mathbf{v}, \mathbf{w}) = \left(\left[\mathbf{v}\right]_S, C\left[\mathbf{w}\right]_S\right), \tag{4}$$

where the inner product on the left is in V and the inner product on the right is the standard inner product on R^n. Using (1) and the fact that C is symmetric, we have

$$(\mathbf{v}, \mathbf{w}) = \left(C\left[\mathbf{v}\right]_S, \left[\mathbf{w}\right]_S\right) \tag{5}$$

(verify). Thus C determines $(\mathbf{v}, \mathbf{w})$ for every $\mathbf{v}$ and $\mathbf{w}$ in V. In summary, we have shown that an inner product on a real finite-dimensional vector space V can be computed by using the standard inner product on R^n, as in (4) or (5).

The matrix C in Theorem 5.2 is called the **matrix of the inner product with respect to the ordered basis** S. If the inner product is as defined in Example 2, then $C = I_n$ (verify).

There is another important property satisfied by the matrix of an inner product. If $\mathbf{u}$ is a nonzero vector in R^n, then $(\mathbf{u}, \mathbf{u}) > 0$; so letting $\mathbf{x} = \left[\mathbf{u}\right]_S$, Equation (2) says that

$$\mathbf{x}^T C\mathbf{x} > 0 \quad \text{for every nonzero } \mathbf{x} \text{ in } R^n.$$

This property of the matrix of an inner product is so important that we specifically identify such matrices. An $n \times n$ symmetric matrix C with the property that $\mathbf{x}^T C\mathbf{x} > 0$ for every nonzero vector $\mathbf{x}$ in R^n is called **positive definite**. A positive definite matrix C is nonsingular, for if C is singular, then the homogeneous system $C\mathbf{x} = \mathbf{0}$ has a nontrivial solution $\mathbf{x}_0$. Then $\mathbf{x}_0^T C\mathbf{x}_0 = 0$, contradicting the requirement that $\mathbf{x}^T C\mathbf{x} > 0$ for any nonzero vector $\mathbf{x}$.

If $C = \left[c_{ij}\right]$ is an $n \times n$ positive definite matrix, then we can use C to define an inner product on V. Using the same notation as before, we define

$$(\mathbf{v}, \mathbf{w}) = \left(\left[\mathbf{v}\right]_S, C\left[\mathbf{w}\right]_S\right) = \sum_{i=1}^{n}\sum_{j=1}^{n} a_i c_{ij} b_j .$$

It is not difficult to show that this defines an inner product on V (verify). The only gap in the preceding discussion is that we still do not know when a symmetric matrix is positive definite, other than trying to verify the definition, which is usually not a fruitful approach. In Section 8.6 (see Theorem 8.11) we provide a characterization of positive definite matrices.

EXAMPLE 7

Let $C = \begin{bmatrix} 2 & 1 \\ 1 & 2 \end{bmatrix}$. In this case we may verify that C is positive definite as follows:

$$\begin{aligned}
\mathbf{x}^T C\mathbf{x} &= \begin{bmatrix} x_1 & x_2 \end{bmatrix} \begin{bmatrix} 2 & 1 \\ 1 & 2 \end{bmatrix} \begin{bmatrix} x_1 \\ x_2 \end{bmatrix} \\
&= 2x_1^2 + 2x_1x_2 + 2x_2^2 \\
&= x_1^2 + x_2^2 + (x_1 + x_2)^2 > 0 \quad \text{if } \mathbf{x} \neq \mathbf{0}
\end{aligned}$$

■

We now define an inner product on P_1 whose matrix with respect to the ordered basis $S = \{t, 1\}$ is C. Thus let $p(t) = a_1t + a_2$ and $q(t) = b_1t + b_2$ be any two vectors in P_1. Let $(p(t), q(t)) = 2a_1b_1 + a_2b_1 + a_1b_2 + 2a_2b_2$. We must verify that $(p(t), p(t)) \geq 0$; that is, $2a_1^2 + 2a_1a_2 + 2a_2^2 \geq 0$. We now have

$$2a_1^2 + 2a_1a_2 + 2a_2^2 = a_1^2 + a_2^2 + (a_1 + a_2)^2 \geq 0.$$

Moreover, if $p(t) = 0$, so that $a_1 = 0$ and $a_2 = 0$, then $(p(t), p(t)) = 0$. Conversely, if $(p(t), p(t)) = 0$, then $a_1 = 0$ and $a_2 = 0$, so $p(t) = 0$. The remaining properties are not difficult to verify.

DEFINITION 5.2

A real vector space that has an inner product defined on it is called an **inner product space**. If the space is finite dimensional, it is called a **Euclidean space**.

If V is an inner product space, then by the **dimension** of V we mean the dimension of V as a real vector space, and a set S is a **basis** for V if S is a basis for the real vector space V. Examples 1 through 6 are inner product spaces. Examples 1 through 5 are Euclidean spaces.

In an inner product space we define the **length** of a vector $\mathbf{u}$ by $\|\mathbf{u}\| = \sqrt{(\mathbf{u}, \mathbf{u})}$. This definition of length seems reasonable, because at least we have $\|\mathbf{u}\| > 0$ if $\mathbf{u} \neq \mathbf{0}$. We can show (see Exercise 7) that $\|\mathbf{0}\| = 0$.

We now prove a result that will enable us to give a worthwhile definition for the cosine of an angle between two nonzero vectors $\mathbf{u}$ and $\mathbf{v}$ in an inner product space V. This result, called the **Cauchy*–Schwarz† inequality**, has many important applications in mathematics. The proof, although not difficult, is one that is not too natural and does call for a clever start.

Theorem 5.3 **Cauchy*–Schwarz† Inequality**

If $\mathbf{u}$ and $\mathbf{v}$ are any two vectors in an inner product space V, then

$$|(\mathbf{u}, \mathbf{v})| \leq \|\mathbf{u}\| \, \|\mathbf{v}\|.$$

AUGUSTIN-LOUIS CAUCHY

KARL HERMANN AMANDUS SCHWARZ

Proof

If $\mathbf{u} = 0$, then $\|\mathbf{u}\| = 0$ and by Exercise 7(b), $(\mathbf{u}, \mathbf{v}) = 0$, so the inequality holds. Now suppose that $\mathbf{u}$ is nonzero. Let r be a scalar and consider the vector $r\mathbf{u} + \mathbf{v}$. Since the inner product of a vector with itself is always nonnegative, we have

$$0 \leq (r\mathbf{u} + \mathbf{v}, r\mathbf{u} + \mathbf{v}) = (\mathbf{u}, \mathbf{u})r^2 + 2r(\mathbf{u}, \mathbf{v}) + (\mathbf{v}, \mathbf{v}) = ar^2 + 2br + c,$$

*Augustin-Louis Cauchy (1789–1857) grew up in a suburb of Paris as a neighbor of several leading mathematicians of the day, attended the École Polytechnique and the École des Ponts et Chaussées, and was for a time a practicing engineer. He was a devout Roman Catholic, with an abiding interest in Catholic charities. He was also strongly devoted to royalty, especially to the Bourbon kings who ruled France after Napoleon's defeat. When Charles X was deposed in 1830, Cauchy voluntarily followed him into exile in Prague.

Cauchy wrote seven books and more than 700 papers of varying quality, touching on all branches of mathematics. He made important contributions to the early theory of determinants, the theory of eigenvalues, the study of ordinary and partial differential equations, the theory of permutation groups, and the foundations of calculus; and he founded the theory of functions of a complex variable.

†Karl Hermann Amandus Schwarz (1843–1921) was born in Poland, but was educated and taught in Germany. He was a protégé of Karl Weierstrass and of Ernst Eduard Kummer, whose daughter he married. His main contributions to mathematics were in the geometric aspects of analysis, such as conformal mappings and minimal surfaces. In connection with the latter, he sought certain numbers associated with differential equations, numbers that have since come to be called eigenvalues. The inequality given in the text was used in the search for these numbers.

where $a = (\mathbf{u}, \mathbf{u})$, $b = (\mathbf{u}, \mathbf{v})$, and $c = (\mathbf{v}, \mathbf{v})$. If we fix $\mathbf{u}$ and $\mathbf{v}$, then $ar^2 + 2br + c = p(r)$ is a quadratic polynomial in r that is nonnegative for all values of r. This means that $p(r)$ has at most one real root, for if it had two distinct real roots, r_1 and r_2, it would be negative between r_1 and r_2 (Figure 5.18). From the quadratic formula, the roots of $p(r)$ are given by

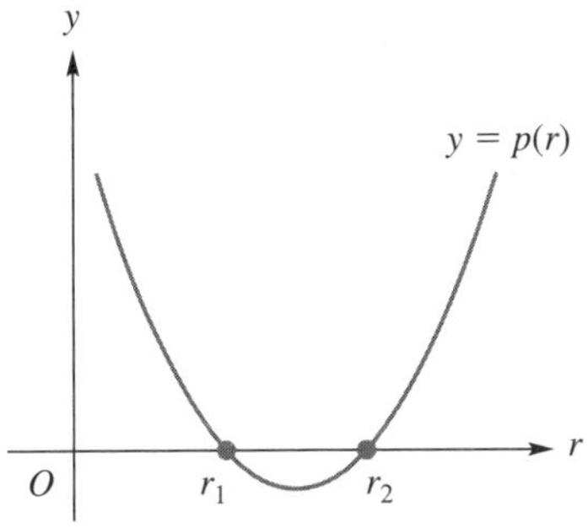

FIGURE 5.18

$$\frac{-b + \sqrt{b^2 - ac}}{a} \quad \text{and} \quad \frac{-b - \sqrt{b^2 - ac}}{a}$$

($a \neq 0$ since $\mathbf{u} \neq \mathbf{0}$). Thus we must have $b^2 - ac \leq 0$, which means that $b^2 \leq ac$. Taking square roots, we have $|b| \leq \sqrt{a}\,\sqrt{c}$. Substituting for a, b, and c, we obtain the desired inequality. ■

Remark The result widely known as the Cauchy–Schwarz inequality (Theorem 5.3) provides a good example of how nationalistic feelings make their way into science. In Russia this result is generally known as Bunyakovsky's[‡] inequality. In France it is often referred to as *Cauchy's inequality*, and in Germany it is frequently called *Schwarz's inequality*. In an attempt to distribute credit for the result among all three contenders, a minority of authors refer to the result as the *CBS inequality*.

EXAMPLE 8

Let

$$\mathbf{u} = \begin{bmatrix} 1 \\ 2 \\ -3 \end{bmatrix} \quad \text{and} \quad \mathbf{v} = \begin{bmatrix} -3 \\ 2 \\ 2 \end{bmatrix}$$

be in the Euclidean space R^3 with the standard inner product. Then $(\mathbf{u}, \mathbf{v}) = -5$, $\|\mathbf{u}\| = \sqrt{14}$, and $\|\mathbf{v}\| = \sqrt{17}$. Therefore, $|(\mathbf{u}, \mathbf{v})| \leq \|\mathbf{u}\|\,\|\mathbf{v}\|$. ■

If $\mathbf{u}$ and $\mathbf{v}$ are any two nonzero vectors in an inner product space V, the Cauchy–Schwarz inequality can be written as

$$-1 \leq \frac{(\mathbf{u}, \mathbf{v})}{\|\mathbf{u}\|\,\|\mathbf{v}\|} \leq 1.$$

It then follows that there is one and only one angle θ such that

$$\cos\theta = \frac{(\mathbf{u}, \mathbf{v})}{\|\mathbf{u}\|\,\|\mathbf{v}\|}, \qquad 0 \leq \theta \leq \pi.$$

We define this angle to be the **angle** between $\mathbf{u}$ and $\mathbf{v}$.

The triangle inequality is an easy consequence of the Cauchy–Schwarz inequality.

[‡] Viktor Yakovlevich Bunyakovsky (1804–1889) was born in Bar, Ukraine. He received a doctorate in Paris in 1825. He carried out additional studies in St. Petersburg and then had a long career there as a professor. Bunyakovsky made important contributions in number theory and also worked in geometry, applied mechanics, and hydrostatics. His proof of the Cauchy–Schwarz inequality appeared in one of his monographs in 1859, 25 years before Schwarz published his proof. He died in St. Petersburg.

Corollary 5.1 Triangle Inequality

If $\mathbf{u}$ and $\mathbf{v}$ are any vectors in an inner product space V, then $\|\mathbf{u}+\mathbf{v}\| \le \|\mathbf{u}\| + \|\mathbf{v}\|$.

Proof

We have

$$\begin{aligned}\|\mathbf{u}+\mathbf{v}\|^2 &= (\mathbf{u}+\mathbf{v}, \mathbf{u}+\mathbf{v}) = (\mathbf{u}, \mathbf{u}) + 2(\mathbf{u}, \mathbf{v}) + (\mathbf{v}, \mathbf{v}) \\ &= \|\mathbf{u}\|^2 + 2(\mathbf{u}, \mathbf{v}) + \|\mathbf{v}\|^2.\end{aligned}$$

The Cauchy–Schwarz inequality states that $(\mathbf{u}, \mathbf{v}) \le |(\mathbf{u}, \mathbf{v})| \le \|\mathbf{u}\|\,\|\mathbf{v}\|$, so

$$\|\mathbf{u}+\mathbf{v}\|^2 \le \|\mathbf{u}\|^2 + 2\|\mathbf{u}\|\,\|\mathbf{v}\| + \|\mathbf{v}\|^2 = (\|\mathbf{u}\| + \|\mathbf{v}\|)^2.$$

Taking square roots, we obtain

$$\|\mathbf{u}+\mathbf{v}\| \le \|\mathbf{u}\| + \|\mathbf{v}\|.$$ ■

We now state the Cauchy–Schwarz inequality for the inner product spaces introduced in several of our examples. In Example 1, if

$$\mathbf{u} = \begin{bmatrix} u_1 \\ u_2 \\ \vdots \\ u_n \end{bmatrix} \quad \text{and} \quad \mathbf{v} = \begin{bmatrix} v_1 \\ v_2 \\ \vdots \\ v_n \end{bmatrix},$$

then

$$|(\mathbf{u}, \mathbf{v})| = \left|\sum_{i=1}^{n} u_i v_i\right| \le \left(\sqrt{\sum_{i=1}^{n} u_i^2}\right)\left(\sqrt{\sum_{i=1}^{n} v_i^2}\right) = \|\mathbf{u}\|\,\|\mathbf{v}\|.$$

In Example 4, if f and g are continuous functions on $[0, 1]$, then

$$|(f, g)| = \left|\int_0^1 f(t)g(t)\,dt\right| \le \left(\sqrt{\int_0^1 f^2(t)\,dt}\right)\left(\sqrt{\int_0^1 g^2(t)\,dt}\right).$$

EXAMPLE 9 Let V be the Euclidean space P_2 with inner product defined as in Example 6. If $p(t) = t + 2$, then the length of $p(t)$ is

$$\|p(t)\| = \sqrt{(p(t), p(t))} = \sqrt{\int_0^1 (t+2)^2\,dt} = \sqrt{\frac{19}{3}}.$$

If $q(t) = 2t - 3$, then to find the cosine of the angle θ between $p(t)$ and $q(t)$, we proceed as follows. First,

$$\|q(t)\| = \sqrt{\int_0^1 (2t-3)^2\,dt} = \sqrt{\frac{13}{3}}.$$

Next,

$$(p(t), q(t)) = \int_0^1 (t+2)(2t-3)\,dt = \int_0^1 (2t^2 + t - 6)\,dt = -\frac{29}{6}.$$

Then

$$\cos\theta = \frac{(p(t), q(t))}{\|p(t)\|\,\|q(t)\|} = \frac{-\frac{29}{6}}{\sqrt{\frac{19}{3}}\sqrt{\frac{13}{3}}} = \frac{-29}{2\sqrt{(19)(13)}}.$$ ■

DEFINITION 5.3

If V is an inner product space, we define the **distance** between two vectors $\mathbf{u}$ and $\mathbf{v}$ in V as $d(\mathbf{u}, \mathbf{v}) = \|\mathbf{u} - \mathbf{v}\|$.

DEFINITION 5.4

Let V be an inner product space. Two vectors $\mathbf{u}$ and $\mathbf{v}$ in V are **orthogonal** if $(\mathbf{u}, \mathbf{v}) = 0$.

EXAMPLE 10

Let V be the Euclidean space R^4 with the standard inner product. If

$$\mathbf{u} = \begin{bmatrix} 1 \\ 0 \\ 0 \\ 1 \end{bmatrix} \quad \text{and} \quad \mathbf{v} = \begin{bmatrix} 0 \\ 2 \\ 3 \\ 0 \end{bmatrix},$$

then $(\mathbf{u}, \mathbf{v}) = 0$, so $\mathbf{u}$ and $\mathbf{v}$ are orthogonal. ■

EXAMPLE 11

Let V be the inner product space P_2 considered in Example 9. The vectors t and $t - \frac{2}{3}$ are orthogonal, since

$$\left(t, t - \frac{2}{3}\right) = \int_0^1 t\left(t - \frac{2}{3}\right) dt = \int_0^1 \left(t^2 - \frac{2t}{3}\right) dt = 0.$$ ■

Of course, the vector $\mathbf{0}_V$ in an inner product space V is orthogonal to every vector in V [see Exercise 7(b)], and two nonzero vectors in V are orthogonal if the angle θ between them is $\pi/2$. Also, the subset of vectors in V orthogonal to a fixed vector in V is a subspace of V (see Exercise 23).

We know from calculus that we can work with any set of coordinate axes for R^3, but that the work becomes less burdensome when we deal with Cartesian coordinates. The comparable notion in an inner product space is that of a basis whose vectors are mutually orthogonal. We now proceed to formulate this idea.

DEFINITION 5.5

Let V be an inner product space. A set S of vectors in V is called **orthogonal** if any two distinct vectors in S are orthogonal. If, in addition, each vector in S is of unit length, then S is called **orthonormal**.

We note here that if $\mathbf{x}$ is a nonzero vector in an inner product space, then we can always find a vector of unit length (called a **unit vector**) in the same direction as $\mathbf{x}$; we let $\mathbf{u} = \frac{1}{\|\mathbf{x}\|}\mathbf{x}$. Then

$$\|\mathbf{u}\| = \sqrt{(\mathbf{u}, \mathbf{u})} = \sqrt{\left(\frac{1}{\|\mathbf{x}\|}\mathbf{x}, \frac{1}{\|\mathbf{x}\|}\mathbf{x}\right)} = \sqrt{\frac{(\mathbf{x}, \mathbf{x})}{\|\mathbf{x}\|\,\|\mathbf{x}\|}} = \sqrt{\frac{\|\mathbf{x}\|^2}{\|\mathbf{x}\|\,\|\mathbf{x}\|}} = 1,$$

and the cosine of the angle between $\mathbf{x}$ and $\mathbf{u}$ is 1, so $\mathbf{x}$ and $\mathbf{u}$ have the same direction.

EXAMPLE 12 If $\mathbf{x}_1 = \begin{bmatrix} 1 \\ 0 \\ 2 \end{bmatrix}$, $\mathbf{x}_2 = \begin{bmatrix} -2 \\ 0 \\ 1 \end{bmatrix}$, and $\mathbf{x}_3 = \begin{bmatrix} 0 \\ 1 \\ 0 \end{bmatrix}$, then $\{\mathbf{x}_1, \mathbf{x}_2, \mathbf{x}_3\}$ is an orthogonal set (verify). The vectors

$$\mathbf{u}_1 = \begin{bmatrix} \dfrac{1}{\sqrt{5}} \\ 0 \\ \dfrac{2}{\sqrt{5}} \end{bmatrix} \quad \text{and} \quad \mathbf{u}_2 = \begin{bmatrix} -\dfrac{2}{\sqrt{5}} \\ 0 \\ \dfrac{1}{\sqrt{5}} \end{bmatrix}$$

are unit vectors in the directions of $\mathbf{x}_1$ and $\mathbf{x}_2$, respectively. Since $\mathbf{x}_3$ is also a unit vector, we conclude that $\{\mathbf{u}_1, \mathbf{u}_2, \mathbf{x}_3\}$ is an orthonormal set. ■

EXAMPLE 13 The natural bases for R^n and R_n are orthonormal sets with respect to the standard inner products on these vector spaces. ■

An important result about orthogonal sets of vectors in an inner product space is the following:

Theorem 5.4 Let $S = \{\mathbf{u}_1, \mathbf{u}_2, \ldots, \mathbf{u}_n\}$ be a finite orthogonal set of nonzero vectors in an inner product space V. Then S is linearly independent.

Proof

Suppose that

$$a_1\mathbf{u}_1 + a_2\mathbf{u}_2 + \cdots + a_n\mathbf{u}_n = \mathbf{0}.$$

Then taking the inner product of both sides with $\mathbf{u}_i$, we have

$$(a_1\mathbf{u}_1 + a_2\mathbf{u}_2 + \cdots + a_i\mathbf{u}_i + \cdots + a_n\mathbf{u}_n, \mathbf{u}_i) = (\mathbf{0}, \mathbf{u}_i) = 0.$$

The left side is

$$a_1(\mathbf{u}_1, \mathbf{u}_i) + a_2(\mathbf{u}_2, \mathbf{u}_i) + \cdots + a_i(\mathbf{u}_i, \mathbf{u}_i) + \cdots + a_n(\mathbf{u}_n, \mathbf{u}_i),$$

and since S is orthogonal, this is $a_i(\mathbf{u}_i, \mathbf{u}_i)$. Thus $a_i(\mathbf{u}_i, \mathbf{u}_i) = 0$. Since $\mathbf{u}_i \neq \mathbf{0}$, $(\mathbf{u}_i, \mathbf{u}_i) \neq 0$, so $a_i = 0$. Repeating this for $i = 1, 2, \ldots, n$, we find that $a_1 = a_2 = \cdots = a_n = 0$, so S is linearly independent. ■

EXAMPLE 14 Let V be the vector space of all continuous real-valued functions on $[-\pi, \pi]$. For f and g in V, we let

$$(f, g) = \int_{-\pi}^{\pi} f(t)g(t)\,dt,$$

which is shown to be an inner product on V (see Example 4). Consider the functions

$$1, \cos t, \sin t, \cos 2t, \sin 2t, \ldots, \cos nt, \sin nt, \ldots, \tag{6}$$

which are clearly in V. The relationships

$$\int_{-\pi}^{\pi} \cos nt\, dt = \int_{-\pi}^{\pi} \sin nt\, dt = \int_{-\pi}^{\pi} \sin nt \cos nt\, dt = 0,$$

$$\int_{-\pi}^{\pi} \cos mt \cos nt\, dt = \int_{-\pi}^{\pi} \sin mt \sin nt\, dt = 0 \quad \text{if } m \neq n$$

demonstrate that $(f, g) = 0$ whenever f and g are distinct functions from (6). Hence every finite subset of functions from (6) is an orthogonal set. Theorem 5.4 then implies that any finite subset of functions from (6) is linearly independent. The functions in (6) were studied by the French mathematician Jean Baptiste Joseph Fourier. We take a closer look at these functions in Section 5.5. ■

Key Terms

Real vector space
Inner product space
Standard inner (dot) product on R^n
Matrix of the inner product with respect to an ordered basis
Positive definite matrix
Euclidean space
Dimension
Basis
Cauchy–Schwarz inequality
CBS inequality
Triangle inequality
Distance between vectors
Orthogonal vectors
Orthogonal set
Unit vectors
Orthonormal set

5.3 Exercises

1. Verify that the standard inner product on R^n satisfies the properties of Definition 5.1.
2. Verify that the function in Example 3 satisfies the remaining three properties of Definition 5.1.
3. Let $V = M_{nn}$ be the real vector space of all $n \times n$ matrices. If A and B are in V, we define $(A, B) = \text{Tr}(B^T A)$, where Tr is the trace function defined in Exercise 43 of Section 1.3. Show that this function is an inner product on V.
4. Verify that the function defined on V in Example 2 is an inner product.
5. Verify that the function defined on R_2 in Example 5 is an inner product.
6. Verify that the function defined on P in Example 6 is an inner product.
7. Let V be an inner product space. Show the following:
   (a) $\|\mathbf{0}\| = 0$.
   (b) $(\mathbf{u}, \mathbf{0}) = (\mathbf{0}, \mathbf{u}) = 0$ for any $\mathbf{u}$ in V.
   (c) If $(\mathbf{u}, \mathbf{v}) = 0$ for all $\mathbf{v}$ in V, then $\mathbf{u} = \mathbf{0}$.
   (d) If $(\mathbf{u}, \mathbf{w}) = (\mathbf{v}, \mathbf{w})$ for all $\mathbf{w}$ in V, then $\mathbf{u} = \mathbf{v}$.
   (e) If $(\mathbf{w}, \mathbf{u}) = (\mathbf{w}, \mathbf{v})$ for all $\mathbf{w}$ in V, then $\mathbf{u} = \mathbf{v}$.

In Exercises 8 and 9, let V be the Euclidean space R_4 with the standard inner product. Compute $(\mathbf{u}, \mathbf{v})$.

8. **(a)** $\mathbf{u} = \begin{bmatrix} 1 & 3 & -1 & 2 \end{bmatrix}, \mathbf{v} = \begin{bmatrix} -1 & 2 & 0 & 1 \end{bmatrix}$
   (b) $\mathbf{u} = \begin{bmatrix} 0 & 0 & 1 & 1 \end{bmatrix}, \mathbf{v} = \begin{bmatrix} 1 & 1 & 0 & 0 \end{bmatrix}$
   (c) $\mathbf{u} = \begin{bmatrix} -2 & 1 & 3 & 4 \end{bmatrix}, \mathbf{v} = \begin{bmatrix} 3 & 2 & 1 & -2 \end{bmatrix}$
9. **(a)** $\mathbf{u} = \begin{bmatrix} 1 & 2 & 3 & 4 \end{bmatrix}, \mathbf{v} = \begin{bmatrix} -1 & 0 & -1 & -1 \end{bmatrix}$
   (b) $\mathbf{u} = \begin{bmatrix} 0 & -1 & 1 & 4 \end{bmatrix}, \mathbf{v} = \begin{bmatrix} 2 & 0 & -8 & 2 \end{bmatrix}$
   (c) $\mathbf{u} = \begin{bmatrix} 0 & 0 & -1 & 2 \end{bmatrix}, \mathbf{v} = \begin{bmatrix} 2 & 3 & -1 & 0 \end{bmatrix}$

In Exercises 10 and 11, use the inner product space of continuous functions on [0, 1] defined in Example 4. Find (f, g) for the following:

10. **(a)** $f(t) = 1 + t,\ g(t) = 2 - t$
    (b) $f(t) = 1,\ g(t) = 3$
    (c) $f(t) = 1,\ g(t) = 3 + 2t$
11. **(a)** $f(t) = 3t,\ g(t) = 2t^2$
    (b) $f(t) = t,\ g(t) = e^t$
    (c) $f(t) = \sin t,\ g(t) = \cos t$

In Exercises 12 and 13, let V be the Euclidean space of Example 3. Compute the length of each given vector.

12. **(a)** $\begin{bmatrix} 1 \\ 3 \end{bmatrix}$ **(b)** $\begin{bmatrix} 3 \\ -1 \end{bmatrix}$ **(c)** $\begin{bmatrix} 1 \\ 0 \end{bmatrix}$

13. **(a)** $\begin{bmatrix} 0 \\ -2 \end{bmatrix}$ **(b)** $\begin{bmatrix} -2 \\ -4 \end{bmatrix}$ **(c)** $\begin{bmatrix} 2 \\ 2 \end{bmatrix}$

In Exercises 14 and 15, let V be the inner product space of Example 6. Find the cosine of the angle between each pair of vectors in V.

14. **(a)** $p(t) = t$, $q(t) = t - 1$

(b) $p(t) = t$, $q(t) = t$

(c) $p(t) = 1$, $q(t) = 2t + 3$

15. **(a)** $p(t) = 1$, $q(t) = 1$

(b) $p(t) = t^2$, $q(t) = 2t^3 - \frac{4}{3}t$

(c) $p(t) = \sin t$, $q(t) = \cos t$

16. Prove the **parallelogram law** for any two vectors in an inner product space:

$$\|\mathbf{u} + \mathbf{v}\|^2 + \|\mathbf{u} - \mathbf{v}\|^2 = 2\|\mathbf{u}\|^2 + 2\|\mathbf{v}\|^2.$$

17. Let V be an inner product space. Show that $\|c\mathbf{u}\| = |c|\,\|\mathbf{u}\|$ for any vector $\mathbf{u}$ and any scalar c.

18. State the Cauchy–Schwarz inequality for the inner product spaces defined in Example 3, Example 5, and Exercise 3.

19. Let V be an inner product space. Prove that if $\mathbf{u}$ and $\mathbf{v}$ are any vectors in V, then $\|\mathbf{u} + \mathbf{v}\|^2 = \|\mathbf{u}\|^2 + \|\mathbf{v}\|^2$ if and only if $(\mathbf{u}, \mathbf{v}) = 0$, that is, if and only if $\mathbf{u}$ and $\mathbf{v}$ are orthogonal. This result is known as the **Pythagorean theorem**.

20. Let $\{\mathbf{u}, \mathbf{v}, \mathbf{w}\}$ be an orthonormal set of vectors in an inner product space V. Compute $\|\mathbf{u} + \mathbf{v} + \mathbf{w}\|^2$.

21. Let V be an inner product space. If $\mathbf{u}$ and $\mathbf{v}$ are vectors in V, show that

$$(\mathbf{u}, \mathbf{v}) = \tfrac{1}{4}\|\mathbf{u} + \mathbf{v}\|^2 - \tfrac{1}{4}\|\mathbf{u} - \mathbf{v}\|^2.$$

22. Let V be the Euclidean space R_4 considered in Exercise 8. Find which of the pairs of vectors listed there are orthogonal.

23. Let V be an inner product space and $\mathbf{u}$ a fixed vector in V. Prove that the set of all vectors in V that are orthogonal to $\mathbf{u}$ is a subspace of V.

24. For each of the inner products defined in Examples 3 and 5, choose an ordered basis S for the vector space and find the matrix of the inner product with respect to S.

25. Let $C = \begin{bmatrix} 3 & -2 \\ -2 & 3 \end{bmatrix}$. Define an inner product on R_2 whose matrix with respect to the natural ordered basis is C.

26. If V is an inner product space, prove that the distance function of Definition 5.3 satisfies the following properties for all vectors $\mathbf{u}$, $\mathbf{v}$, and $\mathbf{w}$ in V:

(a) $d(\mathbf{u}, \mathbf{v}) \geq 0$

(b) $d(\mathbf{u}, \mathbf{v}) = 0$ if and only if $\mathbf{u} = \mathbf{v}$

(c) $d(\mathbf{u}, \mathbf{v}) = d(\mathbf{v}, \mathbf{u})$

(d) $d(\mathbf{u}, \mathbf{v}) \leq d(\mathbf{u}, \mathbf{w}) + d(\mathbf{w}, \mathbf{v})$

In Exercises 27 and 28, let V be the inner product space of Example 4. Compute the distance between the given vectors.

27. **(a)** $\sin t$, $\cos t$ **(b)** t, t^2

28. **(a)** $2t + 3$, $3t^2 - 1$ **(b)** $3t + 1$, 1

In Exercises 29 and 30, which of the given sets of vectors in R^3, with the standard inner product, are orthogonal, orthonormal, or neither?

29. **(a)** $\left\{ \begin{bmatrix} \frac{1}{\sqrt{2}} \\ 0 \\ \frac{1}{\sqrt{2}} \end{bmatrix}, \begin{bmatrix} -\frac{1}{\sqrt{2}} \\ 0 \\ \frac{1}{\sqrt{2}} \end{bmatrix}, \begin{bmatrix} 0 \\ 1 \\ 0 \end{bmatrix} \right\}$

(b) $\left\{ \begin{bmatrix} 1 \\ 1 \\ 0 \end{bmatrix}, \begin{bmatrix} 0 \\ 0 \\ 1 \end{bmatrix}, \begin{bmatrix} 0 \\ 1 \\ 0 \end{bmatrix} \right\}$

(c) $\left\{ \begin{bmatrix} 1 \\ -1 \\ 0 \end{bmatrix}, \begin{bmatrix} 0 \\ 1 \\ 1 \end{bmatrix}, \begin{bmatrix} 0 \\ 0 \\ 1 \end{bmatrix} \right\}$

30. **(a)** $\left\{ \begin{bmatrix} 0 \\ 1 \\ -1 \end{bmatrix}, \begin{bmatrix} 0 \\ 1 \\ 1 \end{bmatrix}, \begin{bmatrix} 2 \\ 0 \\ 0 \end{bmatrix} \right\}$

(b) $\left\{ \begin{bmatrix} -1 \\ 1 \\ 0 \end{bmatrix}, \begin{bmatrix} \frac{1}{\sqrt{3}} \\ \frac{1}{\sqrt{3}} \\ \frac{1}{\sqrt{3}} \end{bmatrix}, \begin{bmatrix} 0 \\ \frac{1}{\sqrt{2}} \\ -\frac{1}{\sqrt{2}} \end{bmatrix} \right\}$

(c) $\left\{ \begin{bmatrix} \frac{2}{3} \\ \frac{2}{3} \\ \frac{1}{3} \end{bmatrix}, \begin{bmatrix} -\frac{2}{3} \\ \frac{1}{3} \\ \frac{2}{3} \end{bmatrix}, \begin{bmatrix} \frac{1}{3} \\ -\frac{2}{3} \\ \frac{2}{3} \end{bmatrix} \right\}$

In Exercises 31 and 32, let V be the inner product space of Example 6.

31. Let $p(t) = 3t + 1$ and $q(t) = at$. For what values of a are $p(t)$ and $q(t)$ orthogonal?

32. Let $p(t) = 3t + 1$ and $q(t) = at + b$. For what values of a and b are $p(t)$ and $q(t)$ orthogonal?

In Exercises 33 and 34, let V be the Euclidean space R^3 with the standard inner product.

33. Let $\mathbf{u} = \begin{bmatrix} 1 \\ 1 \\ -2 \end{bmatrix}$ and $\mathbf{v} = \begin{bmatrix} a \\ -1 \\ 2 \end{bmatrix}$. For what values of a are $\mathbf{u}$ and $\mathbf{v}$ orthogonal?

34. Let $\mathbf{u} = \begin{bmatrix} \frac{1}{\sqrt{2}} \\ 0 \\ \frac{1}{\sqrt{2}} \end{bmatrix}$ and $\mathbf{v} = \begin{bmatrix} a \\ -1 \\ -b \end{bmatrix}$. For what values of a and b is $\{\mathbf{u}, \mathbf{v}\}$ an orthonormal set?

35. Let $A = \begin{bmatrix} 1 & 2 \\ 3 & 4 \end{bmatrix}$. Find a 2×2 matrix $B \neq O$ such that A and B are orthogonal in the inner product space defined in Exercise 3. Can there be more than one matrix B that is orthogonal to A?

36. Let V be the inner product space in Example 4.

(a) If $p(t) = \sqrt{t}$, find $q(t) = a + bt \neq 0$ such that $p(t)$ and $q(t)$ are orthogonal.

(b) If $p(t) = \sin t$, find $q(t) = a + be^t \neq 0$ such that $p(t)$ and $q(t)$ are orthogonal.

37. Let $C = \begin{bmatrix} c_{ij} \end{bmatrix}$ be an $n \times n$ positive definite symmetric matrix and let V be an n-dimensional vector space with ordered basis $S = \{\mathbf{u}_1, \mathbf{u}_2, \ldots, \mathbf{u}_n\}$. For $\mathbf{v} = a_1\mathbf{u}_1 + a_2\mathbf{u}_2 + \cdots + a_n\mathbf{u}_n$ and $\mathbf{w} = b_1\mathbf{u}_1 + b_2\mathbf{u}_2 + \cdots + b_n\mathbf{u}_n$ in V define $(\mathbf{v}, \mathbf{w}) = \sum_{i=1}^{n}\sum_{j=1}^{n} a_i c_{ij} b_j$. Prove that this defines an inner product on V.

38. If A and B are $n \times n$ matrices, show that $(A\mathbf{u}, B\mathbf{v}) = (\mathbf{u}, A^T B\mathbf{v})$ for any vectors $\mathbf{u}$ and $\mathbf{v}$ in Euclidean space R^n with the standard inner product.

39. In the Euclidean space R^n with the standard inner product, prove that $(\mathbf{u}, \mathbf{v}) = \mathbf{u}^T\mathbf{v}$.

40. Consider Euclidean space R^4 with the standard inner product and let

$$\mathbf{u}_1 = \begin{bmatrix} 1 \\ 0 \\ 0 \\ 1 \end{bmatrix} \quad \text{and} \quad \mathbf{u}_2 = \begin{bmatrix} 0 \\ 1 \\ 0 \\ 1 \end{bmatrix}.$$

(a) Prove that the set W consisting of all vectors in R^4 that are orthogonal to both $\mathbf{u}_1$ and $\mathbf{u}_2$ is a subspace of R^4.

(b) Find a basis for W.

41. Let V be an inner product space. Show that if $\mathbf{v}$ is orthogonal to $\mathbf{w}_1, \mathbf{w}_2, \ldots, \mathbf{w}_k$, then $\mathbf{v}$ is orthogonal to every vector in

$$\operatorname{span}\{\mathbf{w}_1, \mathbf{w}_2, \ldots, \mathbf{w}_k\}.$$

42. Suppose that $\{\mathbf{v}_1, \mathbf{v}_2, \ldots, \mathbf{v}_n\}$ is an orthonormal set in R^n with the standard inner product. Let the matrix A be given by $A = \begin{bmatrix} \mathbf{v}_1 & \mathbf{v}_2 & \cdots & \mathbf{v}_n \end{bmatrix}$. Show that A is nonsingular and compute its inverse. Give three different examples of such a matrix in R^2 or R^3.

43. Suppose that $\{\mathbf{v}_1, \mathbf{v}_2, \ldots, \mathbf{v}_n\}$ is an orthogonal set in R^n with the standard inner product. Let A be the matrix whose jth column is $\mathbf{v}_j$, $j = 1, 2, \ldots, n$. Prove or disprove: A is nonsingular.

44. If A is nonsingular, prove that $A^T A$ is positive definite.

45. If C is positive definite, and $\mathbf{x} \neq \mathbf{0}$ is such that $C\mathbf{x} = k\mathbf{x}$ for some scalar k, show that $k > 0$.

46. If C is positive definite, show that its diagonal entries are positive.

47. Let C be positive definite and r any scalar. Prove or disprove: rC is positive definite.

48. If B and C are $n \times n$ positive definite matrices, show that $B + C$ is positive definite.

49. Let S be the set of $n \times n$ positive definite matrices. Is S a subspace of M_{nn}?

50. To compute the standard inner product of a pair of vectors $\mathbf{u}$ and $\mathbf{v}$ in R^n, use the matrix product operation in your software as follows. Let U and V be column matrices for vectors $\mathbf{u}$ and $\mathbf{v}$, respectively. Then $(\mathbf{u}, \mathbf{v}) =$ the product of U^T and V (or V^T and U). Experiment with the vectors in Example 1 and Exercises 8 and 9 (see Exercise 39).

51. Exercise 41 in Section 5.1 can be generalized to R^n, or even R_n in some software. Determine whether this is the case for the software that you use.

52. Exercise 43 in Section 5.1 can be generalized to R^n, or even R_n in some software. Determine whether this is the case for the software that you use.

53. If your software incorporates a computer algebra system that computes definite integrals, then you can compute inner products of functions as in Examples 9 and 14. Use your software to check your results in Exercises 10 and 11.

5.4 Gram*–Schmidt† Process

In this section we prove that for every Euclidean space V we can obtain a basis S for V such that S is an orthonormal set; such a basis is called an **orthonormal basis**, and the method we use to obtain it is the **Gram–Schmidt process**. From our work with the natural bases for R^2, R^3, and in general, R^n, we know that when these bases are present, the computations are kept to a minimum. The reduction in the computational effort is due to the fact that we are dealing with an orthonormal basis. For example, if $S = \{\mathbf{u}_1, \mathbf{u}_2, \dots, \mathbf{u}_n\}$ is a basis for an n-dimensional Euclidean space V, then if $\mathbf{v}$ is any vector in V, we can write $\mathbf{v}$ as

$$\mathbf{v} = c_1\mathbf{u}_1 + c_2\mathbf{u}_2 + \cdots + c_n\mathbf{u}_n.$$

The coefficients $c_1, c_2, \dots, c_n$ are obtained by solving a linear system of n equations in n unknowns.

However, if S is orthonormal, we can produce the same result with much less work. This is the content of the following theorem:

Theorem 5.5 Let $S = \{\mathbf{u}_1, \mathbf{u}_2, \dots, \mathbf{u}_n\}$ be an orthonormal basis for a Euclidean space V and let $\mathbf{v}$ be any vector in V. Then

$$\mathbf{v} = c_1\mathbf{u}_1 + c_2\mathbf{u}_2 + \cdots + c_n\mathbf{u}_n,$$

where

$$c_i = (\mathbf{v}, \mathbf{u}_i), \qquad i = 1, 2, \dots, n.$$

Proof

Exercise 19. ■

JÖRGEN PEDERSEN GRAM

EXAMPLE 1

Let $S = \{\mathbf{u}_1, \mathbf{u}_2, \mathbf{u}_3\}$ be a basis for R^3, where

$$\mathbf{u}_1 = \begin{bmatrix} \frac{2}{3} \\ -\frac{2}{3} \\ \frac{1}{3} \end{bmatrix}, \quad \mathbf{u}_2 = \begin{bmatrix} \frac{2}{3} \\ \frac{1}{3} \\ -\frac{2}{3} \end{bmatrix}, \quad \text{and} \quad \mathbf{u}_3 = \begin{bmatrix} \frac{1}{3} \\ \frac{2}{3} \\ \frac{2}{3} \end{bmatrix}.$$

ERHARD SCHMIDT

*Jörgen Pedersen Gram (1850–1916) was born and educated in Denmark, where he received degrees in Mathematics. In 1875 he began his career at the Hafnia Insurance Company, with whom he was associated until 1910, in ever increasingly important positions. In 1884 he founded his own insurance company while continuing to work for Hafnia. He also did considerable work on mathematical models for maximizing profits in forest management. In addition to his work in actuarial science and probability theory, Gram made many mathematical contributions.

†Erhard Schmidt (1876–1959) taught at several leading German universities and was a student of both Hermann Amandus Schwarz and David Hilbert. He made important contributions to the study of integral equations and partial differential equations and, as part of this study, he introduced the method for finding an orthonormal basis in 1907. In 1908 he wrote a paper on infinitely many linear equations in infinitely many unknowns, in which he founded the theory of Hilbert spaces and in which he again used his method.

Note that S is an orthonormal set. Write the vector

$$\mathbf{v} = \begin{bmatrix} 3 \\ 4 \\ 5 \end{bmatrix}$$

as a linear combination of the vectors in S.

Solution

We have

$$\mathbf{v} = c_1\mathbf{u}_1 + c_2\mathbf{u}_2 + c_3\mathbf{u}_3.$$

Theorem 5.5 shows that c_1, c_2, and c_3 can be derived without having to solve a linear system of three equations in three unknowns. Thus

$$c_1 = (\mathbf{v}, \mathbf{u}_1) = 1, \quad c_2 = (\mathbf{v}, \mathbf{u}_2) = 0, \quad c_3 = (\mathbf{v}, \mathbf{u}_3) = 7,$$

and $\mathbf{v} = \mathbf{u}_1 + 7\mathbf{u}_3$. ■

Theorem 5.6 **Gram–Schmidt Process**

Let V be an inner product space and $W \neq \{\mathbf{0}\}$ an m-dimensional subspace of V. Then there exists an orthonormal basis $T = \{\mathbf{w}_1, \mathbf{w}_2, \ldots, \mathbf{w}_m\}$ for W.

Proof

The proof is constructive; that is, we exhibit the desired basis T. However, we first find an orthogonal basis $T^* = \{\mathbf{v}_1, \mathbf{v}_2, \ldots, \mathbf{v}_m\}$ for W.

Let $S = \{\mathbf{u}_1, \mathbf{u}_2, \ldots, \mathbf{u}_m\}$ be any basis for W. We start by selecting any one of the vectors in S—say, $\mathbf{u}_1$—and call it $\mathbf{v}_1$. Thus $\mathbf{v}_1 = \mathbf{u}_1$. We now look for a vector $\mathbf{v}_2$ in the subspace W_1 of W spanned by $\{\mathbf{u}_1, \mathbf{u}_2\}$ that is orthogonal to $\mathbf{v}_1$. Since $\mathbf{v}_1 = \mathbf{u}_1$, W_1 is also the subspace spanned by $\{\mathbf{v}_1, \mathbf{u}_2\}$. Thus $\mathbf{v}_2 = a_1\mathbf{v}_1 + a_2\mathbf{u}_2$. We determine a_1 and a_2 so that $(\mathbf{v}_2, \mathbf{v}_1) = 0$. Now $0 = (\mathbf{v}_2, \mathbf{v}_1) = (a_1\mathbf{v}_1 + a_2\mathbf{u}_2, \mathbf{v}_1) = a_1(\mathbf{v}_1, \mathbf{v}_1) + a_2(\mathbf{u}_2, \mathbf{v}_1)$. Note that $\mathbf{v}_1 \neq \mathbf{0}$ (Why?), so $(\mathbf{v}_1, \mathbf{v}_1) \neq 0$. Thus

$$a_1 = -a_2\frac{(\mathbf{u}_2, \mathbf{v}_1)}{(\mathbf{v}_1, \mathbf{v}_1)}.$$

We may assign an arbitrary nonzero value to a_2. Thus, letting $a_2 = 1$, we obtain

$$a_1 = -\frac{(\mathbf{u}_2, \mathbf{v}_1)}{(\mathbf{v}_1, \mathbf{v}_1)}.$$

Hence

$$\mathbf{v}_2 = a_1\mathbf{v}_1 + \mathbf{u}_2 = \mathbf{u}_2 - \frac{(\mathbf{u}_2, \mathbf{v}_1)}{(\mathbf{v}_1, \mathbf{v}_1)}\mathbf{v}_1.$$

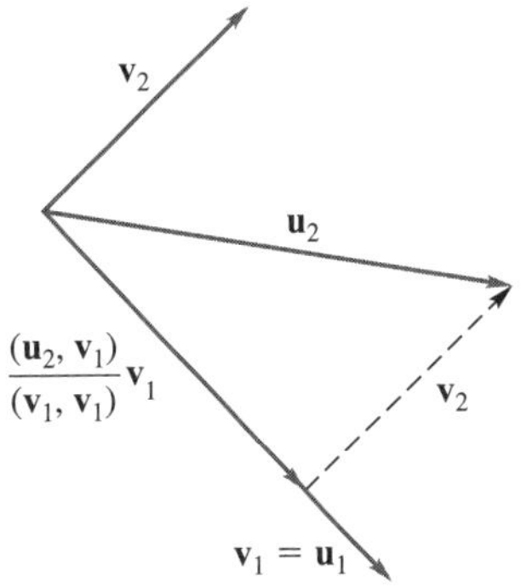

FIGURE 5.19

At this point we have an orthogonal subset $\{\mathbf{v}_1, \mathbf{v}_2\}$ of W (Figure 5.19).

Next, we look for a vector $\mathbf{v}_3$ in the subspace W_2 of W spanned by $\{\mathbf{u}_1, \mathbf{u}_2, \mathbf{u}_3\}$ that is orthogonal to both $\mathbf{v}_1$ and $\mathbf{v}_2$. Of course, W_2 is also the subspace spanned by $\{\mathbf{v}_1, \mathbf{v}_2, \mathbf{u}_3\}$ (Why?). Thus $\mathbf{v}_3 = b_1\mathbf{v}_1 + b_2\mathbf{v}_2 + b_3\mathbf{u}_3$. We try to find b_1, b_2, and b_3 so that $(\mathbf{v}_3, \mathbf{v}_1) = 0$ and $(\mathbf{v}_3, \mathbf{v}_2) = 0$. Now

$$0 = (\mathbf{v}_3, \mathbf{v}_1) = (b_1\mathbf{v}_1 + b_2\mathbf{v}_2 + b_3\mathbf{u}_3, \mathbf{v}_1) = b_1(\mathbf{v}_1, \mathbf{v}_1) + b_3(\mathbf{u}_3, \mathbf{v}_1)$$
$$0 = (\mathbf{v}_3, \mathbf{v}_2) = (b_1\mathbf{v}_1 + b_2\mathbf{v}_2 + b_3\mathbf{u}_3, \mathbf{v}_2) = b_2(\mathbf{v}_2, \mathbf{v}_2) + b_3(\mathbf{u}_3, \mathbf{v}_2).$$

Observe that $\mathbf{v}_2 \neq \mathbf{0}$ (Why?). Solving for b_1 and b_2, we have

$$b_1 = -b_3\frac{(\mathbf{u}_3, \mathbf{v}_1)}{(\mathbf{v}_1, \mathbf{v}_1)} \quad \text{and} \quad b_2 = -b_3\frac{(\mathbf{u}_3, \mathbf{v}_2)}{(\mathbf{v}_2, \mathbf{v}_2)}.$$

We may assign an arbitrary nonzero value to b_3. Thus, letting $b_3 = 1$, we have

$$\mathbf{v}_3 = \mathbf{u}_3 - \frac{(\mathbf{u}_3, \mathbf{v}_1)}{(\mathbf{v}_1, \mathbf{v}_1)}\mathbf{v}_1 - \frac{(\mathbf{u}_3, \mathbf{v}_2)}{(\mathbf{v}_2, \mathbf{v}_2)}\mathbf{v}_2.$$

At this point we have an orthogonal subset $\{\mathbf{v}_1, \mathbf{v}_2, \mathbf{v}_3\}$ of W (Figure 5.20).

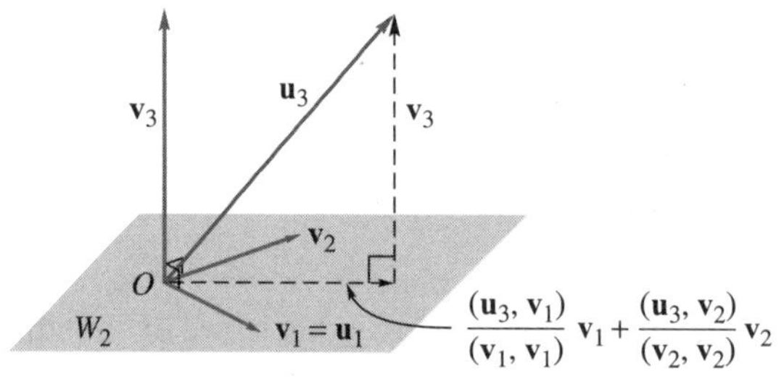

FIGURE 5.20

We next seek a vector $\mathbf{v}_4$ in the subspace W_3 spanned by $\{\mathbf{u}_1, \mathbf{u}_2, \mathbf{u}_3, \mathbf{u}_4\}$, and also by $\{\mathbf{v}_1, \mathbf{v}_2, \mathbf{v}_3, \mathbf{u}_4\}$, which is orthogonal to $\mathbf{v}_1, \mathbf{v}_2, \mathbf{v}_3$. We obtain

$$\mathbf{v}_4 = \mathbf{u}_4 - \frac{(\mathbf{u}_4, \mathbf{v}_1)}{(\mathbf{v}_1, \mathbf{v}_1)}\mathbf{v}_1 - \frac{(\mathbf{u}_4, \mathbf{v}_2)}{(\mathbf{v}_2, \mathbf{v}_2)}\mathbf{v}_2 - \frac{(\mathbf{u}_4, \mathbf{v}_3)}{(\mathbf{v}_3, \mathbf{v}_3)}\mathbf{v}_3.$$

We continue in this manner until we have an orthogonal set

$$T^* = \{\mathbf{v}_1, \mathbf{v}_2, \ldots, \mathbf{v}_m\}$$

of m vectors. By Theorem 5.4 we conclude that T^* is a basis for W. If we now let $\mathbf{w}_i = \dfrac{1}{\|\mathbf{v}_i\|}\mathbf{v}_i$ for $i = 1, 2, \ldots, m$, then $T = \{\mathbf{w}_1, \mathbf{w}_2, \ldots, \mathbf{w}_m\}$ is an orthonormal basis for W. ■

Remark It can be shown that if $\mathbf{u}$ and $\mathbf{v}$ are vectors in an inner product space such that $(\mathbf{u}, \mathbf{v}) = 0$, then $(\mathbf{u}, c\mathbf{v}) = 0$ for any scalar c (Exercise 31). This result can often be used to simplify hand computations in the Gram–Schmidt process. As soon as a vector $\mathbf{v}_i$ is computed in Step 2, multiply it by a proper scalar to clear any fractions that may be present. We shall use this approach in our computational work with the Gram–Schmidt process.

EXAMPLE 2

Let W be the subspace of the Euclidean space R^4 with the standard inner product with basis $S = \{\mathbf{u}_1, \mathbf{u}_2, \mathbf{u}_3\}$, where

$$\mathbf{u}_1 = \begin{bmatrix} 1 \\ 1 \\ 1 \\ 0 \end{bmatrix}, \quad \mathbf{u}_2 = \begin{bmatrix} -1 \\ 0 \\ -1 \\ 1 \end{bmatrix}, \quad \mathbf{u}_3 = \begin{bmatrix} -1 \\ 0 \\ 0 \\ -1 \end{bmatrix}.$$

Transform S to an orthonormal basis $T = \{\mathbf{w}_1, \mathbf{w}_2, \mathbf{w}_3\}$.

Solution

First, let $\mathbf{v}_1 = \mathbf{u}_1$. Then we find that

$$\mathbf{v}_2 = \mathbf{u}_2 - \frac{(\mathbf{u}_2, \mathbf{v}_1)}{(\mathbf{v}_1, \mathbf{v}_1)}\mathbf{v}_1 = \begin{bmatrix} -1 \\ 0 \\ -1 \\ 1 \end{bmatrix} - \left(-\frac{2}{3}\right)\begin{bmatrix} 1 \\ 1 \\ 1 \\ 0 \end{bmatrix} = \begin{bmatrix} -\frac{1}{3} \\ \frac{2}{3} \\ -\frac{1}{3} \\ 1 \end{bmatrix}.$$

Multiplying $\mathbf{v}_2$ by 3 to clear fractions, we get

$$\begin{bmatrix} -1 \\ 2 \\ -1 \\ 3 \end{bmatrix},$$

which we now use as $\mathbf{v}_2$. Next,

$$\begin{aligned} \mathbf{v}_3 &= \mathbf{u}_3 - \frac{(\mathbf{u}_3, \mathbf{v}_1)}{(\mathbf{v}_1, \mathbf{v}_1)}\mathbf{v}_1 - \frac{(\mathbf{u}_3, \mathbf{v}_2)}{(\mathbf{v}_2, \mathbf{v}_2)}\mathbf{v}_2 \\ &= \begin{bmatrix} -1 \\ 0 \\ 0 \\ -1 \end{bmatrix} - \left(-\frac{1}{3}\right)\begin{bmatrix} 1 \\ 1 \\ 1 \\ 0 \end{bmatrix} - \left(-\frac{2}{15}\right)\begin{bmatrix} -1 \\ 2 \\ -1 \\ 3 \end{bmatrix} = \begin{bmatrix} -\frac{4}{5} \\ \frac{3}{5} \\ \frac{1}{5} \\ -\frac{3}{5} \end{bmatrix}. \end{aligned}$$

Multiplying $\mathbf{v}_3$ by 5 to clear fractions, we get

$$\begin{bmatrix} -4 \\ 3 \\ 1 \\ -3 \end{bmatrix},$$

which we now take as $\mathbf{v}_3$. Thus

$$S = \{\mathbf{v}_1, \mathbf{v}_2, \mathbf{v}_3\} = \left\{ \begin{bmatrix} 1 \\ 1 \\ 1 \\ 0 \end{bmatrix}, \begin{bmatrix} -1 \\ 2 \\ -1 \\ 3 \end{bmatrix}, \begin{bmatrix} -4 \\ 3 \\ 1 \\ -3 \end{bmatrix} \right\}$$

is an orthogonal basis for W. Multiplying each vector in S by the reciprocal of its length yields

$$T = \{\mathbf{w}_1, \mathbf{w}_2, \mathbf{w}_3\} = \left\{ \begin{bmatrix} \frac{1}{\sqrt{3}} \\ \frac{1}{\sqrt{3}} \\ \frac{1}{\sqrt{3}} \\ 0 \end{bmatrix}, \begin{bmatrix} -\frac{1}{\sqrt{15}} \\ \frac{2}{\sqrt{15}} \\ -\frac{1}{\sqrt{15}} \\ \frac{3}{\sqrt{15}} \end{bmatrix}, \begin{bmatrix} -\frac{4}{\sqrt{35}} \\ \frac{3}{\sqrt{35}} \\ \frac{1}{\sqrt{35}} \\ -\frac{3}{\sqrt{35}} \end{bmatrix} \right\},$$

which is an orthonormal basis for W. ■

Remark In solving Example 2, as soon as a vector is computed, we multiplied it by an appropriate scalar to eliminate any fractions that may be present. This optional step results in simpler computations when working by hand. Most computer implementations of the Gram–Schmidt process, including those developed with MATLAB, do not clear fractions.

EXAMPLE 3 Let V be the Euclidean space P_3 with the inner product defined in Example 6 of Section 5.3. Let W be the subspace of P_3 having $S = \{t^2, t\}$ as a basis. Find an orthonormal basis for W.

Solution

First, let $\mathbf{u}_1 = t^2$ and $\mathbf{u}_2 = t$. Now let $\mathbf{v}_1 = \mathbf{u}_1 = t^2$. Then

$$\mathbf{v}_2 = \mathbf{u}_2 - \frac{(\mathbf{u}_2, \mathbf{v}_1)}{(\mathbf{v}_1, \mathbf{v}_1)}\mathbf{v}_1 = t - \frac{\frac{1}{4}}{\frac{1}{5}}t^2 = t - \frac{5}{4}t^2,$$

where

$$(\mathbf{v}_1, \mathbf{v}_1) = \int_0^1 t^2t^2\,dt = \int_0^1 t^4\,dt = \frac{1}{5}$$

and

$$(\mathbf{u}_2, \mathbf{v}_1) = \int_0^1 tt^2\,dt = \int_0^1 t^3\,dt = \frac{1}{4}.$$

Since

$$(\mathbf{v}_2, \mathbf{v}_2) = \int_0^1 \left(t - \frac{5}{4}t^2\right)^2 dt = \frac{1}{48},$$

$\left\{\sqrt{5}\,t^2, \sqrt{48}\left(t - \frac{5}{4}t^2\right)\right\}$ is an orthonormal basis for W. If we choose $\mathbf{u}_1 = t$ and $\mathbf{u}_2 = t^2$, then we obtain (verify) the orthonormal basis $\left\{\sqrt{3}\,t, \sqrt{30}\left(t^2 - \frac{1}{2}t\right)\right\}$ for W. ■

In the proof of Theorem 5.6 we have also established the following result: At each stage of the Gram–Schmidt process, the ordered set $\{\mathbf{w}_1, \mathbf{w}_2, \ldots, \mathbf{w}_k\}$ is an orthonormal basis for the subspace spanned by

$$\{\mathbf{u}_1, \mathbf{u}_2, \ldots, \mathbf{u}_k\}, \qquad 1 \le k \le n.$$

Also, the final orthonormal basis T depends upon the order of the vectors in the given basis S. Thus, if we change the order of the vectors in S, we might obtain a different orthonormal basis T for W.

Remark We make one final observation with regard to the Gram–Schmidt process. In our proof of Theorem 5.6 we first obtained an orthogonal basis T^* and then normalized all the vectors in T^* to find the orthonormal basis T. Of course, an alternative course of action is to normalize each vector as it is produced. However, normalizing at the end is simpler for hand computation.

One of the useful consequences of having an orthonormal basis in a Euclidean space V is that an arbitrary inner product on V, when it is expressed in terms of coordinates with respect to the orthonormal basis, behaves like the standard inner product on R^n.

Theorem 5.7 Let V be an n-dimensional Euclidean space, and let $S = \{\mathbf{u}_1, \mathbf{u}_2, \ldots, \mathbf{u}_n\}$ be an orthonormal basis for V. If $\mathbf{v} = a_1\mathbf{u}_1 + a_2\mathbf{u}_2 + \cdots + a_n\mathbf{u}_n$ and $\mathbf{w} = b_1\mathbf{u}_1 + b_2\mathbf{u}_2 + \cdots + b_n\mathbf{u}_n$, then

$$(\mathbf{v}, \mathbf{w}) = a_1b_1 + a_2b_2 + \cdots + a_nb_n.$$

Proof

We first compute the matrix $C = \begin{bmatrix} c_{ij} \end{bmatrix}$ of the given inner product with respect to the ordered basis S. We have

$$c_{ij} = (\mathbf{u}_i, \mathbf{u}_j) = \begin{cases} 1 & \text{if } i = j \\ 0 & \text{if } i \ne j. \end{cases}$$

Hence $C = I_n$, the identity matrix. Now we also know from Equation (2) of Section 5.3 that

$$(\mathbf{v}, \mathbf{w}) = \begin{bmatrix} \mathbf{v} \end{bmatrix}_S^T C \begin{bmatrix} \mathbf{w} \end{bmatrix}_S = \begin{bmatrix} \mathbf{v} \end{bmatrix}_S^T I_n \begin{bmatrix} \mathbf{w} \end{bmatrix}_S = \begin{bmatrix} \mathbf{v} \end{bmatrix}_S^T \begin{bmatrix} \mathbf{w} \end{bmatrix}_S$$

$$= \begin{bmatrix} a_1 & a_2 & \cdots & a_n \end{bmatrix} \begin{bmatrix} b_1 \\ b_2 \\ \vdots \\ b_n \end{bmatrix} = a_1b_1 + a_2b_2 + \cdots + a_nb_n,$$

which establishes the result. ■

The theorem that we just proved has some additional implications. Consider the Euclidean space R_3 with the standard inner product and let W be the subspace with ordered basis $S = \{\begin{bmatrix} 2 & 1 & 1 \end{bmatrix}, \begin{bmatrix} 1 & 1 & 2 \end{bmatrix}\}$. Let $\mathbf{u} = \begin{bmatrix} 5 & 3 & 4 \end{bmatrix}$ be a vector in W. Then

$$\begin{bmatrix} 5 & 3 & 4 \end{bmatrix} = 2\begin{bmatrix} 2 & 1 & 1 \end{bmatrix} + 1\begin{bmatrix} 1 & 1 & 2 \end{bmatrix},$$

so $\begin{bmatrix} 5 & 3 & 4 \end{bmatrix}_S = \begin{bmatrix} 2 \\ 1 \end{bmatrix}$. Now the length of $\mathbf{u}$ is

$$\|\mathbf{u}\| = \sqrt{5^2 + 3^2 + 4^2} = \sqrt{25 + 9 + 16} = \sqrt{50}.$$

We might expect to compute the length of $\mathbf{u}$ by using the coordinate vector with respect to S; that is, $\|\mathbf{u}\| = \sqrt{2^2 + 1^2} = \sqrt{5}$. Obviously, we have the wrong answer. However, let us transform the given basis S for W into an orthonormal basis T for W. Using the Gram–Schmidt process, we find that

$$\left\{\begin{bmatrix} 2 & 1 & 1 \end{bmatrix}, \begin{bmatrix} -\frac{4}{6} & \frac{1}{6} & \frac{7}{6} \end{bmatrix}\right\}$$

is an orthogonal basis for W (verify). It then follows from Exercise 31 that

$$\left\{\begin{bmatrix} 2 & 1 & 1 \end{bmatrix}, \begin{bmatrix} -4 & 1 & 7 \end{bmatrix}\right\}$$

is also an orthogonal basis, so

$$T = \left\{\begin{bmatrix} \frac{2}{\sqrt{6}} & \frac{1}{\sqrt{6}} & \frac{1}{\sqrt{6}} \end{bmatrix}, \begin{bmatrix} -\frac{4}{\sqrt{66}} & \frac{1}{\sqrt{66}} & \frac{7}{\sqrt{66}} \end{bmatrix}\right\}$$

is an orthonormal basis for W. Then the coordinate vector of $\mathbf{u}$ with respect to T is (verify)

$$\begin{bmatrix} \mathbf{u} \end{bmatrix}_T = \begin{bmatrix} \frac{17}{6}\sqrt{6} \\ \frac{1}{6}\sqrt{66} \end{bmatrix}.$$

Computing the length of $\mathbf{u}$ by using these coordinates, we find that

$$\|\mathbf{u}\|_T = \sqrt{\left(\frac{17}{6}\sqrt{6}\right)^2 + \left(\frac{1}{6}\sqrt{66}\right)^2} = \sqrt{\frac{1800}{36}} = \sqrt{50}.$$

It is not difficult to show (Exercise 21) that if T is an orthonormal basis for an inner product space and $\begin{bmatrix} \mathbf{v} \end{bmatrix}_T = \begin{bmatrix} a_1 \\ a_2 \\ \vdots \\ a_n \end{bmatrix}$, then $\|\mathbf{v}\| = \sqrt{a_1^2 + a_2^2 + \cdots + a_n^2}$.

■ *QR*-Factorization

In Section 2.5 we discussed the LU-factorization of a matrix and showed how it leads to a very efficient method for solving a linear system. We now discuss another factorization of a matrix A, called the ***QR*-factorization** of A. This type of factorization is widely used in computer codes to find the eigenvalues of a matrix, to solve linear systems, and to find least squares approximations.

Theorem 5.8 If A is an $m \times n$ matrix with linearly independent columns, then A can be factored as $A = QR$, where Q is an $m \times n$ matrix whose columns form an orthonormal basis for the column space of A and R is an $n \times n$ nonsingular upper triangular matrix.

Proof

Let $\mathbf{u}_1, \mathbf{u}_2, \ldots, \mathbf{u}_n$ denote the linearly independent columns of A, which form a basis for the column space of A. By using the Gram–Schmidt process, we can derive an orthonormal basis $\mathbf{w}_1, \mathbf{w}_2, \ldots, \mathbf{w}_n$ for the column space of A. Recall how this orthonormal basis was obtained. We first constructed an orthogonal basis $\mathbf{v}_1, \mathbf{v}_2, \ldots, \mathbf{v}_n$ as follows: $\mathbf{v}_1 = \mathbf{u}_1$, and then for $i = 2, 3, \ldots, n$, we have

$$\mathbf{v}_i = \mathbf{u}_i - \frac{(\mathbf{u}_i, \mathbf{v}_1)}{(\mathbf{v}_1, \mathbf{v}_1)}\mathbf{v}_1 - \frac{(\mathbf{u}_i, \mathbf{v}_2)}{(\mathbf{v}_2, \mathbf{v}_2)}\mathbf{v}_2 - \cdots - \frac{(\mathbf{u}_i, \mathbf{v}_{i-1})}{(\mathbf{v}_{i-1}, \mathbf{v}_{i-1})}\mathbf{v}_{i-1}. \tag{1}$$

Finally, $\mathbf{w}_i = \dfrac{1}{\|\mathbf{v}_i\|}\mathbf{v}_i$ for $i = 1, 2, 3, \ldots, n$. Now every $\mathbf{u}$-vector can be written as a linear combination of $\mathbf{w}_1, \mathbf{w}_2, \ldots, \mathbf{w}_n$:

$$\begin{aligned}
\mathbf{u}_1 &= r_{11}\mathbf{w}_1 + r_{21}\mathbf{w}_2 + \cdots + r_{n1}\mathbf{w}_n \\
\mathbf{u}_2 &= r_{12}\mathbf{w}_1 + r_{22}\mathbf{w}_2 + \cdots + r_{n2}\mathbf{w}_n \\
&\ \ \vdots \\
\mathbf{u}_n &= r_{1n}\mathbf{w}_1 + r_{2n}\mathbf{w}_2 + \cdots + r_{nn}\mathbf{w}_n.
\end{aligned} \tag{2}$$

From Theorem 5.5, we have

$$r_{ji} = (\mathbf{u}_i, \mathbf{w}_j).$$

Moreover, from Equation (1), we see that $\mathbf{u}_i$ lies in

$$\operatorname{span}\{\mathbf{v}_1, \mathbf{v}_2, \ldots, \mathbf{v}_i\} = \operatorname{span}\{\mathbf{w}_1, \mathbf{w}_2, \ldots, \mathbf{w}_i\}.$$

Since $\mathbf{w}_j$ is orthogonal to span $\{\mathbf{w}_1, \mathbf{w}_2, \ldots, \mathbf{w}_i\}$ for $j > i$, it is orthogonal to $\mathbf{u}_i$. Hence $r_{ji} = 0$ for $j > i$. Let Q be the matrix whose columns are $\mathbf{w}_1, \mathbf{w}_2, \ldots, \mathbf{w}_j$. Let

$$\mathbf{r}_j = \begin{bmatrix} r_{1j} \\ r_{2j} \\ \vdots \\ r_{nj} \end{bmatrix}.$$

Then the equations in (2) can be written in matrix form as

$$A = \begin{bmatrix} \mathbf{u}_1 & \mathbf{u}_2 & \cdots & \mathbf{u}_n \end{bmatrix} = \begin{bmatrix} Q\mathbf{r}_1 & Q\mathbf{r}_2 & \cdots & Q\mathbf{r}_n \end{bmatrix} = QR,$$

where R is the matrix whose columns are $\mathbf{r}_1, \mathbf{r}_2, \ldots, \mathbf{r}_n$. Thus

$$R = \begin{bmatrix} r_{11} & r_{12} & \cdots & r_{1n} \\ 0 & r_{22} & \cdots & r_{2n} \\ 0 & 0 & \cdots & r_{3n} \\ \vdots & \vdots & & \vdots \\ 0 & 0 & \cdots & r_{nn} \end{bmatrix}.$$

We now show that R is nonsingular. Let $\mathbf{x}$ be a solution to the linear system $R\mathbf{x} = \mathbf{0}$. Multiplying this equation by Q on the left, we have

$$Q(R\mathbf{x}) = (QR)\mathbf{x} = A\mathbf{x} = Q\mathbf{0} = \mathbf{0}.$$

As we know from Chapter 1, the homogeneous system $A\mathbf{x} = \mathbf{0}$ can be written as

$$x_1\mathbf{u}_1 + x_2\mathbf{u}_2 + \cdots + x_n\mathbf{u}_n = \mathbf{0},$$

where $x_1, x_2, \ldots, x_n$ are the components of the vector $\mathbf{x}$. Since the columns of A are linearly independent,

$$x_1 = x_2 = \cdots = x_n = 0,$$

so $\mathbf{x}$ must be the zero vector. Then Theorem 2.9 implies that R is nonsingular. In Exercise 36 we ask you to show that the diagonal entries r_{ii} of R are nonzero by first expressing $\mathbf{u}_i$ as a linear combination of $\mathbf{v}_1, \mathbf{v}_2, \ldots, \mathbf{v}_i$ and then computing $r_{ii} = (\mathbf{u}_i, \mathbf{w}_i)$. This provides another proof of the nonsingularity of R. ■

EXAMPLE 4 Find the QR-factorization of

$$A = \begin{bmatrix} 1 & -1 & -1 \\ 1 & 0 & 0 \\ 1 & -1 & 0 \\ 0 & 1 & -1 \end{bmatrix}.$$

Solution

The columns of A are the vectors $\mathbf{u}_1$, $\mathbf{u}_2$, and $\mathbf{u}_3$, respectively, defined in Example 2. In that example we obtained the following orthonormal basis for the column space of A:

$$\mathbf{w}_1 = \begin{bmatrix} \frac{1}{\sqrt{3}} \\ \frac{1}{\sqrt{3}} \\ \frac{1}{\sqrt{3}} \\ 0 \end{bmatrix}, \quad \mathbf{w}_2 = \begin{bmatrix} -\frac{1}{\sqrt{15}} \\ \frac{2}{\sqrt{15}} \\ -\frac{1}{\sqrt{15}} \\ \frac{3}{\sqrt{15}} \end{bmatrix}, \quad \mathbf{w}_3 = \begin{bmatrix} -\frac{4}{\sqrt{35}} \\ \frac{3}{\sqrt{35}} \\ \frac{1}{\sqrt{35}} \\ -\frac{3}{\sqrt{35}} \end{bmatrix}.$$

Then

$$Q = \begin{bmatrix} \frac{1}{\sqrt{3}} & -\frac{1}{\sqrt{15}} & -\frac{4}{\sqrt{35}} \\ \frac{1}{\sqrt{3}} & \frac{2}{\sqrt{15}} & \frac{3}{\sqrt{35}} \\ \frac{1}{\sqrt{3}} & -\frac{1}{\sqrt{15}} & \frac{1}{\sqrt{35}} \\ 0 & \frac{3}{\sqrt{15}} & -\frac{3}{\sqrt{35}} \end{bmatrix} \approx \begin{bmatrix} 0.5774 & -0.2582 & -0.6761 \\ 0.5774 & 0.5164 & 0.5071 \\ 0.5774 & -0.2582 & 0.1690 \\ 0 & 0.7746 & -0.5071 \end{bmatrix}$$

and

$$R = \begin{bmatrix} r_{11} & r_{12} & r_{13} \\ 0 & r_{22} & r_{23} \\ 0 & 0 & r_{33} \end{bmatrix},$$

where $r_{ji} = (\mathbf{u}_i, \mathbf{w}_j)$. Thus

$$R = \begin{bmatrix} \frac{3}{\sqrt{3}} & -\frac{2}{\sqrt{3}} & -\frac{1}{\sqrt{3}} \\ 0 & \frac{5}{\sqrt{15}} & -\frac{2}{\sqrt{15}} \\ 0 & 0 & \frac{7}{\sqrt{35}} \end{bmatrix} \approx \begin{bmatrix} 1.7321 & -1.1547 & -0.5774 \\ 0 & 1.2910 & -0.5164 \\ 0 & 0 & 1.1832 \end{bmatrix}.$$

As you can verify, $A = QR$. ■

Remark State-of-the-art computer implementations (such as in MATLAB) yield an alternative QR-factorization of an $m \times n$ matrix A as the product of an $m \times m$ matrix Q and an $m \times n$ matrix $R = \left[r_{ij}\right]$, where $r_{ij} = 0$ if $i > j$. Thus, if A is 5×3, then

$$R = \begin{bmatrix} * & * & * \\ 0 & * & * \\ 0 & 0 & * \\ 0 & 0 & 0 \\ 0 & 0 & 0 \end{bmatrix}.$$

Key Terms

Orthonormal basis
Gram–Schmidt process
QR-factorization

5.4 Exercises

In this set of exercises, the Euclidean spaces R_n and R^n have the standard inner products on them. Euclidean space P_n has the inner product defined in Example 6 of Section 5.3.

1. Use the Gram–Schmidt process to transform the basis $\left\{\begin{bmatrix}1\\2\end{bmatrix}, \begin{bmatrix}-3\\4\end{bmatrix}\right\}$ for the Euclidean space R^2 into
   (a) an orthogonal basis;
   (b) an orthonormal basis.
2. Use the Gram–Schmidt process to transform the basis $\left\{\begin{bmatrix}1\\0\\1\end{bmatrix}, \begin{bmatrix}-2\\1\\3\end{bmatrix}\right\}$ for the subspace W of Euclidean space R^3 into
   (a) an orthogonal basis;
   (b) an orthonormal basis.
3. Consider the Euclidean space R_4 and let W be the subspace that has

   $$S = \{\begin{bmatrix}1 & 1 & -1 & 0\end{bmatrix}, \begin{bmatrix}0 & 2 & 0 & 1\end{bmatrix}\}$$

   as a basis. Use the Gram–Schmidt process to obtain an orthonormal basis for W.
4. Consider Euclidean space R^3 and let W be the subspace that has basis $S = \left\{\begin{bmatrix}1\\1\\1\end{bmatrix}, \begin{bmatrix}1\\0\\2\end{bmatrix}\right\}$. Use the Gram–Schmidt process to obtain an orthogonal basis for W.

5. Let $S = \{t, 1\}$ be a basis for a subspace W of the Euclidean space P_2. Find an orthonormal basis for W.

6. Repeat Exercise 5 with $S = \{t + 1, t - 1\}$.

7. Let $S = \{t, \sin 2\pi t\}$ be a basis for a subspace W of the inner product space of Example 4 in Section 5.3. Find an orthonormal basis for W.

8. Let $S = \{t, e^t\}$ be a basis for a subspace W of the inner product space of Example 4 in Section 5.3. Find an orthonormal basis for W.

9. Find an orthonormal basis for the Euclidean space R^3 that contains the vectors

$$\begin{bmatrix} \frac{2}{3} \\ -\frac{2}{3} \\ \frac{1}{3} \end{bmatrix} \quad \text{and} \quad \begin{bmatrix} \frac{2}{3} \\ \frac{1}{3} \\ -\frac{2}{3} \end{bmatrix}.$$

10. Use the Gram–Schmidt process to transform the basis

$$\left\{ \begin{bmatrix} 1 \\ 1 \\ 1 \end{bmatrix}, \begin{bmatrix} 0 \\ 1 \\ 1 \end{bmatrix}, \begin{bmatrix} 1 \\ 2 \\ 3 \end{bmatrix} \right\}$$

for the Euclidean space R^3 into an orthonormal basis for R^3.

11. Use the Gram–Schmidt process to construct an orthonormal basis for the subspace W of the Euclidean space R^3 *spanned* by

$$\left\{ \begin{bmatrix} 1 \\ 1 \\ 1 \end{bmatrix}, \begin{bmatrix} 2 \\ 2 \\ 2 \end{bmatrix}, \begin{bmatrix} 0 \\ 0 \\ 1 \end{bmatrix}, \begin{bmatrix} 1 \\ 2 \\ 3 \end{bmatrix} \right\}.$$

12. Use the Gram–Schmidt process to construct an orthonormal basis for the subspace W of the Euclidean space R_3 *spanned* by

$$\{\begin{bmatrix} 1 & -1 & 1 \end{bmatrix}, \begin{bmatrix} -2 & 2 & -2 \end{bmatrix}, \begin{bmatrix} 2 & -1 & 2 \end{bmatrix}, \begin{bmatrix} 0 & 0 & 0 \end{bmatrix}\}.$$

13. Find an orthonormal basis for the subspace of R^3 consisting of all vectors of the form

$$\begin{bmatrix} a \\ a+b \\ b \end{bmatrix}.$$

14. Find an orthonormal basis for the subspace of R_4 consisting of all vectors of the form

$$\begin{bmatrix} a & a+b & c & b+c \end{bmatrix}.$$

15. Find an orthonormal basis for the subspace of R^3 consisting of all vectors $\begin{bmatrix} a \\ b \\ c \end{bmatrix}$ such that $a + b + c = 0$.

16. Find an orthonormal basis for the subspace of R_4 consisting of all vectors $\begin{bmatrix} a & b & c & d \end{bmatrix}$ such that

$$a - b - 2c + d = 0.$$

17. Find an orthonormal basis for the solution space of the homogeneous system

$$\begin{aligned} x_1 + x_2 - x_3 &= 0 \\ 2x_1 + x_2 + 2x_3 &= 0. \end{aligned}$$

18. Find an orthonormal basis for the solution space of the homogeneous system

$$\begin{bmatrix} 1 & 1 & -1 \\ 2 & 1 & 3 \\ 1 & 2 & -6 \end{bmatrix} \begin{bmatrix} x_1 \\ x_2 \\ x_3 \end{bmatrix} = \begin{bmatrix} 0 \\ 0 \\ 0 \end{bmatrix}.$$

19. Prove Theorem 5.5.

20. Let $S = \{\begin{bmatrix} 1 & -1 & 0 \end{bmatrix}, \begin{bmatrix} 1 & 0 & -1 \end{bmatrix}\}$ be a basis for a subspace W of the Euclidean space R_3.

(a) Use the Gram–Schmidt process to obtain an orthonormal basis for W.

(b) Using Theorem 5.5, write $\mathbf{u} = \begin{bmatrix} 5 & -2 & -3 \end{bmatrix}$ as a linear combination of the vectors obtained in part (a).

21. Prove that if T is an orthonormal basis for a Euclidean space and

$$\left[\mathbf{v}\right]_T = \begin{bmatrix} a_1 \\ a_2 \\ \vdots \\ a_n \end{bmatrix},$$

then $\|\mathbf{v}\| = \sqrt{a_1^2 + a_2^2 + \cdots + a_n^2}$.

22. Let W be the subspace of the Euclidean space R^3 with basis

$$S = \left\{ \begin{bmatrix} 1 \\ 0 \\ -2 \end{bmatrix}, \begin{bmatrix} -3 \\ 2 \\ 1 \end{bmatrix} \right\}.$$

Let $\mathbf{v} = \begin{bmatrix} -1 \\ 2 \\ -3 \end{bmatrix}$ be in W.

(a) Find the length of $\mathbf{v}$ directly.

(b) Using the Gram–Schmidt process, transform S into an orthonormal basis T for W.

(c) Find the length of **v** by using the coordinate vector of **v** with respect to T.

23. **(a)** Verify that

$$S = \left\{ \begin{bmatrix} \frac{1}{3} \\ \frac{2}{3} \\ \frac{2}{3} \end{bmatrix}, \begin{bmatrix} \frac{2}{3} \\ \frac{1}{3} \\ -\frac{2}{3} \end{bmatrix}, \begin{bmatrix} \frac{2}{3} \\ -\frac{2}{3} \\ \frac{1}{3} \end{bmatrix} \right\}$$

is an orthonormal basis for the Euclidean space R^3.

(b) Use Theorem 5.5 to find the coordinate vector of $\mathbf{v} = \begin{bmatrix} 15 \\ 3 \\ 3 \end{bmatrix}$ with respect to S.

(c) Find the length of **v** directly and also by using the coordinate vector found in part (b).

24. (***Calculus Required***) Apply the Gram–Schmidt process to the basis $\{1, t, t^2\}$ for the Euclidean space P_2 and obtain an orthonormal basis for P_2.

25. Let V be the Euclidean space of all 2×2 matrices with inner product defined by $(A, B) = \text{Tr}(B^T A)$. (See Exercise 43 in Section 1.3.)

(a) Prove that

$$S = \left\{ \begin{bmatrix} 1 & 0 \\ 0 & 0 \end{bmatrix}, \begin{bmatrix} 0 & 1 \\ 0 & 0 \end{bmatrix}, \begin{bmatrix} 0 & 0 \\ 1 & 0 \end{bmatrix}, \begin{bmatrix} 0 & 0 \\ 0 & 1 \end{bmatrix} \right\}$$

is an orthonormal basis for V.

(b) Use Theorem 5.5 to find the coordinate vector of $\mathbf{v} = \begin{bmatrix} 1 & 2 \\ 3 & 4 \end{bmatrix}$ with respect to S.

26. Let

$$S = \left\{ \begin{bmatrix} 0 & 0 \\ 0 & 1 \end{bmatrix}, \begin{bmatrix} 1 & 1 \\ 0 & 0 \end{bmatrix}, \begin{bmatrix} 1 & 0 \\ 0 & 1 \end{bmatrix} \right\}$$

be a basis for a subspace W of the Euclidean space defined in Exercise 25. Use the Gram–Schmidt process to find an orthonormal basis for W.

27. Repeat Exercise 26 if

$$S = \left\{ \begin{bmatrix} 1 & 0 \\ 0 & 0 \end{bmatrix}, \begin{bmatrix} 0 & 1 \\ 1 & 0 \end{bmatrix}, \begin{bmatrix} 1 & -1 \\ 0 & 1 \end{bmatrix} \right\}.$$

28. Consider the orthonormal basis

$$S = \left\{ \begin{bmatrix} \frac{1}{\sqrt{5}} \\ 0 \\ \frac{2}{\sqrt{5}} \end{bmatrix}, \begin{bmatrix} -\frac{2}{\sqrt{5}} \\ 0 \\ \frac{1}{\sqrt{5}} \end{bmatrix}, \begin{bmatrix} 0 \\ 1 \\ 0 \end{bmatrix} \right\}$$

for R^3. Using Theorem 5.5, write the vector $\begin{bmatrix} 2 \\ -3 \\ 1 \end{bmatrix}$ as a linear combination of the vectors in S.

In Exercises 29 and 30, compute the QR-factorization of A.

29. **(a)** $A = \begin{bmatrix} 1 & 2 \\ -1 & 3 \end{bmatrix}$ **(b)** $A = \begin{bmatrix} 1 & 2 \\ -1 & -2 \\ 1 & 1 \end{bmatrix}$

(c) $A = \begin{bmatrix} 1 & 0 & -1 \\ 2 & -3 & 3 \\ -1 & 2 & 4 \end{bmatrix}$

30. **(a)** $A = \begin{bmatrix} 2 & -1 \\ -1 & 3 \\ 0 & 1 \end{bmatrix}$ **(b)** $A = \begin{bmatrix} 1 & 0 & 2 \\ -1 & 2 & 0 \\ -1 & -2 & 2 \end{bmatrix}$

(c) $A = \begin{bmatrix} 2 & -1 & 1 \\ 1 & 2 & -2 \\ 0 & 1 & -2 \end{bmatrix}$

31. Show that if **u** and **v** are orthogonal vectors in an inner product space, then $(\mathbf{u}, c\mathbf{v}) = 0$ for any scalar c.

32. Let $\mathbf{u}_1, \mathbf{u}_2, \ldots, \mathbf{u}_n$ be vectors in R^n. Show that if **u** is orthogonal to $\mathbf{u}_1, \mathbf{u}_2, \ldots, \mathbf{u}_n$, then **u** is orthogonal to every vector in

$$\text{span}\,\{\mathbf{u}_1, \mathbf{u}_2, \ldots, \mathbf{u}_n\}.$$

33. Let **u** be a fixed vector in R^n. Prove that the set of all vectors in R^n that are orthogonal to **u** is a subspace of R^n.

34. Let $S = \{\mathbf{u}_1, \mathbf{u}_2, \ldots, \mathbf{u}_k\}$ be an orthonormal basis for a subspace W of Euclidean space V that has dimension $n > k$. Discuss how to construct an orthonormal basis for V that includes S.

35. Let $S = \{\mathbf{v}_1, \mathbf{v}_2, \ldots, \mathbf{v}_k\}$ be an orthonormal basis for the Euclidean space V and $\{a_1, a_2, \ldots, a_k\}$ be any set of scalars none of which is zero. Prove that

$$T = \{a_1\mathbf{v}_1, a_2\mathbf{v}_2, \ldots, a_k\mathbf{v}_k\}$$

is an orthogonal basis for V. How should the scalars $a_1, a_2, \ldots, a_k$ be chosen so that T is an orthonormal basis for V?

36. In the proof of Theorem 5.8, show that the diagonal entries r_{ii} are nonzero by first expressing $\mathbf{u}_i$ as a linear combination of $\mathbf{v}_1, \mathbf{v}_2, \ldots, \mathbf{v}_i$ and then computing $r_{ii} = (\mathbf{u}_i, \mathbf{w}_i)$.

37. Show that every nonsingular matrix has a QR-factorization.

38. Determine whether the software that you use has a command to compute an orthonormal set of vectors from a linearly independent set of vectors in R^n. (Assume that the standard inner product is used.) If it does, compare the output from your command with the results in Example 2. To experiment further, use Exercises 2, 7, 8, 11, and 12.

39. Determine whether the software you use has a command to obtain the QR-factorization of a given matrix. If it does, compare the output produced by your command with the results obtained in Example 4. Experiment further with Exercises 29 and 30. Remember the remark following Example 4, which points out that most software in use today will yield a different type of QR-factorization of a given matrix.

5.5 Orthogonal Complements

In Supplementary Exercises 34 and 35 in Chapter 4, we asked you to show that if W_1 and W_2 are subspaces of a vector space V, then $W_1 + W_2$ (the set of all vectors $\mathbf{v}$ in W such that $\mathbf{v} = \mathbf{w}_1 + \mathbf{w}_2$, where $\mathbf{w}_1$ is in W_1 and $\mathbf{w}_2$ is in W_2) is a subspace of V. Moreover, if $V = W_1 + W_2$ and $W_1 \cap W_2 = \{\mathbf{0}\}$, then $V = W_1 \oplus W_2$; that is, V is the direct sum of W_1 and W_2, which means that every vector in V can be written uniquely as $\mathbf{w}_1 + \mathbf{w}_2$, where $\mathbf{w}_1$ is in W_1 and $\mathbf{w}_2$ is in W_2. In this section we show that if V is an inner product space and W is a finite-dimensional subspace of V, then V can be written as a direct sum of W and another subspace of V. This subspace will be used to examine a basic relationship between four vector spaces associated with a matrix.

DEFINITION 5.6 Let W be a subspace of an inner product space V. A vector $\mathbf{u}$ in V is said to be **orthogonal** to W if it is orthogonal to every vector in W. The set of all vectors in V that are orthogonal to all the vectors in W is called the **orthogonal complement** of W in V and is denoted by $W^\perp$. (Read "W perp".)

EXAMPLE 1 Let W be the subspace of R^3 consisting of all multiples of the vector

$$\mathbf{w} = \begin{bmatrix} 2 \\ -3 \\ 4 \end{bmatrix}.$$

Thus $W = \text{span}\,\{\mathbf{w}\}$, so W is a one-dimensional subspace of W. Then a vector $\mathbf{u}$ in R^3 belongs to $W^\perp$ if and only if $\mathbf{u}$ is orthogonal to $c\mathbf{w}$, for any scalar c. Thus, geometrically, $W^\perp$ is the plane with normal $\mathbf{w}$. Using Equations (7) and (8) in optional Section 5.2, $W^\perp$ can also be described as the set of all points $P(x, y, z)$ in R^3 such that

$$2x - 3y + 4z = 0.$$ ■

Observe that if W is a subspace of an inner product space V, then the zero vector of V always belongs to $W^\perp$ (Exercise 24). Moreover, the orthogonal complement of V is the zero subspace, and the orthogonal complement of the zero subspace is V itself (Exercise 25).

Theorem 5.9 Let W be a subspace of an inner product space V. Then the following are true:

(a) $W^\perp$ is a subspace of V.

(b) $W \cap W^\perp = \{\mathbf{0}\}$.

Proof

(a) Let $\mathbf{u}_1$ and $\mathbf{u}_2$ be in $W^\perp$. Then $\mathbf{u}_1$ and $\mathbf{u}_2$ are orthogonal to each vector $\mathbf{w}$ in W. We now have

$$(\mathbf{u}_1 + \mathbf{u}_2, \mathbf{w}) = (\mathbf{u}_1, \mathbf{w}) + (\mathbf{u}_2, \mathbf{w}) = 0 + 0 = 0,$$

so $\mathbf{u}_1 + \mathbf{u}_2$ is in $W^\perp$. Also, let $\mathbf{u}$ be in $W^\perp$ and c be a real scalar. Then for any vector $\mathbf{w}$ in W, we have

$$(c\mathbf{u}, \mathbf{w}) = c(\mathbf{u}, \mathbf{w}) = c\,0 = 0,$$

so $c\mathbf{u}$ is in W. This implies that $W^\perp$ is closed under vector addition and scalar multiplication and hence is a subspace of V.

(b) Let $\mathbf{u}$ be a vector in $W \cap W^\perp$. Then $\mathbf{u}$ is in both W and $W^\perp$, so $(\mathbf{u}, \mathbf{u}) = 0$. From Definition 5.1, it follows that $\mathbf{u} = \mathbf{0}$. ■

In Exercise 26 we ask you to show that if W is a subspace of an inner product space V that is spanned by a set of vectors S, then a vector $\mathbf{u}$ in V belongs to $W^\perp$ if and only if $\mathbf{u}$ is orthogonal to every vector in S. This result can be helpful in finding $W^\perp$, as shown by the next example.

EXAMPLE 2

Let V be the Euclidean space P_3 with the inner product defined in Example 6 of Section 5.3:

$$(p(t), q(t)) = \int_0^1 p(t)q(t)\,dt.$$

Let W be the subspace of P_3 with basis $\{1, t^2\}$. Find a basis for $W^\perp$.

Solution

Let $p(t) = at^3 + bt^2 + ct + d$ be an element of $W^\perp$. Since $p(t)$ must be orthogonal to each of the vectors in the given basis for W, we have

$$(p(t), 1) = \int_0^1 (at^3 + bt^2 + ct + d)\,dt = \frac{a}{4} + \frac{b}{3} + \frac{c}{2} + d = 0,$$

$$(p(t), t^2) = \int_0^1 (at^5 + bt^4 + ct^3 + dt^2)\,dt = \frac{a}{6} + \frac{b}{5} + \frac{c}{4} + \frac{d}{3} = 0.$$

Solving the homogeneous system

$$\begin{aligned} \frac{a}{4} + \frac{b}{3} + \frac{c}{2} + d &= 0 \\ \frac{a}{6} + \frac{b}{5} + \frac{c}{4} + \frac{d}{3} &= 0, \end{aligned}$$

we obtain (verify)

$$a = 3r + 16s, \quad b = -\frac{15}{4}r - 15s, \quad c = r, \quad d = s.$$

Then

$$\begin{aligned} p(t) &= (3r + 16s)t^3 + \left(-\frac{15}{4}r - 15s\right)t^2 + rt + s \\ &= r\left(3t^3 - \frac{15}{4}t^2 + t\right) + s(16t^3 - 15t^2 + 1). \end{aligned}$$

Hence the vectors $3t^3 - \frac{15}{4}t^2 + t$ and $16t^3 - 15t^2 + 1$ span $W^\perp$. Since they are not multiples of each other, they are linearly independent and thus form a basis for $W^\perp$. ■

Theorem 5.10 Let W be a finite-dimensional subspace of an inner product space V. Then

$$V = W \oplus W^\perp.$$

Proof

Let $\dim W = m$. Then W has a basis consisting of m vectors. By the Gram–Schmidt process we can transform this basis to an orthonormal basis. Thus, let $S = \{\mathbf{w}_1, \mathbf{w}_2, \ldots, \mathbf{w}_m\}$ be an orthonormal basis for W. If $\mathbf{v}$ is a vector in V, let

$$\mathbf{w} = (\mathbf{v}, \mathbf{w}_1)\mathbf{w}_1 + (\mathbf{v}, \mathbf{w}_2)\mathbf{w}_2 + \cdots + (\mathbf{v}, \mathbf{w}_m)\mathbf{w}_m \tag{1}$$

and

$$\mathbf{u} = \mathbf{v} - \mathbf{w}. \tag{2}$$

Since $\mathbf{w}$ is a linear combination of vectors in S, $\mathbf{w}$ belongs to W. We next prove that $\mathbf{u}$ lies in $W^\perp$ by showing that $\mathbf{u}$ is orthogonal to every vector in S, a basis for W. For each $\mathbf{w}_i$ in S, we have

$$\begin{aligned} (\mathbf{u}, \mathbf{w}_i) &= (\mathbf{v} - \mathbf{w}, \mathbf{w}_i) = (\mathbf{v}, \mathbf{w}_i) - (\mathbf{w}, \mathbf{w}_i) \\ &= (\mathbf{v}, \mathbf{w}_i) - ((\mathbf{v}, \mathbf{w}_1)\mathbf{w}_1 + (\mathbf{v}, \mathbf{w}_2)\mathbf{w}_2 + \cdots + (\mathbf{v}, \mathbf{w}_m)\mathbf{w}_m, \mathbf{w}_i) \\ &= (\mathbf{v}, \mathbf{w}_i) - (\mathbf{v}, \mathbf{w}_i)(\mathbf{w}_i, \mathbf{w}_i) \\ &= 0, \end{aligned}$$

since $(\mathbf{w}_i, \mathbf{w}_j) = 0$ for $i \neq j$ and $(\mathbf{w}_i, \mathbf{w}_i) = 1$, $1 \leq i \leq m$. Thus $\mathbf{u}$ is orthogonal to every vector in W and so lies in $W^\perp$. Hence

$$\mathbf{v} = \mathbf{w} + \mathbf{u},$$

which means that $V = W + W^\perp$. From part (b) of Theorem 5.9, it follows that

$$V = W \oplus W^\perp.$$ ■

Remark As pointed out at the beginning of this section, we also conclude that the vectors $\mathbf{w}$ and $\mathbf{u}$ defined by Equations (1) and (2) are unique.

Theorem 5.11 If W is a finite-dimensional subspace of an inner product space V, then

$$(W^\perp)^\perp = W.$$

Proof

First, if $\mathbf{w}$ is any vector in W, then $\mathbf{w}$ is orthogonal to every vector $\mathbf{u}$ in $W^{\perp}$, so $\mathbf{w}$ is in $(W^{\perp})^{\perp}$. Hence W is a subspace of $(W^{\perp})^{\perp}$. Conversely, let $\mathbf{v}$ be an arbitrary vector in $(W^{\perp})^{\perp}$. Then, by Theorem 5.10, $\mathbf{v}$ can be written as

$$\mathbf{v} = \mathbf{w} + \mathbf{u},$$

where $\mathbf{w}$ is in W and $\mathbf{u}$ is in $W^{\perp}$. Since $\mathbf{u}$ is in $W^{\perp}$, it is orthogonal to $\mathbf{v}$ and $\mathbf{w}$. Thus

$$0 = (\mathbf{u}, \mathbf{v}) = (\mathbf{u}, \mathbf{w} + \mathbf{u}) = (\mathbf{u}, \mathbf{w}) + (\mathbf{u}, \mathbf{u}) = (\mathbf{u}, \mathbf{u}),$$

or

$$(\mathbf{u}, \mathbf{u}) = 0,$$

which implies that $\mathbf{u} = \mathbf{0}$. Then $\mathbf{v} = \mathbf{w}$, so $\mathbf{v}$ belongs to W. Hence $(W^{\perp})^{\perp} = W$. ■

Remark Since W is the orthogonal complement of $W^{\perp}$, and $W^{\perp}$ is also the orthogonal complement of W, we say that W and $W^{\perp}$ are orthogonal complements.

■ Relations among the Fundamental Subspaces Associated with a Matrix

If A is a given $m \times n$ matrix, we associate the following four fundamental subspaces with A: the null space of A, the row space of A, the null space of A^T, and the column space of A. The following theorem shows that pairs of these four subspaces are orthogonal complements:

Theorem 5.12 If A is a given $m \times n$ matrix, then the following statements are true:

(a) The null space of A is the orthogonal complement of the row space of A.[†]

(b) The null space of A^T is the orthogonal complement of the column space of A.

Proof

(a) Before proving the result, let us verify that the two vector spaces that we wish to show are the same have equal dimensions. If r is the rank of A, then the dimension of the null space of A is $n - r$ (Theorem 4.19). Since the dimension of the row space of A is r, then by Theorem 5.10, the dimension of its orthogonal complement is also $n - r$. Let the vector $\mathbf{x}$ in R^n be in the null space of A. Then $A\mathbf{x} = \mathbf{0}$. Let the vectors $\mathbf{v}_1, \mathbf{v}_2, \ldots, \mathbf{v}_m$ in R^n denote the rows of A. Then the entries in the $m \times 1$ matrix $A\mathbf{x}$ are $\mathbf{v}_1\mathbf{x}, \mathbf{v}_2\mathbf{x}, \ldots, \mathbf{v}_m\mathbf{x}$. Thus we have

$$\mathbf{v}_1\mathbf{x} = 0, \quad \mathbf{v}_2\mathbf{x} = 0, \quad \ldots, \quad \mathbf{v}_m\mathbf{x} = 0. \tag{3}$$

Since $\mathbf{v}_i\mathbf{x} = \mathbf{v}_i \cdot \mathbf{x}$, $i = 1, 2, \ldots, m$, it follows that $\mathbf{x}$ is orthogonal to the vectors $\mathbf{v}_1, \mathbf{v}_2, \ldots, \mathbf{v}_m$, which span the row space of A. It then follows that $\mathbf{x}$ is orthogonal to every vector in the row space of A, so $\mathbf{x}$ lies in the orthogonal complement of the row space of A. Hence the null space of A is contained in the orthogonal complement of the row space of A.

[†]Strictly speaking, the null space of A consists of vectors in R^n—that is, column vectors—whereas the row space of A consists of row vectors—that is, vectors in R_n. Thus the orthogonal complement of the row space should also consist of row vectors. However, by Theorem 4.14 in Section 4.8, R^n and R_n are isomorphic, so an n-vector can be viewed as a row vector or as a column vector.

Conversely, if $\mathbf{x}$ is in the orthogonal complement of the row space of A, then $\mathbf{x}$ is orthogonal to the vectors $\mathbf{v}_1, \mathbf{v}_2, \ldots, \mathbf{v}_m$, so we have the relations given by (3), which imply that $A\mathbf{x} = \mathbf{0}$. Thus $\mathbf{x}$ belongs to the null space of A. Hence the orthogonal complement of the row space of A is contained in the null space of A. It then follows that the null space of A equals the orthogonal complement of the row space of A.

(b) To establish the result, replace A by A^T in part (a) to conclude that the null space of A^T is the orthogonal complement of the row space of A^T. Since the row space of A^T is the column space of A, we have established part (b). ■

In Section 4.9 we briefly discussed the relationship between subspaces associated with a matrix and in Figure 4.33 illustrated their relationships. Using Theorems 5.10 and 5.12, we can say more about the relationships between these subspaces for an $m \times n$ matrix A.

For the row space of A and the null space of A, the following apply:

- Every vector in the row space of A is orthogonal to every vector in the null space of A. [Theorem 5.12(a).]
- The only vector that belongs to both the row space of A and its null space is the zero vector.
- If the rank of A is r, then the dimension of the row space of A is r and the dimension of the null space of A is $n - r$. Thus

$$\text{row space of } A \oplus \text{null space of } A = R^n,$$

and hence every vector in R^n is uniquely expressible as the sum of a vector from the row space of A and a vector from the null space of A. (Theorem 5.10.) This is illustrated in Figure 5.21.

Theorem 5.12(b) tells us that the column space of A and the null space of A^T are related as shown in Figure 5.22.

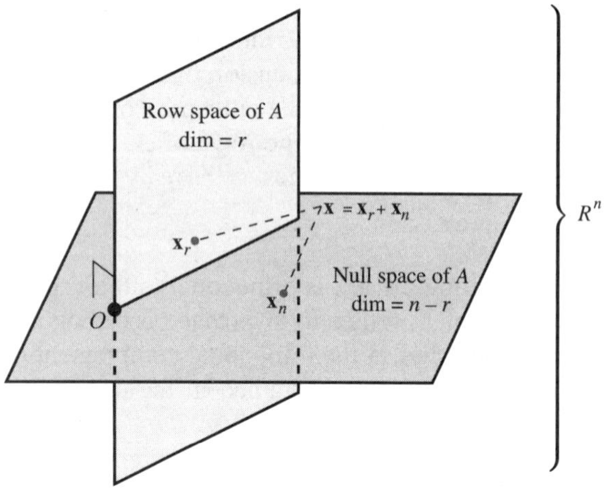

FIGURE 5.21

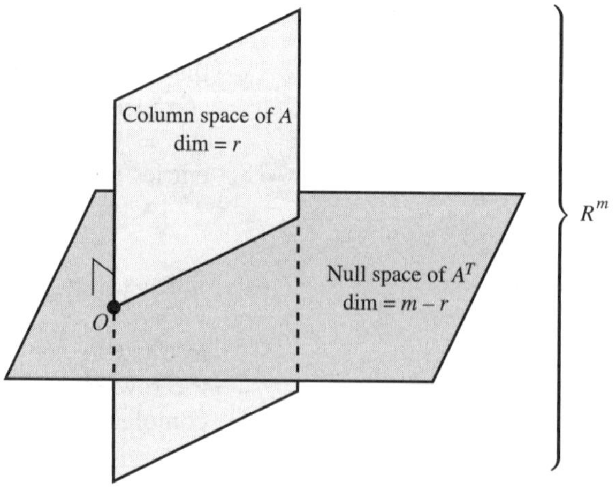

FIGURE 5.22

In order to connect the subspaces represented in Figures 5.21 and 5.22, we use matrix products.

- Let $\mathbf{x}_r$ be in the row space of A. Then $A\mathbf{x}_r = \mathbf{b}$ is a linear combination of the columns of A and hence lies in the column space of A. See Figure 5.23.
- Let $\mathbf{x}_n$ be in the null space of A. Then $A\mathbf{x}_n = \mathbf{0}$, the zero vector in R^n. See Figure 5.23.

For an arbitrary vector $\mathbf{x}$ in R^n, Theorem 5.10 implies that there exist a vector $\mathbf{x}_r$ in the row space of A and a vector $\mathbf{x}_n$ in the null space of A such that

$$\mathbf{x} = \mathbf{x}_r + \mathbf{x}_n.$$

It follows that $A\mathbf{x} = A\mathbf{x}_r = \mathbf{b}$. See Figure 5.23.

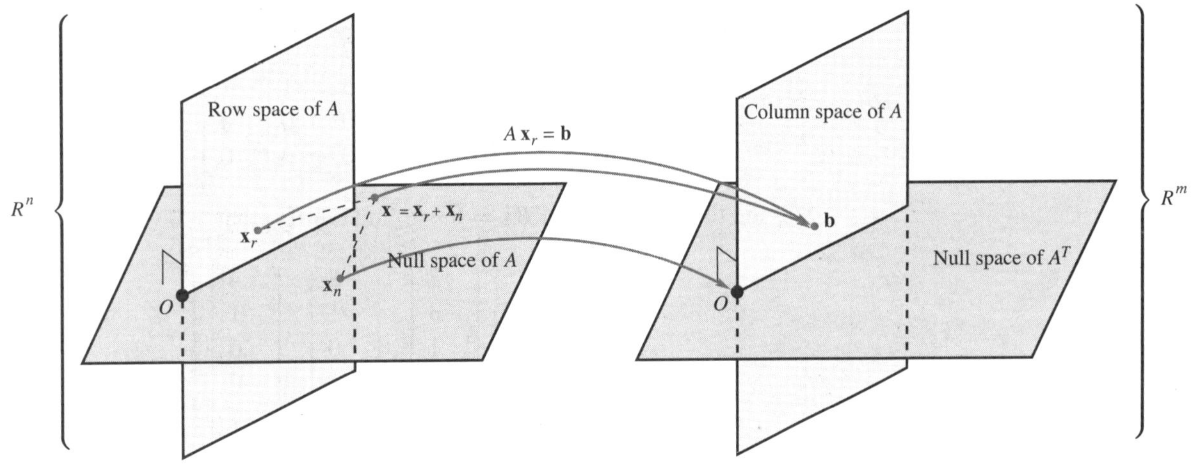

FIGURE 5.23†

In summary, for any $\mathbf{x}$ in R^n, $A\mathbf{x}$ is in the column space of A and hence this case corresponds to a consistent nonhomogeneous linear system $A\mathbf{x} = \mathbf{b}$. In the

†This material is based upon the work of Gilbert Strang*; for further details, see "The Fundamental Theorem of Linear Algebra," *American Mathematical Monthly*, 100 (1993), pp. 848–855.

GILBERT STRANG

*Gilbert Strang (1934–) was born in Chicago, Illinois. He received his undergraduate degree from MIT in 1955 and was a Rhodes Scholar at Balliol College, Oxford. His Ph.D. was from UCLA in 1959, and since then he has taught at MIT. He has been a Sloan Fellow (1966–1967) and a Fairchild Scholar at the California Institute of Technology (1980–1981), and he was named a Fellow of the American Academy of Arts and Sciences in 1985. He is a Professor of Mathematics at MIT and an Honorary Fellow of Balliol College. He was the President of SIAM during 1999 and 2000, and Chair of the Joint Policy Board for Mathematics. He received the von Neumann Medal of the U.S. Association for Computational Mechanics. In 2007, the first Su Buchin Prize from the International Congress of Industrial and Applied Mathematics, and the Haimo Prize from the Mathematical Association of America, were awarded to him for his contributions to teaching around the world. Dr. Strang has been an editor for numerous influential journals, published highly acclaimed books, and authored over 150 papers. His contributions to linear algebra and its applications have had a broad impact on research as well as teaching.

particular case that $\mathbf{x}$ is in the null space of A, the corresponding linear system is the homogeneous system $A\mathbf{x} = \mathbf{0}$.

EXAMPLE 3

Let

$$A = \begin{bmatrix} 1 & -2 & 1 & 0 & 2 \\ 1 & -1 & 4 & 1 & 3 \\ -1 & 3 & 2 & 1 & -1 \\ 2 & -3 & 5 & 1 & 5 \end{bmatrix}.$$

Compute the four fundamental vector spaces associated with A and verify Theorem 5.12.

Solution

We first transform A to reduced row echelon form, obtaining (verify)

$$B = \begin{bmatrix} 1 & 0 & 7 & 2 & 4 \\ 0 & 1 & 3 & 1 & 1 \\ 0 & 0 & 0 & 0 & 0 \\ 0 & 0 & 0 & 0 & 0 \end{bmatrix}.$$

Solving the linear system $B\mathbf{x} = \mathbf{0}$, we find (verify) that

$$S = \left\{ \begin{bmatrix} -7 \\ -3 \\ 1 \\ 0 \\ 0 \end{bmatrix}, \begin{bmatrix} -2 \\ -1 \\ 0 \\ 1 \\ 0 \end{bmatrix}, \begin{bmatrix} -4 \\ -1 \\ 0 \\ 0 \\ 1 \end{bmatrix} \right\}$$

is a basis for the null space of A. Moreover,

$$T = \left\{ \begin{bmatrix} 1 & 0 & 7 & 2 & 4 \end{bmatrix}, \begin{bmatrix} 0 & 1 & 3 & 1 & 1 \end{bmatrix} \right\}$$

is a basis for the row space of A. Since the vectors in S and T are orthogonal, it follows that S is a basis for the orthogonal complement of the row space of A, where we take the vectors in S as row vectors. Next,

$$A^T = \begin{bmatrix} 1 & 1 & -1 & 2 \\ -2 & -1 & 3 & -3 \\ 1 & 4 & 2 & 5 \\ 0 & 1 & 1 & 1 \\ 2 & 3 & -1 & 5 \end{bmatrix}.$$

Solving the linear system $A^T\mathbf{x} = \mathbf{0}$, we find (verify) that

$$S' = \left\{ \begin{bmatrix} 2 \\ -1 \\ 1 \\ 0 \end{bmatrix}, \begin{bmatrix} -1 \\ -1 \\ 0 \\ 1 \end{bmatrix} \right\}$$

is a basis for the null space of A^T. Transforming A^T to reduced row echelon form, we obtain (verify)

$$C = \begin{bmatrix} 1 & 0 & -2 & 1 \\ 0 & 1 & 1 & 1 \\ 0 & 0 & 0 & 0 \\ 0 & 0 & 0 & 0 \\ 0 & 0 & 0 & 0 \end{bmatrix}.$$

Then the nonzero rows of C, read vertically, yield the following basis for the column space of A:

$$T' = \left\{ \begin{bmatrix} 1 \\ 0 \\ -2 \\ 1 \end{bmatrix}, \begin{bmatrix} 0 \\ 1 \\ 1 \\ 1 \end{bmatrix} \right\}.$$

Since the vectors in S' and T' are orthogonal, it follows that S' is a basis for the orthogonal complement of the column space of A. ■

EXAMPLE 4 Find a basis for the orthogonal complement of the subspace W of R_5 spanned by the vectors

$$\mathbf{w}_1 = \begin{bmatrix} 2 & -1 & 0 & 1 & 2 \end{bmatrix}, \quad \mathbf{w}_2 = \begin{bmatrix} 1 & 3 & 1 & -2 & -4 \end{bmatrix},$$
$$\mathbf{w}_3 = \begin{bmatrix} 3 & 2 & 1 & -1 & -2 \end{bmatrix}, \quad \mathbf{w}_4 = \begin{bmatrix} 7 & 7 & 3 & -4 & -8 \end{bmatrix},$$
$$\mathbf{w}_5 = \begin{bmatrix} 1 & -4 & -1 & -1 & -2 \end{bmatrix}.$$

Solution 1

Let $\mathbf{u} = \begin{bmatrix} a & b & c & d & e \end{bmatrix}$ be an arbitrary vector in $W^{\perp}$. Since $\mathbf{u}$ is orthogonal to each of the given vectors spanning W, we have a linear system of five equations in five unknowns, whose coefficient matrix is (verify)

$$A = \begin{bmatrix} 2 & -1 & 0 & 1 & 2 \\ 1 & 3 & 1 & -2 & -4 \\ 3 & 2 & 1 & -1 & -2 \\ 7 & 7 & 3 & -4 & -8 \\ 1 & -4 & -1 & -1 & -2 \end{bmatrix}.$$

Solving the homogeneous system $A\mathbf{x} = \mathbf{0}$, we obtain the following basis for the solution space (verify):

$$S = \left\{ \begin{bmatrix} -\frac{1}{7} \\ -\frac{2}{7} \\ 1 \\ 0 \\ 0 \end{bmatrix}, \begin{bmatrix} 0 \\ 0 \\ 0 \\ -2 \\ 1 \end{bmatrix} \right\}.$$

These vectors, taken in horizontal form, provide a basis for $W^{\perp}$.

Solution 2

Form the matrix whose rows are the given vectors. This matrix is A as shown in Solution 1, so the row space of A is W. By Theorem 5.12, $W^{\perp}$ is the null space of A. Thus we obtain the same basis for $W^{\perp}$ as in Solution 1. ■

The following example will be used to geometrically illustrate Theorem 5.12:

EXAMPLE 5

Let

$$A = \begin{bmatrix} 3 & -1 & 2 \\ 2 & 1 & 3 \\ 7 & 1 & 8 \end{bmatrix}.$$

Transforming A to reduced row echelon form, we obtain

$$\begin{bmatrix} 1 & 0 & 1 \\ 0 & 1 & 1 \\ 0 & 0 & 0 \end{bmatrix},$$

so the row space of A is a two-dimensional subspace of R_3—that is, a plane passing through the origin—with basis $\{(1, 0, 1), (0, 1, 1)\}$. The null space of A is a one-dimensional subspace of R^3 with basis

$$\left\{\begin{bmatrix} -1 \\ -1 \\ 1 \end{bmatrix}\right\}$$

(verify). Since this basis vector is orthogonal to the two basis vectors for the row space of A just given, the null space of A is orthogonal to the row space of A; that is, the null space of A is the orthogonal complement of the row space of A.

Next, transforming A^T to reduced row echelon form, we have

$$\begin{bmatrix} 1 & 0 & 1 \\ 0 & 1 & 2 \\ 0 & 0 & 0 \end{bmatrix},$$

(verify). It follows that

$$\left\{\begin{bmatrix} -1 \\ -2 \\ 1 \end{bmatrix}\right\}$$

is a basis for the null space of A^T (verify). Hence the null space of A^T is a line through the origin. Moreover,

$$\left\{\begin{bmatrix} 1 \\ 0 \\ 1 \end{bmatrix}, \begin{bmatrix} 0 \\ 1 \\ 2 \end{bmatrix}\right\}$$

is a basis for the column space of A^T (verify), so the column space of A^T is a plane through the origin. Since every basis vector for the null space of A^T is orthogonal to every basis vector for the column space of A^T, we conclude that the null space of A^T is the orthogonal complement of the column space of A^T. These results are illustrated in Figure 5.24. ■

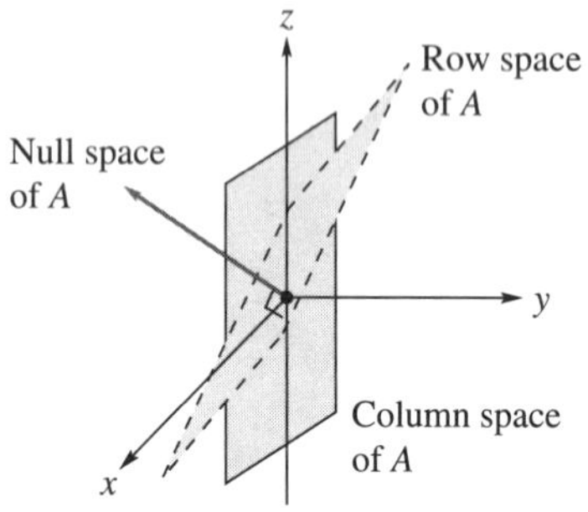

FIGURE 5.24

Projections and Applications

In Theorem 5.10 and in the Remark following the theorem, we have shown that if W is a finite-dimensional subspace of an inner product space V with orthonormal basis $\{\mathbf{w}_1, \mathbf{w}_2, \ldots, \mathbf{w}_m\}$ and $\mathbf{v}$ is any vector in V, then there exist unique vectors $\mathbf{w}$ in W and $\mathbf{u}$ in $W^{\perp}$ such that

$$\mathbf{v} = \mathbf{w} + \mathbf{u}.$$

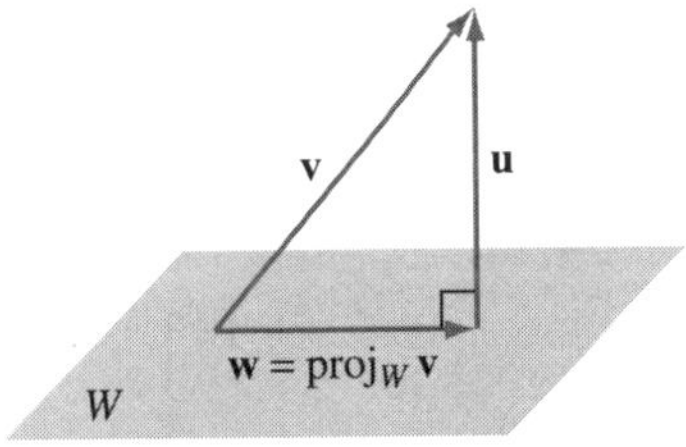

FIGURE 5.25

Moreover, as we saw in Equation (1),

$$\mathbf{w} = (\mathbf{v}, \mathbf{w}_1)\mathbf{w}_1 + (\mathbf{v}, \mathbf{w}_2)\mathbf{w}_2 + \cdots + (\mathbf{v}, \mathbf{w}_m)\mathbf{w}_m,$$

which is called the **orthogonal projection** of $\mathbf{v}$ on W and is denoted by $\text{proj}_W \mathbf{v}$. In Figure 5.25, we illustrate Theorem 5.10 when W is a two-dimensional subspace of R^3 (a plane through the origin).

Often, an orthonormal basis has many fractions, so it is helpful to also have a formula giving $\text{proj}_W \mathbf{v}$ when W has an *orthogonal* basis. In Exercise 29, we ask you to show that if $\{\mathbf{w}_1, \mathbf{w}_2, \ldots, \mathbf{w}_m\}$ is an orthogonal basis for W, then

$$\text{proj}_W \mathbf{v} = \frac{(\mathbf{v}, \mathbf{w}_1)}{(\mathbf{w}_1, \mathbf{w}_1)}\mathbf{w}_1 + \frac{(\mathbf{v}, \mathbf{w}_2)}{(\mathbf{w}_2, \mathbf{w}_2)}\mathbf{w}_2 + \cdots + \frac{(\mathbf{v}, \mathbf{w}_m)}{(\mathbf{w}_m, \mathbf{w}_m)}\mathbf{w}_m.$$

Remark The Gram–Schmidt process described in Theorem 5.6 can be rephrased in terms of projections at each step. Thus, Figure 5.19 (the first step in the Gram–Schmidt process) can be relabeled as follows:

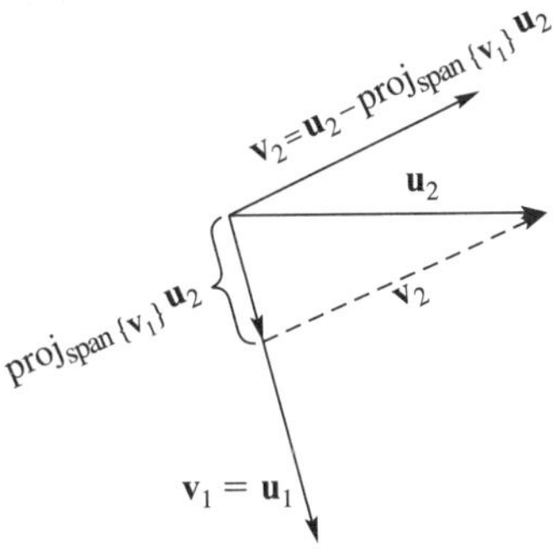

FIGURE 5.19 (relabeled)

EXAMPLE 6

Let W be the two-dimensional subspace of R^3 with orthonormal basis $\{\mathbf{w}_1, \mathbf{w}_2\}$, where

$$\mathbf{w}_1 = \begin{bmatrix} \frac{2}{3} \\ -\frac{1}{3} \\ -\frac{2}{3} \end{bmatrix} \quad \text{and} \quad \mathbf{w}_2 = \begin{bmatrix} \frac{1}{\sqrt{2}} \\ 0 \\ \frac{1}{\sqrt{2}} \end{bmatrix}.$$

Using the standard inner product on R^3, find the orthogonal projection of

$$\mathbf{v} = \begin{bmatrix} 2 \\ 1 \\ 3 \end{bmatrix}$$

on W and the vector $\mathbf{u}$ that is orthogonal to every vector in W.

Solution

From Equation (1), we have

$$\mathbf{w} = \text{proj}_W \mathbf{v} = (\mathbf{v}, \mathbf{w}_1)\mathbf{w}_1 + (\mathbf{v}, \mathbf{w}_2)\mathbf{w}_2 = -1\mathbf{w}_1 + \frac{5}{\sqrt{2}}\mathbf{w}_2 = \begin{bmatrix} \frac{11}{6} \\ \frac{1}{3} \\ \frac{19}{6} \end{bmatrix}$$

and

$$\mathbf{u} = \mathbf{v} - \mathbf{w} = \begin{bmatrix} \frac{1}{6} \\ \frac{2}{3} \\ -\frac{1}{6} \end{bmatrix}.$$

■

It is clear from Figure 5.25 that the distance from $\mathbf{v}$ to the plane W is given by the length of the vector $\mathbf{u} = \mathbf{v} - \mathbf{w}$, that is, by

$$\left\| \mathbf{v} - \text{proj}_W \mathbf{v} \right\|.$$

We prove this result, in general, in Theorem 5.13.

EXAMPLE 7 Let W be the subspace of R^3 defined in Example 6 and let $\mathbf{v} = \begin{bmatrix} 1 \\ 1 \\ 0 \end{bmatrix}$. Find the distance from $\mathbf{v}$ to W.

Solution

We first compute

$$\text{proj}_W \mathbf{v} = (\mathbf{v}, \mathbf{w}_1)\mathbf{w}_1 + (\mathbf{v}, \mathbf{w}_2)\mathbf{w}_2 = \frac{1}{3}\mathbf{w}_1 + \frac{1}{\sqrt{2}}\mathbf{w}_2 = \begin{bmatrix} \frac{13}{18} \\ -\frac{1}{9} \\ \frac{5}{18} \end{bmatrix}.$$

Then

$$\mathbf{v} - \text{proj}_W \mathbf{v} = \begin{bmatrix} 1 \\ 1 \\ 0 \end{bmatrix} - \begin{bmatrix} \frac{13}{18} \\ -\frac{1}{9} \\ \frac{5}{18} \end{bmatrix} = \begin{bmatrix} \frac{5}{18} \\ \frac{10}{9} \\ -\frac{5}{18} \end{bmatrix}$$

and

$$\|\mathbf{v} - \text{proj}_W \mathbf{v}\| = \sqrt{\frac{25}{18^2} + \frac{400}{18^2} + \frac{25}{18^2}} = \frac{15}{18}\sqrt{2} = \frac{5}{6}\sqrt{2},$$

so the distance from $\mathbf{v}$ to W is $\frac{5}{6}\sqrt{2}$. ■

In Example 7, $\|\mathbf{v} - \text{proj}_W \mathbf{v}\|$ represented the distance in 3-space from $\mathbf{v}$ to the plane W. We can generalize this notion of distance from a vector in V to a subspace W of V. We can show that the vector in W that is closest to $\mathbf{v}$ is in fact $\text{proj}_W \mathbf{v}$, so $\|\mathbf{v} - \text{proj}_W \mathbf{v}\|$ represents the distance from $\mathbf{v}$ to W.

Theorem 5.13 Let W be a finite-dimensional subspace of the inner product space V. Then, for vector $\mathbf{v}$ belonging to V, the vector in W closest to $\mathbf{v}$ is $\text{proj}_W \mathbf{v}$. That is, $\|\mathbf{v} - \mathbf{w}\|$, for $\mathbf{w}$ belonging to W, is minimized when $\mathbf{w} = \text{proj}_W \mathbf{v}$.

Proof

Let $\mathbf{w}$ be any vector in W. Then

$$\mathbf{v} - \mathbf{w} = (\mathbf{v} - \text{proj}_W \mathbf{v}) + (\text{proj}_W \mathbf{v} - \mathbf{w}).$$

Since $\mathbf{w}$ and $\text{proj}_W \mathbf{v}$ are both in W, $\text{proj}_W \mathbf{v} - \mathbf{w}$ is in W. By Theorem 5.10, $\mathbf{v} - \text{proj}_W \mathbf{v}$ is orthogonal to every vector in W, so

$$\begin{aligned} \|\mathbf{v} - \mathbf{w}\|^2 &= (\mathbf{v} - \mathbf{w}, \mathbf{v} - \mathbf{w}) \\ &= ((\mathbf{v} - \text{proj}_W \mathbf{v}) + (\text{proj}_W \mathbf{v} - \mathbf{w}), (\mathbf{v} - \text{proj}_W \mathbf{v}) + (\text{proj}_W \mathbf{v} - \mathbf{w})) \\ &= \|\mathbf{v} - \text{proj}_W \mathbf{v}\|^2 + \|\text{proj}_W \mathbf{v} - \mathbf{w}\|^2. \end{aligned}$$

If $\mathbf{w} \neq \text{proj}_W \mathbf{v}$, then $\|\text{proj}_W \mathbf{v} - \mathbf{w}\|^2$ is positive and

$$\|\mathbf{v} - \mathbf{w}\|^2 > \|\mathbf{v} - \text{proj}_W \mathbf{v}\|^2.$$

Thus it follows that $\text{proj}_W \mathbf{v}$ is the vector in W that minimizes $\|\mathbf{v} - \mathbf{w}\|^2$ and hence minimizes $\|\mathbf{v} - \mathbf{w}\|$. ■

In Example 6, $\mathbf{w} = \text{proj}_W \mathbf{v} = \begin{bmatrix} \frac{11}{6} \\ \frac{1}{3} \\ \frac{19}{6} \end{bmatrix}$ is the vector in $W = \text{span}\,\{\mathbf{w}_1, \mathbf{w}_2\}$ that is closest to $\mathbf{v} = \begin{bmatrix} 2 \\ 1 \\ 3 \end{bmatrix}$.

■ Fourier Series (Calculus Required)

In the study of calculus, you most likely encountered functions $f(t)$, which had derivatives of all orders at a point $t = t_0$. Associated with $f(t)$ is its Taylor series, defined by

$$\sum_{k=0}^{\infty} \frac{f^{(k)}(t_0)}{k!}(t - t_0)^k. \tag{4}$$

The expression in (4) is called the **Taylor series of f at t_0** (or **about t_0**, or **centered at t_0**). When $t_0 = 0$, the Taylor series is called a **Maclaurin series**. The coefficients of Taylor and Maclaurin series expansions involve successive derivatives of the given function evaluated at the center of the expansion. If we take the first $n+1$ terms of the series in (4), we obtain a Taylor or Maclaurin polynomial of degree n that approximates the given function.

The function $f(t) = |t|$ does not have a Taylor series expansion at $t_0 = 0$ (a Maclaurin series), because f does not have a derivative at $t = 0$. Thus there is no way to compute the coefficients in such an expansion. The expression in (4) is in terms of the functions $1, t, t^2, \ldots$. However, it is possible to find a series expansion for such a function by using a different type of expansion. One such important expansion involves the set of functions

$$1, \quad \cos t, \quad \sin t, \quad \cos 2t, \quad \sin 2t, \quad \ldots, \quad \cos nt, \quad \sin nt, \ldots,$$

which we discussed briefly in Example 14 of Section 5.3. The French mathematician Jean Baptiste Joseph Fourier* showed that every function f (continuous or not) that is defined on $[-\pi, \pi]$ can be represented by a series of the form

$$\begin{aligned}&\tfrac{1}{2}a_0 + a_1 \cos t + a_2 \cos 2t + \cdots + a_n \cos nt\\&\quad + b_1 \sin t + b_2 \sin 2t + \cdots + b_n \sin nt + \cdots.\end{aligned}$$

It then follows that every function f (continuous or not) that is defined on $[-\pi, \pi]$ can be approximated as closely as we wish by a function of the form

$$\begin{aligned}&\tfrac{1}{2}a_0 + a_1 \cos t + a_2 \cos 2t + \cdots + a_n \cos nt\\&\quad + b_1 \sin t + b_2 \sin 2t + \cdots + b_n \sin nt\end{aligned} \tag{5}$$

for n sufficiently large. The function in (5) is called a **trigonometric polynomial**, and if a_n and b_n are both nonzero, we say that its **degree** is n. The topic of Fourier series is beyond the scope of this book. We limit ourselves to a brief discussion on how to obtain the best approximation of a function by trigonometric polynomials.

JEAN BAPTISTE JOSEPH FOURIER

*Jean Baptiste Joseph Fourier (1768–1830) was born in Auxere, France. His father was a tailor. Fourier received much of his early education in the local military school, which was run by the Benedictine order, and at the age of 19 he decided to study for the priesthood. His strong interest in mathematics, which started developing at the age of 13, continued while he studied for the priesthood. Two years later, he decided not to take his religious vows and became a teacher at the military school where he had studied.

Fourier was active in French politics throughout the French Revolution and the turbulent period that followed. In 1795, he was appointed to a chair at the prestigious École Polytechnique. In 1798, Fourier, as a scientific advisor, accompanied Napoleon in his invasion of Egypt. Upon returning to France, Fourier served for 12 years as prefect of the department of Isére and lived in Grenoble. During this period he did his pioneering work on the theory of heat. In this work he showed that every function can be represented by a series of trigonometric polynomials. Such a series is now called a Fourier series. He died in Paris in 1830.

It is not difficult to show that

$$\int_{-\pi}^{\pi} 1\,dt = 2\pi, \qquad \int_{-\pi}^{\pi} \sin nt\,dt = 0, \qquad \int_{-\pi}^{\pi} \cos nt\,dt = 0,$$

$$\int_{-\pi}^{\pi} \sin nt\,\sin mt\,dt = 0 \quad (n \neq m), \qquad \int_{-\pi}^{\pi} \cos nt\,\cos mt\,dt = 0 \quad (n \neq m),$$

$$\int_{-\pi}^{\pi} \sin nt\,\cos mt\,dt = 0 \quad (n \neq m), \qquad \int_{-\pi}^{\pi} \sin nt\,\sin nt\,dt = \pi,$$

$$\int_{-\pi}^{\pi} \cos nt\,\cos nt\,dt = \pi.$$

(*Hint*: Use your favorite computer algebra system.)

Let V be the vector space of real-valued continuous functions on $[-\pi, \pi]$. If f and g belong to V, then $(f, g) = \int_{-\pi}^{\pi} f(t)g(t)\,dt$ defines an inner product on V, as in Example 14 of Section 5.3. These relations show that the following set of vectors is an orthonormal set in V:

$$\frac{1}{\sqrt{2\pi}}, \quad \frac{1}{\sqrt{\pi}}\cos t, \quad \frac{1}{\sqrt{\pi}}\sin t, \quad \frac{1}{\sqrt{\pi}}\cos 2t, \quad \frac{1}{\sqrt{\pi}}\sin 2t, \quad \ldots,$$

$$\frac{1}{\sqrt{\pi}}\cos nt, \quad \frac{1}{\sqrt{\pi}}\sin nt, \quad \ldots .$$

Now

$$W = \operatorname{span}\left\{\frac{1}{\sqrt{2\pi}}, \frac{1}{\sqrt{\pi}}\cos t, \frac{1}{\sqrt{\pi}}\sin t, \frac{1}{\sqrt{\pi}}\cos 2t, \frac{1}{\sqrt{\pi}}\sin 2t, \quad \ldots, \right.$$
$$\left. \frac{1}{\sqrt{\pi}}\cos nt, \frac{1}{\sqrt{\pi}}\sin nt\right\}$$

is a finite-dimensional subspace of V. Theorem 5.13 implies that the best approximation to a given function f in V by a trigonometric polynomial of degree n is given by $\text{proj}_W f$, the projection of f onto W. This polynomial is called the **Fourier polynomial of degree *n* for *f*.**

EXAMPLE 8

Find Fourier polynomials of degrees 1 and 3 for the function $f(t) = |t|$.

Solution

First, we compute the Fourier polynomial of degree 1. Using Theorem 5.10, we can compute $\text{proj}_W \mathbf{v}$ for $\mathbf{v} = |t|$, as

$$\text{proj}_W |t| = \left(|t|, \frac{1}{\sqrt{2\pi}}\right)\frac{1}{\sqrt{2\pi}} + \left(|t|, \frac{1}{\sqrt{\pi}}\cos t\right)\frac{1}{\sqrt{\pi}}\cos t$$
$$+ \left(|t|, \frac{1}{\sqrt{\pi}}\sin t\right)\frac{1}{\sqrt{\pi}}\sin t.$$

We have

$$\left(|t|, \frac{1}{\sqrt{2\pi}}\right) = \int_{-\pi}^{\pi} |t| \frac{1}{\sqrt{2\pi}}\, dt$$
$$= \frac{1}{\sqrt{2\pi}} \int_{-\pi}^{0} -t\, dt + \frac{1}{\sqrt{2\pi}} \int_{0}^{\pi} t\, dt = \frac{\pi^2}{\sqrt{2\pi}},$$

$$\left(|t|, \frac{1}{\sqrt{\pi}} \cos t\right) = \int_{-\pi}^{\pi} |t| \frac{1}{\sqrt{\pi}} \cos t\, dt$$
$$= \frac{1}{\sqrt{\pi}} \int_{-\pi}^{0} -t \cos t\, dt + \frac{1}{\sqrt{\pi}} \int_{0}^{\pi} t \cos t\, dt$$
$$= -\frac{2}{\sqrt{\pi}} - \frac{2}{\sqrt{\pi}} = -\frac{4}{\sqrt{\pi}},$$

and

$$\left(|t|, \frac{1}{\sqrt{\pi}} \sin t\right) = \int_{-\pi}^{\pi} |t| \frac{1}{\sqrt{\pi}} \sin t\, dt$$
$$= \frac{1}{\sqrt{\pi}} \int_{-\pi}^{0} -t \sin t\, dt + \frac{1}{\sqrt{\pi}} \int_{0}^{\pi} t \sin t\, dt$$
$$= -\sqrt{\pi} + \sqrt{\pi} = 0.$$

Then

$$\operatorname{proj}_W |t| = \frac{\pi^2}{\sqrt{2\pi}} \frac{1}{\sqrt{2\pi}} - \frac{4}{\sqrt{\pi}} \frac{1}{\sqrt{\pi}} \cos t = \frac{\pi}{2} - \frac{4}{\pi} \cos t.$$

Next, we compute the Fourier polynomial of degree 3. By Theorem 5.10,

$$\operatorname{proj}_W |t| = \left(|t|, \frac{1}{\sqrt{2\pi}}\right) \frac{1}{\sqrt{2\pi}}$$
$$+ \left(|t|, \frac{1}{\sqrt{\pi}} \cos t\right) \frac{1}{\sqrt{\pi}} \cos t + \left(|t|, \frac{1}{\sqrt{\pi}} \sin t\right) \frac{1}{\sqrt{\pi}} \sin t$$
$$+ \left(|t|, \frac{1}{\sqrt{\pi}} \cos 2t\right) \frac{1}{\sqrt{\pi}} \cos 2t + \left(|t|, \frac{1}{\sqrt{\pi}} \sin 2t\right) \frac{1}{\sqrt{\pi}} \sin 2t$$
$$+ \left(|t|, \frac{1}{\sqrt{\pi}} \cos 3t\right) \frac{1}{\sqrt{\pi}} \cos 3t + \left(|t|, \frac{1}{\sqrt{\pi}} \sin 3t\right) \frac{1}{\sqrt{\pi}} \sin 3t.$$

We have

$$\int_{-\pi}^{\pi} |t| \frac{1}{\sqrt{\pi}} \cos 2t\, dt = 0, \qquad \int_{-\pi}^{\pi} |t| \frac{1}{\sqrt{\pi}} \sin 2t\, dt = 0,$$

$$\int_{-\pi}^{\pi} |t| \frac{1}{\sqrt{\pi}} \cos 3t\, dt = -\frac{4}{9\sqrt{\pi}}, \qquad \int_{-\pi}^{\pi} |t| \frac{1}{\sqrt{\pi}} \sin 3t\, dt = 0.$$

Hence

$$\text{proj}_W \mathbf{v} = \frac{\pi}{2} - \frac{4}{\pi}\cos t - \frac{4}{9\pi}\cos 3t.$$

Figure 5.26 shows the graphs of f and the Fourier polynomial of degree 1. Figure 5.27 shows the graphs of f and the Fourier polynomial of degree 3. Figure 5.28 shows the graphs of

$$\left||t| - \left(\tfrac{\pi}{2} - \tfrac{4}{\pi}\cos t\right)\right| \quad \text{and} \quad \left||t| - \left(\tfrac{\pi}{2} - \tfrac{4}{\pi}\cos t - \tfrac{4}{9\pi}\cos 3t\right)\right|.$$

Observe how much better the approximation by a Fourier polynomial of degree 3 is. ■

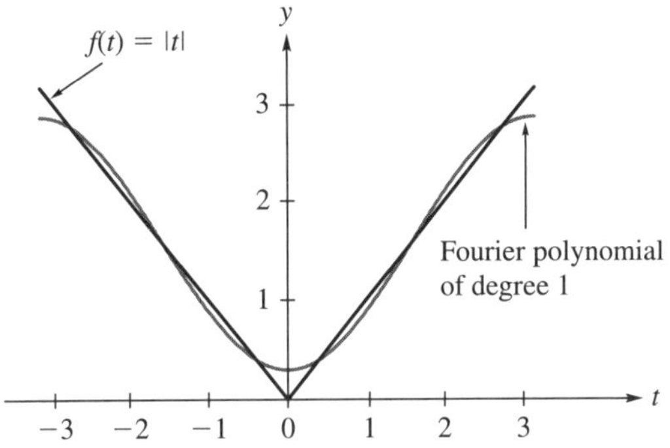

FIGURE 5.26

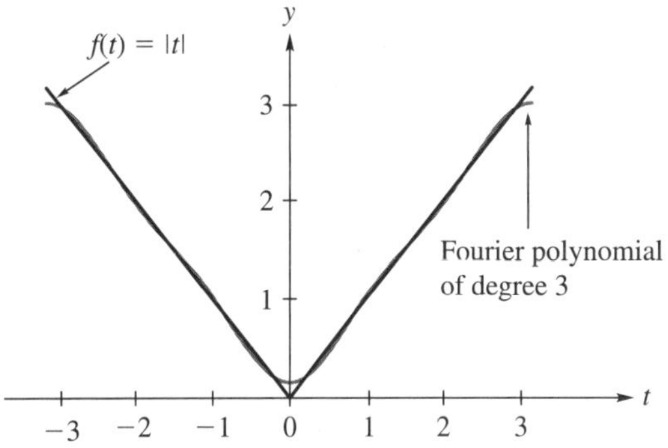

FIGURE 5.27

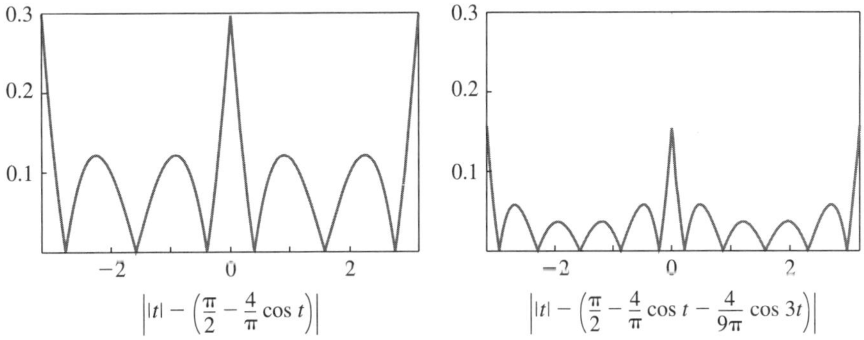

FIGURE 5.28

Fourier series play an important role in the study of heat distribution and in the analysis of sound waves. The study of projections is important in a number of areas in applied mathematics. We illustrate this in Section 5.6 by considering the topic of least squares, which provides a technique for dealing with inconsistent systems.

Key Terms

- Intersection of subspaces
- Direct sum of subspaces
- A vector orthogonal to a subspace
- Orthogonal complement of a subspace
- Null space of the matrix A
- Orthogonal projection
- Orthogonal basis
- Fourier series
- Taylor and Maclaurin series
- Trigonometric polynomial
- Fourier polynomial

5.5 Exercises

1. Let W be the subspace of R^3 spanned by the vector

$$\mathbf{w} = \begin{bmatrix} 2 \\ -3 \\ 1 \end{bmatrix}.$$

(a) Find a basis for $W^\perp$.

(b) Describe $W^\perp$ geometrically. (You may use a verbal or pictorial description.)

2. Let

$$W = \text{span}\left\{\begin{bmatrix} 1 \\ 2 \\ -1 \end{bmatrix}, \begin{bmatrix} -1 \\ 3 \\ 2 \end{bmatrix}\right\}.$$

(a) Find a basis for $W^\perp$.

(b) Describe $W^\perp$ geometrically. (You may use a verbal or pictorial description.)

3. Let W be the subspace of R_5 spanned by the vectors $\mathbf{w}_1$, $\mathbf{w}_2$, $\mathbf{w}_3$, $\mathbf{w}_4$, $\mathbf{w}_5$, where

$$\mathbf{w}_1 = \begin{bmatrix} 2 & -1 & 1 & 3 & 0 \end{bmatrix},$$
$$\mathbf{w}_2 = \begin{bmatrix} 1 & 2 & 0 & 1 & -2 \end{bmatrix},$$
$$\mathbf{w}_3 = \begin{bmatrix} 4 & 3 & 1 & 5 & -4 \end{bmatrix},$$
$$\mathbf{w}_4 = \begin{bmatrix} 3 & 1 & 2 & -1 & 1 \end{bmatrix},$$
$$\mathbf{w}_5 = \begin{bmatrix} 2 & -1 & 2 & -2 & 3 \end{bmatrix}.$$

Find a basis for $W^\perp$.

4. Let W be the subspace of R^4 spanned by the vectors $\mathbf{w}_1$, $\mathbf{w}_2$, $\mathbf{w}_3$, $\mathbf{w}_4$, where

$$\mathbf{w}_1 = \begin{bmatrix} 2 \\ 0 \\ -1 \\ 3 \end{bmatrix}, \quad \mathbf{w}_2 = \begin{bmatrix} 1 \\ 2 \\ 2 \\ -5 \end{bmatrix},$$

$$\mathbf{w}_3 = \begin{bmatrix} 3 \\ 2 \\ 1 \\ -2 \end{bmatrix}, \quad \mathbf{w}_4 = \begin{bmatrix} 7 \\ 2 \\ -1 \\ 4 \end{bmatrix}.$$

Find a basis for $W^\perp$.

5. (***Calculus Required***). Let V be the Euclidean space P_3 with the inner product defined in Example 2. Let W be the subspace of P_3 spanned by $\{t-1, t^2\}$. Find a basis for $W^\perp$.

6. Let V be the Euclidean space P_4 with the inner product defined in Example 2. Let W be the subspace of P_4 spanned by $\{1, t\}$. Find a basis for $W^\perp$.

7. Let W be the plane $3x + 2y - z = 0$ in R^3. Find a basis for $W^\perp$.

8. Let V be the Euclidean space of all 2×2 matrices with the inner product defined by $(A, B) = \text{Tr}(B^T A)$, where Tr is the trace function defined in Exercise 43 of Section 1.3. Let

$$W = \text{span}\left\{\begin{bmatrix} 1 & -1 \\ 2 & 3 \end{bmatrix}, \begin{bmatrix} 1 & 2 \\ 0 & -1 \end{bmatrix}\right\}.$$

Find a basis for $W^\perp$.

In Exercises 9 and 10, compute the four fundamental vector spaces associated with A and verify Theorem 5.12.

9. $A = \begin{bmatrix} 1 & 5 & 3 & 7 \\ 2 & 0 & -4 & -6 \\ 4 & 7 & -1 & 2 \end{bmatrix}$

10. $A = \begin{bmatrix} 2 & -1 & 3 & 4 \\ 0 & -3 & 7 & -2 \\ 1 & 1 & -2 & 3 \\ 1 & 4 & -9 & 5 \end{bmatrix}$

In Exercises 11 through 14, find $\text{proj}_W \mathbf{v}$ *for the given vector* $\mathbf{v}$ *and subspace W.*

11. Let V be the Euclidean space R^3, and W the subspace with basis

$$\begin{bmatrix} \frac{1}{\sqrt{5}} \\ 0 \\ \frac{2}{\sqrt{5}} \end{bmatrix}, \quad \begin{bmatrix} -\frac{2}{\sqrt{5}} \\ 0 \\ \frac{1}{\sqrt{5}} \end{bmatrix}.$$

(a) $\mathbf{v} = \begin{bmatrix} 3 \\ 4 \\ -1 \end{bmatrix}$ **(b)** $\mathbf{v} = \begin{bmatrix} 2 \\ 1 \\ 3 \end{bmatrix}$ **(c)** $\mathbf{v} = \begin{bmatrix} -5 \\ 0 \\ 1 \end{bmatrix}$

12. Let V be the Euclidean space R^4, and W the subspace with basis

$$\begin{bmatrix} 1 & 1 & 0 & 1 \end{bmatrix}, \begin{bmatrix} 0 & 1 & 1 & 0 \end{bmatrix}, \begin{bmatrix} -1 & 0 & 0 & 1 \end{bmatrix}.$$

(a) $\mathbf{v} = \begin{bmatrix} 2 & 1 & 3 & 0 \end{bmatrix}$

(b) $\mathbf{v} = \begin{bmatrix} 0 & -1 & 1 & 0 \end{bmatrix}$

(c) $\mathbf{v} = \begin{bmatrix} 0 & 2 & 0 & 3 \end{bmatrix}$

13. Let V be the vector space of real-valued continuous functions on $[-\pi, \pi]$, and let $W = \text{span}\{1, \cos t, \sin t\}$.

(a) $\mathbf{v} = t$ **(b)** $\mathbf{v} = t^2$ **(c)** $\mathbf{v} = e^t$

14. Let W be the plane in R^3 given by the equation $x + y - 2z = 0$.

(a) $\mathbf{v} = \begin{bmatrix} 0 \\ 2 \\ -1 \end{bmatrix}$ **(b)** $\mathbf{v} = \begin{bmatrix} 3 \\ 1 \\ 4 \end{bmatrix}$ **(c)** $\mathbf{v} = \begin{bmatrix} 1 \\ 0 \\ -1 \end{bmatrix}$

15. Let W be the subspace of R^3 with orthonormal basis $\{\mathbf{w}_1, \mathbf{w}_2\}$, where

$$\mathbf{w}_1 = \begin{bmatrix} 0 \\ 1 \\ 0 \end{bmatrix} \quad \text{and} \quad \mathbf{w}_2 = \begin{bmatrix} \frac{1}{\sqrt{5}} \\ 0 \\ \frac{2}{\sqrt{5}} \end{bmatrix}.$$

Write the vector

$$\mathbf{v} = \begin{bmatrix} 1 \\ 2 \\ -1 \end{bmatrix}$$

as $\mathbf{w} + \mathbf{u}$, with $\mathbf{w}$ in W and $\mathbf{u}$ in $W^\perp$.

16. Let W be the subspace of R^4 with orthonormal basis $\{\mathbf{w}_1, \mathbf{w}_2, \mathbf{w}_3\}$, where

$$\mathbf{w}_1 = \begin{bmatrix} \frac{1}{\sqrt{2}} \\ 0 \\ 0 \\ -\frac{1}{\sqrt{2}} \end{bmatrix}, \quad \mathbf{w}_2 = \begin{bmatrix} 0 \\ 0 \\ 1 \\ 0 \end{bmatrix}, \quad \text{and} \quad \mathbf{w}_3 = \begin{bmatrix} \frac{1}{\sqrt{2}} \\ 0 \\ 0 \\ \frac{1}{\sqrt{2}} \end{bmatrix}.$$

Write the vector

$$\mathbf{v} = \begin{bmatrix} 1 \\ 0 \\ 2 \\ 3 \end{bmatrix}$$

as $\mathbf{w} + \mathbf{u}$ with $\mathbf{w}$ in W and $\mathbf{u}$ in $W^\perp$.

17. Let W be the subspace of continuous functions on $[-\pi, \pi]$ defined in Exercise 13. Write the vector $\mathbf{v} = t - 1$ as $\mathbf{w} + \mathbf{u}$, with $\mathbf{w}$ in W and $\mathbf{u}$ in $W^\perp$.

18. Let W be the plane in R^3 given by the equation $x - y - z = 0$. Write the vector $\mathbf{v} = \begin{bmatrix} 2 \\ -1 \\ 0 \end{bmatrix}$ as $\mathbf{w} + \mathbf{u}$, with $\mathbf{w}$ in W and $\mathbf{u}$ in $W^\perp$.

19. Let W be the subspace of R^3 defined in Exercise 15, and let $\mathbf{v} = \begin{bmatrix} -1 \\ 0 \\ 1 \end{bmatrix}$. Find the distance from $\mathbf{v}$ to W.

20. Let W be the subspace of R^4 defined in Exercise 16, and let $\mathbf{v} = \begin{bmatrix} 1 \\ 2 \\ -1 \\ 0 \end{bmatrix}$. Find the distance from $\mathbf{v}$ to W.

21. Let W be the subspace of continuous functions on $[-\pi, \pi]$ defined in Exercise 13 and let $\mathbf{v} = t$. Find the distance from $\mathbf{v}$ to W.

In Exercises 22 and 23, find the Fourier polynomial of degree 2 for f.

22. (***Calculus Required***) $f(t) = t^2$.

23. (***Calculus Required***) $f(t) = e^t$.

24. Show that if V is an inner product space and W is a subspace of V, then the zero vector of V belongs to $W^\perp$.

25. Let V be an inner product space. Show that the orthogonal complement of V is the zero subspace and the orthogonal complement of the zero subspace is V itself.

26. Show that if W is a subspace of an inner product space V that is spanned by a set of vectors S, then a vector $\mathbf{u}$ in V belongs to $W^\perp$ if and only if $\mathbf{u}$ is orthogonal to every vector in S.

27. Let A be an $m \times n$ matrix. Show that every vector $\mathbf{v}$ in R^n can be written uniquely as $\mathbf{w} + \mathbf{u}$, where $\mathbf{w}$ is in the null space of A and $\mathbf{u}$ is in the column space of A^T.

28. Let V be a Euclidean space, and W a subspace of V. Show that if $\mathbf{w}_1, \mathbf{w}_2, \dots, \mathbf{w}_r$ is a basis for W and $\mathbf{u}_1, \mathbf{u}_2, \dots, \mathbf{u}_s$ is a basis for $W^\perp$, then $\mathbf{w}_1, \mathbf{w}_2, \dots, \mathbf{w}_r, \mathbf{u}_1, \mathbf{u}_2, \dots, \mathbf{u}_s$ is a basis for V, and that $\dim V = \dim W + \dim W^\perp$.

29. Let W be a subspace of an inner product space V and let $\{\mathbf{w}_1, \mathbf{w}_2, \dots, \mathbf{w}_m\}$ be an orthogonal basis for W. Show that if $\mathbf{v}$ is any vector in V, then

$$\text{proj}_W \mathbf{v} = \frac{(\mathbf{v}, \mathbf{w}_1)}{(\mathbf{w}_1, \mathbf{w}_1)}\mathbf{w}_1 + \frac{(\mathbf{v}, \mathbf{w}_2)}{(\mathbf{w}_2, \mathbf{w}_2)}\mathbf{w}_2 + \cdots + \frac{(\mathbf{v}, \mathbf{w}_m)}{(\mathbf{w}_m, \mathbf{w}_m)}\mathbf{w}_m.$$

5.6 Least Squares (Optional)

From Chapter 1 we recall that an $m \times n$ linear system $A\mathbf{x} = \mathbf{b}$ is inconsistent if it has no solution. In the proof of Theorem 4.21 in Section 4.9 we show that $A\mathbf{x} = \mathbf{b}$ is consistent if and only if $\mathbf{b}$ belongs to the column space of A. Equivalently, $A\mathbf{x} = \mathbf{b}$ is inconsistent if and only if $\mathbf{b}$ is *not* in the column space of A. Inconsistent systems do indeed arise in many situations, and we must determine how to deal with them. Our approach is to change the problem so that we do not require the matrix equation $A\mathbf{x} = \mathbf{b}$ to be satisfied. Instead, we seek a vector $\widehat{\mathbf{x}}$ in R^n such that $A\widehat{\mathbf{x}}$ is as close to $\mathbf{b}$ as possible.

If W is the column space of A, then from Theorem 5.13 in Section 5.5, it follows that the vector in W that is closest to $\mathbf{b}$ is $\text{proj}_W \mathbf{b}$. That is, $\|\mathbf{b} - \mathbf{w}\|$, for $\mathbf{w}$ in W, is minimized when $\mathbf{w} = \text{proj}_W \mathbf{b}$. Thus, if we find $\widehat{\mathbf{x}}$ such that $A\widehat{\mathbf{x}} = \text{proj}_W \mathbf{b}$, then we are assured that $\|\mathbf{b} - A\widehat{\mathbf{x}}\|$ will be as small as possible. As shown in the proof of Theorem 5.13, $\mathbf{b} - \text{proj}_W \mathbf{b} = \mathbf{b} - A\widehat{\mathbf{x}}$ is orthogonal to every vector in W. (See Figure 5.29.) It then follows that $\mathbf{b} - A\widehat{\mathbf{x}}$ is orthogonal to each column of A. In terms of a matrix equation, we have

$$A^T(A\widehat{\mathbf{x}} - \mathbf{b}) = \mathbf{0},$$

or equivalently,

$$A^T A\widehat{\mathbf{x}} = A^T\mathbf{b}. \tag{1}$$

Any solution to (1) is called a **least squares solution** to the linear system $A\mathbf{x} = \mathbf{b}$. (**Warning:** In general, $A\widehat{\mathbf{x}} \neq \mathbf{b}$.) Equation (1) is called the **normal system** of equations associated with $A\mathbf{x} = \mathbf{b}$, or simply, the normal system. Observe that if A is nonsingular, a least squares solution to $A\mathbf{x} = \mathbf{b}$ is just the usual solution $\mathbf{x} = A^{-1}\mathbf{b}$ (see Exercise 1).

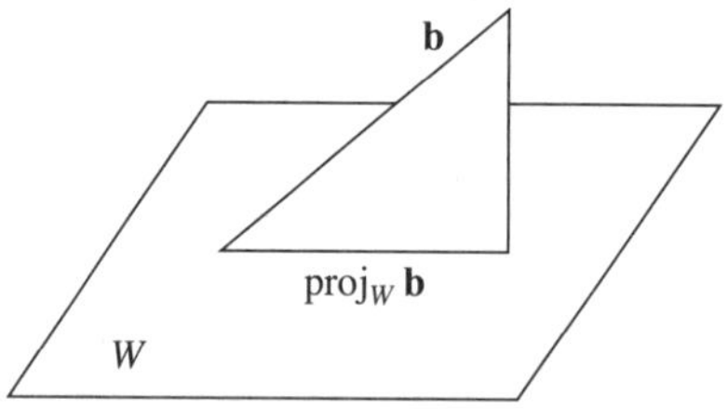

FIGURE 5.29 W = Column space of A.

To compute a least squares solution $\widehat{\mathbf{x}}$ to the linear system $A\mathbf{x} = \mathbf{b}$, we can proceed as follows. Compute $\text{proj}_W \mathbf{b}$ by using Equation (1) in Section 5.5 and then solve $A\widehat{\mathbf{x}} = \text{proj}_W \mathbf{b}$. To compute $\text{proj}_W \mathbf{b}$ requires that we have an orthonormal basis for W, the column space of A. We could first find a basis for W by determining the reduced row echelon form of A^T and taking the transposes of the nonzero rows. Next, apply the Gram–Schmidt process to the basis to find an orthonormal basis for W. The procedure just outlined is theoretically valid when we assume that exact arithmetic is used. However, even small numerical errors, due to, say, roundoff, may adversely affect the results. Thus more sophisticated algorithms are required for numerical applications. We shall not pursue the general case here, but turn our attention to an important special case.

Remark An alternative method for finding $\text{proj}_W \mathbf{b}$ is as follows. Solve Equation (1) for $\widehat{\mathbf{x}}$, the least squares solution to the linear system $A\mathbf{x} = \mathbf{b}$. Then $A\widehat{\mathbf{x}}$ will be $\text{proj}_W \mathbf{b}$.

Theorem 5.14 If A is an $m \times n$ matrix with rank $A = n$, then $A^T A$ is nonsingular and the linear system $A\mathbf{x} = \mathbf{b}$ has a unique least squares solution given by $\widehat{\mathbf{x}} = (A^T A)^{-1} A^T \mathbf{b}$.

Proof

If A has rank n, then the columns of A are linearly independent. The matrix $A^T A$ is nonsingular, provided that the linear system $A^T A\mathbf{x} = \mathbf{0}$ has only the zero solution. Multiplying both sides of $A^T A\mathbf{x} = \mathbf{0}$ by $\mathbf{x}^T$ on the left gives

$$\mathbf{0} = \mathbf{x}^T A^T A\mathbf{x} = (A\mathbf{x})^T (A\mathbf{x}) = (A\mathbf{x}, A\mathbf{x}),$$

when we use the standard inner product on R^n. It follows from Definition 5.1(a) in Section 5.3 that $A\mathbf{x} = \mathbf{0}$. But this implies that we have a linear combination of the linearly independent columns of A that is zero; hence $\mathbf{x} = \mathbf{0}$. Thus $A^T A$ is nonsingular, and Equation (1) has the unique solution $\widehat{\mathbf{x}} = (A^T A)^{-1} A^T \mathbf{b}$. ■

EXAMPLE 1 Determine a least squares solution to $A\mathbf{x} = \mathbf{b}$, where

$$A = \begin{bmatrix} 1 & 2 & -1 & 3 \\ 2 & 1 & 1 & 2 \\ -2 & 3 & 4 & 1 \\ 4 & 2 & 1 & 0 \\ 0 & 2 & 1 & 3 \\ 1 & -1 & 2 & 0 \end{bmatrix}, \qquad \mathbf{b} = \begin{bmatrix} 1 \\ 5 \\ -2 \\ 1 \\ 3 \\ 5 \end{bmatrix}.$$

Solution

Using row reduction, we can show that rank $A = 4$ (verify). Then using Theorem 5.14, we form the normal system $A^T A\widehat{\mathbf{x}} = A^T \mathbf{b}$ (verify),

$$\begin{bmatrix} 26 & 5 & -1 & 5 \\ 5 & 23 & 13 & 17 \\ -1 & 13 & 24 & 6 \\ 5 & 17 & 6 & 23 \end{bmatrix} \widehat{\mathbf{x}} = \begin{bmatrix} 24 \\ 4 \\ 10 \\ 20 \end{bmatrix}.$$

Applying Gaussian elimination, we have the unique least squares solution (verify)

$$\widehat{\mathbf{x}} \approx \begin{bmatrix} 0.9990 \\ -2.0643 \\ 1.1039 \\ 1.8902 \end{bmatrix}.$$

If W is the column space of A, then (verify)

$$\operatorname{proj}_W \mathbf{b} = A\widehat{\mathbf{x}} \approx \begin{bmatrix} 1.4371 \\ 4.8181 \\ -1.8852 \\ 0.9713 \\ 2.6459 \\ 5.2712 \end{bmatrix},$$

which is the vector in W such that $\|\mathbf{b} - \mathbf{y}\|$, $\mathbf{y}$ in W, is minimized. That is,

$$\min_{\mathbf{y} \text{ in } W} \|\mathbf{b} - \mathbf{w}\| = \|\mathbf{b} - A\widehat{\mathbf{x}}\|. \qquad \blacksquare$$

When A is an $m \times n$ matrix whose rank is n, it is computationally more efficient to solve Equation (1) by Gaussian elimination than to determine $(A^T A)^{-1}$ and then form the product $(A^T A)^{-1} A^T \mathbf{b}$. An even better approach is to use the QR-factorization of A, as follows:

Suppose that $A = QR$ is a QR-factorization of A. Substituting this expression for A into Equation (1), we obtain

$$(QR)^T (QR)\widehat{\mathbf{x}} = (QR)^T \mathbf{b},$$

or

$$R^T (Q^T Q) R\widehat{\mathbf{x}} = R^T Q^T \mathbf{b}.$$

Since the columns of Q form an orthonormal set, we have $Q^T Q = I_m$, so

$$R^T R\widehat{\mathbf{x}} = R^T Q^T \mathbf{b}.$$

Since R^T is a nonsingular matrix, we obtain

$$R\widehat{\mathbf{x}} = Q^T \mathbf{b}.$$

Using the fact that R is upper triangular, we readily solve this linear system by back substitution.

EXAMPLE 2 Solve Example 1 by using the QR-factorization of A.

Solution

We use the Gram–Schmidt process, carrying out all computations in MATLAB. We find that Q is given by (verify)

$$Q = \begin{bmatrix} -0.1961 & -0.3851 & 0.5099 & 0.3409 \\ -0.3922 & -0.1311 & -0.1768 & 0.4244 \\ 0.3922 & -0.7210 & -0.4733 & -0.2177 \\ -0.7845 & -0.2622 & -0.1041 & -0.5076 \\ 0 & -0.4260 & 0.0492 & 0.4839 \\ -0.1961 & 0.2540 & -0.6867 & 0.4055 \end{bmatrix}$$

and R is given by (verify)

$$R = \begin{bmatrix} -5.0990 & -0.9806 & 0.1961 & -0.9806 \\ 0 & -4.6945 & -2.8102 & -3.4164 \\ 0 & 0 & -4.0081 & 0.8504 \\ 0 & 0 & 0 & 3.1054 \end{bmatrix}.$$

Then

$$Q^T\mathbf{b} = \begin{bmatrix} 4.7068 \\ -0.1311 \\ 2.8172 \\ 5.8699 \end{bmatrix}.$$

Finally, solving

$$R\widehat{\mathbf{x}} = Q^T\mathbf{b},$$

we find (verify) exactly the same $\widehat{\mathbf{x}}$ as in the solution to Example 1. ■

Remark As we have already observed in our discussion of the QR-factorization of a matrix, if you use a computer program such as MATLAB to find a QR-factorization of A in Example 2, you will find that the program yields Q as a 6×6 matrix whose first four columns agree with the Q found in our solution, and R as a 6×4 matrix whose first four rows agree with the R found in our solution and whose last two rows consist entirely of zeros.

Least squares problems often arise when we try to construct a mathematical model of the form

$$y(t) = x_1 f_1(t) + x_2 f_2(t) + \cdots + x_n f_n(t) \tag{2}$$

to a data set $D = \{(t_i, y_i),\ i = 1, 2, \ldots, m\}$, where $m > n$. Ideally, we would like to determine $x_1, x_2, \ldots, x_n$ such that

$$y_i = x_1 f_1(t_i) + x_2 f_2(t_i) + \cdots + x_n f_n(t_i)$$

for each data point t_i, $i = 1, 2, \ldots, m$. In matrix form we have the linear system $A\mathbf{x} = \mathbf{b}$, where

$$A = \begin{bmatrix} f_1(t_1) & f_2(t_1) & \cdots & f_n(t_1) \\ f_1(t_2) & f_2(t_2) & \cdots & f_n(t_2) \\ \vdots & \vdots & & \vdots \\ f_1(t_m) & f_2(t_m) & \cdots & f_n(t_m) \end{bmatrix}, \tag{3}$$

$$\mathbf{x} = \begin{bmatrix} x_1 & x_2 & \cdots & x_n \end{bmatrix}^T, \quad \text{and} \quad \mathbf{b} = \begin{bmatrix} y_1 & y_2 & \cdots & y_m \end{bmatrix}^T.$$

As is often the case, the system $A\mathbf{x} = \mathbf{b}$ is inconsistent, so we determine a least squares solution $\widehat{\mathbf{x}}$ to $A\mathbf{x} = \mathbf{b}$. If we set $x_i = \hat{x}_i$ in the model equation (2), we say that

$$\hat{y}(t) = \hat{x}_1 f_1(t) + \hat{x}_2 f_2(t) + \cdots + \hat{x}_n f_n(t)$$

gives a least squares model for data set D. In general, $\hat{y}(t_i) \neq y_i$, $i = 1, 2, \ldots, m$. (They may be equal, but there is no guarantee.) Let $e_i = y_i - \hat{y}(t_i)$, $i = 1, 2, \ldots, m$, which represents the error incurred at t_i when $\hat{y}(t)$ is used as a mathematical model for data set D. If

$$\mathbf{e} = \begin{bmatrix} e_1 & e_2 & \cdots & e_m \end{bmatrix}^T,$$

then

$$\mathbf{e} = \mathbf{b} - A\widehat{\mathbf{x}},$$

and Theorem 5.13 in Section 5.5 guarantees that $\|\mathbf{e}\| = \|\mathbf{b} - A\widehat{\mathbf{x}}\|$ is as small as possible. That is,

$$\|\mathbf{e}\|^2 = (\mathbf{e}, \mathbf{e}) = (\mathbf{b} - A\widehat{\mathbf{x}}, \mathbf{b} - A\widehat{\mathbf{x}}) = \sum_{i=1}^{m} \left[y_i - \sum_{j=1}^{n} \hat{x}_j f_j(t_i) \right]^2$$

is minimized. We say that $\widehat{\mathbf{x}}$, the least squares solution, minimizes the sum of the squares of the deviations between the observations y_i and the values $\hat{y}(t_i)$ predicted by the model equation.

EXAMPLE 3 The following data show atmospheric pollutants y_i (relative to an EPA standard) at half-hour intervals t_i:

t_i	1	1.5	2	2.5	3	3.5	4	4.5	5
y_i	−0.15	0.24	0.68	1.04	1.21	1.15	0.86	0.41	−0.08

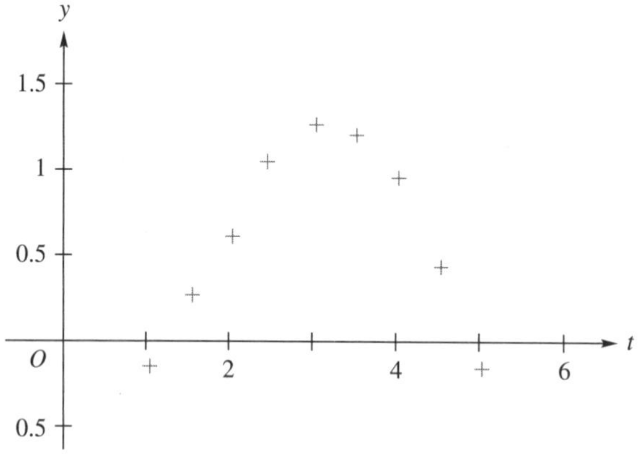

FIGURE 5.30

A plot of these data points, as shown in Figure 5.30, suggests that a quadratic polynomial

$$y(t) = x_1 + x_2 t + x_3 t^2$$

may produce a good model for these data. With $f_1(t) = 1$, $f_2(t) = t$, and

$f_3(t) = t^2$, Equation (3) gives

$$A = \begin{bmatrix} 1 & 1 & 1 \\ 1 & 1.5 & 2.25 \\ 1 & 2 & 4 \\ 1 & 2.5 & 6.25 \\ 1 & 3 & 9 \\ 1 & 3.5 & 12.25 \\ 1 & 4 & 16 \\ 1 & 4.5 & 20.25 \\ 1 & 5 & 25 \end{bmatrix}, \qquad \widehat{\mathbf{x}} = \begin{bmatrix} \hat{x}_1 \\ \hat{x}_2 \\ \hat{x}_3 \end{bmatrix}, \qquad \mathbf{b} = \begin{bmatrix} -0.15 \\ 0.24 \\ 0.68 \\ 1.04 \\ 1.21 \\ 1.15 \\ 0.86 \\ 0.41 \\ -0.08 \end{bmatrix}.$$

The rank of A is 3 (verify), and the normal system is

$$\begin{bmatrix} 9 & 27 & 96 \\ 27 & 96 & 378 \\ 96 & 378 & 1583.25 \end{bmatrix} \begin{bmatrix} \hat{x}_1 \\ \hat{x}_2 \\ \hat{x}_3 \end{bmatrix} = \begin{bmatrix} 5.36 \\ 16.71 \\ 54.65 \end{bmatrix}.$$

Applying Gaussian elimination gives (verify)

$$\widehat{\mathbf{x}} \approx \begin{bmatrix} -1.9317 \\ 2.0067 \\ -0.3274 \end{bmatrix},$$

so we obtain the quadratic polynomial model

$$y(t) = -1.9317 + 2.0067t - 0.3274t^2.$$

Figure 5.31 shows the data set indicated with + and a graph of $y(t)$. We see that $y(t)$ is close to each data point, but is not required to go through the data. ■

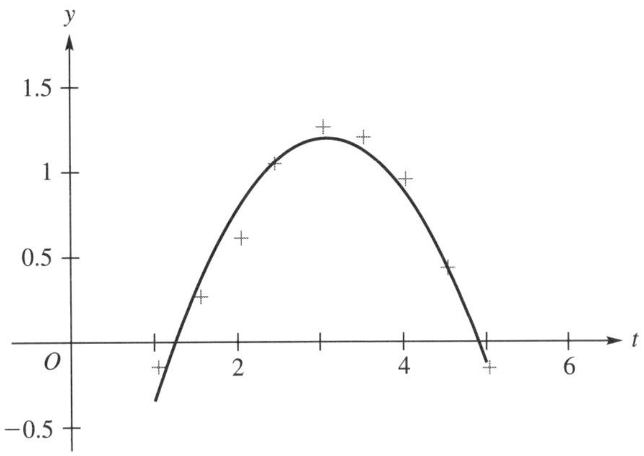

FIGURE 5.31

Key Terms

Inconsistent linear system
Least squares solution
Normal system of equations
Projection
QR-factorization

5.6 Exercises

1. Let A be $n \times n$ and nonsingular. From the normal system of equations in (1), show that the least squares solution to $A\mathbf{x} = \mathbf{b}$ is $\widehat{\mathbf{x}} = A^{-1}\mathbf{b}$.

2. Determine the least squares solution to $A\mathbf{x} = \mathbf{b}$, where

$$A = \begin{bmatrix} 2 & 1 \\ 1 & 0 \\ 0 & -1 \\ -1 & 1 \end{bmatrix} \quad \text{and} \quad \mathbf{b} = \begin{bmatrix} 3 \\ 1 \\ 2 \\ -1 \end{bmatrix}.$$

3. Determine the least squares solution to $A\mathbf{x} = \mathbf{b}$, where

$$A = \begin{bmatrix} 1 & 2 & 1 \\ 1 & 3 & 2 \\ 2 & 5 & 3 \\ 2 & 0 & 1 \\ 3 & 1 & 1 \end{bmatrix} \quad \text{and} \quad \mathbf{b} = \begin{bmatrix} -1 \\ 2 \\ 0 \\ 1 \\ -2 \end{bmatrix}.$$

4. Solve Exercise 2, using QR-factorization of A.

5. Solve Exercise 3, using QR-factorization of A.

6. In the manufacture of product Z, the amount of compound A present in the product is controlled by the amount of ingredient B used in the refining process. In manufacturing a liter of Z, the amount of B used and the amount of A present are recorded. The following data were obtained:

B Used (grams/liter)	2	4	6	8	10
A Present (grams/liter)	3.5	8.2	10.5	12.9	14.6

Determine the least squares line to the data. [In Equation (2), use $f_1(t) = 1$, $f_2(t) = t$.] Also compute $\|\mathbf{e}\|$.

7. In Exercise 6, the least squares line to the data set $D = \{(t_i, y_i),\ i = 1, 2, \ldots, m\}$ is the line $y = \hat{x}_1 + \hat{x}_2 t$, which minimizes

$$E_1 = \sum_{i=1}^{m} [y_i - (x_1 + x_2 t_i)]^2.$$

Similarly, the least squares quadratic (see Example 2) to the data set D is the parabola $y = \hat{x}_1 + \hat{x}_2 t + \hat{x}_3 t^2$, which minimizes

$$E_2 = \sum_{i=1}^{m} \left[y_i - (x_1 + x_2 t_i + x_3 t_i^2)\right]^2.$$

Give a vector space argument to show that $E_2 \leq E_1$.

8. The accompanying table is a sample set of seasonal farm employment data (t_i, y_i) over about a two-year period, where t_i represents months and y_i represents millions of people. A plot of the data is given in the figure. It is decided to develop a least squares mathematical model of the form

$$y(t) = x_1 + x_2 t + x_3 \cos t.$$

Determine the least squares model. Plot the resulting function $y(t)$ and the data set on the same coordinate system.

t_i	y_i	t_i	y_i
3.1	3.7	11.8	5.0
4.3	4.4	13.1	5.0
5.6	5.3	14.3	3.8
6.8	5.2	15.6	2.8
8.1	4.0	16.8	3.3
9.3	3.6	18.1	4.5
10.6	3.6		

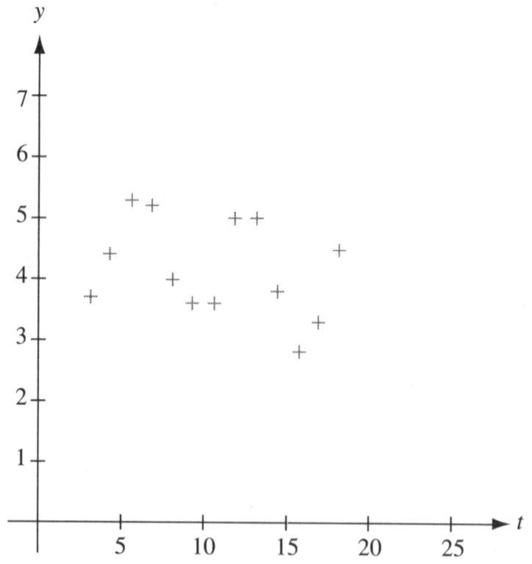

9. Given $A\mathbf{x} = \mathbf{b}$, where

$$A = \begin{bmatrix} 1 & 3 & -3 \\ 2 & 4 & -2 \\ 0 & -1 & 2 \\ 1 & 2 & -1 \end{bmatrix} \quad \text{and} \quad \mathbf{b} = \begin{bmatrix} 1 \\ 0 \\ 0 \\ 1 \end{bmatrix}.$$

(a) Show that rank $A = 2$.

(b) Since rank $A \neq$ number of columns, Theorem 5.14 cannot be used to determine a least squares solution $\widehat{\mathbf{x}}$. Follow the general procedure as discussed prior to Theorem 5.14 to find a least squares solution. Is the solution unique?

10. The following data showing U.S. per capita health care expenditures (in dollars) is available from the National Center for Health Statistics in the 2002 Chartbook on Trends in the Health of Americans.

Year	*Per Capita Expenditures (in \$)*
1960	143
1970	348
1980	1,067
1990	2,738
1995	3,698
1998	4,098
1999	4,302
2000	4,560
2001	4,914
2002	5,317

(a) Determine the line of best fit to the given data.

(b) Predict the per capita expenditure for the years 2008, 2010, and 2015.

11. The following data show the size of the U.S. debt per capita (in dollars). This information was constructed from federal government data on public debt and (estimated) population data.

(a) Determine the line of best fit to the given data.

(b) Predict the debt per capita for the years 2008, 2010, and 2015.

Year	*Debt per Capita (in \$)*
1996	20,070
1997	20,548
1998	20,774
1999	21,182
2000	20,065
2001	20,874
2002	22,274
2003	24,077
2004	25,868
2005	27,430

12. For the data in Exercise 11, find the least squares quadratic polynomial approximation. Compare this model with that obtained in Exercise 11 by computing the error in each case.

13. Gauge is a measure of shotgun bore. Gauge numbers originally referred to the number of lead balls with the diameter equal to that of the gun barrel that could be made from a pound of lead. Thus, a 16-gauge shotgun's bore was smaller than a 12-gauge shotgun's. (Ref., *The World Almanac and Book of Facts 1993*, Pharos Books, NY, 1992, p. 290.) Today, an international agreement assigns millimeter measures to each gauge. The following table gives such information for popular gauges of shotguns:

x	y
Gauge	*Bore Diameter (in mm)*
6	23.34
10	19.67
12	18.52
14	17.60
16	16.81
20	15.90

We would like to develop a model of the form $y = re^{sx}$ for the data in this table. By taking the natural logarithm of both sides in this expression, we obtain

$$\ln y = \ln r + sx. \qquad (*)$$

Let $c_1 = \ln r$ and $c_2 = s$. Substituting the data from the table in Equation $(*)$, we obtain a linear system of six equations in two unknowns.

(a) Show that this system is inconsistent.

(b) Find its least squares solution.

(c) Determine r and s.

(d) Estimate the bore diameter for an 18-gauge shotgun.

14. In some software programs, the command for solving a linear system produces a least squares solution when the coefficient matrix is not square or is nonsingular. Determine whether this is the case in your software. If it is, compare your software's output with the solution given in Example 1. To experiment further, use Exercise 9.

Supplementary Exercises

1. Exercise 33 of Section 5.4 proves that the set of all vectors in R^n that are orthogonal to a fixed vector $\mathbf{u}$ forms a subspace of R^n. For $\mathbf{u} = \begin{bmatrix} 1 \\ -2 \\ 1 \end{bmatrix}$, find an orthogonal basis for the subspace of vectors in R^3 that are orthogonal to $\mathbf{u}$. [*Hint*: Solve the linear system $(\mathbf{u}, \mathbf{v}) = 0$, when $\mathbf{v} = \begin{bmatrix} x_1 \\ x_2 \\ x_3 \end{bmatrix}$.]

2. Use the Gram–Schmidt process to find an orthonormal basis for the subspace of R^4 with basis

$$\left\{ \begin{bmatrix} 1 \\ 0 \\ 0 \\ -1 \end{bmatrix}, \begin{bmatrix} 1 \\ -1 \\ 0 \\ 0 \end{bmatrix}, \begin{bmatrix} 0 \\ 1 \\ 0 \\ 1 \end{bmatrix} \right\}.$$

3. Given the orthonormal basis

$$S = \left\{ \begin{bmatrix} \frac{1}{\sqrt{2}} \\ 0 \\ -\frac{1}{\sqrt{2}} \end{bmatrix}, \begin{bmatrix} 0 \\ 1 \\ 0 \end{bmatrix}, \begin{bmatrix} \frac{1}{\sqrt{2}} \\ 0 \\ \frac{1}{\sqrt{2}} \end{bmatrix} \right\}$$

for R^3, write the vector

$$\mathbf{v} = \begin{bmatrix} 1 \\ 2 \\ 3 \end{bmatrix}$$

as a linear combination of the vectors in S.

4. Use the Gram–Schmidt process to find an orthonormal basis for the subspace of R^4 with basis

$$\left\{ \begin{bmatrix} 1 \\ 0 \\ -1 \\ 0 \end{bmatrix}, \begin{bmatrix} 1 \\ -1 \\ 0 \\ 0 \end{bmatrix}, \begin{bmatrix} 2 \\ 1 \\ 0 \\ 0 \end{bmatrix} \right\}.$$

5. Given vector $\mathbf{v} = \mathbf{i} + 2\mathbf{j} + \mathbf{k}$ and the plane P determined by the vectors

$$\mathbf{w}_1 = \mathbf{i} - 3\mathbf{j} - 2\mathbf{k} \quad \text{and} \quad \mathbf{w}_2 = 3\mathbf{i} - \mathbf{j} - 3\mathbf{k},$$

find the vector in P closest to $\mathbf{v}$ and the distance from $\mathbf{v}$ to P.

6. Find the distance from the point $(2, 3, -1)$ to the plane $3x - 2y + z = 0$. (*Hint*: First find an orthonormal basis for the plane.)

7. Consider the vector space of continuous real-valued functions on $[0, \pi]$ with an inner product defined by $(f, g) = \int_0^\pi f(t)g(t)\,dt$. Show that the collection of functions $\sin nt$, for $n = 1, 2, \ldots,$ is an orthogonal set.

8. Let V be the inner product space defined in Exercise 13 of Section 5.5. In (a) through (c) let W be the subspace spanned by the given orthonormal vectors $\mathbf{w}_1, \mathbf{w}_2, \ldots, \mathbf{w}_n$. Find $\operatorname{proj}_W \mathbf{v}$, for the vector $\mathbf{v}$ in V.

(a) $\mathbf{v} = t + t^2$, $\mathbf{w}_1 = \frac{1}{\sqrt{\pi}} \cos t$, $\mathbf{w}_2 = \frac{1}{\sqrt{\pi}} \sin 2t$

(b) $\mathbf{v} = \sin \frac{1}{2}t$, $\mathbf{w}_1 = \frac{1}{\sqrt{2\pi}}$, $\mathbf{w}_2 = \frac{1}{\sqrt{\pi}} \sin t$

(c) $\mathbf{v} = \cos^2 t$, $\mathbf{w}_1 = \frac{1}{\sqrt{2\pi}}$, $\mathbf{w}_2 = \frac{1}{\sqrt{\pi}} \cos t$, $\mathbf{w}_3 = \frac{1}{\sqrt{\pi}} \cos 2t$

9. Find an orthonormal basis for each null space of A.

(a) $A = \begin{bmatrix} 1 & 0 & 5 & -2 \\ 0 & 1 & -2 & -5 \end{bmatrix}$

(b) $A = \begin{bmatrix} 1 & 0 & 5 & -2 \\ 0 & 1 & -2 & 4 \end{bmatrix}$

10. Find the QR-factorization for each given matrix A.

(a) $\begin{bmatrix} 1 & 0 & 1 \\ -1 & 1 & 2 \\ 2 & 2 & -1 \end{bmatrix}$ (b) $\begin{bmatrix} 2 & 1 \\ -1 & -1 \\ -2 & 3 \end{bmatrix}$

11. Let $W = \operatorname{span} \left\{ \begin{bmatrix} 1 \\ 0 \\ 1 \end{bmatrix}, \begin{bmatrix} 0 \\ 1 \\ 0 \end{bmatrix} \right\}$ in R^3.

(a) Find a basis for the orthogonal complement of W.

(b) Show that vectors $\begin{bmatrix}1\\0\\1\end{bmatrix}$, $\begin{bmatrix}0\\1\\0\end{bmatrix}$, and the basis for the orthogonal complement of W from part (a) form a basis for R^3.

(c) Express each of the given vectors $\mathbf{v}$ as $\mathbf{w}+\mathbf{u}$, where $\mathbf{w}$ is in W and $\mathbf{u}$ is in $W^{\perp}$.

(i) $\mathbf{v}=\begin{bmatrix}1\\0\\0\end{bmatrix}$ **(ii)** $\mathbf{v}=\begin{bmatrix}1\\2\\3\end{bmatrix}$

12. Find the orthogonal complement of the null space of A.

(a) $A=\begin{bmatrix}1&-2&2\\2&3&2\\4&-1&6\end{bmatrix}$

(b) $A=\begin{bmatrix}1&-1&3&2\\1&-4&7&8\\2&1&2&-2\end{bmatrix}$

13. Use the Gram–Schmidt process to find an orthonormal basis for P_2 with respect to the inner product $(f,g)=\int_{-1}^{1} f(t)g(t)\,dt$, starting with the standard basis $\{1,t,t^2\}$ for P_2. The polynomials thus obtained are called the **Legendre* polynomials**.

14. Using the Legendre polynomials from Exercise 13 as an orthonormal basis for P_2 considered as a subspace of the vector space of real-valued continuous functions on $[-1,1]$, find $\text{proj}_{P_2}\mathbf{v}$ for each of the following:

(a) $\mathbf{v}=t^3$ **(b)** $\mathbf{v}=\sin \pi t$ **(c)** $\mathbf{v}=\cos \pi t$

15. Using the Legendre polynomials from Exercise 13 as an orthonormal basis for P_2 considered as a subspace of the vector space of real-valued continuous functions on $[-1,1]$, find the distance from $\mathbf{v}=t^3+1$ to P_2.

16. Let V be the inner product space of real-valued continuous functions on $[-\pi,\pi]$ with inner product defined by $(f,g)=\int_{-\pi}^{\pi} f(t)g(t)\,dt$. (See Exercise 13 in Section 4.5.) Find the distance between $\sin nx$ and $\cos mx$.

17. Let A be an $n\times n$ symmetric matrix, and suppose that R^n has the standard inner product. Prove that if $(\mathbf{u},A\mathbf{u})=(\mathbf{u},\mathbf{u})$ for all $\mathbf{u}$ in R^n, then $A=I_n$.

18. An $n\times n$ symmetric matrix A is **positive semidefinite** if $\mathbf{x}^T A\mathbf{x}\geq 0$ for all $\mathbf{x}$ in R^n. Prove the following:

(a) Every positive definite matrix is positive semidefinite.

(b) If A is singular and positive semidefinite, then A is not positive definite.

(c) A diagonal matrix A is positive semidefinite if and only if $a_{ii}\geq 0$ for $i=1,2,\dots,n$.

19. In Chapter 7 the notion of an orthogonal matrix is discussed. It is shown that an $n\times n$ matrix P is orthogonal if and only if the columns of P, denoted $\mathbf{p}_1,\mathbf{p}_2,\dots,\mathbf{p}_n$, form an orthonormal set in R^n, using the standard inner product. Let P be an orthogonal matrix.

(a) For $\mathbf{x}$ in R^n, prove that $\|P\mathbf{x}\|=\|\mathbf{x}\|$, using Theorem 5.7.

(b) For $\mathbf{x}$ and $\mathbf{y}$ in R^n, prove that the angle between $P\mathbf{x}$ and $P\mathbf{y}$ is the same as that between $\mathbf{x}$ and $\mathbf{y}$.

20. Let A be an $n\times n$ skew symmetric matrix. Prove that $\mathbf{x}^T A\mathbf{x}=0$ for all $\mathbf{x}$ in R^n.

21. Let B be an $m\times n$ matrix with orthonormal columns $\mathbf{b}_1,\mathbf{b}_2,\dots,\mathbf{b}_n$.

(a) Prove that $m\geq n$.

(b) Prove that $B^T B=I_n$.

22. Let $\{\mathbf{u}_1,\dots,\mathbf{u}_k,\mathbf{u}_{k+1},\dots,\mathbf{u}_n\}$ be an orthonormal basis for Euclidean space V, $S=\text{span}\,\{\mathbf{u}_1,\dots,\mathbf{u}_k\}$, and $T=\text{span}\,\{\mathbf{u}_{k+1},\dots,\mathbf{u}_n\}$. For any $\mathbf{x}$ in S and any $\mathbf{y}$ in T, show that $(\mathbf{x},\mathbf{y})=0$.

ADRIEN-MARIE LEGENDRE

*Adrien-Marie Legendre (1752–1833) was born in Paris into a wealthy family and died in Paris. After teaching at the École Militaire in Paris for several years, Legendre won a prestigious prize in 1782 from the Berlin Academy for a paper describing the trajectory of a projectile, taking into account air resistance. He made important contributions to celestial mechanics and number theory and was a member of a commission to standardize weights and measures. He was also the codirector of a major project to produce logarithmic and trigonometric tables. In 1794, Legendre published a basic textbook on geometry that enjoyed great popularity in Europe and the United States for about 100 years. Legendre also developed the method of least squares for fitting a curve to a given set of data.

23. Let V be a Euclidean space and W a subspace of V. Use Exercise 28 in Section 5.5 to show that $(W^{\perp})^{\perp} = W$.

24. Let V be a Euclidean space with basis $S = \{\mathbf{v}_1, \mathbf{v}_2, \ldots, \mathbf{v}_n\}$. Show that if $\mathbf{u}$ is a vector in V that is orthogonal to every vector in S, then $\mathbf{u} = \mathbf{0}$.

25. Show that if A is an $m \times n$ matrix such that AA^T is nonsingular, then rank $A = m$.

26. Let $\mathbf{u}$ and $\mathbf{v}$ be vectors in an inner product space V. If $((\mathbf{u} - \mathbf{v}), (\mathbf{u} + \mathbf{v})) = 0$, show that $\|\mathbf{u}\| = \|\mathbf{v}\|$.

27. Let $S = \{\mathbf{v}_1, \mathbf{v}_2, \ldots, \mathbf{v}_n\}$ be an orthonormal basis for a finite-dimensional inner product space V and let $\mathbf{v}$ and $\mathbf{w}$ be vectors in V with

$$\left[\mathbf{v}\right]_S = \begin{bmatrix} a_1 \\ a_2 \\ \vdots \\ a_n \end{bmatrix} \quad \text{and} \quad \left[\mathbf{w}\right]_S = \begin{bmatrix} b_1 \\ b_2 \\ \vdots \\ b_n \end{bmatrix}.$$

Show that

$$d(\mathbf{v}, \mathbf{w}) = \sqrt{(a_1 - b_1)^2 + (a_2 - b_2)^2 + \cdots + (a_n - b_n)^2}.$$

For Exercises 28 through 33, consider the following information:

*A **vector norm** on R^n is a function that assigns to each vector $\mathbf{v}$ in R^n a nonnegative real number, called the **norm** of $\mathbf{v}$ and denoted by $\|\mathbf{v}\|$, satisfying*

(a) $\|\mathbf{v}\| \geq 0$, *and* $\|\mathbf{v}\| = 0$ *if and only if* $\mathbf{v} = \mathbf{0}$;

(b) $\|c\mathbf{v}\| = |c|\,\|\mathbf{v}\|$ *for any real scalar c and vector $\mathbf{v}$;*

(c) $\|\mathbf{u} + \mathbf{v}\| \leq \|\mathbf{u}\| + \|\mathbf{v}\|$ *for all vectors $\mathbf{u}$ and $\mathbf{v}$ (the triangle inequality).*

*There are three widely used norms in applications of linear algebra, called the **1-norm**, the **2-norm**, and the **∞-norm** and denoted by $\|\ \|_1$, $\|\ \|_2$, $\|\ \|_\infty$, respectively, which are defined as follows: Let*

$$\mathbf{x} = \begin{bmatrix} x_1 \\ x_2 \\ \vdots \\ x_n \end{bmatrix}$$

be a vector in R^n.

$$\|\mathbf{x}\|_1 = |x_1| + |x_2| + \cdots + |x_n|$$

$$\|\mathbf{x}\|_2 = \sqrt{x_1^2 + x_2^2 + \cdots + x_n^2}$$

$$\|\mathbf{x}\|_\infty = \max\{|x_1|, |x_2|, \ldots, |x_n|\}$$

Observe that $\|\mathbf{x}\|_2$ is the length of the vector $\mathbf{x}$ as defined in Section 5.1.

28. For each given vector in R^2, compute $\|\mathbf{x}\|_1$, $\|\mathbf{x}\|_2$, and $\|\mathbf{x}\|_\infty$.

(a) $\begin{bmatrix} 2 \\ 3 \end{bmatrix}$ **(b)** $\begin{bmatrix} 0 \\ -2 \end{bmatrix}$ **(c)** $\begin{bmatrix} -4 \\ -1 \end{bmatrix}$

29. For each given vector in R_3, compute $\|\mathbf{x}\|_1$, $\|\mathbf{x}\|_2$, and $\|\mathbf{x}\|_\infty$.

(a) $\begin{bmatrix} 2 & -2 & 3 \end{bmatrix}$ **(b)** $\begin{bmatrix} 0 & 3 & -2 \end{bmatrix}$

(c) $\begin{bmatrix} 2 & 0 & 0 \end{bmatrix}$

30. Verify that $\|\ \|_1$ is a norm.

31. Verify that $\|\ \|_\infty$ is a norm.

32. Show the following properties:

(a) $\|\mathbf{x}\|_2^2 \leq \|\mathbf{x}\|_1^2$ **(b)** $\dfrac{\|\mathbf{x}\|_1}{n} \leq \|\mathbf{x}\|_\infty \leq \|\mathbf{x}\|_1$

33. Sketch the set of points in R^2 such that

(a) $\|\mathbf{x}\|_1 = 1$; **(b)** $\|\mathbf{x}\|_2 = 1$; **(c)** $\|\mathbf{x}\|_\infty = 1$.

Chapter Review

True or False

1. The distance between the vectors $\mathbf{u}$ and $\mathbf{v}$ is the length of their difference.

2. The cosine of the angle between the vectors $\mathbf{u}$ and $\mathbf{v}$ is given by $\mathbf{u} \cdot \mathbf{v}$.

3. The length of the vector $\mathbf{v}$ is given by $\mathbf{v} \cdot \mathbf{v}$.

4. For any vector $\mathbf{v}$ we can find a unit vector in the same direction.

5. The Cauchy–Schwarz inequality implies that the inner product of a pair of vectors is less than or equal to the product of their lengths.

6. If $\mathbf{v}$ is orthogonal to both $\mathbf{u}$ and $\mathbf{w}$, then $\mathbf{v}$ is orthogonal to every vector in span $\{\mathbf{u}, \mathbf{w}\}$.

7. An othonormal set can contain the zero vector.

8. A linearly independent set in an inner product space is also an orthonormal set.

9. Given any set of vectors, the Gram–Schmidt process will produce an orthonormal set.

10. If

$$W = \text{span}\left\{\begin{bmatrix}1\\0\\1\end{bmatrix}\right\},$$

then $W^{\perp}$ is all vectors of the form $\begin{bmatrix}0\\x\\0\end{bmatrix}$, where x is any real number.

11. If W is a subspace of V, then for $\mathbf{v}$ in V, $\text{proj}_W\mathbf{v}$ is the vector in W closest to $\mathbf{v}$.

12. If W is a subspace of V, then for $\mathbf{v}$ in V, the vector $\mathbf{v} - \text{proj}_W\mathbf{v}$ is orthogonal to W.

Quiz

1. Let

$$\mathbf{u} = \begin{bmatrix}1\\1\\0\end{bmatrix} \quad\text{and}\quad \mathbf{v} = \begin{bmatrix}0\\b\\c\end{bmatrix}.$$

Determine all the unit vectors $\mathbf{v}$ in R^3 so that the angle between $\mathbf{u}$ and $\mathbf{v}$ is 60°.

2. Find all vectors in R^4 that are orthogonal to both

$$\mathbf{v}_1 = \begin{bmatrix}1\\1\\2\\-2\end{bmatrix} \quad\text{and}\quad \mathbf{v}_2 = \begin{bmatrix}2\\1\\1\\2\end{bmatrix}.$$

3. Let V be the vector space of all continuous real-valued functions on the interval $[0, 1]$. For the inner product $(f, g) = \int_0^1 f(t)g(t)\,dt$, determine all polynomials of degree 1 or less that are orthogonal to $f(t) = t + 1$.

4. **(a)** Explain in words what the Cauchy–Schwarz inequality guarantees for a pair of vectors $\mathbf{u}$ and $\mathbf{v}$ in an inner product space.

(b) Briefly explain a primary geometric result derived from the Cauchy–Schwarz inequality.

5. Let $S = \{\mathbf{v}_1, \mathbf{v}_2, \mathbf{v}_3\}$, where

$$\mathbf{v}_1 = \begin{bmatrix}1\\1\\1\\1\end{bmatrix}, \quad \mathbf{v}_2 = \begin{bmatrix}0\\2\\-1\\-1\end{bmatrix}, \quad \mathbf{v}_3 = \begin{bmatrix}-3\\1\\1\\1\end{bmatrix}.$$

(a) Show that S is an orthogonal set.

(b) Determine an orthonormal set T so that span S = span T.

(c) Find a nonzero vector $\mathbf{v}_4$ orthogonal to S and show that $\{\mathbf{v}_1, \mathbf{v}_2, \mathbf{v}_3, \mathbf{v}_4\}$ is a basis for R^4.

6. Let $S = \{\mathbf{u}_1, \mathbf{u}_2, \mathbf{u}_3\}$, where

$$\mathbf{u}_1 = \begin{bmatrix}\frac{1}{3}\\\frac{2}{3}\\\frac{2}{3}\end{bmatrix}, \quad \mathbf{u}_2 = \begin{bmatrix}\frac{2}{3}\\-\frac{2}{3}\\\frac{1}{3}\end{bmatrix}, \quad \mathbf{u}_3 = \begin{bmatrix}\frac{2}{3}\\\frac{1}{3}\\-\frac{2}{3}\end{bmatrix}.$$

(a) Show that S is an orthonormal basis for R^3.

(b) Express

$$\mathbf{w} = \begin{bmatrix}1\\1\\1\end{bmatrix}$$

as a linear combination of the vectors in S.

(c) Let $V = \text{span}\{\mathbf{u}_1, \mathbf{u}_2\}$. Determine the orthogonal projection of

$$\mathbf{w} = \begin{bmatrix}1\\2\\-1\end{bmatrix}$$

onto V and compute the distance from $\mathbf{w}$ to V.

7. Use the Gram–Schmidt process to determine an orthonormal basis for

$$W = \text{span}\left\{\begin{bmatrix}1\\0\\1\\0\end{bmatrix}, \begin{bmatrix}2\\1\\-2\\0\end{bmatrix}, \begin{bmatrix}0\\0\\0\\1\end{bmatrix}\right\}.$$

8. Find a basis for the orthogonal complement of

$$W = \text{span}\left\{\begin{bmatrix}0\\1\\1\\0\end{bmatrix}, \begin{bmatrix}1\\0\\0\\1\end{bmatrix}, \begin{bmatrix}1\\1\\0\\1\end{bmatrix}\right\}.$$

9. Let S be a set of vectors in a finite-dimensional inner product space V. If $W = \text{span}\, S$, and S is linearly dependent, then give a detailed outline of how to develop information for W and $W^{\perp}$.

10. Let V be an inner product space and W a finite-dimensional subspace of V with orthonormal basis $\{\mathbf{w}_1, \mathbf{w}_2, \mathbf{w}_3\}$. For $\mathbf{u}$ and $\mathbf{v}$ in V, prove or disprove that $\text{proj}_W(\mathbf{u} + \mathbf{v}) = \text{proj}_W\mathbf{u} + \text{proj}_W\mathbf{v}$.

Discussion Exercises

1. Let $S = \{\mathbf{v}_1, \mathbf{v}_2, \dots, \mathbf{v}_n\}$ be a basis for a Euclidean space V and $T = \{\mathbf{u}_1, \mathbf{u}_2, \dots, \mathbf{u}_n\}$ an orthonormal basis for V. For a specified vector $\mathbf{w}$ of V provide a discussion that contrasts the work involved in computing $[\mathbf{w}]_S$ and $[\mathbf{w}]_T$.

2. Let $S = \{\mathbf{v}_1, \mathbf{v}_2, \dots, \mathbf{v}_n\}$ be a set of vectors in a Euclidean space V such that span $S = V$. Provide a discussion that outlines a procedure to determine an orthogonal basis for V that uses the set S.

3. Let $A\mathbf{x} = \mathbf{b}$ be a linear system of equations, where A is a nonsingular matrix and assume that we have a QR-factorization of A; $A = QR$.

(a) Explain why $r_{jj} \neq 0$, $j = 1, 2, \dots, n$.

(b) Explain how to solve the linear system using the QR-factorization.

4. Let

$$\mathbf{v} = \begin{bmatrix} v_1 \\ v_2 \end{bmatrix}$$

be a vector in R^2 such that $|v_j| \leq 1$, $j = 1, 2$.

(a) Explain how to determine whether the point corresponding to $\mathbf{v}$ lies within the unit circle centered at the origin.

(b) Suppose that $\mathbf{v}$ and $\mathbf{w}$ are vectors in R^2 that have their corresponding points lying within the unit circle centered at the origin. Explain how to determine whether the circle centered at the midpoint of the line segment connecting the points and having the segment as a diameter lies entirely within the unit circle centered at the origin.

5. Let k be any real number in $(0, 1)$. In the following figures, the function defined on $[0, 1]$ in (a) is denoted by f and in (b) is denoted by g. Both f and g are continuous on $[0, 1]$. Discuss the inner product of f and g using the inner product defined in Example 4 in Section 5.3.

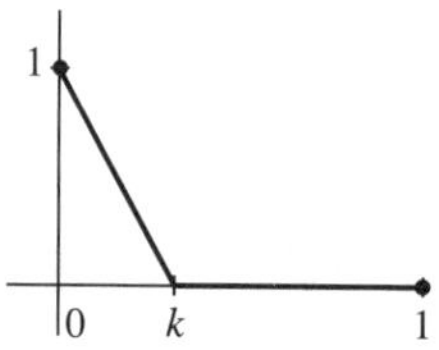

(a) f.

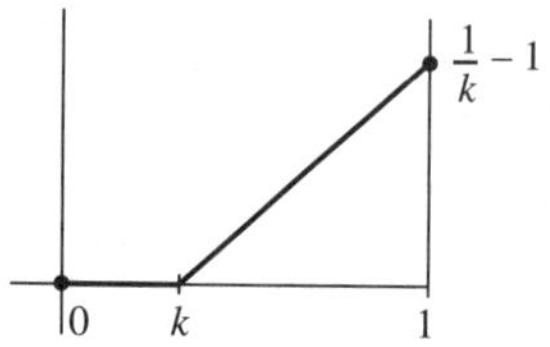

(b) g.

6. Let W be the vector space of all differentiable real-valued functions on the unit interval $[0, 1]$. For f and g in W define

$$(f, g) = \int_0^1 f(x)g(x)\,dx + \int_0^1 f'(x)g'(x)\,dx.$$

Show that (f, g) is an inner product on W.

CHAPTER

6 Linear Transformations and Matrices

6.1 Definition and Examples

As we have noted earlier, much of calculus deals with the study of properties of functions. Indeed, properties of functions are of great importance in every branch of mathematics, and linear algebra is no exception. In Section 1.6, we already encountered functions mapping one vector space into another vector space; these are matrix transformations mapping R^n into R^m. Another example was given by isomorphisms between vector spaces, which we studied in Section 4.8. If we drop some of the conditions that need to be satisfied by a function on a vector space to be an isomorphism, we get another very useful type of function called a linear transformation. Linear transformations play an important role in many areas of mathematics, the physical and social sciences, and economics. A word about notation: In Section 1.6 we denote a matrix transformation by f, and in Section 4.8 we denote an isomorphism by L. In this chapter a function mapping one vector space into another vector space is denoted by L.

DEFINITION 6.1

Let V and W be vector spaces. A function $L: V \to W$ is called a **linear transformation** of V into W if

(a) $L(\mathbf{u}+\mathbf{v}) = L(\mathbf{u}) + L(\mathbf{v})$ for every $\mathbf{u}$ and $\mathbf{v}$ in V.

(b) $L(c\mathbf{u}) = cL(\mathbf{u})$ for any $\mathbf{u}$ in V, and c any real number.

If $V = W$, the linear transformation $L: V \to W$ is also called a **linear operator** on V.

Most of the vector spaces considered henceforth, but not all, are finite-dimensional.

In Definition 6.1, observe that in (a) the $+$ in $\mathbf{u}+\mathbf{v}$ refers to the addition operation in V, whereas the $+$ in $L(\mathbf{u}) + L(\mathbf{v})$ refers to the addition operation in

W. Similarly, in (b) the scalar product $c\mathbf{u}$ is in V, while the scalar product $cL(\mathbf{u})$ is in W.

We have already pointed out in Section 4.8 that an isomorphism is a linear transformation that is one-to-one and onto. Linear transformations occur very frequently, and we now look at some examples. (At this point it might be profitable to review the material of Section A.2.) It can be shown that $L: V \to W$ is a linear transformation if and only if $L(a\mathbf{u} + b\mathbf{v}) = aL(\mathbf{u}) + bL(\mathbf{v})$ for any real numbers a, b and any vectors $\mathbf{u}$, $\mathbf{v}$ in V (see Exercise 6.)

EXAMPLE 1

Let A be an $m \times n$ matrix. In Section 1.6 we defined a matrix transformation as a function $L: R^n \to R^m$ defined by $L(\mathbf{u}) = A\mathbf{u}$. We now show that every matrix transformation is a linear transformation by verifying that properties (a) and (b) in Definition 6.1 hold.

If $\mathbf{u}$ and $\mathbf{v}$ are vectors in R^n, then

$$L(\mathbf{u} + \mathbf{v}) = A(\mathbf{u} + \mathbf{v}) = A\mathbf{u} + A\mathbf{v} = L(\mathbf{u}) + L(\mathbf{v}).$$

Moreover, if c is a scalar, then

$$L(c\mathbf{u}) = A(c\mathbf{u}) = c(A\mathbf{u}) = cL(\mathbf{u}).$$

Hence, every matrix transformation is a linear transformation. ■

For convenience we now summarize the matrix transformations that have already been presented in Section 1.6.

Reflection with respect to the x-axis: $L: R^2 \to R^2$ is defined by

$$L\left(\begin{bmatrix} u_1 \\ u_2 \end{bmatrix}\right) = \begin{bmatrix} u_1 \\ -u_2 \end{bmatrix}.$$

Projection into the xy-plane: $L: R^3 \to R^2$ is defined by

$$L\left(\begin{bmatrix} u_1 \\ u_2 \\ u_3 \end{bmatrix}\right) = \begin{bmatrix} u_1 \\ u_2 \end{bmatrix}.$$

Dilation: $L: R^3 \to R^3$ is defined by $L(\mathbf{u}) = r\mathbf{u}$ for $r > 1$.

Contraction: $L: R^3 \to R^3$ is defined by $L(\mathbf{u}) = r\mathbf{u}$ for $0 < r < 1$.

Rotation counterclockwise through an angle ϕ: $L: R^2 \to R^2$ is defined by

$$L(\mathbf{u}) = \begin{bmatrix} \cos\phi & -\sin\phi \\ \sin\phi & \cos\phi \end{bmatrix}\mathbf{u}.$$

EXAMPLE 2

Let $L: R^3 \to R^3$ be defined by

$$L\left(\begin{bmatrix} u_1 \\ u_2 \\ u_3 \end{bmatrix}\right) = \begin{bmatrix} u_1 + 1 \\ 2u_2 \\ u_3 \end{bmatrix}.$$

To determine whether L is a linear transformation, let

$$\mathbf{u} = \begin{bmatrix} u_1 \\ u_2 \\ u_3 \end{bmatrix} \quad \text{and} \quad \mathbf{v} = \begin{bmatrix} v_1 \\ v_2 \\ v_3 \end{bmatrix}.$$

Then

$$L(\mathbf{u}+\mathbf{v}) = L\left(\begin{bmatrix} u_1 \\ u_2 \\ u_3 \end{bmatrix} + \begin{bmatrix} v_1 \\ v_2 \\ v_3 \end{bmatrix}\right) = L\left(\begin{bmatrix} u_1+v_1 \\ u_2+v_2 \\ u_3+v_3 \end{bmatrix}\right)$$
$$= \begin{bmatrix} (u_1+v_1)+1 \\ 2(u_2+v_2) \\ u_3+v_3 \end{bmatrix}.$$

On the other hand,

$$L(\mathbf{u})+L(\mathbf{v}) = \begin{bmatrix} u_1+1 \\ 2u_2 \\ u_3 \end{bmatrix} + \begin{bmatrix} v_1+1 \\ 2v_2 \\ v_3 \end{bmatrix} = \begin{bmatrix} (u_1+v_1)+2 \\ 2(u_2+v_2) \\ u_3+v_3 \end{bmatrix}.$$

Letting $u_1 = 1$, $u_2 = 3$, $u_3 = -2$, $v_1 = 2$, $v_2 = 4$, and $v_3 = 1$, we see that $L(\mathbf{u}+\mathbf{v}) \neq L(\mathbf{u}) + L(\mathbf{v})$. Hence we conclude that the function L is not a linear transformation. ■

EXAMPLE 3

Let $L\colon R_2 \to R_2$ be defined by

$$L\left(\begin{bmatrix} u_1 & u_2 \end{bmatrix}\right) = \begin{bmatrix} u_1^2 & 2u_2 \end{bmatrix}.$$

Is L a linear transformation?

Solution

Let

$$\mathbf{u} = \begin{bmatrix} u_1 & u_2 \end{bmatrix} \quad \text{and} \quad \mathbf{v} = \begin{bmatrix} v_1 & v_2 \end{bmatrix}.$$

Then

$$\begin{aligned} L(\mathbf{u}+\mathbf{v}) &= L\left(\begin{bmatrix} u_1 & u_2 \end{bmatrix} + \begin{bmatrix} v_1 & v_2 \end{bmatrix}\right) \\ &= L\left(\begin{bmatrix} u_1+v_1 & u_2+v_2 \end{bmatrix}\right) \\ &= \begin{bmatrix} (u_1+v_1)^2 & 2(u_2+v_2) \end{bmatrix}. \end{aligned}$$

On the other hand,

$$\begin{aligned} L(\mathbf{u})+L(\mathbf{v}) &= \begin{bmatrix} u_1^2 & 2u_2 \end{bmatrix} + \begin{bmatrix} v_1^2 & 2v_2 \end{bmatrix} \\ &= \begin{bmatrix} u_1^2+v_1^2 & 2(u_2+v_2) \end{bmatrix}. \end{aligned}$$

Since there are some choices of u and v such that $L(\mathbf{u}+\mathbf{v}) \neq L(\mathbf{u}) + L(\mathbf{v})$, we conclude that L is not a linear transformation. ■

EXAMPLE 4

Let $L\colon P_1 \to P_2$ be defined by

$$L[p(t)] = tp(t).$$

Show that L is a linear transformation.

Solution

Let $p(t)$ and $q(t)$ be vectors in P_1 and let c be a scalar. Then

$$\begin{aligned} L[p(t)+q(t)] &= t[p(t)+q(t)] \\ &= tp(t)+tq(t) \\ &= L[p(t)]+L[q(t)], \end{aligned}$$

and

$$\begin{aligned} L[cp(t)] &= t[cp(t)] \\ &= c[tp(t)] \\ &= cL[p(t)]. \end{aligned}$$

Hence L is a linear transformation. ■

EXAMPLE 5 Let W be the vector space of all real-valued functions and let V be the subspace of all differentiable functions. Let $L\colon V \to W$ be defined by

$$L(f) = f',$$

where f' is the derivative of f. We can show (Exercise 24), using the properties of differentiation, that L is a linear transformation. ■

EXAMPLE 6 Let $V = C[a, b]$ be the vector space of all real-valued functions that are integrable over the interval $[a, b]$. Let $W = R^1$. Define $L\colon V \to W$ by

$$L(f) = \int_a^b f(x)\,dx.$$

We can show (Exercise 25), using the properties of integration, that L is a linear transformation. ■

Since an n-dimensional vector space V is isomorphic to R^n, we can determine whether the set $S = \{\mathbf{v}_1, \mathbf{v}_2, \dots, \mathbf{v}_n\}$ of n vectors in V is a basis for V by checking whether $\{L(\mathbf{v}_1), L(\mathbf{v}_2), \dots, L(\mathbf{v}_n)\}$ is a basis for R^n, where $L\colon V \to R^n$ is an isomorphism. The following example illustrates this approach:

EXAMPLE 7 To find out if $S = \{t^2 + t, t + 1, t - 1\}$ is a basis for P_2, we note that P_2 is a three-dimensional vector space isomorphic to R^3 under the mapping $L\colon P_2 \to R^3$ defined by $L(at^2 + bt + c) = \begin{bmatrix} a \\ b \\ c \end{bmatrix}$. Therefore, S is a basis for P_2 if and only if $T = \{L(t^2 + t), L(t + 1), L(t - 1)\}$ is a basis for R^3. To decide whether this is so, we apply Theorem 4.5. Thus let A be the matrix whose columns are $L(t^2 + t)$, $L(t + 1)$, $L(t - 1)$, respectively. Now

$$L(t^2 + t) = \begin{bmatrix} 1 \\ 1 \\ 0 \end{bmatrix}, \quad L(t + 1) = \begin{bmatrix} 0 \\ 1 \\ 1 \end{bmatrix}, \quad \text{and} \quad L(t - 1) = \begin{bmatrix} 0 \\ 1 \\ -1 \end{bmatrix},$$

so

$$A = \begin{bmatrix} 1 & 0 & 0 \\ 1 & 1 & 1 \\ 0 & 1 & -1 \end{bmatrix}.$$

Since $\det(A) = -2$ (verify), we conclude that T is linearly independent. Hence S is linearly independent, and since $\dim P_2 = 3$, S is a basis for P_2. We could also have reached the same conclusion by computing $\operatorname{rank}(A)$ to be 3 (verify). ■

We now develop some general properties of linear transformations.

Theorem 6.1 Let $L: V \to W$ be a linear transformation. Then

(a) $L(\mathbf{0}_V) = \mathbf{0}_W$.

(b) $L(\mathbf{u} - \mathbf{v}) = L(\mathbf{u}) - L(\mathbf{v})$, for $\mathbf{u}$, $\mathbf{v}$ in V.

Proof

(a) We have

$$\mathbf{0}_V = \mathbf{0}_V + \mathbf{0}_V,$$

so

$$L(\mathbf{0}_V) = L(\mathbf{0}_V + \mathbf{0}_V)$$
$$L(\mathbf{0}_V) = L(\mathbf{0}_V) + L(\mathbf{0}_V).$$

Adding $-L(\mathbf{0}_V)$ to both sides, we obtain

$$L(\mathbf{0}_V) = \mathbf{0}_W.$$

(b) $L(\mathbf{u} - \mathbf{v}) = L(\mathbf{u} + (-1)\mathbf{v}) = L(\mathbf{u}) + L((-1)\mathbf{v})$
$\qquad = L(\mathbf{u}) + (-1)L(\mathbf{v}) = L(\mathbf{u}) - L(\mathbf{v}).$ ■

Remark Example 2 can be solved more easily by observing that

$$L\left(\begin{bmatrix} 0 \\ 0 \\ 0 \end{bmatrix}\right) = \begin{bmatrix} 1 \\ 0 \\ 0 \end{bmatrix},$$

so, by part (a) of Theorem 6.1, L is not a linear transformation.

EXAMPLE 8 Let V be an n-dimensional vector space and $S = \{\mathbf{v}_1, \mathbf{v}_2, \ldots, \mathbf{v}_n\}$ an ordered basis for V. If $\mathbf{v}$ is a vector in V, then $\mathbf{v}$ can be written in one and only one way, as

$$\mathbf{v} = a_1\mathbf{v}_1 + a_2\mathbf{v}_2 + \cdots + a_n\mathbf{v}_n,$$

where $a_1, a_2, \ldots, a_n$ are real numbers, which were called in Section 4.8 the coordinates of $\mathbf{v}$ with respect to S. Recall that in Section 4.8 we defined the coordinate vector of $\mathbf{v}$ with respect to S as

$$\left[\mathbf{v}\right]_S = \begin{bmatrix} a_1 \\ a_2 \\ \vdots \\ a_n \end{bmatrix}.$$

We define $L: V \to R^n$ by

$$L(\mathbf{v}) = \left[\mathbf{v}\right]_S .$$

It is not difficult to show (Exercise 29) that L is a linear transformation. ■

EXAMPLE 9 Let V be an n-dimensional vector space and let $S = \{\mathbf{v}_1, \mathbf{v}_2, \dots, \mathbf{v}_n\}$ and $T = \{\mathbf{w}_1, \mathbf{w}_2, \dots, \mathbf{w}_n\}$ be ordered bases for V. If $\mathbf{v}$ is any vector in V, then Equation (3) in Section 4.8 gives the relationship between the coordinate vector of $\mathbf{v}$ with respect to S and the coordinate vector of $\mathbf{v}$ with respect to T as

$$\left[\mathbf{v}\right]_S = P_{S \leftarrow T} \left[\mathbf{v}\right]_T ,$$

where $P_{S \leftarrow T}$ is the transition matrix from T to S. Let $L: R^n \to R^n$ be defined by

$$L(\mathbf{v}) = P_{S \leftarrow T}\mathbf{v}$$

for $\mathbf{v}$ in R^n. Thus L is a matrix transformation, so it follows that L is a linear transformation. ■

We know from calculus that a function can be specified by a formula which assigns to every member of the domain a unique element of the range. On the other hand, we can also specify a function by listing next to each member of the domain its assigned element of the range. An example of this would be listing the names of all charge account customers of a department store, along with their charge account number. At first glance it appears impossible to describe a linear transformation $L: V \to W$ of a vector space $V \neq \{\mathbf{0}\}$ into a vector space W in this latter manner, since V has infinitely many members in it. However, the next, very useful, theorem tells us that *once we say what a linear transformation L does to a basis for V, then we have completely specified L*. Thus, since in this book we deal mostly with finite-dimensional vector spaces, it is possible to describe L by giving only the images of a finite number of vectors in the domain V.

Theorem 6.2 Let $L: V \to W$ be a linear transformation of an n-dimensional vector space V into a vector space W. Let $S = \{\mathbf{v}_1, \mathbf{v}_2, \dots, \mathbf{v}_n\}$ be a basis for V. If $\mathbf{v}$ is any vector in V, then $L(\mathbf{v})$ is completely determined by $\{L(\mathbf{v}_1), L(\mathbf{v}_2), \dots, L(\mathbf{v}_n)\}$.

Proof

Since $\mathbf{v}$ is in V, we can write $\mathbf{v} = a_1\mathbf{v}_1 + a_2\mathbf{v}_2 + \cdots + a_n\mathbf{v}_n$, where $a_1, a_2, \dots, a_n$ are uniquely determined real numbers. Then

$$\begin{aligned} L(\mathbf{v}) &= L(a_1\mathbf{v}_1 + a_2\mathbf{v}_2 + \cdots + a_n\mathbf{v}_n) \\ &= L(a_1\mathbf{v}_1) + L(a_2\mathbf{v}_2) + \cdots + L(a_n\mathbf{v}_n) \\ &= a_1L(\mathbf{v}_1) + a_2L(\mathbf{v}_2) + \cdots + a_nL(\mathbf{v}_n). \end{aligned}$$

Thus $L(\mathbf{v})$ has been completely determined by the vectors $L(\mathbf{v}_1)$, $L(\mathbf{v}_2)$, $\dots$, $L(\mathbf{v}_n)$. ■

Theorem 6.2 can also be stated in the following useful form: Let $L: V \to W$ and $L': V \to W$ be linear transformations of the n-dimensional vector space V into a vector space W. Let $S = \{\mathbf{v}_1, \mathbf{v}_2, \dots, \mathbf{v}_n\}$ be a basis for V. If $L'(\mathbf{v}_i) = L(\mathbf{v}_i)$ for $i = 1, 2, \dots, n$, then $L'(\mathbf{v}) = L(\mathbf{v})$ for every $\mathbf{v}$ in V; that is, *if L and L' agree on a basis for V, then L and L' are identical linear transformations.*

EXAMPLE 10

Let $L\colon R_4 \to R_2$ be a linear transformation and let $S = \{\mathbf{v}_1, \mathbf{v}_2, \mathbf{v}_3, \mathbf{v}_4\}$ be a basis for R_4, where $\mathbf{v}_1 = \begin{bmatrix} 1 & 0 & 1 & 0 \end{bmatrix}$, $\mathbf{v}_2 = \begin{bmatrix} 0 & 1 & -1 & 2 \end{bmatrix}$, $\mathbf{v}_3 = \begin{bmatrix} 0 & 2 & 2 & 1 \end{bmatrix}$, and $\mathbf{v}_4 = \begin{bmatrix} 1 & 0 & 0 & 1 \end{bmatrix}$. Suppose that

$$L(\mathbf{v}_1) = \begin{bmatrix} 1 & 2 \end{bmatrix}, \quad L(\mathbf{v}_2) = \begin{bmatrix} 0 & 3 \end{bmatrix},$$
$$L(\mathbf{v}_3) = \begin{bmatrix} 0 & 0 \end{bmatrix}, \quad \text{and} \quad L(\mathbf{v}_4) = \begin{bmatrix} 2 & 0 \end{bmatrix}.$$

Let

$$\mathbf{v} = \begin{bmatrix} 3 & -5 & -5 & 0 \end{bmatrix}.$$

Find $L(\mathbf{v})$.

Solution

We first write $\mathbf{v}$ as a linear combination of the vectors in S, obtaining (verify)

$$\mathbf{v} = \begin{bmatrix} 3 & -5 & -5 & 0 \end{bmatrix} = 2\mathbf{v}_1 + \mathbf{v}_2 - 3\mathbf{v}_3 + \mathbf{v}_4.$$

It then follows by Theorem 6.2 that

$$\begin{aligned} L(\mathbf{v}) &= L(2\mathbf{v}_1 + \mathbf{v}_2 - 3\mathbf{v}_3 + \mathbf{v}_4) \\ &= 2L(\mathbf{v}_1) + L(\mathbf{v}_2) - 3L(\mathbf{v}_3) + L(\mathbf{v}_4) = \begin{bmatrix} 4 & 7 \end{bmatrix}. \end{aligned}$$ ■

We already know that if A is an $m \times n$ matrix, then the function $L\colon R^n \to R^m$ defined by $L(\mathbf{x}) = A\mathbf{x}$ for $\mathbf{x}$ in R^n is a linear transformation. In the next example, we show that if $L\colon R^n \to R^m$ is a linear transformation, then L must be of this form. That is, L must be a matrix transformation.

Theorem 6.3 Let $L\colon R^n \to R^m$ be a linear transformation and consider the natural basis $\{\mathbf{e}_1, \mathbf{e}_2, \ldots, \mathbf{e}_n\}$ for R^n. Let A be the $m \times n$ matrix whose jth column is $L(\mathbf{e}_j)$. The matrix A has the following property: If $\mathbf{x} = \begin{bmatrix} x_1 \\ x_2 \\ \vdots \\ x_n \end{bmatrix}$ is any vector in R^n, then

$$L(\mathbf{x}) = A\mathbf{x}. \tag{1}$$

Moreover, A is the only matrix satisfying Equation (1). It is called the **standard matrix representing** L.

Proof

Writing $\mathbf{x}$ as a linear combination of the natural basis for R^n, we have

$$\mathbf{x} = x_1\mathbf{e}_1 + x_2\mathbf{e}_2 + \cdots + x_n\mathbf{e}_n;$$

so by Theorem 6.2,

$$\begin{aligned} L(\mathbf{x}) &= L(x_1\mathbf{e}_1 + x_2\mathbf{e}_2 + \cdots + x_n\mathbf{e}_n) \\ &= x_1L(\mathbf{e}_1) + x_2L(\mathbf{e}_2) + \cdots + x_nL(\mathbf{e}_n). \end{aligned} \tag{2}$$

Since A is the $m \times n$ matrix whose jth column is $L(\mathbf{e}_j)$, we can write Equation (2) in matrix form as

$$L(\mathbf{x}) = A\mathbf{x}.$$

We leave it as an exercise (Exercise 37) to show that A is unique. ■

EXAMPLE 11 Let $L: R^3 \to R^2$ be the linear transformation defined by

$$L\left(\begin{bmatrix} x_1 \\ x_2 \\ x_3 \end{bmatrix}\right) = \begin{bmatrix} x_1 + 2x_2 \\ 3x_2 - 2x_3 \end{bmatrix}.$$

Find the standard matrix representing L.

Solution

Let $\{\mathbf{e}_1, \mathbf{e}_2, \mathbf{e}_3\}$ be the natural basis for R^3. We now compute $L(\mathbf{e}_j)$ for $j = 1, 2, 3$ as follows:

$$L(\mathbf{e}_1) = L\left(\begin{bmatrix} 1 \\ 0 \\ 0 \end{bmatrix}\right) = \begin{bmatrix} 1 \\ 0 \end{bmatrix},$$

$$L(\mathbf{e}_2) = L\left(\begin{bmatrix} 0 \\ 1 \\ 0 \end{bmatrix}\right) = \begin{bmatrix} 2 \\ 3 \end{bmatrix},$$

$$L(\mathbf{e}_3) = L\left(\begin{bmatrix} 0 \\ 0 \\ 1 \end{bmatrix}\right) = \begin{bmatrix} 0 \\ -2 \end{bmatrix}.$$

Hence

$$A = \begin{bmatrix} L(\mathbf{e}_1) & L(\mathbf{e}_2) & L(\mathbf{e}_3) \end{bmatrix} = \begin{bmatrix} 1 & 2 & 0 \\ 0 & 3 & -2 \end{bmatrix}.$$

■

EXAMPLE 12 **(Cryptology)** Cryptology is the technique of coding and decoding messages; it goes back to the time of the ancient Greeks. A simple code is constructed by associating a different number with every letter in the alphabet. For example,

A	B	C	D	⋯	X	Y	Z
↕	↕	↕	↕		↕	↕	↕
1	2	3	4	⋯	24	25	26

Suppose that Mark S. and Susan J. are two undercover agents who want to communicate with each other by using a code because they suspect that their phones are being tapped and their mail is being intercepted. In particular, Mark wants to send Susan the following message:

MEET TOMORROW

Using the substitution scheme just given, Mark sends this message:

13 5 5 20 20 15 13 15 18 18 15 23

A code of this type could be cracked without too much difficulty by a number of techniques, including the analysis of frequency of letters. To make it difficult to crack the code, the agents proceed as follows. First, when they undertook the mission, they agreed on a 3×3 nonsingular matrix, the **encoding matrix**, such as

$$A = \begin{bmatrix} 1 & 2 & 3 \\ 1 & 1 & 2 \\ 0 & 1 & 2 \end{bmatrix}.$$

Mark then breaks the message into four vectors in R^3. (If this cannot be done, we can add extra letters.) Thus we have the vectors

$$\begin{bmatrix} 13 \\ 5 \\ 5 \end{bmatrix}, \quad \begin{bmatrix} 20 \\ 20 \\ 15 \end{bmatrix}, \quad \begin{bmatrix} 13 \\ 15 \\ 18 \end{bmatrix}, \quad \begin{bmatrix} 18 \\ 15 \\ 23 \end{bmatrix}.$$

Mark now defines the linear transformation $L: R^3 \to R^3$ by $L(\mathbf{x}) = A\mathbf{x}$, so the message becomes

$$A \begin{bmatrix} 13 \\ 5 \\ 5 \end{bmatrix} = \begin{bmatrix} 38 \\ 28 \\ 15 \end{bmatrix}, \qquad A \begin{bmatrix} 20 \\ 20 \\ 15 \end{bmatrix} = \begin{bmatrix} 105 \\ 70 \\ 50 \end{bmatrix},$$

$$A \begin{bmatrix} 13 \\ 15 \\ 18 \end{bmatrix} = \begin{bmatrix} 97 \\ 64 \\ 51 \end{bmatrix}, \qquad A \begin{bmatrix} 18 \\ 15 \\ 23 \end{bmatrix} = \begin{bmatrix} 117 \\ 79 \\ 61 \end{bmatrix}.$$

Thus Mark transmits the following message:

$$38 \quad 28 \quad 15 \quad 105 \quad 70 \quad 50 \quad 97 \quad 64 \quad 51 \quad 117 \quad 79 \quad 61$$

Suppose now that Mark receives the message from Susan,

$$77 \quad 54 \quad 38 \quad 71 \quad 49 \quad 29 \quad 68 \quad 51 \quad 33 \quad 76 \quad 48 \quad 40 \quad 86 \quad 53 \quad 52$$

which he wants to decode with the same key matrix A. To decode it, Mark breaks the message into five vectors in R^3:

$$\begin{bmatrix} 77 \\ 54 \\ 38 \end{bmatrix}, \quad \begin{bmatrix} 71 \\ 49 \\ 29 \end{bmatrix}, \quad \begin{bmatrix} 68 \\ 51 \\ 33 \end{bmatrix}, \quad \begin{bmatrix} 76 \\ 48 \\ 40 \end{bmatrix}, \quad \begin{bmatrix} 86 \\ 53 \\ 52 \end{bmatrix}$$

and solves the equation

$$L(\mathbf{x}_1) = \begin{bmatrix} 77 \\ 54 \\ 38 \end{bmatrix} = A\mathbf{x}_1$$

for $\mathbf{x}_1$. Since A is nonsingular,

$$\mathbf{x}_1 = A^{-1} \begin{bmatrix} 77 \\ 54 \\ 38 \end{bmatrix} = \begin{bmatrix} 0 & 1 & -1 \\ 2 & -2 & -1 \\ -1 & 1 & 1 \end{bmatrix} \begin{bmatrix} 77 \\ 54 \\ 38 \end{bmatrix} = \begin{bmatrix} 16 \\ 8 \\ 15 \end{bmatrix}.$$

Similarly,

$$\mathbf{x}_2 = A^{-1}\begin{bmatrix}71\\49\\29\end{bmatrix} = \begin{bmatrix}20\\15\\7\end{bmatrix}, \qquad \mathbf{x}_3 = A^{-1}\begin{bmatrix}68\\51\\33\end{bmatrix} = \begin{bmatrix}18\\1\\16\end{bmatrix},$$

$$\mathbf{x}_4 = A^{-1}\begin{bmatrix}76\\48\\40\end{bmatrix} = \begin{bmatrix}8\\16\\12\end{bmatrix}, \qquad \mathbf{x}_5 = A^{-1}\begin{bmatrix}86\\53\\52\end{bmatrix} = \begin{bmatrix}1\\14\\19\end{bmatrix}.$$

Using our correspondence between letters and numbers, Mark has received the following message:

PHOTOGRAPH PLANS ■

Additional material on cryptology may be found in the references given in Further Readings.

FURTHER READINGS IN CRYPTOLOGY

Elementary presentation

Kohn, Bernice. *Secret Codes and Ciphers*. Englewood Cliffs, N.J.: Prentice-Hall, Inc., 1968 (63 pages).

Advanced presentation

Fisher, James L. *Applications-Oriented Algebra*. New York: T. Harper & Row, Publishers, 1977 (Chapter 9, "Coding Theory").

Garrett, Paul. *Making, Breaking Codes*. Upper Saddle River, N.J.: Prentice Hall, Inc., 2001.

Hardy, Darel W., and Carol L. Walker, *Applied Algebra, Codes, Ciphers and Discrete Algorithms*. Upper Saddle River, N.J.: Prentice Hall, Inc., 2002.

Kahn, David. *The Codebreakers*. New York: The New American Library Inc., 1973.

Key Terms

Linear transformation
Linear operator
Reflection
Projection
Dilation
Contraction
Rotation
Standard matrix representing a linear transformation
Translation

6.1 Exercises

1. Which of the following functions are linear transformations?

(a) $L: R_2 \to R_3$ defined by
$L\left(\begin{bmatrix}u_1 & u_2\end{bmatrix}\right) = \begin{bmatrix}u_1+1 & u_2 & u_1+u_2\end{bmatrix}$

(b) $L: R_2 \to R_3$ defined by
$L\left(\begin{bmatrix}u_1 & u_2\end{bmatrix}\right) = \begin{bmatrix}u_1+u_2 & u_2 & u_1-u_2\end{bmatrix}$

2. Which of the following functions are linear transformations?

(a) $L: R_3 \to R_3$ defined by
$L\left(\begin{bmatrix}u_1 & u_2 & u_3\end{bmatrix}\right) = \begin{bmatrix}u_1 & u_2^2+u_3^2 & u_3^2\end{bmatrix}$

(b) $L: R_3 \to R_3$ defined by
$L\left(\begin{bmatrix}u_1 & u_2 & u_3\end{bmatrix}\right) = \begin{bmatrix}1 & u_3 & u_2\end{bmatrix}$

(c) $L: R_3 \to R_3$ defined by
$L\left(\begin{bmatrix}u_1 & u_2 & u_3\end{bmatrix}\right) = \begin{bmatrix}0 & u_3 & u_2\end{bmatrix}$

3. Which of the following functions are linear transformations? [Here, $p'(t)$ denotes the derivative of $p(t)$ with respect to t.]

(a) $L\colon P_2 \to P_3$ defined by $L(p(t)) = t^3 p'(0) + t^2 p(0)$

(b) $L\colon P_1 \to P_2$ defined by $L(p(t)) = tp(t) + p(0)$

(c) $L\colon P_1 \to P_2$ defined by $L(p(t)) = tp(t) + 1$

4. Which of the following functions are linear transformations?

(a) $L\colon M_{nn} \to M_{nn}$ defined by $L(A) = A^T$

(b) $L\colon M_{nn} \to M_{nn}$ defined by $L(A) = A^{-1}$

5. Which of the following functions are linear transformations?

(a) $L\colon M_{nn} \to R^1$ defined by $L(A) = \det(A)$

(b) $L\colon M_{nn} \to R^1$ defined by $L(A) = \mathrm{Tr}(A)$ (See Exercise 43 in Section 1.3.)

6. Let $L\colon V \to W$ be a mapping of a vector space V into a vector space W. Prove that L is a linear transformation if and only if $L(a\mathbf{u} + b\mathbf{v}) = aL(\mathbf{u}) + bL(\mathbf{v})$ for any real numbers a, b and any vectors $\mathbf{u}$, $\mathbf{v}$ in V.

In Exercises 7 and 8, find the standard matrix representing each given linear transformation.

7. (a) $L\colon R^2 \to R^2$ defined by $L\left(\begin{bmatrix} u_1 \\ u_2 \end{bmatrix}\right) = \begin{bmatrix} -u_1 \\ u_2 \end{bmatrix}$

(b) $L\colon R^2 \to R^2$ defined by $L\left(\begin{bmatrix} u_1 \\ u_2 \end{bmatrix}\right) = \begin{bmatrix} -u_2 \\ u_1 \end{bmatrix}$

(c) $L\colon R^3 \to R^3$ defined by $L\left(\begin{bmatrix} u_1 \\ u_2 \\ u_3 \end{bmatrix}\right) = \begin{bmatrix} u_1 \\ 0 \\ 0 \end{bmatrix}$

8. (a) $L\colon R^2 \to R^2$ defined by $L\left(\begin{bmatrix} u_1 \\ u_2 \end{bmatrix}\right) = \begin{bmatrix} -u_2 \\ -u_1 \end{bmatrix}$

(b) $L\colon R^2 \to R^2$ defined by $L\left(\begin{bmatrix} u_1 \\ u_2 \end{bmatrix}\right) = \begin{bmatrix} u_1 + ku_2 \\ u_2 \end{bmatrix}$

(c) $L\colon R^3 \to R^3$ defined by $L(\mathbf{u}) = k\mathbf{u}$

9. Consider the function $L\colon M_{34} \to M_{24}$ defined by $L(A) = \begin{bmatrix} 2 & 3 & 1 \\ 1 & 2 & -3 \end{bmatrix} A$ for A in M_{34}.

(a) Find $L\left(\begin{bmatrix} 1 & 2 & 0 & -1 \\ 3 & 0 & 2 & 3 \\ 4 & 1 & -2 & 1 \end{bmatrix}\right)$.

(b) Show that L is a linear transformation.

10. Find the standard matrix representing each given linear transformation.

(a) Projection mapping R^3 into the xy-plane

(b) Dilation mapping R^3 into R^3

(c) Reflection with respect to the x-axis mapping R^2 into R^2

11. Find the standard matrix representing each given linear transformation.

(a) $L\colon R^2 \to R^2$ defined by $L\left(\begin{bmatrix} u_1 \\ u_2 \end{bmatrix}\right) = \begin{bmatrix} u_2 \\ u_1 \end{bmatrix}$

(b) $L\colon R^2 \to R^3$ defined by

$$L\left(\begin{bmatrix} u_1 \\ u_2 \end{bmatrix}\right) = \begin{bmatrix} u_1 - 3u_2 \\ 2u_1 - u_2 \\ 2u_2 \end{bmatrix}$$

(c) $L\colon R^3 \to R^3$ defined by

$$L\left(\begin{bmatrix} u_1 \\ u_2 \\ u_3 \end{bmatrix}\right) = \begin{bmatrix} u_1 + 4u_2 \\ -u_3 \\ u_2 + u_3 \end{bmatrix}$$

12. Let $A = \begin{bmatrix} 0 & -1 & 2 \\ -2 & 1 & 3 \\ 1 & 2 & -3 \end{bmatrix}$ be the standard matrix representing the linear transformation $L\colon R^3 \to R^3$.

(a) Find $L\left(\begin{bmatrix} 2 \\ -3 \\ 1 \end{bmatrix}\right)$. **(b)** Find $L\left(\begin{bmatrix} u_1 \\ u_2 \\ u_3 \end{bmatrix}\right)$.

13. Let $L\colon R^3 \to R^2$ be a linear transformation for which we know that

$$L\left(\begin{bmatrix} 1 \\ 0 \\ 0 \end{bmatrix}\right) = \begin{bmatrix} 2 \\ -4 \end{bmatrix},$$

$$L\left(\begin{bmatrix} 0 \\ 1 \\ 0 \end{bmatrix}\right) = \begin{bmatrix} 3 \\ -5 \end{bmatrix}, \quad L\left(\begin{bmatrix} 0 \\ 0 \\ 1 \end{bmatrix}\right) = \begin{bmatrix} 2 \\ 3 \end{bmatrix}.$$

(a) What is $L\left(\begin{bmatrix} 1 \\ -2 \\ 3 \end{bmatrix}\right)$?

(b) What is $L\left(\begin{bmatrix} u_1 \\ u_2 \\ u_3 \end{bmatrix}\right)$?

14. Let $L\colon R_2 \to R_2$ be a linear transformation for which we know that

$$L\left(\begin{bmatrix} 1 & 1 \end{bmatrix}\right) = \begin{bmatrix} 1 & -2 \end{bmatrix},$$
$$L\left(\begin{bmatrix} -1 & 1 \end{bmatrix}\right) = \begin{bmatrix} 2 & 3 \end{bmatrix}.$$

(a) What is $L\left(\begin{bmatrix} -1 & 5 \end{bmatrix}\right)$?

(b) What is $L\left(\begin{bmatrix} u_1 & u_2 \end{bmatrix}\right)$?

15. Let $L: P_2 \to P_3$ be a linear transformation for which we know that $L(1) = 1$, $L(t) = t^2$, $L(t^2) = t^3 + t$.

(a) Find $L(2t^2 - 5t + 3)$. **(b)** Find $L(at^2 + bt + c)$.

16. Let A be a fixed 3×3 matrix; also let $L: M_{33} \to M_{33}$ be defined by $L(X) = AX - XA$, for X in M_{33}. Show that L is a linear transformation.

17. Let $L: R \to R$ be defined by $L(\mathbf{v}) = a\mathbf{v} + b$, where a and b are real numbers. (Of course, $\mathbf{v}$ is a vector in R, which in this case means that $\mathbf{v}$ is also a real number.) Find all values of a and b such that L is a linear transformation.

18. Let V be an inner product space and let $\mathbf{w}$ be a fixed vector in V. Let $L: V \to R$ be defined by $L(\mathbf{v}) = (\mathbf{v}, \mathbf{w})$ for $\mathbf{v}$ in V. Show that L is a linear transformation.

19. Describe the following linear transformations geometrically:

(a) $L\left(\begin{bmatrix} u_1 \\ u_2 \end{bmatrix}\right) = \begin{bmatrix} -u_1 \\ u_2 \end{bmatrix}$

(b) $L\left(\begin{bmatrix} u_1 \\ u_2 \end{bmatrix}\right) = \begin{bmatrix} -u_1 \\ -u_2 \end{bmatrix}$

(c) $L\left(\begin{bmatrix} u_1 \\ u_2 \end{bmatrix}\right) = \begin{bmatrix} -u_2 \\ u_1 \end{bmatrix}$

20. Let $L: P_1 \to P_1$ be a linear transformation for which we know that $L(t + 1) = 2t + 3$ and $L(t - 1) = 3t - 2$.

(a) Find $L(6t - 4)$.

(b) Find $L(at + b)$.

21. Let V and W be vector spaces. Prove that the function $O: V \to W$ defined by $O(\mathbf{v}) = \mathbf{0}_W$ is a linear transformation, which is called the **zero linear transformation**.

22. Let $I: V \to V$ be defined by $I(\mathbf{v}) = \mathbf{v}$, for $\mathbf{v}$ in V. Show that I is a linear transformation, which is called the **identity operator** on V.

23. Let $L: M_{22} \to R^1$ be defined by

$$L\left(\begin{bmatrix} a & b \\ c & d \end{bmatrix}\right) = a + d.$$

Is L a linear transformation?

24. ***(Calculus Required)*** Let W be the vector space of all real-valued functions and let V be the subspace of all differentiable functions. Define $L: V \to W$ by $L(f) = f'$, where f' is the derivative of f. Prove that L is a linear transformation.

25. Let $V = C[a, b]$ be the vector space of all real-valued functions that are integrable over the interval $[a, b]$. Let $W = R^1$. Define $L: V \to W$ by $L(f) = \int_a^b f(x)\,dx$. Prove that L is a linear transformation.

26. Let A be an $n \times n$ matrix and suppose that $L: M_{nn} \to M_{nn}$ is defined by $L(X) = AX$, for X in M_{nn}. Show that L is a linear transformation.

27. Let $L: M_{nn} \to R^1$ be defined by $L(A) = a_{11}a_{22}\cdots a_{nn}$, for an $n \times n$ matrix $A = \begin{bmatrix} a_{ij} \end{bmatrix}$. Is L a linear transformation?

28. Let $T: V \to W$ be the function defined by $T(\mathbf{v}) = \mathbf{v}+\mathbf{b}$, for $\mathbf{v}$ in V, where $\mathbf{b}$ is a fixed nonzero vector in V. T is called a **translation** by vector $\mathbf{v}$. Is T a linear transformation? Explain.

29. Show that the function L defined in Example 8 is a linear transformation.

30. For the linear transformation defined in Example 10, find $L\left(\begin{bmatrix} a & b & c & d \end{bmatrix}\right)$.

31. Let V be an n-dimensional vector space with ordered basis $S = \{\mathbf{v}_1, \mathbf{v}_2, \ldots, \mathbf{v}_n\}$ and let $T = \{\mathbf{w}_1, \mathbf{w}_2, \ldots, \mathbf{w}_n\}$ be an ordered set of vectors in V. Prove that there is a unique linear transformation $L: V \to V$ such that $L(\mathbf{v}_i) = \mathbf{w}_i$ for $i = 1, 2, \ldots, n$. [*Hint*: Let L be a mapping from V into V such that $L(\mathbf{v}_i) = \mathbf{w}_i$; then show how to extend L to be a linear transformation defined on all of V.]

32. Let $L: V \to W$ be a linear transformation from a vector space V into a vector space W. The **image** of a subspace V_1 of V is defined as

$$L(V_1) = \{\mathbf{w} \text{ in } W \mid \mathbf{w} = L(\mathbf{v}) \text{ for some } \mathbf{v} \text{ in } V\}.$$

Show that $L(V_1)$ is a subspace of V.

33. Let L_1 and L_2 be linear transformations from a vector space V into a vector space W. Let $\{\mathbf{v}_1, \mathbf{v}_2, \ldots, \mathbf{v}_n\}$ be a basis for V. Show that if $L_1(\mathbf{v}_i) = L_2(\mathbf{v}_i)$ for $i = 1, 2, \ldots, n$, then $L_1(\mathbf{v}) = L_2(\mathbf{v})$ for any $\mathbf{v}$ in V.

34. Let $L: V \to W$ be a linear transformation from a vector space V into a vector space W. The **preimage** of a subspace W_1 of W is defined as

$$L^{-1}(W_1) = \{\mathbf{v} \text{ in } V \mid L(\mathbf{v}) \text{ is in } W_1\}.$$

Show that $L^{-1}(W_1)$ is a subspace of V.

35. Let $O: R^n \to R^n$ be the zero linear transformation defined by $O(\mathbf{v}) = \mathbf{0}$ for $\mathbf{v}$ in R^n (see Exercise 21). Find the standard matrix representing O.

36. Let $I: R^n \to R^n$ be the identity linear transformation defined by $I(\mathbf{v}) = \mathbf{v}$ for $\mathbf{v}$ in R^n (see Exercise 22). Find the standard matrix representing I.

37. Complete the proof of Theorem 6.3 by showing that the matrix A is unique. (*Hint*: Suppose that there is another matrix B such that $L(\mathbf{x}) = B\mathbf{x}$ for $\mathbf{x}$ in R^n. Consider $L(\mathbf{e}_j)$ for $j = 1, 2, \ldots, n$. Show that $A = B$.)

38. Use the substitution scheme and encoding matrix A of Example 12.

(a) Code the message SEND HIM MONEY.

(b) Decode the message 67 44 41 49 39 19 113 76 62 104 69 55.

39. Use the substitution scheme of Example 12 and the matrix

$$A = \begin{bmatrix} 5 & 3 \\ 2 & 1 \end{bmatrix}.$$

(a) Code the message WORK HARD.

(b) Decode the message

93 36 60 21 159 60 110 43

6.2 Kernel and Range of a Linear Transformation

In this section we study special types of linear transformations; we formulate the notions of one-to-one linear transformations and onto linear transformations. We also develop methods for determining when a linear transformation is one-to-one or onto, and examine some applications of these notions.

DEFINITION 6.2

A linear transformation $L\colon V \to W$ is called **one-to-one** if it is a one-to-one function; that is, if $\mathbf{v}_1 \neq \mathbf{v}_2$ implies that $L(\mathbf{v}_1) \neq L(\mathbf{v}_2)$. An equivalent statement is that L is one-to-one if $L(\mathbf{v}_1) = L(\mathbf{v}_2)$ implies that $\mathbf{v}_1 = \mathbf{v}_2$. (See Figure A.2 in Appendix A.)

EXAMPLE 1

Let $L\colon R^2 \to R^2$ be defined by

$$L\left(\begin{bmatrix} u_1 \\ u_2 \end{bmatrix}\right) = \begin{bmatrix} u_1 + u_2 \\ u_1 - u_2 \end{bmatrix}.$$

To determine whether L is one-to-one, we let

$$\mathbf{v}_1 = \begin{bmatrix} u_1 \\ u_2 \end{bmatrix} \quad \text{and} \quad \mathbf{v}_2 = \begin{bmatrix} v_1 \\ v_2 \end{bmatrix}.$$

Then, if $L(\mathbf{v}_1) = L(\mathbf{v}_2)$, we have

$$\begin{aligned} u_1 + u_2 &= v_1 + v_2 \\ u_1 - u_2 &= v_1 - v_2. \end{aligned}$$

Adding these equations, we obtain $2u_1 = 2v_1$, or $u_1 = v_1$, which implies that $u_2 = v_2$. Hence $\mathbf{v}_1 = \mathbf{v}_2$, and L is one-to-one. ■

EXAMPLE 2

Let $L\colon R^3 \to R^2$ be the linear transformation (a projection) defined by

$$L\left(\begin{bmatrix} u_1 \\ u_2 \\ u_3 \end{bmatrix}\right) = \begin{bmatrix} u_1 \\ u_2 \end{bmatrix}.$$

Since $L\left(\begin{bmatrix} 1 \\ 3 \\ 3 \end{bmatrix}\right) = L\left(\begin{bmatrix} 1 \\ 3 \\ -2 \end{bmatrix}\right)$, yet $\begin{bmatrix} 1 \\ 3 \\ 3 \end{bmatrix} \neq \begin{bmatrix} 1 \\ 3 \\ -2 \end{bmatrix}$, we conclude that L is not one-to-one. ■

We shall now develop some more efficient ways of determining whether a linear transformation is one-to-one.

DEFINITION 6.3 Let $L: V \to W$ be a linear transformation of a vector space V into a vector space W. The **kernel** of L, $\ker L$, is the subset of V consisting of all elements $\mathbf{v}$ of V such that $L(\mathbf{v}) = \mathbf{0}_W$.

We observe that Theorem 6.1 assures us that $\ker L$ is never an empty set, because if $L: V \to W$ is a linear transformation, then $\mathbf{0}_V$ is in $\ker L$.

EXAMPLE 3 Let $L: R^3 \to R^2$ be as defined in Example 2. The vector $\begin{bmatrix} 0 \\ 0 \\ 2 \end{bmatrix}$ is in $\ker L$, since $L\left(\begin{bmatrix} 0 \\ 0 \\ 2 \end{bmatrix}\right) = \begin{bmatrix} 0 \\ 0 \end{bmatrix}$. However, the vector $\begin{bmatrix} 2 \\ -3 \\ 4 \end{bmatrix}$ is not in $\ker L$, since $L\left(\begin{bmatrix} 2 \\ -3 \\ 4 \end{bmatrix}\right) = \begin{bmatrix} 2 \\ -3 \end{bmatrix}$. To find $\ker L$, we must determine all $\mathbf{v}$ in R^3 so that $L(\mathbf{v}) = \mathbf{0}_{R^2}$. That is, we seek $\mathbf{v} = \begin{bmatrix} v_1 \\ v_2 \\ v_3 \end{bmatrix}$ so that

$$L(\mathbf{v}) = L\left(\begin{bmatrix} v_1 \\ v_2 \\ v_3 \end{bmatrix}\right) = \begin{bmatrix} 0 \\ 0 \end{bmatrix} = \mathbf{0}_{R^2}.$$

However, $L(\mathbf{v}) = \begin{bmatrix} v_1 \\ v_2 \end{bmatrix}$. Thus $\begin{bmatrix} v_1 \\ v_2 \end{bmatrix} = \begin{bmatrix} 0 \\ 0 \end{bmatrix}$, so $v_1 = 0$, $v_2 = 0$, and v_3 can be any real number. Hence $\ker L$ consists of all vectors in R^3 of the form $\begin{bmatrix} 0 \\ 0 \\ a \end{bmatrix}$, where a is any real number. It is clear that $\ker L$ consists of the z-axis in (x, y, z) three-dimensional space R^3. ■

An examination of the elements in $\ker L$ allows us to decide whether L is or is not one-to-one.

Theorem 6.4 Let $L: V \to W$ be a linear transformation of a vector space V into a vector space W. Then

(a) $\ker L$ is a subspace of V.

(b) L is one-to-one if and only if $\ker L = \{\mathbf{0}_V\}$.

Proof

(a) We show that if $\mathbf{v}$ and $\mathbf{w}$ are in $\ker L$, then so are $\mathbf{v} + \mathbf{w}$ and $c\mathbf{v}$ for any real number c. If $\mathbf{v}$ and $\mathbf{w}$ are in $\ker L$, then $L(\mathbf{v}) = \mathbf{0}_W$, and $L(\mathbf{w}) = \mathbf{0}_W$.

Then, since L is a linear transformation,

$$L(\mathbf{v}+\mathbf{w}) = L(\mathbf{v}) + L(\mathbf{w}) = \mathbf{0}_W + \mathbf{0}_W = \mathbf{0}_W.$$

Thus $\mathbf{v}+\mathbf{w}$ is in $\ker L$. Also,

$$L(c\mathbf{v}) = cL(\mathbf{v}) = c\,\mathbf{0}_W = \mathbf{0},$$

so $c\mathbf{v}$ is in $\ker L$. Hence $\ker L$ is a subspace of V.

(b) Let L be one-to-one. We show that $\ker L = \{\mathbf{0}_V\}$. Let $\mathbf{v}$ be in $\ker L$. Then $L(\mathbf{v}) = \mathbf{0}_W$. Also, we already know that $L(\mathbf{0}_V) = \mathbf{0}_W$. Then $L(\mathbf{v}) = L(\mathbf{0}_V)$. Since L is one-to-one, we conclude that $\mathbf{v} = \mathbf{0}_V$. Hence $\ker L = \{\mathbf{0}_V\}$.

Conversely, suppose that $\ker L = \{\mathbf{0}_V\}$. We wish to show that L is one-to-one. Let $L(\mathbf{v}_1) = L(\mathbf{v}_2)$ for $\mathbf{v}_1$ and $\mathbf{v}_2$ in V. Then

$$L(\mathbf{v}_1) - L(\mathbf{v}_2) = \mathbf{0}_W,$$

so that $L(\mathbf{v}_1 - \mathbf{v}_2) = \mathbf{0}_W$. This means that $\mathbf{v}_1 - \mathbf{v}_2$ is in $\ker L$, so $\mathbf{v}_1 - \mathbf{v}_2 = \mathbf{0}_V$. Hence $\mathbf{v}_1 = \mathbf{v}_2$, and L is one-to-one. ■

Note that we can also state Theorem 6.4(b) as follows: *L is one-to-one if and only if* $\dim \ker L = 0$.

The proof of Theorem 6.4 has also established the following result, which we state as Corollary 6.1:

Corollary 6.1 If $L(\mathbf{x}) = \mathbf{b}$ and $L(\mathbf{y}) = \mathbf{b}$, then $\mathbf{x} - \mathbf{y}$ belongs to $\ker L$. In other words, any two solutions to $L(\mathbf{x}) = \mathbf{b}$ differ by an element of the kernel of L.

Proof

Exercise 29. ■

EXAMPLE 4 ***(Calculus Required)*** Let $L\colon P_2 \to R$ be the linear transformation defined by

$$L(at^2 + bt + c) = \int_0^1 (at^2 + bt + c)\,dt.$$

(a) Find $\ker L$.

(b) Find $\dim \ker L$.

(c) Is L one-to-one?

Solution

(a) To find $\ker L$, we seek an element $\mathbf{v} = at^2 + bt + c$ in P_2 such that $L(\mathbf{v}) = L(at^2 + bt + c) = \mathbf{0}_R = 0$. Now

$$L(\mathbf{v}) = \left.\frac{at^3}{3} + \frac{bt^2}{2} + ct\right|_0^1 = \frac{a}{3} + \frac{b}{2} + c.$$

Thus $c = -a/3 - b/2$. Then $\ker L$ consists of all polynomials in P_2 of the form $at^2 + bt + (-a/3 - b/2)$, for a and b any real numbers.

(b) To find the dimension of $\ker L$, we obtain a basis for $\ker L$. Any vector in $\ker L$ can be written as

$$at^2 + bt + \left(-\frac{a}{3} - \frac{b}{2}\right) = a\left(t^2 - \frac{1}{3}\right) + b\left(t - \frac{1}{2}\right).$$

Thus the elements $\left(t^2 - \frac{1}{3}\right)$ and $\left(t - \frac{1}{2}\right)$ in P_2 span $\ker L$. Now, these elements are also linearly independent, since they are not constant multiples of each other. Thus $\left\{t^2 - \frac{1}{3}, t - \frac{1}{2}\right\}$ is a basis for $\ker L$, and $\dim \ker L = 2$.

(c) Since $\dim \ker L = 2$, L is not one-to-one. ■

DEFINITION 6.4 If $L: V \to W$ is a linear transformation of a vector space V into a vector space W, then the **range** of L or **image** of V under L, denoted by range L, consists of all those vectors in W that are images under L of vectors in V. Thus $\mathbf{w}$ is in range L if there exists some vector $\mathbf{v}$ in V such that $L(\mathbf{v}) = \mathbf{w}$. The linear transformation L is called **onto** if range $L = W$.

Theorem 6.5 If $L: V \to W$ is a linear transformation of a vector space V into a vector space W, then range L is a subspace of W.

Proof

Let $\mathbf{w}_1$ and $\mathbf{w}_2$ be in range L. Then $\mathbf{w}_1 = L(\mathbf{v}_1)$ and $\mathbf{w}_2 = L(\mathbf{v}_2)$ for some $\mathbf{v}_1$ and $\mathbf{v}_2$ in V. Now

$$\mathbf{w}_1 + \mathbf{w}_2 = L(\mathbf{v}_1) + L(\mathbf{v}_2) = L(\mathbf{v}_1 + \mathbf{v}_2),$$

which implies that $\mathbf{w}_1 + \mathbf{w}_2$ is in range L. Also, if $\mathbf{w}$ is in range L, then $\mathbf{w} = L(\mathbf{v})$ for some $\mathbf{v}$ in V. Then $c\mathbf{w} = cL(\mathbf{v}) = L(c\mathbf{v})$ where c is a scalar, so that $c\mathbf{w}$ is in range L. Hence range L is a subspace of W. ■

EXAMPLE 5 Consider Example 2 of this section again. Is the projection L onto?

Solution

We choose any vector $\mathbf{w} = \begin{bmatrix} c \\ d \end{bmatrix}$ in R^2 and seek a vector $\mathbf{v} = \begin{bmatrix} v_1 \\ v_2 \\ v_3 \end{bmatrix}$ in V such that $L(\mathbf{v}) = \mathbf{w}$. Now $L(\mathbf{v}) = \begin{bmatrix} v_1 \\ v_2 \end{bmatrix}$, so if $v_1 = c$ and $v_2 = d$, then $L(\mathbf{v}) = \mathbf{w}$. Therefore, L is onto and $\dim \text{range } L = 2$. ■

EXAMPLE 6 Consider Example 4 of this section; is L onto?

Solution

Given a vector $\mathbf{w}$ in R, $\mathbf{w} = r$, a real number, can we find a vector $\mathbf{v} = at^2 + bt + c$ in P_2 so that $L(\mathbf{v}) = \mathbf{w} = r$?

Now

$$L(\mathbf{v}) = \int_0^1 (at^2 + bt + c)\,dt = \frac{a}{3} + \frac{b}{2} + c.$$

We can let $a = b = 0$ and $c = r$. Hence L is onto. Moreover, $\dim \text{range } L = 1$. ■

EXAMPLE 7

Let $L\colon R^3 \to R^3$ be defined by

$$L\left(\begin{bmatrix} u_1 \\ u_2 \\ u_3 \end{bmatrix}\right) = \begin{bmatrix} 1 & 0 & 1 \\ 1 & 1 & 2 \\ 2 & 1 & 3 \end{bmatrix} \begin{bmatrix} u_1 \\ u_2 \\ u_3 \end{bmatrix}.$$

(a) Is L onto?
(b) Find a basis for range L.
(c) Find ker L.
(d) Is L one-to-one?

Solution

(a) Given any $\mathbf{w} = \begin{bmatrix} a \\ b \\ c \end{bmatrix}$ in R^3, where a, b, and c are any real numbers, can we find $\mathbf{v} = \begin{bmatrix} v_1 \\ v_2 \\ v_3 \end{bmatrix}$ so that $L(\mathbf{v}) = \mathbf{w}$? We seek a solution to the linear system

$$\begin{bmatrix} 1 & 0 & 1 \\ 1 & 1 & 2 \\ 2 & 1 & 3 \end{bmatrix} \begin{bmatrix} v_1 \\ v_2 \\ v_3 \end{bmatrix} = \begin{bmatrix} a \\ b \\ c \end{bmatrix},$$

and we find the reduced row echelon form of the augmented matrix to be (verify)

$$\left[\begin{array}{ccc|c} 1 & 0 & 1 & a \\ 0 & 1 & 1 & b-a \\ 0 & 0 & 0 & c-b-a \end{array}\right].$$

Thus a solution exists only for $c - b - a = 0$, so L is not onto.

(b) To find a basis for range L, we note that

$$L\left(\begin{bmatrix} v_1 \\ v_2 \\ v_3 \end{bmatrix}\right) = \begin{bmatrix} 1 & 0 & 1 \\ 1 & 1 & 2 \\ 2 & 1 & 3 \end{bmatrix} \begin{bmatrix} v_1 \\ v_2 \\ v_3 \end{bmatrix} = \begin{bmatrix} v_1 + v_3 \\ v_1 + v_2 + 2v_3 \\ 2v_1 + v_2 + 3v_3 \end{bmatrix}$$

$$= v_1 \begin{bmatrix} 1 \\ 1 \\ 2 \end{bmatrix} + v_2 \begin{bmatrix} 0 \\ 1 \\ 1 \end{bmatrix} + v_3 \begin{bmatrix} 1 \\ 2 \\ 3 \end{bmatrix}.$$

This means that

$$\left\{ \begin{bmatrix} 1 \\ 1 \\ 2 \end{bmatrix}, \begin{bmatrix} 0 \\ 1 \\ 1 \end{bmatrix}, \begin{bmatrix} 1 \\ 2 \\ 3 \end{bmatrix} \right\}$$

spans range L. That is, range L is the subspace of R^3 spanned by the columns of the matrix defining L.

The first two vectors in this set are linearly independent, since they are not constant multiples of each other. The third vector is the sum of the first two. Therefore, the first two vectors form a basis for range L, and $\dim \text{range}\, L = 2$.

(c) To find $\ker L$, we wish to find all $\mathbf{v}$ in R^3 so that $L(\mathbf{v}) = \mathbf{0}_{R^3}$. Solving the resulting homogeneous system, we find (verify) that $v_1 = -v_3$ and $v_2 = -v_3$. Thus $\ker L$ consists of all vectors of the form

$$\begin{bmatrix} -a \\ -a \\ a \end{bmatrix} = a \begin{bmatrix} -1 \\ -1 \\ 1 \end{bmatrix},$$

where a is any real number. Moreover, $\dim \ker L = 1$.

(d) Since $\ker L \neq \{\mathbf{0}_{R^3}\}$, it follows from Theorem 6.4(b) that L is not one-to-one. ■

The problem of finding a basis for $\ker L$ always reduces to the problem of finding a basis for the solution space of a homogeneous system; this latter problem has been solved in Section 4.7.

If range L is a subspace of R^m or R_m, then a basis for range L can be obtained by the method discussed in Theorem 4.9 or by the procedure given in Section 4.9. Both approaches are illustrated in the next example.

EXAMPLE 8

Let $L\colon R_4 \to R_3$ be defined by

$$L\left(\begin{bmatrix} u_1 & u_2 & u_3 & u_4 \end{bmatrix}\right) = \begin{bmatrix} u_1 + u_2 & u_3 + u_4 & u_1 + u_3 \end{bmatrix}.$$

Find a basis for range L.

Solution

We have

$$L\left(\begin{bmatrix} u_1 & u_2 & u_3 & u_4 \end{bmatrix}\right) = u_1\begin{bmatrix} 1 & 0 & 1 \end{bmatrix} + u_2\begin{bmatrix} 1 & 0 & 0 \end{bmatrix} + u_3\begin{bmatrix} 0 & 1 & 1 \end{bmatrix} + u_4\begin{bmatrix} 0 & 1 & 0 \end{bmatrix}.$$

Thus

$$S = \left\{\begin{bmatrix} 1 & 0 & 1 \end{bmatrix}, \begin{bmatrix} 1 & 0 & 0 \end{bmatrix}, \begin{bmatrix} 0 & 1 & 1 \end{bmatrix}, \begin{bmatrix} 0 & 1 & 0 \end{bmatrix}\right\}$$

spans range L. To find a subset of S that is a basis for range L, we proceed as in Theorem 4.9 by first writing

$$u_1\begin{bmatrix} 1 & 0 & 1 \end{bmatrix} + u_2\begin{bmatrix} 1 & 0 & 0 \end{bmatrix} + u_3\begin{bmatrix} 0 & 1 & 1 \end{bmatrix} + u_4\begin{bmatrix} 0 & 1 & 0 \end{bmatrix} = \begin{bmatrix} 0 & 0 & 0 \end{bmatrix}.$$

The reduced row echelon form of the augmented matrix of this homogeneous system is (verify)

$$\left[\begin{array}{cccc|c} 1 & 0 & 0 & -1 & 0 \\ 0 & 1 & 0 & 1 & 0 \\ 0 & 0 & 1 & 1 & 0 \end{array}\right].$$

Since the leading 1's appear in columns 1, 2, and 3, we conclude that the first three vectors in S form a basis for range L. Thus

$$\left\{\begin{bmatrix} 1 & 0 & 1 \end{bmatrix}, \begin{bmatrix} 1 & 0 & 0 \end{bmatrix}, \begin{bmatrix} 0 & 1 & 1 \end{bmatrix}\right\}$$

is a basis for range L.

Alternatively, we may proceed as in Section 4.9 to form the matrix whose rows are the given vectors

$$\begin{bmatrix} 1 & 0 & 1 \\ 1 & 0 & 0 \\ 0 & 1 & 1 \\ 0 & 1 & 0 \end{bmatrix}.$$

Transforming this matrix to reduced row echelon form, we get (verify)

$$\begin{bmatrix} 1 & 0 & 0 \\ 0 & 1 & 0 \\ 0 & 0 & 1 \\ 0 & 0 & 0 \end{bmatrix}.$$

Hence $\{[1 \quad 0 \quad 0], [0 \quad 1 \quad 0], [0 \quad 0 \quad 1]\}$ is a basis for range L. ■

To determine if a linear transformation is one-to-one or onto, we must solve a linear system. This is one further demonstration of the frequency with which linear systems must be solved to answer many questions in linear algebra. Finally, from Example 7, where $\dim \ker L = 1$, $\dim \operatorname{range} L = 2$, and $\dim \operatorname{domain} L = 3$, we saw that

$$\dim \ker L + \dim \operatorname{range} L = \dim \operatorname{domain} L.$$

This very important result is always true, and we now prove it in the following theorem:

Theorem 6.6 If $L\colon V \to W$ is a linear transformation of an n-dimensional vector space V into a vector space W, then

$$\dim \ker L + \dim \operatorname{range} L = \dim V.$$

Proof

Let $k = \dim \ker L$. If $k = n$, then $\ker L = V$ (Exercise 42, Section 4.6), which implies that $L(\mathbf{v}) = \mathbf{0}_W$ for every $\mathbf{v}$ in V. Hence $\operatorname{range} L = \{\mathbf{0}_W\}$, $\dim \operatorname{range} L = 0$, and the conclusion holds. Next, suppose that $1 \leq k < n$. We shall prove that $\dim \operatorname{range} L - n - k$. Let $\{\mathbf{v}_1, \mathbf{v}_2, \dots, \mathbf{v}_k\}$ be a basis for $\ker L$. By Theorem 4.11 we can extend this basis to a basis

$$S = \{\mathbf{v}_1, \mathbf{v}_2, \dots, \mathbf{v}_k, \mathbf{v}_{k+1}, \dots, \mathbf{v}_n\}$$

for V. We prove that the set $T = \{L(\mathbf{v}_{k+1}), L(\mathbf{v}_{k+2}), \dots, L(\mathbf{v}_n)\}$ is a basis for range L.

First, we show that T spans range L. Let $\mathbf{w}$ be any vector in range L. Then $\mathbf{w} = L(\mathbf{v})$ for some $\mathbf{v}$ in V. Since S is a basis for V, we can find a unique set of real numbers $a_1, a_2, \dots, a_n$ such that $\mathbf{v} = a_1\mathbf{v}_1 + a_2\mathbf{v}_2 + \cdots + a_n\mathbf{v}_n$. Then

$$\begin{aligned} \mathbf{w} = L(\mathbf{v}) &= L(a_1\mathbf{v}_1 + a_2\mathbf{v}_2 + \cdots + a_k\mathbf{v}_k + a_{k+1}\mathbf{v}_{k+1} + \cdots + a_n\mathbf{v}_n) \\ &= a_1L(\mathbf{v}_1) + a_2L(\mathbf{v}_2) + \cdots + a_kL(\mathbf{v}_k) + a_{k+1}L(\mathbf{v}_{k+1}) + \cdots + a_nL(\mathbf{v}_n) \\ &= a_{k+1}L(\mathbf{v}_{k+1}) + \cdots + a_nL(\mathbf{v}_n) \end{aligned}$$

because $\mathbf{v}_1, \mathbf{v}_2, \dots, \mathbf{v}_k$ are in $\ker L$. Hence T spans range L.

Now we show that T is linearly independent. Suppose that

$$a_{k+1}L(\mathbf{v}_{k+1}) + a_{k+2}L(\mathbf{v}_{k+2}) + \cdots + a_n L(\mathbf{v}_n) = \mathbf{0}_W.$$

Then

$$L(a_{k+1}\mathbf{v}_{k+1} + a_{k+2}\mathbf{v}_{k+2} + \cdots + a_n\mathbf{v}_n) = \mathbf{0}_W.$$

Hence the vector $a_{k+1}\mathbf{v}_{k+1} + a_{k+2}\mathbf{v}_{k+2} + \cdots + a_n\mathbf{v}_n$ is in $\ker L$, and we can write

$$a_{k+1}\mathbf{v}_{k+1} + a_{k+2}\mathbf{v}_{k+2} + \cdots + a_n\mathbf{v}_n = b_1\mathbf{v}_1 + b_2\mathbf{v}_2 + \cdots + b_k\mathbf{v}_k,$$

where $b_1, b_2, \ldots, b_k$ are uniquely determined real numbers. We then have

$$b_1\mathbf{v}_1 + b_2\mathbf{v}_2 + \cdots + b_k\mathbf{v}_k - a_{k+1}\mathbf{v}_{k+1} - a_{k+2}\mathbf{v}_{k+2} - \cdots - a_n\mathbf{v}_n = \mathbf{0}_V.$$

Since S is linearly independent, we find that

$$b_1 = b_2 = \cdots = b_k = a_{k+1} = \cdots = a_n = 0.$$

Hence T is linearly independent and forms a basis for range L.

If $k = 0$, then $\ker L$ has no basis; we let $\{\mathbf{v}_1, \mathbf{v}_2, \ldots, \mathbf{v}_n\}$ be a basis for V. The proof now proceeds as previously. ■

The dimension of $\ker L$ is also called the **nullity** of L. In Section 6.5 we define the rank of L and show that it is equal to $\dim \text{range } L$. With this terminology, the conclusion of Theorem 6.6 is very similar to that of Theorem 4.19. This is not a coincidence, since in the next section we show how to attach a unique $m \times n$ matrix to L, whose properties reflect those of L.

The following example illustrates Theorem 6.6 graphically:

EXAMPLE 9

Let $L\colon R^3 \to R^3$ be the linear transformation defined by

$$L\left(\begin{bmatrix} u_1 \\ u_2 \\ u_3 \end{bmatrix}\right) = \begin{bmatrix} u_1 + u_3 \\ u_1 + u_2 \\ u_2 - u_3 \end{bmatrix}.$$

A vector $\begin{bmatrix} u_1 \\ u_2 \\ u_3 \end{bmatrix}$ is in $\ker L$ if

$$L\left(\begin{bmatrix} u_1 \\ u_2 \\ u_3 \end{bmatrix}\right) = \begin{bmatrix} 0 \\ 0 \\ 0 \end{bmatrix}.$$

We must then find a basis for the solution space of the homogeneous system

$$\begin{aligned} u_1 + \phantom{u_2 -{}} u_3 &= 0 \\ u_1 + u_2 \phantom{{}- u_3} &= 0 \\ u_2 - u_3 &= 0. \end{aligned}$$

We find (verify) that a basis for $\ker L$ is $\left\{\begin{bmatrix} -1 \\ 1 \\ 1 \end{bmatrix}\right\}$, so $\dim \ker L = 1$, and $\ker L$ is a line through the origin.

Next, every vector in range L is of the form

$$\begin{bmatrix} u_1 + u_3 \\ u_1 + u_2 \\ u_2 - u_3 \end{bmatrix},$$

which can be written as

$$u_1 \begin{bmatrix} 1 \\ 1 \\ 0 \end{bmatrix} + u_2 \begin{bmatrix} 0 \\ 1 \\ 1 \end{bmatrix} + u_3 \begin{bmatrix} 1 \\ 0 \\ -1 \end{bmatrix}.$$

Then a basis for range L is

$$\left\{ \begin{bmatrix} 1 \\ 1 \\ 0 \end{bmatrix}, \begin{bmatrix} 0 \\ 1 \\ 1 \end{bmatrix} \right\}$$

(explain), so $\dim \text{range}\, L = 2$ and range L is a plane passing through the origin. These results are illustrated in Figure 6.1. Moreover,

$$\dim R^3 = 3 = \dim \ker L + \dim \text{range}\, L = 1 + 2,$$

verifying Theorem 6.6. ■

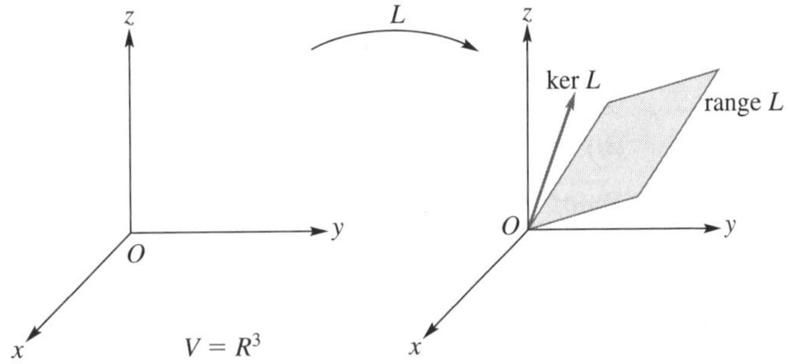

FIGURE 6.1

EXAMPLE 10 Let $L: P_2 \to P_2$ be the linear transformation defined by

$$L(at^2 + bt + c) = (a + 2b)t + (b + c).$$

(a) Find a basis for ker L.
(b) Find a basis for range L.
(c) Verify Theorem 6.6.

Solution

(a) The vector $at^2 + bt + c$ is in ker L if

$$L(at^2 + bt + c) = \mathbf{0},$$

that is, if

$$(a + 2b)t + (b + c) = 0.$$

Then

$$\begin{aligned} a + 2b \quad\quad &= 0 \\ b + c &= 0. \end{aligned}$$

Transforming the augmented matrix of this linear system to reduced row echelon form, we find that a basis for the solution space, and then a basis for $\ker L$, is (verify)

$$\{2t^2 - t + 1\}.$$

(b) Every vector in range L has the form

$$(a + 2b)t + (b + c),$$

so the vectors t and 1 span range L. Since these vectors are also linearly independent, they form a basis for range L.

(c) From (a), $\dim \ker L = 1$, and from (b), $\dim \text{range } L = 2$, so

$$\dim \ker L + \dim \text{range } L = \dim P_2 = 3. \quad \blacksquare$$

We have seen that a linear transformation may be one-to-one and not onto or onto and not one-to-one. However, the following corollary shows that each of these properties implies the other if the vector spaces V and W have the same dimensions:

Corollary 6.2 If $L: V \to W$ is a linear transformation of a vector space V into a vector space W and $\dim V = \dim W$, then the following statements are true:

(a) If L is one-to-one, then it is onto.

(b) If L is onto, then it is one-to-one.

Proof

Exercise 31. ■

A linear transformation $L: V \to W$ of a vector space V into a vector space W is called **invertible** if it is an invertible function—that is, if there exists a unique function $L^{-1}: W \to V$ such that $L \circ L^{-1} = I_W$ and $L^{-1} \circ L = I_V$, where I_V = identity linear transformation on V and I_W = identity linear transformation on W. We now prove the following theorem:

Theorem 6.7 A linear transformation $L: V \to W$ is invertible if and only if L is one-to-one and onto. Moreover, L^{-1} is a linear transformation and $(L^{-1})^{-1} = L$.

Proof

Let L be one-to-one and onto. We define a function $H: W \to V$ as follows. If $\mathbf{w}$ is in W, then since L is onto, $\mathbf{w} = L(\mathbf{v})$ for some $\mathbf{v}$ in V, and since L is one-to-one, $\mathbf{v}$ is unique. Let $H(\mathbf{w}) = \mathbf{v}$; H is a function and $L(H(\mathbf{w})) = L(\mathbf{v}) = \mathbf{w}$, so $L \circ H = I_W$. Also, $H(L(\mathbf{v})) = H(\mathbf{w}) = \mathbf{v}$, so $H \circ L = I_V$. Thus H is an inverse of L. Now H is unique, for if $H_1: W \to V$ is a function such that $L \circ H_1 = I_W$ and $H_1 \circ L = I_V$, then $L(H(\mathbf{w})) = \mathbf{w} = L(H_1(\mathbf{w}))$ for any $\mathbf{w}$ in W. Since L is one-to-one, we conclude that $H(\mathbf{w}) = H_1(\mathbf{w})$. Hence $H = H_1$. Thus $H = L^{-1}$, and L is invertible.

Conversely, let L be invertible; that is, $L \circ L^{-1} = I_W$ and $L^{-1} \circ L = I_V$. We show that L is one-to-one and onto. Suppose that $L(\mathbf{v}_1) = L(\mathbf{v}_2)$ for $\mathbf{v}_1$, $\mathbf{v}_2$ in V. Then $L^{-1}(L(\mathbf{v}_1)) = L^{-1}(L(\mathbf{v}_2))$, so $\mathbf{v}_1 = \mathbf{v}_2$, which means that L is one-to-one. Also, if $\mathbf{w}$ is a vector in W, then $L(L^{-1}(\mathbf{w})) = \mathbf{w}$, so if we let $L^{-1}(\mathbf{w}) = \mathbf{v}$, then $L(\mathbf{v}) = \mathbf{w}$. Thus L is onto.

We now show that L^{-1} is a linear transformation. Let $\mathbf{w}_1$, $\mathbf{w}_2$ be in W, where $L(\mathbf{v}_1) = \mathbf{w}_1$ and $L(\mathbf{v}_2) = \mathbf{w}_2$ for $\mathbf{v}_1$, $\mathbf{v}_2$ in V. Then, since

$$L(a\mathbf{v}_1 + b\mathbf{v}_2) = aL(\mathbf{v}_1) + bL(\mathbf{v}_2) = a\mathbf{w}_1 + b\mathbf{w}_2 \quad \text{for } a, b \text{ real numbers,}$$

we have

$$L^{-1}(a\mathbf{w}_1 + b\mathbf{w}_2) = a\mathbf{v}_1 + b\mathbf{v}_2 = aL^{-1}(\mathbf{w}_1) + bL^{-1}(\mathbf{w}_2),$$

which implies that L^{-1} is a linear transformation.

Finally, since $L \circ L^{-1} = I_W$, $L^{-1} \circ L = I_V$, and inverses are unique, we conclude that $(L^{-1})^{-1} = L$. ■

Remark If $L: V \to V$ is a linear operator that is one-to-one and onto, then L is an isomorphism. See Definition 4.13 in Section 4.8.

EXAMPLE 11

Consider the linear operator $L: R^3 \to R^3$ defined by

$$L\left(\begin{bmatrix} u_1 \\ u_2 \\ u_3 \end{bmatrix}\right) = \begin{bmatrix} 1 & 1 & 1 \\ 2 & 2 & 1 \\ 0 & 1 & 1 \end{bmatrix} \begin{bmatrix} u_1 \\ u_2 \\ u_3 \end{bmatrix}.$$

Since $\ker L = \{\mathbf{0}\}$ (verify), L is one-to-one, and by Corollary 6.2 it is also onto, so it is invertible. To obtain L^{-1}, we proceed as follows. Since $L^{-1}(\mathbf{w}) = \mathbf{v}$, we must solve $L(\mathbf{v}) = \mathbf{w}$ for $\mathbf{v}$. We have

$$L(\mathbf{v}) = L\left(\begin{bmatrix} v_1 \\ v_2 \\ v_3 \end{bmatrix}\right) = \begin{bmatrix} v_1 + v_2 + v_3 \\ 2v_1 + 2v_2 + v_3 \\ v_2 + v_3 \end{bmatrix} = \mathbf{w} = \begin{bmatrix} w_1 \\ w_2 \\ w_3 \end{bmatrix}.$$

We are then solving the linear system

$$\begin{aligned} v_1 + v_2 + v_3 &= w_1 \\ 2v_1 + 2v_2 + v_3 &= w_2 \\ v_2 + v_3 &= w_3 \end{aligned}$$

for v_1, v_2, and v_3. We find that (verify)

$$\begin{bmatrix} v_1 \\ v_2 \\ v_3 \end{bmatrix} = \mathbf{v} = L^{-1}(\mathbf{w}) = L^{-1}\left(\begin{bmatrix} w_1 \\ w_2 \\ w_3 \end{bmatrix}\right) = \begin{bmatrix} w_1 - w_3 \\ -2w_1 + w_2 + w_3 \\ 2w_1 - w_2 \end{bmatrix}.$$

■

The following useful theorem shows that one-to-one linear transformations preserve linear independence of a set of vectors. Moreover, if this property holds, then L is one-to-one.

Theorem 6.8 A linear transformation $L: V \to W$ is one-to-one if and only if the image of every linearly independent set of vectors in V is a linearly independent set of vectors in W.

Proof

Let $S = \{\mathbf{v}_1, \mathbf{v}_2, \ldots, \mathbf{v}_k\}$ be a linearly independent set of vectors in V and let $T = \{L(\mathbf{v}_1), L(\mathbf{v}_2), \ldots, L(\mathbf{v}_k)\}$. Suppose that L is one-to-one; we show that T is linearly independent. Let

$$a_1 L(\mathbf{v}_1) + a_2 L(\mathbf{v}_2) + \cdots + a_k L(\mathbf{v}_k) = \mathbf{0}_W,$$

where $a_1, a_2, \ldots, a_k$ are real numbers. Then

$$L(a_1\mathbf{v}_1 + a_2\mathbf{v}_2 + \cdots + a_k\mathbf{v}_k) = \mathbf{0}_W = L(\mathbf{0}_V).$$

Since L is one-to-one, we conclude that

$$a_1\mathbf{v}_1 + a_2\mathbf{v}_2 + \cdots + a_k\mathbf{v}_k = \mathbf{0}_V.$$

Now S is linearly independent, so $a_1 = a_2 = \cdots = a_k = 0$. Hence T is linearly independent.

Conversely, suppose that the image of any linearly independent set of vectors in V is a linearly independent set of vectors in W. Now $\{\mathbf{v}\}$, where $\mathbf{v} \neq \mathbf{0}_V$, is a linearly independent set in V. Since the set $\{L(\mathbf{v})\}$ is linearly independent, $L(\mathbf{v}) \neq \mathbf{0}_W$, so $\ker L = \{\mathbf{0}_V\}$, which means that L is one-to-one. ■

It follows from this theorem that *if $L: V \to W$ is a linear transformation and* $\dim V = \dim W$, *then L is one-to-one, and thus invertible, if and only if the image of a basis for V under L is a basis for W.* (See Exercise 18.)

We now make one final remark for a linear system $A\mathbf{x} = \mathbf{b}$, where A is $n \times n$. We again consider the linear transformation $L: R^n \to R^n$ defined by $L(\mathbf{x}) = A\mathbf{x}$, for $\mathbf{x}$ in R^n. If A is a nonsingular matrix, then $\dim \operatorname{range} L = \operatorname{rank} A = n$, so $\dim \ker L = 0$. Thus L is one-to-one and hence onto. This means that the given linear system has a unique solution. (Of course, we already knew this result from other considerations.) Now assume that A is singular. Then $\operatorname{rank} A < n$. This means that $\dim \ker L = n - \operatorname{rank} A > 0$, so L is not one-to-one and not onto. Therefore, there exists a vector $\mathbf{b}$ in R^n, for which the system $A\mathbf{x} = \mathbf{b}$ has no solution. Moreover, since A is singular, $A\mathbf{x} = \mathbf{0}$ has a nontrivial solution $\mathbf{x}_0$. If $A\mathbf{x} = \mathbf{b}$ has a solution $\mathbf{y}$, then $\mathbf{x}_0 + \mathbf{y}$ is a solution to $A\mathbf{x} = \mathbf{b}$ (verify). Thus, for A singular, if a solution to $A\mathbf{x} = \mathbf{b}$ exists, then it is not unique.

The following statements are then equivalent:

1. A is nonsingular.
2. $A\mathbf{x} = \mathbf{0}$ has only the trivial solution.
3. A is row (column) equivalent to I_n.
4. The linear system $A\mathbf{x} = \mathbf{b}$ has a unique solution for every vector $\mathbf{b}$ in R^n.
5. A is a product of elementary matrices.
6. $\det(A) \neq 0$.
7. A has rank n.
8. The rows (columns) of A form a linearly independent set of n vectors in R_n (R^n).
9. The dimension of the solution space of $A\mathbf{x} = \mathbf{0}$ is zero.
10. The linear transformation $L: R^n \to R^n$ defined by $L(\mathbf{x}) = A\mathbf{x}$, for $\mathbf{x}$ in R^n, is one-to-one and onto.

We can summarize the conditions under which a linear transformation L of an n-dimensional vector space V into itself (or, more generally, to an n-dimensional vector space W) is invertible by the following equivalent statements:

1. L is invertible.
2. L is one-to-one.
3. L is onto.

Key Terms

One-to-one
Kernel (ker L)
Range
Image of a linear transformation
Onto
Dimension
Nullity
Invertible linear transformation

6.2 Exercises

1. Let $L: R^2 \to R^2$ be the linear transformation defined by $L\left(\begin{bmatrix} u_1 \\ u_2 \end{bmatrix}\right) = \begin{bmatrix} u_1 \\ 0 \end{bmatrix}$.

   (a) Is $\begin{bmatrix} 0 \\ 2 \end{bmatrix}$ in ker L? **(b)** Is $\begin{bmatrix} 2 \\ 2 \end{bmatrix}$ in ker L?

   (c) Is $\begin{bmatrix} 3 \\ 0 \end{bmatrix}$ in range L? **(d)** Is $\begin{bmatrix} 3 \\ 2 \end{bmatrix}$ in range L?

   (e) Find ker L. **(f)** Find range L.

2. Let $L: R^2 \to R^2$ be the linear operator defined by

$$L\left(\begin{bmatrix} u_1 \\ u_2 \end{bmatrix}\right) = \begin{bmatrix} 1 & 2 \\ 2 & 4 \end{bmatrix}\begin{bmatrix} u_1 \\ u_2 \end{bmatrix}.$$

   (a) Is $\begin{bmatrix} 1 \\ 2 \end{bmatrix}$ in ker L? **(b)** Is $\begin{bmatrix} 2 \\ -1 \end{bmatrix}$ in ker L?

   (c) Is $\begin{bmatrix} 3 \\ 6 \end{bmatrix}$ in range L? **(d)** Is $\begin{bmatrix} 2 \\ 3 \end{bmatrix}$ in range L?

   (e) Find ker L.

   (f) Find a set of vectors spanning range L.

3. Let $L: R_4 \to R_2$ be the linear transformation defined by

$$L\left(\begin{bmatrix} u_1 & u_2 & u_3 & u_4 \end{bmatrix}\right) = \begin{bmatrix} u_1 + u_3 & u_2 + u_4 \end{bmatrix}.$$

   (a) Is $\begin{bmatrix} 2 & 3 & -2 & 3 \end{bmatrix}$ in ker L?

   (b) Is $\begin{bmatrix} 4 & -2 & -4 & 2 \end{bmatrix}$ in ker L?

   (c) Is $\begin{bmatrix} 1 & 2 \end{bmatrix}$ in range L?

(d) Is $\begin{bmatrix} 0 & 0 \end{bmatrix}$ in range L?

(e) Find ker L.

(f) Find a set of vectors spanning range L.

4. Let $L\colon R_2 \to R_3$ be the linear transformation defined by $L\left(\begin{bmatrix} u_1 & u_2 \end{bmatrix}\right) = \begin{bmatrix} u_1 & u_1 + u_2 & u_2 \end{bmatrix}$.

(a) Find ker L.

(b) Is L one-to-one?

(c) Is L onto?

5. Let $L\colon R_4 \to R_3$ be the linear transformation defined by

$$L\left(\begin{bmatrix} u_1 & u_2 & u_3 & u_4 \end{bmatrix}\right) = \begin{bmatrix} u_1 + u_2 & u_3 + u_4 & u_1 + u_3 \end{bmatrix}.$$

(a) Find a basis for ker L.

(b) What is dim ker L?

(c) Find a basis for range L.

(d) What is dim range L?

6. Let $L\colon P_2 \to P_3$ be the linear transformation defined by $L(p(t)) = t^2 p'(t)$.

(a) Find a basis for and the dimension of ker L.

(b) Find a basis for and the dimension of range L.

7. Let $L\colon M_{23} \to M_{33}$ be the linear transformation defined by

$$L(A) = \begin{bmatrix} 2 & -1 \\ 1 & 2 \\ 3 & 1 \end{bmatrix} A \quad \text{for } A \text{ in } M_{23}.$$

(a) Find the dimension of ker L.

(b) Find the dimension of range L.

8. Let $L\colon P_2 \to P_1$ be the linear transformation defined by

$$L(at^2 + bt + c) = (a + b)t + (b - c).$$

(a) Find a basis for ker L.

(b) Find a basis for range L.

9. Let $L\colon P_2 \to R_2$ be the linear transformation defined by $L(at^2 + bt + c) = \begin{bmatrix} a & b \end{bmatrix}$.

(a) Find a basis for ker L.

(b) Find a basis for range L.

10. Let $L\colon M_{22} \to M_{22}$ be the linear transformation defined by

$$L(A) = \begin{bmatrix} 1 & 2 \\ 1 & 1 \end{bmatrix} A - A \begin{bmatrix} 1 & 2 \\ 1 & 1 \end{bmatrix}.$$

(a) Find a basis for ker L.

(b) Find a basis for range L.

11. Let $L\colon M_{22} \to M_{22}$ be the linear operator defined by

$$L\left(\begin{bmatrix} a & b \\ c & d \end{bmatrix}\right) = \begin{bmatrix} a+b & b+c \\ a+d & b+d \end{bmatrix}.$$

(a) Find a basis for ker L.

(b) Find a basis for range L.

12. Let $L\colon V \to W$ be a linear transformation.

(a) Show that dim range $L \le \dim V$.

(b) Prove that if L is onto, then $\dim W \le \dim V$.

13. Verify Theorem 6.6 for the following linear transformations:

(a) $L\colon P_2 \to P_2$ defined by $L(p(t)) = tp'(t)$.

(b) $L\colon R_3 \to R_2$ defined by $L\left(\begin{bmatrix} u_1 & u_2 & u_3 \end{bmatrix}\right) = \begin{bmatrix} u_1 + u_2 & u_1 + u_3 \end{bmatrix}$.

(c) $L\colon M_{22} \to M_{23}$ defined by

$$L(A) = A \begin{bmatrix} 1 & 2 & 3 \\ 2 & 1 & 3 \end{bmatrix}, \quad \text{for } A \text{ in } M_{22}.$$

14. Verify Theorem 6.5 for the linear transformation given in Exercise 11.

15. Let A be an $m \times n$ matrix, and consider the linear transformation $L\colon R^n \to R^m$ defined by $L(\mathbf{x}) = A\mathbf{x}$, for $\mathbf{x}$ in R^n. Show that

$$\text{range } L = \text{column space of } A.$$

16. Let $L\colon R^5 \to R^4$ be the linear transformation defined by

$$L\left(\begin{bmatrix} u_1 \\ u_2 \\ u_3 \\ u_4 \\ u_5 \end{bmatrix}\right) = \begin{bmatrix} 1 & 0 & -1 & 3 & -1 \\ 1 & 0 & 0 & 2 & -1 \\ 2 & 0 & -1 & 5 & -1 \\ 0 & 0 & -1 & 1 & 0 \end{bmatrix} \begin{bmatrix} u_1 \\ u_2 \\ u_3 \\ u_4 \\ u_5 \end{bmatrix}.$$

(a) Find a basis for and the dimension of ker L.

(b) Find a basis for and the dimension of range L.

17. Let $L\colon R_3 \to R_3$ be the linear transformation defined by

$$L(\mathbf{e}_1^T) = L\left(\begin{bmatrix} 1 & 0 & 0 \end{bmatrix}\right) = \begin{bmatrix} 3 & 0 & 0 \end{bmatrix},$$
$$L(\mathbf{e}_2^T) = L\left(\begin{bmatrix} 0 & 1 & 0 \end{bmatrix}\right) = \begin{bmatrix} 1 & 1 & 1 \end{bmatrix},$$

and

$$L(\mathbf{e}_3^T) = L\left(\begin{bmatrix} 0 & 0 & 1 \end{bmatrix}\right) = \begin{bmatrix} 2 & 1 & 1 \end{bmatrix}.$$

Is the set

$$\left\{L(\mathbf{e}_1^T), L(\mathbf{e}_2^T), L(\mathbf{e}_3^T)\right\} = \left\{\begin{bmatrix} 3 & 0 & 0 \end{bmatrix}, \begin{bmatrix} 1 & 1 & 1 \end{bmatrix}, \begin{bmatrix} 2 & 1 & 1 \end{bmatrix}\right\}$$

a basis for R_3?

18. Let $L: V \to W$ be a linear transformation, and let $\dim V = \dim W$. Prove that L is invertible if and only if the image of a basis for V under L is a basis for W.

19. Let $L: R^3 \to R^3$ be defined by

$$L\left(\begin{bmatrix}1\\0\\0\end{bmatrix}\right) = \begin{bmatrix}1\\2\\3\end{bmatrix}, \quad L\left(\begin{bmatrix}0\\1\\0\end{bmatrix}\right) = \begin{bmatrix}0\\1\\1\end{bmatrix},$$

$$L\left(\begin{bmatrix}0\\0\\1\end{bmatrix}\right) = \begin{bmatrix}1\\1\\0\end{bmatrix}.$$

(a) Prove that L is invertible.

(b) Find $L^{-1}\left(\begin{bmatrix}2\\3\\4\end{bmatrix}\right)$.

20. Let $L: V \to W$ be a linear transformation, and let $S = \{\mathbf{v}_1, \mathbf{v}_2, \ldots, \mathbf{v}_n\}$ be a set of vectors in V. Prove that if $T = \{L(\mathbf{v}_1), L(\mathbf{v}_2), \ldots, L(\mathbf{v}_n)\}$ is linearly independent, then so is S. What can we say about the converse?

21. Find the dimension of the solution space for the following homogeneous system:

$$\begin{bmatrix}1 & 2 & 1 & 3\\ 2 & 1 & -1 & 2\\ 1 & 0 & 0 & -1\\ 4 & 1 & -1 & 0\end{bmatrix}\begin{bmatrix}x_1\\x_2\\x_3\\x_4\end{bmatrix} = \begin{bmatrix}0\\0\\0\\0\end{bmatrix}.$$

22. Find a linear transformation $L: R_2 \to R_3$ such that $S = \{\begin{bmatrix}1 & -1 & 2\end{bmatrix}, \begin{bmatrix}3 & 1 & -1\end{bmatrix}\}$ is a basis for range L.

23. Let $L: R^3 \to R^3$ be the linear transformation defined by

$$L\left(\begin{bmatrix}u_1\\u_2\\u_3\end{bmatrix}\right) = \begin{bmatrix}1 & 1 & 1\\ 0 & 1 & 2\\ 1 & 2 & 2\end{bmatrix}\begin{bmatrix}u_1\\u_2\\u_3\end{bmatrix}.$$

(a) Prove that L is invertible.

(b) Find $L^{-1}\left(\begin{bmatrix}u_1\\u_2\\u_3\end{bmatrix}\right)$.

24. Let $L: V \to W$ be a linear transformation. Prove that L is one-to-one if and only if $\dim \operatorname{range} L = \dim V$.

25. Let $L: R^4 \to R^6$ be a linear transformation.

(a) If $\dim \ker L = 2$, what is $\dim \operatorname{range} L$?

(b) If $\dim \operatorname{range} L = 3$, what is $\dim \ker L$?

26. Let $L: V \to R^5$ be a linear transformation.

(a) If L is onto and $\dim \ker L = 2$, what is $\dim V$?

(b) If L is one-to-one and onto, what is $\dim V$?

27. Let L be the linear transformation defined in Exercise 24, Section 6.1. Prove or disprove the following:

(a) L is one-to-one.

(b) L is onto.

28. Let L be the linear transformation defined in Exercise 25, Section 6.1. Prove or disprove the following:

(a) L is one-to-one.

(b) L is onto.

29. Prove Corollary 6.1.

30. Let $L: R^n \to R^m$ be a linear transformation defined by $L(\mathbf{x}) = A\mathbf{x}$, for $\mathbf{x}$ in R^n. Prove that L is onto if and only if $\operatorname{rank} A = m$.

31. Prove Corollary 6.2.

6.3 Matrix of a Linear Transformation

In Section 6.2 we saw that if A is an $m \times n$ matrix, then we can define a linear transformation $L: R^n \to R^m$ by $L(\mathbf{x}) = A\mathbf{x}$ for $\mathbf{x}$ in R^n. We shall now develop the following notion: If $L: V \to W$ is a linear transformation of an n-dimensional vector space V into an m-dimensional vector space W, and if we choose ordered bases for V and W, then we can associate a unique $m \times n$ matrix A with L that will enable us to find $L(\mathbf{x})$ for $\mathbf{x}$ in V by merely performing matrix multiplication.

Theorem 6.9 Let $L: V \to W$ be a linear transformation of an n-dimensional vector space V into an m-dimensional vector space W ($n \neq 0$, $m \neq 0$) and let $S = \{\mathbf{v}_1, \mathbf{v}_2, \ldots, \mathbf{v}_n\}$ and $T = \{\mathbf{w}_1, \mathbf{w}_2, \ldots, \mathbf{w}_m\}$ be ordered bases for V and W, respectively. Then the $m \times n$ matrix A whose jth column is the coordinate vector $\left[L(\mathbf{v}_j)\right]_T$ of $L(\mathbf{v}_j)$

with respect to T has the following property:

$$\left[L(\mathbf{x})\right]_T = A\left[\mathbf{x}\right]_S \qquad \text{for every } \mathbf{x} \text{ in } V. \tag{1}$$

Moreover, A is the only matrix with this property.

Proof

We show how to construct the matrix A. Consider the vector $\mathbf{v}_j$ in V for $j = 1, 2, \ldots, n$. Then $L(\mathbf{v}_j)$ is a vector in W, and since T is an ordered basis for W, we can express this vector as a linear combination of the vectors in T in a unique manner. Thus

$$L(\mathbf{v}_j) = c_{1j}\mathbf{w}_1 + c_{2j}\mathbf{w}_2 + \cdots + c_{mj}\mathbf{w}_m. \tag{2}$$

This means that the coordinate vector of $L(\mathbf{v}_j)$ with respect to T is

$$\left[L(\mathbf{v}_j)\right]_T = \begin{bmatrix} c_{1j} \\ c_{2j} \\ \vdots \\ c_{mj} \end{bmatrix}.$$

Recall from Section 4.8 that, to find the coordinate vector $\left[L(\mathbf{v}_j)\right]_T$, we must solve a linear system. We now define an $m \times n$ matrix A by choosing $\left[L(\mathbf{v}_j)\right]_T$ as the jth column of A and show that this matrix satisfies the properties stated in the theorem.

Let $\mathbf{x}$ be any vector in V. Then $L(\mathbf{x})$ is in W. Now let

$$\left[\mathbf{x}\right]_S = \begin{bmatrix} a_1 \\ a_2 \\ \vdots \\ a_n \end{bmatrix} \quad \text{and} \quad \left[L(\mathbf{x})\right]_T = \begin{bmatrix} b_1 \\ b_2 \\ \vdots \\ b_m \end{bmatrix}.$$

This means that $\mathbf{x} = a_1\mathbf{v}_1 + a_2\mathbf{v}_2 + \cdots + a_n\mathbf{v}_n$. Then

$$\begin{aligned} L(\mathbf{x}) &= a_1L(\mathbf{v}_1) + a_2L(\mathbf{v}_2) + \cdots + a_nL(\mathbf{v}_n) \\ &= a_1(c_{11}\mathbf{w}_1 + c_{21}\mathbf{w}_2 + \cdots + c_{m1}\mathbf{w}_m) \\ &\quad + a_2(c_{12}\mathbf{w}_1 + c_{22}\mathbf{w}_2 + \cdots + c_{m2}\mathbf{w}_m) \\ &\quad + \cdots + a_n(c_{1n}\mathbf{w}_1 + c_{2n}\mathbf{w}_2 + \cdots + c_{mn}\mathbf{w}_m) \\ &= (c_{11}a_1 + c_{12}a_2 + \cdots + c_{1n}a_n)\mathbf{w}_1 + (c_{21}a_1 + c_{22}a_2 + \cdots + c_{2n}a_n)\mathbf{w}_2 \\ &\quad + \cdots + (c_{m1}a_1 + c_{m2}a_2 + \cdots + c_{mn}a_n)\mathbf{w}_m. \end{aligned}$$

Now $L(\mathbf{x}) = b_1\mathbf{w}_1 + b_2\mathbf{w}_2 + \cdots + b_m\mathbf{w}_m$. Hence

$$b_i = c_{i1}a_1 + c_{i2}a_2 + \cdots + c_{in}a_n \quad \text{for } i = 1, 2, \ldots, m.$$

Next, we verify Equation (1). We have

$$A\left[\mathbf{x}\right]_S = \begin{bmatrix} c_{11} & c_{12} & \cdots & c_{1n} \\ c_{21} & c_{22} & \cdots & c_{2n} \\ \vdots & \vdots & & \vdots \\ c_{m1} & c_{m2} & \cdots & c_{mn} \end{bmatrix} \begin{bmatrix} a_1 \\ a_2 \\ \vdots \\ a_n \end{bmatrix}$$

$$= \begin{bmatrix} c_{11}a_1 + c_{12}a_2 + \cdots + c_{1n}a_n \\ c_{21}a_1 + c_{22}a_2 + \cdots + c_{2n}a_n \\ \vdots \\ c_{m1}a_1 + c_{m2}a_2 + \cdots + c_{mn}a_n \end{bmatrix} = \begin{bmatrix} b_1 \\ b_2 \\ \vdots \\ b_m \end{bmatrix} = \left[L(\mathbf{x})\right]_T.$$

Finally, we show that $A = \left[c_{ij}\right]$ is the only matrix with this property. Suppose that we have another matrix $\hat{A} = \left[\hat{c}_{ij}\right]$ with the same properties as A, and that $\hat{A} \neq A$. All the elements of A and $\hat{A}$ cannot be equal, so say that the kth columns of these matrices are unequal. Now the coordinate vector of $\mathbf{v}_k$ with respect to the basis S is

$$\left[\mathbf{v}_k\right]_S = \begin{bmatrix} 0 \\ 0 \\ \vdots \\ 1 \\ 0 \\ \vdots \\ 0 \end{bmatrix} \quad \leftarrow k\text{th row.}$$

Then

$$\left[L(\mathbf{v}_k)\right]_T = A \begin{bmatrix} 0 \\ 0 \\ \vdots \\ 1 \\ 0 \\ \vdots \\ 0 \end{bmatrix} = \begin{bmatrix} a_{1k} \\ a_{2k} \\ \vdots \\ a_{mk} \end{bmatrix} = k\text{th column of } A$$

and

$$\left[L(\mathbf{v}_k)\right]_T = \hat{A} \begin{bmatrix} 0 \\ 0 \\ \vdots \\ 1 \\ 0 \\ \vdots \\ 0 \end{bmatrix} = \begin{bmatrix} \hat{a}_{1k} \\ \hat{a}_{2k} \\ \vdots \\ \hat{a}_{mk} \end{bmatrix} = k\text{th column of } \hat{A}.$$

This means that $L(\mathbf{v}_k)$ has two different coordinate vectors with respect to the same ordered basis, which is impossible. Hence the matrix A is unique. ■

We now summarize the procedure given in Theorem 6.9 for computing the matrix of a linear transformation $L\colon V \to W$ with respect to the ordered bases $S = \{\mathbf{v}_1, \mathbf{v}_2, \ldots, \mathbf{v}_n\}$ and $T = \{\mathbf{w}_1, \mathbf{w}_2, \ldots, \mathbf{w}_m\}$ for V and W, respectively.

Step 1. Compute $L(\mathbf{v}_j)$ for $j = 1, 2, \ldots, n$.

Step 2. Find the coordinate vector $\left[L(\mathbf{v}_j)\right]_T$ of $L(\mathbf{v}_j)$ with respect to T. This means that we have to express $L(\mathbf{v}_j)$ as a linear combination of the vectors in T [see Equation (2)], and this requires the solution of a linear system.

Step 3. The matrix A of the linear transformation L with respect to the ordered bases S and T is formed by choosing, for each j from 1 to n, $\left[L(\mathbf{v}_j)\right]_T$ as the jth column of A.

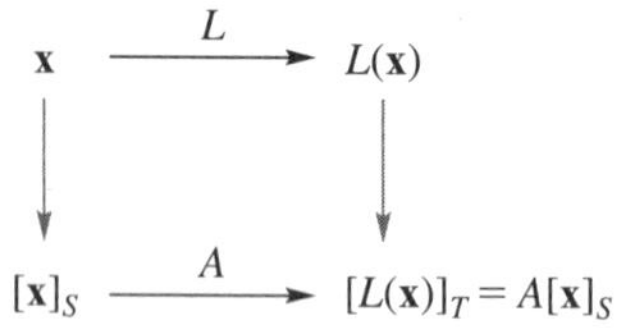

FIGURE 6.2

Figure 6.2 gives a graphical interpretation of Equation (1), that is, of Theorem 6.9. The top horizontal arrow represents the linear transformation L from the n-dimensional vector space V into the m-dimensional vector space W and takes the vector $\mathbf{x}$ in V to the vector $L(\mathbf{x})$ in W. The bottom horizontal line represents the matrix A. Then $\left[L(\mathbf{x})\right]_T$, a coordinate vector in R^m, is obtained simply by multiplying $\left[\mathbf{x}\right]_S$, a coordinate vector in R^n, by the matrix A on the left. We can thus work with matrices rather than with linear transformations.

EXAMPLE 1

(Calculus Required) Let $L\colon P_2 \to P_1$ be defined by $L(p(t)) = p'(t)$, and consider the ordered bases $S = \{t^2, t, 1\}$ and $T = \{t, 1\}$ for P_2 and P_1, respectively.

(a) Find the matrix A associated with L.

(b) If $p(t) = 5t^2 - 3t + 2$, compute $L(p(t))$ directly and then by using A.

Solution

(a) We have

$$L(t^2) = 2t = 2t + 0(1), \qquad \text{so } \left[L(t^2)\right]_T = \begin{bmatrix} 2 \\ 0 \end{bmatrix}.$$

$$L(t) = 1 = 0(t) + 1(1), \qquad \text{so } \left[L(t)\right]_T = \begin{bmatrix} 0 \\ 1 \end{bmatrix}.$$

$$L(1) = 0 = 0(t) + 0(1), \qquad \text{so } \left[L(1)\right]_T = \begin{bmatrix} 0 \\ 0 \end{bmatrix}.$$

In this case, the coordinates of $L(t^2)$, $L(t)$, and $L(1)$ with respect to the T-basis are obtained by observation, since the T-basis is quite simple. Thus

$$A = \begin{bmatrix} 2 & 0 & 0 \\ 0 & 1 & 0 \end{bmatrix}.$$

(b) Since $p(t) = 5t^2 - 3t + 2$, then $L(p(t)) = 10t - 3$. However, we can find $L(p(t))$ by using the matrix A as follows: Since

$$\left[p(t)\right]_S = \begin{bmatrix} 5 \\ -3 \\ 2 \end{bmatrix},$$

then

$$\left[L(p(t))\right]_T = \begin{bmatrix} 2 & 0 & 0 \\ 0 & 1 & 0 \end{bmatrix} \begin{bmatrix} 5 \\ -3 \\ 2 \end{bmatrix} = \begin{bmatrix} 10 \\ -3 \end{bmatrix},$$

which means that $L(p(t)) = 10t - 3$. ■

Remark We observe that Theorem 6.9 states, and Example 1 illustrates, that for a given linear transformation $L: V \to W$, we can obtain the image $L(\mathbf{v})$ of any vector $\mathbf{v}$ in V by a simple matrix multiplication; that is, we multiply the matrix A associated with L by the coordinate vector $\left[\mathbf{v}\right]_S$ of $\mathbf{v}$ with respect to the ordered basis S for V to find the coordinate vector $\left[L(\mathbf{v})\right]_T$ of $L(\mathbf{v})$ with respect to the ordered basis T for W. We then compute $L(\mathbf{v})$, using $\left[L(\mathbf{v})\right]_T$ and T.

EXAMPLE 2

Let $L: P_2 \to P_1$ be defined as in Example 1 and consider the ordered bases $S = \{1, t, t^2\}$ and $T = \{t, 1\}$ for P_2 and P_1, respectively. We then find that the matrix A associated with L is $\begin{bmatrix} 0 & 0 & 2 \\ 0 & 1 & 0 \end{bmatrix}$ (verify). Notice that if we change the order of the vectors in S or T, the matrix may change. ■

EXAMPLE 3

Let $L: P_2 \to P_1$ be defined as in Example 1, and consider the ordered bases $S = \{t^2, t, 1\}$ and $T = \{t + 1, t - 1\}$ for P_2 and P_1, respectively.

(a) Find the matrix A associated with L.
(b) If $p(t) = 5t^2 - 3t + 2$, compute $L(p(t))$.

Solution

(a) We have

$$L(t^2) = 2t.$$

To find the coordinates of $L(t^2)$ with respect to the T-basis, we form

$$L(t^2) = 2t = a_1(t + 1) + a_2(t - 1),$$

which leads to the linear system

$$\begin{aligned} a_1 + a_2 &= 2 \\ a_1 - a_2 &= 0, \end{aligned}$$

whose solution is $a_1 = 1$, $a_2 = 1$ (verify). Hence

$$\left[L(t^2)\right]_T = \begin{bmatrix} 1 \\ 1 \end{bmatrix}.$$

Similarly,

$$L(t) = 1 = \tfrac{1}{2}(t + 1) - \tfrac{1}{2}(t - 1), \qquad \text{so } \left[L(t)\right]_T = \begin{bmatrix} \frac{1}{2} \\ -\frac{1}{2} \end{bmatrix}.$$

$$L(1) = 0 = 0(t + 1) + 0(t - 1), \qquad \text{so } \left[L(1)\right]_T = \begin{bmatrix} 0 \\ 0 \end{bmatrix}.$$

Hence $A = \begin{bmatrix} 1 & \frac{1}{2} & 0 \\ 1 & -\frac{1}{2} & 0 \end{bmatrix}$.

(b) We have

$$\left[L(p(t))\right]_T = \begin{bmatrix} 1 & \frac{1}{2} & 0 \\ 1 & -\frac{1}{2} & 0 \end{bmatrix} \begin{bmatrix} 5 \\ -3 \\ 2 \end{bmatrix} = \begin{bmatrix} \frac{7}{2} \\ \frac{13}{2} \end{bmatrix},$$

so $L(p(t)) = \frac{7}{2}(t+1) + \frac{13}{2}(t-1) = 10t - 3$, which agrees with the result found in Example 1. ■

Notice that the matrices obtained in Examples 1, 2, and 3 are different, even though L is the same in all three examples. In Section 6.5 we discuss the relationship between any two of these three matrices.

The matrix A is called the **representation of L with respect to the ordered bases S and T**. We also say that **A represents L with respect to S and T**. Having A enables us to replace L by A and $\mathbf{x}$ by $\left[\mathbf{x}\right]_S$ to get $A\left[\mathbf{x}\right]_S = \left[L(\mathbf{x})\right]_T$. Thus the result of applying L to $\mathbf{x}$ in V to obtain $L(\mathbf{x})$ in W can be found by multiplying the matrix A by the matrix $\left[\mathbf{x}\right]_S$. That is, we can work with matrices rather than with linear transformations. Physicists and others who deal at great length with linear transformations perform most of their computations with the matrix representations of the linear transformations. Of course, it is easier to work on a computer with matrices than with our abstract definition of a linear transformation. The relationship between linear transformations and matrices is a much stronger one than mere computational convenience. In the next section we show that the set of all linear transformations from an n-dimensional vector space V to an m-dimensional vector space W is a vector space that is isomorphic to the vector space M_{mn} of all $m \times n$ matrices.

We might also mention that if $L\colon R^n \to R^m$ is a linear transformation, then we often use the natural bases for R^n and R^m, which simplifies the task of obtaining a representation of L.

EXAMPLE 4

Let $L\colon R^3 \to R^2$ be defined by

$$L\left(\begin{bmatrix} x_1 \\ x_2 \\ x_3 \end{bmatrix}\right) = \begin{bmatrix} 1 & 1 & 1 \\ 1 & 2 & 3 \end{bmatrix} \begin{bmatrix} x_1 \\ x_2 \\ x_3 \end{bmatrix}.$$

Let

$$\mathbf{e}_1 = \begin{bmatrix} 1 \\ 0 \\ 0 \end{bmatrix}, \quad \mathbf{e}_2 = \begin{bmatrix} 0 \\ 1 \\ 0 \end{bmatrix}, \quad \mathbf{e}_3 = \begin{bmatrix} 0 \\ 0 \\ 1 \end{bmatrix},$$

$$\bar{\mathbf{e}}_1 = \begin{bmatrix} 1 \\ 0 \end{bmatrix}, \quad \text{and} \quad \bar{\mathbf{e}}_2 = \begin{bmatrix} 0 \\ 1 \end{bmatrix}.$$

Then $S = \{\mathbf{e}_1, \mathbf{e}_2, \mathbf{e}_3\}$ and $T = \{\bar{\mathbf{e}}_1, \bar{\mathbf{e}}_2\}$ are the natural bases for R^3 and R^2, respectively.

Now

$$L(\mathbf{e}_1) = \begin{bmatrix} 1 & 1 & 1 \\ 1 & 2 & 3 \end{bmatrix} \begin{bmatrix} 1 \\ 0 \\ 0 \end{bmatrix} = \begin{bmatrix} 1 \\ 1 \end{bmatrix} = 1\overline{\mathbf{e}}_1 + 1\overline{\mathbf{e}}_2, \qquad \text{so } \left[L(\mathbf{e}_1)\right]_T = \begin{bmatrix} 1 \\ 1 \end{bmatrix},$$

$$L(\mathbf{e}_2) = \begin{bmatrix} 1 & 1 & 1 \\ 1 & 2 & 3 \end{bmatrix} \begin{bmatrix} 0 \\ 1 \\ 0 \end{bmatrix} = \begin{bmatrix} 1 \\ 2 \end{bmatrix} = 1\overline{\mathbf{e}}_1 + 2\overline{\mathbf{e}}_2, \qquad \text{so } \left[L(\mathbf{e}_2)\right]_T = \begin{bmatrix} 1 \\ 2 \end{bmatrix},$$

$$L(\mathbf{e}_3) = \begin{bmatrix} 1 & 1 & 1 \\ 1 & 2 & 3 \end{bmatrix} \begin{bmatrix} 0 \\ 0 \\ 1 \end{bmatrix} = \begin{bmatrix} 1 \\ 3 \end{bmatrix} = 1\overline{\mathbf{e}}_1 + 3\overline{\mathbf{e}}_2, \qquad \text{so } \left[L(\mathbf{e}_3)\right]_T = \begin{bmatrix} 1 \\ 3 \end{bmatrix}.$$

In this case, the coordinate vectors of $L(\mathbf{e}_1)$, $L(\mathbf{e}_2)$, and $L(\mathbf{e}_3)$ with respect to the T-basis are readily computed, because T is the natural basis for R^2. Then the representation of L with respect to S and T is

$$A = \begin{bmatrix} 1 & 1 & 1 \\ 1 & 2 & 3 \end{bmatrix}.$$

The reason that A is the same matrix as the one involved in the definition of L is that the natural bases are being used for R^3 and R^2. ■

EXAMPLE 5 Let $L: R^3 \to R^2$ be defined as in Example 4, and consider the ordered bases

$$S = \left\{ \begin{bmatrix} 1 \\ 1 \\ 0 \end{bmatrix}, \begin{bmatrix} 0 \\ 1 \\ 1 \end{bmatrix}, \begin{bmatrix} 0 \\ 0 \\ 1 \end{bmatrix} \right\} \quad \text{and} \quad T = \left\{ \begin{bmatrix} 1 \\ 2 \end{bmatrix}, \begin{bmatrix} 1 \\ 3 \end{bmatrix} \right\}$$

for R^3 and R^2, respectively. Then

$$L\left(\begin{bmatrix} 1 \\ 1 \\ 0 \end{bmatrix} \right) = \begin{bmatrix} 1 & 1 & 1 \\ 1 & 2 & 3 \end{bmatrix} \begin{bmatrix} 1 \\ 1 \\ 0 \end{bmatrix} = \begin{bmatrix} 2 \\ 3 \end{bmatrix}.$$

Similarly,

$$L\left(\begin{bmatrix} 0 \\ 1 \\ 1 \end{bmatrix} \right) = \begin{bmatrix} 2 \\ 5 \end{bmatrix} \quad \text{and} \quad L\left(\begin{bmatrix} 0 \\ 0 \\ 1 \end{bmatrix} \right) = \begin{bmatrix} 1 \\ 3 \end{bmatrix}.$$

To determine the coordinates of the images of the S-basis, we must solve the three linear systems

$$a_1 \begin{bmatrix} 1 \\ 2 \end{bmatrix} + a_2 \begin{bmatrix} 1 \\ 3 \end{bmatrix} = \mathbf{b},$$

where $\mathbf{b} = \begin{bmatrix} 2 \\ 3 \end{bmatrix}, \begin{bmatrix} 2 \\ 5 \end{bmatrix}$, and $\begin{bmatrix} 1 \\ 3 \end{bmatrix}$. This can be done simultaneously, as in Section 4.8, by transforming the partitioned matrix

$$\left[\begin{array}{cc|c|c|c} 1 & 1 & 2 & 2 & 1 \\ 2 & 3 & 3 & 5 & 3 \end{array}\right]$$

to reduced row echelon form, yielding (verify)

$$\left[\begin{array}{cc|c|c|c} 1 & 0 & 3 & 1 & 0 \\ 0 & 1 & -1 & 1 & 1 \end{array}\right].$$

The last three columns of this matrix are the desired coordinate vectors of the image of the S-basis with respect to the T-basis. That is, the last three columns form the matrix A representing L with respect to S and T. Thus

$$A = \begin{bmatrix} 3 & 1 & 0 \\ -1 & 1 & 1 \end{bmatrix}.$$

This matrix, of course, differs from the one that defined L. Thus, although a matrix A may be involved in the definition of a linear transformation L, we cannot conclude that it is necessarily the representation of L that we seek. ■

From Example 5 we see that if $L\colon R^n \to R^m$ is a linear transformation, then a computationally efficient way to obtain a matrix representation A of L with respect to the ordered bases $S = \{\mathbf{v}_1, \mathbf{v}_2, \ldots, \mathbf{v}_n\}$ for R^n and $T = \{\mathbf{w}_1, \mathbf{w}_2, \ldots, \mathbf{w}_m\}$ for R^m is to proceed as follows: Transform the partitioned matrix

$$\left[\begin{array}{cccc|c|c|c|c} \mathbf{w}_1 & \mathbf{w}_2 & \cdots & \mathbf{w}_m & L(\mathbf{v}_1) & L(\mathbf{v}_2) & \cdots & L(\mathbf{v}_n) \end{array}\right]$$

to reduced row echelon form. The matrix A consists of the last n columns of this last matrix.

If $L\colon V \to V$ is a linear operator on an n-dimensional space V, then to obtain a representation of L, we fix ordered bases S and T for V, and obtain a matrix A representing L with respect to S and T. However, it is often convenient in this case to choose $S = T$. To avoid verbosity in this case, we refer to A as the **representation of L with respect to S**. If $L\colon R^n \to R^n$ is a linear operator, then the matrix representing L with respect to the natural basis for R^n has already been discussed in Theorem 6.3 in Section 6.1, where it was called the standard matrix representing L.

Also, we can show readily that the matrix of the identity operator (see Exercise 22 in Section 6.1) on an n-dimensional space, with respect to any basis, is I_n.

Let $I\colon V \to V$ be the identity operator on an n-dimensional vector space V and let $S = \{\mathbf{v}_1, \mathbf{v}_2, \ldots, \mathbf{v}_n\}$ and $T = \{\mathbf{w}_1, \mathbf{w}_2, \ldots, \mathbf{w}_n\}$ be ordered bases for V. It can be shown (Exercise 23) that the matrix of the identity operator with respect to S and T is the transition matrix from the S-basis to the T-basis (see Section 4.8).

If $L\colon V \to V$ is an invertible linear operator and if A is the representation of L with respect to an ordered basis S for V, then A^{-1} is the representation of L^{-1} with respect to S. This fact, which can be proved directly at this point, follows almost trivially in Section 6.4.

Suppose that $L\colon V \to W$ is a linear transformation and that A is the matrix representing L with respect to ordered bases for V and W. Then the problem of finding $\ker L$ reduces to the problem of finding the solution space of $A\mathbf{x} = \mathbf{0}$. Moreover, the problem of finding range L reduces to the problem of finding the column space of A.

We can summarize the conditions under which a linear transformation L of an n-dimensional vector space V into itself (or, more generally, into an n-dimensional vector space W) is invertible by the following equivalent statements:

1. L is invertible.
2. L is one-to-one.
3. L is onto.
4. The matrix A representing L with respect to ordered bases S and T for V and W is nonsingular.

Key Terms

Matrix representing a linear transformation
Ordered basis
Invariant subspace

6.3 Exercises

1. Let $L: R^2 \to R^2$ be defined by

$$L\left(\begin{bmatrix} u_1 \\ u_2 \end{bmatrix}\right) = \begin{bmatrix} u_1 + 2u_2 \\ 2u_1 - u_2 \end{bmatrix}.$$

Let S be the natural basis for R^2 and let

$$T = \left\{\begin{bmatrix} -1 \\ 2 \end{bmatrix}, \begin{bmatrix} 2 \\ 0 \end{bmatrix}\right\}.$$

Find the representation of L with respect to

(a) S; **(b)** S and T; **(c)** T and S; **(d)** T.

(e) Find $L\left(\begin{bmatrix} 1 \\ 2 \end{bmatrix}\right)$ by using the definition of L and also by using the matrices found in parts (a) through (d).

2. Let $L: R_4 \to R_3$ be defined by

$$L\left(\begin{bmatrix} u_1 & u_2 & u_3 & u_4 \end{bmatrix}\right) = \begin{bmatrix} u_1 & u_2 + u_3 & u_3 + u_4 \end{bmatrix}.$$

Let S and T be the natural bases for R_4 and R_3, respectively. Let

$$S' = \{\begin{bmatrix} 1 & 0 & 0 & 1 \end{bmatrix}, \begin{bmatrix} 0 & 0 & 0 & 1 \end{bmatrix}, \begin{bmatrix} 1 & 1 & 0 & 0 \end{bmatrix}, \begin{bmatrix} 0 & 1 & 1 & 0 \end{bmatrix}\}$$

and

$$T' = \{\begin{bmatrix} 1 & 1 & 0 \end{bmatrix}, \begin{bmatrix} 0 & 1 & 0 \end{bmatrix}, \begin{bmatrix} 1 & 0 & 1 \end{bmatrix}\}.$$

(a) Find the representation of L with respect to S and T.

(b) Find the representation of L with respect to S' and T'.

(c) Find $L\left(\begin{bmatrix} 2 & 1 & -1 & 3 \end{bmatrix}\right)$ by using the matrices obtained in parts (a) and (b) and compare this answer with that obtained from the definition for L.

3. Let $L: R^4 \to R^3$ be defined by

$$L\left(\begin{bmatrix} u_1 \\ u_2 \\ u_3 \\ u_4 \end{bmatrix}\right) = \begin{bmatrix} 1 & 0 & 1 & 1 \\ 0 & 1 & 2 & 1 \\ -1 & -2 & 1 & 0 \end{bmatrix} \begin{bmatrix} u_1 \\ u_2 \\ u_3 \\ u_4 \end{bmatrix}.$$

Let S and T be the natural bases for R^4 and R^3, respectively, and consider the ordered bases

$$S' = \left\{\begin{bmatrix} 1 \\ 1 \\ 0 \\ 0 \end{bmatrix}, \begin{bmatrix} 0 \\ 1 \\ 0 \\ 0 \end{bmatrix}, \begin{bmatrix} 0 \\ 0 \\ 1 \\ 1 \end{bmatrix}, \begin{bmatrix} 0 \\ 1 \\ 1 \\ 0 \end{bmatrix}\right\} \quad \text{and}$$

$$T' = \left\{\begin{bmatrix} 1 \\ 0 \\ 1 \end{bmatrix}, \begin{bmatrix} 0 \\ 1 \\ 1 \end{bmatrix}, \begin{bmatrix} 0 \\ 0 \\ 1 \end{bmatrix}\right\}$$

for R^4 and R^3, respectively. Find the representation of L with respect to (a) S and T; (b) S' and T'.

4. Let $L: R^2 \to R^2$ be the linear transformation rotating R^2 counterclockwise through an angle ϕ. Find the representation of L with respect to the natural basis for R^2.

5. Let $L\colon R^3 \to R^3$ be defined by

$$L\left(\begin{bmatrix}1\\0\\0\end{bmatrix}\right)=\begin{bmatrix}1\\1\\0\end{bmatrix},\quad L\left(\begin{bmatrix}0\\1\\0\end{bmatrix}\right)=\begin{bmatrix}2\\0\\1\end{bmatrix},$$
$$L\left(\begin{bmatrix}0\\0\\1\end{bmatrix}\right)=\begin{bmatrix}1\\0\\1\end{bmatrix}.$$

(a) Find the representation of L with respect to the natural basis S for R^3.

(b) Find $L\left(\begin{bmatrix}1\\2\\3\end{bmatrix}\right)$ by using the definition of L and also by using the matrix obtained in part (a).

6. Let $L\colon R^3 \to R^3$ be defined as in Exercise 5. Let $T = \{L(\mathbf{e}_1), L(\mathbf{e}_2), L(\mathbf{e}_3)\}$ be an ordered basis for R^3, and let S be the natural basis for R^3.

(a) Find the representation of L with respect to S and T.

(b) Find $L\left(\begin{bmatrix}1\\2\\3\end{bmatrix}\right)$ by using the matrix obtained in part (a).

7. Let $L\colon R^3 \to R^3$ be the linear transformation represented by the matrix

$$\begin{bmatrix}1&3&1\\1&2&0\\0&1&1\end{bmatrix}$$

with respect to the natural basis for R^3. Find

(a) $L\left(\begin{bmatrix}1\\2\\3\end{bmatrix}\right)$; **(b)** $L\left(\begin{bmatrix}0\\1\\1\end{bmatrix}\right)$.

8. Let $L\colon M_{22} \to M_{22}$ be defined by

$$L(A)=\begin{bmatrix}1&2\\3&4\end{bmatrix}A$$

for A in M_{22}. Consider the ordered bases

$$S=\left\{\begin{bmatrix}1&0\\0&0\end{bmatrix},\begin{bmatrix}0&1\\0&0\end{bmatrix},\begin{bmatrix}0&0\\1&0\end{bmatrix},\begin{bmatrix}0&0\\0&1\end{bmatrix}\right\}$$

and

$$T=\left\{\begin{bmatrix}1&0\\0&1\end{bmatrix},\begin{bmatrix}1&1\\0&0\end{bmatrix},\begin{bmatrix}1&0\\1&0\end{bmatrix},\begin{bmatrix}0&1\\0&0\end{bmatrix}\right\}$$

for M_{22}. Find the representation of L with respect to

(a) S; **(b)** T; **(c)** S and T; **(d)** T and S.

9. ***(Calculus Required)*** Let V be the vector space with basis $S = \{1, t, e^t, te^t\}$ and let $L\colon V \to V$ be a linear operator defined by $L(f) = f' = df/dt$. Find the representation of L with respect to S.

10. Let $L\colon P_1 \to P_2$ be defined by $L(p(t)) = tp(t) + p(0)$. Consider the ordered bases $S = \{t, 1\}$ and $S' = \{t+1, t-1\}$ for P_1, and $T = \{t^2, t, 1\}$ and $T' = \{t^2+1, t-1, t+1\}$ for P_2. Find the representation of L with respect to

(a) S and T **(b)** S' and T'.

(c) Find $L(-3t-3)$ by using the definition of L and the matrices obtained in parts (a) and (b).

11. Let $A=\begin{bmatrix}1&2\\3&4\end{bmatrix}$, and let $L\colon M_{22} \to M_{22}$ be the linear transformation defined by $L(X) = AX - XA$ for X in M_{22}. Let S and T be the ordered bases for M_{22} defined in Exercise 8. Find the representation of L with respect to

(a) S; **(b)** T; **(c)** S and T; **(d)** T and S.

12. Let $L\colon V \to V$ be a linear operator. A nonempty subspace U of V is called **invariant** under L if $L(U)$ is contained in U. Let L be a linear operator with invariant subspace U. Show that if $\dim U = m$ and $\dim V = n$, then L has a representation with respect to a basis S for V of the form $\begin{bmatrix}A&B\\O&C\end{bmatrix}$, where A is $m\times m$, B is $m\times(n-m)$, O is the zero $(n-m)\times m$ matrix, and C is $(n-m)\times(n-m)$.

13. Let $L\colon R^2 \to R^2$ be defined by

$$L\left(\begin{bmatrix}x\\y\end{bmatrix}\right)=\begin{bmatrix}x\\-y\end{bmatrix},$$

a reflection about the x-axis. Consider the natural basis S and the ordered basis

$$T=\left\{\begin{bmatrix}1\\1\end{bmatrix},\begin{bmatrix}-1\\1\end{bmatrix}\right\}$$

for R^2. Find the representation of L with respect to

(a) S; **(b)** T; **(c)** S and T; **(d)** T and S.

14. If $L\colon R_3 \to R_2$ is the linear transformation whose representation with respect to the natural bases for R_3 and R_2 is $\begin{bmatrix}1&-1&2\\2&1&3\end{bmatrix}$, find each of the following:

(a) $L\left(\begin{bmatrix}1&2&3\end{bmatrix}\right)$ **(b)** $L\left(\begin{bmatrix}-1&2&-1\end{bmatrix}\right)$

(c) $L\left(\begin{bmatrix}0&1&2\end{bmatrix}\right)$ **(d)** $L\left(\begin{bmatrix}0&1&0\end{bmatrix}\right)$

(e) $L\left(\begin{bmatrix}0&0&1\end{bmatrix}\right)$

15. If $O: V \to W$ is the zero linear transformation, show that the matrix representation of O with respect to any ordered bases for V and W is the $m \times n$ zero matrix, where $n = \dim V$ and $m = \dim W$.

16. If $I: V \to V$ is the identity linear operator on V defined by $I(\mathbf{v}) = \mathbf{v}$ for $\mathbf{v}$ in V, prove that the matrix representation of I with respect to any ordered basis S for V is I_n, where $\dim V = n$.

17. Let $I: R_2 \to R_2$ be the identity linear operator on R_2. Let $S = \{\begin{bmatrix}1 & 0\end{bmatrix}, \begin{bmatrix}0 & 1\end{bmatrix}\}$ and $T = \{\begin{bmatrix}1 & -1\end{bmatrix}, \begin{bmatrix}2 & 3\end{bmatrix}\}$ be ordered bases for R_2. Find the representation of I with respect to

(a) S; **(b)** T; **(c)** S and T; **(d)** T and S.

18. ***(Calculus Required)*** Let V be the vector space of real-valued continuous functions with basis $S = \{e^t, e^{-t}\}$. Find the representation of the linear operator $L: V \to V$ defined by $L(f) = f'$ with respect to S.

19. Let V be the vector space of real-valued continuous functions with ordered basis $S = \{\sin t, \cos t\}$. Find the representation of the linear operator $L: V \to V$ defined by $L(f) = f'$ with respect to S.

20. ***(Calculus Required)*** Let V be the vector space of real-valued continuous functions with ordered basis $S = \{\sin t, \cos t\}$ and consider $T = \{\sin t - \cos t, \sin t + \cos t\}$, another ordered basis for V. Find the representation of the linear operator $L: V \to V$ defined by $L(f) = f'$ with respect to

(a) S; **(b)** T; **(c)** S and T; **(d)** T and S.

21. Let $L: V \to V$ be a linear operator defined by $L(\mathbf{v}) = c\mathbf{v}$, where c is a fixed constant. Prove that the representation of L with respect to any ordered basis for V is a scalar matrix. (See Section 1.5.)

22. Let the representation of $L: R^3 \to R^2$ with respect to the ordered bases $S = \{\mathbf{v}_1, \mathbf{v}_2, \mathbf{v}_3\}$ and $T = \{\mathbf{w}_1, \mathbf{w}_2\}$ be

$$A = \begin{bmatrix} 1 & 2 & 1 \\ -1 & 1 & 0 \end{bmatrix},$$

where

$$\mathbf{v}_1 = \begin{bmatrix} -1 \\ 1 \\ 0 \end{bmatrix}, \quad \mathbf{v}_2 = \begin{bmatrix} 0 \\ 1 \\ 1 \end{bmatrix}, \quad \mathbf{v}_3 = \begin{bmatrix} 1 \\ 0 \\ 0 \end{bmatrix},$$

$$\mathbf{w}_1 = \begin{bmatrix} 1 \\ 2 \end{bmatrix}, \quad \text{and} \quad \mathbf{w}_2 = \begin{bmatrix} 1 \\ -1 \end{bmatrix}.$$

(a) Compute $\left[L(\mathbf{v}_1)\right]_T$, $\left[L(\mathbf{v}_2)\right]_T$, and $\left[L(\mathbf{v}_3)\right]_T$.

(b) Compute $L(\mathbf{v}_1)$, $L(\mathbf{v}_2)$, and $L(\mathbf{v}_3)$.

(c) Compute $L\left(\begin{bmatrix} 2 \\ 1 \\ -1 \end{bmatrix}\right)$.

23. Let $I: V \to V$ be the identity operator on an n-dimensional vector space V and let $S = \{\mathbf{v}_1, \mathbf{v}_2, \ldots, \mathbf{v}_n\}$ and $T = \{\mathbf{w}_1, \mathbf{w}_2, \ldots, \mathbf{w}_n\}$ be ordered bases for V. Show that the matrix of the identity operator with respect to S and T is the transition matrix from the S-basis to the T-basis. (See Section 4.8.)

6.4 Vector Space of Matrices and Vector Space of Linear Transformations (Optional)

We have already seen in Section 4.2 that the set M_{mn} of all $m \times n$ matrices is a vector space under the operations of matrix addition and scalar multiplication. We now show in this section that the set of all linear transformations of an n-dimensional vector space V into an m-dimensional vector space W is also a vector space U under two suitably defined operations, and we shall examine the relation between U and M_{mn}.

DEFINITION 6.5

Let V and W be two vector spaces of dimensions n and m, respectively. Also, let $L_1: V \to W$ and $L_2: V \to W$ be linear transformations. We define a mapping $L: V \to W$ by $L(\mathbf{x}) = L_1(\mathbf{x}) + L_2(\mathbf{x})$, for $\mathbf{x}$ in V. Of course, the $+$ here is vector addition in W. We shall denote L by $L_1 \boxplus L_2$ and call it the **sum** of L_1 and L_2. Also, if $L_3: V \to W$ is a linear transformation and c is a real number, we define a mapping $H: V \to W$ by $H(\mathbf{x}) = cL_3(\mathbf{x})$ for $\mathbf{x}$ in V. Of course, the operation on the right side is scalar multiplication in W. We denote H by $c \boxdot L_3$ and call it the **scalar multiple** of L_3 by c.

EXAMPLE 1 Let $V = R_3$ and $W = R_2$. Let $L_1\colon R_3 \to R_2$ and $L_2\colon R_3 \to R_2$ be defined by

$$L_1(\mathbf{x}) = L_1\left(\begin{bmatrix} u_1 & u_2 & u_3 \end{bmatrix}\right) = \begin{bmatrix} u_1 + u_2 & u_2 + u_3 \end{bmatrix}$$

and

$$L_2(\mathbf{x}) = L_2\left(\begin{bmatrix} u_1 & u_2 & u_3 \end{bmatrix}\right) = \begin{bmatrix} u_1 + u_3 & u_2 \end{bmatrix}.$$

Then

$$(L_1 \boxplus L_2)(\mathbf{x}) = \begin{bmatrix} 2u_1 + u_2 + u_3 & 2u_2 + u_3 \end{bmatrix}$$

and

$$(3 \boxdot L_1)(\mathbf{x}) = \begin{bmatrix} 3u_1 + 3u_2 & 3u_2 + 3u_3 \end{bmatrix}.$$ ■

We leave it to the reader (see the exercises in this section) to verify that if L, L_1, L_2, and L_3 are linear transformations of V into W and if c is a real number, then $L_1 \boxplus L_2$ and $c \boxdot L_3$ are linear transformations. We also let the reader show that the set U of all linear transformations of V into W is a vector space under the operations $\boxplus$ and $\boxdot$. The linear transformation $O\colon V \to W$ defined by $O(\mathbf{x}) = \mathbf{0}_W$ for $\mathbf{x}$ in V is the zero vector in U. That is, $L \boxplus O = O \boxplus L = L$ for any L in U. Also, if L is in U, then $L \boxplus (-1 \boxdot L) = O$, so we may write $(-1) \boxdot L$ as $-L$. Of course, to say that $S = \{L_1, L_2, \ldots, L_k\}$ is a linearly dependent set in U means merely that there exist k scalars $a_1, a_2, \ldots, a_k$, not all zero, such that

$$(a_1 \boxdot L_1) \boxplus (a_2 \boxdot L_2) \boxplus \cdots \boxplus (a_k \boxdot L_k) = O,$$

where O is the zero linear transformation.

EXAMPLE 2 Let $L_1\colon R_2 \to R_3$, $L_2\colon R_2 \to R_3$, and $L_3\colon R_2 \to R_3$ be defined by

$$\begin{aligned} L_1\left(\begin{bmatrix} u_1 & u_2 \end{bmatrix}\right) &= \begin{bmatrix} u_1 + u_2 & 2u_2 & u_2 \end{bmatrix}, \\ L_2\left(\begin{bmatrix} u_1 & u_2 \end{bmatrix}\right) &= \begin{bmatrix} u_2 - u_1 & 2u_1 + u_2 & u_1 \end{bmatrix}, \\ L_3\left(\begin{bmatrix} u_1 & u_2 \end{bmatrix}\right) &= \begin{bmatrix} 3u_1 & -2u_2 & u_1 + 2u_2 \end{bmatrix}. \end{aligned}$$

Determine whether $S = \{L_1, L_2, L_3\}$ is linearly independent.

Solution

Suppose that

$$(a_1 \boxdot L_1) \boxplus (a_2 \boxdot L_2) \boxplus (a_3 \boxdot L_3) = O,$$

where a_1, a_2, and a_3 are real numbers. Then, for $\mathbf{e}_1^T = \begin{bmatrix} 1 & 0 \end{bmatrix}$, we have

$$((a_1 \boxdot L_1) \boxplus (a_2 \boxdot L_2) \boxplus (a_3 \boxdot L_3))(\mathbf{e}_1^T) = O(\mathbf{e}_1^T) = \begin{bmatrix} 0 & 0 & 0 \end{bmatrix},$$

so

$$\begin{aligned} &a_1 L_1(\mathbf{e}_1^T) + a_2 L_2(\mathbf{e}_1^T) + a_3 L_3(\mathbf{e}_1^T) \\ &\qquad = a_1 \begin{bmatrix} 1 & 2 & 0 \end{bmatrix} + a_2 \begin{bmatrix} -1 & 2 & 1 \end{bmatrix} + a_3 \begin{bmatrix} 3 & 0 & 1 \end{bmatrix} = \begin{bmatrix} 0 & 0 & 0 \end{bmatrix}. \end{aligned}$$

Thus we must solve the homogeneous system

$$\begin{bmatrix} 1 & -1 & 3 \\ 2 & 2 & 0 \\ 0 & 1 & 1 \end{bmatrix} \begin{bmatrix} a_1 \\ a_2 \\ a_3 \end{bmatrix} = \begin{bmatrix} 0 \\ 0 \\ 0 \end{bmatrix},$$

obtaining $a_1 = a_2 = a_3 = 0$ (verify). Hence S is linearly independent. ■

Theorem 6.10 Let U be the vector space of all linear transformations of an n-dimensional vector space V into an m-dimensional vector space W, $n \neq 0$ and $m \neq 0$, under the operations $\boxplus$ and $\boxdot$. Then U is isomorphic to the vector space M_{mn} of all $m \times n$ matrices.

Proof

Let $S = \{\mathbf{v}_1, \mathbf{v}_2, \dots, \mathbf{v}_n\}$ and $T = \{\mathbf{w}_1, \mathbf{w}_2, \dots, \mathbf{w}_m\}$ be ordered bases for V and W, respectively. We define a function $M\colon U \to M_{mn}$ by letting $M(L)$ be the matrix representing L with respect to the bases S and T. We now show that M is an isomorphism.

First, M is one-to-one, for if L_1 and L_2 are two different elements in U, then $L_1(\mathbf{v}_j) \neq L_2(\mathbf{v}_j)$ for some $j = 1, 2, \dots, n$. This means that the jth columns of $M(L_1)$ and $M(L_2)$, which are the coordinate vectors of $L_1(\mathbf{v}_j)$ and $L_2(\mathbf{v}_j)$, respectively, with respect to T, are different, so $M(L_1) \neq M(L_2)$. Hence M is one-to-one.

Next, M is onto. Let $A = \left[a_{ij}\right]$ be a given $m \times n$ matrix; that is, A is an element of M_{mn}. Then we define a function $L\colon V \to W$ by

$$L(\mathbf{v}_i) = \sum_{k=1}^{m} a_{ki}\mathbf{w}_k, \qquad i = 1, 2, \dots, n,$$

and if $\mathbf{x} = c_1\mathbf{v}_1 + c_2\mathbf{v}_2 + \cdots + c_n\mathbf{v}_n$, we define $L(\mathbf{x})$ by

$$L(\mathbf{x}) = \sum_{i=1}^{n} c_i L(\mathbf{v}_i).$$

It is not difficult to show that L is a linear transformation; moreover, the matrix representing L with respect to S and T is $A = \left[a_{ij}\right]$ (verify). Thus $M(L) = A$, so M is onto.

Now let $M(L_1) = A = \left[a_{ij}\right]$ and $M(L_2) = B = \left[b_{ij}\right]$. We show that $M(L_1 \boxplus L_2) = A + B$. First, note that the jth column of $M(L_1 \boxplus L_2)$ is

$$\left[(L_1 \boxplus L_2)(\mathbf{v}_j)\right]_T = \left[L_1(\mathbf{v}_j) + L_2(\mathbf{v}_j)\right]_T = \left[L_1(\mathbf{v}_j)\right]_T + \left[L_2(\mathbf{v}_j)\right]_T.$$

Thus the jth column of $M(L_1 \boxplus L_2)$ is the sum of the jth columns of $M(L_1) = A$ and $M(L_2) = B$. Hence $M(L_1 \boxplus L_2) = A + B$.

Finally, let $M(L) = A$ and c be a real number. Following the idea in the preceding paragraph, we can show that $M(c \boxdot L) = cA$ (verify). Hence U and M_{mn} are isomorphic. ■

This theorem implies that the dimension of U is mn, for $\dim M_{mn} = mn$. Also, it means that when dealing with finite-dimensional vector spaces, we can always replace all linear transformations by their matrix representations and work only with the matrices. Moreover, it should be noted again that matrices lend themselves much more readily than linear transformations to computer implementations.

EXAMPLE 3

Let $A = \begin{bmatrix} 1 & 2 & -1 \\ 2 & -1 & 3 \end{bmatrix}$, and let $S = \{\mathbf{e}_1, \mathbf{e}_2, \mathbf{e}_3\}$ and $T = \{\overline{\mathbf{e}}_1, \overline{\mathbf{e}}_2\}$ be the natural bases for R^3 and R^2, respectively.

(a) Find the unique linear transformation $L\colon R^3 \to R^2$ whose representation with respect to S and T is A.

(b) Let

$$S' = \left\{ \begin{bmatrix} 1 \\ 0 \\ 1 \end{bmatrix}, \begin{bmatrix} 1 \\ 1 \\ 0 \end{bmatrix}, \begin{bmatrix} 0 \\ 1 \\ 1 \end{bmatrix} \right\} \quad \text{and} \quad T' = \left\{ \begin{bmatrix} 1 \\ 3 \end{bmatrix}, \begin{bmatrix} 2 \\ -1 \end{bmatrix} \right\}$$

be ordered bases for R^3 and R^2, respectively. Determine the linear transformation $L\colon R^3 \to R^2$ whose representation with respect to S' and T' is A.

(c) Compute $L\left(\begin{bmatrix} 1 \\ 2 \\ 3 \end{bmatrix}\right)$, using L as determined in part (b).

Solution

(a) Let

$$L(\mathbf{e}_1) = 1\overline{\mathbf{e}}_1 + 2\overline{\mathbf{e}}_2 = \begin{bmatrix} 1 \\ 2 \end{bmatrix},$$

$$L(\mathbf{e}_2) = 2\overline{\mathbf{e}}_1 - 1\overline{\mathbf{e}}_2 = \begin{bmatrix} 2 \\ -1 \end{bmatrix},$$

$$L(\mathbf{e}_3) = -\overline{\mathbf{e}}_1 + 3\overline{\mathbf{e}}_2 = \begin{bmatrix} -1 \\ 3 \end{bmatrix}.$$

Now if $\mathbf{x} = \begin{bmatrix} a_1 \\ a_2 \\ a_3 \end{bmatrix}$ is in R^3, we define $L(\mathbf{x})$ by

$$\begin{aligned} L(\mathbf{x}) &= L(a_1\mathbf{e}_1 + a_2\mathbf{e}_2 + a_3\mathbf{e}_3) \\ &= a_1 L(\mathbf{e}_1) + a_2 L(\mathbf{e}_2) + a_3 L(\mathbf{e}_3), \end{aligned}$$

$$\begin{aligned} L(\mathbf{x}) &= a_1 \begin{bmatrix} 1 \\ 2 \end{bmatrix} + a_2 \begin{bmatrix} 2 \\ -1 \end{bmatrix} + a_3 \begin{bmatrix} -1 \\ 3 \end{bmatrix} \\ &= \begin{bmatrix} a_1 + 2a_2 - a_3 \\ 2a_1 - a_2 + 3a_3 \end{bmatrix}. \end{aligned}$$

Note that

$$L(\mathbf{x}) = \begin{bmatrix} 1 & 2 & -1 \\ 2 & -1 & 3 \end{bmatrix} \begin{bmatrix} a_1 \\ a_2 \\ a_3 \end{bmatrix},$$

so we could have defined L by $L(\mathbf{x}) = A\mathbf{x}$ for $\mathbf{x}$ in R^3. We can do this when the bases S and T are the natural bases.

(b) Let

$$L\left(\begin{bmatrix} 1 \\ 0 \\ 1 \end{bmatrix}\right) = 1\begin{bmatrix} 1 \\ 3 \end{bmatrix} + 2\begin{bmatrix} 2 \\ -1 \end{bmatrix} = \begin{bmatrix} 5 \\ 1 \end{bmatrix},$$

$$L\left(\begin{bmatrix} 1 \\ 1 \\ 0 \end{bmatrix}\right) = 2\begin{bmatrix} 1 \\ 3 \end{bmatrix} - 1\begin{bmatrix} 2 \\ -1 \end{bmatrix} = \begin{bmatrix} 0 \\ 7 \end{bmatrix},$$

$$L\left(\begin{bmatrix} 0 \\ 1 \\ 1 \end{bmatrix}\right) = -1\begin{bmatrix} 1 \\ 3 \end{bmatrix} + 3\begin{bmatrix} 2 \\ -1 \end{bmatrix} = \begin{bmatrix} 5 \\ -6 \end{bmatrix}.$$

Then if $\mathbf{x} = \begin{bmatrix} a_1 \\ a_2 \\ a_3 \end{bmatrix}$, we express $\mathbf{x}$ in terms of the basis S' as

$$\mathbf{x} = b_1\begin{bmatrix} 1 \\ 0 \\ 1 \end{bmatrix} + b_2\begin{bmatrix} 1 \\ 1 \\ 0 \end{bmatrix} + b_3\begin{bmatrix} 0 \\ 1 \\ 1 \end{bmatrix}; \quad \text{hence } \left[\mathbf{x}\right]_{S'} = \begin{bmatrix} b_1 \\ b_2 \\ b_3 \end{bmatrix}.$$

Define $L(\mathbf{x})$ by

$$\begin{aligned} L(\mathbf{x}) &= b_1 L\left(\begin{bmatrix} 1 \\ 0 \\ 1 \end{bmatrix}\right) + b_2 L\left(\begin{bmatrix} 1 \\ 1 \\ 0 \end{bmatrix}\right) + b_3 L\left(\begin{bmatrix} 0 \\ 1 \\ 1 \end{bmatrix}\right) \\ &= b_1\begin{bmatrix} 5 \\ 1 \end{bmatrix} + b_2\begin{bmatrix} 0 \\ 7 \end{bmatrix} + b_3\begin{bmatrix} 5 \\ -6 \end{bmatrix}, \end{aligned}$$

so

$$L(\mathbf{x}) = \begin{bmatrix} 5b_1 + 5b_3 \\ b_1 + 7b_2 - 6b_3 \end{bmatrix}. \tag{1}$$

(c) To find $L\left(\begin{bmatrix} 1 \\ 2 \\ 3 \end{bmatrix}\right)$, we first have (verify)

$$\begin{bmatrix} 1 \\ 2 \\ 3 \end{bmatrix} = 1\begin{bmatrix} 1 \\ 0 \\ 1 \end{bmatrix} + 0\begin{bmatrix} 1 \\ 1 \\ 0 \end{bmatrix} + 2\begin{bmatrix} 0 \\ 1 \\ 1 \end{bmatrix}; \quad \text{hence } \left[\begin{bmatrix} 1 \\ 2 \\ 3 \end{bmatrix}\right]_{S'} = \begin{bmatrix} 1 \\ 0 \\ 2 \end{bmatrix}.$$

Then, using $b_1 = 1$, $b_2 = 0$, and $b_3 = 2$ in Equation (1), we obtain

$$L\left(\begin{bmatrix}1\\2\\3\end{bmatrix}\right) = \begin{bmatrix}15\\-11\end{bmatrix}.$$ ■

The linear transformations obtained in Example 3 depend on the ordered bases for R^2 and R^3. Thus if L is as in part (a), then

$$L\left(\begin{bmatrix}1\\2\\3\end{bmatrix}\right) = \begin{bmatrix}2\\9\end{bmatrix},$$

which differs from the answer obtained in part (c), since the linear transformation in part (b) differs from that in part (a).

Now let V_1 be an n-dimensional vector space, V_2 an m-dimensional vector space, and V_3 a p-dimensional vector space. Let $L_1: V_1 \to V_2$ and $L_2: V_2 \to V_3$ be linear transformations. We can define the composite function

$$L_2 \circ L_1: V_1 \to V_3 \quad \text{by} \quad (L_2 \circ L_1)(\mathbf{x}) = L_2(L_1(\mathbf{x}))$$

for $\mathbf{x}$ in V_1. It follows that $L_2 \circ L_1$ is a linear transformation. If $L: V \to V$, then $L \circ L$ is written as L^2.

EXAMPLE 4

Let $L_1: R^2 \to R^2$ and $L_2: R^2 \to R^2$ be defined by

$$L_1\left(\begin{bmatrix}u_1\\u_2\end{bmatrix}\right) = \begin{bmatrix}u_1\\-u_2\end{bmatrix} \quad \text{and} \quad L_2\left(\begin{bmatrix}u_1\\u_2\end{bmatrix}\right) = \begin{bmatrix}u_2\\u_1\end{bmatrix}.$$

Then

$$(L_1 \circ L_2)\left(\begin{bmatrix}u_1\\u_2\end{bmatrix}\right) = L_1\left(\begin{bmatrix}u_2\\u_1\end{bmatrix}\right) = \begin{bmatrix}u_2\\-u_1\end{bmatrix},$$

while

$$(L_2 \circ L_1)\left(\begin{bmatrix}u_1\\u_2\end{bmatrix}\right) = L_2\left(\begin{bmatrix}u_1\\-u_2\end{bmatrix}\right) = \begin{bmatrix}-u_2\\u_1\end{bmatrix}.$$

Thus $L_1 \circ L_2 \neq L_2 \circ L_1$. ■

Theorem 6.11 Let V_1 be an n-dimensional vector space, V_2 an m-dimensional vector space, and V_3 a p-dimensional vector space with linear transformations L_1 and L_2 such that $L_1: V_1 \to V_2$ and $L_2: V_2 \to V_3$. If the ordered bases P, S, and T are chosen for V_1, V_2, and V_3, respectively, then $M(L_2 \circ L_1) = M(L_2)M(L_1)$.

Proof

Let $M(L_1) = A$ with respect to the P and S ordered bases for V_1 and V_2, respectively, and let $M(L_2) = B$ with respect to the S and T ordered bases for V_2 and V_3, respectively. For any vector $\mathbf{x}$ in V_1, $\left[L_1(\mathbf{x})\right]_S = A\left[\mathbf{x}\right]_P$, and for any vector $\mathbf{y}$ in V_2, $\left[L_2(\mathbf{y})\right]_T = B\left[\mathbf{y}\right]_S$. Then it follows that

$$\begin{aligned}\left[(L_2 \circ L_1)(\mathbf{x})\right]_T &= \left[L_2(L_1(\mathbf{x}))\right]_T \\ &= B\left[L_1(\mathbf{x})\right]_S = B\left(A\left[\mathbf{x}\right]_P\right) = (BA)\left[\mathbf{x}\right]_P.\end{aligned}$$

Since a linear transformation has a unique representation with respect to ordered bases (see Theorem 6.9), we have $M(L_2 \circ L_1) = BA$, and it follows that

$$M(L_2 \circ L_1) = M(L_2)M(L_1).$$

Thus the composite of two linear transformations corresponds to the multiplication of their respective matrix representations. ■

Since AB need not equal BA for matrices A and B, it is thus not surprising that $L_1 \circ L_2$ need not be the same linear transformation as $L_2 \circ L_1$.

If $L: V \to V$ is an invertible linear operator and if A is a representation of L with respect to an ordered basis S for V, then the representation of the identity operator $L \circ L^{-1}$ is the identity matrix I_n. Thus $M(L)M(L^{-1}) = I_n$, which means that A^{-1} is the representation of L^{-1} with respect to S.

Key Terms

Sum of linear transformations
Scalar multiple of a linear transformation
Vector space of linear transformations
Composition of linear transformations

6.4 Exercises

1. Let L_1, L_2, and L be linear transformations of V into W. Prove the following:

(a) $L_1 \boxplus L_2$ is a linear transformation of V into W.

(b) If c is a real number, then $c \boxdot L$ is a linear transformation of V into W.

(c) If A represents L with respect to the ordered bases S and T for V and W, respectively, then cA represents $c \boxdot L$ with respect to S and T, where c is a real number.

2. Let U be the set of all linear transformations of V into W, and let $O: V \to W$ be the zero linear transformation defined by $O(\mathbf{x}) = \mathbf{0}_W$ for all $\mathbf{x}$ in V.

(a) Prove that $O \boxplus L = L \boxplus O = L$ for any L in U.

(b) Show that if L is in U, then

$$L \boxplus ((-1) \boxdot L) = O.$$

3. Let $L_1: R_3 \to R_3$ and $L_2: R_3 \to R_3$ be linear transformations such that

$$L_1(\mathbf{e}_1^T) = \begin{bmatrix} -1 & 2 & 1 \end{bmatrix}, \qquad L_2(\mathbf{e}_1^T) = \begin{bmatrix} 0 & 1 & 3 \end{bmatrix},$$
$$L_1(\mathbf{e}_2^T) = \begin{bmatrix} 0 & 1 & 2 \end{bmatrix}, \quad \text{and} \quad L_2(\mathbf{e}_2^T) = \begin{bmatrix} 4 & -2 & 1 \end{bmatrix},$$
$$L_1(\mathbf{e}_3^T) = \begin{bmatrix} -1 & 1 & 3 \end{bmatrix}, \qquad L_2(\mathbf{e}_3^T) = \begin{bmatrix} 0 & 2 & 2 \end{bmatrix},$$

where $S = \{\mathbf{e}_1^T, \mathbf{e}_2^T, \mathbf{e}_3^T\}$ is the natural basis for R_3. Find the following:

(a) $(L_1 \boxplus L_2)\left(\begin{bmatrix} u_1 & u_2 & u_3 \end{bmatrix}\right)$

(b) $(L_1 \boxplus L_2)\left(\begin{bmatrix} 2 & 1 & -3 \end{bmatrix}\right)$

(c) The representation of $L_1 \boxplus L_2$ with respect to S

(d) $(-2 \boxdot L_1)\left(\begin{bmatrix} u_1 & u_2 & u_3 \end{bmatrix}\right)$

(e) $(-2 \boxdot L_1) \boxplus (4 \boxdot L_2)\left(\begin{bmatrix} 2 & 1 & -3 \end{bmatrix}\right)$

(f) The representation of $(-2 \boxdot L_1) \boxplus (4 \boxdot L_2)$ with respect to S

4. Verify that the set U of all linear transformations of V into W is a vector space under the operations $\boxplus$ and $\boxdot$.

5. Let $L_i: R_3 \to R_3$ be a linear transformation defined by

$$L_i(\mathbf{x}) = A_i\mathbf{x}, \qquad i = 1, 2,$$

where

$$A_1 = \begin{bmatrix} 1 & 1 & 1 \\ 2 & 2 & 2 \\ 3 & 3 & 3 \end{bmatrix} \quad \text{and} \quad A_2 = \begin{bmatrix} 1 & 1 & 2 \\ 2 & 2 & 4 \\ -2 & -2 & -4 \end{bmatrix}.$$

(a) For $\mathbf{x} = \begin{bmatrix} 3 & -2 & 1 \end{bmatrix}^T$, find $(L_1 \boxplus L_2)(\mathbf{x})$.

(b) Determine $\ker L_1$, $\ker L_2$, and $\ker L_1 \cap \ker L_2$ (intersection of kernels).

(c) Determine $\ker(L_1 \boxplus L_2)$.

(d) What is the relationship between $\ker L_1 \cap \ker L_2$ and $\ker(L_1 \boxplus L_2)$?

6. Let V_1, V_2, and V_3 be vector spaces of dimensions n, m, and p, respectively. Also let $L_1: V_1 \to V_2$ and $L_2: V_2 \to V_3$ be linear transformations. Prove that $L_2 \circ L_1: V_1 \to V_3$ is a linear transformation.

7. ***(Calculus Required)*** Let $L_1\colon P_1 \to P_2$ be the linear transformation defined by $L_1(p(t)) = tp(t)$ and let $L_2\colon P_2 \to P_3$ be the linear transformation defined by $L_2(p(t)) = t^2p'(t)$. Let $R = \{t+1, t-1\}$, $S = \{t^2, t-1, t+2\}$, and $T = \{t^3, t^2-1, t, t+1\}$ be ordered bases for P_1, P_2, and P_3, respectively.

(a) Find the representation C of $L_2 \circ L_1$ with respect to R and T.

(b) Compute the representation A of L_1 with respect to R and S and the representation B of L_2 with respect to S and T. Verify that BA is the matrix C obtained in part (a).

8. Let L_1, L_2, and S be as in Exercise 3. Find the following:

(a) $(L_1 \circ L_2)\left(\begin{bmatrix} u_1 & u_2 & u_3 \end{bmatrix}\right)$

(b) $(L_2 \circ L_1)\left(\begin{bmatrix} u_1 & u_2 & u_3 \end{bmatrix}\right)$

(c) The representation of $L_1 \circ L_2$ with respect to S

(d) The representation of $L_2 \circ L_1$ with respect to S

9. If $\begin{bmatrix} 1 & 4 & -1 \\ 2 & 1 & 3 \\ 1 & -1 & 2 \end{bmatrix}$ is the representation of a linear operator $L\colon R^3 \to R^3$ with respect to ordered bases S and T for R^3, find the representation with respect to S and T of

(a) $2 \boxdot L$; **(b)** $2 \boxdot L \boxplus L \circ L$.

10. Let L_1, L_2, and L_3 be linear transformations of R_3 into R_2 defined by

$$\begin{aligned} L_1\left(\begin{bmatrix} u_1 & u_2 & u_3 \end{bmatrix}\right) &= \begin{bmatrix} u_1+u_2 & u_1-u_3 \end{bmatrix}, \\ L_2\left(\begin{bmatrix} u_1 & u_2 & u_3 \end{bmatrix}\right) &= \begin{bmatrix} u_1-u_2 & u_3 \end{bmatrix}, \quad \text{and} \\ L_3\left(\begin{bmatrix} u_1 & u_2 & u_3 \end{bmatrix}\right) &= \begin{bmatrix} u_1 & u_2+u_3 \end{bmatrix}. \end{aligned}$$

Prove that $S = \{L_1, L_2, L_3\}$ is a linearly independent set in the vector space U of all linear transformations of R_3 into R_2.

11. Find the dimension of the vector space U of all linear transformations of V into W for each of the following:

(a) $V = R^2$, $W = R^3$ **(b)** $V = P_2$, $W = P_1$

(c) $V = M_{21}$, $W = M_{32}$ **(d)** $V = R_3$, $W = R_4$

12. Repeat Exercise 11 for each of the following:

(a) $V = W$ is the vector space with basis $\{\sin t, \cos t\}$.

(b) $V = W$ is the vector space with basis $\{1, t, e^t, te^t\}$.

(c) V is the vector space *spanned* by $\{1, t, 2t\}$, and W is the vector space with basis $\{t^2, t, 1\}$.

13. Let $A = \begin{bmatrix} a_{ij} \end{bmatrix}$ be a given $m \times n$ matrix, and let V and W be given vector spaces of dimensions n and m, respectively. Let $S = \{\mathbf{v}_1, \mathbf{v}_2, \ldots, \mathbf{v}_n\}$ be an ordered basis for V, and let $T = \{\mathbf{w}_1, \mathbf{w}_2, \ldots, \mathbf{w}_m\}$ be an ordered basis for W. Define a function $L\colon V \to W$ by

$$L(\mathbf{v}_i) = \sum_{k=1}^{m} a_{ki}\mathbf{w}_k, \qquad i = 1, 2, \ldots, n,$$

and if $\mathbf{x} = c_1\mathbf{v}_1 + c_2\mathbf{v}_2 + \cdots + c_n\mathbf{v}_n$, we define $L(\mathbf{x})$ by

$$L(\mathbf{x}) = \sum_{i=1}^{n} c_i L(\mathbf{v}_i).$$

(a) Show that L is a linear transformation.

(b) Show that A represents L with respect to S and T.

14. Let $A = \begin{bmatrix} 1 & 2 & -2 \\ 3 & 4 & -1 \end{bmatrix}$. Let S be the natural basis for R^3 and T be the natural basis for R^2.

(a) Find the linear transformation $L\colon R^3 \to R^2$ determined by A.

(b) Find $L\left(\begin{bmatrix} u_1 \\ u_2 \\ u_3 \end{bmatrix}\right)$. **(c)** Find $L\left(\begin{bmatrix} 1 \\ 2 \\ 3 \end{bmatrix}\right)$.

15. Let A be as in Exercise 14. Consider the ordered bases $S = \{t^2, t, 1\}$ and $T = \{t, 1\}$ for P_2 and P_1, respectively.

(a) Find the linear transformation $L\colon P_2 \to P_1$ determined by A.

(b) Find $L(at^2 + bt + c)$. **(c)** Find $L(2t^2 - 5t + 4)$.

16. Find two linear transformations $L_1\colon R^2 \to R^2$ and $L_2\colon R^2 \to R^2$ such that $L_2 \circ L_1 \neq L_1 \circ L_2$.

17. Find a linear transformation $L\colon R^2 \to R^2$, $L \neq I$, the identity operator, such that $L^2 = L \circ L = I$.

18. Find a linear transformation $L\colon R^2 \to R^2$, $L \neq O$, the zero transformation, such that $L^2 = O$.

19. Find a linear transformation $L\colon R^2 \to R^2$, $L \neq I$, $L \neq O$, such that $L^2 = L$.

20. Let $L\colon R^3 \to R^3$ be the linear transformation defined in Exercise 19 of Section 6.2. Find the matrix representing L^{-1} with respect to the natural basis for R^3.

21. Let $L\colon R^3 \to R^3$ be the linear transformation defined in Exercise 23 of Section 6.2. Find the matrix representing L^{-1} with respect to the natural basis for R^3.

22. Let $L\colon R^3 \to R^3$ be the invertible linear transformation represented by

$$A = \begin{bmatrix} 2 & 0 & 4 \\ -1 & 1 & -2 \\ 2 & 3 & 3 \end{bmatrix}$$

with respect to an ordered basis S for R^3. Find the representation of L^{-1} with respect to S.

23. Let $L: V \to V$ be a linear transformation represented by a matrix A with respect to an ordered basis S for V. Show that A^2 represents $L^2 = L \circ L$ with respect to S. Moreover, show that if k is a positive integer, then A^k represents $L^k = L \circ L \circ \cdots \circ L$ (k times) with respect to S.

24. Let $L: P_1 \to P_1$ be the invertible linear transformation represented by

$$A = \begin{bmatrix} 4 & 2 \\ 3 & -1 \end{bmatrix}$$

with respect to an ordered basis S for P_1. Find the representation of L^{-1} with respect to S.

6.5 Similarity

In Section 6.3 we saw how the matrix representing a linear transformation of an n-dimensional vector space V into an m-dimensional vector space W depends upon the ordered bases we choose for V and W. We now see how this matrix changes when the bases for V and W are changed. For simplicity, in this section we represent the transition matrices $P_{S \leftarrow S'}$ and $Q_{T \leftarrow T'}$ as P and Q, respectively.

Theorem 6.12 Let $L: V \to W$ be a linear transformation of an n-dimensional vector space V into an m-dimensional vector space W. Let $S = \{\mathbf{v}_1, \mathbf{v}_2, \ldots, \mathbf{v}_n\}$ and $S' = \{\mathbf{v}'_1, \mathbf{v}'_2, \ldots, \mathbf{v}'_n\}$ be ordered bases for V, with transition matrix P from S' to S; let $T = \{\mathbf{w}_1, \mathbf{w}_2, \ldots, \mathbf{w}_m\}$ and $T' = \{\mathbf{w}'_1, \mathbf{w}'_2, \ldots, \mathbf{w}'_m\}$ be ordered bases for W with transition matrix Q from T' to T. If A is the representation of L with respect to S and T, then $Q^{-1}AP$ is the representation of L with respect to S' and T'.

Proof

Recall Section 4.8, where the transition matrix was first introduced. If P is the transition matrix from S' to S, and $\mathbf{x}$ is a vector in V, then

$$\left[\mathbf{x}\right]_S = P\left[\mathbf{x}\right]_{S'}, \tag{1}$$

where the jth column of P is the coordinate vector $\left[\mathbf{v}'_j\right]_S$ of $\mathbf{v}'_j$ with respect to S. Similarly, if Q is the transition matrix from T' to T and $\mathbf{y}$ is a vector in W, then

$$\left[\mathbf{y}\right]_T = Q\left[\mathbf{y}\right]_{T'}, \tag{2}$$

where the jth column of Q is the coordinate vector $\left[\mathbf{w}'_j\right]_T$ of $\mathbf{w}'_j$ with respect to T. If A is the representation of L with respect to S and T, then

$$\left[L(\mathbf{x})\right]_T = A\left[\mathbf{x}\right]_S \tag{3}$$

for $\mathbf{x}$ in V. Substituting $\mathbf{y} = L(\mathbf{x})$ in (2), we have $\left[L(\mathbf{x})\right]_T = Q\left[L(\mathbf{x})\right]_{T'}$. Now, using first (3) and then (1) in this last equation, we obtain

$$Q\left[L(\mathbf{x})\right]_{T'} = AP\left[\mathbf{x}\right]_{S'},$$

so

$$\left[L(\mathbf{x})\right]_{T'} = Q^{-1}AP\left[\mathbf{x}\right]_{S'}.$$

This means that $Q^{-1}AP$ is the representation of L with respect to S' and T'. ■

Theorem 6.12 can be illustrated as follows: Consider Figure 6.3(a), where $I_V: V \to V$ is the identity operator on V, and $I_W: W \to W$ is the identity operator on W.

Let $\left[S\right]_{\text{coord}}$ be the set consisting of the coordinate vectors of each of the vectors in S, and let $\left[T\right]_{\text{coord}}$, $\left[S'\right]_{\text{coord}}$, and $\left[T'\right]_{\text{coord}}$ be defined similarly. In terms of matrix representations, Figure 6.3(a) becomes Figure 6.3(b).

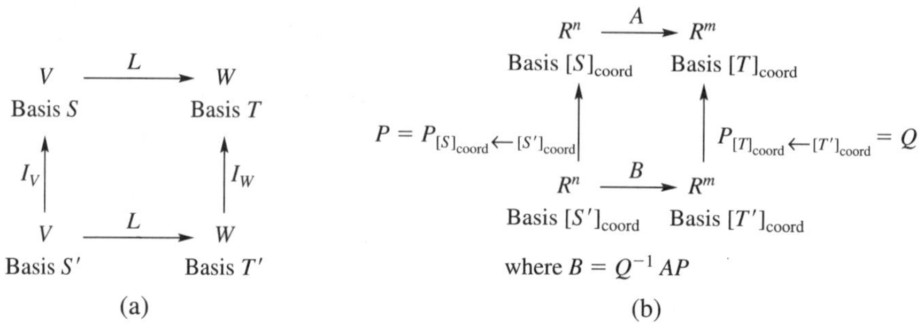

FIGURE 6.3

Using Theorem 2.13, we see that *two representations of a linear transformation with respect to different pairs of bases are equivalent.*

EXAMPLE 1

Let $L: R^3 \to R^2$ be defined by

$$L\left(\begin{bmatrix} u_1 \\ u_2 \\ u_3 \end{bmatrix}\right) = \begin{bmatrix} u_1 + u_3 \\ u_2 - u_3 \end{bmatrix}.$$

Consider the ordered bases

$$S = \left\{\begin{bmatrix} 1 \\ 0 \\ 0 \end{bmatrix}, \begin{bmatrix} 0 \\ 1 \\ 0 \end{bmatrix}, \begin{bmatrix} 0 \\ 0 \\ 1 \end{bmatrix}\right\} \quad \text{and} \quad S' = \left\{\begin{bmatrix} 1 \\ 1 \\ 0 \end{bmatrix}, \begin{bmatrix} 0 \\ 1 \\ 1 \end{bmatrix}, \begin{bmatrix} 0 \\ 0 \\ 1 \end{bmatrix}\right\}$$

for R^3, and

$$T = \left\{\begin{bmatrix} 1 \\ 0 \end{bmatrix}, \begin{bmatrix} 0 \\ 1 \end{bmatrix}\right\} \quad \text{and} \quad T' = \left\{\begin{bmatrix} 1 \\ 1 \end{bmatrix}, \begin{bmatrix} 1 \\ 3 \end{bmatrix}\right\}$$

for R^2. We can establish (verify) that $A = \begin{bmatrix} 1 & 0 & 1 \\ 0 & 1 & -1 \end{bmatrix}$ is the representation of L with respect to S and T.

The transition matrix P from S' to S is the matrix whose jth column is the coordinate vector of the jth vector in the basis S' with respect to S. Thus $P = \begin{bmatrix} 1 & 0 & 0 \\ 1 & 1 & 0 \\ 0 & 1 & 1 \end{bmatrix}$, and the transition matrix Q from T' to T is $Q = \begin{bmatrix} 1 & 1 \\ 1 & 3 \end{bmatrix}$.

Now $Q^{-1} = \begin{bmatrix} \frac{3}{2} & -\frac{1}{2} \\ -\frac{1}{2} & \frac{1}{2} \end{bmatrix}$. (We could also obtain Q^{-1} as the transition matrix from T to T'.) Then the representation of L with respect to S' and T' is

$$B = Q^{-1}AP = \begin{bmatrix} 1 & \frac{3}{2} & 2 \\ 0 & -\frac{1}{2} & -1 \end{bmatrix}.$$

On the other hand, we can compute the representation of L with respect to S' and T' directly. We have

$$L\left(\begin{bmatrix} 1 \\ 1 \\ 0 \end{bmatrix}\right) = \begin{bmatrix} 1 \\ 1 \end{bmatrix} = 1\begin{bmatrix} 1 \\ 1 \end{bmatrix} + 0\begin{bmatrix} 1 \\ 3 \end{bmatrix}, \quad \text{so} \quad \left[L\left(\begin{bmatrix} 1 \\ 1 \\ 0 \end{bmatrix}\right)\right]_{T'} = \begin{bmatrix} 1 \\ 0 \end{bmatrix};$$

$$L\left(\begin{bmatrix} 0 \\ 1 \\ 1 \end{bmatrix}\right) = \begin{bmatrix} 1 \\ 0 \end{bmatrix} = \frac{3}{2}\begin{bmatrix} 1 \\ 1 \end{bmatrix} - \frac{1}{2}\begin{bmatrix} 1 \\ 3 \end{bmatrix}, \quad \text{so} \quad \left[L\left(\begin{bmatrix} 0 \\ 1 \\ 1 \end{bmatrix}\right)\right]_{T'} = \begin{bmatrix} \frac{3}{2} \\ -\frac{1}{2} \end{bmatrix};$$

$$L\left(\begin{bmatrix} 0 \\ 0 \\ 1 \end{bmatrix}\right) = \begin{bmatrix} 1 \\ -1 \end{bmatrix} = 2\begin{bmatrix} 1 \\ 1 \end{bmatrix} - 1\begin{bmatrix} 1 \\ 3 \end{bmatrix}, \quad \text{so} \quad \left[L\left(\begin{bmatrix} 0 \\ 0 \\ 1 \end{bmatrix}\right)\right]_{T'} = \begin{bmatrix} 2 \\ -1 \end{bmatrix}.$$

Then the representation of L with respect to S' and T' is

$$\begin{bmatrix} 1 & \frac{3}{2} & 2 \\ 0 & -\frac{1}{2} & -1 \end{bmatrix},$$

which agrees with our earlier result. ■

Taking $V = W$ in Theorem 6.11, we obtain an important result, which we state as Corollary 6.3.

Corollary 6.3 Let $L: V \to V$ be a linear operator on an n-dimensional vector space. Let $S = \{\mathbf{v}_1, \mathbf{v}_2, \ldots, \mathbf{v}_n\}$ and $S' = \{\mathbf{v}'_1, \mathbf{v}'_2, \ldots, \mathbf{v}'_n\}$ be ordered bases for V with transition matrix P from S' to S. If A is the representation of L with respect to S, then $P^{-1}AP$ is the representation of L with respect to S'. ■

We may define the **rank** of a linear transformation $L: V \to W$, rank L, as the rank of any matrix representing L. This definition makes sense, since if A and B represent L, then A and B are equivalent (see Section 2.4); by Section 4.9 we know that equivalent matrices have the same rank.

We can now restate Theorem 6.6 as follows: If $L: V \to W$ is a linear transformation, then

$$\text{nullity } L + \text{rank } L = \dim V.$$

If $L: R^n \to R^m$ is a linear transformation defined by $L(\mathbf{x}) = A\mathbf{x}$, for $\mathbf{x}$ in R^n, where A is an $m \times n$ matrix, then it follows that nullity L = nullity A. (See Section 4.7.)

Theorem 6.13 Let $L: V \to W$ be a linear transformation. Then rank $L = \dim \text{range } L$.

Proof

Let $n = \dim V$, $m = \dim W$, and $r = \dim \operatorname{range} L$. Then, from Theorem 6.6, $\dim \ker L = n - r$. Let $\mathbf{v}_{r+1}, \mathbf{v}_{r+2}, \ldots, \mathbf{v}_n$ be a basis for $\ker L$. By Theorem 4.11, there exist vectors $\mathbf{v}_1, \mathbf{v}_2, \ldots, \mathbf{v}_r$ in V such that $S = \{\mathbf{v}_1, \mathbf{v}_2, \ldots, \mathbf{v}_r, \mathbf{v}_{r+1}, \ldots, \mathbf{v}_n\}$ is a basis for V. The vectors $\mathbf{w}_1 = L(\mathbf{v}_1)$, $\mathbf{w}_2 = L(\mathbf{v}_2)$, $\ldots$, $\mathbf{w}_r = L(\mathbf{v}_r)$ form a basis for range L. (They clearly span range L, and there are r of them, so Theorem 4.12 applies.) Again by Theorem 4.11, there exist vectors $\mathbf{w}_{r+1}, \mathbf{w}_{r+2}, \ldots, \mathbf{w}_m$ in W such that $T = \{\mathbf{w}_1, \mathbf{w}_2, \ldots, \mathbf{w}_r, \mathbf{w}_{r+1}, \mathbf{w}_{r+2}, \ldots, \mathbf{w}_m\}$ is a basis for W. Let A denote the $m \times n$ matrix that represents L with respect to S and T. The columns of A are (Theorem 6.9)

$$\left[L(\mathbf{v}_i)\right]_T = \left[\mathbf{w}_i\right]_T = \mathbf{e}_i, \qquad i = 1, 2, \ldots, r$$

and

$$\left[L(\mathbf{v}_j)\right]_T = \left[\mathbf{0}_W\right]_T = \mathbf{0}_{R^m}, \qquad j = r+1, r+2, \ldots, n.$$

Hence

$$A = \begin{bmatrix} I_r & O \\ O & O \end{bmatrix}.$$

Therefore

$$\operatorname{rank} L = \operatorname{rank} A = r = \dim \operatorname{range} L. \qquad \blacksquare$$

DEFINITION 6.6

If A and B are $n \times n$ matrices, we say that B is **similar** to A if there is a nonsingular matrix P such that $B = P^{-1}AP$.

We can show readily (Exercise 1) that B is similar to A if and only if A is similar to B. Thus we replace the statements "A is similar to B" and "B is similar to A" by "A and B are similar."

By Corollary 6.3 we then see that any two representations of a linear operator $L: V \to V$ are similar. Conversely, let $A = \left[a_{ij}\right]$ and $B = \left[b_{ij}\right]$ be similar $n \times n$ matrices and let V be an n-dimensional vector space. (We may take V as R^n.) We wish to show that A and B represent the same linear transformation $L: V \to V$ with respect to different bases. Since A and B are similar, $B = P^{-1}AP$ for some nonsingular matrix $P = \left[p_{ij}\right]$. Let $S = \{\mathbf{v}_1, \mathbf{v}_2, \ldots, \mathbf{v}_n\}$ be an ordered basis for V; from the proof of Theorem 6.10, we know that there exists a linear transformation $L: V \to V$, which is represented by A with respect to S. Then

$$\left[L(\mathbf{x})\right]_S = A\left[\mathbf{x}\right]_S. \tag{4}$$

We wish to prove that B also represents L with respect to some basis for V. Let

$$\mathbf{w}_j = \sum_{i=1}^{n} p_{ij}\mathbf{v}_i. \tag{5}$$

We first show that $T = \{\mathbf{w}_1, \mathbf{w}_2, \ldots, \mathbf{w}_n\}$ is also a basis for V. Suppose that

$$a_1\mathbf{w}_1 + a_2\mathbf{w}_2 + \cdots + a_n\mathbf{w}_n = \mathbf{0}.$$

Then, from (5), we have

$$a_1\left(\sum_{i=1}^{n} p_{i1}\mathbf{v}_i\right)+a_2\left(\sum_{i=1}^{n} p_{i2}\mathbf{v}_i\right)+\cdots+a_n\left(\sum_{i=1}^{n} p_{in}\mathbf{v}_i\right)=\mathbf{0},$$

which can be rewritten as

$$\left(\sum_{j=1}^{n} p_{1j}a_j\right)\mathbf{v}_1+\left(\sum_{j=1}^{n} p_{2j}a_j\right)\mathbf{v}_2+\cdots+\left(\sum_{j=1}^{n} p_{nj}a_j\right)\mathbf{v}_n=\mathbf{0}. \tag{6}$$

Since S is linearly independent, each of the coefficients in (6) is zero. Thus

$$\sum_{j=1}^{n} p_{ij}a_j=0, \qquad i=1,2,\ldots,n,$$

or equivalently,

$$P\mathbf{a}=\mathbf{0},$$

where $\mathbf{a}=\begin{bmatrix}a_1 & a_2 & \cdots & a_n\end{bmatrix}^T$. This is a homogeneous system of n equations in the n unknowns $a_1, a_2, \ldots, a_n$, whose coefficient matrix is P. Since P is nonsingular, the only solution is the trivial one. Hence $a_1=a_2=\cdots=a_n=0$, and T is linearly independent. Moreover, Equation (5) implies that P is the transition matrix from T to S (see Section 4.8). Thus

$$\left[\mathbf{x}\right]_S=P\left[\mathbf{x}\right]_T. \tag{7}$$

In (7), replace $\mathbf{x}$ by $L(\mathbf{x})$, giving

$$\left[L(\mathbf{x})\right]_S=P\left[L(\mathbf{x})\right]_T.$$

Using (4), we have

$$A\left[\mathbf{x}\right]_S=P\left[L(\mathbf{x})\right]_T,$$

and by (7),

$$AP\left[\mathbf{x}\right]_T=P\left[L(\mathbf{x})\right]_T.$$

Hence

$$\left[L(\mathbf{x})\right]_T=P^{-1}AP\left[\mathbf{x}\right]_T,$$

which means that the representation of L with respect to T is $B=P^{-1}AP$. We can summarize these results in the following theorem:

Theorem 6.14 Let V be any n-dimensional vector space and let A and B be any $n\times n$ matrices. Then A and B are similar if and only if A and B represent the same linear transformation $L\colon V\to V$ with respect to two ordered bases for V. ■

EXAMPLE 2 Let $L\colon R_3\to R_3$ be defined by

$$L\left(\begin{bmatrix}u_1 & u_2 & u_3\end{bmatrix}\right)=\begin{bmatrix}2u_1-u_3 & u_1+u_2-u_3 & u_3\end{bmatrix}.$$

Let $S = \{[1 \quad 0 \quad 0], [0 \quad 1 \quad 0], [0 \quad 0 \quad 1]\}$ be the natural basis for R_3. The representation of L with respect to S is

$$A = \begin{bmatrix} 2 & 0 & -1 \\ 1 & 1 & -1 \\ 0 & 0 & 1 \end{bmatrix}.$$

Now consider the ordered basis

$$S' = \{[1 \quad 0 \quad 1], [0 \quad 1 \quad 0], [1 \quad 1 \quad 0]\}$$

for R_3. The transition matrix P from S' to S is

$$P = \begin{bmatrix} 1 & 0 & 1 \\ 0 & 1 & 1 \\ 1 & 0 & 0 \end{bmatrix}; \quad \text{moreover, } P^{-1} = \begin{bmatrix} 0 & 0 & 1 \\ -1 & 1 & 1 \\ 1 & 0 & -1 \end{bmatrix}.$$

Then the representation of L with respect to S' is

$$B = P^{-1}AP = \begin{bmatrix} 1 & 0 & 0 \\ 0 & 1 & 0 \\ 0 & 0 & 2 \end{bmatrix}.$$

The same result can be calculated directly (verify). The matrices A and B are similar. ■

Observe that the matrix B obtained in Example 2 is diagonal. We can now ask a number of related questions. First, given $L: V \to V$, when can we choose a basis S for V such that the representation of L with respect to S is diagonal? How do we choose such a basis? In Example 2 we apparently pulled our basis S' "out of the air." If we cannot choose a basis giving a representation of L that is diagonal, can we choose a basis giving a matrix that is close in appearance to a diagonal matrix? What do we gain from having such simple representations? First, we already know from Section 6.4 that if A represents $L: V \to V$ with respect to some ordered basis S for V, then A^k represents $L \circ L \circ \cdots \circ L = L^k$ with respect to S; now, if A is similar to B, then B^k also represents L^k. Of course, if B is diagonal, then it is a trivial matter to compute B^k: The diagonal elements of B^k are those of B raised to the kth power. We shall also find that if A is similar to a diagonal matrix, then we can easily solve a homogeneous linear system of differential equations with constant coefficients. The answers to these questions are taken up in detail in Chapter 7.

Similar matrices enjoy other nice properties. For example, if A and B are similar, then $\text{Tr}(A) = \text{Tr}(B)$ (Exercise 8). (See Exercise 43 in Section 1.3 for a definition of trace.) Also, if A and B are similar, then A^k and B^k are similar for any positive integer k (Exercise 6).

We obtain one final result on similar matrices.

Theorem 6.15 If A and B are similar $n \times n$ matrices, then rank A = rank B.

Proof

We know from Theorem 6.14 that A and B represent the same linear transformation $L: R^n \to R^n$ with respect to different bases. Hence rank A = rank L = rank B. ■

Key Terms

Transition matrix
Equivalent representations
Rank of a linear transformation
Similar matrices

6.5 Exercises

1. Let A, B, and C be square matrices. Show that

(a) A is similar to A.

(b) If B is similar to A, then A is similar to B.

(c) If C is similar to B and B is similar to A, then C is similar to A.

2. Let L be the linear transformation defined in Exercise 2, Section 6.3.

(a) Find the transition matrix P from S' to S.

(b) Find the transition matrix from S to S' and verify that it is P^{-1}.

(c) Find the transition matrix Q from T' to T.

(d) Find the representation of L with respect to S' and T'.

(e) What is the dimension of range L?

3. Do Exercise 1(d) of Section 6.3, using transition matrices.

4. Do Exercise 8(b) of Section 6.3, using transition matrices.

5. Let $L: R^2 \to R^2$ be defined by

$$L\left(\begin{bmatrix} u_1 \\ u_2 \end{bmatrix}\right) = \begin{bmatrix} u_1 \\ -u_2 \end{bmatrix}.$$

(a) Find the representation of L with respect to the natural basis S for R^2.

(b) Find the representation of L with respect to the ordered basis

$$T = \left\{\begin{bmatrix} 1 \\ -1 \end{bmatrix}, \begin{bmatrix} 1 \\ 2 \end{bmatrix}\right\}.$$

(c) Verify that the matrices obtained in parts (a) and (b) are similar.

(d) Verify that the ranks of the matrices obtained in parts (a) and (b) are equal.

6. Show that if A and B are similar matrices, then A^k and B^k are similar for any positive integer k. (*Hint*: If $B = P^{-1}AP$, find $B^2 = BB$, and so on.)

7. Show that if A and B are similar, then A^T and B^T are similar.

8. Prove that if A and B are similar, then $\text{Tr}(A) = \text{Tr}(B)$. (*Hint*: See Exercise 43 in Section 1.3 for a definition of trace.)

9. Let $L: R_3 \to R_2$ be the linear transformation whose representation is

$$A = \begin{bmatrix} 2 & -1 & 3 \\ 3 & 1 & 0 \end{bmatrix}$$

with respect to the ordered bases

$$S = \{\begin{bmatrix} 1 & 0 & -1 \end{bmatrix}, \begin{bmatrix} 0 & 2 & 0 \end{bmatrix}, \begin{bmatrix} 1 & 2 & 3 \end{bmatrix}\}$$

and

$$T = \{\begin{bmatrix} 1 & -1 \end{bmatrix}, \begin{bmatrix} 2 & 0 \end{bmatrix}\}.$$

Find the representation of L with respect to the natural bases for R_3 and R_2.

10. Let $L: R^3 \to R^3$ be the linear transformation whose representation with respect to the natural basis for R^3 is $A = \begin{bmatrix} a_{ij} \end{bmatrix}$. Let

$$P = \begin{bmatrix} 0 & 1 & 1 \\ 1 & 0 & 1 \\ 1 & 1 & 0 \end{bmatrix}.$$

Find a basis T for R^3 with respect to which $B = P^{-1}AP$ represents L. (*Hint*: See the solution of Example 2.)

11. Let A and B be similar. Show that

(a) If A is nonsingular, then B is nonsingular.

(b) If A is nonsingular, then A^{-1} and B^{-1} are similar.

12. Do Exercise 13(b) of Section 6.3, using transition matrices.

13. Do Exercise 17(b) of Section 6.3, using transition matrices.

14. Do Exercise 10(b) of Section 6.3, using transition matrices.

15. Do Exercise 20(b) of Section 6.3, using transition matrices.

16. Prove that A and O are similar if and only if $A = O$.

17. Show that if A and B are similar matrices, then $\det(A) = \det(B)$.

6.6 Introduction to Homogeneous Coordinates (Optional)

The mathematics underlying computer graphics is closely connected to matrix multiplication. In either 2-space or 3-space we perform rotations, reflections, or scaling operations (dilations or contractions) by means of a matrix transformation, where the transformation is defined in terms of the standard matrix A that is 2×2 or 3×3, respectively. The effect of the transformation can be described as a function f that operates on a *picture*, viewed as a set of data, to produce an *image*:

$$f(\textit{picture}) = \textit{image}.$$

When the data representing the picture are properly arranged, the operation of the function f is executed by a matrix multiplication using the matrix A:

$$f(\textit{picture}) = A * \textit{picture} = \textit{image}.$$

Unfortunately, a general transformation includes not only rotations, reflections, and scalings, but also translations, or shifts (which "transport" the picture without distortion from where it used to be to a new location). Such a general transformation cannot be expressed using the associated standard matrix of corresponding size 2×2 or 3×3. To use matrix multiplication seamlessly in dealing with translations, we introduce the notion of homogeneous coordinates, discussed in this section, and obtain an extended representation of the data representing the *picture*.

2D and 3D Transformations

In our earlier discussion of matrix transformations and computer graphics in Sections 1.6 and 1.7, we used elementary two-dimensional graphics to illustrate the principles involved. Examples of such transformations from R^2 into R^2, called **2D transformations**, included rotations, reflections, scalings, and shears. Associated with each transformation is a 2×2 matrix A representing the matrix transformation (see Example 1), and the image of the vector $\mathbf{x}$ is the matrix product $A\mathbf{x}$. Composite transformations are performed by successive matrix multiplications. To obtain a single matrix that would perform a sequence of such transformations, we need only compute the corresponding product of the associated matrices.

EXAMPLE 1 Matrices associated with 2D transformations:

- Identity transformation: $\begin{bmatrix} 1 & 0 \\ 0 & 1 \end{bmatrix}$
- Reflection with respect to the x-axis: $\begin{bmatrix} 1 & 0 \\ 0 & -1 \end{bmatrix}$
- Reflection with respect to the y-axis: $\begin{bmatrix} -1 & 0 \\ 0 & 1 \end{bmatrix}$

- Reflection with respect to the line $y = x$: $\begin{bmatrix} 0 & 1 \\ 1 & 0 \end{bmatrix}$
- Rotation counterclockwise through a positive angle θ: $\begin{bmatrix} \cos\theta & -\sin\theta \\ \sin\theta & \cos\theta \end{bmatrix}$
- Scaling by h in the x-direction and by k in the y-direction: $\begin{bmatrix} h & 0 \\ 0 & k \end{bmatrix}$
- Shear in the x-direction by the factor k: $\begin{bmatrix} 1 & k \\ 0 & 1 \end{bmatrix}$
- Shear in the y-direction by the factor k: $\begin{bmatrix} 1 & 0 \\ k & 1 \end{bmatrix}$ ■

For matrix transformations from R^3 into R^3, called **3D transformations**, the corresponding matrix A is 3×3. As in the case of 2D transformations, the image of the vector $\mathbf{x}$ under a 3D transformation is the matrix product $A\mathbf{x}$; composite transformations are performed by successive matrix multiplications. Example 2 shows matrices associated with several 3D transformations.

EXAMPLE 2

Matrices associated with 3D transformations:

- Identity transformation: $\begin{bmatrix} 1 & 0 & 0 \\ 0 & 1 & 0 \\ 0 & 0 & 1 \end{bmatrix}$
- Scaling by h in the x-direction, by k in the y-direction, and by m in the z-direction: $\begin{bmatrix} h & 0 & 0 \\ 0 & k & 0 \\ 0 & 0 & m \end{bmatrix}$
- Rotation counterclockwise through a positive angle θ about the x-axis:
$$R_x = \begin{bmatrix} 1 & 0 & 0 \\ 0 & \cos\theta & -\sin\theta \\ 0 & \sin\theta & \cos\theta \end{bmatrix}$$
- Rotation counterclockwise through a positive angle β about the y-axis:
$$R_y = \begin{bmatrix} \cos\beta & 0 & \sin\beta \\ 0 & 1 & 0 \\ -\sin\beta & 0 & \cos\beta \end{bmatrix}$$
- Rotation counterclockwise through a positive angle γ about the z-axis:
$$R_z = \begin{bmatrix} \cos\gamma & -\sin\gamma & 0 \\ \sin\gamma & \cos\gamma & 0 \\ 0 & 0 & 1 \end{bmatrix}$$
■

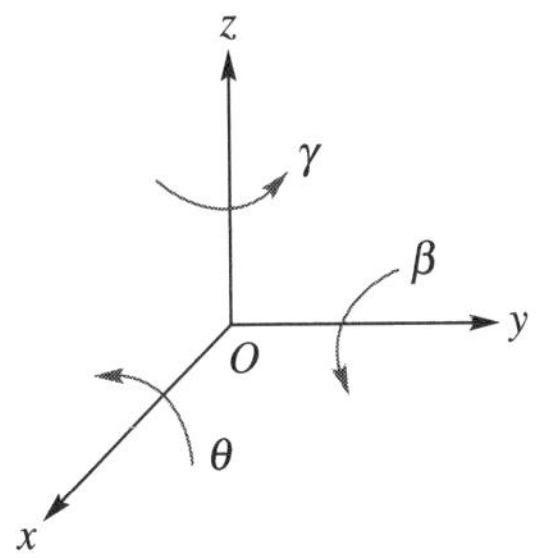

FIGURE 6.4

The matrices for the rotations given in Example 2 are consistent with a right-handed coordinate system. We illustrate this in Figure 6.4. Since the 2D and 3D transformations are matrix transformations, they are linear transformations.

Translations

The translation of a point, vector, or object defined by a set of points is performed by adding the same quantity Δx to each x-coordinate, the same quantity Δy to each y-coordinate, and in three dimensions the same quantity Δz to each z-coordinate. We emphasize that Δx, Δy, and Δz are not required to be equal in magnitude and can be positive, negative, or zero. We illustrate this in Figure 6.5 for a point in R^3, where the coordinates of the translated point are

$$(x^*, y^*, z^*) = (x + \Delta x, y + \Delta y, z + \Delta z).$$

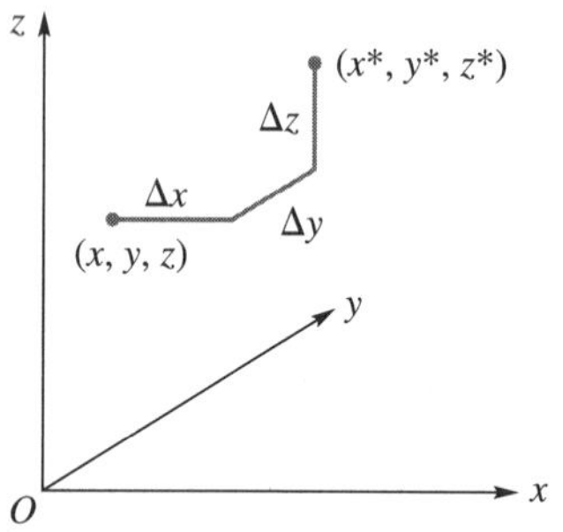

FIGURE 6.5

The point (x, y, z) can also be interpreted as the vector

$$\mathbf{v} = \begin{bmatrix} x \\ y \\ z \end{bmatrix}.$$

If we define the vector

$$\mathbf{t} = \begin{bmatrix} t_x \\ t_y \\ t_z \end{bmatrix} = \begin{bmatrix} \Delta x \\ \Delta y \\ \Delta z \end{bmatrix},$$

then the translation of the vector $\mathbf{v}$ by the vector $\mathbf{t}$ is the vector

$$\mathbf{v}^* = \mathbf{v} + \mathbf{t} = \begin{bmatrix} x + t_x \\ y + t_y \\ z + t_z \end{bmatrix}.$$

Figure 6.6 shows

$$\mathbf{v} = \begin{bmatrix} 5 \\ 7 \\ 10 \end{bmatrix}, \quad \mathbf{t} = \begin{bmatrix} -3 \\ 4 \\ -5 \end{bmatrix}, \quad \mathbf{v}^* = \begin{bmatrix} 2 \\ 11 \\ 5 \end{bmatrix}.$$

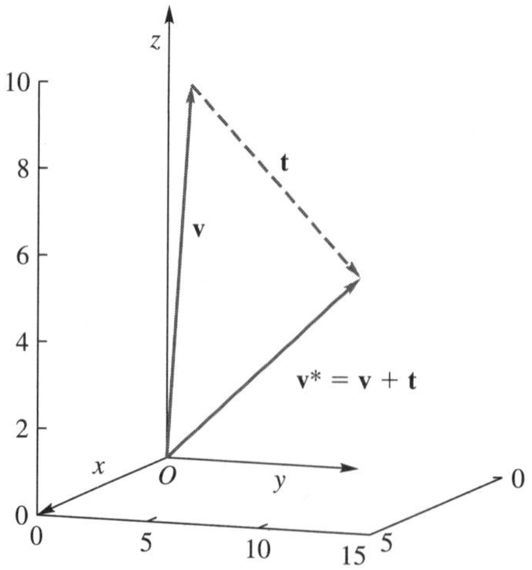

FIGURE 6.6

A line segment connects two points P_0 and P_1 and can be viewed as the set of points given by the expression $sP_1 + (1 - s)P_0$, where s is a parameter with $0 \le s \le 1$. If we let $\mathbf{r}_0$ be the vector from the origin to P_0 and $\mathbf{r}_1$ the vector from the origin to P_1, then the linear combination

$$s\mathbf{r}_1 + (1 - s)\mathbf{r}_0, \quad 0 \le s \le 1$$

sweeps out the line segment between P_0 and P_1. We illustrate this in 2-space in Figure 6.7, where the asterisks, $*$, lie on the line segment between P_0 and P_1.

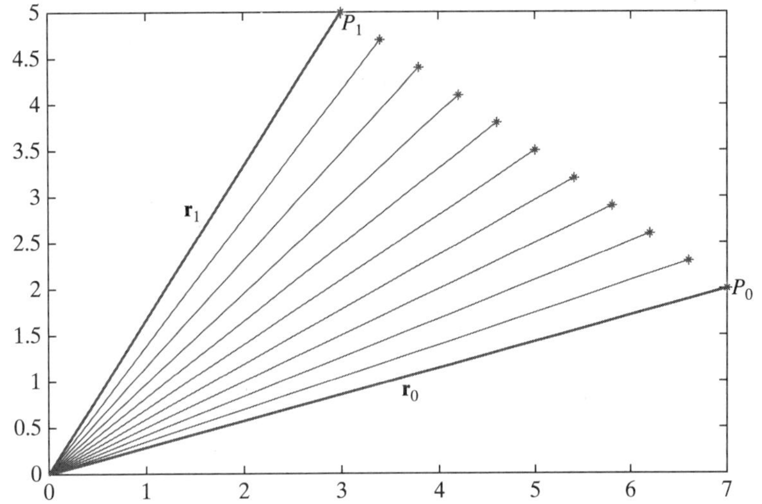

FIGURE 6.7

The translation of a line segment preserves its orientation and length. To carry out the translation of a line segment, we simply translate its endpoints. In 2-space, as shown in Figure 6.7, if $P_0 = (x_0, y_0)$, $P_1 = (x_1, y_1)$, and the line segment connecting these points is to be translated by

$$\mathbf{t} = \begin{bmatrix} t_x \\ t_y \end{bmatrix} = \begin{bmatrix} \Delta x \\ \Delta y \end{bmatrix},$$

then the endpoints of the translated line segment are $P_0^* = (x_0 + \Delta x, y_0 + \Delta y)$ and $P_1^* = (x_1 + \Delta x, y_1 + \Delta y)$. The corresponding vector expression is $s\mathbf{r}_1 + (1 - s)\mathbf{r}_0 + \mathbf{t}$, $0 \le s \le 1$. See Figure 6.8.

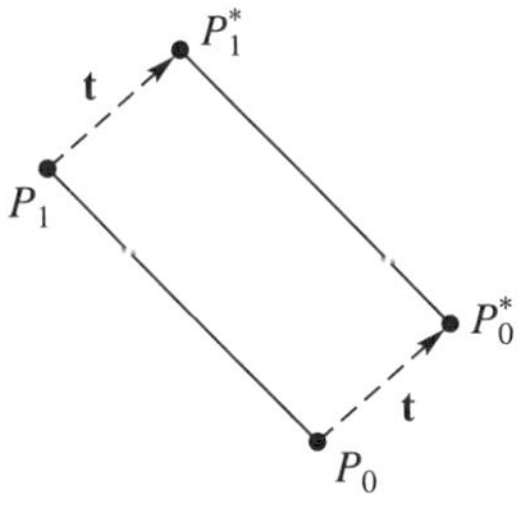

FIGURE 6.8

Example 3 demonstrates the use of translations with other 2D transformations from the list given in Example 1.

EXAMPLE 3

Let $\mathbf{v} = \begin{bmatrix} -4 \\ 6 \end{bmatrix}$ and $\mathbf{t} = \begin{bmatrix} 2 \\ 5 \end{bmatrix}$.

(a) Reflect $\mathbf{v}$ with respect to the line $y = x$ and then translate the resulting vector by $\mathbf{t}$.

(b) Scale the vector $\mathbf{v}$ by a factor of $-\frac{1}{2}$, translate the result by the vector $\mathbf{t}$, and rotate the result about the origin by an angle of $\frac{\pi}{6}$ radians.

Solution

(a) The matrix defining the reflection with respect to the line $y = x$ is

$$A = \begin{bmatrix} 0 & 1 \\ 1 & 0 \end{bmatrix}.$$

Then the desired vector, $\mathbf{w}$, is given by

$$\mathbf{w} = A\mathbf{v} + \mathbf{t} = \begin{bmatrix} 8 \\ 1 \end{bmatrix}.$$

(Verify.) See Figure 6.9(a).

(b) The matrix defining the scaling by $-\frac{1}{2}$ is

$$A = \begin{bmatrix} -\frac{1}{2} & 0 \\ 0 & -\frac{1}{2} \end{bmatrix},$$

and the matrix defining the rotation is

$$R = \begin{bmatrix} \cos\frac{\pi}{6} & -\sin\frac{\pi}{6} \\ \sin\frac{\pi}{6} & \cos\frac{\pi}{6} \end{bmatrix}.$$

Then the final vector $\mathbf{w}$ is computed as $\mathbf{w} = R(A\mathbf{v} + \mathbf{t})$. (Verify.) See Figure 6.9(b). ■

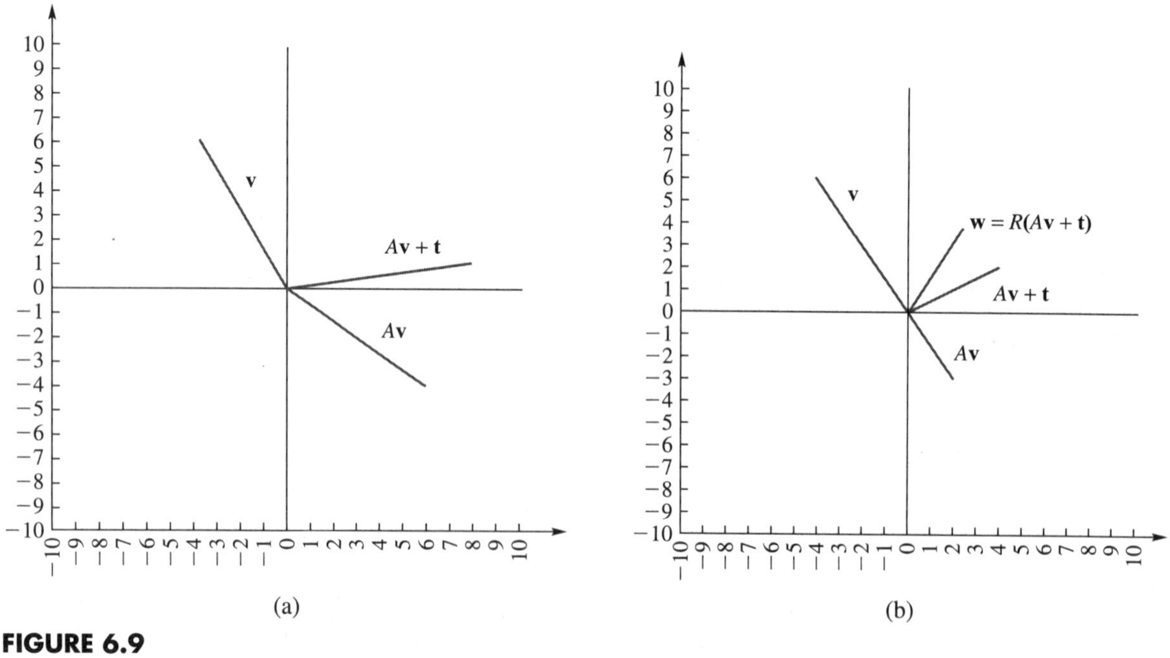

FIGURE 6.9

In Example 3(b), if we require only the scaling followed by the rotation, we could compute the matrix product $B = RA$ and obtain the desired vector, $\mathbf{w}$, as $\mathbf{w} = B\mathbf{v}$.

EXAMPLE 4 Figure 6.10(a) shows an object called boxhead, which is described as a set of vertices connected by line segments as shown. To translate boxhead by the vector

$$\mathbf{t} = \begin{bmatrix} -3 \\ 5 \end{bmatrix},$$

we translate each vertex of the object and reconnect the new vertices in the same order. However, we can show the translation as a sequence of steps of the figure, as follows. Let B be the $2 \times n$ matrix of vertices of boxhead. Then to translate boxhead by the vector $\mathbf{t}$, we add $\mathbf{t}$ to each column of B, giving the new vertices in a $2 \times n$ matrix B^*. If $T = \begin{bmatrix} \mathbf{t} & \mathbf{t} & \cdots & \mathbf{t} \end{bmatrix}$, which contains n repetitions of the vector $\mathbf{t}$, then $B^* = B + T$. To make boxhead with vertices in B move along a path to a final image of boxhead with vertices in B^*, we use the parametric expression $sB^* + (1 - s)B$ for $0 \leq s \leq 1$. Figure 6.10(b) shows boxhead's successive movements along the translation path for $s = 0.2, 0.4, 0.6, 0.8, 1.0$. For each value of s, we plot and connect the vertices given by the parametric expression. ■

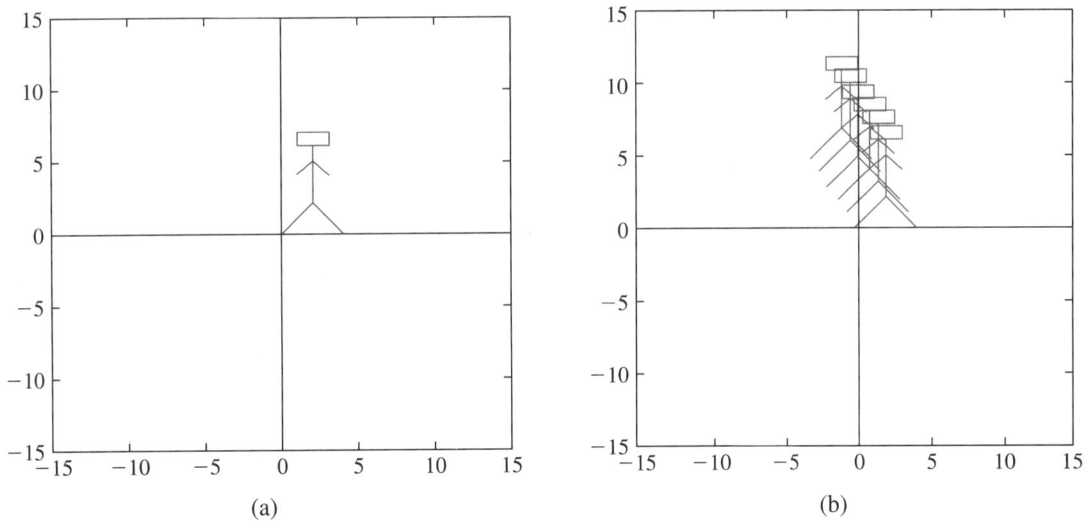

FIGURE 6.10

Unfortunately, a translation cannot be performed by a matrix multiplication using 2×2 matrices. This makes combining matrix transformations, like those in Examples 1 and 2, with translations awkward. The use of successive matrix transformations can be done by multiplying the corresponding matrices (in the proper order, since matrix multiplication is not commutative) prior to applying the transformation to the vector or object. However, when translations are interspersed, this convenience is lost. While the preceding examples were in the plane, that is, in 2-space, the corresponding behavior occurs in 3-space.

Homogeneous Coordinates

In order to have scalings, projections, and rotations interact smoothly with translations, we change the vector space in which we work.* As noted in Example 1, 2D transformations take a vector **v** from R^2 and compute its image as another vector in R^2 by multiplying it by a 2×2 matrix. Similarly, from Example 2, 3D transformations perform a matrix multiplication by using a 3×3 matrix. In order to use matrix multiplication to perform translations and avoid addition, as illustrated in Examples 3 and 4, we adjoin another component to vectors and border matrices with another row and column. This change is said to use **homogeneous coordinates**. To use homogeneous coordinates, we make the following identifications: A vector $\begin{bmatrix} x \\ y \end{bmatrix}$ in R^2 is identified with the vector $\begin{bmatrix} x \\ y \\ 1 \end{bmatrix}$ in R^3. The first two components are the same, and the third component is 1. Similarly, a vector

$$\begin{bmatrix} x \\ y \\ z \end{bmatrix}$$

in R^3 is identified with a vector

$$\begin{bmatrix} x \\ y \\ z \\ 1 \end{bmatrix}$$

in R^4. If

$$\begin{bmatrix} x \\ y \\ 1 \end{bmatrix}$$

is considered a point in homogeneous coordinates, then we merely plot the ordered pair (x, y). Similarly, for the point

$$\begin{bmatrix} x \\ y \\ z \\ 1 \end{bmatrix}$$

in homogeneous coordinates, we plot the ordered triple (x, y, z). Each of the matrices A associated with the matrix transformations in Example 1 is identified with a 3×3 matrix of the form

$$\begin{bmatrix} A & \begin{bmatrix} 0 \\ 0 \end{bmatrix} \\ \begin{bmatrix} 0 & 0 \end{bmatrix} & \begin{bmatrix} 1 \end{bmatrix} \end{bmatrix} = \begin{bmatrix} a_{11} & a_{12} & 0 \\ a_{21} & a_{22} & 0 \\ 0 & 0 & 1 \end{bmatrix}.$$

*We shall see that by using four-dimensional vectors, we can perform transformations in 3-space in an easy fashion.

EXAMPLE 5

(a) When using homogeneous coordinates for R^2, a reflection with respect to the y-axis is identified with the 3×3 matrix

$$\begin{bmatrix} -1 & 0 & 0 \\ 0 & 1 & 0 \\ 0 & 0 & 1 \end{bmatrix}.$$

(b) When using homogeneous coordinates for R^2, a rotation by an angle θ is identified with the 3×3 matrix

$$\begin{bmatrix} \cos\theta & -\sin\theta & 0 \\ \sin\theta & \cos\theta & 0 \\ 0 & 0 & 1 \end{bmatrix}.$$

■

Each of the matrices A associated with the matrix transformations from R^3 into R^3 in Example 2 is identified with a 4×4 matrix of the form

$$\begin{bmatrix} & A & & \begin{bmatrix} 0 \\ 0 \\ 0 \end{bmatrix} \\ \begin{bmatrix} 0 & 0 & 0 \end{bmatrix} & & & \begin{bmatrix} 1 \end{bmatrix} \end{bmatrix} = \begin{bmatrix} a_{11} & a_{12} & a_{13} & 0 \\ a_{21} & a_{22} & a_{23} & 0 \\ a_{31} & a_{32} & a_{33} & 0 \\ 0 & 0 & 0 & 1 \end{bmatrix}.$$

EXAMPLE 6

(a) When using homogeneous coordinates for R^3, a scaling by h, k, and m in the respective directions x, y, and z is identified with the 4×4 matrix

$$\begin{bmatrix} h & 0 & 0 & 0 \\ 0 & k & 0 & 0 \\ 0 & 0 & m & 0 \\ 0 & 0 & 0 & 1 \end{bmatrix}.$$

(b) When using homogeneous coordinates for R^3, a rotation about the z-axis through the angle γ is identified with the 4×4 matrix

$$\begin{bmatrix} \cos\gamma & -\sin\gamma & 0 & 0 \\ \sin\gamma & \cos\gamma & 0 & 0 \\ 0 & 0 & 1 & 0 \\ 0 & 0 & 0 & 1 \end{bmatrix}.$$

■

A translation in R^2 of $\mathbf{v} = \begin{bmatrix} x \\ y \end{bmatrix}$ by $\mathbf{t} = \begin{bmatrix} \Delta x \\ \Delta y \end{bmatrix}$ when using homogeneous coordinates is performed as

$$\begin{bmatrix} x \\ y \\ 1 \end{bmatrix} + \begin{bmatrix} \Delta x \\ \Delta y \\ 1 \end{bmatrix} = \begin{bmatrix} x + \Delta x \\ y + \Delta y \\ 1 \end{bmatrix}.$$

(Note that the third entries are *not* added.) This translation can be performed by matrix multiplication, using the 3×3 matrix

$$\begin{bmatrix} 1 & 0 & \Delta x \\ 0 & 1 & \Delta y \\ 0 & 0 & 1 \end{bmatrix}.$$

We have

$$\begin{bmatrix} 1 & 0 & \Delta x \\ 0 & 1 & \Delta y \\ 0 & 0 & 1 \end{bmatrix} \begin{bmatrix} x \\ y \\ 1 \end{bmatrix} = \begin{bmatrix} x + \Delta x \\ y + \Delta y \\ 1 \end{bmatrix}.$$

Similarly, a translation in R^3 by

$$\begin{bmatrix} \Delta x \\ \Delta y \\ \Delta z \\ 1 \end{bmatrix},$$

using homogeneous coordinates, is performed by multiplying by the matrix

$$\begin{bmatrix} 1 & 0 & 0 & \Delta x \\ 0 & 1 & 0 & \Delta y \\ 0 & 0 & 1 & \Delta z \\ 0 & 0 & 0 & 1 \end{bmatrix}.$$

Thus, homogeneous coordinates provide an easy way to use successive matrix multiplications to perform the composition of scalings, projections, rotations, and translations in both R^2 and R^3.

EXAMPLE 7 The composition of operations in Example 3(b), using homogeneous coordinates, can be expressed as $\mathbf{w} = R(T(A\mathbf{v})) = RTA\mathbf{v}$, where

$$T = \begin{bmatrix} 1 & 0 & 2 \\ 0 & 1 & 5 \\ 0 & 0 & 1 \end{bmatrix}$$

is the matrix determined by the translation of the vector

$$\mathbf{t} = \begin{bmatrix} 2 \\ 5 \\ 1 \end{bmatrix},$$

which is in homogeneous form. Here $\mathbf{v}$, A, and R must be expressed in homogeneous form, also. (Construct the matrix RTA and verify this result.) ■

EXAMPLE 8 Let S be the unit square in the plane with the lower left corner at the origin. S is to be rotated by 45° ($\pi/4$ radians) counterclockwise and then translated by the vector $\mathbf{t} = \begin{bmatrix} 1 \\ 1 \end{bmatrix}$. To determine the 3×3 matrix in homogeneous form that performs the composition of these transformations, we compute the rotation matrix R and multiply it (on the left) by the translation matrix T. We have

$$M = TR = \begin{bmatrix} 1 & 0 & 1 \\ 0 & 1 & 1 \\ 0 & 0 & 1 \end{bmatrix} \begin{bmatrix} \cos(\pi/4) & -\sin(\pi/4) & 0 \\ \sin(\pi/4) & \cos(\pi/4) & 0 \\ 0 & 0 & 1 \end{bmatrix}$$

$$= \begin{bmatrix} \sqrt{2}/2 & -\sqrt{2}/2 & 1 \\ \sqrt{2}/2 & \sqrt{2}/2 & 1 \\ 0 & 0 & 1 \end{bmatrix}.$$

If S is described by the coordinates of its vertices in homogeneous form arranged as columns of the matrix Q, where

$$Q = \begin{bmatrix} 0 & 1 & 1 & 0 \\ 0 & 0 & 1 & 1 \\ 1 & 1 & 1 & 1 \end{bmatrix},$$

then the image of the unit square is given by the matrix product MQ, which is (verify)

$$\begin{bmatrix} 1 & 1+\sqrt{2}/2 & 1 & 1-\sqrt{2}/2 \\ 1 & 1+\sqrt{2}/2 & 1+\sqrt{2} & 1+\sqrt{2}/2 \\ 1 & 1 & 1 & 1 \end{bmatrix}. \tag{1}$$

Figure 6.11 shows the result of the transformation. ■

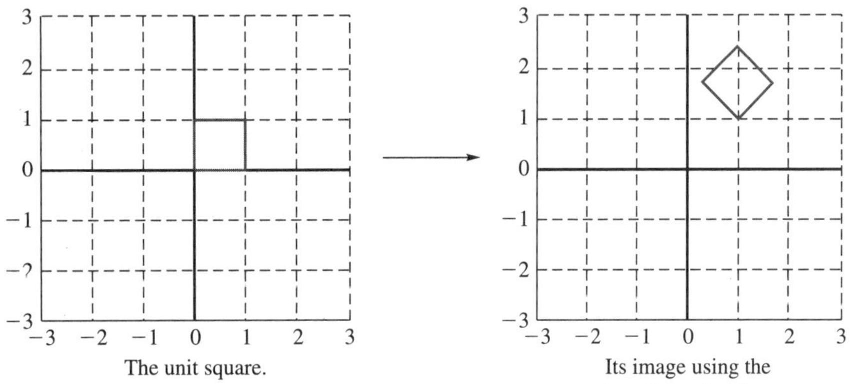

The unit square.

Its image using the transformation by matrix *M*.

FIGURE 6.11

EXAMPLE 9 Figure 6.12 shows an object called boxdog, which is described as a set of vertices connected by line segments in 3-space. Successive frames show composite images of boxdog rotated about the z-axis, then reflected with respect to the xy-plane, and finally translated by the vector

$$\mathbf{t} = \begin{bmatrix} 0 \\ 0 \\ -5 \end{bmatrix}.$$

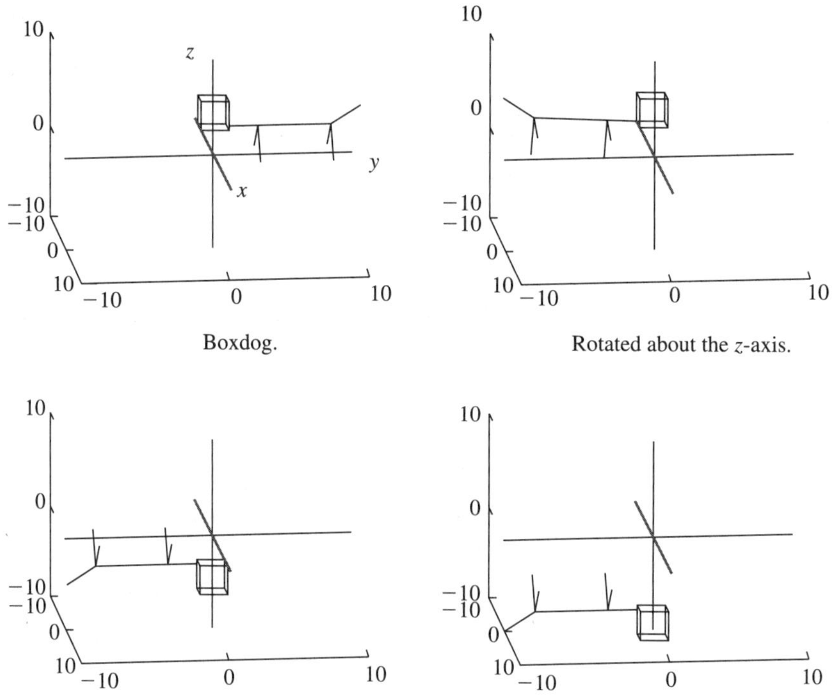

Boxdog. Rotated about the z-axis.

Reflected through the xy-plane. Translated by vector $\begin{bmatrix} 0 \\ 0 \\ -5 \end{bmatrix}$.

FIGURE 6.12

The sequence of frames in Figure 6.12 was obtained by using homogeneous coordinate representations for vectors in 3-space and multiplying the appropriate transformation matrices R, F, and T, which perform rotation, reflection, and translation, where

$$R = \begin{bmatrix} -1 & 0 & 0 & 0 \\ 0 & -1 & 0 & 0 \\ 0 & 0 & 1 & 0 \\ 0 & 0 & 0 & 1 \end{bmatrix}, \qquad F = \begin{bmatrix} 1 & 0 & 0 & 0 \\ 0 & 1 & 0 & 0 \\ 0 & 0 & -1 & 0 \\ 0 & 0 & 0 & 1 \end{bmatrix},$$

$$T = \begin{bmatrix} 1 & 0 & 0 & 0 \\ 0 & 1 & 0 & 0 \\ 0 & 0 & 1 & -5 \\ 0 & 0 & 0 & 1 \end{bmatrix}.$$

■

Key Terms

Image
2D transformations
Reflection
Rotation
Scaling
Shear
3D transformations
Translation
Homogeneous coordinates
Geometric sweep
Screw transformation

6.6 Exercises

1. Refer to Example 8.
   (a) Determine the matrix M in homogeneous form that would first perform the translation and then the rotation.
   (b) Sketch the unit square S in one coordinate plane and its image by the transformation represented by the matrix M in a second coordinate plane.
   (c) Are the images shown in Figure 6.11 and that determined in part (b) the same? Explain why or why not.

2. Let the columns of

$$S = \begin{bmatrix} 0 & 2 & 4 & 0 \\ 0 & 3 & 4 & 0 \\ 1 & 1 & 1 & 1 \end{bmatrix}$$

be the homogeneous form of the coordinates of the vertices of a triangle in the plane. Note that the first and last column are the same, indicating that the figure is a closed region.
   (a) In a coordinate plane, sketch the triangle determined by S. Connect its vertices with straight line segments.
   (b) The triangle determined by S is to be scaled by $\frac{1}{2}$ in both the x- and y-directions and then translated by $\mathbf{t} = \begin{bmatrix} -1 \\ -1 \end{bmatrix}$. Determine the 3×3 matrix M in homogeneous form that represents the composite transformation that first performs the scaling and then the translation.
   (c) Use the matrix M from part (b) to determine the image of the triangle and sketch it in a coordinate plane.
   (d) Determine the matrix Q in homogeneous form that would first perform the translation and then the scaling.
   (e) Use the matrix Q from part (d) to determine the image of the triangle and sketch it in a coordinate plane.
   (f) Are the images from parts (c) and (e) the same? Explain why or why not.

3. Let the columns of

$$S = \begin{bmatrix} 0 & 2 & 4 & 0 \\ 0 & 3 & 4 & 0 \\ 1 & 1 & 1 & 1 \end{bmatrix}$$

be the homogeneous form of the coordinates of the vertices of a triangle in the plane. Note that the first and last columns are the same, indicating that the figure is a closed region.
   (a) In a coordinate plane, sketch the triangle determined by S. Connect its vertices with straight line segments.
   (b) The triangle determined by S is to be rotated by 30° counterclockwise and then rotated again by 90° counterclockwise. Determine the 3×3 matrix M in homogeneous form that represents the composite transformation of these two rotations.
   (c) Use the matrix M from part (b) to determine the image of the triangle and sketch it in a coordinate plane.
   (d) Determine the matrix Q in homogeneous form that would first perform the rotation through 90° followed by the rotation through 30°.
   (e) Use the matrix Q from part (d) to determine the image of the triangle and sketch it in a coordinate plane.
   (f) Are the images from parts (c) and (e) the same? Explain why or why not.

4. A plane figure S is to be translated by $\mathbf{t} = \begin{bmatrix} -2 \\ 3 \end{bmatrix}$ and then the resulting figure translated by $\mathbf{v} = \begin{bmatrix} 4 \\ -1 \end{bmatrix}$.
   (a) Determine the 3×3 matrix M in homogeneous form that will perform this composition of translations.
   (b) Can the transformations be reversed? That is, is it possible to determine a matrix P that will return the image from part (a) to the original position? If it is possible, determine the matrix P. Explain your construction of P.

5. Let A be the 3×3 matrix in homogeneous form that reflects a plane figure about the x-axis and let B be the 3×3 matrix in homogeneous form that translates a plane figure by the vector $\mathbf{t} = \begin{bmatrix} 4 \\ -2 \end{bmatrix}$. Will the image be the same regardless of the order in which the transformations are performed? Explain your answer.

6. Let A be the 3×3 matrix in homogeneous form that translates a plane figure by $\mathbf{t} = \begin{bmatrix} -3 \\ -2 \end{bmatrix}$ and let B be the 3×3 matrix in homogeneous form that translates a plane figure by the vector $\mathbf{v} = \begin{bmatrix} 1 \\ 3 \end{bmatrix}$. Will the image be the same regardless of the order in which the transformations are performed? Explain your answer.

7. Determine the matrix in homogeneous form that produced the image of the rectangle depicted in the following figure:

Original figure

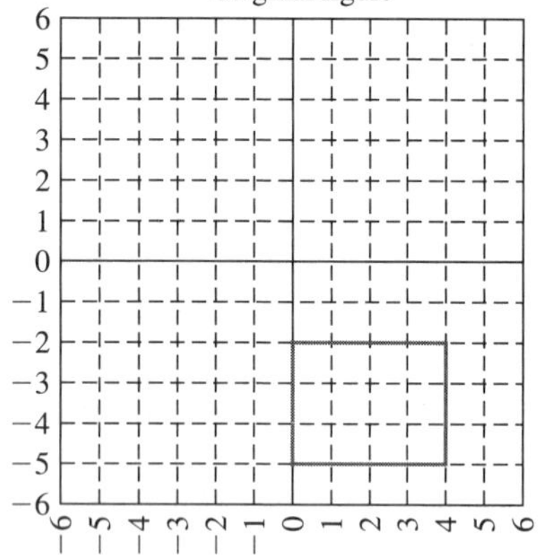

Image

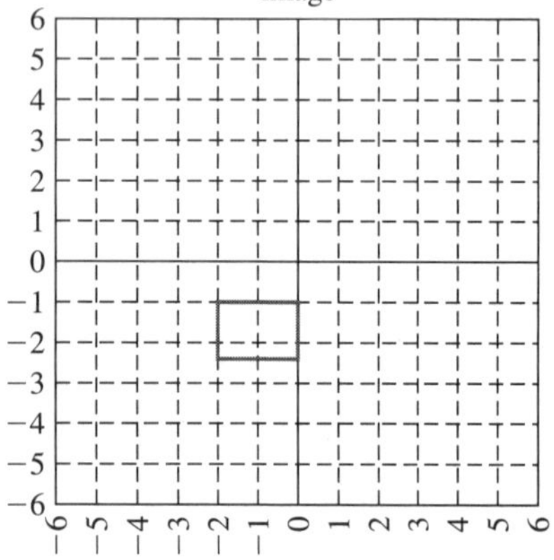

8. Determine the matrix in homogeneous form that produced the image of the triangle depicted in the following figure:

Original figure

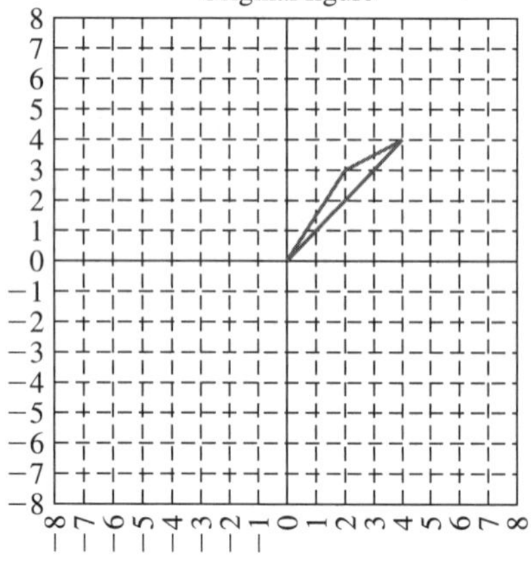

Image

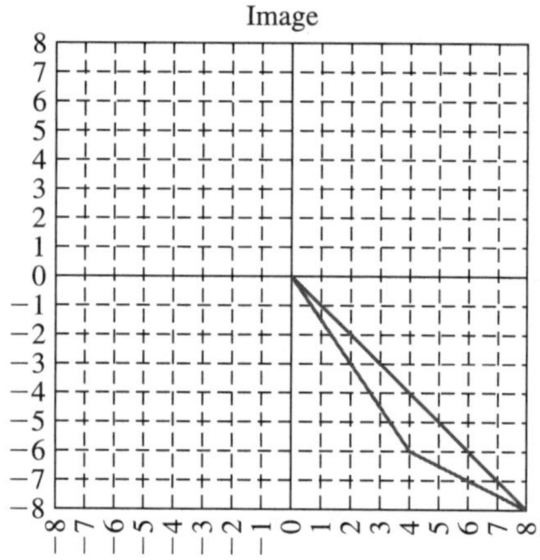

9. Determine the matrix in homogeneous form that produced the image of the semicircle depicted in the following figure:

Original figure

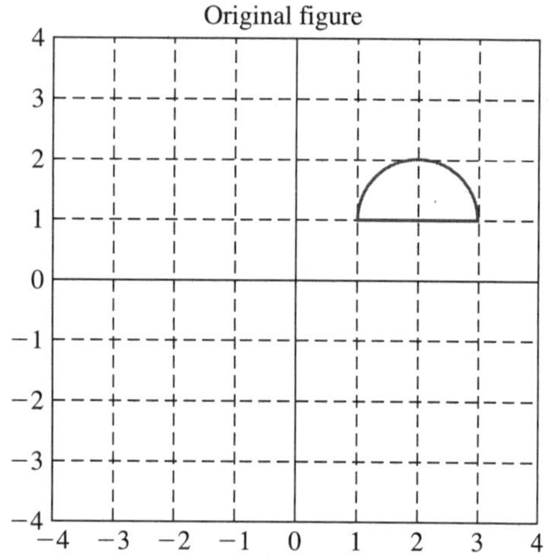

Image

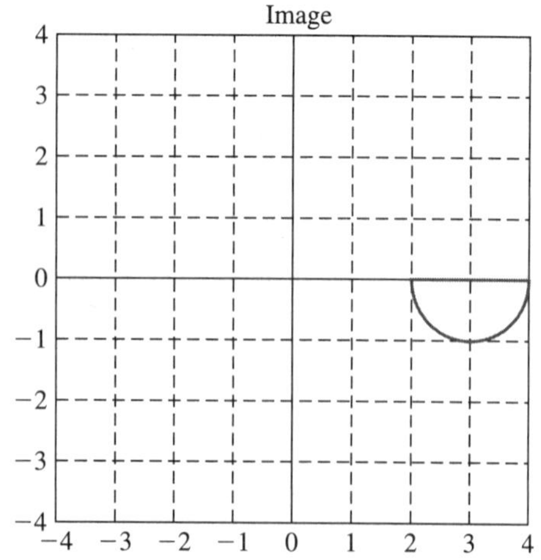

10. Determine the matrix in homogeneous form that produced the image of the semicircle depicted in the following figure:

Original figure

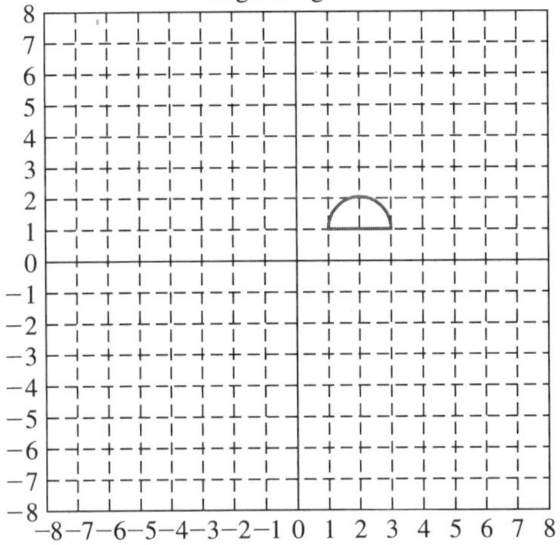

Image

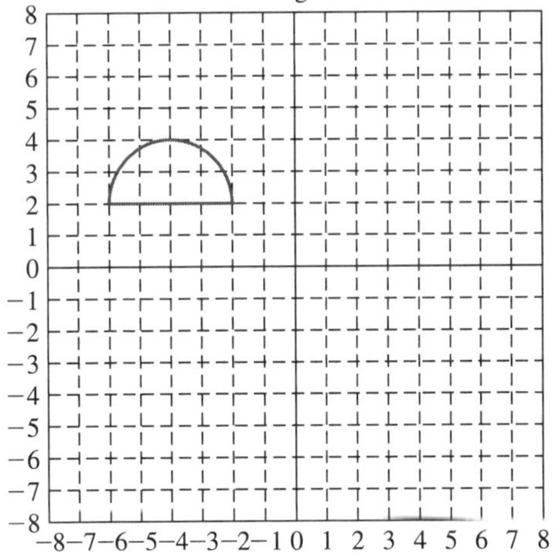

11. Determine the matrix in homogeneous form that produced the image of the semicircle depicted in the following figure:

Original figure

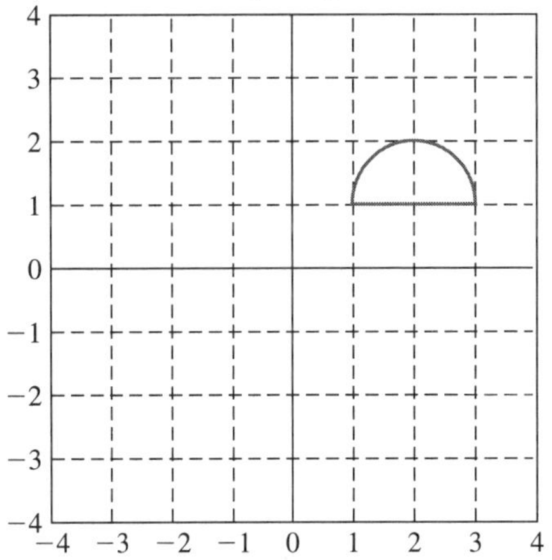

Image

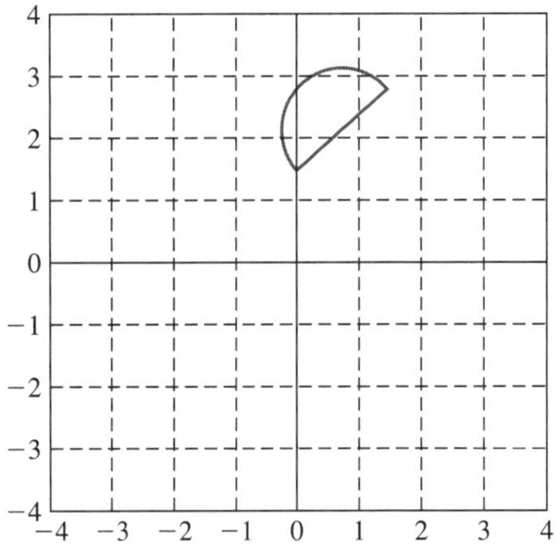

12. Determine the matrix in homogeneous form that produced the image of the semicircle depicted in the following figure:

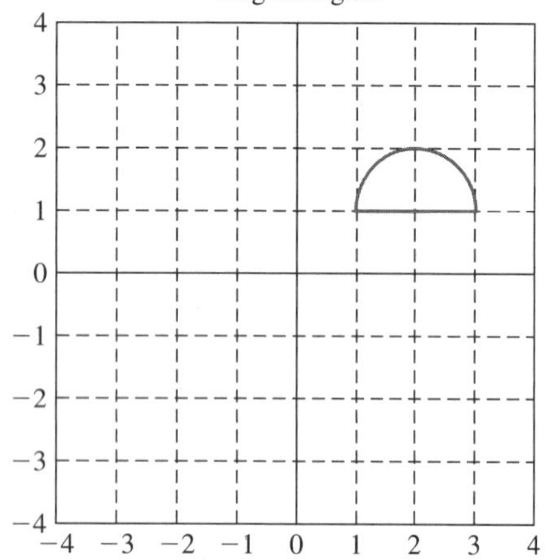

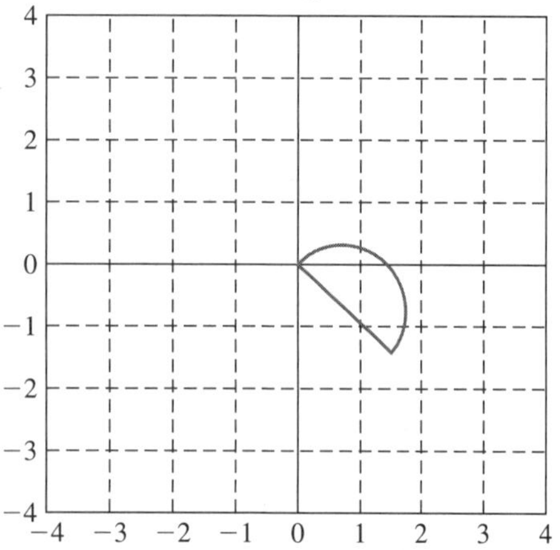

13. The semicircle depicted as the original figure in Exercise 12 is to be transformed by a rotation of 90° counterclockwise about the point $(1, 1)$.

(a) Sketch the image of the semicircle.

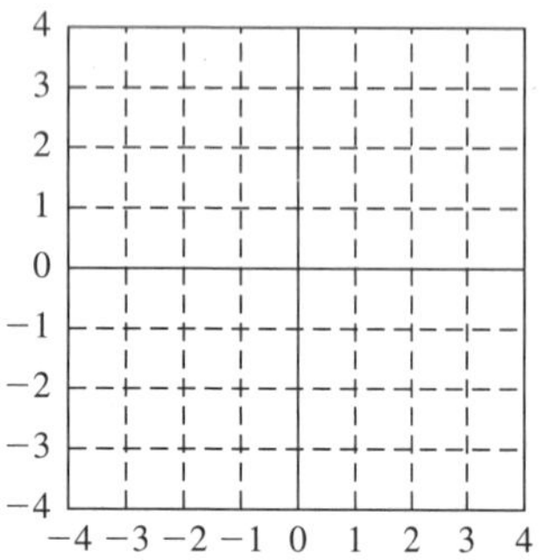

(b) Using transformations, we must first translate the semicircle to the origin, rotate it, and then translate it back to its original position. Determine the 3×3 matrix M in homogeneous form that will perform this transformation.

(c) Verify that your matrix M is correct by computing the image of the endpoints of the horizontal diameter of the semicircle.

14. **Sweeps** In calculus a surface of revolution is generated by rotating a curve $y = f(x)$ defined on an interval $[a, b]$ around an axis of revolution. For example, $y = x^2$ on $[0, 1]$ rotated about the x-axis generates the surface shown in Figure 6.13. We say that the surface is "swept out" by the curve rotating about the x-axis.

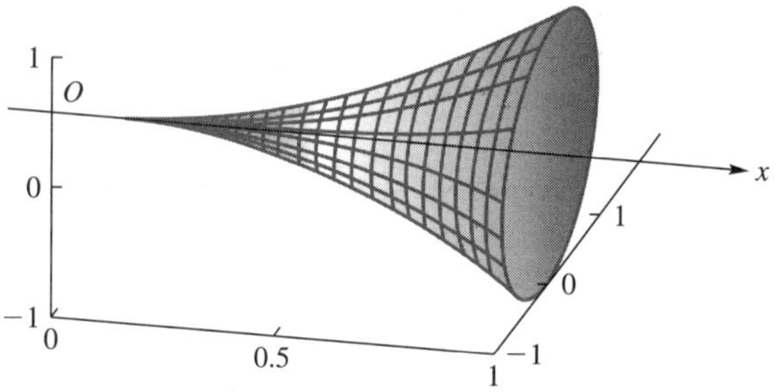

FIGURE 6.13

A continuous sequence of composite transformations of a shape is said to produce a *geometric sweep*. The set of transformations employed is called a *geometric sweep transformation*. In Figure 6.14 we show the swept surface of a triangular tube. The original triangle with vertices $A(3, 0, 0)$, $B(0, 0, 5)$, and $C(-6, 0, 0)$ is translated to sweep out a tube that terminates with the triangle whose vertices are $D(6, 6, 3)$, $E(6, 6, 8)$, and $F(-3, 6, 3)$.

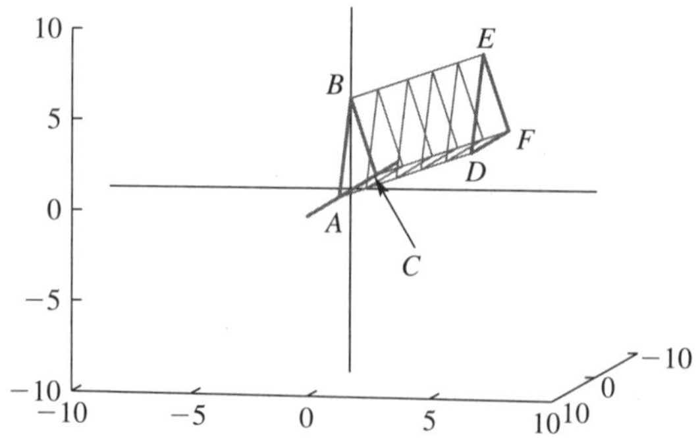

FIGURE 6.14

The figure shows several intermediate states of the sweep process. The tube is generated by translating triangle

ABC by the vector

$$\mathbf{t} = \begin{bmatrix} t_x \\ t_y \\ t_z \end{bmatrix} = \begin{bmatrix} 3 \\ 6 \\ 3 \end{bmatrix}.$$

The terminal face of the tube is triangle *DEF*. The triangular tube in Figure 6.14 was swept out by the geometric sweep transformation, which is given by the following matrix W:

$$W = \begin{bmatrix} 1 & 0 & 0 & s_{j+1}t_x \\ 0 & 1 & 0 & s_{j+1}t_y \\ 0 & 0 & 1 & s_{j+1}t_z \\ 0 & 0 & 0 & 1 \end{bmatrix}.$$

The matrix expression that performs the steps of the sweep, where T_0 represents triangle *ABC* and T_5 represents triangle *DEF*, is given by

$$T_{j+1} = WT_j, \qquad (*)$$
where $j = 0, 1, 2, 3, 4$ and $s_{j+1} = (0.2)(j + 1)$.

The values of the parameter s_j control the position of the intermediate triangles in the sweep from triangle *ABC* to triangle *DEF*. The number of steps in the sweep (here, there are five) is determined by the range of the index j with the requirements $s_1 < s_2 < \cdots < s_5 = 1$.

(a) Figure 6.15 shows the swept surface we generate by translating the triangle *ABC* along the y-axis. Point D in Figure 6.15 has coordinates (3, 10, 0). Construct the matrix W corresponding to this geometric sweep transformation.

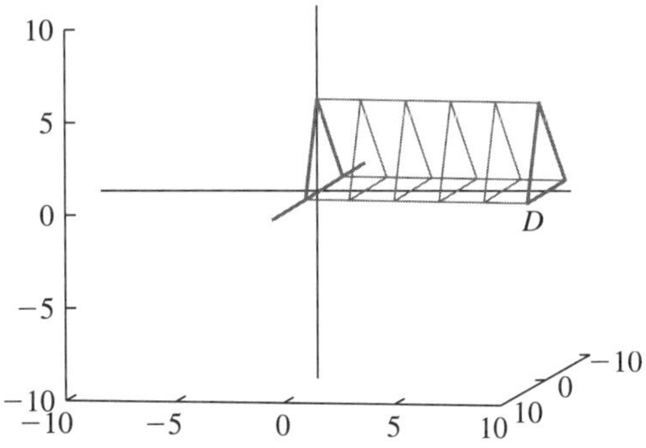

FIGURE 6.15

(b) We can generate a twisted sweep by including a parameterized rotation at each step of the sweep. If we wanted to rotate face T_0 through an angle θ about the y-axis as we sweep to obtain face T_5, then we parameterize the rotation in a manner similar to that used for the sweep translation matrix that appears in Equation (∗). Such a rotation matrix has the form

$$\begin{bmatrix} \cos(s_{j+1}\theta) & 0 & \sin(s_{j+1}\theta) & 0 \\ 0 & 1 & 0 & 0 \\ -\sin(s_{j+1}\theta) & 0 & \cos(s_{j+1}\theta) & 0 \\ 0 & 0 & 0 & 1 \end{bmatrix},$$

where $j = 0, 1, 2, 3, 4$ and $s_{j+1} = (0.2)(j+1)$. The result of the composite transformation of the translation sweep and the parameterized rotation, using $\theta = \pi/4$, is shown in Figure 6.16. Figure 6.16(a) shows the steps of the sweep, and Figure 6.16(b) emphasizes the tube effect.

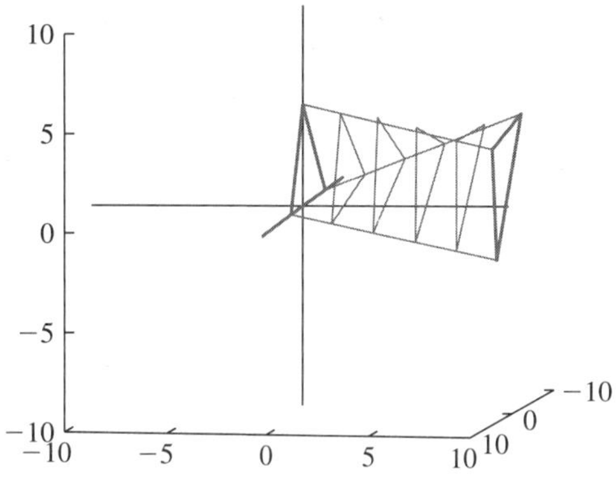

(a)

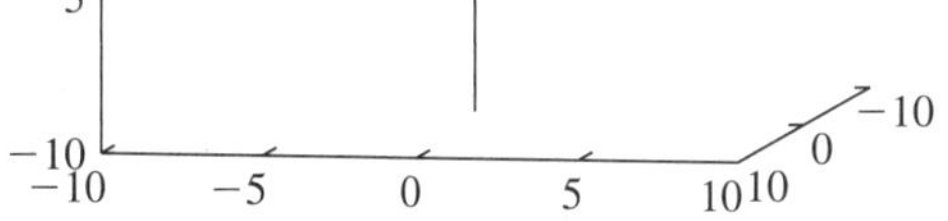

(b)

FIGURE 6.16

Determine the composite parameterized sweep matrix that will perform the sweep with a twist, as shown in Figure 6.16.

(c) The sweep process can be used with any type of transformation. Determine the parameterized sweep matrix that will perform the sweep with scaling of $\frac{1}{2}$ in the z-direction, as shown in Figure 6.17.

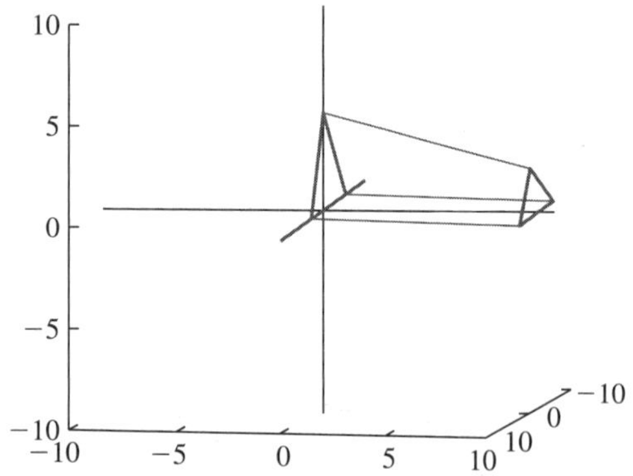

FIGURE 6.17

15. (***Screw Transformation***) A screw transformation is a three-dimensional transformation that consists of a translation together with a rotation about an axis parallel to the direction of the translation. (There is nothing comparable in two dimensions.) If the rotation is through an angle θ, then the translation is $k\theta$, where k is a fixed value. In order to show the screw effect, we parameterize the angle values from 0 to θ in the form $\theta_j = s_j\theta$ for $j = 0, 1, \ldots, n$, where $s_j = (j/n)\theta$. We then apply successive composite transformations of the rotation and the translation. In Figure 6.18(a) we take a vector $\mathbf{v}$, rotate it about the z-axis in steps of θ_j, and translate it along the z-axis in steps of $k\theta_j$, where $k = -1$. The angle $\theta = 8\pi$. To view the process, we leave an asterisk to indicate the endpoint of the images of vector $\mathbf{v}$. Figure 6.18(b) shows the generation of the track through an angle of 2π, and Figure 6.18(c) displays the screw effect of this transformation.

(a) Determine the parameterized matrix for a step of the screw transformation illustrated in Figure 6.19.

(b) In Figure 6.19 we have included, with the parameterized matrix for a step of the screw transformation, another transformation. Discuss the type of transformation included that would generate such behavior.

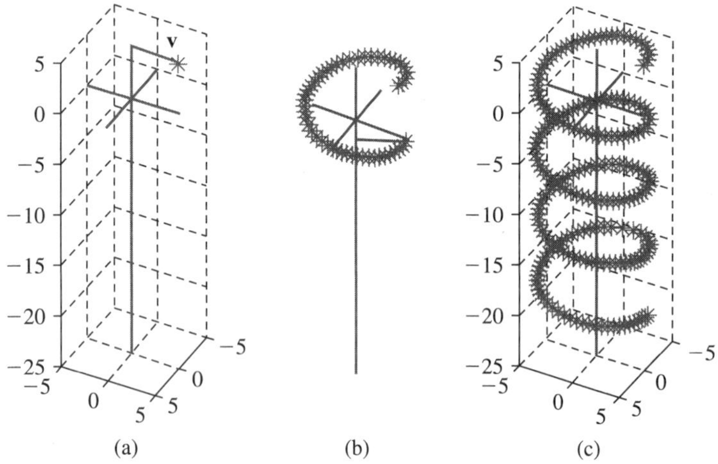

FIGURE 6.18

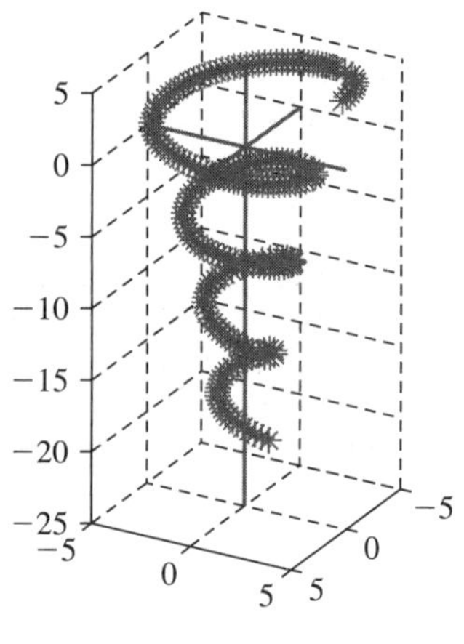

FIGURE 6.19

Supplementary Exercises

1. For an $n \times n$ matrix A, the trace of A, $\text{Tr}(A)$, is defined as the sum of the diagonal entries of A. (See Exercise 43 in Section 1.3.) Prove that the trace defines a linear transformation from M_{nn} to the vector space of all real numbers.

2. Let $L\colon M_{nm} \to M_{mn}$ be the function defined by $L(A) = A^T$ (the transpose of A), for A in V. Is L a linear transformation? Justify your answer.

3. Let V be the vector space of all $n \times n$ matrices and let $L\colon V \to V$ be the function defined by

$$L(A) = \begin{cases} A^{-1} & \text{if } A \text{ is nonsingular} \\ O & \text{if } A \text{ is singular} \end{cases}$$

for A in V. Is L a linear transformation? Justify your answer.

4. Let $L: R_3 \to R_3$ be a linear transformation for which we know that

$$L\left(\begin{bmatrix}1 & 0 & 1\end{bmatrix}\right) = \begin{bmatrix}1 & 2 & 3\end{bmatrix},$$
$$L\left(\begin{bmatrix}0 & 1 & 2\end{bmatrix}\right) = \begin{bmatrix}1 & 0 & 0\end{bmatrix}, \quad \text{and}$$
$$L\left(\begin{bmatrix}1 & 1 & 0\end{bmatrix}\right) = \begin{bmatrix}1 & 0 & 1\end{bmatrix}.$$

(a) What is $L\left(\begin{bmatrix}4 & 1 & 0\end{bmatrix}\right)$?

(b) What is $L\left(\begin{bmatrix}0 & 0 & 0\end{bmatrix}\right)$?

(c) What is $L\left(\begin{bmatrix}u_1 & u_2 & u_3\end{bmatrix}\right)$?

5. Let $L: P_1 \to P_1$ be a linear transformation defined by

$$L(t-1) = t+2 \quad \text{and} \quad L(t+1) = 2t+1.$$

(a) What is $L(5t+1)$?

(b) What is $L(at+b)$?

6. Let $L: P_2 \to P_2$ be a linear transformation defined by

$$L(at^2+bt+c) = (a+c)t^2 + (b+c)t.$$

(a) Is $t^2 - t - 1$ in $\ker L$?

(b) Is $t^2 + t - 1$ in $\ker L$?

(c) Is $2t^2 - t$ in range L?

(d) Is $t^2 - t + 2$ in range L?

(e) Find a basis for $\ker L$.

(f) Find a basis for range L.

7. Let $L: P_3 \to P_3$ be the linear transformation defined by

$$L(at^3+bt^2+ct+d) = (a-b)t^3 + (c-d)t.$$

(a) Is $t^3 + t^2 + t - 1$ in $\ker L$?

(b) Is $t^3 - t^2 + t - 1$ in $\ker L$?

(c) Is $3t^3 + t$ in range L?

(d) Is $3t^3 - t^2$ in range L?

(e) Find a basis for $\ker L$.

(f) Find a basis for range L.

8. Let $L: M_{22} \to M_{22}$ be the linear transformation defined by $L(A) = A^T$.

(a) Find a basis for $\ker L$.

(b) Find a basis for range L.

9. Let $L: M_{22} \to M_{22}$ be defined by

$$L(A) = A^T.$$

Let

$$S = \left\{\begin{bmatrix}1 & 0\\0 & 0\end{bmatrix}, \begin{bmatrix}0 & 1\\0 & 0\end{bmatrix}, \begin{bmatrix}0 & 0\\1 & 0\end{bmatrix}, \begin{bmatrix}0 & 0\\0 & 1\end{bmatrix}\right\}$$

and

$$T = \left\{\begin{bmatrix}1 & 1\\0 & 0\end{bmatrix}, \begin{bmatrix}0 & 1\\0 & 0\end{bmatrix}, \begin{bmatrix}0 & 0\\1 & 1\end{bmatrix}, \begin{bmatrix}1 & 0\\0 & 1\end{bmatrix}\right\}$$

be bases for M_{22}. Find the matrix of L with respect to

(a) S; **(b)** S and T; **(c)** T and S; **(d)** T.

10. For the vector space V and linear transformation L of Exercise 24 in Section 6.1, find a basis for $\ker L$.

11. Let V be the vector space of real-valued continuous functions on $[0, 1]$ and let $L: V \to R$ be given by $L(f) = f(0)$, for f in V.

(a) Show that L is a linear transformation.

(b) Describe the kernel of L and give examples of polynomials, quotients of polynomials, and trigonometric functions that belong to $\ker L$.

(c) If we redefine L by $L(f) = f\left(\frac{1}{2}\right)$, is it still a linear transformation? Explain.

12. ***(Calculus Required)*** Let $L: P_1 \to R$ be the linear transformation defined by

$$L(p(t)) = \int_0^1 p(t)\,dt.$$

(a) Find a basis for $\ker L$.

(b) Find a basis for range L.

(c) Verify Theorem 6.6 for L.

13. Let $L: P_2 \to P_2$ be the linear transformation defined by

$$L(at^2+bt+c) = (a+2c)t^2 + (b-c)t + (a-c).$$

Let $S = \{1, t, t^2\}$ and $T = \{t^2-1, t, t-1\}$ be ordered bases for P_2.

(a) Find the matrix of L with respect to S and T.

(b) If $p(t) = 2t^2 - 3t + 1$, compute $L(p(t))$, using the matrix obtained in part (a).

14. Let $L: P_1 \to P_1$ be a linear transformation that is represented by the matrix

$$A = \begin{bmatrix}2 & -3\\1 & 2\end{bmatrix}$$

with respect to the basis $S = \{p_1(t), p_2(t)\}$, where

$$p_1(t) = t-2 \quad \text{and} \quad p_2(t) = t+1.$$

(a) Compute $L(p_1(t))$ and $L(p_2(t))$.

(b) Compute $\left[L(p_1(t))\right]_S$ and $\left[L(p_2(t))\right]_S$.

(c) Compute $L(t+2)$.

15. Let $L: P_3 \to P_3$ be defined by

$$L(at^3 + bt^2 + ct + d) = 3at^2 + 2bt + c.$$

Find the matrix of L with respect to the basis $S = \{t^3, t^2, t, 1\}$ for P_3.

16. Consider R^n as an inner product space with the standard inner product and let $L: R^n \to R^n$ be a linear operator. Prove that for any vector $\mathbf{u}$ in R^n, $\|L(\mathbf{u})\| = \|\mathbf{u}\|$ if and only if

$$(L(\mathbf{u}), L(\mathbf{v})) = (\mathbf{u}, \mathbf{v})$$

for any vectors $\mathbf{u}$ and $\mathbf{v}$ in R^n. Such a linear operator is said to **preserve inner products**.

17. Let $L_1: V \to V$ and $L_2: V \to V$ be linear transformations on a vector space V. Prove that

$$(L_1 + L_2)^2 = L_1^2 + 2L_1 \circ L_2 + L_2^2$$

if and only if $L_1 \circ L_2 = L_2 \circ L_1$. (See Exercise 23 in Section 6.4.)

18. Let $\mathbf{u}$ and $\mathbf{v}$ be nonzero vectors in R^n. In Section 5.3 we defined the angle between $\mathbf{u}$ and $\mathbf{v}$ to be the angle θ such that

$$\cos\theta = \frac{(\mathbf{u}, \mathbf{v})}{\|\mathbf{u}\|\,\|\mathbf{v}\|}, \qquad 0 \le \theta \le \pi.$$

A linear operator $L: R^n \to R^n$ is called **angle preserving** if the angle between $\mathbf{u}$ and $\mathbf{v}$ is the same as that between $L(\mathbf{u})$ and $L(\mathbf{v})$. Prove that if L is inner product preserving (see Exercise 16), then it is angle preserving.

19. Let $L: R^n \to R^n$ be a linear operator that preserves inner products (see Exercise 16), and let the $n \times n$ matrix A represent L with respect to some ordered basis S for R^n.

(a) Prove that $\ker L = \{\mathbf{0}\}$.

(b) Prove that $AA^T = I_n$. (*Hint*: Use Supplementary Exercise 17 in Chapter 5.)

20. Let $L: V \to W$ be a linear transformation. If $\{\mathbf{v}_1, \mathbf{v}_2, \ldots, \mathbf{v}_k\}$ spans V, show that $\{L(\mathbf{v}_1), L(\mathbf{v}_2), \ldots, L(\mathbf{v}_k)\}$ spans range L.

21. Let V be an n-dimensional vector space and $S = \{\mathbf{v}_1, \mathbf{v}_2, \ldots, \mathbf{v}_n\}$ a basis for V. Define $L: R^n \to V$ as follows: If

$$\mathbf{w} = \begin{bmatrix} a_1 \\ a_2 \\ \vdots \\ a_n \end{bmatrix}$$

is a vector in R^n, let

$$L(\mathbf{w}) = a_1\mathbf{v}_1 + a_2\mathbf{v}_2 + \cdots + a_n\mathbf{v}_n.$$

Show that

(a) L is a linear transformation;

(b) L is one-to-one;

(c) L is onto.

22. Let V be an n-dimensional vector space. The vector space of all linear transformations from V into R^1 is called the **dual space** and is denoted by V^*. Prove that $\dim V = \dim V^*$. What does this imply?

23. If A and B are nonsingular, show that AB and BA are similar.

Chapter Review

True or False

1. If $L: V \to W$ is a linear transformation, then for $\mathbf{v}_1$ and $\mathbf{v}_2$ in V, we have $L(\mathbf{v}_1 - \mathbf{v}_2) = L(\mathbf{v}_1) - L(\mathbf{v}_2)$.

2. If $L: R^2 \to R^2$ is a linear transformation defined by

$$L\left(\begin{bmatrix} u_1 \\ u_2 \end{bmatrix}\right) = \begin{bmatrix} u_1 \\ 0 \end{bmatrix},$$

then L is one-to-one.

3. Let $L: V \to W$ be a linear transformation. If $\mathbf{v}_1$ and $\mathbf{v}_2$ are in $\ker L$, then so is $\operatorname{span}\{\mathbf{v}_1, \mathbf{v}_2\}$.

4. If $L: V \to W$ is a linear transformation, then for any vector $\mathbf{w}$ in W there is a vector $\mathbf{v}$ in V such that $L(\mathbf{v}) = \mathbf{w}$.

5. If $L: R^4 \to R^3$ is a linear transformation, then it is possible that $\dim \ker L = 1$ and $\dim \operatorname{range} L = 2$.

6. A linear transformation is invertible if and only if it is onto and one-to-one.

7. Similar matrices represent the same linear transformation with respect to different ordered bases.

8. If a linear transformation $L: V \to W$ is onto, then the image of V is all of W.

9. If a linear transformation $L: R^3 \to R^3$ is onto, then L is invertible.

10. If $L: V \to W$ is a linear transformation, then the image of a linearly independent set of vectors in V is a linearly independent set in W.

11. Similar matrices have the same determinant.

12. The determinant of 3×3 matrices defines a linear transformation from M_{33} to R^1.

Quiz

1. Let $L: M_{22} \to R$ be defined by

$$L\left(\begin{bmatrix} a & b \\ c & d \end{bmatrix}\right) = a - d + b - c.$$

Is L a linear transformation? Explain.

2. Let k be a fixed real scalar.

   (a) Show that $L_k: R^2 \to R^2$ defined by

$$L_k\left(\begin{bmatrix} a \\ b \end{bmatrix}\right) = \begin{bmatrix} a \\ ka + b \end{bmatrix}$$

   is a linear transformation.

   (b) Find the standard matrix representing L_k.

3. Let $L: R^3 \to R^3$ be the linear transformation defined by $L(\mathbf{x}) = A\mathbf{x}$, where A is a 3×3 matrix with no zero entries and all of its entries different.

   (a) Find a matrix A so that $\begin{bmatrix} 1 \\ 0 \\ 1 \end{bmatrix}$ is in $\ker L$.

   (b) Can A be chosen nonsingular? Explain.

4. Let $L: R^3 \to R^3$ be the linear transformation defined by

$$L\left(\begin{bmatrix} 1 \\ 1 \\ 0 \end{bmatrix}\right) = \begin{bmatrix} 0 \\ 1 \\ 1 \end{bmatrix}, \quad L\left(\begin{bmatrix} 1 \\ 2 \\ 1 \end{bmatrix}\right) = \begin{bmatrix} 2 \\ 1 \\ 2 \end{bmatrix},$$

$$L\left(\begin{bmatrix} 0 \\ 1 \\ 2 \end{bmatrix}\right) = \begin{bmatrix} 2 \\ 0 \\ 0 \end{bmatrix}.$$

Find $L\left(\begin{bmatrix} 3 \\ 1 \\ -5 \end{bmatrix}\right)$.

5. Let $L: R^2 \to R^2$ be the linear transformation defined by

$$L\left(\begin{bmatrix} x_1 \\ x_2 \end{bmatrix}\right) = \begin{bmatrix} x_2 - x_1 \\ 2x_1 + x_2 \end{bmatrix},$$

let S be the natural basis for R^2, and

$$T = \left\{\begin{bmatrix} 1 \\ 1 \end{bmatrix}, \begin{bmatrix} 2 \\ 1 \end{bmatrix}\right\}.$$

Find the representation of L with respect to T and S.

6. Let $L: R^2 \to R^3$ be the linear transformation defined by

$$L\left(\begin{bmatrix} x_1 \\ x_2 \end{bmatrix}\right) = \begin{bmatrix} x_1 - x_2 \\ x_1 + x_2 \\ 2x_1 \end{bmatrix}.$$

Let S be the natural basis for R^2 and

$$S' = \left\{\begin{bmatrix} 1 \\ 2 \end{bmatrix}, \begin{bmatrix} 0 \\ 1 \end{bmatrix}\right\}$$

be another basis for R^2, while T is the natural basis for R^3 and

$$T' = \left\{\begin{bmatrix} 1 \\ 1 \\ 0 \end{bmatrix}, \begin{bmatrix} 1 \\ 2 \\ 1 \end{bmatrix}, \begin{bmatrix} 2 \\ 0 \\ 0 \end{bmatrix}\right\}$$

is another basis for R^3.

   (a) Find the representation A of L with respect to S and T.

   (b) Find the transition matrix P from S' to S.

   (c) Find the transition matrix Q from T' to T.

   (d) Find the representation B of L with respect to S' and T'.

Discussion Exercises

1. Let $L: V \to W$ be a linear transformation. Explain the meaning of the following statement:

   The action of the linear transformation L is completely determined by its action on a basis for V.

2. Let $L: V \to W$ be a linear transformation such that the $m \times n$ matrix A represents L with respect to particular ordered bases for V and W. Explain the connection between the null space of A and properties of L.

3. Let $L: V \to W$ be a linear transformation that is represented by a matrix A and also by a matrix B, $A \neq B$. Explain why this is possible.

4. A translation in R^2 is a mapping T from R^2 to R^2 defined by $T(\mathbf{v}) = \mathbf{v} + \mathbf{b}$, where $\mathbf{v}$ is any vector in R^2 and $\mathbf{b}$ is a fixed vector in R^2.

   (a) Explain why T is not a linear transformation when $\mathbf{b} \neq \mathbf{0}$.

(b) Let S be the set of all points on the unit circle centered at the origin. Carefully describe the result of applying the translation T on S for each of the following choices of $\mathbf{b}$:

(i) $\mathbf{b} = \begin{bmatrix} 2 \\ 0 \end{bmatrix}$ **(ii)** $\mathbf{b} = \begin{bmatrix} 0 \\ -3 \end{bmatrix}$ **(iii)** $\mathbf{b} = \begin{bmatrix} -2 \\ 4 \end{bmatrix}$

5. The transformation $T: R^2 \to R^2$ defined by $T(\mathbf{v}) = A\mathbf{v} + \mathbf{b}$, where A is a specified 2×2 matrix, $\mathbf{v}$ is any vector in R^2, and $\mathbf{b}$ is a fixed vector in R^2, is called an **affine transformation**.

(a) Let $A = \begin{bmatrix} 3 & 0 \\ 0 & 3 \end{bmatrix}$ and $\mathbf{b} = \begin{bmatrix} 1 \\ 2 \end{bmatrix}$. Carefully describe $T(\mathbf{v})$.

(b) Let $A = \begin{bmatrix} 3 & 0 \\ 0 & 0 \end{bmatrix}$ and $\mathbf{b} = \begin{bmatrix} 1 \\ 2 \end{bmatrix}$. Carefully describe $T(\mathbf{v})$.

(c) Determine when an affine transformation is a linear transformation.

(d) Explain geometric differences between a translation and an affine transformation for $\mathbf{b} \neq \mathbf{0}$.

6. Any mapping of a vector space V into a vector space W that is not a linear transformation is called a **nonlinear transformation**. Let $V = W = M_{nn}$ and let A be any vector in V. Determine which of the following transformations are nonlinear:

(a) $T(A) = A^T$ **(b)** $T(A) = A^2$

(c) $T(A) = A + I_n$

(d) $T(A) = P^{-1}AP$ for P a fixed nonsingular matrix in M_{nn}.

(e) $T(A) = BAC$ for B and C fixed matrices in M_{nn}

(f) $T(A) = \begin{cases} O_n, & \text{if } A \text{ is singular} \\ A^{-1}, & \text{if } A \text{ is nonsingular} \end{cases}$

7. Suppose that $T: R^n \to R^n$ is a nonlinear transformation. (See Discussion Exercise 6.) Explain why the action of T cannot be completely described in terms of multiplication by an $n \times n$ matrix.

8. Let V be the set of points inside or on the unit circle centered at the origin in R^2. Let $\mathbf{v} = (a, b)$ be any point in V and

$T(\mathbf{v})$ = the point on the circumference of the unit circle closest to $\mathbf{v}$.

Explain why T is neither a linear nor a nonlinear transformation.

9. Explain why an affine transformation (see Exercise 5) maps a line segment to a line segment.

10. From Discussion Exercise 9, affine transformations seem simple enough. However, compositions of different affine transformations can lead to surprisingly complex patterns, which are referred to as **fractals**. To physically construct a simple fractal, proceed as follows:

(a) Take a strip of paper about 12 inches long and 1 inch wide. Fold it in half; then open it up so that there is a right angle between the halves. See Step 1 in Figure 6.20.

(b) Next, fold the original strip in half as before, and then fold it in half again in the same direction. Open it up, keeping the angles between the pieces a right angle; this is Step 2.

(c) Repeat the folding in half successively, in the same direction, and opening, maintaining a right angle between the pieces to get Step 3.

(d) Perform Step 4.

The figure shows a number of steps. After a large number of steps, the resulting figure resembles a dragon. Hence this fractal is called a **dragon curve**.

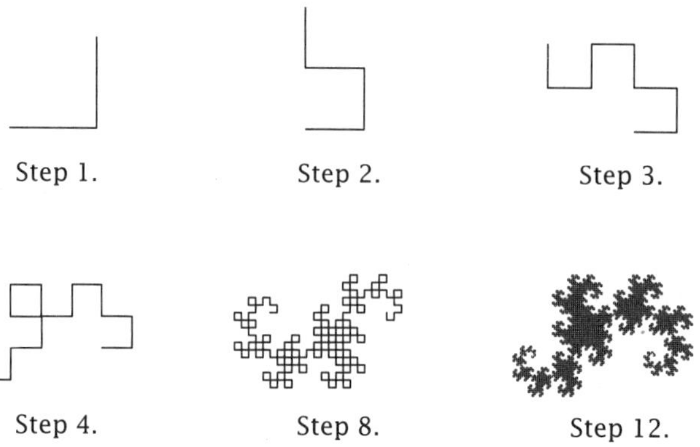

FIGURE 6.20

11. To use linear algebra in place of the paper folding in Discussion Exercise 10, we use the following approach: Let the original strip of paper be represented by a line segment that connects the point $A(0, 0)$ to the point $B\left(\frac{\sqrt{2}}{2}, \frac{\sqrt{2}}{2}\right)$. The first fold replaces this segment by the legs of an isosceles right triangle that has hypotenuse AB. (See Figure 6.21.) We view the image of the segment AB as the union of two line segments, each an image of AB.

(a) Construct an affine transformation T_1 that maps AB to CD. (*Hint*: The length of AB is 1, while CD has length $\frac{\sqrt{2}}{2}$ and is rotated $-45°$.)

(b) Construct an affine transformation T_2 that maps AB to DE. (*Hint*: The length of AB is 1, while DE has length $\frac{\sqrt{2}}{2}$, is rotated $45°$, and has undergone a translation.)

(c) Thus Step 1 is the union of $T_1(AB)$ and $T_2(AB)$. In a similar fashion, Step 2 is the union of $T_1(CD)$, $T_2(CD)$, $T_1(DE)$, and $T_2(DE)$. Both transformations T_1 and T_2 are applied to each segment of a step of the dragon curve to obtain the next step as a union of the images. Suppose that 10 steps of this process are performed. Determine the total number of line segments that must be drawn to get Step 10.[†]

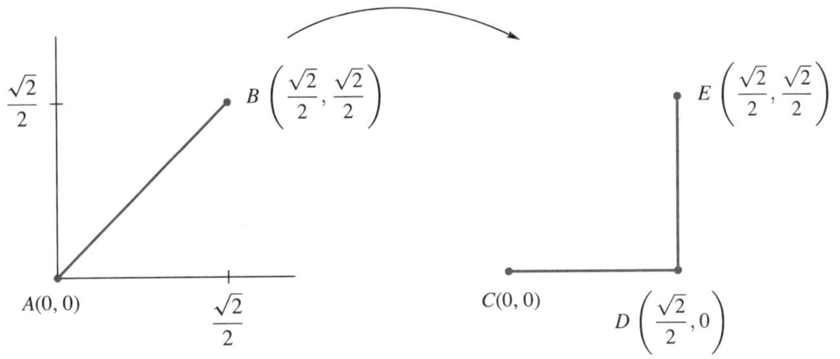

FIGURE 6.21

[†]The orientation of the dragon curve can vary, depending on the choice of the original segment AB. A web search on the dragon curve will yield additional information. More information on fractals is available on the internet.

CHAPTER

7 Eigenvalues and Eigenvectors

7.1 Eigenvalues and Eigenvectors

■ Definitions and Examples

Let $L: V \to V$ be a linear transformation of an n-dimensional vector space V into itself (a linear operator on V). Then L maps a vector $\mathbf{v}$ in V to another vector $L(\mathbf{v})$ in V. A question that arises in a wide variety of applications is that of determining whether $L(\mathbf{v})$ can be a multiple of $\mathbf{v}$. If V is R^n or C^n, then this question becomes one of determining whether $L(\mathbf{v})$ can be parallel to $\mathbf{v}$. Note that if $\mathbf{v} = \mathbf{0}$, then $L(\mathbf{0}) = \mathbf{0}$, so $L(\mathbf{0})$ is a multiple of $\mathbf{0}$ and is then parallel to $\mathbf{0}$. Thus we need consider only nonzero vectors in V.

EXAMPLE 1

Let $L: R^2 \to R^2$ be a reflection with respect to the x-axis, defined by

$$L(\mathbf{v}) = L\left(\begin{bmatrix} x_1 \\ x_2 \end{bmatrix}\right) = \begin{bmatrix} x_1 \\ -x_2 \end{bmatrix},$$

which we considered in Example 4 of Section 1.6. See Figure 7.1. If we want

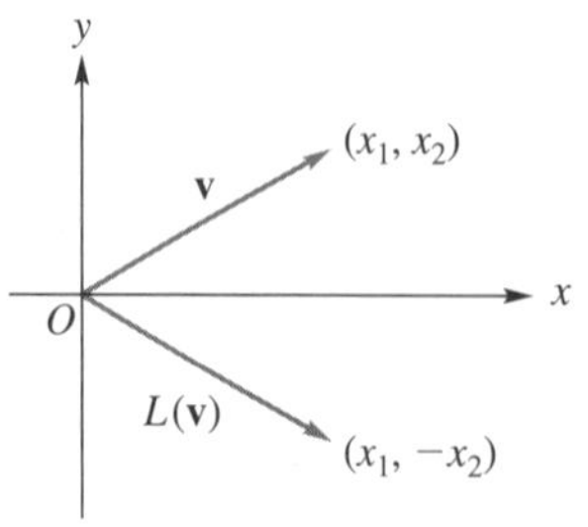

FIGURE 7.1

$L(\mathbf{v})$ to be parallel to a nonzero vector $\mathbf{v}$, we must have

$$L\left(\begin{bmatrix} x_1 \\ x_2 \end{bmatrix}\right) = \lambda \begin{bmatrix} x_1 \\ x_2 \end{bmatrix},$$

where λ is a scalar. Thus

$$\lambda \begin{bmatrix} x_1 \\ x_2 \end{bmatrix} = \begin{bmatrix} x_1 \\ -x_2 \end{bmatrix}$$

or

$$\begin{aligned} \lambda x_1 \quad &= \quad x_1 \\ \lambda x_2 &= -x_2. \end{aligned}$$

Since $\mathbf{v}$ is not the zero vector, both x_1 and x_2 cannot be zero. If $x_1 \neq 0$, then from the first equation it follows that $\lambda = 1$, and from the second equation we conclude that $x_2 = 0$. Thus $\mathbf{v} = \begin{bmatrix} r \\ 0 \end{bmatrix}$, $r \neq 0$, which represents any vector along the x-axis. If $x_2 \neq 0$, then from the second equation it follows that $\lambda = -1$, and from the first equation we have $x_1 = 0$. Thus $\mathbf{v} = \begin{bmatrix} 0 \\ s \end{bmatrix}$, $s \neq 0$, which represents any vector along the y-axis. Hence, for any vector $\mathbf{v}$ along the x-axis or along the y-axis, $L(\mathbf{v})$ will be parallel to $\mathbf{v}$. ■

EXAMPLE 2

Let $L: R^2 \to R^2$ be the linear operator defined by

$$L\left(\begin{bmatrix} x \\ y \end{bmatrix}\right) = \begin{bmatrix} \cos\phi & -\sin\phi \\ \sin\phi & \cos\phi \end{bmatrix} \begin{bmatrix} x \\ y \end{bmatrix},$$

a counterclockwise rotation through the angle ϕ, $0 \leq \phi < 2\pi$, as defined in Example 8 of Section 1.6. See Figure 7.2. It follows that if $\phi \neq 0$ and $\phi \neq \pi$, then for every vector $\mathbf{v} = \begin{bmatrix} x \\ y \end{bmatrix}$, $L(\mathbf{v})$ is oriented in a direction different from that of $\mathbf{v}$, so $L(\mathbf{v})$ and $\mathbf{v}$ are never parallel. If $\phi = 0$, then $L(\mathbf{v}) = \mathbf{v}$ (verify), which means that $L(\mathbf{v})$ and $\mathbf{v}$ are in the same direction and are thus parallel. If $\phi = \pi$, then $L(\mathbf{v}) = -\mathbf{v}$ (verify), so $L(\mathbf{v})$ and $\mathbf{v}$ are in opposite directions and are thus parallel. ■

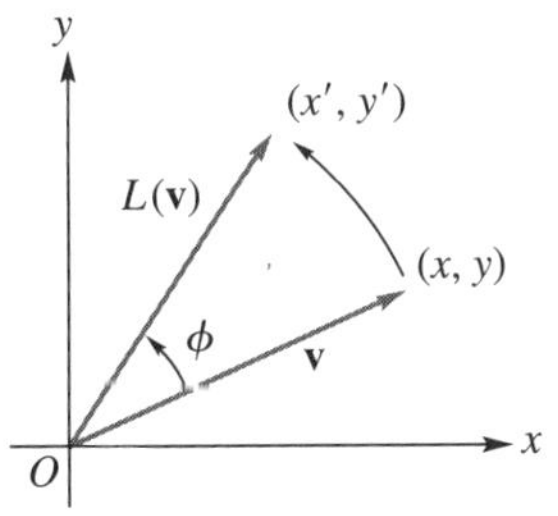

FIGURE 7.2

As we can see from Examples 1 and 2, the problem of determining all vectors $\mathbf{v}$ in an n-dimensional vector space V that are mapped by a given linear operator $L: V \to V$ to a multiple of $\mathbf{v}$ does not seem simple. We now formulate some terminology to study this important problem.

DEFINITION 7.1

Let $L: V \to V$ be a linear transformation of an n-dimensional vector space V into itself (a linear operator on V). The number λ is called an **eigenvalue** of L if there exists a *nonzero* vector $\mathbf{x}$ in V such that

$$L(\mathbf{x}) = \lambda \mathbf{x}. \tag{1}$$

Every nonzero vector $\mathbf{x}$ satisfying this equation is then called an **eigenvector** of L **associated with the eigenvalue** $\boldsymbol{\lambda}$. The word *eigenvalue* is a hybrid (*eigen* in German means *proper*). Eigenvalues are also called **proper values**, **characteristic**

values, and **latent values**; and eigenvectors are also called **proper vectors**, and so on, accordingly.

Remark In Definition 7.1, the number λ can be real or complex and the vector $\mathbf{x}$ can have real or complex components.

Note that if we do not require that $\mathbf{x}$ be nonzero in Definition 7.1, then *every* number λ would be an eigenvalue, since $L(\mathbf{0}) = \mathbf{0} = \lambda\mathbf{0}$. Such a definition would be of no interest. This is why we insist that $\mathbf{x}$ be nonzero.

EXAMPLE 3

Let $L: V \to V$ be the linear operator defined by $L(\mathbf{x}) = 2\mathbf{x}$. We can see that the only eigenvalue of L is $\lambda = 2$ and that every nonzero vector in V is an eigenvector of L associated with the eigenvalue $\lambda = 2$. ■

Example 3 shows that an eigenvalue λ can have associated with it many different eigenvectors. In fact, if $\mathbf{x}$ is an eigenvector of L associated with the eigenvalue λ [i.e., $L(\mathbf{x}) = \lambda\mathbf{x}$], then

$$L(r\mathbf{x}) = rL(\mathbf{x}) = r(\lambda\mathbf{x}) = \lambda(r\mathbf{x}),$$

for any real number r. Thus, if $r \neq 0$, then $r\mathbf{x}$ is also an eigenvector of L associated with λ so that eigenvectors are never unique.

EXAMPLE 4

Let $L: R^2 \to R^2$ be the linear operator defined by

$$L\left(\begin{bmatrix} a_1 \\ a_2 \end{bmatrix}\right) = \begin{bmatrix} -a_2 \\ a_1 \end{bmatrix}.$$

To find eigenvalues of L and associated eigenvectors, we proceed as follows: We need to find a number λ such that

$$L\left(\begin{bmatrix} a_1 \\ a_2 \end{bmatrix}\right) = \lambda \begin{bmatrix} a_1 \\ a_2 \end{bmatrix}.$$

Then

$$\begin{bmatrix} -a_2 \\ a_1 \end{bmatrix} = \lambda \begin{bmatrix} a_1 \\ a_2 \end{bmatrix},$$

so

$$\begin{aligned} -a_2 &= \lambda a_1 \\ a_1 &= \lambda a_2, \end{aligned}$$

so

$$-a_2 = \lambda^2 a_2.$$

If $a_2 \neq 0$, then $\lambda^2 = -1$. Hence

$$\lambda = i \quad \text{and} \quad \lambda = -i.$$

This means that there is no vector $\begin{bmatrix} a_1 \\ a_2 \end{bmatrix}$ in R^2 such that $L\left(\begin{bmatrix} a_1 \\ a_2 \end{bmatrix}\right)$ is parallel to $\begin{bmatrix} a_1 \\ a_2 \end{bmatrix}$. If we now consider L as defined previously to map C^2 into C^2, then L has the eigenvalue $\lambda = i$ with associated eigenvector $\begin{bmatrix} i \\ 1 \end{bmatrix}$ (verify), and the eigenvalue $\lambda = -i$ with associated eigenvector $\begin{bmatrix} -i \\ 1 \end{bmatrix}$ (verify). ■

EXAMPLE 5 Let $L\colon R^2 \to R^2$ be the linear operator defined by

$$L\left(\begin{bmatrix} a_1 \\ a_2 \end{bmatrix}\right) = \begin{bmatrix} a_2 \\ a_1 \end{bmatrix}.$$

It follows that

$$L\left(\begin{bmatrix} r \\ r \end{bmatrix}\right) = 1\begin{bmatrix} r \\ r \end{bmatrix} \quad \text{and} \quad L\left(\begin{bmatrix} r \\ -r \end{bmatrix}\right) = -1\begin{bmatrix} r \\ -r \end{bmatrix}.$$

Thus any vector of the form $\begin{bmatrix} r \\ r \end{bmatrix}$, where r is any nonzero real number—such as $\mathbf{x}_1 = \begin{bmatrix} 1 \\ 1 \end{bmatrix}$—is an eigenvector of L associated with the eigenvalue $\lambda = 1$; any vector of the form $\begin{bmatrix} r \\ -r \end{bmatrix}$, where r is any nonzero real number, such as $\mathbf{x}_2 = \begin{bmatrix} 1 \\ -1 \end{bmatrix}$, is an eigenvector of L associated with the eigenvalue $\lambda = -1$ (see Figure 7.3). ■

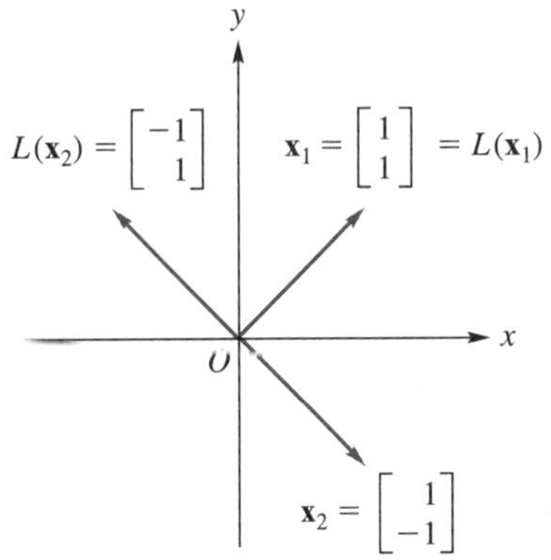

FIGURE 7.3

EXAMPLE 6 Let $L\colon R^2 \to R^2$ be counterclockwise rotation through the angle ϕ, $0 \le \phi < 2\pi$, as defined in Example 2. It follows from our discussion in Example 2 that $\phi = 0$ and $\phi = \pi$ are the only angles for which L has eigenvalues. Thus, if $\phi = 0$, then $\lambda = 1$ is the only eigenvalue of L and every nonzero vector in R^2 is an eigenvector of L associated with the eigenvalue $\lambda = 1$. The geometric approach used in Example 2 shows that the linear operator L has no real eigenvalues and associated eigenvectors when $\phi = \pi/4$. Let us now proceed algebraically and consider the linear operator as defined in Example 2, but now mapping C^2 into C^2. In this case, we find that

$$\lambda = \frac{\sqrt{2}}{2} + \frac{\sqrt{2}}{2}i$$

is an eigenvalue of L with associated eigenvector $\begin{bmatrix} i \\ 1 \end{bmatrix}$ (verify) and

$$\lambda = \frac{\sqrt{2}}{2} - \frac{\sqrt{2}}{2}i$$

is an eigenvalue of L with associated eigenvector $\begin{bmatrix} i \\ -1 \end{bmatrix}$ (verify). ■

EXAMPLE 7

Let $L\colon R_2 \to R_2$ be defined by $L\left(\begin{bmatrix} x_1 & x_2 \end{bmatrix}\right) = \begin{bmatrix} 0 & x_2 \end{bmatrix}$. We can then see that

$$L\left(\begin{bmatrix} r & 0 \end{bmatrix}\right) = \begin{bmatrix} 0 & 0 \end{bmatrix} = 0\begin{bmatrix} r & 0 \end{bmatrix},$$

so a vector of the form $\begin{bmatrix} r & 0 \end{bmatrix}$, where r is any nonzero real number (such as $\begin{bmatrix} 2 & 0 \end{bmatrix}$), is an eigenvector of L associated with the eigenvalue $\lambda = 0$. Also,

$$L\left(\begin{bmatrix} 0 & r \end{bmatrix}\right) = \begin{bmatrix} 0 & r \end{bmatrix} = 1\begin{bmatrix} 0 & r \end{bmatrix},$$

so a vector of the form $\begin{bmatrix} 0 & r \end{bmatrix}$, where r is any nonzero real number such as $\begin{bmatrix} 0 & 1 \end{bmatrix}$, is an eigenvector of L associated with the eigenvalue $\lambda = 1$. ■

By definition, the zero vector cannot be an eigenvector. However, Example 7 shows that the scalar zero can be an eigenvalue.

EXAMPLE 8

(Calculus Required) Although we introduced this chapter with the requirement that V be an n-dimensional vector space, the notions of eigenvalues and eigenvectors can be considered for infinite-dimensional vector spaces. In this example we look at such a situation.

Let V be the vector space of all real-valued functions of a single variable that have derivatives of all orders. Let $L\colon V \to V$ be the linear operator defined by

$$L(f) = f'.$$

Then the problem presented in Definition 7.1 can be stated as follows: Can we find a number λ and a function $f \neq 0$ in V so that

$$L(f) = \lambda f? \tag{2}$$

If $y = f(x)$, then (2) can be written as

$$\frac{dy}{dx} = \lambda y. \tag{3}$$

Equation (3) states that the quantity y is one whose rate of change, with respect to x, is proportional to y itself. Examples of physical phenomena in which a quantity satisfies (3) include growth of human population, growth of bacteria and other organisms, investment problems, radioactive decay, carbon dating, and concentration of a drug in the body.

For each number λ (an eigenvalue of L) we obtain, by using calculus, an associated eigenvector given by

$$f(x) = Ke^{\lambda x},$$

where K is an arbitrary nonzero constant. ■

Equation (3) is a simple example of a differential equation. The subject of differential equations is a major area in mathematics. In Section 8.4 we provide a brief introduction to homogeneous linear systems of differential equations.

Let L be a linear transformation of an n-dimensional vector space V into itself. If $S = \{\mathbf{x}_1, \mathbf{x}_2, \ldots, \mathbf{x}_n\}$ is a basis for V, then there is an $n \times n$ matrix A that represents L with respect to S (see Section 6.3). To determine an eigenvalue λ of L and an eigenvector $\mathbf{x}$ of L associated with the eigenvalue λ, we solve the equation

$$L(\mathbf{x}) = \lambda\mathbf{x}.$$

Using Theorem 6.9, we see that an equivalent matrix equation is

$$A\left[\mathbf{x}\right]_S = \lambda\left[\mathbf{x}\right]_S.$$

This formulation allows us to use techniques for solving linear systems in R^n to determine eigenvalue–eigenvector pairs of L.

EXAMPLE 9

Let $L\colon P_2 \to P_2$ be a linear operator defined by

$$L(at^2 + bt + c) = -bt - 2c.$$

The eigen-problem for L can be formulated in terms of a matrix representing L with respect to a specific basis for P_2. Find the corresponding matrix eigen-problem for each of the bases $S = \{1 - t, 1 + t, t^2\}$ and $T = \{t - 1, 1, t^2\}$ for P_2.

Solution

To find the matrix A that represents L with respect to the basis S, we compute (verify)

$$L(1-t) = t - 2 = -\tfrac{3}{2}(1-t) - \tfrac{1}{2}(1+t) + 0t^2, \qquad \text{so } \left[L(1-t)\right]_S = \begin{bmatrix} -\frac{3}{2} \\ -\frac{1}{2} \\ 0 \end{bmatrix},$$

$$L(1+t) = -t - 2 = -\tfrac{1}{2}(1-t) - \tfrac{3}{2}(1+t) + 0t^2, \qquad \text{so } \left[L(1+t)\right]_S = \begin{bmatrix} -\frac{1}{2} \\ -\frac{3}{2} \\ 0 \end{bmatrix},$$

$$L(t^2) = 0 = 0(1-t) + 0(1+t) + 0t^2, \qquad \text{so } \left[L(t^2)\right]_S = \begin{bmatrix} 0 \\ 0 \\ 0 \end{bmatrix}.$$

Then

$$A = \begin{bmatrix} -\frac{3}{2} & -\frac{1}{2} & 0 \\ -\frac{1}{2} & -\frac{3}{2} & 0 \\ 0 & 0 & 0 \end{bmatrix},$$

and the matrix eigen-problem for L with respect to S is that of finding a number λ and a nonzero vector $\mathbf{x}$ in R^3 or C^3 so that

$$A\mathbf{x} = \lambda\mathbf{x}.$$

In a similar fashion we can show that the matrix B which represents L with respect to the basis T is

$$B = \begin{bmatrix} -1 & 0 & 0 \\ 1 & -2 & 0 \\ 0 & 0 & 0 \end{bmatrix}$$

(verify), and the corresponding matrix eigen-problem for L with respect to T is

$$B\mathbf{x} = \lambda\mathbf{x}.$$

Thus the matrix eigen-problem for L depends on the basis selected for V. We show in Section 7.2 that the eigenvalues of L will not depend on the matrix representing L. ■

As we have seen in Example 9, the eigen-problem for a linear transformation can be expressed in terms of a matrix representing L. We now formulate the notions of eigenvalue and eigenvector for *any* square matrix. If A is an $n \times n$ matrix, we can consider, as in Section 6.1, the linear operator $L\colon R^n \to R^n$ ($C^n \to C^n$) defined by $L(\mathbf{x}) = A\mathbf{x}$ for $\mathbf{x}$ in R^n (C^n). If λ is a scalar (real or complex), and $\mathbf{x} \neq \mathbf{0}$ a vector in R^n (C^n) such that

$$A\mathbf{x} = \lambda\mathbf{x}, \tag{4}$$

then we say that λ is an **eigenvalue** of A and $\mathbf{x}$ is an **eigenvector** of A **associated with** $\boldsymbol{\lambda}$. That is, λ is an eigenvalue of L and $\mathbf{x}$ is an eigenvector of L associated with λ.

Remark Although we began this chapter with the problem of finding the eigenvalues and associated eigenvectors of a linear operator, from now on we emphasize the problem of finding the eigenvalues and associated eigenvectors of an $n \times n$ matrix.

■ Computing Eigenvalues and Eigenvectors

Thus far we have found the eigenvalues and associated eigenvectors of a given linear transformation by inspection, geometric arguments, or very simple algebraic approaches. In the following example, we compute the eigenvalues and associated eigenvectors of a matrix by a somewhat more systematic method.

EXAMPLE 10

Let $A = \begin{bmatrix} 1 & 1 \\ -2 & 4 \end{bmatrix}$. We wish to find the eigenvalues of A and their associated eigenvectors. Thus we wish to find all numbers λ and all nonzero vectors $\mathbf{x} = \begin{bmatrix} x_1 \\ x_2 \end{bmatrix}$ that satisfy Equation (4):

$$\begin{bmatrix} 1 & 1 \\ -2 & 4 \end{bmatrix} \begin{bmatrix} x_1 \\ x_2 \end{bmatrix} = \lambda \begin{bmatrix} x_1 \\ x_2 \end{bmatrix}, \tag{5a}$$

which yields

$$\begin{aligned} x_1 + x_2 &= \lambda x_1 \\ -2x_1 + 4x_2 &= \lambda x_2 \end{aligned} \quad \text{or} \quad \begin{aligned} (\lambda - 1)x_1 - x_2 &= 0 \\ 2x_1 + (\lambda - 4)x_2 &= 0. \end{aligned} \tag{5b}$$

This homogeneous system of two equations in two unknowns has a nontrivial solution if and only if the determinant of the coefficient matrix is zero. Thus

$$\begin{vmatrix} \lambda - 1 & -1 \\ 2 & \lambda - 4 \end{vmatrix} = 0.$$

This means that

$$\lambda^2 - 5\lambda + 6 = 0 = (\lambda - 3)(\lambda - 2),$$

and so $\lambda_1 = 2$ and $\lambda_2 = 3$ are the eigenvalues of A. That is, Equation (5b) will have a nontrivial solution only when $\lambda_1 = 2$ or $\lambda_2 = 3$. To find all eigenvectors of A associated with $\lambda_1 = 2$, we substitute $\lambda_1 = 2$ in Equation (5a):

$$\begin{bmatrix} 1 & 1 \\ -2 & 4 \end{bmatrix} \begin{bmatrix} x_1 \\ x_2 \end{bmatrix} = 2 \begin{bmatrix} x_1 \\ x_2 \end{bmatrix},$$

which yields

$$\begin{aligned} x_1 + x_2 &= 2x_1 \\ -2x_1 + 4x_2 &= 2x_2 \end{aligned} \quad \text{or} \quad \begin{aligned} (2 - 1)x_1 - x_2 &= 0 \\ 2x_1 + (2 - 4)x_2 &= 0 \end{aligned}$$

$$\text{or} \quad \begin{aligned} x_1 - x_2 &= 0 \\ 2x_1 - 2x_2 &= 0. \end{aligned}$$

Note that we could have obtained this last homogeneous system by merely substituting $\lambda_1 = 2$ in (5b). All solutions to this last system are given by

$$\begin{aligned} x_1 &= x_2 \\ x_2 &= \text{any number } r. \end{aligned}$$

Hence all eigenvectors associated with the eigenvalue $\lambda_1 = 2$ are given by $\begin{bmatrix} r \\ r \end{bmatrix}$, where r is any nonzero number. In particular, for $r = 1$, $\mathbf{x}_1 = \begin{bmatrix} 1 \\ 1 \end{bmatrix}$ is an eigenvector associated with $\lambda_1 = 2$. Similarly, substituting $\lambda_2 = 3$ in Equation (5b), we obtain

$$\begin{aligned} (3 - 1)x_1 - x_2 &= 0 \\ 2x_1 + (3 - 4)x_2 &= 0 \end{aligned} \quad \text{or} \quad \begin{aligned} 2x_1 - x_2 &= 0 \\ 2x_1 - x_2 &= 0. \end{aligned}$$

All solutions to this last homogeneous system are given by

$$\begin{aligned} x_1 &= \tfrac{1}{2}x_2 \\ x_2 &= \text{any number } s. \end{aligned}$$

Hence all eigenvectors associated with the eigenvalue $\lambda_2 = 3$ are given by $\begin{bmatrix} \frac{s}{2} \\ s \end{bmatrix}$, where s is any nonzero number. In particular, for $s = 2$, $\mathbf{x}_2 = \begin{bmatrix} 1 \\ 2 \end{bmatrix}$ is an eigenvector associated with the eigenvalue $\lambda_2 = 3$. ■

We now use the method followed in Example 10 as our standard for finding the eigenvalues and associated eigenvectors of a given matrix. We first state some terminology.

DEFINITION 7.2

Let $A = \begin{bmatrix} a_{11} & a_{12} & \cdots & a_{1n} \\ a_{21} & a_{22} & \cdots & a_{2n} \\ \vdots & \vdots & & \vdots \\ a_{n1} & a_{n2} & \cdots & a_{nn} \end{bmatrix}$ be an $n \times n$ matrix. Then the determinant of the matrix

$$\lambda I_n - A = \begin{bmatrix} \lambda - a_{11} & -a_{12} & \cdots & -a_{1n} \\ -a_{21} & \lambda - a_{22} & \cdots & -a_{2n} \\ \vdots & \vdots & & \vdots \\ -a_{n1} & -a_{n2} & \cdots & \lambda - a_{nn} \end{bmatrix}$$

is called the **characteristic polynomial** of A. The equation

$$p(\lambda) = \det(\lambda I_n - A) = 0$$

is called the **characteristic equation** of A.

Recall from Chapter 3 that each term in the expansion of the determinant of an $n \times n$ matrix is a product of n entries of the matrix, containing exactly one entry from each row and exactly one entry from each column. Thus, if we expand $\det(\lambda I_n - A)$, we obtain a polynomial of degree n. The expression involving λ^n in the characteristic polynomial of A comes from the product

$$(\lambda - a_{11})(\lambda - a_{22}) \cdots (\lambda - a_{nn}),$$

and so the coefficient of λ^n is 1. We can then write

$$\det(\lambda I_n - A) = p(\lambda) = \lambda^n + a_1\lambda^{n-1} + a_2\lambda^{n-2} + \cdots + a_{n-1}\lambda + a_n.$$

Note that if we let $\lambda = 0$ in $\det(\lambda I_n - A)$, as well as in the expression on the right, then we get $\det(-A) = a_n$, and thus the constant term of the characteristic polynomial of A is $a_n = (-1)^n \det(A)$.

EXAMPLE 11

Let $A = \begin{bmatrix} 1 & 2 & -1 \\ 1 & 0 & 1 \\ 4 & -4 & 5 \end{bmatrix}$. The characteristic polynomial of A is

$$p(\lambda) = \det(\lambda I_3 - A) = \begin{vmatrix} \lambda - 1 & -2 & 1 \\ -1 & \lambda & -1 \\ -4 & 4 & \lambda - 5 \end{vmatrix} = \lambda^3 - 6\lambda^2 + 11\lambda - 6$$

(verify). ■

We now connect the characteristic polynomial of a matrix with its eigenvalues in the following theorem:

Theorem 7.1 Let A be an $n \times n$ matrix. The eigenvalues of A are the roots of the characteristic polynomial of A.

Proof

Let $\mathbf{x}$ in R^n be an eigenvector of A associated with the eigenvalue λ. Then

$$A\mathbf{x} = \lambda\mathbf{x} \quad \text{or} \quad A\mathbf{x} = (\lambda I_n)\mathbf{x} \quad \text{or} \quad (\lambda I_n - A)\mathbf{x} = \mathbf{0}.$$

This is a homogeneous system of n equations in n unknowns; a nontrivial solution exists if and only if $\det(\lambda I_n - A) = 0$. Hence λ is a root of the characteristic polynomial of A.

Conversely, if λ is a root of the characteristic polynomial of A, then $\det(\lambda I_n - A) = 0$, so the homogeneous system $(\lambda I_n - A)\mathbf{x} = \mathbf{0}$ has a nontrivial solution. Hence λ is an eigenvalue of A. ■

Thus, to find the eigenvalues of a given matrix A, we must find the roots of its characteristic polynomial $p(\lambda)$. There are many methods for finding approximations to the roots of a polynomial, some of them more effective than others. Two results that are sometimes useful in this connection are as follows: (1) The product of all the roots of the polynomial

$$p(\lambda) = \lambda^n + a_1\lambda^{n-1} + \cdots + a_{n-1}\lambda + a_n$$

is $(-1)^n a_n$; and (2) If $a_1, a_2, \ldots, a_n$ are integers, then $p(\lambda)$ cannot have a rational root that is not already an integer. Thus we need to try only the integer factors of a_n as possible rational roots of $p(\lambda)$. Of course, $p(\lambda)$ might well have irrational roots or complex roots.

To minimize the computational effort, and as a convenience to the reader, *most of the characteristic polynomials to be solved in the rest of this chapter have only integer roots*, and each of these roots is a factor of the constant term of the characteristic polynomial of A. The corresponding eigenvectors are obtained by substituting for λ in the matrix equation

$$(\lambda I_n - A)\mathbf{x} = \mathbf{0} \tag{6}$$

and solving the resulting homogeneous system. The solution to these types of problems has been studied in Section 4.7.

EXAMPLE 12 Compute the eigenvalues and associated eigenvectors of the matrix A defined in Example 11.

Solution

In Example 11 we found the characteristic polynomial of A to be

$$p(\lambda) = \lambda^3 - 6\lambda^2 + 11\lambda - 6.$$

The possible integer roots of $p(\lambda)$ are ± 1, ± 2, ± 3, and ± 6. By substituting these values in $p(\lambda)$, we find that $p(1) = 0$, so $\lambda = 1$ is a root of $p(\lambda)$. Hence $(\lambda - 1)$ is a factor of $p(\lambda)$. Dividing $p(\lambda)$ by $(\lambda - 1)$, we obtain

$$p(\lambda) = (\lambda - 1)(\lambda^2 - 5\lambda + 6) \quad \text{(verify)}.$$

Factoring $\lambda^2 - 5\lambda + 6$, we have

$$p(\lambda) = (\lambda - 1)(\lambda - 2)(\lambda - 3).$$

The eigenvalues of A are then $\lambda_1 = 1$, $\lambda_2 = 2$, and $\lambda_3 = 3$. To find an eigenvector $\mathbf{x}_1$ associated with $\lambda_1 = 1$, we substitute $\lambda = 1$ in (6) to get

$$\begin{bmatrix} 1-1 & -2 & 1 \\ -1 & 1 & -1 \\ -4 & 4 & 1-5 \end{bmatrix} \begin{bmatrix} x_1 \\ x_2 \\ x_3 \end{bmatrix} = \begin{bmatrix} 0 \\ 0 \\ 0 \end{bmatrix}$$

or

$$\begin{bmatrix} 0 & -2 & 1 \\ -1 & 1 & -1 \\ -4 & 4 & -4 \end{bmatrix} \begin{bmatrix} x_1 \\ x_2 \\ x_3 \end{bmatrix} = \begin{bmatrix} 0 \\ 0 \\ 0 \end{bmatrix}.$$

The vector $\begin{bmatrix} -\frac{r}{2} \\ \frac{r}{2} \\ r \end{bmatrix}$ is a solution for any number r. Thus

$$\mathbf{x}_1 = \begin{bmatrix} -1 \\ 1 \\ 2 \end{bmatrix}$$

is an eigenvector of A associated with $\lambda_1 = 1$ (r was taken as 2).

To find an eigenvector $\mathbf{x}_2$ associated with $\lambda_2 = 2$, we substitute $\lambda = 2$ in (6), obtaining

$$\begin{bmatrix} 2-1 & -2 & 1 \\ -1 & 2 & -1 \\ -4 & 4 & 2-5 \end{bmatrix} \begin{bmatrix} x_1 \\ x_2 \\ x_3 \end{bmatrix} = \begin{bmatrix} 0 \\ 0 \\ 0 \end{bmatrix}$$

or

$$\begin{bmatrix} 1 & -2 & 1 \\ -1 & 2 & -1 \\ -4 & 4 & -3 \end{bmatrix} \begin{bmatrix} x_1 \\ x_2 \\ x_3 \end{bmatrix} = \begin{bmatrix} 0 \\ 0 \\ 0 \end{bmatrix}.$$

The vector $\begin{bmatrix} -\frac{r}{2} \\ \frac{r}{4} \\ r \end{bmatrix}$ is a solution for any number r. Thus $\mathbf{x}_2 = \begin{bmatrix} -2 \\ 1 \\ 4 \end{bmatrix}$ is an eigenvector of A associated with $\lambda_2 = 2$ (r was taken as 4).

To find an eigenvector $\mathbf{x}_3$ associated with $\lambda_3 = 3$, we substitute $\lambda = 3$ in (6), obtaining

$$\begin{bmatrix} 3-1 & -2 & 1 \\ -1 & 3 & -1 \\ -4 & 4 & 3-5 \end{bmatrix} \begin{bmatrix} x_1 \\ x_2 \\ x_3 \end{bmatrix} = \begin{bmatrix} 0 \\ 0 \\ 0 \end{bmatrix}$$

or

$$\begin{bmatrix} 2 & -2 & 1 \\ -1 & 3 & -1 \\ -4 & 4 & -2 \end{bmatrix} \begin{bmatrix} x_1 \\ x_2 \\ x_3 \end{bmatrix} = \begin{bmatrix} 0 \\ 0 \\ 0 \end{bmatrix}.$$

The vector $\begin{bmatrix} -\frac{r}{4} \\ \frac{r}{4} \\ r \end{bmatrix}$ is a solution for any number r. Thus

$$\mathbf{x}_3 = \begin{bmatrix} -1 \\ 1 \\ 4 \end{bmatrix}$$

is an eigenvector of A associated with $\lambda_3 = 3$ (r was taken as 4). ■

EXAMPLE 13

Compute the eigenvalues and associated eigenvectors of

$$A = \begin{bmatrix} 0 & 0 & 3 \\ 1 & 0 & -1 \\ 0 & 1 & 3 \end{bmatrix}.$$

Solution

The characteristic polynomial of A is

$$p(\lambda) = \det(\lambda I_3 - A) = \begin{vmatrix} \lambda - 0 & 0 & -3 \\ -1 & \lambda - 0 & 1 \\ 0 & -1 & \lambda - 3 \end{vmatrix} = \lambda^3 - 3\lambda^2 + \lambda - 3$$

(verify). We find that $\lambda = 3$ is a root of $p(\lambda)$. Dividing $p(\lambda)$ by $(\lambda - 3)$, we get $p(\lambda) = (\lambda - 3)(\lambda^2 + 1)$. The eigenvalues of A are then

$$\lambda_1 = 3, \quad \lambda_2 = i, \quad \lambda_3 = -i.$$

To compute an eigenvector $\mathbf{x}_1$ associated with $\lambda_1 = 3$, we substitute $\lambda = 3$ in (6), obtaining

$$\begin{bmatrix} 3-0 & 0 & -3 \\ -1 & 3-0 & 1 \\ 0 & -1 & 3-3 \end{bmatrix} \begin{bmatrix} x_1 \\ x_2 \\ x_3 \end{bmatrix} = \begin{bmatrix} 0 \\ 0 \\ 0 \end{bmatrix}.$$

We find that the vector $\begin{bmatrix} r \\ 0 \\ r \end{bmatrix}$ is a solution for any number r (verify). Letting $r = 1$, we conclude that

$$\mathbf{x}_1 = \begin{bmatrix} 1 \\ 0 \\ 1 \end{bmatrix}$$

is an eigenvector of A associated with $\lambda_1 = 3$. To obtain an eigenvector $\mathbf{x}_2$ associated with $\lambda_2 = i$, we substitute $\lambda = i$ in (6), which yields

$$\begin{bmatrix} i-0 & 0 & -3 \\ -1 & i-0 & 1 \\ 0 & -1 & i-3 \end{bmatrix} \begin{bmatrix} x_1 \\ x_2 \\ x_3 \end{bmatrix} = \begin{bmatrix} 0 \\ 0 \\ 0 \end{bmatrix}.$$

We find that the vector $\begin{bmatrix} (-3i)r \\ (-3+i)r \\ r \end{bmatrix}$ is a solution for any number r (verify). Letting $r = 1$, we conclude that

$$\mathbf{x}_2 = \begin{bmatrix} -3i \\ -3+i \\ 1 \end{bmatrix}$$

is an eigenvector of A associated with $\lambda_2 = i$. Similarly, we find that

$$\mathbf{x}_3 = \begin{bmatrix} 3i \\ -3-i \\ 1 \end{bmatrix}$$

is an eigenvector of A associated with $\lambda_3 = -i$. ■

EXAMPLE 14 Let L be the linear operator on P_2 defined in Example 9. Using the matrix B obtained there representing L with respect to the basis $\{t-1, 1, t^2\}$ for P_2, find the eigenvalues and associated eigenvectors of L.

Solution

The characteristic polynomial of

$$B = \begin{bmatrix} -1 & 0 & 0 \\ 1 & -2 & 0 \\ 0 & 0 & 0 \end{bmatrix}$$

is $p(\lambda) = \lambda(\lambda+2)(\lambda+1)$ (verify), so the eigenvalues of L are $\lambda_1 = 0$, $\lambda_2 = -2$, and $\lambda_3 = -1$. Associated eigenvectors are (verify)

$$\mathbf{x}_1 = \begin{bmatrix} 0 \\ 0 \\ 1 \end{bmatrix}, \quad \mathbf{x}_2 = \begin{bmatrix} 1 \\ 1 \\ 0 \end{bmatrix}, \quad \mathbf{x}_3 = \begin{bmatrix} 0 \\ 1 \\ 0 \end{bmatrix}.$$

These are the coordinate vectors of the eigenvectors of L, so the corresponding eigenvectors of L are

$$0(t-1)+0(1)+1(t^2)=t^2$$

$$1(t-1)+1(1)+0(t^2)=t$$

and

$$0(t-1)+1(1)+0(t^2)=1,$$

respectively. ■

The procedure for finding the eigenvalues and associated eigenvectors of a matrix is as follows:

Step 1. Determine the roots of the characteristic polynomial

$$p(\lambda)=\det(\lambda I_n - A).$$

These are the eigenvalues of A.

Step 2. For each eigenvalue λ, find all the nontrivial solutions to the homogeneous system $(\lambda I_n - A)\mathbf{x}=\mathbf{0}$. These are the eigenvectors of A associated with the eigenvalue λ.

The characteristic polynomial of a matrix may have some complex roots, and it may, as seen in Example 13, even have no real roots. However, in the important case of symmetric matrices, all the roots of the characteristic polynomial are real. We prove this in Section 7.3 (Theorem 7.6).

Eigenvalues and eigenvectors satisfy many important and interesting properties. For example, if A is an upper (lower) triangular matrix, then the eigenvalues of A are the elements on the main diagonal of A (Exercise 11). Other properties are developed in the exercises for this section.

It must be pointed out that the method for finding the eigenvalues of a linear transformation or matrix by obtaining the real roots of the characteristic polynomial is not practical for $n > 4$, since it involves evaluating a determinant. Efficient numerical methods for finding eigenvalues and associated eigenvectors are studied in numerical analysis courses.

Warning When finding the eigenvalues and associated eigenvectors of a matrix A, do not make the common mistake of first transforming A to reduced row echelon form B and then finding the eigenvalues and eigenvectors of B. To see quickly how this approach fails, consider the matrix A defined in Example 10. Its eigenvalues are $\lambda_1 = 2$ and $\lambda_2 = 3$. Since A is a nonsingular matrix, when we transform it to reduced row echelon form B, we have $B = I_2$. The eigenvalues of I_2 are $\lambda_1 = 1$ and $\lambda_2 = 1$.

Key Terms

Eigenvalue
Eigenvector
Proper value
Characteristic value
Latent value
Characteristic polynomial
Characteristic equation
Roots of the characteristic polynomial

7.1 Exercises

1. Let $L: R^2 \to R^2$ be counterclockwise rotation through an angle π. Find the eigenvalues and associated eigenvectors of L.

2. Let $L: P_1 \to P_1$ be the linear operator defined by $L(at+b) = bt + a$. Using the matrix representing L with respect to the basis $\{1, t\}$ for P_1, find the eigenvalues and associated eigenvectors of L.

3. Let $L: P_2 \to P_2$ be the linear operator defined by

$$L(at^2+bt+c) = c - at^2.$$

Using the matrix representing L with respect to the basis $\{t^2+1, t, 1\}$ for P_2, find the eigenvalues and associated eigenvectors of L.

4. Let $L: R_3 \to R_3$ be defined by

$$L\left(\begin{bmatrix} a_1 & a_2 & a_3 \end{bmatrix}\right) = \begin{bmatrix} 2a_1+3a_2 & -a_2+4a_3 & 3a_3 \end{bmatrix}.$$

Using the natural basis for R_3, find the eigenvalues and associated eigenvectors of L.

5. Find the characteristic polynomial of each of the following matrices:

(a) $\begin{bmatrix} 2 & 1 \\ -1 & 3 \end{bmatrix}$ **(b)** $\begin{bmatrix} 1 & 2 & 1 \\ 0 & 1 & 2 \\ -1 & 3 & 2 \end{bmatrix}$

(c) $\begin{bmatrix} 4 & -1 & 3 \\ 0 & 2 & 1 \\ 0 & 0 & 3 \end{bmatrix}$ **(d)** $\begin{bmatrix} 4 & 2 \\ 3 & 3 \end{bmatrix}$

6. Find the characteristic polynomial, the eigenvalues, and associated eigenvectors of each of the following matrices:

(a) $\begin{bmatrix} 1 & 1 \\ 1 & 1 \end{bmatrix}$ **(b)** $\begin{bmatrix} 1 & 0 & 0 \\ -1 & 3 & 0 \\ 3 & 2 & -2 \end{bmatrix}$

(c) $\begin{bmatrix} 0 & 1 & 2 \\ 0 & 0 & 3 \\ 0 & 0 & 0 \end{bmatrix}$ **(d)** $\begin{bmatrix} 2 & 1 & 2 \\ 2 & 2 & -2 \\ 3 & 1 & 1 \end{bmatrix}$

7. Find the characteristic polynomial, the eigenvalues, and associated eigenvectors of each of the following matrices:

(a) $\begin{bmatrix} 1 & -1 \\ 2 & 4 \end{bmatrix}$ **(b)** $\begin{bmatrix} 2 & -2 & 3 \\ 0 & 3 & -2 \\ 0 & -1 & 2 \end{bmatrix}$

(c) $\begin{bmatrix} 2 & 2 & 3 \\ 1 & 2 & 1 \\ 2 & -2 & 1 \end{bmatrix}$ **(d)** $\begin{bmatrix} -2 & -2 & 3 \\ 0 & 3 & -2 \\ 0 & -1 & 2 \end{bmatrix}$

8. Find all the eigenvalues and associated eigenvectors of each of the following matrices:

(a) $\begin{bmatrix} 1 & 4 \\ 1 & -2 \end{bmatrix}$ **(b)** $\begin{bmatrix} 0 & -9 \\ 1 & 0 \end{bmatrix}$

(c) $\begin{bmatrix} 4 & 2 & -4 \\ 1 & 5 & -4 \\ 0 & 0 & 6 \end{bmatrix}$ **(d)** $\begin{bmatrix} 0 & -1 & 0 \\ 1 & 0 & 0 \\ 0 & 1 & 0 \end{bmatrix}$

9. Find the characteristic polynomial, the eigenvalues, and associated eigenvectors of each of the following matrices:

(a) $\begin{bmatrix} 0 & 1 \\ -1 & 0 \end{bmatrix}$ **(b)** $\begin{bmatrix} -2 & -4 & -8 \\ 1 & 0 & 0 \\ 0 & 1 & 0 \end{bmatrix}$

(c) $\begin{bmatrix} 2-i & 2i & 0 \\ 1 & 0 & 0 \\ 0 & 1 & 0 \end{bmatrix}$ **(d)** $\begin{bmatrix} 5 & 2 \\ -1 & 3 \end{bmatrix}$

10. Find all the eigenvalues and associated eigenvectors of each of the following matrices:

(a) $\begin{bmatrix} -1 & -1+i \\ 1 & 0 \end{bmatrix}$ **(b)** $\begin{bmatrix} i & 1 & 0 \\ 1 & i & 0 \\ 0 & 0 & 1 \end{bmatrix}$

(c) $\begin{bmatrix} 0 & -1 & 0 \\ 1 & 0 & 0 \\ 0 & 1 & 0 \end{bmatrix}$ **(d)** $\begin{bmatrix} 0 & 0 & -9 \\ 0 & 1 & 0 \\ 1 & 0 & 0 \end{bmatrix}$

11. Prove that if A is an upper (lower) triangular matrix, then the eigenvalues of A are the elements on the main diagonal of A.

12. Prove that A and A^T have the same eigenvalues. What, if anything, can we say about the associated eigenvectors of A and A^T?

13. Let

$$A = \begin{bmatrix} 1 & 2 & 3 & 4 \\ 0 & -1 & 3 & 2 \\ 0 & 0 & 3 & 3 \\ 0 & 0 & 0 & 2 \end{bmatrix}$$

represent the linear transformation $L: M_{22} \to M_{22}$ with respect to the basis

$$S = \left\{ \begin{bmatrix} 1 & 0 \\ 0 & 0 \end{bmatrix}, \begin{bmatrix} 0 & 1 \\ 0 & 0 \end{bmatrix}, \begin{bmatrix} 0 & 0 \\ 1 & 0 \end{bmatrix}, \begin{bmatrix} 0 & 0 \\ 0 & 1 \end{bmatrix} \right\}.$$

Find the eigenvalues and associated eigenvectors of L.

14. Let $L: V \to V$ be a linear operator, where V is an n-dimensional vector space. Let λ be an eigenvalue of L.

Prove that the subset of V consisting of $\mathbf{0}_V$ and all eigenvectors of L associated with λ is a subspace of V. This subspace is called the **eigenspace** associated with λ.

15. Let λ be an eigenvalue of the $n \times n$ matrix A. Prove that the subset of R^n (C^n) consisting of the zero vector and all eigenvectors of A associated with λ is a subspace of R^n (C^n). This subspace is called the **eigenspace** associated with λ. (This result is a corollary to the result in Exercise 14.)

16. In Exercises 14 and 15, why do we have to include $\mathbf{0}_V$ in the set of all eigenvectors associated with λ?

In Exercises 17 and 18, find a basis for the eigenspace (see Exercise 15) associated with λ for each given matrix.

17. **(a)** $\begin{bmatrix} 0 & 0 & 1 \\ 0 & 1 & 0 \\ 1 & 0 & 0 \end{bmatrix}, \lambda = 1$

(b) $\begin{bmatrix} 2 & 1 & 0 \\ 1 & 2 & 1 \\ 0 & 1 & 2 \end{bmatrix}, \lambda = 2$

18. **(a)** $\begin{bmatrix} 3 & 0 & 0 \\ -2 & 3 & -2 \\ 2 & 0 & 5 \end{bmatrix}, \lambda = 3$

(b) $\begin{bmatrix} 4 & 2 & 0 & 0 \\ 3 & 3 & 0 & 0 \\ 0 & 0 & 2 & 5 \\ 0 & 0 & 0 & 2 \end{bmatrix}, \lambda = 2$

19. Let $A = \begin{bmatrix} 0 & -4 & 0 \\ 1 & 0 & 0 \\ 0 & 1 & 0 \end{bmatrix}$.

(a) Find a basis for the eigenspace associated with the eigenvalue $\lambda_1 = 2i$.

(b) Find a basis for the eigenspace associated with the eigenvalue $\lambda_2 = -2i$.

20. Let $A = \begin{bmatrix} 2 & 2 & 3 & 4 \\ 0 & 2 & 3 & 2 \\ 0 & 0 & 1 & 1 \\ 0 & 0 & 0 & 1 \end{bmatrix}$.

(a) Find a basis for the eigenspace associated with the eigenvalue $\lambda_1 = 1$.

(b) Find a basis for the eigenspace associated with the eigenvalue $\lambda_2 = 2$.

21. Prove that if λ is an eigenvalue of a matrix A with associated eigenvector $\mathbf{x}$, and k is a positive integer, then λ^k is an eigenvalue of the matrix

$$A^k = A \cdot A \cdot \cdots \cdot A \quad (k \text{ factors})$$

with associated eigenvector $\mathbf{x}$.

22. Let

$$A = \begin{bmatrix} 1 & 4 \\ 1 & -2 \end{bmatrix}$$

be the matrix of Exercise 8(a). Find the eigenvalues and eigenvectors of A^2 and verify Exercise 21.

23. Prove that if $A^k = O$ for some positive integer k [i.e., if A is a nilpotent matrix (see Supplementary Exercise 22 in Chapter 1)], then 0 is the only eigenvalue of A. (*Hint*: Use Exercise 21.)

24. Let A be an $n \times n$ matrix.

(a) Show that $\det(A)$ is the product of all the roots of the characteristic polynomial of A.

(b) Show that A is singular if and only if 0 is an eigenvalue of A.

(c) Also prove the analogous statement for a linear transformation: If $L: V \to V$ is a linear transformation, show that L is not one-to-one if and only if 0 is an eigenvalue of L.

(d) Show that if A is nilpotent (see Supplementary Exercise 22 in Chapter 1), then A is singular.

25. Let $L: V \to V$ be an invertible linear operator and let λ be an eigenvalue of L with associated eigenvector $\mathbf{x}$.

(a) Show that $1/\lambda$ is an eigenvalue of L^{-1} with associated eigenvector $\mathbf{x}$.

(b) State and prove the analogous statement for matrices.

26. Let A be an $n \times n$ matrix with eigenvalues λ_1 and λ_2, where $\lambda_1 \neq \lambda_2$. Let S_1 and S_2 be the eigenspaces associated with λ_1 and λ_2, respectively. Explain why the zero vector is the only vector that is in both S_1 and S_2.

27. Let λ be an eigenvalue of A with associated eigenvector $\mathbf{x}$. Show that $\lambda + r$ is an eigenvalue of $A + rI_n$ with associated eigenvector $\mathbf{x}$. Thus, adding a scalar multiple of the identity matrix to A merely shifts the eigenvalues by the scalar multiple.

28. Let A be an $n \times n$ matrix and consider the linear operator on R^n defined by $L(\mathbf{u}) = A\mathbf{u}$, for $\mathbf{u}$ in R^n. A subspace W of R^n is called **invariant** under L if for any $\mathbf{w}$ in W, $L(\mathbf{w})$ is also in W. Show that an eigenspace of A is invariant under L.

29. Let A and B be $n \times n$ matrices such that $A\mathbf{x} = \lambda\mathbf{x}$ and $B\mathbf{x} = \mu\mathbf{x}$. Show that

(a) $(A + B)\mathbf{x} = (\lambda + \mu)\mathbf{x}$;

(b) $(AB)\mathbf{x} = (\lambda\mu)\mathbf{x}$.

30. The **Cayley*–Hamilton† theorem** states that a matrix satisfies its characteristic equation; that is, if A is an $n \times n$ matrix with characteristic polynomial

$$p(\lambda) = \lambda^n + a_1\lambda^{n-1} + \cdots + a_{n-1}\lambda + a_n,$$

then

$$A^n + a_1A^{n-1} + \cdots + a_{n-1}A + a_nI_n = O.$$

The proof and applications of this result, unfortunately, lie beyond the scope of this book. Verify the Cayley–Hamilton theorem for the following matrices:

(a) $\begin{bmatrix} 1 & 2 & 3 \\ 2 & -1 & 5 \\ 3 & 2 & 1 \end{bmatrix}$ **(b)** $\begin{bmatrix} 1 & 2 & 3 \\ 0 & 2 & 2 \\ 0 & 0 & -3 \end{bmatrix}$

(c) $\begin{bmatrix} 3 & 3 \\ 2 & 4 \end{bmatrix}$

31. Let A be an $n \times n$ matrix whose characteristic polynomial is

$$p(\lambda) = \lambda^n + a_1\lambda^{n-1} + \cdots + a_{n-1}\lambda + a_n.$$

If A is nonsingular, show that

$$A^{-1} = -\frac{1}{a_n}(A^{n-1} + a_1A^{n-2} + \cdots + a_{n-2}A + a_{n-1}I_n).$$

[*Hint*: Use the Cayley–Hamilton theorem (Exercise 30).]

32. Let

$$A = \begin{bmatrix} a & b \\ c & d \end{bmatrix}.$$

Prove that the characteristic polynomial $p(\lambda)$ of A is given by

$$p(\lambda) = \lambda^2 - \mathrm{Tr}(A)\lambda + \det(A),$$

where $\mathrm{Tr}(A)$ denotes the trace of A (see Exercise 43 in Section 1.3).

33. Show that if A is a matrix all of whose columns add up to 1, then $\lambda = 1$ is an eigenvalue of A. (*Hint*: Consider the product $A^T\mathbf{x}$, where $\mathbf{x}$ is a vector all of whose entries are 1, and use Exercise 12.)

ARTHUR CAYLEY

WILLIAM ROWAN HAMILTON

*Arthur Cayley (1821–1895) was born in Richmond, Surrey, England, into an established and talented family, and died in Cambridge. As a youngster he showed considerable talent in mathematics, and his teachers persuaded his father to let him go to Cambridge instead of entering the family business as a merchant. At Cambridge he distinguished himself in his studies and published a number of papers as an undergraduate. After graduation, he accepted a Fellowship at Trinity College in Cambridge, but left to study law. Although he spent 14 years working as a successful lawyer, he was still able to spend considerable time working on mathematics and published nearly 200 papers during this period. In 1863, he was appointed to a Professorship at Cambridge University. His prolific work in matrix theory and other areas of linear algebra was of fundamental importance to the development of the subject. He also made many important contributions to abstract algebra, geometry, and other areas of modern mathematics.

† William Rowan Hamilton (1805–1865) was born and died in Dublin, Ireland. He was a child prodigy who had learned Greek, Latin, and Hebrew by the age of 5. His extraordinary mathematical talent became apparent by the age of 12. He studied at Trinity College in Dublin, where he was an outstanding student. As an undergraduate he was appointed Royal Astronomer of Ireland, Director of the Dunsink Observatory, and Professor of Astronomy. In his early papers he made significant contributions to the field of optics. At the age of 30 he was knighted after having his mathematical theory of conical refraction in optics confirmed experimentally. Hamilton is also known for his discovery of the algebra of quaternions, a set consisting of quadruples satisfying certain algebraic properties. He spent the last 20 years of his life working on quaternions and their applications.

34. Show that if A is an $n \times n$ matrix whose kth row is the same as the kth row of I_n, then 1 is an eigenvalue of A.

35. Let A be a square matrix.

(a) Suppose that the homogeneous system $A\mathbf{x} = \mathbf{0}$ has a nontrivial solution $\mathbf{x} = \mathbf{u}$. Show that $\mathbf{u}$ is an eigenvector of A.

(b) Suppose that 0 is an eigenvalue of A and $\mathbf{v}$ is an associated eigenvector. Show that the homogeneous system $A\mathbf{x} = \mathbf{0}$ has a nontrivial solution.

36. Determine whether your software has a command for finding the characteristic polynomial of a matrix A. If it does, compare the output from your software with the results in Examples 10 and 13. Software output for a characteristic polynomial often is just the set of coefficients of the polynomial with the powers of λ omitted. Carefully determine the order in which the coefficients are listed. Experiment further with the matrices in Exercises 5 and 6.

37. If your software has a command for finding the characteristic polynomial of a matrix A (see Exercise 36), it probably has another command for finding the roots of polynomials. Investigate the use of these commands in your software. The roots of the characteristic polynomial of A are the eigenvalues of A.

38. Assuming that your software has the commands discussed in Exercises 36 and 37, apply them to find the eigenvalues of $A = \begin{bmatrix} 0 & 1 \\ -1 & 0 \end{bmatrix}$. If your software is successful, the results should be $\lambda = i, -i$, where $i = \sqrt{-1}$. (See Appendix B.) (**Caution:** Some software does not handle complex roots and may not permit complex elements in a matrix. Determine the situation for the software you use.)

39. Most linear algebra software has a command for automatically finding the eigenvalues of a matrix. Determine the command available in your software. Test its behavior on Examples 12 and 13. Often, such a command uses techniques that are different than finding the roots of the characteristic polynomial. Use the documentation accompanying your software to find the method used. (**Warning:** It may involve ideas from Section 7.3 or more sophisticated procedures.)

40. Following the ideas in Exercise 39, determine the command in your software for obtaining the eigenvectors of a matrix. Often, it is a variation of the eigenvalue command. Test it on the matrices in Examples 10 and 12. These examples cover the types of cases for eigenvectors that you will encounter frequently in this course.

7.2 Diagonalization and Similar Matrices

If $L: V \to V$ is a linear operator on an n-dimensional vector space V, as we have already seen in Section 7.1 then we can find the eigenvalues of L and associated eigenvectors by using a matrix representing L with respect to a basis for V. The computational steps involved depend upon the matrix selected to represent L. An ideal situation would be the following one: Suppose that L is represented by a matrix A with respect to a certain basis for V. Find a basis for V with respect to which L is represented by a diagonal matrix D whose eigenvalues are the same as the eigenvalues of A. Of course, this is a very desirable situation, since the eigenvalues of D are merely its entries on the main diagonal. Now recall from Theorem 6.14 in Section 6.5 that A and D represent the same linear operator $L: V \to V$ with respect to two bases for V if and only if they are similar—that is, if and only if there exists a nonsingular matrix P such that $D = P^{-1}AP$. In this section we examine the type of linear transformations and matrices for which this situation is possible. For convenience, we work only with matrices all of whose entries and eigenvalues are real numbers.

Remark The need to compute powers of a matrix A (see Section 1.5) arises frequently in applications. (See the discussion of Fibonacci numbers in Section 1.5, and recursion relations in the exercises of Section 1.5 and Section 8.1.) If the matrix A is similar to a diagonal matrix D, then $D = P^{-1}AP$, for an appropriate matrix P. It follows that $A = PDP^{-1}$ (verify) and that $A^k = PD^kP^{-1}$ (verify).

Since D is diagonal, so is D^k, and its diagonal entries are d_{jj}^k. Hence A^k is easy to compute. In Section 8.4 we see another instance in which problems involving matrices that are similar to diagonal matrices can be solved quite efficiently.

DEFINITION 7.3 Let $L: V \to V$ be a linear operator on an n-dimensional vector space V. We say that L is **diagonalizable**, or can be **diagonalized**, if there exists a basis S for V such that L is represented with respect to S by a diagonal matrix D.

EXAMPLE 1 In Example 2 of Section 6.5 we considered the linear transformation $L: R_3 \to R_3$ defined by

$$L\left(\begin{bmatrix} u_1 & u_2 & u_3 \end{bmatrix}\right) = \begin{bmatrix} 2u_1 - u_3 & u_1 + u_2 - u_3 & u_3 \end{bmatrix}.$$

In that example we used the basis

$$S' = \left\{\begin{bmatrix} 1 & 0 & 1 \end{bmatrix}, \begin{bmatrix} 0 & 1 & 0 \end{bmatrix}, \begin{bmatrix} 1 & 1 & 0 \end{bmatrix}\right\}$$

for R_3 and showed that the representation of L with respect to S' is

$$B = \begin{bmatrix} 1 & 0 & 0 \\ 0 & 1 & 0 \\ 0 & 0 & 2 \end{bmatrix}.$$

Hence L is a diagonalizable linear transformation. ■

We next show that similar matrices have the same eigenvalues.

Theorem 7.2 Similar matrices have the same eigenvalues.

Proof

Let A and B be similar. Then $B = P^{-1}AP$, for some nonsingular matrix P. We prove that A and B have the same characteristic polynomials, $p_A(\lambda)$ and $p_B(\lambda)$, respectively. We have

$$\begin{aligned} p_B(\lambda) &= \det(\lambda I_n - B) = \det(\lambda I_n - P^{-1}AP) \\ &= \det(P^{-1}\lambda I_n P - P^{-1}AP) = \det(P^{-1}(\lambda I_n - A)P) \\ &= \det(P^{-1})\det(\lambda I_n - A)\det(P) \\ &= \det(P^{-1})\det(P)\det(\lambda I_n - A) \\ &= \det(\lambda I_n - A) = p_A(\lambda). \end{aligned} \tag{1}$$

Since $p_A(\lambda) = p_B(\lambda)$, it follows that A and B have the same eigenvalues. ■

Note that in the proof of Theorem 7.2 we have used the facts that the product of $\det(P^{-1})$ and $\det(P)$ is 1 and that determinants are numbers, so their order as factors in multiplication does not matter.

Let $L: V \to V$ be a diagonalizable linear operator on an n-dimensional vector space V and let $S = \{\mathbf{x}_1, \mathbf{x}_2, \ldots, \mathbf{x}_n\}$ be a basis for V such that L is represented with respect to S by a diagonal matrix

$$D = \begin{bmatrix} \lambda_1 & 0 & \cdots & 0 & 0 \\ 0 & \lambda_2 & \cdots & 0 & 0 \\ \vdots & \vdots & & \vdots & \vdots \\ 0 & 0 & \cdots & \lambda_{n-1} & 0 \\ 0 & 0 & \cdots & 0 & \lambda_n \end{bmatrix},$$

where $\lambda_1, \lambda_2, \ldots, \lambda_n$ are real scalars. Now recall that if D represents L with respect to S, then the jth column of D is the coordinate vector $\left[L(\mathbf{x}_j)\right]_S$ of $L(\mathbf{x}_j)$ with respect to S. Thus we have

$$\left[L(\mathbf{x}_j)\right]_S = \begin{bmatrix} 0 \\ 0 \\ \vdots \\ 0 \\ \lambda_j \\ 0 \\ \vdots \\ 0 \end{bmatrix} \quad \leftarrow j\text{th row},$$

which means that

$$L(\mathbf{x}_j) = 0\mathbf{x}_1 + 0\mathbf{x}_2 + \cdots + 0\mathbf{x}_{j-1} + \lambda_j\mathbf{x}_j + 0\mathbf{x}_{j+1} + \cdots + 0\mathbf{x}_n = \lambda_j\mathbf{x}_j.$$

Conversely, let $S = \{\mathbf{x}_1, \mathbf{x}_2, \ldots, \mathbf{x}_n\}$ be a basis for V such that

$$\begin{aligned} L(\mathbf{x}_j) = \lambda_j\mathbf{x}_j &= 0\mathbf{x}_1 + 0\mathbf{x}_2 + \cdots + 0\mathbf{x}_{j-1} \\ &\quad + \lambda_j\mathbf{x}_j + 0\mathbf{x}_{j+1} + \cdots + 0\mathbf{x}_n \quad \text{for } j = 1, 2, \ldots, n. \end{aligned}$$

We now find the matrix representing L with respect to S. The jth column of this matrix is

$$\left[L(\mathbf{x}_j)\right]_S = \begin{bmatrix} 0 \\ 0 \\ \vdots \\ 0 \\ \lambda_j \\ 0 \\ \vdots \\ 0 \end{bmatrix}.$$

Hence

$$D = \begin{bmatrix} \lambda_1 & 0 & \cdots & 0 & 0 \\ 0 & \lambda_2 & \cdots & 0 & 0 \\ \vdots & \vdots & & \vdots & \vdots \\ 0 & 0 & \cdots & \lambda_{n-1} & 0 \\ 0 & 0 & \cdots & 0 & \lambda_n \end{bmatrix},$$

a diagonal matrix, represents L with respect to S, so L is diagonalizable.

We can now state the following theorem, whose proof has just been given:

Theorem 7.3 Let $L: V \to V$ be a linear operator on an n-dimensional vector space V. Then L is diagonalizable if and only if V has a basis S of eigenvectors of L. Moreover, if D is the diagonal matrix representing L with respect to S, then the entries on the main diagonal of D are the eigenvalues of L. ■

In terms of matrices, Theorem 7.3 can be stated as follows:

Theorem 7.4 An $n \times n$ matrix A is similar to a diagonal matrix D if and only if A has n linearly independent eigenvectors. Moreover, the elements on the main diagonal of D are the eigenvalues of A. ■

Remark If a matrix A is similar to a diagonal matrix, we say that A is **diagonalizable** or can be diagonalized.

To use Theorem 7.4, we need show only that there is a set of n eigenvectors of A that are linearly independent.

EXAMPLE 2

Let $A = \begin{bmatrix} 1 & 1 \\ -2 & 4 \end{bmatrix}$. In Example 10 of Section 7.1 we found that the eigenvalues of A are $\lambda_1 = 2$ and $\lambda_2 = 3$, with associated eigenvectors

$$\mathbf{x}_1 = \begin{bmatrix} 1 \\ 1 \end{bmatrix} \quad \text{and} \quad \mathbf{x}_2 = \begin{bmatrix} 1 \\ 2 \end{bmatrix},$$

respectively. Since

$$S = \left\{ \begin{bmatrix} 1 \\ 1 \end{bmatrix}, \begin{bmatrix} 1 \\ 2 \end{bmatrix} \right\}$$

is linearly independent (verify), A can be diagonalized. From Theorem 7.4, we conclude that A is similar to $D = \begin{bmatrix} 2 & 0 \\ 0 & 3 \end{bmatrix}$. ■

EXAMPLE 3

Let $A = \begin{bmatrix} 1 & 1 \\ 0 & 1 \end{bmatrix}$. Can A be diagonalized?

Solution

Since A is upper triangular, its eigenvalues are the entries on its main diagonal (Exercise 11 in Section 7.1). Thus, the eigenvalues of A are $\lambda_1 = 1$ and $\lambda_2 = 1$.

We now find eigenvectors of A associated with $\lambda_1 = 1$. Equation (6) of Section 7.1, $(\lambda I_n - A)\mathbf{x} = \mathbf{0}$, becomes, with $\lambda = 1$, the homogeneous system

$$\begin{aligned} (1-1)x_1 - x_2 &= 0 \\ (1-1)x_2 &= 0. \end{aligned}$$

The vector $\begin{bmatrix} r \\ 0 \end{bmatrix}$, for any number r, is a solution. Thus all eigenvectors of A are multiples of the vector $\begin{bmatrix} 1 \\ 0 \end{bmatrix}$. Since A does not have two linearly independent eigenvectors, it cannot be diagonalized. ■

If an $n \times n$ matrix A is similar to a diagonal matrix D, then $P^{-1}AP = D$ for some nonsingular matrix P. We now discuss how to construct such a matrix P. We have $AP = PD$. Let

$$D = \begin{bmatrix} \lambda_1 & 0 & \cdots & 0 & 0 \\ 0 & \lambda_2 & \cdots & 0 & 0 \\ \vdots & \vdots & & \vdots & \vdots \\ 0 & 0 & \cdots & \lambda_{n-1} & 0 \\ 0 & 0 & \cdots & 0 & \lambda_n \end{bmatrix},$$

and let $\mathbf{x}_j$, $j = 1, 2, \ldots, n$, be the jth column of P. Note that the jth column of AP is $A\mathbf{x}_j$, and the jth column of PD is $\lambda_j \mathbf{x}_j$. (See Exercise 46 in Section 1.3.) Thus we have

$$A\mathbf{x}_j = \lambda_j \mathbf{x}_j,$$

which means that λ_j is an eigenvalue of A and $\mathbf{x}_j$ is an associated eigenvector.

Conversely, if $\lambda_1, \lambda_2, \ldots, \lambda_n$ are n eigenvalues of an $n \times n$ matrix A and $\mathbf{x}_1, \mathbf{x}_2, \ldots, \mathbf{x}_n$ are associated eigenvectors forming a linearly independent set, we let P be the matrix whose jth column is $\mathbf{x}_j$. Then rank $P = n$, so by Corollary 4.7, P is nonsingular. Since $A\mathbf{x}_j = \lambda_j \mathbf{x}_j$, $j = 1, 2, \ldots, n$, we have $AP = PD$, or $P^{-1}AP = D$, which means that A is diagonalizable. Thus, if n eigenvectors $\mathbf{x}_1, \mathbf{x}_2, \ldots, \mathbf{x}_n$ of the $n \times n$ matrix A form a linearly independent set, we can diagonalize A by letting P be the matrix whose jth column is $\mathbf{x}_j$, and we find that $P^{-1}AP = D$, a diagonal matrix whose entries on the main diagonal are the associated eigenvalues of A. Of course, the order of columns of P determines the order of the diagonal entries of D.

EXAMPLE 4

Let A be as in Example 2. The eigenvalues of A are $\lambda_1 = 2$ and $\lambda_2 = 3$, and associated eigenvectors are

$$\mathbf{x}_1 = \begin{bmatrix} 1 \\ 1 \end{bmatrix} \quad \text{and} \quad \mathbf{x}_2 = \begin{bmatrix} 1 \\ 2 \end{bmatrix},$$

respectively. Thus

$$P = \begin{bmatrix} 1 & 1 \\ 1 & 2 \end{bmatrix} \quad \text{and} \quad P^{-1} = \begin{bmatrix} 2 & -1 \\ -1 & 1 \end{bmatrix} \quad \text{(verify)}.$$

Hence

$$P^{-1}AP = \begin{bmatrix} 2 & -1 \\ -1 & 1 \end{bmatrix} \begin{bmatrix} 1 & 1 \\ -2 & 4 \end{bmatrix} \begin{bmatrix} 1 & 1 \\ 1 & 2 \end{bmatrix} = \begin{bmatrix} 2 & 0 \\ 0 & 3 \end{bmatrix}.$$

On the other hand, if we let $\lambda_1 = 3$ and $\lambda_2 = 2$, then

$$\mathbf{x}_1 = \begin{bmatrix} 1 \\ 2 \end{bmatrix} \quad \text{and} \quad \mathbf{x}_2 = \begin{bmatrix} 1 \\ 1 \end{bmatrix};$$

$$P = \begin{bmatrix} 1 & 1 \\ 2 & 1 \end{bmatrix} \quad \text{and} \quad P^{-1} = \begin{bmatrix} -1 & 1 \\ 2 & -1 \end{bmatrix},$$

and

$$P^{-1}AP = \begin{bmatrix} -1 & 1 \\ 2 & -1 \end{bmatrix} \begin{bmatrix} 1 & 1 \\ -2 & 4 \end{bmatrix} \begin{bmatrix} 1 & 1 \\ 2 & 1 \end{bmatrix} = \begin{bmatrix} 3 & 0 \\ 0 & 2 \end{bmatrix}.$$

■

The following useful theorem identifies a large class of matrices that can be diagonalized:

Theorem 7.5 If the roots of the characteristic polynomial of an $n \times n$ matrix A are all different from each other (i.e., distinct), then A is diagonalizable.

Proof

Let $\{\lambda_1, \lambda_2, \dots, \lambda_n\}$ be the set of distinct eigenvalues of A, and let $S = \{\mathbf{x}_1, \mathbf{x}_2, \dots, \mathbf{x}_n\}$ be a set of associated eigenvectors. We wish to prove that S is linearly independent.

Suppose that S is linearly dependent. Then Theorem 4.7 implies that some vector $\mathbf{x}_j$ is a linear combination of the preceding vectors in S. We can assume that $S_1 = \{\mathbf{x}_1, \mathbf{x}_2, \dots, \mathbf{x}_{j-1}\}$ is linearly independent, for otherwise, one of the vectors in S_1 is a linear combination of the preceding ones, and we can choose a new set S_2, and so on. We thus have that S_1 is linearly independent and that

$$\mathbf{x}_j = a_1\mathbf{x}_1 + a_2\mathbf{x}_2 + \cdots + a_{j-1}\mathbf{x}_{j-1}, \tag{2}$$

where $a_1, a_2, \dots, a_{j-1}$ are scalars. This means that

$$\begin{aligned} A\mathbf{x}_j &= A(a_1\mathbf{x}_1 + a_2\mathbf{x}_2 + \cdots + a_{j-1}\mathbf{x}_{j-1}) \\ &= a_1A\mathbf{x}_1 + a_2A\mathbf{x}_2 + \cdots + a_{j-1}A\mathbf{x}_{j-1}. \end{aligned} \tag{3}$$

Since $\lambda_1, \lambda_2, \dots, \lambda_j$ are eigenvalues and $\mathbf{x}_1, \mathbf{x}_2, \dots, \mathbf{x}_j$ are associated eigenvectors, we know that $A\mathbf{x}_i = \lambda_i\mathbf{x}_i$ for $i = 1, 2, \dots, n$. Substituting in (3), we have

$$\lambda_j\mathbf{x}_j = a_1\lambda_1\mathbf{x}_1 + a_2\lambda_2\mathbf{x}_2 + \cdots + a_{j-1}\lambda_{j-1}\mathbf{x}_{j-1}. \tag{4}$$

Multiplying (2) by λ_j, we get

$$\lambda_j\mathbf{x}_j = \lambda_j a_1\mathbf{x}_1 + \lambda_j a_2\mathbf{x}_2 + \cdots + \lambda_j a_{j-1}\mathbf{x}_{j-1}. \tag{5}$$

Subtracting (4) from (3), we have

$$\begin{aligned} \mathbf{0} &= \lambda_j\mathbf{x}_j - \lambda_j\mathbf{x}_j \\ &= a_1(\lambda_1 - \lambda_j)\mathbf{x}_1 + a_2(\lambda_2 - \lambda_j)\mathbf{x}_2 + \cdots + a_{j-1}(\lambda_{j-1} - \lambda_j)\mathbf{x}_{j-1}. \end{aligned}$$

Since S_1 is linearly independent, we must have

$$a_1(\lambda_1 - \lambda_j) = 0, \quad a_2(\lambda_2 - \lambda_j) = 0, \quad \dots, \quad a_{j-1}(\lambda_{j-1} - \lambda_j) = 0.$$

Now $(\lambda_1 - \lambda_j) \neq 0$, $(\lambda_2 - \lambda_j) \neq 0, \dots, (\lambda_{j-1} - \lambda_j) \neq 0$, since the λ's are distinct, which implies that

$$a_1 = a_2 = \cdots = a_{j-1} = 0.$$

This means that $\mathbf{x}_j = \mathbf{0}$, which is impossible if $\mathbf{x}_j$ is an eigenvector. Hence S is linearly independent, so A is diagonalizable. ■

Remark In the proof of Theorem 7.5 we have actually established the following somewhat stronger result: Let A be an $n \times n$ matrix and let $\lambda_1, \lambda_2, \ldots, \lambda_k$ be k distinct eigenvalues of A with associated eigenvectors $\mathbf{x}_1, \mathbf{x}_2, \ldots, \mathbf{x}_k$. Then $\mathbf{x}_1, \mathbf{x}_2, \ldots, \mathbf{x}_k$ are linearly independent (Exercise 25).

If all the roots of the characteristic polynomial of A are not all distinct, then A may or may not be diagonalizable. The characteristic polynomial of A can be written as the product of n factors, each of the form $\lambda - \lambda_0$, where λ_0 is a root of the characteristic polynomial. Now the eigenvalues of A are the roots of the characteristic polynomial of A. Thus the characteristic polynomial can be written as

$$(\lambda - \lambda_1)^{k_1}(\lambda - \lambda_2)^{k_2} \cdots (\lambda - \lambda_r)^{k_r},$$

where $\lambda_1, \lambda_2, \ldots, \lambda_r$ are the distinct eigenvalues of A, and $k_1, k_2, \ldots, k_r$ are integers whose sum is n. The integer k_i is called the **multiplicity** of λ_i. Thus in Example 3, $\lambda = 1$ is an eigenvalue of

$$A = \begin{bmatrix} 1 & 1 \\ 0 & 1 \end{bmatrix}$$

of multiplicity 2. It can be shown that A can be diagonalized if and only if, for each eigenvalue λ of multiplicity k, we can find k linearly independent eigenvectors. This means that the solution space of the homogeneous system $(\lambda I_n - A)\mathbf{x} = \mathbf{0}$ has dimension k. It can also be shown that if λ is an eigenvalue of A of multiplicity k, then we can never find more than k linearly independent eigenvectors associated with λ.

EXAMPLE 5 Let

$$A = \begin{bmatrix} 0 & 0 & 1 \\ 0 & 1 & 2 \\ 0 & 0 & 1 \end{bmatrix}.$$

Then the characteristic polynomial of A is $p(\lambda) = \lambda(\lambda - 1)^2$ (verify), so the eigenvalues of A are $\lambda_1 = 0$, $\lambda_2 = 1$, and $\lambda_3 = 1$; thus $\lambda_2 = 1$ is an eigenvalue of multiplicity 2. We now consider the eigenvectors associated with the eigenvalues $\lambda_2 = \lambda_3 = 1$. They are computed by solving the homogeneous system $(1I_3 - A)\mathbf{x} = \mathbf{0}$ [Equation (6) in Section 7.1]:

$$\begin{bmatrix} 1 & 0 & -1 \\ 0 & 0 & -2 \\ 0 & 0 & 0 \end{bmatrix} \begin{bmatrix} x_1 \\ x_2 \\ x_3 \end{bmatrix} = \begin{bmatrix} 0 \\ 0 \\ 0 \end{bmatrix}.$$

The solutions are the vectors of the form

$$\begin{bmatrix} 0 \\ r \\ 0 \end{bmatrix},$$

where r is any number, so the dimension of the solution space of $(1I_3 - A)\mathbf{x} = \mathbf{0}$ is 1 (Why?), and we cannot find two linearly independent eigenvectors. Thus A cannot be diagonalized. ■

EXAMPLE 6

Let $A = \begin{bmatrix} 0 & 0 & 0 \\ 0 & 1 & 0 \\ 1 & 0 & 1 \end{bmatrix}$. The characteristic polynomial of A is $p(\lambda) = \lambda(\lambda - 1)^2$ (verify), so the eigenvalues of A are $\lambda_1 = 0$, $\lambda_2 = 1$, and $\lambda_3 = 1$; thus $\lambda_2 = 1$ is again an eigenvalue of multiplicity 2. Now we consider the eigenvectors associated with the eigenvalues $\lambda_2 = \lambda_3 = 1$. They are computed by solving the homogeneous system $(1I_3 - A)\mathbf{x} = \mathbf{0}$ [Equation (6) in Section 7.1]:

$$\begin{bmatrix} 1 & 0 & 0 \\ 0 & 0 & 0 \\ -1 & 0 & 0 \end{bmatrix} \begin{bmatrix} x_1 \\ x_2 \\ x_3 \end{bmatrix} = \begin{bmatrix} 0 \\ 0 \\ 0 \end{bmatrix}.$$

The solutions are the vectors of the form $\begin{bmatrix} 0 \\ r \\ s \end{bmatrix}$ for any numbers r and s. Thus $\mathbf{x}_2 = \begin{bmatrix} 0 \\ 1 \\ 0 \end{bmatrix}$ and $\mathbf{x}_3 = \begin{bmatrix} 0 \\ 0 \\ 1 \end{bmatrix}$ are eigenvectors.

Next, we look for an eigenvector associated with $\lambda_1 = 0$. We have to solve the homogeneous system

$$\begin{bmatrix} 0 & 0 & 0 \\ 0 & -1 & 0 \\ -1 & 0 & -1 \end{bmatrix} \begin{bmatrix} x_1 \\ x_2 \\ x_3 \end{bmatrix} = \begin{bmatrix} 0 \\ 0 \\ 0 \end{bmatrix}.$$

The solutions are the vectors of the form $\begin{bmatrix} r \\ 0 \\ -r \end{bmatrix}$ for any number r. Thus $\mathbf{x}_1 = \begin{bmatrix} 1 \\ 0 \\ -1 \end{bmatrix}$ is an eigenvector associated with $\lambda_1 = 0$. Now $S = \{\mathbf{x}_1, \mathbf{x}_2, \mathbf{x}_3\}$ is linearly independent, so A can be diagonalized. ■

Thus an $n \times n$ matrix will fail to be diagonalizable only if it does not have n linearly independent eigenvectors.

Now define the **characteristic polynomial** of a linear operator $L: V \to V$ as the characteristic polynomial of any matrix representing L; by Theorem 7.2 all representations of L will give the same characteristic polynomial. It follows that a scalar λ is an eigenvalue of L if and only if λ is a root of the characteristic polynomial of L.

EXAMPLE 7

In Example 14 of Section 7.1, we derived the matrix eigen-problem for the linear operator $L: P_2 \to P_2$, defined in Example 9 of that section, by $L(at^2 + bt + c) = -bt - 2c$ by using the matrix

$$B = \begin{bmatrix} -1 & 0 & 0 \\ 1 & -2 & 0 \\ 0 & 0 & 0 \end{bmatrix},$$

which represents L with respect to the basis $\{t-1, 1, t^2\}$ for P_2. We computed the characteristic polynomial of B to be $p(\lambda) = \lambda(\lambda+2)(\lambda+1)$, so this is also the characteristic polynomial of L. Since the eigenvalues are distinct, it follows that B and L are diagonalizable. Of course, any other matrix representing L could be used in place of B. ■

Key Terms

Diagonalizable
Diagonalized
Similar matrices
Characteristic polynomial
Eigenvalues/eigenvectors
Distinct eigenvalues
Multiplicity of an eigenvalue

7.2 Exercises

1. Let $L: P_2 \to P_2$ be the linear operator defined by $L(p(t)) = p'(t)$ for $p(t)$ in P_2. Is L diagonalizable? If it is, find a basis S for P_2 with respect to which L is represented by a diagonal matrix.

2. Let $L: P_1 \to P_1$ be the linear operator defined by
$$L(at+b) = -bt - a.$$
Find, if possible, a basis for P_1 with respect to which L is represented by a diagonal matrix.

3. Let $L: P_2 \to P_2$ be the linear operator defined by
$$L(at^2+bt+c) = at^2 - c.$$
Find, if possible, a basis for P_2 with respect to which L is represented by a diagonal matrix.

4. ***(Calculus Required)*** Let V be the vector space of continuous functions with basis $\{\sin t, \cos t\}$, and let $L: V \to V$ be defined as $L(g(t)) = g'(t)$. Is L diagonalizable?

5. Let $L: P_2 \to P_2$ be the linear operator defined by
$$L(at^2+bt+c) = (2a+b+c)t^2 + (2c-3b)t + 4c.$$
Find the eigenvalues and eigenvectors of L. Is L diagonalizable?

6. Which of the following matrices are diagonalizable?

(a) $\begin{bmatrix} 1 & 4 \\ 1 & -2 \end{bmatrix}$ **(b)** $\begin{bmatrix} 1 & 0 \\ -2 & 1 \end{bmatrix}$

(c) $\begin{bmatrix} 1 & 1 & -2 \\ 4 & 0 & 4 \\ 1 & -1 & 4 \end{bmatrix}$ **(d)** $\begin{bmatrix} 1 & 2 & 3 \\ 0 & -1 & 2 \\ 0 & 0 & 2 \end{bmatrix}$

7. Which of the following matrices are diagonalizable?

(a) $\begin{bmatrix} 3 & 1 & 0 \\ 0 & 3 & 1 \\ 0 & 0 & 3 \end{bmatrix}$ **(b)** $\begin{bmatrix} -2 & 2 \\ 5 & 1 \end{bmatrix}$

(c) $\begin{bmatrix} 2 & 0 & 3 \\ 0 & 1 & 0 \\ 0 & 1 & 2 \end{bmatrix}$ **(d)** $\begin{bmatrix} 2 & 3 & 3 & 5 \\ 3 & 2 & 2 & 3 \\ 0 & 0 & 2 & 2 \\ 0 & 0 & 0 & 2 \end{bmatrix}$

8. Find a 2×2 nondiagonal matrix whose eigenvalues are 2 and -3, and associated eigenvectors are $\begin{bmatrix} -1 \\ 2 \end{bmatrix}$ and $\begin{bmatrix} 1 \\ 1 \end{bmatrix}$, respectively.

9. Find a 3×3 nondiagonal matrix whose eigenvalues are -2, -2, and 3, and associated eigenvectors are $\begin{bmatrix} 1 \\ 0 \\ 1 \end{bmatrix}$, $\begin{bmatrix} 0 \\ 1 \\ 1 \end{bmatrix}$, and $\begin{bmatrix} 1 \\ 1 \\ 1 \end{bmatrix}$, respectively.

10. For each of the following matrices find, if possible, a nonsingular matrix P such that $P^{-1}AP$ is diagonal:

(a) $\begin{bmatrix} 4 & 2 & 3 \\ 2 & 1 & 2 \\ -1 & -2 & 0 \end{bmatrix}$ **(b)** $\begin{bmatrix} 1 & 1 & 2 \\ 0 & 1 & 0 \\ 0 & 1 & 3 \end{bmatrix}$

(c) $\begin{bmatrix} 1 & 2 & 3 \\ 0 & 1 & 0 \\ 2 & 1 & 2 \end{bmatrix}$ **(d)** $\begin{bmatrix} 0 & -1 \\ 2 & 3 \end{bmatrix}$

11. For each of the following matrices find, if possible, a nonsingular matrix P such that $P^{-1}AP$ is diagonal:

(a) $\begin{bmatrix} 3 & -2 & 1 \\ 0 & 2 & 0 \\ 0 & 0 & 0 \end{bmatrix}$ **(b)** $\begin{bmatrix} 2 & 2 & 2 \\ 2 & 2 & 2 \\ 2 & 2 & 2 \end{bmatrix}$

(c) $\begin{bmatrix} 3 & 0 & 0 \\ 2 & 3 & 0 \\ 0 & 0 & 3 \end{bmatrix}$ **(d)** $\begin{bmatrix} 1 & 0 & 1 \\ 0 & 1 & 0 \\ 0 & 1 & 2 \end{bmatrix}$

12. Let A be a 2×2 matrix whose eigenvalues are 3 and 4, and associated eigenvectors are $\begin{bmatrix} -1 \\ 1 \end{bmatrix}$ and $\begin{bmatrix} 2 \\ 1 \end{bmatrix}$, respectively. Without computation, find a diagonal matrix D that is similar to A, and a nonsingular matrix P such that $P^{-1}AP = D$.

13. Let A be a 3×3 matrix whose eigenvalues are -3, 4, and 4, and associated eigenvectors are

$$\begin{bmatrix} -1 \\ 0 \\ 1 \end{bmatrix}, \quad \begin{bmatrix} 0 \\ 0 \\ 1 \end{bmatrix}, \quad \text{and} \quad \begin{bmatrix} 0 \\ 1 \\ 1 \end{bmatrix},$$

respectively. Without computation, find a diagonal matrix D that is similar to A, and a nonsingular matrix P such that $P^{-1}AP = D$.

14. Which of the following matrices are similar to a diagonal matrix?

(a) $\begin{bmatrix} 2 & 3 & 0 \\ 0 & 1 & 0 \\ 0 & 0 & 2 \end{bmatrix}$ **(b)** $\begin{bmatrix} 2 & 3 & 1 \\ 0 & 1 & 0 \\ 0 & 0 & 2 \end{bmatrix}$

(c) $\begin{bmatrix} -3 & 0 \\ 1 & 2 \end{bmatrix}$ **(d)** $\begin{bmatrix} 1 & 1 & 0 \\ 2 & 2 & 0 \\ 3 & 3 & 3 \end{bmatrix}$

15. Show that each of the following matrices is diagonalizable and find a diagonal matrix similar to each given matrix:

(a) $\begin{bmatrix} 4 & 2 \\ 3 & 3 \end{bmatrix}$ **(b)** $\begin{bmatrix} 3 & 2 \\ 6 & 4 \end{bmatrix}$

(c) $\begin{bmatrix} 2 & -2 & 3 \\ 0 & 3 & -2 \\ 0 & -1 & 2 \end{bmatrix}$ **(d)** $\begin{bmatrix} 0 & -2 & 1 \\ 1 & 3 & -1 \\ 0 & 0 & 1 \end{bmatrix}$

16. Show that none of the following matrices is diagonalizable:

(a) $\begin{bmatrix} 1 & 1 \\ 0 & 1 \end{bmatrix}$ **(b)** $\begin{bmatrix} 2 & 0 & 0 \\ 3 & 2 & 0 \\ 0 & 0 & 5 \end{bmatrix}$

(c) $\begin{bmatrix} 10 & 11 & 3 \\ -3 & -4 & -3 \\ -8 & -8 & -1 \end{bmatrix}$ **(d)** $\begin{bmatrix} 2 & 3 & 3 & 5 \\ 3 & 2 & 2 & 3 \\ 0 & 0 & 1 & 1 \\ 0 & 0 & 0 & 1 \end{bmatrix}$

17. A matrix A is called **defective** if A has an eigenvalue λ of multiplicity $m > 1$ for which the associated eigenspace has a basis of fewer than m vectors; that is, the dimension of the eigenspace associated with λ is less than m. Use the eigenvalues of the following matrices to determine which matrices are defective:

(a) $\begin{bmatrix} 8 & 7 \\ 0 & 8 \end{bmatrix}$, $\lambda = 8, 8$

(b) $\begin{bmatrix} 3 & 0 & 0 \\ -2 & 3 & -2 \\ 2 & 0 & 5 \end{bmatrix}$, $\lambda = 3, 3, 5$

(c) $\begin{bmatrix} 3 & 3 & 3 \\ 3 & 3 & 3 \\ -3 & -3 & -3 \end{bmatrix}$, $\lambda = 0, 0, 3$

(d) $\begin{bmatrix} 0 & 0 & 1 & 0 \\ 0 & 0 & 0 & -1 \\ 1 & 0 & 0 & 0 \\ 0 & -1 & 0 & 0 \end{bmatrix}$, $\lambda = 1, 1, -1, -1$

18. Let $D = \begin{bmatrix} 2 & 0 \\ 0 & -2 \end{bmatrix}$. Compute D^9.

19. Let $A = \begin{bmatrix} 3 & -5 \\ 1 & -3 \end{bmatrix}$. Compute A^9. (*Hint*: Find a matrix P such that $P^{-1}AP$ is a diagonal matrix D and show that $A^9 = PD^9P^{-1}$.)

20. Let $A = \begin{bmatrix} a & b \\ c & d \end{bmatrix}$. Find necessary and sufficient conditions for A to be diagonalizable.

21. Let A and B be nonsingular $n \times n$ matrices. Prove that AB and BA have the same eigenvalues.

22. **(*Calculus Required*)** Let V be the vector space of continuous functions with basis $\{e^t, e^{-t}\}$. Let $L\colon V \to V$ be defined by $L(g(t)) = g'(t)$ for $g(t)$ in V. Show that L is diagonalizable.

23. Prove that if A is diagonalizable, then **(a)** A^T is diagonalizable, and **(b)** A^k is diagonalizable, where k is a positive integer.

24. Show that if A is nonsingular and diagonalizable, then A^{-1} is diagonalizable.

25. Let $\lambda_1, \lambda_2, \ldots, \lambda_k$ be distinct eigenvalues of an $n \times n$ matrix A with associated eigenvectors $\mathbf{x}_1, \mathbf{x}_2, \ldots, \mathbf{x}_k$. Prove that $\mathbf{x}_1, \mathbf{x}_2, \ldots, \mathbf{x}_k$ are linearly independent. (*Hint*: See the proof of Theorem 7.5.)

26. Let A and B be nonsingular $n \times n$ matrices. Prove that AB^{-1} and $B^{-1}A$ have the same eigenvalues.

27. Show that if a matrix A is similar to a diagonal matrix D, then $\mathrm{Tr}(A) = \mathrm{Tr}(D)$, where $\mathrm{Tr}(A)$ is the trace of A. [*Hint*: See Exercise 43, Section 1.3, where part (c) establishes $\mathrm{Tr}(AB) = \mathrm{Tr}(BA)$.]

28. Let A be an $n \times n$ matrix and let $B = P^{-1}AP$ be similar to A. Show that if $\mathbf{x}$ is an eigenvector of A associated with the eigenvalue λ of A, then $P^{-1}\mathbf{x}$ is an eigenvector of B associated with the eigenvalue λ of the matrix B.

7.3 Diagonalization of Symmetric Matrices

In this section we consider the diagonalization of a symmetric matrix (i.e., a matrix A for which $A = A^T$). We restrict our attention to symmetric matrices, because they are easier to handle than general matrices and because they arise in many applied problems. One of these applications is discussed in Section 8.6. By definition, all entries of a symmetric matrix are real numbers.

Theorem 7.5 assures us that an $n \times n$ matrix A is diagonalizable if it has n distinct eigenvalues; if this is not so, then A may fail to be diagonalizable. However, every symmetric matrix can be diagonalized; that is, if A is symmetric, there exists a nonsingular matrix P such that $P^{-1}AP = D$, where D is a diagonal matrix. Moreover, P has some noteworthy properties that we remark on. We thus turn to the study of symmetric matrices in this section.

We first prove that all the roots of the characteristic polynomial of a symmetric matrix are real. Appendix B.2 contains additional examples of matrices with complex eigenvalues. A review of complex arithmetic appears in Appendix B.1.

Theorem 7.6 All the roots of the characteristic polynomial of a symmetric matrix are real numbers.

Proof

We give two proofs of this result. They both require some facts about complex numbers, which are covered in Appendix B.1. The first proof requires fewer of these facts, but is more computational and longer. Let $\lambda = a + bi$ be any root of the characteristic polynomial of A. We shall prove that $b = 0$, so that λ is a real number. Now

$$\det(\lambda I_n - A) = 0 = \det((a + bi)I_n - A).$$

This means that the homogeneous system

$$((a + bi)I_n - A)(\mathbf{x} + \mathbf{y}i) = \mathbf{0} = \mathbf{0} + \mathbf{0}i \tag{1}$$

has a nontrivial solution $\mathbf{x} + \mathbf{y}i$, where $\mathbf{x}$ and $\mathbf{y}$ are vectors in R^n that are not both the zero vector. Carrying out the multiplication in (1), we obtain

$$(aI_n\mathbf{x} - A\mathbf{x} - bI_n\mathbf{y}) + i(aI_n\mathbf{y} + bI_n\mathbf{x} - A\mathbf{y}) = \mathbf{0} + \mathbf{0}i. \tag{2}$$

Setting the real and imaginary parts equal to $\mathbf{0}$, we have

$$\begin{aligned} aI_n\mathbf{x} - A\mathbf{x} - bI_n\mathbf{y} &= \mathbf{0} \\ aI_n\mathbf{y} - A\mathbf{y} + bI_n\mathbf{x} &= \mathbf{0}. \end{aligned} \tag{3}$$

Forming the inner products of both sides of the first equation in (3) with $\mathbf{y}$ and of both sides of the second equation of (3) with $\mathbf{x}$, we have

$$\begin{aligned} (\mathbf{y}, aI_n\mathbf{x} - A\mathbf{x} - bI_n\mathbf{y}) &= (\mathbf{y}, \mathbf{0}) = 0 \\ (aI_n\mathbf{y} - A\mathbf{y} + bI_n\mathbf{x}, \mathbf{x}) &= (\mathbf{0}, \mathbf{x}) = 0, \end{aligned}$$

or

$$\begin{aligned} a(\mathbf{y}, I_n\mathbf{x}) - (\mathbf{y}, A\mathbf{x}) - b(\mathbf{y}, I_n\mathbf{y}) &= 0 \\ a(I_n\mathbf{y}, \mathbf{x}) - (A\mathbf{y}, \mathbf{x}) + b(I_n\mathbf{x}, \mathbf{x}) &= 0. \end{aligned} \tag{4}$$

Now, by Equation (3) in Section 5.3, we see that $(I_n\mathbf{y}, \mathbf{x}) = (\mathbf{y}, I_n^T\mathbf{x}) = (\mathbf{y}, I_n\mathbf{x})$ and that $(A\mathbf{y}, \mathbf{x}) = (\mathbf{y}, A^T\mathbf{x}) = (\mathbf{y}, A\mathbf{x})$. Note that we have used the facts that $I_n^T = I_n$ and that, since A is symmetric, we have $A^T = A$. Subtracting the two equations in (4), we now get

$$-b(\mathbf{y}, I_n\mathbf{y}) - b(I_n\mathbf{x}, \mathbf{x}) = 0, \tag{5}$$

or

$$-b[(\mathbf{y}, \mathbf{y}) + (\mathbf{x}, \mathbf{x})] = 0. \tag{6}$$

Since $\mathbf{x}$ and $\mathbf{y}$ are not both the zero vector, $(\mathbf{x}, \mathbf{x}) > 0$ or $(\mathbf{y}, \mathbf{y}) > 0$. From (6), we conclude that $b = 0$. Hence every root of the characteristic polynomial of A is a real number.

Alternative Proof

Let λ be any root of the characteristic polynomial of A. We will prove that λ is real by showing that $\lambda = \overline{\lambda}$, its complex conjugate. We have

$$A\mathbf{x} = \lambda\mathbf{x}.$$

Multiplying both sides of this equation by $\overline{\mathbf{x}}^T$ on the left, we obtain

$$\overline{\mathbf{x}}^T A\mathbf{x} = \overline{\mathbf{x}}^T \lambda\mathbf{x}.$$

Taking the conjugate transpose of both sides yields

$$\overline{\mathbf{x}}^T \overline{A}^T \mathbf{x} = \overline{\lambda}\,\overline{\mathbf{x}}^T\mathbf{x},$$

or

$$\overline{\mathbf{x}}^T A\mathbf{x} = \overline{\lambda}\,\overline{\mathbf{x}}^T\mathbf{x} \qquad (\text{since } A = A^T)$$
$$\lambda\,\overline{\mathbf{x}}^T\mathbf{x} = \overline{\lambda}\,\overline{\mathbf{x}}^T\mathbf{x},$$

so

$$(\lambda - \overline{\lambda})(\overline{\mathbf{x}}^T\mathbf{x}) = 0.$$

Since $\mathbf{x} \neq \mathbf{0}$, $\overline{\mathbf{x}}^T\mathbf{x} \neq 0$. Hence $\lambda - \overline{\lambda} = 0$ or $\lambda = \overline{\lambda}$. ■

Now that we have established this result, we know that complex numbers do not enter into the study of the diagonalization problem for symmetric matrices.

Theorem 7.7 If A is a symmetric matrix, then eigenvectors that belong to distinct eigenvalues of A are orthogonal.

Proof

Let $\mathbf{x}_1$ and $\mathbf{x}_2$ be eigenvectors of A that are associated with the distinct eigenvalues λ_1 and λ_2 of A. We then have

$$A\mathbf{x}_1 = \lambda_1\mathbf{x}_1 \quad \text{and} \quad A\mathbf{x}_2 = \lambda_2\mathbf{x}_2.$$

Now, using Equation (3) of Section 5.3 and the fact that $A^T = A$, since A is symmetric, we have

$$\begin{aligned}\lambda_1(\mathbf{x}_1, \mathbf{x}_2) &= (\lambda_1\mathbf{x}_1, \mathbf{x}_2) = (A\mathbf{x}_1, \mathbf{x}_2)\\ &= (\mathbf{x}_1, A^T\mathbf{x}_2) = (\mathbf{x}_1, A\mathbf{x}_2)\\ &= (\mathbf{x}_1, \lambda_2\mathbf{x}_2) = \lambda_2(\mathbf{x}_1, \mathbf{x}_2).\end{aligned}$$

Thus

$$\lambda_1(\mathbf{x}_1, \mathbf{x}_2) = \lambda_2(\mathbf{x}_1, \mathbf{x}_2),$$

and subtracting, we obtain

$$\begin{aligned} 0 &= \lambda_1(\mathbf{x}_1, \mathbf{x}_2) - \lambda_2(\mathbf{x}_1, \mathbf{x}_2) \\ &= (\lambda_1 - \lambda_2)(\mathbf{x}_1, \mathbf{x}_2). \end{aligned}$$

Since $\lambda_1 \neq \lambda_2$, we conclude that $(\mathbf{x}_1, \mathbf{x}_2) = 0$. ■

EXAMPLE 1

Let $A = \begin{bmatrix} 0 & 0 & -2 \\ 0 & -2 & 0 \\ -2 & 0 & 3 \end{bmatrix}$. The characteristic polynomial of A is

$$p(\lambda) = (\lambda + 2)(\lambda - 4)(\lambda + 1)$$

(verify), so the eigenvalues of A are $\lambda_1 = -2$, $\lambda_2 = 4$, $\lambda_3 = -1$. Associated eigenvectors are the nontrivial solutions of the homogeneous system [Equation (6) in Section 7.1]

$$\begin{bmatrix} \lambda & 0 & 2 \\ 0 & \lambda + 2 & 0 \\ 2 & 0 & \lambda - 3 \end{bmatrix} \begin{bmatrix} x_1 \\ x_2 \\ x_3 \end{bmatrix} = \begin{bmatrix} 0 \\ 0 \\ 0 \end{bmatrix}.$$

For $\lambda_1 = -2$, we find that $\mathbf{x}_1$ is any vector of the form $\begin{bmatrix} 0 \\ r \\ 0 \end{bmatrix}$, where r is any nonzero number (verify). Thus we may take $\mathbf{x}_1 = \begin{bmatrix} 0 \\ 1 \\ 0 \end{bmatrix}$. For $\lambda_2 = 4$, we find that $\mathbf{x}_2$ is any vector of the form $\begin{bmatrix} -\frac{r}{2} \\ 0 \\ r \end{bmatrix}$, where r is any nonzero number (verify). Thus we may take $\mathbf{x}_2 = \begin{bmatrix} -1 \\ 0 \\ 2 \end{bmatrix}$. For $\lambda_3 = -1$, we find that $\mathbf{x}_3$ is any vector of the form $\begin{bmatrix} 2r \\ 0 \\ r \end{bmatrix}$, where r is any nonzero number (verify). Thus we may take $\mathbf{x}_3 = \begin{bmatrix} 2 \\ 0 \\ 1 \end{bmatrix}$. It is clear that $\{\mathbf{x}_1, \mathbf{x}_2, \mathbf{x}_3\}$ is orthogonal and linearly independent. Thus A is similar to $D = \begin{bmatrix} -2 & 0 & 0 \\ 0 & 4 & 0 \\ 0 & 0 & -1 \end{bmatrix}$. ■

If A can be diagonalized, then there exists a nonsingular matrix P such that $P^{-1}AP$ is diagonal. Moreover, the columns of P are eigenvectors of A. Now, if the eigenvectors of A form an orthogonal set S, as happens when A is symmetric and the eigenvalues of A are distinct, then since any nonzero scalar multiple of an eigenvector of A is also an eigenvector of A, we can normalize S to obtain an orthonormal set $T = \{\mathbf{x}_1, \mathbf{x}_2, \ldots, \mathbf{x}_n\}$ of eigenvectors of A. Let the jth column of P be the eigenvector $\mathbf{x}_j$, and we now see what type of matrix P must be. We can write P as a partitioned matrix in the form $P = \begin{bmatrix} \mathbf{x}_1 & \mathbf{x}_2 & \cdots & \mathbf{x}_n \end{bmatrix}$. Then

$$P^T = \begin{bmatrix} \mathbf{x}_1^T \\ \mathbf{x}_2^T \\ \vdots \\ \mathbf{x}_n^T \end{bmatrix},$$

where $\mathbf{x}_i^T$ is the transpose of the $n \times 1$ matrix (or vector) $\mathbf{x}_i$. We find that the (i, j) entry in $P^T P$ is $(\mathbf{x}_i, \mathbf{x}_j)$. Since $(\mathbf{x}_i, \mathbf{x}_j) = 1$ if $i = j$ and $(\mathbf{x}_i, \mathbf{x}_j) = 0$ if $i \neq j$, we have $P^T P = I_n$, which means that $P^T = P^{-1}$. Such matrices are important enough to have a special name.

DEFINITION 7.4 A real square matrix A is called **orthogonal** if $A^{-1} = A^T$. Of course, we can also say that A is orthogonal if $A^T A = I_n$.

EXAMPLE 2 Let

$$A = \begin{bmatrix} \frac{2}{3} & -\frac{2}{3} & \frac{1}{3} \\ \frac{2}{3} & \frac{1}{3} & -\frac{2}{3} \\ \frac{1}{3} & \frac{2}{3} & \frac{2}{3} \end{bmatrix}.$$

It is easy to check that $A^T A = I_n$, so A is an orthogonal matrix. ■

EXAMPLE 3 Let A be the matrix defined in Example 1. We already know that the set of eigenvectors

$$\left\{ \begin{bmatrix} 0 \\ 1 \\ 0 \end{bmatrix}, \begin{bmatrix} -1 \\ 0 \\ 2 \end{bmatrix}, \begin{bmatrix} 2 \\ 0 \\ 1 \end{bmatrix} \right\}$$

is orthogonal. If we normalize these vectors, we find that

$$T = \left\{ \begin{bmatrix} 0 \\ 1 \\ 0 \end{bmatrix}, \begin{bmatrix} -\frac{1}{\sqrt{5}} \\ 0 \\ \frac{2}{\sqrt{5}} \end{bmatrix}, \begin{bmatrix} \frac{2}{\sqrt{5}} \\ 0 \\ \frac{1}{\sqrt{5}} \end{bmatrix} \right\}$$

is an orthonormal basis for R^3. A matrix P such that $P^{-1}AP$ is diagonal is the matrix whose columns are the vectors in T. Thus

$$P = \begin{bmatrix} 0 & -\frac{1}{\sqrt{5}} & \frac{2}{\sqrt{5}} \\ 1 & 0 & 0 \\ 0 & \frac{2}{\sqrt{5}} & \frac{1}{\sqrt{5}} \end{bmatrix}.$$

We leave it to the reader to verify that P is an orthogonal matrix and that

$$P^{-1}AP = P^TAP = \begin{bmatrix} -2 & 0 & 0 \\ 0 & 4 & 0 \\ 0 & 0 & -1 \end{bmatrix}.$$

■

The following theorem is not difficult to prove:

Theorem 7.8 The $n \times n$ matrix A is orthogonal if and only if the columns (rows) of A form an orthonormal set.

Proof

Exercise 5. ■

If A is an orthogonal matrix, then we can show that $\det(A) = \pm 1$ (Exercise 8). We now look at some of the geometric properties of orthogonal matrices. If A is an orthogonal $n \times n$ matrix, let $L: R^n \to R^n$ be the linear operator defined by $L(\mathbf{x}) = A\mathbf{x}$, for $\mathbf{x}$ in R^n (recall Chapter 6). If $\det(A) = 1$ and $n = 2$, it then follows that L is a counterclockwise rotation. It can also be shown that if $\det(A) = -1$, then L is a reflection about the x-axis followed by a counterclockwise rotation (see Exercise 32).

Again, let A be an orthogonal $n \times n$ matrix and let $L: R^n \to R^n$ be defined by $L(\mathbf{x}) = A\mathbf{x}$ for $\mathbf{x}$ in R^n. We now compute $(L(\mathbf{x}), L(\mathbf{y}))$ for any vectors $\mathbf{x}$, $\mathbf{y}$ in R^n, using the standard inner product on R^n. We have

$$(L(\mathbf{x}), L(\mathbf{y})) = (A\mathbf{x}, A\mathbf{y}) = (\mathbf{x}, A^TA\mathbf{y}) = (\mathbf{x}, A^{-1}A\mathbf{y}) = (\mathbf{x}, I_n\mathbf{y}) = (\mathbf{x}, \mathbf{y}), \quad (7)$$

where we have used Equation (3) in Section 5.3. This means that L preserves the inner product of two vectors and, consequently, L preserves length. (Why?) It also follows that if θ is the angle between vectors $\mathbf{x}$ and $\mathbf{y}$ in R^n, then the angle between $L(\mathbf{x})$ and $L(\mathbf{y})$ is also θ. A linear transformation satisfying Equation (7), $(L(\mathbf{x}), L(\mathbf{y})) = (\mathbf{x}, \mathbf{y})$, is called an **isometry** (from the Greek meaning *equal length*). Conversely, let $L: R^n \to R^n$ be an isometry, so that $(L(\mathbf{x}), L(\mathbf{y})) = (\mathbf{x}, \mathbf{y})$ for any $\mathbf{x}$ and $\mathbf{y}$ in R^n. Let A be the standard matrix representing L. Then $L(\mathbf{x}) = A\mathbf{x}$. We now have

$$(\mathbf{x}, \mathbf{y}) = (L(\mathbf{x}), L(\mathbf{y})) = (A\mathbf{x}, A\mathbf{y}) = (\mathbf{x}, A^TA\mathbf{y}).$$

Since this holds for all $\mathbf{x}$ in R^n, then by Exercise 7(e) in Section 5.3, we conclude that $A^TA\mathbf{y} = \mathbf{y}$ for any $\mathbf{y}$ in R^n. It follows that $A^TA = I_n$ (Exercise 36), so A is an orthogonal matrix. Other properties of orthogonal matrices and isometries are examined in the exercises. (See also Supplementary Exercises 16 and 18 in Chapter 6.)

We now turn to the general situation for a symmetric matrix; even if A has eigenvalues whose multiplicities are greater than one, it turns out that we can still diagonalize A. We omit the proof of the next theorem. For a proof, see J. M. Ortega, *Matrix Theory: A Second Course*, New York: Plenum Press, 1987.

Theorem 7.9 If A is a symmetric $n \times n$ matrix, then there exists an orthogonal matrix P such that $P^{-1}AP = P^TAP = D$, a diagonal matrix. The eigenvalues of A lie on the main diagonal of D. ■

It can be shown (see the book by Ortega cited previously) that if a symmetric matrix A has an eigenvalue λ of multiplicity k, then the solution space of the homogeneous system $(\lambda I_n - A)\mathbf{x} = \mathbf{0}$ [Equation (6) in Section 7.1] has dimension k. This means that there exist k linearly independent eigenvectors of A associated with the eigenvalue λ. By the Gram–Schmidt process, we can choose an orthonormal basis for this solution space. Thus we obtain a set of k orthonormal eigenvectors associated with the eigenvalue λ. Since eigenvectors associated with distinct eigenvalues are orthogonal, if we form the set of all eigenvectors, we get an orthonormal set. Hence the matrix P whose columns are the eigenvectors is orthogonal.

EXAMPLE 4

Let

$$A = \begin{bmatrix} 0 & 2 & 2 \\ 2 & 0 & 2 \\ 2 & 2 & 0 \end{bmatrix}.$$

The characteristic polynomial of A is

$$p(\lambda) = (\lambda + 2)^2(\lambda - 4)$$

(verify), so its eigenvalues are

$$\lambda_1 = -2, \quad \lambda_2 = -2, \quad \text{and} \quad \lambda_3 = 4.$$

That is, -2 is an eigenvalue of multiplicity 2. To find eigenvectors associated with -2, we solve the homogeneous system $(-2I_3 - A)\mathbf{x} = \mathbf{0}$:

$$\begin{bmatrix} -2 & -2 & -2 \\ -2 & -2 & -2 \\ -2 & -2 & -2 \end{bmatrix} \begin{bmatrix} x_1 \\ x_2 \\ x_3 \end{bmatrix} = \begin{bmatrix} 0 \\ 0 \\ 0 \end{bmatrix}. \tag{8}$$

A basis for the solution space of (8) consists of the eigenvectors

$$\mathbf{x}_1 = \begin{bmatrix} -1 \\ 1 \\ 0 \end{bmatrix} \quad \text{and} \quad \mathbf{x}_2 = \begin{bmatrix} -1 \\ 0 \\ 1 \end{bmatrix}$$

(verify). Now $\mathbf{x}_1$ and $\mathbf{x}_2$ are not orthogonal, since $(\mathbf{x}_1, \mathbf{x}_2) \neq 0$. We can use the Gram–Schmidt process to obtain an orthonormal basis for the solution space of (8) (the eigenspace associated with -2) as follows: Let $\mathbf{y}_1 = \mathbf{x}_1$ and

$$\mathbf{y}_2 = \mathbf{x}_2 - \left(\frac{\mathbf{x}_2 \cdot \mathbf{y}_1}{\mathbf{y}_1 \cdot \mathbf{y}_1}\right)\mathbf{x}_1 = \begin{bmatrix} -\frac{1}{2} \\ -\frac{1}{2} \\ 1 \end{bmatrix}.$$

To eliminate fractions, we let

$$\mathbf{y}_2^* = 2\mathbf{y}_2 = \begin{bmatrix} -1 \\ -1 \\ 2 \end{bmatrix}.$$

The set $\{\mathbf{y}_1, \mathbf{y}_2^*\}$ is an orthogonal set of vectors. Normalizing, we obtain

$$\mathbf{z}_1 = \frac{1}{\|\mathbf{y}_1\|}\mathbf{y}_1 = \frac{1}{\sqrt{2}}\begin{bmatrix} -1 \\ 1 \\ 0 \end{bmatrix} \quad \text{and} \quad \mathbf{z}_2 = \frac{1}{\|\mathbf{y}_2^*\|}\mathbf{y}_2^* = \frac{1}{\sqrt{6}}\begin{bmatrix} -1 \\ -1 \\ 2 \end{bmatrix}.$$

The set $\{\mathbf{z}_1, \mathbf{z}_2\}$ is an orthonormal basis for the eigenspace associated with $\lambda = -2$. Now we find a basis for the eigenspace associated with $\lambda = 4$ by solving the homogeneous system $(4I_3 - A)\mathbf{x} = \mathbf{0}$:

$$\begin{bmatrix} 4 & -2 & -2 \\ -2 & 4 & -2 \\ -2 & -2 & 4 \end{bmatrix}\begin{bmatrix} x_1 \\ x_2 \\ x_3 \end{bmatrix} = \begin{bmatrix} 0 \\ 0 \\ 0 \end{bmatrix}. \tag{9}$$

A basis for this eigenspace consists of the vector

$$\mathbf{x}_3 = \begin{bmatrix} 1 \\ 1 \\ 1 \end{bmatrix}$$

(verify). Normalizing this vector, we have the eigenvector

$$\mathbf{z}_3 = \frac{1}{\sqrt{3}}\begin{bmatrix} 1 \\ 1 \\ 1 \end{bmatrix}$$

as a basis for the eigenspace associated with $\lambda = 4$. Since eigenvectors associated with distinct eigenvalues are orthogonal, we observe that $\mathbf{z}_3$ is orthogonal to both $\mathbf{z}_1$ and $\mathbf{z}_2$. Thus the set $\{\mathbf{z}_1, \mathbf{z}_2, \mathbf{z}_3\}$ is an orthonormal basis for R^3 consisting of eigenvectors of A. The matrix P is the matrix whose jth column is $\mathbf{z}_j$:

$$P = \begin{bmatrix} -\frac{1}{\sqrt{2}} & -\frac{1}{\sqrt{6}} & \frac{1}{\sqrt{3}} \\ \frac{1}{\sqrt{2}} & -\frac{1}{\sqrt{6}} & \frac{1}{\sqrt{3}} \\ 0 & \frac{2}{\sqrt{6}} & \frac{1}{\sqrt{3}} \end{bmatrix}.$$

We leave it to the reader to verify that

$$P^{-1}AP = P^TAP = \begin{bmatrix} -2 & 0 & 0 \\ 0 & -2 & 0 \\ 0 & 0 & 4 \end{bmatrix}.$$

■

EXAMPLE 5 Let

$$A = \begin{bmatrix} 1 & 2 & 0 & 0 \\ 2 & 1 & 0 & 0 \\ 0 & 0 & 1 & 2 \\ 0 & 0 & 2 & 1 \end{bmatrix}.$$

Either by straightforward computation or by Exercise 16 in Section 3.2, we find that the characteristic polynomial of A is

$$p(\lambda) = (\lambda + 1)^2(\lambda - 3)^2,$$

so its eigenvalues are

$$\lambda_1 = -1, \quad \lambda_2 = -1, \quad \lambda_3 = 3, \quad \text{and} \quad \lambda_4 = 3.$$

We now compute associated eigenvectors and the orthogonal matrix P. The eigenspace associated with the eigenvalue -1, of multiplicity 2, is the solution space of the homogeneous system $(-1I_4 - A)\mathbf{x} = \mathbf{0}$, namely,

$$\begin{bmatrix} -2 & -2 & 0 & 0 \\ 2 & 2 & 0 & 0 \\ 0 & 0 & -2 & -2 \\ 0 & 0 & -2 & -2 \end{bmatrix} \begin{bmatrix} x_1 \\ x_2 \\ x_3 \\ x_4 \end{bmatrix} = \begin{bmatrix} 0 \\ 0 \\ 0 \\ 0 \end{bmatrix},$$

which is the set of all vectors of the form

$$\begin{bmatrix} r \\ -r \\ s \\ -s \end{bmatrix} = r\begin{bmatrix} 1 \\ -1 \\ 0 \\ 0 \end{bmatrix} + s\begin{bmatrix} 0 \\ 0 \\ 1 \\ -1 \end{bmatrix},$$

where r and s are taken as any real numbers. Thus the eigenvectors

$$\begin{bmatrix} 1 \\ -1 \\ 0 \\ 0 \end{bmatrix} \quad \text{and} \quad \begin{bmatrix} 0 \\ 0 \\ 1 \\ -1 \end{bmatrix}$$

form a basis for the eigenspace associated with -1, and the dimension of this eigenspace is 2. Note that the eigenvectors

$$\begin{bmatrix} 1 \\ -1 \\ 0 \\ 0 \end{bmatrix} \quad \text{and} \quad \begin{bmatrix} 0 \\ 0 \\ 1 \\ -1 \end{bmatrix}$$

happen to be orthogonal. Since we are looking for an orthonormal basis for this eigenspace, we take

$$\mathbf{x}_1 = \begin{bmatrix} \frac{1}{\sqrt{2}} \\ -\frac{1}{\sqrt{2}} \\ 0 \\ 0 \end{bmatrix} \quad \text{and} \quad \mathbf{x}_2 = \begin{bmatrix} 0 \\ 0 \\ \frac{1}{\sqrt{2}} \\ -\frac{1}{\sqrt{2}} \end{bmatrix}$$

as eigenvectors associated with λ_1 and λ_2, respectively. Then $\{\mathbf{x}_1, \mathbf{x}_2\}$ is an orthonormal basis for the eigenspace associated with -1. The eigenspace associated with the eigenvalue 3, of multiplicity 2, is the solution space of the homogeneous system $(3I_4 - A)\mathbf{x} = \mathbf{0}$, namely,

$$\begin{bmatrix} 2 & -2 & 0 & 0 \\ -2 & 2 & 0 & 0 \\ 0 & 0 & 2 & -2 \\ 0 & 0 & -2 & 2 \end{bmatrix} \begin{bmatrix} x_1 \\ x_2 \\ x_3 \\ x_4 \end{bmatrix} = \begin{bmatrix} 0 \\ 0 \\ 0 \\ 0 \end{bmatrix},$$

which is the set of all vectors of the form

$$\begin{bmatrix} r \\ r \\ s \\ s \end{bmatrix} = r \begin{bmatrix} 1 \\ 1 \\ 0 \\ 0 \end{bmatrix} + s \begin{bmatrix} 0 \\ 0 \\ 1 \\ 1 \end{bmatrix},$$

where r and s are taken as any real numbers. Thus the eigenvectors

$$\begin{bmatrix} 1 \\ 1 \\ 0 \\ 0 \end{bmatrix} \quad \text{and} \quad \begin{bmatrix} 0 \\ 0 \\ 1 \\ 1 \end{bmatrix}$$

form a basis for the eigenspace associated with 3, and the dimension of this eigenspace is 2. Since these eigenvectors are orthogonal, we normalize them and let

$$\mathbf{x}_3 = \begin{bmatrix} \frac{1}{\sqrt{2}} \\ \frac{1}{\sqrt{2}} \\ 0 \\ 0 \end{bmatrix} \quad \text{and} \quad \mathbf{x}_4 = \begin{bmatrix} 0 \\ 0 \\ \frac{1}{\sqrt{2}} \\ \frac{1}{\sqrt{2}} \end{bmatrix}$$

be eigenvectors associated with λ_3 and λ_4, respectively. Then $\{\mathbf{x}_3, \mathbf{x}_4\}$ is an orthonormal basis for the eigenspace associated with 3. Now eigenvectors associated with distinct eigenvalues are orthogonal, so $\{\mathbf{x}_1, \mathbf{x}_2, \mathbf{x}_3, \mathbf{x}_4\}$ is an orthonormal basis for R^4. The matrix P is the matrix whose jth column is $\mathbf{x}_j$, $j = 1, 2, 3, 4$. Thus

$$P = \begin{bmatrix} \frac{1}{\sqrt{2}} & 0 & \frac{1}{\sqrt{2}} & 0 \\ -\frac{1}{\sqrt{2}} & 0 & \frac{1}{\sqrt{2}} & 0 \\ 0 & \frac{1}{\sqrt{2}} & 0 & \frac{1}{\sqrt{2}} \\ 0 & -\frac{1}{\sqrt{2}} & 0 & \frac{1}{\sqrt{2}} \end{bmatrix}.$$

We leave it to the reader to verify that P is an orthogonal matrix and that

$$P^{-1}AP = P^TAP = \begin{bmatrix} -1 & 0 & 0 & 0 \\ 0 & -1 & 0 & 0 \\ 0 & 0 & 3 & 0 \\ 0 & 0 & 0 & 3 \end{bmatrix}.$$

■

The procedure for diagonalizing a matrix A is as follows:

Step 1. Form the characteristic polynomial $p(\lambda) = \det(\lambda I_n - A)$ of A.

Step 2. Find the roots of the characteristic polynomial of A.

Step 3. For each eigenvalue λ_j of A of multiplicity k_j, find a basis for the solution space of $(\lambda_j I_n - A)\mathbf{x} = \mathbf{0}$ (the eigenspace associated with λ_j). If the dimension of the eigenspace is less than k_j, then A is not diagonalizable. We thus determine n linearly independent eigenvectors of A. In Section 4.7 we solved the problem of finding a basis for the solution space of a homogeneous system.

Step 4. Let P be the matrix whose columns are the n linearly independent eigenvectors determined in Step 3. Then $P^{-1}AP = D$, a diagonal matrix whose diagonal elements are the eigenvalues of A that correspond to the columns of P.

If A is an $n \times n$ symmetric matrix, we know that we can find an orthogonal matrix P such that $P^{-1}AP$ is diagonal. Conversely, suppose that A is a matrix for which we can find an orthogonal matrix P such that $P^{-1}AP = D$ is a diagonal matrix. What type of matrix is A? Since $P^{-1}AP = D$, $A = PDP^{-1}$. Also, $P^{-1} = P^T$, since P is orthogonal. Then

$$A^T = (PDP^T)^T = (P^T)^T D^T P^T = PDP^T = A,$$

which means that A is symmetric.

■ Application: The Image of the Unit Circle by a Symmetric Matrix

In Example 5 of Section 1.7 we showed that the image of the unit circle by a matrix transformation whose associated matrix is diagonal was an ellipse centered at the origin with major and minor axes parallel to the coordinate axes; that is, in standard position. Here, we investigate the image of the unit circle by a matrix transformation whose associated matrix A is symmetric. We show the fundamental role played by eigenvalues and associated eigenvectors of A in determining both the size and orientation of the image.

Let A be a 2×2 symmetric matrix. Then by Theorem 7.9 there exists an orthogonal matrix $P = \begin{bmatrix} \mathbf{p}_1 & \mathbf{p}_2 \end{bmatrix}$ such that

$$P^TAP = D = \begin{bmatrix} \lambda_1 & 0 \\ 0 & \lambda_2 \end{bmatrix},$$

where λ_1 and λ_2 are the eigenvalues of A with associated eigenvectors $\mathbf{p}_1$ and $\mathbf{p}_2$, respectively. Moreover, $\{\mathbf{p}_1, \mathbf{p}_2\}$ is an orthonormal set. It follows that $A = PDP^T$ (verify).

Any point on the unit circle in R^2 is represented by a vector (or point)

$$\mathbf{v} = \begin{bmatrix} \cos t \\ \sin t \end{bmatrix}, \qquad 0 \le t < 2\pi.$$

Hence the image of the unit circle consists of all vectors (or points)

$$A\mathbf{v} = (PDP^T)\mathbf{v}.$$

It is convenient to view the image points as those obtained from a composition of matrix transformations; that is,

$$A\mathbf{v} = P(D(P^T\mathbf{v})). \tag{10}$$

This corresponds to three successive matrix transformations.

To determine the form or shape of this image, we use the following fact:

If Q is any 2×2 orthogonal matrix, then there is a real number φ such that

$$Q = \begin{bmatrix} \cos\varphi & -\sin\varphi \\ \sin\varphi & \cos\varphi \end{bmatrix} \quad \text{or} \quad Q = \begin{bmatrix} \cos\varphi & \sin\varphi \\ \sin\varphi & -\cos\varphi \end{bmatrix}.$$

In the first case Q is a rotation matrix; see Section 1.7. In the second case Q performs a reflection (about the x-axis, the y-axis, the line $y = x$, or the line $y = -x$) or a rotation followed by a reflection about the x-axis.

With this information and the result of Example 5 in Section 1.7, we can determine the geometric form of the image of the unit circle. Using the composite form displayed in Equation (10), we have the following actions:

P^T takes the unit circle to another unit circle,
D takes this unit circle to an ellipse in standard position, and (11)
P rotates or reflects the ellipse.

Thus the image of the unit circle by a symmetric matrix is an ellipse with center at the origin, but possibly with its axes not parallel to the coordinate axes (see Figure 7.4).

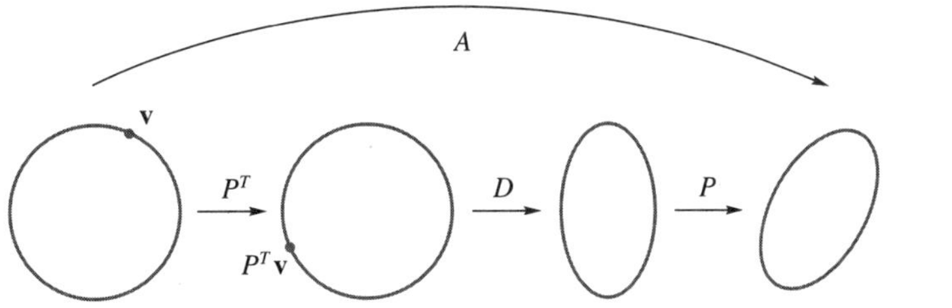

FIGURE 7.4 Unit circle. The image; an ellipse.

Next, we show how the eigenvalues and associated eigenvectors of A completely determine the elliptical image. The unit vectors $\mathbf{i} = \begin{bmatrix} 1 \\ 0 \end{bmatrix}$ and $\mathbf{j} = \begin{bmatrix} 0 \\ 1 \end{bmatrix}$ are on the unit circle. Then $P\mathbf{i} = \mathbf{p}_1$ and $P\mathbf{j} = \mathbf{p}_2$. (Verify.) But $\mathbf{p}_1$ and $\mathbf{p}_2$ are also on the unit circle, since P is an orthogonal matrix. Hence

$$P^{-1}\mathbf{p}_1 = P^T\mathbf{p}_1 = \mathbf{i} \quad \text{and} \quad P^{-1}\mathbf{p}_2 = P^T\mathbf{p}_2 = \mathbf{j}.$$

In (10), let $\mathbf{v} = \mathbf{p}_1$; then we have

$$A\mathbf{p}_1 = P(D(P^T\mathbf{p}_1)) = P(D(\mathbf{i})) = P(\lambda_1\mathbf{i}) = \lambda_1 P\mathbf{i} = \lambda_1\mathbf{p}_1 \tag{12}$$

and also $A\mathbf{p}_2 = \lambda_2\mathbf{p}_2$. But, of course, we knew this, because λ_1 and λ_2 are the eigenvalues of A with associated eigenvectors $\mathbf{p}_1$ and $\mathbf{p}_2$. However, this sequence of steps shows that eigenvectors of A on the original unit circle become multiples of themselves in the elliptical image. Moreover, since $\mathbf{p}_1$ and $\mathbf{p}_2$ are orthogonal, so are $A\mathbf{p}_1$ and $A\mathbf{p}_2$ (Why?), and these are the axes of the elliptical image. We display this graphically in Figure 7.5. It follows that the elliptical image is completely determined by the eigenvalues and associated eigenvectors of the matrix A.

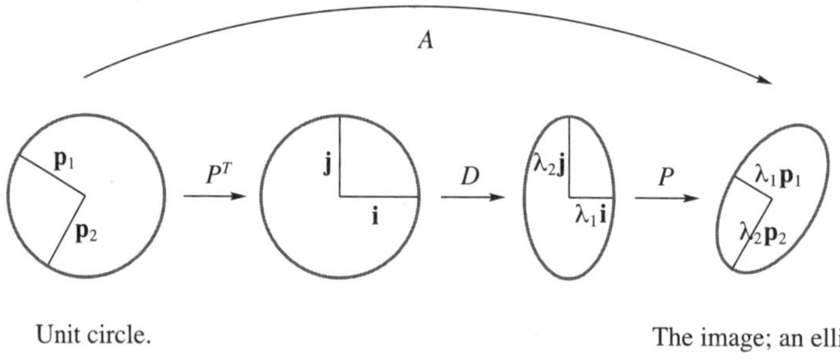

FIGURE 7.5

EXAMPLE 6 Let $A = \begin{bmatrix} 3 & 4 \\ 4 & 3 \end{bmatrix}$. Then the eigenvalues of A are $\lambda_1 = -1$ and $\lambda_2 = 7$ with associated orthonormal eigenvectors

$$\mathbf{p}_1 = \begin{bmatrix} \frac{1}{\sqrt{2}} \\ -\frac{1}{\sqrt{2}} \end{bmatrix} \quad \text{and} \quad \mathbf{p}_2 = \begin{bmatrix} \frac{1}{\sqrt{2}} \\ \frac{1}{\sqrt{2}} \end{bmatrix}.$$

(Verify.) Figure 7.6 shows the unit circle with $\mathbf{p}_1$ and $\mathbf{p}_2$ displayed as solid line segments and the elliptical image of the unit circle with axes displayed as dashed line segments. ■

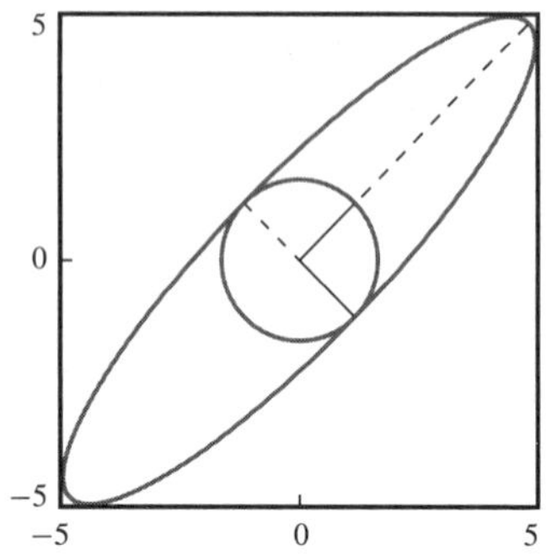

FIGURE 7.6

The preceding results and example show the following:

1. The eigenvalues determine the stretching of axes.
2. Eigenvectors determine the orientation of the images of the axes.

So, indeed, the eigenvalues and associated eigenvectors completely determine the image. These results generalize to $n \times n$ symmetric matrices. The image of the unit n-ball is an n-dimensional ellipse. For $n = 3$, the image is an ellipsoid.

Some remarks about nonsymmetric matrices are in order at this point. Theorem 7.5 assures us that an $n \times n$ matrix A is diagonalizable if all the roots of its characteristic polynomial are distinct. We also studied examples, in Section 7.2, of nonsymmetric matrices that had repeated eigenvalues which were diagonalizable (see Example 7) and others that were not diagonalizable (see Examples 3 and 6). There are some striking differences between the symmetric and nonsymmetric cases, which we now summarize. If A is nonsymmetric, then the roots of its characteristic polynomial need not all be real numbers; if an eigenvalue λ has multiplicity k, then the solution space of $(\lambda I_n - A)\mathbf{x} = \mathbf{0}$ may have dimension less than k; if the roots of the characteristic polynomial of A are all real, it is possible that A will not have n linearly independent eigenvectors; eigenvectors associated with distinct eigenvalues need not be orthogonal. Thus, in Example 6 of Section 7.2, the eigenvectors $\mathbf{x}_1$ and $\mathbf{x}_3$ associated with the eigenvalues $\lambda_1 = 0$ and $\lambda_3 = 1$ are not orthogonal. If a matrix A cannot be diagonalized, then we can often find a matrix B similar to A that is "nearly diagonal." The matrix B is said to be in **Jordan canonical form**. The study of such matrices lies beyond the scope of this book, but they are studied in advanced books on linear algebra (e.g., K. Hoffman and R. Kunze, *Linear Algebra*, 2d ed., Englewood Cliffs, NJ: Prentice-Hall, 1971); they play a key role in many applications of linear algebra.

It should be noted that in many applications we need find only a diagonal matrix D that is similar to the given matrix A; that is, we do not explicitly have to know the matrix P such that $P^{-1}AP = D$.

Eigenvalue problems arise in all applications involving vibrations; they occur in aerodynamics, elasticity, nuclear physics, mechanics, chemical engineering, biology, differential equations, and so on. Many of the matrices to be diagonalized in applied problems either are symmetric or all the roots of their characteristic polynomial are real. Of course, the methods for finding eigenvalues that have been presented in this chapter are not recommended for matrices of large order because of the need to evaluate determinants.

Key Terms

Symmetric matrix
Diagonalization
Eigenvalues/eigenvectors
Distinct eigenvalues
Orthogonal matrix
Orthonormal set
Gram–Schmidt process

7.3 Exercises

1. Verify that

$$P = \begin{bmatrix} \frac{2}{3} & -\frac{2}{3} & \frac{1}{3} \\ \frac{2}{3} & \frac{1}{3} & -\frac{2}{3} \\ \frac{1}{3} & \frac{2}{3} & \frac{2}{3} \end{bmatrix}$$

is an orthogonal matrix.

2. Find the inverse of each of the following orthogonal matrices:

(a) $A = \begin{bmatrix} 1 & 0 & 0 \\ 0 & \cos\phi & \sin\phi \\ 0 & -\sin\phi & \cos\phi \end{bmatrix}$

(b) $B = \begin{bmatrix} 1 & 0 & 0 \\ 0 & \frac{1}{\sqrt{2}} & -\frac{1}{\sqrt{2}} \\ 0 & -\frac{1}{\sqrt{2}} & -\frac{1}{\sqrt{2}} \end{bmatrix}$

3. Show that if A and B are orthogonal matrices, then AB is an orthogonal matrix.

4. Show that if A is an orthogonal matrix, then A^{-1} is orthogonal.

5. Prove Theorem 7.8.

6. Verify Theorem 7.8 for the matrices in Exercise 2.

7. Verify that the matrix P in Example 3 is an orthogonal matrix and that

$$P^{-1}AP = P^TAP = \begin{bmatrix} -2 & 0 & 0 \\ 0 & 4 & 0 \\ 0 & 0 & -1 \end{bmatrix}.$$

8. Show that if A is an orthogonal matrix, then $\det(A) = \pm 1$.

9. **(a)** Verify that the matrix

$$\begin{bmatrix} \cos\phi & -\sin\phi \\ \sin\phi & \cos\phi \end{bmatrix}$$

is orthogonal.

(b) Prove that if A is an orthogonal 2×2 matrix, then there exists a real number ϕ such that either

$$A = \begin{bmatrix} \cos\phi & -\sin\phi \\ \sin\phi & \cos\phi \end{bmatrix}$$

or

$$A = \begin{bmatrix} \cos\phi & \sin\phi \\ \sin\phi & -\cos\phi \end{bmatrix}.$$

10. For the orthogonal matrix

$$A = \begin{bmatrix} \frac{1}{\sqrt{2}} & -\frac{1}{\sqrt{2}} \\ -\frac{1}{\sqrt{2}} & -\frac{1}{\sqrt{2}} \end{bmatrix},$$

verify that $(A\mathbf{x}, A\mathbf{y}) = (\mathbf{x}, \mathbf{y})$ for any vectors $\mathbf{x}$ and $\mathbf{y}$ in R^2.

11. Let A be an $n \times n$ orthogonal matrix, and let $L: R^n \to R^n$ be the linear operator associated with A; that is, $L(\mathbf{x}) = A\mathbf{x}$ for $\mathbf{x}$ in R^n. Let θ be the angle between vectors $\mathbf{x}$ and $\mathbf{y}$ in R^n. Prove that the angle between $L(\mathbf{x})$ and $L(\mathbf{y})$ is also θ.

12. A linear operator $L: V \to V$, where V is an n-dimensional Euclidean space, is called **orthogonal** if $(L(\mathbf{x}), L(\mathbf{y})) = (\mathbf{x}, \mathbf{y})$. Let S be an orthonormal basis for V, and let the matrix A represent the orthogonal linear operator L with respect to S. Prove that A is an orthogonal matrix.

13. Let $L: R^2 \to R^2$ be the linear operator performing a counterclockwise rotation through $\pi/4$ and let A be the matrix representing L with respect to the natural basis for R^2. Prove that A is orthogonal.

14. Let A be an $n \times n$ matrix and let $B = P^{-1}AP$ be similar to A. Prove that if $\mathbf{x}$ is an eigenvector of A associated with the eigenvalue λ of A, then $P^{-1}\mathbf{x}$ is an eigenvector of B associated with the eigenvalue λ of B.

In Exercises 15 through 20, diagonalize each given matrix and find an orthogonal matrix P such that $P^{-1}AP$ is diagonal.

15. $A = \begin{bmatrix} 2 & 2 \\ 2 & 2 \end{bmatrix}$

16. $A = \begin{bmatrix} 0 & 0 & 1 \\ 0 & 0 & 0 \\ 1 & 0 & 0 \end{bmatrix}$

17. $A = \begin{bmatrix} 0 & 0 & 0 \\ 0 & 2 & 2 \\ 0 & 2 & 2 \end{bmatrix}$

18. $A = \begin{bmatrix} 0 & 0 & 0 & 0 \\ 0 & 0 & 0 & 0 \\ 0 & 0 & 0 & 1 \\ 0 & 0 & 1 & 0 \end{bmatrix}$

19. $A = \begin{bmatrix} 0 & -1 & -1 \\ -1 & 0 & -1 \\ -1 & -1 & 0 \end{bmatrix}$

20. $A = \begin{bmatrix} -1 & 2 & 2 \\ 2 & -1 & 2 \\ 2 & 2 & -1 \end{bmatrix}$

In Exercises 21 through 28, diagonalize each given matrix.

21. $A = \begin{bmatrix} 2 & 1 \\ 1 & 2 \end{bmatrix}$

22. $A = \begin{bmatrix} 2 & 2 & 0 & 0 \\ 2 & 2 & 0 & 0 \\ 0 & 0 & 2 & 2 \\ 0 & 0 & 2 & 2 \end{bmatrix}$

23. $A = \begin{bmatrix} 1 & 1 & 0 \\ 1 & 1 & 0 \\ 0 & 0 & 1 \end{bmatrix}$

24. $A = \begin{bmatrix} 1 & 0 & 0 \\ 0 & 3 & -2 \\ 0 & -2 & 3 \end{bmatrix}$

25. $A = \begin{bmatrix} 1 & 0 & 0 \\ 0 & 1 & 1 \\ 0 & 1 & 1 \end{bmatrix}$

26. $A = \begin{bmatrix} 0 & 0 & 0 & 1 \\ 0 & 0 & 0 & 0 \\ 0 & 0 & 0 & 0 \\ 1 & 0 & 0 & 0 \end{bmatrix}$

27. $A = \begin{bmatrix} 1 & -1 & 2 \\ -1 & 1 & 2 \\ 2 & 2 & 2 \end{bmatrix}$

28. $A = \begin{bmatrix} -3 & 0 & -1 \\ 0 & -2 & 0 \\ -1 & 0 & -3 \end{bmatrix}$

29. Prove Theorem 7.9 for the 2×2 case by studying the two possible cases for the roots of the characteristic polynomial of A.

30. Let $L: V \to V$ be an orthogonal linear operator (see Exercise 12), where V is an n-dimensional Euclidean space. Show that if λ is an eigenvalue of L, then $|\lambda| = 1$.

31. Let $L: R^2 \to R^2$ be defined by

$$L\left(\begin{bmatrix} x \\ y \end{bmatrix}\right) = \begin{bmatrix} \frac{1}{\sqrt{2}} & \frac{1}{\sqrt{2}} \\ \frac{1}{\sqrt{2}} & -\frac{1}{\sqrt{2}} \end{bmatrix} \begin{bmatrix} x \\ y \end{bmatrix}.$$

Show that L is an isometry of R^2.

32. Let $L: R^2 \to R^2$ be defined by $L(\mathbf{x}) = A\mathbf{x}$, for $\mathbf{x}$ in R^2, where A is an orthogonal matrix.

(a) Prove that if $\det(A) = 1$, then L is a counterclockwise rotation.

(b) Prove that if $\det(A) = -1$, then L is a reflection about the x-axis, followed by a counterclockwise rotation.

33. Let $L: R^n \to R^n$ be a linear operator.

(a) Prove that if L is an isometry, then $\|L(\mathbf{x})\| = \|\mathbf{x}\|$, for $\mathbf{x}$ in R^n.

(b) Prove that if L is an isometry and θ is the angle between vectors $\mathbf{x}$ and $\mathbf{y}$ in R^n, then the angle between $L(\mathbf{x})$ and $L(\mathbf{y})$ is also θ.

34. Let $L: R^n \to R^n$ be a linear operator defined by $L(\mathbf{x}) = A\mathbf{x}$ for $\mathbf{x}$ in R^n. Prove that if L is an isometry, then L^{-1} is an isometry.

35. Let $L: R^n \to R^n$ be a linear operator and $S = \{\mathbf{v}_1, \mathbf{v}_2, \dots, \mathbf{v}_n\}$ an orthonormal basis for R^n. Prove that L is an isometry if and only if $T = \{L(\mathbf{v}_1), L(\mathbf{v}_2), \dots, L(\mathbf{v}_n)\}$ is an orthonormal basis for R^n.

36. Show that if $A^T A\mathbf{y} = \mathbf{y}$ for all $\mathbf{y}$ in R^n, then $A^T A = I_n$.

37. Show that if A is an orthogonal matrix, then A^T is also orthogonal.

38. Let A be an orthogonal matrix. Show that cA is orthogonal if and only if $c = \pm 1$.

39. Assuming that the software you use has a command for eigenvalues and eigenvectors (see Exercises 39 and 40 in Section 7.1), determine whether a set of orthonormal eigenvectors is returned when the input matrix A is symmetric. (See Theorem 7.9.) Experiment with the matrices in Examples 4 and 5.

40. If the answer to Exercise 39 is no, you can use the Gram–Schmidt procedure to obtain an orthonormal set of eigenvectors. (See Exercise 38 in Section 5.4.) Experiment with the matrices in Examples 4 and 5 if necessary.

Supplementary Exercises

1. Find the eigenvalues and associated eigenvectors for each of the matrices in Supplementary Exercise 2 in Chapter 3. Which of these matrices are similar to a diagonal matrix?

2. Let

$$A = \begin{bmatrix} 1 & 0 & -4 \\ 0 & 5 & 4 \\ -4 & 4 & 3 \end{bmatrix}.$$

(a) Find the eigenvalues and associated eigenvectors of A.

(b) Is A similar to a diagonal matrix? If so, find a nonsingular matrix P such that $P^{-1}AP$ is diagonal. Is P unique? Explain.

(c) Find the eigenvalues of A^{-1}.

(d) Find the eigenvalues and associated eigenvectors of A^2.

3. Let A be any $n \times n$ real matrix.

(a) Prove that the coefficient of λ^{n-1} in the characteristic polynomial of A is given by $-\operatorname{Tr}(A)$ (see Exercise 43 in Section 1.3).

(b) Prove that $\operatorname{Tr}(A)$ is the sum of the eigenvalues of A.

(c) Prove that the constant coefficient of the characteristic polynomial of A is $\pm$ times the product of the eigenvalues of A.

4. Prove or disprove: Every nonsingular matrix is similar to a diagonal matrix.

5. Let $p(x) = a_0 + a_1 x + a_2 x^2 + \cdots + a_k x^k$ be a polynomial in x. Show that the eigenvalues of matrix $p(A) = a_0 I_n + a_1 A + a_2 A^2 + \cdots + a_k A^k$ are $p(\lambda_i)$, $i = 1, 2, \dots, n$, where λ_i are the eigenvalues of A.

6. Let $p_1(\lambda)$ be the characteristic polynomial of A_{11}, and $p_2(\lambda)$ the characteristic polynomial of A_{22}. What is the characteristic polynomial of each of the following partitioned matrices?

(a) $A = \begin{bmatrix} A_{11} & O \\ O & A_{22} \end{bmatrix}$ **(b)** $A = \begin{bmatrix} A_{11} & A_{21} \\ O & A_{22} \end{bmatrix}$

(*Hint*: See Exercises 16 and 19 in Section 3.2.)

7. Let $L: P_1 \to P_1$ be the linear operator defined by $L(at + b) = \dfrac{a+b}{2}t$. Let $S = \{2 - t, 3 + t\}$ be a basis for P_1.

(a) Find $\left[L(2 - t)\right]_S$ and $\left[L(3 + t)\right]_S$.

(b) Find a matrix A representing L with respect to S.

(c) Find the eigenvalues and associated eigenvectors of A.

(d) Find the eigenvalues and associated eigenvectors of L.

(e) Describe the eigenspace for each eigenvalue of L.

8. Let $V = M_{22}$ and let $L: V \to V$ be the linear operator defined by $L(A) = A^T$, for A in V. Let $S = \{A_1, A_2, A_3, A_4\}$, where

$$A_1 = \begin{bmatrix} 1 & 0 \\ 0 & 0 \end{bmatrix}, \quad A_2 = \begin{bmatrix} 0 & 1 \\ 0 & 0 \end{bmatrix},$$

$$A_3 = \begin{bmatrix} 0 & 0 \\ 1 & 0 \end{bmatrix}, \quad \text{and} \quad A_4 = \begin{bmatrix} 0 & 0 \\ 0 & 1 \end{bmatrix},$$

be a basis for V.

(a) Find $\left[L(A_i)\right]_S$ for $i = 1, 2, 3, 4$.

(b) Find the matrix B representing L with respect to S.

(c) Find the eigenvalues and associated eigenvectors of B.

(d) Find the eigenvalues and associated eigenvectors of L.

(e) Show that one of the eigenspaces is the set of all 2×2 symmetric matrices and that the other is the set of all 2×2 skew symmetric matrices.

9. ***(Calculus Required)*** Let V be the *real* vector space of trigonometric polynomials of the form $a + b\sin x + c\cos x$. Let $L: V \to V$ be the linear operator defined by $L(\mathbf{v}) = \dfrac{d}{dx}\left[\mathbf{v}\right]$. Find the eigenvalues and associated eigenvectors of L. (*Hint*: Use the basis $S = \{1, \sin x, \cos x\}$ for V.)

10. Let V be the *complex* vector space (see Appendix B.2) of trigonometric polynomials

$$a + b\sin x + c\cos x.$$

For L as defined in Exercise 9, find the eigenvalues and associated eigenvectors.

11. Prove that if the matrix A is similar to a diagonal matrix, then A is similar to A^T.

Chapter Review

True or False

1. If $\mathbf{x}$ is an eigenvector of A, then so is $k\mathbf{x}$ for any scalar k.

2. Zero is never an eigenvalue of a matrix.

3. The roots of the characteristic polynomial of a matrix are its eigenvalues.

4. Given that $\lambda = 4$ is an eigenvalue of A, then an associated eigenvector is a nontrivial solution of the homogeneous system $(4I_n - A)\mathbf{x} = \mathbf{0}$.

5. If an $n \times n$ matrix A is real, then all of its eigenvalues are real.

6. If a 3×3 matrix A has eigenvalues $\lambda = 1, -1, 3$, then A is diagonalizable.

7. If $\mathbf{x}$ and $\mathbf{y}$ are eigenvectors of A associated with the eigenvalue λ, then for any nonzero vector $\mathbf{w}$ in span$\{\mathbf{x}, \mathbf{y}\}$, $A\mathbf{w} = \lambda\mathbf{w}$.

8. If P is nonsingular and D is diagonal, then the eigenvalues of $A = P^{-1}DP$ are the diagonal entries of D.

9. If A is 3×3 and has eigenvectors that form a basis for R^3, then A is diagonalizable.

10. The eigenvalues of A^2 are the squares of the eigenvalues of A.

11. Every $n \times n$ matrix has n eigenvalues, so it is diagonalizable.

12. If A is similar to an upper triangular matrix U, then the eigenvalues of A are the diagonal entries of U.

13. Every symmetric matrix is diagonalizable.

14. The inverse of an orthogonal matrix is its transpose.

15. If A is 4×4 and orthogonal, then the inner product of any two different columns of A is zero.

16. If A is 4×4 and orthogonal, then its columns are a basis for R^4.

17. If A is 4×4 and symmetric, then we can find eigenvectors of A that are a basis for R^4.

18. Let $L: R^3 \to R^3$ be represented by the matrix A. If $\mathbf{x}$ is an eigenvector of A, then $L(\mathbf{x})$ and $\mathbf{x}$ are parallel.

19. If A is diagonalizable and has eigenvalues with absolute value less than 1, then $\lim_{k\to\infty} A^k = O$.

20. If A is orthogonal, then $|\det(A)| = 1$.

Quiz

1. Find the eigenvalues and associated eigenvectors of

$$A = \begin{bmatrix} -1 & -2 \\ 4 & 5 \end{bmatrix}.$$

2. Let $L: P_2 \to P_2$ be defined by $L(at^2 + bt + c) = ct + b$.
   (a) Find the matrix representing L with respect to the basis $S = \{t^2, 2 + t, 2 - t\}$.
   (b) Find the eigenvalues and associated eigenvectors of L.
3. The characteristic polynomial of A is $p(\lambda) = \lambda^3 - 3\lambda^2 + 4$. Find its eigenvalues.
4. If A has eigenvalue $\lambda = 2$ and associated eigenvector $\mathbf{x} = \begin{bmatrix} 1 \\ -1 \end{bmatrix}$, find an eigenvalue and associated eigenvector of $B = A^3 - A + 3I_2$.
5. For

$$A = \begin{bmatrix} 2 & 1 & 1 \\ 0 & 2 & 0 \\ 0 & -1 & 1 \end{bmatrix},$$

find a basis for the eigenspace associated with $\lambda = 2$.
6. Find a diagonal matrix similar to

$$A = \begin{bmatrix} 2 & 1 & -3 \\ 3 & 0 & 3 \\ -1 & 1 & 0 \end{bmatrix}.$$

7. Is $A = \begin{bmatrix} 1 & 1 & 1 \\ 0 & 1 & 1 \\ 0 & 0 & 0 \end{bmatrix}$ diagonalizable? Explain.
8. Let A be a 3×3 matrix whose columns satisfy $\text{col}_j(A)^T \text{col}_i(A) = 0$ for $i \neq j$. Is A an orthogonal matrix? Explain.
9. Let $\mathbf{x} = \begin{bmatrix} 1 \\ -1 \\ 2 \end{bmatrix}$ and $\mathbf{y} = \begin{bmatrix} 3 \\ 3 \\ 0 \end{bmatrix}$.
   (a) Find a nonzero vector $\mathbf{z}$ orthogonal to both $\mathbf{x}$ and $\mathbf{y}$.
   (b) Let $A = \begin{bmatrix} \mathbf{x} & \mathbf{y} & \mathbf{z} \end{bmatrix}$. Compute $A^T A$. Describe the resulting matrix and explain the meaning of each of its entries.
   (c) Use the information from part (b) to form an orthogonal matrix related to A. Explain your procedure.
   (d) Form a conjecture to complete the following: If A is an $n \times n$ matrix whose columns are mutually orthogonal, then $A^T A$ is ______.
   (e) Prove your conjecture.
10. Prove or disprove: If we interchange two rows of a square matrix, then the eigenvalues are unchanged.
11. Let A be a 3×3 matrix with first row $\begin{bmatrix} k & 0 & 0 \end{bmatrix}$ for some nonzero real number k. Prove that k is an eigenvalue of A.
12. Let $A = \begin{bmatrix} 9 & -1 & -2 \\ -1 & 9 & -2 \\ -2 & -2 & 6 \end{bmatrix}$.
   (a) Show that $\lambda = 4$ and $\lambda = 10$ are eigenvalues of A and find bases for the associated eigenspaces.
   (b) Find an orthogonal matrix P so that $P^T A P$ is diagonal.

Discussion Exercises

1. Given a particular matrix A and the scalar 5, discuss how to determine whether 5 is an eigenvalue of A.
2. Given a particular matrix A and a vector $\mathbf{x} \neq \mathbf{0}$, discuss how to determine whether $\mathbf{x}$ is an eigenvector of A.
3. Suppose that the 5×5 matrix A has five linearly independent eigenvectors given by the set $S = \{\mathbf{x}_1, \mathbf{x}_2, \mathbf{x}_3, \mathbf{x}_4, \mathbf{x}_5\}$.
   (a) Explain why S is a basis for R^5.
   (b) Explain how to obtain a nonsingular matrix P so that $P^{-1}AP$ is diagonal. Can there be more than one such matrix P? Explain.
   (c) Discuss what is known about the eigenvalues of A.
   (d) Is the matrix A nonsingular? Explain.
   (e) Explain how to determine an orthonormal basis for R^5, using the set S.
4. In the description of an eigenspace (see Exercise 14 in Section 7.1), explain why we explicitly insist that the zero vector be included.
5. Let

$$S = \{\mathbf{x}_1, \mathbf{x}_2\} = \left\{ \begin{bmatrix} 1 \\ 1 \end{bmatrix}, \begin{bmatrix} 1 \\ 2 \end{bmatrix} \right\}$$

and $\lambda_1 = 4$, $\lambda_2 = 1$, where $A\mathbf{x}_1 = \lambda_1 \mathbf{x}_1$ and $A\mathbf{x}_2 = \lambda_2 \mathbf{x}_2$. Also, let V_1 be the eigenspace associated with λ_1, and V_2 the eigenspace associated with λ_2. Discuss the validity of the following statement and cite reasons for your conclusion:

> If $\mathbf{y} \neq \mathbf{0}$ is a vector in R^2, then $\mathbf{y}$ belongs to either V_1 or V_2.

6. Let $p(\lambda)$ be the characteristic polynomial of the 4×4 matrix A. Using algebra, you compute the roots of $p(\lambda) = 0$ to be r_1, r_2, r_3, and r_4. Discuss how to determine whether your calculations of the roots is correct by using the matrix A.

7. Let $A = I_2$ and $B = \begin{bmatrix} 1 & 1 \\ 0 & 1 \end{bmatrix}$. Discuss the validity of the following statement and cite reasons for your conclusion:

> If A and B have the same trace, determinant, rank, and eigenvalues, then the matrices are similar.

8. Let A be an $n \times n$ matrix with distinct eigenvectors $\mathbf{x}_1$ and $\mathbf{x}_2$. Discuss conditions under which any nonzero vector in span$\{\mathbf{x}_1, \mathbf{x}_2\}$ is an eigenvector of A.

9. Let A be a 3×3 symmetric matrix with eigenvalues λ_j, $j = 1, 2, 3$, such that $|\lambda_j| < 1$ for each j. Discuss the behavior of the sequence of matrices $A, A^2, A^3, \ldots, A^n$, as $n \to \infty$.

10. Let $A = \begin{bmatrix} 1 & 1 \\ 1 & 0 \end{bmatrix}$.

(a) Find the eigenvalues and associated eigenvectors of A.

(b) Determine a diagonal matrix D similar to A.

(c) In Section 1.5 we briefly discussed Fibonacci numbers. Discuss how to use the matrix D of part (b) to determine the nth Fibonacci number.

CHAPTER

8 Applications of Eigenvalues and Eigenvectors (Optional)

8.1 Stable Age Distribution in a Population; Markov Processes

In this section we deal with two applications of eigenvalues and eigenvectors. These applications find use in a wide variety of everyday situations, including harvesting of animal resources and planning of mass transportation systems.

■ Stable Age Distribution in a Population

Consider a population of animals that can live to a maximum age of n years (or any other time unit). Suppose that the number of males in the population is always a fixed percentage of the female population. Thus, in studying the growth of the entire population, we can ignore the male population and concentrate our attention on the female population. We divide the female population into $n+1$ age groups as follows:

$$x_i^{(k)} = \text{number of females of age } i \text{ who are alive at time } k,\ 0 \le i \le n;$$
$$f_i = \text{fraction of females of age } i \text{ who will be alive a year later;}$$
$$b_i = \text{average number of females born to a female of age } i.$$

Let

$$\mathbf{x}^{(k)} = \begin{bmatrix} x_0^{(k)} \\ x_1^{(k)} \\ \vdots \\ x_n^{(k)} \end{bmatrix} \qquad (k \ge 0)$$

denote the age distribution vector at time k.

The number of females in the first age group (age zero) at time $k+1$ is merely the total number of females born from time k to time $k+1$. There are $x_0^{(k)}$ females in the first age group at time k and each of these females, on the average, produces b_0 female offspring, so the first age group produces a total of $b_0 x_0^{(k)}$ females. Similarly, the $x_1^{(k)}$ females in the second age group (age 1) produce a total of $b_1 x_1^{(k)}$ females. Thus

$$x_0^{(k+1)} = b_0 x_0^{(k)} + b_1 x_1^{(k)} + \cdots + b_n x_n^{(k)}. \tag{1}$$

The number $x_1^{(k+1)}$ of females in the second age group at time $k+1$ is the number of females from the first age group at time k who are alive a year later. Thus

$$x_1^{(k+1)} = \begin{pmatrix}\text{fraction of females in}\\ \text{first age group who are}\\ \text{alive a year later}\end{pmatrix} \times \begin{pmatrix}\text{number of females in}\\ \text{first age group}\end{pmatrix},$$

or

$$x_1^{(k+1)} = f_0 x_0^{(k)},$$

and, in general,

$$x_j^{(k+1)} = f_{j-1} x_{j-1}^{(k)} \qquad (1 \le j \le n). \tag{2}$$

We can write (1) and (2), using matrix notation, as

$$\mathbf{x}^{(k+1)} = A\mathbf{x}^{(k)} \qquad (k \ge 1), \tag{3}$$

where

$$A = \begin{bmatrix} b_0 & b_1 & b_2 & \cdots & b_{n-1} & b_n \\ f_0 & 0 & 0 & \cdots & 0 & 0 \\ 0 & f_1 & 0 & \cdots & 0 & 0 \\ \vdots & \vdots & \vdots & & \vdots & \vdots \\ 0 & 0 & 0 & \cdots & f_{n-1} & 0 \end{bmatrix}.$$

We can use Equation (3) to try to determine a distribution of the population by age groups at time $k+1$ so that the number of females in each age group at time $k+1$ will be a fixed multiple of the number in the corresponding age group at time k. That is, if λ is the multiplier, we want

$$\mathbf{x}^{(k+1)} = \lambda \mathbf{x}^{(k)}$$

or

$$A\mathbf{x}^{(k)} = \lambda \mathbf{x}^{(k)}.$$

Thus λ is an eigenvalue of A and $\mathbf{x}^{(k)}$ is a corresponding eigenvector. If $\lambda = 1$, the number of females in each age group will be the same, year after year. If we can find an eigenvector $\mathbf{x}^{(k)}$ corresponding to the eigenvalue $\lambda = 1$, we say that we have a **stable age distribution**.

EXAMPLE 1

Consider a beetle that can live to a maximum age of two years and whose population dynamics are represented by the matrix

$$A = \begin{bmatrix} 0 & 0 & 6 \\ \frac{1}{2} & 0 & 0 \\ 0 & \frac{1}{3} & 0 \end{bmatrix}.$$

We find that $\lambda = 1$ is an eigenvalue of A with corresponding eigenvector

$$\begin{bmatrix} 6 \\ 3 \\ 1 \end{bmatrix}.$$

Thus, if the numbers of females in the three groups are proportional to 6:3:1, we have a stable age distribution. That is, if we have 600 females in the first age group, 300 in the second, and 100 in the third, then, year after year, the number of females in each age group will remain the same. ■

Remark Population growth problems of the type considered in Example 1 have applications to animal harvesting.

■ Markov Processes

A **Markov*** **chain**, or **Markov process**, is a process in which the probability of the system being in a particular state at a given observation period depends only on its state at the immediately preceding observation period.

Suppose that the system has n possible states. For each $i = 1, 2, \ldots, n$, and $j = 1, 2, \ldots, n$, let t_{ij} be the probability that if the system is in state j at a certain observation period, it will be in state i at the next observation period; t_{ij} is called a **transition probability**. Moreover, t_{ij} applies to every period; that is, it does not change with time.

Since t_{ij} is a probability, we must have

$$0 \leq t_{ij} \leq 1 \qquad (1 \leq i, j \leq n).$$

ANDREI ANDREYEVICH MARKOV

Also, if the system is in state j at a certain observation period, then it must be in one of the n states (it may remain in state j) at the next observation period. Thus we have

$$t_{1j} + t_{2j} + \cdots + t_{nj} = 1. \tag{4}$$

It is convenient to arrange the transition probabilities as the $n \times n$ matrix $T = \left[t_{ij}\right]$, which is called the **transition matrix** of the Markov process. Other names for a transition matrix are **Markov matrix**, **stochastic matrix**, and **probability**

*Andrei Andreyevich Markov (1856–1922) was born in Ryazan, Russia, and died in St. Petersburg, Russia. After graduating from St. Petersburg University, he became a professor of mathematics at that institution in 1893. At that time, he became involved in liberal movements and expressed his opposition to the tsarist regime. His early contributions in mathematics were in number theory and analysis. He turned to probability theory and later developed the field that is now known as Markov chains to analyze the structure of literary texts. Today, Markov chains are widely used in many applications, such as modern physics, the fluctuation of stock prices, and genetics.

matrix. We see that the entries in each column of T are nonnegative and, from Equation (4), add up to 1.

We shall now use the transition matrix of the Markov process to determine the probability of the system being in any of the n states at future times.

Let

$$\mathbf{x}^{(k)} = \begin{bmatrix} p_1^{(k)} \\ p_2^{(k)} \\ \vdots \\ p_n^{(k)} \end{bmatrix} \qquad (k \geq 0)$$

denote the **state vector** of the Markov process at the observation period k, where $p_j^{(k)}$ is the probability that the system is in state j at the observation period k. The state vector $\mathbf{x}^{(0)}$, at the observation period 0, is called the **initial state vector**.

It follows from the basic properties of probability theory that if T is the transition matrix of a Markov process, then the state vector $\mathbf{x}^{(k+1)}$, at the $(k+1)$th observation period, can be determined from the state vector $\mathbf{x}^{(k)}$, at the kth observation period, as

$$\mathbf{x}^{(k+1)} = T\mathbf{x}^{(k)}. \tag{5}$$

From (5), we have

$$\begin{aligned} \mathbf{x}^{(1)} &= T\mathbf{x}^{(0)} \\ \mathbf{x}^{(2)} &= T\mathbf{x}^{(1)} = T(T\mathbf{x}^{(0)}) = T^2\mathbf{x}^{(0)} \\ \mathbf{x}^{(3)} &= T\mathbf{x}^{(2)} = T(T^2\mathbf{x}^{(0)}) = T^3\mathbf{x}^{(0)}, \end{aligned}$$

and, in general,

$$\mathbf{x}^{(n)} = T^n\mathbf{x}^{(0)}.$$

Thus the transition matrix and the initial state vector completely determine every other state vector.

For certain types of Markov processes, as the number of observation periods increases, the state vectors converge to a fixed vector. In this case, we say that the Markov process has reached **equilibrium**. The fixed vector is called the **steady-state vector**. Markov processes are generally used to determine the behavior of a system in the long run—for example, the share of the market that a certain manufacturer can expect to retain on a somewhat permanent basis. Thus, the question of whether or not a Markov process reaches equilibrium is quite important. To identify a class of Markov processes that reach equilibrium, we need several additional notions.

The vector

$$\mathbf{u} = \begin{bmatrix} u_1 \\ u_2 \\ \vdots \\ u_n \end{bmatrix}$$

is called a **probability vector** if $u_i \geq 0$ $(1 \leq i \leq n)$ and

$$u_1 + u_2 + \cdots + u_n = 1.$$

A Markov process is called **regular** if its transition matrix T has the property that all the entries in some power of T are positive. It can be shown that a regular Markov process always reaches equilibrium and the steady-state vector is a probability vector. We can find the steady-state vector by obtaining the limit of the successive powers $T^n\mathbf{x}$ for an arbitrary probability vector $\mathbf{x}$. Observe that if $\mathbf{u}$ is the steady-state vector of a Markov process with transition matrix T, then $T\mathbf{u} = \mathbf{u}$, so that $\lambda = 1$ is an eigenvalue of T with associated eigenvector $\mathbf{u}$ (a probability vector).

EXAMPLE 2

Suppose that the weather in a certain city is either rainy or dry. As a result of extensive record-keeping, it has been determined that the probability of a rainy day following a dry day is $\frac{1}{3}$, and the probability of a rainy day following a rainy day is $\frac{1}{2}$. Let state D be a dry day and state R be a rainy day. Then the transition matrix of this Markov process is

$$T = \begin{array}{c} \begin{array}{cc} \text{D} & \text{R} \end{array} \\ \begin{bmatrix} \frac{2}{3} & \frac{1}{2} \\ \frac{1}{3} & \frac{1}{2} \end{bmatrix} \end{array} \begin{array}{c} \\ \text{D} \\ \text{R} \end{array}.$$

Since all the entries in T are positive, we are dealing with a regular Markov process, so the process reaches equilibrium. Suppose that when we begin our observations (day 0), it is dry, so the initial state vector is

$$\mathbf{x}^{(0)} = \begin{bmatrix} 1 \\ 0 \end{bmatrix},$$

a probability vector. Then the state vector on day 1 (the day after we begin our observations) is

$$\mathbf{x}^{(1)} = T\mathbf{x}^{(0)} = \begin{bmatrix} 0.67 & 0.5 \\ 0.33 & 0.5 \end{bmatrix} \begin{bmatrix} 1 \\ 0 \end{bmatrix} = \begin{bmatrix} 0.67 \\ 0.33 \end{bmatrix},$$

where for convenience we have written $\frac{2}{3}$ and $\frac{1}{3}$ as 0.67 and 0.33, respectively. Moreover, to simplify matters, the output of calculations is recorded to three digits of accuracy. Thus the probability of no rain on day 1 is 0.67, and the probability of rain on that day is 0.33. Similarly,

$$\mathbf{x}^{(2)} = T\mathbf{x}^{(1)} = \begin{bmatrix} 0.67 & 0.5 \\ 0.33 & 0.5 \end{bmatrix} \begin{bmatrix} 0.67 \\ 0.33 \end{bmatrix} = \begin{bmatrix} 0.614 \\ 0.386 \end{bmatrix}$$

$$\mathbf{x}^{(3)} = T\mathbf{x}^{(2)} = \begin{bmatrix} 0.67 & 0.5 \\ 0.33 & 0.5 \end{bmatrix} \begin{bmatrix} 0.614 \\ 0.386 \end{bmatrix} = \begin{bmatrix} 0.604 \\ 0.396 \end{bmatrix}$$

$$\mathbf{x}^{(4)} = T\mathbf{x}^{(3)} = \begin{bmatrix} 0.67 & 0.5 \\ 0.33 & 0.5 \end{bmatrix} \begin{bmatrix} 0.604 \\ 0.396 \end{bmatrix} = \begin{bmatrix} 0.603 \\ 0.397 \end{bmatrix}$$

$$\mathbf{x}^{(5)} = T\mathbf{x}^{(4)} = \begin{bmatrix} 0.67 & 0.5 \\ 0.33 & 0.5 \end{bmatrix} \begin{bmatrix} 0.603 \\ 0.397 \end{bmatrix} = \begin{bmatrix} 0.603 \\ 0.397 \end{bmatrix}.$$

From the fourth day on, the state vector is always the same,

$$\begin{bmatrix} 0.603 \\ 0.397 \end{bmatrix},$$

so this is the steady-state vector.

This means that from the fourth day on, it is dry about 60% of the time, and it rains about 40% of the time.

The steady-state vector can also be found as follows: Since $\lambda = 1$ is an eigenvalue of T, we find an associated eigenvector $\mathbf{u} = \begin{bmatrix} u_1 \\ u_2 \end{bmatrix}$ by solving the equation

$$T\mathbf{u} = \mathbf{u}$$

or

$$(I_2 - T)\mathbf{u} = \mathbf{0}.$$

From the infinitely many solutions that can be obtained by solving the resulting homogeneous system, we determine a unique solution $\mathbf{u}$ by requiring that its components add up to 1 (since $\mathbf{u}$ is a probability vector). In this case we have to solve the homogeneous system

$$\begin{aligned} \tfrac{1}{3}u_1 - \tfrac{1}{2}u_2 &= 0 \\ -\tfrac{1}{3}u_1 + \tfrac{1}{2}u_2 &= 0. \end{aligned}$$

Then $u_2 = \frac{2}{3}u_1$. Substituting in the equation

$$u_1 + u_2 = 1,$$

we get $u_1 = 0.6$ and $u_2 = 0.4$. ■

Key Terms

Stable age distribution	Markov matrix	Equilibrium
Markov chain	Stochastic matrix	Steady-state vector
Markov process	Probability matrix	Probability vector
Regular Markov process	Transition probability	Initial state vector

8.1 Exercises

1. Consider a living organism that can live to a maximum age of 2 years and whose matrix is

$$A = \begin{bmatrix} 0 & 0 & 8 \\ \frac{1}{4} & 0 & 0 \\ 0 & \frac{1}{2} & 0 \end{bmatrix}.$$

Find a stable age distribution.

2. Consider a living organism that can live to a maximum age of 2 years and whose matrix is

$$A = \begin{bmatrix} 0 & 4 & 0 \\ \frac{1}{4} & 0 & 0 \\ 0 & \frac{1}{2} & 0 \end{bmatrix}.$$

Find a stable age distribution.

3. Which of the following can be transition matrices of a Markov process?

(a) $\begin{bmatrix} 0.3 & 0.7 \\ 0.4 & 0.6 \end{bmatrix}$ **(b)** $\begin{bmatrix} 0.2 & 0.3 & 0.1 \\ 0.8 & 0.5 & 0.7 \\ 0.0 & 0.2 & 0.2 \end{bmatrix}$

(c) $\begin{bmatrix} 0.55 & 0.33 \\ 0.45 & 0.67 \end{bmatrix}$ **(d)** $\begin{bmatrix} 0.3 & 0.4 & 0.2 \\ 0.2 & 0.0 & 0.8 \\ 0.1 & 0.3 & 0.6 \end{bmatrix}$

4. Which of the following are probability vectors?

(a) $\begin{bmatrix} \frac{1}{2} \\ \frac{1}{3} \\ \frac{2}{3} \end{bmatrix}$ **(b)** $\begin{bmatrix} 0 \\ 1 \\ 0 \end{bmatrix}$

(c) $\begin{bmatrix} \frac{1}{4} \\ \frac{1}{6} \\ \frac{1}{3} \\ \frac{1}{4} \end{bmatrix}$ **(d)** $\begin{bmatrix} \frac{1}{5} \\ \frac{2}{5} \\ \frac{1}{10} \\ \frac{2}{10} \end{bmatrix}$

5. Consider the transition matrix

$$T = \begin{bmatrix} 0.7 & 0.4 \\ 0.3 & 0.6 \end{bmatrix}.$$

(a) If $\mathbf{x}^{(0)} = \begin{bmatrix} 1 \\ 0 \end{bmatrix}$, compute $\mathbf{x}^{(1)}$, $\mathbf{x}^{(2)}$, and $\mathbf{x}^{(3)}$ to three decimal places.

(b) Show that T is regular and find its steady-state vector.

6. Consider the transition matrix

$$T = \begin{bmatrix} 0 & 0.2 & 0.0 \\ 0 & 0.3 & 0.3 \\ 1 & 0.5 & 0.7 \end{bmatrix}.$$

(a) If

$$\mathbf{x}^{(0)} = \begin{bmatrix} 0 \\ 1 \\ 0 \end{bmatrix},$$

compute $\mathbf{x}^{(1)}$, $\mathbf{x}^{(2)}$, $\mathbf{x}^{(3)}$, and $\mathbf{x}^{(4)}$ to three decimal places.

(b) Show that T is regular and find its steady-state vector.

7. Which of the following transition matrices are regular?

(a) $\begin{bmatrix} 0 & \frac{1}{2} \\ 1 & \frac{1}{2} \end{bmatrix}$ **(b)** $\begin{bmatrix} \frac{1}{2} & 0 & 0 \\ 0 & 1 & \frac{1}{2} \\ \frac{1}{2} & 0 & \frac{1}{2} \end{bmatrix}$

(c) $\begin{bmatrix} 1 & \frac{1}{3} & 0 \\ 0 & \frac{1}{3} & 1 \\ 0 & \frac{1}{3} & 0 \end{bmatrix}$ **(d)** $\begin{bmatrix} \frac{1}{4} & \frac{3}{5} & \frac{1}{2} \\ \frac{1}{2} & 0 & 0 \\ \frac{1}{4} & \frac{2}{5} & \frac{1}{2} \end{bmatrix}$

8. Show that each of the following transition matrices reaches a state of equilibrium:

(a) $\begin{bmatrix} \frac{1}{2} & 1 \\ \frac{1}{2} & 0 \end{bmatrix}$ **(b)** $\begin{bmatrix} 0.4 & 0.2 \\ 0.6 & 0.8 \end{bmatrix}$

(c) $\begin{bmatrix} \frac{1}{3} & 1 & \frac{1}{2} \\ \frac{1}{3} & 0 & \frac{1}{4} \\ \frac{1}{3} & 0 & \frac{1}{4} \end{bmatrix}$ **(d)** $\begin{bmatrix} 0.3 & 0.1 & 0.4 \\ 0.2 & 0.4 & 0.0 \\ 0.5 & 0.5 & 0.6 \end{bmatrix}$

9. Find the steady-state vector of each of the following regular matrices:

(a) $\begin{bmatrix} \frac{1}{3} & \frac{1}{2} \\ \frac{2}{3} & \frac{1}{2} \end{bmatrix}$ **(b)** $\begin{bmatrix} 0.3 & 0.1 \\ 0.7 & 0.9 \end{bmatrix}$

(c) $\begin{bmatrix} \frac{1}{4} & \frac{1}{2} & \frac{1}{3} \\ 0 & \frac{1}{2} & \frac{2}{3} \\ \frac{3}{4} & 0 & 0 \end{bmatrix}$ **(d)** $\begin{bmatrix} 0.4 & 0.0 & 0.1 \\ 0.2 & 0.5 & 0.3 \\ 0.4 & 0.5 & 0.6 \end{bmatrix}$

10. (***Psychology***) A behavioral psychologist places a rat each day in a cage with two doors, A and B. The rat can go through door A, where it receives an electric shock, or through door B, where it receives some food. A record is made of the door through which the rat passes. At the start of the experiment, on a Monday, the rat is equally likely to go through door A as through door B. After going through door A and receiving a shock, the probability of going through the same door on the next day is 0.3. After going through door B and receiving food, the probability of going through the same door on the next day is 0.6.

(a) Write the transition matrix for the Markov process.

(b) What is the probability of the rat going through door A on Thursday (the third day after starting the experiment)?

(c) What is the steady-state vector?

11. (***Sociology***) A study has determined that the occupation of a boy, as an adult, depends upon the occupation of his father and is given by the following transition matrix where P = professional, F = farmer, and L = laborer:

		Father's occupation		
		P	F	L
Son's occupation	P	0.8	0.3	0.2
	F	0.1	0.5	0.2
	L	0.1	0.2	0.6

Thus the probability that the son of a professional will also be a professional is 0.8, and so on.

(a) What is the probability that the grandchild of a professional will also be a professional?

(b) In the long run, what proportion of the population will be farmers?

12. (***Genetics***) Consider a plant that can have red flowers (R), pink flowers (P), or white flowers (W), depending upon the genotypes RR, RW, and WW. When we cross each of these genotypes with a genotype RW, we obtain the transition matrix

Flowers of parent plant

$$\begin{array}{ll} & \quad\;\; \text{R} \qquad \text{P} \qquad \text{W} \\ \text{Flowers of offspring plant} \begin{array}{c} \text{R} \\ \text{P} \\ \text{W} \end{array} & \begin{bmatrix} 0.5 & 0.25 & 0.0 \\ 0.5 & 0.50 & 0.5 \\ 0.0 & 0.25 & 0.5 \end{bmatrix} \end{array}$$

Suppose that each successive generation is produced by crossing only with plants of RW genotype. When the process reaches equilibrium, what percentage of the plants will have red, pink, or white flowers?

13. (***Mass Transit***) A new mass transit system has just gone into operation. The transit authority has made studies that predict the percentage of commuters who will change to mass transit (M) or continue driving their automobile (A). The following transition matrix shows the results:

This year

$$\begin{array}{ll} & \quad\;\; \text{M} \qquad \text{A} \\ \text{Next year} \quad \begin{array}{c} \text{M} \\ \text{A} \end{array} & \begin{bmatrix} 0.7 & 0.2 \\ 0.3 & 0.8 \end{bmatrix}. \end{array}$$

Suppose that the population of the area remains constant, and that initially 30% of the commuters use mass transit and 70% use their automobiles.

(a) What percentage of the commuters will be using the mass transit system after 1 year? After 2 years?

(b) What percentage of the commuters will be using the mass transit system in the long run?

8.2 Spectral Decomposition and Singular Value Decomposition

Theorem 7.9 tells us that an $n \times n$ symmetric matrix A can be expressed as the matrix product

$$A = PDP^T, \tag{1}$$

where D is a diagonal matrix and P is an orthogonal matrix. The diagonal entries of D are the eigenvalues of A, $\lambda_1, \lambda_2, \ldots, \lambda_n$, and the columns of P are associated orthonormal eigenvectors $\mathbf{x}_1, \mathbf{x}_2, \ldots, \mathbf{x}_n$. The expression in (1) is called the **spectral decomposition** of A. It is helpful to write (1) in the following form:

$$A = \begin{bmatrix} \mathbf{x}_1 & \mathbf{x}_2 & \cdots & \mathbf{x}_n \end{bmatrix} \begin{bmatrix} \lambda_1 & 0 & \cdots & 0 & 0 \\ 0 & \lambda_2 & \cdots & 0 & 0 \\ \vdots & \vdots & & \vdots & \vdots \\ 0 & 0 & \cdots & \lambda_{n-1} & 0 \\ 0 & 0 & \cdots & 0 & \lambda_n \end{bmatrix} \begin{bmatrix} \mathbf{x}_1^T \\ \mathbf{x}_2^T \\ \vdots \\ \mathbf{x}_n^T \end{bmatrix}. \tag{2}$$

The expression in (2) can be used to gain information about quadratic forms as shown in Section 8.6, and the nature of the eigenvalues of A can be utilized in other applications. However, we can use (1) and (2) to reveal aspects of the information contained in the matrix A. In particular, we can express A as a linear combination of simple symmetric matrices that are fundamental building blocks of the total information within A. The product DP^T can be computed and gives

$$DP^T = \begin{bmatrix} \lambda_1 & 0 & \cdots & 0 & 0 \\ 0 & \lambda_2 & \cdots & 0 & 0 \\ \vdots & \vdots & & \vdots & \vdots \\ 0 & 0 & \cdots & \lambda_{n-1} & 0 \\ 0 & 0 & \cdots & 0 & \lambda_n \end{bmatrix} \begin{bmatrix} \mathbf{x}_1^T \\ \mathbf{x}_2^T \\ \vdots \\ \mathbf{x}_n^T \end{bmatrix} = \begin{bmatrix} \lambda_1\mathbf{x}_1^T \\ \lambda_2\mathbf{x}_2^T \\ \vdots \\ \lambda_n\mathbf{x}_n^T \end{bmatrix},$$

and hence (2) becomes

$$A = \begin{bmatrix} \mathbf{x}_1 & \mathbf{x}_2 & \cdots & \mathbf{x}_n \end{bmatrix} \begin{bmatrix} \lambda_1\mathbf{x}_1^T \\ \lambda_2\mathbf{x}_2^T \\ \vdots \\ \lambda_n\mathbf{x}_n^T \end{bmatrix}. \tag{3}$$

A careful analysis of the product on the right side of (3) reveals that we can express A as a linear combination of the matrices $\mathbf{x}_j\mathbf{x}_j^T$, and the coefficients are the eigenvalues of A. That is,

$$A = \sum_{j=1}^{n} \lambda_j \mathbf{x}_j \mathbf{x}_j^T = \lambda_1\mathbf{x}_1\mathbf{x}_1^T + \lambda_2\mathbf{x}_2\mathbf{x}_2^T + \cdots + \lambda_n\mathbf{x}_n\mathbf{x}_n^T. \tag{4}$$

A formal proof of (4) is quite tedious. In Example 1 we show the process for obtaining (4) in the case $n = 2$. The steps involved reveal the pattern that can be followed for the general case.

EXAMPLE 1

Let A be a 2×2 symmetric matrix with eigenvalues λ_1 and λ_2 and associated orthonormal eigenvectors $\mathbf{x}_1$ and $\mathbf{x}_2$. Let $P = \begin{bmatrix} \mathbf{x}_1 & \mathbf{x}_2 \end{bmatrix}$ and, to make the manipulations easier to see, let

$$\mathbf{x}_1 = \begin{bmatrix} a \\ b \end{bmatrix} \quad \text{and} \quad \mathbf{x}_2 = \begin{bmatrix} c \\ d \end{bmatrix}.$$

Since A is symmetric, it is diagonalizable by using an orthogonal matrix. We have

$$P^TAP = D = \begin{bmatrix} \lambda_1 & 0 \\ 0 & \lambda_2 \end{bmatrix} \quad \text{and so} \quad A = P\begin{bmatrix} \lambda_1 & 0 \\ 0 & \lambda_2 \end{bmatrix}P^T.$$

Next, perform the following matrix and algebraic operations:

$$
\begin{aligned}
A &= \begin{bmatrix} a & c \\ b & d \end{bmatrix} \begin{bmatrix} \lambda_1 a & \lambda_1 b \\ \lambda_2 c & \lambda_2 d \end{bmatrix} = \begin{bmatrix} \lambda_1 a^2 + \lambda_2 c^2 & \lambda_1 ab + \lambda_2 cd \\ \lambda_1 ab + \lambda_2 cd & \lambda_1 b^2 + \lambda_2 d^2 \end{bmatrix} \\
&= \begin{bmatrix} \lambda_1 a^2 & \lambda_1 ab \\ \lambda_1 ab & \lambda_1 b^2 \end{bmatrix} + \begin{bmatrix} \lambda_2 c^2 & \lambda_2 cd \\ \lambda_2 cd & \lambda_2 d^2 \end{bmatrix} \\
&= \lambda_1 \begin{bmatrix} a^2 & ab \\ ab & b^2 \end{bmatrix} + \lambda_2 \begin{bmatrix} c^2 & cd \\ cd & d^2 \end{bmatrix} \\
&= \lambda_1 \begin{bmatrix} a\begin{bmatrix} a & b \end{bmatrix} \\ b\begin{bmatrix} a & b \end{bmatrix} \end{bmatrix} + \lambda_2 \begin{bmatrix} c\begin{bmatrix} c & d \end{bmatrix} \\ d\begin{bmatrix} c & d \end{bmatrix} \end{bmatrix} \\
&= \lambda_1 \begin{bmatrix} a \\ b \end{bmatrix} \begin{bmatrix} a & b \end{bmatrix} + \lambda_2 \begin{bmatrix} c \\ d \end{bmatrix} \begin{bmatrix} c & d \end{bmatrix} \\
&= \lambda_1 \mathbf{x}_1 \mathbf{x}_1^T + \lambda_2 \mathbf{x}_2 \mathbf{x}_2^T.
\end{aligned}
$$

■

The expression in (4) is equivalent to the spectral decomposition in (1), but displays the eigenvalue and eigenvector information in a different form. Example 2 illustrates the spectral decomposition in both forms.

EXAMPLE 2

Let

$$A = \begin{bmatrix} 2 & 1 & 0 \\ 1 & 2 & 0 \\ 0 & 0 & -1 \end{bmatrix}.$$

To determine the spectral representation of A, we first obtain its eigenvalues and eigenvectors. We find that A has three distinct eigenvalues $\lambda_1 = 1$, $\lambda_2 = 3$, and $\lambda_3 = -1$ and that associated eigenvectors are, respectively (verify),

$$\begin{bmatrix} 1 \\ -1 \\ 0 \end{bmatrix}, \quad \begin{bmatrix} 1 \\ 1 \\ 0 \end{bmatrix}, \quad \text{and} \quad \begin{bmatrix} 0 \\ 0 \\ 1 \end{bmatrix}.$$

Since the eigenvalues are distinct, we are assured that the corresponding eigenvectors form an orthogonal set. (See Theorem 7.7.) Normalizing these vectors, we obtain eigenvectors of unit length that are an orthonormal set:

$$\frac{1}{\sqrt{2}}\begin{bmatrix} 1 \\ -1 \\ 0 \end{bmatrix}, \quad \frac{1}{\sqrt{2}}\begin{bmatrix} 1 \\ 1 \\ 0 \end{bmatrix}, \quad \text{and} \quad \begin{bmatrix} 0 \\ 0 \\ 1 \end{bmatrix}.$$

Then the spectral representation of A is

$$A = \lambda_1 \mathbf{x}_1 \mathbf{x}_1^T + \lambda_2 \mathbf{x}_2 \mathbf{x}_2^T + \lambda_3 \mathbf{x}_3 \mathbf{x}_3^T$$

$$= 1\left(\frac{1}{\sqrt{2}}\right)^2 \begin{bmatrix} 1 \\ -1 \\ 0 \end{bmatrix} \begin{bmatrix} 1 & -1 & 0 \end{bmatrix} + 3\left(\frac{1}{\sqrt{2}}\right)^2 \begin{bmatrix} 1 \\ 1 \\ 0 \end{bmatrix} \begin{bmatrix} 1 & 1 & 0 \end{bmatrix}$$

$$+ (-1) \begin{bmatrix} 0 \\ 0 \\ 1 \end{bmatrix} \begin{bmatrix} 0 & 0 & 1 \end{bmatrix}$$

$$= 1\left(\frac{1}{2}\right) \begin{bmatrix} 1 & -1 & 0 \\ -1 & 1 & 0 \\ 0 & 0 & 0 \end{bmatrix} + 3\left(\frac{1}{2}\right) \begin{bmatrix} 1 & 1 & 0 \\ 1 & 1 & 0 \\ 0 & 0 & 0 \end{bmatrix} + (-1) \begin{bmatrix} 0 & 0 & 0 \\ 0 & 0 & 0 \\ 0 & 0 & 1 \end{bmatrix}$$

$$= 1 \begin{bmatrix} \frac{1}{2} & -\frac{1}{2} & 0 \\ -\frac{1}{2} & \frac{1}{2} & 0 \\ 0 & 0 & 0 \end{bmatrix} + 3 \begin{bmatrix} \frac{1}{2} & \frac{1}{2} & 0 \\ \frac{1}{2} & \frac{1}{2} & 0 \\ 0 & 0 & 0 \end{bmatrix} + (-1) \begin{bmatrix} 0 & 0 & 0 \\ 0 & 0 & 0 \\ 0 & 0 & 1 \end{bmatrix}. \quad \blacksquare$$

In Example 2 the eigenvalues of A are distinct, so the associated eigenvectors form an orthogonal set. If a symmetric matrix has an eigenvalue that is repeated, then the linearly independent eigenvectors associated with the repeated eigenvalue need not form an orthogonal set. However, we can apply the Gram–Schmidt process to the linearly independent eigenvectors associated with a repeated eigenvalue to obtain a set of orthogonal eigenvectors.

We note that (4) expresses the symmetric matrix A as a linear combination of matrices, $\mathbf{x}_j \mathbf{x}_j^T$, which are $n \times n$, since $\mathbf{x}_j$ is $n \times 1$ and $\mathbf{x}_j^T$ is $1 \times n$. The matrix $\mathbf{x}_j \mathbf{x}_j^T$ has a simple construction, as shown in Figure 8.1.

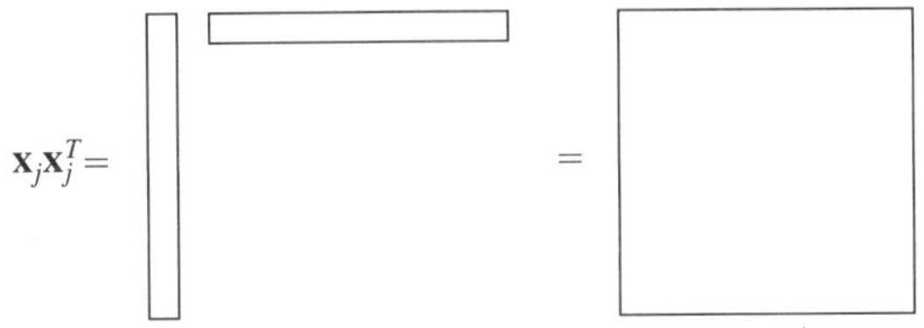

FIGURE 8.1

We call $\mathbf{x}_j \mathbf{x}_j^T$ an **outer product**. (See Supplementary Exercises in Chapter 1.) It can be shown that each row is a multiple of $\mathbf{x}_j^T$. Hence the reduced row echelon form of $\mathbf{x}_j \mathbf{x}_j^T$ [denoted $\mathbf{rref}(\mathbf{x}_j \mathbf{x}_j^T)$] has one nonzero row and thus has rank one. We interpret this in (4) to mean that each outer product $\mathbf{x}_j \mathbf{x}_j^T$ contributes just one piece of information to the construction of matrix A. Thus we can say that the spectral decomposition certainly reveals basic information about the matrix A.

The spectral decomposition in (4) expresses A as a linear combination of outer products of the eigenvectors of A with coefficients that are the corresponding eigenvalues. Since each outer product has rank one, we could say that they have "equal value" in building the matrix A. However, the magnitude of the eigenvalue

λ_j determines the "weight" given to the information contained in the outer product $\mathbf{x}_j\mathbf{x}_j^T$. Intuitively, if we label the eigenvalues so that $|\lambda_1| \geq |\lambda_2| \geq \cdots \geq |\lambda_n|$, then the beginning terms in the sum

$$A = \sum_{j=1}^{n} \lambda_j \mathbf{x}_j \mathbf{x}_j^T = \lambda_1 \mathbf{x}_1 \mathbf{x}_1^T + \lambda_2 \mathbf{x}_2 \mathbf{x}_2^T + \cdots + \lambda_n \mathbf{x}_n \mathbf{x}_n^T$$

contribute more information than the later terms, which correspond to the smaller eigenvalues. Note that if any eigenvalue is zero, then its eigenvector contributes no information to the construction of A. This is the case for a singular symmetric matrix.

EXAMPLE 3

From Example 2, we have that the eigenvalues are ordered as $|3| \geq |-1| \geq |1|$. Thus the contribution of the eigenvectors corresponding to eigenvalues -1 and 1 can be considered equal, whereas that corresponding to 3 is dominant. Rewriting the spectral decomposition, using the terms in the order of the magnitude of the eigenvalues, we obtain the following:

$$\begin{aligned} A &= 3\left(\frac{1}{\sqrt{2}}\right)^2 \begin{bmatrix} 1 \\ 1 \\ 0 \end{bmatrix} \begin{bmatrix} 1 & 1 & 0 \end{bmatrix} + (-1) \begin{bmatrix} 0 \\ 0 \\ 1 \end{bmatrix} \begin{bmatrix} 0 & 0 & 1 \end{bmatrix} \\ &\quad + 1\left(\frac{1}{\sqrt{2}}\right)^2 \begin{bmatrix} 1 \\ -1 \\ 0 \end{bmatrix} \begin{bmatrix} 1 & -1 & 0 \end{bmatrix} \\ &= 3\left(\frac{1}{2}\right) \begin{bmatrix} 1 & 1 & 0 \\ 1 & 1 & 0 \\ 0 & 0 & 0 \end{bmatrix} + (-1) \begin{bmatrix} 0 & 0 & 0 \\ 0 & 0 & 0 \\ 0 & 0 & 1 \end{bmatrix} + 1\left(\frac{1}{2}\right) \begin{bmatrix} 1 & -1 & 0 \\ -1 & 1 & 0 \\ 0 & 0 & 0 \end{bmatrix}. \end{aligned}$$

■

Looking at the terms of the partial sums in Example 3 individually, we have the following matrices:

$$\begin{bmatrix} \frac{3}{2} & \frac{3}{2} & 0 \\ \frac{3}{2} & \frac{3}{2} & 0 \\ 0 & 0 & 0 \end{bmatrix}$$

$$\begin{bmatrix} \frac{3}{2} & \frac{3}{2} & 0 \\ \frac{3}{2} & \frac{3}{2} & 0 \\ 0 & 0 & 0 \end{bmatrix} + \begin{bmatrix} 0 & 0 & 0 \\ 0 & 0 & 0 \\ 0 & 0 & -1 \end{bmatrix} = \begin{bmatrix} \frac{3}{2} & \frac{3}{2} & 0 \\ \frac{3}{2} & \frac{3}{2} & 0 \\ 0 & 0 & -1 \end{bmatrix}$$

$$\begin{bmatrix} \frac{3}{2} & \frac{3}{2} & 0 \\ \frac{3}{2} & \frac{3}{2} & 0 \\ 0 & 0 & -1 \end{bmatrix} + \begin{bmatrix} \frac{1}{2} & -\frac{1}{2} & 0 \\ -\frac{1}{2} & \frac{1}{2} & 0 \\ 0 & 0 & 0 \end{bmatrix} = \begin{bmatrix} 2 & 1 & 0 \\ 1 & 2 & 0 \\ 0 & 0 & -1 \end{bmatrix} = A.$$

This suggests that we can approximate the information in the matrix A, using the partial sums of the spectral decomposition in (4). In fact, this type of development is the foundation of a number of approximation procedures in mathematics.

■ Application: The Image of the Unit Circle by a 2 × 2 Matrix

Let A be any 2×2 real matrix. Then we know that A has a singular value decomposition $A = USV^T$, where $U = \begin{bmatrix} \mathbf{u}_1 & \mathbf{u}_2 \end{bmatrix}$ and $V = \begin{bmatrix} \mathbf{v}_1 & \mathbf{v}_2 \end{bmatrix}$ are 2×2 orthogonal matrices. The matrix S is a 2×2 diagonal matrix with the singular values of A as its diagonal entries. The analysis of the form of the image of the unit circle in this case is nearly the same as in Section 7.3, except that we use singular vectors of A in place of eigenvectors. The image is again an ellipse. We have the following important equivalent relations:

$A = USV^T$ is equivalent to $AV = US$ (explain).

$AV = US$ is equivalent to $A\begin{bmatrix} \mathbf{v}_1 & \mathbf{v}_2 \end{bmatrix} = \begin{bmatrix} \mathbf{u}_1 & \mathbf{u}_2 \end{bmatrix}\begin{bmatrix} s_{11} & 0 \\ 0 & s_{22} \end{bmatrix}$.

$A\begin{bmatrix} \mathbf{v}_1 & \mathbf{v}_2 \end{bmatrix} = \begin{bmatrix} \mathbf{u}_1 & \mathbf{u}_2 \end{bmatrix}\begin{bmatrix} s_{11} & 0 \\ 0 & s_{22} \end{bmatrix}$ is equivalent to $A\mathbf{v}_1 = s_{11}\mathbf{u}_1$ and $A\mathbf{v}_2 = s_{22}\mathbf{u}_2$.

Thus the image of the columns of V are scalar multiples of the columns of $\mathbf{U}$. This is reminiscent of the analogous results for eigenvalues and eigenvectors in Section 7.3. Since V is orthogonal, its columns are orthonormal vectors and $V^T = V^{-1}$, so $V^T\mathbf{v}_1 = \mathbf{i}$ and $V^T\mathbf{v}_2 = \mathbf{j}$. Figure 8.2 is analogous to Figure 7.5 in Section 7.3, with eigenvectors replaced by singular vectors.

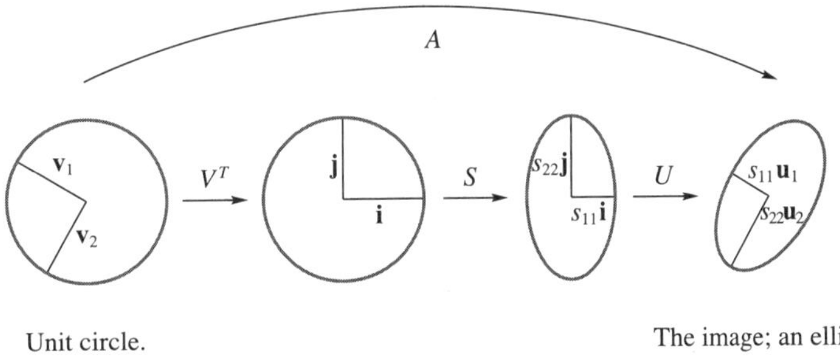

FIGURE 8.2

EXAMPLE 4

Let

$$A = \begin{bmatrix} 1 & 1 \\ -2 & 4 \end{bmatrix}.$$

Using MATLAB, we find that the singular value decomposition is $A = USV^T$, where

$$U = \begin{bmatrix} -0.1091 & -0.9940 \\ -0.9940 & 0.1091 \end{bmatrix}, \quad S = \begin{bmatrix} 4.4966 & 0 \\ 0 & 1.3343 \end{bmatrix},$$

and

$$V = \begin{bmatrix} 0.4179 & -0.9085 \\ -0.9085 & -0.4179 \end{bmatrix}.$$

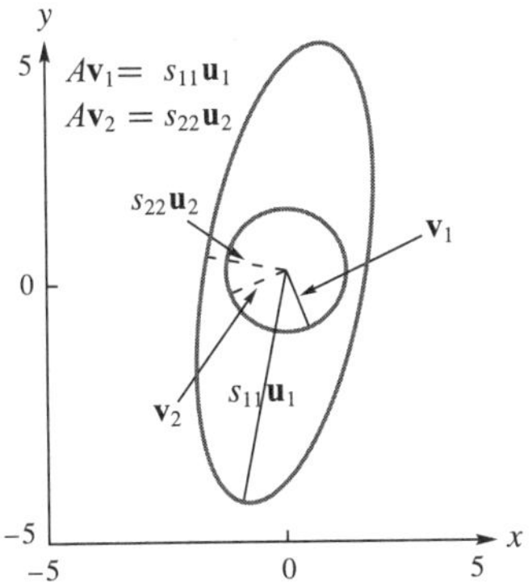

FIGURE 8.3

Figure 8.3 shows the unit circle with the vectors $\mathbf{v}_1$ and $\mathbf{v}_2$ displayed and the elliptical image with its major and minor axes shown as $s_{11}\mathbf{u}_1$ and $s_{22}\mathbf{u}_2$, respectively. ■

■ Application: Symmetric Images

Suppose that we have a large symmetric matrix A of information that must be transmitted quickly, but we really need only most of the information content. This might be the case if the symmetric matrix represented information from a pattern or a photo. Let us also assume that we can compute eigenvalues and associated eigenvectors of the matrix with relative ease. Using the spectral decomposition of the matrix, we can approximate the image by using **partial sums*** rather than all the eigenvalues and eigenvectors. To do this, we consider the eigenvalues as weights for the information contained in the eigenvectors and label the eigenvalues so that

$$|\lambda_1| \geq |\lambda_2| \geq \cdots \geq |\lambda_n|.$$

Then $A \approx \lambda_1 \mathbf{x}_1 \mathbf{x}_1^T + \lambda_2 \mathbf{x}_2 \mathbf{x}_2^T + \cdots + \lambda_k \mathbf{x}_k \mathbf{x}_k^T$, where $k \leq n$. This scheme uses the information associated with the larger eigenvalues first and then adjoins it to that associated with the smaller eigenvalues.

EXAMPLE 5 Consider the geometric pattern shown in Figure 8.4, which we digitize by using a 1 for a black block and 0 for a white block. Let A be the 9×9 matrix of corresponding zeros and ones in Figure 8.5.

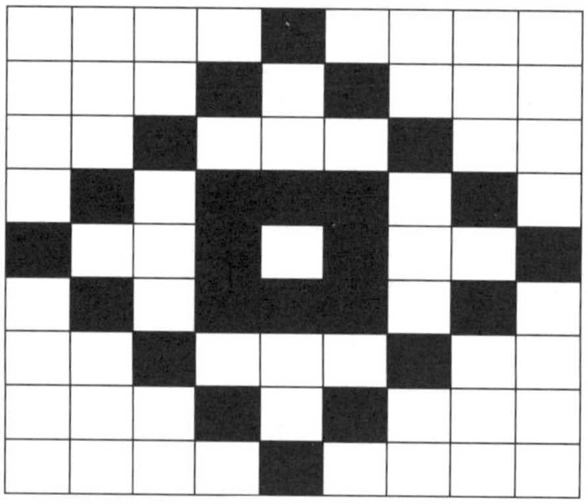

FIGURE 8.4

0	0	0	0	1	0	0	0	0
0	0	0	1	0	1	0	0	0
0	0	1	0	0	0	1	0	0
0	1	0	1	1	1	0	1	0
1	0	0	1	0	1	0	0	1
0	1	0	1	1	1	0	1	0
0	0	1	0	0	0	1	0	0
0	0	0	1	0	1	0	0	0
0	0	0	0	1	0	0	0	0

FIGURE 8.5

We can approximate the geometric pattern by using partial sums of the spectral representation. With MATLAB, we can show that the eigenvalues are, in order of magnitude to tenths, 3.7, 2, −2, 1.1, −0.9, 0, 0, 0, 0. Rather than display the approximations digitally, we show a pictorial representation where an entry is a black square if the corresponding numerical entry is greater than or equal to $\frac{1}{2}$ in absolute value; otherwise, it is shown as a white square. We show the first two partial sums in Figures 8.6 and 8.7, respectively. The third partial sum reveals Figure 8.7 again, and the fourth partial sum is the original pattern of Figure 8.4. If we had to transmit enough information to build a black-and-white approximation to the pattern, then we could send just the first four eigenvalues and their associated eigenvectors; that is, $4 + (4)9 = 40$ numerical values. This is a significant savings, compared with transmitting all of the matrix A, which requires 81 values. ■

*When fewer than the n terms in (4) are combined, we call this a partial sum of the spectral representation.

$A \approx 3.7\mathbf{p}_1\mathbf{p}_1^* \quad \Rightarrow$

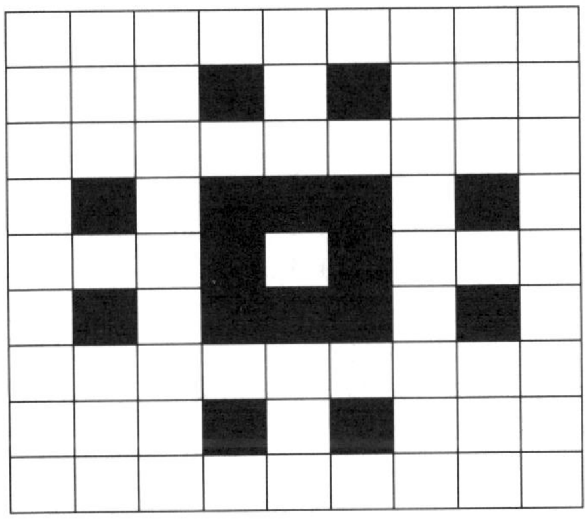

FIGURE 8.6

$A \approx 3.7\mathbf{x}_1\mathbf{x}_1^* + 2\mathbf{x}_2\mathbf{x}_2^* \quad \Rightarrow$

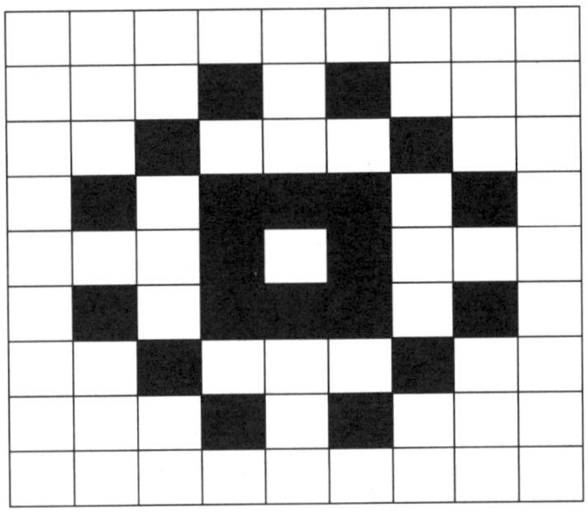

FIGURE 8.7

While the utility of a symmetric matrix is somewhat limited in terms of the representation of information, a generalization of a spectral style decomposition to arbitrary size matrices dramatically extends the usefulness of the approximation procedure by partial sums. We indicate this more general decomposition next.

We state the following result, whose proof can be found in the references listed at the end of this section:

The Singular Value Decomposition of a Matrix

Let A be an $m \times n$ real matrix. Then there exist orthogonal matrices U of size $m \times m$ and V of size $n \times n$ such that

$$A = USV^T, \tag{5}$$

where S is an $m \times n$ matrix with nondiagonal entries all zero and $s_{11} \geq s_{22} \geq \cdots \geq s_{pp} \geq 0$, where $p = \min\{m, n\}$.

The diagonal entries of S are called the **singular values** of A, the columns of U are called the **left singular vectors** of A, and the columns of V are called the **right singular vectors** of A. The singular value decomposition of A in (5) can be expressed as the following linear combination:

$$A = \mathrm{col}_1(U)s_{11}\mathrm{col}_1(V)^T + \mathrm{col}_2(U)s_{22}\mathrm{col}_2(V)^T + \cdots + \mathrm{col}_p(U)s_{pp}\mathrm{col}_p(V)^T, \tag{6}$$

which has the same form as the spectral representation of a symmetric matrix as given in Equation (4).

To determine the matrices U, S, and V in the singular value decomposition given in (5), we start as follows: An $n \times n$ symmetric matrix related to A is $A^T A$. By Theorem 7.9 there exists an orthogonal $n \times n$ matrix V such that

$$V(A^T A)V^T = D,$$

where D is a diagonal matrix whose diagonal entries $\lambda_1, \lambda_2, \ldots, \lambda_n$ are the eigenvalues of $A^T A$. If $\mathbf{v}_j$ denotes column j of V, then $(A^T A)\mathbf{v}_j = \lambda_j \mathbf{v}_j$. Multiply both sides of this expression on the left by $\mathbf{v}_j^T$; then we can rearrange the expression as

$$\mathbf{v}_j^T(A^T A)\mathbf{v}_j = \lambda_j \mathbf{v}_j^T \mathbf{v}_j \quad \text{or} \quad (A\mathbf{v}_j)^T(A\mathbf{v}_j) = \lambda_j \mathbf{v}_j^T \mathbf{v}_j \quad \text{or} \quad \|A\mathbf{v}_j\|^2 = \lambda_j \|\mathbf{v}_j\|^2.$$

Since the length of a vector is nonnegative, the last expression implies that $\lambda_j \geq 0$. Hence each eigenvalue of $A^T A$ is nonnegative. If necessary, renumber the eigenvalues of $A^T A$ so that $\lambda_1 \geq \lambda_2 \geq \cdots \geq \lambda_n$; then define $s_{jj} = \sqrt{\lambda_j}$. We note that since V is an orthogonal matrix, each of its columns is a unit vector; that is, $\|\mathbf{v}_j\| = 1$. Hence $s_{jj} = \|A\mathbf{v}_j\|$. (Verify.) Thus the singular values of A are the square roots of the eigenvalues of $A^T A$.

Finally, we determine the $m \times m$ orthogonal matrix U. Given the matrix equation in (5), let us see what the columns of U should look like.

- Since U is to be orthogonal, its columns must be an orthonormal set; hence they are linearly independent $m \times 1$ vectors.
- The matrix S has the form (block diagonal)

$$S = \left[\begin{array}{cccc|c} s_{11} & 0 & \cdots & 0 & \\ 0 & s_{22} & \cdots & 0 & O_{p,\,n-p} \\ \vdots & \vdots & & \vdots & \\ 0 & 0 & \cdots & s_{pp} & \\ \hline & O_{m-p,\,p} & & & O_{m-p,\,n-p} \end{array}\right],$$

where $O_{r,\,s}$ denotes an $r \times s$ matrix of zeros.

- From (5), $AV = US$, so

$$A\begin{bmatrix}\mathbf{v}_1 & \mathbf{v}_2 & \cdots & \mathbf{v}_n\end{bmatrix} = \begin{bmatrix}\mathbf{u}_1 & \mathbf{u}_2 & \cdots & \mathbf{u}_m\end{bmatrix} \left[\begin{array}{cccc|c} s_{11} & 0 & \cdots & 0 & \\ 0 & s_{22} & \cdots & 0 & O_{p,\,n-p} \\ \vdots & \vdots & & \vdots & \\ 0 & 0 & \cdots & s_{pp} & \\ \hline & O_{m-p,\,p} & & & O_{m-p,\,n-p} \end{array}\right].$$

This implies that we need to require that $A\mathbf{v}_j = s_{jj}\mathbf{u}_j$ for $j = 1, 2, \ldots, p$. (Verify.)

- However, U must have m orthonormal columns, and $m \geq p$. In order to construct U, we need an orthonormal basis for R^m whose first p vectors are $\mathbf{u}_j = \dfrac{1}{s_{jj}}A\mathbf{v}_j$. In Theorem 4.11 we showed how to extend any linearly independent subset of a vector space to a basis. We use that technique here to obtain the remaining $m - p$ columns of U. (This is necessary only if $m > p$.) Since these $m - p$ columns are not unique, matrix U is not unique. (Neither is V if any of the eigenvalues of $A^T A$ are repeated.)

It can be shown that the preceding construction gives matrices U, S, and V, so that $A = USV^T$. We illustrate the process in Example 6.

EXAMPLE 6

To find the singular value decomposition of $A = \begin{bmatrix} 2 & -4 \\ 2 & 2 \\ -4 & 0 \\ 1 & 4 \end{bmatrix}$, we follow the steps outlined previously. First, we compute $A^T A$, and then compute its eigenvalues and eigenvectors. We obtain (verify)

$$A^T A = \begin{bmatrix} 25 & 0 \\ 0 & 36 \end{bmatrix},$$

and since it is diagonal, we know that its eigenvalues are its diagonal entries. It follows that $\begin{bmatrix} 1 \\ 0 \end{bmatrix}$ is an eigenvector associated with eigenvalue 25 and that $\begin{bmatrix} 0 \\ 1 \end{bmatrix}$ is an eigenvector associated with eigenvalue 36. (Verify.) We label the eigenvalues in decreasing magnitude as $\lambda_1 = 36$ and $\lambda_2 = 25$ with corresponding eigenvectors $\mathbf{v}_1 = \begin{bmatrix} 0 \\ 1 \end{bmatrix}$ and $\mathbf{v}_2 = \begin{bmatrix} 1 \\ 0 \end{bmatrix}$, respectively. Hence

$$V = \begin{bmatrix} \mathbf{v}_1 & \mathbf{v}_2 \end{bmatrix} = \begin{bmatrix} 0 & 1 \\ 1 & 0 \end{bmatrix} \quad \text{and} \quad s_{11} = 6 \quad \text{and} \quad s_{22} = 5.$$

It follows that

$$S = \begin{bmatrix} 6 & 0 \\ 0 & 5 \\ \hdashline \multicolumn{2}{c}{O_2} \end{bmatrix}. \quad (O_2 \text{ is the } 2 \times 2 \text{ zero matrix.})$$

Next, we determine the matrix U, starting with the first two columns:

$$\mathbf{u}_1 = \frac{1}{s_{11}}A\mathbf{v}_1 = \frac{1}{6}\begin{bmatrix} -4 \\ 2 \\ 0 \\ 4 \end{bmatrix} \quad \text{and} \quad \mathbf{u}_2 = \frac{1}{s_{22}}A\mathbf{v}_2 = \frac{1}{5}\begin{bmatrix} 2 \\ 2 \\ -4 \\ 1 \end{bmatrix}.$$

The remaining two columns are found by extending the linearly independent vectors in $\{\mathbf{u}_1, \mathbf{u}_2\}$ to a basis for R^4 and then applying the Gram–Schmidt process. We proceed as follows: We first compute the reduced row echelon form (verify):

$$\mathbf{rref}\left(\begin{bmatrix}\mathbf{u}_1 & \mathbf{u}_2 & I_4\end{bmatrix}\right) = \begin{bmatrix} 1 & 0 & 0 & 0 & \frac{3}{8} & \frac{3}{2} \\ 0 & 1 & 0 & 0 & -\frac{3}{2} & 0 \\ 0 & 0 & 1 & 0 & \frac{3}{4} & 1 \\ 0 & 0 & 0 & 1 & \frac{3}{8} & -\frac{1}{2} \end{bmatrix}.$$

This tells us that the set $\{\mathbf{u}_1, \mathbf{u}_2, \mathbf{e}_1, \mathbf{e}_2\}$ is a basis for R^4. (Explain why.) Next, we apply the Gram–Schmidt process to this set to find the orthogonal matrix U. Recording the results to six decimals, we obtain

$$U = \begin{bmatrix} & & 0.628932 & 0 \\ \mathbf{u}_1 & \mathbf{u}_2 & 0.098933 & 0.847998 \\ & & 0.508798 & 0.317999 \\ & & 0.579465 & -0.423999 \end{bmatrix}.$$

Thus we have the singular value decomposition of A, and

$$A = USV^T = \begin{bmatrix} & & 0.628932 & 0 \\ \mathbf{u}_1 & \mathbf{u}_2 & 0.098933 & 0.847998 \\ & & 0.508798 & 0.317999 \\ & & 0.579465 & -0.423999 \end{bmatrix} \begin{bmatrix} 6 & 0 \\ 0 & 5 \\ \hdashline \multicolumn{2}{c}{O_2} \end{bmatrix} \begin{bmatrix}\mathbf{v}_1 & \mathbf{v}_2\end{bmatrix}^T$$

$$= \begin{bmatrix} -\frac{2}{3} & \frac{2}{5} & 0.628932 & 0 \\ \frac{1}{3} & \frac{2}{5} & 0.098933 & 0.847998 \\ 0 & -\frac{4}{5} & 0.508798 & 0.317999 \\ \frac{2}{3} & \frac{1}{5} & 0.579465 & -0.423999 \end{bmatrix} \begin{bmatrix} 6 & 0 \\ 0 & 5 \\ 0 & 0 \\ 0 & 0 \end{bmatrix} \begin{bmatrix} 0 & 1 \\ 1 & 0 \end{bmatrix}^T. \quad \blacksquare$$

The singular decomposition of a matrix A has been called one of the most useful tools in terms of the information that it reveals. For example, previously in Section 4.9, we indicated that we could compute the rank A by determining the number of nonzero rows in the reduced row echelon form of A. An implicit assumption in this statement is that all the computational steps in the row operations would use exact arithmetic. Unfortunately, in most computing environments, when we perform row operations, exact arithmetic is not used. Rather, floating point arithmetic, which is a model of exact arithmetic, is used. Thus we may lose accuracy because of the accumulation of small errors in the arithmetic steps. In some cases this loss of accuracy is enough to introduce doubt into the computation of rank. The following two results, which we state without proof, indicate how the singular value decomposition of a matrix can be used to compute its rank.

Theorem 8.1 Let A be an $m \times n$ matrix and let B and C be nonsingular matrices of sizes $m \times m$ and $n \times n$, respectively. Then

$$\operatorname{rank} BA = \operatorname{rank} A = \operatorname{rank} AC. \quad \blacksquare$$

Corollary 8.1 The rank of A is the number of nonzero singular values of A. (Multiple singular values are counted according to their multiplicity.)

Proof

Exercise 7. ■

Because matrices U and V of a singular value decomposition are orthogonal, it can be shown that most of the errors due to the use of floating point arithmetic occur in the computation of the singular values. The size of the matrix A and characteristics of the floating point arithmetic are often used to determine a threshold value below which singular values are considered zero. It has been argued that singular values and singular value decomposition give us a computationally reliable way to compute the rank of A.

In addition to determining rank, the singular value decomposition of a matrix provides orthonormal bases for the fundamental subspaces associated with a linear system of equations. We state without proof the following:

Theorem 8.2 Let A be an $m \times n$ real matrix of rank r with singular value decomposition USV^T. Then the following statements are true:

(a) The first r columns of U are an orthonormal basis for the column space of A.

(b) The first r columns of V are an orthonormal basis for the row space of A.

(c) The last $n - r$ columns of V are an orthonormal basis for the null space of A. ■

The singular value decomposition is also a valuable tool in the computation of least-squares approximations and can be used to approximate images that have been digitized. For examples of these applications, refer to the following references.

REFERENCES

Hern, T., and C. Long. "Viewing Some Concepts and Applications in Linear Algebra," in *Visualization in Teaching and Learning Mathematics*. MAA Notes Number 19, The Mathematical Association of America, 1991, pp. 173–190.

Hill, D. *Experiments in Computational Matrix Algebra*. New York: Random House, 1988.

Hill, D. "Nuggets of Information," in *Using MATLAB in the Classroom*. Upper Saddle River, NJ: Prentice Hall, 1993, pp. 77–93.

Kahaner, D., C. Moler, and S. Nash. *Numerical Methods and Software*. Upper Saddle River, NJ: Prentice Hall, 1989.

Stewart, G. *Introduction to Matrix Computations*. New York: Academic Press, 1973.

Key Terms

Diagonal matrix
Orthogonal matrix
Spectral decomposition
Outer product
Singular value decomposition
Singular values
Singular vectors (left/right)
Symmetric image
Gram–Schmidt process
Rank
Orthonormal basis

8.2 Exercises

1. Find the singular values of each of the following matrices:

(a) $A = \begin{bmatrix} 5 & 0 \\ 0 & 0 \\ 0 & -1 \end{bmatrix}$ **(b)** $A = \begin{bmatrix} 1 & 1 \\ 1 & 1 \end{bmatrix}$

(c) $A = \begin{bmatrix} 1 & 2 \\ 0 & 1 \\ -2 & 1 \end{bmatrix}$

(d) $A = \begin{bmatrix} 1 & 0 & 1 & -1 \\ 0 & 1 & -1 & 1 \end{bmatrix}$

2. Determine the singular value decomposition of

$$A = \begin{bmatrix} 1 & -4 \\ -2 & 2 \\ 2 & 4 \end{bmatrix}.$$

3. Determine the singular value decomposition of

$$A = \begin{bmatrix} 1 & 1 \\ -1 & 1 \end{bmatrix}.$$

4. Determine the singular value decomposition of

$$A = \begin{bmatrix} 1 & 0 & 1 \\ 1 & 1 & -1 \\ -1 & 1 & 0 \\ 0 & 1 & 1 \end{bmatrix}.$$

5. There is a MATLAB command, **svd**, that computes the singular value decomposition of a matrix A. Used in the form **svd(A)**, the output is a list of the singular values of A. In the form **[U, S, V] = svd(A)**, we get matrices U, S, and V such that $A = USV^T$. Use this command to determine the singular values of each of the following matrices:

(a) $\begin{bmatrix} 5 & 2 \\ 8 & 5 \end{bmatrix}$ **(b)** $\begin{bmatrix} 3 & 12 & -10 \\ 1 & -1 & 5 \\ -2 & -8 & 5 \end{bmatrix}$

(c) $\begin{bmatrix} 1 & 5 & 9 \\ 2 & 6 & 10 \\ 3 & 7 & 11 \\ 4 & 8 & 12 \end{bmatrix}$

6. An $m \times n$ matrix A is said to have **full rank** if rank $A =$ minimum $\{m, n\}$. The singular value decomposition lets us measure how close A is to not having full rank. If any singular value is zero, then A does not have full rank. If $s_{\min}$ is the smallest singular value of A and $s_{\min} \neq 0$, then the distance from A to the set of matrices with rank $r = \min\{m, n\} - 1$ is $s_{\min}$. Determine the distance from each of the given matrices to the matrices of the same size with rank $\min\{m, n\} - 1$. (Use MATLAB to find the singular values.)

(a) $\begin{bmatrix} 1 & 3 & 2 \\ 4 & 1 & 0 \\ 2 & -5 & -4 \\ 1 & 2 & 1 \end{bmatrix}$ **(b)** $\begin{bmatrix} 1 & 2 & 3 \\ 1 & 0 & 1 \\ 3 & 6 & 9 \end{bmatrix}$

(c) $\begin{bmatrix} 1 & 0 & 0 & -1 & 0 \\ -1 & 1 & 1 & 0 & 0 \\ 0 & 0 & 1 & -2 & -1 \\ 0 & 1 & 0 & 1 & 1 \end{bmatrix}$

7. Prove Corollary 8.1, using Theorem 8.1 and the singular value decomposition.

Exercises 8 and 9 use supplemental MATLAB *commands constructed for this book. See Section 9.9.*

8. To use MATLAB to illustrate geometrically the approximation of a picture composed of black and white squares, like that in Example 5, proceed as follows. We have put a four-letter word into a matrix by using a 1 for a black block and 0 for a white block. The objective is to determine the word, using as few as possible singular value and singular vector terms to construct a partial sum from Equation (6). Follow the directions appearing on the screen in the routine **picgen**. Use the following MATLAB commands:

load svdword1 ← This will load a word encoded in a matrix, using 0's and 1's.

picgen(svdword1) ← This will show the partial sum images as determined by the singular value and singular vectors that appear in Equation (6).

9. Follow the procedure given in Exercise 8, using these commands:

load svdword2

picgen(svdword2)

(*Hint*: The letters in this case have been written on an angle.)

8.3 Dominant Eigenvalue and Principal Component Analysis

Dominant Eigenvalue

In Section 8.2 we discussed and illustrated the spectral decomposition of a symmetric matrix. We showed that if A is an $n \times n$ symmetric matrix, then we can express A as a linear combination of matrices of rank one, using the eigenvalues of A and associated eigenvectors as follows. Let $\lambda_1, \lambda_2, \ldots, \lambda_n$ be the eigenvalues of A and $\mathbf{x}_1, \mathbf{x}_2, \ldots, \mathbf{x}_n$ a set of associated orthonormal eigenvectors. Then the spectral decomposition of A is given by

$$A = \lambda_1 \mathbf{x}_1 \mathbf{x}_1^T + \lambda_2 \mathbf{x}_2 \mathbf{x}_2^T + \cdots + \lambda_n \mathbf{x}_n \mathbf{x}_n^T. \tag{1}$$

Furthermore, if we label the eigenvalues so that

$$|\lambda_1| \geq |\lambda_2| \geq \cdots \geq |\lambda_n|,$$

then we can construct approximations to the matrix A, using partial sums of the spectral decomposition. We illustrated such approximations by using matrices of zeros and ones that corresponded to pictures represented by matrices of black and white blocks. As remarked in Section 8.2, the terms using eigenvalues of largest magnitude in the partial sums in (1) contributed a large part of the "information" represented by the matrix A. In this section we investigate two other situations where the largest eigenvalue and its corresponding eigenvector can supply valuable information.

DEFINITION 8.1 If A is a real $n \times n$ matrix with real eigenvalues $\lambda_1, \lambda_2, \ldots, \lambda_n$, then an eigenvalue of largest magnitude is called a **dominant eigenvalue** of A.

EXAMPLE 1 Let

$$A = \begin{bmatrix} 9 & 6 & -14 \\ -2 & 1 & 2 \\ 6 & 6 & -11 \end{bmatrix}.$$

The eigenvalues of A are 3, 1, and -5 (verify). Thus the dominant eigenvalue of A is -5, since $|-5| > 1$ and $|-5| > 3$. ■

Remark Observe that λ_j is a dominant eigenvalue of A, provided that $|\lambda_j| \geq |\lambda_i|$, $i = 1, 2, \ldots, j-1, j+1, \ldots, n$. A matrix can have more than one dominant eigenvalue. For example, the matrix

$$\begin{bmatrix} 4 & 2 & -1 \\ 0 & -2 & 7 \\ 0 & 0 & -4 \end{bmatrix}$$

has both 4 and -4 as dominant eigenvalues.

Let A be a real $n \times n$ matrix with real eigenvalues $\lambda_1, \lambda_2, \ldots, \lambda_n$, where $|\lambda_1| > |\lambda_i|$, $i = 2, \ldots, n$. Then A has a unique dominant eigenvalue. Furthermore, suppose that A is diagonalizable with associated linearly independent eigenvectors $\mathbf{x}_1, \mathbf{x}_2, \ldots, \mathbf{x}_n$. Hence $S = \{\mathbf{x}_1, \mathbf{x}_2, \ldots, \mathbf{x}_n\}$ is a basis for R^n, and every vector $\mathbf{x}$ in R^n is expressible as a linear combination of the vectors in S. Let

$$\mathbf{x} = c_1\mathbf{x}_1 + c_2\mathbf{x}_2 + \cdots + c_n\mathbf{x}_n \quad \text{with } c_1 \neq 0$$

and compute the sequence of vectors $A\mathbf{x}$, $A^2\mathbf{x}$, $A^3\mathbf{x}, \ldots, A^k\mathbf{x}, \ldots$. We obtain the following:

$$\begin{aligned}
A\mathbf{x} &= c_1A\mathbf{x}_1 + c_2A\mathbf{x}_2 + \cdots + c_nA\mathbf{x}_n = c_1\lambda_1\mathbf{x}_1 + c_2\lambda_2\mathbf{x}_2 + \cdots + c_n\lambda_n\mathbf{x}_n \\
A^2\mathbf{x} &= c_1\lambda_1A\mathbf{x}_1 + c_2\lambda_2A\mathbf{x}_2 + \cdots + c_n\lambda_nA\mathbf{x}_n = c_1\lambda_1^2\mathbf{x}_1 + c_2\lambda_2^2\mathbf{x}_2 + \cdots + c_n\lambda_n^2\mathbf{x}_n \\
A^3\mathbf{x} &= c_1\lambda_1^2A\mathbf{x}_1 + c_2\lambda_2^2A\mathbf{x}_2 + \cdots + c_n\lambda_n^2A\mathbf{x}_n = c_1\lambda_1^3\mathbf{x}_1 + c_2\lambda_2^3\mathbf{x}_2 + \cdots + c_n\lambda_n^3\mathbf{x}_n \\
&\ \vdots \\
A^k\mathbf{x} &= c_1\lambda_1^k\mathbf{x}_1 + c_2\lambda_2^k\mathbf{x}_2 + \cdots + c_n\lambda_n^k\mathbf{x}_n \\
&\ \vdots
\end{aligned}$$

We have

$$\begin{aligned}
A^k\mathbf{x} &= \lambda_1^k\left(c_1\mathbf{x}_1 + c_2\frac{\lambda_2^k}{\lambda_1^k}\mathbf{x}_2 + \cdots + c_n\frac{\lambda_n^k}{\lambda_1^k}\mathbf{x}_n\right) \\
&= \lambda_1^k\left(c_1\mathbf{x}_1 + c_2\left(\frac{\lambda_2}{\lambda_1}\right)^k\mathbf{x}_2 + \cdots + c_n\left(\frac{\lambda_n}{\lambda_1}\right)^k\mathbf{x}_n\right),
\end{aligned}$$

and since λ_1 is the dominant eigenvalue of A, $\left|\dfrac{\lambda_i}{\lambda_1}\right| < 1$ for $i > 1$. Hence, as $k \to \infty$, it follows that

$$A^k\mathbf{x} \to \lambda_1^k c_1\mathbf{x}_1. \tag{2}$$

Using this result, we can make the following observations: For a real diagonalizable matrix with all real eigenvalues and a unique dominant eigenvalue λ_1, we have

(a) $A^k\mathbf{x}$ approaches the zero vector for any vector $\mathbf{x}$, provided that $|\lambda_1| < 1$.
(b) The sequence of vectors $A^k\mathbf{x}$ does not converge, provided that $|\lambda_1| > 1$.
(c) If $|\lambda_1| = 1$, then the limit of the sequence of vectors $A^k\mathbf{x}$ is an eigenvector associated with λ_1.

In certain **iterative** processes in numerical linear algebra, sequences of vectors of the form $A^k\mathbf{x}$ arise frequently. In order to determine the convergence of the sequence, it is important to determine whether or not the dominant eigenvalue is smaller than 1. Rather than compute the eigenvalues, it is often easier to determine an upper bound on the dominant eigenvalue. We next investigate one such approach.

In the Supplementary Exercises 28–33 in Chapter 5 we defined the **1-norm** of a vector $\mathbf{x}$ in R^n as the sum of the absolute values of its entries; that is,

$$\|\mathbf{x}\|_1 = |x_1| + |x_2| + \cdots + |x_n|.$$

For an $n \times n$ matrix A, we extend this definition and define the **1-norm** of the matrix A to be the maximum of the 1-norms of its columns. We denote the 1-norm of the matrix A by $\|A\|_1$, and it follows that

$$\|A\|_1 = \max_{j=1,2,\ldots,n} \{\|\mathrm{col}_j(A)\|_1\}.$$

EXAMPLE 2

Let

$$A = \begin{bmatrix} 9 & 6 & -14 \\ -2 & 1 & 2 \\ 6 & 6 & -11 \end{bmatrix}$$

as in Example 1. It follows that $\|A\|_1 = \max\{17, 13, 27\} = 27$. ■

Theorem 8.3 For an $n \times n$ matrix A, the absolute value of the dominant eigenvalue of A is less than or equal to $\|A\|_1$.

Proof

Let $\mathbf{x}$ be any n-vector. Then the product $A\mathbf{x}$ can be expressed as a linear combination of the columns of A in the form

$$A\mathbf{x} = x_1\mathrm{col}_1(A) + x_2\mathrm{col}_2(A) + \cdots + x_n\mathrm{col}_n(A).$$

We proceed by using properties of a norm:

$$\begin{aligned}
\|A\mathbf{x}\|_1 &= \|x_1\mathrm{col}_1(A) + x_2\mathrm{col}_2(A) + \cdots + x_n\mathrm{col}_n(A)\|_1 \\
&\quad \text{(Compute the 1-norm of each side.)} \\
&\le \|x_1\mathrm{col}_1(A)\|_1 + \|x_2\mathrm{col}_2(A)\|_1 + \cdots + \|x_n\mathrm{col}_n(A)\|_1 \\
&\quad \text{(Use the triangle inequality of the 1-norm.)} \\
&= |x_1|\|\mathrm{col}_1(A)\|_1 + |x_2|\|\mathrm{col}_2(A)\|_1 + \cdots + |x_n|\|\mathrm{col}_n(A)\|_1 \\
&\quad \text{(Use the scalar multiple of a norm.)} \\
&\le |x_1|\|A\|_1 + |x_2|\|A\|_1 + \cdots + |x_n|\|A\|_1 \\
&\quad \text{(Use } \|\mathrm{col}_j(A)\| \le \|A\|_1\text{.)} \\
&= \|\mathbf{x}\|_1\|A\|_1.
\end{aligned}$$

Next, suppose that $\mathbf{x}$ is the eigenvector corresponding to the dominant eigenvalue λ of A and recall that $A\mathbf{x} = \lambda\mathbf{x}$. Then we have

$$\|A\mathbf{x}\|_1 = \|\lambda\mathbf{x}\|_1 = |\lambda|\|\mathbf{x}\|_1 \le \|\mathbf{x}\|_1\|A\|_1,$$

and since $\mathbf{x}$ is an eigenvector, we have $\|\mathbf{x}\|_1 \neq 0$, so

$$|\lambda| \le \|A\|_1.$$

Hence the absolute value of the dominant eigenvalue is "bounded above" by the matrix 1-norm of A. ■

EXAMPLE 3

Let

$$A = \begin{bmatrix} 9 & 6 & -14 \\ -2 & 1 & 2 \\ 6 & 6 & -11 \end{bmatrix}$$

as in Example 1. Then

$$|\text{dominant eigenvalue of } A| \leq \|A\|_1 = \max\{17, 13, 27\} = 27.$$

From Example 1, we know that |dominant eigenvalue of A| = 5. ■

Theorem 8.4 If $\|A\|_1 < 1$, then the sequence of vectors $A^k\mathbf{x}$ approaches the zero vector for any vector $\mathbf{x}$.

Proof

Exercise 9. ■

In Section 8.1 we saw sequences of vectors of the form $T^n\mathbf{x}$ that arise in the analysis of Markov processes. If T is a transition matrix (also called a Markov matrix), then $\|T\|_1 = 1$. Then from Theorem 8.3 we know that the absolute value of the dominant eigenvalue of T is less than or equal to 1. However, we have the following stronger result:

Theorem 8.5 If T is a transition matrix of a Markov process, then the dominant eigenvalue of T is 1.

Proof

Let $\mathbf{x}$ be the n-vector of all ones. Then $T^T\mathbf{x} = \mathbf{x}$ (verify), so 1 is an eigenvalue of T^T. Since a matrix and its transpose have the same eigenvalues, $\lambda = 1$ is also an eigenvalue of T. Now by Theorem 8.3 and the statement in the paragraph preceding this theorem, we conclude that the dominant eigenvalue of the transition matrix T is 1. ■

The preceding results about the dominant eigenvalue were very algebraic in nature. We now turn to a graphical look at the effect of the dominant eigenvalue and an associated eigenvector.

From (2) we see that the sequence of vectors $A\mathbf{x}$, $A^2\mathbf{x}$, $A^3\mathbf{x}$, ... approaches a scalar multiple of an eigenvector associated with the dominant eigenvalue. Geometrically, we can say that the sequence of vectors $A\mathbf{x}$, $A^2\mathbf{x}$, $A^3\mathbf{x}$, ... approaches a line in n-space that is parallel to an eigenvector associated with the dominant eigenvalue. Example 4 illustrates this observation in R^2.

EXAMPLE 4

Let L be a linear transformation from R^2 to R^2 that is represented by the matrix

$$A = \begin{bmatrix} 7 & -2 \\ 4 & 1 \end{bmatrix}$$

with respect to the natural basis for R^2. For a 2-vector $\mathbf{x}$ we compute the terms $A^k\mathbf{x}$, $k = 1, 2, \ldots, 7$. Since we are interested in the direction of this set of vectors in R^2 and for ease in displaying these vectors in the plane R^2, we first scale each of the vectors $A^k\mathbf{x}$ to be a unit vector. Setting

$$\mathbf{x} = \begin{bmatrix} 0.2 \\ 0.5 \end{bmatrix}$$

and computing the set of vectors, we get the following:

$\mathbf{x}$	$A\mathbf{x}$	$A^2\mathbf{x}$	$A^3\mathbf{x}$	$A^4\mathbf{x}$	$A^5\mathbf{x}$	$A^6\mathbf{x}$	$A^7\mathbf{x}$
0.2	0.2941	0.0688	−0.7654	−0.9397	−0.8209	−0.7667	−0.7402
0.5	0.9558	0.9976	0.6436	−0.3419	−0.5711	−0.6420	−0.6724

Here, we have shown only four decimal digits. Figure 8.8 shows these vectors in R^2, where $\mathbf{x}$ is labeled with 0 and the vectors $A^k\mathbf{x}$ are labeled with the value of k. ■

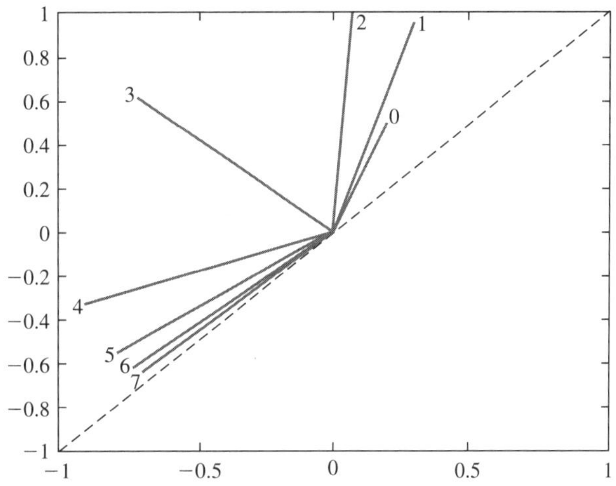

FIGURE 8.8

An eigenvector associated with a dominant eigenvalue is shown as a dashed line segment. The choice of the vector $\mathbf{x}$ is *almost* arbitrary, in the sense that $\mathbf{x}$ cannot be an eigenvector associated with an eigenvalue that is not a dominant eigenvalue, since in that case the sequence $A^k\mathbf{x}$ would always be in the direction of that eigenvector.

EXAMPLE 5

For the linear transformation in Example 4, we compute successive images of the unit circle; that is, $A^k \times$ (unit circle). (See Example 5 of Section 1.7 for a special case.) The first image is an ellipse, and so are the successive images for $k = 2, 3, \ldots$. Figure 8.9 displays five images (where each point displayed in the graphs is the terminal point of a vector that has been scaled to be a unit vector in R^2), and again we see the alignment of the images in the direction of an eigenvector associated with the dominant eigenvalue. This eigenvector is shown as a dashed line segment. ■

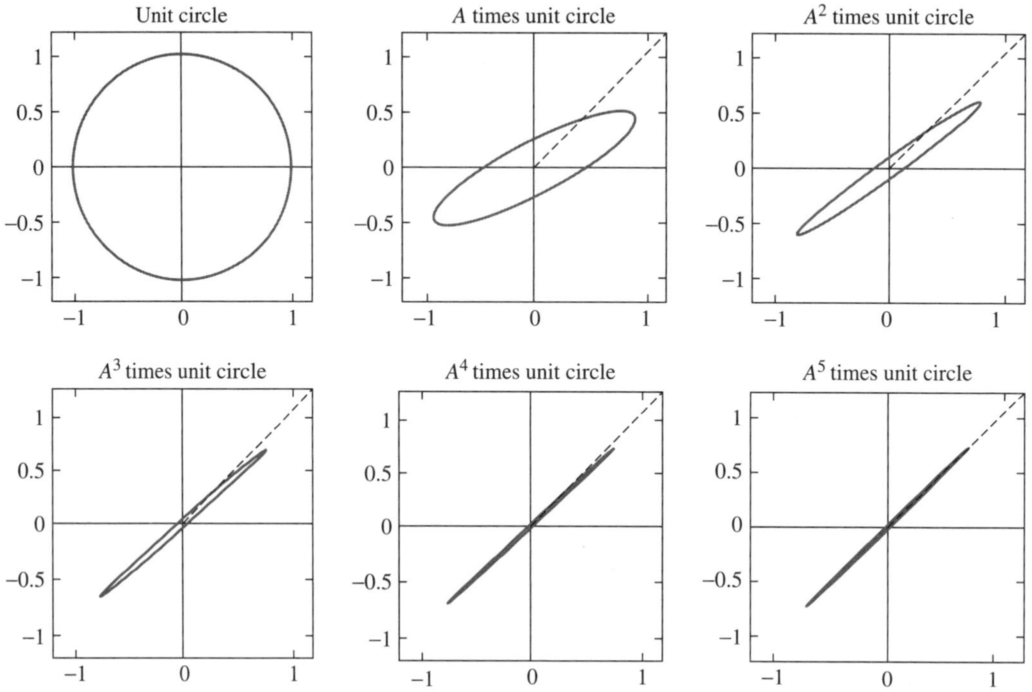

FIGURE 8.9

The sequence of vectors $A\mathbf{x}$, $A^2\mathbf{x}$, $A^3\mathbf{x}$, ... also forms the basis for the numerical method called the **power method** for estimating the dominant eigenvalue of a matrix A. Details of this method can be found in D. R. Hill and B. Kolman, *Modern Matrix Algebra*, Upper Saddle River, NJ: Prentice Hall, 2001, as well as in numerical analysis and numerical linear algebra texts.

■ Principal Component Analysis

The second application that involves the dominant eigenvalue and its eigenvector is taken from applied multivariate statistics and is called **principal component analysis**, often abbreviated **PCA**. To provide a foundation for this topic, we briefly discuss some selected terminology from statistics and state some results that involve a matrix that is useful in statistical analysis.

Multivariate statistics concerns the analysis of data in which several variables are measured on a number of subjects, patients, objects, items, or other entities of interest. The goal of the analysis is to understand the relationships between the variables: how they vary separately, how they vary together, and how to develop an algebraic model that expresses the interrelationships of the variables.

The sets of observations of the variables, the data, are represented by a matrix. Let x_{jk} indicate the particular value of the kth variable that is observed on the jth item. We let n be the number of items being observed and p the number of variables measured. Such data are organized and represented by a rectangular matrix X given by

$$X = \begin{bmatrix} x_{11} & x_{12} & \cdots & x_{1k} & \cdots & x_{1p} \\ x_{21} & x_{22} & \cdots & x_{2k} & \cdots & x_{2p} \\ \vdots & \vdots & & \vdots & & \vdots \\ x_{j1} & x_{j2} & \cdots & x_{jk} & \cdots & x_{jp} \\ \vdots & \vdots & & \vdots & & \vdots \\ x_{n1} & x_{n2} & \cdots & x_{nk} & \cdots & x_{np} \end{bmatrix},$$

a **multivariate data matrix**. The matrix X contains all the observations on all of the variables. Each column represents the data for a different variable, and linear combinations of the set of observations are formed by the matrix product $X\mathbf{c}$, where $\mathbf{c}$ is a $p \times 1$ matrix. Useful algebraic models are derived in this way by imposing some optimization criteria for the selection of the entries of the coefficient vector $\mathbf{c}$.

In a single-variable case where the matrix X is $n \times 1$, such as exam scores, the data are often summarized by calculating the arithmetic average, or sample mean, and a measure of spread, or variation. Such summary calculations are referred to as **descriptive statistics**. In this case, for

$$X = \begin{bmatrix} x_1 \\ x_2 \\ \vdots \\ x_n \end{bmatrix},$$

the

$$\textbf{sample mean} = \overline{x} = \frac{1}{n}\sum_{j=1}^{n} x_j$$

and the

$$\textbf{sample variance} = s^2 = \frac{1}{n}\sum_{j=1}^{n}(x_j - \overline{x})^2.$$

In addition, the square root of the sample variance is known as the **sample standard deviation**.

EXAMPLE 6 If the matrix

$$X = \begin{bmatrix} 97 & 92 & 90 & 87 & 85 & 83 & 83 & 78 & 72 & 71 & 70 & 65 \end{bmatrix}^T$$

is the set of scores out of 100 for an exam in linear algebra, then the associated descriptive statistics are $\overline{x} \approx 81$, $s^2 \approx 90.4$, and the standard deviation $s \approx 9.5$. ■

These descriptive statistics are also applied to the set of observations of each of the variables in a multivariate data matrix. We next define these, together with statistics that provide a measure of the relationship between pairs of variables:

$$\text{Sample mean for the } k\text{th variable} = \overline{x}_k = \frac{1}{n}\sum_{j=1}^{n} x_{jk}, \quad k = 1, 2, \ldots, p.$$

$$\text{Sample variance for the } k\text{th variable} = s_k^2 = \frac{1}{n}\sum_{j=1}^{n} (x_{jk} - \overline{x}_k)^2, \quad k = 1, 2, \ldots, p.$$

Remark The sample variance is often defined with a divisor of $n - 1$ rather than n, for theoretical reasons, especially in the case where n, the number of samples, is small. In many multivariate statistics texts, there is a notational convention employed to distinguish between the two versions. For simplicity in our brief excursion into multivariate statistics, we will use the expression given previously.

Presently, we shall introduce a matrix which contains statistics that relate pairs of variables. For convenience of matrix notation, we shall use the alternative notation s_{kk} for the variance of the kth variable; that is,

$$s_{kk} = s_k^2 = \frac{1}{n}\sum_{j=1}^{n} (x_{jk} - \overline{x}_k)^2, \quad k = 1, 2, \ldots, p.$$

A measure of the linear association between a pair of variables is provided by the notion of **sample covariance**. The measure of association between the ith and kth variables in the multivariate data matrix X is given by

$$\text{Sample covariance} = s_{ik} = \frac{1}{n}\sum_{j=1}^{n} (x_{ji} - \overline{x}_i)(x_{jk} - \overline{x}_k), \quad \begin{aligned} i &= 1, 2, \ldots, p, \\ k &= 1, 2, \ldots, p, \end{aligned}$$

which is the average product of the deviations from their respective sample means. It follows that $s_{ik} = s_{ki}$, for all i and k, and that for $i = k$, the sample covariance is just the variance, $s_k^2 = s_{kk}$.

We next organize the descriptive statistics associated with a multivariate data matrix into matrices:

$$\text{Matrix of sample means} = \overline{\mathbf{x}} = \begin{bmatrix} \overline{x}_1 \\ \overline{x}_2 \\ \vdots \\ \overline{x}_p \end{bmatrix}.$$

$$\text{Matrix of sample variances and covariances} = S_n = \begin{bmatrix} s_{11} & s_{12} & \cdots & s_{1p} \\ s_{21} & s_{22} & \cdots & s_{2p} \\ \vdots & \vdots & & \vdots \\ s_{p1} & s_{p2} & \cdots & s_{pp} \end{bmatrix}.$$

The matrix S_n is a symmetric matrix whose diagonal entries are the sample variances and the subscript n is a notational device to remind us that the divisor n was used to compute the variances and covariances. The matrix S_n is often called the **covariance matrix**, for simplicity.

EXAMPLE 7

A selection of six receipts from a supermarket was collected to investigate the nature of food sales. On each receipt was the cost of the purchases and the number of items purchased. Let the first variable be the cost of the purchases rounded to whole dollars, and the second variable the number of items purchased. The corresponding multivariate data matrix is

$$X = \begin{bmatrix} 39 & 21 \\ 59 & 28 \\ 18 & 10 \\ 21 & 13 \\ 14 & 13 \\ 22 & 10 \end{bmatrix}.$$

Determine the sample statistics given previously, recording numerical values to one decimal place and using this approximation in subsequent calculations.

Solution

We find that the sample means are

$$\overline{x}_1 \approx 28.8 \quad \text{and} \quad \overline{x}_2 \approx 15.8,$$

and thus we take the matrix of sample means as

$$\overline{\mathbf{x}} = \begin{bmatrix} 28.8 \\ 15.8 \end{bmatrix}.$$

The variances are

$$s_{11} \approx 243.1 \quad \text{and} \quad s_{22} \approx 43.1,$$

while the covariances are

$$s_{12} = s_{21} \approx 97.8.$$

Hence we take the covariance matrix as

$$S_n = \begin{bmatrix} 243.1 & 97.8 \\ 97.8 & 43.1 \end{bmatrix}.$$

■

In a more general setting the multivariate data matrix X is a matrix whose entries are random variables. In this setting the matrices of descriptive statistics are computed using probability distributions and expected value. We shall not consider this case, but just note that the vector of means and the covariance matrix can be computed in an analogous fashion. In particular, the covariance matrix is symmetric, as it is for the "sample" case illustrated previously.

We now state several results that indicate how to use information about the covariance matrix to define a set of new variables. These new variables are linear combinations of the original variables represented by the columns of the data matrix X. The technique is called **principal component analysis**, PCA, and is among the oldest and most widely used of multivariate techniques. The new variables are derived in decreasing order of importance so that the first, called the **first principal component**, accounts for as much as possible of the variation in the original data. The second new variable, called the **second principal component**, accounts for another, but smaller, portion of the variation, and so on. For a situation involving p variables, p components are required to account for all the variation, but often, much of the variation can be accounted for by a small number of principal components. Thus, PCA has as its goals the interpretation of the variation and data reduction.

The description of PCA given previously is analogous to the use of the spectral decomposition of a symmetric matrix in the application to symmetric images discussed in Section 8.2. In fact, we use the eigenvalues and associated orthonormal eigenvectors of the covariance matrix S_n to construct the principal components and derive information about them. We have the following result, which we state without proof:

Theorem 8.6 Let S_n be the $p \times p$ covariance matrix associated with the multivariate data matrix X. Let the eigenvalues of S_n be λ_j, $j = 1, 2, \ldots, p$, where $\lambda_1 \geq \lambda_2 \geq \cdots \geq \lambda_p \geq 0$, and let the associated orthonormal eigenvectors be $\mathbf{u}_j$, $j = 1, 2, \ldots, p$. Then the ith principal component $\mathbf{y}_i$ is given by the linear combination of the columns of X, where the coefficients are the entries of the eigenvector $\mathbf{u}_i$; that is,

$$\mathbf{y}_i = i\text{th principal component} = X\mathbf{u}_i.$$

In addition, the variance of $\mathbf{y}_i$ is λ_i, and the covariance of $\mathbf{y}_i$ and $\mathbf{y}_k$, $i \neq k$, is zero. (If some of the eigenvalues are repeated, then the choices of the associated eigenvectors are not unique; hence the principal components are not unique.) ■

Theorem 8.7 Under the hypotheses of Theorem 8.6, the total variance of X given by $\sum_{i=1}^{p} s_{ii}$ is the same as the sum of the eigenvalues of the covariance matrix S_n.

Proof

Exercise 18. ■

This result implies that

$$\begin{pmatrix} \text{Proportion of the} \\ \text{total variance due} \\ \text{to the } k\text{th principal} \\ \text{component} \end{pmatrix} = \frac{\lambda_k}{\lambda_1 + \lambda_2 + \cdots + \lambda_p}, \qquad k = 1, 2, \ldots, p. \tag{3}$$

Thus we see that if $\lambda_1 > \lambda_2$, then λ_1 is the dominant eigenvalue of the covariance matrix. Hence the first principal component is a new variable that "explains," or accounts for, more of the variation than any other principal component. If a large percentage of the total variance for a data matrix with a large number p of columns can be attributed to the first few principal components, then these new variables can replace the original p variables without significant loss of information. Thus we can achieve a significant reduction in data.

EXAMPLE 8

Compute the first principal component for the data matrix X given in Example 7.

Solution

The covariance matrix S_n is computed in Example 7, so we determine its eigenvalues and associated orthonormal eigenvectors. (Here, we record the numerical values to only four decimal places.) We obtain the eigenvalues

$$\lambda_1 = 282.9744 \quad \text{and} \quad \lambda_2 = 3.2256$$

and associated eigenvectors

$$\mathbf{u}_1 = \begin{bmatrix} 0.9260 \\ 0.3775 \end{bmatrix} \quad \text{and} \quad \mathbf{u}_2 = \begin{bmatrix} 0.3775 \\ -0.9260 \end{bmatrix}.$$

Then, using Theorem 8.7, we find that the first principal component is

$$\mathbf{y}_1 = 0.9260\text{col}_1(X) + 0.3775\text{col}_2(X),$$

and it follows that $\mathbf{y}_1$ accounts for the proportion

$$\frac{\lambda_1}{\lambda_1 + \lambda_2} \quad \text{(about 98.9\%)}$$

of the total variance of X (verify). ■

EXAMPLE 9

Suppose that we have a multivariate data matrix X with three columns, which we denote as $\mathbf{x}_1$, $\mathbf{x}_2$, and $\mathbf{x}_3$, and the covariance matrix (recording values to only four decimal places) is

$$S_n = \begin{bmatrix} 3.6270 & 2.5440 & 0 \\ 2.5440 & 6.8070 & 0 \\ 0 & 0 & 1 \end{bmatrix}.$$

Determine the principal components $\mathbf{y}_1$, $\mathbf{y}_2$, and $\mathbf{y}_3$.

Solution

We find that the eigenvalues and associated orthonormal eigenvectors are

$$\lambda_1 = 8.2170, \quad \lambda_2 = 2.2170, \quad \text{and} \quad \lambda_3 = 1,$$

$$\mathbf{u}_1 = \begin{bmatrix} 0.4848 \\ 0.8746 \\ 0 \end{bmatrix}, \quad \mathbf{u}_2 = \begin{bmatrix} -0.8746 \\ 0.4848 \\ 0 \end{bmatrix}, \quad \text{and} \quad \mathbf{u}_3 = \begin{bmatrix} 0 \\ 0 \\ 1 \end{bmatrix}.$$

Thus the principal components are

$$\begin{aligned} \mathbf{y}_1 &= 0.4848\mathbf{x}_1 + 0.8746\mathbf{x}_2 \\ \mathbf{y}_2 &= -0.8746\mathbf{x}_1 + 0.4848\mathbf{x}_2 \\ \mathbf{y}_3 &= \mathbf{x}_3. \end{aligned}$$

Then it follows from (3) that $\mathbf{y}_1$ accounts for 78.61% of the total variance, while $\mathbf{y}_2$ and $\mathbf{y}_3$ account for 19.39% and 8.75%, respectively. ■

There is much more information concerning the influence of and relationship between variables that can be derived from the computations associated with PCA. For more information on PCA, see the references below.

We now make several observations about the geometric nature of PCA. The fact that the covariance matrix S_n is symmetric means that we can find an orthogonal matrix U consisting of eigenvectors of S_n such that $U^T S_n U = D$, a diagonal matrix. The geometric consequence of this result is that the p original variables are rotated to p new orthogonal variables, called the principal components. Moreover, these principal components are linear combinations of the original variables. (The orthogonality follows from the fact that the covariance of $\mathbf{y}_i$ and $\mathbf{y}_k$, $i \neq k$, is zero.) Hence the computation of the principal components amounts to transforming a coordinate system that consists of axes that may not be mutually perpendicular to a new coordinate system with mutually perpendicular axes. The new coordinate system, the principal components, represents the original variables in a more ordered and convenient way. An orthogonal coordinate system makes it possible to easily use projections to derive further information about the relationships between variables. For details, refer to the following references:

REFERENCES

Johnson, Richard A., and Dean W. Wichern. *Applied Multivariate Statisical Analysis*, 5th ed. Upper Saddle River, NJ: Prentice Hall, 2002.

Jolliffe, I. T. *Principal Component Analysis*. New York: Springer-Verlag, 1986.

Wickens, Thomas D. *The Geometry of Multivariate Statistics*. Hillsdale, NJ: Lawrence Erlbaum Associates, 1995.

For an interesting application to trade routes in geography, see Philip D. Straffin, "Linear Algebra in Geography: Eigenvectors of Networks," *Mathematics Magazine*, vol. 53, no. 5, Nov. 1980, pp. 269–276.

■ Searching with Google: Using the Dominant Eigenvalue

In Section 1.2, after Example 6, we introduced the connectivity matrix A used by the software that drives Google's search engine. Matrix A has entries that are either 0 or 1, with $a_{ij} = 1$ if website j links to website i; otherwise, $a_{ij} = 0$.

EXAMPLE 10

A company with seven employees encourages the use of websites for a variety of business reasons. Each employee has a website, and certain employees include links to coworkers' sites. For this small company, their connectivity matrix is as follows:

$A =$

	E_1	E_2	E_3	E_4	E_5	E_6	E_7
E_1	0	1	0	1	1	0	0
E_2	1	0	1	1	0	1	0
E_3	1	0	0	0	0	0	1
E_4	0	0	1	0	1	1	0
E_5	1	1	0	0	0	1	0
E_6	0	1	1	1	0	0	1
E_7	1	0	0	0	0	0	1

Here, we have assigned the names E_k, $k = 1, 2, \ldots, 7$ to designate the employees. We see from the column labeled E_3 that this employee links to the sites of coworkers E_2, E_4, and E_6. ■

Upon inspecting the connectivity matrix in Example 10, we might try to assign a rank to an employee website by merely counting the number of sites that are linked to it. But this strategy does not take into account the rank of the websites that link to a given site.

There are many applications that use the ranking of objects, teams, or people in associated order of importance. One approach to the ranking strategy is to create a connectivity matrix and compute its dominant eigenvalue and associated eigenvector. For a wide class of such problems, the entries of the eigenvector can be taken to be all positive and scaled so that the sum of their squares is 1. In such cases, if the kth entry is largest, then the kth item that is being ranked is considered the most important; that is, it has the highest rank. The other items are ranked according to the size of the corresponding entry of the eigenvector.

EXAMPLE 11

For the connectivity matrix in Example 10, an eigenvector associated with the dominant eigenvalue is

$$\mathbf{v} = \begin{bmatrix} 0.4261 \\ 0.4746 \\ 0.2137 \\ 0.3596 \\ 0.4416 \\ 0.4214 \\ 0.2137 \end{bmatrix}.$$

It follows that $\max\{v_1, v_2, \ldots, v_7\} = 0.4746$; hence employee number 2 has the highest-ranked website, followed by that of number 5, and then number 1. Notice that the site for employee 6 was referenced more times than that of employee 1 or employee 5, but is considered lower in rank. ■

In carrying out a Google search, the ranking of websites is a salient feature that determines the order of the sites returned to a query. The strategy for ranking uses the basic idea that the rank of a site is higher if other highly ranked sites link to it. In order to implement this strategy for the huge connectivity matrix that is a part of Google's ranking mechanism, a variant of the dominant eigenvalue/eigenvector idea of Example 10 is used. In their algorithm the Google team determines the rank of a site so that it is proportional to the sum of the ranks of all sites that link to it. This approach generates a large eigenvalue/eigenvector problem that uses the connectivity matrix in a more general fashion than that illustrated in Example 11.

REFERENCES

Moler, Cleve. "The World's Largest Matrix Computation: Google's PageRank Is an Eigenvector of a Matrix of Order 2.7 Billion." *MATLAB News and Notes*, October 2002, pp. 12–13.

Wilf, Herbert S. "Searching the Web with Eigenvectors." *The UMAP Journal*, 23(2), 2002, pp. 101–103.

Key Terms

Symmetric matrix	Principal component analysis (PCA)	First principal component
Orthonormal eigenvectors	Multivariate data matrix	Second principal component
Dominant eigenvalue	Descriptive statistics	Total variance
Iterative process	Sample mean	Google
1-norm	Sample variance	Connectivity matrix
Markov process	Sample standard deviation	
Power method	Covariance matrix	

8.3 Exercises

1. Find the dominant eigenvalue of each of the following matrices:

(a) $\begin{bmatrix} 1 & 1 \\ -2 & 4 \end{bmatrix}$ **(b)** $\begin{bmatrix} 1 & 2 & -1 \\ 1 & 0 & 1 \\ 4 & -4 & 5 \end{bmatrix}$

2. Find the dominant eigenvalue of each of the following matrices:

(a) $\begin{bmatrix} 4 & 2 \\ 3 & 3 \end{bmatrix}$ **(b)** $\begin{bmatrix} 2 & 1 & 2 \\ 2 & 2 & -2 \\ 3 & 1 & 1 \end{bmatrix}$

3. Find the 1-norm of each of the following matrices:

(a) $\begin{bmatrix} 3 & -5 \\ 2 & 2 \end{bmatrix}$ **(b)** $\begin{bmatrix} 4 & 1 & 0 \\ 2 & 5 & 0 \\ -4 & -4 & 7 \end{bmatrix}$

(c) $\begin{bmatrix} 2 & -1 & 1 & 0 \\ 4 & -2 & 2 & 3 \\ 1 & 0 & 2 & 6 \\ -3 & 4 & 8 & 1 \end{bmatrix}$

4. Find the 1-norm of each of the following matrices:

(a) $\begin{bmatrix} 4 & -1 \\ 1 & 0 \end{bmatrix}$ **(b)** $\begin{bmatrix} -2 & 0 & 0 \\ -2 & 3 & -1 \\ 3 & -2 & 2 \end{bmatrix}$

(c) $\begin{bmatrix} 1 & 2 & -3 & 4 \\ 4 & 1 & 2 & -3 \\ -3 & 4 & 1 & 2 \\ 2 & -3 & 4 & 1 \end{bmatrix}$

5. Determine a bound on the absolute value of the dominant eigenvalue for each of the matrices in Exercise 1.

6. Determine a bound on the absolute value of the dominant eigenvalue for each of the matrices in Exercise 2.

7. Prove that if A is symmetric, then $\|A\|_1 = \|A^T\|_1$.

8. Determine a matrix A for which $\|A\|_1 = \|A^T\|_1$, but A is not symmetric.

9. Prove Theorem 8.4.

10. Explain why $\|A\|_1$ can be greater than 1 and the sequence of vectors $A^k\mathbf{x}$ can still approach the zero vector.

11. Let $X = \begin{bmatrix} 56 & 62 & 59 & 73 & 75 \end{bmatrix}^T$ be the weight in ounces of scoops of birdseed obtained by the same person using the same scoop. Find the sample mean, the variation, and standard deviation of these data.

12. Let $X = \begin{bmatrix} 5400 & 4900 & 6300 & 6700 \end{bmatrix}^T$ be the estimates in dollars for the cost of replacing a roof on the same home. Find the sample mean, the variation, and standard deviation of these data.

13. For the five most populated cities in the United States in 2002, we have the following crime information: For violent offenses known to police per 100,000 residents, the number of robberies appears in column 1 of the data matrix X, and the number of aggravated assaults in column 2. (Values are rounded to the nearest whole number.)

$$X = \begin{bmatrix} 337 & 425 \\ 449 & 847 \\ 631 & 846 \\ 550 & 617 \\ 582 & 647 \end{bmatrix}$$

Determine the vector of sample means and the covariance matrix. (Data taken from TIME Almanac 2006, Information Please LLC, Pearson Education, Boston, MA.)

14. For the five most populated cities in the United States in 2002, we have the following crime information: For property crimes known to police per 100,000 residents, the number of burglaries appears in column 1 of the data matrix X and the number of motor vehicle thefts in column 2. (Values are rounded to the nearest whole number.)

$$X = \begin{bmatrix} 372 & 334 \\ 662 & 891 \\ 869 & 859 \\ 1319 & 1173 \\ 737 & 873 \end{bmatrix}$$

Determine the vector of sample means and the covariance matrix. (Data taken from TIME Almanac 2006, Information Please LLC, Pearson Education, Boston, MA.)

15. For the data in Exercise 13, determine the first principal component.

16. For the data in Exercise 14, determine the first principal component.

17. In Section 5.3 we defined a positive definite matrix as a square symmetric matrix C such that $\mathbf{y}^T C\mathbf{y} > 0$ for every nonzero vector $\mathbf{y}$ in R^n. Prove that any eigenvalue of a positive definite matrix is positive.

18. Let S_n be a covariance matrix satisfying the hypotheses of Theorem 8.6. To prove Theorem 8.7, proceed as follows:

(a) Show that the trace of S_n is the total variance. (See Section 1.3, Exercise 43, for the definition of trace.)

(b) Show that there exists an orthogonal matrix P such that $P^T S_n P = D$, a diagonal matrix.

(c) Show that the trace of S_n is equal to the trace of D.

(d) Complete the proof.

8.4 Differential Equations

A **differential equation** is an equation that involves an unknown function and its derivatives. An important, simple example of a differential equation is

$$\frac{d}{dt}x(t) = rx(t),$$

where r is a constant. The idea here is to find a function $x(t)$ that will satisfy the given differential equation. This differential equation is discussed further subsequently. Differential equations occur often in all branches of science and engineering; linear algebra is helpful in the formulation and solution of differential equations. In this section we provide only a brief survey of the approach; books on differential equations deal with the subject in much greater detail, and several suggestions for further reading are given at the end of this chapter.

Homogeneous Linear Systems

We consider the **first-order homogeneous linear system** of differential equations,

$$\begin{aligned} x_1'(t) &= a_{11}x_1(t) + a_{12}x_2(t) + \cdots + a_{1n}x_n(t) \\ x_2'(t) &= a_{21}x_1(t) + a_{22}x_2(t) + \cdots + a_{2n}x_n(t) \\ &\vdots \\ x_n'(t) &= a_{n1}x_1(t) + a_{n2}x_2(t) + \cdots + a_{nn}x_n(t), \end{aligned} \tag{1}$$

where the a_{ij} are known constants. We seek functions $x_1(t), x_2(t), \ldots, x_n(t)$ defined and differentiable on the real line and satisfying (1).

We can write (1) in matrix form by letting

$$\mathbf{x}(t) = \begin{bmatrix} x_1(t) \\ x_2(t) \\ \vdots \\ x_n(t) \end{bmatrix}, \qquad A = \begin{bmatrix} a_{11} & a_{12} & \cdots & a_{1n} \\ a_{21} & a_{22} & \cdots & a_{2n} \\ \vdots & \vdots & & \vdots \\ a_{n1} & a_{n2} & \cdots & a_{nn} \end{bmatrix},$$

and defining

$$\mathbf{x}'(t) = \begin{bmatrix} x_1'(t) \\ x_2'(t) \\ \vdots \\ x_n'(t) \end{bmatrix}.$$

Then (1) can be written as

$$\mathbf{x}'(t) = A\mathbf{x}(t). \tag{2}$$

We shall often write (2) more briefly as

$$\mathbf{x}' = A\mathbf{x}.$$

With this notation, an n-vector function

$$\mathbf{x}(t) = \begin{bmatrix} x_1(t) \\ x_2(t) \\ \vdots \\ x_n(t) \end{bmatrix}$$

satisfying (2) is called a **solution** to the given system.

We leave it to the reader to verify that if $\mathbf{x}^{(1)}(t), \mathbf{x}^{(2)}(t), \ldots, \mathbf{x}^{(n)}(t)$ are all solutions to (2), then any linear combination

$$\mathbf{x}(t) = b_1\mathbf{x}^{(1)}(t) + b_2\mathbf{x}^{(2)}(t) + \cdots + b_n\mathbf{x}^{(n)}(t) \tag{3}$$

is also a solution to (2).

It can be shown (Exercise 4) that the set of all solutions to the homogeneous linear system of differential equations (2) is a subspace of the vector space of differentiable real-valued n-vector functions.

A set of vector functions $\{\mathbf{x}^{(1)}(t), \mathbf{x}^{(2)}(t), \ldots, \mathbf{x}^{(n)}(t)\}$ is said to be a **fundamental system** for (1) if every solution to (1) can be written in the form (3). In this case, the right side of (3), where $b_1, b_2, \ldots, b_n$ are arbitrary constants, is said to be the **general solution** to (2).

It can be shown (see the book by Boyce and DiPrima or the book by Cullen cited in Further Readings) that any system of the form (2) has a fundamental system (in fact, infinitely many).

In general, differential equations arise in the course of solving physical problems. Typically, once a general solution to the differential equation has been obtained, the physical constraints of the problem impose certain definite values on the arbitrary constants in the general solution, giving rise to a **particular solution**. An important particular solution is obtained by finding a solution $\mathbf{x}(t)$ to Equation (2) such that $\mathbf{x}(0) = \mathbf{x}_0$, an **initial condition**, where $\mathbf{x}_0$ is a given vector. This problem is called an **initial value problem**. If the general solution (3) is known, then the initial value problem can be solved by setting $t = 0$ in (3) and determining the constants $b_1, b_2, \ldots, b_n$ so that

$$\mathbf{x}_0 = b_1\mathbf{x}^{(1)}(0) + b_2\mathbf{x}^{(2)}(0) + \cdots + b_n\mathbf{x}^{(n)}(0).$$

It is readily seen that this is actually an $n \times n$ linear system with unknowns $b_1, b_2, \ldots, b_n$. This linear system can also be written as

$$C\mathbf{b} = \mathbf{x}_0, \tag{4}$$

where

$$\mathbf{b} = \begin{bmatrix} b_1 \\ b_2 \\ \vdots \\ b_n \end{bmatrix}$$

and C is the $n \times n$ matrix whose columns are $\mathbf{x}^{(1)}(0), \mathbf{x}^{(2)}(0), \ldots, \mathbf{x}^{(n)}(0)$, respectively. It can be shown (see the book by Boyce and DiPrima or the book by Cullen cited in Further Readings) that if $\mathbf{x}^{(1)}(t), \mathbf{x}^{(2)}(t), \ldots, \mathbf{x}^{(n)}(t)$ form a fundamental system for (1), then C is nonsingular, so (4) always has a unique solution.

EXAMPLE 1

The simplest system of the form (1) is the single equation

$$\frac{dx}{dt} = ax, \tag{5}$$

where a is a constant. From calculus, the solutions to this equation are of the form

$$x = be^{at}; \tag{6}$$

that is, this is the general solution to (5). To solve the initial value problem

$$\frac{dx}{dt} = ax, \qquad x(0) = x_0,$$

we set $t = 0$ in (6) and obtain $b = x_0$. Thus the solution to the initial value problem is

$$x = x_0 e^{at}.$$ ■

The system (2) is said to be **diagonal** if the matrix A is diagonal. Then (1) can be rewritten as

$$\begin{aligned} x_1'(t) &= a_{11}x_1(t) \\ x_2'(t) &= \qquad\quad a_{22}x_2(t) \\ &\vdots \\ x_n'(t) &= \qquad\qquad\qquad a_{nn}x_n(t). \end{aligned} \tag{7}$$

This system is easy to solve, since the equations can be solved separately. Applying the results of Example 1 to each equation in (7), we obtain

$$\begin{aligned} x_1(t) &= b_1 e^{a_{11}t} \\ x_2(t) &= b_2 e^{a_{22}t} \\ &\vdots \\ x_n(t) &= b_n e^{a_{nn}t}, \end{aligned} \tag{8}$$

where $b_1, b_2, \ldots, b_n$ are arbitrary constants. Writing (8) in vector form yields

$$\mathbf{x}(t) = \begin{bmatrix} b_1 e^{a_{11}t} \\ b_2 e^{a_{22}t} \\ \vdots \\ b_n e^{a_{nn}t} \end{bmatrix} = b_1 \begin{bmatrix} 1 \\ 0 \\ 0 \\ \vdots \\ 0 \end{bmatrix} e^{a_{11}t} + b_2 \begin{bmatrix} 0 \\ 1 \\ 0 \\ \vdots \\ 0 \end{bmatrix} e^{a_{22}t} + \cdots + b_n \begin{bmatrix} 0 \\ 0 \\ \vdots \\ 0 \\ 1 \end{bmatrix} e^{a_{nn}t}.$$

This implies that the vector functions

$$\mathbf{x}^{(1)}(t) = \begin{bmatrix} 1 \\ 0 \\ 0 \\ \vdots \\ 0 \end{bmatrix} e^{a_{11}t}, \quad \mathbf{x}^{(2)}(t) = \begin{bmatrix} 0 \\ 1 \\ 0 \\ \vdots \\ 0 \end{bmatrix} e^{a_{22}t}, \quad \ldots, \quad \mathbf{x}^{(n)}(t) = \begin{bmatrix} 0 \\ 0 \\ \vdots \\ 0 \\ 1 \end{bmatrix} e^{a_{nn}t}$$

form a fundamental system for the diagonal system (7).

EXAMPLE 2

The diagonal system

$$\begin{bmatrix} x_1' \\ x_2' \\ x_3' \end{bmatrix} = \begin{bmatrix} 3 & 0 & 0 \\ 0 & -2 & 0 \\ 0 & 0 & 4 \end{bmatrix} \begin{bmatrix} x_1 \\ x_2 \\ x_3 \end{bmatrix} \tag{9}$$

can be written as three equations:

$$\begin{aligned} x_1' &= 3x_1 \\ x_2' &= -2x_2 \\ x_3' &= 4x_3. \end{aligned}$$

Solving these equations, we obtain

$$x_1 = b_1 e^{3t}, \qquad x_2 = b_2 e^{-2t}, \qquad x_3 = b_3 e^{4t},$$

where b_1, b_2, and b_3 are arbitrary constants. Thus

$$\mathbf{x}(t) = \begin{bmatrix} b_1 e^{3t} \\ b_2 e^{-2t} \\ b_3 e^{4t} \end{bmatrix} = b_1 \begin{bmatrix} 1 \\ 0 \\ 0 \end{bmatrix} e^{3t} + b_2 \begin{bmatrix} 0 \\ 1 \\ 0 \end{bmatrix} e^{-2t} + b_3 \begin{bmatrix} 0 \\ 0 \\ 1 \end{bmatrix} e^{4t}$$

is the general solution to (9), and the functions

$$\mathbf{x}^{(1)}(t) = \begin{bmatrix} 1 \\ 0 \\ 0 \end{bmatrix} e^{3t}, \qquad \mathbf{x}^{(2)}(t) = \begin{bmatrix} 0 \\ 1 \\ 0 \end{bmatrix} e^{-2t}, \qquad \mathbf{x}^{(3)}(t) = \begin{bmatrix} 0 \\ 0 \\ 1 \end{bmatrix} e^{4t}$$

form a fundamental system for (9). ■

If the system (2) is not diagonal, then it cannot be solved as simply as the system in the preceding example. However, there is an extension of this method that yields the general solution in the case where A is diagonalizable. Suppose that A is diagonalizable and P is a nonsingular matrix such that

$$P^{-1}AP = D, \tag{10}$$

where D is diagonal. Then, multiplying the given system

$$\mathbf{x}' = A\mathbf{x}$$

on the left by P^{-1}, we obtain

$$P^{-1}\mathbf{x}' = P^{-1}A\mathbf{x}.$$

Since $P^{-1}P = I_n$, we can rewrite the last equation as

$$P^{-1}\mathbf{x}' = (P^{-1}AP)(P^{-1}\mathbf{x}). \tag{11}$$

Temporarily, let

$$\mathbf{u} = P^{-1}\mathbf{x}. \tag{12}$$

Since P^{-1} is a constant matrix,

$$\mathbf{u}' = P^{-1}\mathbf{x}'. \tag{13}$$

Therefore, substituting (10), (12), and (13) into (11), we obtain

$$\mathbf{u}' = D\mathbf{u}. \tag{14}$$

Equation (14) is a diagonal system and can be solved by the methods just discussed. Before proceeding, however, let us recall from Theorem 7.4 that

$$D = \begin{bmatrix} \lambda_1 & 0 & \cdots & 0 \\ 0 & \lambda_2 & \cdots & 0 \\ \vdots & \vdots & & \vdots \\ 0 & 0 & \cdots & \lambda_n \end{bmatrix},$$

where $\lambda_1, \lambda_2, \ldots, \lambda_n$ are the eigenvalues of A, and that the columns of P are linearly independent eigenvectors of A associated, respectively, with $\lambda_1, \lambda_2, \ldots, \lambda_n$. From the discussion just given for diagonal systems, the general solution to (14) is

$$\mathbf{u}(t) = b_1\mathbf{u}^{(1)}(t) + b_2\mathbf{u}^{(2)}(t) + \cdots + b_n\mathbf{u}^{(n)}(t) = \begin{bmatrix} b_1e^{\lambda_1 t} \\ b_2e^{\lambda_2 t} \\ \vdots \\ b_ne^{\lambda_n t} \end{bmatrix},$$

where

$$\mathbf{u}^{(1)}(t) = \begin{bmatrix} 1 \\ 0 \\ 0 \\ \vdots \\ 0 \end{bmatrix} e^{\lambda_1 t}, \quad \mathbf{u}^{(2)}(t) = \begin{bmatrix} 0 \\ 1 \\ 0 \\ \vdots \\ 0 \end{bmatrix} e^{\lambda_2 t}, \quad \ldots, \quad \mathbf{u}^{(n)}(t) = \begin{bmatrix} 0 \\ 0 \\ \vdots \\ 0 \\ 1 \end{bmatrix} e^{\lambda_n t} \tag{15}$$

and $b_1, b_2, \ldots, b_n$ are arbitrary constants. From Equation (12), $\mathbf{x} = P\mathbf{u}$, so the general solution to the given system $\mathbf{x}' = A\mathbf{x}$ is

$$\mathbf{x}(t) = P\mathbf{u}(t) = b_1 P\mathbf{u}^{(1)}(t) + b_2 P\mathbf{u}^{(2)}(t) + \cdots + b_n P\mathbf{u}^{(n)}(t). \tag{16}$$

However, since the constant vectors in (15) are the columns of the identity matrix and $PI_n = P$, (16) can be rewritten as

$$\mathbf{x}(t) = b_1\mathbf{p}_1 e^{\lambda_1 t} + b_2\mathbf{p}_2 e^{\lambda_2 t} + \cdots + b_n\mathbf{p}_n e^{\lambda_n t}, \tag{17}$$

where $\mathbf{p}_1, \mathbf{p}_2, \ldots, \mathbf{p}_n$ are the columns of P, and therefore eigenvectors of A associated with $\lambda_1, \lambda_2, \ldots, \lambda_n$, respectively.

We summarize this discussion in the following theorem:

Theorem 8.8 If the $n \times n$ matrix A has n linearly independent eigenvectors $\mathbf{p}_1, \mathbf{p}_2, \ldots, \mathbf{p}_n$ associated with the eigenvalues $\lambda_1, \lambda_2, \ldots, \lambda_n$, respectively, then the general solution to the homogeneous linear system of differential equations

$$\mathbf{x}' = A\mathbf{x}$$

is given by Equation (17). ■

EXAMPLE 3 For the system

$$\mathbf{x}' = \begin{bmatrix} 1 & -1 \\ 2 & 4 \end{bmatrix}\mathbf{x},$$

the matrix

$$A = \begin{bmatrix} 1 & -1 \\ 2 & 4 \end{bmatrix}$$

has eigenvalues $\lambda_1 = 2$ and $\lambda_2 = 3$ with respective associated eigenvectors (verify)

$$\mathbf{p}_1 = \begin{bmatrix} 1 \\ -1 \end{bmatrix} \quad \text{and} \quad \mathbf{p}_2 = \begin{bmatrix} 1 \\ -2 \end{bmatrix}.$$

These eigenvectors are automatically linearly independent, since they are associated with distinct eigenvalues (proof of Theorem 7.5). Hence the general solution to the given system is

$$\mathbf{x}(t) = b_1 \begin{bmatrix} 1 \\ -1 \end{bmatrix} e^{2t} + b_2 \begin{bmatrix} 1 \\ -2 \end{bmatrix} e^{3t}.$$

In terms of components, this can be written as

$$\begin{aligned} x_1(t) &= b_1 e^{2t} + b_2 e^{3t} \\ x_2(t) &= -b_1 e^{2t} - 2b_2 e^{3t}. \end{aligned}$$ ■

EXAMPLE 4 Consider the following homogeneous linear system of differential equations:

$$\mathbf{x}' = \begin{bmatrix} x_1' \\ x_2' \\ x_3' \end{bmatrix} = \begin{bmatrix} 0 & 1 & 0 \\ 0 & 0 & 1 \\ 8 & -14 & 7 \end{bmatrix} \begin{bmatrix} x_1 \\ x_2 \\ x_3 \end{bmatrix}.$$

The characteristic polynomial of A is (verify)

$$p(\lambda) = \lambda^3 - 7\lambda^2 + 14\lambda - 8,$$

or

$$p(\lambda) = (\lambda - 1)(\lambda - 2)(\lambda - 4),$$

so the eigenvalues of A are $\lambda_1 = 1$, $\lambda_2 = 2$, and $\lambda_3 = 4$. Associated eigenvectors are (verify)

$$\begin{bmatrix} 1 \\ 1 \\ 1 \end{bmatrix}, \quad \begin{bmatrix} 1 \\ 2 \\ 4 \end{bmatrix}, \quad \begin{bmatrix} 1 \\ 4 \\ 16 \end{bmatrix},$$

respectively. The general solution is then given by

$$\mathbf{x}(t) = b_1 \begin{bmatrix} 1 \\ 1 \\ 1 \end{bmatrix} e^t + b_2 \begin{bmatrix} 1 \\ 2 \\ 4 \end{bmatrix} e^{2t} + b_3 \begin{bmatrix} 1 \\ 4 \\ 16 \end{bmatrix} e^{4t},$$

where b_1, b_2, and b_3 are arbitrary constants. ■

EXAMPLE 5

For the linear system of Example 4, solve the initial value problem determined by the **initial conditions** $x_1(0) = 4$, $x_2(0) = 6$, and $x_3(0) = 8$.

Solution

We write our general solution in the form $\mathbf{x} = P\mathbf{u}$ as

$$\mathbf{x}(t) = \begin{bmatrix} 1 & 1 & 1 \\ 1 & 2 & 4 \\ 1 & 4 & 16 \end{bmatrix} \begin{bmatrix} b_1 e^t \\ b_2 e^{2t} \\ b_3 e^{4t} \end{bmatrix}.$$

Now

$$\mathbf{x}(0) = \begin{bmatrix} 4 \\ 6 \\ 8 \end{bmatrix} = \begin{bmatrix} 1 & 1 & 1 \\ 1 & 2 & 4 \\ 1 & 4 & 16 \end{bmatrix} \begin{bmatrix} b_1 e^0 \\ b_2 e^0 \\ b_3 e^0 \end{bmatrix},$$

or

$$\begin{bmatrix} 1 & 1 & 1 \\ 1 & 2 & 4 \\ 1 & 4 & 16 \end{bmatrix} \begin{bmatrix} b_1 \\ b_2 \\ b_3 \end{bmatrix} = \begin{bmatrix} 4 \\ 6 \\ 8 \end{bmatrix}. \tag{18}$$

Solving (18) by Gauss–Jordan reduction, we obtain (verify)

$$b_1 = \tfrac{4}{3}, \qquad b_2 = 3, \qquad b_3 = -\tfrac{1}{3}.$$

Therefore, the solution to the initial value problem is

$$\mathbf{x}(t) = \tfrac{4}{3} \begin{bmatrix} 1 \\ 1 \\ 1 \end{bmatrix} e^t + 3 \begin{bmatrix} 1 \\ 2 \\ 4 \end{bmatrix} e^{2t} - \tfrac{1}{3} \begin{bmatrix} 1 \\ 4 \\ 16 \end{bmatrix} e^{4t}.$$

■

We now recall several facts from Chapter 7. If A does not have distinct eigenvalues, then we may or may not be able to diagonalize A. Let λ be an eigenvalue of A of multiplicity k. Then A can be diagonalized if and only if the dimension of the eigenspace associated with λ is k—that is, if and only if the rank of the matrix $(\lambda I_n - A)$ is $n - k$ (verify). If the rank of $(\lambda I_n - A)$ is $n - k$, then we can find k linearly independent eigenvectors of A associated with λ.

EXAMPLE 6 Consider the linear system

$$\mathbf{x}' = A\mathbf{x} = \begin{bmatrix} 1 & 0 & 0 \\ 0 & 3 & -2 \\ 0 & -2 & 3 \end{bmatrix} \mathbf{x}.$$

The eigenvalues of A are $\lambda_1 = \lambda_2 = 1$ and $\lambda_3 = 5$ (verify). The rank of the matrix

$$(1I_3 - A) = \begin{bmatrix} 0 & 0 & 0 \\ 0 & -2 & 2 \\ 0 & 2 & -2 \end{bmatrix}$$

is 1, and the linearly independent eigenvectors

$$\begin{bmatrix} 1 \\ 0 \\ 0 \end{bmatrix} \quad \text{and} \quad \begin{bmatrix} 0 \\ 1 \\ 1 \end{bmatrix}$$

are associated with the eigenvalue 1 (verify). The eigenvector

$$\begin{bmatrix} 0 \\ 1 \\ -1 \end{bmatrix}$$

is associated with the eigenvalue 5 (verify). The general solution to the given system is then

$$\mathbf{x}(t) = b_1 \begin{bmatrix} 1 \\ 0 \\ 0 \end{bmatrix} e^t + b_2 \begin{bmatrix} 0 \\ 1 \\ 1 \end{bmatrix} e^t + b_3 \begin{bmatrix} 0 \\ 1 \\ -1 \end{bmatrix} e^{5t},$$

where b_1, b_2, and b_3 are arbitrary constants. ■

If we cannot diagonalize A as in the examples, we have a considerably more difficult situation. Methods for dealing with such problems are discussed in more advanced books (see Further Readings).

■ Application—A Diffusion Process

The following example is a modification of an example presented by Derrick and Grossman in *Elementary Differential Equations with Applications* (see Further Readings):

EXAMPLE 7

Consider two adjoining cells separated by a permeable membrane, and suppose that a fluid flows from the first cell to the second one at a rate (in milliliters per minute) that is numerically equal to three times the volume (in milliliters) of the fluid in the first cell. It then flows out of the second cell at a rate (in milliliters per minute) that is numerically equal to twice the volume in the second cell. Let $x_1(t)$ and $x_2(t)$ denote the volumes of the fluid in the first and second cells at time t, respectively. Assume that, initially, the first cell has 40 milliliters of fluid, while the second one has 5 milliliters of fluid. Find the volume of fluid in each cell at time t. See Figure 8.10.

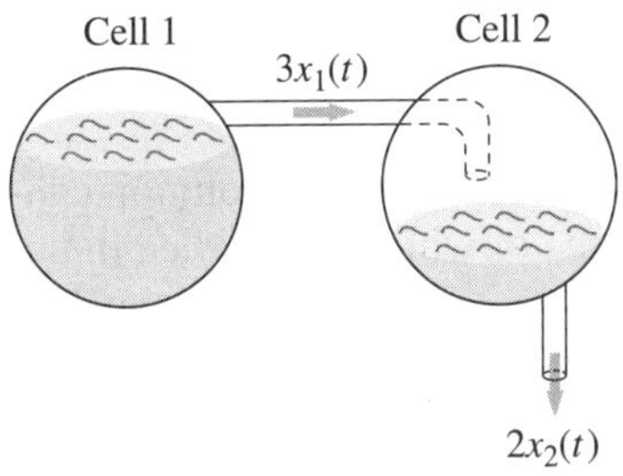

FIGURE 8.10

Solution

The change in volume of the fluid in each cell is the difference between the amount flowing in and the amount flowing out. Since no fluid flows into the first cell, we have

$$\frac{dx_1(t)}{dt} = -3x_1(t),$$

where the minus sign indicates that the fluid is flowing out of the cell. The flow $3x_1(t)$ from the first cell flows into the second cell. The flow out of the second cell is $2x_2(t)$. Thus the change in volume of the fluid in the second cell is given by

$$\frac{dx_2(t)}{dt} = 3x_1(t) - 2x_2(t).$$

We have then obtained the linear system

$$\frac{dx_1(t)}{dt} = -3x_1(t)$$

$$\frac{dx_2(t)}{dt} = 3x_1(t) - 2x_2(t),$$

which can be written in matrix form as

$$\begin{bmatrix} x_1'(t) \\ x_2'(t) \end{bmatrix} = \begin{bmatrix} -3 & 0 \\ 3 & -2 \end{bmatrix} \begin{bmatrix} x_1(t) \\ x_2(t) \end{bmatrix}.$$

The eigenvalues of the matrix

$$A = \begin{bmatrix} -3 & 0 \\ 3 & -2 \end{bmatrix}$$

are (verify) $\lambda_1 = -3$, $\lambda_2 = -2$, and respective associated eigenvectors are (verify)

$$\begin{bmatrix} 1 \\ -3 \end{bmatrix}, \quad \begin{bmatrix} 0 \\ 1 \end{bmatrix}.$$

Hence the general solution is given by

$$\mathbf{x}(t) = \begin{bmatrix} x_1(t) \\ x_2(t) \end{bmatrix} = b_1 \begin{bmatrix} 1 \\ -3 \end{bmatrix} e^{-3t} + b_2 \begin{bmatrix} 0 \\ 1 \end{bmatrix} e^{-2t}.$$

Using the initial conditions, we find that (verify)

$$b_1 = 40, \qquad b_2 = 125.$$

Thus the volume of fluid in each cell at time t is given by

$$\begin{aligned} x_1(t) &= 40e^{-3t} \\ x_2(t) &= -120e^{-3t} + 125e^{-2t}. \end{aligned}$$ ■

It should be pointed out that for many differential equations the solution cannot be written as a formula. Numerical methods, some of which are studied in numerical analysis, exist for obtaining numerical solutions to differential equations; computer codes for some of these methods are widely available.

FURTHER READINGS

Boyce, W. E. *Elementary Differential Equations*, 9th ed. New York: John Wiley & Sons, Inc., 2007.

Cullen, C. G. *Linear Algebra and Differential Equations*, 2d ed. Boston: PWS-Kent, 1991.

Derrick, W. R., and S. I. Grossman. *Elementary Differential Equations*, 4th ed. Reading, MA: Addison-Wesley, 1997.

Dettman, J. H. *Introduction to Linear Algebra and Differential Equations*. New York: Dover, 1986.

Goode, S. W. *Differential Equations and Linear Algebra*, 2d ed. Upper Saddle River, NJ: Prentice Hall, Inc., 2000.

Rabenstein, A. L. *Elementary Differential Equations with Linear Algebra*, 4th ed. Philadelphia: W. B. Saunders, 1992.

Key Terms

Differential equation
Homogeneous system of differential equations
Initial value problem
Initial condition
Fundamental system
Particular solution
Diffusion process
General solution
Eigenvectors

8.4 Exercises

1. Consider the linear system of differential equations

$$\begin{bmatrix} x_1' \\ x_2' \\ x_3' \end{bmatrix} = \begin{bmatrix} -3 & 0 & 0 \\ 0 & 4 & 0 \\ 0 & 0 & 2 \end{bmatrix} \begin{bmatrix} x_1 \\ x_2 \\ x_3 \end{bmatrix}.$$

(a) Find the general solution.

(b) Find the solution to the initial value problem determined by the initial conditions $x_1(0) = 3$, $x_2(0) = 4$, $x_3(0) = 5$.

2. Consider the linear system of differential equations

$$\begin{bmatrix} x_1' \\ x_2' \\ x_3' \end{bmatrix} = \begin{bmatrix} 1 & 0 & 0 \\ 0 & -2 & 1 \\ 0 & 0 & 3 \end{bmatrix} \begin{bmatrix} x_1 \\ x_2 \\ x_3 \end{bmatrix}.$$

(a) Find the general solution.

(b) Find the solution to the initial value problem determined by the initial conditions $x_1(0) = 2$, $x_2(0) = 7$, $x_3(0) = 20$.

3. Find the general solution to the linear system of differential equations

$$\begin{bmatrix} x_1' \\ x_2' \\ x_3' \end{bmatrix} = \begin{bmatrix} 4 & 0 & 0 \\ 3 & -5 & 0 \\ 2 & 1 & 2 \end{bmatrix} \begin{bmatrix} x_1 \\ x_2 \\ x_3 \end{bmatrix}.$$

4. Prove that the set of all solutions to the homogeneous linear system of differential equations $\mathbf{x}' = A\mathbf{x}$, where A is $n \times n$, is a subspace of the vector space of all differentiable real-valued n-vector functions. This subspace is called the **solution space** of the given linear system.

5. Find the general solution to the linear system of differential equations

$$\begin{bmatrix} x_1' \\ x_2' \\ x_3' \end{bmatrix} = \begin{bmatrix} 5 & 0 & 0 \\ 0 & -4 & 3 \\ 0 & 3 & 4 \end{bmatrix} \begin{bmatrix} x_1 \\ x_2 \\ x_3 \end{bmatrix}.$$

6. Find the general solution to the linear system of differential equations

$$\begin{bmatrix} x_1' \\ x_2' \end{bmatrix} = \begin{bmatrix} 3 & -2 \\ -2 & 3 \end{bmatrix} \begin{bmatrix} x_1 \\ x_2 \end{bmatrix}.$$

7. Find the general solution to the linear system of differential equations

$$\begin{bmatrix} x_1' \\ x_2' \\ x_3' \end{bmatrix} = \begin{bmatrix} -2 & -2 & 3 \\ 0 & -2 & 2 \\ 0 & 2 & 1 \end{bmatrix} \begin{bmatrix} x_1 \\ x_2 \\ x_3 \end{bmatrix}.$$

8. Find the general solution to the linear system of differential equations

$$\begin{bmatrix} x_1' \\ x_2' \\ x_3' \end{bmatrix} = \begin{bmatrix} 1 & 1 & 2 \\ 0 & 1 & 0 \\ 0 & 1 & 3 \end{bmatrix} \begin{bmatrix} x_1 \\ x_2 \\ x_3 \end{bmatrix}.$$

9. Consider two competing species that live in the same forest, and let $x_1(t)$ and $x_2(t)$ denote the respective populations of the species at time t. Suppose that the initial populations are $x_1(0) = 500$ and $x_2(0) = 200$. If the growth rates of the species are given by

$$\begin{aligned} x_1'(t) &= -3x_1(t) + 6x_2(t) \\ x_2'(t) &= x_1(t) - 2x_2(t), \end{aligned}$$

what is the population of each species at time t?

10. Suppose that we have a system consisting of two interconnected tanks, each containing a brine solution. Tank A contains $x(t)$ pounds of salt in 200 gallons of brine, and tank B contains $y(t)$ pounds of salt in 300 gallons of brine. The mixture in each tank is kept uniform by constant stirring. When $t = 0$, brine is pumped from tank A to tank B at 20 gallons/minute and from tank B to tank A at 20 gallons/minute. Find the amount of salt in each tank at time t if $x(0) = 10$ and $y(0) = 40$.

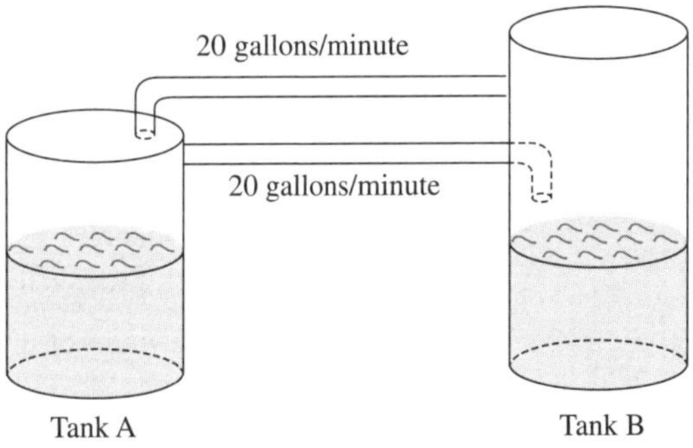

8.5 Dynamical Systems

In Section 8.4 we studied how to solve homogeneous linear systems of differential equations for which an initial condition had been specified. We called such systems initial value problems and wrote them in the form

$$\mathbf{x}'(t) = A\mathbf{x}(t), \qquad \mathbf{x}(0) = \mathbf{x}_0, \tag{1}$$

where

$$\mathbf{x}(t) = \begin{bmatrix} x_1(t) \\ x_2(t) \\ \vdots \\ x_n(t) \end{bmatrix}, \qquad A = \begin{bmatrix} a_{11} & a_{12} & a_{13} & \cdots & a_{1n} \\ a_{21} & a_{22} & a_{23} & \cdots & a_{2n} \\ a_{31} & a_{32} & a_{33} & \cdots & a_{3n} \\ \vdots & \vdots & \vdots & & \vdots \\ a_{n1} & a_{n2} & a_{n3} & \cdots & a_{nn} \end{bmatrix},$$

and $\mathbf{x}_0$ is a specified vector of constants. In the case that A was diagonalizable, we used the eigenvalues and eigenvectors of A to construct a particular solution to (1).

In this section we focus our attention on the case $n = 2$, and for ease of reference, we use x and y instead of x_1 and x_2. Such homogeneous linear systems of differential equations can be written in the form

$$\begin{aligned} \frac{dx}{dt} &= ax + by \\ \frac{dy}{dt} &= cx + dy, \end{aligned} \tag{2}$$

where a, b, c, and d are real constants, or

$$\mathbf{x}'(t) = \frac{d}{dt}\begin{bmatrix} x \\ y \end{bmatrix} = A\begin{bmatrix} x \\ y \end{bmatrix} = A\mathbf{x}(t), \tag{3}$$

where $A = \begin{bmatrix} a & b \\ c & d \end{bmatrix}$. For the systems (2) and (3), we try to describe properties of the solution based on the differential equation itself. This area is called the **qualitative theory** of differential equations and was studied extensively by J. H. Poincaré.*

The systems (2) and (3) are called **autonomous** differential equations, since the rates of change $\frac{dx}{dt}$ and $\frac{dy}{dt}$ explicitly depend on only the values of x and y, not on the independent variable t. For our purposes, we shall call the independent variable t **time**, and then (2) and (3) are said to be time-independent systems. Using this convention, the systems in (2) and (3) provide a model for the change of x and y as time goes by. Hence such systems are called **dynamical systems**. We use this terminology throughout our discussion in this section.

JULES HENRI POINCARÉ

*Jules Henri Poincaré (1854–1912) was born in Nancy, France, to a well-to-do family, many of whose members played key roles in the French government. As a youngster, he was clumsy and absent-minded, but showed great talent in mathematics. In 1873, he entered the École Polytéchnique, and in 1879, he received his doctorate from the University of Paris. He then began a university career, finally joining the University of Paris in 1881, where he remained until his death. Poincaré is considered the last of the universalists in mathematics—that is, someone who can work in all branches of mathematics, both pure and applied. His doctoral dissertation dealt with the existence of solutions to differential equations. In applied mathematics, he made contributions to the fields of optics, electricity, elasticity, thermodynamics, quantum mechanics, the theory of relativity, and cosmology. In pure mathematics, he was one of the principal creators of algebraic topology and made numerous contributions to algebraic geometry, analytic functions, and number theory. He was the first person to think of chaos in connection with his work in astronomy. In his later years, he wrote several books popularizing mathematics. In some of these books he dealt with the psychological processes involved in mathematical discovery and with the aesthetic aspects of mathematics.

A qualitative analysis of dynamical systems probes such questions as the following:

- Are there any constant solutions?
- If there are constant solutions, do nearby solutions move toward or away from the constant solution?
- What is the behavior of solutions as $t \to \pm\infty$?
- Are there any solutions that oscillate?

Each of these questions has a geometric flavor. Hence we introduce a helpful device for studying the behavior of dynamical systems. If we regard t, time, as a parameter, then $x = x(t)$ and $y = y(t)$ will represent a curve in the xy-plane. Such a curve is called a **trajectory**, or an **orbit**, of the systems (2) and (3). The xy-plane is called the **phase plane** of the dynamical system.

EXAMPLE 1 The system

$$\mathbf{x}'(t) = A\mathbf{x}(t) = \begin{bmatrix} 0 & 1 \\ -1 & 0 \end{bmatrix} \mathbf{x}(t) \qquad \text{or} \qquad \begin{aligned} \frac{dx}{dt} &= y \\ \frac{dy}{dt} &= -x \end{aligned}$$

has the general solution*

$$\begin{aligned} x &= b_1 \sin(t) + b_2 \cos(t) \\ y &= b_1 \cos(t) - b_2 \sin(t). \end{aligned} \tag{4}$$

It follows that the trajectories† satisfy $x^2 + y^2 = c^2$, where $c^2 = b_1^2 + b_2^2$. Hence the trajectories of this dynamical system are circles in the phase plane centered at the origin. We note that if an initial condition $\mathbf{x}(0) = \begin{bmatrix} k_1 \\ k_2 \end{bmatrix}$ is specified, then setting $t = 0$ in (4) gives the linear system

$$\begin{aligned} b_1 \sin(0) + b_2 \cos(0) &= k_1 \\ b_1 \cos(0) - b_2 \sin(0) &= k_2. \end{aligned}$$

It follows that the solution is $b_2 = k_1$ and $b_1 = k_2$, so the corresponding particular solution to the initial value problem $\mathbf{x}'(t) = A\mathbf{x}(t)$, $\mathbf{x}_0 = \begin{bmatrix} k_1 \\ k_2 \end{bmatrix}$ determines the trajectory $x^2 + y^2 = k_1^2 + k_2^2$; that is, a circle centered at the origin with radius $\sqrt{k_1^2 + k_2^2}$. ■

---

*We verify this later in the section.

†We can obtain the trajectories directly by noting that we can eliminate t to get $\frac{dy}{dx} = \frac{-x}{y}$. Then separating the variables gives $y\,dy = -x\,dx$, and upon integrating, we get $\frac{y^2}{2} = -\frac{x^2}{2} + k^2$, or equivalently, $x^2 + y^2 = c^2$.

A sketch of the trajectories of a dynamical system in the phase plane is called a **phase portrait**. A phase portrait usually contains the sketches of a few trajectories and an indication of the direction in which the curve is traversed. Arrowheads are placed on the trajectory to indicate the direction of motion of a point (x, y) as t increases. The direction is indicated by the **velocity vector**

$$\mathbf{v} = \begin{bmatrix} \dfrac{dx}{dt} \\ \dfrac{dy}{dt} \end{bmatrix}.$$

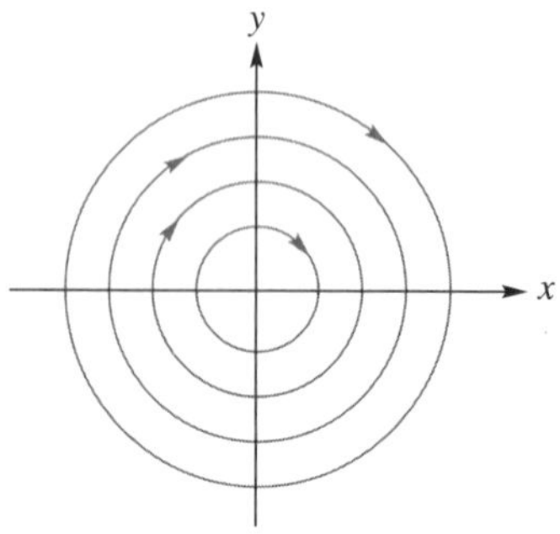

FIGURE 8.11

For the dynamical system in Example 1, we have $\mathbf{v} = \begin{bmatrix} y \\ -x \end{bmatrix}$. Thus in the phase plane for $x > 0$ and $y > 0$, the vector $\mathbf{v}$ is oriented downward to the right; hence these trajectories are traversed clockwise, as shown in Figure 8.11. (**Warning:** In other dynamical systems not all trajectories are traversed in the same direction.)

One of the questions posed earlier concerned the existence of constant solutions. For a dynamical system (which we regard as a model depicting the change of x and y as time goes by) to have a constant solution, both $\dfrac{dx}{dt}$ and $\dfrac{dy}{dt}$ must be zero. That is, the system doesn't change. It follows that points in the phase plane that correspond to constant solutions are determined by solving

$$\frac{dx}{dt} = ax + by = 0$$

$$\frac{dy}{dt} = cx + dy = 0,$$

which leads to the homogeneous linear system

$$A\mathbf{x} = \begin{bmatrix} a & b \\ c & d \end{bmatrix} \begin{bmatrix} x \\ y \end{bmatrix} = \begin{bmatrix} 0 \\ 0 \end{bmatrix}.$$

We know that one solution to this linear system is $x = 0$, $y = 0$ and that there exist other solutions if and only if $\det(A) = 0$. In Example 1,

$$A = \begin{bmatrix} 0 & 1 \\ -1 & 0 \end{bmatrix} \quad \text{and} \quad \det(A) = 1.$$

Thus for the dynamical system in Example 1, the only point in the phase plane that corresponds to a constant solution is $x = 0$ and $y = 0$, the origin.

DEFINITION 8.2 A point in the phase plane at which both $\dfrac{dx}{dt}$ and $\dfrac{dy}{dt}$ are zero is called an **equilibrium point**, or **fixed point**, of the dynamical system.

The behavior of trajectories near an equilibrium point serves as a way to characterize different types of equilibrium points. If trajectories through all points near an equilibrium point converge to the equilibrium point, then we say that the equilibrium point is **stable**, or an **attractor**. This is the case for the origin shown in

Figure 8.12, where the trajectories are all straight lines heading into the origin. The dynamical system whose phase portrait is shown in Figure 8.12 is

$$\frac{dx}{dt} = -x$$
$$\frac{dy}{dt} = -y,$$

which we discuss later in Example 3. Another situation is shown in Figure 8.11, where, again, the only equilibrium point is the origin. In this case, trajectories through points near the equilibrium point stay a small distance away. In such a case, the equilibrium point is called **marginally stable**. At other times, nearby trajectories tend to move away from an equilibrium point. In such cases, we say that the equilibrium point is **unstable**, or a **repelling** point. (See Figure 8.13.) In addition, we can have equilibrium points where nearby trajectories on one side move toward it and on the other side move away from it. Such an equilibrium point is called a **saddle point**. (See Figure 8.14.)

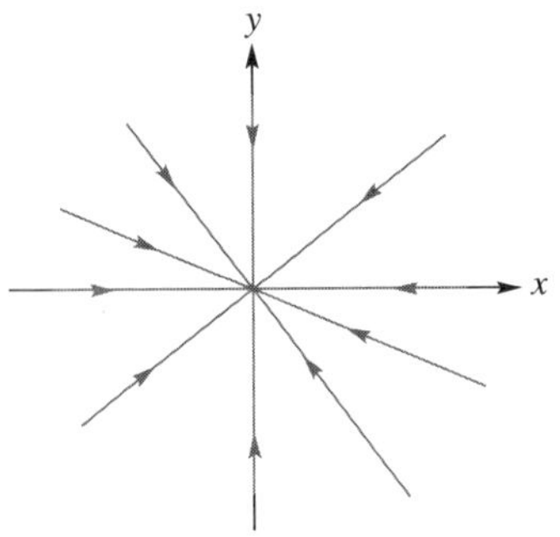

FIGURE 8.12 A stable equilibrium point at (0, 0).

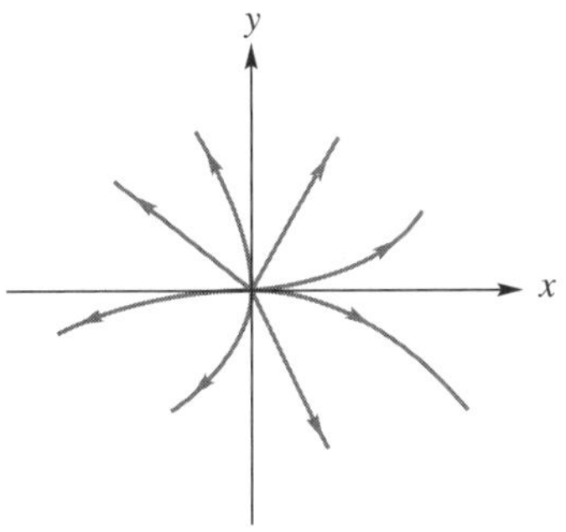

FIGURE 8.13 An unstable equilibrium point at (0, 0).

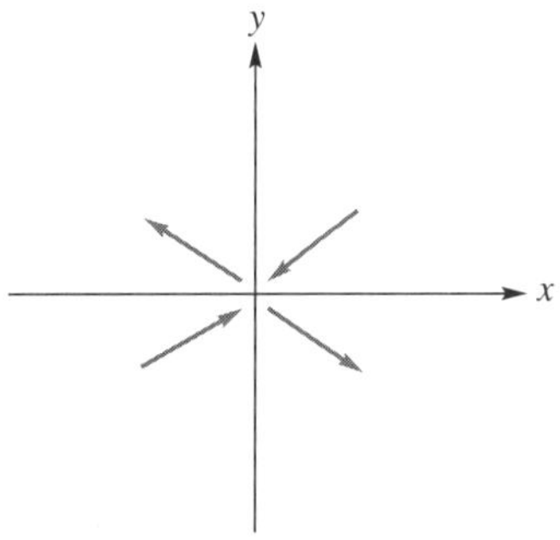

FIGURE 8.14 A saddle point at (0, 0).

From the developments in Section 8.4, we expect that the eigenvalues and eigenvectors of the coefficient matrix $A = \begin{bmatrix} a & b \\ c & d \end{bmatrix}$ of the dynamical system will determine features of the phase portrait of the system. From Equation (17) in Section 8.4, we have

$$\mathbf{x}(t) = \begin{bmatrix} x \\ y \end{bmatrix} = b_1\mathbf{p}_1e^{\lambda_1 t} + b_2\mathbf{p}_2e^{\lambda_2 t}, \tag{5}$$

where λ_1 and λ_2 are the eigenvalues of A, and $\mathbf{p}_1$ and $\mathbf{p}_2$ are associated eigenvectors. We also require that both λ_1 and λ_2 be nonzero.* Hence A is nonsingular. (Explain why.) And so the only equilibrium point is $x = 0$, $y = 0$, the origin.

*It can be shown that if both eigenvalues of A are zero, then all solutions to (2) as given in (5) are either constants or constants and straight lines. In addition, we can show that if one eigenvalue of A is zero and the other nonzero, then there is a line of equilibrium points. See Further Readings at the end of this section.

To show how we use the eigen information from A to determine the phase portrait, we treat the case of complex eigenvalues separately from the case of real eigenvalues.

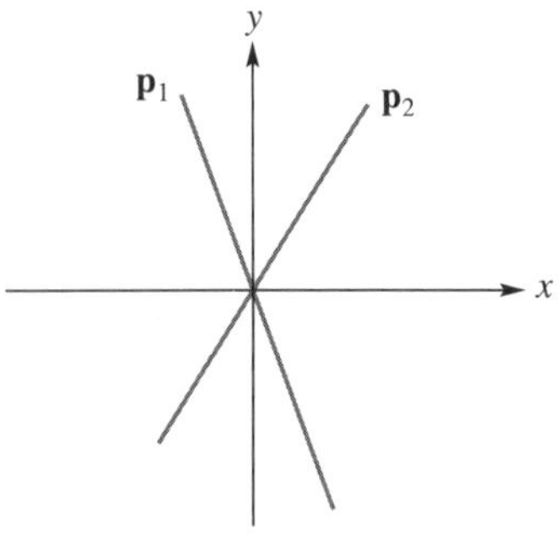

FIGURE 8.15

Case λ_1 and λ_2 Real

For real eigenvalues (and eigenvectors), the phase plane interpretation of Equation (5) is that $\mathbf{x}(t)$ is in span $\{\mathbf{p}_1, \mathbf{p}_2\}$. Hence $\mathbf{p}_1$ and $\mathbf{p}_2$ are trajectories. It follows that the eigenvectors $\mathbf{p}_1$ and $\mathbf{p}_2$ determine lines or rays through the origin in the phase plane, and a phase portrait for this case has the general form shown in Figure 8.15. To complete the portrait, we need more than the special trajectories corresponding to the eigen directions. These other trajectories depend on the values of λ_1 and λ_2.

Eigenvalues negative and distinct: $\lambda_1 < \lambda_2 < 0$

From (5), as $t \to \infty$, $\mathbf{x}(t)$ gets small. Hence all the trajectories tend toward the equilibrium point at the origin as $t \to \infty$. See Example 2 and Figure 8.16.

EXAMPLE 2

Determine the phase plane portrait of the dynamical system

$$\mathbf{x}'(t) = A\mathbf{x}(t) = \begin{bmatrix} -2 & -2 \\ 1 & -5 \end{bmatrix} \mathbf{x}(t).$$

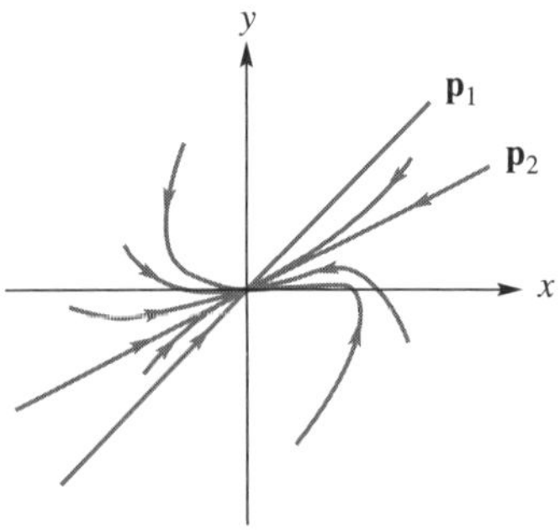

FIGURE 8.16

We begin by finding the eigenvalues and associated eigenvectors of A. We find (verify)

$$\lambda_1 = -4, \quad \lambda_2 = -3, \quad \text{and} \quad \mathbf{p}_1 = \begin{bmatrix} 1 \\ 1 \end{bmatrix}, \quad \mathbf{p}_2 = \begin{bmatrix} 2 \\ 1 \end{bmatrix}.$$

It follows that

$$\mathbf{x}(t) = \begin{bmatrix} x \\ y \end{bmatrix} = b_1\mathbf{p}_1 e^{-4t} + b_2\mathbf{p}_2 e^{-3t},$$

and as $t \to \infty$, $\mathbf{x}(t)$ gets small. It is helpful to rewrite this expression in the form

$$\mathbf{x}(t) = \begin{bmatrix} x \\ y \end{bmatrix} = b_1\mathbf{p}_1 e^{-4t} + b_2\mathbf{p}_2 e^{-3t} = e^{-3t}(b_1\mathbf{p}_1 e^{-t} + b_2\mathbf{p}_2).$$

As long as $b_2 \neq 0$, the term $b_1\mathbf{p}_1 e^{-t}$ is negligible in comparison to $b_2\mathbf{p}_2$. This implies that, as $t \to \infty$, all trajectories, except those starting on $\mathbf{p}_1$, will align themselves in the direction of $\mathbf{p}_2$ as they get close to the origin. Hence the phase portrait appears like that given in Figure 8.16. The origin is an attractor. ■

Eigenvalues positive and distinct: $\lambda_1 > \lambda_2 > 0$

From (5), as $t \to \infty$, $\mathbf{x}(t)$ gets large. Hence all the trajectories tend to go away from the equilibrium point at the origin. The phase portrait for such dynamical systems is like that in Figure 8.16, except that all the arrowheads are reversed, indicating motion away from the origin. In this case $(0, 0)$ is called an **unstable** equilibrium point.

Both eigenvalues negative, but equal: $\lambda_1 = \lambda_2 < 0$

All trajectories go to a stable equilibrium at the origin, but they may bend differently than the trajectories depicted in Figure 8.16. Their behavior depends upon the number of linearly independent eigenvectors of matrix A. If there are two linearly independent eigenvectors, then $\mathbf{x}(t) = e^{\lambda_1 t}(b_1\mathbf{p}_1 + b_2\mathbf{p}_2)$, which is a multiple of the constant vector $b_1\mathbf{p}_1 + b_2\mathbf{p}_2$. Thus it follows that all the trajectories are lines through the origin, and since $\lambda_1 < 0$, motion along them is toward the origin. See Figure 8.17. We illustrate this in Example 3.

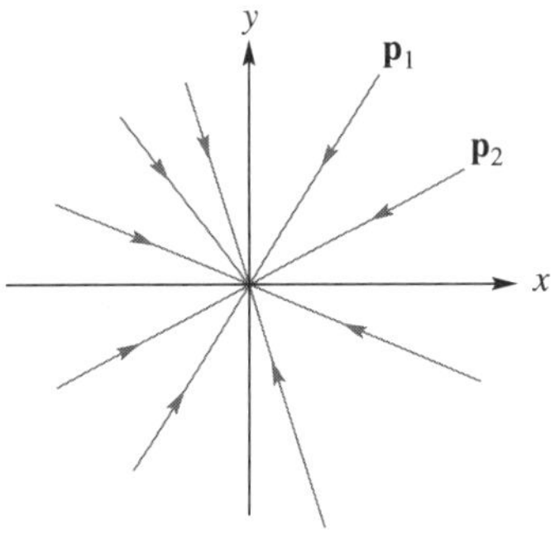

FIGURE 8.17

EXAMPLE 3

The matrix A of the dynamical system

$$\mathbf{x}'(t) = A\mathbf{x}(t) = \begin{bmatrix} -1 & 0 \\ 0 & -1 \end{bmatrix}\mathbf{x}(t)$$

has eigenvalues $\lambda_1 = \lambda_2 = -1$ and corresponding eigenvectors

$$\mathbf{p}_1 = \begin{bmatrix} 1 \\ 0 \end{bmatrix} \quad \text{and} \quad \mathbf{p}_2 = \begin{bmatrix} 0 \\ 1 \end{bmatrix} \qquad \text{(verify).}$$

It follows that

$$\mathbf{x}(t) = e^{-t}(b_1\mathbf{p}_1 + b_2\mathbf{p}_2) = e^{-t}\begin{bmatrix} b_1 \\ b_2 \end{bmatrix},$$

so $x = b_1e^{-t}$ and $y = b_2e^{-t}$. If $b_1 \neq 0$, then $y = \dfrac{b_2}{b_1}x$. If $b_1 = 0$, then we are on the trajectory in the direction of $\mathbf{p}_2$. It follows that all trajectories are straight lines through the origin, as in Figures 8.12 and 8.17. ■

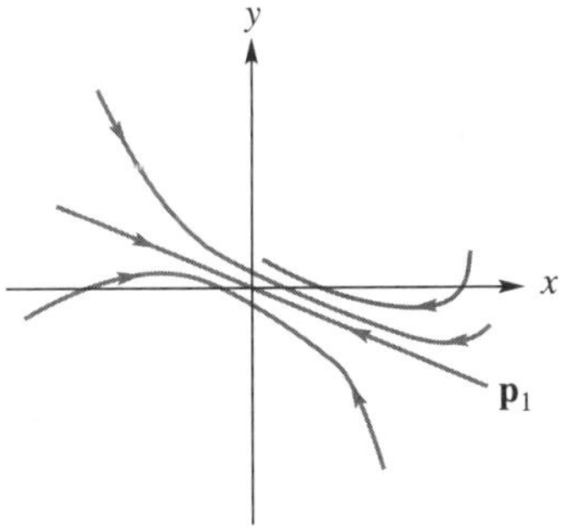

FIGURE 8.18

If there is only one linearly independent eigenvector, then it can be shown that all trajectories passing through points not on the eigenvector align themselves to be tangent to the eigenvector at the origin. We will not develop this case, but the phase portrait is similar to Figure 8.18. In the case that the eigenvalues are positive and equal, the phase portraits for these two cases are like Figures 8.17 and 8.18 with the arrowheads reversed.

One positive eigenvalue and one negative eigenvalue: $\lambda_1 < 0 < \lambda_2$

From (5), as $t \to \infty$, one of the terms is increasing while the other term is decreasing. This causes a trajectory that is not in the direction of an eigenvector to head toward the origin, but bend away as t gets larger. The origin in this case is called a **saddle point**. The phase portrait resembles Figure 8.19.

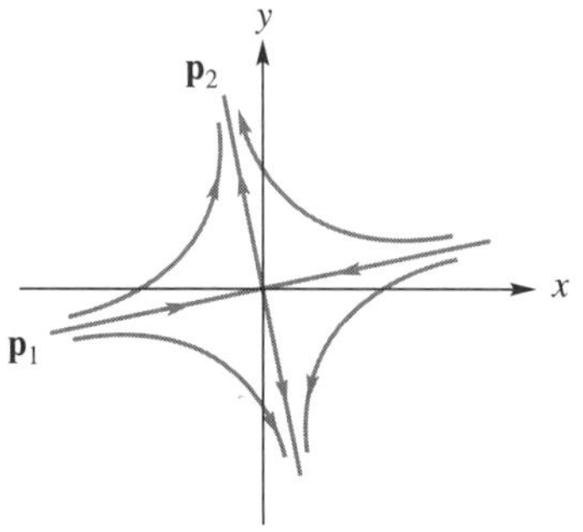

FIGURE 8.19

EXAMPLE 4 Determine the phase plane portrait of the dynamical system

$$\mathbf{x}'(t) = A\mathbf{x}(t) = \begin{bmatrix} 1 & -1 \\ -2 & 0 \end{bmatrix} \mathbf{x}(t).$$

We begin by finding the eigenvalues and associated eigenvectors of A. We find (verify)

$$\lambda_1 = -1, \quad \lambda_2 = 2, \quad \text{and} \quad \mathbf{p}_1 = \begin{bmatrix} 1 \\ 2 \end{bmatrix}, \quad \mathbf{p}_2 = \begin{bmatrix} 1 \\ -1 \end{bmatrix}.$$

It follows that the origin is a saddle point and that we have

$$\mathbf{x}(t) = \begin{bmatrix} x \\ y \end{bmatrix} = b_1\mathbf{p}_1e^{-t} + b_2\mathbf{p}_2e^{2t}.$$

We see that if $b_1 \neq 0$ and $b_2 = 0$, then the motion is in the direction of eigenvector $\mathbf{p}_1$ and toward the origin. Similarly, if $b_1 = 0$ and $b_2 \neq 0$, then the motion is in the direction of $\mathbf{p}_2$, but away from the origin. If we look at the components of the original system and eliminate t, we obtain (verify)

$$\frac{dy}{dx} = \frac{2x}{y - x}.$$

This expression tells us the slope along trajectories in the phase plane. Inspecting this expression, we see the following (explain):

- All trajectories crossing the y-axis have horizontal tangents.
- As a trajectory crosses the line $y = x$, it has a vertical tangent.
- Whenever a trajectory crosses the x-axis, it has slope -2.

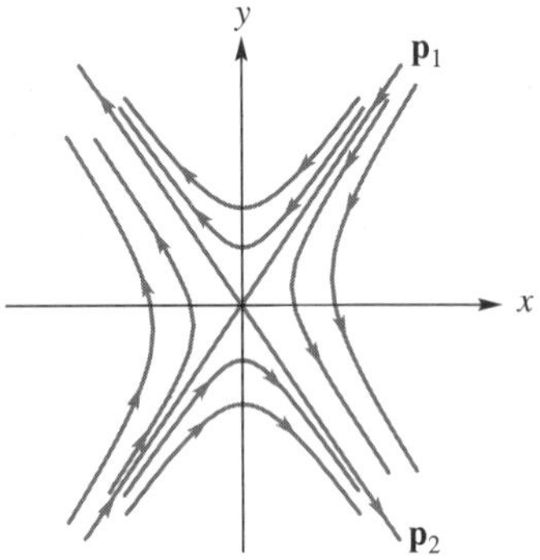

FIGURE 8.20

Using the general form of a saddle point, shown in Figure 8.19, we can produce quite an accurate phase portrait for this dynamical system, shown in Figure 8.20. ■

Both Eigenvalues Complex Numbers

For real 2×2 matrices A, the characteristic equation $\det(\lambda I_2 - A) = 0$ is a quadratic polynomial. If the roots of this quadratic equation λ_1 and λ_2 are complex numbers, then they are conjugates of one another. (See Appendix B.2.) If $\lambda_1 = \alpha + \beta i$, where α and β are real numbers with $\beta \neq 0$, then $\lambda_2 = \overline{\lambda_1} = \alpha - \beta i$. Hence in Equation (5) we have the exponential of a complex number:

$$e^{\lambda_1 t} = e^{(\alpha+\beta i)t} = e^{\alpha t} e^{\beta i t}.$$

The term $e^{\alpha t}$ is a standard exponential function, but $e^{\beta i t}$ is quite different, since $i = \sqrt{-1}$. Fortunately, there is a simple way to express such an exponential function in terms of more manageable functions. We use Euler's identity,

$$e^{i\theta} = \cos(\theta) + i\sin(\theta),$$

which we state without proof. By Euler's identity, we have

$$e^{\lambda_1 t} = e^{(\alpha+\beta i)t} = e^{\alpha t} e^{\beta i t} = e^{\alpha t}(\cos(\beta t) + i\sin(\beta t)),$$

and

$$\begin{aligned} e^{\lambda_2 t} = e^{(\alpha-\beta i)t} &= e^{\alpha t} e^{-\beta i t} \\ &= e^{\alpha t}(\cos(-\beta t) + i\sin(-\beta t)) \\ &= e^{\alpha t}(\cos(\beta t) - i\sin(\beta t)). \end{aligned}$$

It can be shown that the system given in (5) can be written so that the components $x(t)$ and $y(t)$ are linear combinations of $e^{\alpha t}\cos(\beta t)$ and $e^{\alpha t}\sin(\beta t)$ with real coefficients. The behavior of the trajectories can now be analyzed by considering the sign of α, since $\beta \neq 0$.

Complex eigenvalues: $\lambda_1 = \alpha + \beta i$ and $\lambda_2 = \alpha - \beta i$ with $\alpha = 0$, $\beta \neq 0$

For this case, $x(t)$ and $y(t)$ are linear combinations of $\cos(\beta t)$ and $\sin(\beta t)$. It can be shown that the trajectories are ellipses whose major and minor axes are determined by the eigenvectors. (See Example 1 for a particular case.) The motion is periodic, since the trajectories are closed curves. The origin is a marginally stable equilibrium point, since the trajectories through points near the origin do not move very far away. (See Figure 8.11.)

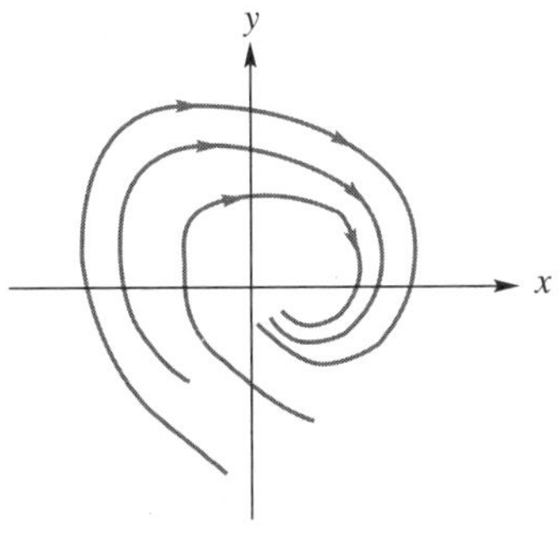

FIGURE 8.21

Complex eigenvalues: $\lambda_1 = \alpha + \beta i$ and $\lambda_2 = \alpha - \beta i$ with $\alpha \neq 0$, $\beta \neq 0$

For this case, $x(t)$ and $y(t)$ are linear combinations of $e^{\alpha t}\cos(\beta t)$ and $e^{\alpha t}\sin(\beta t)$. It can be shown that the trajectories are spirals. If $\alpha > 0$, then the spiral goes outward, away from the origin. Thus the origin is an unstable equilibrium point. If $\alpha < 0$, then the spiral goes inward, toward the origin, and so the origin is a stable equilibrium point. The phase portrait is a collection of spirals as shown in Figure 8.21, with arrowheads appropriately affixed.

The dynamical system in (2) may appear quite special. However, the experience gained from the qualitative analysis we have performed is the key to understanding the behavior of nonlinear dynamical systems of the form

$$\frac{dx}{dt} = f(x, y)$$

$$\frac{dy}{dt} = g(x, y).$$

The extension of the phase plane and phase portraits to such nonlinear systems is beyond the scope of this brief introduction. Such topics are part of courses on differential equations and can be found in the books listed in Further Readings.

FURTHER READINGS

Boyce, W. E. *Elementary Differential Equations*, 9th ed. New York: John Wiley & Sons, Inc., 2007.

Campbell, S. L. *An Introduction to Differential Equations and Their Applications*, 2d ed. Belmont, CA: Wadsworth Publishing Co., 1990.

Farlow, S. J. *An Introduction to Differential Equations and Their Applications*. New York: McGraw-Hill, Inc., 1994.

Key Terms

Qualitative theory
Autonomous differential equation
Dynamical system
Phase portrait
Equilibrium point
Fixed point
Attractor
Marginally stable
Unstable point
Repelling point
Saddle point
Eigenvalues
Complex eigenvalues

8.5 Exercises

For each of the dynamical systems in Exercises 1 through 10, determine the nature of the equilibrium point at the origin, and describe the phase portrait.

1. $\mathbf{x}'(t) = \begin{bmatrix} -1 & 0 \\ 0 & -3 \end{bmatrix} \mathbf{x}(t)$

2. $\mathbf{x}'(t) = \begin{bmatrix} 2 & 0 \\ 0 & 1 \end{bmatrix} \mathbf{x}(t)$

3. $\mathbf{x}'(t) = \begin{bmatrix} -1 & 2 \\ 0 & -1 \end{bmatrix} \mathbf{x}(t)$

4. $\mathbf{x}'(t) = \begin{bmatrix} -2 & 0 \\ 3 & 1 \end{bmatrix} \mathbf{x}(t)$

5. $\mathbf{x}'(t) = \begin{bmatrix} 1 & 1 \\ 3 & -1 \end{bmatrix} \mathbf{x}(t)$

6. $\mathbf{x}'(t) = \begin{bmatrix} -1 & -1 \\ 1 & -1 \end{bmatrix} \mathbf{x}(t)$

7. $\mathbf{x}'(t) = \begin{bmatrix} -2 & 1 \\ 2 & -3 \end{bmatrix} \mathbf{x}(t)$

8. $\mathbf{x}'(t) = \begin{bmatrix} -2 & 1 \\ -1 & -2 \end{bmatrix} \mathbf{x}(t)$

9. $\mathbf{x}'(t) = \begin{bmatrix} 3 & -13 \\ 1 & -3 \end{bmatrix} \mathbf{x}(t)$

10. $\mathbf{x}'(t) = \begin{bmatrix} 3 & 2 \\ 2 & 3 \end{bmatrix} \mathbf{x}(t)$

8.6 Real Quadratic Forms

In your precalculus and calculus courses you have seen that the graph of the equation

$$ax^2 + 2bxy + cy^2 = d, \tag{1}$$

where a, b, c, and d are real numbers, is a **conic section** centered at the origin of a rectangular Cartesian coordinate system in two-dimensional space. Similarly, the graph of the equation

$$ax^2 + 2dxy + 2exz + by^2 + 2fyz + cz^2 = g, \tag{2}$$

where a, b, c, d, e, f, and g are real numbers, is a **quadric surface** centered at the origin of a rectangular Cartesian coordinate system in three-dimensional space. If a conic section or quadric surface is not centered at the origin, its equations are more complicated than those given in (1) and (2).

The identification of the conic section or quadric surface that is the graph of a given equation often requires the rotation and translation of the coordinate axes. These methods can best be understood as an application of eigenvalues and eigenvectors of matrices, discussed in Sections 8.7 and 8.8.

The expressions on the left sides of Equations (1) and (2) are examples of quadratic forms. Quadratic forms arise in statistics, mechanics, and in other areas of physics; in quadratic programming; in the study of maxima and minima of functions of several variables; and in other applied problems. In this section we use our results on eigenvalues and eigenvectors of matrices to give a brief treatment of real quadratic forms in n variables. In Section 8.7 we apply these results to the classification of the conic sections, and in Section 8.8 to the classification of the quadric surfaces.

DEFINITION 8.3

If A is a symmetric matrix, then the function $g\colon R^n \to R^1$ (a real-valued function on R^n) defined by

$$g(\mathbf{x}) = \mathbf{x}^T A\mathbf{x},$$

where

$$\mathbf{x} = \begin{bmatrix} x_1 \\ x_2 \\ \vdots \\ x_n \end{bmatrix},$$

is called a **real quadratic form in the n variables** $x_1, x_2, \dots, x_n$. The matrix A is called the **matrix of the quadratic form g**. We shall also denote the quadratic form by $g(\mathbf{x})$.

EXAMPLE 1

Write the left side of (1) as the quadratic form in the variables x and y.

Solution

Let

$$\mathbf{x} = \begin{bmatrix} x \\ y \end{bmatrix} \quad \text{and} \quad A = \begin{bmatrix} a & b \\ b & c \end{bmatrix}.$$

Then the left side of (1) is the quadratic form

$$g(\mathbf{x}) = \mathbf{x}^T A\mathbf{x}.$$ ■

EXAMPLE 2 Write the left side of (2) as the quadratic form in the variables x, y, and z.

Solution

Let

$$\mathbf{x} = \begin{bmatrix} x \\ y \\ z \end{bmatrix} \quad \text{and} \quad A = \begin{bmatrix} a & d & e \\ d & b & f \\ e & f & c \end{bmatrix}.$$

Then the left side of (2) is the quadratic form

$$g(\mathbf{x}) = \mathbf{x}^T A\mathbf{x}.$$ ■

EXAMPLE 3 The following expressions are quadratic forms:

(a) $3x^2 - 5xy - 7y^2 = \begin{bmatrix} x & y \end{bmatrix} \begin{bmatrix} 3 & -\frac{5}{2} \\ -\frac{5}{2} & -7 \end{bmatrix} \begin{bmatrix} x \\ y \end{bmatrix}$

(b) $3x^2 - 7xy + 5xz + 4y^2 - 4yz - 3z^2 = \begin{bmatrix} x & y & z \end{bmatrix} \begin{bmatrix} 3 & -\frac{7}{2} & \frac{5}{2} \\ -\frac{7}{2} & 4 & -2 \\ \frac{5}{2} & -2 & -3 \end{bmatrix} \begin{bmatrix} x \\ y \\ z \end{bmatrix}$ ■

Suppose now that $g(\mathbf{x}) = \mathbf{x}^T A\mathbf{x}$ is a quadratic form. To simplify the quadratic form, we change from the variables $x_1, x_2, \dots, x_n$ to the variables $y_1, y_2, \dots, y_n$, where we assume that the old variables are related to the new variables by $\mathbf{x} = P\mathbf{y}$ for some orthogonal matrix P. Then

$$g(\mathbf{x}) = \mathbf{x}^T A\mathbf{x} = (P\mathbf{y})^T A(P\mathbf{y}) = \mathbf{y}^T (P^T AP)\mathbf{y} = \mathbf{y}^T B\mathbf{y},$$

where $B = P^T AP$. We shall let you verify that if A is a symmetric matrix, then $P^T AP$ is also symmetric (Exercise 25). Thus

$$h(\mathbf{y}) = \mathbf{y}^T B\mathbf{y}$$

is another quadratic form and $g(\mathbf{x}) = h(\mathbf{y})$.

This situation is important enough to formulate the following definitions:

DEFINITION 8.4 If A and B are $n \times n$ matrices, we say that B is **congruent** to A if $B = P^T AP$ for a nonsingular matrix P.

In light of Exercise 26, "A is congruent to B" and "B is congruent to A" can both be replaced by "A and B are congruent."

DEFINITION 8.5 Two quadratic forms g and h with matrices A and B, respectively, are said to be **equivalent** if A and B are congruent.

The congruence of matrices and equivalence of forms are more general concepts than the notion of similarity, since the matrix P is required only to be nonsingular (not necessarily orthogonal). We consider here the more restrictive situation, with P orthogonal.

EXAMPLE 4

Consider the quadratic form in the variables x and y defined by

$$g(\mathbf{x}) = 2x^2 + 2xy + 2y^2 = \begin{bmatrix} x & y \end{bmatrix} \begin{bmatrix} 2 & 1 \\ 1 & 2 \end{bmatrix} \begin{bmatrix} x \\ y \end{bmatrix}. \tag{3}$$

We now change from the variables x and y to the variables x' and y'. Suppose that the old variables are related to the new variables by the equations

$$x = \frac{1}{\sqrt{2}}x' - \frac{1}{\sqrt{2}}y' \quad \text{and} \quad y = \frac{1}{\sqrt{2}}x' + \frac{1}{\sqrt{2}}y', \tag{4}$$

which can be written in matrix form as

$$\mathbf{x} = \begin{bmatrix} x \\ y \end{bmatrix} = \begin{bmatrix} \frac{1}{\sqrt{2}} & -\frac{1}{\sqrt{2}} \\ \frac{1}{\sqrt{2}} & \frac{1}{\sqrt{2}} \end{bmatrix} \begin{bmatrix} x' \\ y' \end{bmatrix} = P\mathbf{y},$$

where the orthogonal (hence nonsingular) matrix

$$P = \begin{bmatrix} \frac{1}{\sqrt{2}} & -\frac{1}{\sqrt{2}} \\ \frac{1}{\sqrt{2}} & \frac{1}{\sqrt{2}} \end{bmatrix} \quad \text{and} \quad \mathbf{y} = \begin{bmatrix} x' \\ y' \end{bmatrix}.$$

We shall soon see why and how this particular matrix P was selected. Substituting in (3), we obtain

$$\begin{aligned}
g(\mathbf{x}) &= \mathbf{x}^T A\mathbf{x} = (P\mathbf{y})^T A(P\mathbf{y}) = \mathbf{y}^T P^T A P\mathbf{y} \\
&= \begin{bmatrix} x' & y' \end{bmatrix} \begin{bmatrix} \frac{1}{\sqrt{2}} & -\frac{1}{\sqrt{2}} \\ \frac{1}{\sqrt{2}} & \frac{1}{\sqrt{2}} \end{bmatrix}^T \begin{bmatrix} 2 & 1 \\ 1 & 2 \end{bmatrix} \begin{bmatrix} \frac{1}{\sqrt{2}} & -\frac{1}{\sqrt{2}} \\ \frac{1}{\sqrt{2}} & \frac{1}{\sqrt{2}} \end{bmatrix} \begin{bmatrix} x' \\ y' \end{bmatrix} \\
&= \begin{bmatrix} x' & y' \end{bmatrix} \begin{bmatrix} 3 & 0 \\ 0 & 1 \end{bmatrix} \begin{bmatrix} x' \\ y' \end{bmatrix} = h(\mathbf{y}) \\
&= 3x'^2 + y'^2.
\end{aligned}$$

Thus the matrices

$$\begin{bmatrix} 2 & 1 \\ 1 & 2 \end{bmatrix} \quad \text{and} \quad \begin{bmatrix} 3 & 0 \\ 0 & 1 \end{bmatrix}$$

are congruent, and the quadratic forms g and h are equivalent. ■

We now turn to the question of how to select the matrix P.

Theorem 8.9 **Principal Axes Theorem**

Any quadratic form in n variables $g(\mathbf{x}) = \mathbf{x}^T A\mathbf{x}$ is equivalent by means of an orthogonal matrix P to a quadratic form, $h(\mathbf{y}) = \lambda_1 y_1^2 + \lambda_2 y_2^2 + \cdots + \lambda_n y_n^2$, where

$$\mathbf{y} = \begin{bmatrix} y_1 \\ y_2 \\ \vdots \\ y_n \end{bmatrix}$$

and $\lambda_1, \lambda_2, \ldots, \lambda_n$ are the eigenvalues of the matrix A of g.

Proof

If A is the matrix of g, then since A is symmetric, we know by Theorem 7.9 that A can be diagonalized by an orthogonal matrix. This means that there exists an orthogonal matrix P such that $D = P^{-1}AP$ is a diagonal matrix. Since P is orthogonal, $P^{-1} = P^T$, so $D = P^T AP$. Moreover, the elements on the main diagonal of D are the eigenvalues $\lambda_1, \lambda_2, \ldots, \lambda_n$ of A, which are real numbers. The quadratic form h with matrix D is given by

$$h(\mathbf{y}) = \lambda_1 y_1^2 + \lambda_2 y_2^2 + \cdots + \lambda_n y_n^2;$$

g and h are equivalent. ■

EXAMPLE 5

Consider the quadratic form g in the variables x, y, and z defined by

$$g(\mathbf{x}) = 2x^2 + 4y^2 + 6yz - 4z^2.$$

Determine a quadratic form h of the form in Theorem 8.9 to which g is equivalent.

Solution

The matrix of g is

$$A = \begin{bmatrix} 2 & 0 & 0 \\ 0 & 4 & 3 \\ 0 & 3 & -4 \end{bmatrix},$$

and the eigenvalues of A are

$$\lambda_1 = 2, \quad \lambda_2 = 5, \quad \text{and} \quad \lambda_3 = -5 \qquad \text{(verify)}.$$

Let h be the quadratic form in the variables x', y', and z' defined by

$$h(\mathbf{y}) = 2x'^2 + 5y'^2 - 5z'^2.$$

Then g and h are equivalent by means of some orthogonal matrix. Note that $\hat{h}(\mathbf{y}) = -5x'^2 + 2y'^2 + 5z'^2$ is also equivalent to g. ■

Observe that to apply Theorem 8.9 to diagonalize a given quadratic form, as shown in Example 5, we do not need to know the eigenvectors of A (nor the matrix P); we require only the eigenvalues of A.

To understand the significance of Theorem 8.9, we consider quadratic forms in two and three variables. As we have already observed at the beginning of this section, the graph of the equation

$$g(\mathbf{x}) = \mathbf{x}^T A\mathbf{x} = 1,$$

where $\mathbf{x}$ is a vector in R^2 and A is a symmetric 2×2 matrix, is a conic section centered at the origin of the xy-plane. From Theorem 8.9 it follows that there is a Cartesian coordinate system in the xy-plane with respect to which the equation of this conic section is

$$ax'^2 + by'^2 = 1,$$

where a and b are real numbers. Similarly, the graph of the equation

$$g(\mathbf{x}) = \mathbf{x}^T A\mathbf{x} = 1,$$

where $\mathbf{x}$ is a vector in R^3 and A is a symmetric 3×3 matrix, is a quadric surface centered at the origin of the xyz Cartesian coordinate system. From Theorem 8.9 it follows that there is a Cartesian coordinate system in 3-space with respect to which the equation of the quadric surface is

$$ax'^2 + by'^2 + cz'^2 = 1,$$

where a, b, and c are real numbers. The principal axes of the conic section or quadric surface lie along the new coordinate axes, and this is the reason for calling Theorem 8.9 the **Principal Axes Theorem**.

EXAMPLE 6

Consider the conic section whose equation is

$$g(\mathbf{x}) = 2x^2 + 2xy + 2y^2 = 9.$$

From Example 4 it follows that this conic section can also be described by the equation

$$h(\mathbf{y}) = 3x'^2 + y'^2 = 9,$$

which can be rewritten as

$$\frac{x'^2}{3} + \frac{y'^2}{9} = 1.$$

The graph of this equation is an ellipse (Figure 8.22) whose major axis is along the y'-axis. The major axis is of length 6; the minor axis is of length $2\sqrt{3}$. We now note that there is a very close connection between the eigenvectors of the matrix of (3) and the location of the x'- and y'-axes.

Since $\mathbf{x} = P\mathbf{y}$, we have $\mathbf{y} = P^{-1}\mathbf{x} = P^T\mathbf{x} = P\mathbf{x}$. ($P$ is orthogonal and, in this example, also symmetric.) Thus

$$x' = \frac{1}{\sqrt{2}}x + \frac{1}{\sqrt{2}}y \quad \text{and} \quad y' = -\frac{1}{\sqrt{2}}x + \frac{1}{\sqrt{2}}y.$$

This means that, in terms of the x- and y-axes, the x'-axis lies along the vector

$$\mathbf{x}_1 = \begin{bmatrix} \frac{1}{\sqrt{2}} \\ \frac{1}{\sqrt{2}} \end{bmatrix},$$

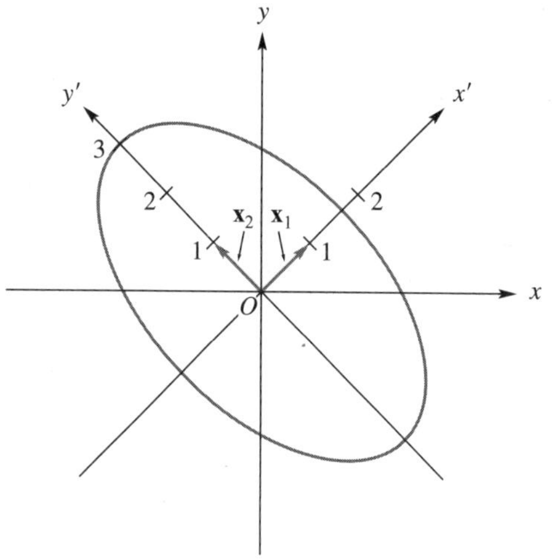

FIGURE 8.22

and the y'-axis lies along the vector

$$\mathbf{x}_2 = \begin{bmatrix} -\frac{1}{\sqrt{2}} \\ \frac{1}{\sqrt{2}} \end{bmatrix}.$$

Now $\mathbf{x}_1$ and $\mathbf{x}_2$ are the columns of the matrix

$$P = \begin{bmatrix} \frac{1}{\sqrt{2}} & -\frac{1}{\sqrt{2}} \\ \frac{1}{\sqrt{2}} & \frac{1}{\sqrt{2}} \end{bmatrix},$$

which in turn are eigenvectors of the matrix of (3). Thus the x'- and y'-axes lie along the eigenvectors of the matrix of (3). (See Figure 8.22.) ■

The situation described in Example 6 is true in general. That is, the principal axes of a conic section or quadric surface lie along the eigenvectors of the matrix of the quadratic form.

Let $g(\mathbf{x}) = \mathbf{x}^T A\mathbf{x}$ be a quadratic form in n variables. Then we know that g is equivalent to the quadratic form

$$h(\mathbf{y}) = \lambda_1 y_1^2 + \lambda_2 y_2^2 + \cdots + \lambda_n y_n^2,$$

where $\lambda_1, \lambda_2, \ldots, \lambda_n$ are eigenvalues of the symmetric matrix A of g, and hence are all real. We can label the eigenvalues so that all the positive eigenvalues of A, if any, are listed first, followed by all the negative eigenvalues, if any, followed by the zero eigenvalues, if any. Thus let $\lambda_1, \lambda_2, \ldots, \lambda_p$ be positive, $\lambda_{p+1}, \lambda_{p+2}, \ldots, \lambda_r$ be negative, and $\lambda_{r+1}, \lambda_{r+2}, \ldots, \lambda_n$ be zero. We now define the diagonal matrix

H whose entries on the main diagonal are

$$\frac{1}{\sqrt{\lambda_1}},\ \frac{1}{\sqrt{\lambda_2}},\ \ldots,\ \frac{1}{\sqrt{\lambda_p}},\ \frac{1}{\sqrt{-\lambda_{p+1}}},\ \frac{1}{\sqrt{-\lambda_{p+2}}},\ \ldots,\ \frac{1}{\sqrt{-\lambda_r}},\ 1,\ 1,\ \ldots,\ 1,$$

with $n-r$ ones. Let D be the diagonal matrix whose entries on the main diagonal are $\lambda_1, \lambda_2, \ldots, \lambda_p, \lambda_{p+1}, \ldots, \lambda_r, \lambda_{r+1}, \ldots, \lambda_n$; A and D are congruent. Let $D_1 = H^T DH$ be the matrix whose diagonal elements are $1, 1, \ldots, 1, -1, \ldots, -1, 0, 0, \ldots, 0$ (p ones, $r-p-1$ negative ones, and $n-r$ zeros); D and D_1 are then congruent. From Exercise 26 it follows that A and D_1 are congruent. In terms of quadratic forms, we have established Theorem 8.10.

Theorem 8.10 A quadratic form $g(\mathbf{x}) = \mathbf{x}^T A\mathbf{x}$ in n variables is equivalent to a quadratic form

$$h(\mathbf{y}) = y_1^2 + y_2^2 + \cdots + y_p^2 - y_{p+1}^2 - y_{p+2}^2 - \cdots - y_r^2. \qquad ■$$

It is clear that the rank of the matrix D_1 is r, the number of nonzero entries on its main diagonal. Now it can be shown (see J. L. Goldberg, *Matrix Theory with Applications*, New York: McGraw-Hill, Inc., 1991) that congruent matrices have equal ranks. Since the rank of D_1 is r, the rank of A is also r. We also refer to r as the **rank** of the quadratic form g whose matrix is A. It can be shown (see the book by Goldberg cited earlier) that the number p of positive terms in the quadratic form h of Theorem 8.10 is unique; that is, no matter how we simplify the given quadratic form g to obtain an equivalent quadratic form, the latter will always have p positive terms. Hence the quadratic form h in Theorem 8.10 is unique; it is often called the **canonical form** of a quadratic form in n variables. The difference between the number of positive eigenvalues and the number of negative eigenvalues is $s = p - (r - p) = 2p - r$ and is called the **signature** of the quadratic form. Thus, if g and h are equivalent quadratic forms, then they have equal ranks and signatures. However, it can also be shown (see the book by Goldberg cited earlier) that if g and h have equal ranks and signatures, then they are equivalent.

EXAMPLE 7 Consider the quadratic form in x_1, x_2, x_3, given by

$$g(\mathbf{x}) = 3x_2^2 + 8x_2x_3 - 3x_3^2 = \mathbf{x}^T A\mathbf{x} = \begin{bmatrix} x_1 & x_2 & x_3 \end{bmatrix} \begin{bmatrix} 0 & 0 & 0 \\ 0 & 3 & 4 \\ 0 & 4 & -3 \end{bmatrix} \begin{bmatrix} x_1 \\ x_2 \\ x_3 \end{bmatrix}.$$

The eigenvalues of A are (verify)

$$\lambda_1 = 5, \quad \lambda_2 = -5, \quad \text{and} \quad \lambda_3 = 0.$$

In this case A is congruent to

$$D = \begin{bmatrix} 5 & 0 & 0 \\ 0 & -5 & 0 \\ 0 & 0 & 0 \end{bmatrix}.$$

If we let

$$H = \begin{bmatrix} \frac{1}{\sqrt{5}} & 0 & 0 \\ 0 & \frac{1}{\sqrt{5}} & 0 \\ 0 & 0 & 1 \end{bmatrix},$$

then

$$D_1 = H^T D H = \begin{bmatrix} 1 & 0 & 0 \\ 0 & -1 & 0 \\ 0 & 0 & 0 \end{bmatrix}$$

and A are congruent, and the given quadratic form is equivalent to the canonical form

$$h(\mathbf{y}) = y_1^2 - y_2^2.$$

The rank of g is 2, and since $p = 1$, the signature $s = 2p - r = 0$. ■

For a final application of quadratic forms we consider positive definite symmetric matrices. We recall that in Section 5.3, a symmetric $n \times n$ matrix A was called positive definite if $\mathbf{x}^T A\mathbf{x} > 0$ for every nonzero vector $\mathbf{x}$ in R^n.

If A is a symmetric matrix, then $g(\mathbf{x}) = \mathbf{x}^T A\mathbf{x}$ is a quadratic form, and by Theorem 8.9, g is equivalent to h, where

$$h(\mathbf{y}) = \lambda_1 y_1^2 + \lambda_2 y_2^2 + \cdots + \lambda_p y_p^2 + \lambda_{p+1} y_{p+1}^2 + \lambda_{p+2} y_{p+2}^2 + \cdots + \lambda_r y_r^2.$$

Now A is positive definite if and only if $h(\mathbf{y}) > 0$ for each $\mathbf{y} \neq \mathbf{0}$. However, this can happen if and only if all summands in $h(\mathbf{y})$ are positive and $r = n$. These remarks have established the following theorem:

Theorem 8.11 A symmetric matrix A is positive definite if and only if all the eigenvalues of A are positive. ■

A quadratic form is then called **positive definite** if its matrix is positive definite.

Key Terms

Conic section
Quadric surface
Quadratic form
Matrix of a quadratic form
Congruent matrices
Equivalent quadratic forms
Principal axes theorem
Canonical form
Signature
Positive definite matrix

8.6 Exercises

In Exercises 1 and 2, write each quadratic form as $\mathbf{x}^T A\mathbf{x}$, where A is a symmetric matrix.

1. **(a)** $-3x^2 + 5xy - 2y^2$

(b) $2x_1^2 + 3x_1x_2 - 5x_1x_3 + 7x_2x_3$

(c) $3x_1^2 + x_2^2 - 2x_3^2 + x_1x_2 - x_1x_3 - 4x_2x_3$

2. **(a)** $x_1^2 - 3x_2^2 + 4x_3^2 - 4x_1x_2 + 6x_2x_3$

(b) $4x^2 - 6xy - 2y^2$

(c) $-2x_1x_2 + 4x_1x_3 + 6x_2x_3$

In Exercises 3 and 4, for each given symmetric matrix A, find a diagonal matrix D that is congruent to A.

3. **(a)** $A = \begin{bmatrix} -1 & 0 & 0 \\ 0 & 1 & 1 \\ 0 & 1 & 1 \end{bmatrix}$

(b) $A = \begin{bmatrix} 1 & 1 & 1 \\ 1 & 1 & 1 \\ 1 & 1 & 1 \end{bmatrix}$

(c) $A = \begin{bmatrix} 0 & 2 & 2 \\ 2 & 0 & 2 \\ 2 & 2 & 0 \end{bmatrix}$

4. **(a)** $A = \begin{bmatrix} 3 & 4 & 0 \\ 4 & -3 & 0 \\ 0 & 0 & 5 \end{bmatrix}$

(b) $A = \begin{bmatrix} 2 & 1 & 1 \\ 1 & 2 & 1 \\ 1 & 1 & 2 \end{bmatrix}$

(c) $A = \begin{bmatrix} 0 & 0 & 1 \\ 0 & 1 & 0 \\ 1 & 0 & 0 \end{bmatrix}$

In Exercises 5 through 10, find a quadratic form of the type in Theorem 8.9 that is equivalent to the given quadratic form.

5. $2x^2 - 4xy - y^2$

6. $x_1^2 + x_2^2 + x_3^2 + 2x_2x_3$

7. $2x_1x_3$

8. $2x_2^2 + 2x_3^2 + 4x_2x_3$

9. $-2x_1^2 - 4x_2^2 + 4x_3^2 - 6x_2x_3$

10. $6x_1x_2 + 8x_2x_3$

In Exercises 11 through 16, find a quadratic form of the type in Theorem 8.10 that is equivalent to the given quadratic form.

11. $2x^2 + 4xy + 2y^2$

12. $x_1^2 + x_2^2 + x_3^2 + 2x_1x_2$

13. $2x_1^2 + 4x_2^2 + 4x_3^2 + 10x_2x_3$

14. $2x_1^2 + 3x_2^2 + 3x_3^2 + 4x_2x_3$

15. $-3x_1^2 + 2x_2^2 + 2x_3^2 + 4x_2x_3$

16. $-3x_1^2 + 5x_2^2 + 3x_3^2 - 8x_1x_3$

17. Let $g(\mathbf{x}) = 4x_2^2 + 4x_3^2 - 10x_2x_3$ be a quadratic form in three variables. Find a quadratic form of the type in Theorem 8.10 that is equivalent to g. What is the rank of g? What is the signature of g?

18. Let $g(\mathbf{x}) = 3x_1^2 - 3x_2^2 - 3x_3^2 + 4x_2x_3$ be a quadratic form in three variables. Find a quadratic form of the type in Theorem 8.10 that is equivalent to g. What is the rank of g? What is the signature of g?

19. Find all quadratic forms $g(\mathbf{x}) = \mathbf{x}^T A\mathbf{x}$ in two variables of the type described in Theorem 8.10. What conics do the equations $\mathbf{x}^T A\mathbf{x} = 1$ represent?

20. Find all quadratic forms $g(\mathbf{x}) = \mathbf{x}^T A\mathbf{x}$ in two variables of rank 1 of the type described in Theorem 8.10. What conics do the equations $\mathbf{x}^T A\mathbf{x} = 1$ represent?

In Exercises 21 and 22, which of the given quadratic forms in three variables are equivalent?

21. $g_1(\mathbf{x}) = x_1^2 + x_2^2 + x_3^2 + 2x_1x_2$
$g_2(\mathbf{x}) = 2x_2^2 + 2x_3^2 + 2x_2x_3$
$g_3(\mathbf{x}) = 3x_2^2 - 3x_3^2 + 8x_2x_3$
$g_4(\mathbf{x}) = 3x_2^2 + 3x_3^2 - 4x_2x_3$

22. $g_1(\mathbf{x}) = x_2^2 + 2x_1x_3$
$g_2(\mathbf{x}) = 2x_1^2 + 2x_2^2 + x_3^2 + 2x_1x_2 + 2x_1x_3 + 2x_2x_3$
$g_3(\mathbf{x}) = 2x_1x_2 + 2x_1x_3 + 2x_2x_3$
$g_4(\mathbf{x}) = 4x_1^2 + 3x_2^2 + 4x_3^2 + 10x_1x_3$

In Exercises 23 and 24, which of the given matrices are positive definite?

23. **(a)** $\begin{bmatrix} 2 & -1 \\ -1 & 2 \end{bmatrix}$ **(b)** $\begin{bmatrix} 2 & 1 \\ 1 & 2 \end{bmatrix}$

(c) $\begin{bmatrix} 3 & 1 & 0 \\ 1 & 3 & 0 \\ 0 & 0 & 3 \end{bmatrix}$ **(d)** $\begin{bmatrix} 1 & 0 & 0 \\ 0 & 2 & 0 \\ 0 & 0 & -3 \end{bmatrix}$

(e) $\begin{bmatrix} 2 & 2 \\ 2 & 2 \end{bmatrix}$

24. **(a)** $\begin{bmatrix} 0 & -1 \\ -1 & 0 \end{bmatrix}$ **(b)** $\begin{bmatrix} 1 & 1 \\ 1 & 1 \end{bmatrix}$

(c) $\begin{bmatrix} 0 & 0 & 0 \\ 0 & 1 & 2 \\ 0 & 2 & 1 \end{bmatrix}$ **(d)** $\begin{bmatrix} 7 & 4 & 4 \\ 4 & 7 & 4 \\ 4 & 4 & 7 \end{bmatrix}$

(e) $\begin{bmatrix} 2 & 0 & 0 & 0 \\ 0 & 1 & 0 & 0 \\ 0 & 0 & 3 & 4 \\ 0 & 0 & 4 & -3 \end{bmatrix}$

25. Prove that if A is a symmetric matrix, then $P^T AP$ is also symmetric.

26. If A, B, and C are $n \times n$ symmetric matrices, prove the following:

(a) A is congruent to A.

(b) If B is congruent to A, then A is congruent to B.

(c) If C is congruent to B and B is congruent to A, then C is congruent to A.

27. Prove that if A is symmetric, then A is congruent to a diagonal matrix D.

28. Let $A = \begin{bmatrix} a & b \\ b & d \end{bmatrix}$ be a 2×2 symmetric matrix. Prove that A is positive definite if and only if $\det(A) > 0$ and $a > 0$.

29. Prove that a symmetric matrix A is positive definite if and only if $A = P^T P$ for a nonsingular matrix P.

8.7 Conic Sections

In this section we discuss the classification of the conic sections in the plane. A **quadratic equation** in the variables x and y has the form

$$ax^2 + 2bxy + cy^2 + dx + ey + f = 0, \tag{1}$$

where a, b, c, d, e, and f are real numbers. The graph of Equation (1) is a **conic section**, a curve so named because it is produced by intersecting a plane with a right circular cone that has two nappes. In Figure 8.23 we show that a plane cuts the cone in a circle, ellipse, parabola, or hyperbola. Degenerate cases of the conic sections are a point, a line, a pair of lines, or the empty set.

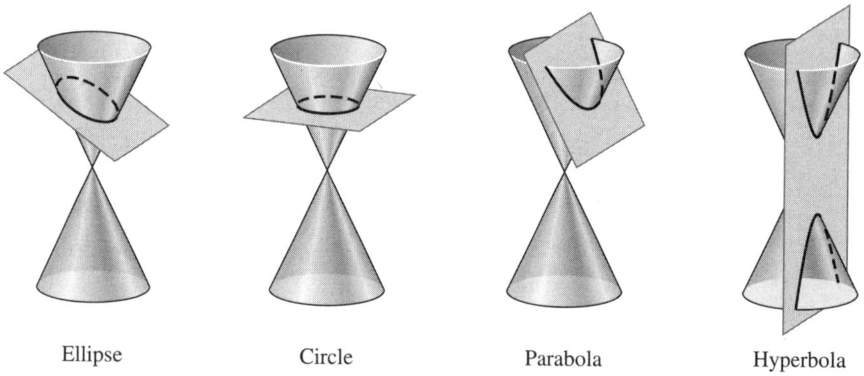

FIGURE 8.23 Nondegenerate conic sections.

The nondegenerate conics are said to be in **standard position** if their graphs and equations are as given in Figure 8.24. The equation is said to be in **standard form**.

EXAMPLE 1

Identify the graph of each given equation.

(a) $4x^2 + 25y^2 - 100 = 0$

(b) $9y^2 - 4x^2 = -36$

(c) $x^2 + 4y = 0$

(d) $y^2 = 0$

(e) $x^2 + 9y^2 + 9 = 0$

(f) $x^2 + y^2 = 0$

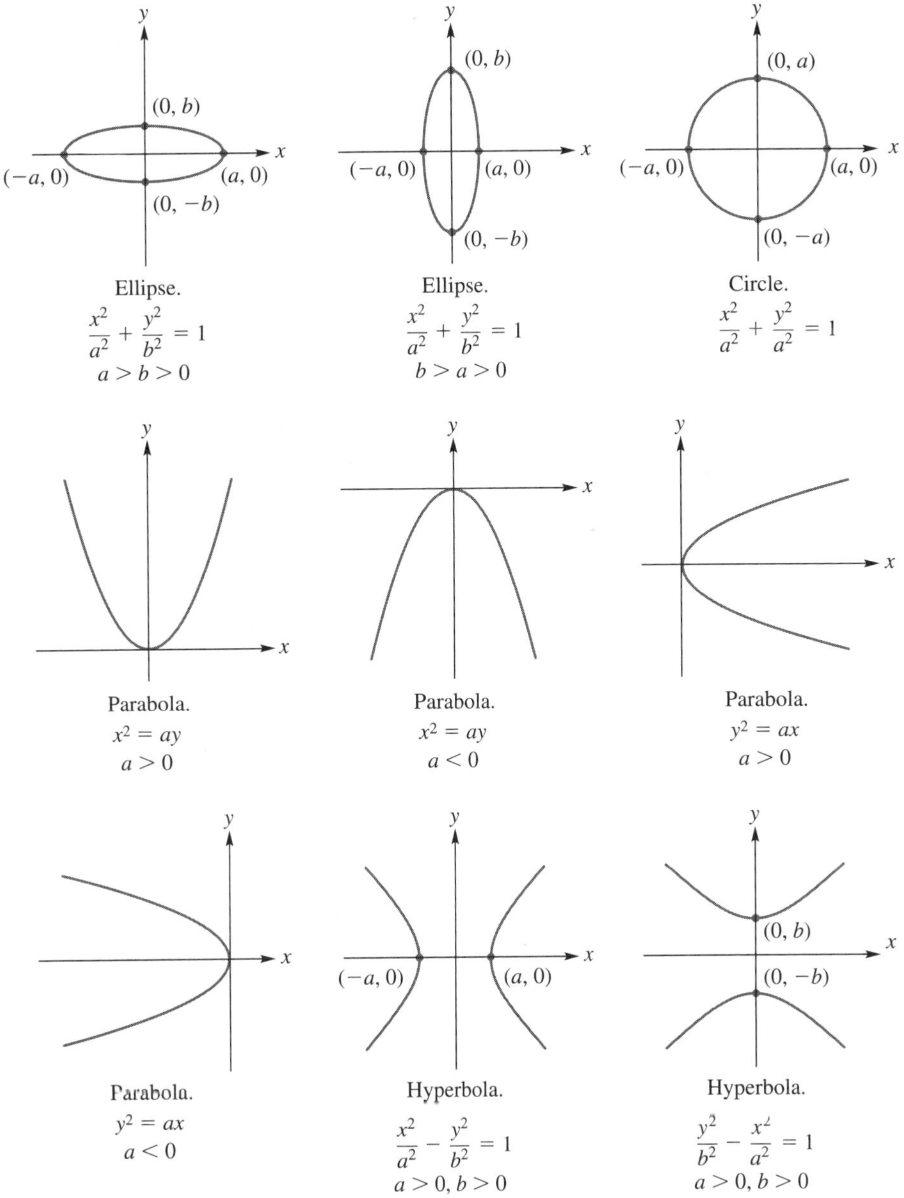

FIGURE 8.24 Conic sections in standard position.

Solution

(a) We rewrite the given equation as

$$\frac{4}{100}x^2 + \frac{25}{100}y^2 = \frac{100}{100},$$

or

$$\frac{x^2}{25} + \frac{y^2}{4} = 1,$$

whose graph is an ellipse in standard position with $a = 5$ and $b = 2$. Thus the x-intercepts are $(5, 0)$ and $(-5, 0)$, and the y-intercepts are $(0, 2)$ and $(0, -2)$.

(b) Rewriting the given equation as

$$\frac{x^2}{9} - \frac{y^2}{4} = 1,$$

we see that its graph is a hyperbola in standard position with $a = 3$ and $b = 2$. The x-intercepts are $(3, 0)$ and $(-3, 0)$.

(c) Rewriting the given equation as

$$x^2 = -4y,$$

we see that its graph is a parabola in standard position with $a = -4$, so it opens downward.

(d) Every point satisfying the given equation must have a y-coordinate equal to zero. Thus the graph of this equation consists of all the points on the x-axis.

(e) Rewriting the given equation as

$$x^2 + 9y^2 = -9,$$

we conclude that there are no points in the plane whose coordinates satisfy the given equation.

(f) The only point satisfying the equation is the origin $(0, 0)$, so the graph of this equation is the single point consisting of the origin. ■

We next turn to the study of conic sections whose graphs are not in standard position. First, notice that the equations of the conic sections whose graphs are in standard position do not contain an xy-term (called a **cross-product term**). If a cross-product term appears in the equation, the graph is a conic section that has been rotated from its standard position [see Figure 8.25(a)]. Also, notice that none of the equations in Figure 8.24 contain an x^2-term and an x-term or a y^2-term and a y-term. If either of these cases occurs and there is no xy-term in the equation, the graph is a conic section that has been translated from its standard position [see Figure 8.25(b)]. On the other hand, if an xy-term is present, the graph is a conic section that has been rotated and possibly also translated [see Figure 8.25(c)].

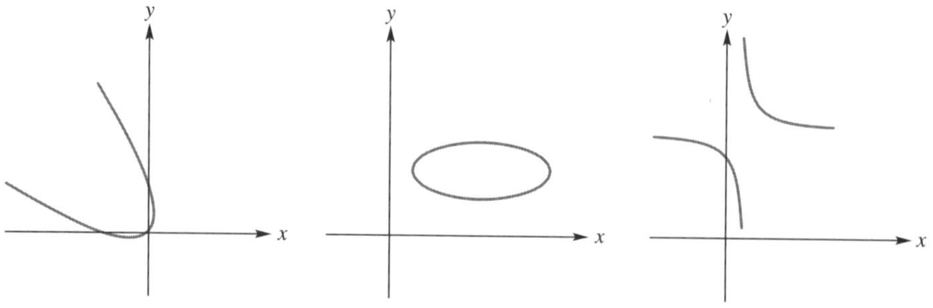

FIGURE 8.25

(a) A parabola that has been rotated.

(b) An ellipse that has been translated.

(c) A hyperbola that has been rotated and translated.

To identify a nondegenerate conic section whose graph is not in standard position, we proceed as follows:

1. If a cross-product term is present in the given equation, rotate the xy-coordinate axes by means of an orthogonal linear transformation so that in the resulting equation the xy-term no longer appears.
2. If an xy-term is not present in the given equation, but an x^2-term and an x-term, or a y^2-term and a y-term appear, translate the xy-coordinate axes by completing the square so that the graph of the resulting equation will be in standard position with respect to the origin of the new coordinate system.

Thus, if an xy-term appears in a given equation, we first rotate the xy-coordinate axes and then, if necessary, translate the rotated axes. In the next example, we deal with the case requiring only a translation of axes.

EXAMPLE 2 Identify and sketch the graph of the equation

$$x^2 - 4y^2 + 6x + 16y - 23 = 0. \tag{2}$$

Also, write its equation in standard form.

Solution

Since there is no cross-product term, we need only translate axes. Completing the squares in the x- and y-terms, we have

$$\begin{aligned} x^2 + 6x + 9 - 4(y^2 - 4y + 4) - 23 &= 9 - 16 \\ (x+3)^2 - 4(y-2)^2 &= 23 + 9 - 16 = 16. \end{aligned} \tag{3}$$

Letting

$$x' = x + 3 \quad \text{and} \quad y' = y - 2,$$

we can rewrite Equation (3) as

$$x'^2 - 4y'^2 = 16,$$

or in standard form as

$$\frac{x'^2}{16} - \frac{y'^2}{4} = 1. \tag{4}$$

If we translate the xy-coordinate system to the $x'y'$-coordinate system, whose origin is at $(-3, 2)$, then the graph of Equation (4) is a hyperbola in standard position with respect to the $x'y'$-coordinate system (see Figure 8.26). ■

We now turn to the problem of identifying the graph of Equation (1), where we assume that $b \neq 0$; that is, a cross-product term is present. This equation can be written in matrix form as

$$\mathbf{x}^T A\mathbf{x} + B\mathbf{x} + f = 0, \tag{5}$$

where

$$\mathbf{x} = \begin{bmatrix} x \\ y \end{bmatrix}, \quad A = \begin{bmatrix} a & b \\ b & c \end{bmatrix}, \quad \text{and} \quad B = \begin{bmatrix} d & e \end{bmatrix}.$$

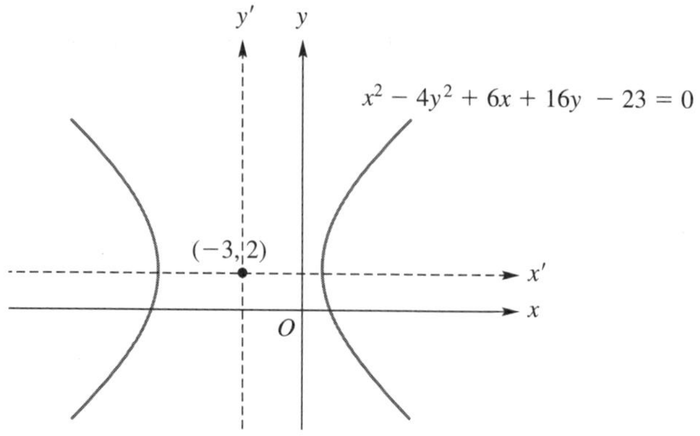

FIGURE 8.26

Since A is a symmetric matrix, we know from Section 7.3 that it can be diagonalized by an orthogonal matrix P. Thus

$$P^T A P = \begin{bmatrix} \lambda_1 & 0 \\ 0 & \lambda_2 \end{bmatrix},$$

where λ_1 and λ_2 are the eigenvalues of A and the columns of P are $\mathbf{x}_1$ and $\mathbf{x}_2$, orthonormal eigenvectors of A associated with λ_1 and λ_2, respectively.

Letting

$$\mathbf{x} = P\mathbf{y}, \quad \text{where} \quad \mathbf{y} = \begin{bmatrix} x' \\ y' \end{bmatrix},$$

we can rewrite Equation (5) as

$$(P\mathbf{y})^T A(P\mathbf{y}) + B(P\mathbf{y}) + f = 0$$
$$\mathbf{y}^T (P^T A P)\mathbf{y} + B(P\mathbf{y}) + f = 0$$

or

$$\begin{bmatrix} x' & y' \end{bmatrix} \begin{bmatrix} \lambda_1 & 0 \\ 0 & \lambda_2 \end{bmatrix} \begin{bmatrix} x' \\ y' \end{bmatrix} + B(P\mathbf{y}) + f = 0 \tag{6}$$

or

$$\lambda_1 x'^2 + \lambda_2 y'^2 + d'x' + e'y' + f = 0. \tag{7}$$

Equation (7) is the resulting equation for the given conic section, and it has no cross-product term.

As discussed in Section 8.6, the x' and y' coordinate axes lie along the eigenvectors $\mathbf{x}_1$ and $\mathbf{x}_2$, respectively. Since P is an orthogonal matrix, $\det(P) = \pm 1$, and if necessary, we can interchange the columns of P (the eigenvectors $\mathbf{x}_1$ and $\mathbf{x}_2$ of A) or multiply a column of P by -1 so that $\det(P) = 1$. As noted in Section 7.3, it then follows that P is the matrix of a counterclockwise rotation of R^2 through an angle θ that can be determined as follows: If

$$\mathbf{x}_1 = \begin{bmatrix} x_{11} \\ x_{21} \end{bmatrix},$$

then

$$\theta = \tan^{-1}\left(\frac{x_{21}}{x_{11}}\right),$$

a result that is frequently developed in a calculus course.

EXAMPLE 3 Identify and sketch the graph of the equation

$$5x^2 - 6xy + 5y^2 - 24\sqrt{2}\,x + 8\sqrt{2}\,y + 56 = 0. \tag{8}$$

Write the equation in standard form.

Solution

Rewriting the given equation in matrix form, we obtain

$$\begin{bmatrix} x & y \end{bmatrix}\begin{bmatrix} 5 & -3 \\ -3 & 5 \end{bmatrix}\begin{bmatrix} x \\ y \end{bmatrix} + \begin{bmatrix} -24\sqrt{2} & 8\sqrt{2} \end{bmatrix}\begin{bmatrix} x \\ y \end{bmatrix} + 56 = 0.$$

We now find the eigenvalues of the matrix

$$A = \begin{bmatrix} 5 & -3 \\ -3 & 5 \end{bmatrix}.$$

Thus

$$\begin{aligned} |\lambda I_2 - A| &= \begin{bmatrix} \lambda - 5 & 3 \\ 3 & \lambda - 5 \end{bmatrix} \\ &= (\lambda - 5)(\lambda - 5) - 9 = \lambda^2 - 10\lambda + 16 \\ &= (\lambda - 2)(\lambda - 8), \end{aligned}$$

so the eigenvalues of A are

$$\lambda_1 = 2, \quad \lambda_2 = 8.$$

Associated eigenvectors are obtained by solving the homogeneous system

$$(\lambda I_2 - A)\mathbf{x} = \mathbf{0}.$$

Thus, for $\lambda_1 = 2$, we have

$$\begin{bmatrix} -3 & 3 \\ 3 & -3 \end{bmatrix}\mathbf{x} = \mathbf{0},$$

so an eigenvector of A associated with $\lambda_1 = 2$ is

$$\begin{bmatrix} 1 \\ 1 \end{bmatrix}.$$

For $\lambda_2 = 8$, we have

$$\begin{bmatrix} 3 & 3 \\ 3 & 3 \end{bmatrix}\mathbf{x} = \mathbf{0},$$

so an eigenvector of A associated with $\lambda_2 = 8$ is

$$\begin{bmatrix} -1 \\ 1 \end{bmatrix}.$$

Normalizing these eigenvectors, we obtain the orthogonal matrix

$$P = \begin{bmatrix} \frac{1}{\sqrt{2}} & -\frac{1}{\sqrt{2}} \\ \frac{1}{\sqrt{2}} & \frac{1}{\sqrt{2}} \end{bmatrix}.$$

Then

$$P^T A P = \begin{bmatrix} 2 & 0 \\ 0 & 8 \end{bmatrix}.$$

Letting $\mathbf{x} = P\mathbf{y}$, we write the transformed equation for the given conic section, Equation (6), as

$$2x'^2 + 8y'^2 - 16x' + 32y' + 56 = 0$$

or

$$x'^2 + 4y'^2 - 8x' + 16y' + 28 = 0.$$

To identify the graph of this equation, we need to translate axes, so we complete the squares, which yields

$$(x' - 4)^2 + 4(y' + 2)^2 + 28 = 16 + 16$$

$$(x' - 4)^2 + 4(y' + 2)^2 = 4$$

$$\frac{(x' - 4)^2}{4} + \frac{(y' + 2)^2}{1} = 1. \tag{9}$$

Letting $x'' = x' - 4$ and $y'' = y' + 2$, we find that Equation (9) becomes

$$\frac{x''^2}{4} + \frac{y''^2}{1} = 1, \tag{10}$$

whose graph is an ellipse in standard position with respect to the $x''y''$-coordinate axes, as shown in Figure 8.27, where the origin of the $x''y''$-coordinate system is in the $x'y'$-coordinate system at $(4, -2)$, which is $(3\sqrt{2}, \sqrt{2})$ in the xy-coordinate

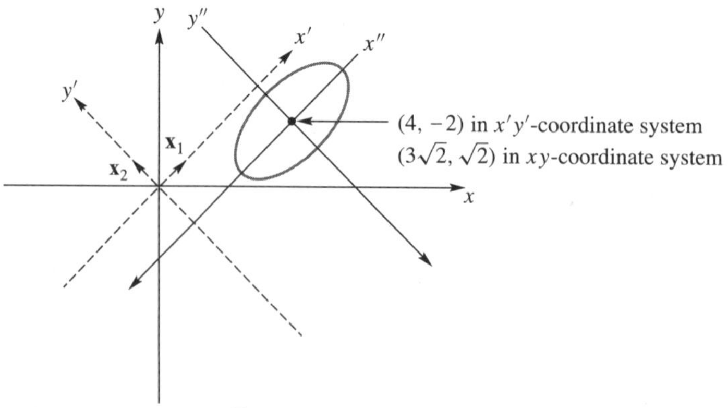

FIGURE 8.27 $5x^2 - 6xy + 5y^2 - 24\sqrt{2}x + 8\sqrt{2}y + 56 = 0.$

system. Equation (10) is the standard form of the equation of the ellipse. Since the eigenvector

$$\mathbf{x}_1 = \begin{bmatrix} \frac{1}{\sqrt{2}} \\ \frac{1}{\sqrt{2}} \end{bmatrix},$$

the xy-coordinate axes have been rotated through the angle θ, where

$$\theta = \tan^{-1}\left(\frac{\frac{1}{\sqrt{2}}}{\frac{1}{\sqrt{2}}}\right) = \tan^{-1}1,$$

so $\theta = 45°$. The x'- and y'-axes lie along the respective eigenvectors

$$\mathbf{x}_1 = \begin{bmatrix} \frac{1}{\sqrt{2}} \\ \frac{1}{\sqrt{2}} \end{bmatrix} \quad \text{and} \quad \mathbf{x}_2 = \begin{bmatrix} -\frac{1}{\sqrt{2}} \\ \frac{1}{\sqrt{2}} \end{bmatrix},$$

as shown in Figure 8.27. ■

The graph of a given quadratic equation in x and y can be identified from the equation that is obtained after rotating axes, that is, from Equation (6) or (7). The identification of the conic section given by these equations is shown in Table 8.1.

TABLE 8.1 Identification of the Conic Sections

λ_1, λ_2 Both Nonzero		***Exactly One of λ_1, λ_2 Is Zero***
$\lambda_1\lambda_2 > 0$	$\lambda_1\lambda_2 < 0$	
Ellipse	Hyperbola	Parabola

Key Terms

Quadratic equation
Conic section
Circle
Ellipse
Parabola
Hyperbola
Standard position
Standard form
Rotation of axes

8.7 Exercises

In Exercises 1 through 10, identify the graph of each equation.

1. $x^2 + 9y^2 - 9 = 0$

2. $x^2 = 2y$

3. $25y^2 - 4x^2 = 100$

4. $y^2 - 16 = 0$

5. $3x^2 - y^2 = 0$

6. $y = 0$

7. $4x^2 + 4y^2 - 9 = 0$

8. $-25x^2 + 9y^2 + 225 = 0$

9. $4x^2 + y^2 = 0$

10. $9x^2 + 4y^2 + 36 = 0$

In Exercises 11 through 18, translate axes to identify the graph of each equation and write each equation in standard form.

11. $x^2 + 2y^2 - 4x - 4y + 4 = 0$

12. $x^2 - y^2 + 4x - 6y - 9 = 0$

13. $x^2 + y^2 - 8x - 6y = 0$

14. $x^2 - 4x + 4y + 4 = 0$

15. $y^2 - 4y = 0$

16. $4x^2 + 5y^2 - 30y + 25 = 0$

17. $x^2 + y^2 - 2x - 6y + 10 = 0$

18. $2x^2 + y^2 - 12x - 4y + 24 = 0$

In Exercises 19 through 24, rotate axes to identify the graph of each equation and write each equation in standard form.

19. $x^2 + xy + y^2 = 6$

20. $xy = 1$

21. $9x^2 + y^2 + 6xy = 4$

22. $x^2 + y^2 + 4xy = 9$

23. $4x^2 + 4y^2 - 10xy = 0$

24. $9x^2 + 6y^2 + 4xy - 5 = 0$

In Exercises 25 through 30, identify the graph of each equation and write each equation in standard form.

25. $9x^2 + y^2 + 6xy - 10\sqrt{10}\,x + 10\sqrt{10}\,y + 90 = 0$

26. $5x^2 + 5y^2 - 6xy - 30\sqrt{2}\,x + 18\sqrt{2}\,y + 82 = 0$

27. $5x^2 + 12xy - 12\sqrt{13}\,x = 36$

28. $6x^2 + 9y^2 - 4xy - 4\sqrt{5}\,x - 18\sqrt{5}\,y = 5$

29. $x^2 - y^2 + 2\sqrt{3}\,xy + 6x = 0$

30. $8x^2 + 8y^2 - 16xy + 33\sqrt{2}\,x - 31\sqrt{2}\,y + 70 = 0$

8.8 Quadric Surfaces

In Section 8.7 conic sections were used to provide geometric models for quadratic forms in two variables. In this section we investigate quadratic forms in three variables and use particular surfaces called quadric surfaces as geometric models. Quadric surfaces are often studied and sketched in analytic geometry and calculus. Here we use Theorems 8.9 and 8.10 to develop a classification scheme for quadric surfaces.

A **second-degree polynomial equation** in three variables x, y, and z has the form

$$ax^2 + by^2 + cz^2 + 2dxy + 2exz + 2fyz + gx + hy + iz = j, \qquad (1)$$

where coefficients a through j are real numbers with $a, b, \ldots, f$ not all zero. Equation (1) can be written in matrix form as

$$\mathbf{x}^T A\mathbf{x} + B\mathbf{x} = j, \qquad (2)$$

where

$$A = \begin{bmatrix} a & d & e \\ d & b & f \\ e & f & c \end{bmatrix}, \quad B = \begin{bmatrix} g & h & i \end{bmatrix}, \quad \text{and} \quad \mathbf{x} = \begin{bmatrix} x \\ y \\ z \end{bmatrix}.$$

We call $\mathbf{x}^T A\mathbf{x}$ the **quadratic form (in three variables) associated with the second-degree polynomial** in (1). As in Section 8.6, the symmetric matrix A is called the matrix of the quadratic form.

The graph of (1) in R^3 is called a **quadric surface**. As in the case of the classification of conic sections in Section 8.7, the classification of (1) as to the type of surface represented depends on the matrix A. Using the ideas in Section

8.7, we have the following strategies to determine a simpler equation for a quadric surface:

1. If A is not diagonal, then a rotation of axes is used to eliminate any cross-product terms xy, xz, or yz.
2. If $B = \begin{bmatrix} g & h & i \end{bmatrix} \neq \mathbf{0}$, then a translation of axes is used to eliminate any first-degree terms.

The resulting equation will have the standard form

$$\lambda_1 x''^2 + \lambda_2 y''^2 + \lambda_3 z''^2 = k,$$

or in matrix form,

$$\mathbf{y}^T C \mathbf{y} = k, \tag{3}$$

where $\mathbf{y} = \begin{bmatrix} x'' \\ y'' \\ z'' \end{bmatrix}$, k is some real constant, and C is a diagonal matrix with diagonal entries $\lambda_1, \lambda_2, \lambda_3$, which are the eigenvalues of A.

We now turn to the classification of quadric surfaces.

DEFINITION 8.6

Let A be an $n \times n$ symmetric matrix. The **inertia** of A, denoted $\text{In}(A)$, is an ordered triple of numbers

$$(\text{pos, neg, zer}),$$

where pos, neg, and zer are the number of positive, negative, and zero eigenvalues of A, respectively.

EXAMPLE 1

Find the inertia of each of the following matrices:

$$A_1 = \begin{bmatrix} 2 & 2 \\ 2 & 2 \end{bmatrix}, \quad A_2 = \begin{bmatrix} 2 & 1 \\ 1 & 2 \end{bmatrix}, \quad A_3 = \begin{bmatrix} 0 & 2 & 2 \\ 2 & 0 & 2 \\ 2 & 2 & 0 \end{bmatrix}.$$

Solution

We determine the eigenvalues of each of the matrices. It follows that (verify)

$\det(\lambda I_2 - A_1) = \lambda(\lambda - 4) = 0,$ so $\lambda_1 = 0, \lambda_2 = 4$, and $\text{In}(A_1) = (1, 0, 1)$.

$\det(\lambda I_2 - A_2) = (\lambda - 1)(\lambda - 3) = 0,$ so $\lambda_1 = 1, \lambda_2 = 3$, and $\text{In}(A_2) = (2, 0, 0)$.

$\det(\lambda I_3 - A_3) = (\lambda + 2)^2(\lambda - 4) = 0,$ so $\lambda_1 = \lambda_2 = -2, \lambda_3 = 4$, and $\text{In}(A_3) = (1, 2, 0)$. ■

From Section 8.6, the signature of a quadratic form $\mathbf{x}^T A\mathbf{x}$ is the difference between the number of positive eigenvalues and the number of negative eigenvalues of A. In terms of inertia, the signature of $\mathbf{x}^T A\mathbf{x}$ is $s = \text{pos} - \text{neg}$.

In order to use inertia for classification of quadric surfaces (or conic sections), we assume that the eigenvalues of an $n \times n$ symmetric matrix A of a quadratic form in n variables are denoted by

$$\begin{aligned} \lambda_1 \geq \cdots \geq \lambda_{\text{pos}} &> 0 \\ \lambda_{\text{pos}+1} \leq \cdots \leq \lambda_{\text{pos}+\text{neg}} &< 0 \\ \lambda_{\text{pos}+\text{neg}+1} = \cdots = \lambda_n &= 0. \end{aligned}$$

The largest positive eigenvalue is denoted by λ_1 and the smallest one by λ_{pos}. We also assume that $\lambda_1 > 0$ and $j \geq 0$ in (2), which eliminates redundant and impossible cases. For example, if

$$A = \begin{bmatrix} -1 & 0 & 0 \\ 0 & -2 & 0 \\ 0 & 0 & -3 \end{bmatrix}, \quad B = \begin{bmatrix} 0 & 0 & 0 \end{bmatrix}, \quad \text{and} \quad j = 5,$$

then the second-degree polynomial is $-x^2 - 2y^2 - 3z^2 = 5$, which has an empty solution set. That is, the surface represented has no points. However, if $j = -5$, then the second-degree polynomial is $-x^2 - 2y^2 - 3z^2 = -5$, which is identical to $x^2 + 2y^2 + 3z^2 = 5$. The assumptions $\lambda_1 > 0$ and $j \geq 0$ avoid such a redundant representation.

EXAMPLE 2

Consider a quadratic form in two variables with matrix A, and assume that $\lambda_1 > 0$ and $f \geq 0$ in Equation (1) of Section 8.7. Then there are only three possible cases for the inertia of A, which we summarize as follows:

1. $\text{In}(A) = (2, 0, 0)$; then the quadratic form represents an ellipse.
2. $\text{In}(A) = (1, 1, 0)$; then the quadratic form represents a hyperbola.
3. $\text{In}(A) = (1, 0, 1)$; then the quadratic form represents a parabola.

This classification is identical to that given in Table 8.2 later in this section, taking the assumptions into account. ■

Note that the classification of the conic sections in Example 2 does not distinguish between special cases within a particular geometric class. For example, both $y = x^2$ and $x = y^2$ have inertia $(1, 0, 1)$.

Before classifying quadric surfaces, using inertia, we present the quadric surfaces in the standard forms met in analytic geometry and calculus. (In the following, a, b, and c are positive unless otherwise designated.)

Ellipsoid (See Figure 8.28.)

$$\frac{x^2}{a^2} + \frac{y^2}{b^2} + \frac{z^2}{c^2} = 1$$

The special case $a = b = c$ is a **sphere**.

Elliptic Paraboloid (See Figure 8.29.)

$$z = \frac{x^2}{a^2} + \frac{y^2}{b^2}, \qquad y = \frac{x^2}{a^2} + \frac{z^2}{c^2}, \qquad x = \frac{y^2}{b^2} + \frac{z^2}{c^2}$$

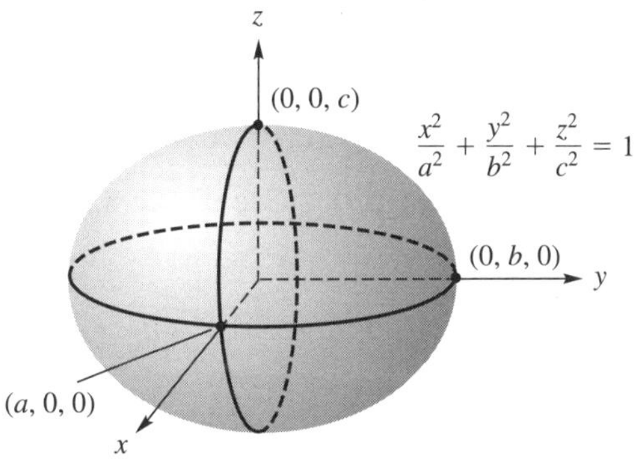

FIGURE 8.28 Ellipsoid.

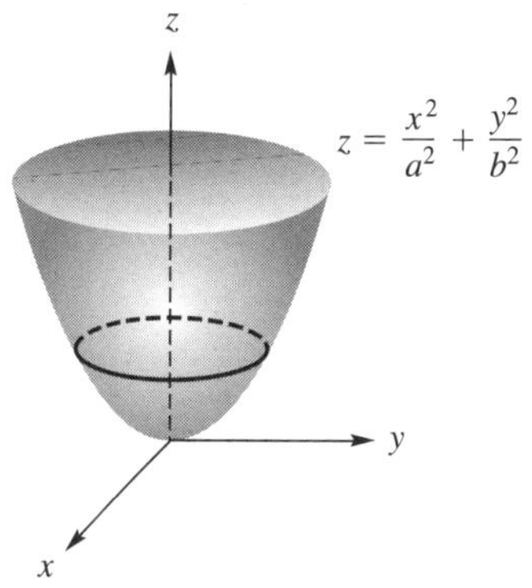

FIGURE 8.29 Elliptic paraboloid.

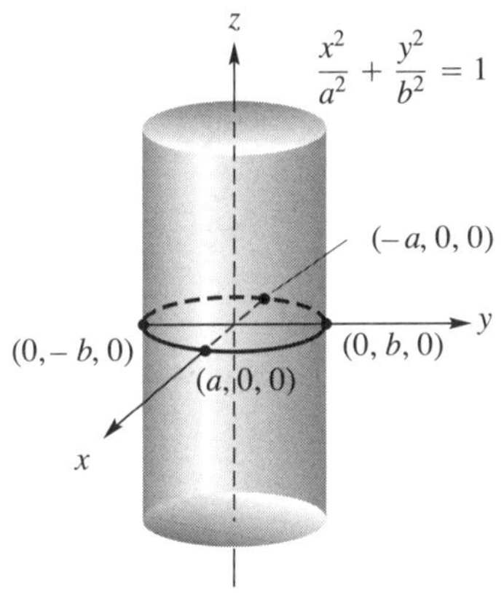

FIGURE 8.30 Elliptic cylinder.

A degenerate case of a parabola is a line, so a degenerate case of an elliptic paraboloid is an **elliptic cylinder** (see Figure 8.30), which is given by

$$\frac{x^2}{a^2}+\frac{y^2}{b^2}=1, \qquad \frac{x^2}{a^2}+\frac{z^2}{c^2}=1, \qquad \frac{y^2}{b^2}+\frac{z^2}{c^2}=1.$$

Hyperboloid of One Sheet (See Figure 8.31.)

$$\frac{x^2}{a^2}+\frac{y^2}{b^2}-\frac{z^2}{c^2}=1, \qquad \frac{x^2}{a^2}-\frac{y^2}{b^2}+\frac{z^2}{c^2}=1, \qquad -\frac{x^2}{a^2}+\frac{y^2}{b^2}+\frac{z^2}{c^2}=1$$

A degenerate case of a hyperbola is a pair of lines through the origin; hence a degenerate case of a hyperboloid of one sheet is a **cone** (see Figure 8.32), which is given by

$$\frac{x^2}{a^2}+\frac{y^2}{b^2}-\frac{z^2}{c^2}=0, \qquad \frac{x^2}{a^2}-\frac{y^2}{b^2}+\frac{z^2}{c^2}=0, \qquad -\frac{x^2}{a^2}+\frac{y^2}{b^2}+\frac{z^2}{c^2}=0.$$

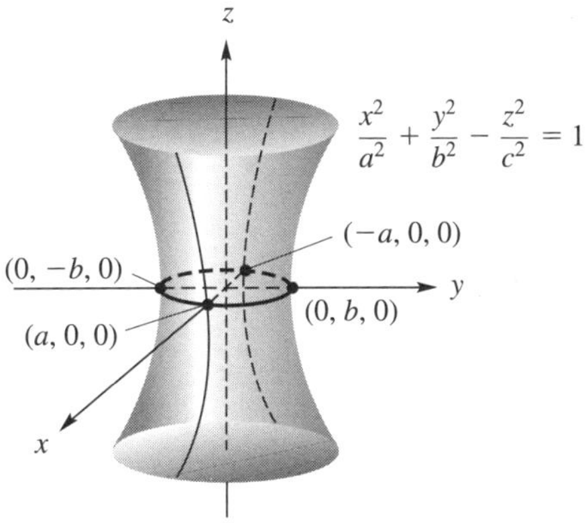

FIGURE 8.31 Hyperboloid of one sheet.

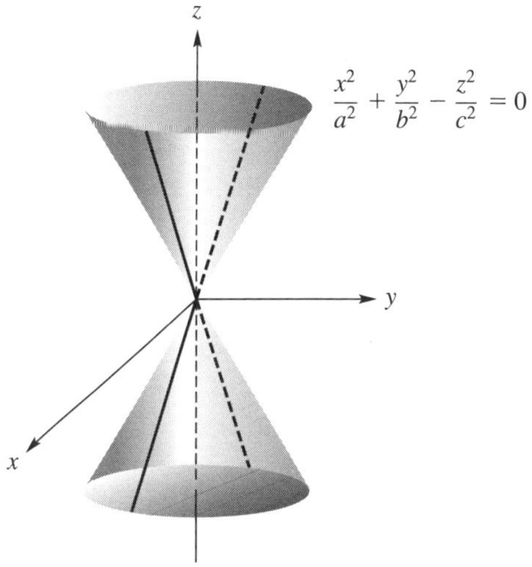

FIGURE 8.32 Cone.

Hyperboloid of Two Sheets (See Figure 8.33.)

$$\frac{x^2}{a^2} - \frac{y^2}{b^2} - \frac{z^2}{c^2} = 1, \qquad -\frac{x^2}{a^2} - \frac{y^2}{b^2} + \frac{z^2}{c^2} = 1, \qquad -\frac{x^2}{a^2} + \frac{y^2}{b^2} - \frac{z^2}{c^2} = 1$$

Hyperbolic Paraboloid (See Figure 8.34.)

$$\pm z = \frac{x^2}{a^2} - \frac{y^2}{b^2}, \qquad \pm y = \frac{x^2}{a^2} - \frac{z^2}{b^2}, \qquad \pm x = \frac{y^2}{a^2} - \frac{z^2}{b^2}$$

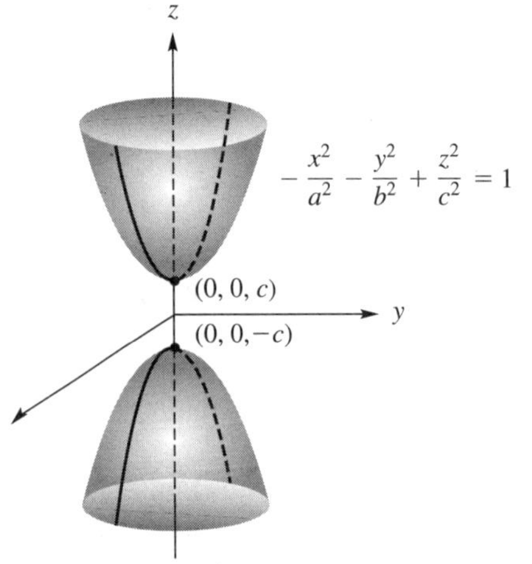

FIGURE 8.33 Hyperboloid of two sheets.

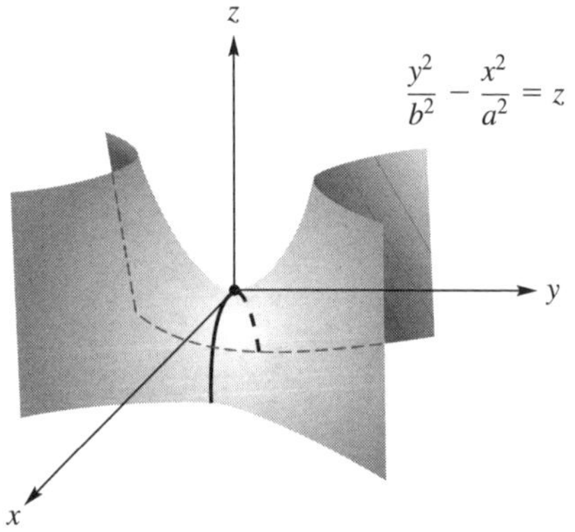

FIGURE 8.34 Hyperbolic paraboloid.

A degenerate case of a parabola is a line, so a degenerate case of a hyperbolic paraboloid is a hyperbolic cylinder (see Figure 8.35), which is given by

$$\frac{x^2}{a^2} - \frac{y^2}{b^2} = \pm 1, \qquad \frac{x^2}{a^2} - \frac{z^2}{b^2} = \pm 1, \qquad \frac{y^2}{a^2} - \frac{z^2}{b^2} = \pm 1.$$

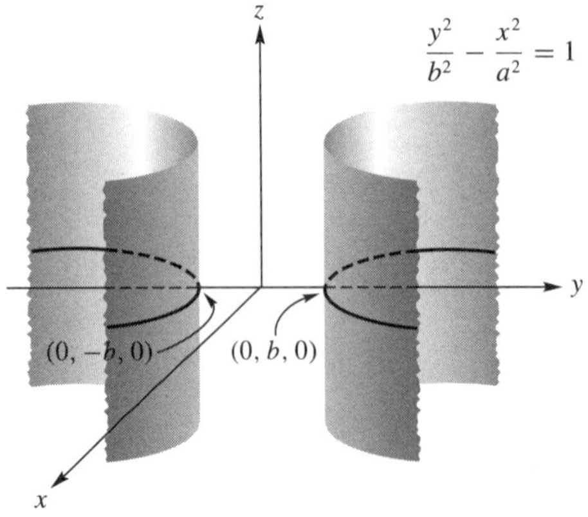

FIGURE 8.35 Hyperbolic cylinder.

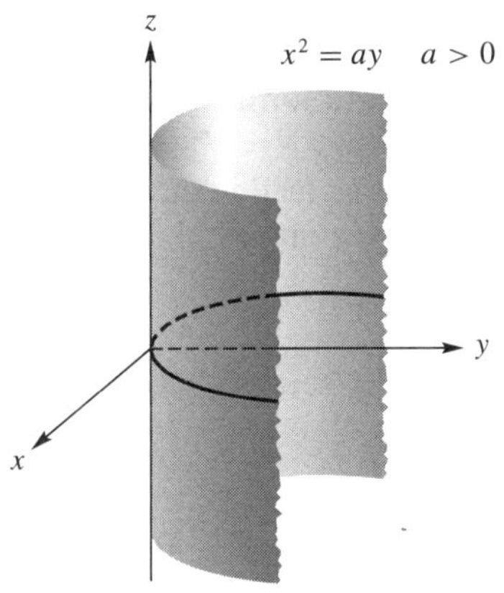

FIGURE 8.36 Parabolic cylinder.

Parabolic Cylinder (See Figure 8.36.) One of a or b is not zero.

$$x^2 = ay + bz, \qquad y^2 = ax + bz, \qquad z^2 = ax + by$$

For a quadratic form in three variables with matrix A, under the assumptions $\lambda_1 > 0$ and $j \geq 0$ in (2), there are exactly six possibilities for the inertia of A. We present these in Table 8.2. As with the conic section classification of Example 2, the classification of quadric surfaces in Table 8.2 does not distinguish between special cases within a particular geometric class.

TABLE 8.2 Identification of the Quadric Surfaces

$\text{In}(A) = (3, 0, 0)$	Ellipsoid
$\text{In}(A) = (2, 0, 1)$	Elliptic paraboloid
$\text{In}(A) = (2, 1, 0)$	Hyperboloid of one sheet
$\text{In}(A) = (1, 2, 0)$	Hyperboloid of two sheets
$\text{In}(A) = (1, 1, 1)$	Hyperbolic paraboloid
$\text{In}(A) = (1, 0, 2)$	Parabolic cylinder

EXAMPLE 3

Classify the quadric surface represented by the quadratic form $\mathbf{x}^T A\mathbf{x} = 3$, where

$$A = \begin{bmatrix} 0 & 2 & 2 \\ 2 & 0 & 2 \\ 2 & 2 & 0 \end{bmatrix} \quad \text{and} \quad \mathbf{x} = \begin{bmatrix} x \\ y \\ z \end{bmatrix}.$$

Solution

From Example 1 we have $\text{In}(A) = (1, 2, 0)$, and hence the quadric surface is a hyperboloid of two sheets. ■

EXAMPLE 4

Classify the quadric surface given by

$$2x^2 + 4y^2 - 4z^2 + 6yz - 5x + 3y = 2.$$

Solution

Rewrite the second-degree polynomial as a quadratic form in three variables to identify the matrix A of the quadratic form. This gives

$$A = \begin{bmatrix} 2 & 0 & 0 \\ 0 & 4 & 3 \\ 0 & 3 & -4 \end{bmatrix}.$$

Its eigenvalues are $\lambda_1 = 5$, $\lambda_2 = 2$, and $\lambda_3 = -5$ (verify). Thus $\text{In}(A) = (2, 1, 0)$, and hence the quadric surface is a hyperboloid of one sheet. ■

It is much easier to classify a quadric surface than to transform it to the standard forms that are used in analytic geometry and calculus. The algebraic steps followed to produce an equation in standard form from a second-degree polynomial equation (1) require, in general, a rotation and translation of axes, as mentioned earlier. The rotation requires both the eigenvalues and eigenvectors of the matrix A of the quadratic form. The eigenvectors of A are used to form an orthogonal matrix P so that $\det(P) = 1$, and hence the change of variables $\mathbf{x} = P\mathbf{y}$ represents a rotation. The resulting associated quadratic form is that obtained in the principal axes theorem, Theorem 8.9; that is, all cross-product terms are eliminated. We illustrate this with the next example.

EXAMPLE 5 For the quadric surface in Example 4,

$$\mathbf{x}^T A\mathbf{x} + \begin{bmatrix} -5 & 3 & 0 \end{bmatrix} \mathbf{x} = 2,$$

determine the rotation so that all cross-product terms are eliminated.

Solution

The eigenvalues of

$$A = \begin{bmatrix} 2 & 3 & 0 \\ 0 & 4 & 3 \\ 0 & 3 & -4 \end{bmatrix}$$

are

$$\lambda_1 = 5, \quad \lambda_2 = 2, \quad \lambda_3 = -5$$

and associated eigenvectors are (verify),

$$\mathbf{v}_1 = \begin{bmatrix} 0 \\ 3 \\ 1 \end{bmatrix}, \quad \mathbf{v}_2 = \begin{bmatrix} 1 \\ 0 \\ 0 \end{bmatrix}, \quad \mathbf{v}_3 = \begin{bmatrix} 0 \\ 1 \\ -3 \end{bmatrix}.$$

The eigenvectors $\mathbf{v}_i$ are mutually orthogonal, since they correspond to distinct eigenvalues of a symmetric matrix (see Theorem 7.7). We normalize the eigenvectors as

$$\mathbf{u}_1 = \frac{1}{\sqrt{10}} \begin{bmatrix} 0 \\ 3 \\ 1 \end{bmatrix}, \qquad \mathbf{u}_2 = \mathbf{v}_2, \qquad \mathbf{u}_3 = \frac{1}{\sqrt{10}} \begin{bmatrix} 0 \\ 1 \\ -3 \end{bmatrix}$$

and define $P = \begin{bmatrix} \mathbf{u}_1 & \mathbf{u}_2 & \mathbf{u}_3 \end{bmatrix}$. Then $\det(P) = 1$ (verify), so we let $\mathbf{x} = P\mathbf{y}$ and obtain the representation

$$(P\mathbf{y})^T A(P\mathbf{y}) + \begin{bmatrix} -5 & 3 & 0 \end{bmatrix} P\mathbf{y} = 2$$
$$\mathbf{y}^T(P^T AP)\mathbf{y} + \begin{bmatrix} -5 & 3 & 0 \end{bmatrix} P\mathbf{y} = 2.$$

Since $P^T AP = D$, and letting $\mathbf{y} = \begin{bmatrix} x' \\ y' \\ z' \end{bmatrix}$, we have

$$\mathbf{y}^T D\mathbf{y} + \begin{bmatrix} -5 & 3 & 0 \end{bmatrix} P\mathbf{y} = 2,$$

$$\mathbf{y}^T \begin{bmatrix} 5 & 0 & 0 \\ 0 & 2 & 0 \\ 0 & 0 & -5 \end{bmatrix} \mathbf{y} + \begin{bmatrix} \frac{9}{\sqrt{10}} & -5 & \frac{3}{\sqrt{10}} \end{bmatrix} \mathbf{y} = 2$$

[if det(P) $\neq 1$, we redefine P by reordering its columns until we get its determinant to be 1], or

$$5x'^2 + 2y'^2 - 5z'^2 + \frac{9}{\sqrt{10}}x' - 5y' + \frac{3}{\sqrt{10}}z' = 2. \quad \blacksquare$$

To complete the transformation to standard form, we introduce a change of variable to perform a translation that eliminates any first-degree terms. Algebraically, we complete the square in each of the three variables.

EXAMPLE 6 Continue with Example 5 to eliminate the first-degree terms.

Solution

The last expression for the quadric surface in Example 5 can be written as

$$5x'^2 + \frac{9}{\sqrt{10}}x' + 2y'^2 - 5y' - 5z'^2 + \frac{3}{\sqrt{10}}z' = 2.$$

Completing the square in each variable, we have

$$5\left(x'^2 + \frac{9}{5\sqrt{10}}x' + \frac{81}{1000}\right) + 2\left(y'^2 - \frac{5}{2}y' + \frac{25}{16}\right) - 5\left(z'^2 - \frac{3}{5\sqrt{10}}z' + \frac{9}{1000}\right)$$

$$= 5\left(x' + \frac{9}{10\sqrt{10}}\right)^2 + 2\left(y' - \frac{5}{4}\right)^2 - 5\left(z' - \frac{3}{10\sqrt{10}}\right)^2$$

$$= 2 + \frac{405}{1000} + \frac{50}{16} - \frac{45}{1000} = \frac{5485}{1000}.$$

Letting

$$x'' = x' + \frac{9}{10\sqrt{10}}, \qquad y'' = y' - \frac{5}{4}, \qquad z'' = z' - \frac{3}{10\sqrt{10}},$$

we can write the equation of the quadric surface as

$$5x''^2 + 2y''^2 - 5z''^2 = \frac{5485}{1000} = 5.485.$$

This can be written in standard form as

$$\frac{x''^2}{\frac{5.485}{5}} + \frac{y''^2}{\frac{5.485}{2}} - \frac{z''^2}{\frac{5.485}{5}} = 1. \quad \blacksquare$$

Key Terms

Second-degree polynomial
Quadratic form
Quadric surface
Symmetric matrix
Inertia
Classification of quadric surfaces
Ellipse
Hyperbola
Parabola
Ellipsoid
Elliptic cylinder
Elliptic paraboloid
Hyperboloid of one sheet
Hyperboloid of two sheets
Cone
Hyperbolic paraboloid
Degenerate case
Parabolic cylinder

8.8 Exercises

In Exercises 1 through 14, use inertia to classify the quadric surface given by each equation.

1. $x^2 + y^2 + 2z^2 - 2xy - 4xz - 4yz + 4x = 8$
2. $x^2 + 3y^2 + 2z^2 - 6x - 6y + 4z - 2 = 0$
3. $z = 4xy$
4. $x^2 + y^2 + z^2 + 2xy = 4$
5. $x^2 - y = 0$
6. $2xy + z = 0$
7. $5y^2 + 20y + z - 23 = 0$
8. $x^2 + y^2 + 2z^2 - 2xy + 4xz + 4yz = 16$
9. $4x^2 + 9y^2 + z^2 + 8x - 18y - 4z - 19 = 0$
10. $y^2 - z^2 - 9x - 4y + 8z - 12 = 0$
11. $x^2 + 4y^2 + 4x + 16y - 16z - 4 = 0$
12. $4x^2 - y^2 + z^2 - 16x + 8y - 6z + 5 = 0$
13. $x^2 - 4z^2 - 4x + 8z = 0$
14. $2x^2 + 2y^2 + 4z^2 + 2xy - 2xz - 2yz + 3x - 5y + z = 7$

In Exercises 15 through 28, classify the quadric surface given by each equation and determine its standard form.

15. $x^2 + 2y^2 + 2z^2 + 2yz = 1$
16. $x^2 + y^2 + 2z^2 - 2xy + 4xz + 4yz = 16$
17. $2xz - 2z - 4y - 4z + 8 = 0$
18. $x^2 + 3y^2 + 3z^2 - 4yz = 9$
19. $x^2 + y^2 + z^2 + 2xy = 8$
20. $-x^2 - y^2 - z^2 + 4xy + 4xz + 4yz = 3$
21. $2x^2 + 2y^2 + 4z^2 - 4xy - 8xz - 8yz + 8x = 15$
22. $4x^2 + 4y^2 + 8z^2 + 4xy - 4xz - 4yz + 6x - 10y + 2z = \frac{9}{4}$
23. $2y^2 + 2z^2 + 4yz + \frac{16}{\sqrt{2}}x + 4 = 0$
24. $x^2 + y^2 - 2z^2 + 2xy + 8xz + 8yz + 3x + z = 0$
25. $-x^2 - y^2 - z^2 + 4xy + 4xz + 4yz + \frac{3}{\sqrt{2}}x - \frac{3}{\sqrt{2}}y = 6$
26. $2x^2 + 3y^2 + 3z^2 - 2yz + 2x + \frac{1}{\sqrt{2}}y + \frac{1}{\sqrt{2}}z = \frac{3}{8}$
27. $x^2 + y^2 - z^2 - 2x - 4y - 4z + 1 = 0$
28. $-8x^2 - 8y^2 + 10z^2 + 32xy - 4xz - 4yz = 24$

APPENDIX

A Preliminaries

In this appendix, which can be consulted as the need arises, we present the basic ideas of sets and functions that are used in Chapters 4, 5, 6, and 7.

A.1 Sets

A **set** is a collection, class, aggregate, or family of objects, which are called **elements**, or **members**, of the set. A set will be denoted by a capital letter, and an element of a set by a lowercase letter. A set S is specified either by describing all the elements of S, or by stating a property that determines, unequivocally, whether an element is or is not an element of S. Let $S = \{1, 2, 3\}$ be the set of all positive integers < 4. Then a real number belongs to S if it is a positive integer < 4. Thus S has been described in both ways. Sets A and B are said to be **equal** if each element of A belongs to B and if each element of B belongs to A. We write $A = B$. Thus $\{1, 2, 3\} = \{3, 2, 1\} = \{2, 1, 3\}$, and so on. If A and B are sets such that every element of A belongs to B, then A is said to be a **subset** of B. The set of all rational numbers is a subset of the set of all real numbers; the set $\{1, 3\}$ is a subset of $\{1, 2, 3\}$; the set of all isosceles triangles is a subset of the set of all triangles. We can see that every set is a subset of itself. The **empty set** is the set that has no elements in it. The set of all real numbers whose squares equal -1 is empty, because the square of a real number is never negative.

A.2 Functions

A **function** f from a set S into a set T is a rule that assigns to each element s of S a unique element t of T. We denote the function f by $f: S \to T$ and write $t = f(s)$. Functions constitute the basic ingredient of calculus and other branches of mathematics, and the reader has dealt extensively with them. The set S is called the **domain** of f; the set T is called the **codomain** of f; the subset $f(S)$ of T consisting of all the elements $f(s)$, for s in S, is called the **range** of f, or the **image** of S under f. As examples of functions, we consider the following:

1. Let $S = T =$ the set of all real numbers. Let $f: S \to T$ be defined by the rule $f(s) = s^2$, for s in S.
2. Let $S =$ the set of all real numbers and let $T =$ the set of all nonnegative real numbers. Let $f: S \to T$ be defined by the rule $f(s) = s^2$, for s in S.
3. Let $S =$ three-dimensional space, where each point is described by x-, y-, and z-coordinates (x, y, z). Let $T =$ the (x, y)-plane as a subset of S. Let $f: S \to T$ be defined by the rule $f((x, y, z)) = (x, y, 0)$. To see what f does, we take a point (x, y, z) in S, draw a line from (x, y, z) perpendicular to T, the (x, y)-plane, and find the point of intersection $(x, y, 0)$ of this line with the (x, y)-plane. This point is the image of (x, y, z) under f; f is called a **projection function** (Figure A.1).

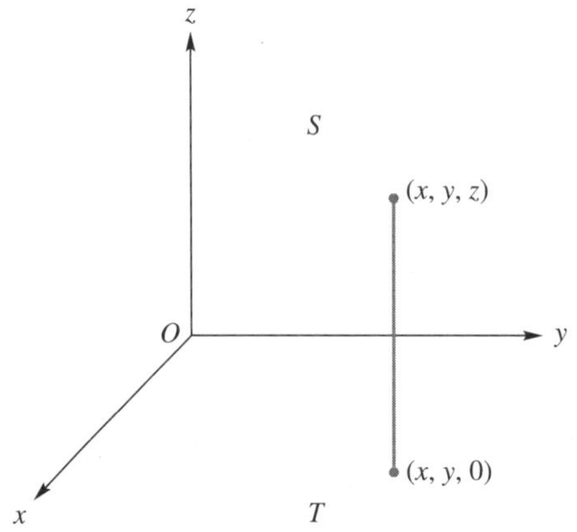

FIGURE A.1

4. Let $S = T =$ the set of all real numbers. Let $f: S \to T$ be defined by the rule $f(s) = 2s + 1$, for s in S.
5. Let $S =$ the x-axis in the (x, y)-plane and let $T =$ the (x, y)-plane. Let $f: S \to T$ be defined by the rule $f((s, 0)) = (s, 1)$, for s in S.
6. Let $S =$ the set of all real numbers. Let $T =$ the set of all positive real numbers. Let $f: S \to T$ be defined by the rule $f(s) = e^s$, for s in S.

There are two properties of functions that we need to distinguish. A function $f: S \to T$ is called **one-to-one** if $f(s_1) \neq f(s_2)$ whenever s_1 and s_2 are distinct elements of S. That is, f is one-to-one if two different elements of S cannot be sent by f to the same element of T. An equivalent statement is that if $f(s_1) = f(s_2)$, then we must have $s_1 = s_2$ (see Figure A.2). A function $f: S \to T$ is called **onto** if the range of f is all of T—that is, if for any given t in T there is at least one s in S such that $f(s) = t$ (see Figure A.3).

We now examine the listed functions:

1. f is not one-to-one, for if $f(s_1) = f(s_2)$, it need not follow that

$$s_1 = s_2 \qquad [f(2) = f(-2) = 4].$$

Since the range of f is the set of nonnegative real numbers, f is not onto. Thus if $t = -4$, then there is no s such that $f(s) = -4$.

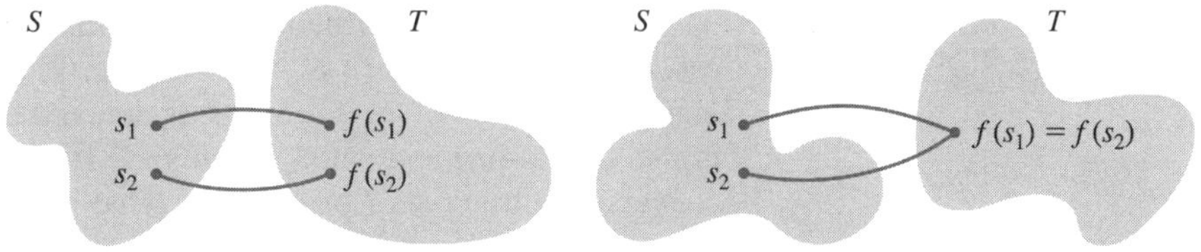

(a) *f* is one-to-one. (b) *f* is not one-to-one.

FIGURE A.2

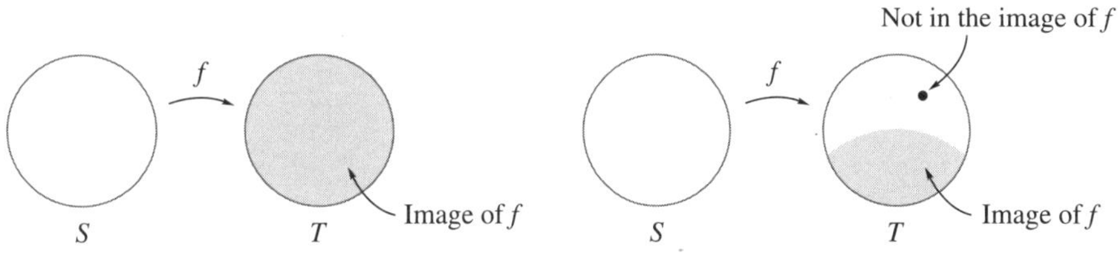

(a) *f* is onto. (b) *f* is not onto.

FIGURE A.3

2. f is not one-to-one, but is onto. For if t is a given nonnegative real number, then $s = \sqrt{t}$ is in S and $f(s) = t$. Note that the codomain makes a difference here. The *formulas* are the same in 1 and 2, but the *functions* are not.
3. f is not one-to-one, for if $f((a_1, a_2, a_3)) = f((b_1, b_2, b_3))$, then $(a_1, a_2, 0) = (b_1, b_2, 0)$, so $a_1 = b_1$ and $a_2 = b_2$. However, b_3 need not equal a_3. The range of f is T; that is, f is onto. For let $(x_1, x_2, 0)$ be any element of T. Can we find an element (a_1, a_2, a_3) of S such that $f((a_1, a_2, a_3)) = (x_1, x_2, 0)$? We merely let $a_1 = x_1$, $a_2 = x_2$, and $a_3 =$ any real number we wish—say, $a_3 = 5$.
4. f is one-to-one, for if $f(s_1) = f(s_2)$, then $2s_1 + 1 = 2s_2 + 1$, which means that $s_1 = s_2$. Also, f is onto, for given a real number t, we seek a real number s so that $f(s) = t$; that is, we need to solve $2s + 1 = t$ for s, which we can do, obtaining $s = \frac{1}{2}(t - 1)$.
5. f is one-to one, but f is not onto, because not every element in T has 1 for its y-coordinate.
6. f is one-to-one and onto, because $e^{s_1} \neq e^{s_2}$ if $s_1 \neq s_2$, and for any positive t we can always solve $t = e^s$, obtaining $s = \ln t$.

If $f: S \to T$ and $g: T \to U$ are functions, then we can define a new function $g \circ f$, by $(g \circ f)(s) = g(f(s))$, for s in S. The function $g \circ f: S \to U$ is called the **composite** of f and g. Thus, if f and g are the functions 4 and 6 in the preceding list of functions, then $g \circ f$ is defined by $(g \circ f)(s) = g(f(s)) = e^{2s+1}$, and $f \circ g$ is defined by $(f \circ g)(s) = f(g(s)) = 2e^s + 1$. The function $i: S \to S$ defined by $i(s) = s$, for s in S, is called the **identity function** on S. A function $f: S \to T$ for which there is a function $g: T \to S$ such that $g \circ f = i_S =$ identity function on S and $f \circ g = i_T =$ identity function on T is called an **invertible function**, and g is called an **inverse** of f. It can be shown that a function can have at most one inverse. It is not difficult to show—and we do so in Chapter 6 for a special case—that a function $f: S \to T$ is invertible if and only if it is one-to-one and onto. The inverse of f, if it exists, is denoted by f^{-1}. If f is invertible, then f^{-1} is also one-to-one and onto. "If and only if" means that both the statement and its converse are true (see Appendix C). That is, if $f: S \to T$ is invertible, then f is one-to-one and onto; if $f: S \to T$ is one-to-one and onto, then f is invertible. Functions 4 and 6 are invertible; the inverse of function 4 is $g: T \to S$ defined by $g(t) = \frac{1}{2}(t - 1)$ for t in T; the inverse of function 6 is $g: T \to S$, defined by $g(t) = \ln t$.

APPENDIX

B Complex Numbers

B.1 Complex Numbers

Complex numbers are usually introduced in an algebra course to "complete" the solution to the quadratic equation

$$ax^2 + bx + c = 0.$$

In using the quadratic formula

$$x = \frac{-b \pm \sqrt{b^2 - 4ac}}{2a},$$

the case in which $b^2 - 4ac < 0$ is not resolved unless we can cope with the square roots of negative numbers. In the sixteenth century, mathematicians and scientists justified this "completion" of the solution of quadratic equations by intuition. Naturally, a controversy arose, with some mathematicians denying the existence of these numbers and others using them along with real numbers. The use of complex numbers did not lead to any contradictions, and the idea proved to be an important milestone in the development of mathematics.

A **complex number** c is of the form $c = a + bi$, where a and b are real numbers and where $i = \sqrt{-1}$; a is called the **real part** of c, and b is called the **imaginary part** of c. The term *imaginary part* arose from the mysticism surrounding the beginnings of complex numbers; however, these numbers are as "real" as the real numbers.

EXAMPLE 1

(a) $5 - 3i$ has real part 5 and imaginary part -3;

(b) $-6 + \sqrt{2}\,i$ has real part -6 and imaginary part $\sqrt{2}$. ■

The symbol $i = \sqrt{-1}$ has the property that $i^2 = -1$, and we can deduce the following relationships:

$$i^3 = -i, \quad i^4 = 1, \quad i^5 = i, \quad i^6 = -1, \quad i^7 = -i, \quad \dots .$$

These results will be handy for simplifying operations involving complex numbers.

We say that two complex numbers $c_1 = a_1 + b_1 i$ and $c_2 = a_2 + b_2 i$ are **equal** if their real and imaginary parts are equal, that is, if $a_1 = a_2$ and $b_1 = b_2$. Of course, every real number a is a complex number with its imaginary part zero: $a = a + 0i$.

Operations on Complex Numbers

If $c_1 = a_1 + b_1 i$ and $c_2 = a_2 + b_2 i$ are complex numbers, then their **sum** is

$$c_1 + c_2 = (a_1 + a_2) + (b_1 + b_2)i,$$

and their **difference** is

$$c_1 - c_2 = (a_1 - a_2) + (b_1 - b_2)i.$$

In words, to form the sum of two complex numbers, add the real parts and add the imaginary parts. The **product** of c_1 and c_2 is

$$\begin{aligned} c_1 c_2 &= (a_1 + b_1 i) \cdot (a_2 + b_2 i) = a_1 a_2 + (a_1 b_2 + b_1 a_2)i + b_1 b_2 i^2 \\ &= (a_1 a_2 - b_1 b_2) + (a_1 b_2 + b_1 a_2)i. \end{aligned}$$

A special case of multiplication of complex numbers occurs when c_1 is real. In this case, we obtain the simple result

$$c_1 c_2 = c_1 \cdot (a_2 + b_2 i) = c_1 a_2 + c_1 b_2 i.$$

If $c = a + bi$ is a complex number, then the **conjugate** of c is the complex number $\overline{c} = a - bi$. It is not difficult to show that if c and d are complex numbers, then the following basic properties of complex arithmetic hold:

1. $\overline{\overline{c}} = c$.
2. $\overline{c + d} = \overline{c} + \overline{d}$.
3. $\overline{cd} = \overline{c}\,\overline{d}$.
4. c is a real number if and only if $c = \overline{c}$.
5. $c\overline{c}$ is a nonnegative real number and $c\overline{c} = 0$ if and only if $c = 0$.

We prove property 4 here and leave the others as exercises. Let $c = a + bi$ so that $\overline{c} = a - bi$. If $c = \overline{c}$, then $a + bi = a - bi$, so $b = 0$ and c is real. On the other hand, if c is real, then $c = a$ and $\overline{c} = a$, so $c = \overline{c}$.

EXAMPLE 2

Let $c_1 = 5 - 3i$, $c_2 = 4 + 2i$, and $c_3 = -3 + i$.

(a) $c_1 + c_2 = (5 - 3i) + (4 + 2i) = 9 - i$

(b) $c_2 - c_3 = (4 + 2i) - (-3 + i) = (4 - (-3)) + (2 - 1)i = 7 + i$

(c) $c_1 c_2 = (5 - 3i) \cdot (4 + 2i) = 20 + 10i - 12i - 6i^2 = 26 - 2i$

(d) $c_1 \overline{c_3} = (5 - 3i) \cdot \overline{(-3 + i)} = (5 - 3i) \cdot (-3 - i)$
$= -15 - 5i + 9i + 3i^2$
$= -18 + 4i$

(e) $3c_1 + 2\bar{c}_2 = 3(5 - 3i) + 2\overline{(4 + 2i)} = (15 - 9i) + 2(4 - 2i)$
$= (15 - 9i) + (8 - 4i) = 23 - 13i$

(f) $c_1\bar{c}_1 = (5 - 3i)\overline{(5 - 3i)} = (5 - 3i)(5 + 3i) = 34$ ■

When we consider systems of linear equations with complex coefficients, we need to divide complex numbers to complete the solution process and derive a reasonable form for the solution. Let $c_1 = a_1 + b_1 i$ and $c_2 = a_2 + b_2 i$. If $c_2 \neq 0$—that is, if $a_2 \neq 0$ or $b_2 \neq 0$—then we can **divide** c_1 by c_2:

$$\frac{c_1}{c_2} = \frac{a_1 + b_1 i}{a_2 + b_2 i}.$$

To conform to our practice of expressing a complex number in the form real part + imaginary part $\cdot\, i$, we must simplify the foregoing expression for c_1/c_2. To simplify this complex fraction, we multiply the numerator and the denominator by the conjugate of the denominator. Thus, dividing c_1 by c_2 gives the complex number

$$\frac{c_1}{c_2} = \frac{a_1 + b_1 i}{a_2 + b_2 i} = \frac{(a_1 + b_1 i)(a_2 - b_2 i)}{(a_2 + b_2 i)(a_2 - b_2 i)} = \frac{a_1 a_2 + b_1 b_2}{a_2^2 + b_2^2} - \frac{a_1 b_2 + a_2 b_1}{a_2^2 + b_2^2}\, i.$$

EXAMPLE 3

Let $c_1 = 2 - 5i$ and $c_2 = -3 + 4i$. Then

$$\frac{c_1}{c_2} = \frac{2 - 5i}{-3 + 4i} = \frac{(2 - 5i)(-3 - 4i)}{(-3 + 4i)(-3 - 4i)} = \frac{-26 + 7i}{(-3)^2 + (4)^2} = -\frac{26}{25} + \frac{7}{25}\, i.$$ ■

Finding the reciprocal of a complex number is a special case of division of complex numbers. If $c = a + bi$, $c \neq 0$, then

$$\frac{1}{c} = \frac{1}{a + bi} = \frac{a - bi}{(a + bi)(a - bi)} = \frac{a - bi}{a^2 + b^2}$$
$$= \frac{a}{a^2 + b^2} - \frac{b}{a^2 + b^2}\, i.$$

EXAMPLE 4

(a) $\dfrac{1}{2 + 3i} = \dfrac{2 - 3i}{(2 + 3i)(2 - 3i)} = \dfrac{2 - 3i}{2^2 + 3^2} = \dfrac{2}{13} - \dfrac{3}{13}\, i$

(b) $\dfrac{1}{i} = \dfrac{-i}{i(-i)} = \dfrac{-i}{-i^2} = \dfrac{-i}{-(-1)} = -i$ ■

Summarizing, we can say that complex numbers are mathematical objects for which addition, subtraction, multiplication, and division are defined in such a way that these operations on real numbers can be derived as special cases. In fact, it can be shown that complex numbers form a mathematical system that is called a **field**.

■ Geometric Representation of Complex Numbers

A complex number $c = a + bi$ may be regarded as an ordered pair (a, b) of real numbers. This ordered pair of real numbers corresponds to a point in the plane. Such a correspondence naturally suggests that we represent $a + bi$ as a point in the **complex plane**, where the horizontal axis is used to represent the real part of c and the vertical axis is used to represent the imaginary part of c. To simplify matters, we call these the **real axis** and the **imaginary axis**, respectively (see Figure B.1).

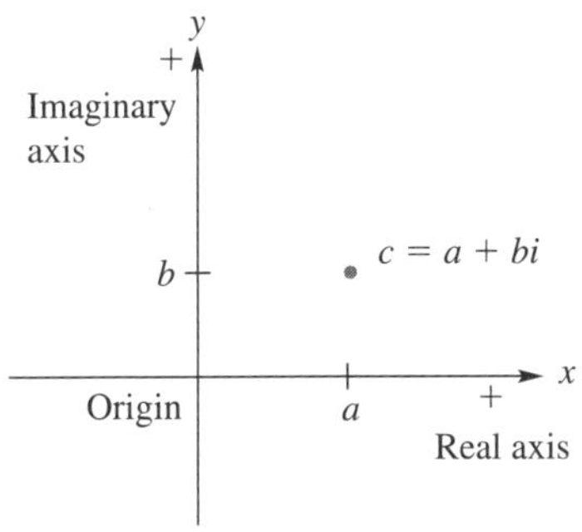

FIGURE B.1 Complex plane.

EXAMPLE 5

Plot the complex numbers $c = 2 - 3i$, $d = 1 + 4i$, $e = -3$, and $f = 2i$ in the complex plane.

Solution

See Figure B.2. ■

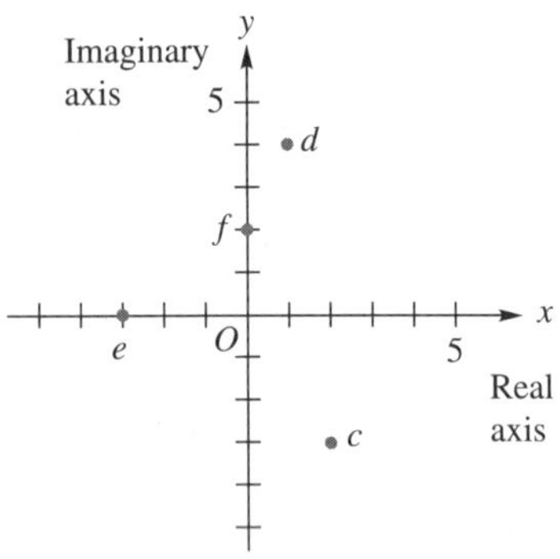

FIGURE B.2

The rules concerning inequality of real numbers, such as less than and greater than, *do not apply to complex numbers.* There is no way to arrange the complex numbers according to size. However, using the geometric representation from the complex plane, we can attach a notion of size to a complex number by measuring its distance from the origin. The distance from the origin to $c = a + bi$ is called the **absolute value**, or **modulus**, of the complex number and is denoted by $|c| = |a+bi|$. Using the formula for the distance between ordered pairs of real numbers, we get

$$|c| = |a + bi| = \sqrt{a^2 + b^2}.$$

It follows that $c\,\overline{c} = |c|^2$ (verify).

EXAMPLE 6

Referring to Example 5, we note that $|c| = \sqrt{13}$; $|d| = \sqrt{17}$; $|e| = 3$; $|f| = 2$. ■

A different, but related interpretation of a complex number is obtained if we associate with $c = a + bi$ the vector OP, where O is the origin $(0, 0)$ and P is the point (a, b). There is an obvious correspondence between this representation and vectors in the plane discussed in calculus, which we reviewed in Section 4.1. Using a vector representation, addition and subtraction of complex numbers can be viewed as the corresponding vector operations. These are represented in Figures 4.5, 4.6, and 4.9. We will not pursue the manipulation of complex numbers by vector operations here, but such a point of view is important for the development and study of complex variables.

■ Matrices with Complex Entries

If the entries of a matrix are complex numbers, we can perform the matrix operations of addition, subtraction, multiplication, and scalar multiplication in a manner completely analogous to that for real matrices. We verify the validity of these operations, using properties of complex arithmetic and just imitating the proofs for real matrices presented in the text. We illustrate these concepts in the following example:

EXAMPLE 7

Let

$$A = \begin{bmatrix} 4+i & -2+3i \\ 6+4i & -3i \end{bmatrix}, \quad B = \begin{bmatrix} 2-i & 3-4i \\ 5+2i & -7+5i \end{bmatrix},$$

$$C = \begin{bmatrix} 1+2i & i \\ 3-i & 8 \\ 4+2i & 1-i \end{bmatrix}.$$

(a)
$$A + B = \begin{bmatrix} (4+i)+(2-i) & (-2+3i)+(3-4i) \\ (6+4i)+(5+2i) & (-3i)+(-7+5i) \end{bmatrix}$$

$$= \begin{bmatrix} 6 & 1-i \\ 11+6i & -7+2i \end{bmatrix}$$

(b) $B - A = \begin{bmatrix} (2-i)-(4+i) & (3-4i)-(-2+3i) \\ (5+2i)-(6+4i) & (-7+5i)-(-3i) \end{bmatrix}$

$= \begin{bmatrix} -2-2i & 5-7i \\ -1-2i & -7+8i \end{bmatrix}$

(c) $CA = \begin{bmatrix} 1+2i & i \\ 3-i & 8 \\ 4+2i & 1-i \end{bmatrix} \begin{bmatrix} 4+i & -2+3i \\ 6+4i & -3i \end{bmatrix}$

$= \begin{bmatrix} (1+2i)(4+i)+(i)(6+4i) & (1+2i)(-2+3i)+(i)(-3i) \\ (3-i)(4+i)+(8)(6+4i) & (3-i)(-2+3i)+(8)(-3i) \\ (4+2i)(4+i)+(1-i)(6+4i) & (4+2i)(-2+3i)+(1-i)(-3i) \end{bmatrix}$

$= \begin{bmatrix} -2+15i & -5-i \\ 61+31i & -3-13i \\ 24+10i & -17+5i \end{bmatrix}$

(d) $(2+i)B = \begin{bmatrix} (2+i)(2-i) & (2+i)(3-4i) \\ (2+i)(5+2i) & (2+i)(-7+5i) \end{bmatrix}$

$= \begin{bmatrix} 5 & 10-5i \\ 8+9i & -19+3i \end{bmatrix}$ ■

Just as we can compute the conjugate of a complex number, we can compute the **conjugate of a matrix** by computing the conjugate of each entry of the matrix. We denote the conjugate of a matrix A by $\overline{A}$ and write

$$\overline{A} = \left[\overline{a_{ij}}\right].$$

EXAMPLE 8 Referring to Example 7, we find that

$$\overline{A} = \begin{bmatrix} 4-i & -2-3i \\ 6-4i & 3i \end{bmatrix} \quad \text{and} \quad \overline{B} = \begin{bmatrix} 2+i & 3+4i \\ 5-2i & -7-5i \end{bmatrix}.$$ ■

The following properties of the conjugate of a matrix hold:

1. $\overline{\overline{A}} = A$.
2. $\overline{A+B} = \overline{A} + \overline{B}$.
3. $\overline{AB} = \overline{A}\,\overline{B}$.
4. For any real number k, $\overline{kA} = k\,\overline{A}$.
5. For any complex number c, $\overline{cA} = \overline{c}\,\overline{A}$.
6. $(\overline{A})^T = \overline{A^T}$.
7. If A is nonsingular, then $(\overline{A})^{-1} = \overline{A^{-1}}$.

We prove properties 5 and 6 here and leave the others as exercises. First, property 5 is proved as follows: If c is complex, the (i, j) entry of $\overline{cA}$ is

$$\overline{ca_{ij}} = \overline{c}\,\overline{a_{ij}},$$

which is the (i, j) entry of $\overline{c}\,\overline{A}$. Next, we prove property 6: The (i, j) entry of $(\overline{A})^T$ is $\overline{a_{ji}}$, which is the (i, j) entry of $\overline{A^T}$.

■ Special Types of Complex Matrices

As we have already seen, certain types of real matrices satisfy some important properties. The same situation applies to complex matrices, and we now discuss several of these types of matrices.

An $n \times n$ complex matrix A is called **Hermitian**† if

$$\overline{A^T} = A.$$

This is equivalent to saying that $\overline{a_{ji}} = a_{ij}$ for all i and j. Every real symmetric matrix is Hermitian [Exercise 11(c)], so we may consider Hermitian matrices as the analogs of real symmetric matrices.

EXAMPLE 9

The matrix

$$A = \begin{bmatrix} 2 & 3+i \\ 3-i & 5 \end{bmatrix}$$

is Hermitian, since

$$\overline{A^T} = \overline{\begin{bmatrix} 2 & 3-i \\ 3+i & 5 \end{bmatrix}} = \begin{bmatrix} 2 & 3+i \\ 3-i & 5 \end{bmatrix} = A.$$

■

An $n \times n$ complex matrix A is called **unitary** if

$$(\overline{A^T})A = A(\overline{A^T}) = I_n.$$

This is equivalent to saying that $\overline{A^T} = A^{-1}$. Every real orthogonal matrix is unitary [Exercise 12(a)], so we may consider unitary matrices as the analogs of real orthogonal matrices.

EXAMPLE 10

The matrix

$$A = \begin{bmatrix} \frac{1}{\sqrt{3}} & \frac{1+i}{\sqrt{3}} \\ \frac{1-i}{\sqrt{3}} & -\frac{1}{\sqrt{3}} \end{bmatrix}$$

is unitary, since (verify)

$$(\overline{A^T})A = \begin{bmatrix} \frac{1}{\sqrt{3}} & \frac{1+i}{\sqrt{3}} \\ \frac{1-i}{\sqrt{3}} & -\frac{1}{\sqrt{3}} \end{bmatrix} \begin{bmatrix} \frac{1}{\sqrt{3}} & \frac{1+i}{\sqrt{3}} \\ \frac{1-i}{\sqrt{3}} & -\frac{1}{\sqrt{3}} \end{bmatrix} = I_2$$

and, similarly, $A(\overline{A^T}) = I_2$. ■

CHARLES HERMITE

†Charles Hermite (1822–1901) was born to a well-to-do middle-class family in Dieuze, Lorraine, France and died in Paris. He studied at the École Polytechnique for only one year and continued his mathematical studies on his own with the encouragement of several of the leading mathematicians of the day, who recognized his extraordinary abilities at a young age. He did not like geometry but did make many important contributions in number theory, algebra, and linear algebra. One of his two major contributions was to show that the general fifth-degree polynomial can be solved by using a special type of function called an elliptic function. His second major contribution was to show that the number e (the base for the system of natural logarithms) is transcendental; that is, e is not the root of any polynomial equation with integer coefficients.

There is one more type of complex matrix that is important. An $n \times n$ complex matrix is called **normal** if

$$(\overline{A^T})\,A = A\,(\overline{A^T}).$$

EXAMPLE 11

The matrix

$$A = \begin{bmatrix} 5-i & -1+i \\ -1-i & 3-i \end{bmatrix}$$

is normal, since (verify)

$$(\overline{A^T})\,A = A\,(\overline{A^T}) = \begin{bmatrix} 28 & -8+8i \\ -8-8i & 12 \end{bmatrix}.$$

Moreover, A is not Hermitian, since $\overline{A^T} \neq A$ (verify). ■

■ Complex Numbers and Roots of Polynomials

A polynomial of degree n with real coefficients has n complex roots, some, all, or none of which may be real numbers. Thus the polynomial $f_1(x) = x^4 - 1$ has the roots i, $-i$, 1, and -1; the polynomial $f_2(x) = x^2 - 1$ has the roots 1 and -1; and the polynomial $f_3(x) = x^2 + 1$ has the roots i and $-i$.

B.1 Exercises

1. Let $c_1 = 3+4i$, $c_2 = 1-2i$, and $c_3 = -1+i$. Compute each of the following and simplify as much as possible:
   - **(a)** $c_1 + c_2$
   - **(b)** $c_3 - c_1$
   - **(c)** $c_1 c_2$
   - **(d)** $c_2 \overline{c_3}$
   - **(e)** $4c_3 + \overline{c_2}$
   - **(f)** $(-i) \cdot c_2$
   - **(g)** $\overline{3c_1 - ic_2}$
   - **(h)** $c_1 c_2 c_3$
2. Write in the form $a + bi$.
   - **(a)** $\dfrac{1+2i}{3-4i}$
   - **(b)** $\dfrac{2-3i}{3-i}$
   - **(c)** $\dfrac{(2+i)^2}{i}$
   - **(d)** $\dfrac{1}{(3+2i)(1+i)}$
3. Represent each complex number as a point and as a vector in the complex plane.
   - **(a)** $4+2i$
   - **(b)** $-3+i$
   - **(c)** $3-2i$
   - **(d)** $i(4+i)$
4. Find the modulus of each complex number in Exercise 3.
5. If $c = a + bi$, then we can denote the real part of c by $\text{Re}(c)$ and the imaginary part of c by $\text{Im}(c)$.
   - **(a)** For any complex numbers $c_1 = a_1 + b_1 i$, $c_2 = a_2 + b_2 i$, prove that $\text{Re}(c_1 + c_2) = \text{Re}(c_1) + \text{Re}(c_2)$ and $\text{Im}(c_1 + c_2) = \text{Im}(c_1) + \text{Im}(c_2)$.
   - **(b)** For any real number k, prove that $\text{Re}(kc) = k\,\text{Re}(c)$ and $\text{Im}(kc) = k\,\text{Im}(c)$.
   - **(c)** Is part (b) true if k is a complex number?
   - **(d)** Prove or disprove:

     $$\text{Re}(c_1 c_2) = \text{Re}(c_1) \cdot \text{Re}(c_2).$$
6. Sketch, in the complex plane, the vectors corresponding to c and $\overline{c}$ if $c = 2+3i$ and $c = -1+4i$. Geometrically, we can say that $\overline{c}$ is the reflection of c with respect to the real axis. (See also Example 4 in Section 1.6.)
7. Let

   $$A = \begin{bmatrix} 2+2i & -1+3i \\ -2 & 1-i \end{bmatrix},$$

   $$B = \begin{bmatrix} 2i & 1+2i \\ 0 & 3-i \end{bmatrix}, \qquad C = \begin{bmatrix} 2+i \\ -i \end{bmatrix}.$$

   Compute each of the following and simplify each entry as $a + bi$:
   - **(a)** $A + B$
   - **(b)** $(1-2i)C$
   - **(c)** AB
   - **(d)** BC
   - **(e)** $A - 2I_2$
   - **(f)** $\overline{B}$
   - **(g)** $A\overline{C}$
   - **(h)** $(\overline{A+B})C$
8. Let A and B be $m \times n$ complex matrices, and let C be an $n \times n$ nonsingular matrix.
   - **(a)** Prove that $\overline{A+B} = \overline{A} + \overline{B}$.
   - **(b)** Prove that for any real number k, $\overline{kA} = k\overline{A}$.
   - **(c)** Prove that $(\overline{C})^{-1} = \overline{C^{-1}}$.

9. If $A = \begin{bmatrix} 0 & i \\ i & 0 \end{bmatrix}$, compute A^2, A^3, and A^4. Give a general rule for A^n, n a positive integer.

10. Which of the following matrices are Hermitian, unitary, or normal?

(a) $\begin{bmatrix} 3 & 2+i \\ 2-i & 4 \end{bmatrix}$ **(b)** $\begin{bmatrix} 2 & 1-i \\ 3+i & -2 \end{bmatrix}$

(c) $\begin{bmatrix} \frac{1-i}{2} & \frac{1+i}{2} \\ \frac{1+i}{2} & \frac{1-i}{2} \end{bmatrix}$ **(d)** $\begin{bmatrix} 1 & -1 \\ 1 & 1 \end{bmatrix}$

(e) $\begin{bmatrix} 1 & 3-i & 4-i \\ 3+i & -2 & 2+i \\ 4+i & 2-i & 3 \end{bmatrix}$

(f) $\begin{bmatrix} 3 & \frac{3-i}{2} & \frac{4-i}{2} \\ \frac{3-i}{2} & -2 & 2+i \\ \frac{4-i}{2} & 2-i & 5 \end{bmatrix}$

(g) $\begin{bmatrix} 3+2i & -1 \\ -i & 2+i \end{bmatrix}$ **(h)** $\begin{bmatrix} i & i \\ -i & 1 \end{bmatrix}$

(i) $\begin{bmatrix} 1 & 0 & 0 \\ 0 & \frac{1+i}{\sqrt{3}} & \frac{1}{\sqrt{3}} \\ 0 & -\frac{1}{\sqrt{3}} & \frac{1-i}{\sqrt{3}} \end{bmatrix}$

(j) $\begin{bmatrix} 4+7i & -2-i \\ 1-2i & 3+4i \end{bmatrix}$

11. (a) Prove that the diagonal entries of a Hermitian matrix must be real.

(b) Prove that every Hermitian matrix A can be written as $A = B + iC$, where B is real and symmetric and C is real and skew symmetric (see Definition 1.9). [*Hint*: Consider $B = (A + \overline{A})/2$ and $C = (A - \overline{A})/2i$.]

(c) Prove that every real symmetric matrix is Hermitian.

12. (a) Show that every real orthogonal matrix is unitary.

(b) Show that if A is a unitary matrix, then A^T is unitary.

(c) Show that if A is a unitary matrix, then A^{-1} is unitary.

13. Let A be an $n \times n$ complex matrix.

(a) Show that A can be written as $B + iC$, where B and C are Hermitian.

(b) Show that A is normal if and only if

$$BC = CB.$$

[*Hint*: Consider $B = (A + \overline{A^T})/2$ and $C = (A - \overline{A^T})/2i$.]

14. (a) Prove that every Hermitian matrix is normal.

(b) Prove that every unitary matrix is normal.

(c) Find a 2×2 normal matrix that is neither Hermitian nor unitary.

15. An $n \times n$ complex matrix A is called **skew Hermitian** if

$$\overline{A^T} = -A.$$

Show that a matrix $A = B + iC$, where B and C are real matrices, is skew Hermitian if and only if B is skew symmetric and C is symmetric.

16. Find all the roots.

(a) $x^2 + x + 1 = 0$

(b) $x^3 + 2x^2 + x + 2 = 0$

(c) $x^5 + x^4 - x - 1 = 0$

17. Let $p(x)$ denote a polynomial and let A be a square matrix. Then $p(A)$ is called a **matrix polynomial**, or a **polynomial in the matrix** A. For $p(x) = 2x^2 + 5x - 3$, compute $p(A) = 2A^2 + 5A - 3I_n$ for each of the following:

(a) $A = \begin{bmatrix} -3 & 0 \\ 0 & -3 \end{bmatrix}$ **(b)** $A = \begin{bmatrix} 1 & 2 \\ 0 & 1 \end{bmatrix}$

(c) $A = \begin{bmatrix} 0 & i \\ i & 0 \end{bmatrix}$ **(d)** $A = \begin{bmatrix} 1 & i \\ 0 & 0 \end{bmatrix}$

18. Let $p(x) = x^2 + 1$.

(a) Determine two different 2×2 matrices A of the form kI_2 that satisfy $p(A) = O$.

(b) Verify that $p(A) = O$, for $A = \begin{bmatrix} 1 & 2 \\ -1 & -1 \end{bmatrix}$.

19. Find all the 2×2 matrices A of the form kI_2 that satisfy $p(A) = O$ for $p(x) = x^2 - x - 2$.

20. In Supplementary Exercise 4 in Chapter 1, we introduced the concept of a square root of a matrix with real entries. We can generalize the notion of a square root of a matrix if we permit complex entries.

(a) Compute a complex square root of

$$A = \begin{bmatrix} -1 & 0 \\ 0 & 0 \end{bmatrix}.$$

(b) Compute a complex square root of

$$A = \begin{bmatrix} -2 & 2 \\ 2 & -2 \end{bmatrix}.$$

B.2 Complex Numbers in Linear Algebra

The primary goal of this appendix is to provide an easy transition to complex numbers in linear algebra. This is of particular importance in Chapter 7, where complex eigenvalues and eigenvectors arise naturally for matrices with real entries. Hence we restate only the main theorems in the complex case and provide a discussion and examples of the major ideas needed to accomplish this transition. It will soon be evident that the increased computational effort of complex arithmetic becomes quite tedious if done by hand.

■ Solving Linear Systems with Complex Entries

The results and techniques dealing with the solution of linear systems that we developed in Chapter 2 carry over directly to linear systems with complex coefficients. We shall illustrate row operations and echelon forms for such systems with Gauss–Jordan reduction, using complex arithmetic.

EXAMPLE 1

Solve the following linear system by Gauss–Jordan reduction:

$$\begin{aligned}(1+i)x_1 + (2+i)x_2 &= 5\\(2-2i)x_1 + \qquad\quad ix_2 &= 1+2i.\end{aligned}$$

Solution

We form the augmented matrix and use elementary row operations to transform it to reduced row echelon form. For the augmented matrix $\left[A \mid B\right]$,

$$\left[\begin{array}{cc|c} 1+i & 2+i & 5 \\ 2-2i & i & 1+2i \end{array}\right],$$

multiply the first row by $1/(1+i)$ to obtain

$$\left[\begin{array}{cc|c} 1 & \frac{3}{2}-\frac{1}{2}i & \frac{5}{2}-\frac{5}{2}i \\ 2-2i & i & 1+2i \end{array}\right].$$

We now add $-(2-2i)$ times the first row to the second row to get

$$\left[\begin{array}{cc|c} 1 & \frac{3}{2}-\frac{1}{2}i & \frac{5}{2}-\frac{5}{2}i \\ 0 & -2+5i & 1+12i \end{array}\right].$$

Multiply the second row by $1/(-2+5i)$ to obtain

$$\left[\begin{array}{cc|c} 1 & \frac{3}{2}-\frac{1}{2}i & \frac{5}{2}-\frac{5}{2}i \\ 0 & 1 & 2-i \end{array}\right],$$

which is in row echelon form. To get to reduced row echelon form, we add $-\left(\frac{3}{2}-\frac{1}{2}i\right)$ times the second row to the first row to obtain

$$\left[\begin{array}{cc|c} 1 & 0 & 0 \\ 0 & 1 & 2-i \end{array}\right].$$

Hence the solution is $x_1 = 0$ and $x_2 = 2 - i$. ■

If you carry out the arithmetic for the row operations in the preceding example, you will feel the burden of the complex arithmetic, even though there were just two equations in two unknowns. Gaussian elimination with back substitution can also be used on linear systems with complex coefficients.

EXAMPLE 2

Suppose that the augmented matrix of a linear system has been transformed to the following matrix in row echelon form:

$$\left[\begin{array}{ccc|c} 1 & 0 & 1+i & -1 \\ 0 & 1 & 3i & 2+i \\ 0 & 0 & 1 & 2i \end{array}\right].$$

The back substitution procedure gives us

$$\begin{aligned} x_3 &= 2i \\ x_2 &= 2 + i - 3i(2i) = 2 + i + 6 = 8 + i \\ x_1 &= -1 - (1+i)(2i) = -1 - 2i + 2 = 3 - 2i. \end{aligned}$$ ■

We can alleviate the tedium of complex arithmetic by using computers to solve linear systems with complex entries. However, we must still pay a high price, because the execution time will be approximately twice as long as that for the same size linear system with all real entries. We illustrate this by showing how to transform an $n \times n$ linear system with complex coefficients to a $2n \times 2n$ linear system with only real coefficients.

EXAMPLE 3

Consider the linear system

$$\begin{aligned} (2+i)x_1 + \ \ (1+i)x_2 &= 3 + 6i \\ (3-i)x_1 + (2-2i)x_2 &= 7 - i. \end{aligned}$$

If we let $x_1 = a_1 + b_1 i$ and $x_2 = a_2 + b_2 i$, with a_1, b_1, a_2, and b_2 real numbers, then we can write this system in matrix form as

$$\begin{bmatrix} 2+i & 1+i \\ 3-i & 2-2i \end{bmatrix} \begin{bmatrix} a_1 + b_1 i \\ a_2 + b_2 i \end{bmatrix} = \begin{bmatrix} 3+6i \\ 7-i \end{bmatrix}.$$

We first rewrite the given linear system as

$$\left(\begin{bmatrix} 2 & 1 \\ 3 & 2 \end{bmatrix} + i \begin{bmatrix} 1 & 1 \\ -1 & -2 \end{bmatrix} \right) \left(\begin{bmatrix} a_1 \\ a_2 \end{bmatrix} + i \begin{bmatrix} b_1 \\ b_2 \end{bmatrix} \right) = \begin{bmatrix} 3 \\ 7 \end{bmatrix} + i \begin{bmatrix} 6 \\ -1 \end{bmatrix}.$$

Multiplying, we have

$$\left(\begin{bmatrix} 2 & 1 \\ 3 & 2 \end{bmatrix} \begin{bmatrix} a_1 \\ a_2 \end{bmatrix} - \begin{bmatrix} 1 & 1 \\ -1 & -2 \end{bmatrix} \begin{bmatrix} b_1 \\ b_2 \end{bmatrix} \right)$$

$$+ i \left(\begin{bmatrix} 2 & 1 \\ 3 & 2 \end{bmatrix} \begin{bmatrix} b_1 \\ b_2 \end{bmatrix} + \begin{bmatrix} 1 & 1 \\ -1 & -2 \end{bmatrix} \begin{bmatrix} a_1 \\ a_2 \end{bmatrix} \right) = \begin{bmatrix} 3 \\ 7 \end{bmatrix} + i \begin{bmatrix} 6 \\ -1 \end{bmatrix}.$$

The real and imaginary parts on both sides of the equation must agree, respectively, and so we have

$$\begin{bmatrix} 2 & 1 \\ 3 & 2 \end{bmatrix}\begin{bmatrix} a_1 \\ a_2 \end{bmatrix} - \begin{bmatrix} 1 & 1 \\ -1 & -2 \end{bmatrix}\begin{bmatrix} b_1 \\ b_2 \end{bmatrix} = \begin{bmatrix} 3 \\ 7 \end{bmatrix}$$

and

$$\begin{bmatrix} 2 & 1 \\ 3 & 2 \end{bmatrix}\begin{bmatrix} b_1 \\ b_2 \end{bmatrix} + \begin{bmatrix} 1 & 1 \\ -1 & -2 \end{bmatrix}\begin{bmatrix} a_1 \\ a_2 \end{bmatrix} = \begin{bmatrix} 6 \\ -1 \end{bmatrix}.$$

This leads to the linear system

$$\begin{aligned} 2a_1 + a_2 - b_1 - b_2 &= 3 \\ 3a_1 + 2a_2 + b_1 + 2b_2 &= 7 \\ a_1 + a_2 + 2b_1 + b_2 &= 6 \\ -a_1 - 2a_2 + 3b_1 + 2b_2 &= -1, \end{aligned}$$

which can be written as

$$\begin{bmatrix} 2 & 1 & -1 & -1 \\ 3 & 2 & 1 & 2 \\ 1 & 1 & 2 & 1 \\ -1 & -2 & 3 & 2 \end{bmatrix}\begin{bmatrix} a_1 \\ a_2 \\ b_1 \\ b_2 \end{bmatrix} = \begin{bmatrix} 3 \\ 7 \\ 6 \\ -1 \end{bmatrix}.$$

This linear system of four equations in four unknowns is now solved as in Chapter 2. The solution is (verify) $a_1 = 1$, $a_2 = 2$, $b_1 = 2$, and $b_2 = -1$. Thus $x_1 = 1 + 2i$ and $x_2 = 2 - i$ is the solution to the given linear system. ■

■ Determinants of Complex Matrices

The definition of a determinant and all the properties derived in Chapter 3 apply to matrices with complex entries. The following example is an illustration:

EXAMPLE 4 Let A be the coefficient matrix of Example 3. Compute $|A|$.

Solution

$$\begin{aligned} \begin{vmatrix} 2+i & 1+i \\ 3-i & 2-2i \end{vmatrix} &= (2+i)(2-2i) - (3-i)(1+i) \\ &= (6-2i) - (4+2i) \\ &= 2-4i \end{aligned}$$ ■

■ Complex Vector Spaces

A **complex vector space** is defined exactly as is a real vector space in Definition 4.4, except that the scalars in properties 5 through 8 are permitted to be complex numbers. The terms *complex* vector space and *real* vector space emphasize the set from which the scalars are chosen. It happens that, in order to satisfy the closure property of scalar multiplication [Definition 4.4(b)] in a complex vector space, we must, in most examples, consider vectors that involve complex numbers.

Most of the real vector spaces of Chapter 4 have complex vector space analogs.

EXAMPLE 5

(a) Consider C^n, the set of all $n \times 1$ matrices

$$\begin{bmatrix} a_1 \\ a_2 \\ \vdots \\ a_n \end{bmatrix}$$

with complex entries. Let the operation $\oplus$ be matrix addition and let the operation $\odot$ be multiplication of a matrix by a complex number. We can verify that C^n is a complex vector space by using the properties of matrices established in Section 1.4 and the properties of complex arithmetic established in Section B.1. (Note that if the operation $\odot$ is taken as multiplication of a matrix by a real number, then C^n is a real vector space whose vectors have complex components.)

(b) The set of all $m \times n$ matrices, with complex entries with matrix addition as $\oplus$ and multiplication of a matrix by a complex number as $\odot$, is a complex vector space (verify). We denote this vector space by C_{mn}.

(c) The set of polynomials, with complex coefficients with polynomial addition as $\oplus$ and multiplication of a polynomial by a complex constant as $\odot$, forms a complex vector space. Verification follows the pattern of Example 6 in Section 4.2.

(d) The set of complex-valued continuous functions defined on the interval $[a, b]$ (i.e., all functions of the form $f(t) = f_1(t) + if_2(t)$, where f_1 and f_2 are real-valued continuous functions on $[a, b]$), with $\oplus$ defined by $(f \oplus g)(t) = f(t) + g(t)$ and $\odot$ defined by $(c \odot f)(t) = cf(t)$ for a complex scalar c, forms a complex vector space. The corresponding real vector space is given in Example 7 in Section 4.2 for the interval $(-\infty, \infty)$. ■

A **complex vector subspace** W of a complex vector space V is defined as in Definition 4.5, but with real scalars replaced by complex ones. The analog of Theorem 4.3 can be proved to show that a nonempty subset W of a complex vector space V is a complex vector subspace if and only if the following conditions hold:

(a) If **u** and **v** are any vectors in W, then $\mathbf{u} \oplus \mathbf{v}$ is in W.

(b) If c is any complex number and **u** is any vector in W, then $c \odot \mathbf{u}$ is in W.

EXAMPLE 6

(a) Let W be the set of all vectors in C_{31} of the form

$$\begin{bmatrix} a \\ 0 \\ b \end{bmatrix},$$

where a and b are complex numbers. It follows that

$$\begin{bmatrix} a \\ 0 \\ b \end{bmatrix} \oplus \begin{bmatrix} d \\ 0 \\ e \end{bmatrix} = \begin{bmatrix} a+d \\ 0 \\ b+e \end{bmatrix}$$

belongs to W and, for any complex scalar c,

$$c \odot \begin{bmatrix} a \\ 0 \\ b \end{bmatrix} = \begin{bmatrix} ca \\ 0 \\ cb \end{bmatrix}$$

belongs to W. Hence W is a complex vector subspace of C_{31}.

(b) Let W be the set of all vectors in C_{mn} having only real entries. If $A = \begin{bmatrix} a_{ij} \end{bmatrix}$ and $B = \begin{bmatrix} b_{ij} \end{bmatrix}$ belong to W, then so will $A \oplus B$, because if a_{ij} and b_{ij} are real, then so is their sum. However, if c is any complex scalar and A belongs to W, then $c \odot A = cA$ can have entries ca_{ij} that need not be real numbers. It follows that $c \odot A$ need not belong to W, so W is not a complex vector subspace. ■

■ Linear Independence and Basis in Complex Vector Spaces

The notions of linear combinations, spanning sets, linear dependence, linear independence, and basis are unchanged for complex vector spaces, except that we use complex scalars. (See Sections 4.4 and 4.6.)

EXAMPLE 7

Let V be the complex vector space C^3. Let

$$\mathbf{v}_1 = \begin{bmatrix} 1 \\ i \\ 0 \end{bmatrix}, \quad \mathbf{v}_2 = \begin{bmatrix} i \\ 0 \\ 1+i \end{bmatrix}, \quad \text{and} \quad \mathbf{v}_3 = \begin{bmatrix} 1 \\ 1 \\ 1 \end{bmatrix}.$$

(a) Determine whether $\mathbf{v} = \begin{bmatrix} -1 \\ -3+3i \\ -4+i \end{bmatrix}$ is a linear combination of $\mathbf{v}_1$, $\mathbf{v}_2$, and $\mathbf{v}_3$.

(b) Determine whether $\{\mathbf{v}_1, \mathbf{v}_2, \mathbf{v}_3\}$ spans C^3.

(c) Determine whether $\{\mathbf{v}_1, \mathbf{v}_2, \mathbf{v}_3\}$ is a linearly independent subset of C^3.

(d) Is $\{\mathbf{v}_1, \mathbf{v}_2, \mathbf{v}_3\}$ a basis for C^3?

Solution

(a) We proceed as in Example 8 of Section 4.4. We form a linear combination of $\mathbf{v}_1$, $\mathbf{v}_2$, and $\mathbf{v}_3$, with unknown coefficients a_1, a_2, and a_3, respectively, and set it equal to $\mathbf{v}$:

$$a_1\mathbf{v}_1 + a_2\mathbf{v}_2 + a_3\mathbf{v}_3 = \mathbf{v}.$$

If we substitute the vectors $\mathbf{v}_1$, $\mathbf{v}_2$, $\mathbf{v}_3$, and $\mathbf{v}$ into this expression, we obtain (verify) the linear system

$$\begin{aligned} a_1 + \qquad ia_2 + a_3 &= -1 \\ ia_1 \qquad\qquad + a_3 &= -3+3i \\ (1+i)a_2 + a_3 &= -4+i. \end{aligned}$$

We next investigate the consistency of this linear system by using elementary row operations to transform its augmented matrix to either row echelon or reduced row echelon form. A row echelon form is (verify)

$$\left[\begin{array}{ccc:c} 1 & i & 1 & -1 \\ 0 & 1 & 1-i & -3+4i \\ 0 & 0 & 1 & -3 \end{array}\right],$$

which implies that the system is consistent; hence $\mathbf{v}$ is a linear combination of $\mathbf{v}_1$, $\mathbf{v}_2$, and $\mathbf{v}_3$. In fact, back substitution gives (verify) $a_1 = 3$, $a_2 = i$, and $a_3 = -3$.

(b) Let $\mathbf{v} = \begin{bmatrix} c_1 \\ c_2 \\ c_3 \end{bmatrix}$ be an arbitrary vector of C^3. We form the linear combination

$$a_1\mathbf{v}_1 + a_2\mathbf{v}_2 + a_3\mathbf{v}_3 = \mathbf{v}$$

and solve for a_1, a_2, and a_3. The resulting linear system is

$$\begin{aligned} a_1 + \quad & ia_2 + a_3 = c_1 \\ ia_1 \quad & \quad\;\; + a_3 = c_2 \\ & (1+i)a_2 + a_3 = c_3. \end{aligned}$$

Transforming the augmented matrix to row echelon form, we obtain (verify)

$$\left[\begin{array}{ccc:c} 1 & i & 1 & c_1 \\ 0 & 1 & 1-i & c_2 - ic_1 \\ 0 & 0 & 1 & -c_3 + (1+i)(c_2 - ic_1) \end{array}\right].$$

Hence we can solve for a_1, a_2, a_3 for any choice of complex numbers c_1, c_2, c_3, which implies that $\{\mathbf{v}_1, \mathbf{v}_2, \mathbf{v}_3\}$ spans C^3.

(c) Proceeding as in Example 7 of Section 4.5, we form the equation

$$a_1\mathbf{v}_1 + a_2\mathbf{v}_2 + a_3\mathbf{v}_3 = \mathbf{0}$$

and solve for a_1, a_2, and a_3. The resulting homogeneous system is

$$\begin{aligned} a_1 + \quad & ia_2 + a_3 = 0 \\ ia_1 \quad & \quad\;\; + a_3 = 0 \\ & (1+i)a_2 + a_3 = 0. \end{aligned}$$

Transforming the augmented matrix to row echelon form, we obtain (verify)

$$\left[\begin{array}{ccc:c} 1 & i & 1 & 0 \\ 0 & 1 & 1-i & 0 \\ 0 & 0 & 1 & 0 \end{array}\right],$$

and hence the only solution is $a_1 = a_2 = a_3 = 0$, showing that $\{\mathbf{v}_1, \mathbf{v}_2, \mathbf{v}_3\}$ is linearly independent.

(d) Yes, because $\mathbf{v}_1$, $\mathbf{v}_2$, and $\mathbf{v}_3$ span C^3 [part (b)] and they are linearly independent [part (c)]. ■

Just as for a real vector space, the questions of spanning sets, linearly independent or linearly dependent sets, and basis in a complex vector space are resolved by using an appropriate linear system. The definition of the dimension of a complex vector space is the same as that given in Definition 4.11. In discussing the dimension of a complex vector space such as C^n, we must adjust our intuitive picture. For example, C^1 consists of all complex multiples of a single nonzero vector. This collection can be put into one-to-one correspondence with the complex numbers themselves—that is, with all the points in the complex plane (see Figure B.1). Just as the elements of a two-dimensional real vector space can be put into a one-to-one correspondence with the points of R^2 (see Section 4.1), a complex vector space of dimension one has a geometric model that is in one-to-one correspondence with a geometric model of a two-dimensional real vector space. Similarly, a complex vector space of dimension two is the same, geometrically, as a four-dimensional real vector space.

Complex Inner Products

Let V be a complex vector space. An **inner product** on V is a function that assigns, to each ordered pair of vectors $\mathbf{u}$, $\mathbf{v}$ in V, a complex number $(\mathbf{u}, \mathbf{v})$ satisfying the following conditions:

(a) $(\mathbf{u}, \mathbf{v}) \geq 0$; $(\mathbf{u}, \mathbf{u}) = 0$ if and only if $\mathbf{u} = \mathbf{0}_V$
(b) $\overline{(\mathbf{v}, \mathbf{u})} = (\mathbf{u}, \mathbf{v})$ for any $\mathbf{u}$, $\mathbf{v}$ in V
(c) $(\mathbf{u} + \mathbf{v}, \mathbf{w}) = (\mathbf{u}, \mathbf{w}) + (\mathbf{v}, \mathbf{w})$ for any $\mathbf{u}$, $\mathbf{v}$, $\mathbf{w}$ in V
(d) $(c\mathbf{u}, \mathbf{v}) = c(\mathbf{u}, \mathbf{v})$ for any $\mathbf{u}$, $\mathbf{v}$ in V, and c a complex scalar

Remark Observe how similar this definition is to Definition 5.1 of a real inner product.

EXAMPLE 8

We can define the **standard inner product** on C^n by defining $(\mathbf{u}, \mathbf{v})$ for

$$\mathbf{u} = \begin{bmatrix} u_1 \\ u_2 \\ \vdots \\ u_n \end{bmatrix} \quad \text{and} \quad \mathbf{v} = \begin{bmatrix} v_1 \\ v_2 \\ \vdots \\ v_n \end{bmatrix} \quad \text{in } C^n \text{ as}$$

$$(\mathbf{u}, \mathbf{v}) = u_1\overline{v}_1 + u_2\overline{v}_2 + \cdots + u_n\overline{v}_n,$$

which can also be expressed as $(\mathbf{u}, \mathbf{v}) = \mathbf{u}^T\,\overline{\mathbf{v}}$.

Thus, if

$$\mathbf{u} = \begin{bmatrix} 1-i \\ 2 \\ -3+2i \end{bmatrix} \quad \text{and} \quad \mathbf{v} = \begin{bmatrix} 3+2i \\ 3-4i \\ -3i \end{bmatrix}$$

are vectors in C^3, then

$$\begin{aligned}(\mathbf{u}, \mathbf{v}) &= (1-i)\overline{(3+2i)} + 2\overline{(3-4i)} + (-3+2i)\overline{(-3i)} \\ &= (1-5i) + (6+8i) + (-6-9i) \\ &= 1-6i.\end{aligned}$$

■

A complex vector space that has a complex inner product defined on it is called a **complex inner product space**. If V is a complex inner product space, then we can define the length of a vector $\mathbf{u}$ in V exactly as in the real case:

$$\|\mathbf{u}\| = \sqrt{(\mathbf{u}, \mathbf{u})}.$$

Moreover, the vectors $\mathbf{u}$ and $\mathbf{v}$ in V are said to be **orthogonal** if $(\mathbf{u}, \mathbf{v}) = 0$.

Complex Eigenvalues and Eigenvectors

In the case of complex matrices, we have the following analogs of the theorems presented in Section 7.3, which show the role played by the special matrices discussed in Section 7.3:

Theorem B.1 If A is a Hermitian matrix, then the eigenvalues of A are all real. Moreover, eigenvectors belonging to distinct eigenvalues are orthogonal (complex analog of Theorems 7.6 and 7.7). ■

Theorem B.2 If A is a Hermitian matrix, then there exists a unitary matrix U such that $U^{-1}AU = D$, a diagonal matrix. The eigenvalues of A lie on the main diagonal of D (complex analog of Theorem 7.9). ■

In Section 7.3 we proved that if A is a real symmetric matrix, then there exists an orthogonal matrix P such that $P^{-1}AP = D$, a diagonal matrix; and conversely, if there is an orthogonal matrix P such that $P^{-1}AP$ is a diagonal matrix, then A is a symmetric matrix. For complex matrices, the situation is more complicated. The converse of Theorem B.2 is not true. That is, if A is a matrix for which there exists a unitary matrix U such that $U^{-1}AU = D$, a diagonal matrix, then A need not be a Hermitian matrix. The correct statement involves normal matrices. (See Section B.1.) The following result can be established:

Theorem B.3 If A is a normal matrix, then there exists a unitary matrix U such that $U^{-1}AU = D$, a diagonal matrix. Conversely, if A is a matrix for which there exists a unitary matrix U such that $U^{-1}AU = D$, a diagonal matrix, then A is a normal matrix. ■

B.2 Exercises

1. Solve by using Gauss–Jordan reduction.

(a) $\begin{aligned}(1+2i)x_1 + (-2+i)x_2 &= 1-3i\\ (2+i)x_1 + (-1+2i)x_2 &= -1-i\end{aligned}$

(b) $\begin{aligned}2ix_1 - (1-i)x_2 &= 1+i\\ (1-i)x_1 + x_2 &= 1-i\end{aligned}$

(c) $\begin{aligned}(1+i)x_1 - x_2 &= -2+i\\ 2ix_1 + (1-i)x_2 &= i\end{aligned}$

2. Transform the given augmented matrix of a linear system to row echelon form and solve by back substitution.

(a) $\left[\begin{array}{ccc|c} 2 & i & 0 & 1-i \\ 0 & 3i & -2+i & 4 \\ 0 & 0 & 2+i & 2-i \end{array}\right]$

(b) $\left[\begin{array}{ccc|c} i & 2 & 1+i & 3i \\ 0 & 1-i & 0 & 2+i \\ 0 & 0 & 3 & 6-3i \end{array}\right]$

3. Solve by Gaussian elimination with back substitution.

(a) $\begin{aligned}ix_1 + (1+i)x_2 \phantom{{}+ x_3} &= i\\ (1-i)x_1 + x_2 - ix_3 &= 1\\ ix_2 + x_3 &= 1\end{aligned}$

(b) $$\begin{aligned} x_1 + ix_2 + (1-i)x_3 &= 2+i \\ ix_1 \quad + (1+i)x_3 &= -1+i \\ 2ix_2 - \quad x_3 &= 2-i \end{aligned}$$

4. Compute the determinant and simplify as much as possible.

(a) $\begin{vmatrix} 1+i & -1 \\ 2i & 1+i \end{vmatrix}$

(b) $\begin{vmatrix} 2-i & 1+i \\ 1+2i & -(1-i) \end{vmatrix}$

(c) $\begin{vmatrix} 1+i & 2 & 2-i \\ i & 0 & 3+i \\ -2 & 1 & 1+2i \end{vmatrix}$

(d) $\begin{vmatrix} 2 & 1-i & 0 \\ 1+i & -1 & i \\ 0 & -i & 2 \end{vmatrix}$

5. Find the inverse of each of the following matrices, if possible:

(a) $\begin{bmatrix} i & 2 \\ 1+i & -i \end{bmatrix}$ **(b)** $\begin{bmatrix} 2 & i & 3 \\ 1+i & 0 & 1-i \\ 2 & 1 & 2+i \end{bmatrix}$

6. Determine whether the following subsets W of C_{22} are complex vector subspaces:

(a) W is the set of all 2×2 complex matrices with zeros on the main diagonal.

(b) W is the set of all 2×2 complex matrices that have diagonal entries with real part equal to zero.

(c) W is the set of all symmetric 2×2 complex matrices.

7. (a) Prove or disprove: The set W of all $n \times n$ Hermitian matrices is a complex vector subspace of C_{nn}.

(b) Prove or disprove: The set W of all $n \times n$ Hermitian matrices is a real vector subspace of the real vector space of all $n \times n$ complex matrices.

8. Prove or disprove: The set W of all $n\times n$ unitary matrices is a complex vector subspace of C_{nn}.

9. Let $W = \text{span}\{\mathbf{v}_1, \mathbf{v}_2, \mathbf{v}_3\}$, where

$$\mathbf{v}_1 = \begin{bmatrix} -1+i \\ 2 \\ 1 \end{bmatrix}, \quad \mathbf{v}_2 = \begin{bmatrix} 1 \\ 1+i \\ i \end{bmatrix},$$

$$\mathbf{v}_3 = \begin{bmatrix} -5+2i \\ -1-3i \\ 2-3i \end{bmatrix}.$$

(a) Does $\mathbf{v} = \begin{bmatrix} i \\ 0 \\ 0 \end{bmatrix}$ belong to W?

(b) Is the set $\{\mathbf{v}_1, \mathbf{v}_2, \mathbf{v}_3\}$ linearly independent or linearly dependent?

10. Let $\{\mathbf{v}_1, \mathbf{v}_2, \mathbf{v}_3\}$ be a basis for a complex vector space V. Determine whether or not $\mathbf{w}$ is in span $\{\mathbf{w}_1, \mathbf{w}_2\}$.

(a) $$\begin{aligned} \mathbf{w}_1 &= i\mathbf{v}_1 + (1-i)\mathbf{v}_2 + 2\mathbf{v}_3 \\ \mathbf{w}_2 &= (2+i)\mathbf{v}_1 + 2i\mathbf{v}_2 + (3-i)\mathbf{v}_3 \\ \mathbf{w} &= (-2-3i)\mathbf{v}_1 + (3-i)\mathbf{v}_2 + (-2-2i)\mathbf{v}_3 \end{aligned}$$

(b) $$\begin{aligned} \mathbf{w}_1 &= 2i\mathbf{v}_1 + \mathbf{v}_2 + (1-i)\mathbf{v}_3 \\ \mathbf{w}_2 &= 3i\mathbf{v}_1 + (1+i)\mathbf{v}_2 + 3\mathbf{v}_3 \\ \mathbf{w} &= (2+3i)\mathbf{v}_1 + (2+i)\mathbf{v}_2 + (4-2i)\mathbf{v}_3 \end{aligned}$$

11. Find the eigenvalues and associated eigenvectors of the following complex matrices:

(a) $A = \begin{bmatrix} 1 & 1 \\ -1 & 1 \end{bmatrix}$ **(b)** $A = \begin{bmatrix} 1 & i \\ -i & 1 \end{bmatrix}$

(c) $A = \begin{bmatrix} 2 & 0 & 0 \\ 0 & 2 & i \\ 0 & -i & 2 \end{bmatrix}$

12. For each of the parts in Exercise 11, find a matrix P such that $P^{-1}AP = D$, a diagonal matrix. For part (c), find three different matrices P that diagonalize A.

13. (a) Prove that if A is Hermitian, then the eigenvalues of A are real.

(b) Verify that A in Exercise 11(c) is Hermitian.

(c) Are the eigenvectors associated with an eigenvalue of a Hermitian matrix guaranteed to be real vectors? Explain.

14. Prove that an $n \times n$ complex matrix A is unitary if and only if the columns (rows) of A form an orthonormal set with respect to the standard inner product on C^n. (*Hint*: See Theorem 7.8.)

15. Show that if A is a skew Hermitian matrix (see Exercise 15 in Section B.1) and λ is an eigenvalue of A, then the real part of λ is zero.

APPENDIX

C Introduction to Proofs

C.1 Logic

Logic is the discipline that deals with the methods of reasoning. It provides rules and techniques for determining whether a given argument is valid. Logical reasoning is used in mathematics to prove theorems.

A **statement** or **proposition** is a declarative sentence that is either true or false, but not both.

EXAMPLE 1

Which of the following are statements?

(a) Nero is dead.

(b) Every rectangle is a square.

(c) Do you speak English?

(d) $x^2 \geq 0$, for every real number x.

(e) $4 + 3 = 7$.

(f) Call me tomorrow.

(g) $x^2 - 3x + 2 = 0$.

Solution

(a) (b), (d), and (e) are statements. (a), (d), and (e) are true, while (b) is false; (c) is a question, not a statement; (f) is a command, not a statement; (g) is a declarative sentence that is true for some values of x and false for others. ■

If p is a statement, then the statement "not p" is called the **negation** of p and is denoted by $\sim p$. The negation of p is also called the **opposite** of p. The statement $\sim p$ is true when p is false. The truth value of $\sim p$ relative to p is given in Table C.1, which is called a **truth table**.

TABLE C.1

p	$\sim p$
T	F
F	T

EXAMPLE 2 Give the negation of each of the following statements:

(a) p: $2 + 3 > 1$
(b) q: It is cold.

Solution

(a) $\sim p$: $2 + 3$ is not greater than 1. That is, $\sim p$: $2 + 3 \leq 1$. Since p is true, $\sim p$ is false.
(b) $\sim q$: It is not cold. ■

The statements p and q can be combined by a number of logical connectives to form compound statements. We look at the most important logical connectives.

Let p and q be statements.

1. The statement "p and q" is denoted by $p \wedge q$ and is called the **conjunction** of p and q. The statement $p \wedge q$ is true only when both p and q are true. The truth table giving the truth values of $p \wedge q$ is given in Table C.2.
2. The statement "p or q" is denoted by $p \vee q$ and is called the **disjunction** of p and q. The statement $p \vee q$ is true only when either p or q or both are true. The truth table giving the truth values of $p \vee q$ is given in Table C.3.

TABLE C.2

p	q	$p \wedge q$
T	T	T
T	F	F
F	T	F
F	F	F

TABLE C.3

p	q	$p \vee q$
T	T	T
T	F	T
F	T	T
F	F	F

EXAMPLE 3 Form the conjunction of the statements p: $2 < 3$ and q: $-5 > -8$.

Solution

$p \wedge q$: $2 < 3$ and $-5 > -8$, a true statement. ■

EXAMPLE 4 Form the disjunction of the statements p: -2 is a negative integer and q: $\sqrt{3}$ is a rational number.

Solution

$p \vee q$: -2 is a negative integer or $\sqrt{3}$ is a rational number, a true statement. (Why?) ■

The connective *or* is more complicated than the connective *and*, because it is used in two different ways in the English language. When we say "I left for Paris on Monday or I left for Paris on Tuesday," we have a disjunction of the statements p: I left for Paris on Monday and q: I left for Paris on Tuesday. Of course, exactly one of the two possibilities occurred; both could not have occurred. Thus the connective *or* is being used in an *exclusive* sense. On the other hand, consider the disjunction "I failed French or I passed mathematics." In this case, at least one of the two possibilities could have occurred, but both possibilities could have occurred. Thus, the connective *or* is being used in an *inclusive* sense. In mathematics and computer science, we always use the connective *or* in the inclusive sense.

Two statements are **equivalent** if they have the same truth values. This means that in the course of a proof or computation, we can always replace a given statement by an equivalent statement. Thus,

Multiplying by 1 is equivalent to multiplying by $\frac{5}{5}$ or by $\frac{x-3}{x-3}$, $x \neq 3$.

Dividing by 2 is equivalent to multiplying by $\frac{1}{2}$.

Equivalent statements are used heavily in constructing proofs, as we indicate later.

If p and q are statements, the compound statement "if p then q," denoted by $p \Longrightarrow q$, is called a **conditional statement**, or an **implication**. The statement p is called the **antecedent** or **hypothesis**, and the statement q is called the **consequent** or **conclusion**. The connective *if ... then* is denoted by the symbol $\Longrightarrow$.

EXAMPLE 5

The following are implications:

(a) If $\underbrace{\text{two lines are parallel}}_{p}$, then $\underbrace{\text{the lines do not intersect}}_{q}$.

(b) If $\underbrace{\text{I am hungry}}_{p}$, then $\underbrace{\text{I will eat}}_{q}$. ■

The conditional statement $p \Longrightarrow q$ is true whenever the hypothesis is false or the conclusion is true. Thus, the truth table giving the truth values of $p \Longrightarrow q$ is shown in Table C.4.

TABLE C.4

p	q	$p \Longrightarrow q$
T	T	T
T	F	F
F	T	T
F	F	T

A conditional statement can appear disguised in various forms. Each of the following is equivalent to $p \Longrightarrow q$:

p implies q;
q, if p;
p only if q;
p is sufficient for q;
q is necessary for p.

One of the primary objectives in mathematics is to show that the implication $p \Longrightarrow q$ is true; that is, we want to show that if p is true, then q must be true.

If $p \implies q$ is an implication, then the **contrapositive** of $p \implies q$ is the implication $\sim q \implies \sim p$. The truth table giving its truth values is shown in Table C.5, which we observe is exactly the same as Table C.4, the truth table for the conditional statement $p \implies q$.

TABLE C.5

p	q	$\sim q \implies \sim p$
T	T	T
T	F	F
F	T	T
F	F	T

TABLE C.6

q	p	$q \implies p$
T	T	T
T	F	F
F	T	T
F	F	T

If $p \implies q$ is an implication, then the **converse** of $p \implies q$ is the implication $q \implies p$. The truth table giving its truth values is shown in Table C.6. Observe that the converse of $p \implies q$ is obtained by interchanging the hypothesis and conclusion.

EXAMPLE 6

Form the contrapositive and converse of the given implication.

(a) If two different lines are parallel, then the lines do not intersect.

(b) If the numbers a and b are positive, then ab is positive.

(c) If $n + 1$ is odd, then n is even.

Solution

(a) Contrapositive: If two different lines intersect, then they are not parallel. The given implication and the contrapositive are true.

Converse: If two different lines do not intersect, then they are parallel. In this case, the given implication and the converse are true.

(b) Contrapositive: If ab is not positive, then a and b are not both positive. The given implication and the contrapositive are true.

Converse: If ab is positive, then a and b are positive. In this case, the given implication is true, but the converse is false (take $a = -1$ and $b = -2$).

(c) Contrapositive: If n is odd, then $n + 1$ is even. The given implication and the contrapositive are true.

Converse: If n is even, then $n + 1$ is odd. In this case, the given implication and the converse are true. ■

TABLE C.7

p	q	$p \iff q$
T	T	T
T	F	F
F	T	F
F	F	T

If p and q are statements, the compound statement "p if and only if q," denoted by $p \iff q$, is called a **biconditional**. The connective *if and only if* is denoted by the symbol $\iff$. The truth values of $p \iff q$ are given in Table C.7. Observe that $p \iff q$ is true only when both p and q are true or when both are false. The biconditional $p \iff q$ can also be stated as p is necessary and sufficient for q.

EXAMPLE 7

Each of the following is a biconditional statement:

(a) $a > b$ if and only if $a - b > 0$.

(b) An integer n is prime if and only if its only divisors are 1 and itself. ■

EXAMPLE 8

Is the following biconditional a true statement? $3 > 2$ if and only if $0 < 3 - 2$.

Solution

Let p be the statement $3 > 2$ and let q be the statement $0 < 3 - 2$. Since both p and q are true, we conclude that $p \Longleftrightarrow q$ is true. ■

A convenient way to think of a biconditional is as follows: $p \Longleftrightarrow q$ is true exactly when p and q are equivalent. It is also not difficult to show that, to prove $p \Longleftrightarrow q$, we must show that *both* $p \Longrightarrow q$ and $q \Longrightarrow p$ are true.

We soon turn to a brief introduction to techniques of proof. First, we present in Table C.8 a number of equivalences that are useful in this regard. Thus, in any proof we may replace any statement by its equivalent statement.

TABLE C.8

	Statement	*Equivalent Statement*
(a)	$\sim(\sim p)$	p
(b)	$\sim(p \vee q)$	$(\sim p) \wedge (\sim q)$
(c)	$\sim(p \wedge q)$	$(\sim p) \vee (\sim q)$
(d)	$(p \Longrightarrow q)$	$(\sim p) \vee q$
(e)	$(p \Longrightarrow q)$	$\sim q \Longrightarrow \sim p$
(f)	$(p \Longleftrightarrow q)$	$(p \Longrightarrow q) \wedge (q \Longrightarrow p)$
(g)	$\sim(p \Longrightarrow q)$	$p \wedge \sim q$
(h)	$\sim(p \Longleftrightarrow q)$	$(p \wedge \sim q) \vee (q \wedge \sim p)$
(i)	$(p \Longrightarrow q)$	$((p \wedge (\sim q)) \Longrightarrow c$, where c is a statement that is always false

Finally, in Table C.9, we present a number of implications that are always true. Some of these are useful in techniques of proof.

TABLE C.9

(a)	$(p \wedge q) \Longrightarrow p$
(b)	$(p \wedge q) \Longrightarrow q$
(c)	$p \Longrightarrow (p \vee q)$
(d)	$q \Longrightarrow (p \vee q)$
(e)	$\sim p \Longrightarrow (p \Longrightarrow q)$
(f)	$\sim(p \Longrightarrow q) \Longrightarrow p$
(g)	$(p \wedge (p \Longrightarrow q)) \Longrightarrow q$
(h)	$(\sim p \wedge (p \vee q)) \Longrightarrow q$
(i)	$(\sim q \wedge (p \Longrightarrow q)) \Longrightarrow \sim p$
(j)	$((p \wedge q) \wedge (q \wedge r)) \Longrightarrow (p \Longrightarrow r)$

C.2 Techniques of Proof

In this section we discuss techniques for constructing proofs of conditional statements $p \implies q$. To prove $p \implies q$, we must show that whenever p is true it follows that q is true, by a logical argument in the language of mathematics. The construction of this logical argument may be quite elusive; the logical argument itself is what we call the proof. Conceptually, the proof that $p \implies q$ is a sequence of steps that logically connect p to q. Each step in the "connection" must be justified or have a reason for its validity, which is usually a previous definition, a property or axiom that is known to be true, a previously proven theorem or solved problem, or even a previously verified step in the current proof. Thus we connect p and q by logically building blocks of known (or accepted) facts. Often, it is not clear what building blocks (facts) to use and exactly how to get started on a fruitful path. In many cases, *the first step of the proof is crucial.* Unfortunately, we have no explicit guidelines in this area, other than to recommend a careful reading of the hypothesis p and conclusion q in order to clearly understand them. Only in this way can we begin to seek relationships (connections) between them. At any stage in a proof, we can replace a statement that needs to be derived by an equivalent statement.

The construction of a proof requires the building of a step-by-step connection (a logical bridge) between p and q. If we let $b_1, b_2, \ldots, b_n$ represent logical building blocks, then, conceptually, our proof appears as

$$p \implies b_1 \implies b_2 \implies \cdots \implies b_n \implies q,$$

where each conditional statement must be justified. This approach is known as a **direct proof**. We illustrate this in Example 1.

EXAMPLE 1

Prove: If m and n are even integers, then $m+n$ is even.

Solution

Let p be the statement "m and n are even integers" and let q be the statement "$m+n$ is even." We start by assuming that p is true and then ask what facts we know which can lead us to q. Since both p and q involve even numbers, we try the following:

$$p \implies \underbrace{m = 2k,\ n = 2j}_{b_1}, \text{ for some integers } k \text{ and } j.$$

Since q deals with the sum of $m+n$, we try to form this sum in b_2:

$$b_1 \implies \underbrace{m+n = 2k+2j = 2(k+j)}_{b_2}.$$

Observe that b_2 implies that the sum $m+n$ is a multiple of 2. Hence $m+n$ is even. This is just q, so we have $b_2 \implies q$. In summary,

$$p \implies b_1 \implies b_2 \implies q. \tag{1}$$

■

In Example 1 we proceeded forward to build a bridge from p to q. We call this **forward building**. Alternatively, we could also start with q and ask what fact b_n will lead us to q, what fact b_{n-1} will lead us to b_n, and so on, until we reach p [see Expression (1)]. Such a logical bridge is called **backward building**. The two techniques can be combined; build forward a few steps, build backwards a few steps, and try to logically join the two ends.

In practice, the construction of proofs is an art and must be learned in part from observation and experience. The choice of intermediate steps and the methods for deriving them are creative activities that cannot be precisely described.

Another proof technique replaces the original statement $p \implies q$ by an equivalent statement and then proves the new statement. Such a procedure is called an **indirect method of proof**. One indirect method uses the equivalence between $p \implies q$ and its contrapositive $\sim q \implies \sim p$ [Table C.8(e)]. When the proof of the contrapositive is done directly, we call this **proof by contrapositive**. Unfortunately, there is no way to predict in advance that an indirect method of proof by contrapositive may be successful. Sometimes, the appearance of the word not in the conclusion $\sim q$ is a *suggestion* to try this method. There are no guarantees that it will work. We illustrate the use of proof by contrapositive in Example 2.

EXAMPLE 2

Let n be an integer. Prove that if n^2 is odd, then n is odd.

Solution

Let p: n^2 is odd and q: n is odd. We have to prove that $p \implies q$ is true. Instead, we prove the contrapositive $\sim q \implies \sim p$. Thus, suppose that n is not odd, so that n is even. Then $n = 2k$, where k is an integer. We have $n^2 = (2k)^2 = 4k^2 = 2(2k^2)$, so n^2 is even. We have thus shown that if n is even, then n^2 is even, which is the contrapositive of the given statement. Hence the given statement has been established by the method of proof by contrapositive. ■

A second indirect method of proof, called **proof by contradiction**, uses the equivalence between the conditional $p \implies q$ and the statement $((p \wedge (\sim q)) \implies c$, where c is a statement that is always false [Table C.8(i)]. We can see why this method works by referring to Table C.4. The conditional $p \implies q$ is false only when the hypothesis p is true and the conclusion q is false. The method of proof by contradiction starts with the assumption that p is true. We would like to show that q is also true, so we assume that q is false and then attempt to build a logical bridge to a statement that is known to be always false. When this is done, we say that we have reached a contradiction, so our additional hypothesis that q is false must be incorrect. Therefore, q must be true. If we are unable to build a bridge to some statement that is always false, then we cannot conclude that q is false. Possibly, we were not clever enough to build a correct bridge. As with proof by contrapositive, we cannot tell in advance that the method of proof by contradiction will be successful.

EXAMPLE 3

Show that $\sqrt{2}$ is irrational.

Solution

Let p: $x = \sqrt{2}$ and q: x is irrational. We assume that p is true and need to show that q is true. We try proof by contradiction. Thus we also assume that $\sim q$ is true, so we have assumed that $x = \sqrt{2}$ and x is rational. Then

$$x = \sqrt{2} = \frac{n}{d},$$

where n and d are integers having no common factors; that is, $\frac{n}{d}$ is in lowest terms. Then

$$2 = \frac{n^2}{d^2}, \quad \text{so} \quad 2d^2 = n^2.$$

This implies that n is even, since the square of an odd number is odd. Thus, $n = 2k$, for some integer k, so

$$2d^2 = n^2 = (2k)^2.$$

Hence $2d^2 = 4k^2$, and therefore $d^2 = 2k^2$, an even number. Hence d is even. We have now concluded that both n and d are even, which implies that they have a common factor of 2, contradicting the fact that $\frac{n}{d}$ is in lowest terms. Thus our assumption $\sim q$ is invalid, and it follows that q is true. ■

As a final observation, we note that many mathematical results make a statement that is true for all objects of a certain type. For example, the statement "Let m and n be integers; prove that $n^2 = m^2$ if and only if $m = n$ or $m = -n$" is actually saying, "For all integers m and n, $n^2 = m^2$ if and only if $m = n$ or $m = -n$." To prove this result, we must make sure that all the steps in the proof are valid for every integer m and n. We cannot prove the result for specific values of m and n. On the other hand, to disprove a result claiming that a certain property holds for all objects of a certain type, we need find only one instance in which the property does not hold. Such an instance is called a **counterexample**.

EXAMPLE 4

Prove or disprove the statement that if x and y are real numbers, then

$$(x^2 = y^2) \iff (x = y).$$

Solution

Let $x = -2$ and $y = 2$. Then $x^2 = y^2$, that is, $(-2)^2 = (2)^2$, but $x \neq y$, so the biconditional is false. That is, we have disproved the result by producing a counterexample. Many other counterexamples could be used equally well. ■

GLOSSARY FOR LINEAR ALGEBRA

Additive inverse of a matrix: The additive inverse of an $m \times n$ matrix A is an $m \times n$ matrix B such that $A + B = O$. Such a matrix B is the negative of A, denoted $-A$, which is equal to $(-1)A$.

Adjoint: For an $n \times n$ matrix $A = \left[a_{ij} \right]$ the adjoint of A, denoted adj A is the transpose of the matrix formed by replacing each entry by its cofactor A_{ij}; that is, adj $A = \left[A_{ji} \right]$.

Angle between vectors: For nonzero vectors $\mathbf{u}$ and $\mathbf{v}$ in R^n the angle θ between $\mathbf{u}$ and $\mathbf{v}$ is determined from the expression

$$\cos(\theta) = \frac{\mathbf{u} \cdot \mathbf{v}}{\|\mathbf{u}\| \|\mathbf{v}\|}.$$

Augmented matrix: For the linear system $A\mathbf{x} = \mathbf{b}$, the augmented matrix is formed by adjoining to the coefficient matrix A the right side vector $\mathbf{b}$. We expressed the augmented matrix as $\left[A \;\vdots\; \mathbf{b} \right]$.

Back substitution: If $U = \left[u_{ij} \right]$ is an upper triangular matrix all of whose diagonal entries are not zero, then the linear system $U\mathbf{x} = \mathbf{b}$ can be solved by back substitution. The process starts with the last equation and computes

$$x_n = \frac{b_n}{u_{nn}};$$

we use the next to last equation and compute

$$x_{n-1} = \frac{b_{n-1} - u_{n-1\,n} x_n}{u_{n-1\,n-1}};$$

continuing in this fashion using the jth equation we compute

$$x_j = \frac{b_j - \sum_{k=n}^{j+1} u_{jk} x_k}{u_{jj}}.$$

Basis: A set of vectors $S = \{\mathbf{v}_1, \mathbf{v}_2, \ldots, \mathbf{v}_k\}$ from a vector space V is called a basis for V provided S spans V and S is a linearly independent set.

Cauchy–Schwarz inequality: For vectors $\mathbf{v}$ and $\mathbf{u}$ in R^n, the Cauchy–Schwarz inequality says that the absolute value of the dot product of $\mathbf{v}$ and $\mathbf{u}$ is less than or equal to the product of the lengths of $\mathbf{v}$ and $\mathbf{u}$; that is, $|\mathbf{v} \cdot \mathbf{u}| \leq \|\mathbf{v}\| \|\mathbf{u}\|$.

Characteristic equation: For a square matrix A, its characteristic equation is given by $f(t) = \det(A - tI) = 0$.

Characteristic polynomial: For a square matrix A, its characteristic polynomial is given by $f(t) = \det(A - tI)$.

Closure properties: Let V be a given set, with members that we call vectors, and two operations, one called vector addition, denoted $\oplus$, and the second called scalar multiplication, denoted $\odot$. We say that V is closed under $\oplus$, provided for $\mathbf{u}$ and $\mathbf{v}$ in V, $\mathbf{u} \oplus \mathbf{v}$ is a member of V. We say that V is closed under $\odot$, provided for any real number k, $k \odot \mathbf{u}$ is a member of V.

Coefficient matrix: A linear system of m equations in n unknowns has the form

$$\begin{aligned} a_{11}x_1 + a_{12}x_2 + \cdots + a_{1n}x_n &= b_1 \\ a_{21}x_1 + a_{22}x_2 + \cdots + a_{2n}x_n &= b_2 \\ &\vdots \\ a_{m1}x_1 + a_{m2}x_2 + \cdots + a_{mn}x_n &= b_m. \end{aligned}$$

The matrix

$$A = \begin{bmatrix} a_{11} & a_{12} & \cdots & a_{1n} \\ a_{21} & a_{22} & \cdots & a_{2n} \\ \vdots & \vdots & & \vdots \\ a_{m1} & a_{m2} & \cdots & a_{mn} \end{bmatrix}$$

is called the coefficient matrix of the linear system.

Cofactor: For an $n \times n$ matrix $A = \left[a_{ij} \right]$ the cofactor A_{ij} of a_{ij} is defined as $A_{ij} = (-1)^{i+j} \det(M_{ij})$, where M_{ij} is the ij-minor of A.

Column rank: The column rank of a matrix A is the dimension of the column space of A or equivalently the number of linearly independent columns of A.

Column space: The column space of a real $m \times n$ matrix A is the subspace of R^m spanned by the columns of A.

Complex vector space: A complex vector space V is a set, with members that we call vectors, and two operations: one called vector addition, denoted $\oplus$, and the second called scalar multiplication, denoted $\odot$. We require that V be closed under $\oplus$, that is, for $\mathbf{u}$ and $\mathbf{v}$ in V, $\mathbf{u} \oplus \mathbf{v}$ is a member of V; in addition we require that V be closed under $\odot$, that is, for any complex number k, $k \odot \mathbf{u}$ is a member of V. There are 8 other properties that must be satisfied before V with the two operations $\oplus$ and $\odot$ is called a complex vector space.

Complex vector subspace: A subset W of a complex vector space V that is closed under addition and scalar multiplication is called a complex subspace of V.

Components of a vector: The components of a vector $\mathbf{v}$ in R^n are its entries;

$$\mathbf{v} = \begin{bmatrix} v_1 \\ v_2 \\ \vdots \\ v_n \end{bmatrix}.$$

Composite linear transformation: Let L_1 and L_2 be linear transformations with $L_1\colon V \to W$ and $L_2\colon W \to U$. Then the composition $L_2 \circ L_1\colon V \to U$ is a linear transformation and for $\mathbf{v}$ in V, we compute $(L_2 \circ L_1)(\mathbf{v}) = L_2(L_1(v))$.

Computation of a determinant via reduction to triangular form: For an $n \times n$ matrix A the determinant of A, denoted $\det(A)$ or $|A|$, can be computed with the aid of elementary row operations as follows. Use elementary row operations on A, keeping track of the operations used, to obtain an upper triangular matrix. Using the changes in the determinant as the result of applying a row operation and the fact that the determinant of an upper triangular matrix is the product of its diagonal entries, we can obtain an appropriate expression for $\det(A)$.

Consistent linear system: A linear system $A\mathbf{x} = \mathbf{b}$ is called consistent if the system has at least one solution.

Coordinates: The coordinates of a vector $\mathbf{v}$ in a vector space V with ordered basis $S = \{\mathbf{v}_1, \mathbf{v}_2, \ldots, \mathbf{v}_n\}$ are the coefficients $c_1, c_2, \ldots, c_n$ such that $\mathbf{v} = c_1\mathbf{v}_1 + c_2\mathbf{v}_2 + \cdots + c_n\mathbf{v}_n$. We denote the coordinates of $\mathbf{v}$ relative to the basis S by $\left[\mathbf{v}\right]_S$ and write

$$\left[\mathbf{v}\right]_S = \begin{bmatrix} c_1 \\ c_2 \\ \vdots \\ c_n \end{bmatrix}.$$

Cross product: The cross product of a pair of vectors $\mathbf{u}$ and $\mathbf{v}$ from R^3 is denoted $\mathbf{u} \times \mathbf{v}$, and is computed as the determinant

$$\begin{vmatrix} \mathbf{i} & \mathbf{j} & \mathbf{k} \\ u_1 & u_2 & u_3 \\ v_1 & v_2 & v_3 \end{vmatrix},$$

where $\mathbf{i}$, $\mathbf{j}$, and $\mathbf{k}$ are the unit vectors in the x-, y-, and z-directions, respectively.

Defective matrix: A square matrix A is called defective if it has an eigenvalue of multiplicity $m > 1$ for which the associated eigenspace has a basis with fewer than m vectors.

Determinant: For an $n \times n$ matrix A the determinant of A, denoted $\det(A)$ or $|A|$, is a scalar that is computed as the sum of all possible products of n entries of A each with its appropriate sign, with exactly one entry from each row and exactly one entry from each column.

Diagonal matrix: A square matrix $A = \left[a_{ij}\right]$ is called diagonal provided $a_{ij} = 0$ whenever $i \neq j$.

Diagonalizable: A square matrix A is called diagonalizable provided it is similar to a diagonal matrix D; that is, there exists a nonsingular matrix P such that $P^{-1}AP = D$.

Difference of matrices: The difference of the $m \times n$ matrices A and B is denoted $A - B$ and is equal to the sum $A + (-1)B$. The difference $A - B$ is the $m \times n$ matrix whose entries are the difference of corresponding entries of A and B.

Difference of vectors: The difference of the vectors $\mathbf{v}$ and $\mathbf{w}$ in a vector space V is denoted $\mathbf{v} - \mathbf{w}$, which is equal to the sum $\mathbf{v} + (-1)\mathbf{w}$. If $V = R^n$, then $\mathbf{v} - \mathbf{w}$ is computed as the difference of corresponding entries.

Dilation: The linear transformation $L\colon R^n \to R^n$ given by $L(\mathbf{v}) = k\mathbf{v}$, for $k > 1$, is called a dilation.

Dimension: The dimension of a nonzero vector space V is the number of vectors in a basis for V. The dimension of the vector space $\{\mathbf{0}\}$ is defined as zero.

Distance between points (or vectors): The distance between the points $(u_1, u_2, \ldots, u_n)$ and $(v_1, v_2, \ldots, v_n)$ is the length of the vector $\mathbf{u} - \mathbf{v}$, where $\mathbf{u} = (u_1, u_2, \ldots, u_n)$ and $\mathbf{v} = (v_1, v_2, \ldots, v_n)$, and is given by

$$\|\mathbf{u} - \mathbf{v}\| = \sqrt{(u_1 - v_1)^2 + (u_2 - v_2)^2 + \cdots + (u_n - v_n)^2}.$$

Thus we see that the distance between vectors in R^n is also $\|\mathbf{u} - \mathbf{v}\|$.

Dot product: For vectors $\mathbf{v}$ and $\mathbf{w}$ in R^n the dot product of $\mathbf{v}$ and $\mathbf{w}$ is also called the standard inner product or just the inner product of $\mathbf{v}$ and $\mathbf{w}$. The dot product of $\mathbf{v}$ and $\mathbf{w}$ in R^n is denoted $\mathbf{v} \cdot \mathbf{w}$ and is computed as

$$\mathbf{v} \cdot \mathbf{w} = v_1w_1 + v_2w_2 + \cdots + v_nw_n.$$

Eigenspace: The set of all eigenvectors of a square matrix A associated with a specified eigenvalue λ of A, together with the zero vector, is called the eigenspace associated with the eigenvalue λ.

Eigenvalue: An eigenvalue of an $n \times n$ matrix A is a scalar λ for which there exists a nonzero n-vector $\mathbf{x}$ such that $A\mathbf{x} = \lambda\mathbf{x}$. The vector $\mathbf{x}$ is an eigenvector associated with the eigenvalue λ.

Eigenvector: An eigenvector of an $n \times n$ matrix A is a nonzero n-vector $\mathbf{x}$ such that $A\mathbf{x}$ is a scalar multiple of $\mathbf{x}$; that is, there exists some scalar λ such that $A\mathbf{x} = \lambda\mathbf{x}$. The scalar is an eigenvalue of the matrix A.

Elementary row operations: An elementary row operation on a matrix is any of the following three operations: (1) an interchange of rows, (2) multiplying a row by a nonzero scalar, and (3) replacing a row by adding a scalar multiple of a different row to it.

Equal matrices: The $m \times n$ matrices A and B are equal provided corresponding entries are equal; that is, $A = B$ if $a_{ij} = b_{ij}$, $i = 1, 2, \ldots, m$, $j = 1, 2, \ldots, n$.

Equal vectors: Vectors **v** and **w** in R^n are equal provided corresponding entries are equal; that is, $\mathbf{v} = \mathbf{w}$ if their corresponding components are equal.

Finite-dimensional vector space: A vector space V that has a basis that is a finite subset of V is said to be finite dimensional.

Forward substitution: If $L = \left[l_{ij} \right]$ is a lower triangular matrix all of whose diagonal entries are not zero, then the linear system $L\mathbf{x} = \mathbf{b}$ can be solved by forward substitution. The process starts with the first equation and computes

$$x_1 = \frac{b_1}{l_{11}};$$

next we use the second equation and compute

$$x_2 = \frac{b_2 - l_{21}x_1}{l_{22}};$$

continuing in this fashion using the jth equation we compute

$$x_j = \frac{b_j - \displaystyle\sum_{k=1}^{j-1} l_{jk}x_k}{l_{jj}}.$$

Fundamental vector spaces associated with a matrix: If A is an $m \times n$ matrix there are four fundamental subspaces associated with A: (1) the null space of A, a subspace of R^n; (2) the row space of A, a subspace of R^n; (3) the null space of A^T, a subspace of R^m; and (4) the column space of A, a subspace of R^m.

Gaussian elimination: For the linear system $A\mathbf{x} = \mathbf{b}$ form the augmented matrix $\left[A \mid \mathbf{b} \right]$. Compute the row echelon form of the augmented matrix; then the solution can be computed using back substitution.

Gauss–Jordan reduction: For the linear system $A\mathbf{x} = \mathbf{b}$ form the augmented matrix $\left[A \mid \mathbf{b} \right]$. Compute the reduced row echelon form of the augmented matrix; then the solution can be computed using back substitution.

General solution: The general solution of a consistent linear system $A\mathbf{x} = \mathbf{b}$ is the set of all solutions to the system. If $\mathbf{b} = \mathbf{0}$, then the general solution is just the set of all solutions to the homogeneous system $A\mathbf{x} = \mathbf{0}$, denoted $\mathbf{x}_h$. If $\mathbf{b} \neq \mathbf{0}$, then the general solution of the nonhomogeneous system consists of a particular solution of $A\mathbf{x} = \mathbf{b}$, denoted $\mathbf{x}_p$, together with $\mathbf{x}_h$; that is, the general solution is expressed as $\mathbf{x}_p + \mathbf{x}_h$.

Gram–Schmidt process: The Gram–Schmidt process converts a basis for a subspace into an orthonormal basis for the same subspace.

Hermitian matrix: An $n \times n$ complex matrix A is called Hermitian provided $\overline{A}^T = A$.

Homogeneous system: A homogeneous system is a linear system in which the right side of each equation is zero. We denote a homogeneous system by $A\mathbf{x} = \mathbf{0}$.

Identity matrix: The $n \times n$ identity matrix, denoted I_n, is a diagonal matrix with diagonal entries of all 1s.

Inconsistent linear system: A linear system $A\mathbf{x} = \mathbf{b}$ that has no solution is called inconsistent.

Infinite-dimensional vector space: A vector space V for which there is no finite subset of vectors that form a basis for V is said to be infinite dimensional.

Inner product: For vectors **v** and **w** in R^n the inner product of **v** and **w** is also called the dot product or standard inner product of **v** and **w**. The inner product of **v** and **w** in R^n is denoted $\mathbf{v} \cdot \mathbf{w}$ and is computed as

$$\mathbf{v} \cdot \mathbf{w} = v_1 w_1 + v_2 w_2 + \cdots + v_n w_n.$$

Invariant subspace: A subspace W of a vector space V is said to be invariant under the linear transformation $L\colon V \to V$, provided $L(\mathbf{v})$ is in W for all vectors **v** in W.

Inverse linear transformation: See invertible linear transformation.

Inverse of a matrix: An $n \times n$ matrix A is said to have an inverse provided there exists an $n \times n$ matrix B such that $AB = BA = I$. We call B the inverse of A and denote it as A^{-1}. In this case, A is also called nonsingular.

Invertible linear transformation: A linear transformation $L\colon V \to W$ is called invertible if there exists a linear transformation, denoted L^{-1}, such that $L^{-1}(L(\mathbf{v})) = \mathbf{v}$, for all vectors **v** in V and $L(L^{-1}(\mathbf{w})) = \mathbf{w}$, for all vectors **w** in W.

Isometry: An isometry is a linear transformation L that preserves the distance between pairs of vectors; that is, $\|L(\mathbf{v}) - L(\mathbf{u})\| = \|\mathbf{v} - \mathbf{u}\|$, for all vectors **u** and **v**. Since an isometry preserves distances, it also preserves lengths; that is, $\|L(\mathbf{v})\| = \|\mathbf{v}\|$, for all vectors **v**.

Length (or magnitude) of a vector: The length of a vector **v** in R^n is denoted $\|\mathbf{v}\|$ and is computed as the expression

$$\sqrt{v_1^2 + v_2^2 + \cdots + v_n^2}.$$

For a vector **v** in a vector space V on which an inner product (dot product) is defined, the length of **v** is computed as $\|\mathbf{v}\| = \sqrt{\mathbf{v} \cdot \mathbf{v}}$.

Linear combination: A linear combination of vectors $\mathbf{v}_1, \mathbf{v}_2, \ldots, \mathbf{v}_k$ from a vector space V is an expression of the form $c_1\mathbf{v}_1 + c_2\mathbf{v}_2 + \cdots + c_k\mathbf{v}_k$, where the $c_1, c_2, \ldots, c_k$ are scalars. A linear combination of the $m \times n$ matrices $A_1, A_2, \ldots, A_k$ is given by $c_1 A_1 + c_2 A_2 + \cdots + c_k A_k$.

Linear operator: A linear operator is a linear transformation L from a vector space to itself; that is, $L\colon V \to V$.

Linear system: A system of m linear equations in n unknowns $x_1, x_2, \ldots, x_n$ is a set of linear equations in the n unknowns. We express a linear system in matrix form as $A\mathbf{x} = \mathbf{b}$, where A is the matrix of coefficients, $\mathbf{x}$ is the vector of unknowns, and $\mathbf{b}$ is the vector of right sides of the linear equations. (See coefficient matrix.)

Linear transformation: A linear transformation $L\colon V \to W$ is a function assigning a unique vector $L(\mathbf{v})$ in W to each vector $\mathbf{v}$ in V such that two properties are satisfied: (1) $L(\mathbf{u}+\mathbf{v}) = L(\mathbf{u}) + L(\mathbf{v})$, for every $\mathbf{u}$ and $\mathbf{v}$ in V, and (2) $L(k\mathbf{v}) = kL(\mathbf{v})$, for every $\mathbf{v}$ in V and every scalar k.

Linearly dependent: A set of vectors $S = \{\mathbf{v}_1, \mathbf{v}_2, \ldots, \mathbf{v}_n\}$ is called linearly dependent provided there exists a linear combination $c_1\mathbf{v}_1 + c_2\mathbf{v}_2 + \cdots + c_n\mathbf{v}_n$ that produces the zero vector when not all the coefficients are zero.

Linearly independent: A set of vectors $S = \{\mathbf{v}_1, \mathbf{v}_2, \ldots, \mathbf{v}_n\}$ is called linearly independent provided the only linear combination $c_1\mathbf{v}_1 + c_2\mathbf{v}_2 + \cdots + c_n\mathbf{v}_n$ that produces the zero vector is when all the coefficients are zero, that is, only when $c_1 = c_2 = \cdots = c_n = 0$.

Lower triangular matrix: A square matrix with zero entries above its diagonal entries is called lower triangular.

LU-factorization (or LU-decomposition): An LU-factorization of a square matrix A expresses A as the product of a lower triangular matrix L and an upper triangular matrix U; that is, $A = LU$.

Main diagonal of a matrix: The main diagonal of an $n \times n$ matrix A is the set of entries $a_{11}, a_{22}, \ldots, a_{nn}$.

Matrix: An $m \times n$ matrix A is a rectangular array of mn entries arranged in m rows and n columns.

Matrix addition: For $m \times n$ matrices $A = \left[a_{ij}\right]$ and $B = \left[b_{ij}\right]$, the addition of A and B is performed by adding corresponding entries; that is, $A + B = \left[a_{ij}\right] + \left[b_{ij}\right]$. This is also called the sum of the matrices A and B.

Matrix representing a linear transformation: Let $L\colon V \to W$ be a linear transformation from an n-dimensional space V to an m-dimensional space W. For a basis $S = \{\mathbf{v}_1, \mathbf{v}_2, \ldots, \mathbf{v}_n\}$ in V and a basis $T = \{\mathbf{w}_1, \mathbf{w}_2, \ldots, \mathbf{w}_m\}$ in W there exists an $m \times n$ matrix A, with column j of $A = \left[L(\mathbf{v}_j)\right]_T$ such that the coordinates of $L(\mathbf{x})$, for any x in V, with respect to the T basis can be computed as $\left[L(\mathbf{x})\right]_T = A\left[\mathbf{x}\right]_S$. We say A is the matrix representing the linear transformation L.

Matrix transformation: For an $m \times n$ matrix A the function f defined by $f(\mathbf{u}) = A\mathbf{u}$ for $\mathbf{u}$ in R^n is called the matrix transformation from R^n to R^m defined by the matrix A.

Minor: Let $A = \left[a_{ij}\right]$ be an $n \times n$ matrix and M_{ij} the $(n-1) \times (n-1)$ submatrix of A obtained by deleting the ith row and jth column of A. The determinant $\det(M_{ij})$ is called the minor of a_{ij}.

Multiplicity of an eigenvalue: The multiplicity of an eigenvalue λ of a square matrix A is the number of times λ is a root of the characteristic polynomial of A.

Natural (or standard) basis: The natural basis for R^n is the set of vectors $\mathbf{e}_j$ = column j (or, equivalently, row j) of the $n \times n$ identity matrix, $j = 1, 2, \ldots, n$.

Negative of a vector: The negative of a vector $\mathbf{u}$ is a vector $\mathbf{w}$ such that $\mathbf{u} + \mathbf{w} = \mathbf{0}$, the zero vector. We denote the negative of $\mathbf{u}$ as $-\mathbf{u} = (-1)\mathbf{u}$.

Nonhomogeneous system: A linear system $A\mathbf{x} = \mathbf{b}$ is called nonhomogeneous provided the vector $\mathbf{b}$ is not the zero vector.

Nonsingular (or invertible) matrix: An $n \times n$ matrix A is called nonsingular provided there exists an $n \times n$ matrix B such that $AB = BA = I$. We call B the inverse of A and denote it as A^{-1}.

Nontrivial solution: A nontrivial solution of a linear system $A\mathbf{x} = \mathbf{b}$ is any vector $\mathbf{x}$ containing at least one nonzero entry such that $A\mathbf{x} = \mathbf{b}$.

Normal matrix: An $n \times n$ complex matrix A is called normal provided $\left(\overline{A}^T\right) A = A\left(\overline{A}^T\right)$.

n-space: The set of all n-vectors is called n-space. For vectors whose entries are real numbers we denote n-space as R^n. For a special case see 2-space.

Nullity: The nullity of the matrix A is the dimension of the null space of A.

n-vector: A $1 \times n$ or an $n \times 1$ matrix is called an n-vector. When n is understood, we refer to n-vectors merely as vectors.

One-to-one: A function $f\colon S \to T$ is said to be one-to-one provided $f(s_1) \neq f(s_2)$ whenever s_1 and s_2 are distinct elements of S. A linear transformation $L\colon V \to W$ is called one-to-one provided L is a one-to-one function.

Onto: A function $f\colon S \to T$ is said to be onto provided for each member t of T there is some member s in S so that $f(s) = t$. A linear transformation $L\colon V \to W$ is called onto provided range $L = W$.

Ordered basis: A set of vectors $S = \{\mathbf{v}_1, \mathbf{v}_2, \ldots, \mathbf{v}_k\}$ in a vector space V is called an ordered basis for V provided S is a basis for V and if we reorder the vectors in S, this new ordering of the vectors in S is considered a different basis for V.

Orthogonal basis: A basis for a vector space V that is also an orthogonal set is called an orthogonal basis for V.

Orthogonal complement: The orthogonal complement of a set S of vectors in a vector space V is the set of all vectors in V that are orthogonal to all vectors in S.

Orthogonal matrix: A square matrix P is called orthogonal provided $P^{-1} = P^T$.

Orthogonal projection: For a vector $\mathbf{v}$ in a vector space V, the orthogonal projection of $\mathbf{v}$ onto a subspace W of V with orthonormal basis $\{\mathbf{w}_1, \mathbf{w}_2, \ldots, \mathbf{w}_k\}$ is the vector $\mathbf{w}$ in W, where

$\mathbf{w} = (\mathbf{v} \cdot \mathbf{w}_1)\mathbf{w}_1 + (\mathbf{v} \cdot \mathbf{w}_2)\mathbf{w}_2 + \cdots + (\mathbf{v} \cdot \mathbf{w}_k)\mathbf{w}_k$. Vector $\mathbf{w}$ is the vector in W that is closest to $\mathbf{v}$.

Orthogonal set: A set of vectors $S = \{\mathbf{w}_1, \mathbf{w}_2, \ldots, \mathbf{w}_k\}$ from a vector space V on which an inner product is defined is an orthogonal set provided none of the vectors is the zero vector and the inner product of any two different vectors is zero.

Orthogonal vectors: A pair of vectors is called orthogonal provided their dot (inner) product is zero.

Orthogonally diagonalizable: A square matrix A is said to be orthogonally diagonalizable provided there exists an orthogonal matrix P such that $P^{-1}AP$ is a diagonal matrix. That is, A is similar to a diagonal matrix using an orthogonal matrix P.

Orthonormal basis: A basis for a vector space V that is also an orthonormal set is called an orthonormal basis for V.

Orthonormal set: A set of vectors $S = \{\mathbf{w}_1, \mathbf{w}_2, \ldots, \mathbf{w}_k\}$ from a vector space V on which an inner product is defined is an orthonormal set provided each vector is a unit vector and the inner product of any two different vectors is zero.

Parallel vectors: Two nonzero vectors are said to be parallel if one is a scalar multiple of the other.

Particular solution: A particular solution of a consistent linear system $A\mathbf{x} = \mathbf{b}$ is a vector $\mathbf{x}_p$ containing no arbitrary constants such that $A\mathbf{x}_p = \mathbf{b}$.

Partitioned matrix: A matrix that has been partitioned into submatrices by drawing horizontal lines between rows and/or vertical lines between columns is called a partitioned matrix. There are many ways to partition a matrix.

Perpendicular (or orthogonal) vectors: A pair of vectors is said to be perpendicular or orthogonal provided their dot product is zero.

Pivot: When using row operations on a matrix A, a pivot is a nonzero entry of a row that is used to zero-out entries in the column in which the pivot resides.

Positive definite: Matrix A is positive definite provided A is symmetric and all of its eigenvalues are positive.

Powers of a matrix: For a square matrix A and nonnegative integer k, the kth power of A, denoted A^k, is the product of A with itself k times; $A^k = A \cdot A \cdot \cdots \cdot A$, where there are k factors.

Projection: The projection of a point P in a plane onto a line L in the same plane is the point Q obtained by intersecting the line L with the line through P that is perpendicular to L. The linear transformation $L\colon R^3 \to R^2$ defined by $L(x, y, z) = (x, y)$ is called a projection of R^3 onto R^2. (See also orthogonal projection.)

Range: The range of a function $f\colon S \to T$ is the set of all members t of T such that there is a member s in S with $f(s) = t$. The range of a linear transformation $L\colon V \to W$ is the set of all vectors in W that are images under L of vectors in V.

Rank: Since row rank A = column rank A, we just refer to the rank of the matrix A as rank A. Equivalently, rank A = the number of linearly independent rows (columns) of A = the number of leading 1s in the reduced row echelon form of A.

Real vector space: A real vector space V is a set, with members that we call vectors and two operations: one is called vector addition, denoted $\oplus$, and the second called scalar multiplication, denoted $\odot$. We require that V be closed under $\oplus$; that is, for $\mathbf{u}$ and $\mathbf{v}$ in V, $\mathbf{u} \oplus \mathbf{v}$ is a member of V. In addition we require that V be closed under $\odot$; that is, for any real number k, $k \odot \mathbf{u}$ is a member of V. There are 8 other properties that must be satisfied before V with the two operations $\oplus$ and $\odot$ is called a vector space.

Reduced row echelon form: A matrix is said to be in reduced row echelon form provided it satisfies the following properties: (1) All zero rows, if there are any, appear as bottom rows. (2) The first nonzero entry in a nonzero row is a 1; it is called a leading 1. (3) For each nonzero row, the leading 1 appears to the right and below any leading 1s in preceding rows. (4) If a column contains a leading 1, then all other entries in that column are zero.

Reflection: The linear transformation $L\colon R^2 \to R^2$ given by $L(x, y) = (x, -y)$ is called a reflection with respect to the x-axis. Similarly, $L(x, y) = (-x, y)$ is called a reflection with respect to the y-axis.

Roots of the characteristic polynomial: For a square matrix A, the roots of its characteristic polynomial $f(t) = \det(A - tI)$ are the eigenvalues of A.

Rotation: The linear transformation $L\colon R^2 \to R^2$ given by

$$L\left(\begin{bmatrix} x \\ y \end{bmatrix}\right) = \begin{bmatrix} \cos(\theta) & -\sin(\theta) \\ \sin(\theta) & \cos(\theta) \end{bmatrix} \begin{bmatrix} x \\ y \end{bmatrix}$$

is called a counterclockwise rotation in the plane by the angle θ.

Row echelon form: A matrix is said to be in row echelon form provided it satisfies the following properties: (1) All zero rows, if there are any, appear as bottom rows. (2) The first nonzero entry in a nonzero row is a 1; it is called a leading 1. (3) For each nonzero row, the leading 1 appears to the right and below any leading 1s in preceding rows.

Row equivalent: The $m \times n$ matrices A and B are row equivalent provided there exists a set of row operations that when performed on A yield B.

Row rank: The row rank of matrix A is the dimension of the row space of A or, equivalently, the number of linearly independent rows of A.

Row space: The row space of a real $m \times n$ matrix A is the subspace of R^n spanned by the rows of A.

Scalar matrix: Matrix A is a scalar matrix provided A is a diagonal matrix with equal diagonal entries.

Scalar multiple of a matrix: For an $m \times n$ matrix $A = \left[a_{ij}\right]$ and scalar r, the scalar multiple of A by r gives the $m \times n$ matrix $rA = \left[ra_{ij}\right]$.

Scalar multiple of a vector: If $\mathbf{v}$ is in real vector space V, then for any real number k, a scalar, the scalar multiple of $\mathbf{v}$ by k is denoted $k\mathbf{v}$. If $V = R^n$, then $k\mathbf{v} = (kv_1, kv_2, \ldots, kv_n)$.

Scalars: In a real vector space V the scalars are real numbers and are used when we form scalar multiples $k\mathbf{v}$, where $\mathbf{v}$ is in V. Also, when we form linear combinations of vectors the coefficients are scalars.

Shear: A shear in the x-direction is defined by the matrix transformation

$$L(\mathbf{u}) = \begin{bmatrix} 1 & k \\ 0 & 1 \end{bmatrix} \begin{bmatrix} u_1 \\ u_2 \end{bmatrix},$$

where k is a scalar. Similarly, a shear in the y-direction is given by

$$L(\mathbf{u}) = \begin{bmatrix} 1 & 0 \\ k & 1 \end{bmatrix} \begin{bmatrix} u_1 \\ u_2 \end{bmatrix}.$$

Similar matrices: Matrices A and B are similar provided there exists a nonsingular matrix P such that $A = P^{-1}BP$.

Singular (or noninvertible) matrix: A matrix A that has no inverse matrix is said to be singular. Any square matrix whose reduced row echelon form is not the identity matrix is singular.

Skew symmetric matrix: A square real matrix A such that $A = -A^T$ is called skew symmetric.

Solution space: The solution space of an $m \times n$ real homogeneous system $A\mathbf{x} = \mathbf{0}$ is the set W of all n-vectors $\mathbf{x}$ such that A times $\mathbf{x}$ gives the zero vector. W is a subspace of R^n.

Solution to a homogeneous system: A solution to a homogeneous system $A\mathbf{x} = \mathbf{0}$ is a vector $\mathbf{x}$ such that A times $\mathbf{x}$ gives the zero vector.

Solution to a linear system: A solution to a linear system $A\mathbf{x} = \mathbf{b}$ is any vector $\mathbf{x}$ such that A times $\mathbf{x}$ gives the vector $\mathbf{b}$.

Span: The span of a set $W = \{\mathbf{w}_1, \mathbf{w}_2, \ldots, \mathbf{w}_k\}$, denoted by span W, from a vector space V is the set of all possible linear combinations of the vectors $\mathbf{w}_1, \mathbf{w}_2, \ldots, \mathbf{w}_k$. Span W is a subspace of V.

Square matrix: A matrix with the same number of rows as columns is called a square matrix.

Standard inner product: For vectors $\mathbf{v}$ and $\mathbf{w}$ in R^n the standard inner product of $\mathbf{v}$ and $\mathbf{w}$ is also called the dot product of $\mathbf{v}$ and $\mathbf{w}$, denoted $\mathbf{v} \cdot \mathbf{w} = v_1w_1 + v_2w_2 + \cdots + v_nw_n$.

Submatrix: A matrix obtained from a matrix A by deleting rows and/or columns is called a submatrix of A.

Subspace: A subset W of a vector space V that is closed under addition and scalar multiplication is called a subspace of V.

Sum of vectors: The sum of two vectors is also called vector addition. In R^n adding corresponding components of the vectors performs the sum of two vectors. In a vector space V, $\mathbf{u} \oplus \mathbf{v}$ is computed using the definition of the operation $\oplus$.

Summation notation: A compact notation to indicate the sum of a set $\{a_1, a_2, \ldots, a_n\}$; the sum of a_1 through a_n is denoted in summation notation as $\sum_{i=1}^{n} a_i$.

Symmetric matrix: A square real matrix A such that $A = A^T$ is called symmetric.

Transition matrix: Let $S = \{\mathbf{v}_1, \mathbf{v}_2, \ldots, \mathbf{v}_n\}$ and $T = \{\mathbf{w}_1, \mathbf{w}_2, \ldots, \mathbf{w}_n\}$ be bases for an n-dimensional vector space V. The transition matrix from the T-basis to the S-basis is an $n \times n$ matrix, denoted $P_{S\leftarrow T}$, that converts the coordinates of a vector $\mathbf{v}$ relative to the T-basis into the coordinates of $\mathbf{v}$ relative to the S-basis; $\left[\mathbf{v}\right]_S = P_{S\leftarrow T}\left[\mathbf{v}\right]_T$.

Translation: Let $T\colon V \to V$ be defined by $T(\mathbf{v}) = \mathbf{v} + \mathbf{b}$ for all $\mathbf{v}$ in V and any fixed vector $\mathbf{b}$ in V. We call this the translation by the vector $\mathbf{b}$.

Transpose of a matrix: The transpose of an $m \times n$ matrix A is the $n \times m$ matrix obtained by forming columns from each row of A. The transpose of A is denoted A^T.

Trivial solution: The trivial solution of a homogeneous system $A\mathbf{x} = \mathbf{0}$ is the zero vector.

2-space: The set of all 2-vectors is called 2-space. For vectors whose entries are real numbers we denote 2-space as R^2.

Unit vector: A vector of length 1 is called a unit vector.

Unitary matrix: An $n \times n$ complex matrix A is called unitary provided $A^{-1} = \overline{A}^T$.

Upper triangular matrix: A square matrix with zero entries below its diagonal entries is called upper triangular.

Vector: The generic name for any member of a vector space. (See also 2-vector and n-vector.)

Vector addition: The sum of two vectors is also called vector addition. In R^n adding corresponding components of the vectors performs vector addition.

Zero matrix: A matrix with all zero entries is called the zero matrix.

Zero polynomial: A polynomial all of whose coefficients are zero is called the zero polynomial.

Zero subspace: The subspace consisting of exactly the zero vector of a vector space is called the zero subspace.

Zero vector: A vector with all zero entries is called the zero vector.

ANSWERS TO ODD-NUMBERED EXERCISES

CHAPTER 1

Section 1.1, p. 8

1. $x = 4, y = 2.$

3. $x = -4, y = 2, z = 10.$

5. $x = 2, y = -1, z = -2.$

7. $x = -20, y = \left(\frac{1}{4}\right) z + 8, z =$ any real number.

9. This linear system has no solution. It is inconsistent.

11. $x = 5, y = 1.$

13. This linear system has no solution. It is inconsistent.

15. **(a)** $t = 10.$ **(b)** One value is $t = 3.$

(c) The choice $t = 3$ in part (b) was arbitrary. Any choice for t, other than $t = 10$, makes the system inconsistent. Hence there are infinitely many ways to choose a value for t in part (b).

17. **(a)** $t = 0.$ **(b)** $t = 1.$ **(c)** Any $t \neq 0.$

21. $x = 2, y = 1, z = 0.$

23. There is no such value of r.

27. Zero, infinitely many, zero.

29. No points of intersection:

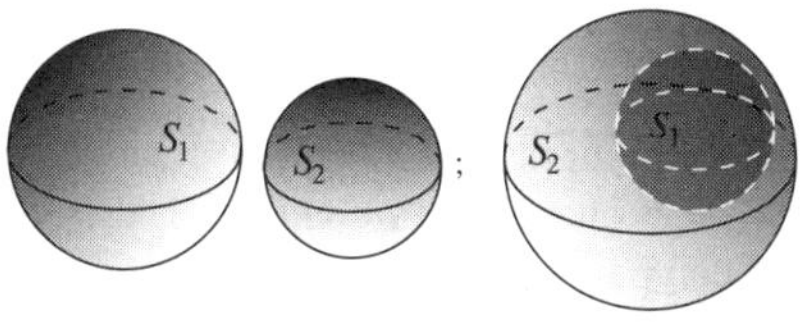

One point of intersection:

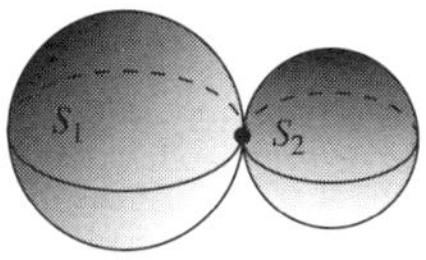

Infinitely many points of intersection:

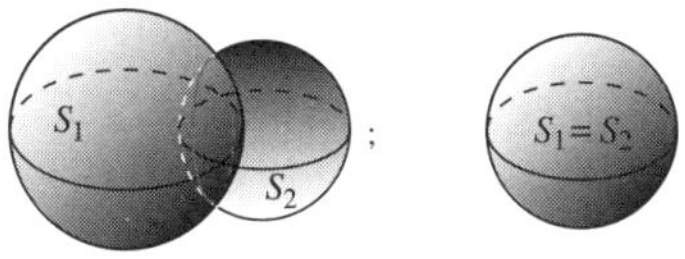

Intersection is a circle   Intersection is S_1 $(=S_2)$

31. 1.5 tons of regular and 2.5 tons of special plastic.

33. 20 tons of 2-minute developer and a total of 40 tons of 6-minute and 9-minute developer.

35. \$7000, \$14,000, \$3000.

Section 1.2, p. 19

1. **(a)** $-3, -5, 4.$ **(b)** $4, 5.$ **(c)** $2, 6, -1.$

3. **(a)**

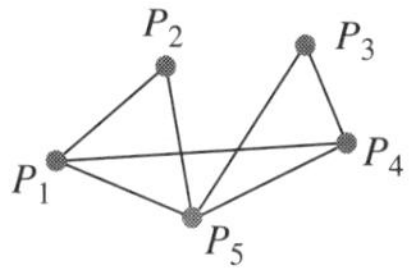

(b)

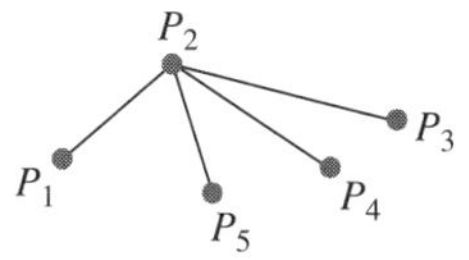

5. $a = 0, b = 2, c = 1, d = 2.$

7. **(a)** $\begin{bmatrix} 1 & 4 \\ 10 & 18 \end{bmatrix}.$

(b) $3(2A) = 6A = \begin{bmatrix} 6 & 12 & 18 \\ 12 & 6 & 24 \end{bmatrix}.$

(c) $3A + 2A = 5A = \begin{bmatrix} 5 & 10 & 15 \\ 10 & 5 & 20 \end{bmatrix}.$

(d) $2(D + F) = 2D + 2F = \begin{bmatrix} -2 & 6 \\ 8 & 14 \end{bmatrix}.$

(e) $(2 + 3)D = 2D + 3D = \begin{bmatrix} 15 & -10 \\ 10 & 20 \end{bmatrix}.$

(f) Impossible.

9. **(a)** $\begin{bmatrix} 2 & 4 \\ 4 & 2 \\ 6 & 8 \end{bmatrix}.$ **(b)** Impossible.

(c) $\begin{bmatrix} 1 & -4 \\ 2 & 1 \\ 3 & -2 \end{bmatrix}.$ **(d)** Impossible.

(e) $(-A)^T = -(A^T) = \begin{bmatrix} -1 & -2 \\ -2 & -1 \\ -3 & -4 \end{bmatrix}.$

(f) Impossible.

11. No. **13.** Zero.

15. The entries are symmetric about the main diagonal.

19. **(a)** True. **(b)** True. **(c)** True.

21. $\frac{1}{2}(\mathbf{t} + \mathbf{b}).$

Section 1.3, p. 30

1. **(a)** 2. **(b)** 1. **(c)** 4. **(d)** 1.

3. ± 2. **5.** $x = 4,\ y = -6$.

7. 1. **9.** $\pm\frac{\sqrt{2}}{2}$.

11. **(a)** $\begin{bmatrix} 14 & 8 \\ 16 & 9 \end{bmatrix}$. **(b)** $\begin{bmatrix} 1 & 2 & 3 \\ 4 & 5 & 10 \\ 7 & 8 & 17 \end{bmatrix}$.

(c) $\begin{bmatrix} 7 & 10 & -2 \\ 19 & 6 & 31 \end{bmatrix}$.

(d) Impossible. **(e)** $\begin{bmatrix} 19 & -8 \\ 32 & 30 \end{bmatrix}$.

13. **(a)** $\begin{bmatrix} -2 & 12 \\ 2 & 17 \\ 10 & 13 \end{bmatrix}$. **(b)** $\begin{bmatrix} 8 & 12 \\ 12 & -1 \end{bmatrix}$.

(c) $\begin{bmatrix} 11 & 4 \\ 27 & 19 \end{bmatrix}$.

(d) Impossible. **(e)** Impossible.

15. **(a)** $\begin{bmatrix} 1 & 2 \\ 2 & 1 \\ 3 & 4 \end{bmatrix}$. **(b)** $(A^T)^T = A$.

(c) $\begin{bmatrix} 14 & 16 \\ 8 & 9 \end{bmatrix}$. **(d)** Same as (c).

(e) $(C + E)^T B = C^T B + E^T B = \begin{bmatrix} 28 & 14 \\ 8 & 8 \\ 38 & 17 \end{bmatrix}$.

(f) $A(2B) = 2(AB) = \begin{bmatrix} 28 & 16 \\ 32 & 18 \end{bmatrix}$.

17. **(a)** 4. **(b)** 13. **(c)** 3. **(d)** 12.

19. $AB = \begin{bmatrix} -4 & 7 \\ 0 & 5 \end{bmatrix}$; $BA = \begin{bmatrix} -1 & 2 \\ 9 & 2 \end{bmatrix}$.

21. **(a)** $\begin{bmatrix} 6 \\ 25 \\ 10 \\ 25 \end{bmatrix}$. **(b)** $\begin{bmatrix} 12 \\ 11 \\ 17 \\ 20 \end{bmatrix}$.

23. $2\begin{bmatrix} 2 \\ -1 \\ 5 \end{bmatrix} + 1\begin{bmatrix} -3 \\ 2 \\ -1 \end{bmatrix} + 4\begin{bmatrix} 4 \\ 3 \\ -2 \end{bmatrix}$.

27. There are infinitely many choices. For example, $r = 1$, $s = 0$; or $r = 0$, $s = 2$; or $r = 10$, $s = -18$.

31.
$$\begin{aligned} -2x_1 - x_2 \qquad\quad + 4x_4 &= 5 \\ -3x_1 + 2x_2 + 7x_3 + 8x_4 &= 3 \\ x_1 \qquad\qquad\quad + 2x_4 &= 4 \\ 3x_1 \qquad + x_3 + 3x_4 &= 6. \end{aligned}$$

33. $\begin{bmatrix} 2 & 3 & 0 \\ 0 & 3 & 1 \\ 2 & -1 & 0 \end{bmatrix}\begin{bmatrix} x_1 \\ x_2 \\ x_3 \end{bmatrix} = \begin{bmatrix} 0 \\ 0 \\ 0 \end{bmatrix}$.

35. The linear systems are equivalent. That is, they have the same solutions.

37. **(a)** $\begin{aligned} 2x_1 + x_2 &= 4 \\ 3x_2 &= 2. \end{aligned}$ **(b)** $\begin{aligned} x_1 \qquad + 3x_3 + x_4 &= 2 \\ 2x_1 + x_2 + 4x_3 + 3x_4 &= 5 \\ -x_1 + 2x_2 + 5x_3 + 4x_4 &= 8. \end{aligned}$

39. **(a)** $\mathbf{x} = \begin{bmatrix} 0 \\ 0 \\ 0 \end{bmatrix}$ is the only solution.

(b) $\mathbf{x} = \begin{bmatrix} 2 \\ 2 \\ 2 \\ 2 \end{bmatrix}$ is a solution; another solution is $\begin{bmatrix} -9 \\ 11 \\ 1 \\ 1 \end{bmatrix}$.

49. AB gives the total cost of producing each kind of product in each city:

	Salt Lake City	Chicago
Chair	38	44
Table	67	78

Section 1.4, p. 40

9. One such pair is $A = \begin{bmatrix} 1 & 1 \\ 2 & 2 \end{bmatrix}$ and $B = \begin{bmatrix} 1 & 1 \\ -1 & -1 \end{bmatrix}$.
Another pair is $A = \begin{bmatrix} 2 & 6 \\ 4 & 12 \end{bmatrix}$ and $B = \begin{bmatrix} -3 & -3 \\ 1 & 1 \end{bmatrix}$.
Yet another pair is $A = O$ and $B =$ any 2×2 matrix with at least one nonzero element. There are infinitely many such pairs.

11. There are many such pairs of matrices. For example, $A = \begin{bmatrix} 1 & 1 \\ 0 & 1 \end{bmatrix}$ and $B = \begin{bmatrix} 1 & -1 \\ 0 & 1 \end{bmatrix}$ or $A = \begin{bmatrix} 1 & 1 \\ 1 & 2 \end{bmatrix}$ and $B = \begin{bmatrix} 2 & -1 \\ -1 & 1 \end{bmatrix}$. Note also that for $A = k\begin{bmatrix} 1 & 0 \\ 0 & 1 \end{bmatrix}$ and $B = \left(\frac{1}{k}\right)\begin{bmatrix} 1 & 0 \\ 0 & 1 \end{bmatrix}$, $k \neq 0, \pm 1$, we have $A \neq B$ and $AB = \begin{bmatrix} 1 & 0 \\ 0 & 1 \end{bmatrix}$.

21. **(a)** $3A = 3\begin{bmatrix} \cos 0^\circ & \sin 0^\circ \\ \cos 1^\circ & \sin 1^\circ \\ \vdots & \vdots \\ \cos 359^\circ & \sin 359^\circ \end{bmatrix}$.

(b) The ordered pairs obtained from A are $(x_i, y_i) = (3\cos i, 3\sin i)$, where $i = 0, 1, \ldots, 359$. Since

$$x_i^2 + y_i^2 = 9,$$

we conclude that the point (x_i, y_i) lies on the circle $x^2 + y^2 = 9$.

23. $r = 2$. **25.** $s = r^2$. **31.** $k = \pm\sqrt{\frac{1}{6}}$.

Section 1.5, p. 52

7. **(a)** $\begin{bmatrix} -7 & 2 & 1 \\ 8 & 1 & 5 \\ 1 & -1 & 4 \end{bmatrix}$. **(b)** $\begin{bmatrix} 2 & 0 & 1 \\ 1 & 1 & 1 \\ 2 & 0 & 3 \end{bmatrix}$.

(c) $\begin{bmatrix} -8 & 0 & 0 \\ -5 & 25 & 100 \\ -13 & 25 & 100 \end{bmatrix}$.

15. $B = \begin{bmatrix} 1 & 3 \\ 3 & 1 \end{bmatrix}$ is such that $AB = BA$. There are infinitely many such matrices B.

33. **(a)** $A^{-1} = \begin{bmatrix} -\frac{2}{13} & \frac{3}{13} \\ \frac{5}{13} & -\frac{1}{13} \end{bmatrix}$. **(b)** $A^{-1} = \begin{bmatrix} -\frac{1}{3} & \frac{2}{3} \\ \frac{2}{3} & -\frac{1}{3} \end{bmatrix}$.

35. $\begin{bmatrix} 11 & 19 \\ 7 & 0 \end{bmatrix}$. **37.** $\mathbf{x} = \begin{bmatrix} 15 \\ 9 \end{bmatrix}$. **39.** $\mathbf{x} = \begin{bmatrix} 2 \\ 1 \end{bmatrix}$.

41. **(a)** $\begin{bmatrix} \frac{6}{13} \\ \frac{11}{13} \end{bmatrix}$. **(b)** $\begin{bmatrix} \frac{8}{13} \\ \frac{19}{13} \end{bmatrix}$.

43. Possible answer: $\begin{bmatrix} 1 & 2 \\ -3 & 4 \end{bmatrix}$ and $\begin{bmatrix} -1 & -2 \\ 3 & -4 \end{bmatrix}$.

59. **(a)** $\mathbf{w}_1 = \begin{bmatrix} 5 \\ 1 \end{bmatrix}$, $\mathbf{w}_2 = \begin{bmatrix} 19 \\ 5 \end{bmatrix}$, $\mathbf{w}_3 = \begin{bmatrix} 65 \\ 19 \end{bmatrix}$, $u_2 = 5$, $u_3 = 19$, $u_4 = 65$.

(b) $\mathbf{w}_{n-1} = A^{n-1}\mathbf{w}_0$.

Section 1.6, p. 62

1.

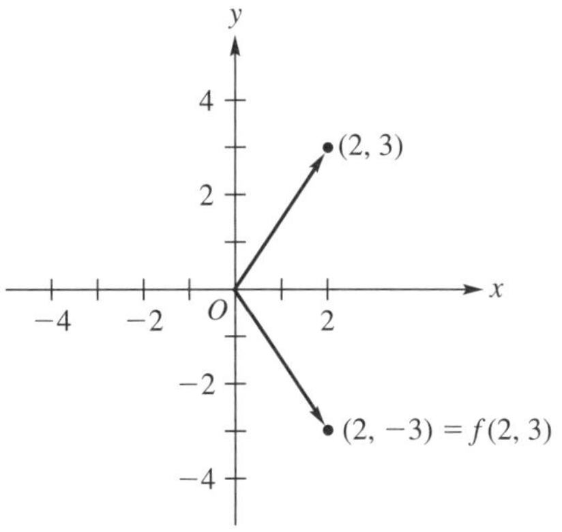

3.

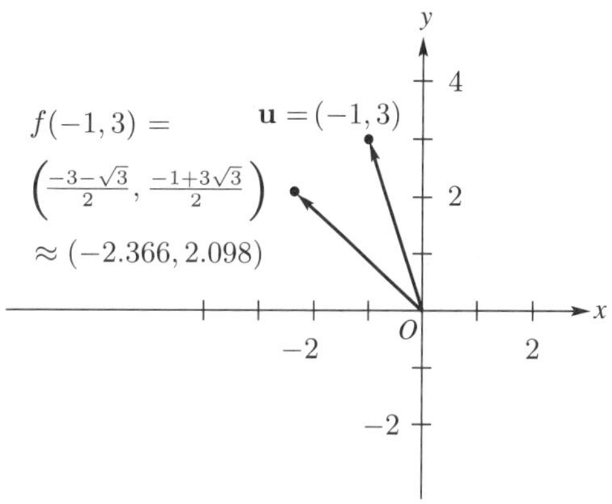

5.

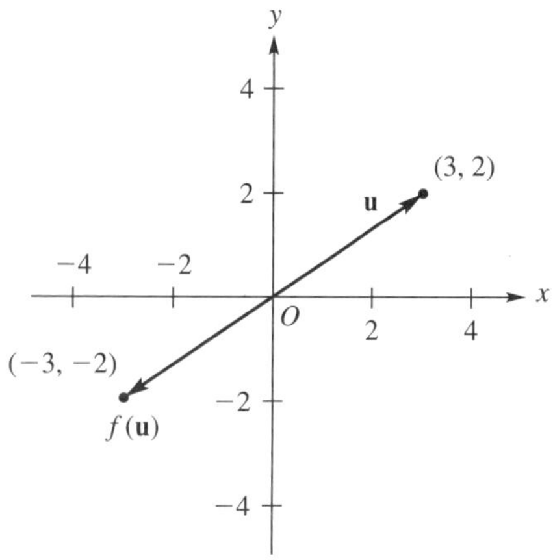

7.

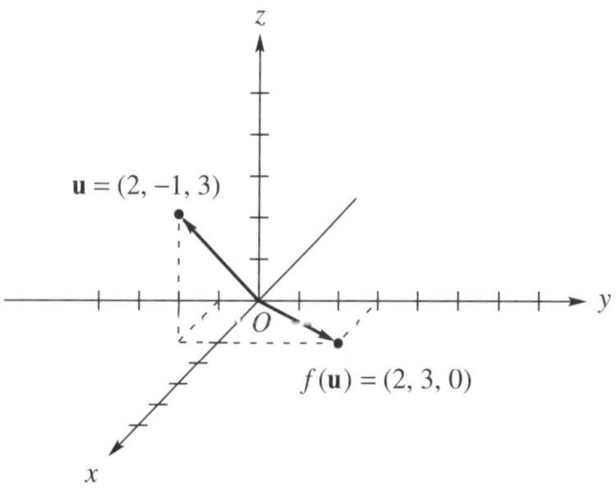

9. Yes. **11.** Yes. **13.** No.

15. **(a)** Reflection about the y-axis.

(b) Rotate counterclockwise through $\frac{\pi}{2}$.

17. **(a)** Projection onto the x-axis.

(b) Projection onto the y-axis.

19. **(a)** Counterclockwise rotation by 60°.

(b) Clockwise rotation by 30°.

(c) 12.

Section 1.7, p. 70

1.

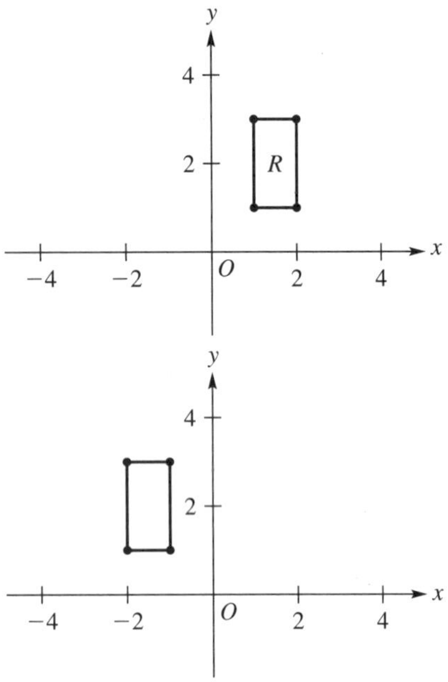

3.

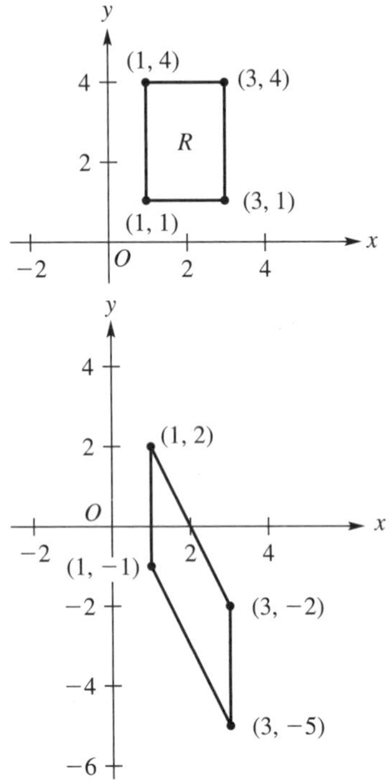

5.

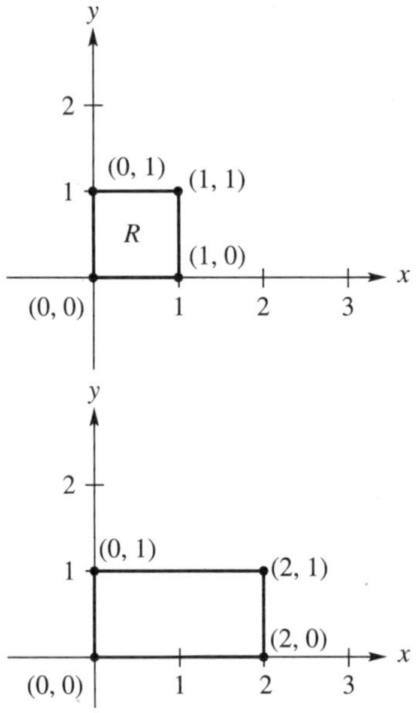

7. (−10, 15), (3, 12), (−5, 2).

9.

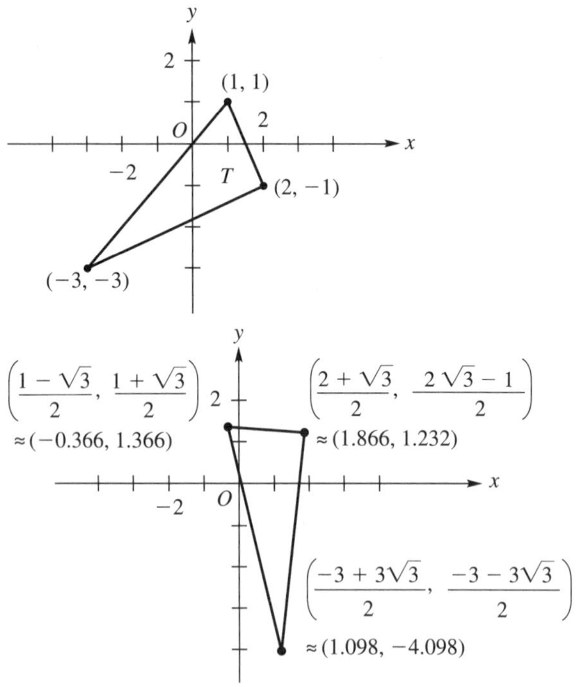

11. The image of the vertices of T under L consists of the points $(-9, -18)$, $(0, 0)$, and $(3, 6)$. Thus the image of T under L is a line segment.

13.

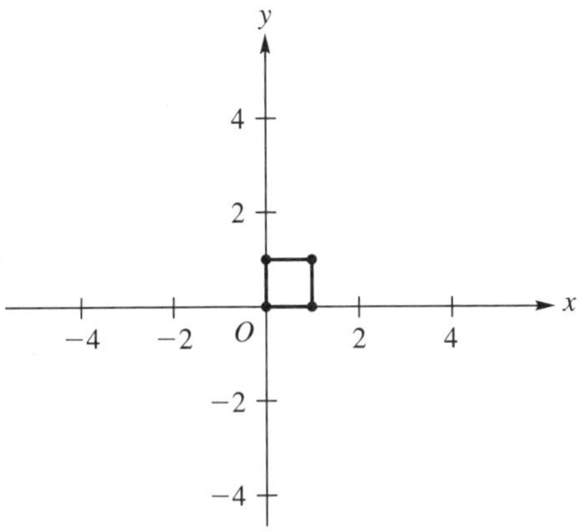

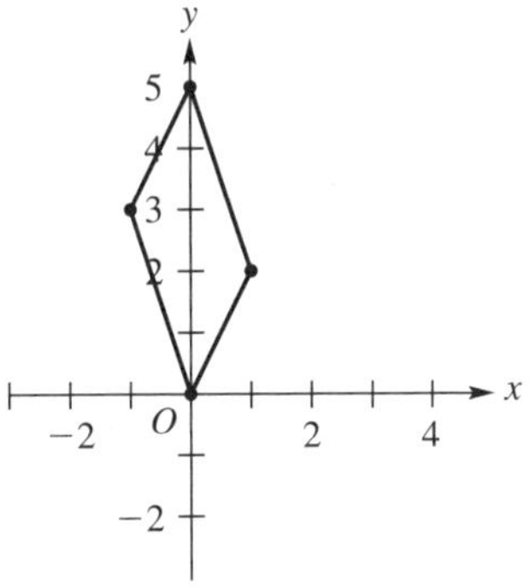

15. (a) Possible answer: First perform f_1 (90° counterclockwise rotation) then f_3.

(b) Possible answer: Perform f_1 (−135° counterclockwise rotation).

17. $\cos(\theta_1 - \theta_2) = \cos\theta_1 \cos\theta_2 + \sin\theta_1 \sin\theta_2$
$\sin(\theta_1 - \theta_2) = \sin\theta_1 \cos\theta_2 - \cos\theta_1 \sin\theta_2$.

Section 1.8, p. 79

1. A has correlation coefficient 0.93.
B has correlation coefficient 0.76.
C has correlation coefficient 0.97.
D has correlation coefficient 0.88.

3.

Distance	200	200	400	400	500	500	500	700	700
Amplitude	12.6	19.9	9.3	9.5	7.9	7.8	8.0	6.0	6.4

Correlation coefficient = −0.8482.
Angle in radians = 2.5834.
Angle in degrees = 148.0161°.
Moderately negatively correlated.

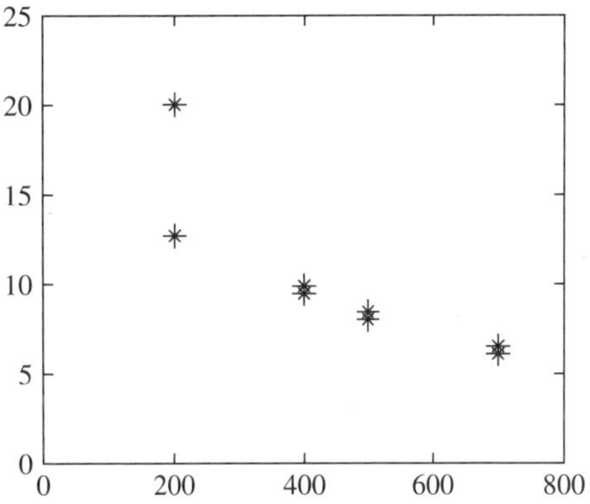

Supplementary Exercises, p. 80

1. (a) 3. **(b)** 6. **(c)** 10. **(d)** $\frac{n}{2}(n+1)$.

3. $\begin{bmatrix}1 & 0\\0 & 1\end{bmatrix}, \begin{bmatrix}-1 & 0\\0 & -1\end{bmatrix}, \begin{bmatrix}1 & b\\0 & -1\end{bmatrix}, \begin{bmatrix}-1 & b\\0 & 1\end{bmatrix}$,
where b is any real number.

23. $\mathbf{w} = \begin{bmatrix}1 & 1 & 1 & 1\end{bmatrix}^T$.

Chapter Review, p. 83

True/False

1. F. **2.** F. **3.** T. **4.** T. **5.** T.
6. T. **7.** T. **8.** T. **9.** T. **10.** T.

Quiz

1. $\begin{bmatrix}2\\-4\end{bmatrix}$. **2.** $r = 0$. **3.** $a = b = 4$.

4. $a = 2$, $b = 10$, $c =$ any real number.

5. $\begin{bmatrix}3\\r\end{bmatrix}$, where r is any real number.

CHAPTER 2

Section 2.1, p. 94

1. (a) Possible answer:
$\mathbf{r}_1 \to -\mathbf{r}_1$
$\mathbf{r}_2 \to \mathbf{r}_2 - 2\mathbf{r}_1$
$\mathbf{r}_3 \to \mathbf{r}_3 - 2\mathbf{r}_1$
$\mathbf{r}_2 \to \frac{1}{3}\mathbf{r}_2$
$\mathbf{r}_3 \to \mathbf{r}_3 - 2\mathbf{r}_2$
$\mathbf{r}_3 \to -3\mathbf{r}_3$

$$\begin{bmatrix}1 & -2 & 5\\0 & 1 & -\frac{4}{3}\\0 & 0 & 1\end{bmatrix}$$

(b) Possible answer:
$\mathbf{r}_2 \to \mathbf{r}_2 - 3\mathbf{r}_1$
$\mathbf{r}_3 \to \mathbf{r}_3 - 5\mathbf{r}_1$
$\mathbf{r}_4 \to \mathbf{r}_4 + 2\mathbf{r}_1$

$$\begin{bmatrix}1 & 1 & -1\\0 & 1 & 2\\0 & 0 & 0\\0 & 0 & 0\end{bmatrix}$$

3. (a) $\mathbf{r}_2 \to \mathbf{r}_2 + 2\mathbf{r}_3$, $\mathbf{r}_1 \to \mathbf{r}_1 - 4\mathbf{r}_3$, $\mathbf{r}_1 \to \mathbf{r}_1 - 2\mathbf{r}_2$ $\begin{bmatrix} 1 & 0 & 0 \\ 0 & 1 & 0 \\ 0 & 0 & 1 \end{bmatrix}$

(b) $\mathbf{r}_2 \to \mathbf{r}_2 + 4\mathbf{r}_3$, $\mathbf{r}_1 \to \mathbf{r}_1 - 5\mathbf{r}_3$, $\mathbf{r}_1 \to \mathbf{r}_1 - 3\mathbf{r}_3$ $\begin{bmatrix} 1 & 4 & 0 & 0 \\ 0 & 0 & 1 & 0 \\ 0 & 0 & 0 & 1 \\ 0 & 0 & 0 & 0 \end{bmatrix}$

5. (a) $\mathbf{r}_2 \to \mathbf{r}_2 + 2\mathbf{r}_1$, $\mathbf{r}_3 \to \mathbf{r}_3 - 3\mathbf{r}_1$, $\mathbf{r}_3 \to \mathbf{r}_3 - 2\mathbf{r}_2$ $\begin{bmatrix} 1 & 0 & -2 \\ 0 & 1 & 5 \\ 0 & 0 & 0 \end{bmatrix}$

(b) $\mathbf{r}_2 \to \mathbf{r}_2 + \mathbf{r}_1$, $\mathbf{r}_4 \to \mathbf{r}_4 + 2\mathbf{r}_1$, $\mathbf{r}_2 \to \frac{1}{2}\mathbf{r}_2$, $\mathbf{r}_3 \to \mathbf{r}_3 - \mathbf{r}_2$, $\mathbf{r}_4 \to \mathbf{r}_4 - 7\mathbf{r}_2$, $\mathbf{r}_3 \to 2\mathbf{r}_3$, $\mathbf{r}_4 \to \mathbf{r}_4 - \frac{1}{2}\mathbf{r}_3$, $\mathbf{r}_2 \to \mathbf{r}_2 + \frac{1}{2}\mathbf{r}_3$, $\mathbf{r}_1 \to \mathbf{r}_1 - \mathbf{r}_3$ $\begin{bmatrix} 1 & 0 & 0 \\ 0 & 1 & 0 \\ 0 & 0 & 1 \\ 0 & 0 & 0 \end{bmatrix}$

7. (a) N **(b)** REF **(c)** RREF.

11. (a) Possible answer: $\begin{bmatrix} 1 & 0 & 0 & 0 \\ -1 & 1 & 0 & 0 \\ 0 & \frac{1}{2} & 1 & 0 \\ 2 & -\frac{1}{2} & 3 & 1 \end{bmatrix}$

(b) $I_4 = \begin{bmatrix} 1 & 0 & 0 & 0 \\ 0 & 1 & 0 & 0 \\ 0 & 0 & 1 & 0 \\ 0 & 0 & 0 & 1 \end{bmatrix}$

13. $\begin{bmatrix} 1 & 0 \\ 0 & 1 \end{bmatrix}$.

Section 2.2, p. 113

1. (a) $x = 8, y = 1, z = 4$.

(b) $x = -21 + 3t, y = t, z = 5, w = 1$.

3. (a) $x = 2 - s, y = s, z = -3 - t, w = t$.

(b) $x = 3, y = 0, z = 1$.

5. (a) $x = 1, y = 2, z = -2$.

(b) $x = 1, y = 2, z = -2$.

7. (a) $x = -1, y = 4, z = -3$.

(b) $x = y = z = 0$.

(c) $x = r, y = -2r, z = r$, where r = any real number.

(d) $x = -2r, y = r, z = 0$, where r = any real number.

9. (a) $x = 1, y = 2, z = 2$.

(b) $x = y = z = 0$.

11. $\mathbf{x} = \begin{bmatrix} r \\ r \end{bmatrix}, r \neq 0$. **13.** $\mathbf{x} = \begin{bmatrix} -\frac{1}{2}r \\ \frac{1}{2}r \\ r \end{bmatrix}, r \neq 0$.

15. (a) $a = \pm\sqrt{3}$. **(b)** $a \neq \pm\sqrt{3}$.

(c) There is no value of a such that this system has infinitely many solutions.

17. (a) $a = -3$. **(b)** $a \neq \pm 3$. **(c)** $a = 3$.

21. $x = -2 + r, y = 2 - 2r, z = r$, where r is any real number.

23. $c - b - a = 0$. **27.** $-a + b - c = 0$.

31. $2x^2 + 2x + 1$.

33. $T_1 = 36.25°C, T_2 = 36.25°C, T_3 = 28.75°C, T_4 = 28.75°C$.

35. Radius = 37.

39. One solution is $2C_2H_6 + 7O_2 \to 4CO_2 + 6H_2O$.

41. $\begin{bmatrix} 1 - i \\ 2 \\ -i \end{bmatrix}$. **43.** $\begin{bmatrix} i \\ 2i \\ -1 \end{bmatrix}$.

Section 2.3, p. 124

3. (a) $\begin{bmatrix} 1 & -4 & 0 & 0 \\ 0 & 1 & 0 & 0 \\ 0 & 0 & 1 & 0 \\ 0 & 0 & 0 & 1 \end{bmatrix}$.

(b) $\begin{bmatrix} 1 & 0 & 0 & 0 \\ 0 & 0 & 1 & 0 \\ 0 & 1 & 0 & 0 \\ 0 & 0 & 0 & 1 \end{bmatrix}$. **(c)** $\begin{bmatrix} 1 & 0 & 0 & 0 \\ 0 & 1 & 0 & 0 \\ 0 & 0 & 4 & 0 \\ 0 & 0 & 0 & 1 \end{bmatrix}$.

5. (a) $C = \begin{bmatrix} 1 & 0 & 0 \\ 0 & 1 & 0 \\ 0 & 0 & 1 \end{bmatrix}$.

(b) $B = \begin{bmatrix} -1 & 1 & 1 \\ 2 & -1 & -2 \\ 2 & -1 & -1 \end{bmatrix}$.

(c) A and B are inverses of each other.

7. $A^{-1} = \begin{bmatrix} -2 & \frac{3}{2} \\ 1 & -\frac{1}{2} \end{bmatrix}$.

9. (a) Singular. **(b)** $A^{-1} = \begin{bmatrix} \frac{1}{2} & -\frac{1}{4} \\ \frac{1}{6} & \frac{1}{12} \end{bmatrix}$.

(c) $A^{-1} = \begin{bmatrix} 0 & 1 & -1 \\ 2 & -2 & -1 \\ -1 & 1 & 1 \end{bmatrix}$. **(d)** Singular.

11. **(a)** $A^{-1} = \begin{bmatrix} 1 & 0 & -1 \\ 1 & -1 & 2 \\ -1 & 1 & -1 \end{bmatrix}$.

(b) $A^{-1} = \begin{bmatrix} \frac{7}{3} & -\frac{1}{3} & -\frac{1}{3} & -\frac{2}{3} \\ \frac{4}{9} & -\frac{1}{9} & -\frac{4}{9} & \frac{1}{9} \\ -\frac{1}{9} & -\frac{2}{9} & \frac{1}{9} & \frac{2}{9} \\ -\frac{5}{3} & \frac{2}{3} & \frac{2}{3} & \frac{1}{3} \end{bmatrix}$.

(c) Singular.

(d) $A^{-1} = \begin{bmatrix} \frac{3}{2} & -1 & \frac{1}{2} \\ \frac{1}{2} & 0 & -\frac{1}{2} \\ -\frac{3}{2} & 1 & \frac{1}{2} \end{bmatrix}$.

(e) Singular.

13. $A = \begin{bmatrix} 1 & 0 \\ 3 & 1 \end{bmatrix} \begin{bmatrix} 1 & 0 \\ 0 & -2 \end{bmatrix} \begin{bmatrix} 1 & 2 \\ 0 & 1 \end{bmatrix}$.

15. $A = \begin{bmatrix} \frac{1}{2} & -1 \\ -\frac{1}{2} & 2 \end{bmatrix}$.

17. (a) and (b).

19. A^{-1} exists for $a \neq 0$. Then

$$A^{-1} = \begin{bmatrix} 0 & 1 & 0 \\ 1 & -1 & 0 \\ -\dfrac{2}{a} & \dfrac{1}{a} & \dfrac{1}{a} \end{bmatrix}.$$

Section 2.4, p. 129

3. **(a)** $\begin{bmatrix} 1 & 0 & 0 & 0 \\ 0 & 1 & 0 & 0 \\ 0 & 0 & 0 & 0 \end{bmatrix}$. **(b)** $\begin{bmatrix} 1 & 0 & 0 \\ 0 & 1 & 0 \\ 0 & 0 & 1 \\ 0 & 0 & 0 \end{bmatrix}$.

(c) $\begin{bmatrix} 1 & 0 & 0 & 0 \\ 0 & 1 & 0 & 0 \\ 0 & 0 & 1 & 0 \\ 0 & 0 & 0 & 0 \\ 0 & 0 & 0 & 0 \end{bmatrix}$. **(d)** I_3.

7. $B = \begin{bmatrix} 1 & 0 & 0 & 0 \\ 0 & 1 & 0 & 0 \\ 0 & 0 & 0 & 0 \\ 0 & 0 & 0 & 0 \end{bmatrix}$, $P = \begin{bmatrix} -1 & 1 & 0 & 0 \\ 2 & -1 & 0 & 0 \\ -4 & 0 & 1 & 0 \\ 0 & 0 & 0 & 1 \end{bmatrix}$,

$$Q = \begin{bmatrix} 1 & 0 & -1 & 2 \\ 0 & 1 & 1 & 5 \\ 0 & 0 & 1 & 0 \\ 0 & 0 & 0 & 1 \end{bmatrix}.$$

Section 2.5, p. 136

1. $\mathbf{x} = \begin{bmatrix} 1 \\ 2 \\ 1 \end{bmatrix}$. **3.** $\mathbf{x} = \begin{bmatrix} 1 \\ 0 \\ 2 \\ -4 \end{bmatrix}$.

5. $L = \begin{bmatrix} 1 & 0 & 0 \\ 2 & 1 & 0 \\ 2 & -2 & 1 \end{bmatrix}$, $U = \begin{bmatrix} 2 & 3 & 4 \\ 0 & -1 & 2 \\ 0 & 0 & -2 \end{bmatrix}$,

$\mathbf{x} = \begin{bmatrix} 4 \\ -2 \\ 1 \end{bmatrix}$.

7. $L = \begin{bmatrix} 1 & 0 & 0 \\ 0.5 & 1 & 0 \\ 0.25 & -1.5 & 1 \end{bmatrix}$, $U = \begin{bmatrix} 4 & 2 & 3 \\ 0 & -1 & 3.5 \\ 0 & 0 & 5.5 \end{bmatrix}$,

$\mathbf{x} = \begin{bmatrix} 2 \\ -2 \\ -1 \end{bmatrix}$.

9. $L = \begin{bmatrix} 1 & 0 & 0 & 0 \\ 0.5 & 1 & 0 & 0 \\ -1 & 0.2 & 1 & 0 \\ 2 & -0.4 & 2 & 1 \end{bmatrix}$,

$U = \begin{bmatrix} 2 & 1 & 0 & -4 \\ 0 & -0.5 & 0.25 & 1 \\ 0 & 0 & 0.2 & 2 \\ 0 & 0 & 0 & 2 \end{bmatrix}$, $\mathbf{x} = \begin{bmatrix} 0.5 \\ 2 \\ -2 \\ 1.5 \end{bmatrix}$.

Supplementary Exercises, p. 137

1. $B = \begin{bmatrix} 1 & 0 & 0 \\ 0 & 1 & 0 \\ 0 & 0 & 1 \end{bmatrix}$. **3.** $a = \pm 1$.

5. **(a)** Multiply the jth row of B by $\frac{1}{k}$.

(b) Interchange the ith and jth rows of B.

(c) Add $-k$ times the jth row of B to its ith row.

7. $\dfrac{1}{2}\begin{bmatrix} 1 & 1 & -1 \\ -1 & 1 & 1 \\ 1 & -1 & 1 \end{bmatrix}$.

9. **(a)** The results must be identical, since an inverse is unique.

(b) The instructor computes AA_1 and AA_2. If the result is I_{10}, then the answer submitted by the student is correct.

11. $s \neq \frac{5}{4}$. **19.** $r = 3, s = 1, t = 3, p = -2$.

Chapter Review, p. 138

True/False

1. F. **2.** T. **3.** F. **4.** T. **5.** T.

6. T. **7.** T. **8.** T. **9.** T. **10.** F.

Quiz

1. $\begin{bmatrix} 1 & 0 & 2 \\ 0 & 1 & 3 \\ 0 & 0 & 0 \end{bmatrix}$.

2. (a) No. The entry above the leading one in the second row is not zero.

(b) Infinitely many.

(c) No. If we transform A to reduced row echelon form, we do not obtain I_4.

(d) $\begin{bmatrix} -6+2r+7s \\ r \\ -3s \\ s \end{bmatrix}$, where r and s are any real numbers.

3. $k = 6$.

4. $\begin{bmatrix} 0 \\ 0 \\ 0 \end{bmatrix}$. **5.** $\begin{bmatrix} -\frac{1}{2} & \frac{1}{2} & \frac{1}{2} \\ 1 & -1 & 0 \\ -\frac{1}{2} & \frac{3}{2} & -\frac{1}{2} \end{bmatrix}$.

6. $P = A^{-1}$, $Q = B$.

7. Diagonal, zero, or symmetric.

CHAPTER 3

Section 3.1, p. 145

1. (a) 5. **(b)** 7. **(c)** 4.

3. (a) Even. **(b)** Odd. **(c)** Even.

5. (a) $-$. **(b)** $+$. **(c)** $-$.

7. (a) 9.

(b) Number of inversions in 416235 is 6; number of inversions in 436215 is 9.

9. (a) 0. **(b)** 0.

11. (a) 9. **(b)** 0. **(c)** 144.

13. (a) $t^2 - 3t - 4$. **(b)** $t^3 - 4t^2 + 3t$.

15. (a) $t = 4, t = -1$. **(b)** $t = 1, t = 0, t = 3$.

Section 3.2, p. 154

1. (a) 3. **(b)** 2. **(c)** 24. **(d)** 29.

(e) 4. **(f)** -30.

3. 3. **5.** 8.

7. (a) 2. **(b)** -120.

(c) $(t-1)(t-2)(t-3) = t^3 - 6t^2 + 11t - 6$.

(d) $t^2 - 2t - 11$.

23. 32. **25.** (b) is nonsingular.

27. The system has a nontrivial solution.

Section 3.3, p. 164

1. (a) 1. **(b)** 7. **(c)** 2. **(d)** 10.

3. (a) -2. **(b)** 9. **(c)** -2. **(d)** -1.

5. (a) 3. **(d)** 29. **(e)** 4.

7. (a) 4. **(c)** -30. **(f)** 0.

11. (a) $t = 0, t = 5$. **(b)** $t = 1, t = 4$.

15. (a) 6. **(b)** $(3, -6), (-1, 2), (13, -14)$.

(c) 24.

17. $\frac{41}{2}$.

Section 3.4, p. 169

3. (a) $\begin{bmatrix} 24 & -42 & -30 \\ 19 & -2 & -30 \\ -4 & 32 & 30 \end{bmatrix}$. **(b)** 150.

5. (a) $A^{-1} = \begin{bmatrix} 1 & 0 & -1 \\ 1 & -1 & 2 \\ -1 & 1 & -1 \end{bmatrix}$.

(b) $A^{-1} = \begin{bmatrix} \frac{7}{3} & -\frac{1}{3} & -\frac{1}{3} & -\frac{2}{3} \\ \frac{4}{9} & -\frac{1}{9} & -\frac{4}{9} & \frac{1}{9} \\ -\frac{1}{9} & -\frac{2}{9} & \frac{1}{9} & \frac{2}{9} \\ -\frac{5}{3} & \frac{2}{3} & \frac{2}{3} & \frac{1}{3} \end{bmatrix}$.

(c) Singular. **(d)** $A^{-1} = \begin{bmatrix} \frac{3}{2} & -1 & \frac{1}{2} \\ \frac{1}{2} & 0 & -\frac{1}{2} \\ -\frac{3}{2} & 1 & \frac{1}{2} \end{bmatrix}$.

(e) Singular.

7. (a) $-\frac{1}{28}\begin{bmatrix} -30 & -5 & 9 & 46 \\ -32 & 4 & 4 & 36 \\ -12 & -2 & -2 & 24 \\ 16 & -2 & -2 & -32 \end{bmatrix}$.

(b) $\begin{bmatrix} \frac{3}{14} & -\frac{3}{7} & \frac{1}{7} \\ \frac{1}{7} & \frac{5}{7} & -\frac{4}{7} \\ -\frac{1}{14} & \frac{1}{7} & \frac{2}{7} \end{bmatrix}$. **(c)** $\begin{bmatrix} \frac{2}{9} & -\frac{1}{9} \\ \frac{1}{6} & \frac{1}{6} \end{bmatrix}$.

9. $\begin{bmatrix} \dfrac{d}{ad-bc} & \dfrac{-b}{ad-bc} \\ \dfrac{-c}{ad-bc} & \dfrac{a}{ad-bc} \end{bmatrix}$.

11. $\begin{bmatrix} \frac{1}{4} & 0 & 0 \\ 0 & -\frac{1}{3} & 0 \\ 0 & 0 & \frac{1}{2} \end{bmatrix}$.

Section 3.5, p. 172

1. $x_1 = -2, x_2 = 0, x_3 = 1$.

3. $x_3 = \frac{4}{5}$.

5. $x_1 = x_2 = x_3 = 0$.

7. Since $\det(A) = 0$, we cannot use Cramer's rule.

Supplementary Exercises, p. 174

1. **(a)** 5. **(b)** 4. **(c)** 36. **(d)** 5.

Chapter Review, p. 174

True/False

1. F. **2.** T. **3.** F. **4.** T.
5. T. **6.** F. **7.** F. **8.** T.
9. T. **10.** F. **11.** T. **12.** F.

Quiz

1. -54. **3.** -1. **4.** -2. **6.** 19.

7. $A^{-1} = \begin{bmatrix} -\frac{1}{2} & \frac{5}{2} & 1 \\ 1 & -3 & -1 \\ -1 & 4 & 1 \end{bmatrix}$.

8. $x_1 = \frac{11}{7}, x_2 = -\frac{4}{7}, x_3 = -\frac{5}{7}$.

CHAPTER 4

Section 4.1, p. 187

1.

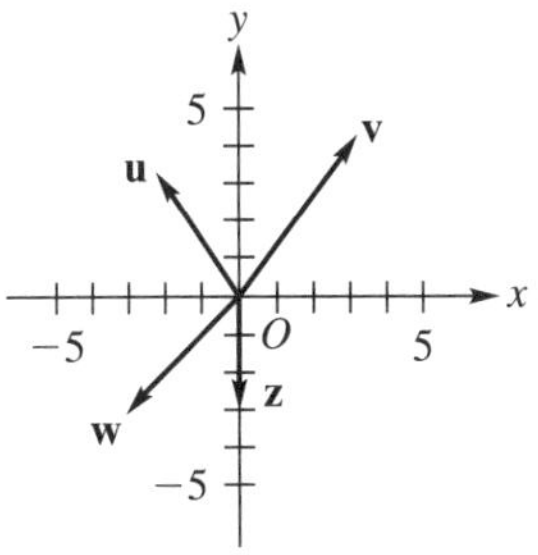

3. Tail $(-1, -4)$.

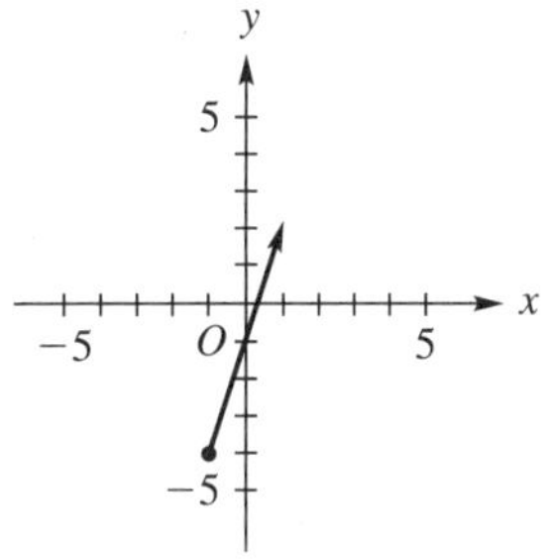

5. $a = 3, b = -1$.

7. **(a)** $\begin{bmatrix} 2 \\ 3 \end{bmatrix}$. **(b)** $\begin{bmatrix} -1 \\ 3 \\ -1 \end{bmatrix}$.

9. **(a)** $\begin{bmatrix} 4 \\ 3 \end{bmatrix}$. **(b)** $\begin{bmatrix} 2 \\ 3 \\ 7 \end{bmatrix}$.

11. **(a)** $\mathbf{u}+\mathbf{v} = \begin{bmatrix} 0 \\ 8 \end{bmatrix}$, $\mathbf{u}-\mathbf{v} = \begin{bmatrix} 4 \\ -2 \end{bmatrix}$,
$2\mathbf{u} = \begin{bmatrix} 4 \\ 6 \end{bmatrix}$, $3\mathbf{u}-2\mathbf{v} = \begin{bmatrix} 10 \\ -1 \end{bmatrix}$.

(b) $\mathbf{u}+\mathbf{v} = \begin{bmatrix} 3 \\ 5 \end{bmatrix}$, $\mathbf{u}-\mathbf{v} = \begin{bmatrix} -3 \\ 1 \end{bmatrix}$,
$2\mathbf{u} = \begin{bmatrix} 0 \\ 6 \end{bmatrix}$, $3\mathbf{u}-2\mathbf{v} = \begin{bmatrix} -6 \\ 5 \end{bmatrix}$.

(c) $\mathbf{u}+\mathbf{v} = \begin{bmatrix} 5 \\ 8 \end{bmatrix}$, $\mathbf{u}-\mathbf{v} = \begin{bmatrix} -1 \\ 4 \end{bmatrix}$,
$2\mathbf{u} = \begin{bmatrix} 4 \\ 12 \end{bmatrix}$, $3\mathbf{u}-2\mathbf{v} = \begin{bmatrix} 0 \\ 14 \end{bmatrix}$.

13. **(a)** $\begin{bmatrix} 1 \\ 5 \\ 3 \end{bmatrix}$. **(b)** $\begin{bmatrix} -4 \\ -3 \\ -1 \end{bmatrix}$.

(c) $\begin{bmatrix} 1 \\ 6 \\ 2 \end{bmatrix}$. **(d)** $\begin{bmatrix} -7 \\ 1 \\ 13 \end{bmatrix}$.

15. **(a)** $r = \frac{1}{2}, s = \frac{3}{2}$.

(b) $r = -2, s = 1, t = -1$.

(c) $r = -2, s = 6$.

17. Impossible.

19. Possible answer: $c_1 = -2, c_2 = -1, c_3 = 1$.

Section 4.2, p. 196

1. **(b)** No; $0 \odot p(t) = 0$, which does not have exact degree 2.
3. **(a)** Yes. **(b)** Yes. **(c)** $O = \begin{bmatrix} 0 & 0 \\ 0 & 0 \end{bmatrix}$.

   (d) Yes. The negative of $\begin{bmatrix} a & b \\ 2b & d \end{bmatrix}$ is $\begin{bmatrix} -a & -b \\ -2b & -d \end{bmatrix}$.

   (e) Yes. It satisfies all the properties in Definition 4.4
7. Properties (3), (4), (b), (5), (6), and (7).
9. Properties (5), (6), and (8).
11. Property (8).
17. No.

Section 4.3, p. 205

1. Yes. Properties (a) and (b) of Theorem 4.3 are satisfied.
3. No. A scalar multiple of a vector in W may not lie in W.
5. (b), (c), and (d).
7. (b) and (c).
9. (a) and (c).
13. Yes.
15. (a) and (c).
17. (a) and (b).
19. (b), (c), and (e).
33. (b) and (c).
37. (a) and (c).
39. **(a)** Possible answer: $x = 2 + 2t$, $y = -3 + 5t$, $z = 1 + 4t$.

    (b) Possible answer: $x = -3 + 8t$, $y = -2 + 7t$, $z = -2 + 6t$.

Section 4.4, p. 215

1. **(a)** Possible answers:
   $$\left\{ \begin{bmatrix} 1 \\ 0 \\ 0 \end{bmatrix}, \begin{bmatrix} 1 \\ 1 \\ 0 \end{bmatrix}, \begin{bmatrix} 1 \\ 1 \\ 1 \end{bmatrix}, \begin{bmatrix} 0 \\ 0 \\ 1 \end{bmatrix}, \begin{bmatrix} 0 \\ 1 \\ 1 \end{bmatrix}, \begin{bmatrix} 1 \\ 1 \\ 1 \end{bmatrix} \right\}.$$

   (b) Possible answers:
   $$\left\{ \begin{bmatrix} 1 & 0 \\ 0 & 0 \end{bmatrix}, \begin{bmatrix} 0 & 1 \\ 0 & 0 \end{bmatrix}, \begin{bmatrix} 0 & 0 \\ 1 & 0 \end{bmatrix}, \begin{bmatrix} 0 & 0 \\ 0 & 1 \end{bmatrix}, \begin{bmatrix} 1 & 0 \\ 0 & 0 \end{bmatrix}, \begin{bmatrix} 1 & 1 \\ 0 & 0 \end{bmatrix}, \begin{bmatrix} 1 & 1 \\ 1 & 0 \end{bmatrix}, \begin{bmatrix} 1 & 1 \\ 1 & 1 \end{bmatrix} \right\}.$$

   (c) Possible answers: $\{t^2, t+1, t-1\}$, $\{t^2+t, t^2-t, t+1\}$.
3. **(a)** Yes. **(b)** No. **(c)** Yes. **(d)** No.
7. (a) and (d).
9. No.
11. Possible answer: $\left\{ \begin{bmatrix} -1 \\ -1 \\ 1 \\ 0 \end{bmatrix} \right\}$.
15. $$\left\{ \begin{bmatrix} 1 & 0 & 0 \\ 0 & 0 & 0 \\ 0 & 0 & 0 \end{bmatrix}, \begin{bmatrix} 0 & 0 & 1 \\ 0 & 0 & 0 \\ 0 & 0 & 0 \end{bmatrix}, \begin{bmatrix} 0 & 0 & 0 \\ 0 & 1 & 0 \\ 0 & 0 & 0 \end{bmatrix}, \begin{bmatrix} 0 & 0 & 0 \\ 0 & 0 & 0 \\ 1 & 0 & 0 \end{bmatrix}, \begin{bmatrix} 0 & 0 & 0 \\ 0 & 0 & 0 \\ 0 & 0 & 1 \end{bmatrix} \right\}.$$

Section 4.5, p. 226

3. Yes.
5. No.
7. Yes.
9. No.
11. (a) and (c) are linearly dependent; (b) is linearly independent.

    (a) $\begin{bmatrix} 3 & 6 & 6 \end{bmatrix} = 2\begin{bmatrix} 1 & 1 & 0 \end{bmatrix} + 1\begin{bmatrix} 0 & 2 & 3 \end{bmatrix} + 1\begin{bmatrix} 1 & 2 & 3 \end{bmatrix}$.

    (c) $\begin{bmatrix} 0 & 0 & 0 \end{bmatrix} = 0\begin{bmatrix} 1 & 1 & 0 \end{bmatrix} + 0\begin{bmatrix} 0 & 2 & 3 \end{bmatrix} + 0\begin{bmatrix} 1 & 2 & 3 \end{bmatrix}$.
13. (a) and (b) are linearly independent; (c) is linearly dependent: $t + 13 = 3(2t^2 + t + 1) - 2(3t^2 + t - 5)$.
15. (b) is linearly dependent:
    $$\begin{bmatrix} 1 \\ 2 \\ -2 \end{bmatrix} = 2\begin{bmatrix} 1 \\ 1 \\ -1 \end{bmatrix} + \begin{bmatrix} 0 \\ 1 \\ 1 \end{bmatrix} - \begin{bmatrix} 1 \\ 1 \\ 1 \end{bmatrix}.$$
17. For $c \neq \pm 2$.

Section 4.6, p. 242

1. (a) and (d).
3. (a) and (d).
5. (c).
7. (a) is a basis for R^3 and
   $$\begin{bmatrix} 2 \\ 1 \\ 3 \end{bmatrix} = \frac{3}{2}\begin{bmatrix} 1 \\ 1 \\ 1 \end{bmatrix} + \frac{1}{2}\begin{bmatrix} 1 \\ 2 \\ 3 \end{bmatrix} - \frac{3}{2}\begin{bmatrix} 0 \\ 1 \\ 0 \end{bmatrix}.$$
9. (a) forms a basis.
   $$5t^2 - 3t + 8 = 5(t^2 + t) - 8(t - 1).$$
11. $\left\{ \begin{bmatrix} 1 \\ 2 \\ 2 \end{bmatrix}, \begin{bmatrix} 3 \\ 2 \\ 1 \end{bmatrix} \right\}$; $\dim W = 2$.
13. $\{t^3 + t^2 - 2t + 1, t^2 + 1\}$, $\dim W = 2$.

15. For $a \neq -1, 0, 1$.

17. $S = \left\{ \begin{bmatrix} 1 & 0 & 0 \\ 0 & 0 & 0 \\ 0 & 0 & 0 \end{bmatrix}, \begin{bmatrix} 0 & 0 & 0 \\ 0 & 1 & 0 \\ 0 & 0 & 0 \end{bmatrix}, \begin{bmatrix} 0 & 0 & 0 \\ 0 & 0 & 0 \\ 0 & 0 & 1 \end{bmatrix} \right\}$.

19. **(a)** $\left\{ \begin{bmatrix} 1 \\ 1 \\ 0 \end{bmatrix}, \begin{bmatrix} 0 \\ 1 \\ 1 \end{bmatrix} \right\}$. **(b)** $\left\{ \begin{bmatrix} 1 \\ 1 \\ 0 \end{bmatrix}, \begin{bmatrix} 0 \\ 0 \\ 1 \end{bmatrix} \right\}$.

(c) $\left\{ \begin{bmatrix} 1 \\ 0 \\ 2 \end{bmatrix}, \begin{bmatrix} 0 \\ 1 \\ 1 \end{bmatrix} \right\}$.

21. $\{t^2 + 2, t - 3\}$.

23. **(a)** 3. **(b)** 2.

25. **(a)** 2. **(b)** 1. **(c)** 2. **(d)** 2.

27. **(a)** 4. **(b)** 3. **(c)** 3. **(d)** 4.

29. $\{t^3 + t, t^2 - t, t^3, 1\}$. **31.** 2.

33. Possible answer: the set of all vectors of the form $\begin{bmatrix} a \\ b \\ 0 \\ 0 \end{bmatrix}$, where a and b are real numbers.

49. $\left\{ \begin{bmatrix} 1 & 0 & 0 \\ 0 & 0 & 0 \\ 0 & 0 & -1 \end{bmatrix}, \begin{bmatrix} 0 & 1 & 0 \\ 0 & 0 & 0 \\ 0 & 0 & 0 \end{bmatrix}, \begin{bmatrix} 0 & 0 & 1 \\ 0 & 0 & 0 \\ 0 & 0 & 0 \end{bmatrix}, \begin{bmatrix} 0 & 0 & 0 \\ 1 & 0 & 0 \\ 0 & 0 & 0 \end{bmatrix}, \begin{bmatrix} 0 & 0 & 0 \\ 0 & 1 & 0 \\ 0 & 0 & -1 \end{bmatrix}, \begin{bmatrix} 0 & 0 & 0 \\ 0 & 0 & 1 \\ 0 & 0 & 0 \end{bmatrix}, \begin{bmatrix} 0 & 0 & 0 \\ 0 & 0 & 0 \\ 1 & 0 & 0 \end{bmatrix}, \begin{bmatrix} 0 & 0 & 0 \\ 0 & 0 & 0 \\ 0 & 1 & 0 \end{bmatrix} \right\}$; $\dim W = 8$.

Section 4.7, p. 251

1. **(a)** $\mathbf{x} = \begin{bmatrix} \frac{1}{2}r + s \\ r \\ s \end{bmatrix}$, where r and s are any real numbers.

(b) $\mathbf{x} = r \begin{bmatrix} \frac{1}{2} \\ 1 \\ 0 \end{bmatrix} + s \begin{bmatrix} 1 \\ 0 \\ 1 \end{bmatrix}$.

(c)

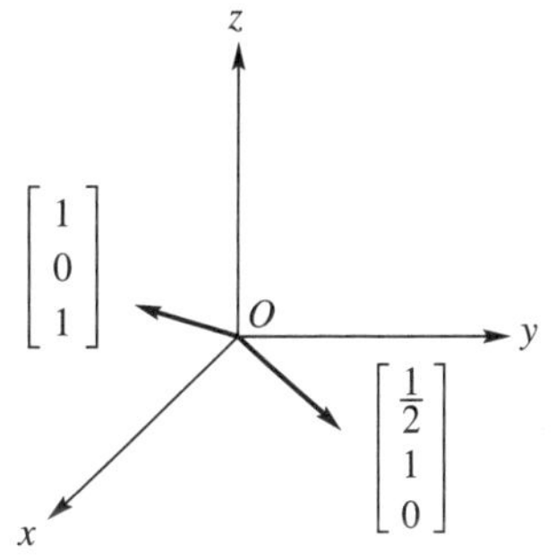

3. $\left\{ \begin{bmatrix} 0 \\ -1 \\ 0 \\ 1 \end{bmatrix}, \begin{bmatrix} 2 \\ -3 \\ 1 \\ 0 \end{bmatrix} \right\}$; dimension = 2.

5. $\left\{ \begin{bmatrix} 1 \\ -\frac{8}{3} \\ -\frac{4}{3} \\ 1 \end{bmatrix} \right\}$; dimension = 1.

7. $\left\{ \begin{bmatrix} -2 \\ 1 \\ 0 \\ 0 \\ 0 \end{bmatrix}, \begin{bmatrix} -3 \\ 0 \\ 1 \\ 1 \\ 0 \end{bmatrix} \right\}$; dimension = 2.

9. $\left\{ \begin{bmatrix} -2 \\ \frac{1}{2} \\ 0 \\ 0 \\ 1 \end{bmatrix}, \begin{bmatrix} -1 \\ 1 \\ 0 \\ 1 \\ 0 \end{bmatrix} \right\}$; dimension = 2.

11. $\left\{ \begin{bmatrix} -3 \\ 2 \\ 0 \\ 1 \end{bmatrix}, \begin{bmatrix} 5 \\ -4 \\ 1 \\ 0 \end{bmatrix} \right\}$.

13. $\left\{ \begin{bmatrix} -1 \\ 1 \end{bmatrix} \right\}$.

15. $\left\{ \begin{bmatrix} 1 \\ -2 \\ 1 \end{bmatrix} \right\}$.

17. $\lambda = 3$ or -4. **19.** $\lambda = 0$ or 1.

21. $\mathbf{x}_p = \begin{bmatrix} -2 \\ 0 \end{bmatrix}$, $\mathbf{x}_h = r \begin{bmatrix} -2 \\ 1 \end{bmatrix}$, where r is any real number.

Section 4.8, p. 267

1. $\begin{bmatrix} 3 \\ -2 \end{bmatrix}$. **3.** $\begin{bmatrix} 2 \\ -1 \end{bmatrix}$. **5.** $\begin{bmatrix} 1 \\ -1 \\ 0 \\ 2 \end{bmatrix}$.

7. $\begin{bmatrix} 0 \\ 3 \end{bmatrix}$. **9.** $4t - 3$. **11.** $\begin{bmatrix} -1 & 0 \\ 9 & 7 \end{bmatrix}$.

13. **(b)** $\left[\mathbf{v}\right]_S = \begin{bmatrix} 10 \\ -8 \end{bmatrix}$.

(c) $\lambda_1 = -0.3$.

(d) $\lambda_2 = 0.25$.

(e) $\mathbf{v} = 10\lambda_1^n \mathbf{v}_1 - 8\lambda_2^n \mathbf{v}_2$.

(f) The sequence approaches the zero vector.

15. **(a)** $\left[\mathbf{v}\right]_T = \begin{bmatrix} -7 \\ 4 \end{bmatrix}$; $\left[\mathbf{w}\right]_T = \begin{bmatrix} 7 \\ -1 \end{bmatrix}$.

(b) $P_{S \leftarrow T} = \begin{bmatrix} 1 & 2 \\ -1 & -1 \end{bmatrix}$.

(c) $\left[\mathbf{v}\right]_S = \begin{bmatrix} 1 \\ 3 \end{bmatrix}$; $\left[\mathbf{w}\right]_S = \begin{bmatrix} 5 \\ -6 \end{bmatrix}$.

(d) Same as (c).

(e) $Q_{T \leftarrow S} = \begin{bmatrix} -1 & -2 \\ 1 & 1 \end{bmatrix}$.

(f) Same as (a).

17. **(a)** $\left[\mathbf{v}\right]_T = \begin{bmatrix} 3 \\ 2 \\ -7 \end{bmatrix}$; $\left[\mathbf{w}\right]_T = \begin{bmatrix} 2 \\ 3 \\ -3 \end{bmatrix}$.

(b) $P_{S \leftarrow T} = \begin{bmatrix} 2 & 1 & 0 \\ 1 & -\frac{2}{5} & \frac{3}{5} \\ 0 & \frac{2}{5} & \frac{2}{5} \end{bmatrix}$.

(c) $\left[\mathbf{v}\right]_S = \begin{bmatrix} 8 \\ -2 \\ -2 \end{bmatrix}$; $\left[\mathbf{w}\right]_S = \begin{bmatrix} 7 \\ -1 \\ 0 \end{bmatrix}$.

(d) Same as (c).

(e) $Q_{T \leftarrow S} = \begin{bmatrix} \frac{1}{3} & \frac{1}{3} & -\frac{1}{2} \\ \frac{1}{3} & -\frac{2}{3} & 1 \\ -\frac{1}{3} & \frac{2}{3} & \frac{3}{2} \end{bmatrix}$.

(f) $\left[\mathbf{v}\right]_T = Q_{T \leftarrow S}\left[\mathbf{v}\right]_S = \begin{bmatrix} 3 \\ 2 \\ -7 \end{bmatrix}$.

$\left[\mathbf{w}\right]_T = Q_{T \leftarrow S}\left[\mathbf{w}\right]_S = \begin{bmatrix} 2 \\ 3 \\ -3 \end{bmatrix}$; same as (a).

19. **(a)** $\left[\mathbf{v}\right]_T = \begin{bmatrix} 1 \\ 1 \\ 1 \\ 0 \end{bmatrix}$; $\left[\mathbf{w}\right]_T = \begin{bmatrix} 2 \\ -2 \\ 1 \\ -1 \end{bmatrix}$.

(b) $P_{S \leftarrow T} = \begin{bmatrix} 1 & 0 & 0 & 1 \\ \frac{1}{3} & \frac{2}{3} & -\frac{2}{3} & 0 \\ \frac{1}{3} & -\frac{1}{3} & \frac{1}{3} & 0 \\ -\frac{1}{3} & \frac{1}{3} & \frac{2}{3} & 0 \end{bmatrix}$.

(c) $\left[\mathbf{v}\right]_S = \begin{bmatrix} 1 \\ \frac{1}{3} \\ \frac{1}{3} \\ \frac{2}{3} \end{bmatrix}$; $\left[\mathbf{w}\right]_S = \begin{bmatrix} 1 \\ -\frac{4}{3} \\ \frac{5}{3} \\ -\frac{2}{3} \end{bmatrix}$.

(d) Same as (c).

(e) $Q_{T \leftarrow S} = \begin{bmatrix} 0 & 1 & 2 & 0 \\ 0 & 1 & 0 & 1 \\ 0 & 0 & 1 & 1 \\ 1 & -1 & -2 & 0 \end{bmatrix}$.

(f) Same as (a).

21. $\begin{bmatrix} 1 \\ 1 \end{bmatrix}$. **23.** $\begin{bmatrix} 1 \\ 1 \\ -1 \end{bmatrix}$.

25. $S = \{t + 1, 5t - 2\}$. **27.** $S = \{-t + 5, t - 3\}$.

Section 4.9, p. 282

1. A possible basis is $\{\mathbf{v}_1, \mathbf{v}_2, \mathbf{v}_3\}$, where

$\mathbf{v}_1 = \begin{bmatrix} 1 \\ 0 \\ 0 \end{bmatrix}$, $\mathbf{v}_2 = \begin{bmatrix} 0 \\ 1 \\ 0 \end{bmatrix}$, and $\mathbf{v}_3 = \begin{bmatrix} 0 \\ 0 \\ 1 \end{bmatrix}$.

(a) $\begin{bmatrix} 3 \\ 4 \\ 12 \end{bmatrix} = 3\mathbf{v}_1 + 4\mathbf{v}_2 + 12\mathbf{v}_3$.

(b) $\begin{bmatrix} 3 \\ 2 \\ 2 \end{bmatrix} = 3\mathbf{v}_1 + 2\mathbf{v}_2 + 2\mathbf{v}_3$.

(c) $\begin{bmatrix} 1 \\ 2 \\ 6 \end{bmatrix} = \mathbf{v}_1 + 2\mathbf{v}_2 + 6\mathbf{v}_3$.

3. A possible answer is $\left\{\begin{bmatrix} 1 & 0 \\ 0 & 0 \end{bmatrix}, \begin{bmatrix} 0 & 1 \\ 0 & 0 \end{bmatrix}, \begin{bmatrix} 0 & 0 \\ 1 & 0 \end{bmatrix},\right.$

$\left.\begin{bmatrix} 0 & 0 \\ 0 & 1 \end{bmatrix}\right\}$.

5. **(a)** $\{\begin{bmatrix}1 & 0 & -1\end{bmatrix}, \begin{bmatrix}0 & 1 & 0\end{bmatrix}\}$.

(b) $\{\begin{bmatrix}1 & 2 & -1\end{bmatrix}, \begin{bmatrix}1 & 9 & -1\end{bmatrix}\}$.

7. **(a)** $\left\{\begin{bmatrix}1\\0\\0\\0\end{bmatrix}, \begin{bmatrix}0\\1\\0\\\frac{1}{5}\end{bmatrix}, \begin{bmatrix}0\\0\\1\\\frac{3}{5}\end{bmatrix}\right\}$.

(b) $\left\{\begin{bmatrix}1\\1\\3\\2\end{bmatrix}, \begin{bmatrix}-2\\-1\\2\\1\end{bmatrix}, \begin{bmatrix}0\\0\\5\\3\end{bmatrix}\right\}$.

9. **(a)** 3. **(b)** 5.

13. **(a)** rank $= 2$, nullity $= 2$.

(b) rank $= 3$, nullity $= 1$.

15. B and C are equivalent; A, D, and E are equivalent.

17. Neither.

19. (b).

21. **(a)** Yes. **(b)** No.

23. (b).

25. **(a)** 2. **(b)** 3.

27. Has a unique solution.

29. Linearly dependent.

31. Yes.

33. For $c \neq \pm 2$.

35. **(a)** Linearly dependent.

(b) Linearly independent.

Supplementary Exercises, p. 285

1. **(b)** $k = 0$.

3. **(a)** No. **(b)** Yes. **(c)** Yes.

11. $a = 1$ or $a = 2$.

13. $k \neq 1, -1$.

17. **(a)** $\mathbf{b} = \begin{bmatrix}a\\b\\c\end{bmatrix}$, where $b + c - 3a = 0$.

(b) Any $\mathbf{b}$.

25. **(a)** $T = \left\{\begin{bmatrix}2\\2\\1\end{bmatrix}, \begin{bmatrix}3\\1\\2\end{bmatrix}, \begin{bmatrix}5\\4\\4\end{bmatrix}\right\}$.

(b) $S = \left\{\begin{bmatrix}\frac{6}{5}\\-\frac{2}{5}\\\frac{13}{5}\end{bmatrix}, \begin{bmatrix}\frac{2}{5}\\\frac{1}{5}\\-\frac{4}{5}\end{bmatrix}, \begin{bmatrix}-\frac{1}{5}\\\frac{2}{5}\\-\frac{3}{5}\end{bmatrix}\right\}$.

Chapter Review, p. 288

True/False

1. T.	**2.** T.	**3.** F.	**4.** F.	**5.** T.
6. F.	**7.** T.	**8.** T.	**9.** T.	**10.** F.
11. F.	**12.** T.	**13.** F.	**14.** T.	**15.** T.
16. T.	**17.** T.	**18.** T.	**19.** F.	**20.** F.
21. T.				

Quiz

1. No. Property 1 in Definition 4.4 is not satisfied.

2. No. Properties 5–8 in Definition 4.4 are not satisfied.

3. Yes. All the properties in Definition 4.4 are satisfied.

Basis: $\left\{\begin{bmatrix}1\\0\\-1\end{bmatrix}, \begin{bmatrix}0\\1\\0\end{bmatrix}\right\}$; dimension $= 2$.

4. No. Property (b) in Definition 4.4 is not satisfied.

5. If $p(t)$ and $q(t)$ are in W and c is a scalar, then

$$(p + q)(0) = p(0) + q(0) = 0 + 0 = 0,$$

and

$$(cp)(0) = c[p(0)] = c0 = 0.$$

Basis $= \{t^2, t\}$.

6. No. S is linearly dependent.

7. Possible answer: Basis $= \left\{\begin{bmatrix}1\\2\\0\end{bmatrix}, \begin{bmatrix}1\\0\\1\end{bmatrix}, \begin{bmatrix}3\\0\\0\end{bmatrix}\right\}$.

8. $\left\{\begin{bmatrix}-1\\0\\1\\0\end{bmatrix}, \begin{bmatrix}1\\-1\\0\\1\end{bmatrix}\right\}$.

9. $\{\begin{bmatrix}1 & 0 & 2\end{bmatrix}, \begin{bmatrix}0 & 1 & -2\end{bmatrix}\}$.

10. Dimension of null space $= n - \text{rank}\, A = 3 - 2 = 1$.

11. $\mathbf{x}_p = \begin{bmatrix}-\frac{1}{3}\\\frac{7}{6}\\0\\0\end{bmatrix}$, $\mathbf{x}_h = r\begin{bmatrix}-1\\-\frac{1}{4}\\\frac{3}{2}\\1\end{bmatrix}$, where r is any real number.

CHAPTER 5

Section 5.1, p. 297

1. **(a)** 1. **(b)** 0. **(c)** $\sqrt{5}$.

3. **(a)** 1. **(b)** $\sqrt{2}$.

5. **(a)** $\sqrt{74}$. **(b)** $\sqrt{58}$.

7. $c = \pm\sqrt{5}$.

9. **(a)** $\dfrac{-14}{\sqrt{5}\sqrt{41}}$. **(b)** $\dfrac{-6}{\sqrt{5}\sqrt{41}}$.

11. **(a)** 1, 0, 0.

(b) $\dfrac{1}{\sqrt{14}}, \dfrac{3}{\sqrt{14}}, \dfrac{2}{\sqrt{14}}$.

(c) $\dfrac{-1}{\sqrt{14}}, \dfrac{-2}{\sqrt{14}}, \dfrac{-3}{\sqrt{14}}$.

(d) $\dfrac{4}{\sqrt{29}}, \dfrac{-3}{\sqrt{29}}, \dfrac{2}{\sqrt{29}}$.

17. **(a)** $\mathbf{v}_1$ and $\mathbf{v}_4$, $\mathbf{v}_1$ and $\mathbf{v}_6$, $\mathbf{v}_3$ and $\mathbf{v}_4$, $\mathbf{v}_3$ and $\mathbf{v}_6$, $\mathbf{v}_4$ and $\mathbf{v}_5$, $\mathbf{v}_5$ and $\mathbf{v}_6$.

(b) $\mathbf{v}_1$ and $\mathbf{v}_5$, $\mathbf{v}_4$ and $\mathbf{v}_6$.

(c) $\mathbf{v}_1$ and $\mathbf{v}_3$, $\mathbf{v}_3$ and $\mathbf{v}_5$.

19. (b).

21.

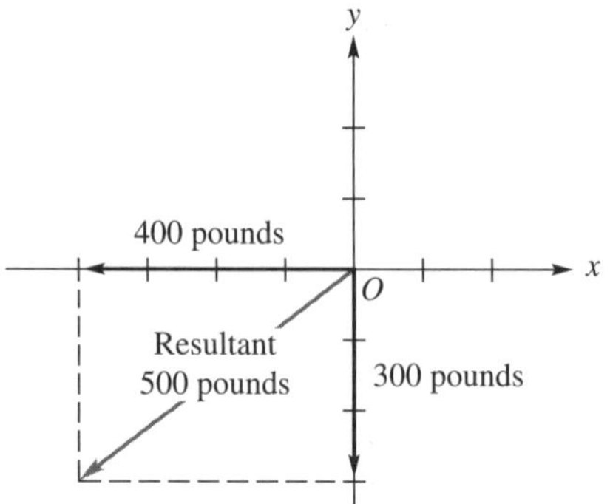

25. $\frac{5}{2}$. **27.** $a = -2, b = 2$.

Section 5.2, p. 306

1. **(a)** $-15\mathbf{i} - 2\mathbf{j} + 9\mathbf{k}$. **(b)** $-3\mathbf{i} + 3\mathbf{j} + 3\mathbf{k}$.

(c) $7\mathbf{i} + 5\mathbf{j} - \mathbf{k}$. **(d)** $0\mathbf{i} + 0\mathbf{j} + 0\mathbf{k}$.

5. $(\mathbf{u} \times \mathbf{v}) \cdot \mathbf{w} = 24$.

13. $\frac{3}{2}\sqrt{10}$. **15.** 1. **17.** (a).

19. **(a)** $x - z + 2 = 0$. **(b)** $3x + y - 14z + 47 = 0$.

21. $4x - 4y + z + 16 = 0$.

23. $x = -2 + 2t, y = 5 - 3t, z = -3 + 4t$.

25. $\left\{\begin{bmatrix} \frac{2}{3} \\ 1 \\ 0 \end{bmatrix}, \begin{bmatrix} \frac{5}{3} \\ 0 \\ 1 \end{bmatrix}\right\}$.

27. **(a)** $-15\mathbf{i} - 2\mathbf{j} + 9\mathbf{k}$. **(b)** $-3\mathbf{i} + 3\mathbf{j} + 3\mathbf{k}$.

(c) $7\mathbf{i} + 5\mathbf{j} - \mathbf{k}$. **(d)** $3\mathbf{i} - 8\mathbf{j} - \mathbf{k}$.

Section 5.3, p. 317

9. **(a)** -8. **(b)** 0. **(c)** 1.

11. **(a)** $\frac{3}{2}$. **(b)** 1. **(c)** $\frac{1}{2}\sin^2 1$.

13. **(a)** 2. **(b)** $2\sqrt{5}$. **(c)** $2\sqrt{2}$.

15. **(a)** 1. **(b)** 0. **(c)** $\dfrac{2\sin^2 1}{\sqrt{4 - \sin^2 2}}$.

25. If $\mathbf{u} = \begin{bmatrix} u_1 & u_2 \end{bmatrix}$ and $\mathbf{v} = \begin{bmatrix} v_1 & v_2 \end{bmatrix}$, then $(\mathbf{u}, \mathbf{v}) = 3u_1v_1 - 2u_1v_2 - 2u_2v_1 + 3u_2v_2$.

27. **(a)** $\sqrt{1 - \sin^2 1}$. **(b)** $\sqrt{\frac{1}{30}}$.

29. **(a)** Orthonormal. **(b)** Neither.

(c) Neither.

31. $a = 0$. **33.** $a = 5$.

35. $B = \begin{bmatrix} b_{11} & b_{12} \\ b_{21} & b_{22} \end{bmatrix}$ with $b_{11} + 3b_{21} + 2b_{12} + 4b_{22} = 0$.

Section 5.4, p. 329

1. **(a)** $\left\{\begin{bmatrix} 1 \\ 2 \end{bmatrix}, \begin{bmatrix} -4 \\ 2 \end{bmatrix}\right\}$.

(b) $\left\{\frac{1}{\sqrt{5}}\begin{bmatrix} 1 \\ 2 \end{bmatrix}, \frac{1}{\sqrt{5}}\begin{bmatrix} -2 \\ 1 \end{bmatrix}\right\}$.

3. $\left\{\frac{1}{\sqrt{3}}\begin{bmatrix} 1 & 1 & -1 & 0 \end{bmatrix}, \frac{1}{\sqrt{33}}\begin{bmatrix} -2 & 4 & 2 & 3 \end{bmatrix}\right\}$.

5. $\left\{\sqrt{3}\,t, 2 - 3t\right\}$.

7. $\left\{\sqrt{3}\,t, \dfrac{\sin 2\pi t + \left(\frac{3}{2\pi}\right)t}{\sqrt{\frac{1}{2} - \frac{3}{4\pi^2}}}\right\}$.

9. $\left\{\begin{bmatrix} \frac{2}{3} \\ -\frac{2}{3} \\ \frac{1}{3} \end{bmatrix}, \begin{bmatrix} \frac{2}{3} \\ \frac{1}{3} \\ -\frac{2}{3} \end{bmatrix}, \begin{bmatrix} \frac{1}{3} \\ \frac{2}{3} \\ \frac{2}{3} \end{bmatrix}\right\}$.

11. Possible answer:

$\left\{\frac{1}{\sqrt{3}}\begin{bmatrix} 1 \\ 1 \\ 1 \end{bmatrix}, \frac{1}{\sqrt{6}}\begin{bmatrix} -1 \\ -1 \\ 2 \end{bmatrix}, \frac{1}{\sqrt{2}}\begin{bmatrix} -1 \\ 1 \\ 0 \end{bmatrix}\right\}$.

13. $\left\{\begin{bmatrix} \frac{1}{\sqrt{2}} \\ \frac{1}{\sqrt{2}} \\ 0 \end{bmatrix}, \begin{bmatrix} -\frac{1}{\sqrt{6}} \\ \frac{1}{\sqrt{6}} \\ \frac{2}{\sqrt{6}} \end{bmatrix}\right\}$.

15. $\left\{\begin{bmatrix} \frac{1}{\sqrt{2}} \\ 0 \\ -\frac{1}{\sqrt{2}} \end{bmatrix}, \begin{bmatrix} -\frac{1}{\sqrt{6}} \\ \frac{2}{\sqrt{6}} \\ -\frac{1}{\sqrt{6}} \end{bmatrix}\right\}$.

17. $\left\{\frac{1}{\sqrt{26}}\begin{bmatrix} -3 \\ 4 \\ 1 \end{bmatrix}\right\}$.

23. **(b)** $\begin{bmatrix} 9 \\ 9 \\ 9 \end{bmatrix}$. **(c)** $\|\mathbf{v}\| = 9\sqrt{3}$.

25. **(b)** $\left[\mathbf{v}\right]_S = \begin{bmatrix} 1 \\ 2 \\ 3 \\ 4 \end{bmatrix}$.

27. Possible answer:

$$\left\{\begin{bmatrix} 1 & 0 \\ 0 & 0 \end{bmatrix}, \frac{1}{\sqrt{2}}\begin{bmatrix} 0 & 1 \\ 1 & 0 \end{bmatrix}, \frac{1}{\sqrt{6}}\begin{bmatrix} 0 & -1 \\ 1 & 2 \end{bmatrix}\right\}.$$

29. **(a)** $Q = \begin{bmatrix} \frac{1}{\sqrt{2}} & \frac{1}{\sqrt{2}} \\ -\frac{1}{\sqrt{2}} & \frac{1}{\sqrt{2}} \end{bmatrix} \approx \begin{bmatrix} 0.7071 & 0.7071 \\ -0.7071 & 0.7071 \end{bmatrix}$,

$$R = \begin{bmatrix} \sqrt{2} & -\frac{1}{\sqrt{2}} \\ 0 & \frac{5}{\sqrt{2}} \end{bmatrix} \approx \begin{bmatrix} 1.4142 & 0.7071 \\ 0 & 3.5355 \end{bmatrix}.$$

(b) $Q = \begin{bmatrix} \frac{\sqrt{3}}{3} & \frac{1}{\sqrt{6}} \\ -\frac{\sqrt{3}}{3} & -\frac{1}{\sqrt{6}} \\ \frac{\sqrt{3}}{3} & -\frac{2}{\sqrt{6}} \end{bmatrix} \approx \begin{bmatrix} 0.5774 & 0.4082 \\ -0.5774 & -0.4082 \\ 0.5774 & -0.8165 \end{bmatrix}$,

$$R = \begin{bmatrix} \sqrt{3} & \frac{5}{\sqrt{3}} \\ 0 & \frac{2}{\sqrt{6}} \end{bmatrix} \approx \begin{bmatrix} 1.7321 & 2.8868 \\ 0 & 0.8165 \end{bmatrix}.$$

(c) $Q = \begin{bmatrix} \frac{1}{\sqrt{6}} & \frac{4}{\sqrt{21}} & -\frac{1}{\sqrt{14}} \\ \frac{2}{\sqrt{6}} & -\frac{1}{\sqrt{21}} & \frac{2}{\sqrt{14}} \\ -\frac{1}{\sqrt{6}} & \frac{2}{\sqrt{21}} & \frac{3}{\sqrt{14}} \end{bmatrix}$

$$\approx \begin{bmatrix} 0.4082 & 0.8729 & -0.2673 \\ 0.8165 & -0.2182 & 0.5345 \\ -0.4082 & 0.4364 & 0.8018 \end{bmatrix},$$

$$R = \begin{bmatrix} \frac{6}{\sqrt{6}} & -\frac{8}{\sqrt{6}} & \frac{1}{\sqrt{6}} \\ 0 & \frac{7}{\sqrt{21}} & \frac{1}{\sqrt{21}} \\ 0 & 0 & \frac{19}{\sqrt{14}} \end{bmatrix}$$

$$\approx \begin{bmatrix} 2.4495 & -3.2660 & 0.4082 \\ 0 & 1.5275 & 0.2182 \\ 0 & 0 & 5.0780 \end{bmatrix}.$$

Section 5.5, p. 348

1. **(a)** $\left\{\begin{bmatrix} \frac{3}{2} \\ 1 \\ 0 \end{bmatrix}, \begin{bmatrix} -\frac{1}{2} \\ 0 \\ 1 \end{bmatrix}\right\}$.

(b) The set of all points $P(x, y, z)$ such that $2x - 3y + z = 0$. $W^{\perp}$ is the plane whose normal is $\mathbf{w}$.

3. $\left\{\begin{bmatrix} -\frac{17}{5} & \frac{6}{5} & 5 & 1 & 0 \end{bmatrix}, \begin{bmatrix} \frac{8}{5} & \frac{1}{5} & -3 & 0 & 1 \end{bmatrix}\right\}$.

5. $\left\{\frac{45}{14}t^3 - \frac{55}{14}t^2 + t,\ \frac{130}{7}t^3 - \frac{120}{7}t^2 + 1\right\}$.

7. $\begin{bmatrix} -3 \\ -2 \\ 1 \end{bmatrix}$.

9. Null space of A has basis $\left\{\begin{bmatrix} 2 \\ -1 \\ 1 \\ 0 \end{bmatrix}, \begin{bmatrix} 3 \\ -2 \\ 0 \\ 1 \end{bmatrix}\right\}$.

Basis for row space of A is $\left\{\begin{bmatrix} 1 & 0 & -2 & -3 \end{bmatrix}, \begin{bmatrix} 0 & 1 & 1 & 2 \end{bmatrix}\right\}$.

Null space of A^T has basis $\left\{\begin{bmatrix} -\frac{7}{5} \\ -\frac{13}{10} \\ 1 \end{bmatrix}\right\}$.

Basis for column space of A^T is $\left\{\begin{bmatrix} 1 \\ 0 \\ \frac{7}{5} \end{bmatrix}, \begin{bmatrix} 0 \\ 1 \\ \frac{13}{10} \end{bmatrix}\right\}$.

11. **(a)** $\begin{bmatrix} 3 \\ 0 \\ -1 \end{bmatrix}$. **(b)** $\begin{bmatrix} 2 \\ 0 \\ 3 \end{bmatrix}$. **(c)** $\begin{bmatrix} -5 \\ 0 \\ 1 \end{bmatrix}$.

13. **(a)** $2\sin t$.

(b) $\dfrac{\pi^2}{3} - 4\cos t$.

(c) $\left(\dfrac{e^{\pi} - e^{-\pi}}{2\pi}\right) + \left(\dfrac{e^{-\pi} - e^{\pi}}{2\pi}\right)\cos t + \left(\dfrac{e^{\pi} - e^{-\pi}}{2\pi}\right)\sin t$.

15. $\mathbf{w} = \begin{bmatrix} -\frac{1}{5} \\ 2 \\ -\frac{2}{5} \end{bmatrix}, \mathbf{u} = \begin{bmatrix} \frac{6}{5} \\ 0 \\ -\frac{3}{5} \end{bmatrix}$.

17. $\mathbf{w} = 2\sin t - 1, \mathbf{u} = t - 1 - [2\sin t - 1] = t - 2\sin t$.

19. $\frac{3}{5}\sqrt{5}$.

21. $\sqrt{\dfrac{2\pi^3}{3} - 4\pi} \approx 2.847$.

23. $\text{proj}_W e^t = \dfrac{1}{2\pi}\left(e^{\pi} - e^{-\pi}\right) + \dfrac{1}{\pi}\left(-\dfrac{1}{2}e^{\pi} + \dfrac{1}{2}e^{-\pi}\right)\cos t + \dfrac{1}{\pi}\left(\dfrac{1}{2}e^{\pi} - \dfrac{1}{2}e^{-\pi}\right)\sin t + \dfrac{1}{\pi}\left(\dfrac{1}{5}e^{\pi} - \dfrac{1}{5}e^{-\pi}\right)\cos 2t + \dfrac{1}{\pi}\left(-\dfrac{2}{5}e^{\pi} + \dfrac{2}{5}e^{-\pi}\right)\sin 2t$.

Section 5.6, p. 356

3. $\widehat{\mathbf{x}} \approx \begin{bmatrix} -1.5333 \\ -1.8667 \\ 4.2667 \end{bmatrix}$.

9. **(b)** $\widehat{\mathbf{x}} = \begin{bmatrix} -\frac{5}{11} \\ \frac{4}{11} \\ 0 \end{bmatrix}$. The solution is unique.

11. **(a)** $y = 752x - 1{,}482{,}059.80$.

(b)

Year	Predicted value rounded to whole dollars
2008	27,956
2010	29,460
2015	33,220

13. **(b)** $\ln y = -0.0272x + 3.2709$.

(c) $r = 26.3350, s = -0.0272$. **(d)** 16.14 mm.

Supplementary Exercises, p. 358

1. $\left\{\begin{bmatrix} 2 \\ 1 \\ 0 \end{bmatrix}, \begin{bmatrix} -\frac{1}{5} \\ \frac{2}{5} \\ 1 \end{bmatrix}\right\}$.

3. $\mathbf{v} = -\sqrt{2}\begin{bmatrix} \frac{1}{\sqrt{2}} \\ 0 \\ -\frac{1}{\sqrt{2}} \end{bmatrix} + 2\begin{bmatrix} 0 \\ 1 \\ 0 \end{bmatrix} + 2\sqrt{2}\begin{bmatrix} \frac{1}{\sqrt{2}} \\ 0 \\ \frac{1}{\sqrt{2}} \end{bmatrix}$.

5. Vector in P closest to $\mathbf{v}$ is $\dfrac{1}{122}\begin{bmatrix} 59 \\ 271 \\ 50 \end{bmatrix}$; distance is $\dfrac{9}{\sqrt{122}}$.

9. **(a)** Possible answer: $\left\{\dfrac{1}{\sqrt{30}}\begin{bmatrix} -5 \\ 2 \\ 1 \\ 0 \end{bmatrix}, \dfrac{1}{\sqrt{30}}\begin{bmatrix} 2 \\ 5 \\ 0 \\ 1 \end{bmatrix}\right\}$.

(b) Possible answer: $\left\{\dfrac{1}{\sqrt{30}}\begin{bmatrix} -5 \\ 2 \\ 1 \\ 0 \end{bmatrix}, \dfrac{1}{\sqrt{255}}\begin{bmatrix} -5 \\ -14 \\ 3 \\ 5 \end{bmatrix}\right\}$.

11. **(a)** $\left\{\begin{bmatrix} -1 \\ 0 \\ 1 \end{bmatrix}\right\}$.

(c) (i) $\mathbf{w} = \dfrac{1}{2}\begin{bmatrix} 1 \\ 0 \\ 1 \end{bmatrix}, \mathbf{u} = -\dfrac{1}{2}\begin{bmatrix} -1 \\ 0 \\ 1 \end{bmatrix}$;

(ii) $\mathbf{w} = \begin{bmatrix} 2 \\ 2 \\ 2 \end{bmatrix}, \mathbf{u} = \begin{bmatrix} -1 \\ 0 \\ 1 \end{bmatrix}$.

13. $\left\{\dfrac{1}{\sqrt{2}}, \sqrt{\dfrac{3}{2}}\,t, \sqrt{\dfrac{5}{8}}\,(3t^2 - 1)\right\}$.

15. $\sqrt{\dfrac{8}{175}} \approx 0.214$.

29. **(a)** 7, 4.1231, 3.

(b) 5, 3.6055, 3.

(c) 2, 2, 2.

33.

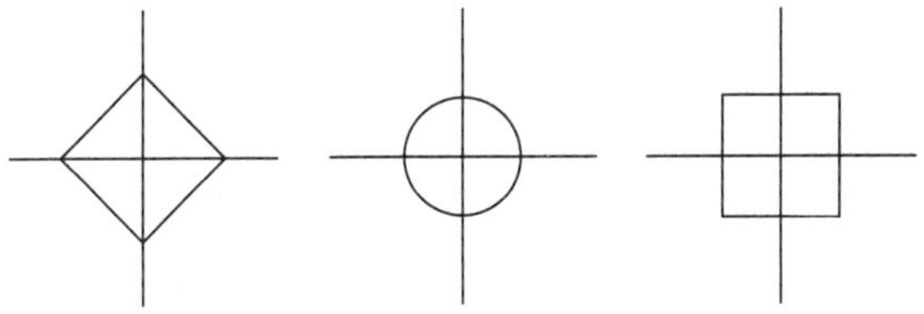

(a) $\|\mathbf{x}\|_1 = 1$ (b) $\|\mathbf{x}\|_2 = 1$ (c) $\|\mathbf{x}\|_\infty = 1$

Chapter Review, p. 360

True/False

1. T. **2.** F. **3.** F. **4.** F.

5. T. **6.** T. **7.** F. **8.** F.

9. F. **10.** F. **11.** T. **12.** T.

Quiz

1. $b = \frac{\sqrt{2}}{2}, c = \pm\frac{\sqrt{2}}{2}$.

2. $r\begin{bmatrix} 1 \\ -3 \\ 1 \\ 0 \end{bmatrix} + s\begin{bmatrix} -4 \\ 6 \\ 0 \\ 1 \end{bmatrix}$, where r and s are any real numbers.

3. $a + bt$, where $a = -\frac{5}{9}b$.

4. **(b)** The cosine of the angle between $\mathbf{u}$ and $\mathbf{v}$ lies between -1 and 1.

5. **(b)** $\left\{\frac{1}{2}\mathbf{v}_1, \frac{1}{\sqrt{6}}\mathbf{v}_2, \frac{1}{\sqrt{12}}\mathbf{v}_3\right\}$.

(c) $\mathbf{v}_4 = \begin{bmatrix} 0 \\ 0 \\ 1 \\ -1 \end{bmatrix}$.

6. **(b)** $\mathbf{w} = \frac{5}{3}\mathbf{u}_1 + \frac{1}{3}\mathbf{u}_2 + \frac{1}{3}\mathbf{u}_3$. **(c)** $\begin{bmatrix} -\frac{1}{3} & \frac{4}{3} & \frac{1}{3} \end{bmatrix}^T$; $\frac{2}{3}\sqrt{6}$.

7. $\left\{\begin{bmatrix} \frac{1}{\sqrt{2}} \\ 0 \\ \frac{1}{\sqrt{2}} \\ 0 \end{bmatrix}, \begin{bmatrix} \frac{2}{3} \\ \frac{1}{3} \\ -\frac{2}{3} \\ 0 \end{bmatrix}, \begin{bmatrix} 0 \\ 0 \\ 0 \\ 1 \end{bmatrix}\right\}$.

8. $\begin{bmatrix} -1 \\ 0 \\ 0 \\ 1 \end{bmatrix}$.

9. *Hint*: To start, consider the construction of a basis for W_1.

CHAPTER 6

Section 6.1, p. 372

1. (b).

3. (a) and (b).

5. (b)

7. **(a)** $\begin{bmatrix} -1 & 0 \\ 0 & 1 \end{bmatrix}$. **(b)** $\begin{bmatrix} 0 & -1 \\ 1 & 0 \end{bmatrix}$.

(c) $\begin{bmatrix} 1 & 0 & 0 \\ 0 & 0 & 0 \\ 0 & 0 & 0 \end{bmatrix}$.

9. **(a)** $\begin{bmatrix} 15 & 5 & 4 & 8 \\ -5 & -1 & 10 & 2 \end{bmatrix}$.

11. **(a)** $A = \begin{bmatrix} 0 & 1 \\ 1 & 0 \end{bmatrix}$.

(b) $A = \begin{bmatrix} 1 & -3 \\ 2 & -1 \\ 0 & 2 \end{bmatrix}$.

(c) $A = \begin{bmatrix} 1 & 4 & 0 \\ 0 & 0 & -1 \\ 0 & 1 & 1 \end{bmatrix}$.

13. **(a)** $\begin{bmatrix} 2 \\ 15 \end{bmatrix}$. **(b)** $\begin{bmatrix} 2u_1 + 3u_2 + 2u_3 \\ -4u_1 - 5u_2 + 3u_3 \end{bmatrix}$.

15. **(a)** $2t^3 - 5t^2 + 2t + 3$.

(b) $at^3 + bt^2 + at + c$.

17. a = any real number, $b = 0$.

19. **(a)** Reflection about the y-axis.

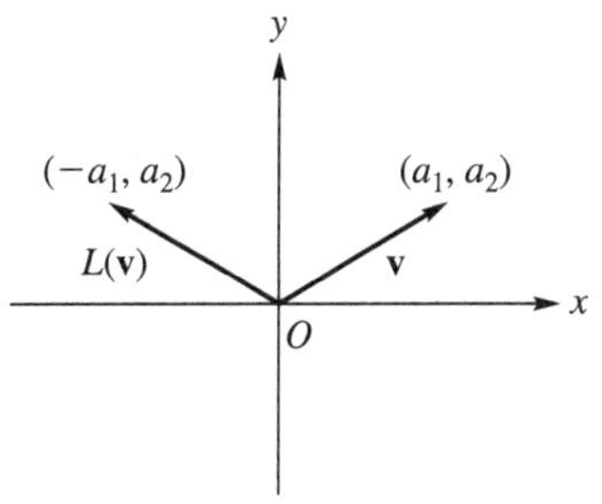

(b) Reflection about the origin.

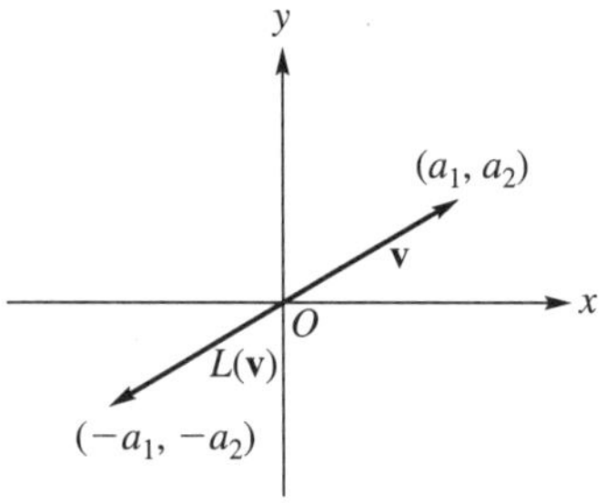

(c) Counterclockwise rotation through $\pi/2$ radians.

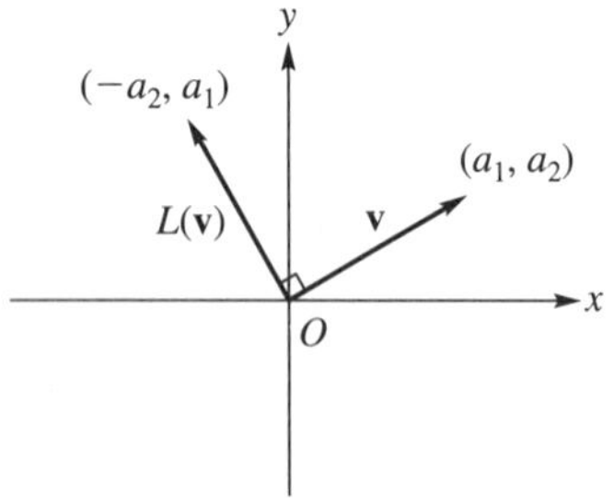

23. Yes. **27.** No. **35.** The $n \times n$ zero matrix.

Section 6.2, p. 387

1. **(a)** Yes. **(b)** No. **(c)** Yes. **(d)** No.

(e) All vectors $\begin{bmatrix} 0 \\ a \end{bmatrix}$, where a is any real number; that is, the y-axis.

(f) All vectors $\begin{bmatrix} a \\ 0 \end{bmatrix}$, where a is any real number; that is, the x-axis.

3. **(a)** No. **(b)** Yes. **(c)** Yes. **(d)** Yes.

(e) All vectors of the form $\begin{bmatrix} -r & -s & r & s \end{bmatrix}$, where r and s are real numbers.

(f) $\{\begin{bmatrix} 1 & 0 \end{bmatrix}, [0, 1]\}$.

5. **(a)** $\{\begin{bmatrix} 1 & -1 & -1 & 1 \end{bmatrix}\}$. **(b)** 1.

(c) $\{\begin{bmatrix} 1 & 0 & 1 \end{bmatrix}, \begin{bmatrix} 1 & 0 & 0 \end{bmatrix}, \begin{bmatrix} 0 & 1 & 1 \end{bmatrix}\}$. **(d)** 3.

7. **(a)** 0. **(b)** 6.

9. **(a)** $\{1\}$. **(b)** $\{\begin{bmatrix} 1 & 0 \end{bmatrix}, \begin{bmatrix} 0 & 1 \end{bmatrix}\}$.

11. **(a)** $\ker L = \left\{\begin{bmatrix} 0 & 0 \\ 0 & 0 \end{bmatrix}\right\}$, so $\ker L$ has no basis.

(b) $\left\{\begin{bmatrix} 1 & 0 \\ 1 & 0 \end{bmatrix}, \begin{bmatrix} 1 & 1 \\ 0 & 1 \end{bmatrix}, \begin{bmatrix} 0 & 1 \\ 0 & 0 \end{bmatrix}, \begin{bmatrix} 0 & 0 \\ 1 & 1 \end{bmatrix}\right\}$.

13. **(a)** $\dim \ker L = 1$, $\dim \operatorname{range} L = 2$.

(b) $\dim \ker L = 1$, $\dim \operatorname{range} L = 2$.

(c) $\dim \ker L = 0$, $\dim \operatorname{range} L = 4$.

17. No.

19. **(b)** $L^{-1}\left(\begin{bmatrix} 2 \\ 3 \\ 4 \end{bmatrix}\right) = \begin{bmatrix} \frac{3}{2} \\ -\frac{1}{2} \\ \frac{1}{2} \end{bmatrix}$. **21.** 1.

23. **(b)** $\begin{bmatrix} -2u_1 - u_3 \\ -2u_1 - u_2 + 2u_3 \\ -2u_1 + u_2 - u_3 \end{bmatrix}$.

25. **(a)** 2. **(b)** 1.

Section 6.3, p. 397

1. **(a)** $\begin{bmatrix} 1 & 2 \\ 2 & -1 \end{bmatrix}$. **(b)** $\begin{bmatrix} 1 & -\frac{1}{2} \\ 1 & \frac{3}{4} \end{bmatrix}$.

(c) $\begin{bmatrix} 3 & 2 \\ -4 & 4 \end{bmatrix}$. **(d)** $\begin{bmatrix} -2 & 2 \\ \frac{1}{2} & 2 \end{bmatrix}$.

(e) $\begin{bmatrix} 5 \\ 0 \end{bmatrix}$.

3. **(a)** $\begin{bmatrix} 1 & 0 & 1 & 1 \\ 0 & 1 & 2 & 1 \\ -1 & -2 & 1 & 0 \end{bmatrix}$.

(b) $\begin{bmatrix} 1 & 0 & 2 & 1 \\ 1 & 1 & 3 & 3 \\ -5 & -3 & -4 & -5 \end{bmatrix}$.

5. **(a)** $\begin{bmatrix} 1 & 2 & 1 \\ 1 & 0 & 0 \\ 0 & 1 & 1 \end{bmatrix}$. **(b)** $\begin{bmatrix} 8 \\ 1 \\ 5 \end{bmatrix}$.

7. **(a)** $\begin{bmatrix} 10 \\ 5 \\ 5 \end{bmatrix}$. **(b)** $\begin{bmatrix} 4 \\ 2 \\ 2 \end{bmatrix}$.

9. $\begin{bmatrix} 0 & 1 & 0 & 0 \\ 0 & 0 & 0 & 0 \\ 0 & 0 & 1 & 1 \\ 0 & 0 & 0 & 1 \end{bmatrix}$.

11. **(a)** $\begin{bmatrix} 0 & -3 & 2 & 0 \\ -2 & -3 & 0 & 2 \\ 3 & 0 & 3 & -3 \\ 0 & 3 & -2 & 0 \end{bmatrix}$.

(b) $\begin{bmatrix} 0 & 3 & -2 & 3 \\ 0 & -9 & -2 & -6 \\ 0 & 3 & 6 & 0 \\ 0 & 4 & 0 & 3 \end{bmatrix}$.

(c) $\begin{bmatrix} 0 & 3 & -2 & 0 \\ -3 & -6 & 1 & 3 \\ 3 & 0 & 3 & -3 \\ 1 & 3 & -1 & -1 \end{bmatrix}$.

(d) $\begin{bmatrix} 0 & -3 & 2 & -3 \\ 0 & -5 & -2 & -3 \\ 0 & 3 & 6 & 0 \\ 0 & 3 & -2 & 3 \end{bmatrix}$.

13. (a) $\begin{bmatrix} 1 & 0 \\ 0 & -1 \end{bmatrix}$. (b) $\begin{bmatrix} 0 & -1 \\ -1 & 0 \end{bmatrix}$.

(c) $\begin{bmatrix} \frac{1}{2} & -\frac{1}{2} \\ -\frac{1}{2} & -\frac{1}{2} \end{bmatrix}$. (d) $\begin{bmatrix} 1 & -1 \\ -1 & -1 \end{bmatrix}$.

17. (a) $\begin{bmatrix} 1 & 0 \\ 0 & 1 \end{bmatrix}$. (b) $\begin{bmatrix} 1 & 0 \\ 0 & 1 \end{bmatrix}$.

(c) $\begin{bmatrix} \frac{3}{5} & -\frac{2}{5} \\ \frac{1}{5} & \frac{1}{5} \end{bmatrix}$. (d) $\begin{bmatrix} 1 & 2 \\ -1 & 3 \end{bmatrix}$.

19. $\begin{bmatrix} 0 & -1 \\ 1 & 0 \end{bmatrix}$.

Section 6.4, p. 405

3. (a) $\begin{bmatrix} -u_1 + 4u_2 - u_3 & 3u_1 - u_2 + 3u_3 & 4u_1 + 3u_2 + 5u_3 \end{bmatrix}$.

(b) $\begin{bmatrix} 5 & -4 & -4 \end{bmatrix}$.

(c) $\begin{bmatrix} -1 & 4 & -1 \\ 3 & -1 & 3 \\ 4 & 3 & 5 \end{bmatrix}$.

(d) $\begin{bmatrix} 2u_1 + 2u_3 & -4u_1 - 2u_2 - 2u_3 & -2u_1 - 4u_2 - 6u_3 \end{bmatrix}$.

(e) $\begin{bmatrix} 14 & -28 & 14 \end{bmatrix}$.

(f) $\begin{bmatrix} 2 & 16 & 2 \\ 0 & -10 & 6 \\ 10 & 0 & 2 \end{bmatrix}$.

5. (a) $\begin{bmatrix} 5 \\ 10 \\ 0 \end{bmatrix}$.

(b) $\ker L_1$ = all vectors of the form $\begin{bmatrix} -r-s \\ r \\ s \end{bmatrix}$,

$\ker L_2$ = all vectors of the form $\begin{bmatrix} -r-2s \\ r \\ s \end{bmatrix}$,

$\ker L_1 \cap \ker L_2$ = all vectors of the form $\begin{bmatrix} -r \\ r \\ 0 \end{bmatrix}$.

(c) All vectors of the form $\begin{bmatrix} -r \\ r \\ 0 \end{bmatrix}$.

(d) They are the same.

7. (a) $C = \begin{bmatrix} 2 & 2 \\ 1 & -1 \\ -1 & 1 \\ 1 & -1 \end{bmatrix}$.

(b) $A = \begin{bmatrix} 1 & 1 \\ \frac{2}{3} & -\frac{2}{3} \\ \frac{1}{3} & -\frac{1}{3} \end{bmatrix}$, $B = \begin{bmatrix} 2 & 0 & 0 \\ 0 & 1 & 1 \\ 0 & -1 & -1 \\ 0 & 1 & 1 \end{bmatrix}$.

9. (a) $\begin{bmatrix} 2 & 8 & -2 \\ 4 & 2 & 6 \\ 2 & -2 & 4 \end{bmatrix}$. (b) $\begin{bmatrix} 10 & 17 & 7 \\ 11 & 8 & 13 \\ 3 & -1 & 4 \end{bmatrix}$.

11. (a) 6. (b) 6. (c) 12. (d) 12.

15. (a) $L(t^2) = t + 3$, $L(t) = 2t + 4$, $L(1) = -2t - 1$.

(b) $(a + 2b - 2c)t + (3a + 4b - c)$.

(c) $-16t - 18$.

17. Possible answers: $L\left(\begin{bmatrix} u_1 \\ u_2 \end{bmatrix}\right) = \begin{bmatrix} u_2 \\ u_1 \end{bmatrix}$; $L\left(\begin{bmatrix} u_1 \\ u_2 \end{bmatrix}\right) = \begin{bmatrix} -u_1 \\ -u_2 \end{bmatrix}$.

19. Possible answers: $L\left(\begin{bmatrix} u_1 \\ u_2 \end{bmatrix}\right) = \begin{bmatrix} u_1 \\ 0 \end{bmatrix}$; $L\left(\begin{bmatrix} u_1 \\ u_2 \end{bmatrix}\right) = \begin{bmatrix} \dfrac{u_1 + u_2}{2} \\ \dfrac{u_1 + u_2}{2} \end{bmatrix}$.

21. $\begin{bmatrix} 2 & 0 & -1 \\ -2 & -1 & 2 \\ 1 & 1 & -1 \end{bmatrix}$.

Section 6.5, p. 413

3. $\begin{bmatrix} -2 & 2 \\ \frac{1}{2} & 2 \end{bmatrix}$.

5. (a) $A = \begin{bmatrix} 1 & 0 \\ 0 & -1 \end{bmatrix}$. (b) $B = \begin{bmatrix} \frac{1}{3} & \frac{4}{3} \\ \frac{2}{3} & -\frac{1}{3} \end{bmatrix}$.

(c) $P = \begin{bmatrix} 1 & 1 \\ -1 & 2 \end{bmatrix}$.

9. $\begin{bmatrix} \frac{13}{2} & \frac{1}{2} & -\frac{3}{2} \\ -\frac{5}{2} & \frac{1}{2} & -\frac{1}{2} \end{bmatrix}$.

13. $\begin{bmatrix} 1 & 0 \\ 0 & 1 \end{bmatrix}$.

15. $\begin{bmatrix} 0 & -1 \\ 1 & 0 \end{bmatrix}$.

Section 6.6, p. 425

1. (a) $M = \begin{bmatrix} \sqrt{2}/2 & -\sqrt{2}/2 & 0 \\ \sqrt{2}/2 & \sqrt{2}/2 & \sqrt{2} \\ 0 & 0 & 1 \end{bmatrix}$.

(b)

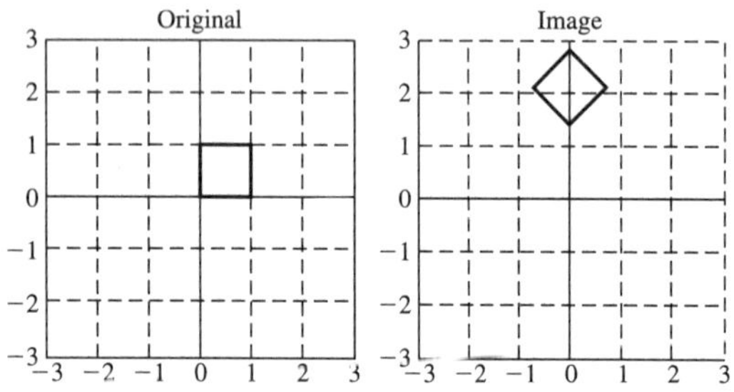

(c) No.

3. (a)

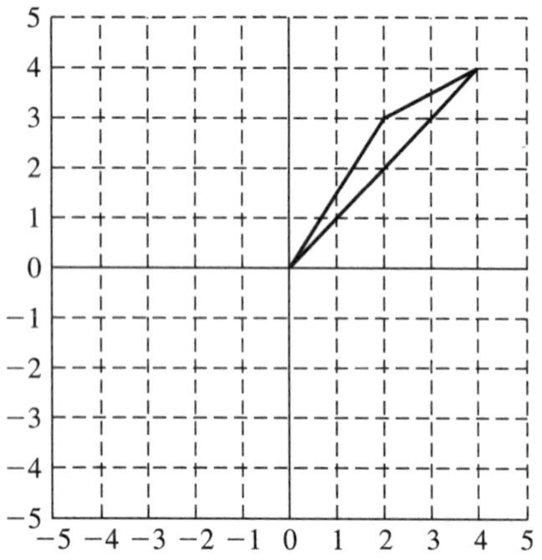

(b) $M = \begin{bmatrix} -1/2 & -\sqrt{3}/2 & 0 \\ \sqrt{3}/2 & 1/2 & 0 \\ 0 & 0 & 1 \end{bmatrix}$.

(c)

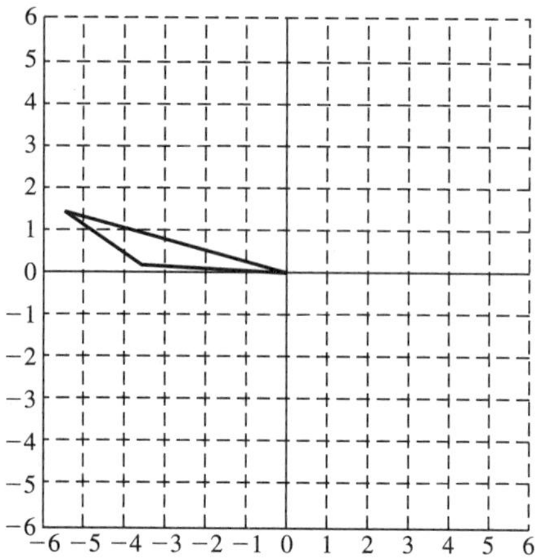

(d) $Q = \begin{bmatrix} -1/2 & -\sqrt{3}/2 & 0 \\ \sqrt{3}/2 & 1/2 & 0 \\ 0 & 0 & 1 \end{bmatrix}$.

(e) It is the same as in part (c).

(f) They are the same since $M = Q$. Rotations are commutative.

5. $A = \begin{bmatrix} 1 & 0 & 0 \\ 0 & -1 & 0 \\ 0 & 0 & 1 \end{bmatrix}$ and $B = \begin{bmatrix} 1 & 0 & 4 \\ 0 & 1 & -2 \\ 0 & 0 & 1 \end{bmatrix}$. The images will not be the same since $AB \neq BA$.

7. $M = \begin{bmatrix} -\frac{1}{2} & 0 & 0 \\ 0 & \frac{1}{2} & 0 \\ 0 & 0 & 1 \end{bmatrix}$.

9. $M = \begin{bmatrix} 1 & 0 & 1 \\ 0 & -1 & 1 \\ 0 & 0 & 1 \end{bmatrix}$.

11. $M = \begin{bmatrix} \sqrt{2}/2 & -\sqrt{2}/2 & 0 \\ \sqrt{2}/2 & \sqrt{2}/2 & 0 \\ 0 & 0 & 1 \end{bmatrix}$.

13. (a)

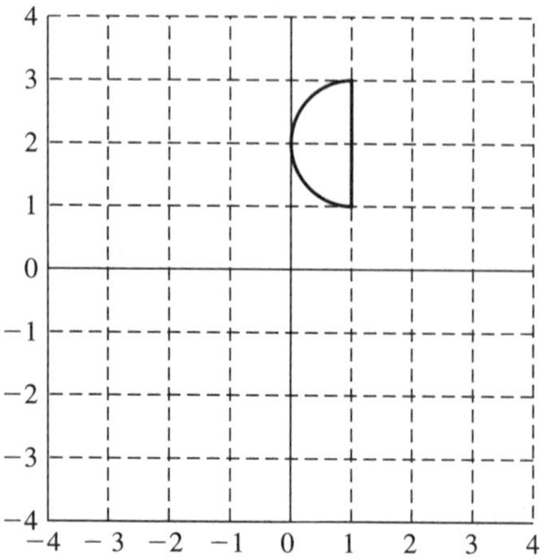

(b) $M = \begin{bmatrix} 0 & -1 & 2 \\ 1 & 0 & 0 \\ 0 & 0 & 1 \end{bmatrix}$.

(c) $M\begin{bmatrix} 1 \\ 1 \\ 1 \end{bmatrix} = \begin{bmatrix} 1 \\ 1 \\ 1 \end{bmatrix}$ and $M\begin{bmatrix} 3 \\ 1 \\ 1 \end{bmatrix} = \begin{bmatrix} 1 \\ 3 \\ 1 \end{bmatrix}$.

15. (a) $\begin{bmatrix} \cos\theta_j & -\sin\theta_j & 0 & 0 \\ \sin\theta_j & \cos\theta_j & 0 & 0 \\ 0 & 0 & 1 & k\theta_j \\ 0 & 0 & 0 & 1 \end{bmatrix}$.

(b) It appears that the length of the vector being "screwed" is decreasing as we move down the z-axis. Thus a scaling in the x- and y-directions by a factor smaller than 1 is included.

Supplementary Exercises, p. 430

3. No.

5. (a) $8t+7$. **(b)** $\frac{1}{2}(3a+b)t + \frac{1}{2}(3a-b)$.

7. (a) No. **(b)** No. **(c)** Yes. **(d)** No.

(e) $\{t^3+t^2, t+1\}$. **(f)** $\{t^3, t\}$.

9. (a) $\begin{bmatrix} 1 & 0 & 0 & 0 \\ 0 & 0 & 1 & 0 \\ 0 & 1 & 0 & 0 \\ 0 & 0 & 0 & 1 \end{bmatrix}$.

(b) $\begin{bmatrix} 1 & 1 & 0 & -1 \\ -1 & -1 & 1 & 1 \\ 0 & 1 & 0 & 0 \\ 0 & -1 & 0 & 1 \end{bmatrix}$.

(c) $\begin{bmatrix} 1 & 0 & 0 & 1 \\ 0 & 0 & 1 & 0 \\ 1 & 1 & 0 & 0 \\ 0 & 0 & 1 & 1 \end{bmatrix}$.

(d) $\begin{bmatrix} 2 & 1 & -1 & 0 \\ -2 & -1 & 2 & 0 \\ 1 & 1 & 0 & 0 \\ -1 & -1 & 1 & 1 \end{bmatrix}$.

11. (b) $\ker L =$ the set of all continuous functions f on $[0, 1]$ such that $f(0) = 0$.

(c) Yes.

13. (a) $\begin{bmatrix} 2 & 0 & 1 \\ 0 & 1 & 2 \\ -1 & 0 & -2 \end{bmatrix}$. **(b)** $4t^2 - 4t + 1$.

15. $\begin{bmatrix} 0 & 0 & 0 & 0 \\ 3 & 0 & 0 & 0 \\ 0 & 2 & 0 & 0 \\ 0 & 0 & 1 & 0 \end{bmatrix}$.

Chapter Review, p. 432

True/False

1. T. **2.** F. **3.** T. **4.** F. **5.** F.
6. T. **7.** T. **8.** T. **9.** T. **10.** F.

Quiz

1. Yes. **2. (b)** $\begin{bmatrix} 1 & 0 \\ k & 1 \end{bmatrix}$. **3. (b)** No.

4. $\begin{bmatrix} -4 \\ 3 \\ 4 \end{bmatrix}$. **5.** $\begin{bmatrix} 0 & -1 \\ 3 & 5 \end{bmatrix}$.

6. (a) $A = \begin{bmatrix} 1 & -1 \\ 1 & 1 \\ 2 & 0 \end{bmatrix}$. **(b)** $P = \begin{bmatrix} 1 & 0 \\ 2 & 1 \end{bmatrix}$.

(c) $Q = \begin{bmatrix} 1 & 1 & 2 \\ 1 & 2 & 0 \\ 0 & 1 & 0 \end{bmatrix}$. **(d)** $B = \begin{bmatrix} -1 & 1 \\ 2 & 0 \\ -1 & -1 \end{bmatrix}$.

CHAPTER 7

Section 7.1, p. 450

1. The only eigenvalue of L is $\lambda = -1$. Every nonzero vector in R^2 is an eigenvector of L associated with λ.

3. The eigenvalues are $\lambda_1 = 1, \lambda_2 = -1, \lambda_3 = 0$. Associated eigenvectors are $1, t^2$, and t, respectively.

5. (a) $p(\lambda) = \lambda^2 - 5\lambda + 7$.

(b) $p(\lambda) = \lambda^3 - 4\lambda^2 + 7$.

(c) $p(\lambda) = (\lambda-4)(\lambda-2)(\lambda-3) = \lambda^3 - 9\lambda^2 + 26\lambda - 24$.

(d) $p(\lambda) = \lambda^2 - 7\lambda + 6$.

7. (a) $p(\lambda) = \lambda^2 - 5\lambda + 6$. The eigenvalues are $\lambda_1 = 2$ and $\lambda_2 = 3$. Associated eigenvectors are

$$\mathbf{x}_1 = \begin{bmatrix} 1 \\ -1 \end{bmatrix} \quad \text{and} \quad \mathbf{x}_2 = \begin{bmatrix} 1 \\ -2 \end{bmatrix}.$$

(b) $p(\lambda) = \lambda^3 - 7\lambda^2 + 14\lambda - 8$. The eigenvalues are $\lambda_1 = 1, \lambda_2 = 2$, and $\lambda_3 = 4$. Associated eigenvectors are

$$\mathbf{x}_1 = \begin{bmatrix} -1 \\ 1 \\ 1 \end{bmatrix}, \quad \mathbf{x}_2 = \begin{bmatrix} 1 \\ 0 \\ 0 \end{bmatrix}, \quad \text{and}$$

$$\mathbf{x}_3 = \begin{bmatrix} 7 \\ -4 \\ 2 \end{bmatrix}.$$

(c) $p(\lambda) = \lambda^3 - 5\lambda^2 + 2\lambda + 8$. The eigenvalues are $\lambda_1 = -1$, $\lambda_2 = 2$, and $\lambda_3 = 4$. Associated eigenvectors are

$$\mathbf{x}_1 = \begin{bmatrix} 1 \\ 0 \\ -1 \end{bmatrix}, \quad \mathbf{x}_2 = \begin{bmatrix} -2 \\ -3 \\ 2 \end{bmatrix}, \quad \text{and}$$

$$\mathbf{x}_3 = \begin{bmatrix} 8 \\ 5 \\ 2 \end{bmatrix}.$$

(d) $p(\lambda) = \lambda^3 - 3\lambda^2 - 6\lambda + 8$. The eigenvalues are $\lambda_1 = -2$, $\lambda_2 = 4$, and $\lambda_3 = 1$. Associated eigenvectors are

$$\mathbf{x}_1 = \begin{bmatrix} 1 \\ 0 \\ 0 \end{bmatrix}, \quad \mathbf{x}_2 = \begin{bmatrix} 7 \\ -12 \\ 6 \end{bmatrix}, \quad \text{and}$$

$$\mathbf{x}_3 = \begin{bmatrix} 1 \\ 3 \\ 3 \end{bmatrix}.$$

9. (a) $p(\lambda) = \lambda^2 + 1$. The eigenvalues are $\lambda_1 = i$ and $\lambda_2 = -i$. Associated eigenvectors are

$$\mathbf{x}_1 = \begin{bmatrix} 1 \\ i \end{bmatrix} \quad \text{and} \quad \mathbf{x}_2 = \begin{bmatrix} 1 \\ -i \end{bmatrix}.$$

(b) $p(\lambda) = \lambda^3 + 2\lambda^2 + 4\lambda + 8$. The eigenvalues are $\lambda_1 = -2$, $\lambda_2 = 2i$, and $\lambda_3 = -2i$. Associated eigenvectors are

$$\mathbf{x}_1 = \begin{bmatrix} 4 \\ -2 \\ 1 \end{bmatrix}, \quad \mathbf{x}_2 = \begin{bmatrix} -4 \\ 2i \\ 1 \end{bmatrix}, \quad \text{and}$$

$$\mathbf{x}_3 = \begin{bmatrix} -4 \\ -2i \\ 1 \end{bmatrix}.$$

(c) $p(\lambda) = \lambda^3 + (-2 + i)\lambda^2 - 2i\lambda$. The eigenvalues are $\lambda_1 = 0$, $\lambda_2 = -i$, and $\lambda_3 = 2$. Associated eigenvectors are

$$\mathbf{x}_1 = \begin{bmatrix} 0 \\ 0 \\ 1 \end{bmatrix}, \quad \mathbf{x}_2 = \begin{bmatrix} -1 \\ -i \\ 1 \end{bmatrix}, \quad \text{and} \quad \mathbf{x}_3 = \begin{bmatrix} 4 \\ 2 \\ 1 \end{bmatrix}.$$

(d) $p(\lambda) = \lambda^2 - 8\lambda + 17$. The eigenvalues are $\lambda_1 = 4 + i$ and $\lambda_2 = 4 - i$. Associated eigenvectors are

$$\mathbf{x}_1 = \begin{bmatrix} 2 \\ -1 + i \end{bmatrix} \quad \text{and} \quad \mathbf{x}_2 = \begin{bmatrix} 2 \\ -1 - i \end{bmatrix}.$$

13. The characteristic polynomial of A is

$$p(\lambda) = (\lambda - 1)(\lambda + 1)(\lambda - 3)(\lambda - 2).$$

The eigenvalues of L are $\lambda_1 = 1$, $\lambda_2 = -1$, $\lambda_3 = 3$, and $\lambda_4 = 2$. Associated eigenvectors are

$$\begin{bmatrix} 1 & 0 \\ 0 & 0 \end{bmatrix}, \quad \begin{bmatrix} 1 & -1 \\ 0 & 0 \end{bmatrix},$$

$$\begin{bmatrix} 9 & 3 \\ 4 & 0 \end{bmatrix}, \quad \text{and} \quad \begin{bmatrix} -29 & -7 \\ -9 & 3 \end{bmatrix}.$$

17. (a) $\left\{ \begin{bmatrix} 1 \\ 0 \\ 1 \end{bmatrix}, \begin{bmatrix} 0 \\ 1 \\ 0 \end{bmatrix} \right\}$. **(b)** $\left\{ \begin{bmatrix} -1 \\ 0 \\ 1 \end{bmatrix} \right\}$.

19. (a) $\left\{ \begin{bmatrix} -4 \\ 2i \\ 1 \end{bmatrix} \right\}$. **(b)** $\left\{ \begin{bmatrix} -4 \\ -2i \\ 1 \end{bmatrix} \right\}$.

Section 7.2, p. 461

1. L is not diagonalizable. The eigenvalues of L are $\lambda_1 = \lambda_2 = \lambda_3 = 0$. The set of associated eigenvectors does not form a basis for P_2.

3. $\{t^2, t, 1\}$.

5. The eigenvalues of L are $\lambda_1 = 2$, $\lambda_2 = -3$, and $\lambda_3 = 4$. Associated eigenvectors are t^2, $t^2 - 5t$, and $9t^2 + 4t + 14$. L is diagonalizable.

7. (a) Not diagonalizable. **(b)** Diagonalizable.

(c) Not diagonalizable.

(d) Not diagonalizable.

9. $\begin{bmatrix} 3 & 5 & -5 \\ 5 & 3 & -5 \\ 5 & 5 & -7 \end{bmatrix}$.

11. (a) $P = \begin{bmatrix} 1 & 2 & 1 \\ 0 & 1 & 0 \\ 0 & 0 & -3 \end{bmatrix}$.

(b) $P = \begin{bmatrix} -1 & -1 & 1 \\ 1 & 0 & 1 \\ 0 & 1 & 1 \end{bmatrix}$.

(c) Not possible.

(d) Not possible.

13. $D = \begin{bmatrix} -3 & 0 & 0 \\ 0 & 4 & 0 \\ 0 & 0 & 4 \end{bmatrix}$, $P = \begin{bmatrix} -1 & 0 & 0 \\ 0 & 0 & 1 \\ 1 & 1 & 1 \end{bmatrix}$.

15. **(a)** $D = \begin{bmatrix} 6 & 0 \\ 0 & 1 \end{bmatrix}$. **(b)** $D = \begin{bmatrix} 0 & 0 \\ 0 & 7 \end{bmatrix}$.

(c) $D = \begin{bmatrix} 2 & 0 & 0 \\ 0 & 4 & 0 \\ 0 & 0 & 1 \end{bmatrix}$.

(d) $D = \begin{bmatrix} 1 & 0 & 0 \\ 0 & 2 & 0 \\ 0 & 0 & 1 \end{bmatrix}$.

17. **(a)** Defective.
(b) Not defective.
(c) Not defective.
(d) Not defective.

19. $\begin{bmatrix} 768 & -1280 \\ 256 & -768 \end{bmatrix}$.

Section 7.3, p. 475

7. $P^T P = I_3$.

9. **(a)** If B is the given matrix, verify that $B^T B = I_2$.

15. A is similar to $D = \begin{bmatrix} 0 & 0 \\ 0 & 4 \end{bmatrix}$ and $P = \begin{bmatrix} \frac{1}{\sqrt{2}} & \frac{1}{\sqrt{2}} \\ -\frac{1}{\sqrt{2}} & \frac{1}{\sqrt{2}} \end{bmatrix}$.

17. A is similar to $D = \begin{bmatrix} 0 & 0 & 0 \\ 0 & 0 & 0 \\ 0 & 0 & 4 \end{bmatrix}$ and

$$P = \begin{bmatrix} 1 & 0 & 0 \\ 0 & -\frac{1}{\sqrt{2}} & \frac{1}{\sqrt{2}} \\ 0 & \frac{1}{\sqrt{2}} & \frac{1}{\sqrt{2}} \end{bmatrix}.$$

19. A is similar to $D = \begin{bmatrix} -2 & 0 & 0 \\ 0 & 1 & 0 \\ 0 & 0 & 1 \end{bmatrix}$ and

$$P = \begin{bmatrix} \frac{1}{\sqrt{3}} & -\frac{1}{\sqrt{2}} & -\frac{1}{\sqrt{6}} \\ \frac{1}{\sqrt{3}} & \frac{1}{\sqrt{2}} & -\frac{1}{\sqrt{6}} \\ \frac{1}{\sqrt{3}} & 0 & \frac{2}{\sqrt{6}} \end{bmatrix}.$$

21. A is similar to $D = \begin{bmatrix} 3 & 0 \\ 0 & 1 \end{bmatrix}$.

23. A is similar to $D = \begin{bmatrix} 1 & 0 & 0 \\ 0 & 2 & 0 \\ 0 & 0 & 0 \end{bmatrix}$.

25. A is similar to $D = \begin{bmatrix} 1 & 0 & 0 \\ 0 & 0 & 0 \\ 0 & 0 & 2 \end{bmatrix}$.

27. A is similar to $D = \begin{bmatrix} 2 & 0 & 0 \\ 0 & -2 & 0 \\ 0 & 0 & 4 \end{bmatrix}$.

Supplementary Exercises, p. 477

1. **(a)** $\lambda_1 = 1, \lambda_2 = 1, \lambda_3 = 4$;

associated eigenvectors: $\begin{bmatrix} 1 \\ 0 \\ 0 \end{bmatrix}, \begin{bmatrix} 0 \\ 0 \\ 1 \end{bmatrix}, \begin{bmatrix} 4 \\ 3 \\ 0 \end{bmatrix}$.

(b) $\lambda_1 = 3, \lambda_2 = 4, \lambda_3 = -1$;

associated eigenvectors: $\begin{bmatrix} 0 \\ 0 \\ 1 \end{bmatrix}, \begin{bmatrix} 1 \\ -1 \\ 0 \end{bmatrix}, \begin{bmatrix} 2 \\ 3 \\ 0 \end{bmatrix}$.

(c) $\lambda_1 = 1, \lambda_2 = 2, \lambda_3 = 3$;

associated eigenvectors: $\begin{bmatrix} 1 \\ 1 \\ 1 \end{bmatrix}, \begin{bmatrix} 1 \\ 2 \\ 4 \end{bmatrix}, \begin{bmatrix} 1 \\ 3 \\ 9 \end{bmatrix}$.

(d) $\lambda_1 = -3, \lambda_2 = 1, \lambda_3 = -1$;

associated eigenvectors: $\begin{bmatrix} 1 \\ -3 \\ 9 \end{bmatrix}, \begin{bmatrix} 1 \\ 1 \\ 1 \end{bmatrix}, \begin{bmatrix} -1 \\ 1 \\ -1 \end{bmatrix}$.

7. **(a)** $\begin{bmatrix} -\frac{3}{10} \\ \frac{1}{5} \end{bmatrix}, \begin{bmatrix} -\frac{6}{5} \\ \frac{4}{5} \end{bmatrix}$.

(b) $A = \dfrac{1}{10}\begin{bmatrix} -3 & -12 \\ 2 & 8 \end{bmatrix}$.

(c) The eigenvalues are $\lambda_1 = 0, \lambda_2 = \frac{1}{2}$. Associated eigenvectors are

$$\mathbf{x}_1 = \begin{bmatrix} -4 \\ 1 \end{bmatrix} \quad \text{and} \quad \mathbf{x}_2 = \begin{bmatrix} -3 \\ 2 \end{bmatrix}.$$

(d) The eigenvalues are $\lambda_1 = 0$ and $\lambda_2 = \frac{1}{2}$. Associated eigenvectors are $p_1(t) = 5t - 5$ and $p_2(t) = 5t$.

(e) The eigenspace for $\lambda_1 = 0$ is the subspace of P_1 with basis $\{5t - 5\}$. The eigenspace for $\lambda_2 = \frac{1}{2}$ is the subspace of P_1 with basis $\{5t\}$.

9. The only eigenvalue is $\lambda_1 = 0$ and an associated eigenvector is $p_1(x) = 1$.

Chapter Review, p. 478

True/False

1. T.	**2.** F.	**3.** T.	**4.** T.	**5.** F.
6. T.	**7.** T.	**8.** T.	**9.** T.	**10.** T.
11. F.	**12.** T.	**13.** T.	**14.** T.	**15.** T.
16. T.	**17.** T.	**18.** T.	**19.** T.	**20.** T.

Quiz

1. $\lambda_1 = 1, \mathbf{x}_1 = \begin{bmatrix} 1 \\ -1 \end{bmatrix}$.

$\lambda_2 = 3, \mathbf{x}_2 = \begin{bmatrix} -1 \\ 2 \end{bmatrix}$.

2. (a) $A = \begin{bmatrix} 0 & 0 & 0 \\ 0 & \frac{5}{4} & \frac{3}{4} \\ 0 & -\frac{3}{4} & -\frac{5}{4} \end{bmatrix}$.

(b) $\lambda_1 = 0, \mathbf{x}_1 = \begin{bmatrix} 1 \\ 0 \\ 0 \end{bmatrix}$,

$\lambda_2 = 1, \mathbf{x}_2 = \begin{bmatrix} 0 \\ -3 \\ 1 \end{bmatrix}$,

$\lambda_3 = -1, \mathbf{x}_3 = \begin{bmatrix} 0 \\ 1 \\ -3 \end{bmatrix}$.

3. $\lambda_1 = -1, \lambda_2 = 2$, and $\lambda_3 = 2$.

4. $\lambda = 9, \mathbf{x}$.

5. $\left\{ \begin{bmatrix} 1 \\ 0 \\ 0 \end{bmatrix}, \begin{bmatrix} 0 \\ 1 \\ -1 \end{bmatrix} \right\}$.

6. $D = \begin{bmatrix} -3 & 0 & 0 \\ 0 & 2 & 0 \\ 0 & 0 & 3 \end{bmatrix}$.

7. No. **8.** No.

9. (a) Possible answer: $\begin{bmatrix} -2 \\ 2 \\ 2 \end{bmatrix}$.

(b) Possible answer: $P = \begin{bmatrix} \frac{1}{\sqrt{6}} & -\frac{1}{\sqrt{2}} & \frac{1}{\sqrt{3}} \\ \frac{1}{\sqrt{6}} & \frac{1}{\sqrt{2}} & \frac{1}{\sqrt{3}} \\ \frac{2}{\sqrt{6}} & 0 & -\frac{1}{\sqrt{3}} \end{bmatrix}$

12. (a) $\lambda_1 = 4, \left\{ \begin{bmatrix} 1 \\ 1 \\ 2 \end{bmatrix} \right\}$.

$\lambda_2 = 10, \left\{ \begin{bmatrix} -1 \\ 1 \\ 0 \end{bmatrix}, \begin{bmatrix} -2 \\ 0 \\ 1 \end{bmatrix} \right\}$.

CHAPTER 8

Section 8.1, p. 486

1. $\begin{bmatrix} 8 \\ 2 \\ 1 \end{bmatrix}$. **3.** (b) and (c).

5. (a) $\mathbf{x}^{(1)} = \begin{bmatrix} 0.7 \\ 0.3 \end{bmatrix}, \mathbf{x}^{(2)} = \begin{bmatrix} 0.61 \\ 0.39 \end{bmatrix}, \mathbf{x}^{(3)} = \begin{bmatrix} 0.583 \\ 0.417 \end{bmatrix}$.

(b) Since all entries of T are positive, it is regular; $\mathbf{u} = \begin{bmatrix} 0.571 \\ 0.429 \end{bmatrix}$.

7. (a) and (d).

9. (a) $\begin{bmatrix} \frac{3}{7} \\ \frac{4}{7} \end{bmatrix}$. **(b)** $\begin{bmatrix} \frac{1}{8} \\ \frac{7}{8} \end{bmatrix}$.

(c) $\begin{bmatrix} \frac{4}{11} \\ \frac{4}{11} \\ \frac{3}{11} \end{bmatrix}$. **(d)** $\begin{bmatrix} \frac{1}{11} \\ \frac{4}{11} \\ \frac{6}{11} \end{bmatrix}$.

11. (a) 0.69.

(b) 20.7 percent of the population will be farmers.

13. (a) 35 percent, 37.5 percent.

(b) 40 percent.

Section 8.2, p. 500

1. (a) 5, 1. **(b)** 2, 0.

(c) $\sqrt{5}, \sqrt{6}$. **(d)** $0, 0, 1, \sqrt{5}$.

3. $A = USV^T = \begin{bmatrix} \frac{1}{\sqrt{2}} & \frac{1}{\sqrt{2}} \\ -\frac{1}{\sqrt{2}} & \frac{1}{\sqrt{2}} \end{bmatrix} \begin{bmatrix} \sqrt{2} & 0 \\ 0 & \sqrt{2} \end{bmatrix} \begin{bmatrix} 1 & 0 \\ 0 & 1 \end{bmatrix}^T$.

5. (a) 10.8310, 0.8310.

(b) 18.9245, 3.8400, 0.3440.

(c) 25.4368, 1.7226, 0.

Section 8.3, p. 514

1. **(a)** 3. **(b)** 3.
3. **(a)** 7. **(b)** 10. **(c)** 13.
5. **(a)** 5. **(b)** 7.
11. Sample mean = 65, variation = 58, standard deviation ≈ 7.6158.
13. Sample means $\approx \begin{bmatrix} 509.8 \\ 676.4 \end{bmatrix}$.

    Covariance matrix $\approx \begin{bmatrix} 11{,}014.96 & 9{,}822.88 \\ 9{,}822.88 & 25{,}092.64 \end{bmatrix}$.
15. First principal component $\approx \begin{bmatrix} 951.0 \\ 1{,}345.5 \\ 1{,}716.3 \\ 826.7 \\ 746.9 \end{bmatrix}$.

Section 8.4, p. 524

1. **(a)** $\mathbf{x}(t) = \begin{bmatrix} x_1(t) \\ x_2(t) \\ x_3(t) \end{bmatrix} = \begin{bmatrix} b_1 e^{-3t} \\ b_2 e^{4t} \\ b_3 e^{2t} \end{bmatrix}$

   $= b_1 \begin{bmatrix} 1 \\ 0 \\ 0 \end{bmatrix} e^{-3t} + b_2 \begin{bmatrix} 0 \\ 1 \\ 0 \end{bmatrix} e^{4t} + b_3 \begin{bmatrix} 0 \\ 0 \\ 1 \end{bmatrix} e^{2t}$.

   (b) $\begin{bmatrix} 3e^{-3t} \\ 4e^{4t} \\ 5e^{2t} \end{bmatrix} = 3 \begin{bmatrix} 1 \\ 0 \\ 0 \end{bmatrix} e^{-3t} + 4 \begin{bmatrix} 0 \\ 1 \\ 0 \end{bmatrix} e^{4t} + 5 \begin{bmatrix} 0 \\ 0 \\ 1 \end{bmatrix} e^{2t}$.
3. $\mathbf{x}(t) = b_1 \begin{bmatrix} 6 \\ 2 \\ 7 \end{bmatrix} e^{4t} + b_2 \begin{bmatrix} 0 \\ 7 \\ -1 \end{bmatrix} e^{-5t} + b_3 \begin{bmatrix} 0 \\ 0 \\ 1 \end{bmatrix} e^{2t}$.
5. $\mathbf{x}(t) = b_1 \begin{bmatrix} 1 \\ 0 \\ 0 \end{bmatrix} e^{5t} + b_2 \begin{bmatrix} 0 \\ 1 \\ 3 \end{bmatrix} e^{5t} + b_3 \begin{bmatrix} 0 \\ -3 \\ 1 \end{bmatrix} e^{-5t}$.
7. $\mathbf{x}(t) = b_1 \begin{bmatrix} 1 \\ 0 \\ 0 \end{bmatrix} e^{-2t} + b_2 \begin{bmatrix} 1 \\ 1 \\ 2 \end{bmatrix} e^{2t} + b_3 \begin{bmatrix} -7 \\ -2 \\ 1 \end{bmatrix} e^{-3t}$.
9. $\mathbf{x}(t) = \begin{bmatrix} 440 + 60e^{-5t} \\ 220 - 20e^{-5t} \end{bmatrix}$.

Section 8.5, p. 534

1. The origin is a stable equilibrium. The phase portrait shows all trajectories tending toward the origin.
3. The origin is a stable equilibrium. The phase portrait shows all trajectories tending toward the origin with those passing through points not on the eigenvector aligning themselves to be tangent to the eigenvector at the origin.
5. The origin is a saddle point. The phase portrait shows trajectories not in the direction of an eigenvector heading toward the origin, but bending away as $t \to \infty$.
7. The origin is a stable equilibrium. The phase portrait shows all trajectories tending toward the origin.
9. The origin is called marginally stable.

Section 8.6, p. 542

1. **(a)** $\begin{bmatrix} x & y \end{bmatrix} \begin{bmatrix} -3 & \frac{5}{2} \\ \frac{5}{2} & -2 \end{bmatrix} \begin{bmatrix} x \\ y \end{bmatrix}$.

   (b) $\begin{bmatrix} x_1 & x_2 & x_3 \end{bmatrix} \begin{bmatrix} 2 & \frac{3}{2} & -\frac{5}{2} \\ \frac{3}{2} & 0 & \frac{7}{2} \\ -\frac{5}{2} & \frac{7}{2} & 0 \end{bmatrix} \begin{bmatrix} x_1 \\ x_2 \\ x_3 \end{bmatrix}$.

   (c) $\begin{bmatrix} x_1 & x_2 & x_3 \end{bmatrix} \begin{bmatrix} 3 & \frac{1}{2} & -\frac{1}{2} \\ \frac{1}{2} & 1 & -2 \\ -\frac{1}{2} & -2 & -2 \end{bmatrix} \begin{bmatrix} x_1 \\ x_2 \\ x_3 \end{bmatrix}$.
3. **(a)** $\begin{bmatrix} -1 & 0 & 0 \\ 0 & 2 & 0 \\ 0 & 0 & 0 \end{bmatrix}$.

   (b) $\begin{bmatrix} 3 & 0 & 0 \\ 0 & 0 & 0 \\ 0 & 0 & 0 \end{bmatrix}$.

   (c) $\begin{bmatrix} 4 & 0 & 0 \\ 0 & -2 & 0 \\ 0 & 0 & -2 \end{bmatrix}$.
5. $3x'^2 - 2y'^2$.
7. $y_1^2 - y_3^2$.
9. $-2y_1^2 + 5y_2^2 - 5y_3^2$.
11. y'^2.
13. $y_1^2 + y_2^2 - y_3^2$.
15. $y_1^2 - y_2^2$.
17. $y_1^2 - y_2^2$, rank = 2, signature = 0.
21. g_1, g_2, and g_4.
23. (a), (b), and (c).

Section 8.7, p. 551

1. Ellipse.

3. Hyperbola.

5. Two intersecting lines.

7. Circle.

9. Point.

11. Ellipse; $\frac{x'^2}{2} + y'^2 = 1$.

13. Circle; $\frac{x'^2}{5^2} + \frac{y'^2}{5^2} = 1$.

15. Pair of parallel lines; $y' = 2$, $y' = -2$; $y'^2 = 4$.

17. Point $(1, 3)$; $x'^2 + y'^2 = 0$.

19. Possible answer: ellipse; $\frac{x'^2}{12} + \frac{y'^2}{4} = 1$.

21. Possible answer: pair of parallel lines $y' = \frac{2}{\sqrt{10}}$ and $y' = -\frac{2}{\sqrt{10}}$.

23. Possible answer: two intersecting lines $y' = 3x'$ and $y' = -3x'$; $9x'^2 - y'^2 = 0$.

25. Possible answer: parabola; $y''^2 = -4x''$.

27. Possible answer: hyperbola; $\frac{x''^2}{4} - \frac{y''^2}{9} = 1$.

29. Possible answer: hyperbola; $\frac{x''^2}{\frac{9}{8}} - \frac{y''^2}{\frac{9}{8}} = 1$.

Section 8.8, p. 560

1. Hyperboloid of one sheet.

3. Hyperbolic paraboloid.

5. Parabolic cylinder.

7. Parabolic cylinder.

9. Ellipsoid.

11. Elliptic paraboloid.

13. Hyperbolic paraboloid.

15. Ellipsoid; $x'^2 + y'^2 + \frac{z'^2}{\frac{1}{3}} = 1$.

17. Hyperbolic paraboloid; $\frac{x''^2}{4} - \frac{y''^2}{4} = z''$.

19. Elliptic paraboloid; $\frac{x'^2}{4} + \frac{y'^2}{8} = 1$.

21. Hyperboloid of one sheet; $\frac{x''^2}{2} + \frac{y''^2}{4} - \frac{z''}{4} = 1$.

23. Parabolic cylinder; $x''^2 = \frac{4}{\sqrt{2}} y''$.

25. Hyperboloid of two sheets; $\frac{x''^2}{\frac{7}{4}} - \frac{y''^2}{\frac{7}{4}} - \frac{z''^2}{\frac{7}{4}} = 1$.

27. Cone; $x''^2 + y''^2 - z''^2 = 0$.

APPENDIX B

Section B.1, p. A-11

1. **(a)** $4 + 2i$. **(b)** $-4 - 3i$. **(c)** $11 - 2i$.
(d) $-3 + i$. **(e)** $-3 + 6i$. **(f)** $-2 - i$.
(g) $7 - 11i$. **(h)** $-9 + 13i$.

3. **(a)**

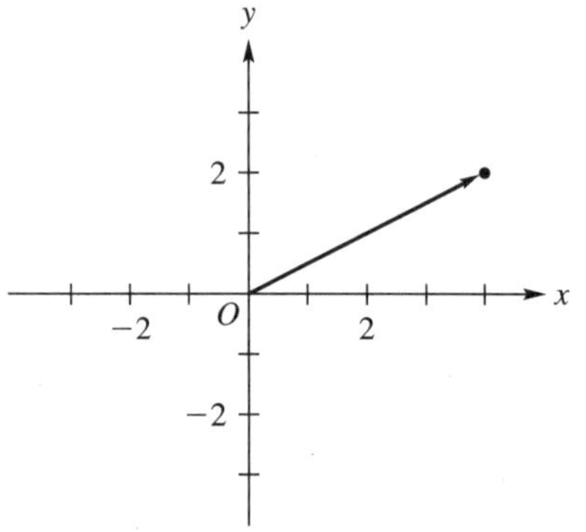

(b)

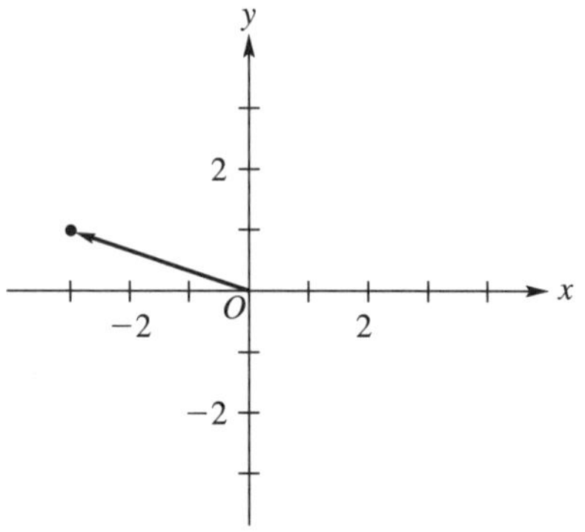

(c)

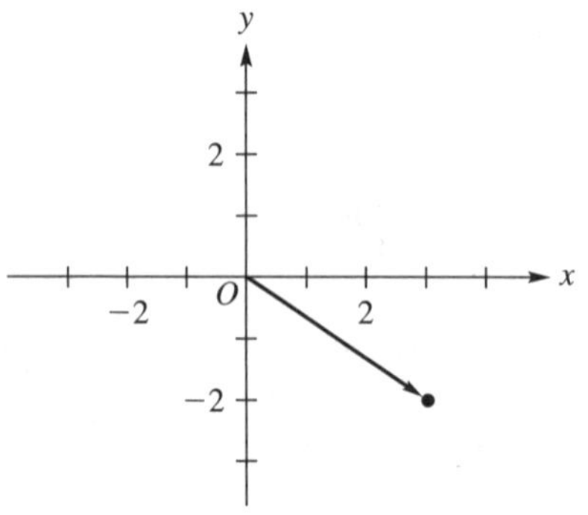

(d)

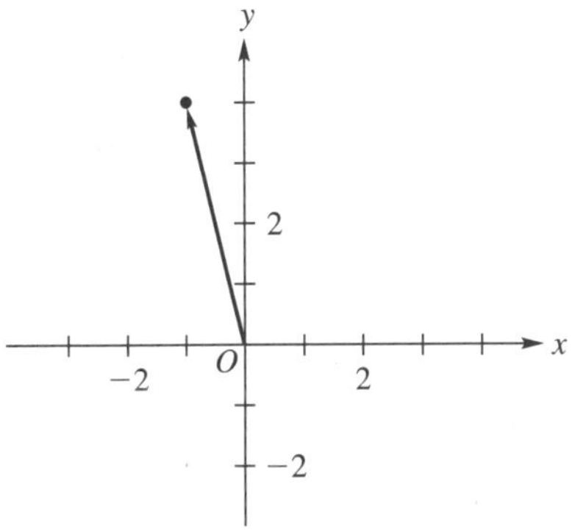

7. (a) $\begin{bmatrix} 2+4i & 5i \\ -2 & 4-2i \end{bmatrix}$. (b) $\begin{bmatrix} 4-3i \\ -2-i \end{bmatrix}$.

(c) $\begin{bmatrix} -4+4i & -2+16i \\ -4i & -8i \end{bmatrix}$.

(d) $\begin{bmatrix} 3i \\ -1-3i \end{bmatrix}$. (e) $\begin{bmatrix} 2i & -1+3i \\ -2 & -1-i \end{bmatrix}$.

(f) $\begin{bmatrix} -2i & 1-2i \\ 0 & 3+i \end{bmatrix}$. (g) $\begin{bmatrix} 3+i \\ -3+3i \end{bmatrix}$.

(h) $\begin{bmatrix} 3-6i \\ -2-6i \end{bmatrix}$.

9. $A^2 = \begin{bmatrix} -1 & 0 \\ 0 & -1 \end{bmatrix}$, $A^3 = \begin{bmatrix} 0 & -i \\ -i & 0 \end{bmatrix}$,

$A^4 = \begin{bmatrix} 1 & 0 \\ 0 & 1 \end{bmatrix}$, $A^{4n} = I_2$, $A^{4n+1} = A$,

$A^{4n+2} = A^2 = -I_2$, $A^{4n+3} = A^3 = -A$.

17. (a) $\begin{bmatrix} 0 & 0 \\ 0 & 0 \end{bmatrix}$. (b) $\begin{bmatrix} 4 & 18 \\ 0 & 4 \end{bmatrix}$.

(c) $\begin{bmatrix} -5 & 5i \\ 5i & 5 \end{bmatrix}$. (d) $\begin{bmatrix} 4 & 7i \\ 0 & -3 \end{bmatrix}$.

19. $\begin{bmatrix} 2 & 0 \\ 0 & 2 \end{bmatrix}$, $\begin{bmatrix} -1 & 0 \\ 0 & -1 \end{bmatrix}$.

Section B.2, p. A-20

1. (a) No solution. (b) No solution.

(c) $x_1 = \frac{3}{4} + \frac{5}{4}i$, $x_2 = \frac{3}{2} + i$.

3. (a) $x_1 = i$, $x_2 = 1$, $x_3 = 1 - i$.

(b) $x_1 = 0$, $x_2 = -i$, $x_3 = i$.

5. (a) $\frac{1}{5}\begin{bmatrix} 2+i & 2-4i \\ 3-i & -2-i \end{bmatrix}$.

(b) $\frac{1}{6}\begin{bmatrix} i & 1-3i & 1 \\ -2-3i & 2i & 3+2i \\ 1 & 2i & -i \end{bmatrix}$.

9. (a) Yes. (b) Linearly independent.

11. (a) The eigenvalues are $\lambda_1 = 1+i$, $\lambda_2 = 1-i$. Associated eigenvectors are

$$\mathbf{x}_1 = \begin{bmatrix} -i \\ 1 \end{bmatrix} \quad \text{and} \quad \mathbf{x}_2 = \begin{bmatrix} i \\ 1 \end{bmatrix}.$$

(b) The eigenvalues are $\lambda_1 = 0$, $\lambda_2 = 2$. Associated eigenvectors are

$$\mathbf{x}_1 = \begin{bmatrix} -i \\ 1 \end{bmatrix} \quad \text{and} \quad \mathbf{x}_2 = \begin{bmatrix} i \\ 1 \end{bmatrix}.$$

(c) The eigenvalues are $\lambda_1 = 1$, $\lambda_2 = 2$, $\lambda_3 = 3$. Associated eigenvectors are

$$\mathbf{x}_1 = \begin{bmatrix} 0 \\ -i \\ 1 \end{bmatrix}, \quad \mathbf{x}_2 = \begin{bmatrix} 1 \\ 0 \\ 0 \end{bmatrix}, \quad \text{and} \quad \mathbf{x}_3 = \begin{bmatrix} 0 \\ i \\ 1 \end{bmatrix}.$$

Indexes

Photo Credits

Chapter 1
Page 82
Alston Householder: Oak Ridge National Laboratory

Chapter 2
Page 97
Bettmann/Getty Images;
Historical image collection by Bildagentur-online/Alamy Stock Photo

Chapter 3
Page 170
Photo Researchers, Inc/Alamy Stock Photo

Chapter 4
Page 178
Georgios Kollidas/Shutterstock

Chapter 5
Page 312
Juulijs/Fotolia ; akg-images
Page 320
Jörgen Pedersen Gram: Institut Mittag-Leffler, The Royal Swedish Academy of Sciences
imagno/akg-images
Page 337
Gilbert Strang: Gilbert Strang
Page 344
Heritage Images/Hulton Archive/Getty Images
Page 359
Classic Image/Alamy Stock Photo

Chapter 7
Page 452
Bettmann/Getty Images;
Hulton-Deutsch Collection/Corbis/Corbis via Getty Images
Page 483
Juulijs/Fotolia

Chapter 8
Page 526
Popperfoto/Getty Images

Appendix B
Page 624
Photo Researchers, Inc/Alamy Stock Photo

读者意见反馈

为收集对教材的意见建议，进一步完善教材编写并做好服务工作，读者可将对本教材的意见建议通过如下渠道反馈至我社。

咨询电话 400-810-0598

反馈邮箱 hepsci@pub.hep.cn

通信地址 北京市朝阳区惠新东街4号富盛大厦1座

高等教育出版社理科事业部

邮政编码 100029